# BIOLOGY
## The Unity and Diversity of Life

### SIXTH EDITION

## CECIE STARR
Belmont, California

## RALPH TAGGART
Michigan State University

**Wadsworth Publishing Company**
Belmont, California
A Division of Wadsworth, Inc.

BIOLOGY PUBLISHER: *Jack C. Carey*

EDITORIAL ASSISTANT: *Kathryn Shea*

ART DIRECTOR AND DESIGNER: *Stephen Rapley*

PRODUCTION EDITOR: *Mary Forkner Douglas*

COPY EDITOR: *Carolyn McGovern*

PRODUCTION COORDINATOR: *Jerry Holloway*

MANUFACTURING: *Randy Hurst*

MARKETING: *Todd Armstrong, Karen Culver*

EDITORIAL PRODUCTION: *Scott Alkire, John Douglas, Kathy Hart, Gloria Joyce, Ed Serdziak, Karen Stough*

PERMISSIONS: *Marion Hansen*

PHOTO RESEARCH: *Marion Hansen, Stuart Kenter*

ARTISTS: *Susan Breitbard, Lewis Calver, Joan Carol, Raychel Ciemma, Robert Demerest, Ron Erwin, Enid Hatton, Darwin Hennings, Vally Hennings, Joel Ito, Robin Jensen, Keith Kasnot, Julie Leech, Laszlo Mezoly, Leonard Morgan, Palay/Beaubois (Phoebe Gloeckner, Lynne Larson, Betsy Palay), Victor Royer, Jeanne Schreiber, Kevin Somerville, John Waller, Judy Waller, Jennifer Wardrip*

DESIGN CONSULTANT: *Gary Head*

COVER DESIGN: *Stephen Rapley*

COVER PHOTOGRAPH: *© Thomas D. Mangelsen*

COMPOSITOR: *G&S Typesetters, Inc.: Bill M. Grosskopf, Merry Finley, Pat Molenaar, Beverly Zigal, Maurine Zook*

COLOR SEPARATOR: *H&S Graphics, Inc.: Tom Andersen, Nancy Dean, Roger Tillander, Marty O'Dean, Dennis Schnell*

PRINTING: *R. R. Donnelley & Sons Company/Willard*

2 3 4 5 6 7 8 9 10     96 95 94 93 92

**Library of Congress Cataloging in Publication Data**
Starr, Cecie.
    Biology: the unity and diversity of life / Cecie Starr, Ralph Taggart—6th ed.
        p.    cm.
    Includes bibliographical references and index.
    ISBN 0-534-16566-4
    1. Biology.   I. Taggart, Ralph.   II. Title.
QH308.2.S72  1992
574—dc20                    91-44997
                                    CIP

# PREFACE

Ask people at random to comment on the photograph on the cover of this book and you might hear something like, *What sweet little birds!* (as we did).

Those "sweet little birds" are highly specialized predators, the African bee-eaters of the family Meropidae. They swoop preferentially after bees and wasps in midair, using their long bill to catch safely, hold onto, and crush the stinging types. Then they WHAP WHAP WHAP the crushed body against the side of a branch or some other hard surface until the stinger protrudes and its venom drips out. After tossing the body several times in the air, the bee-eater gulps it down, head first, stinger last.

All textbooks for survey courses in biology include such descriptions of events in the natural world. All put on a parade of representatives from the five great kingdoms of organisms. They all describe the structure and function of those organisms at different levels of biological organization. The descriptions are useful, in that they help students build a working vocabulary about the parts and processes of life.

And yet, if textbooks are to convey accurately the nature of biological science, they must be more than a collection of observations. They must help students become familiar with the approach that biologists take to answering questions about what it is they see. With this approach, for example, the photograph of bee-eaters may open a door of inquiry, with all sorts of questions tumbling out.

> . . . I've read about the aggressive African bees and what happened after they were introduced to South America. If the bees are so aggressive, why don't they defend themselves? Are they color-blind? How else could such brilliant-blue birds sneak up on them in midair? Maybe bees don't see blue. (Come to think of it, they must not be color-blind—I see them all the time on yellow flowers.) How could I test this idea? If it's true, then bees should pass up blue flowers in favor of yellow ones. How might I set up an experiment to test this prediction? For one thing, I'd better use odorless plastic flowers . . . .

Biologists ask questions, make educated guesses (hypotheses) about possible answers, then devise ways to rigorously test predictions that will hold true if the hypotheses are correct. In broad outline, their approach is that simple. Yet it has proved to be one of the most useful of all tools for explaining the world around us. Students can use the biological approach to satisfy curiosity about bees and bee-eaters. They can use this approach to pick their way logically through today's environmental, medical, and social landmines. Finally, they can use it to understand the past and predict possible futures for ourselves and all other organisms.

## OBJECTIVES FOR THIS EDITION

*Biology: The Unity and Diversity of Life* has been evolving for eighteen years. More than 2 million students have used this book, and each revision becomes more refined in response to their experience with it. More than 1,500 dedicated teachers and researchers have shared insights with us during the years of refinements. They are our guardians of reading level and depth of coverage, of currency and accuracy.

As with previous editions, we mapped out major objectives to guide us in our approach to the sixth edition: First, write in a clear, engaging style, without being patronizing. Second, give enough examples of problem solving and experiments to provide familiarity with a scientific approach to interpreting the world. Third, identify the key concepts and select topics that reflect current research in all major fields, then present this material in light of two major themes in biology— evolution and energy flow. Fourth, use interesting, informative applications to stimulate student interest. Fifth, create easy-to-follow line art and select informative photographs.

## REVISION HIGHLIGHTS

### Writing Style

Years ago, we thought students could be enticed into the biological sciences with lively writing, memorable analogies, and engaging bits of natural history. However, the prevailing view was that such an approach somehow would be inappropriate for a textbook dealing with biological science. And so, for the past fifteen years, we focused primarily on making the writing clear and the science accurate.

Today, students often pick up biology textbooks with apprehension. If the words do not engage them, they sometimes end up hating the book *and* the subject. Instructors still ask for a scientifically accurate book—but now they also ask for one that puts the life back in.

We could not be more pleased. Because we devoted so many years to writing about biology with confidence, precision, and objectivity, we knew where the writing in this new edition *could* be loosened up. Interrupting a description of, say, the mechanisms of mitosis with a dithering of words will do the struggling student no good. Plunking humorous anecdotes into a chapter on the correlation between geologic and organismic evolution trivializes a magnificent story. Taking up valuable reading time with bits of natural history is pointless—unless those bits lead students to the big concepts.

By contrast, it certainly is appropriate to liven up a paragraph on, say, the structure of the nuclear envelope (page 61), the role of mitosis in growth and development (page 146), and the functions of skin (page 621). When you look through this new edition, you will see stunning line art and photographs. Don't let them distract you from the line-by-line judgment calls made with respect to the writing in every chapter. Improving and livening up the writing was our major objective.

## Vignettes

What authors say in a preface sometimes bears no apparent resemblance to what they did in the book. As corroboration of what we did with the writing, look at the chapter introductions, each a short story that leads into the chapter's key concepts. Some provide glimpses into the natural history of an organism, others show how biological science applies to human affairs, and many do both.

If the Chapter 9 vignette rings true, it is because Beverly McMillan sat quietly beside the Alagnak at dawn, watching life end for a female salmon. If the body language of a bulldog making his contribution to cardiology provides a light touch (Chapter 38), this is Margaret Warner's offering. Remembering a journal article from her graduate school days, she rummaged through her university's archives and found those photographs for us. If you wonder whether a confrontation with a tornado will work in class, this is Fred Delcomyn's story (the Chapter 32 vignette) and it works for him. If the Chapter 24 vignette on daisies as supermarkets seems accurate yet refreshing, Edward Ross has been thinking about this for a long time.

## Applications

This new edition features a greatly increased number of applications, all indexed on the back endpapers for easy reference. Some examples enrich the text. Others are boxed illustrations or *Commentaries* that provide in-depth information for the interested student but do not interrupt the text flow.

Many examples convey the importance of biological science in general, as when basic concepts in population ecology are applied to the prospects and problems of 5.4 billion of us now living on this planet (pages 797 and 875). Others bring home the impact of biological research on individual lives, as when students are asked to think about the effects of crack cocaine (page 562), anabolic steroids (page 637) or the implications of human gene therapy (page 257), which to a limited extent is already under way.

Starting in February, 1992, Wadsworth will be publishing an *Annual Newsletter* on important new applications that may be used to supplement those already incorporated into the sixth edition.

## Doing Science

Earlier editions included many examples of biologists at work as a way to help students develop their own understanding of critical thinking. The entire chapter on DNA structure and function has been especially successful in this respect. So have the descriptions of experimental evidence for the concepts being discussed, as in the chapter on plant growth and development. (See also the index entry, Experiments.) This edition builds on our base of science in action.

At John Alcock's suggestion, we added *Doing Science* essays. In one, students will follow Molly Lutcavage's line of questioning in her studies of the leatherback turtle, a species on the brink of extinction (page 712). In one of the essays that John drafted himself, they will see how DNA fingerprinting was used to help explain self-sacrificing behavior—not among insects, but among a fascinating group of mammals (page 918).

## Illustrations

One of us (Cecie Starr) has for eighteen years been obsessed with writing and creating illustrations simultaneously. It takes her almost as much time to research, develop, and integrate art with the text as to write and rewrite manuscripts. The obsession extends to positioning art and text references on the same two-page spread, no page-flipping required. Chapter 8, an obvious example, shows how layouts make it easier to study glycolysis, the Krebs cycle, electron transport phosphorylation, and anaerobic pathways.

Icons (pictorial representations) next to the main art show students where pathways or structures occur in a cell, multicelled body, or some other system. Zoom-sequence illustrations, from the macroscopic to microscopic, serve a similar purpose. Simple color-coded diagrams help students interpret micrographs.

Careful use of color helps students track information on hard-to-visualize topics. Throughout the book, for instance, proteins are color-coded green, carbohydrates pink, lipids yellow and gold, DNA blue, and RNA orange. Full-color anatomical paintings help give students a sense of the splendid internal complexity of organisms.

Often we incorporated written summaries *within* diagrams to make concepts easier to grasp. Where possible, we broke down information into a series of steps that are far less threatening than one large, complex diagram. Students find this approach useful, particularly with respect to art on mitosis, meiosis, and protein synthesis. It works just as effectively for such topics as antibody-mediated immunity and neural functioning.

Consider also the pedagogical impact of illustration size, as Starr did for every page. One photograph (page 885) conveys the magnitude of tropical rain forest destruction; a small photograph of a patch of burning trees could never do this. Probably few students gasp in wonder over a diagram of biomes— but ask them to use that diagram to interpret the spectacular photograph preceding Chapter 1. Pieced together from thousands of satellite images, it reveals the sweep of the Sahara, the collective green of boreal forests, and other features of the earth's surface.

## STUDY AIDS

New to this edition is a *list of key concepts* following the vignette for each chapter. We increased the number of summary statements of concepts within the text itself to help keep readers on track. Several end-of-chapter study aids reinforce the key concepts. Each chapter has a *summary* in list form, *review questions*, a *self-quiz*, *selected key terms*, and *recommended readings*. Page numbers tie each review question and key term to the relevant text page.

Numerous *genetics problems* help students grasp the principles of inheritance. The *glossary* includes pronunciation guides and origins of words, when such information will make formidable words less so. The *index* is comprehensive; students find a door to the text more quickly through finer divisions of topics. The first appendix has *metric-English conversion charts*. The second is a *classification scheme* that students can use for reference purposes. The third has *detailed answers* to the genetics problems; and the fourth, *answers* to self-quizzes. The final appendix shows structural formulas for *major metabolic pathways* for interested students and instructors who prefer the added detail.

The appendixes and glossary are printed on paper of different tints to preclude frustrating searches for where one ends and the next starts.

## SUPPLEMENTS

Twenty supplements are available. *Full-color transparencies* and *35mm slides* of almost all illustrations from the book are labeled with large, boldface type. A *Test Items* booklet has 5,000 questions by outstanding test writers. Questions are available in electronic form on IBM, Apple IIe, and Macintosh.

An *Instructor's Resource Manual* has, for each chapter, an outline, objectives, list of boldface or italic terms, and a detailed lecture outline. It also includes suggestions for lecture presentations, classroom and laboratory demonstrations, suggested discussion questions, research paper topics, and annotations for filmstrips and videos. *Lecture outlines* in the Instructor's Resource Manual are available on a data disk for those who wish to modify the material. A *Videodisc Correlation Directory and Barcode Guide* correlates the text with popular videodiscs. Software—*HyperCard Stacks for Videodiscs*—correlates the text with the same videodiscs.

A new, active-oriented *Study Guide and Workbook* asks students to respond to almost all questions by writing in the guide. Questions are arranged by chapter section. Each chapter also has a set of critical thinking questions. The *chapter objectives* of the Study Guide are available on disk as part of the testing file for those who wish to modify or select portions of the material. An *electronic study guide* consists of multiple-choice questions different from those in the test-item booklet. Students get feedback on why their answer is correct or incorrect. A 100-page *Answer Booklet* has answers to the book's end-of-chapter review questions.

A special version of *STELLA II*, a software tool for developing critical thinking skills, is available to users of the book, together with a workbook.

Approximately 400 *flashcards* with 1,000 glossary items are available. There are four anthologies. *Contemporary Readings in Biology* has articles on applications of interest to students. *Science and the Human Spirit: Contexts for Writing and Learning* helps students learn to write effectively about biology. *Ethical Issues in the New Reproductive Technologies* discusses some major issues of our time. *The Game of Science* gives students a realistic view of what science is and what scientists do.

A new *Laboratory Manual* has 38 experiments and exercises. It now contains hundreds of labeled photographs, and all illustrations are in full color. Many experiments are divided into distinct parts that can be assigned individually, depending on time available. All have objectives, discussion (introduction, background, and relevance), a list of materials for each part of an experiment, procedural steps, pre-lab questions, and post-lab questions. An *Instructor's Manual* accompanies the Laboratory Manual. It covers quantities, procedures for preparing reagents, time requirements for each portion of the exercise, hints to make the lab a success, and vendors of materials with item numbers.

# IMPROVEMENTS IN CONTENT AND ORGANIZATION

Although the following paragraphs are by no means inclusive, they convey the magnitude of the revision.

**INTRODUCTION** We streamlined the first two chapters of the preceding edition. Now one chapter provides an overview of key biological concepts and a revised treatment of scientific methods, livened up by John Alcock. Simple examples introduce the pertinent points of evolution by natural selection, but the history of evolutionary thought now is the stage-setting chapter for the evolution unit (III).

**UNIT I. CELLULAR BASIS OF LIFE** More concise writing and greatly improved art make the chapters on cell structure and biochemistry more accessible. Chapter 2 has a new *Commentary* on radioisotopes. Chapter 3 includes an improved discussion of carbohydrates, a new *Commentary* on cholesterol and atherosclerosis, and a better diagram of hemoglobin structure. Chapter 4 has a simpler description of the cytomembrane system. Notice the cell icons in the illustrations of organelles. We updated the classification of membrane proteins (Chapter 5). Better diagrams of freeze-fracturing accompany the *Doing Science* essay (page 78). Chapter 6 presents a simpler overview of basic metabolism.

The plant in the zoom-sequence of chloroplast structure (Chapter 7) is now the same species used in Chapter 28, which continues the story by showing translocation. David Fisher helped develop these illustrations. Chapter 8 is reorganized—first the aerobic pathway, then anaerobic pathways. To keep text concepts uncluttered, details of ATP formation in Chapters 7 and 8 are presented in boxed illustrations.

**UNIT II. PRINCIPLES OF INHERITANCE** Chapters 9 through 11 already are effective in the classroom. We sharpened the writing, included new examples (such as Labrador coat color), but left the organization much the same. Robert Robbins suggested the vignette for Chapter 10 and the *Commentary* on HeLa cells (page 147). Chromosomal inheritance and human genetics are combined in one chapter (12). Morgan's fruit fly experiments are in an optional, boxed illustration. Students should enjoy the new *Commentary* on sex determination (page 186).

The Chapter 13 vignette provides background for the Watson-Crick story and reminds students that science proceeds as a community effort (more or less). The organization of DNA in chromosomes is now described and illustrated in this chapter (page 212). The Chapter 14 vignette actually makes the idea of reading about protein synthesis nonthreatening. A simple overview and improved art make the chapter easier to follow. Early studies of gene function are described in an optional boxed illustration. Gene mutation is now introduced in this chapter, with a *Commentary* on its role in evolution.

Chapter 15 (gene regulation) is updated, with better delineation between prokaryotic and eukaryotic mechanisms. The vignette on control of cell division sets the stage for the *Commentary* on cancer. Daniel Fairbanks and Lisa Starr made solid contributions to Chapter 16, which provides a reorganized, updated, and more accurate picture of recombinant DNA technology and genetic engineering. The *Doing Science* essay gives interested students simple descriptions of gel electrophoresis of DNA and DNA sequencing methods.

**UNIT III. PRINCIPLES OF EVOLUTION** The evolution and diversity units now immediately follow the genetics unit. We overhauled the content and added spectacular illustrations (see page 312). Chapter 17 provides the historical background. We polished the chapter on microevolution (18) and added an in-depth look at a current study of speciation (page 288).

Chapter 19 has crisper descriptions of the evidence for the origin of the earth and life. Events and mechanisms underlying large-scale evolutionary patterns and rates of change are described succinctly. Macroevolutionary patterns dominating each major geologic era are sketched out. Figure 19.16 graphically emphasizes a central concept—that changes in the environment have been a profound force in the evolution of life.

Aaron Bauer wrote a new chapter on systematics (20) and, amazingly, made cladistics understandable (page 324). We moved the case study on human evolution (Chapter 21) here to make the unit self-contained. It includes a *Doing Science* essay on mitochondrial DNA and recent human ancestry. A section on classification outlines the five-kingdom scheme.

In Chapter 20 and elsewhere, we remind students that boundaries between taxa are not real; we impose them on a continuum of evolutionary lines. Taxonomists take the impositions seriously, and possibly our decision to classify the red, brown, and green algae as plants will make some of them cranky. However, we did not make the decision lightly. It reflects the overwhelming preference of hundreds of teachers who responded to a questionnaire on this issue.

**UNIT IV. EVOLUTION AND DIVERSITY**  We reworked the diversity unit extensively. Introductory texts (our earlier editions included) tend to slight the microbial world. Notice the expanded, richly illustrated coverage of viruses, bacteria, and protistans in Chapter 22. The chapter also has *Commentaries* on infectious diseases (page 352), eukaryotic origins (page 361), and the beginnings of multicellularity (page 369). Fungi now have their own chapter (23) that conveys the diversity in this often-ignored kingdom.

We clarified the Chapter 24 survey of evolutionary trends among plants. More applications are woven into descriptions of the major divisions, as on page 388. Once again, Eugene Kozloff guided us through the maze of invertebrates and helped refine the chapter (25). Vertebrates are described in a separate chapter (26). In both chapters, icons serve as effective roadmaps.

**UNIT V. PLANT STRUCTURE AND FUNCTION**  We made the writing easier to follow, added applications, and improved the art and page layouts. Chapter 28 has better coverage of root nodules (page 489) and superior diagrams for absorption (491), transpiration (492), and translocation (496).

The vignette in Chapter 29 gives new meaning to the word chocolate, the *Commentary* (page 506) provides vivid examples of pollination, and a *Doing Science* essay asking why some plants produce so many flowers (page 512) reminds students of what it means to think critically.

**UNIT VI. ANIMAL STRUCTURE AND FUNCTION**  Extensive rewriting and many more applications make this inherently complex unit approachable. The new art speaks for itself.

We updated tissue classification and micrographs and explained homeostasis with tangible examples (Chapter 31). Neurobiologists helped update Chapter 32, and our teacher reviewers helped make it accessible. Chapter 33 better describes the evolution of nervous systems and the neural wiring of vertebrates.

The endocrine chapter (34) has less abstract examples and art. Chapter 35 provides a more accurate picture of sensory function. It has new material on echolocation, pain, and vision, including a *Commentary* on eye disorders (612).

We packed Chapter 36 with applications that should hold student interest. The sections on muscle function and energy metabolism make better sense. An integrative diagram at the start of Chapters 37, 38, and 39 helps students visualize how systems are integrated. We expanded the material on human nutrition and included a *Commentary* on eating disorders (650). Chapters 38 through 42 underwent major reorganization and updating. Whether assigned or not, Chapter 43 on human reproduction and development is one that students read closely, and we took special care to provide them with accurate and current information.

**UNIT VII. ECOLOGY AND BEHAVIOR**  We worked closely with Robert Colwell and George Cox to reorganize and update the ecology chapters. Growth equations in Chapter 44 are described more clearly and the section on human population growth is expanded. Chapter 45 has refined definitions for habitat, niche, and species richness. Charles Krebs suggested an update for the Canadian lynx-hare story (pages 811–812). Jane Lubchenko's study of predation and competition is included (page 817). There is a new *Commentary* on species introduction (page 818).

The Chapter 46 vignette on a major environmental issue may leave students with a sense that things *can* change when we put our minds to it. A new carbon cycle diagram (page 837) leads into the *Commentary* on global warming (838). Chapter 47 provides a tighter overview of factors shaping climate, hence ecosystems. Photographs are large enough to show biome features. The text on lake ecosystems and intertidal zonation is more straightforward. Ernest Benfield provided material on stream ecosystems (864). Tropical reefs are now illustrated in this chapter (868). The Chapter 48 vignette describes tropical rain forests, then the text conveys the magnitude and pace of their destruction (874 and 884). Our friend Tyler Miller, Jr., helped us update this important chapter.

John Alcock's interest in teaching students how to think critically is evident in his two chapters (49 and 50), starting with the vignette on nest-building behavior. He updated and reorganized both.

## Advisors for the Sixth Edition

BRENGELMANN, GEORGE, *University of Washington*
DELCOMYN, FRED, *University of Illinois, Urbana*
DENNISTON, KATHERINE, *Towson State University*
FISHER, DAVID, *University of Hawaii, Manoa*
FONDACARO, JOSEPH, *Marion Merrell Dow*
FORTNEY, SUSAN, *Johnson Space Center*
FRYE, BERNARD, *University of Texas, Arlington*
HESS, WILLIAM, *Brigham Young University*
HOHAM, RONALD, *Colgate University*
JACKSON, JOHN, *North Hennepin Community College*
KENDRICK, BRYCE, *University of Waterloo*
LASSITER, WILLIAM, *University of North Carolina, Chapel Hill*
MARR, ELEANOR, *Dutchess Community College*
MILLER, G. TYLER, *Pittsboro, North Carolina*
MUCH, DAVID, *Muhlenberg College*
ROBBINS, ROBERT, *National Science Foundation*
ROSE, GREIG, *West Valley College*
SLOBODA, ROGER, *Dartmouth College*
WARNER, MARGARET, *Indiana University, Krennert Institute*
WEISS, MARK, *Wayne State University*
WENDEROTH, MARY PAT, *University of Washington*
WOLFE, STEPHEN, *Emeritus, University of California, Davis*

## Reviewers for the Sixth Edition

ABBAS, ABDUL, *Brigham Womens Hospital*
ADAMS, THOMAS, *Michigan State University*
ANDERSON, DEBRA, *Oregon Health Sciences University*
ARMSTRONG, PETER, *University of California, Davis*
ARNOLD, STEVAN, *University of Chicago*
ATCHISON, GARY, *Iowa State University*
BAJER, ANDREW, *University of Oregon*
BAKKEN, AIMEE, *University of Washington*
BAPTISTA, LUIS, *California Academy of Sciences*
BARKWORTH, MARY, *Utah State University*
BATZING, J. L., *State University of New York, Cortland*
BECK, CHARLES, *University of Michigan*
BEECHER, MICHAEL, *University of Washington*
BEEVERS, HARRY, *University of California, Santa Cruz*
BENACERRAF, BARUJ, *Harvard School of Medicine*
BENDER, KRISTEN, *California State University, Long Beach*
BENES, ELINOR, *California State University, Sacramento*
BIRKEY, C. WILLIAM, *Ohio State University*
BISHOP, VERNON, *University of Texas, San Antonio*
BOHR, D. F., *University of Michigan Medical School*
BONNER, JAMES, *California Institute of Technology*
BOOHAR, RICHARD, *University of Nebraska, Lincoln*
BOTTRELL, CLYDE, *Tarrant County Junior College South*
BRADBURY, JACK, *University of California, San Diego*
BRINKLEY, B., *University of Alabama, Birmingham*
BRINSON, MARK, *East Carolina University*
BROMAGE, TIM, *City University of New York*
BROOKS, VIRGINIA, *Oregon Health Sciences University*
BROWN, ARTHUR, *University of Arkansas*
BUCKNER, VIRGINIA, *Johnson County Community College*
BURNES, E. R., *University of Arkansas, Little Rock*
BURTON, HAROLD, *State University of New York, Buffalo*
CABOT, JOHN, *State University of New York, Stony Brook*
CALVIN, CLYDE, *Portland State University*
CARLSON, ALBERT, *State University of New York, Stony Brook*
CASE, CHRISTINE, *Skyline College*
CASE, TED, *University of California, San Diego*
CASEY, KENNETH, *Veterans Administration Medical Center*
CENTANNI, RUSSELL, *Boise State University*
CHAMPION, REBECCA, *Kennesaw State College*
CHAPMAN, DAVID, *University of California, Los Angeles*

CHARLESWORTH, BRIAN, *University of Chicago*
CHERNOFF, BARRY, *Field Museum of Natural History*
CHISZAR, DAVID, *University of Colorado*
CHRISTENSEN, A. KENT, *University of Michigan Medical School*
CHRISTIAN, DONALD, *University of Minnesota*
CLAIRBORNE, JAMES, *Georgia Southern University*
CLARK, NANCY, *University of Connecticut*
CONNELL, MARY, *Appalachian State University*
COQUELIN, ARTHUR, *Texas Tech University*
COTTER, DAVID, *Georgia College*
COYNE, JERRY, *University of Chicago*
DANIELS, JUDY, *Washtenau Community College*
DAVIS, DAVID GALE, *University of Alabama*
DAVIS, JERRY, *University of Wisconsin*
DEMPSEY, JEROME, *University of Wisconsin*
DENGLER, NANCY, *University of Toronto*
DENISON, WILLIAM, *Oregon State University*
DICKSON, KATHRYN, *California State University, Fullerton*
DLUZEN, DEAN, *University of Illinois, Urbana*
DOYLE, PATRICK, *Middle Tennessee State University*
DULING, BRIAN, *University of Virginia, Charlottesville*
ECK, GERALD, *University of Washington*
EDLIN, GORDON, *University of Hawaii, Manoa*
ENDLER, JOHN, *University of California, Santa Barbara*
ENGLISH, DARREL, *Northern Arizona University*
ERICKSON, GINA, *Highline Community College*
ERWIN, CINDY, *City College of San Francisco*
ESTEP, DAN, *University of Georgia*
EWALD, PAUL, *Amherst College*
FALK, RICHARD, *University of California, Davis*
FINNEGAN, D. J., *University of Edinburgh*
FISHER, DONALD, *Washington State University*
FLESCH, DAVID, *Mansfield University*
FLESSA, KARL, *University of Arizona*
FOYER, CHRISTINE, *Laboratoire du Metabolisme*
FROEHLICH, JEFFREY, *University of New Mexico*
FUNK, FRED, *Northern Arizona University*
GAINES, MICHAEL, *University of Kansas*
GENUTH, SAUL, *Case Western Reserve University*
GHOLZ, HENRY, *University of Florida*
GIBSON, ARTHUR, *University of California, Los Angeles*
GIBSON, THOMAS, *San Diego State University*
GIDDAY, JEFFREY, *University of Virginia School of Medicine*
GOFF, CHRISTOPHER, *Haverford College*
GOODMAN, MAURICE, *University of Massachusetts Medical School*
GORDON, ALBERT, *University of Washington*
GOSZ, JAMES, *University of New Mexico*
GRAHAM, LINDA, *University of Wisconsin*
GRANT, BRUCE, *College of William and Mary*
GREEN, GARETH, *Johns Hopkins University*
GREENE, HARRY, *University of California, Berkeley*
GREGG, KATHERINE, *West Virginia Wesleyan College*
HANKEN, JAMES, *University of Colorado*
HANRATTY, PAMELA, *University of Southern Mississippi*
HARDIN, JOYCE, *Hendrix College*
HARLEY, JOHN, *Eastern Kentucky University*
HARRIS, JAMES, *Utah Valley Community College*
HARTNEY, KRISTINE BEHRENTS, *California State University, Fullerton*
HASSAN, ASLAM, *University of Illinois, Urbana*
HELLER, LOIS JANE, *University of Minnesota*
HERTZ, PAUL, *Barnard College*
HEWITSON, WALTER, *Bridgewater State College*
HILDEBRAND, MILTON, *University of California, Davis*
HILFER, S. ROBERT, *Temple University*
HINCK, LARRY, *Arkansas State University*
HODGSON, RONALD, *Central Michigan University*
HOLLINGER, TOM, *University of Florida, Gainesville*
HOLMES, KENNETH, *University of Illinois, Urbana*
HOSICK, HOWARD, *Washington State University*
JAKOBSON, ERIC, *University of Illinois, Urbana*
JENSEN, STEVEN, *Southwest Missouri State University*
JOHNSON, LEONARD, *University of Tennessee School of Medicine*
JOHNSON, TED, *St. Olaf College*

JONES, PATRICIA, *Stanford University*
JUILLERAT, FLORENCE, *Indiana University–Purdue University*
KAYE, GORDON, *Albany Medical School*
KEIM, MARY, *Seminole College*
KELLY, DOUGLAS, *University of Southern California*
KELSEN, STEVEN, *Temple University Hospital*
KEYES, JACK, *Linfield College, Portland*
KIGER, JOHN, *University of California, Davis*
KIMBALL, JOHN, *Tufts University*
KIRK, HELEN, *University of Western Ontario*
KNUTTGEN, HAROLD, *Boston University*
KREBS, CHARLES, *University of British Columbia*
KURIS, ARMAND, *University of California, Santa Barbara*
KUTCHAI, HOWARD, *University of Virginia Medical School*
LATIES, GEORGE, *University of California, Los Angeles*
LATTA, VIRGINIA, *Jefferson State Junior College*
LEFEVRE, GEORGE, *California State University, Northridge*
LEVY, MATTHEW, *School of Medicine, City University of New York*
LEWIS, LARRY, *Bradford University*
LINDE, RANDY, *Palo Alto Medical Foundation*
LINDSEY, JERRI, *Tarrant County Junior College*
LITTLE, ROBERT, *Medical College of Georgia*
LOCKE, MICHAEL, *University of Western Ontario*
MACKLIN, MONICA, *Northeastern State University*
MADIGAN, MICHAEL, *Southern Illinois University, Carbondale*
MAJUMDAR, S. K., *Lafayette College*
MALLOCH, DAVID, *University of Toronto*
MANN, ALAN, *University of Pennsylvania*
MARGULIES, MAURICE, *Rockville, Maryland*
MARGULIS, LYNN, *University of Massachusetts, Amherst*
MARTIN, JAMES, *Reynolds Community College*
MATHEWS, ROBERT, *University of Georgia*
MATSON, RONALD, *Kennesaw State College*
MATTHAI, WILLIAM, *Tarrant County Junior College*
MAXSON, LINDA, *Pennsylvania State University*
MAXWELL, JOYCE, *California State University, Northridge*
MCCLINTIC, J. ROBERT, *California State University, Fresno*
MCEDWARD, LARRY, *University of Florida*
MCKEAN, HEATHER, *Eastern Washington State University*
MCKEE, DOROTHY, *Auburn University, Montgomery*
MCNABB, ANNE, *Virginia Polytechnic Institute and State University*
MCREYNOLDS, JOHN, *University of Michigan Medical School*
MERTENS, THOMAS, *Ball State University*
MEYER, NANCY, *University of Michigan, Dearborn*
MIMMS, CHARLES, *University of Georgia*
MITZNER, WAYNE, *Johns Hopkins University*
MOCK, DOUG, *University of Washington*
MOHRMAN, DAVID, *University of Minnesota*
MOISES, HYLAND, *University of Michigan Medical School*
MOORE-LANDECKER, ELIZABETH, *Glassboro State University*
MORBECK, MARY ELLEN, *University of Arizona*
MORRISON, WILLIAM, *Shippensburg University*
MORTON, DAVID, *Frostburg State University*
MOUNT, DAVID, *University of Arizona*
MURPHY, RICHARD, *University of Virginia Medical School*
MURRISH, DAVID, *State University of New York, Binghamton*
MYERS, NORMAN, *Oxford University, England*
MYRES, BRIAN, *Cypress College*
NAGLE, JAMES, *Drew University*
NEMEROFSKY, ARNOLD, *State University of New York, New Paltz*
NICHOLS-KIRK, HELEN, *University of Western Ontario*
NORRIS, DAVID, *University of Colorado*
O'BRIEN, ELINOR, *Boston College*
OJANLATVA, ANSA, *Sacramento, California*
OLSON, MERLE, *University of Texas Health Science Center*
ORR, CLIFTON, *University of Arkansas*
ORR, ROBERT, *California Academy of Sciences*
PAI, ANNA, *Montclair State College*
PALMBLAD, IVAN, *Utah State University*
PAPPENFUSS, HERBERT, *Boise State University*
PARSONS, THOMAS, *University of Toronto*
PAULY, JOHN, *University of Arkansas for Medical Sciences*
PECHENIK, JAN, *Tufts University*

PERRY, JAMES, *Frostburg State University*
PETERSON, GARY, *South Dakota State University*
PIERCE, CARL, *Washington University*
PIKE, CARL, *Franklin and Marshall College*
PLEASANTS, BARBARA, *Iowa State University*
POWELL, FRANK, *University of California, San Diego*
RALPH, CHARLES, *Colorado State University*
RAWN, CARROLL, *Seton Hall University*
REEVE, MARIAN, *Emeritus, Merritt Community College*
RICKETT, JOHN, *University of Arkansas, Little Rock*
RIEDER, CONLY, *Wadsworth Center for Laboratories and Research*
ROMANO, FRANK, *Jacksonville State University*
ROSEN, FRED, *Harvard University School of Medicine*
ROSS, EDWARD, *California Academy of Sciences*
ROSS, GORDON, *University of California Medical Center, Los Angeles*
ROSS, IAN, *University of California, Santa Barbara*
ROST, THOMAS, *University of California, Davis*
RUIBAL, RUDOLFO, *University of California, Riverside*
SACHS, GEORGE, *University of California, Los Angeles*
SACKETT, JAMES, *University of California, Los Angeles*
SALISBURY, FRANK, *Utah State University*
SCHAPIRO, HARRIET, *San Diego State University*
SCHECKLER, STEPHEN, *Virginia Polytechnic Institute and State University*
SCHIMEL, DAVID, *NASA Ames Research Center*
SCHLESINGER, WILLIAM, *Duke University*
SCHMID, RUDI, *University of California, Berkeley*
SCHMOYER, IRVIN, *Muhlenberg College*
SCHNERMANN, JURGEN, *University of Michigan School of Medicine*
SEARLES, RICHARD, *Duke University*
SEHGAL, PREM, *East Carolina University*
SHARP, ROGER, *University of Nebraska, Omaha*
SHEPHERD, JOHN, *Mayo Medical University*
SHERMAN, PAUL, *Cornell University*
SHOEMAKER, DAVID, *Emory University School of Medicine*
SHONTZ, NANCY, *Grand Valley State University*
SHOPPER, MARILYN, *Johnson County Community College*
SILK, WENDY, *University of California, Davis*
SLATKIN, MONTGOMERY, *University of California, Berkeley*
SMILES, MICHAEL, *State University of New York, Farmingdale*
SMITH, RALPH, *University of California, Berkeley*
SOLOMON, TRAVIS, *Kansas City V.A. Medical Center*
STARR, LISA, *Scripps Clinic and Research Foundation*
STEIN-TAYLOR, JANET, *University of Illinois, Chicago*
STEINERT, KATHLEEN, *Bellevue Community College*
STITT, JOHN, *John B. Pierce Foundation Laboratory*
SULLIVAN, LAWRENCE, *University of Kansas*
SUMMERS, GERALD, *University of Missouri*
SUNDBERG, MARSHALL, *Louisiana State University*
SWABY, JAMES, *United States Air Force Academy*
TERHUNE, JERRY, *Jefferson Community College, University of Kentucky*
THAMES, MARC, *Medical College of Virginia*
TIFFANY, LOIS, *Iowa State University*
TIZARD, IAN, *Texas A & M University*
TORSTVEIT, ELINOR, *Concordia College*
TRAMMELL, JAMES, JR., *Arapahoe Community College*
TUTTLE, JEREMY, *University of Virginia, Charlottesville*
VALENTINE, JAMES, *University of California, Santa Barbara*
VALTIN, HEINZ, *Dartmouth Medical School*
WAALAND, ROBERT, *University of Washington*
WADE, MICHAEL, *University of Chicago*
WALSH, BRUCE, *University of Arizona*
WARING, RICHARD, *Oregon State University*
WARMBRODT, ROBERT, *University of Maryland*
WEIGL, ANN, *Winston-Salem State University*
WEISBRODT, NORMAN, *University of Texas Medical School, Houston*
WELKIE, GEORGE, *Utah State University*
WHEELIS, MARK, *University of California, Davis*
WHIPP, BRIAN, *University of California, Los Angeles*
WHITTOW, G. CAUSEY, *University of Hawaii School of Medicine*
WINICUR, SANDRA, *Indiana University, South Bend*
WISE, ROBERT, *Francis Scott Key Medical Center*
WOMBLE, MARK, *University of Michigan Medical School*
ZIHLMAN, ADRIENNE, *University of California, Santa Cruz*

## A COMMUNITY EFFORT

One, two, or a smattering of authors can write accurately and often very well about their field of interest, but it takes more than this to deal with the full breadth of biological sciences. For us, it takes an educational network that extends through the United States and on into Canada, England, Germany, France, Sweden, Australia, and elsewhere. We continually track down respected researchers, teachers, and photographers. For a few rarified topics, we invite resource manuscripts from specialists who have never turned their back on the call to teach a new generation of students. We integrate such material into our own manuscript, rewriting or graphically shaping it according to our strong convictions about what an introductory book must be.

John Alcock, Aaron Bauer, Rob Colwell, George Cox, Daniel Fairbanks, Eugene Kozloff, Bill Parson, Cleon Ross, and Sam Sweet have been exceptional in their commitment to our efforts. They have responded with grace to phone calls and faxes, to queries concerning the odd fact.

Many reviewers were content specialists, others were diary reviewers who evaluated the fifth edition's effectiveness in their own classrooms. Collectively, their comments helped us shape the revision. Our advisors assisted us in evaluating reviewer comments and in suggesting improvements in the new manuscripts and art. Katherine Denniston, Pamela Hanratty, William Hess, John Jackson, Bernard Frye, Greig Rose, and Nancy Meyer advised us on every chapter for level and clarity. Jackson and Denniston helped us develop new self-quiz questions.

If the writing now seems fresher, this is largely because of our creative interaction with Beverly McMillan, the developmental editor for this edition.

She also researched material and drafted several vignettes, *Doing Science* essays, and *Commentaries*. We treasure her as a friend and gifted writer.

Dick Greenberg, Jack Carey, Kathie Head, Stephen Rapley, and Randy Hurst at Wadsworth never let the users of this book down. This time they assembled the best production team and manufacturers in the business, starting with Mary Douglas, who has our unequivocal respect and friendship. Mary has the talent, toughness, sensitivity, compulsiveness, and oblique sense of humor required to shepherd text and art manuscripts of this complexity.

Because of Jerry Holloway and Kathryn Shea, we kept smiling instead of bashing our head against the wall. Marion Hansen would have liked to bash our head against the wall, but instead she persevered through one more edition and, with Stuart Kenter, collected exquisite photographs. Carolyn McGovern, Gloria Joyce, and Ed Serdziak took care of complicated editorial functions. Todd Armstrong, Karen Culver, and Debbie Dennis have been most supportive. Ryan Carey was the captive student reader. Barbara Odone, John Douglas, and Verbal Clark kept the paper flowing.

Besides designing a memorable cover for the book, Stephen Rapley worked out its interior design in consultations with Gary Head. Raychel Ciemma, Bob Demerest, Darwin Hennings, Vally Hennings, Len Morgan, and Betsy Palay did much of the outstanding new art. Susan Breitbard, Natalie Hill, Carole Lawson, Joan Olson, Jill Turney, and Kathryn Werhane worked directly with us as resident artists. Tom Anderson once again was responsive to picky requests on color separations. G&S is in a league by itself.

Jack Carey, would you have believed eighteen years ago that you would be publisher of such a widely used textbook in biology? You charted our course, and made it happen.

# CONTENTS IN BRIEF

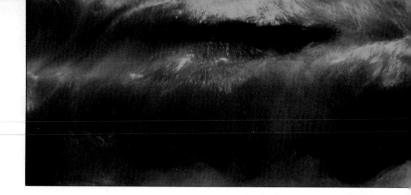

# DETAILED TABLE OF CONTENTS

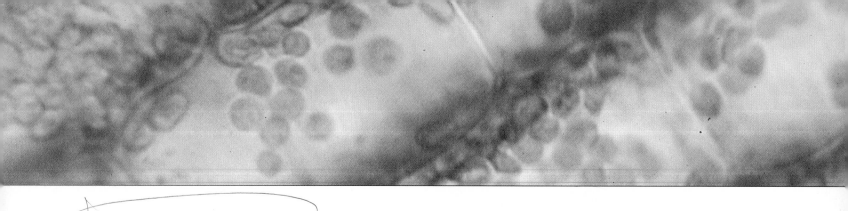

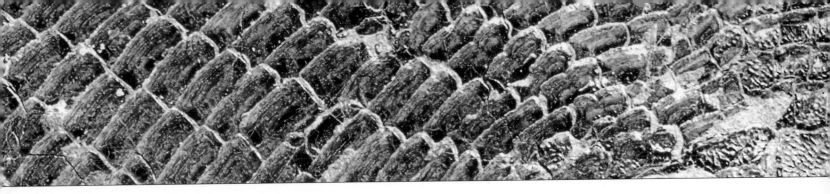

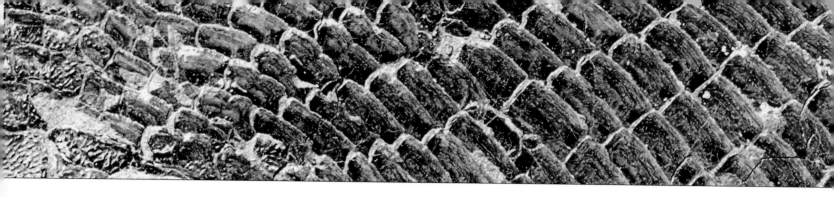

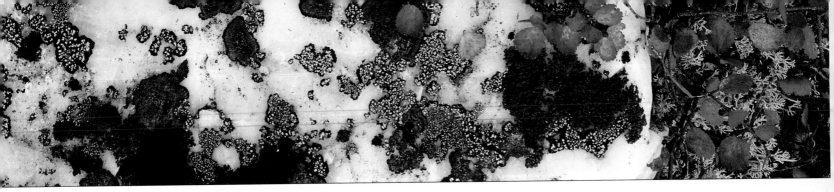

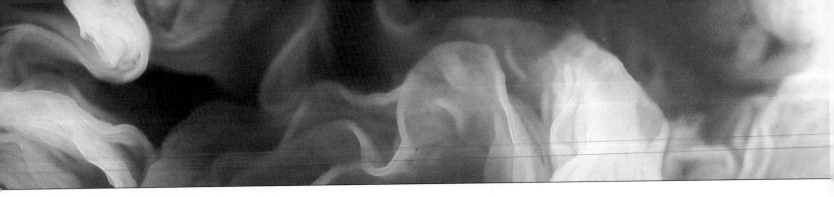

Current configurations of the earth's oceans and land masses—
the geologic stage upon which life's drama continues to unfold.
Thousands of separate images were pieced together to create
this remarkable photograph of our planet.

# 1 METHODS AND CONCEPTS IN BIOLOGY

## Biology Revisited

Buried somewhere in that mass of nerve tissue just above and behind your eyes are memories of first encounters with the living world. Still in residence are sensations of discovering your own two hands and feet, your family, the change of seasons, the smell of rain-drenched earth and grass. In that brain are traces of early introductions to a great disorganized parade of insects, flowers, friends, and furred things, mostly living, sometimes dead. There, too, are memories of questions—*"What is life?"* and, inevitably, *"What is death?"* There are memories of answers, some satisfying, others less so.

Observing, asking questions, accumulating answers—in this manner you have acquired a store of knowledge about the world of life. As you have grown older, experience and education have been refining your questions, and no doubt the answers are more difficult to come by. What *is* life? What defines the living state? The answers you get may vary, depending, for example, on whether they come from a physician, a court of law, or the parents of a severely injured girl who is being maintained by mechanical life support systems because her brain no longer functions at all.

Yet despite the changing character of your questions, the world of living things persists much as it did before. New leaves still unfurl during the spring rains. Animals are born, they grow, reproduce, and die even as new individuals of their kind replace them. The world of life has not changed in these respects. It is just that your perceptions about them have deepened.

It is scarcely appropriate, then, for a book to claim that it is your introduction to biology—the study of life—when you have been studying life ever since awareness of the world began penetrating your brain. The subject is the same familiar world you have already thought about for many years. That is why this book claims only to be biology *revisited*, in ways that may help carry your thoughts about life to deeper, more organized levels of understanding.

Let us return to the question, What is life? Offhandedly, you might reply that you know it when you see it. Yet there is no simple answer, for the question opens up a story that has been unfolding in countless directions for several billion years! To biologists, "life" is an outcome of ancient events that led to the assembly of nonliving materials into the first organized, living cells. "Life" is a way of capturing and using energy and materials. "Life" is a way of sensing and responding to specific changes in the environment. "Life" is a capacity to reproduce; it is a capacity to follow programs of growth and development. And "life" evolves, meaning that details in the body plan and functions of each kind of organism can change through successive generations.

Clearly, a short description only hints at the meaning of life. Deeper insight requires wide-ranging study of life's characteristics.

Throughout this book you will encounter many different examples of living things—how they are constructed, how they function, where they live, what they do. The examples provide evidence in support of certain concepts which, when taken together, will give you a sense of what "life" is. The next few pages will provide you with a brief overview of those basic concepts. As you continue your reading in subsequent chapters, you may find it useful now and then to return to this overview as a way of reinforcing your grasp of details.

**Figure 1.1** Think back on all you have known and seen, and this is the foundation for your deeper probes into the world of life.

**1.** All organisms are alike in these respects: Their structure, organization, and interactions arise from the properties of matter and energy. They obtain and use energy and materials from their environment, and they make controlled responses to changing conditions. They grow and reproduce, and instructions for traits that they pass on from one generation to the next reside in their DNA.

**2.** Organisms show great diversity in their structure, function, and behavior, largely as a result of evolution by means of natural selection.

**3.** The theories of science are based on systematic observations, hypotheses, predictions, and relentless testing. The external world, not internal conviction, is the testing ground for scientific theories.

## SHARED CHARACTERISTICS OF LIFE

### DNA and Biological Organization

Picture a frog on a rock, busily croaking. Without even thinking about it, you know that the frog is alive and the rock is not. At a much deeper level, however, the difference between them blurs. Frogs, rocks, and all other living or nonliving things are composed of the same particles (protons, electrons, and neutrons). The particles are organized into atoms, in every case according to the same physical laws. At the heart of those laws is something called **energy**—a capacity to make things happen, to do work. Energetic interactions bind atom to atom in predictable patterns, giving rise to the structured bits of matter we call molecules. Energetic interactions among molecules hold a rock together—and they hold a frog together.

It takes a special type of molecule called deoxyribonucleic acid, or **DNA**, to set living things apart from the nonliving world. No chunk of granite or quartz has it. DNA molecules contain the instructions for assembling each new organism from carbon, hydrogen, and a few

3

With blueprints and some energy, you can put together highly ordered patterns from a disordered pile of just two kinds of ceramic tiles:

or

other kinds of "lifeless" molecules. By analogy, think of what you can do with just two kinds of ceramic tiles in a crafts kit. With a little effort, you can glue the tiles together according to the kit's directions, so that you can produce many organized patterns of tiles (Figure 1.2). Similarly, the organization of life emerges from lifeless matter with DNA "directions," some raw materials, and energy.

Look carefully at Figure 1.3, which outlines the levels of organization in nature. The quality of "life" actually emerges at the level of cells. A *cell* is the basic living unit. This means it has the capacity to maintain itself as an independent unit and to reproduce, given appropriate sources of energy and raw materials. Amoebas and many other single-celled organisms lead such independent lives.

A *multicelled organism* is more complex, with specialized cells typically arranged into tissues, organs, and often organ systems. Its cells depend on the integrated activities of one another, but each generally retains the capacity for independent existence. How do we know this? Individual cells that have been removed from humans and other multicelled organisms can be kept alive under controlled laboratory conditions.

The next, more inclusive level of organization is the *population:* a group of single-celled or multicelled organisms of the same kind occupying a given area. A congregation of penguins at a rookery in Antarctica is an example. Moving on, the populations of whales, seals, fishes, and all other organisms living in the same area as the penguins make up a *community.*

The next level, the *ecosystem,* includes the community *and* its physical and chemical environment. The most inclusive level of organization is the *biosphere.* The biosphere includes all regions of the earth's waters, crust, and atmosphere in which organisms live.

**The structure and organization of nonliving *and* living things arise from the fundamental properties of matter and energy.**

**The structure and organization *unique* to living things starts with instructions contained in DNA molecules.**

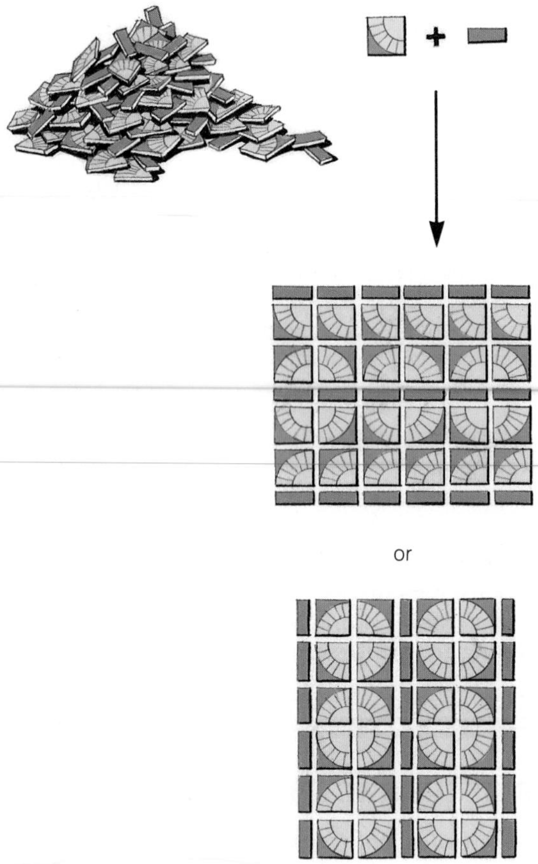

**Figure 1.2** Emergence of organized patterns from disorganized beginnings. Two ceramic tile patterns are shown here. You probably can visualize other possible patterns using the same two kinds of tiles. Similarly, the organization characteristic of life emerges from pools of simple building blocks, given energy sources and specific DNA "blueprints."

## Metabolism

You never, ever will find a rock engaged in metabolic activities. Only living cells can do this. **Metabolism** refers to the cell's capacity to (1) *extract and transform energy* from its surroundings and (2) *use energy* and so maintain itself, grow, and reproduce. In essence, metabolism means "energy transfers" within the cell.

A growing rice plant nicely illustrates this aspect of life. Like other plants, it has cells that engage in *photosynthesis*. The cells convert sunlight energy to chemical energy, which is then parceled out to the tasks of building sugars, starch, and other good things from simple raw materials in the environment. (Chemical energy is remarkable stuff. Cells use it to build large molecules out of smaller bits. They also use it to split molecules apart and liberate various bits.) In photosynthesis, energy from sunlight drives the attachment of a bit of phosphate to a certain molecule, which thereby becomes known as ATP. ATP is a generous molecule. It readily transfers chemical energy to other molecules that function as metabolic workers (enzymes), building blocks, or energy reserves.

In rice plants, energy reserves are especially concentrated in starchy seeds—rice grains—from which more rice plants may grow. The energy reserves in countless trillions of rice grains also provide energy for billions of rice-eating humans around the world. How? In rice plants, humans, and most other organisms, stored chemical energy can be tapped for use by way of another metabolic process, called *aerobic respiration*. Later chapters will describe the splendid metabolic jugglings of photosynthesis and aerobic respiration. For now, the point to keep in mind is this:

**Living things show metabolic activity: Their cells acquire and use energy to stockpile, tear down, build, and eliminate materials in ways that promote survival and reproduction.**

## Interdependency Among Organisms

With few exceptions, a flow of energy from the sun maintains the great pattern of organization in nature. Plants and some other photosynthetic organisms are the entry point for this flow—the food "producers" for the world of life. Animals are "consumers." Directly or indirectly, they feed on the energy stored in plant parts. For example, zebras tap directly into the stored energy when they nibble on grass, and lions tap into it indirectly when they nibble on zebras. Certain bacteria and fungi are "decomposers." When they feed on the tissues or remains of other organisms, they break down complex molecules to simple raw materials—which can be recycled back to the producers.

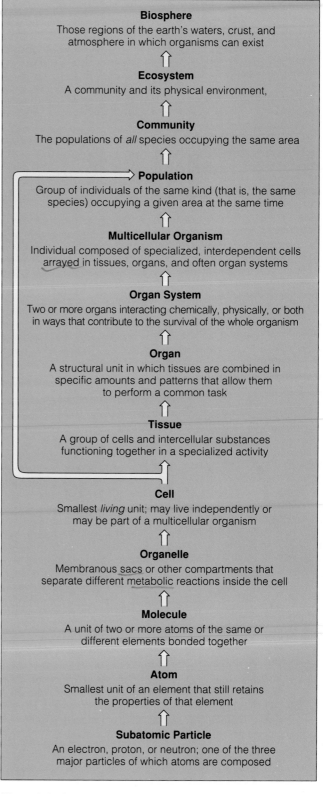

**Biosphere**
Those regions of the earth's waters, crust, and atmosphere in which organisms can exist

⇑

**Ecosystem**
A community and its physical environment,

⇑

**Community**
The populations of *all* species occupying the same area

⇑

**Population**
Group of individuals of the same kind (that is, the same species) occupying a given area at the same time

⇑

**Multicellular Organism**
Individual composed of specialized, interdependent cells arrayed in tissues, organs, and often organ systems

⇑

**Organ System**
Two or more organs interacting chemically, physically, or both in ways that contribute to the survival of the whole organism

⇑

**Organ**
A structural unit in which tissues are combined in specific amounts and patterns that allow them to perform a common task

⇑

**Tissue**
A group of cells and intercellular substances functioning together in a specialized activity

⇑

**Cell**
Smallest *living* unit; may live independently or may be part of a multicellular organism

⇑

**Organelle**
Membranous sacs or other compartments that separate different metabolic reactions inside the cell

⇑

**Molecule**
A unit of two or more atoms of the same or different elements bonded together

⇑

**Atom**
Smallest unit of an element that still retains the properties of that element

⇑

**Subatomic Particle**
An electron, proton, or neutron; one of the three major particles of which atoms are composed

**Figure 1.3** Simplified picture of the levels of organization in nature, starting with the subatomic particles that serve as the fundamental building blocks of all organisms.

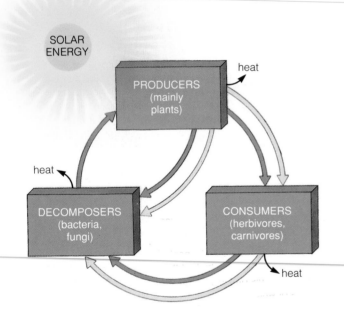

Figure 1.4 is a generalized picture of energy flow and the cycling of materials through the world of life. Figure 1.5 is a specific example of the resulting interdependencies among organisms.

Energy flows to, within, and from single cells and multicelled organisms. It flows within and between populations, communities, and ecosystems. As you will see, interactions among organisms are part of the cycling of carbon and other substances on a global scale. They also have profound influence on the earth's energy "budget." Understand the extent of those interactions and you will gain insight into the greenhouse effect, acid rain, and many other modern-day problems.

**All organisms are part of webs of organization in nature, in that they depend directly or indirectly on one another for energy and raw materials.**

**Figure 1.4** Energy flow and the cycling of materials in the biosphere. Here, grasses of the African savanna are producers that provide energy directly for zebras (herbivores) and indirectly for lions and vultures (carnivores). The wastes and remains of all these organisms are energy sources for decomposers, which cycle nutrients back to the producers.

## Homeostasis

It is often said that only living organisms "respond" to the environment. Yet a rock also responds to the environment, as when it yields to gravity and tumbles downhill or when it changes shape slowly under the battering of wind, rain, or tides. The real difference is this: *Organisms have the cellular means to sense environmental changes and make controlled responses to them.* They do so with the help of diverse **receptors**, which are molecules and structures that can detect specific information about the environment. When cells receive information from receptors, their activities become adjusted in ways that bring about an appropriate response.

a

b

c

d

Your body, for example, can withstand only so much heat or cold. It must rid itself of harmful substances. Certain foods must be available to it, in certain amounts. Yet temperatures shift, harmful substances may be encountered, and food is sometimes plentiful and sometimes scarce.

Even so, your body usually can adjust to the variations and so maintain internal operating conditions for its cells. **Homeostasis** refers to a state in which conditions in this "internal environment" are being maintained within a tolerable range. Homeostasis, too, is a common attribute of living things.

Think about what happens after you eat and simple sugar molecules make their way into your bloodstream. Certain cells detect the rising level of sugar in your blood and cause molecules of insulin, a hormone, to be released. Most of your body's cells have receptors for insulin, which prods the cells into taking up sugar molecules. With this uptake, the blood sugar level returns to normal. Now suppose you can't eat when you should and your blood sugar level falls. Then, a different hormone prods cells in your liver and elsewhere to dig into

**Figure 1.5** How organisms interact through their requirements for energy and raw materials. This example makes the point, even though the cast of characters seems of a most improbable sort.

First we have the adult male elephant of the African savanna (**a**). It stands almost two stories high at the shoulder and weighs more than eight tons. This grazing animal eats large quantities of plants, the remains of which leave its body as droppings of considerable size (**b**). Appearances to the contrary, locked in the droppings are substantial stores of unused nutrients. With resource availability being what it is, even waste products from one kind of organism are food for another.

And so we next have little dung beetles rushing to the scene almost simultaneously with the uplifting of the elephant's tail. With great precision they carve out fragments of the dung into round balls (**c**). The dung balls are rolled off and buried underground in burrows, where they serve as compact food supplies. In these balls the beetles lay eggs, a reproductive behavior that assures the forthcoming offspring of a food supply (**d**). Also assured is an uncluttered environment. If the dung were to remain aboveground, it would dry out and pile up beneath the hot African sun. Instead, the surface of the land is tidied up, the beetle has its resource, and the remains of the dung are left to decay in burrows—there to enrich the soil that nourishes the plants that sustain (among others) the elephants.

**Figure 1.6** "The insect"—a continuum of developmental stages, with new adaptive properties emerging at each stage. Shown here: the development of a silkworm moth, from egg (**a**) to larval stage (**b**), to pupal form (**c**), to emergence of the splendid adult form (**d, e**).

a

b

their storehouses of energy-rich molecules. Those molecules are broken down into simple sugars, which are released into the bloodstream—and again the blood sugar level returns to normal.

> **All organisms respond to changing conditions through use of homeostatic controls, which help maintain their internal operating conditions.**

### Reproduction

We humans tend to think we enter the world rather abruptly and are destined to leave it the same way. Yet we and all other organisms are more than this. *We are part of an immense, ongoing journey that began billions of years ago.* Think about the first cell of a new human individual, which is produced when a sperm joins with an egg. The cell would not even exist if the sperm and egg had not been formed earlier, according to DNA instructions that were passed down through countless generations. With time-tested DNA instructions, a new human body develops in ways that will prepare it, ultimately, for *reproduction*. With reproduction, the journey of life continues.

If someone asked you to think of a moth, would you simply picture a winged insect? What of the tiny fertilized egg deposited on a branch by a female moth (Figure 1.6)? The egg contains all the instructions necessary to become an adult. By those instructions, the egg first develops into a caterpillar, a larval form adapted for rapid feeding and growth. The caterpillar eats and increases in size until an internal "alarm clock" goes off. Then its body enters a so-called pupal stage of development, which requires wholesale remodeling. Some cells

die, while other cells multiply and become organized in different patterns. Now the adult moth emerges. It is equipped with organs in which eggs or sperm develop. Its wings are brightly colored and flutter at a frequency that can attract a potential mate. In short, the adult stage is adapted for reproduction.

None of these stages is "the insect." "The insect" is a series of organized stages from one fertilized egg to the next, each vital for the ultimate production of new moths. The instructions for each stage were written into moth DNA long before each moment of reproduction—and so the ancient moth story continues.

> **Each organism arises through *reproduction* (the production of offspring by one or more parents).**
>
> **Each organism is part of a reproductive continuum that extends back through countless generations.**

### Mutation and Adapting to Change

The word *inheritance* refers to the transmission, from parents to offspring, of structural and functional patterns characteristic of each kind of organism. In living cells, hereditary instructions are encoded in molecules of DNA. Those instructions have two striking qualities. They assure that offspring will resemble their parents—and they also permit *variations* in the details of their traits. By "trait" we mean some aspect of an organism's body, functioning, or behavior. For example, having five fingers on each hand is a human trait. Yet some humans are born with six fingers on each hand instead of five! Variations in traits arise through **mutations**, which are changes in the structure or number of DNA molecules.

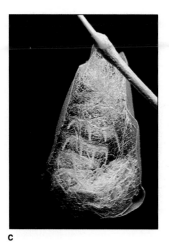

c

d

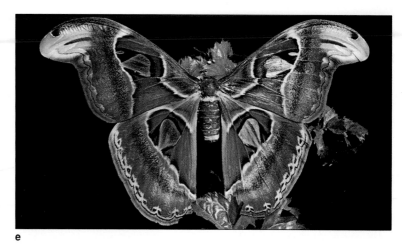

e

Many mutations are harmful, for the separate bits of information in DNA are part of a coordinated whole. A single mutation in a tiny segment of human DNA may lead to a genetic disorder such as hemophilia, in which blood cannot clot properly in response to a cut or bruise. Yet on rare occasions, a mutation may prove to be harmless, even beneficial, under prevailing conditions. One type of mutation in light-colored moths leads to dark-colored offspring. What happens when a dark moth rests on a soot-covered tree? Bird predators simply do not see it. If most trees are soot-covered (as in industrial regions), light moths are more likely to be seen and eaten—so the dark form has a better chance of living long enough to reproduce. Under such conditions, the mutated form of the trait is more adaptive.

An **adaptive trait** simply is one that helps an organism survive and reproduce under a given set of environmental conditions.

---

In all organisms, DNA is the molecule of inheritance: Its instructions for reproducing traits are passed on from parents to offspring.

Mutations introduce variations in heritable traits.

Although most mutations are harmful, some give rise to variations in form, function, or behavior that turn out to be adaptive under prevailing conditions.

---

## LIFE'S DIVERSITY

### Five Kingdoms, Millions of Species

Until now, we have focused on the unity of life—on characteristics shared by all organisms. Superimposed on this shared heritage is immense diversity. Many mil-lions of different kinds of organisms, or **species**, inhabit the earth. And many millions more existed in the past and became extinct. Early attempts to make sense of life's diversity led to a classification scheme in which each species was assigned a two-part name. The first part designates the **genus** (plural, genera). It encompasses all the species having perceived similarities to one another. The second part designates a particular species within that genus.

For instance, *Quercus alba* is the scientific name of the white oak. *Quercus rubra* is the name of the red oak. (Once the genus name has been spelled out, subsequent uses of it in the same document can be abbreviated—for example, to *Q. rubra*.)

Life's diversity is classified further by using more inclusive groupings. For example, similar genera are placed in the same *family*, similar families into the same *order*, then similar orders into the same *class*. Similar classes are placed into a *division* or *phylum* (plural, phyla). In turn, phyla are assigned to a *kingdom*. Today, most biologists recognize the following five kingdoms:

| | |
|---|---|
| **Monera** | *Bacteria. Single cells of relatively little internal complexity. Producers or decomposers.* |
| **Protista** | *Protistans. Single cells of considerable internal complexity. Producers or consumers.* |
| **Fungi** | *Fungi. Mostly multicelled. Decomposers.* |
| **Plantae** | *Plants. Mostly multicelled. Mostly producers.* |
| **Animalia** | *Animals. Multicelled. Consumers.* |

**Figure 1.7** Representatives of the five kingdoms of life.

**Kingdom Monera. (a)** A bacterium, seen with the aid of a microscope. The single-celled bacteria making up this kingdom live nearly everywhere, including in or on other organisms. The ones in your gut and on your skin outnumber the cells making up your body.

**Kingdom Protista. (b)** A parasitic trichonomad, from a termite gut. This kingdom of single-celled organisms has poorly defined boundaries; many lineages seem to have evolutionary connections with plants, fungi, and animals.

**Kingdom Fungi. (c)** A stinkhorn fungus. Many fungi are major decomposers. Even a single elm tree can shed 400 pounds of leaves in one season. Without decomposers, communities would gradually be buried in their own garbage.

**Kingdom Plantae. (d)** A grove of California coast redwoods. Like nearly all members of the plant kingdom, redwoods produce their own food through photosynthesis. **(e)** Flower of a plant called a composite. Its intricate pattern guides bees to the flower's nectar. The bees get food and the plant gets pollinated. Many organisms are locked in such mutually beneficial interactions.

**Kingdom Animalia. (f)** Male bighorn sheep competing for females. Like all animals, they cannot produce their own food and so must eat other organisms. Like most animals, they move about far more than the adult organisms of other kingdoms (most of which do not move about at all).

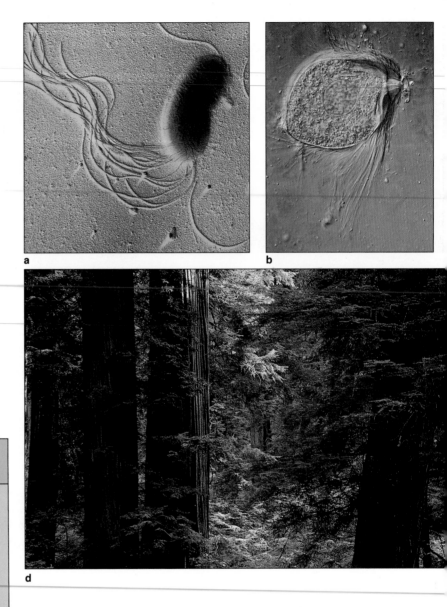

a             b

d

| Table 1.1 Characteristics of Organisms in All Five Kingdoms |
| --- |
| 1. Complex structural organization based on instructions contained in DNA molecules. |
| 2. Directly or indirectly, dependence on other organisms for energy and material resources. |
| 3. Metabolic activity by the single cell or multiple cells composing the body. |
| 4. Use of homeostatic controls that maintain favorable operating conditions in the body despite changing conditions in the environment. |
| 5. Reproductive capacity, by which the instructions for heritable traits are passed from parents to offspring. |
| 6. Diversity in body form, in the functions of various body parts, and in behavior. Such traits are adaptations to changing conditions in the environment. |
| 7. The capacity to evolve, based ultimately on variations in traits that arise through mutations in DNA. |

Figure 1.7 shows a few representatives of the five kingdoms. When looking at this figure, look also at Table 1.1, which summarizes the main characteristics of life that have been described so far in this chapter. *Every living organism in all five kingdoms displays these characteristics.*

**An Evolutionary View of Diversity**

If organisms are so much alike in so many ways, what could possibly account for their diversity? In biology, a key explanation is called evolution by means of natural selection.

c

e

f

By way of example, suppose a DNA mutation gives rise to a different form of a trait in one member of a population—say, black moth wings instead of white. Suppose black wings prove to be adaptive in concealing the moth from predators. Because the black-winged moth has an advantage over a white moth right next to it on a soot-covered tree trunk, it lives to reproduce. So do its black-winged offspring—and so do *their* offspring.

The variant form of this trait is now popping up with greater frequency. In time it may even become the more common form, so that what was once a population of mostly white-winged moths now consists mostly of dark-winged moths. **Evolution** is taking place—the character of the population is changing through successive generations.

Long ago, Charles Darwin used pigeons to explain how evolution might occur. Domesticated pigeons show great variation in their traits. Darwin pointed out that pigeon breeders who wish to promote certain traits, such as black tail feathers with curly edges, will "select" individual pigeons having the most black and the most curl in their tail feathers. By permitting only those birds to mate, they will foster the desired traits and eliminate others from their captive population.

Thus Darwin used *artificial* selection as a model for natural selection.

**Figure 1.8** A few examples of the more than 300 varieties of domesticated pigeons. Such forms have been derived, by selective breeding, from the wild rock dove (**a**).

Figure 1.8 shows a few of the variations in size, feather color, and other traits of pigeons that resulted from *artificial* selection. In later chapters, we will look at the mechanisms and consequences of *natural* selection. For now, these are the points to remember:

**1.** Members of a population vary in form, function, and behavior, and much of this variation is heritable.

**2.** Some forms of heritable traits are more adaptive than others; they improve chances of surviving and reproducing. Thus individuals with adaptive traits tend to make up more of the reproductive base in each new generation.

**3.** **Natural selection** is simply a measure of the difference in survival and reproduction that has occurred among individuals that differ from one another in one or more traits.

**4.** Any population *evolves* when some forms of traits increase in frequency and others decrease or disappear over the generations. In this manner, variations have accumulated in different lines of organisms. Life's diversity is the sum total of those variations.

## THE NATURE OF BIOLOGICAL INQUIRY

### On Scientific Methods

Species evolve in myriad directions, like branches growing in many directions on a single tree of life. Today we say this with confidence, but it was not always so. Awareness of evolution developed over centuries, as naturalists and travelers collected specimens of living and extinct organisms, then asked questions about the similarities and differences among them. Darwin and others proposed only tentative answers to those questions. Evidence supporting some of their answers came much later, after generations of scientists devised a staggering number of ingenious ways to test them.

If you have not had much exposure to the way scientists think and the methods they use to track down answers, you might believe that "doing science" is a mysterious ritual. There they are, exquisitely trained persons pondering terribly complex problems and doing experiments late into the night. Becoming a practicing scientist generally does require special training—but thinking scientifically does not.

For example, any reasonably alert person might wonder about some of the organisms illustrated in this chapter. Why does the silkworm moth (*Hyalophora cecropia*) have such distinct, boldly patterned wings? Why don't all trees grow as tall as redwoods? Why do zebras have striped coats? Why don't dung beetles simply burrow under a dung pat instead of rolling a ball of the stuff great distances before putting it underground?

There is no such thing as a single "scientific method" of investigating such questions. However, the following list is a good starting point for understanding how a scientist might proceed with such an investigation:

**1.** Identify a problem or ask a question about some aspect of the natural world.

**2.** Develop one or more **hypotheses**, or educated guesses, about what the solution or answer might be. This might involve sorting through what has been learned already about related phenomena.

**3.** Think about what predictably will occur or will be observed if the hypothesis is correct. This is sometimes called the "if-then" process. (*If* gravity pulls objects toward the earth, *then* it should be possible to observe apples falling down, not up, from a tree.)

**4.** Devise ways to *test* the accuracy of predictions drawn from the hypothesis. This typically involves making observations, developing models, and performing experiments.

**5.** If the tests do not turn out as expected, check to see what might have gone wrong. (Maybe a substance being tested was tainted, maybe a dial was set incorrectly or a relevant factor overlooked. Or maybe the hypothesis just isn't a good one.)

**6.** Repeat or devise new tests—the more the better. Hypotheses that have been supported by many different tests are more likely to be correct.

**7.** Objectively report the results from tests and the conclusions drawn from them.

Using this list as a guide, let's return to the question posed earlier about the silkworm moth's wing pattern. Is it a mating "flag"? If so, it would help the male and female moths identify each other, rather than dallying with members of the wrong species and producing defective offspring. Or does the pattern camouflage the moths from bird predators? Maybe it blends with plants that the moths rest on during the day. (Moths fly at night, not during daylight hours.)

These actually are two plausible explanations, based on what scientists have already learned about mate discrimination and camouflaging among insects in general. Returning to still another question, do dung beetles roll away their prize ball so other dung beetles won't get it? Or do they bury the ball in distant concealed places where predators are less likely to find the beetle larvae that will grow inside it?

In nearly all cases, questions about what causes a natural phenomenon have more than one possible answer. In science, *alternative hypotheses are the rule, not the exception*.

## Testing Alternative Hypotheses

Identifying which of two or more hypotheses may be correct depends on tests. The trick here is to use each hypothesis as a guide for producing testable predictions. *If* the moth wing pattern functions in mate discrimination (the hypothesis), *then* it follows logically that mating should occur only at times of day when moths can see each other's wing patterns (the prediction).

Scientists do not use "prediction" as fortunetellers do, to "look into the future." They use it as a statement of what you should be able to observe in nature, if you were to go looking for it. Start from the observation that moths mate at night. If their wing pattern helps potential mates identify each other, then there won't be any moths mating on moonless nights. They won't be able to see the patterns.

Suppose you test this prediction by stealthily watching moths on a moonless night. To your surprise, you see moths mating just as often in total darkness as in moonlight. Here is evidence of a mistaken prediction—and, by extension, a mistaken hypothesis.

Whether test results support or undermine your hypothesis, you should repeat the test several times. Doing so will provide insight into the reliability of your findings. It's safe to say that most respectable conclusions in science rest on numerous studies, carried out by people who tried to test many alternative hypotheses.

**By testing a prediction, you test the underlying hypothesis.**

## The Role of Experiments

The preceding example shows how you might test a prediction by simple observation. You also might be able to test it by **experiments**. These are tests in which nature is manipulated as a way to reveal its secrets. Generally, scientists try to design experiments in such a way that the results will clearly show that a hypothesis is mistaken. Why? It is often easy to *disprove* a hypothesis but almost impossible to *prove* one. (An infinite number of experiments would have to be performed to show that it holds true under all possible conditions.)

Suppose you come up with another hypothesis about the wing pattern of *H. cecropia*. You propose that individuals with *altered* color patterns should have difficulty securing a mate. To test this, you sit out night after night, waiting for a peculiarly patterned moth to fly by. None does, so you decide to do an experiment. You *paint* the wings of a group of moths.

In a fair test of your prediction, you do more than put the group of painted moths in a cage with unaltered ones to observe the outcome. You also have a **control group**, which is used to evaluate possible side effects of the manipulation of the experimental group. Ideally, members of a control group should be *identical* to those of an experimental group in every respect—*except* for the key factor, or **variable**, that is under study. The variable here is wing color pattern. You also make sure that the number of individuals in both groups is large enough to give you more confidence that the experimental outcome will not be due to chance alone. Two or three moths will tell you nothing, as you will discover later in the book, in a section on probability (pages 169 and 170).

How can you be sure there are no other variables between the two groups that might influence the outcome of an experiment? After all, maybe paint fumes are as repulsive to a potential mate as the painted pattern. Maybe you rough up the moths when you handle them and somehow make them less desirable than those in the control group. Maybe the paint weighs enough to change the flutter frequency of the wings.

And so you decide that the control group also must be painted, using the same kind of paint, the same kind of brushes, and the same amount of handling. But for this group, you *duplicate* the natural wing color pattern as you paint. Thus your experimental group and control group are identical except for the variable under study. If only those moths with altered wing patterns turn out to be unlucky in love, then your control group will help substantiate your hypothesis.

**Experiments are tests in which nature is manipulated.** They require careful design of a set of controls to evaluate possible side effects of the manipulation.

### About the Word "Theory"

More than a century ago, Darwin unveiled his ideas about the evolution of species and ushered in one of the most dramatic of all scientific revolutions. The core of his thinking—that life's diversity is the outcome of evolution by natural selection—became popularly known as "the theory of evolution." We will be looking closely at Darwin's scientific work in a later chapter, but here let's focus on the word *theory.*

In science, a **theory** is a related set of hypotheses which, taken together, form a broad-ranging explanation about some fundamental aspect of the natural world. A scientific theory differs from a scientific hypothesis in its *breadth of application.* Darwin's theory fits this description—it is a big, encompassing "Aha!" explanation that, in a few intellectual strokes, makes sense of a huge number of observable phenomena. Think about it. His theory explains what has caused most of the diversity among many millions of different living things!

There are many other major theories in biology. One explains what causes all offspring to resemble their parents, no matter what the species. Another explains what caused the mass extinction of the dinosaurs about 65 million years ago. We will be looking at these and other theories throughout the book. None is bigger in scope than Darwin's, but all do a good job of attempting to make sense of the natural world.

Like hypotheses, theories are accepted or rejected on the basis of tests. For example, several competing theories about evolution were pushed vigorously in Darwin's time, but since then they have been tested and essentially rejected. After thousands of different tests, Darwin's theory still stands, with only some modification. Today, most biologists accept the modified theory as correct—but they still keep their eyes open for new evidence that might call it into question.

Scientists admire tested theories for good reason. A tested theory serves as a general frame of reference for additional research, and research is what scientists do. If Darwin's theory is correct, then the diverse attributes of living things exist because, at least in the past, they helped individuals leave descendants. Therefore, when biologists look at a moth's wings, an immediate question comes to mind—"I wonder how that wing color helps that moth leave descendants." When using Darwin's theory as a guide to developing plausible hypotheses, they will probably focus on possible answers that relate directly or indirectly to reproductive success—as we saw in our earlier example.

A scientific theory is an explanation about the cause or causes of a broad range of related phenomena. Like hypotheses, theories are open to tests, revision, and tentative acceptance or rejection.

### Uncertainty in Science

Isn't anything ever "for sure" in science? Are there no comfortable, final conclusions? In an ultimate sense, no. Scientists must be content with *relative* certainty about whether an idea is correct or not. When a theory or hypothesis withstands exhaustive testing by many independent researchers, that "relative certainty" can be very great. Even so, there is always a chance that one of the tests has hidden flaws, which would invalidate the results. That is why scientific papers include a section on the methods employed for the tests they describe. This allows other scientists to check the procedures used, even to the point of duplicating the research.

Knowing that others will scrutinize your ideas and the methods used to test them has a wonderful effect. It forces you to try to remain objective—even if you believe fiercely in what you propose.

In short, individual scientists must keep asking themselves: "Are there tests or observations that will show my ideas to be incorrect?" They are expected to put aside pride or bias by testing their ideas, even in ways that might prove them wrong. Even if an individual scientist doesn't (or won't) do this, *others will*—for science proceeds as a community that is both cooperative and competitive. Ideas are shared, with the understanding that it is just as important to expose errors as it is to applaud insights.

**The fact that scientists can and do change their mind when presented with new evidence is a *strength* of their profession, not a weakness.**

### The Limits of Science

The call for objectivity strengthens the theories that do emerge from scientific studies. Yet it also puts limits on the kinds of studies that can be carried out. Beyond the realm of scientific analysis, certain events remain unexplained. Why do we exist, for what purpose? Why does any one of us have to die at a particular moment and not another?

Answers to such questions are *subjective*. This means they come from within, as an outcome of all the experiences and mental connections that shape our consciousness. Because individuals differ so enormously in this regard, subjective answers do not readily lend themselves to scientific analysis.

This is not to say that subjective answers are without value. No human society can function without a shared commitment to standards for making judgments, however subjective those judgments might be. Moral, aesthetic, economic, and philosophical standards vary from one society to the next. But all guide their members in deciding what is important and good, and what is not. All attempt to give meaning to what we do.

Every so often, scientists stir up controversy when they explain part of the world that was previously considered beyond natural explanation—that is, belonging to the "supernatural." This is sometimes true when moral codes are interwoven with religious narratives, which grew out of observations by ancestors. Exploring some longstanding view of the world from a scientific perspective may be misinterpreted as questioning morality, even though the two are not remotely synonymous.

For example, centuries ago Nicolaus Copernicus studied the movements of planets and stated that the earth circles the sun. Today the statement seems obvious. Back then, it was heresy. The prevailing belief was that the Creator had made the earth (and, by extension, humankind) the immovable center of the universe! Not long afterward a respected professor, Galileo Galilei, studied the Copernican model of the solar system. He thought it was a good one and said so. He was forced to retract his statement publicly, on his knees, and to put the earth back as the fixed center of things. (Word has it that when he stood up he muttered, "But it moves nevertheless.")

Today, as then, society has its sets of standards. Today, as then, those standards may be called into question when a new, natural explanation runs counter to supernatural belief. When this happens it doesn't mean that scientists as a group are less moral, less lawful, less sensitive, or less caring than any other group. Their work, however, is guided by one additional standard: *The external world, not internal conviction, must be the testing ground for scientific beliefs.*

**Systematic observations, hypotheses, predictions, tests—in all these ways, science differs from systems of belief that are based on faith, force, authority, or simple consensus.**

## SUMMARY

1. All organisms are alike in the following characteristics:
    a. Their structure, organization, and interactions arise from the basic properties of matter and energy.
    b. They rely on metabolic and homeostatic processes.
    c. They have the capacity for growth, development, and reproduction.
    d. Their heritable instructions are encoded in DNA.

2. There are many millions of different kinds of organisms. Each distinct kind of organism is called a species. Distinct species resembling one another more than they resemble other species are grouped into the same genus, and so on with increasingly inclusive groupings into family, order, class, phylum (or division), and kingdom.

3. Diversity among organisms arises through mutations that introduce changes in the DNA. These changes lead to heritable variation in the form, functioning, or behavior of individual offspring.

4. Individuals in a population vary in their heritable traits, and the variations influence their ability to survive and reproduce. Under prevailing conditions, certain varieties of a given trait may be more adaptive than others. They will be "selected" and others eliminated

through successive generations. The changing frequencies of different traits change the character of the population over time; it evolves. These points are central to the principle of evolution by natural selection.

5. There are many scientific methods of gathering information. These are key terms associated with scientific inquiry:

a. Theory: An explanation of a broad range of related phenomena. An example is the theory of how processes of natural selection bring about evolution.

b. Hypothesis: A possible explanation of a specific phenomenon. Sometimes called an educated guess.

c. Prediction: A claim about what an observer can expect to see in nature *if* a particular theory or hypothesis is correct.

d. Test: The attempt to secure actual observations in order to determine whether they match the predicted or expected observations.

e. Conclusion: A statement about whether and to what extent a particular theory or hypothesis can be accepted or rejected, based on the tests of predictions derived from it.

6. The external world, not internal conviction, is the testing ground for scientific theories.

## Review Questions

1. Why is it difficult to give a simple definition of life? (For this and subsequent chapters, *italic numbers* following review questions indicate the pages on which the answers may be found.) *2*

2. What does *adaptive* mean? Give some examples of environmental conditions to which plants and animals must be adapted. *9*

3. Study Figure 1.3. Then, on your own, arrange and define the levels of biological organization. What concept ties this organization to the history of life, from the time of origin to the present? *5*

4. In what fundamental ways are all organisms alike? *9–10*

5. What is metabolic activity? *4–5*

6. What are the "instructions" contained in DNA? What is a mutation? Why are most mutations likely to be harmful? *3–4*

7. Outline the one-way flow of energy and the cycling of materials through the biosphere. *6*

8. What does evolution mean? *11*

9. Witnesses in a court of law are asked to "swear to tell the truth, the whole truth, and nothing but the truth." What are some of the problems inherent in the question? Can you think of a better alternative?

10. Design a test to support or refute the following hypothesis: The body fat in rabbits appears yellow in certain mutant individuals—but only when those mutants also eat leafy plants containing a yellow pigment molecule called xanthophyll.

## Self-Quiz *(Answers in Appendix IV)*

1. The complex patterns of structural organization characteristic of life are based on instructions contained in _____ .

2. Directly or indirectly, all living organisms depend on one another for _____ .

3. _____ is the ability of organisms to extract and transform energy from the environment and use it during maintenance, growth, and reproduction.

4. _____ means maintaining the body's internal operating conditions within a tolerable range even when environmental conditions change.

5. Diverse structural, functional, and behavioral traits are considered to be _____ to changing conditions in the environment.

6. The capacity to evolve is based on variations in traits, which orginally arise through _____ .

7. Organisms show _____ , the ability to transmit instructions for heritable traits from parents to offspring.

8. That each of us has great-great-great-great-grandmothers and grandfathers is an example of a unique property of life known as _____ .

    a. metabolism       c. reproduction
    b. homeostasis     d. organization

9. A scientific approach to explaining various aspects of the natural world includes all of the following except _____ .

    a. hypothesis     c. faith and simple consensus
    b. testing        d. systematic observations

10. A related set of hypotheses that collectively explain some aspect of the natural world is a scientific _____ .

    a. prediction     d. authority
    b. test          e. observation
    c. theory

## Selected Key Terms

| | | |
|---|---|---|
| adaptive trait *9* | experiment *13* | photosynthesis *5* |
| aerobic respiration *5* | Fungi *9* | Plantae *9* |
| Animalia *9* | genus *9* | population *4* |
| biosphere *4* | homeostasis *7* | prediction *13* |
| cell *4* | hypothesis *13* | Protista *9* |
| community *4* | inheritance *8* | reproduction *8* |
| control group *14* | metabolism *4* | species *9* |
| DNA *3* | Monera *9* | test *13* |
| ecosystem *4* | multicelled | theory *14* |
| energy *3* | organism *4* | variable *14* |
| evolution *11* | mutation *8* | |

FACING PAGE: *Living cells of a green alga* (Elodea), *with their chemical factories called chloroplasts.*

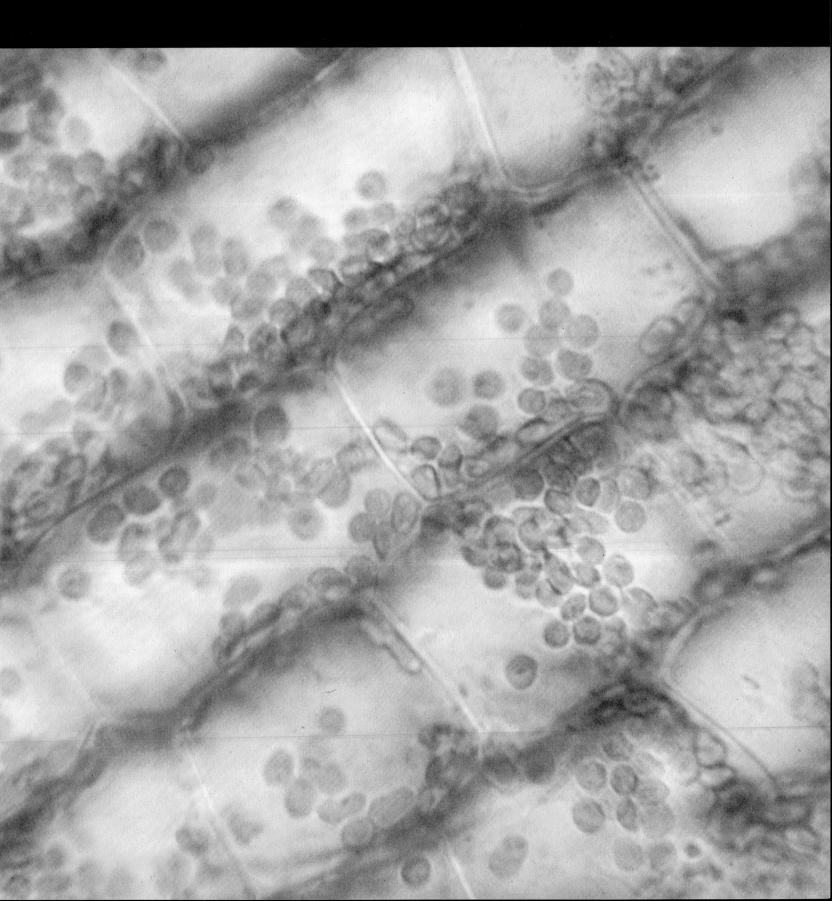

## The Chemistry In and Around You

As you read this page, thousands of chemical reactions are proceeding inside you in ways that keep your body running smoothly. Whether it is daytime or the middle of the night, streams of sunlight are reaching half of the earth's surface. And countless plants are converting energy contained in the sun's rays into forms that can be used for assembling the carbohydrates and other building blocks of roots, stems, and leaves. Your very life depends on breathing in the oxygen released into the atmosphere during chemical reactions such as these, the first of which occurred more than 2.5 billion years ago.

In the past two centuries—a mere blip of evolutionary time—we have managed to discover what chemical substances are made of, how they can be transformed into different substances, and what it takes to accomplish the transformations. Some of the products of these discoveries are synthetic fabrics, fertilizers, vaccines, antibiotics, and the plastic components of refrigerators, computers, television sets, jet planes, and cars.

Our chemical "magic" brings us great benefits *and* monumental problems. For example, fertilizer applications and other agricultural practices maintain food supplies for the 5.4 billion people on earth. Without them, much of the human population would starve to death. But weeds don't understand that the fertilizers are for crop plants, and plant-eating insects don't understand that the crops are for us, not them. Each year they gobble up or ruin about 45 percent of what we grow. In 1945 we began battling back with synthetic compounds that kill weeds as well as insects, worms, rodents, and other animals. In 1988 alone, Americans managed to spread more than a billion pounds of herbicides and pesticides through homes, gardens, offices, industries, and farmlands (Figure 2.1).

Among the insect killers we find the carbamates, organophosphates (including malathion), and halogenated compounds (including chlordane). Most are neurotoxins, meaning they block vital communication signals between nerve cells. Some remain active for days, others for weeks or years. Unfortunately, they kill many of nature's insect eaters, including dragonflies and birds, as well as the targeted pests. Over time, the pests build up resistance to the pesticides, for reasons that will become apparent in later chapters. And only 10 percent of the insecticides being sold today have been assessed for potential health hazards.

Consider that *we* inhale pesticides, ingest them with food, or absorb them through skin. After entering the body, many pesticides can cause headaches, rashes, asthma, and bronchitis in susceptible people and can increase their vulnerability to chronic infections. Susceptibility depends on genetic makeup, overall health, nutritional habits, and concurrent exposure to other

**Figure 2.1** Cropduster with its rain of pesticides.

toxic substances. Exposure to certain pesticides can trigger hives, pain in the joints, and other moderate allergic reactions in about 11 million Americans. They can trigger severe, life-threatening immune reactions in another 5 million.

Maintaining our food supplies, industries, and health depends on chemistry—and so do our chances of reducing harmful side effects of its application. You owe it to yourself and others to gain greater understanding of chemical substances and their interactions. By demystifying the "magic" of chemistry, you will be better equipped to assess the benefits and risks of its application to the world of life.

KEY CONCEPTS

**1.** Atoms give up, acquire, or share their electrons with other atoms in specific ways. These interactions are the basis for the structural organization and activities of all living things.

**2.** Chemical bonds are unions between the electron structures of different atoms. The most common bonds in biological molecules are ionic bonds, hydrogen bonds, and covalent bonds.

**3.** Life depends on the properties of water, including its temperature-stabilizing effects, cohesiveness, and capacity to dissolve many substances.

**4.** Even though cells continuously produce and use hydrogen ions ($H^+$) during chemical reactions, they have the means to maintain the $H^+$ concentration within narrow limits.

## ORGANIZATION OF MATTER

"Matter" is anything that occupies space and has mass. It includes the solids, liquids, and gases around you and within your body. All of these forms of matter are made of one or more **elements**, or fundamental substances that cannot be broken down to a different substance by ordinary chemical means. About ninety-two elements occur naturally on earth. It takes only four kinds—hydrogen, carbon, nitrogen, and oxygen—to make up most of the human body.

Some elements are vital for normal body functioning, even though they are present in what might seem to be insignificant amounts. Collectively, these so-called **trace elements** represent less than 0.01 percent of all the atoms in any organism. Copper is an example. Carefully dry out and analyze the tissues of a maple tree and you will find they are only about 0.006 percent copper. But if the tree has a copper deficiency, its leaf buds will die, yellow or dead spots will form on its leaves, and its overall growth will be stunted.

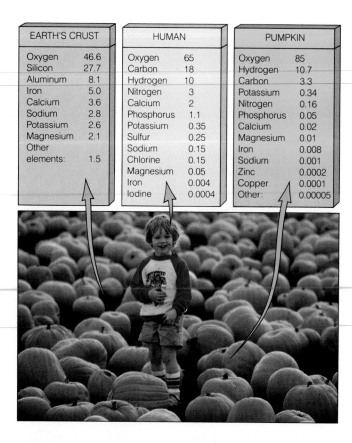

| EARTH'S CRUST | | HUMAN | | PUMPKIN | |
|---|---|---|---|---|---|
| Oxygen | 46.6 | Oxygen | 65 | Oxygen | 85 |
| Silicon | 27.7 | Carbon | 18 | Hydrogen | 10.7 |
| Aluminum | 8.1 | Hydrogen | 10 | Carbon | 3.3 |
| Iron | 5.0 | Nitrogen | 3 | Potassium | 0.34 |
| Calcium | 3.6 | Calcium | 2 | Nitrogen | 0.16 |
| Sodium | 2.8 | Phosphorus | 1.1 | Phosphorus | 0.05 |
| Potassium | 2.6 | Potassium | 0.35 | Calcium | 0.02 |
| Magnesium | 2.1 | Sulfur | 0.25 | Magnesium | 0.01 |
| Other | | Sodium | 0.15 | Iron | 0.008 |
| elements: | 1.5 | Chlorine | 0.15 | Sodium | 0.001 |
| | | Magnesium | 0.05 | Zinc | 0.0002 |
| | | Iron | 0.004 | Copper | 0.0001 |
| | | Iodine | 0.0004 | Other: | 0.00005 |

**Figure 2.2** Comparison of the proportions of different elements in the earth's crust, the human body, and a pumpkin as percentages of the total weight of each.

| Table 2.1 | Atomic Number and Mass Number of Elements Common in Living Things | | |
|---|---|---|---|
| Element | Symbol | Atomic Number | Most Common Mass Number |
| Hydrogen | H | 1 | 1 |
| Carbon | C | 6 | 12 |
| Nitrogen | N | 7 | 14 |
| Oxygen | O | 8 | 16 |
| Sodium | Na | 11 | 23 |
| Magnesium | Mg | 12 | 24 |
| Phosphorus | P | 15 | 31 |
| Sulfur | S | 16 | 32 |
| Chlorine | Cl | 17 | 35 |
| Potassium | K | 19 | 39 |
| Calcium | Ca | 20 | 40 |
| Iron | Fe | 26 | 56 |
| Iodine | I | 53 | 127 |

Take a look at Figure 2.2, which gives an example of how organisms differ in the proportions of their constituent elements, relative to nonliving material.

By international agreement, a one- or two-letter chemical symbol stands for each element, regardless of the element's name in different languages. What we call *nitrogen* is called *azoto* in Italian and *stickstoff* in German. But the symbol for this element is always N. Similarly, the symbol for the element sodium is always Na (from the Latin *natrium*). Table 2.1 lists the chemical symbols of elements that are common in living things.

In substances called **compounds**, two or more elements are combined in fixed proportions. Water is a compound; it is always 11.9 percent hydrogen and 88.1 percent oxygen, no matter where you find it.

### The Structure of Atoms

Look at some water in a glass, then imagine your eyes probing ever deeper into its underlying structure. First you would discover molecules composed of hydrogen and oxygen. By definition, a **molecule** is a unit of two or more atoms (of the same or different elements) bonded together. For any compound, it is the smallest unit that maintains the compound's elemental composition. In turn, each kind of **atom** is the smallest unit of matter that is unique to a particular element. Below the level of atoms, you always find the same types of particles. The particles are protons, neutrons, and electrons, the universal building blocks of atoms.

Protons and neutrons make up the atom's core region, or nucleus. Protons have a positive charge ($p^+$); neutrons are electrically neutral. Electrons, which have a negative charge ($e^-$), are attracted to the nucleus. They move rapidly around it and occupy most of the atom's volume. An atom has just as many protons as electrons, so it has no *net* charge, overall.

The number of protons in the nucleus, called the **atomic number**, differs for each element. The hydrogen atom has only one proton; its atomic number is 1. The carbon atom has six protons; its atomic number is 6. Table 2.1 lists other examples. The total number of protons *and* neutrons in the nucleus is the **mass number**. For example, the mass number of a carbon atom with six protons and six neutrons is 12. (The relative masses of atoms are also called atomic weights. This term is not precise—mass is not quite the same thing as weight—but its use continues.)

As you will see, knowing an atom's atomic number and mass number tells us something about how it will interact (if at all) with other atoms. Knowing those values gives us an idea of whether that atom can give up, acquire, or share its electrons with other atoms. *Such electron activity is the basis for the flow of materials and energy through the living world.*

# *Commentary*

## Dating Fossils, Tracking Chemicals, and Saving Lives—Some Uses of Radioisotopes

In the winter of 1896, the physicist Henri Becquerel tucked a heavily wrapped rock of uranium into a desk drawer, on top of an unexposed photographic plate. A few days later, he opened the drawer and discovered a faint image of the rock on the plate—apparently caused by energy emitted from the rock. One of his coworkers, Marie Curie, gave the name "radioactivity" to the phenomenon.

As we now know, radioisotopes are unstable atoms, with too many protons or neutrons. The instability causes them to capture or emit electrons or some other particle. This spontaneous process, called radioactive decay, continues until the original isotope has changed to a new, stable isotope, one that is not radioactive.

### Radioactive Dating

Each type of radioisotope has a characteristic number of protons and neutrons, and it decays spontaneously at a characteristic rate into a different isotope. The *half-life* is the time it takes for half the nuclei in any given amount of a radioactive element to decay into another isotope. The half-life cannot be modified by temperature, pressure, chemical reactions, or any other environmental factor. That is why radioactive dating is such a reliable method of determining the age of rock layers in the earth—hence the age of fossils they may contain. To determine a rock's age, we can compare the amount of one of its radioisotopes with the amount of the decay product for that isotope.

(a) Fossilized frond of a tree fern, one of many species that lived more than 250 million years ago. (b) Fossilized sycamore leaf that dropped 50 million years ago.

For example, $^{40}$potassium has a half-life of 1.3 billion years and decays to $^{40}$argon, a stable isotope. The age of anything that contains $^{40}$potassium can be determined by measuring the ratio of $^{40}$argon to $^{40}$potassium. In this way, researchers have dated fossils that are millions, even billions, of years old (Figures *a* and *b*). Radioactive dating

| Main Radioisotopes Used in Dating | | | |
|---|---|---|---|
| Radioisotope (unstable) | Stable Product | Half-Life (years) | Useful Range (years) |
| $^{87}$rubidium → $^{87}$strontium | | 49 billion | 100 million |
| $^{232}$thorium → $^{208}$lead | | 14 billion | 200 million |
| $^{238}$uranium → $^{206}$lead | | 4.5 billion | 100 million |
| $^{40}$potassium → $^{40}$argon | | 1.3 billion | 100 million |
| $^{235}$uranium → $^{207}$lead | | 704 million | 100,000 |
| $^{14}$carbon → $^{14}$nitrogen | | 5,730 | 0–60,000 |

a

b

### Isotopes

Although all atoms of an element have the same number of protons, they may vary slightly in how many neutrons they have. Atoms having the same atomic number but a different mass number are **isotopes**. Thus "a carbon atom" might be carbon 12 (containing six protons, six neutrons), carbon 13 (six protons, seven neutrons), or carbon 14 (six protons, eight neutrons). These can be written as $^{12}$C, $^{13}$C, and $^{14}$C. All isotopes of an element have the same number of electrons, so they interact with other atoms the same way. Accordingly, cells can use any carbon isotope for a metabolic reaction.

You have probably heard of radioactive isotopes, or **radioisotopes**. They are unstable isotopes that tend to break apart (decay) into more stable atoms. The *Commentary* describes some of the ways in which radioisotopes are used in research, in medicine, and in establishing the age of fossil-containing rocks.

All atoms of an element have the same number of electrons and protons, but they can vary slightly in the number of neutrons. The variant forms are isotopes.

with $^{238}$uranium, which has a half-life of 4.5 billion years, indicates that the earth formed 4.6 billion years ago. The list above Figure *a* shows the useful ranges of the main radioisotopes used in dating methods.

## Tracking Chemicals

Emissions from radioisotopes can be detected by a scintillation counter and other devices. This means that isotopes can be used as *tracers*. They can be used to identify the pathways or destination of a substance that has been introduced into a cell, the human body, an ecosystem, or some other "system."

For example, because all isotopes of an element have the same number of electrons, they all interact with other atoms in the same way. Accordingly, cells can use isotopes of carbon for a given metabolic reaction. Carbon happens to be a key building block for photosynthesis. By putting

plant cells in a medium enriched in $^{14}$carbon, researchers identified the steps by which plants take up carbon and incorporate it into newly forming carbohydrates. Tracers also are helping us increase crop production by providing insights into how plants use synthetic fertilizers and naturally occurring nutrients.

What about medical applications? As one example, the thyroid is the only gland in the body to take up iodine. A tiny amount of the radioisotope $^{123}$iodine can be injected into a patient's bloodstream, then the thyroid can be scanned with a scintillation counter. This is called a radioisotope scan. Figure *c* shows examples of what these scans may reveal.

## Saving Lives

In nuclear medicine, radioisotopes are used to diagnose and to treat diseases. Patients with irregular heartbeats use pacemakers, which are powered by energy emitted from $^{238}$plutonium. (This otherwise dangerous radioisotope is sealed in a case to prevent its emissions from damaging body tissues.) With PET (positron-emission tomography), radioisotopes provide diagnostic information about abnormalities in the metabolic functions of specific tissues. The radioisotopes are incorporated in glucose or some other biological molecule, then they are injected into a patient, who is moved into a PET scanner. Cells in certain tissues absorb the glucose. The radioisotopes give off energy that can be used to produce a vivid image of the variations in metabolic activity among different cells (Figure *d*).

Finally, some cancer treatments make use of the fact that radioisotopes can damage or destroy living cells. In radiation therapy, localized cancers are deliberately bombarded with $^{226}$radium or $^{60}$cobalt.

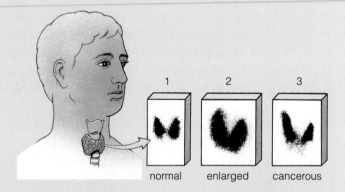

**c** Scans of human thyroid glands after $^{123}$iodine was injected into the bloodstream. The thyroid normally takes up iodine (including radioisotopes) and uses it in hormone production. (1) Uptake by a normal gland. (2) Enlarged gland of a patient with a thyroid disorder. (3) Cancerous thyroid gland.

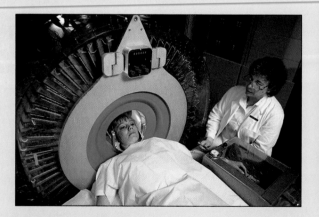
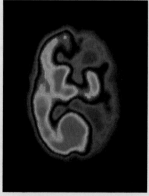

**d** Patient being moved into a PET scanner. The inset shows a vivid image of a brain scan of a child with a severe neurological disorder. The different colors signify differences in metabolic activity in one half of the brain; the other half shows no activity.

# BONDS BETWEEN ATOMS

Let's turn now to the nature of reactions among atoms. In case you are not familiar with such reactions, take a moment to review Figure 2.3, which summarizes a few conventions used in describing them.

## The Nature of Chemical Bonds

A **chemical bond** is a union between the electron structures of atoms. In other words, *it is an energy relationship.* Usually an atom gives up, gains, or shares one or more electrons with another atom. Some atoms do this rather easily, but others do not. Such differences in bonding behavior arise through differences in the number and arrangement of electrons in the atoms of each kind of element.

Picture three narcissistic actresses arriving at the Academy Awards wearing the same bright-red designer dress. Each has a compulsion to be in the limelight but dreads being photographed next to the others. Two might maneuver themselves *near* the center of attention while scooting away from each other to some extent; but by unspoken agreement, all three never, ever stay in the same place at the same time.

Electrons behave roughly the same way. They are attracted to the protons of a nucleus but repelled by other electrons that may be present. They spend as much time as possible near the nucleus and as far away from each other by moving in different *orbitals,* which are regions of space around the nucleus in which electrons are likely to be at any instant. Each orbital has enough room for two electrons, at most.

In all atoms, the orbital closest to a nucleus is ball-shaped (Figure 2.4). In a hydrogen atom, a single electron occupies that orbital. In a helium atom, two electrons occupy it. Electrons in the closest orbital to the nucleus are said to be at the *lowest energy level.*

Atoms larger than helium have two electrons in the first orbital. They also have other electrons that occupy different orbitals. On the average, those other electrons are farther away from the nucleus, and they are said to be at *higher energy levels.*

A simple although not quite accurate way to think about electron orbitals is to imagine them occupying the

**Figure 2.4** Very simplified model of atomic structure, using a hydrogen atom and a helium atom as examples. The nucleus, a core region, consists of some number of protons and (except for hydrogen) neutrons. Electrons occur in orbitals around the nucleus.

**Figure 2.3** Chemical bookkeeping.

Symbols for elements are used in writing *formulas*, which identify the composition of compounds. (For example, water has the formula $H_2O$. The subscript indicates two hydrogen atoms are present for every oxygen atom.) Symbols and formulas are used in *chemical equations*: representations of reactions among atoms and molecules.

In written chemical reactions, an arrow means "yields." Substances entering a reaction (*reactants*) are to the left of the arrow. Products of the reaction are to the right. For example, the overall process of photosynthesis is often written this way:

$$6CO_2 \ + \ 6H_2O \longrightarrow C_6H_{12}O_6 \ + \ 6O_2$$

| $6CO_2$ | $6H_2O$ | $C_6H_{12}O_6$ | $6O_2$ |
|---|---|---|---|
| 6 carbons | 12 hydrogens | 6 carbons | 12 oxygens |
| 12 oxygens | 6 oxygens | 12 hydrogens | |
| | | 6 oxygens | |

Notice there are as many atoms of each element to the right of the arrow as there are to the left, even though they are combined in different forms. Atoms taking part in chemical reactions may be rearranged but they are never destroyed. The *law of conservation of mass* states that the total mass of all materials entering a reaction equals the total mass of all the products.

When thinking about cellular reactions, keep in mind that no atoms are lost, so the equations you use to represent them must be balanced in this manner.

Both the reactants and products can be expressed in moles. A "mole" is a certain number of atoms or molecules of any substance, just as "a dozen" can refer to any twelve cats, roses, and so forth. Its weight (in grams) equals the total atomic weight of the atoms that compose the substance.

For example, the atomic weight of carbon is 12. Hence one mole of carbon weighs 12 grams. A mole of oxygen (atomic weight 16) weighs 16 grams. Can you show why a mole of water ($H_2O$) weighs 18 grams, and why a mole of glucose ($C_6H_{12}O_6$) weighs 180 grams?

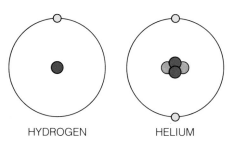

HYDROGEN          HELIUM

○ electron
● proton
◐ neutron

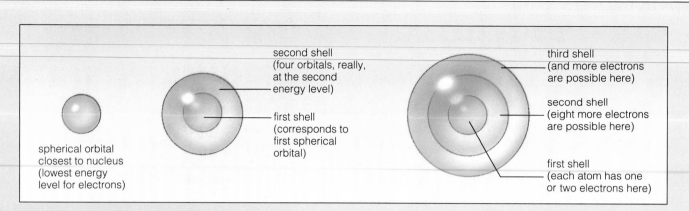

**a** The shell model of electron distribution in atoms.

**Figure 2.5** Arrangement of electrons in atoms. (**a**) In every atom, one or at most two electrons occupy a ball-shaped volume of space (an orbital) close to the nucleus. At this scale, the nucleus would be an invisible speck at the ball's center. This orbital is at the lowest energy level.

At the next (higher) energy level, there can be as many as eight more electrons—two in each of the four orbitals shown in (**b**). As you can see, the orbital shapes get tricky. For our purposes, we can ignore the shapes and simply think of the total number of orbitals at a given energy level as being somewhere inside a "shell" of the sort sketched in (**a**).

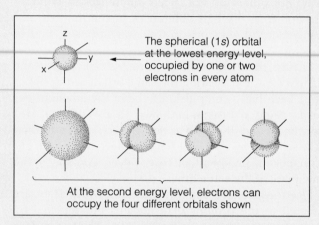

The spherical (1s) orbital at the lowest energy level, occupied by one or two electrons in every atom

At the second energy level, electrons can occupy the four different orbitals shown

**b** Shapes of electron orbitals that occur at the first and second energy levels, which correspond to the first and second "shells" in **a**.

**Table 2.2  Electron Distribution for a Few Elements**

| Element | Chemical Symbol | Atomic Number | Electron Distribution |  |  |
|---|---|---|---|---|---|
|  |  |  | First Shell | Second Shell | Third Shell |
| Hydrogen | H | 1 | 1 | — | — |
| Helium | He | 2 | 2 | — | — |
| Carbon | C | 6 | 2 | 4 | — |
| Nitrogen | N | 7 | 2 | 5 | — |
| Oxygen | O | 8 | 2 | 6 | — |
| Neon | Ne | 10 | 2 | 8 | — |
| Sodium | Na | 11 | 2 | 8 | 1 |
| Magnesium | Mg | 12 | 2 | 8 | 2 |
| Phosphorus | P | 15 | 2 | 8 | 5 |
| Sulfur | S | 16 | 2 | 8 | 6 |
| Chlorine | Cl | 17 | 2 | 8 | 7 |

space inside hollow *shells* around the nucleus. The shell closest to the nucleus has one orbital, so it can hold no more than two electrons. The next shell can have as many as eight electrons, two in each of four orbitals. Successive shells can have still more electrons, as Figure 2.5 shows.

Hydrogen, the simplest atom, has one electron in its first (and only) shell. Notice in Figure 2.6 that sodium, with eleven electrons, has a lone electron in its outermost shell. In other words, electrons only *partly fill* the highest occupied shell of either atom. *Such atoms tend to react with other atoms.* As you can see from Table 2.2, atoms that tend to enter into reactions include not only hydrogen but also carbon, nitrogen, and oxygen. These atoms are the main building blocks of organisms.

**An atom tends to react with other atoms when its outermost shell is only partly filled with electrons.**

## Ionic Bonding

Sometimes the balance between the protons and electrons of an atom is disturbed so much that one or more electrons are knocked out of the atom, pulled away from it, or added to it. When an atom loses or gains one or more electrons, it becomes positively or negatively charged. In this state, it is an **ion**.

When an atom loses or gains electrons, another atom of the right kind must be nearby to accept or donate the electrons. Since one loses and one gains electrons, *both* become ionized. Depending on the surroundings, the two ions can go their separate ways or stay together through the mutual attraction of their opposite charges. An association of two oppositely charged ions is an **ionic bond**. NaCl, the table salt we sprinkle on food, has ions of sodium ($Na^+$) and chloride ($Cl^-$) linked together this way. Figure 2.7 shows their arrangement.

**An ion is an atom or molecule that has gained or lost one or more electrons, and so has acquired an overall positive or negative charge.**

**In an ionic bond, a positive and a negative ion are linked by the mutual attraction of opposite charges.**

## Covalent Bonding

Often, an attraction between two atoms is not quite enough for one atom to pull electrons completely away from the other. The atoms end up sharing electrons, in what is called a **covalent bond**. We can use a line to represent a single covalent bond between two atoms, as in

H—H. If we want to focus on the number of electrons being shared, we can use a dot to represent each one:

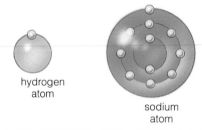

In a double covalent bond, two atoms share two pairs of electrons. An example is the $O_2$ molecule, or O=O. In a triple covalent bond (such as N≡N), two atoms share three pairs of electrons.

Covalent bonds may be nonpolar or polar. In a *nonpolar* covalent bond, both atoms exert the same pull on shared electrons. The word nonpolar implies there is no difference between the two "ends" (or poles) of the bond. An example is the H—H molecule. Its hydrogen atoms, with one proton each, attract the shared electrons equally.

**Figure 2.6** Distribution of electrons (yellow dots) for hydrogen and sodium atoms. Each atom has a lone electron (and room for more) in its outermost shell. Atoms having such partly filled shells tend to enter into reactions with other atoms.

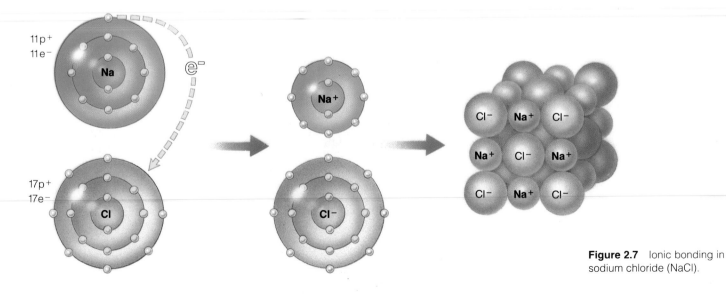

**Figure 2.7** Ionic bonding in sodium chloride (NaCl).

In a *polar* covalent bond, atoms of different elements (which have different numbers of protons) do not exert the same pull on electrons. The more attractive atom ends up with a slight negative charge (it is "electronegative"). But this is balanced out by the other atom, which ends up with a slight positive charge. In other words, the two atoms together have no *net* charge, but the *distribution* of charge within the bond is asymmetric.

A water molecule (H—O—H) has polar covalent bonds. Its electrons are less attracted to the hydrogens than to the oxygen (which has more protons).

---

**In a covalent bond, atoms share electrons.**

**If electrons are shared equally, the bond is nonpolar. If they are not shared equally, the bond is polar (electronegative at one end and electropositive at the other).**

---

Covalent bonds provide strong structural links between atoms in the carbohydrates, lipids, proteins, and other building blocks of cell architecture. In DNA, a double-stranded molecule, they link atoms together in each strand.

### Hydrogen Bonding

In a **hydrogen bond**, an atom of a molecule interacts weakly with a neighboring hydrogen atom that is already taking part in a polar covalent bond. (The hydrogen, which has a slight positive charge, is attracted to the slight negative charge of the other atom.) Hydrogen bonds can form between two different molecules, as shown in Figure 2.8. They also can form between two different regions of the same molecule where it twists back on itself.

Hydrogen bonds are common in large biological molecules. For example, many occur between the two strands of a DNA molecule, along the lines shown in Figure 2.8b. Individually, those bonds are easily broken, but collectively they help stabilize DNA structure. In later chapters, we will look at the energy it takes to break these and other bonds. Here, we will turn next to the hydrogen bonds between water molecules. These bonds impart some structure to liquid water—and they are responsible for many of water's life-sustaining properties.

---

**In a hydrogen bond, an atom or molecule interacts weakly with a neighboring hydrogen atom that is already taking part in a polar covalent bond.**

---

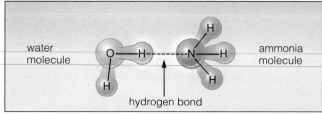

**a**

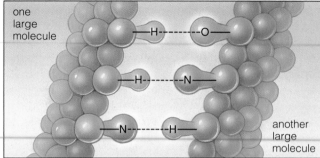

**b**

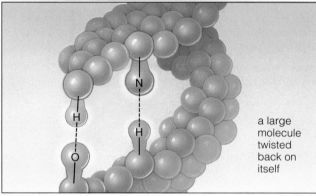

**c**

**Figure 2.8**  Some examples of hydrogen bonds. These are easier to break than covalent bonds. But where they occur in profusion, their collective action helps give water and other substances some of their characteristic properties. And collectively they help stabilize the shape of large molecules.

## PROPERTIES OF WATER

Of all the planets in the solar system, only the earth is bathed in liquid water (Figure 2.9). Water covers about three-fourths of its surface. Life apparently originated in water. Many kinds of organisms still live in it, and the ones that don't carry an abundance of water around with them, in cells and in tissue spaces. The water molecules inside and outside cells are absolutely crucial for life. They are required for many important reactions, and they have major roles in the shape and internal molecular organization of cells.

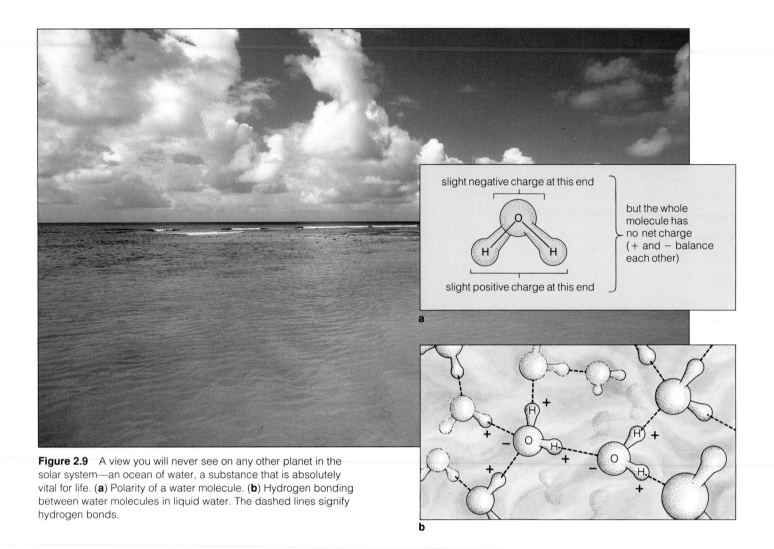

**Figure 2.9** A view you will never see on any other planet in the solar system—an ocean of water, a substance that is absolutely vital for life. (**a**) Polarity of a water molecule. (**b**) Hydrogen bonding between water molecules in liquid water. The dashed lines signify hydrogen bonds.

## Polarity of the Water Molecule

Even though a water molecule carries no net charge, it interacts weakly with many other substances. How? Remember that the water molecule carries *partial* charges. As a result of its electron arrangements and bond angles, the whole molecule is slightly electronegative at the oxygen "end." And it is slightly electropositive at the "end" where the two hydrogens are positioned (Figure 2.9b).

Their polarity allows water molecules to interact with one another. Figure 2.9c shows how their hydrogen atoms form weak hydrogen bonds with oxygen atoms of neighboring molecules in liquid water. The same type of interaction is possible between water and many different polar substances. Polar substances are **hydrophilic** (water-loving). This means they are attracted to one end or the other of a water molecule and may form weak hydrogen bonds with it.

By contrast, nonpolar substances are **hydrophobic** (water-dreading). This means they tend to be repelled by water. Shake a bottle containing water and salad oil (a hydrophobic substance), then put it on a counter. The

oil and water hold little attraction for each other. Gradually, hydrogen bonds reunite the water molecules (they replace bonds that were broken when you shook the bottle). As they do, the oil molecules are pushed aside and forced to cluster in droplets or in a film at the water's surface.

**In a hydrophilic interaction, polar substances form weak hydrogen bonds with water.**

**In a hydrophobic interaction, nonpolar groups cluster together in water. The clustering is not a true bond. The surrounding water molecules simply hydrogen-bond with one another and push out the nonpolar groups.**

Hydrogen bonds and hydrophobic interactions underlie three properties of water that are biologically important. These are its resistance to changes in temperature, its internal cohesion, and its capacity to dissolve many substances.

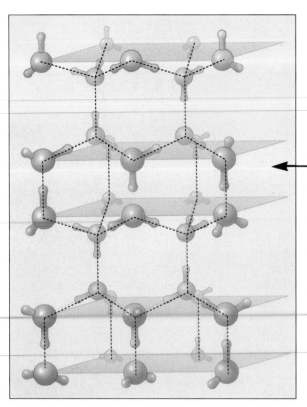

**Figure 2.10** Hydrogen bonding between water molecules in ice. Below 0°C, each water molecule becomes locked by four hydrogen bonds into a crystal lattice. In this bonding pattern, molecules are spaced farther apart than in liquid water at room temperature. When water is liquid, constant molecular motion usually prevents the maximum number of hydrogen bonds from forming.

## Temperature-Stabilizing Effects

Have you ever plunged into a solar-heated swimming pool well after sundown during a hot summer month? Even if the evening air has cooled off considerably, the water is still comfortably warm. Why doesn't the water temperature drop as fast as that of the surrounding air?

The *temperature* of a substance is a measure of how fast its molecules are moving. When water becomes heated, its molecules cannot move faster until hydrogen bonds among them are broken. It takes considerable heat to break those countless bonds and keep them from re-forming among molecules in liquid water. That is why it takes longer to boil a pot of water than it does to boil, say, alcohol or most other liquids.

When you heat the water to its boiling point, of course, its molecules have absorbed enough energy to separate from one another and you start to see steam (water vapor) rising from the pot. This is an example of **evaporation**, the process by which water molecules escape from a fluid surface and enter the surrounding air. Keep in mind that each water molecule in the liquid has *absorbed* some of the excess energy. When those molecules break free and depart in large numbers, they carry that energy away and the surface temperature of the water is lowered.

You cool off on hot, dry days through evaporative water loss from water that sweat glands secrete to your skin surface. (Sweat is 99 percent water.) You can lose quite a bit of water, given that you have more than 2½ million sweat glands distributed over nearly all of your body parts. Oasis plants of hot, dry deserts don't have sweat glands, but they too rely on evaporative water loss. Few plants elsewhere can do this—they would be severely stressed by such ongoing water losses. But the springs of oases provide constant replacements.

Water also resists freezing. At room temperature and down to about 0°C, hydrogen bonds between its molecules constantly break and form again, permitting some freedom of movement and keeping the water liquid. Below 0°C, however, the molecules become locked in the extended bonding pattern of ice. Because of this pattern, which is shown in Figure 2.10, ice is less dense than liquid water and is able to float on it. During winter freezes, sheets of ice typically form on surfaces of ponds, lakes, and streams. They act like a blanket that holds in the water's heat.

## Cohesive Properties

Swimming in a pool on a hot summer night can be deliciously refreshing, as long as you don't mind the large night-flying bugs that accidentally hit the water's surface and float about on it. Like other substances with cohesive properties, liquid water resists rupturing when stretched

a                                    b

c

**Figure 2.11** Evidence of water's cohesive properties. (**a, b**) A kingfisher plunging into water after a fish dinner. Notice how the water molecules stay put as a continuous sheet at the stream-air interface in the first photograph. Notice how they form many droplets when the kingfisher forces them away from the surface. The water's appearance in both cases results from hydrogen bonding among water molecules. The hydrogen bonds resist breaking apart when put under tension. (**c**) A water strider "walking" on water.

(placed under tension). Hydrogen bonds impart cohesion to water.

Where air and water meet at the pool surface, hydrogen bonds exert a constant inward pull on the uppermost water molecules. Their collective action imparts a high surface tension—insects have a tough time penetrating it, as do leaves and other bits of debris that land on the water (Figure 2.11).

In large land-dwelling plants, cohesion helps pull up water through narrow cellular pipelines that extend from the roots through stems and up to leaves. Hydrogen bonds hold the water molecules together in narrow columns along the route. Water evaporates from the leaf cells—and more water molecules are "pulled" into leaf cells as replacements (Chapter 28). The cohesion is strong enough to pull water even to the top of mature redwoods and other impressively tall trees.

### Solvent Properties

Because of the polar nature of its molecules, water is an excellent solvent for ions and polar molecules. A *solvent* is any fluid in which one or more substances can be dissolved. The dissolved substances themselves are called *solutes*. What does "dissolved" actually mean? Consider what happens when you pour some table salt into a glass of water. The salt crystals eventually separate into $Na^+$ and $Cl^-$ ions. Water molecules cluster around each

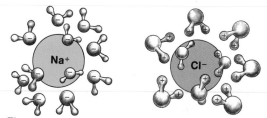

**Figure 2.12** Spheres of hydration around charged ions.

positively charged ion with their "negative" ends pointing toward it. They also cluster around each negatively charged ion with their "positive" ends pointing toward it (Figure 2.12).

These clusterings are "spheres of hydration." They shield charged ions and keep them from interacting, so the ions can remain dispersed in water. *A substance is "dissolved" in water when spheres of hydration form around its individual ions or molecules.* This happens to solutes in cells, in the sap of maple trees, in your blood, and in the body fluids of all other organisms.

**Cell structure and function depend on three properties of water: its internal cohesion, temperature-stabilizing effects, and capacity to dissolve many substances.**

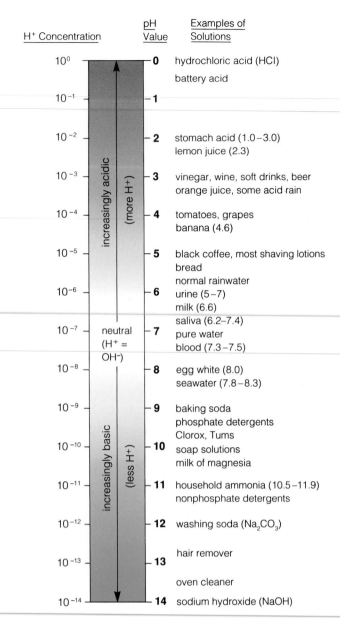

| H⁺ Concentration | pH Value | Examples of Solutions |

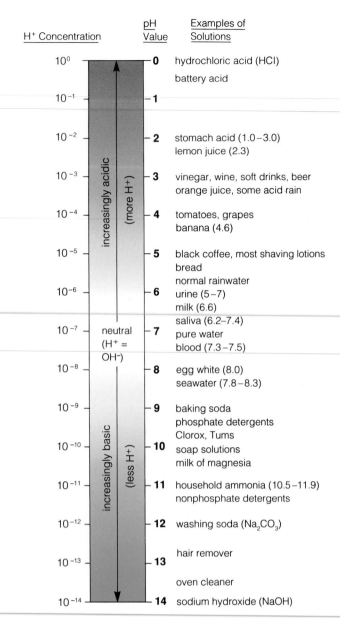

**Figure 2.13** The pH scale, in which a liter of fluid is assigned a number according to the number of hydrogen ions present. The most useful part of the scale ranges from 0 (most acidic) to 14 (most basic), with 7 representing neutrality.

A change of only 1 on the pH scale means a tenfold change in hydrogen ion concentration. For example, the gastric juice in your stomach is ten times more acidic than vinegar, and vinegar is ten times more acidic than tomatoes.

## ACIDS, BASES, AND SALTS

### Acids and Bases

Some substances release one or more protons when they dissolve in water. Such "naked" protons are also known as **hydrogen ions** ($H^+$). A substance that releases $H^+$ in water is an **acid**.

Think about what happens when you inhale the fragrance of some fried chicken, then chew and swallow it, thereby sending the chicken on its way to the fluid contained in your saclike stomach. Cells in the stomach lining are stimulated into secreting hydrochloric acid (HCl), which separates into $H^+$ and $Cl^-$. The stomach fluid becomes more acidic, and a good thing, too. The increased acidity switches on enzymes that can digest the chicken proteins. It also helps kill bacteria that may have lurked in or on the chicken. Of course, if you stuff yourself with too much fried food, you may end up with a truly "acid stomach"—and then you might reach for an antacid tablet.

Milk of magnesia is one kind of antacid. It contains magnesium hydroxide [$Mg(OH)_2$]. This substance can separate into a magnesium ion ($Mg^{++}$) and a **hydroxide ion** ($OH^-$). The hydroxide ion may then *combine with* one of those excess hydrogen ions in the potent fluid inside your stomach and so help settle things down. Any substance that releases $OH^-$ in water is a **base**.

### The pH Scale

The **pH scale** is used to measure the concentration of free (unbound) hydrogen ions in different solutions. As Figure 2.13 shows, the scale ranges from 0 (most acidic) to 14 (most basic). The midpoint, 7, represents a neutral solution in which $H^+$ and $OH^-$ concentrations are the same.

Pure water, with a pH of 7, is a neutral solution. Acidic solutions, such as hydrochloric acid and lemon juice, have a pH below 7. (That is, they have more $H^+$ than $OH^-$ ions.) The opposite is true of alkaline solutions, which have a pH above 7. *The greater the $H^+$ concentration, the lower the pH value.* For example, a tenfold increase in the $H^+$ concentration of a fluid corresponds to a decrease by one unit of the pH scale.

### Buffers

The life of each cell in each organism depends on ongoing chemical reactions—which happen to be extremely sensitive to even slight shifts in pH. Various mechanisms control the pH inside cells so that it hovers close to a neutral 7. Even though chemical reactions are con-

**Figure 2.14** Sulfur dioxide emissions from a coal-burning power plant. Special camera filters revealed these otherwise invisible emissions. Together with other airborne pollutants, sulfur dioxides dissolve in atmospheric water to form acidic solutions. They are a major component of acid rain.

tinually using up and producing hydrogen ions, the cellular pH normally will not swing abruptly.

Control mechanisms also maintain the pH of blood and tissue fluids that bathe all living cells in your body. The pH of this "extracellular fluid" is a little more alkaline than the fluid inside cells; it generally ranges between 7.35 and 7.45. Later in the book, we will consider how the lungs and kidneys help control the body's overall acid-to-base balance. Here, simply keep in mind that the balancing acts depend largely on buffer molecules. A **buffer** is any molecule that can combine with hydrogen ions, release them, or both, and so help stabilize pH.

Different kinds of buffers can sponge up or release hydrogen ions when conditions dictate. Bicarbonate ($HCO_3^-$) in blood is a good example. It can combine with hydrogen ions to form carbonic acid ($H_2CO_3$). In doing so, it decreases the acidity of blood:

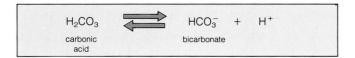

$$H_2CO_3 \rightleftarrows HCO_3^- + H^+$$

carbonic acid          bicarbonate

Conversely, when the $H^+$ concentration declines, bicarbonate can release hydrogen ions and help increase blood's acidity. The buffering action normally helps stabilize the pH of blood. This is important, because if the pH value shoots above or drops below the optimal range, death may follow.

So far, we have been talking about pH values inside cells and multicelled organisms. The pH of the outside environment is often much higher or lower. Cells of sphagnum mosses grow in peat bogs, where the pH is 3.2 to 4.6 (highly acidic). Some roundworms live in places where the pH is 3.4. River water ranges between 6.8 and 8.6. Industrial wastes are sometimes

so acidic they affect the pH of rain (Figure 2.14 and Chapter 48).

**Buffer molecules help maintain cellular pH near 7. They also help maintain the pH of extracellular fluid within the range of about 7.35 to 7.45. Environmental pH may be notably above or below that range.**

### Dissolved Salts

We have mentioned salts in passing but have not yet defined them. A **salt** is an ionic compound, formed when an acid reacts with a base. Sodium chloride can form this way:

$$HCl + NaOH \longrightarrow NaCl + H_2O$$

hydrochloric acid          sodium hydroxide, a base          salt

Many salts dissolve into ions that play vital roles in cells. For example, ions of potassium ($K^+$) and sodium take part in the "messages" traveling through the nervous system. Calcium ions ($Ca^{++}$) take part in cell movements, cell division, nerve cell function, muscle contraction, and blood clotting.

**Salts are ionic compounds that form when an acid reacts with a base, and that usually dissociate into positively and negatively charged ions in water.**

## WATER AND BIOLOGICAL ORGANIZATION

The organization of almost all large biological molecules is influenced by their interaction with water. Proteins, one of the topics of the next chapter, are a prime example. Depending on a variety of factors, the protein surface can be positively or negatively charged overall. The charged regions and polar groups attract water molecules. They also attract ions—which in turn attract water. The result is an electrically charged "cushion" of ions and water around the protein's surface:

Through such interactions with water and ions, the protein remains dispersed in cellular fluid rather than randomly settling against some cell structure. Why is it important to prevent settling? Many chemical reactions are played out on specific molecular regions of proteins—*they are the stages for life-sustaining tasks performed inside cells.*

Many more examples could be given of interactions between the polar water molecule and other substances characteristic of the cellular world. For now, the point to keep in mind is this:

**The properties of water profoundly influence the organization and behavior of substances that make up cells as well as the cellular environment.**

## SUMMARY

1. Matter is composed of elements, the atoms of which are composed of protons, neutrons, and electrons. An atom has a *net* electric charge of zero. An ion is an atom or compound that has gained or lost one or more electrons and so has acquired an overall positive or negative charge.

2. All atoms of an element have the same number of protons but they can vary slightly in the number of neutrons (they are isotopes). Radioisotopes decay spontaneously into atoms of different types.

3. Electrons occupy orbitals in shells around the nucleus. An orbital can hold no more than two electrons. Atoms of hydrogen, carbon, nitrogen, and oxygen (the main elements of biological molecules) have an unfilled orbital in their outermost shell and tend to form bonds with other elements.

4. The following chemical bonds are common in cells:
   a. Ionic bond: A positive ion and a negative ion remain together by the mutual attraction of opposite charges.
   b. Nonpolar covalent bond: Atoms share one or more electrons equally; there is no difference in charge between the two poles of the bond.
   c. Polar covalent bond: Atoms share one or more electrons unequally, causing a slight difference in charge between the two poles of the bond.
   d. Hydrogen bond: An atom of a molecule interacts weakly with a hydrogen atom that is already taking part in a polar covalent bond.

5. Acids release hydrogen ions ($H^+$) in solution; bases combine with hydrogen ions. Hydrogen ions are the same thing as free (unbound) protons.

6. Cellular pH is commonly kept close to neutral, meaning the $H^+$ and $OH^-$ concentrations are nearly equal. Reactions between an acid and a base produce salts—ionic compounds with vital roles in cell functions.

7. A water molecule shows polarity. Due to its electron arrangements, one end carries a partial negative charge and the other, a partial positive charge.
   a. Other polar molecules are attracted to water (they are hydrophilic).
   b. Nonpolar molecules are repelled by water (they are hydrophobic).

8. The properties of water influence the organization and behavior of substances that make up cells and cellular environments.
   a. Hydrogen bonding between its individual molecules gives water its temperature-stabilizing and cohesive properties.
   b. The polarity of the water molecules allows ions and polar molecules to dissolve readily in water. It also influences the shapes of large molecules in cells through hydrophobic and hydrophilic interactions.

## Review Questions

1. Define element, atom, molecule, and compound. What are the six main elements (and their symbols) in most organisms? *19–20*

2. Define an atom, an ion, and an isotope. *20, 21, 25*

3. Explain the differences among covalent, ionic, and hydrogen bonds. *25, 26*

4. What is the difference between a hydrophilic and a hydrophobic interaction? Is a film of oil on water an outcome of bonding between the molecules making up the oil? *27*

5. Define an acid, a base, and a salt. On a pH scale from 0 to 14, what is the acid range? Why are buffers important in cells? *30–31*

6. What type of bond is associated with the temperature-stabilizing, cohesive, and solvent properties of water? Is that bond also important in hydrophobic interactions? *27–29*

## Self-Quiz *(Answers in Appendix IV)*

1. Atoms are constructed of protons, neutrons, and _____ .

2. An _____ has a net charge of zero; an _____ has gained or lost one or more electrons, and so has become negatively or positively charged.
   a. ion; ion
   b. ion; atom
   c. atom; atom
   d. atom; ion

3. Interactions between atoms as they give up, acquire, or share _____ help determine the organization and activities of living things.

4. _____ are atoms of the same element that vary only in the number of neutrons they possess.

5. The main chemical elements found in biological molecules are:
   a. hydrogen, sulfur, nitrogen, oxygen
   b. phosphorus, hydrogen, carbon, oxygen
   c. carbon, oxygen, hydrogen, nitrogen
   d. carbon, oxygen, nitrogen, sulfur

6. Orbitals within shells around the nucleus of an atom can each hold no more than _____ electrons.
   a. one
   b. two
   c. three
   d. four

7. Electrons are shared unequally in a(n) _____ bond.
   a. nonpolar covalent
   b. ionic
   c. hydrogen
   d. polar covalent

8. Polar substances are _____; nonpolar substances are _____.
   a. hydrophilic; also hydrophilic
   b. hydrophilic; hydrophobic
   c. hydrophobic; also hydrophobic
   d. hydrophobic; hydrophilic

9. Which characterizes the internal pH of most cells?
   a. high concentration of $H^+$
   b. nearly equal concentration of $H^+$ and $OH^-$
   c. high concentration of $OH^-$
   d. both b and c are correct

10. A(n) _____ can combine with hydrogen ions or release them in response to changes in cellular pH.
    a. acid
    b. salt
    c. base
    d. buffer

11. Match these chemistry concepts appropriately:
    _____ water molecule polarity
    _____ common bonds in biological molecules
    _____ cellular pH
    _____ hydrogen bonds between water molecules
    _____ salt

    a. close to neutral
    b. temperature-stabilizing and cohesive properties
    c. permits ions and polar molecules to dissolve more easily
    d. produced by reaction between acid and base
    e. ionic, covalent, and hydrogen

## Selected Key Terms

| | | |
|---|---|---|
| acid *30* | hydrogen ion ($H^+$) *30* | molecule *20* |
| atom *20* | hydrophilic | nonpolar covalent |
| atomic number *20* | substance *27* | bond *25* |
| base *30* | hydrophobic | pH scale *30* |
| buffer *31* | substance *27* | polar covalent |
| chemical bond *23* | hydroxide ion | bond *26* |
| compound *20* | ($OH^-$) *30* | radioisotope *21* |
| element *19* | ion *25* | salt *31* |
| evaporation *28* | ionic bond *25* | solute *29* |
| hydrogen bond *26* | mass number *20* | solvent *29* |

## Readings

Lehninger, A. 1982. *Principles of Biochemistry*. New York: Worth. Classic reference book in the field.

Miller, G. T. 1991. *Chemistry: A Contemporary Approach*. Third edition. Belmont, California: Wadsworth. Simple introduction to basic chemical concepts and their application to everyday life.

# 3 CARBON COMPOUNDS IN CELLS

## Ancient Carbon Treasures

More than 400 million years ago, immense geologic forces were changing the contours of the earth's surface. Vast seas drained away as the seafloor rose slowly beneath them, and they left behind sediments loaded with the nutrient-rich remains of untold generations of marine organisms. Think of it! For perhaps a billion years, simple green plants had been confined to the seas—yet now they were at the threshold of a sunlit, nutrient-rich, uncrowded world. Some pioneer species made the most of the promising new environment. Within a mere 50 million years, their descendants included ferns, broadleafed shrubs, and huge trees, some more than twelve stories tall (Figure 3.1).

Between 360 million and 280 million years ago, however, sea levels swung dramatically, and the continents were submerged and drained no less than fifty times. When the seas moved out, swamp forests with large, scaly-barked trees gradually carpeted the wet lowlands. When the seas moved in, the forests were submerged again. The organic mess left behind eventually became transformed into the vast coal deposits of Britain and of the Appalachian Mountains of North America.

*Coal* is 55 to 95 percent carbon; the rest is mostly hydrogen, oxygen, nitrogen, and sulfur. It is a legacy of photosynthesis in those ancient swamp forests. Countless numbers of plants converted the energy of sunlight into chemical energy, which became stored in glucose and other carbon-containing compounds, and then into plant tissues. When the seas made their incursions, the tissue remains were submerged and became buried in sediments that protected them from decay. Gradually, the sediments compressed the saturated, undecayed remains into what we now call *peat*. As more sediments accumulated, the increased heat and pressure squeezed out some of the hydrogen, sulfur, and other elements from the peat. And so the peat became even more compact, with a higher percentage of carbon—it became coal, one of the "fossil fuels" we use as energy resources.

It took a fantastic amount of photosynthesis, burial, and compaction to form each major seam of coal in the earth. It has taken humans only a few centuries to deplete much of the known coal deposits. Often you will hear about our annual "production rates" for coal or some other fossil fuel. But how much do we really produce each year? None. We simply *extract* it from the earth. Given the millions of years required for its formation, coal is, for all intents and purposes, a nonrenewable resource.

With this bit of history as background, we turn to the substances that hold onto energy as it flows through the living world. Those substances are the so-called

a

**Figure 3.1** (a) Reconstruction of the kinds of plants that formed the vast swamp forests of the Carboniferous Period, about 50 million years ago. Burial and compaction of the remains of those forests produced the world's deposits of coal, a carbon-rich fossil fuel (b).

biological molecules, which range from complex carbohydrates to proteins, lipids, and nucleic acids. Carbon atoms, linked together, form the structural backbone of every one of them. Those molecules are the foundation for the structure and function of every cell, every organism; they were the foundation for every chunk of coal. Today your body uses them as building materials, as "worker" molecules such as enzymes, and as storehouses of energy that can drive all of our activities—from eating a leaf of lettuce to adding coal to the furnace fire.

b

KEY CONCEPTS

**1.** Carbon atoms linked together into chains and rings serve as the skeletons of organic compounds—skeletons to which hydrogen, oxygen, and other atoms are attached.

**2.** Cells are able to assemble simple sugars, fatty acids, amino acids, and nucleotides. Those four families of small organic molecules serve as energy sources or as building blocks for the large molecules characteristic of life—the complex carbohydrates, lipids, proteins, and nucleic acids.

**3.** Carbohydrates include glucose and other simple sugars. They also include polysaccharides, which may consist of hundreds or thousands of simple sugars linked into straight or branched chains. What we call fats are only one of several types of lipids. Carbohydrates and lipids serve as energy reserves and building blocks.

**4.** Some proteins have structural roles, but others have functional roles. One class of proteins, the enzymes, makes metabolic reactions proceed much, much faster than they would on their own. The nucleic acids called DNA and RNA are the basis of inheritance and cell reproduction.

## PROPERTIES OF CARBON COMPOUNDS

By far, the three most abundant elements in living things are oxygen, hydrogen, and carbon. Much of the oxygen and hydrogen is linked together in the form of water. But significant amounts of those two elements also are linked to carbon, the most important structural element in the body.

Carbon's central role in the molecules of life arises from its bonding properties. *Each carbon atom can form as many as four covalent bonds with other carbon atoms as well as with other elements.* In cells, carbon atoms that are linked in chains or rings form the backbones, or skeletons, for diverse compounds. The backbones occur in strandlike, globular, and sheetlike molecules, some of which contain

**Figure 3.2** Carbon compounds. There is a Tinkertoy quality to carbon compounds, in that a single carbon atom can be the start of truly diverse molecules assembled from "straight-stick" covalent bonds. Start with a hydrocarbon, which consists only of carbon and hydrogen. Methane ($CH_4$) is the simplest hydrocarbon. If you were to strip one hydrogen from methane, the result would be a methyl group, which occurs in fats, oils, and waxes:

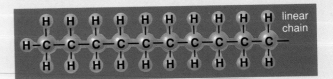

To such a linear chain, you could add branches:

Now imagine that two methane molecules are each stripped of a hydrogen atom and bonded together. If the resulting structure were to lose a hydrogen atom, you would end up with an ethyl group:

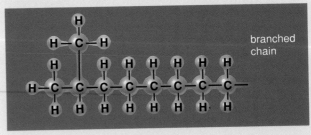

You could go on building a continuous chain, with all the carbon atoms arranged in a line:

You might even have the chains coiled back on themselves into rings, which may be represented in any of these ways:

| GROUP: | STRUCTURE: | OCCURS IN: |
|---|---|---|
| Methyl (—CH₃) | | fats, oils, waxes |
| Hydroxyl (—OH) | | sugars |
| Aldehyde | | sugars |
| Ketone | | sugars |
| Carboxyl (—COOH) | | sugars, fats, amino acids |
| Amino (—NH₂) | | amino acids, proteins |
| Phosphate (—Ⓟ) | | phosphate compounds (e.g, ATP) |

**Figure 3.3** Major functional groups that confer distinctive properties upon carbon compounds.

thousands, even millions, of atoms. Figure 3.2 shows some of the carbon bonding arrangements.

The carbon compounds assembled in cells are *organic* molecules. The term distinguishes them from the simple *inorganic* compounds, such as water and carbon dioxide, which have no carbon chains or rings.

### Families of Small Organic Molecules

Compounds having no more than twenty or so carbon atoms are considered to be small organic molecules. The four main families of small organic molecules found in cells are *simple sugars, fatty acids, amino acids, and nucleotides.* Usually the compounds are present in cellular fluid. Cells use them as energy sources and as building blocks for large molecules, or "macromolecules." The main macromolecules are *polysaccharides* (one of three classes of carbohydrates), *lipids, proteins,* and *nucleic acids.* We will survey their characteristics shortly.

### Functional Groups

The structure and behavior of organic compounds depend on the properties of their **functional groups,** which are atoms covalently bonded to the carbon backbone. For example, butter and other fats have "methyl groups" (—CH$_3$). Nonpolar covalent bonds link the hydrogen atoms to carbon in a methyl group (Figure 3.3). Water cannot form hydrogen bonds with nonpolar groups—and that is why butter does not dissolve in water. Neither does wax, for the same reason. Alcohols, which include sugars, have "hydroxyl groups" (—OH) attached to the carbon backbone. Water *can* form hydrogen bonds at hydroxyl groups—and that is why sugars dissolve in water. Figure 3.3 shows the main functional groups that characterize organic compounds.

Because its atoms share electrons equally, the carbon backbone of an organic compound is a stable structure that does not break down easily. By comparison, functional groups are much more prone to take part in chemical reactions. They can influence the electron arrangements of neighboring atoms and so affect the structure and reactivity of the molecule as a whole.

**The characteristic behavior of organic compounds depends largely on the type of functional groups that are attached to their carbon backbone.**

## Condensation and Hydrolysis

Small organic compounds are linked together into macromolecules with the help of enzymes. *Enzymes* are a special class of proteins that speed up reactions between specific substances. We will study these remarkable proteins in later chapters. For now, it is enough to know that when enzymes go to work on a target molecule, they usually are recognizing specific functional groups and bringing about specific changes in the structure of those groups.

Think about a **condensation reaction**, which results in the covalent linkage of small molecules and, often, the formation of water. Enzyme action causes an H atom to be split away from one molecule and an —OH group to be split away from another. A covalent bond forms between the two molecules at the exposed sites (Figure 3.4a). The parts that were split off are now free ions (H$^+$ and OH$^-$), and they may combine to form a water molecule.

Repeated condensation reactions can produce a polymer. A *polymer* is a molecule composed of anywhere from three to millions of relatively small subunits, which may or may not be identical. The individual units incorporated in polymers are often called *monomers*.

Now think about **hydrolysis**. As you can see from Figure 3.4b, this process is like condensation in reverse.

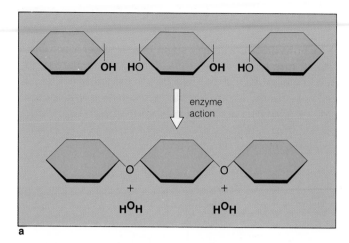

a

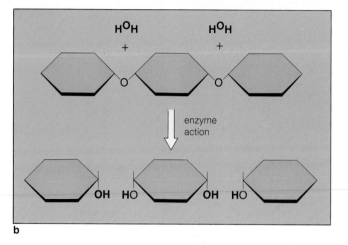

b

**Figure 3.4** (a) Condensation. In this generalized example, three monomers become covalently bonded to form a larger molecule, and two water molecules are released during the reaction. (b) Hydrolysis. In this example, two covalent bonds of a molecule are split, and the H$^+$ and OH$^-$ ions derived from water molecules become attached to the molecular fragments.

During hydrolysis, enzyme action splits a molecule into two or more parts by breaking covalent bonds. At the same time, H$^+$ and OH$^-$ derived from a water molecule are attached to the exposed bonding sites. Hydrolysis reactions that break apart starch and other polymers are common in cells. The released subunits can be used as building blocks or energy sources.

**Condensation is the covalent linkage of small molecules in a reaction that can also involve the formation of water.**

**Hydrolysis is the cleavage of a molecule into two or more parts by reaction with water.**

# CARBOHYDRATES

A **carbohydrate** is a simple sugar or a polymer assembled from a number of sugar units. Carbohydrates are probably the most abundant molecules in the world of life. All cells use them as transportable packets of quick energy, storage forms of energy, and structural materials. We recognize three classes of carbohydrates: the monosaccharides, oligosaccharides, and polysaccharides.

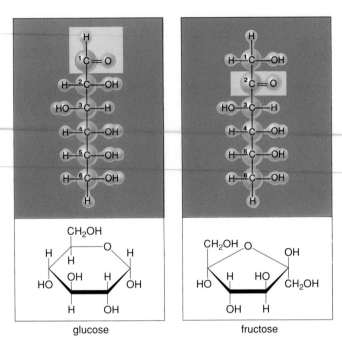

glucose          fructose

**Figure 3.5** Straight-chain and ring forms of glucose and fructose. (For reference purposes, sometimes the carbon atoms of sugars are numbered in sequence, starting at the end of the molecule closest to the aldehyde or ketone group.)

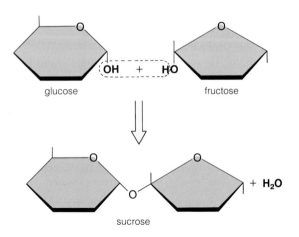

**Figure 3.6** Condensation of two monosaccharides (glucose and fructose) into a disaccharide (sucrose).

## Monosaccharides

The simplest type of carbohydrate is the sugar monomer, or **monosaccharide**. ("Saccharide" comes from a Greek word meaning sugar.) Sugar monomers are soluble in water, and most are sweet-tasting. As Figure 3.3 shows, they have two or more —OH groups and either an aldehyde or a ketone group. The most common ones have a backbone of five or six carbon atoms and tend to form ring structures when dissolved in cellular fluid.

Ribose and deoxyribose, which occur in RNA and DNA respectively, are five-carbon sugars. Figure 3.5 shows the structure of glucose, a six-carbon sugar ($C_6H_{12}O_6$). You will be encountering glucose repeatedly in this book. It is the main energy source for most organisms. And it is the precursor, or "parent" molecule, of many other compounds. Sucrose, fructose, and other specialized sugar molecules are derived from glucose. Large carbohydrates, including starch, are assembled from many glucose units.

You also will be encountering three other important compounds derived from sugar monomers. Glycerol (a sugar acid) is a component of fats. Vitamin C (a sugar acid) is essential in human nutrition. Glucose-6-phosphate (a sugar phosphate) is a premier entrant into major reaction pathways, including aerobic respiration.

## Oligosaccharides

An **oligosaccharide** is a short chain of two or more sugar monomers. Sucrose, lactose, and maltose are examples. Each belongs to the subclass called *disaccharides*, meaning they each consist of two covalently joined monomers. A sucrose molecule is simply a glucose monomer joined to a fructose monomer (Figure 3.6). Sucrose is probably the most plentiful sugar in nature. It is the form in which carbohydrates are transported to and from different parts of leafy plants. We make table sugar by extracting and crystallizing sucrose from plants such as sugar cane. Lactose (a glucose and a galactose unit) is present in milk. Maltose (two glucose units) is present in germinating seeds. Oligosaccharides with three or more sugar monomers are usually attached as short side chains to proteins and other large molecules. Some of these side chains have roles in cell membrane function and in immunity, which are topics of later chapters.

## Polysaccharides

A **polysaccharide** is a straight or branched chain of hundreds or thousands of sugar monomers, of the same or different kinds. The most common polysaccharides—starch, cellulose, glycogen, and chitin—are all polymers of glucose.

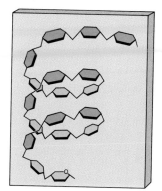

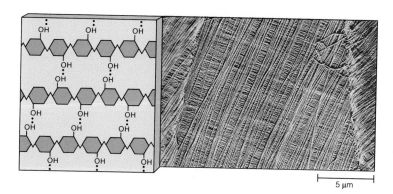

**Figure 3.7** Oxygen bridges between the glucose subunits of amylose, a form of starch. The boxed inset depicts the coiling of an amylose molecule, which is stabilized by hydrogen bonds.

**Figure 3.8** Structure of cellulose, which is composed of glucose subunits. Neighboring cellulose molecules hydrogen-bond together at —OH groups to form a fine strand. Such strands may twist together and then coil up as cellulose threads, of the sort shown in this micrograph of the cell wall of *Cladophora*, a green alga.

Starch is a storage form for sugar, and it can be readily hydrolyzed when cells require sugar units for energy or for building programs. Cellulose, a fiberlike structural material in the cell walls of plants, is tough and insoluble in water. It has been likened to the steel rods in reinforced concrete; it can withstand enormous weight and stress.

Given that both starch and cellulose consist of glucose units, why do they have such different properties? The answer is that adjacent glucose units are linked together in a different way in each substance. In starch, the linkages allow chains of glucose to twist into a coil, with many —OH groups facing outward (Figure 3.7). In cellulose, the chains stretch out side by side, and they hydrogen-bond to one another at —OH groups (Figure 3.8). The bonds stabilize the chains in tight bundles that resist breakdown. Only termites and a few other organisms have enzymes that can digest cellulose.

Glycogen, a highly branched polysaccharide, is the animal's equivalent of starch. Liver and muscle tissues are notable for their stores of glycogen. When blood sugar levels fall, liver cells break down glycogen and the glucose units are released to the blood. Similarly, when muscle cells are being given a workout, they tap into their glycogen supplies. The glycogen gives the cell quick access to energy. Glucose units can be released simultaneously from the ends of glycogen's numerous branches, a few of which are shown in Figure 3.9.

Many animals and fungi have cells that secrete chitin, a polysaccharide with nitrogen atoms attached to

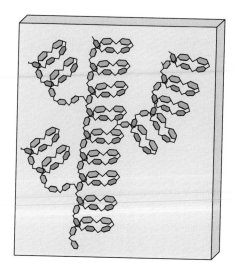

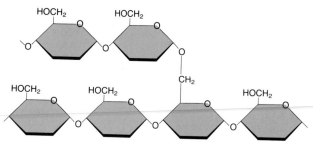

**Figure 3.9** Branched structure of glycogen, a form in which sugars are stored in some animal tissues.

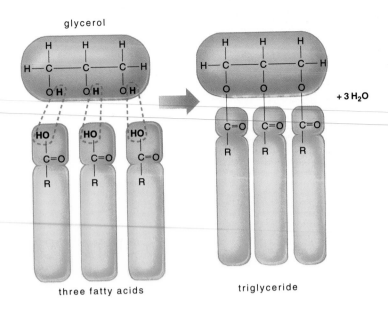

glycerol

three fatty acids     triglyceride

$+3 H_2O$

**Figure 3.11** Condensation of fatty acids into a triglyceride. Here, the R signifies the "rest" of the carbon chain in each fatty acid molecule.

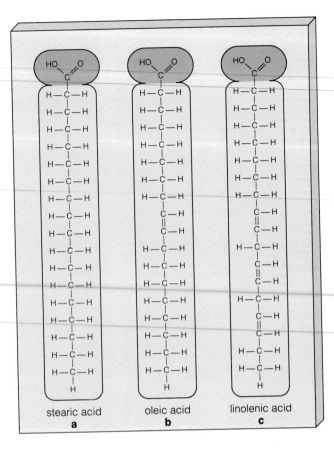

stearic acid
a

oleic acid
b

linolenic acid
c

**Figure 3.10** (*Above*) Structural formulas for representative fatty acids. (**a**) Stearic acid is fully saturated. (**b**) Oleic acid, with its double bond in the carbon backbone, is unsaturated. (**c**) Linolenic acid, with three double bonds, is one of the "polyunsaturated" fatty acids.

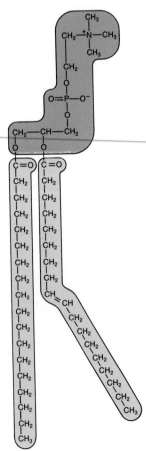

**Figure 3.12** Structural formula of a typical phospholipid found in animal cell membranes. The hydrophilic head is shown in orange, and the hydrophobic tails in gold.

the glucose backbone. Chitin is the main structural material in external skeletons and other hard body parts of many insects and crustaceans, including crabs. Chitin is the reason fresh mushrooms, which contain a great deal of water, are firm as well as soft to the touch. Chitin is the main structural material in the cell walls of most fungal species, not only the mushroom-producing ones.

## LIPIDS

**Lipids** are greasy or oily compounds that show little tendency to dissolve in water, but they dissolve in nonpolar solvents such as ether. Like polysaccharides and proteins, lipids can be broken down by hydrolysis reactions. Some lipids function in energy storage. Others are structural materials in membranes, coatings, and other cell structures. Here we will focus on two types: lipids with and without fatty acid components.

### Lipids With Fatty Acids

A "fatty acid" is a long, water-insoluble chain of mostly carbon and hydrogen, with a —COOH group at one end (Figure 3.10). When a fatty acid is part of a more complex lipid molecule, it is usually stretched out like a flexible tail. Let's look briefly at three common lipids having fatty acid tails: the glycerides, phospholipids, and waxes.

**Glycerides.** The substances commonly called fats and oils are composed of glycerides. A **glyceride** molecule has one to three fatty acid tails attached to a backbone of

**Figure 3.13** Penguins of the Antarctic, one of several types of animals that have a very thick, insulative layer of tryglycerides under the skin. Penguins also have a large, pear-shaped oil gland where their tail joins the body. These birds use their face and bill to spread the oil over their feathers. The oily coating keeps the feathers watertight and dry. And a good thing, too. Penguins may spend half a year in the open ocean without going ashore. They would become waterlogged and die in a few hours without their oil coating.

glycerol (Figure 3.11). Monoglycerides have one tail, diglycerides have two, and triglycerides have three. Glycerides are abundant lipids and a rich source of stored energy. Gram for gram, triglycerides yield more than twice as much energy as carbohydrates. Some animals, including seals and penguins, have a very thick layer of triglycerides beneath the skin (Figure 3.13). The layer serves as insulation against the near-freezing temperatures of their surroundings.

*Saturated fats*, including butter and lard, tend to be solids at room temperature. "Saturated" means all the carbon atoms in the fatty acid tails are joined by single C—C bonds and as many hydrogen atoms as possible are linked to them. Figure 3.10a shows this bonding scheme. The tails of adjacent molecules snuggle together in parallel array.

*Unsaturated fats,* or oils, tend to be liquid at room temperature. In this case, one or more double bonds occur between the carbon atoms in the fatty acid tails (Figure 3.10b and c). Oils are liquid because the double bonds create kinks that disrupt packing between tails.

Some amount of unsaturated fats is important in nutrition. In one study, immature rats were placed on a fat-free diet. The rats did not grow normally, their hair fell off, their skin turned scaly, and they died young. The abnormal conditions never developed in other rats that were fed small amounts of linoleic acid and linolenic acid, two unsaturated fats required in the diets of rats (and humans).

**Phospholipids**. In contrast to triglycerides, the phospholipids seldom serve as energy storage molecules. Rather, they are the main structural material of cell membranes. A **phospholipid** has two fatty acid tails attached to a glycerol backbone. It also has a hydrophilic "head," composed of a small polar group and a phosphate group, that dissolves in water (Figure 3.12). As you will see in Chapter 5, cell membranes have two layers of lipids

**Figure 3.14** Beads of water on the waxy cuticle of cherries.

pressed against each other. One layer has its molecular heads dissolved in the fluid inside the cell, the other layer has its heads dissolved in the fluid surroundings, and all the tails are sandwiched between them.

**Waxes**. The lipids called **waxes** have long-chain fatty acids linked to long-chain alcohols or to carbon rings. Wax secretions form coatings that help keep the skin and hair of various animals protected, lubricated, and pliable. Waxes secreted from special glands in waterfowl and other birds help make feathers water-repellant (Figure 3.13). In many plants, waxes are embedded in a matrix of lipid polymers (cutin or suberin). For example, waxes and cutin form the *cuticle*, a covering on the surface of aboveground plant parts that helps restrict water loss. A waxy cuticle gives cherries, apples, and many other fruits a shiny appearance (Figure 3.14).

### Lipids Without Fatty Acids

Lipids that have no fatty acid tails are less abundant than the ones described so far, but many have roles in membrane structure and in controls over metabolism.

## Cholesterol Invasions of Your Arteries

Arteriosclerosis is a condition in which the blood vessels called arteries thicken and lose elasticity. Conditions worsen with *atherosclerosis*, the buildup of lipid deposits in the wall of arteries and the subsequent shrinking of arterial diameter. The lipid culprits are saturated fats and cholesterol.

Your body requires certain amounts of cholesterol, but the liver normally manufactures enough to meet the body's demands. Many of the foods you eat also contain cholesterol, and it ends up circulating in the blood along with cholesterol produced by the liver. If you habitually eat too much cholesterol-rich food, you may end up with high levels of cholesterol in the blood. If you have a certain heritable (genetic) disorder, the same thing might happen no matter what kinds of food you eat.

Cholesterol does not float freely in the blood. When it leaves the liver, it is bound to *low-density lipoproteins* (LDLs), which the liver also produces. The LDLs in turn bind to receptor molecules on cells in different parts of the body. So bound, they can be engulfed and their cholesterol cargo used for a range of cell activities. Other molecules, called *high-density lipoproteins* (HDLs), cart cholesterol back to the liver, where it can be metabolized.

Cells of some people with high blood cholesterol do not seem to have enough LDL receptors, so they cannot remove as much LDL from the blood. LDL levels increase, and so does the risk of atherosclerosis. LDLs, with their cholesterol cargo, seem to have a penchant for infiltrating arterial walls. Abnormal cells and cell products multiply at the infiltration sites, then calcium salts and a fibrous net form over the whole mass. The result is an atherosclerotic plaque. You can see such plaques in Figure *a*. Blood clots may form where they occur and narrow or block the arteries, leading to a heart attack (Chapter 38).

In contrast, HDLs seem to attract cholesterol out of arterial walls and transport it to the liver. Atherosclerosis is uncommon in rats, which have mostly HDLs. It is common in humans, who in general have mostly LDLs.

People who want to reduce their LDL levels are usually advised to restrict their intake of saturated fats. At the same time, they are encouraged to add fish oils and unsaturated vegetable oils (such as olive oil) to their diet. Some research suggests that HDL levels are higher in people who exercise regularly. They also appear to be higher in people who do not smoke cigarettes or other forms of tobacco, and who drink little or no alcohol.

**a** Plaques in arteries of a heart patient.

Some are long, water-insoluble chains. Others, including the steroids, have ring structures. All **steroids** have the same backbone of four carbon rings, but they vary in the number, position, and type of functional groups attached to it:

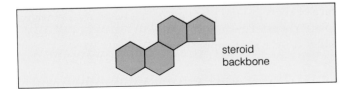

steroid backbone

You have probably heard of cholesterol. This steroid is an important component of animal cell membranes. It also is used in the synthesis of vitamin D, which is necessary for the proper development of bones and teeth. In excess amounts, however, cholesterol contributes to atherosclerosis. This is a disorder in which lipids become deposited in the walls of blood vessels (see *Commentary*). Plant tissues have no cholesterol, but they do contain steroids called phytosterols.

Many *hormones* are steroids. Among them are testosterone and estrogens, two major kinds of sex hormones. Hormones help regulate the body's growth, development, and reproduction, as well as its everyday functions. Bodybuilders and athletes sometimes use certain hormonelike steroids to increase their muscle mass. Unfortunately, use of those substances also can result in pronounced behavioral disorders and other abnormalities (Chapter 36).

# PROTEINS

**Proteins** are polymers of amino acids. An **amino acid** is a small organic molecule having a hydrogen atom, an amino group, a carboxyl group, and one or more atoms called its R group. All four parts are covalently bonded to the same carbon atom. Under cellular conditions, the amino and the carboxyl parts are ionized (charged), as shown here:

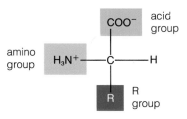

Proteins are assembled from only twenty different kinds of amino acids. (Figure 3.15 shows some of them.) Yet proteins are the most diverse of all biological molecules. Among their ranks are the enzymes, which make specific reactions proceed faster than they would on their own. Many molecules concerned with cell movements and transport of cell substances are proteins. So are many hormones and substances called antibodies, which help defend the body against disease-causing agents. Still other proteins are structural materials, the stuff of muscles, bone and cartilage, hoof and claw.

## Protein Structure

**Primary Structure.** Cells build proteins by stringing amino acids together, one after the other, with *peptide bonds*. These are simply covalent bonds that form between the amino group of one amino acid and the carboxyl group of another, as shown in Figure 3.16. Three or more amino acids linked together are called a **polypeptide chain**.

Different sequences of amino acids occur in different proteins. For example, the two polypeptide chains making up the protein insulin always have the amino acid

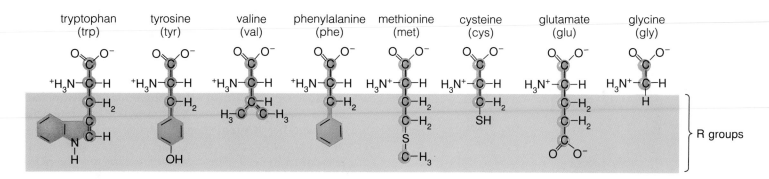

**Figure 3.15** Structural formulas for eight of the twenty common amino acids. The R groups are highlighted by the green-shaded box.

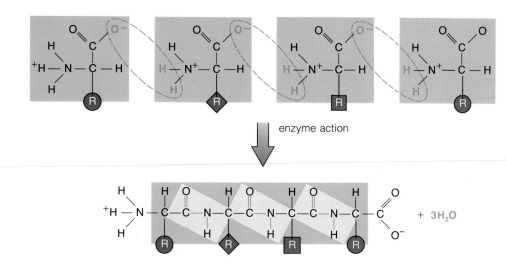

**Figure 3.16** Condensation of a polypeptide chain from four amino acids.

enzyme action

+ 3H₂O

Figure 3.17 Linear sequence of amino acids in cattle insulin, as determined by Frederick Sanger in 1953. This protein is composed of two polypeptide chains, linked by disulfide bridges (—S—S—).

Figure 3.18 Hydrogen bonds (dotted lines) in a polypeptide chain. Such bonds can give rise to a coiled chain (a) or sheetlike array of chains (b).

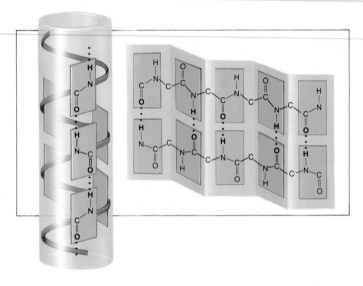

sequence shown in Figure 3.17. The specific sequence of amino acids in a polypeptide chain is the protein's *primary structure*.

**Three-Dimensional Structure.** The sequence of amino acids influences the shape that a protein can assume, what its function will be, and how it will interact with other substances. It does so in two major ways. First, oxygen and other atoms of specific amino acids in the sequence affect the patterns of hydrogen bonding between different amino acids along the chain. Second, R groups in the sequence interact and determine how the chain bends and twists into its three-dimensional shape.

In Figure 3.16, notice the peptide groups, which are indicated by the tan-shaded squares. Because of the way electrons are shared in each square, the atoms linked by the red bonds tend to be positioned rigidly in the same plane (the square). Only the atoms outside the squares have some freedom in how they become oriented. These

bonding patterns impose limits on the protein structures possible. In many cases, hydrogen bonds between every third amino acid hold the chain in a helical coil (Figure 3.18a). The structure of hemoglobin, an oxygen-carrying protein, includes coils like this. In other cases, the chain is extended, and hydrogen bonds hold two or more chains side by side in a sheetlike structure (Figure 3.18b). This bonding pattern occurs in silk proteins. The term *secondary structure* refers to the helical or extended pattern brought about by hydrogen bonds at regular intervals along a polypeptide chain.

Most helically coiled chains become further folded into a characteristic shape when one R group interacts with another R group some distance away, with the polypeptide backbone itself, or with other substances present in the cell. The term *tertiary structure* refers to the folding that arises through interactions among R groups of a polypeptide chain. Figure 3.19a shows one of the diverse shapes achieved through such interactions. Some proteins have *quaternary structure*, meaning they incorporate two or more polypeptide chains.

Proteins can be fiberlike, globular, or some combination of the two. Hemoglobin, shown in Figure 3.19b, has a globular shape, overall. Keratin, the structural material of hair and fur, is fibrous (Figure 3.20). So is collagen, the most common animal protein. Skin, bone, tendons, cartilage, blood vessels, heart valves, corneas—these and other components of the animal body depend on the strength inherent in collagen.

The amino acid sequence of a polypeptide chain dictates the final three-dimensional structure of a protein—and that structure dictates how the protein will interact with other cell substances.

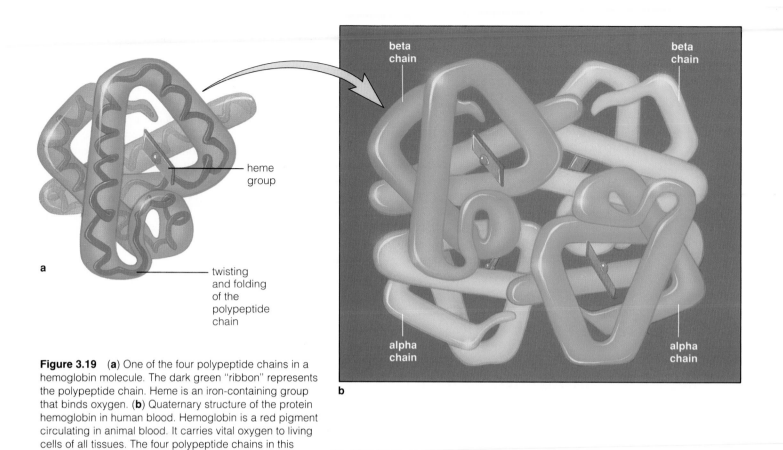

heme
group

twisting
and folding
of the
polypeptide
chain

a

b

beta
chain

beta
chain

alpha
chain

alpha
chain

**Figure 3.19** (**a**) One of the four polypeptide chains in a hemoglobin molecule. The dark green "ribbon" represents the polypeptide chain. Heme is an iron-containing group that binds oxygen. (**b**) Quaternary structure of the protein hemoglobin in human blood. Hemoglobin is a red pigment circulating in animal blood. It carries vital oxygen to living cells of all tissues. The four polypeptide chains in this molecule are held together tightly by numerous weak bonds.

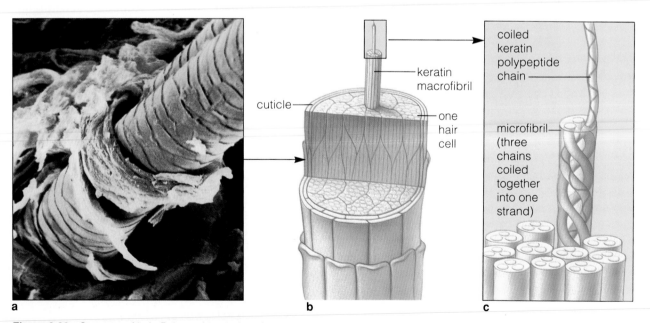

cuticle

keratin
macrofibril

one
hair
cell

coiled
keratin
polypeptide
chain

microfibril
(three
chains
coiled
together
into one
strand)

a

b

c

**Figure 3.20** Structure of hair. Polypeptide chains of the protein keratin are synthesized inside hair cells, which are derived from epidermal cells of the skin. The chains become organized into fine fibers (microfibrils), which become bundled together into larger, cablelike fibers (macrofibrils) that eventually fill the cells. The dead, flattened cells form a tubelike cuticle around the developing hair shaft.

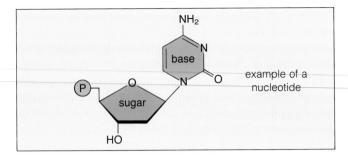

**Figure 3.21** Generalized structure of a nucleotide.

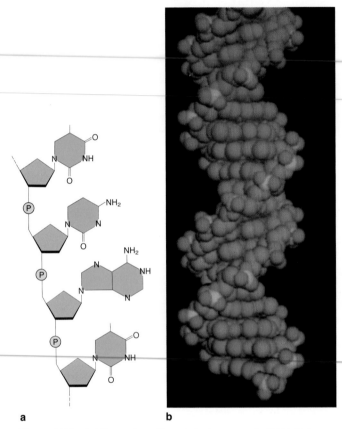

**a**　　　　**b**

**Figure 3.22** (**a**) Examples of bonds between nucleotides in a nucleic acid molecule. (**b**) Model of a segment of DNA, a molecule that is central to maintaining and reproducing the cell. The nucleotide bases are shown in blue.

## Lipoproteins and Glycoproteins

Many proteins are normally combined with other types of molecules. **Lipoproteins**, for example, have both lipid and protein components. Lipoproteins circulate in the blood, where they transport fats and cholesterol (see *Commentary*). Most **glycoproteins** are proteins to which oligosaccharides are covalently bonded. Some of the oligosaccharides are linear chains, others are branched. Glycoproteins make up nearly all of the proteins on the outer surface of animal cells, most of the protein products secreted from cells, and most of the proteins found in blood.

## Protein Denaturation

The hydrogen bonds and other interactions that hold a protein in its normal, three-dimensional shape are relatively weak. They also are sensitive to pH and temperature. When the $H^+$ concentration shifts or temperatures increase above 60°C, those interactions can be disrupted. **Denaturation** refers to the loss of a molecule's three-dimensional shape through disruption of the weak bonds responsible for it. When protein molecules have undergone denaturation, their polypeptide chains have unwound or have changed shape, so the proteins are no longer functional.

Think about the "egg white" of an uncooked chicken egg, which is a concentrated solution of the protein albumin. When you cook an egg, heat does not affect the strong covalent bonds of albumin's primary structure—but it destroys the weaker bonds responsible for its secondary and tertiary structure. Although denaturation can be reversed for some kinds of proteins when normal conditions are restored, albumin isn't one of them. There is no way to uncook a cooked egg.

## NUCLEOTIDES AND NUCLEIC ACIDS

The small organic compounds called nucleotides are central to life. Each **nucleotide** contains three components: a five-carbon sugar (ribose or deoxyribose), a nitrogen-containing base that has either a single-ring or double-ring structure, and a phosphate group. The three components of a nucleotide are hooked together as shown in Figure 3.21.

Three kinds of nucleotides or nucleotide-based molecules are the adenosine phosphates, the nucleotide coenzymes, and the nucleic acids. We will explore the structure and function of these molecules in later chapters.

*Adenosine phosphates* are relatively small molecules that function as chemical messengers within and between cells, and as energy carriers. Cyclic adenosine monophosphate (cAMP) is a chemical messenger. The nucleotide called adenosine triphosphate (ATP) serves as a carrier of chemical energy from one reaction site to another in cells.

*Nucleotide coenzymes* transport hydrogen atoms and electrons from one reaction site to another in cells. The hydrogens and electrons are necessary in metabolism. Nicotinamide adenine dinucleotide ($NAD^+$) and flavin adenine dinucleotide (FAD) are two of the major coenzymes.

*Nucleic acids* are large, single or double strands of nucleotides. Adjacent nucleotides in the strand are connected by phosphate bridges, and their bases stick out to the side (Figure 3.22a). The sequence in which the four kinds of bases follow one another in the strand varies among nucleic acids.

Ribonucleic acids (RNAs) and deoxyribonucleic acids (DNAs) are built according to the plan just outlined. RNA is a single nucleotide strand, most often. DNA is usually a double-stranded molecule that twists helically, like a spiral staircase (Figure 3.22b). The two strands are held together by hydrogen bonds that form between them. You will be reading more about these molecules in chapters to come. For now, it is enough to know the following: (1) Genetic instructions are encoded in the sequence of bases in DNA, and (2) RNA molecules function in the processes by which genetic instructions are used to build proteins.

## SUMMARY

1. Structurally, carbon atoms are the starting point for the large organic compounds in cells—the polysaccharides, lipids, proteins, and nucleic acids. Each carbon atom can form up to four covalent bonds with other atoms. Organic compounds are those with a backbone of carbon atoms covalently linked into a chain or ring structure.

2. Cells assemble larger organic compounds from simple sugars, fatty acids, amino acids, and nucleotides. By definition, all of these are small organic molecules that include no more than twenty or so carbon atoms.

3. The structure and behavior of organic compounds depend on the properties of functional groups, which are atoms or clusters of atoms bonded covalently to the carbon backbone. Unlike the relatively stable carbon backbone, the functional groups tend to enter into chemical reactions.

4. Organic compounds are put together and split apart by enzyme-mediated reactions. In condensation reactions, small molecules become covalently linked and water forms. In hydrolysis, a water molecule is used in a reaction that splits a molecule into two or more parts.

5. Table 3.1 on the next page summarizes the main categories of biological molecules that have been described in this chapter. Included in this table are the most common classes of molecules within each category. We will have occasion to return to their nature and roles in diverse life processes.

## Review Questions

1. Four main families of small organic molecules are used in cells for the assembly of carbohydrates, lipids, proteins, and nucleic acids (the large biological molecules). What are they? *36*

2. Identify which of the following is the carbohydrate, fatty acid, amino acid, and polypeptide: *38, 40, 43–44*
   a. $^+NH_3$—CHR—$COO^-$      c. (glycine)$_{20}$
   b. $C_6H_{12}O_6$      d. $CH_3(CH_2)_{16}COOH$

3. Is this statement true or false? Enzymes are proteins, but not all proteins are enzymes. *43*

4. Describe the four levels of protein structure. How do the side groups of a protein molecule influence its interactions with other substances? Give an example of what happens when the bonds holding a protein together are disrupted. *43–44, 46*

5. Distinguish between the following:
   a. monosaccharide, polysaccharide *38*
   b. peptide bond, polypeptide *43*
   c. glycerol, fatty acid *40–41*
   d. nucleotide, nucleic acid *46–47*

## Self-Quiz   *(Answers in Appendix IV)*

1. The backbone of organic compounds is formed by the chemical bonding of _____ atoms into chains and rings.

**Table 3.1  Summary of the Main Carbon Compounds in Living Things**

| Category | Main Subcategories | Some Examples and Their Functions | |
|---|---|---|---|
| **CARBOHYDRATES**<br><br>*contain an aldehyde or a ketone group, and one or more hydroxyl groups* | **Monosaccharides** | Glucose | Structural roles, energy source |
| | **Oligosaccharides** | Sucrose (a disaccharide) | Form of sugar transported in plants |
| | **Polysaccharides** | Starch<br>Cellulose | Energy storage<br>Structural roles |
| **LIPIDS**<br><br>*are largely hydrocarbon, generally do not dissolve in water but dissolve in nonpolar substances* | **Lipids with fatty acids:**<br><br>*Glycerides*: one, two, or three fatty acid tails attached to glycerol backbone | Fats (e.g., butter)<br>Oils (e.g., corn oil) | Energy storage |
| | *Phospholipids*: phosphate group, another polar group, and (often) two fatty acids attached to glycerol backbone | Phosphatidylcholine | Key component of cell membranes |
| | *Waxes*: long-chain fatty acid tails attached to alcohol | Waxes in cutin | Water retention by plants |
| | **Lipids with no fatty acids:**<br><br>*Steroids*: four carbon rings; the number, position, and type of functional groups vary | Cholesterol | Component of animal cell membranes; can be rearranged into other steroids (e.g., vitamin D, sex hormones) |
| **PROTEINS**<br><br>*are polypeptides (up to several thousand amino acids, covalently linked)* | **Fibrous proteins:**<br><br>Individual polypeptide chains, often linked into tough, water-insoluble molecules | Keratin<br>Collagen | Structural element of hair, nails<br>Structural element of bones and cartilage |
| | **Globular proteins:**<br><br>One or more polypeptide chains folded and linked into globular shapes; many roles in cell activities | Enzymes<br>Hemoglobin<br>Insulin<br>Antibodies | Increase in rates of reactions<br>Oxygen transport<br>Control of glucose metabolism<br>Tissue defense |
| **NUCLEOTIDES**<br><br>*are units (or chains) having a five-carbon sugar, phosphate, and a nitrogen-containing base* | **Adenosine phosphates** | ATP | Energy carrier |
| | **Nucleotide coenzymes** | $NAD^+$, $NADP^+$ | Transport of protons ($H^+$) and electrons from one reaction site to another |
| | **Nucleic acids**<br><br>Chains of thousands to millions of nucleotides | DNA, RNAs | Storage, transmission, translation of genetic information |

2. Each carbon atom can form up to _____ bonds with other atoms.
   a. four
   b. six
   c. eight
   d. sixteen

3. The four types of large organic molecules characteristic of life are the _____, _____, _____, and _____.

4. All of the following *except* _____ belong to the four families of small organic molecules that serve as building blocks for large biological molecules or as energy sources.
   a. fatty acids
   b. simple sugars
   c. lipids
   d. nucleotides
   e. amino acids

5. Which of the following would *not* be included in the family of carbohydrates?
   a. glucose molecules
   b. simple sugars
   c. fats
   d. polysaccharides

6. Increasing the rate of metabolic reactions is the role of functional proteins known as _____.
   a. DNA
   b. amino acids
   c. fatty acids
   d. enzymes

7. Nucleic acids, the basis of inheritance and of cell reproduction, include _____.
   a. polysaccharides
   b. DNA and RNA
   c. proteins
   d. simple sugars

8. Which of the following best describes the role of functional groups?
   a. assembling large organic compounds from smaller ones
   b. determining the structure and behavior of organic compounds
   c. splitting molecules into two or more parts
   d. speeding up metabolic reactions

9. In _____ reactions, small molecules become covalently linked, and water can also form.
   a. symbiotic
   b. hydrolysis
   c. condensation
   d. ionic

10. Match each type of molecule with the correct description.
   _____ long chain of amino acids            a. carbohydrate
   _____ energy carrier                       b. phospholipid
   _____ glycerol, fatty acids, phosphate     c. protein
   _____ long chain of nucleotides            d. DNA
   _____ one or more sugar monomers           e. ATP

## Selected Key Terms

adenosine phosphate *47*
amino acid *43*
carbohydrate *38*
condensation reaction *37*
denaturation *46*
enzyme *37*
fatty acid *36*
functional group *36*
glyceride *40*

glycoprotein *46*
hydrolysis *37*
inorganic molecule *36*
lipid *40*
lipoprotein *46*
monomer *37*
monosaccharide *38*
nucleic acid *47*
nucleotide *46*
nucleotide coenzyme *47*
oligosaccharide *38*

organic molecule *36*
peptide bond *43*
phospholipid *41*
polymer *37*
polypeptide chain *43*
polysaccharide *38*
protein *43*
saturated fat *41*
steroid *42*
unsaturated fat *41*
wax *41*

## Readings

Hegstrom, R., and D. Kondepudi. January 1990. "The Handedness of the Universe." *Scientific American* 262 (1): 108–115.

Lehninger, A. 1982. *Principles of Biochemistry.* New York: Worth.

*Scientific American.* "The Molecules of Life." October 1985. This entire issue is devoted to articles on current insights into DNA, proteins, and other biological molecules. Excellent illustrations.

## Pastures of the Seas

Drifting through the surface waters of the world ocean are vast populations of living, single cells, busily engaged in photosynthesis. You can't see them without a microscope—a row of 7 million cells of one species would be less than a quarter-inch long—yet are they abundant! In some parts of the world, a cup of seawater may hold 24 million cells of one species, and that doesn't even include all the cells of *other* species.

Many of the photosynthetic drifters are protistans, such as the exquisitely shelled diatoms shown in Figure 4.1a. Others are bacteria, and still others are single-celled members of the plant kingdom. Together they are the pastures of the seas, grazed upon by microscopic animals that in turn are food for squids, fishes, and other predators. The pastures "bloom" in spring, when the waters have become warmer and enriched with nutrients, churned up from the deep by winter currents. Then, populations burgeon as their cellular members divide again and again. Some of those populations double in size not once, not twice, but *seven times* in a single day.

Biologists have known about these drifting populations for more than a century. They gave them the not-quite-accurate name "phytoplankton" (*phyto-* for plant, *plankton* for drifting). But no one suspected that the number of cells and their distribution were truly mind-boggling—until satellites started sending back photographs from space. For example, satellite images of the surface waters of the North Atlantic revealed springtime blooms that stretch from North Carolina to Spain. Those huge blooms also extend downward, several hundred feet beneath the ocean's surface!

Collectively, those single microscopic cells have enormous impact on the world's climate. They use about half of the carbon dioxide that is released each year during fossil fuel burning and other human activities. If they did not do this, atmospheric concentrations of carbon dioxide would be building up more rapidly to levels that may warm the planet, by way of the greenhouse effect described in Chapter 46. Warming can lead to changes in sea level and patterns of rainfall that can flood coastal areas and cause global shifts in the areas suitable for food production.

All cells of the phytoplankton have the same dependency on sunlight, carbon dioxide, and traces of phosphorus, nitrogen, and other nutrients. They all are sensitive to changes in the chemical composition of seawater. Even so, we daily dump industrial wastes, fertilizers and other chemicals of agricultural runoff, and raw sewage into the world ocean. How do the drifting cells tolerate that noxious chemical brew? How much more *can* they tolerate?

With these questions we start thinking about the way cells are put together and how they respond to changing conditions in their world. As our example of the photosynthetic drifters makes clear, tiny cells have significance beyond their capacity to survive and keep busy. Viewed collectively, cells work together in keeping you and all other kinds of multicelled organisms alive—and untold numbers work in ways that shape the physical character of the biosphere.

**Figure 4.1** (**a**) Representative diatoms—members of the floating pastures of the seas, the marine phytoplankton. (**b, c**) Satellite images that reveal the concentration of chlorophyll at the ocean surface. (Chlorophyll, a type of light-trapping pigment, occurs in nearly all photosynthetic organisms.) In these images, red indicates high chlorophyll concentrations, purple indicates low. During winter (**b**), phytoplankton are abundant in a few coastal areas only. During the spring (**c**), the pastures spread across the entire North Atlantic.

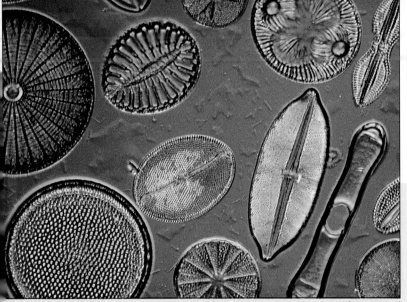

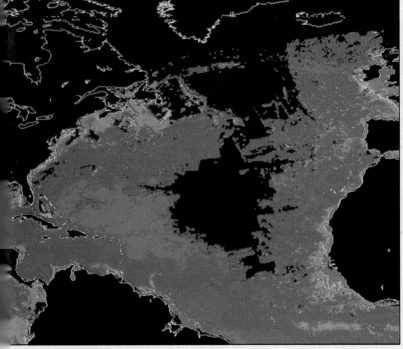

**b** *(Above)* Winter  **c** *(Below)* Spring

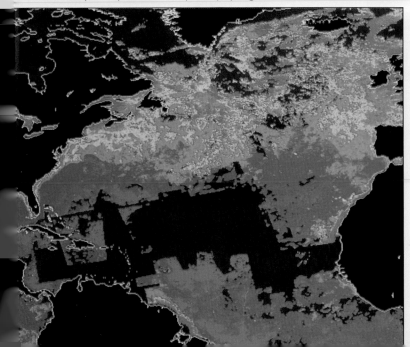

KEY CONCEPTS

**1.** Cells are the smallest units that still retain the characteristics of life, including complex organization, metabolic activity, and reproductive behavior.

**2.** All cells have a plasma membrane, and it surrounds an inner region of cytoplasm. The plasma membrane keeps events within the cell separate from the surrounding environment, so that the events proceed in organized, controlled ways.

**3.** Eukaryotic cells have a nucleus and other organelles (membrane-bound compartments) within the cytoplasm. The membranes of these organelles separate different chemical reactions in the space of the cytoplasm and so allow the reactions to proceed in orderly fashion. Prokaryotic cells (bacteria) do not have comparable organelles.

## GENERALIZED PICTURE OF THE CELL

### Emergence of the Cell Theory

Early in the seventeenth century, Galileo Galilei arranged two glass lenses in a cylinder. With this instrument he happened to look at an insect, and afterward he described the stunning geometric patterns of its tiny eyes. Thus Galileo, who was not a biologist, was the first to record a biological observation made through a microscope. The study of the cellular basis of life was about to begin. First in Italy, then in France and England, biologists set out to explore a world whose existence had not even been suspected.

At mid-century Robert Hooke, "Curator of Instruments" for the Royal Society of England, was at the forefront of the studies. When Hooke first turned one of his microscopes to a thinly sliced piece of cork from a mature tree, he observed tiny, empty compartments. He gave them the Latin name *cellulae* (meaning small rooms); hence the origin of the biological term "cell." They were actually walls of dead cells, which is what cork is made of. But Hooke did not think of them as being dead because he

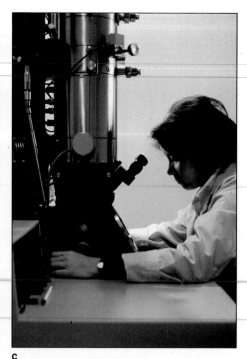

**Figure 4.2**  Microscopy then and now. (**a**) Robert Hooke's compound microscope and his drawing of dead cork cells. (**b**) Anton van Leeuwenhoek, microscope in hand. (**c**) An electron microscope in a modern laboratory.

did not know that cells could be alive. He also noted that cells in other plant materials contained "juices." Figure 4.2 shows Hooke's microscope.

Given the simplicity of their instruments, it is amazing that the pioneers in microscopy saw as much as they did. Antony van Leeuwenhoek, a shopkeeper, had great skill in constructing lenses and possibly the keenest vision. He even observed a single bacterium—a type of organism so small it would not be seen again for another two centuries! Yet this was mostly an age of exploration, not of interpretation. Once the limits of their simple instruments had been reached, the early microscopists had to give up interest in cell structure without being able to explain what they had seen.

Then, in the 1820s, improvements in lens design brought cells into sharper focus. Robert Brown, a botanist, observed a spherelike structure in every plant cell he examined. He called the structure a "nucleus." By 1839, the zoologist Theodor Schwann reported the presence of cells in animal tissues. He began working with Matthias Schleiden, a botanist who had concluded that all plant tissues are composed of cells and that the nucleus is somehow paramount in a cell's development. Both investigators proposed that each cell develops as an independent unit, even though its life is influenced by the organism as a whole. Schwann distilled the mean-

ing of the new observations into what became known as the first two generalizations of the **cell theory:**

**All organisms are composed of one or more cells.**

**The cell is the basic *living* unit of organization for all organisms.**

Yet a question remained: Where do cells come from? A decade later the physiologist Rudolf Virchow completed studies of growth and reproduction (their division into two cells). He reached the following conclusion, which became the third generalization of the cell theory:

**All cells arise from preexisting cells.**

Not only was a cell viewed as the smallest living unit, the continuity of life was now seen to arise directly from the growth and division of single cells. Within each tiny cell, events were going on that had profound implications for all levels of biological organization!

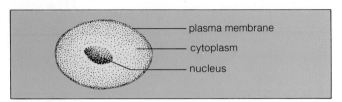

*[hɛrɛdɪtɛri]—mental or physical quality, or a disease is hereditary, ↑ it is passed to a child from the genes of his/her parents.*

*heredity [həˈrɛdəti]- the process of passing on a mental or physical quality from a parent's genes to a child.*

## Basic Aspects of Cell Structure and Function

Cells vary in size, shape, and complexity. However, they are alike in a few basic features. They all have a *plasma membrane* surrounding an inner region, the *cytoplasm*. Their hereditary material, DNA, is not scattered haphazardly through the cell interior. Bacterial cells have DNA physically concentrated in a part of the cytoplasm designated the nucleoid. All other cell types have DNA organized in a *nucleus:*

*[handwritten margin note, illegible]*

These features, which are the structural basis of all cellular events, can be defined in this way:

**1. Plasma membrane.** This is the cell's outermost membrane, composed of two lipid layers in which proteins are embedded. It separates internal events from the environment so that they can proceed in organized, controlled ways. The membrane does not totally isolate the interior; many substances move across it. It also has receptors for outside information that can alter cell behavior.

**2. Nucleus.** This membrane-bound compartment contains hereditary instructions (DNA) and other molecules that function in how the instructions are read, modified, and dispersed.

**3. Cytoplasm.** The cytoplasm is everything enclosed by the plasma membrane, except for the nucleus. In all but bacterial cells, it has compartments in which specific metabolic reactions occur. It includes particles and filaments bathed in a semifluid substance. The filaments form a skeleton that imparts shape and permits movement.

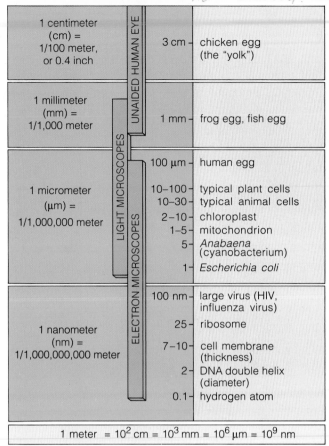

**Figure 4.3** Units of measure used in microscopy. The micrometer is used in describing whole cells or large cell structures. The nanometer is used in describing cell ultrastructures and large organic molecules.

## Cell Size and Cell Shape

Can any cell be seen with the unaided human eye? There are a few, including the "yolks" of bird eggs, cells in the red part of a watermelon, and the fish eggs we call caviar. Generally, however, cells are too small to be observed without microscopes; they are measured in micrometers. A micrometer is only *one-millionth* of a meter long. Your red blood cells are about 6 to 8 micrometers across—and a string of about 2,000 would only be as long as your thumbnail is wide! As Figures 4.3 through 4.5 indicate, light microscopes reveal details down to about 0.2 micrometer. Electron microscopes are used for details smaller than this.

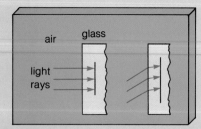

**a** *Refraction of light rays (The angle of entry and the molecular structure of the glass determine how much the rays will bend)*

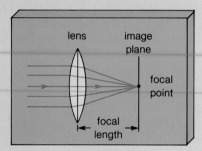

**b** *Focusing of light rays*

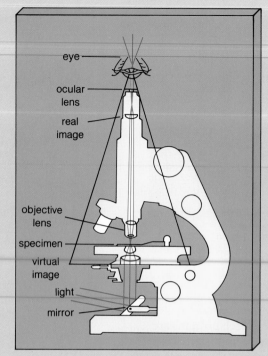

**c** *Compound light microscope*

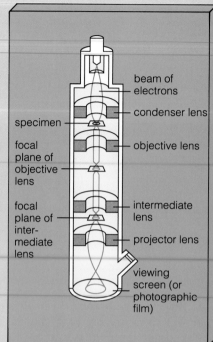

**d** *Transmission electron microscope*

**Figure 4.4** Microscopes—gateways to the cell.

**Light Microscopes**. (**a**) Light microscopy relies on the bending, or refraction, of light rays. Light rays pass straight through the center of a curved lens. The farther they are from the center, the more they bend. (**b, c**) The *compound light microscope* is a two-lens system. All rays coming from the object being viewed are channeled through the system of lenses to a single place where they can be seen by the human eye.

If you wish to observe a *living* cell, it must be small or thin enough for light to pass through. Also, structures inside cells can be seen only if they differ in color and contrast from their surroundings—but most are nearly colorless and optically uniform in density. Specimens can be stained (exposed to dyes that react with some cell structures but not others), but staining usually alters the structures and kills the cells. Finally, dead cells begin to break down at once, so they must be preserved (fixed) before staining. Most observations have been made of dead, fixed, or stained cells. Largely transparent living cells can be observed through the *phase-contrast microscope*. Here, small differences in the way different structures refract light are converted to larger variations in brightness.

No matter how good a glass or quartz lens system may be, when magnification exceeds 2,000× (when the image diameter is 2,000 times as large as the object's diameter), cell structures appear large but are not clearer. By analogy, when you hold a

magnifying glass close to a newspaper photograph, you see only black dots. You cannot see a detail as small as or smaller than a dot; the dot would cover it up. In microscopy, something like dot size intervenes to limit *resolution* (the property that determines whether small objects close together can be seen as separate things). That limiting factor is the physical size of wavelengths of visible light.

Light comes in different wavelengths, or colors. The red wavelengths are about 750 nanometers and violet wavelengths are about 400 nanometers; all other colors fall in between. If an object is smaller than about one-half the wavelength, light rays passing by it will overlap so much that the object won't be visible. The best light microscopes resolve detail only to about 200 nanometers.

**Transmission Electron Microscopes**. Electrons are usually thought of as particles, but they also behave like waves. The wavelengths of electrons used in electron microscopes are about 0.005 nanometer—about 100,000 times shorter than those of visible light! Ordinary lenses cannot be used to focus such accelerated streams of electrons, because glass scatters them. But each electron carries an electric charge, which responds to magnetic force. A magnetic field can divert electrons along defined paths and channel them to a focal point. Magnetic lenses are used in *transmission electron microscopes* (**d**).

Electrons must travel in a vacuum, otherwise they would be randomly scattered by molecules in the air. Cells can't live in a

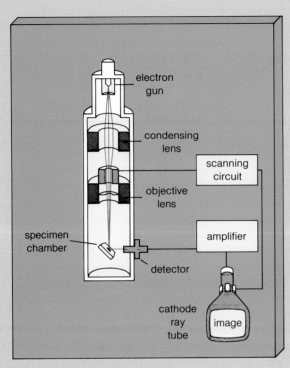

**e** *Scanning electron microscope*

vacuum, so living cells cannot be observed at this higher magnification. In addition, specimens must be sliced extremely thin so that electron scattering corresponds to the density of different structures. (The more dense the structure, the greater the scattering and the darker the area in the final image formed.) Specimen fixation is crucial. Fine cell structures are the first to fall apart when cells die, and artifacts (structures that do not really exist in cells) may result. Because most cell materials are somewhat transparent to electrons, they must be stained with heavy metal "dyes," which can create more artifacts.

With *high-voltage electron microscopes*, electrons can be made ten times more energetic than with the standard electron microscope. With the energy boost, intact cells several micrometers thick can be penetrated. The image produced is something like an x-ray plate and reveals the three-dimensional internal organization of cells (see, for example, Figure 4.21).

**Scanning Electron Microscopes.** (**e**) With a *scanning electron microscope*, a narrow electron beam is played back and forth across a specimen's surface, which has been coated with a thin metal layer. Electron energy triggers the emission of secondary electrons in the metal. Equipment similar to a television camera detects the emission patterns, and an image is formed. Scanning electron microscopy does not approach the high resolution of transmission instruments. However, its images have fantastic depth.

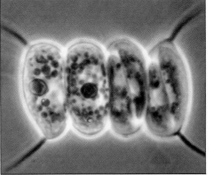

As on other micrographs in this book, the short bar provides a reference for size. Each micrometer (μm) is only 1/1,000,000 of a meter.

**a** Light micrograph (phase-contrast).

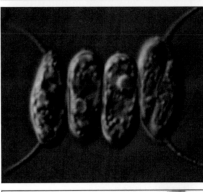

**b** Light micrograph (Nomarski process).

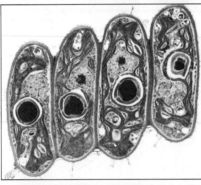

**c** Transmission electron micrograph, thin section.

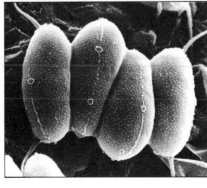

**d** Scanning electron micrograph.

**Figure 4.5** Comparison of how different types of microscopes reveal cellular details. The specimen is the green alga *Scenedesmus;* the magnification is the same in all cases. (**a**) Phase-contrast and (**b**) Nomarski techniques create optical contrasts without staining the cells; both have enhanced the value of light microscopes. (**c**) Details of a cell's internal structure show up best with transmission electron microscopy. (**d**) Scanning electron microscopy provides a three-dimensional view of surface features.

Why are most cells so small? There is a physical constraint on increases in cell size, called the **surface-to-volume ratio**. Simply put, *as a cell expands in diameter, its volume increases more rapidly than its surface area*. Figure 4.6 illustrates the relationship: Volume increases with the cube of the diameter, but surface area increases only with the square.

For example, suppose we figure out a way to make a round cell grow four times in diameter. Its cytoplasmic volume increases (4 × 4 × 4), or sixty-four times. But the surface area of its plasma membrane increases by only sixteen times (4 × 4). To survive, cells must constantly exchange materials with their surroundings. The greater the volume of cytoplasm, the more plasma membrane is required to handle the increased traffic. Unfortunately for our rotund cell, each unit of plasma membrane must now serve four times as much cytoplasm as before! Past a certain point, the inward flow of nutrients and outward flow of wastes (some toxic) is not fast enough, and we have a dead cell on our hands.

A very large, round cell would have the added problem of moving nutrients and wastes through its large volume of cytoplasm. By contrast, small or skinny cells don't have this problem. The random motion of molecules easily distributes substances through their small or stretched-out volume of cytoplasm. This molecular motion, called diffusion, is an important topic of the next chapter.

So you can see why most cells are small—or long and thin, or have outfoldings and infoldings that increase their membrane surface relative to the volume of cytoplasm. *The smaller or more stretched out or frilly-surfaced the cell, the more efficiently materials can cross its plasma membrane and diffuse through the cytoplasm.*

Surface-to-volume constraints also influence multicelled body plans. Some algae grow as delicate strands. Their cells are attached end to end, and each interacts directly with the environment. Other algae and a few protistans are sheetlike, with all cells at or near the body surface. In massive plants and animals, transport systems move materials to and from the millions, billions, even trillions of cells packed together in their tissues. That is the point of having an incessantly pumping heart and an elaborate network of blood vessels inside your body. This efficient circulatory system quickly delivers materials from the environment to the doorstep of all living cells and sweeps away wastes. Its "highways" cut through the volume of tissue and so shrink the distance that would otherwise have to be traversed by diffusion.

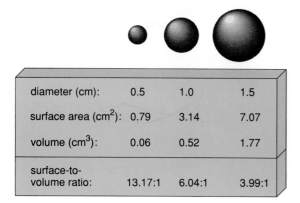

| diameter (cm): | 0.5 | 1.0 | 1.5 |
|---|---|---|---|
| surface area (cm²): | 0.79 | 3.14 | 7.07 |
| volume (cm³): | 0.06 | 0.52 | 1.77 |
| surface-to-volume ratio: | 13.17:1 | 6.04:1 | 3.99:1 |

**Figure 4.6** Relationship between the surface area and volume when a sphere is enlarged. Notice that as the diameter increases, the volume increases more rapidly than the surface area.

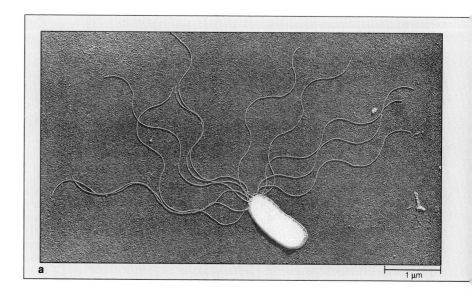

a

1 μm

# PROKARYOTIC CELLS— THE BACTERIA

Let's now turn to the characteristics of specific cell types, starting with bacteria. Bacteria are the smallest cells and, in structural terms, the simplest to think about. For most bacteria, a rigid or semirigid cell wall surrounds the plasma membrane. The wall, formed by secretions from the bacterium, supports the cell and imparts shape to it (Figure 4.7). Beneath the wall, the plasma membrane controls the movement of substances into and out of the cytoplasm. Bacterial cells have only a small volume of cytoplasm, although many ribosomes are dispersed through it.

A **ribosome** consists of two molecular subunits, each composed of RNA and protein molecules. In all cells, not just bacteria, ribosomes serve as workbenches for making proteins. At the ribosomal surface, enzymes speed the construction of polypeptide chains. Each new protein consists of one or more of those chains (page 44).

Bacterial cells are said to be prokaryotic because they do not have a nucleus. The DNA is concentrated in an irregularly shaped region of cytoplasm called the *nucleoid*. The word *prokaryotic* means "before the nucleus," and it implies that some forms of bacteria existed on earth before the evolution of cells having a nucleus. Chapters 19 and 22 provide closer looks at the evolution, structure, and functioning of bacteria. Here our focus will be on the nucleated cells, the eukaryotes.

# EUKARYOTIC CELLS

## Function of Organelles

Outside the realm of bacteria, all cells of all organisms—from the diatoms of phytoplankton to polar bears and peach trees and puffball mushrooms—are eukaryotic. Only eukaryotic cells have a profusion of organelles. **Organelles** are membranous sacs, envelopes, and other compartmented portions of the cytoplasm. The most conspicuous organelle is the nucleus. (Hence the name *eukaryotic*, which means "true nucleus.")

No chemical apparatus in the world can match the eukaryotic cell for the sheer number of chemical reactions that proceed in so small a space. Many of the reactions are incompatible. For example, a starch molecule can be put together by some reactions and taken apart by others—but a cell would gain nothing if the reactions proceeded at the same time on the same molecule. Yet reactions proceed smoothly in the cytoplasm, and they do so largely for these reasons:

Organelles physically separate chemical reactions (many of which are incompatible) into different regions of the cytoplasm.

Organelles separate different reactions in time, as when molecules are produced in one organelle, then used later in other reaction sequences.

**Figure 4.7** Bacterial body plans. (**a**) Surface view of *Pseudomonas marginalis*, which is equipped with bacterial flagella. Like other types of flagella, these structures are used in propelling the cell through its environment. (**b**) Sketch and transmission electron micrograph of *Escherichia coli*.

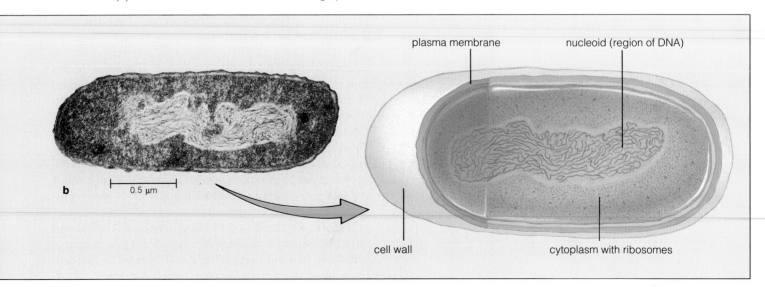

plasma membrane · nucleoid (region of DNA)

b · 0.5 μm

cell wall · cytoplasm with ribosomes

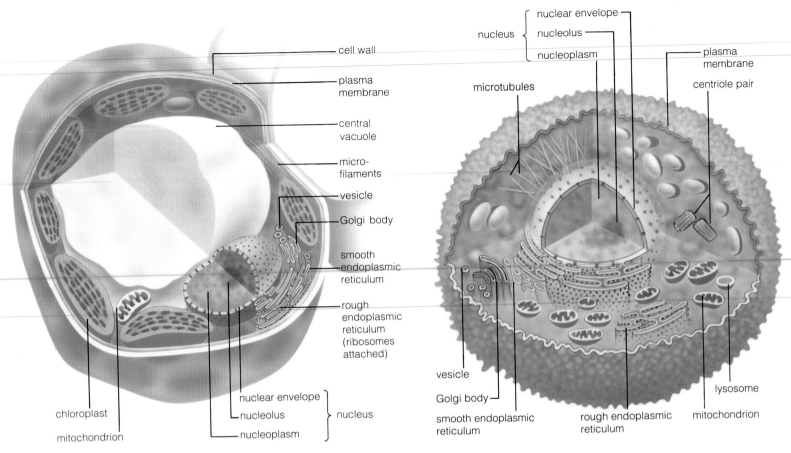

**Figure 4.8** Generalized sketch of a plant cell, showing the types of organelles that may be present.

**Figure 4.9** Generalized sketch of an animal cell, showing the types of organelles that may be present.

## Typical Components of Eukaryotic Cells

In general, eukaryotic cells contain the following organelles, each with specific functions:

| | |
|---|---|
| nucleus | *physical isolation and organization of DNA* |
| endoplasmic reticulum | *modification of polypeptide chains into mature proteins; lipid synthesis* |
| Golgi bodies | *further modification, sorting, and shipping of proteins and lipids for secretion or for use in cell* |
| lysosomes | *digestion (breakdown) within the cell* |
| transport vesicles | *transport of a variety of materials to and from organelles and plasma membrane* |
| mitochondria | *ATP formation* |

Besides the organelles just listed, eukaryotic cells have many thousands of *ribosomes*, either "free" in the cytoplasm or attached to certain membranes. They also have a *cytoskeleton*, an internal network of protein filaments. The cytoskeleton imparts shape to a cell and keeps its internal parts structurally organized. It also underlies movements of cell structures and organelles—and often of the entire cell through the environment.

Only photosynthetic cells of plants contain *chloroplasts*. Inside these organelles, sunlight energy is converted to forms that are used in the synthesis of biological molecules. One or more *central vacuoles* often occupy most of the space inside fungal and plant cells. A *cell wall* surrounds the plasma membrane of many protistan, fungal, and plant cells.

Figures 4.8 through 4.11 show where organelles and structures might be located in a typical plant and animal cell. Keep in mind that calling a cell "typical" is like calling a squid or cactus a "typical" animal or plant. Mind-boggling variations exist on the basic plan.

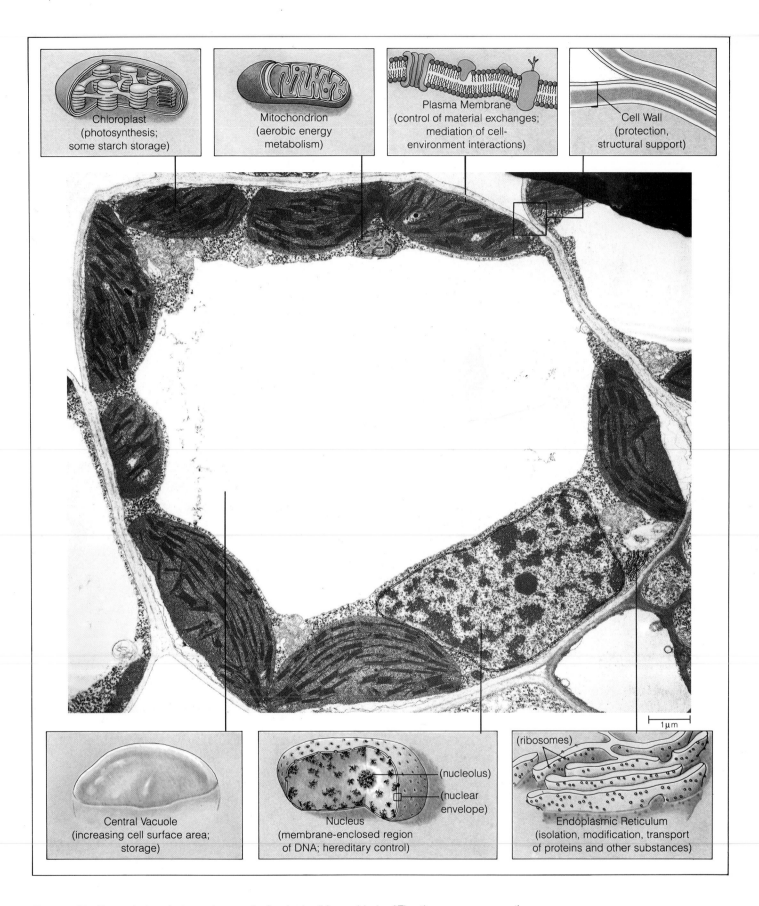

**Figure 4.10** Transmission electron micrograph of a plant cell from a blade of Timothy grass, cross-section.

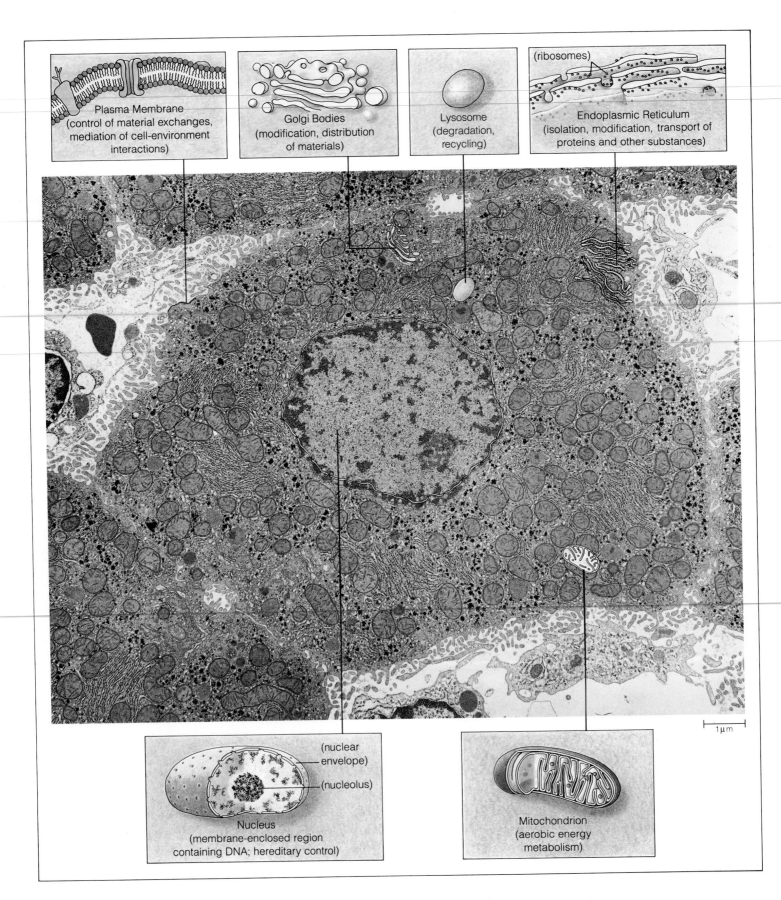

Plasma Membrane
(control of material exchanges, mediation of cell-environment interactions)

Golgi Bodies
(modification, distribution of materials)

Lysosome
(degradation, recycling)

(ribosomes)

Endoplasmic Reticulum
(isolation, modification, transport of proteins and other substances)

1 µm

(nuclear envelope)

(nucleolus)

Nucleus
(membrane-enclosed region containing DNA; hereditary control)

Mitochondrion
(aerobic energy metabolism)

**Figure 4.11** Transmission electron micrograph of an animal cell from a rat liver, cross-section.

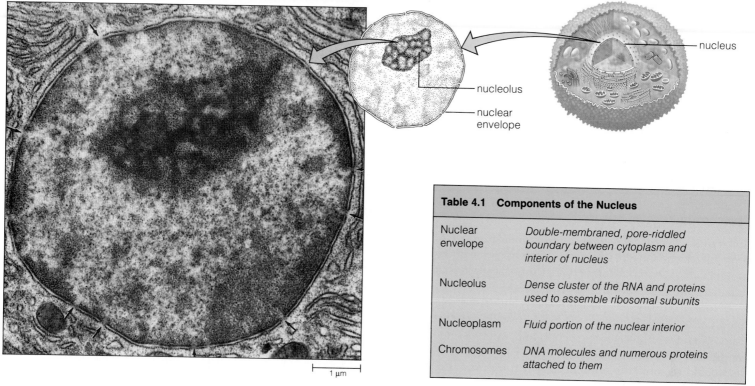

**Figure 4.12** Transmission electron micrograph of the nucleus from a pancreatic cell. Arrows point to pores in the nuclear envelope.

| Table 4.1 | Components of the Nucleus |
|---|---|
| Nuclear envelope | *Double-membraned, pore-riddled boundary between cytoplasm and interior of nucleus* |
| Nucleolus | *Dense cluster of the RNA and proteins used to assemble ribosomal subunits* |
| Nucleoplasm | *Fluid portion of the nuclear interior* |
| Chromosomes | *DNA molecules and numerous proteins attached to them* |

## THE NUCLEUS

There would be no cells whatsoever without carbohydrates, lipids, proteins, and nucleic acids. It takes a special class of proteins called enzymes to build and use those molecules. Thus, *cell structure and function begin with proteins—and instructions for building the proteins themselves are contained in DNA.*

A membrane-bound compartment, the **nucleus**, isolates the DNA in eukaryotic cells. Figure 4.12 shows its structure. Table 4.1 lists its components.

The nucleus serves two functions. First, it helps control access to the instructions contained in DNA. Second, the nucleus keeps the DNA separate from all the substances and metabolic machinery in the cytoplasm. This makes it easier to package up the DNA when the time comes for a cell to divide. The DNA can be sorted efficiently into parcels, one for each new cell that forms.

### Nuclear Envelope

Imagine a golf ball sheathed in a double layer of Saran Wrap. The outermost part of the nucleus, the **nuclear envelope**, is something like that; it has a two-membrane structure. (As with cell membranes in general, each of the two membranes of the nuclear envelope is actually two layers of lipid molecules, studded with a variety of

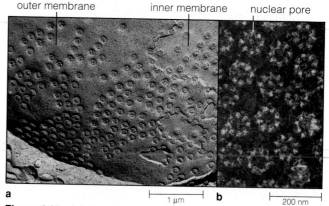

**Figure 4.13** (a) Electron micrograph of a freeze-fractured nucleus (page 78). The two membranes that form the nuclear envelope are positioned like sheets, one atop the other. Each membrane has a lipid bilayer structure, as described on page 76. (b) Closer look at the pores that cross both membranes.

proteins.) Ribosomes dot the side of the membrane facing the cytoplasm.

As Figures 4.12 and 4.13 indicate, pores occur at regular intervals over the entire nuclear envelope. Observations show that the pores serve as passageways for the controlled exchange of specific substances between the nucleus and cytoplasm.

## Nucleolus

As eukaryotic cells grow, two or more dense masses of irregular size and shape appear in the nucleus. Each mass is a **nucleolus** (plural, nucleoli). Nucleoli are sites where the protein and RNA subunits of ribosomes are assembled. The subunits are then shipped out of the nucleus, into the cytoplasm, where they come together as intact ribosomes. Figure 4.12 shows a nucleolus in a nondividing cell.

## Chromosomes

Eukaryotic DNA is threadlike, with a great number of proteins attached to it like beads on a string. Some of the proteins are enzymes. Many others form a scaffold that organizes the DNA during cell division.

Before a cell divides, its DNA molecules are duplicated (both new cells get all the DNA instructions this way). Then the duplicated molecules fold and twist into condensed structures, proteins and all. Early microscopists could see only the condensed structures, and they called them **chromosomes** ("colored bodies"). Today, we call DNA and its proteins a chromosome regardless of whether it is in threadlike or condensed form.

The point is this: *Chromosomes do not always look the same during the life of a cell.* We will consider different aspects of "the chromosome" in chapters to come, and it will help to keep this point—and the following sketch—in mind:

unduplicated, uncondensed chromosome (a DNA double helix + proteins)

duplicated uncondensed chromosome (two DNA double helices + proteins)

duplicated, condensed chromosome

## THE CYTOMEMBRANE SYSTEM

The polypeptide chains of proteins are assembled in the cytoplasm. What happens to the newly formed chains? Many are dissolved in the cytoplasm, and others enter the cytomembrane system. The **cytomembrane system** includes the *endoplasmic reticulum, Golgi bodies, lysosomes,*

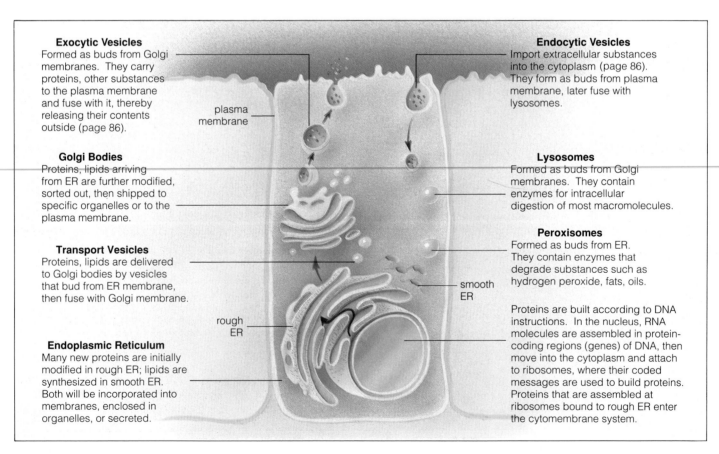

**Exocytic Vesicles**
Formed as buds from Golgi membranes. They carry proteins, other substances to the plasma membrane and fuse with it, thereby releasing their contents outside (page 86).

**Golgi Bodies**
Proteins, lipids arriving from ER are further modified, sorted out, then shipped to specific organelles or to the plasma membrane.

**Transport Vesicles**
Proteins, lipids are delivered to Golgi bodies by vesicles that bud from ER membrane, then fuse with Golgi membrane.

**Endoplasmic Reticulum**
Many new proteins are initially modified in rough ER; lipids are synthesized in smooth ER. Both will be incorporated into membranes, enclosed in organelles, or secreted.

plasma membrane

rough ER

smooth ER

**Endocytic Vesicles**
Import extracellular substances into the cytoplasm (page 86). They form as buds from plasma membrane, later fuse with lysosomes.

**Lysosomes**
Formed as buds from Golgi membranes. They contain enzymes for intracellular digestion of most macromolecules.

**Peroxisomes**
Formed as buds from ER. They contain enzymes that degrade substances such as hydrogen peroxide, fats, oils.

Proteins are built according to DNA instructions. In the nucleus, RNA molecules are assembled in protein-coding regions (genes) of DNA, then move into the cytoplasm and attach to ribosomes, where their coded messages are used to build proteins. Proteins that are assembled at ribosomes bound to rough ER enter the cytomembrane system.

**Figure 4.14** Cytomembrane system. Endoplasmic reticulum, transport vesicles, Golgi bodies, and endocytic vesicles are components of the secretory pathway of this system (upward-directed arrows).

and a variety of *vesicles*. As Figures 4.14 and 4.15 indicate, many proteins as well as lipids take on their final form and are distributed by way of this system.

## Endoplasmic Reticulum and Ribosomes

**Endoplasmic reticulum**, or ER, is a membrane with rough and smooth regions, owing largely to the presence or absence of ribosomes on the surface facing the cytoplasm. In animal cells, the ER begins at the nuclear envelope and curves through the cytoplasm.

*Rough ER* has many ribosomes attached. Often it is arranged as stacked, flattened sacs of the sort shown in Figure 4.16a. Polypeptide chains are assembled on the ribosomes. Only the newly forming chains that have a "signal" enter the sacs (the rest join the cytoplasmic pool of proteins). The signal is a sequence of about fifteen to twenty specific amino acids. As the chains pass through the rough ER, enzymes attach oligosaccharides to them. They are destined for membranes, for secretion outside the cell, or for delivery to other organelles.

Many kinds of cells specialize in secreting proteins, and rough ER is notably abundant in them. In your own body, for example, some cells of the pancreas produce and secrete proteins (enzymes) that end up in your small intestine, where they function in digestion. Proteins also are produced in quantities by immature egg cells that grow rapidly in size (frog and other amphibian eggs are like this).

*Smooth ER* is free of ribosomes and curves through the cytoplasm like connecting pipes (Figure 4.16b). It is the main site of lipid synthesis in many cells. Smooth ER is highly developed in seeds and in animal cells that secrete steroid hormones. Drugs and some harmful

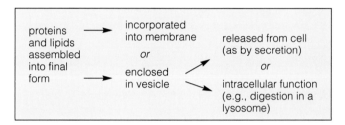

**Figure 4.15** Destination of proteins and lipids that take on their final form in the cytomembrane system.

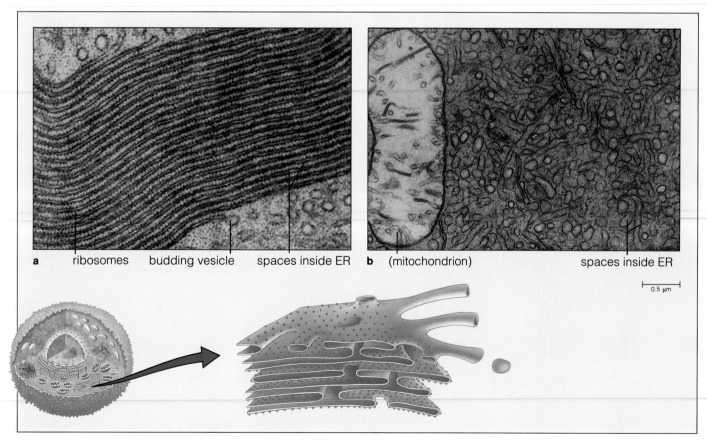

a   ribosomes   budding vesicle   spaces inside ER   b   (mitochondrion)   spaces inside ER

0.5 μm

**Figure 4.16** Endoplasmic reticulum. (**a**) Rough ER, showing how the membrane surface facing the cytoplasm is studded with ribosomes. (**b**) Smooth ER, in cross-section.

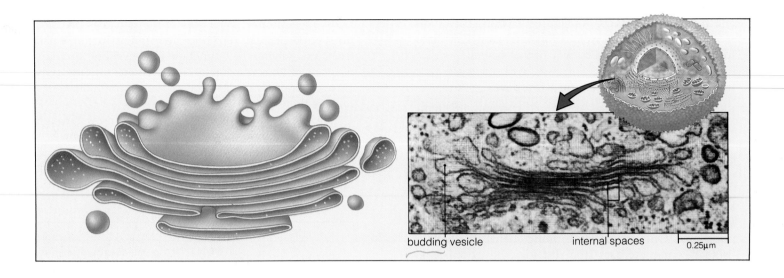

**Figure 4.17** Electron micrograph and sketch of a Golgi body.

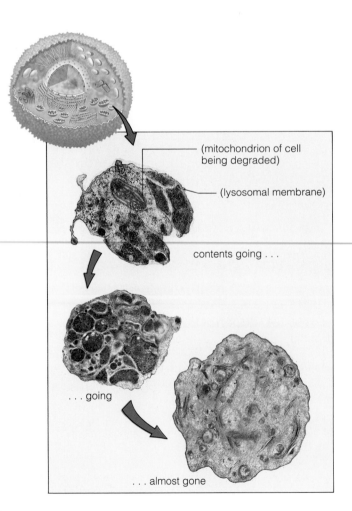

**Figure 4.18** Digestion of organelles from a destroyed cell, as seen in a lysosome.

by-products of metabolism are inactivated in the smooth ER of liver cells. One type of smooth ER (the sarcoplasmic reticulum) occurs in skeletal muscle; it stores and releases calcium ions that play a role in muscle contraction.

## Golgi Bodies

In **Golgi bodies**, many proteins and lipids undergo final modification, then they are sorted out and packaged for specific destinations. Here, polysaccharides that were attached to each protein in the ER become trimmed and embellished in specific ways. The final results allow components of the Golgi membrane to "recognize" differences among many products and to form vesicles with special mailing tags around them.

All eukaryotic cells have one or more Golgi bodies. In outward appearance, a Golgi body resembles a stack of pancakes—usually eight or less, and usually curled at the edges (Figure 4.17). Each "pancake" is a flattened compartment. The topmost pancakes bulge at the edges, then the bulges break away as vesicles. Some secretory cells concentrate and store products in such vesicles until the cell is signaled to release them. Figure 4.14 shows where the Golgi bodies fit in the secretory pathway.

## Assorted Vesicles

In animals, some of the vesicles budding from Golgi bodies becomes **lysosomes**, the main organelles of digestion inside the cell. Forty or so enzymes in these membrane bags can break down every polysaccharide, nucleic acid, and protein, along with some lipids. (Comparable enzymes are in central vacuoles of plant cells.) Lysosomes fuse with vesicles that carry a variety of substances or damaged cell parts to be degraded (Figure 4.18). They also can destroy bacteria and foreign particles.

Some vesicles that bud from the ER are called *peroxisomes*. They contain enzymes that use oxygen to break down fatty acids and amino acids. Hydrogen peroxide, a potentially harmful substance, is a product of the reactions. Another enzyme converts the hydrogen peroxide to water and oxygen or uses it to break down alcohol. If you drink alcohol, nearly half of it is degraded in peroxisomes of your liver and kidney cells.

Another type of vesicle, the *glyoxysome*, is abundant in lipid-rich seeds, such as those of peanut plants. Its enzymes help convert stored fats and oils to the sugars necessary for rapid, early growth of the plant.

## MITOCHONDRIA

As mentioned earlier in the book, the energy of ATP drives nearly all cell activities. Eukaryotic cells with high demands for ATP rely heavily on the **mitochondrion** (plural, mitochondria). This organelle specializes in liberating the energy stored in sugars and using it to form *many* ATP molecules. Mitochondria extract far more energy from sugars and fats than can be done by any other means, and they do so with the help of oxygen. When you breathe in, you are taking in oxygen for your mitochondria.

Some cell types have only a sprinkling of mitochondria. The cells with the greatest demands for ATP may have more than a thousand. Not surprisingly, mitochondria are especially profuse in muscle cells, parts of nerve cells, and cells that specialize in absorbing or secreting substances.

In size and certain biochemical features, mitochondria resemble bacteria. As described on page 361, they may have originated from ancient bacteria that had been engulfed by predatory cells. In brief, by this scenario, the bacteria managed to escape digestion. They even thrived and reproduced in the cytoplasm of the predatory cell and *its* descendants. As they became permanent, protected residents, they lost many structures and functions necessary for independent life. In time the stripped-down bacterial descendants became mitochondria.

In shape, mitochondria can resemble balls, lumpy potatoes, tubes, or threads. But their shapes change. Depending on chemical conditions in the cell, mitochondria grow and branch out, even fuse with one another and divide in two. Each mitochondrion has an outer membrane facing the cytoplasm and an inner membrane, usually with many deep, inward folds called cristae (Figure 4.19). The double-membrane system creates two compartments. Hydrogen ions move from one compartment to the other in ways that cause ATP to form. This important process is described more fully in Chapter 8.

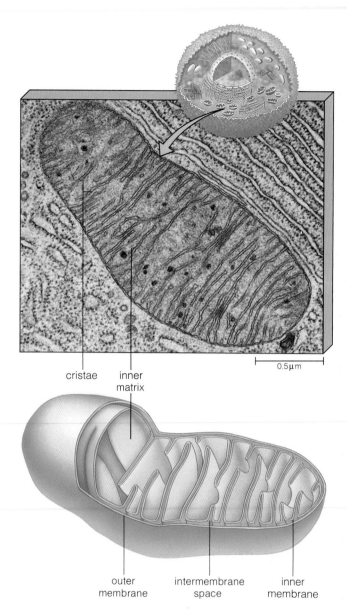

cristae   inner
          matrix

outer          intermembrane        inner
membrane           space          membrane

**Figure 4.19** Micrograph and generalized sketch of a mitochondrion.

## SPECIALIZED PLANT ORGANELLES

### Chloroplasts and Other Plastids

Most plant cells contain one or more "plastids," a category of organelles specialized for photosynthesis and storage. Three kinds are common:

chloroplasts   *with photosynthetic pigments and starch-storing capacity*

chromoplasts   *with pigments that often may function in pollination and seed dispersal by visually attracting animals*

amyloplasts   *with starch-storing capacity; no pigments*

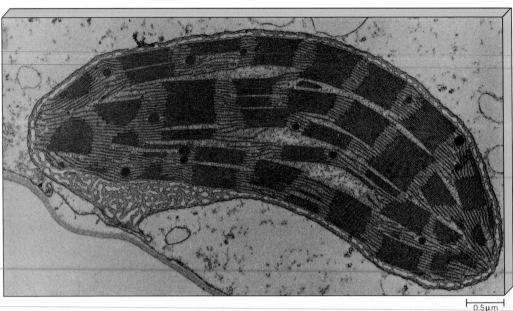

**Figure 4.20** Micrograph and generalized sketch of a chloroplast. A semifluid substance (the stroma) surrounds an elaborate system of membrane compartments. Commonly, many of the compartments are organized as stacks of flattened disks (grana, singular, granum). Like mitochondria, chloroplasts have a double-membrane envelope.

0.5μm

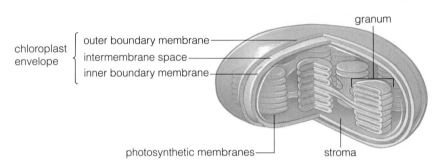

chloroplast envelope { outer boundary membrane — intermembrane space — inner boundary membrane —

granum

photosynthetic membranes —

stroma

The **chloroplast** functions in photosynthesis. It has a double-membrane envelope that surrounds a semifluid substance, *stroma*.

Within the chloroplast's stroma is an elaborate system of flattened membrane compartments. These interconnect with one another, and it is common to see many disk-shaped compartments organized into stacks (Figure 4.20). Each stack is a *granum* (plural, grana). Pigments, enzymes, and other molecules of the membrane system trap sunlight energy and have roles in ATP formation.

In the stroma, enzymes speed the assembly of sugars, starch, and other products of photosynthesis. Clusters of new starch molecules ("starch grains") are often stored briefly inside the chloroplast.

Chloroplasts often are oval or disk-shaped and may be green, yellow-green, or golden-brown. Their color depends on the kinds and amounts of light-absorbing pigment molecules in their membranes. Chlorophyll, a green pigment, is an example. Chlorophyll is present in all chloroplasts, but it may be masked by other pigments, as it is in brown algae.

Chromoplasts store red or brown pigments that give flower petals, fruits, and some roots (such as carrots) their characteristic colors. The colorless amyloplasts, which occur in plant parts exposed to little (if any) sunlight, are often storage sites for starch grains. They are abundant in cells of stems, potato tubers, and many seeds.

**Central Vacuoles**

Mature, living plant cells often have a large, fluid-filled **central vacuole** (Figure 4.10). This organelle usually occupies 50 to 90 percent of the cell interior, so there is only a narrow zone of cytoplasm between the vacuole and the plasma membrane.

A central vacuole can store amino acids, sugars, ions, and toxic wastes. It also increases cell surface area. During growth, cell walls enlarge under the force of water pressure that builds up inside the vacuole. The cell also enlarges, but its cytoplasm is "stretched out" between the vacuole and wall. The improved surface-to-volume ratio enhances mineral absorption.

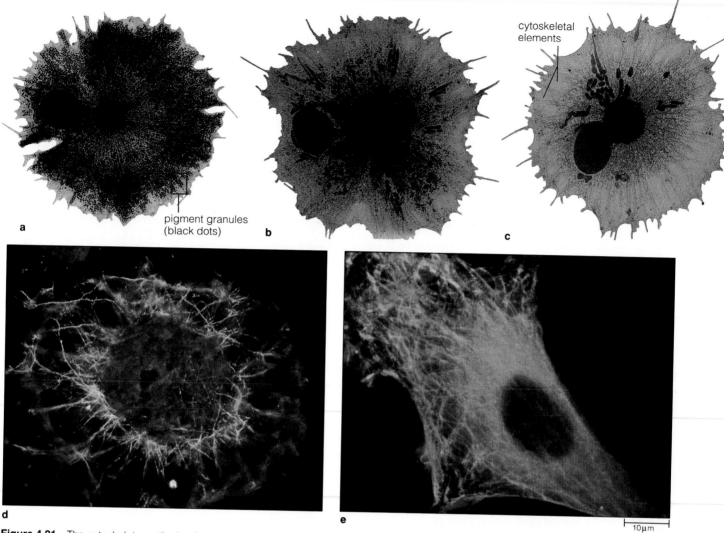

cytoskeletal elements

pigment granules (black dots)

a    b    c

d    e

10μm

**Figure 4.21** The cytoskeleton—the basis of a cell's shape, internal organization, and capacity for motion.

**(a-c)** High-voltage electron micrographs of a pigment-containing cell from a squirrelfish. Amphibian skin and fish scales often have pigment-containing cells that collectively cause color changes. By changing color, the animal often can blend better with its surroundings and so escape the attention of predators.

The color darkens when, in response to signals from the nervous and endocrine systems, pigment granules are moved to the periphery of these cells (**a**). The color lightens when granules become condensed near the center of the cells (**b**). Granules are moved rapidly along abundant tracks of microtubules and other cytoskeletal elements (**c**).

**(d)** Cytoskeleton of a plant cell (African blood lily), made visible with *fluorescence microscopy.* By this process, cells take up molecules that bind only to certain types of proteins—*after* those molecules have been labeled with fluorescent dyes. The glow from the bound molecules marks the location of different proteins. Here, the green filaments are microtubules, composed of the protein tubulin. They are outside the nucleus, in which chromosomes (stained purple) are clustered.

**(e)** Cytoskeleton of a fibroblast, a cell that gives rise to certain connective tissues in animals. Fluorescence microscopy reveals the location of three different proteins. In this composite of three images, actin (blue) and vinculin (red) are associated with microfilaments. Tubulin (green) is associated with microtubules.

## THE CYTOSKELETON

### Components of the Cytoskeleton

Each cell type has a characteristic shape and internal organization made possible by its own tiny **cytoskeleton**. This interconnected system of bundled fibers, slender threads, and lattices extends from the nucleus all the way to the plasma membrane (Figure 4.21).

Some parts of the cytoskeleton are *transient;* they appear and disappear at different times in the life of a cell. Before a cell divides, for instance, some fibers assemble into a "spindle" structure that attaches to chromosomes and moves them about, then they disassemble when the task is done. But many parts of the cytoskeleton are *permanent.* These include the filaments in skeletal muscle cells, which are the basis of contraction. Other

**Figure 4.22** Three major classes of protein fibers making up the cytoskeleton of eukaryotic cells. None occurs in prokaryotic cells. Microtubules consist of globular protein subunits (tubulins) linked in parallel rows. The ones involved in cell movements are typically assembled and disassembled within minutes.

Intermediate filaments are unique to specialized types of animal cells. For example, some help form "spot welds" that help hold adjacent cells together in certain tissues. The intermediate filaments in different cell types are composed of different protein subunits.

The protein actin is the key subunit of microfilaments. Actin is present in all eukaryotic cells. Often it is the most abundant cytoplasmic protein. In contractile cells, it functions in association with thick fibers composed of myosin. Myosin is another type of microfilament (Chapter 36).

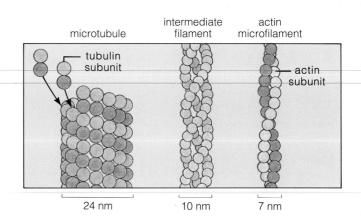

**Figure 4.23** (**a**) Scanning electron micrograph showing the hairlike cilia on the surface of *Paramecium*. (**b**) The 9 + 2 array of microtubules in a cilium or flagellum.

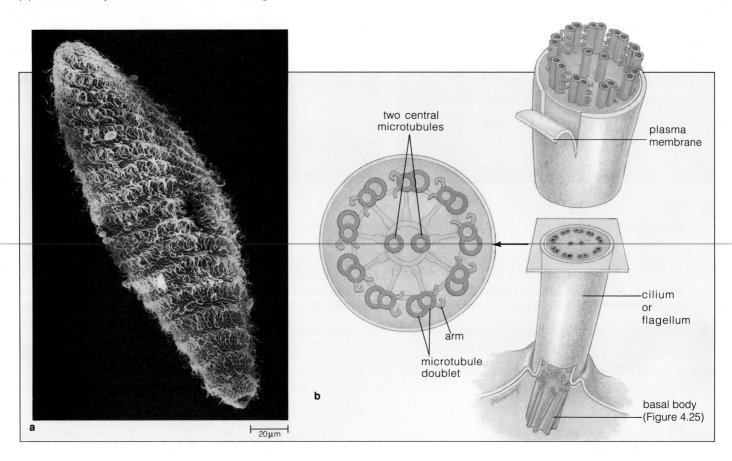

examples are flagella and cilia, two kinds of motile structures that will be described shortly.

*Microtubules* and *intermediate filaments* are the main cytoskeletal elements. *Microfilaments* occur in some specialized animal cells. All three types are assembled from protein subunits (Figure 4.22).

## Flagella and Cilia

Microtubular structures propel many free-living eukaryotic cells through their surroundings. These structures are **flagella** (singular, flagellum). Certain protistans have one or more flagella. Dinoflagellates, a large contingent in marine phytoplankton, have a pair of them. Human sperm cells have one flagellum that makes up most of their length.

**Cilia** (singular, cilium), a similar kind of microtubular structure, typically are arrayed at the cell surface (Figure 4.23). Many free-living cells use cilia for propulsion. Cells with fixed positions in some tissues use cilia for stirring up their surroundings. For example, airborne bacteria and other particles are drummed out of your lungs when many thousands of cells lining the air tubes beat their cilia in coordination.

Cilia are shorter and more numerous than flagella, but both have the same organization. Nine pairs of microtubules ring two central microtubules, in what is called a *9 + 2 array.* An extension of the plasma membrane surrounds the array. Figure 4.24 describes how interactions between microtubules cause the cilium or flagellum to bend, this being the basis of propulsion.

## MTOCs and Centrioles

As each new cell develops, microtubules influence what its shape will be. We know, for example, that cellulose strands maintain the shape of plant cell walls. But a temporary "scaffold" of microtubules laid down earlier by those cells during their growth guides the placement of cellulose deposits in the newly forming walls. So the question becomes this: *What organizes the microtubules?*

The organization and orientation of microtubules depend on the number, type, and location of small collections of proteins and other substances in the cytoplasm. Each such collection is a **microtubule organizing center** (MTOC). In most animal cells, a prominent MTOC near the nucleus also includes a pair of centrioles. **Centrioles** are small cylinders composed of triplet microtubules (Figure 4.25). While DNA is being duplicated before cell division, centrioles also are duplicated; a new one grows at right angles to the parent structure.

Centrioles play a vital role in ciliated or flagellated cells. When such cells are first forming, a centriole moves away from the nucleus and through the cytoplasm, then becomes positioned near the plasma membrane. There,

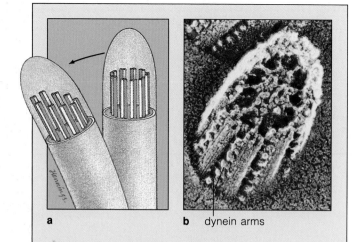

**a**    **b** dynein arms

**Figure 4.24** Movement of cilia and flagella. (**a**) In an unbent cilium or flagellum, all microtubule doublets extend the same distance into the tip. With bending, the doublets on the outside of the arc are displaced farthest from the tip. This relationship shows that microtubule doublets slide over each other, rather than contracting, when the cilium or flagellum bends.

(**b**) Many clawlike "arms" extend at regular intervals along the length of each microtubule doublet in the outer ring. The arms are composed of dynein, an ATP-hydrolyzing enzyme. All arms protrude in the same direction—toward the next doublet in the ring. When dynein binds and splits an ATP molecule, the angle of the arm changes with respect to the doublet in front of it. The arm bends and is strongly attracted to the doublet in front of it. On contact, the arm "unbends" with great force and causes its neighbor to move. The dynein releases its hold, after which it can grab another ATP molecule and attach to a new binding site on the doublet.

In other words, dynein arms on one doublet swing back and forth like tiny oars, displacing the neighboring doublet with each oarlike arc. The displacement produces the doublet sliding responsible for bending the flagellum or cilium.

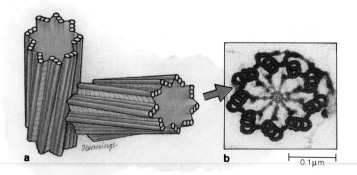

**a**    **b**    0.1μm

**Figure 4.25** (**a**) A pair of centrioles, which occur near the nucleus of many cells. Centrioles apparently help organize the cytoskeleton. In many species they become basal bodies, which give rise to the microtubular core of cilia and flagella. (**b**) Electron micrograph of a basal body, thin section, from a protistan (*Saccinobacculus*).

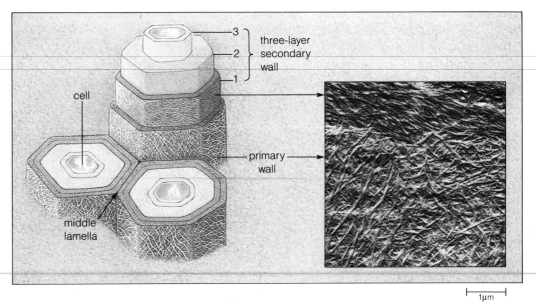

**Figure 4.26** Primary and secondary walls of plant cells as they would appear in partial cross-section. The micrograph shows primary and secondary walls of a plant cell. The strands are largely cellulose.

it gives rise to the microtubules that form the core structure of a cilium or flagellum (Figure 4.23). After giving rise to the microtubules, the centrioles remain attached to the motile structure as a **basal body**.

At one time, centrioles also were thought to direct cell division. Yet division proceeds as usual in flowering plants, conifers, and other species that have no centrioles. It also proceeded even when the centrioles near the nucleus of a fertilized mouse egg were experimentally removed. However, those centrioles may govern the *plane* of cell division.

During development, each cell must divide at a prescribed angle relative to the other cells. The successive division planes influence the shape of the developing embryo and the adult form. Intriguingly, a mouse embryo did result from repeated divisions of the centriole-deprived cells. But the divisions followed a disorganized pattern, so the embryo was deformed!

## CELL SURFACE SPECIALIZATIONS

### Cell Walls

For many cells, surface deposits outside the plasma membrane form coats, capsules, sheaths, or walls. **Cell walls** occur among bacteria, protistans, fungi, and plants. Animal cells do not produce walls, although some secrete products to the surface layer of tissues in which those cells occur.

Most cell walls have carbohydrate frameworks. They generally provide support and resist mechanical pressure, as when they keep plant cells from stretching too much while the cells expand with incoming water during growth. Even the most solid-looking walls have microscopic spaces that make them porous, so water and solutes can move to and from the plasma membrane.

In new plant cells, cellulose strands are bundled together and added to a developing *primary cell wall*. After the main growth phase, many types of plant cells also deposit materials to form an inner, rigid *secondary cell wall*, which often consists of several layers (Figure 4.26). Cutin, suberin, and waxes commonly are embedded in cell walls at or near the plant's surface. They play a protective role and help reduce water loss.

### Extracellular Matrix and Cell Junctions

In multicelled organisms, each living cell must interact with its physical surroundings, but it also must interact with its cellular neighbors. At the surface of organs, body cavities, or the body itself, cells must link tightly together so that the interior of the organism (or organ) is not indiscriminately exposed to the outside world. In all tissues, cells of the same type must recognize one another and physically stick together. Finally, in tissues where cells must act in coordinated fashion, the cells must share channels to exchange signals, nutrients, or both. Heart muscle cells are an example.

**Extracellular Matrix.** In animals, a meshwork of macromolecules called the *extracellular matrix* holds the cells of many tissues together. The shape of the matrix influences how cells of a given tissue will divide and what their shape will be. Its composition influences cell metabolism. Components of the matrix often include collagen and other fibrous proteins, glycoproteins, and specialized polysaccharides that form a jellylike or watery "ground substance." Nutrients, hormones, and other molecules readily diffuse from cell to cell through the ground substance. Much of the body weight of vertebrates consists of extracellular matrix material, as in bone.

In multicelled plants, adjacent cells are cemented together at their primary walls (Figure 4.26). The cementing material (the middle lamella) has an abundance of pectin compounds. Home cooks use some of those pectins to "bind" jams and jellies.

**Cell Junctions in Animals.** Cell-to-cell junctions are common in animal tissues. They are illustrated in Chapter 31, but here we can mention the three most common types. *Tight junctions* occur between cells of epithelial tissues, which line the body's outer surface, inner cavities, and organs. They form tight seals that keep molecules from freely crossing the epithelium and entering deeper tissues. Such seals keep stomach acids from leaking into other tissues, for example.

*Adhering junctions* are like spot welds at the plasma membranes of two adjacent cells. They help hold cells together in tissues that are subject to stretching, such as epithelium of the skin, heart, and stomach.

At *gap junctions,* small, open channels directly link the cytoplasm of adjacent cells. In heart muscle and smooth muscle, gap junctions provide rapid communication between cells. In liver and other tissues, they allow small molecules and ions to pass directly from one cell to the next.

**Cell Junctions in Plants.** In land plants, living cells are linked wall-to-wall, not membrane-to-membrane. However, channels called *plasmodesmata* (singular, plasmo-

desma) extend across adjacent walls and connect the cytoplasm of neighboring cells. There can be 1,000 to 100,000 plasmodesmata penetrating the walls of a cell. The total number affects the rate at which nutrients and other substances are transported between adjacent cells.

We will be returning to such cell-to-cell interactions. For now, the point to remember is this: *In multicelled organisms, coordinated cell activities depend on specialized forms of linkage and communication between cells.*

## SUMMARY

1. The cell theory has three main points. First, all living things are made of one or more cells. Second, each cell is the basic living unit. It either can exist independently or has the potential to do so. Third, a new cell arises only from cells that already exist.

2. At the minimum, all cells have a nucleus (or nucleoid, in the case of bacteria), cytoplasm, and a plasma membrane.

3. The plasma membrane separates internal cellular events from the environment so that they can proceed in organized ways. It also acts as a boundary for exchanges between the cell's interior and its surroundings.

4. Eukaryotic cells have a variety of organelles—membranous compartments concerned with acquiring and using energy, building molecules, and tearing down molecules in controlled, specialized ways.

5. Organelle membranes separate different chemical reactions in the space of the cytoplasm and so allow them to proceed in orderly fashion.

6. Table 4.2 on the next page summarizes the organelles and other structures found in both prokaryotic and eukaryotic cells.

### Review Questions

1. State the three principles of the cell theory. *52*

2. All cells share three features: a nucleus (or nucleoid), cytoplasm, and a plasma membrane. Describe the functions of each. *53*

3. Why is it highly improbable that you will ever encounter a predatory two-ton living cell on the sidewalk? *56*

4. Suppose you want to observe details of the surface of an insect's compound eye. Would you benefit most from a compound light microscope, transmission electron microscope, or scanning electron microscope? *54–55*

5. Are all cells microscopic? Is the micrometer used in describing whole cells or extremely small cell structures? Is the nanometer used in describing whole cells or cell ultrastructure? *53*

6. Eukaryotic cells generally contain these organelles: nucleus, endoplasmic reticulum, Golgi bodies, and mitochondria. Describe the function of each. *61–65*

7. What are the components of the cytomembrane system? Sketch their general arrangement, from the nuclear envelope to the plasma

**Table 4.2  Summary of Typical Components of Prokaryotic and Eukaryotic Cells**

| Cell Component | Function | Prokaryotic | Eukaryotic | | | |
| --- | --- | --- | --- | --- | --- | --- |
| | | Moneran | Protistan | Fungus | Plant | Animal |
| Cell wall | Protection, structural support | ✓* | ✓* | ✓ | ✓ | none |
| Plasma membrane | Regulation of substances moving into and out of cell | ✓ | ✓ | ✓ | ✓ | ✓ |
| Nucleus | Physical isolation and organization of DNA | none | ✓ | ✓ | ✓ | ✓ |
| DNA | Encoding of hereditary information | ✓ | ✓ | ✓ | ✓ | ✓ |
| RNA | Transcription, translation of DNA messages into specific proteins | ✓ | ✓ | ✓ | ✓ | ✓ |
| Nucleolus | Assembly of ribosomal subunits | none | ✓ | ✓ | ✓ | ✓ |
| Ribosome | Protein synthesis | ✓ | ✓ | ✓ | ✓ | ✓ |
| Endoplasmic reticulum | Modification of many proteins into mature form; lipid synthesis | none | ✓ | ✓ | ✓ | ✓ |
| Golgi body | Final modification of proteins, lipids; sorting and packaging them for shipment inside cell or for export | none | ✓ | ✓ | ✓ | ✓ |
| Lysosome | Intracellular digestion | none | ✓ | ✓* | ✓* | ✓ |
| Mitochondrion | ATP formation | ** | ✓ | ✓ | ✓ | ✓ |
| Photosynthetic pigment | Light-energy conversion | ✓* | ✓* | none | ✓ | none |
| Chloroplast | Photosynthesis, some starch storage | none | ✓* | none | ✓ | none |
| Central vacuole | Increasing cell surface area, storage | none | none | ✓* | ✓ | none |
| Cytoskeleton | Cell shape, internal organization, basis of cellular motion | none | ✓* | ✓* | ✓* | ✓ |
| Complex flagellum, cilium | Movement | none | ✓* | ✓* | ✓* | ✓ |

*Known to occur in at least some groups.
**Aerobic reactions do occur in many groups, but mitochondria are not involved.

membrane, and describe the role of each in the flow of materials between these two boundary layers. *62–65*

8. Lysosomes dismantle and dispose of malfunctioning organelles and foreign particles. Can you describe how? *64*

9. Describe the structure and function of chloroplasts and mitochondria. Mention the ways in which they are similar. *65–66*

10. Is this statement true or false? All chloroplasts are plastids, but not all plastids are chloroplasts. *65*

11. What are the functions of the central vacuole in mature, living plant cells? *66*

12. What is a cytoskeleton? How do you suppose it might aid in cell functioning? *67–69*

13. Are all components of the cytoskeleton permanent? *67*

14. What gives rise to the microtubular array of cilia and flagella? Distinguish between a centriole and a basal body. *69–70*

15. Cell walls occur among which organisms: bacteria, protistans, plants, fungi, or animals? Are cell walls solid or porous? 70

16. In plants, is a secondary cell wall deposited inside or outside the surface of the primary cell wall? Do all plant cells have secondary walls? 70

17. What are some functions of the extracellular matrix in animal tissues? 70–71

18. In multicelled organisms, coordinated interactions depend on linkages and communication between adjacent cells. What types of junctions occur between adjacent animal cells? Plant cells? 71

19. With a sheet of paper, cover the Table 4.2 column entitled Function. Can you now name the primary functions of the cell structures listed in this table?

20. Having done the preceding exercise, can you now write a paragraph describing the differences between prokaryotic and eukaryotic cells?

## Self-Quiz *(Answers in Appendix IV)*

1. _____ are the smallest units that have complex organization, show metabolic activity, reproduce, and exhibit other characteristics of life.

2. The plasma membrane _____
   a. surrounds an inner region of cytoplasm
   b. separates the nucleus from the cytoplasm
   c. separates internal cell events from the environment
   d. acts as a nucleus in prokaryotic cells
   e. only a and c are correct

3. Which is *not* a key point of the cell theory?
   a. all living things are made of one or more cells
   b. the cell is the basic living unit
   c. no cell can exist unless at least one other cell is present
   d. new cells arise only from cells that already exist

4. Unlike eukaryotic cells, prokaryotic cells _____
   a. do not have a plasma membrane
   b. have RNA, not DNA
   c. do not have a nucleus
   d. all of the above

5. Organelles _____
   a. are membrane-bound compartments
   b. are typical of eukaryotic cells, not prokaryotic cells
   c. separate chemical reactions in time and space
   d. all of the above are functions of organelles

6. Plant cells but not animal cells have _____ .
   a. mitochondria
   b. a plasma membrane
   c. ribosomes
   d. a cell wall

7. Eukaryotic DNA is contained within the _____ .
   a. central vacuole
   b. nucleus
   c. lysosome
   d. Golgi body
   e. b and d are correct

8. The cytomembrane system does *not* include:
   a. ER
   b. transport vesicles
   c. Golgi bodies
   d. plastids
   e. all of the above are parts of the system

9. The _____ is responsible for cell shape, internal structural organization, and cell movement.

10. Match each organelle with its correct function.
    _____ protein synthesis
    _____ movement
    _____ intracellular digestion
    _____ modification of new proteins
    _____ lipid synthesis
    _____ photosynthesis
    _____ ATP formation
    _____ ribosome subunit assembly

    a. mitochondrion
    b. chloroplast
    c. ribosome
    d. smooth ER
    e. rough ER
    f. nucleolus
    g. lysosome
    h. flagellum

## Selected Key Terms

basal body 70
cell 52
cell theory 52
cell wall 70
central vacuole 66
centriole 69
chloroplast 65
chromosome 62
cilium 69
cytomembrane system 62
cytoplasm 53
cytoskeleton 67
endoplasmic reticulum (ER) 63
eukaryotic cell 57
flagellum 69
Golgi body 64
intermediate filament 71

lysosome 64
microfilament 71
microtubule 71
microtubule organizing center (MTOC) 69
mitochondrion 58, 65
nuclear envelope 61
nucleoid 57
nucleolus 62
nucleus 53, 61
organelle 57
plasma membrane 53
plastid 65
prokaryotic cell 57
ribosome 57
surface-to-volume ratio 56

## Readings

Alberts, B., et al. 1989. *Molecular Biology of the Cell.* Second edition. New York: Garland.

Bloom, W., and D. Fawcett. 1986. *A Textbook of Histology.* Eleventh edition. Philadelphia: Saunders. Outstanding reference book on cell structure.

deDuve, C. 1985. *A Guided Tour of the Living Cell.* New York: Freeman. Beautifully illustrated introduction to the cell; two short volumes.

Kessel, R., and C. Shih. 1974. *Scanning Electron Microscopy in Biology.* New York: Springer-Verlag. Stunning micrographs.

Rothman, J. September 1985. "The Compartmental Organization of the Golgi Apparatus." *Scientific American* 253(3):74–89. Describes the three specialized compartments of Golgi bodies.

Weber, K., and M. Osborn. October 1985. "The Molecules of the Cell Matrix." *Scientific American* 253(4):110–120. Summarizes techniques used to study the cytoskeleton.

Weibe, H. 1978. "The Significance of Plant Vacuoles." *Bioscience* 28: 327–331.

# 5 MEMBRANE STRUCTURE AND FUNCTION

## Water, Water Everywhere

Not one living thing on earth can survive without water—the stuff that bathes cells inside and out, donates its molecules to metabolic reactions, and dissolves vital ions and keeps them from settling in a heap. But "water" in one place may not be the same as "water" in another. Seawater is much saltier than lake water (many more ions are dissolved in it), and some seas are saltier than others. And not all places on earth have a continuous supply of the stuff. Water obviously is available year-round in seas and lakes, but it comes and goes in brief, seasonal pulses in deserts and it stays frozen in the far north except for slushy surface thaws during the brief summer months.

If all goes well, each organism holds on to enough water and dissolved ions—not too little, not too much—to maintain the structure and metabolic activities of its living cells. Both single-celled and multicelled organisms have mechanisms by which they conserve the volume and ionic composition of the fluids inside them.

But who is to say that life consistently goes well? Think about a hard-shelled goose barnacle drifting offshore on a log, more or less at the mercy of ocean currents. For better or worse, the log is its home. A stalk at one end of this marine organism is attached to the wood, and featherlike appendages at the other end extend into the water, collecting microscopic bits of food (Figure 5.1a). The salty fluid inside the living cells that make up the barnacle body is in balance with the surrounding seawater. But suppose the log drifts by chance into the dilute waters from a melting glacier (Figure 5.1b). The balance is upset, and the goose barnacle has no way to deal with this. Its cells become horrifically stressed, and it is likely that the barnacle may die.

The same thing can happen to burrowing worms and other soft-bodied organisms that live between the high and low tide marks along a rocky shore. During an unpredictably heavy storm, when the seawater becomes dilute with runoff from the land, the balance between

**Figure 5.1** Goose barnacles, which live attached to logs and other floating objects in the seas. Their featherlike appendages extend from the shell and comb the water for edibles. The cells of these animals are vulnerable to changes in salt concentration. If their log home were to drift into glacial meltwaters—which are low in salinity—the animals would die. In the photograph to the right, the dark blue water has a high salt concentration; the lighter blue water is flowing out from the glacier in the background.

a

the fluid in their cells and the surrounding water is similarly upset. At such times, the resulting deaths in the intertidal zone reach catastrophic proportions.

With this example we begin to see the cell for what it is: a tiny, organized bit of life in a world that is, by comparison, unorganized and sometimes harsh. *No matter what goes on outside, the cell must bring in certain substances, release or keep out others, and conduct its internal activities with great precision.* For this bit of life, the bastion against disorganization is the **plasma membrane**—a thin, seemingly flimsy surface layering of little more than lipids and proteins, dotted here and there with guarded passageways. Across this membrane, materials are exchanged between the cytoplasm and the surroundings. Then, within the cytoplasm of eukaryotic cells, exchanges are made across **internal cell membranes**, which form the compartments called organelles. This most fundamental of all cell structures, the membrane, will be our focus here.

b

KEY CONCEPTS

**1.** Cell membranes are composed mainly of phospholipids and proteins. The phospholipids form a double layer (bilayer) that gives the membrane its basic structure and serves as a barrier to water-soluble substances. The proteins carry out most membrane functions.

**2.** The plasma membrane helps control the types and amounts of substances moving into and out of cells. It also has built-in mechanisms for receiving chemical signals from the outside and for chemically recognizing other cells.

**3.** Internal cell membranes form compartments, which separate the many different metabolic reactions that proceed in the space of the cytoplasm.

**4.** Simple diffusion is the natural, unassisted movement of a solute from one region to another region where it is not as concentrated. Many small molecules having no net charge diffuse across cell membranes. Membrane transport proteins assist ions and many large molecules across.

**5.** Some transport proteins are open channels for water-soluble substances; some have molecular gates that open in controlled ways. Other transport proteins (carriers) bind solutes and shunt them across the membrane, either passively or actively (after receiving an energy boost).

## FLUID MEMBRANES IN A LARGELY FLUID WORLD

### The Lipid Bilayer

Fluid bathes the outer surface of living cells and fills most of their interior. You might think that the plasma membrane would have to be a rather solid structure, given that both sides of it are immersed in fluid, yet the plasma membrane is fluid, too! When you puncture a cell with a fine needle, the cell does not lose cytoplasm when the needle is withdrawn. Instead, the cell surface seems to flow over and seal the puncture. How can a fluid cell membrane remain distinct from fluid surroundings? The answer is found in the properties of lipid molecules.

Lipid molecules cluster spontaneously when surrounded by water. Consider the phospholipids, the most abundant type of lipid in cell membranes. A **phospholipid**, recall, has a hydrophilic (water-loving) head and two fatty acid tails, which are largely hydrophobic (water-dreading):

- polar (hydrophilic) head

- nonpolar (hydrophobic) tails

When many phospholipid molecules are immersed in water, hydrophobic interactions may force them together into two layers, with all the fatty acid tails sandwiched between the hydrophilic heads. This arrangement is called a **lipid bilayer**, and it is the structural basis of all cell membranes:

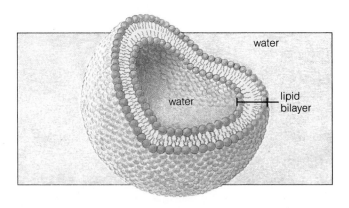

water

water

lipid bilayer

Because lipid bilayers minimize the number of hydrophobic groups exposed to water, the fatty acid tails do not have to spend a lot of energy fighting the water molecules, so to speak. Thus the reason a punctured plasma membrane tends to seal itself is that the puncture is energetically unfavorable (it leaves too many hydrophobic groups exposed).

Ordinarily, of course, few cells are ever punctured with fine needles. But the self-sealing behavior of their membrane lipids is useful for more than damage control. This behavior underlies the formation of vesicles, as described briefly in the preceding chapter. When vesicles bud off ER or Golgi membranes, for example, hydrophobic interactions with water molecules of the surrounding fluids push lipid molecules together and so close off the breakaway sites. The same thing happens when some of the vesicles fuse with the plasma membrane, and when pinched-off bits of the plasma membrane become vesicles that move into the cytoplasm. We will return to this membrane behavior later in the chapter.

## Fluid Mosaic Model of Membrane Structure

Three types of lipids are common in cell membranes: phospholipids, glycolipids, and sterols, all of which are largely hydrophobic with a hydrophilic head. The phospholipids differ in their head regions. They also differ in the length and degree of saturation of their fatty acid tails. (Fully saturated fatty acids have no double bonds in the carbon backbone; unsaturated fatty acids have one or more.) Figure 5.2 shows phosphatidylcholine, a phospholipid that is common to cell membranes.

Glycolipids are similar in structure but they have one or more sugar monomers attached at the head end. The steroid cholesterol is abundant in animal cell membranes but nonexistent in plant cell membranes, which have phytosterols.

Within a lipid bilayer, individual lipid molecules show quite a bit of movement. They move sideways, they spin

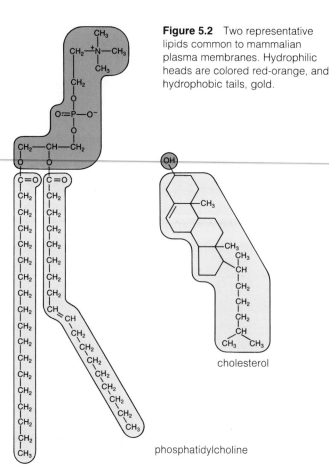

**Figure 5.2** Two representative lipids common to mammalian plasma membranes. Hydrophilic heads are colored red-orange, and hydrophobic tails, gold.

cholesterol

phosphatidylcholine

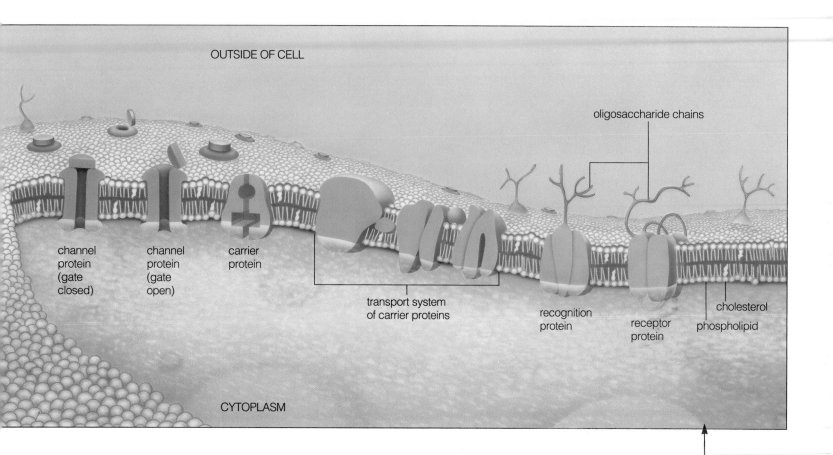

OUTSIDE OF CELL

oligosaccharide chains

channel protein (gate closed)

channel protein (gate open)

carrier protein

transport system of carrier proteins

recognition protein

receptor protein

cholesterol

phospholipid

CYTOPLASM

about their long axis, and their tails flex back and forth. The movements help keep adjacent lipids from packing into a solid layer. Packing is also disrupted by the presence of cholesterol and by lipids with short tails (which cannot interact as strongly as long-tailed neighbors do) or unsaturated tails (which tend to kink at the sites of double bonds).

So far, you may have the impression that a cell membrane consists only of lipids. Actually, a variety of proteins are nestled in the lipid bilayer and positioned at its two surfaces. In other words, *the membrane is a mosaic of lipids and proteins.* Most proteins of the plasma membrane are *glycoproteins,* meaning they have sugars covalently bonded to them. Usually the sugars are short-chain oligosaccharides. They always extend outward into the extracellular fluid, never into the cytoplasm. Although certain proteins are tethered to specific locations, many are not so restricted; their position can shift in the plane of the membrane. Taken together, the molecular movements and packing variations help account for the fact that a membrane behaves more like a *fluid* than a solid.

The lipids and proteins of one of the membrane's surfaces differ in number, kind, and arrangement from those of the other surface. In other words, the membrane has an *asymmetrical structure.* As you will see, this

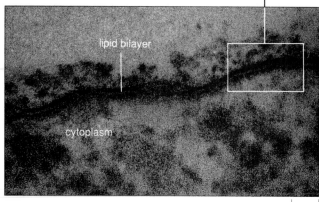

lipid bilayer

cytoplasm

0.02µm

**Figure 5.3** Views of the plasma membrane, based on the fluid mosaic model of membrane structure. The micrographs show the plasma membrane of a eukaryotic cell, thin section.

asymmetry reflects differences in the tasks carried out at the two membrane surfaces.

Figure 5.3 shows a model of membrane structure. It is a recent version of a model put together in 1972 by S. J. Singer and G. Nicolson. Evidence favoring this model comes from many sources, including freeze-fracture microscopy, as described next in the *Doing Science* essay.

## Discovering Details About Membrane Structure

Cells come in a spectacular array of shapes and sizes, and they perform a spectacular array of different tasks. Particularly in multicelled organisms, specialized cells engage in a kind of export-import business, selectively shipping off materials that they have manufactured for use by other cells and just as selectively receiving substances that neighboring or distant cells of other types have produced. The shipping and receiving functions are related to the structure of the plasma membrane.

If you wanted to deduce the structure of a plasma membrane, you would probably begin with attempts to identify its molecular components. Your first challenge is to secure a large sample of plasma membrane that has not been contaminated with internal cell membranes, such as those of the nucleus. Using red blood cells will simplify your task. Such cells are abundant, they are easy to collect—and structurally they are simple. As red blood cells develop, the nucleus becomes inactivated and is expelled from the cell body. At maturity, each cell is little more than a sack of hemoglobin, ribosomes, and a few other parts that will keep the cell functional for its life span of about 120 days.

By placing a sample of red blood cells in a test tube containing distilled water, you can separate the plasma membranes from the rest of the cell components. Red

blood cells are hypotonic in such a solution. This means there are more solutes (and fewer water molecules) inside the cell body than outside. Because water can move across the plasma membrane, it tends to follow its concentration gradient and diffuse into the cell body (by osmosis).

Red blood cells have no mechanisms for actively taking in or expelling water, so they tend to swell when placed in hypotonic solutions. In time the cells in your sample burst and their contents spill out. This gives you a mixture of membranes, hemoglobin, and other cell parts in the water.

How do you separate the bits of membrane? The trick here is to place a tube containing a special solution of cell parts in a *centrifuge*, a device that spins test tubes at high speed. Each component of a cell has its own molecular composition, and this gives it a characteristic density. As the centrifuge spins at the appropriate speed, components with the greatest density will move toward the bottom of the tube. Other components will take up layered positions above them according to their relative densities.

If you have done the centrifugation properly, one layer in the solution inside the tube will contain only shreds of cell membrane. You can carefully draw off the layer and examine it microscopically to verify that your membrane sample is not contaminated.

Through standard chemical analysis, you find that the plasma membrane is composed of lipids and proteins. Today we know how those molecules are organized in a plasma membrane, but in the past there were two competing models of membrane structure. In the "protein coat" model, the membrane was composed of a bilayer of lipids,

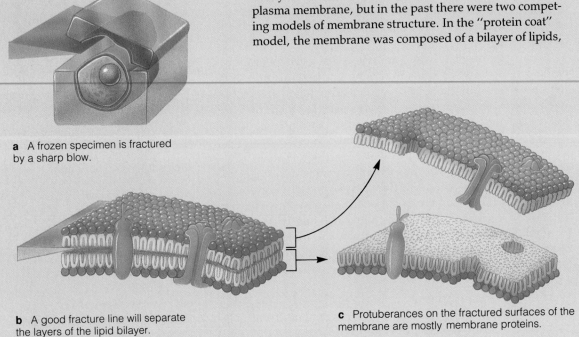

**a** A frozen specimen is fractured by a sharp blow.

**b** A good fracture line will separate the layers of the lipid bilayer.

**c** Protuberances on the fractured surfaces of the membrane are mostly membrane proteins.

coated on both surfaces with a layer of proteins. In the "fluid mosaic" model, proteins were largely embedded within the bilayer.

How would you test the protein coat model? One way would be to predict and then measure the amount of protein present in a given sample of membranes from red blood cells. First you calculate how much protein would be required to coat the inner and outer surface of a known number of cells. Then you separate a membrane sample into its lipid and protein fractions. By measuring the *observed* ratio of proteins to lipids, you can match the results against the ratio *predicted* on the basis of the model.

Such tests have been done on membrane samples, and they reveal that there is far too little protein to cover both surfaces of a lipid bilayer. The tests provide evidence against the protein coat model, at least for red blood cells.

What would be a direct test of the fluid mosaic model? You could employ freeze-fracturing and freeze-etching techniques. As Figures *a–e* show, these are special methods of preparing cells for electron microscopy. First a sample of cells is immersed in liquid nitrogen, an extremely cold fluid. The cells freeze instantly, after which they can be struck with a microscopically small chisel. A properly directed blow will fracture the cell in such a way that one layer of the lipid bilayer separates from the other. Those preparations can then be inspected under extremely high magnifications.

(If the *protein coat* model were correct, what do you suppose one such layer would reveal upon examination?)

What you actually see is not a perfectly smooth, pure lipid layer. Freeze-fractured cell membranes reveal many bumps and other irregularities in lipid layers. The bumps are the proteins that are incorporated whole, directly in the bilayer—just as the fluid mosaic model requires.

(**a**) Freeze-fracturing and freeze-etching. In the freeze-fracture step, specimens being prepared for electron microscopy are rapidly frozen, then fractured by a sharp blow from the edge of a fine blade. (**b, c**) Fractured membranes commonly split down the middle of the lipid bilayer. Typically, one inner surface is studded with particles and depressions, and the other is a complementary pattern of depressions and particles. The particles are membrane proteins.

(**d**) Sometimes specimens are freeze-etched: more ice is evaporated from the fracture face to expose the outer membrane surface. (**e**) In a process called metal shadowing, the fractured surface is coated with a layer of carbon and heavy metal such as platinum. The coating is thin enough to replicate details of the exposed specimen surface. The metal replica, not the specimen itself, is used for micrographs. (**f**) The micrograph shows part of a replica of a red blood cell, prepared by freeze-fracturing and freeze-etching.

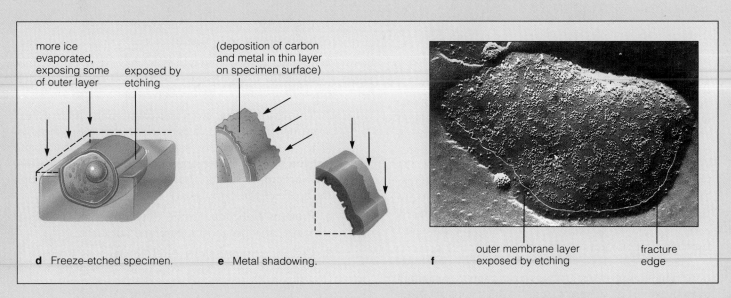

more ice evaporated, exposing some of outer layer

exposed by etching

(deposition of carbon and metal in thin layer on specimen surface)

**d**  Freeze-etched specimen.

**e**  Metal shadowing.

**f**

outer membrane layer exposed by etching

fracture edge

*Fluid mosaic model of membrane structure:* **A cell membrane is an asymmetrical mosaic of lipids *and* proteins. The membrane shows fluid behavior because of movements and packing variations among its lipids and proteins.**

Keep in mind that the fluid mosaic model is only a starting point for discussing membrane structure and function. Membranes vary not only in their composition but also in their fluidity. Consider one effect of cholesterol, an important lipid in animal cell membranes. Cholesterol has a rigid ring structure (page 42). When its rings interact with the tails of neighboring phospholipids, they immobilize parts of their neighbors' tails and so make the membrane less fluid at that particular spot. Yet at high concentration, cholesterol actually helps keep the membrane fluid by preventing saturated fatty acid tails from packing together.

Why is maintaining membrane fluidity so important? Cell survival depends on it. For example, we know that a bacterial or yeast cell doesn't have many mechanisms for dealing with cold spells. When temperatures fall, membranes tend to become less fluid, and the functioning of their enzymes and other proteins suffers. In response to the cold, however, those cells rapidly synthesize lipids that have double bonds in their tails—and the infusion of more kinky lipids into the membrane helps keep it from stiffening up.

### Functions of Membrane Proteins

Whereas the lipid bilayer serves as the overall structural framework for a cell membrane, proteins carry out most membrane functions. Three categories of proteins are crucial in this respect:

>Membrane transport proteins
>Recognition proteins
>Receptor proteins

As is the case for other biological molecules, each of these proteins has certain numbers and kinds of atoms linked to its carbon backbone. The specific arrangement and location of those atoms are the basis for chemical interactions with other substances. For example, when a substance weakly binds with some of those atoms, the binding might induce changes in the protein's shape. The physical changes can "bump" the bound substance from one side of the membrane to the other. This is a very simplified description of a membrane transport function, but you get the idea.

There are two classes of proteins with transport functions: channel proteins and carrier proteins. A **channel protein** serves as a pore through which ions or other water-soluble substances move from one side of the membrane to the other. Some channels are perpetually open. Others, being equipped with molecular "gates," open and close in controlled ways that permit or block the passage of specific ions. A **carrier protein** binds specific substances and changes its shape in ways that shunt the substances across the membrane. As you will see shortly, some carrier proteins do this rather passively. Others use energy as they actively pump substances in a specific direction.

A **recognition protein** is like a molecular fingerprint at the cell surface; it identifies the cell as being of a certain type. When a multicelled embryo is developing, recognition proteins help guide the ordering of cells into tissues. Later, they function in cell-to-cell interactions. The oligosaccharide chains of certain recognition proteins help form a "sugar coat" at the outer surface of the plasma membrane. Some proteins of the coat promote adhesion between cells. Others function in cell-to-cell recognition and in coordinating cell behavior within tissues.

A **receptor protein** is like a switch that turns on or off when particular substances bind to it. Receptor proteins have binding sites for hormones or other substances that can trigger alterations in a cell's metabolism or behavior. For example, the binding of the hormone somatotropin to a receptor turns on enzymes that crank up the machinery for cell growth and division. Different cell types have different combinations of receptors. In vertebrates, some receptors are widespread among the body's cells. Others are restricted to only a few cell types. Most commonly, the receptors are arrayed at the plasma membrane; some are located inside the cell.

To sum up, all cell membranes have the following characteristics in common:

**1.** Cell membranes are composed of lipids (especially phospholipids) and proteins.

**2.** In membranes, lipid molecules have their hydrophilic heads at the two outer faces of a bilayer and their hydrophobic tails sandwiched in between.

**3.** The lipid bilayer gives cell membranes their overall *structure* and serves as a barrier between two solutions. (The plasma membrane separates the fluids inside and outside the cell. The membranes of organelles or other cellular compartments separate different solutions within the space of the cytoplasm.)

**4.** Proteins embedded in the bilayer or positioned at its surfaces carry out most membrane *functions*.

**5.** Cell membranes show fluid behavior as a result of movements and packing variations among their lipid and protein components.

# DIFFUSION

With the preceding overview in mind, we are almost ready to explore the details of membrane function. First, however, let's consider the natural, unassisted movements of water and solutes. Many membrane properties involve mechanisms that work with or against these movements.

## Gradients Defined

"Concentration" refers to the number of molecules (or ions) of a substance in a given volume of fluid. In the absence of other forces, molecules of a given type move down their **concentration gradient**. They tend to move from a region of greater concentration to a region where they are less concentrated. They are driven to do so because they are constantly colliding with one another millions of times a second. Random collisions do send the molecules back and forth, but the *net* movement is outward from the region of greater concentration. Similarly, for any defined volume, gradients can also exist between two regions that differ in pressure, temperature, or net electric charge.

When your heart contracts, it generates fluid pressure that is greatest in the first artery leaving the heart and lowest in the last vein leading back into it. When you slip while skiing and find yourself sitting in a snowbank, heat energy flows down a thermal gradient and is transferred away from your body to the snow. Every moment of your life, you depend on electrical gradients that are associated largely with differences in the concentration of sodium ions inside and outside the plasma membrane of your nerve cells. When sodium ions are permitted to flow down their concentration gradient, they create an electric current across the membrane. Such currents are the "messages" that travel through your nervous system. As you will see throughout this book, gradients are central to a variety of dynamic processes.

## Simple Diffusion

The random movement of like molecules or ions down a concentration gradient is called **simple diffusion**. Simple diffusion accounts for the movements of many small molecules across cell membranes.

The direction in which a substance diffuses depends on its own concentration gradient, not on any others. In other words, each substance diffuses *independently* of other substances that may be present. Suppose you put a few drops of red food coloring into one end of a pan filled with water. At first all of the dye molecules remain at that end, but many molecules start careening toward the other end. Even though collisions are also sending

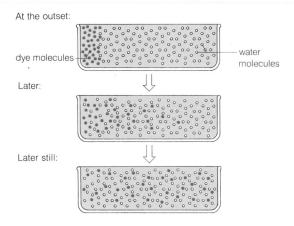

**Figure 5.4** Diagram of the diffusion of dye molecules in one direction and water molecules in the opposite direction in a pan of water.

some dye molecules back, *more* molecules are leaving the concentrated region, and the net movement is down the gradient. For the same reason, water moves in the opposite direction, to the region where water molecules are less concentrated in the pan (Figure 5.4).

Even when dye molecules and water molecules are dispersed evenly in the pan, collisions still occur at random. But now there are no more gradients, hence there is no *net* movement in either direction. The molecules are said to be at dynamic equilibrium.

The *rate* of diffusion depends on several factors. For example, diffusion proceeds faster when the concentration gradient is steep. It also proceeds faster at higher temperatures, because heat energy causes molecules to move more rapidly (hence to collide more frequently). Diffusion rates are also affected by molecular size. Other factors being equal, smaller molecules move faster than large ones do.

## Bulk Flow

Diffusion rates are often enhanced by **bulk flow**. This is the tendency of all the different substances in a fluid to move together in the same direction in response to a pressure gradient.

For example, in large multicelled plants and animals, nutrients and other substances do not diffuse slowly across large tissue masses to reach the plasma membrane of individual cells. They move rapidly by bulk flow through transport tubes, which occur in the respiratory and circulatory systems of animals and in the vascular systems of plants. In effect, bulk flow "shrinks" the diffusion distances for molecules that must move between the environment and interior cells. It also shrinks the diffusion distance between two body regions, as between leaves and roots.

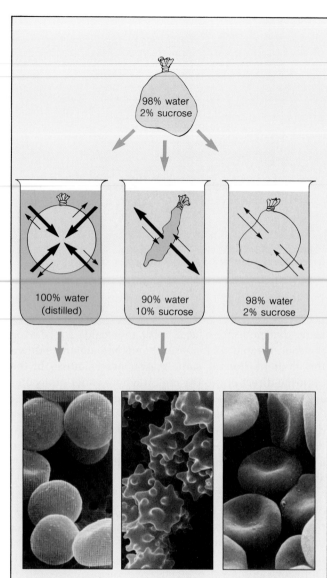

**Hypotonic**
Water diffuses
inward

**Hypertonic**
Water diffuses
outward

**Isotonic**
No net change
in water movement

**Figure 5.5** Effects of osmosis in different environments. The sketches show why it is important for cells to be matched to solute levels in their environment. (In each sketched container, arrow width represents the relative amount of water movement.)

The micrographs correspond to the sketches. They show the kinds of shapes that might be seen in human red blood cells placed in *hypotonic* solutions (influx of water into the cell), *hypertonic* solutions (outward flow of water from the cell), and *isotonic* solutions (internal and external solute concentrations are matched, no net movement of water).

Red blood cells have no special mechanisms for actively taking in or expelling water molecules. Hence they would swell or shrivel up, if solute levels in their environment were to change.

## OSMOSIS

### Osmosis Defined

A membrane is "differentially permeable" when some substances but not others can pass through it. **Osmosis** is the movement of water across any differentially permeable membrane in response to solute concentration gradients, a pressure gradient, or both. (A *solute* is any dissolved substance.)

Imagine you have a plastic bag that acts like a membrane, in that it is permeable to water molecules but impermeable to larger molecules. You fill the bag with water containing a small amount of table sugar, then you put it in a container of distilled water. "Distilled" means the water is nearly free of solutes. Inside the bag, the concentration of water molecules per unit volume is less than it is outside—so water moves into the bag (Figure 5.5). However, only the water is moving across the membrane—the sugar molecules are too large to do so. Thus, the sugar inside the bag cannot follow *its* concentration gradient and move out. Soon the bag swells with water, and eventually it may spring a leak or burst.

Suppose you had immersed the bag in water having more dissolved sugar than did the solution inside the bag. The net movement of water would have been outward, and the bag would have shriveled.

Finally, suppose you had used distilled water inside and outside the bag. Because the solute concentrations would be the same, there would be no water concentration gradient—and there would be no net movement of water in either direction.

Osmosis is the passive movement of water across a differentially permeable membrane in response to solute concentration gradients, a pressure gradient, or both.

### Tonicity

Osmotic movements across cell membranes are affected by **tonicity**—that is, the relative concentrations of solutes in two fluids. In this case, we are talking about the extracellular fluid and the fluid portion of cytoplasm. When solute concentrations are equal in both fluids, or *isotonic*, there is no net osmotic movement of water in either direction. When the solute concentrations are not equal, one fluid is *hypotonic* (has less solutes) and the other is *hypertonic* (has more solutes). Water molecules tend to move from a hypotonic fluid to a hypertonic one.

If cells did not have mechanisms for adjusting to differences in tonicity, they would shrivel or burst like the plastic bag described above. For example, many solutes are dissolved in your bloodstream and in the cytoplasm

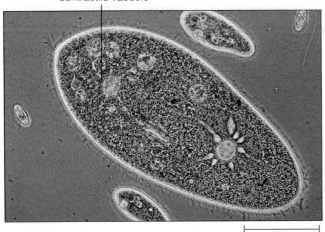

contractile vacuole

**Figure 5.6** Osmosis and *Paramecium*, a single-celled protozoan that lives in fresh water (a hypotonic environment). Because the cell interior is hypertonic relative to the surroundings, water tends to move into the cell by osmosis. If the influx were left unchecked, the cell would become bloated and the plasma membrane would rupture. However, the excess is expelled by an energy-requiring transport process that is based on a specialized organelle called the contractile vacuole. Tubelike extensions of this organelle extend through the cytoplasm. Water enters these extensions and collects in a central vacuolar space. When filled, the vacuole contracts and the water is forced into a small pore that empties to the outside.

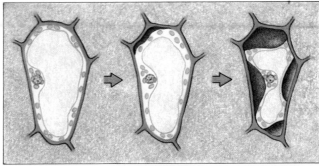

a   b   c

**Figure 5.7** Effects of osmosis on plant cells. Most land plants grow in hypotonic soil, so water tends to move into their cells by osmosis. The cells have fairly rigid walls outside the plasma membrane; when water is absorbed, pressure increases against the wall. *Turgor pressure* refers to the internal pressure on a cell wall resulting from the inward osmotic movement of water.

If turgor pressure becomes high enough to counter the effects of cytoplasmic solutes, water also will be squeezed back out. When the *outward* flow equals the *inward* flow, turgor pressure is constant, cell walls cannot collapse and the soft plant parts stay erect. When soil solutes reach high concentrations, however, the net water movement is outward and wilting occurs.

(**a**) At the start of this experiment, 10 grams of salt (NaCl) in 60 milliliters of water are added to a pot containing tomato plants. (**b**) The plants start collapsing after about 5 minutes. (**c**) Wilting is severe in less than 30 minutes. The corresponding sketches show progressive plasmolysis (shrinking of cytoplasm away from the cell walls).

of red blood cells. But suppose you immerse a red blood cell in a hypotonic solution. That cell contains many large organic molecules, which cannot cross the plasma membrane. Although those molecules cannot move out, water can move in. Internal pressure builds up, but red blood cells have no mechanism for disposing of such a large volume of excess water. The cell continues to swell until the membrane ruptures. Then the cell undergoes *lysis*—it becomes grossly "leaky"—and is destroyed.

If the red blood cell had been placed in a hypertonic solution, water would have moved out of the cytoplasm and the cell would have shriveled. This particular type of cell maintains its volume only when solute concentrations stay much the same on both sides of the plasma membrane. Other cell types can live in hypotonic or hypertonic environments, as Figure 5.6 illustrates.

## Water Potential

The soil in which most land plants grow is hypotonic relative to the cells in those plants. Thus water tends to move by osmosis into the plant. When individual cells absorb the water, pressure increases against the cell wall, which is fairly rigid. This internal fluid pressure on a cell wall is called **turgor pressure**.

Pressure will build up inside any walled cell when water moves inward by osmosis, but water will also be squeezed back out when turgor pressure is great enough to counter the effects of internal solutes. Both forces have the potential to cause the directional movement of water. The sum of these two opposing forces is called the **water potential**.

Turgor pressure is constant when the outward flow of water from a plant cell equals the rate of inward diffusion. There is enough pressure to keep cell walls from collapsing, and the soft parts of the plant body are maintained in erect positions. When the environment becomes so dry that the movement of water into the plant dwindles or stops, water moves out of the cells and soft parts of the plant body wilt. Wilting also occurs when solutes reach high concentrations in the environment (Figure 5.7).

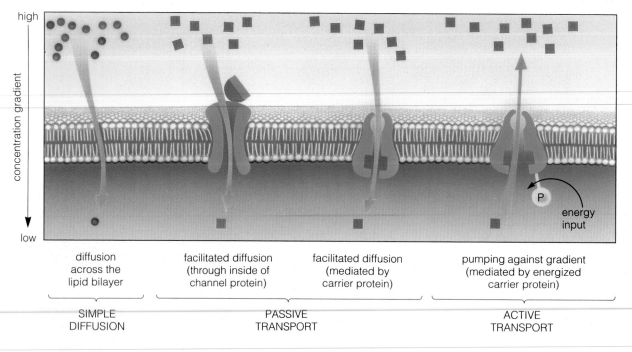

<table>
diffusion across the lipid bilayer | facilitated diffusion (through inside of channel protein) | facilitated diffusion (mediated by carrier protein) | pumping against gradient (mediated by energized carrier protein)
</table>

SIMPLE DIFFUSION | PASSIVE TRANSPORT | ACTIVE TRANSPORT

**Figure 5.8** Overview of active and passive transport mechanisms.

## MOVEMENT OF WATER AND SOLUTES ACROSS CELL MEMBRANES

### The Available Routes

We are now ready to consider how substances move into and out of cells, with or without assistance from membrane proteins. Only small, electrically neutral molecules readily diffuse across the bilayer. Examples are oxygen, carbon dioxide, and ethanol, all of which have no net charge:

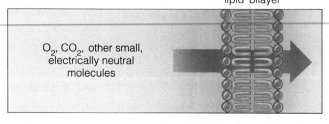

In contrast, glucose and other large, water-soluble molecules that are electrically neutral almost never diffuse freely across the lipid bilayer. Neither do positively or negatively charged ions, no matter how small they are:

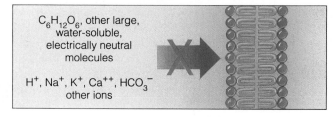

Different types of proteins move these substances across the membrane by active and passive transport mechanisms. In **active transport**, the protein becomes activated into moving a solute against its concentration gradient. In **passive transport**, the protein does *not* require an energy boost. The solute simply moves through the protein's interior, following its concentration gradient.

Figure 5.8 provides an overview of how substances move across cell membranes. Through a combination of simple diffusion, passive transport, and active transport, cells or organelles are supplied with numerous raw materials and they are rid of wastes, at controlled rates. These mechanisms help maintain pH and volume inside the cell or organelle within functional ranges.

### Facilitated Diffusion

A membrane protein with passive transport functions is highly selective about which solutes it will assist across the membrane. A protein that transports glucose, for instance, will not transport amino acids. And the protein *only* helps a particular solute move in the direction that simple diffusion would take it (down its concentration gradient). That is why this transport mechanism is called **facilitated diffusion**.

Membrane proteins involved in facilitated diffusion may have an interior space that can be opened up to one side of the membrane at a time (Figure 5.9). Probably a specific array of hydrophilic groups project into that

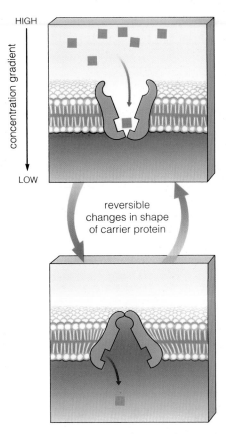

**Figure 5.9** One model of facilitated diffusion across a cell membrane. The carrier protein shown here can exist in two configurational states. In one state, the binding sites for a solute are exposed to the fluid outside the lipid bilayer of the membrane. In the other state, the same binding sites are exposed to cytoplasmic fluid. The transition from one state to the other is reversible and depends on the direction of the solute concentration gradient across the membrane. The changes in shape are induced when the solute is bound in place.

space. When water-soluble molecules bind with the groups, the binding seems to trigger a change in the protein shape. This change permits the solute to move through the hydrophilic interior. While the solute makes its passage, the protein closes in behind it and returns to its original shape.

### Active Transport

A membrane protein that *actively* moves specific solutes into and out of cells undergoes changes in shape, somewhat like the changes induced in the proteins responsible for facilitated diffusion. In this case, however, an energy boost leads to a *series* of changes, which cause the protein to pump solutes across the membrane (Figure 5.10). Most often, ATP donates energy to the protein.

One active transport system, the *sodium-potassium pump,* helps maintain high concentrations of potassium and low concentrations of sodium inside the cell. Another, the *calcium pump,* helps keep calcium concentrations at least a thousand times lower inside the

**Figure 5.10** (*Below*) Simplified picture of an active transport system in animal cell membranes. In this example, transport of one kind of solute across the membrane is coupled with transport of another kind in the opposite direction. The transport protein receives an energy boost from ATP and so undergoes changes in its shape that are necessary for the transport process. The net result of active transport is that a solute is moved across the membrane against its concentration gradient, an event that is coupled to the expenditure of energy.

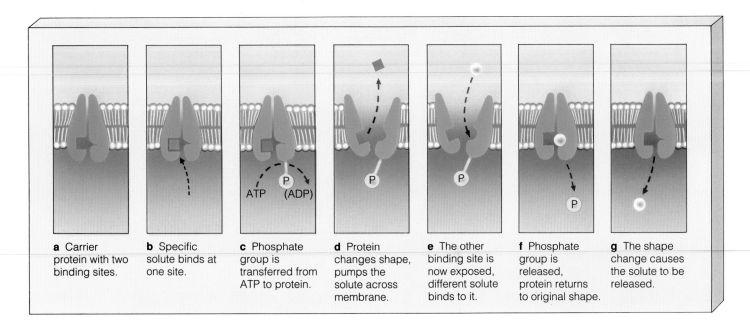

**a** Carrier protein with two binding sites.

**b** Specific solute binds at one site.

**c** Phosphate group is transferred from ATP to protein.

**d** Protein changes shape, pumps the solute across membrane.

**e** The other binding site is now exposed, different solute binds to it.

**f** Phosphate group is released, protein returns to original shape.

**g** The shape change causes the solute to be released.

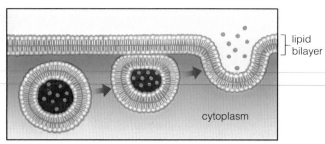

**a** Exocytosis

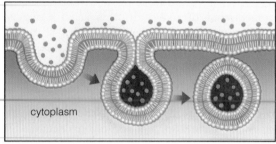

**b** Endocytosis

**Figure 5.11** (**a**) Fusion of a vesicle with the plasma membrane during exocytosis. (**b**) Formation of a vesicle during endocytosis.

cell than outside. The features that all active transport systems hold in common can be summarized this way:

**1.** In **active transport,** small ions, small charged molecules, and large molecules are pumped across a cell membrane, against their concentration gradient.

**2.** Certain proteins spanning the lipid bilayer are the active transport systems. They are highly selective about which solutes they will bind and transport.

**3.** The proteins act when a specific solute is bound in place and when they receive an energy boost, as from ATP.

## Exocytosis and Endocytosis

Exocytosis and endocytosis are processes by which small regions of plasma membrane or organelle membranes pinch off and form transport vesicles around substances (Figure 5.11). In **exocytosis**, vesicles in the cytoplasm travel to the plasma membrane, fuse with it, and have their contents released to the cell's surroundings. Vesicles derived from Golgi bodies and released from secretory cells are an example; these are illustrated in Figure 4.14. In **endocytosis**, a region of the plasma membrane encloses particles at or near the cell surface, then pinches off to form a vesicle that moves into the cytoplasm.

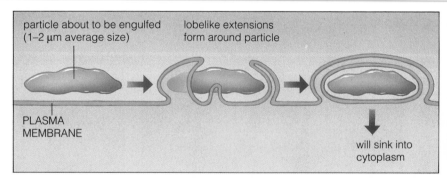

**a** Phagocytosis

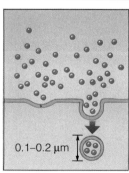

**b** Pinocytosis

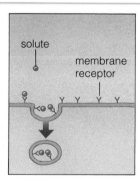

**c** Receptor-mediated endocytosis

**Figure 5.12** Mechanisms of endocytosis.

The amoeba relies on endocytosis. This single-celled organism is phagocytic—a "cell eater." When it encounters a chemically "tasty" particle or cell, one or two lobes form at its surface. The lobelike extensions curve back and form a compartment around the particle, and this compartment becomes an endocytic vesicle (Figure 5.12). Many endocytic vesicles fuse with lysosomes. As mentioned on page 64, lysosomes are filled with digestive enzymes. Following fusion, the contents of the endocytic vesicle are digested. Phagocytic white blood cells rely on endocytosis when they destroy harmful agents such as bacteria.

Endocytosis also transports liquid droplets into animal cells. (Sometimes this process is called pinocytosis, which means "cell drinking.") A depression forms at the surface of the plasma membrane and dimples inward around extracellular fluid. An endocytic vesicle forms and moves inside the cytoplasm, where it fuses with lysosomes.

In *receptor-mediated endocytosis*, specific molecules are brought into the cell through involvement of specialized regions of the plasma membrane that form coated pits. Each pit, or shallow depression, is coated on its cytoplasmic side with a dense lattice of proteins (Figure 5.13). The pit appears to be lined with surface receptors that are specific (in the example shown here) for lipoprotein particles. When lipoproteins are bound to the receptors, the pit sinks into the cytoplasm and forms an endocytic vesicle.

Figure 5.14 shows the possible destinations of vesicles that form by way of exocytosis and endocytosis.

indentation on extracellular side of plasma membrane

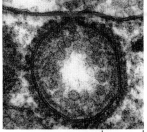
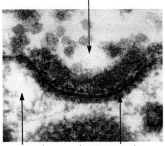

**a** cytoplasm  plasma membrane

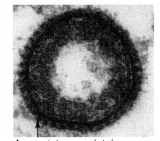

**c**       0.1 µm

**b** proteins of lipoprotein cytoskeleton particles bound to plasma membrane receptors

**d** vesicle completely formed, moving into cytoplasm

**Figure 5.13** A closer look at receptor-mediated endocytosis and the formation of a transport vesicle. The electron micrographs show part of the plasma membrane of an immature egg from a hen. The shallow indentation in (**a**) is an example of a *coated pit*. On the cytoplasmic side of the pit, proteins (clathrin) form a dense lattice. On the extracellular side of the membrane, the pit is thought to be lined with surface receptors that are specific for lipoprotein particles (**b**). In (**c**), the pit has deepened and has become rounded. In (**d**), the coated vesicle is fully formed, with lipoproteins (and protein receptors) enclosed within it.

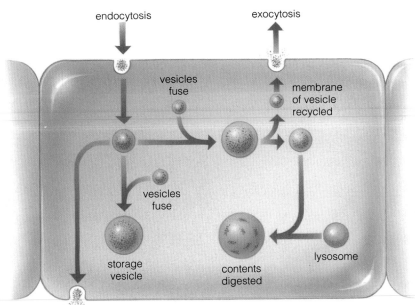

endocytosis        exocytosis

vesicles fuse

membrane of vesicle recycled

vesicles fuse

storage vesicle

contents digested

lysosome

substance released at opposing cell surface

**Figure 5.14** Possible fates of substances in vesicles formed by endocytosis. Compare Figure 4.14, which shows the formation of vesicles in the cytomembrane system.

# SUMMARY

## Membrane Structure

1. Living cells are bathed in fluid of one sort or another, and their interior is also fluid. The plasma membrane is the boundary between the external fluid world and the cytoplasm. Internal cell membranes are boundaries between different cytoplasmic regions.

2. The lipid bilayer is the basic structure of all cell membranes. It has two layers of lipids (phospholipids especially), with the hydrophobic tails of the molecules sandwiched in between the hydrophilic heads. The membrane is rather fluid because of packing variations and rapid movements among the individual molecules.

3. Most membrane functions are carried out by diverse proteins embedded in the bilayer or weakly bonded to one of its surfaces.

## Membrane Functions

1. Three classes of proteins are crucial for membrane functions. These are membrane transport proteins, recognition proteins, and receptor proteins. The following points summarize their functions.

2. *Control of substances moving into and out of cells.* Transport proteins passively or actively move specific solutes across the lipid bilayer portion of the plasma membrane. The selective movements of solutes affect metabolism, cell volume and cellular pH, nutrient stockpiling, and the removal of harmful substances.

3. *Compartmentalization of internal cellular space.* Membranes within the cytoplasm form compartments (organelles) in which specialized activities occur. The nucleus, chloroplasts, and the inner and outer mitochondrial membranes are examples.

4. *Signal reception.* Some receptor proteins bind signaling molecules such as hormones, and when they do, they trigger alterations in cell behavior or metabolism.

5. *Cell-to-cell recognition.* In multicelled animals, recognition proteins help cells of like type identify and adhere to one another during tissue formation and, at later stages of development, during tissue interactions.

6. *Transport of macromolecules and particles.* Through exocytosis and endocytosis, cells eject or take in large molecules or particles across the plasma membrane.

## Movement of Water and Solutes Across Membranes

1. Different types of solutes move across the membrane, in some cases with and in some cases against the solute concentration gradient.

2. Simple diffusion is a natural, unassisted movement of solutes from a region of high concentration to a region of lower concentration.

3. Osmosis is the movement of water across a differentially permeable membrane in response to solute concentration gradients, a pressure gradient, or both. When conditions are isotonic (equal concentrations of solutes across the membrane), there is no osmotic movement in either direction. Water tends to move from hypotonic fluids (with lower concentrations of solutes) to hypertonic fluids (with higher concentrations of solutes).

4. Channel and carrier proteins carry out membrane transport functions. Open or gated channel proteins serve as pores through which water-soluble substances cross the membrane. Carrier proteins bind solutes and so shunt them across. Some carrier proteins do this passively, others require an energy boost to pump solutes across.

5. The facilitated diffusion of a solute through channel proteins or carrier proteins is called passive transport. The solute moves passively through the protein interior, in the direction that its concentration gradient takes it.

6. The pumping of solutes across membranes by energized carrier proteins is called active transport.

## Review Questions

1. Describe the fluid mosaic model of plasma membranes. What makes the membrane fluid? What parts constitute the mosaic? *76–80*

2. List the structural features that all cell membranes have in common. *80*

3. Describe some functions of membrane proteins. *80*

4. How does diffusion work? *81*

5. What is osmosis, and what causes it? *82*

6. Explain the difference between active and passive transport mechanisms. *84*

7. What types of substances can readily diffuse across the lipid bilayer of a cell membrane? What types of substances must be actively or passively transported across? *84*

8. Can you explain the difference between exocytosis and endocytosis? *86–87*

1. Substances move into and out of the cell by crossing the _____.

2. Cell membranes consist mainly of a _____.
   a. carbohydrate bilayer and proteins
   b. protein bilayer and phospholipids
   c. phospholipid bilayer and proteins
   d. nucleic acid bilayer and proteins

3. Most membrane functions are carried out by _____ associated with the bilayer.
   a. proteins
   b. phospholipids
   c. nucleic acids
   d. hormones

4. When a cell is placed in a hypotonic solution, _____.
   a. water tends to move into the cell
   b. water tends to move out of the cell
   c. water does not move into or out of the cell
   d. exocytosis will have to occur

5. _____ cannot diffuse across the lipid bilayer of the plasma membrane.
   a. ethanol
   b. ions
   c. oxygen
   d. carbon dioxide

6. Three classes of proteins essential for membrane functions are _____, _____, and _____.

7. _____ are membrane proteins that bind hormones and other substances that can trigger changes in cell behavior or metabolism.
   a. gated channel protein
   b. receptor protein
   c. carrier protein
   d. recognition protein

8. The passive movement of a substance through channel proteins as it follows its concentration gradient across a cell membrane is an example of _____.
   a. osmosis
   b. active transport
   c. diffusion
   d. facilitated diffusion

9. _____ help cells of like type identify and adhere to one another during tissue development.
   a. open channel proteins
   b. energized carrier proteins
   c. recognition proteins
   d. receptor proteins

10. Match each membrane structure/function with the appropriate description.
    _____ substance expelled from cell when vesicle fuses with plasma membrane
    _____ water moves across membrane due to concentration or pressure gradients
    _____ natural, unassisted movement of solutes down a concentration gradient
    _____ facilitated diffusion of solutes via channel or carrier proteins
    _____ substances move into cell by invagination of plasma membrane
    _____ carry out membrane transport functions
    _____ energized carrier protein pumps solutes across membrane

    a. passive transport
    b. channel and carrier proteins
    c. active transport
    d. simple diffusion
    e. endocytosis
    f. osmosis
    g. exocytosis

active transport *84*
bulk flow *81*
calcium pump *85*
carrier protein *80*
channel protein *80*
concentration gradient *81*
endocytosis *86*
exocytosis *86*
facilitated diffusion *84*
fluid mosaic model *80*
hypertonic *82*
hypotonic *82*
internal cell membrane *75*
isotonic *82*
lipid bilayer *76*
lysis *83*
osmosis *82*
passive transport *84*
phospholipid *76*
plasma membrane *75*
recognition protein *80*
receptor protein *80*
receptor-mediated endocytosis *87*
simple diffusion *81*
sodium-potassium pump *85*
solute *82*
tonicity *82*
turgor pressure *83*
water potential *83*

Alberts, B. et al. 1989. *Molecular Biology of the Cell*. Second edition. New York: Garland.

Bretscher, M. October 1985. "The Molecules of the Cell Membrane." *Scientific American* 253(4):100–108. Fairly recent description of the structure and function of the plasma membrane.

Dautry-Varsat, A., and H. Lodish. May 1984. "How Receptors Bring Proteins and Particles Into Cells." *Scientific American* 250(5): 52–58. Describes receptor-mediated endocytosis.

Singer, S., and G. Nicolson. 1972. "The Fluid Mosaic Model of the Structure of Cell Membranes." *Science* 175:720–731.

Unwin, N., and R. Henderson. February 1984. "The Structure of Proteins in Biological Membranes." *Scientific American* 250(2): 78–94.

# 6 GROUND RULES OF METABOLISM

## The Old Man of the Woods

Deep in a Michigan forest, on a moist and rotting trunk of a fallen tree, a quiet celebration of life is about to unfold. The huge trunk is jarring in the midst of such luxuriant greenery, an unsettling reminder of the death that comes to all organisms with intricately specialized cells. Yet through a patch of decaying bark a mushroom erupts. Soon it has grown several inches tall. It is dirty brown and ugly to boot, with a cap that is shaggy, cracked, and lumpy with gray-black scales (Figure 6.1). Mushroom hunters love it. Rather than using its scientific name, *Strobilomyces floccopus*, they commonly call it the "old man of the woods" and welcome its presence at the dinner table.

Suppose you come across an old man of the woods and decide to dissect it gently. You see almost immediately that it consists of a highly ordered array of parts. It has a stalk, which had grown tall and strong enough to support the cap some distance above the tree trunk. Inside the cap, hundreds of tiny tubes point downward. Even tinier reproductive bodies called spores have been growing inside the tubes and still are sequestered there.

If you had left the mushroom on the trunk, those spores would have been released from the tubes, and air currents would have dispersed them to good (or bad) places for germination.

As you probe the rotting wood where the mushroom had been growing, you realize that the old man of the woods is more than a stalk and a cap. You have yanked the mushroom from the rest of the fungal body—slender filaments threading every which way through its substrate. Collectively the filaments represent the mycelium, the part of the fungus devoted to securing food energy from the surroundings. Like other fungi, the old man of the woods produces and secretes enzymes to the outside of its body. Those enzymes digest large organic molecules in the tree's once-living tissues. The digested molecular bits are small enough for fungal cells to absorb—and to use as energy sources and raw materials.

Thus in death, the fallen tree had become the basis of new life. *The energy and materials derived from its organic molecules had helped another organism grow, maintain its highly organized state, make spores, and so serve as a bridge to a new generation of life.*

You might have focused your attention on a living tree in the forest, a robin, a squirrel, an earthworm, or any other organism besides a fungus. You would have arrived at the same understanding of the metabolic connections between energy and the living state. Metabolism—the controlled acquisition and use of energy in the synthesis and breakdown of organic compounds—happens only in living organisms. It depends on specific enzymes and organized arrays of cell parts, the construction of which depends ultimately on instructions encoded in DNA.

This chapter is our starting point for exploring the nature of energy and how it can be used to do cellular work. It will serve as our background for considering specific mechanisms by which organisms acquire and release the energy that keeps them alive—the central topics of the next two chapters.

**Figure 6.1** Reproductive structures of the old man of the woods, assembled with energy and materials that had been stored previously in another organism in the web of life.

## KEY CONCEPTS

**1.** Cells have the controlled capacity to trap and use energy for stockpiling, breaking apart, building, and eliminating substances in ways that contribute to survival and reproduction. This capacity is called metabolism.

**2.** The complex organization characteristic of life is maintained by a steady input of energy. To stay alive, cells must replace the energy that they inevitably lose as heat by metabolic reactions. Directly or indirectly, the sun is the source of energy replacements for nearly all organisms.

**3.** In biosynthetic pathways, organic molecules are assembled and energy becomes stored in them. In degradative pathways, organic molecules are broken apart and energy is released.

**4.** Enzymes (which speed reaction rates) take part in nearly all metabolic pathways. So does ATP, an organic compound that transfers energy from one reaction site to another in the cell.

By definition, **metabolism** is the controlled capacity to acquire and use energy for stockpiling, breaking apart, building, and eliminating substances in ways that contribute to survival and reproduction.

You see outward signs of metabolic activity whenever you look at a living cell through a microscope. Most obviously, the cell is moving about. Through its movements, it is identifying and taking in raw materials suspended in the water droplet on the slide. To power those tiny movements, the cell is extracting energy from food molecules stored away earlier. Even as you watch it, the cell is using energy and materials as it builds and maintains its membranes, its stores of chemical compounds, its DNA, its pools of enzymes. It is alive, it is growing, it may divide in two. Multiply this activity by *65 trillion cells* and you have an inkling of the metabolic activity going on in your own body as you sit quietly, observing that single cell!

## ENERGY AND LIFE

**Energy** is a capacity to do work. You use energy when you run or sleep or think about something bad, dangerous, or pleasant. In all cases, some cells of your muscles, brain, and other body parts are being put to work. Each cell is using energy, as when it is contracting or when it is producing and secreting hormone molecules that can alter your behavior.

You (or the cell) cannot create your own energy from scratch, however. You must get it from someplace else. That is the message of the **first law of thermodynamics:**

**The total amount of energy in the universe remains constant. More energy cannot be created; existing energy cannot be destroyed. It can only be converted from one form to another.**

Consider what this law means. The universe has only so much energy, distributed in a variety of forms. One form can be converted to another, as when corn plants absorb sunlight energy and convert it to the chemical energy of starch. By eating corn, you can extract and convert its energy to other forms, such as mechanical energy for your movements. Energy conversions of all kinds are notable for this reason: Whenever one takes place, a little energy escapes to the surroundings as heat. (Your body steadily gives off about as much heat as a

100-watt light bulb because of ongoing conversions in your cells.) As Figure 6.2 suggests, however, none of the energy vanishes. It just ends up someplace else.

This brings us to another important concept, the *quality* of energy available. Energy concentrated in a starch molecule is high quality, since it lends itself to conversions. Heat energy spread out in the atmosphere is low quality since, for all practical purposes, it can't be gathered up and converted to other forms. Heat can be transferred from a hot object to a cooler one, but it cannot on its own be transferred in the opposite direction.

The amount of low-quality energy is increasing in the grand scheme of things. That is the point of the **second law of thermodynamics:**

**The spontaneous direction of energy flow is from high-quality to low-quality forms. With each conversion, some energy is randomly dispersed in a form that is not as readily available to do work.**

Without energy to maintain it, any organized system tends to become disorganized over time. As Figure 6.2 indicates, "system" means all the matter in a specified region—a plant, a strand of DNA, a galaxy, and so on. **Entropy** is a measure of the degree of randomness or disorder of these and all other systems. The second law says that, in any process that happens spontaneously, the total entropy of the system and its surroundings must increase.

Think about the Egyptian pyramids—originally organized, presently crumbling, and many thousands of years from now, dust. The *ultimate* destination of the pyramids and everything else in the universe is a state of maximum entropy. Why? Through the eons, high-quality energy that can be tapped to maintain order in systems ultimately will be converted to energy of the lowest quality—perhaps uniformly dispersed heat—and how will that ever be gathered up to do work? Billions of years from now, energy conversions as we know them may never happen again.

Can it be that life is one glorious pocket of resistance to the rather depressing flow toward complete entropy? After all, every time a new organism grows, atoms become linked together into precise arrays and energy becomes more concentrated and organized, not less so! We see order everywhere, in patterned butterfly wings, in the petaled symmetry of flowers, in the structure of DNA. Yet a simple example will show that the second law does indeed apply to life on earth.

Think of each living cell as a speck of order in a universe that is tending toward disorder. The cell stays that

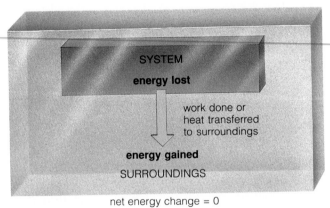

net energy change = 0

**Figure 6.2** The nature of energy. According to the first law of thermodynamics, the total energy content of any system and its surroundings remains constant. "System" means all matter within a specific region, such as a plant, a DNA molecule, or a galaxy. The "surroundings" are everything in the universe *except* the system.

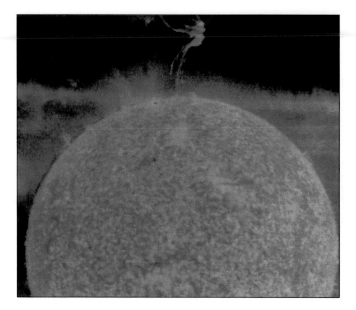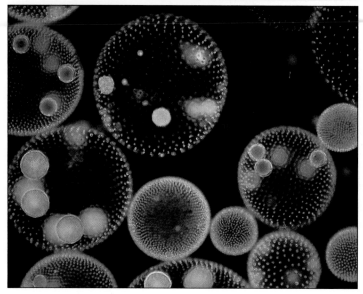

**Figure 6.3** All events large and small, from the birth of stars to the death of a microorganism, are governed by laws of thermodynamics. Shown here, eruptions on the sun's surface and, to the right, *Volvox*—each sphere a colony of microscopically small single cells able to capture sunlight energy that indirectly drives their life processes.

way as long as it continually taps into energy from an outside source. The primary source of energy for life on earth is the sun—which is steadily losing energy (Figure 6.3). Plants capture some sunlight energy, and they convert it to other forms of energy such as that of glucose and other organic molecules. Some of the stored energy gets transferred to organisms that feed, directly or indirectly, on plants. At every energy transfer along the way, however, some energy is lost, usually as heat. This adds more heat to the universal pool.

Overall, then, energy is still flowing in one direction. *The world of life maintains a high degree of organization only because it is being resupplied with energy lost from someplace else.*

**There is a steady flow of sunlight energy into the interconnected web of life, and this compensates for the steady flow of energy leaving it.**

## THE NATURE OF METABOLISM

### Energy Changes in Metabolic Reactions

The next two chapters focus mainly on the chemical reactions of photosynthesis and aerobic respiration, the main pathways of energy flow in the world of life. Those reactions will make more sense if you keep the following three concepts in mind:

**1.** The substances present at the end of a reaction (the products) may have *less* or *more* energy than did the starting substances (the reactants).

**2.** Most reactions are reversible. Besides proceeding in the forward direction (from reactants to end products), they can proceed in the reverse direction (from end products back to reactants).

**3.** A reversible reaction tends to approach equilibrium, a state in which it proceeds at about the same rate in both directions.

**Energies of Cellular Substances.** When we speak of the energy of a cellular substance, we really are speaking of the usable energy that is released when the substance is broken down to simpler materials. Chemical energies are often measured in terms of kilocalories per mole. A *kilocalorie* is the same thing as a thousand calories—the amount of energy needed to heat 1,000 grams of water from 14.5°C to 15.5°C at standard pressure. Because energy can be converted from one form to another, bond energies can be expressed in kilocalories even though the energy is used for something other than heating water.

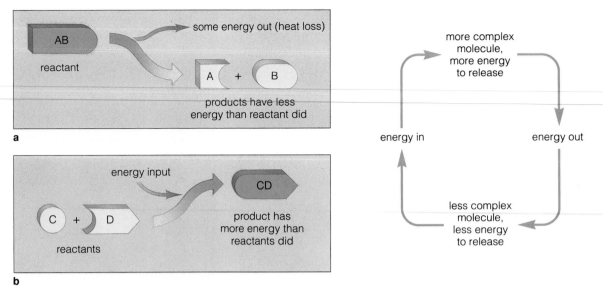

**Figure 6.4** Energy changes in metabolic reactions. (**a**) In exergonic reactions, products have less energy than the reactants did. Some energy released by the reactions is harnessed to do cellular work. The breakdown of glucose into carbon dioxide and water is an example (page 120). (**b**) In endergonic reactions, products have more energy than the reactants did. An energy input drives these reactions. An example is photosynthesis, in which sunlight energy drives the linkage of carbon dioxide and water into sugars and other compounds of higher energy content.

**Energy Losses and Gains**. Why is it important to think about chemical energies? Just as you might use firewood as a source of energy, so do cells use glucose and other molecules. You light a match to wood; cells use ATP like a "match" to start breaking the covalent bonds of glucose. The bonds occur between carbon, hydrogen, and oxygen atoms of glucose ($C_6H_{12}O_6$). When a glucose molecule is fully broken down, all that is left are six molecules each of carbon dioxide ($CO_2$) and water ($H_2O$). These are small molecules, with new covalent bonds among their atoms. The new bonds are stronger than the covalent bonds between the carbon atoms in glucose. Thus carbon dioxide and water are more stable compounds than glucose. Said another way, the overall breakdown reactions *release* energy—as much as 686 kilocalories per mole of glucose—just as a fire releases heat energy.

Glucose breakdown is a good example of a reaction in which the products end up with less energy than the reactants had. Reactions that show a net *loss* in energy are said to be *exergonic*, meaning "energy out" (Figure 6.4a). Your body runs on a good deal of energy released during exergonic reactions. On the average, the breakdown of food molecules in your body releases about 2,000 to 2,800 kilocalories each day.

Conversely, many other reactions simply won't proceed on their own without an energy boost. The energy-requiring reactions in which starch and other large molecules are assembled from smaller, energy-poor ones are examples of this. Reactions that show a net *gain* in energy are said to be *endergonic*, meaning "energy in" (Figure 6.4b).

**Reversible Reactions**. Like most chemical reactions, the ones that proceed in living cells are "reversible." In other words, once product molecules have formed, they can be converted back to reactant molecules.

To understand how this works, recall that living cells are bathed in fluid and are largely fluid themselves. Molecules in this fluid move about constantly and bump into each other, and most often they react spontaneously in regions where they are most concentrated. The more concentrated they are, the more their random movements put them on collision courses. When they collide, one molecule might cause the other to split apart or change its shape, or one might join up with the other.

As long as the concentration of reactant molecules is high enough, a chemical reaction proceeds in the forward direction to product molecules. But when the concentration of product builds up, some number of product molecules will revert to reactants. Reversible reactions are indicated by arrows running in the "forward" and "reverse" directions (Figure 6.5).

**Dynamic Equilibrium**. Unless other events in the cell (or body) keep it from doing so, a reaction that is reversible approaches **dynamic equilibrium**. Then, the forward and reverse reactions proceed at equal rates (Figure 6.6). There is no *net* change in the concentrations of reac-

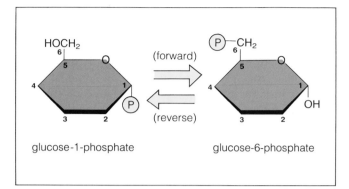

**Figure 6.5** A reversible reaction. In most cells, phosphate can become attached to a glucose molecule in several different ways. With high concentrations of glucose-1-phosphate, the reaction shown here tends to run in the forward direction; with high concentrations of glucose-6-phosphate, it runs in reverse. (The "1" and "6" identify the particular carbon atom of the glucose ring to which phosphate is attached.)

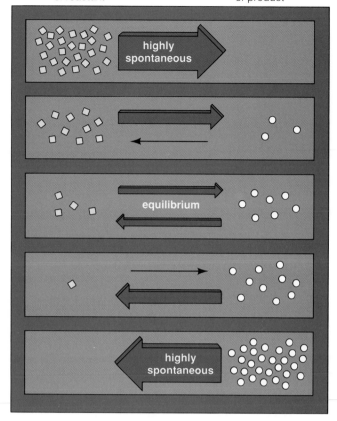

**Figure 6.6** Chemical equilibrium. With high concentrations of reactant molecules, reactions generally proceed most strongly in the forward direction. With high concentrations of product molecules, they proceed most strongly in reverse. At chemical equilibrium, the *rates* of the forward and reverse reactions are equal.

tants and products even though molecules are still being converted from one form to the other. It's like a party with as many people wandering in as wandering out of two adjoining rooms. The total number in each room stays the same—say, thirty in one and ten in the other—even though the mix of people in each room continually changes.

How many reactant molecules and how many product molecules will there be at equilibrium? That depends on the substances involved. *Every reaction has its own characteristic ratio of products to reactants at equilibrium.*

In one reaction, glucose-1-phosphate is rearranged into glucose-6-phosphate. As Figure 6.5 shows, these are simply glucose molecules with a phosphate group attached to one of their six carbon atoms. Suppose we allow the reaction to proceed when the concentrations of both substances are the same. At this point the forward reaction occurs 19 times faster than the reverse reaction does. Said another way, the forward reaction is producing more molecules in the same amount of time. Only when there finally are 19 molecules of glucose-6-phosphate for every molecule of glucose-1-phosphate will the forward and reverse reactions proceed at the same rate. For this example, then, the characteristic ratio of products to reactants at equilibrium is 19:1.

You may be thinking that which way a reaction proceeds is a remote issue, of concern only to laboratory chemists. As you will now see, this is really a central issue in our study of what it takes to be alive.

## Metabolic Pathways

Cells never stop building up and tearing down substances until they are dead. This incessant activity requires control over the directions in which different reactions proceed. Why? Cells use only so many molecules of different substances at a given time, and they have only so much internal space to hold any excess. If cells were to produce more than they could use, store, or secrete, the excess might cause problems.

Think about what happens to a baby born with a mutation that can give rise to the genetic disorder phenylketonuria (PKU). In this disorder, a series of reactions is blocked and a substance (phenylalanine) reaches high concentrations in the baby's body. The excess enters into reactions that produce phenylketones, and the accumulation of those substances leads to severe mental retardation. The symptoms can be prevented if affected individuals are detected soon enough. They are immediately placed on a diet that makes it difficult for phenylalanine to accumulate in the body.

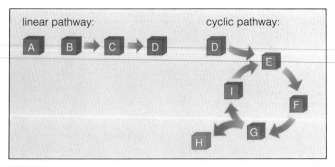

**Figure 6.7** Linear and cyclic metabolic pathways.

Normally, cells maintain, increase, and decrease the concentrations of substances by coordinating a variety of metabolic pathways. A **metabolic pathway** is an orderly sequence of reactions, the steps of which are quickened with the help of specific enzymes. Most sequences are linear; some are circular (Figure 6.7). Branches often link different pathways, with products of one pathway serving as reactants for others.

Overall, the main metabolic pathways are either degradative or biosynthetic. In **degradative pathways**, carbohydrates, lipids, and proteins are broken down in stepwise reactions that lead to products of lower energy. Often, some of the energy released is used to do cellular work. In **biosynthetic pathways**, small molecules are assembled into larger molecules such as polysaccharides, proteins, and nucleic acids.

The participants in metabolic pathways can be defined in the following way:

| | |
|---|---|
| **reactants** | *substances able to enter into a reaction; also called substrates or precursors* |
| **intermediates** | *compounds formed between the start and the end of a metabolic pathway* |
| **enzymes** | *proteins that catalyze (speed up) reactions* |
| **cofactors** | *small molecules and metal ions that help enzymes or that carry atoms or electrons from one reaction site to another* |
| **energy carriers** | *mainly ATP, which readily donates energy to diverse reactions* |
| **end products** | *the substances present at the conclusion of a metabolic pathway* |

Let's take a closer look at some of these substances and at their vital roles in metabolism.

## ENZYMES

A cup of sugar left undisturbed for twenty years changes very little. But when you eat sugar, it rapidly undergoes chemical change. Enzymes secreted by some of your cells account for the difference in the rate of change. **Enzymes** are molecules with enormous catalytic power, meaning that they greatly enhance the rate at which specific reactions approach equilibrium. The vast majority are proteins, although a few forms of RNA also have catalytic properties (page 301).

### Characteristics of Enzymes and Their Substrates

Enzymes have four characteristics in common. *First*, enzymes do not make anything happen that would not eventually happen on its own. They just make it happen faster—at least a million times faster, usually. *Second*, enzyme molecules are not permanently altered or used up in the reactions they catalyze; they can be used over and over again. *Third*, each type of enzyme is highly selective about its substrates.

**Substrates** are specific molecules that an enzyme can chemically recognize, bind briefly to itself, and modify in a specific way. Thrombin, an enzyme involved in blood clotting, helps break a specific peptide bond in a specific protein. It chemically recognizes a bond between two amino acids (arginine and glycine) and speeds its breakdown:

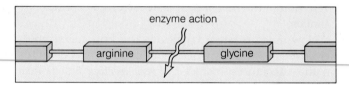

*Fourth*, an enzyme can recognize both the reactants and the products of a given reaction as its substrates. In other words, the same enzyme can catalyze a reaction in both the forward and reverse directions.

### Enzyme Structure and Function

Substrates interact with an enzyme at one or more regions called active sites. An **active site** is a surface region where an enzyme molecule is folded in the shape of a crevice and where a particular reaction is catalyzed. Figure 6.8 shows an example.

As long ago as 1890, Emil Fischer thought the shape of some region of the enzyme's surface must match a complementary region on its substrate, like a lock precisely matching its key. Today we know the match is not so rigid. In Daniel Koshland's **induced-fit model**, an active site almost *but not quite* matches its substrate when

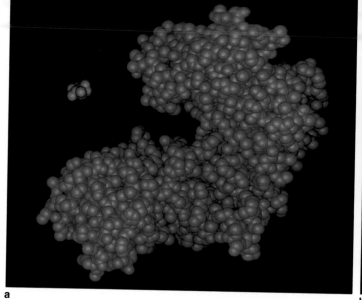

a

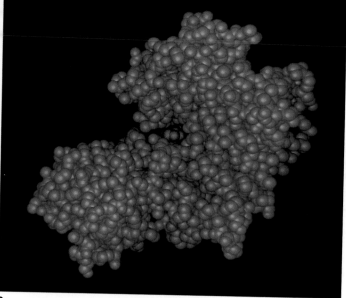

b

**Figure 6.8** Model of the induced fit between the enzyme hexokinase (blue) and its bound substrate (a glucose molecule, shown in red).

(**a**) The cleft into which the glucose is heading is the enzyme's active site. (**b**) In this enzyme-substrate complex, notice how the enzyme shape is altered temporarily: the upper and lower parts now close in around the substrate.

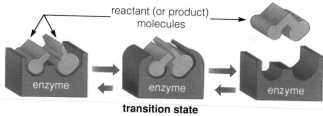

**transition state**
(tightest binding but least stable)

**Figure 6.9** Induced-fit model of enzyme-substrate interactions. Only when the substrate is bound in place is the enzyme's active site complementary to it. The most precise fit occurs during a transition state that precedes the reaction. An enzyme-substrate complex is short-lived, partly because only weak bonds hold it together.

first making contact with it. The enzyme-substrate interaction is enough to strain certain bonds within the substrate molecule. The strained bonds are easier to break, and this paves the way for the formation of the new bonding arrangements characteristic of the product molecules.

Figure 6.9 is a simple picture of how an enzyme catalyzes a reaction between two substrate molecules. When the molecules fit most precisely into the active site, they are in an activated condition called the "transition state." Now the reaction between them proceeds spontaneously, just as a boulder pushed up and over the crest of a hill rolls down on its own.

In the cellular world, reactants typically do not enter the transition state without an energy "push." Simply put, the reactant molecules must collide with some minimum amount of energy. For any given reaction, the minimum amount of energy needed to bring all the reactant molecules to the transition state is called the **activation energy**. That amount is like a hill over which the molecules must be pushed (Figure 6.10).

An enzyme increases the rate of a given reaction by *lowering* the required activation energy. How? Among other things, weak but extensive bonding at the active site puts the enzyme's substrates in orientations that promote reaction. In contrast, reactants colliding on their own do so from random directions, so mutually attractive chemical groups may not make contact and reaction may not occur.

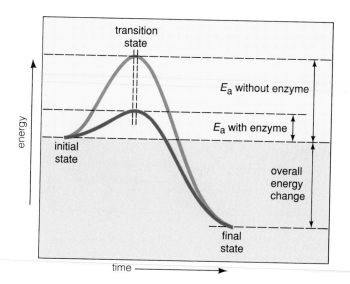

**Figure 6.10** Energy hill diagram showing the effect of enzyme action. An enzyme greatly enhances the rate at which a reaction proceeds because it lowers the required activation energy ($E_a$). In other words, not as much collision energy is needed to boost reactant molecules to the crest of the energy hill (transition state).

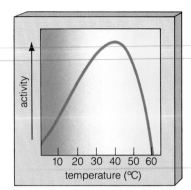

**Figure 6.11** Effect of increases in temperature on enzyme activity.

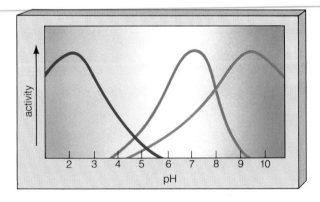

**Figure 6.12** Visible effects of environmental temperature on enzyme activity. In Siamese cats, the fur on the ears and paws contains more dark-brown pigment (melanin) than the rest of the body. A heat-sensitive enzyme controlling melanin production is less active in warmer body regions, and this results in lighter fur in those regions.

**Figure 6.13** Effect of pH on enzyme activity. The brown line charts the activity of an enzyme that is fully functional in neutral solutions. The red line charts the activity of one that is functional in basic solutions; the purple line, in acidic solutions.

An enzyme enhances the rate of a given reaction by lowering the activation energy required for that reaction.

### Effects of Temperature and pH on Enzymes

Each type of enzyme functions best within a certain temperature range (Figures 6.11 and 6.12). When the temperature becomes too high, reaction rates decrease sharply. The increased thermal energy disrupts weak bonds holding the enzyme in its three-dimensional shape, and denaturation occurs (page 46). Because the active site becomes altered, substrates cannot bind to it.

Brief exposure to temperatures that are much higher than an organism typically encounters can destroy enzymes and adversely affect metabolism. This happens during fevers that are high enough to be extremely dangerous. Thus, when the internal body temperature of a human reaches 44°C (112°F), death generally follows.

Also, most enzymes function best within a limited range of pH, which commonly is near pH 7. Higher or lower pH values generally disrupt the enzyme's three-dimensional shape and its function (Figure 6.13). Pepsin is one of the exceptions; this enzyme functions in the extremely acidic fluid of the stomach.

Enzymes function only within limited ranges of temperature and pH.

### Control of Enzyme Activity

Through controls over enzymes of different pathways, cells direct the flow of nutrients, wastes, and other substances in suitable ways. Some of the controls govern the number of enzyme molecules available. For example, certain mechanisms work to accelerate or slow down the production of enzymes. Others stimulate or inhibit the activity of enzymes already formed.

When you eat too much sugar, for example, enzymes in your liver cells act on the excess, converting it first to glucose and then to glycogen or fat. When your body uses up glucose and needs more, enzymes break down glycogen to release its glucose subunits. In this case, a hormone called glucagon acts as a control over enzyme activity. It stimulates the key enzyme in the pathway by which glycogen is degraded, and it inhibits the enzyme that catalyzes glycogen formation.

**Inhibitors**, which can bind with enzymes and interfere with their function, are one type of control mechanism. A certain inhibitor molecule acts on trypsin, a

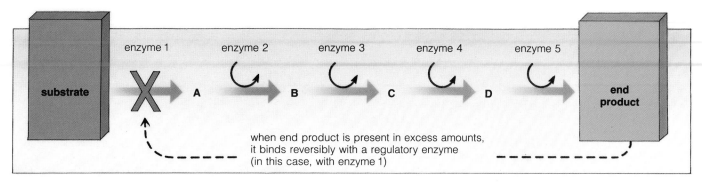

enzyme 1 enzyme 2 enzyme 3 enzyme 4 enzyme 5

substrate  A  B  C  D  end product

when end product is present in excess amounts,
it binds reversibly with a regulatory enzyme
(in this case, with enzyme 1)

**Figure 6.14** Example of feedback inhibition of a metabolic pathway. Here, the end product binds to the first enzyme in the pathway leading to its formation. When the product concentration drops, fewer molecules are around to inhibit the regulatory enzyme in the pathway, so production can rise again. Feedback inhibition allows concentrations of substances to be adjusted quickly to the cell's requirements.

protein-digesting enzyme. Cells in the pancreas produce and secrete trypsin into the small intestine. Prior to secretion, trypsin is kept isolated in vesicles, in inactive form. Any molecules that do escape packaging are shut down by the inhibitor. Without this safeguard, trypsin could be unleashed against the proteins of tissues and blood vessels in the pancreas. In *acute pancreatitis*, this enzyme and others are activated prematurely, sometimes with fatal results.

Some enzymes are governed by **allosteric control**. Besides the active site, "allosteric" enzymes have control sites where specific substances can bind and alter enzyme activity. For example, when a cell produces tryptophan molecules faster than it uses them, the unused molecules bind to and inhibit an allosteric enzyme. The enzyme happens to catalyze a step necessary in the synthesis of tryptophan.

Control of tryptophan production is a form of **feedback inhibition**. The output of the process works in a way that inhibits further output. Figure 6.14 illustrates this control mechanism.

---

Enzymes act only on specific substrates, and controls over their activity are central to the directed flow of substrates into, through, and out of the cell.

---

## COFACTORS

Some enzymes speed up the transfer of electrons, atoms, or functional groups from one substrate to another. Many must be assisted by nonprotein components called **cofactors,** either to help catalyze the reaction or to serve fleetingly as the transfer agents.

Cofactors include some large organic molecules that function as *coenzymes.* Among them are **NAD$^+$** (nicotinamide adenine dinucleotide) and **FAD** (flavin adenine dinucleotide). Both types have roles in the breakdown of glucose and other carbohydrates. When enzymes dismantle a glucose molecule, unbound protons (H$^+$) and electrons become available. These are picked up by NAD$^+$, FAD, or both and turned over to other reaction sites. When carrying their cargo, the two coenzymes are abbreviated NADH and FADH$_2$ respectively.

**NADP$^+$** (nicotinamide adenine dinucleotide phosphate) is a coenzyme with a central role in photosynthesis. It also serves as a link between some of the main degradative and biosynthetic pathways (for example, the ones by which fatty acids are assembled). Discerning enzymes recognize the phosphate "tag" on NADP$^+$ molecules and allow them to transfer protons and electrons to specific reaction sites. When carrying this cargo, the coenzyme is often abbreviated NADPH.

Some *metal ions* also serve as cofactors. They include ferrous iron (Fe$^{++}$), which is a component of cytochrome molecules. The cytochromes are carrier proteins bound in cell membranes, such as the membranes of chloroplasts and mitochondria.

## ELECTRON TRANSFERS IN METABOLIC PATHWAYS

If you were to throw some glucose into a woodfire, its atoms would quickly let go of one another and combine with oxygen in the atmosphere, forming CO$_2$ and H$_2$O. The energy that had been stored in glucose would be lost as heat. Cells make better use of a glucose molecule; they do not "burn" it all at once and so waste its stored energy. Instead, atoms are plucked away from glucose in controlled steps, so that *intermediate* molecules form

**Figure 6.15** (a) When hydrogen and oxygen are made to react (say, by an electric spark), energy is released as heat. (b) In cells, the same type of reaction is made to occur in many small steps that allow much of the released energy to be harnessed in usable form. These "steps" are electron transfers, often between molecules that operate together as an electron transport system.

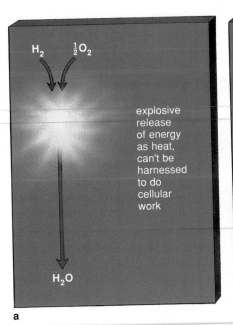

a

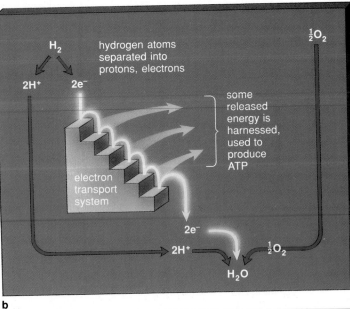

b

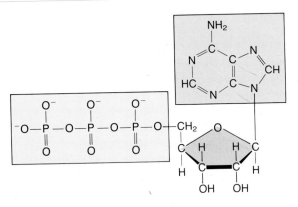

**Figure 6.16** Structural formula for adenosine triphosphate, or ATP. The triphosphate group is shaded in gold, the sugar ribose in pink, and the adenine portion in blue.

along the route from $C_6H_{12}O_6$ to the $CO_2$ and $H_2O$. At each step, a specific enzyme lowers the activation energy for an intermediate. And at each step, only *some* energy is released.

Remember that a chemical bond is a union of the electron structures of two atoms. Breaking a chemical bond puts electrons up for grabs. In chloroplasts and mitochondria, such liberated electrons are sent through **electron transport systems**. The systems consist of enzymes and cofactors, bound in a cell membrane, that transfer electrons in a highly organized sequence. One molecule "donates" electrons, and the next in line "accepts" them. Each time a donor gives up electrons, it is said to be "oxidized." Each time an acceptor acquires

electrons, it is "reduced." *Oxidation-reduction* merely means an electron transfer.

Electron transfers can occur after an atom (or a molecule) absorbs enough energy to boost one or more of its electrons to a shell farther from its nucleus (page 24). An excited electron quickly returns to the lowest energy level available to it—and it gives off energy when it does this. *Electron transport systems "intercept" excited electrons and make use of the energy they release.*

By analogy, think of an electron transport system as a staircase (Figure 6.15). The electrons at the top of the staircase have the most energy. They drop down the staircase, one step at a time (they are transferred from one electron carrier to another). With each drop, some energy being released can be harnessed to do work—for example, to make hydrogen ions move in ways that establish pH and electric gradients across membranes. Such gradients, as you will discover, are central in ATP formation.

Electron transfers may seem rather abstract. The *Commentary* makes the idea more concrete by describing unusual but striking effects of electron transfers.

## ATP: THE UNIVERSAL ENERGY CARRIER

### Structure and Function of ATP

Sunlight is the primary energy source for the web of life. Before the sun's energy can be used in cell activities, it must be transformed into the chemical energy of **ATP** (adenosine triphosphate). Similarly, cells cannot directly

## You Light Up My Life—Visible Effects of Electron Transfers

Travel the ocean at night and you may see evidence of comb jellies near the water's surface. These relatives of jellyfishes give good displays of *bioluminescence*—which simply means luminescent flashes in body tissues (Figure *a*). Such flashes also occur in several groups of bacteria, fungi, fishes, and insects, including fireflies. The insects use flashes as signals that may attract a mate.

Highly fluorescent substances called luciferins can be prodded into giving up electrons. Once they have been released, the electrons drop to a lower energy level—and light is emitted when they do. Enzymes called luciferases catalyze the reactions.

Several years ago, the biochemist Keith Wood "stole" the soft, yellowish light that flashes in fireflies. Wood and his coworkers isolated the firefly gene that contains the instructions for building luciferase molecules. Just recently they stole more lights from the Jamaican click beetle, otherwise known as the kittyboo. The kittyboo has four luciferase genes. One codes for green flashes, another for greenish-yellow, and the others for yellow and orange. Biochemists have been inserting the luciferase genes into a variety of organisms, including *Escherichia coli*. Figure *b* shows four bacterial cells, glowing with the borrowed light.

The ability to transfer luciferase genes into *E. coli* may prove useful in studying the mechanisms by which particular genes are turned on and off. Imagine a light bulb that couldn't be seen when you turned on the switch. What biochemists are doing is effectively substituting a light bulb for a gene. They are interested in the switches, not the bulb. When light flashes in a modified *E. coli* cell, they have observable evidence that a *switch* controlling a gene has been activated.

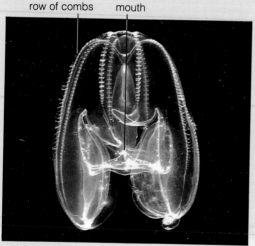

row of combs    mouth

**a** Bioluminescence in a comb jelly

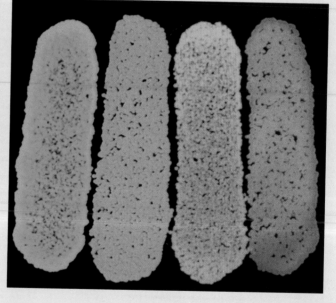

**b** Bioluminescence imparted to four bacterial cells

use the energy that is released during the breakdown of carbohydrates or other large organic molecules. They must first use the stored energy to produce ATP.

How does ATP act as an energy carrier from one reaction to another? As Figure 6.16 shows, ATP is composed of adenine (a nitrogen-containing compound), ribose (a five-carbon sugar), and a triphosphate (three phosphate groups), all linked together by covalent bonds.

Under cellular conditions, one of the phosphates can be split off easily by hydrolysis, and a good deal of usable energy is released when this happens. Said another way, ATP is an unstable molecule that can readily break down to more stable products.

Many hundreds of different enzymes can couple ATP hydrolysis to many hundreds of different metabolic reactions. Energy released during ATP hydrolysis provides energy for biosynthesis, active transport across cell membranes, and molecular displacements, such as those underlying muscle contraction. In fact, ATP molecules are like the coins of a nation—they are a common currency of energy in all cells.

**ATP directly or indirectly delivers energy to or picks up energy from almost all metabolic pathways.**

### The ATP/ADP Cycle

In the **ATP/ADP cycle**, an energy input drives the linkage of **ADP** (adenosine diphosphate) and a phosphate group (or inorganic phosphate) into ATP. Then the ATP donates a phosphate group elsewhere and reverts back to ADP:

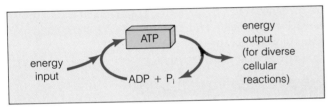

Adding phosphate to a molecule is called **phosphorylation**. What is important about it? When a molecule becomes phosphorylated by ATP, its store of energy generally increases *and it becomes primed to enter a specific reaction.*

With the ATP/ADP cycle, cells have a renewable means of conserving energy and transferring it to specific reactions. The ATP turnover is breathtaking. Even if you were bedridden for twenty-four hours, your cells would turn over approximately 40 kilograms (88 pounds) of ATP molecules simply for routine maintenance!

### SUMMARY

1. Cells acquire and use energy to synthesize, accumulate, break down, and rid themselves of substances in controlled ways. These activities, which sustain cell growth, maintenance, and reproduction, are called metabolism.

2. Cellular use of energy conforms to two laws of thermodynamics. According to the first law, energy can be converted from one form to another but the total amount in the universe never changes. According to the second law, with each energy conversion, some energy is dispersed in a form that is not as readily available to do work.

3. Cells lose energy during metabolic reactions, but in nearly all cases, they replace it with energy derived directly or indirectly from the sun.

4. A metabolic pathway is a stepwise sequence of reactions in a cell. In *biosynthetic* pathways, organic molecules are assembled and energy becomes stored in them. In *degradative* pathways, organic molecules are broken apart and energy is released.

5. The following substances take part in metabolic reactions:

a. Reactants or substrates: the substances that enter a reaction.

b. Enzymes: proteins that serve as catalysts (they speed up reactions). Enzymes do not change the expected outcome of a reaction. They only change the rate at which the reaction proceeds.

c. Cofactors: coenzymes (including $NAD^+$) and metal ions that help catalyze reactions or carry functional groups stripped from substrates.

d. Energy carriers: mainly ATP, which readily donates energy to other molecules. Most biosynthetic pathways run directly or indirectly on ATP energy.

e. End products: the substances formed at the end of a metabolic pathway.

6. If left undisturbed, most metabolic reactions approach a state of dynamic equilibrium. Then, there is no further net change in the concentrations of reactants and products.

7. Enzymes do not change what the concentration ratio of reactant to product molecules will be at equilibrium. They only increase the rate at which a reaction approaches equilibrium (by lowering the required activation energy).

8. During many electron transfers (oxidation-reduction reactions), energy is released that can be used to do work—for example, to make ATP.

1. State the first and second laws of thermodynamics. Which law deals with the *quality* of available energy, and which deals with the *quantity*? Give some examples of high-quality energy. *92–93*

2. Does the living state violate the second law of thermodynamics? In other words, how does the world of living things maintain a high degree of organization, even though there is a universal trend toward disorganization? *92–93*

3. In metabolic reactions, does equilibrium imply equal concentrations of reactants and products? Can you think of a cellular event that might keep a reaction from approaching equilibrium? *94*

4. Describe an enzyme and its role in metabolic reactions. *96–98*

5. What are the three molecular components of ATP? What is the function of ATP, and why is phosphorylation of a molecule by ATP so important? *101–102*

6. What is an oxidation-reduction reaction? What is its function in cells? *100*

7. Enzymes _____ .
   a. enhance reaction rates
   b. are affected by pH
   c. act on specific substrates
   d. all of the above are correct

8. All electron transport systems involve _____ and _____ .
   a. enzymes, cofactors
   b. electron transfers, released energy
   c. chloroplasts, mitochondria
   d. both a and b are correct

9. The main energy carriers in cells are _____ .
   a. NAD$^+$ molecules
   b. cofactors
   c. ATP molecules
   d. enzymes

10. Match each substance with its correct description.
    _____ a coenzyme or metal ion
    _____ substance formed at end of a metabolic pathway
    _____ mainly ATP
    _____ substance entering a reaction
    _____ protein that catalyzes a reaction
    
    a. reactant
    b. enzyme
    c. cofactor
    d. energy carrier
    e. end product

## Self-Quiz *(Answers in Appendix IV)*

1. A cell's capacity to acquire and use energy for building and breaking apart molecules is called _____ .

2. Two laws of _____ govern how cells acquire, convert, and transfer energy during metabolic reactions.

3. The ultimate source of energy for nearly all organisms on earth is _____ .
   a. food
   b. water
   c. the sun
   d. ATP

4. Which of the following statements is *not* true of dynamic equilibrium?
   a. The concentration of product is equal to the concentration of reactant.
   b. The rate of forward reaction is equal to the rate of reverse reaction.
   c. There is no further change in the concentrations of reactant and product.
   d. It is unchanged by enzyme activity.

5. In biosynthetic pathways, _____ .
   a. organic molecules are simply broken apart
   b. energy is not required for the reactions
   c. organic molecules are assembled
   d. energy is stored in organic molecules
   e. both c and d are correct

6. Which of the following is *not* true? Metabolic pathways _____ .
   a. occur in stepwise series of chemical reactions
   b. are speeded up by enzyme activity
   c. may degrade or assemble molecules
   d. overcome the second law of thermodynamics

## Selected Key Terms

| | |
|---|---|
| activation energy *97* | enzyme *96* |
| active site *96* | exergonic reaction *94* |
| ADP *102* | FAD *99* |
| allosteric control *99* | feedback inhibition *99* |
| ATP *100* | first law of thermodynamics *92* |
| biosynthetic pathway *96* | intermediate compound *96* |
| coenzyme *99* | kilocalorie *93* |
| cofactor *99* | metabolic pathway *96* |
| degradative pathway *96* | metabolism *91* |
| dynamic equilibrium *94* | NAD$^+$ *99* |
| electron transport system *100* | NADP$^+$ *99* |
| end product *96* | oxidation-reduction *100* |
| endergonic reaction *94* | phosphorylation *102* |
| energy *92* | reactant *96* |
| energy carrier *96* | second law of thermodynamics *92* |
| entropy *92* | substrate *96* |

## Readings

Atkins, P. 1984. *The Second Law.* New York: Freeman.

Doolittle, R. 1985. "Proteins." *Scientific American* 253(4):88–99.

Fenn, J. 1982. *Engines, Energy, and Entropy.* New York: Freeman. Deceptively simple introduction to thermodynamics; good analogies. Paperback.

Fersht, A. 1985. *Enzyme Structure and Mechanism.* Second edition. New York: Freeman.

## Sunlight, Rain, and Cellular Work

Just before dawn in the Midwest the air is dry and motionless. The heat that has scorched the land for weeks still rises from the earth and hangs in the air of a new day. There are no clouds in sight. There is no promise of rain. For hundreds of miles, crops stretch out, withered or dead. All the marvels of modern agriculture can't save them now. In the absence of one vital resource—water—life in each cell of those many thousands of plants has ceased.

In Los Angeles, a student reading the morning newspaper complains that the Midwest drought will probably cause a hike in food prices. In Washington, D.C., economists calculate the crop failures in terms of decreased tonnage available for domestic consumption and export; government officials brood about what that

means to the nation's balance of payments. In Ethiopia, a child with bloated belly and spindly legs waits passively for death. Even if food donations were to reach her now, it would be too late. Deprived of food resources too long, the cells of her body will never grow normally again.

You are about to explore the ways in which cells trap and use energy. At first the cellular pathways may seem to be far removed from the world of your interests. *Yet the food molecules on which you and nearly all other organisms depend cannot be built or used without those pathways and the raw materials—including water—required for their operation.*

We will return repeatedly to this point in later chapters, when we address such concerns as human

**Figure 7.1** Links between photosynthesis and aerobic respiration—the main energy-acquiring and energy-releasing pathways in the world of life.

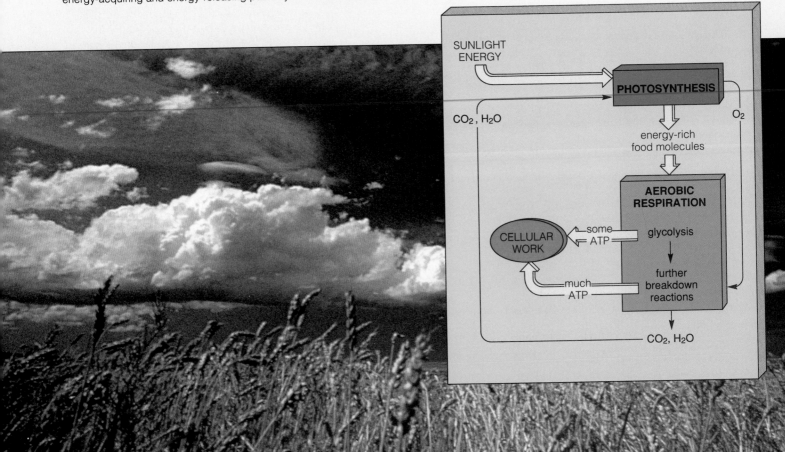

nutrition, human population growth and environmental limits on agriculture, genetic engineering of new crop plants, and the effects of pollution on food production. Here, our point of departure is the *source* of food, which isn't a farm or a supermarket or a refrigerator. What we call "food" was put together somewhere in the world by living cells from organic compounds. Given that the basic structure of every organic compound is a carbon backbone, the questions become these:

1. *Where does the carbon come from in the first place?*

2. *Where does the energy come from to drive the linkage of carbon and other atoms into organic compounds?*

3. *How does the energy inherent in those compounds become available to do cellular work?*

The answers vary, depending on whether you are talking about autotrophic or heterotrophic organisms.

**Autotrophs** obtain carbon and energy from the physical environment. They are "self-nourishing" (which is what autotroph means). Their carbon source is carbon dioxide ($CO_2$), a gaseous substance all around us in the air and dissolved in water. Only the *photosynthetic* autotrophs can get energy from sunlight. Plants, some protistans, and some bacteria fall in this category. A few kinds of bacteria are *chemosynthetic* autotrophs; they get energy by stripping electrons from sulfur or some other inorganic substance.

**Heterotrophs** are not self-nourishing. They feed on autotrophs, each other, and organic wastes. They must get carbon and energy from organic compounds *already built* by autotrophs. Animals, fungi, many protistans, and most bacteria are heterotrophs.

It follows, from the above, that carbon and energy enter the web of life primarily by **photosynthesis**. Energy stored in organic compounds as a result of photosynthesis can be released by several different pathways, all of which begin with the same breakdown reactions (*glycolysis*). Of these, the pathway called **aerobic respiration** releases the most energy. Figure 7.1 shows the links between photosynthesis and aerobic respiration—the focus of this chapter and the next.

KEY CONCEPTS

**1.** Organic compounds, with their carbon backbones, are the key building blocks and energy stores for life. Plants and other photosynthetic autotrophs assemble their own organic compounds. They use carbon dioxide from the air as a source for the carbon, and they trap sunlight energy to drive the synthesis reactions.

**2.** Photosynthesis is the main biosynthetic pathway by which carbon and energy enter the web of life. Animals and other heterotrophs must obtain their carbon and energy from organic compounds already built by the autotrophs.

**3.** In plants, photosynthesis proceeds at clusters of pigment molecules and electron transport systems present in the membranes of organelles called chloroplasts.

**4.** During the light-dependent reactions, pigment molecules absorb light energy and are compelled to give up electrons to transport systems. Movement of electrons through the systems sets up electrochemical gradients across membranes, and these gradients drive the formation of ATP and NADPH.

**5.** During the actual synthesis reactions, which proceed independently of sunlight, ATP provides energy to drive the joining of carbon and oxygen (from carbon dioxide) with hydrogen and electrons (from NADPH). The reactions produce sugar phosphates, which are used to form sucrose, starch, and other end products of photosynthesis.

## PHOTOSYNTHESIS

### Simplified Picture of Photosynthesis

Photosynthesis proceeds in two stages, each with its own set of reactions. In the *light-dependent* reactions, sunlight energy is absorbed and converted to chemical energy, which is transferred to ATP and NADPH. (Here you may wish to refer to pages 99 and 100.) In the *light-independent* reactions, sugars and other compounds are assembled with the help of ATP and NADPH. Photosynthesis is typically summarized this way:

$$2H_2O + CO_2 \xrightarrow{\text{sunlight}} O_2 + (CH_2O) + H_2O$$

As you can see from this equation, hydrogen atoms obtained from water molecules end up in newly synthesized compounds that are based on some number of $(CH_2O)$ units. For instance, for the reactions leading to a new glucose molecule $(C_6H_{12}O_6)$, you would have to multiply everything by six to get the six carbons, twelve hydrogens, and six oxygens in it:

$$12H_2O + 6CO_2 \xrightarrow{\text{sunlight}} 6O_2 + C_6H_{12}O_6 + 6H_2O$$

Here, glucose is shown as an end product in order to keep the chemical bookkeeping simple. The reactions

a

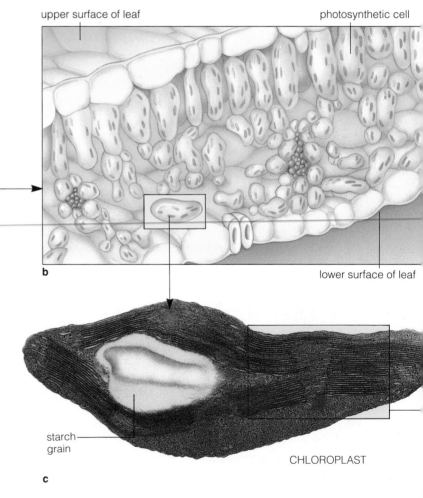

upper surface of leaf

photosynthetic cell

b

lower surface of leaf

starch grain

CHLOROPLAST

c

**Figure 7.2** Functional zones of a chloroplast from the leaf of a sow thistle (*Sonchus*) shown in **a**. The light-dependent reactions of photosynthesis occur at thylakoid membranes, and they lead to ATP and NADPH formation. The light-independent reactions occur in the stroma. They lead to production of sugars and other carbon-containing molecules. (**b**) Section through a sow thistle leaf, showing chloroplast-containing cells. (**c**) Chloroplast in cross-section. (**d**) Two of the grana. (**e**) Where photosynthetic reactions of the sort described in this chapter occur.

don't really end with glucose, however. Newly formed glucose and other simple sugars are linked at once into sucrose, starch, and other carbohydrates—the true end products of photosynthesis.

## Chloroplast Structure and Function

Let's take a look at how photosynthesis proceeds in the chloroplasts of leafy plants. As indicated earlier in Figure 4.20, this organelle has a double-membrane envelope around its semifluid interior, the *stroma*. An elaborate membrane system weaves through the stroma. It takes the form of flattened channels and disklike compartments organized into stacks, called *grana* (singular, *granum*). This is the **thylakoid membrane** system of many kinds of plants.

The spaces inside the thylakoid disks and channels connect with one another. They form a continuous compartment where hydrogen ions can be accumulated and used in ways that help produce ATP. The stroma is where enzymes speed the assembly of the actual products of photosynthesis, including starch molecules (Figure 7.2).

If you could line up 2,000 chloroplasts, one after another, the lineup would be no wider than a dime. Imagine all the chloroplasts in just one lettuce leaf, each a tiny factory for producing sugars and starch—and you get an idea of the magnitude of metabolic events required to feed you and all other organisms living together on this planet.

## LIGHT-DEPENDENT REACTIONS

Three events unfold during the **light-dependent reactions**, the initial stage of photosynthesis. *First*, pigments absorb light energy and give up electrons. *Second*, electron and hydrogen transfers lead to ATP and NADPH formation. *Third*, the pigments that gave up electrons in the first place get electron replacements.

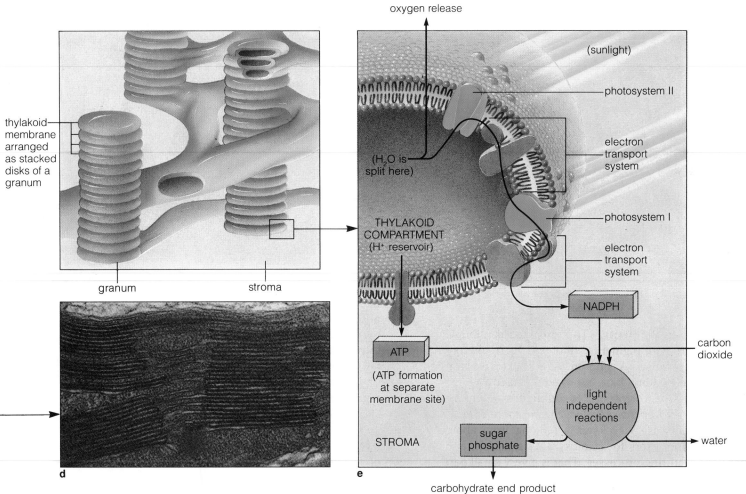

oxygen release

thylakoid membrane arranged as stacked disks of a granum

granum

stroma

d

(sunlight)

photosystem II

electron transport system

photosystem I

electron transport system

(H₂O is split here)

THYLAKOID COMPARTMENT (H⁺ reservoir)

NADPH

carbon dioxide

ATP

(ATP formation at separate membrane site)

light independent reactions

STROMA

sugar phosphate

water

e

carbohydrate end product (e.g., sucrose, starch, cellulose)

**Photosystems.** A thylakoid membrane has many thousands of **photosystems**, which are organized clusters of 200 to 300 light-trapping pigment molecules. The pigments are bound to membrane proteins. Over 90 percent of the pigments in a photosystem simply "harvest" sunlight. When they absorb photon energy, one of their electrons gets boosted to a higher energy level (page 23). The electron returns almost at once to a lower level, and the extra energy is released when it does:

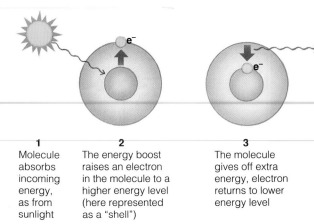

| 1 | 2 | 3 |
|---|---|---|
| Molecule absorbs incoming energy, as from sunlight | The energy boost raises an electron in the molecule to a higher energy level (here represented as a "shell") | The molecule gives off extra energy, electron returns to lower energy level |

The released energy hops from one molecule to another in the photosystem. With each hop, a little of that energy is usually lost (as heat). The energy remaining corresponds to longer and longer wavelengths, compared to the original photon energy. Only a few chlorophyll molecules in a photosystem can respond to the longest of those wavelengths. They act like a sink, or *trap*, for all the energy being harvested by all the other pigments. Those chlorophylls alone give up the electrons used in photosynthesis.

When energy flows into an energy trap, an electron is excited and is rapidly transferred to an acceptor molecule embedded in the thylakoid membrane. Thus, *the first event of photosynthesis is the light-activated transfer of an electron from a special chlorophyll molecule to an acceptor molecule.* It is over in less than a billionth of a second.

### How ATP and NADPH Form in Chloroplasts

Electrons expelled from a chlorophyll molecule pass through one or two **electron transport systems** of the sort described on page 100. Each system is a series of molecules bound in the thylakoid membrane (Figure 7.2d). Each molecule accepts and then donates electrons to the next molecule in line. The electron flow sets up gradients across the membrane, and these drive the attachment of phosphate to ADP, so forming ATP. We call this *photo*phosphorylation, because the pathway depends on an earlier input of light energy. Such pathways are cyclic or noncyclic.

**Cyclic Pathway.** A special chlorophyll molecule, P700, dominates the type of pigment cluster called *photosystem I*. This molecule absorbs wavelengths of about 700 nanometers. In the simplest ATP-producing pathway, electrons "travel in a circle" from a P700 chlorophyll, through a transport system, then back to P700:

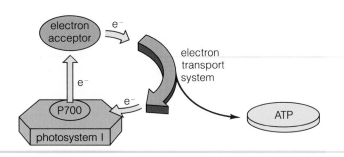

This pathway is called **cyclic photophosphorylation**. It is probably the oldest means of ATP production. Early photosynthetic autotrophs were no larger than existing bacteria, so their body-building programs could scarcely have been enormous. They could have used ATP alone to build organic compounds even though such reactions are rather inefficient. (ATP only carries energy to sites where organic compounds are built. The electrons and hydrogen atoms required must be obtained by other means.) However, energy from the cyclic pathway would not have sustained the evolution of larger photosynthesizers, including leafy plants.

**Noncyclic Pathway.** Today, leafy plants rely mostly on **noncyclic photophosphorylation**. In this ATP-producing pathway, electrons move through two photosystems and two electron transport systems embedded in the thylakoid membrane. But the electrons do not move in a circle through these membrane sites. At the end of the transport system that contains P700, they are picked up by a coenzyme, NADP$^+$, that delivers hydrogen as well as the electrons to a reaction site in the stroma. There, the hydrogen and electrons are used *directly* in the synthesis of organic compounds!

Figure 7.6 is a diagram of the sequence of events in the noncyclic pathway. The reactions begin at a photosystem distinguished by a chlorophyll molecule (P680) that absorbs wavelengths of about 680 nanometers. The photosystem having this type of molecule is designated *photosystem II*. When P680 absorbs light energy, it gives up an electron to an acceptor molecule. From there, the electron moves through a transport system—and then to chlorophyll P700 of photosystem I.

The excited electron has not yet returned to the lowest available energy level. When the P700 absorbs light energy, electrons are boosted even higher and passed to a second transport system. Transport systems, recall, are like steps on an energy staircase—and this boost places electrons at the top of a higher staircase. There is enough energy left at the bottom of this staircase to attach two electrons and a hydrogen ion ($H^+$) to $NADP^+$, the result being NADPH.

Thus, in the noncyclic pathway, electrons flow in one direction to NADPH. In the meantime, the P680 molecule that gives up electrons in the first place is getting replacements—from water. Inside the thylakoid compartment, water molecules are being split into oxygen, unbound protons (that is, hydrogen ions), and electrons. Photon energy indirectly drives this reaction sequence, which is called **photolysis** (Figure 7.6).

Oxygen atoms split from water molecules are by-products of the noncyclic pathway. Oxygen has been accumulating ever since this pathway evolved, more than 3.5 billion years ago. It profoundly changed the earth's atmosphere. And it made possible aerobic respiration, the most efficient pathway for extracting energy from organic compounds. The emergence of the noncyclic pathway ultimately allowed you and all other animals to be around today, breathing the oxygen that helps keep your cells alive.

**With cyclic photophosphorylation, ATP alone forms.**

**With noncyclic photophosphorylation, ATP *and* NADPH form, and oxygen is released as a by-product.**

**After the noncyclic pathway evolved, oxygen accumulated in the atmosphere and made aerobic respiration possible.**

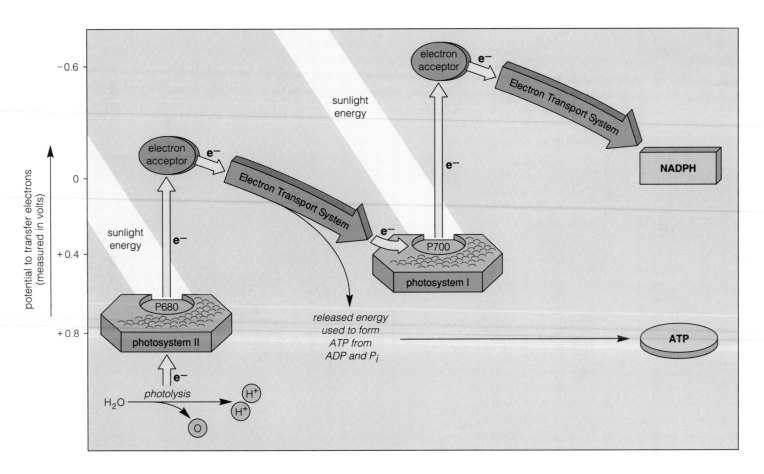

**Figure 7.6** Noncyclic photophosphorylation, which yields NADPH as well as ATP. Electrons derived from the splitting of water molecules (photolysis) travel through two photosystems, which work together in boosting the electrons to an energy level high enough to lead to NADPH formation. Figure 7.7 provides a closer look at the mechanism by which ATP forms.

## A Closer Look at ATP Formation

So far, you have seen what happens to the electrons and oxygen released when water molecules are split at the start of the noncyclic pathway. What happens to the hydrogen ions? They accumulate inside the thylakoid compartment of the chloroplast. Hydrogen ions also accumulate there when electron transport systems are operating. (This is true of both the cyclic and noncyclic pathways.) This accumulation sets up concentration and electric gradients across the thylakoid membrane—and the energy inherent in those gradients is tapped to form ATP.

Figure 7.7 shows how the combined force of the gradients propels hydrogen ions out of the thylakoid compartment and into the stroma. They flow through channel proteins that have enzyme activity. Through enzyme action, inorganic phosphate ($P_i$) becomes linked to ADP, and so ATP is formed.

The idea that the concentration and electric gradients across a membrane drive ATP formation is known as the **chemiosmotic theory**.

## LIGHT-INDEPENDENT REACTIONS

The **light-independent reactions** are the "synthesis" part of photosynthesis. They are the reactions by which carbohydrates are put together in the stroma of chloroplasts. ATP provides energy and NADPH provides hydrogen and electrons for the reactions. The air around photosynthetic cells provides carbon and oxygen (in the form of carbon dioxide).

The reactions are called "light-independent" because they do not depend directly on sunlight. They *can* proceed as long as ATP and NADPH are available from the light-dependent reactions. But ATP and NADPH normally are produced during daylight, so the light-independent reactions occur mostly during the day.

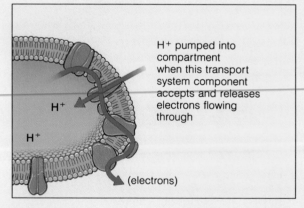

**Figure 7.7** The chemiosmotic theory of ATP formation, as it occurs in chloroplasts of leafy plants.

thylakoid membrane system in chloroplast

stroma
thylakoid membrane
thylakoid compartment

O₂

P680

2H₂O

H⁺  H⁺
H⁺
H⁺

(electrons)

**a**

H⁺ pumped into compartment when this transport system component accepts and releases electrons flowing through

H⁺

H⁺

(electrons)

**b**

(**a**) The start of the noncyclic pathway of photosynthesis, water molecules are split into oxygen, "naked" protons (or hydrogen ions), and electrons. (The reaction sequence is called photolysis.)

The oxygen is released as an end product and the electrons are sent through transport systems. The hydrogen ions accumulate inside the thylakoid compartment of the chloroplast, as sketched above.

Hydrogen ions also accumulate in the compartment when electron transport systems are operating. (This is true of both the cyclic and noncyclic pathway.) When certain molecules of the transport system accept electrons, they also pick up hydrogen ions from the stroma and release them inside the compartment (**b**).

Through photolysis and electron transport, hydrogen ions become more concentrated in the thylakoid compartment than in the stroma. The lopsided distribution of those positively

## Calvin-Benson Cycle

The heart of the light-independent reactions is the **Calvin-Benson cycle**, which is named after its discoverers, Melvin Calvin and Andrew Benson. During this cyclic pathway, carbon is captured from carbon dioxide, and a sugar phosphate forms in reactions that require ATP and NADPH. The cycle ends with regeneration of RuBP (ribulose bisphosphate)—a compound that is required to capture the carbon in the first place.

As Figure 7.8 shows, the reactions begin with **carbon dioxide fixation**. Carbon dioxide diffuses into leaves and is present in the spaces between photosynthetic cells. Those cells have enzymes that capture carbon dioxide and attach it to *RuBP*, which has a backbone of five carbon atoms. The attachment produces an unstable, six-carbon intermediate that splits at once into two molecules of *PGA* (phosphoglycerate).

Each PGA receives a phosphate group from ATP. The resulting intermediate receives hydrogen and electrons from NADPH to form *PGAL* (phosphoglyceraldehyde). It takes six carbon dioxide and six RuBP molecules to produce twelve PGAL. Most of the PGAL becomes rearranged into new RuBP molecules—which can be used to fix more carbon. But two PGAL are joined together to form a *sugar phosphate*. Such sugars have phosphate

groups attached that prime them for further reaction. Glucose-6-phosphate is an example. Its name simply means a phosphate group is attached to the sixth carbon atom of the glucose molecule (Figure 6.5).

The Calvin-Benson cycle yields enough RuBP to replace the ones used at the start of carbon dioxide fixation. The ADP, NADP$^+$, and phosphate leftovers are sent back to the light-dependent reaction sites, where they are converted once more to NADPH and ATP. The sugar phosphate formed in the cycle can serve as a building block for the plant's main carbohydrates, including sucrose, starch, and cellulose. Synthesis of those com-

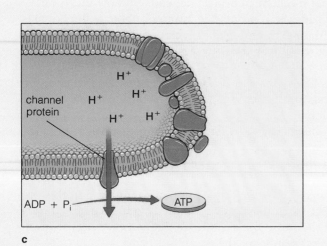

charged ions also creates a difference in electric charge across the thylakoid membrane. An electric gradient as well as a concentration gradient has been established.

**(c)** The combined force of the concentration and electric gradients propels hydrogen ions out of the compartment and into the stroma. The ions flow through channel proteins (called ATP synthases). The proteins span the membrane and have built-in enzyme machinery. The ion flow drives the enzyme machinery by which ADP combines with inorganic phosphate to form ATP.

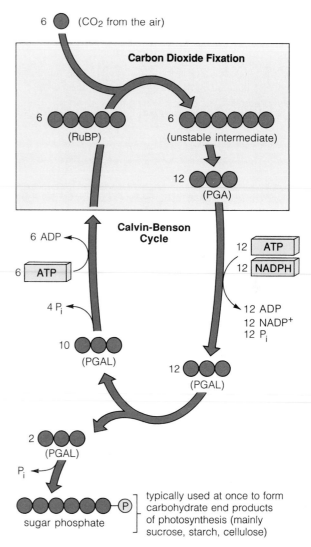

**Figure 7.8** Summary of the light-independent reactions of photosynthesis. The carbon atoms of the different molecules are depicted in red. All of the intermediates have one or two phosphate groups; for simplicity we show only the one on the "end product" (sugar phosphate). The phosphate groups that have been detached from molecules during the reactions are designated P$_i$, meaning "inorganic phosphate."

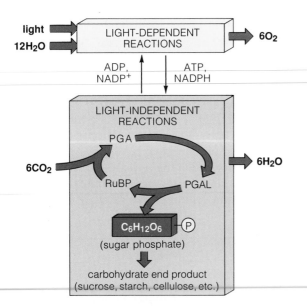

**Figure 7.9** Summary of the main reactants, intermediates, and products of photosynthesis corresponding to the equation:

$$12H_2O + 6CO_2 \xrightarrow{\text{sunlight}} 6O_2 + C_6H_{12}O_6 + 6H_2O$$

pounds by different metabolic pathways marks the conclusion of the light-independent reactions.

Figure 7.9 summarizes the main reactants, intermediates, and products of photosynthesis, from start to finish. Beginning with the light-dependent reactions, photolysis of every two water molecules yields four electrons and one $O_2$ molecule. For every four electrons, three ATP and two NADPH molecules are produced. *Each turn of the Calvin-Benson cycle requires three ATP and two NADPH molecules, as well as a $CO_2$ molecule.* It takes six turns of the cycle to get the six $CO_2$ molecules required for each six-carbon sugar formed.

---

**During the Calvin-Benson cycle, carbon is "captured" from carbon dioxide, a sugar phosphate forms in reactions requiring ATP and NADPH, and RuBP (needed to capture the carbon) is regenerated.**

---

### How Autotrophs Use Intermediates and Products of Photosynthesis

So here we are, with sugar phosphates formed in a microscopic speck of an organelle. Visualize millions of such specks in, say, a corn plant. Where do the sugar

phosphates go from here? Although a fraction of them are used at once as fuel to provide energy for cellular work, almost all are used as building blocks in the synthesis of sucrose, starch, and cellulose.

Of all the carbohydrates produced by photosynthesis, sucrose is the most easily transportable. Conducting tissues carry it from sites in the leaf to living cells in all parts of the corn plant body. Starch is the main storage form of carbohydrate in the leaves, stems, and roots.

What happens in other plants besides corn? Some of the sucrose produced in leaves of potato plants is converted and stored as starch in underground stem regions called tubers (the "potatoes"). In sugar beets, onions, and sugarcane, sucrose itself is the main storage form. Photosynthetic autotrophs also use intermediates and products of photosynthesis when they assemble lipids and amino acids. Indeed, some green algae use more than 90 percent of the carbon fixed when they construct proteins and lipids. These plants have a brief life cycle, and they live in places where sunlight and water are plentiful. They put most photosynthetic products into rapid growth and reproduction instead of diverting them to storage forms.

### C4 Plants

Plant growth depends on photosynthesis, and photosynthesis depends on efficient carbon dioxide fixation. That mechanism in turn depends on how much carbon dioxide is concentrated around photosynthetic cells. Although there is plenty of carbon dioxide in the air, it is not always abundantly available to the cells *inside the leaves* of land plants.

Leaves have a waxy covering that retards moisture loss, and water escapes mainly through tiny passages (stomata) across the leaf's surface. On hot, dry days the stomata are closed and the plant conserves water. This means, however, that carbon dioxide can't enter the leaf. At the same time, oxygen builds up inside the leaf as a by-product of photosynthesis. The stage is set for a wasteful process called "photorespiration." In this process, oxygen instead of carbon dioxide becomes attached to the RuBP used in the Calvin-Benson cycle, with different results for the plant:

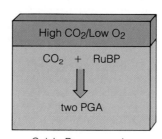

Calvin-Benson cycle predominates

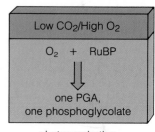

photorespiration predominates

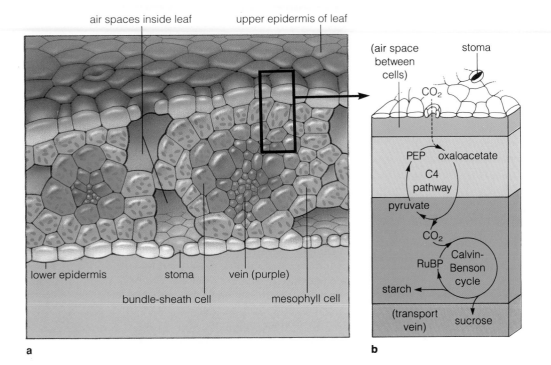

air spaces inside leaf     upper epidermis of leaf

lower epidermis    stoma    vein (purple)

bundle-sheath cell     mesophyll cell

a

(air space between cells)

stoma

$CO_2$

PEP   oxaloacetate

C4 pathway

pyruvate

$CO_2$

RuBP   Calvin-Benson cycle

starch

(transport vein)    sucrose

b

**Figure 7.10** C4 pathway. (**a**) This is the internal structure of a leaf from corn (*Zea mays*), a typical C4 plant. Notice how the photosynthetic bundle-sheath cells (dark green) surround the veins and are in turn surrounded by photosynthetic mesophyll cells (lighter green). (**b**) C4 plants have a carbon-fixing system that *precedes* the Calvin-Benson cycle.

Formation of sugar phosphates depends on PGA. When photorespiration wins out, less PGA forms and the plant's capacity for growth suffers.

Many plants, including crabgrass, sugarcane, and corn, continue to fix carbon dioxide even when their stomata close on hot, dry days. In these plants, the first compound formed by carbon dioxide fixation is the four-carbon oxaloacetate. Hence their name, "C4" plants. (Three-carbon PGA is produced in "C3" plants.)

C4 plants fix carbon dioxide not once but twice, in two different types of photosynthetic cells. Carbon dioxide that has diffused into their leaves is fixed initially in *mesophyll cells*, forming oxaloacetate. This is a temporary fix, so to speak. The oxaloacetate is quickly transferred to *bundle-sheath cells*, which form a layer around every vein in the leaf. In those cells, the carbon dioxide is released, fixed again, and used to build carbohydrates (Figure 7.10). Thus carbon dioxide accumulates in the bundle-sheath cells, and its concentration can remain high even when the Calvin-Benson cycle is operating.

When the temperature climbs, stomata close and oxygen accumulates in the leaves of any type of plant. C3 plants photorespire even more than usual in hot weather, and their growth slows. By contrast, C4 plants do not photorespire as much, even in hot weather. Because of their special carbon-fixing system, the carbon dioxide concentration still increases in the bundle-sheath cells—so oxygen loses out in the competition for RuBP.

C4 species are abundant in regions with the highest temperatures during the growing season. For example, 80 percent of all native species in Florida are C4 plants—compared to 0 percent in Manitoba, Canada. Kentucky bluegrass and other C3 species have the advantage in regions where temperatures drop below 25°C; they are better adapted to cold. When we mix C3 and C4 species from different regions in our gardens, one or the other kind will have the advantage during at least some part of the year. That is why a lawn of Kentucky bluegrass can thrive during cool spring weather in San Diego, only to be overwhelmed during the hot summer months by a C4 plant—crabgrass.

## CHEMOSYNTHESIS

Photosynthesis so dominates the energy-trapping pathways that sometimes it is easy to overlook other, less common routes. The *chemosynthetic autotrophs* obtain energy not from sunlight but rather from the oxidation of inorganic substances (ammonium ions, iron or sulfur compounds, and so on). That is their source of energy for building organic compounds.

For example, some bacteria that live in soil use ammonia ($NH_3$) molecules as an energy source, stripping them of protons and electrons. Nitrite ions ($NO_2^-$) and nitrate ions ($NO_3^-$) are the remnants of their activities. Compared with ammonium ions, nitrite and nitrate ions are readily washed out of the soil, so the action of the so-called nitrifying bacteria can lower soil fertility. When farmers are ready to plant their fields, they sometimes add chemicals to the soil that inhibit bacterial metabolism.

We will return later to the environmental effects of chemosynthetic autotrophs. In this unit, we turn next to pathways by which energy is released from carbohydrates and other biological molecules—the chemical legacy of autotrophs and, ultimately, of sunlight.

## Energy From Artificial Chloroplasts?

If innovative researchers are successful, motorists may one day fill their gas tanks with an abundant, nonpolluting hydrogen fuel. As motorists stand at the pump, they may also worry less about the greenhouse effect and global warming. When carried out on a large scale, the same technology that produces the fuel could gobble up carbon dioxide—a "greenhouse gas" that is accumulating to alarming levels in the atmosphere.

Such rewards might emerge from efforts to create artificial assemblies of the molecules required for photosynthesis. Such assemblies in principle will convert energy from sunlight into molecules such as methane and hydrogen. Methane is a type of natural gas, and hydrogen is a potential liquid fuel for use in automobiles or for generating electricity. Most importantly, sunlight is an abundant, "no-cost" energy source.

In attempting to devise a workable artificial system, scientists have focused on events in the photosystems of chloroplast membranes. In those natural systems, photons absorbed by chlorophyll molecules provide the energy to boost electrons to a higher energy level. An artificial system must incorporate natural or manmade molecules that can absorb light, donate electrons, and receive excited electrons. Such a system must also maintain the one-way flow of electrons. And it must keep the process going in a way that permits released energy to be captured in a useful chemical form—much as it is in the NADPH formed by the noncyclic pathway of photosynthesis.

In the past few years, researchers have put together molecular assemblies that actually mimic the light absorption and electron excitation in photosystems. For example, chemists at Arizona State University have synthesized a promising, five-part molecule of natural compounds, including some derived from chlorophyll and from electron transport systems. The artificial assembly maintains an energetic state long enough for useful products to be obtained. Another candidate is a chain of inorganic molecules that absorb light like chlorophyll. The chain is embedded in a matrix of a porous mineral (zeolite). The tiny pores of zeolite can be aligned to form molecule-sized tunnels, which can keep the photosynthetic apparatus physically oriented in the proper way.

While theoretical and technical problems remain, the promise of artificial photosynthesis is simply too great to be dismissed. We may soon learn the intricate details of chloroplast function. The human population—so inextricably dependent on green plants—will then owe chloroplasts even more.

## SUMMARY

1. Plants and other photosynthetic autotrophs use sunlight (as an energy source) and carbon dioxide (as the carbon source) for building organic compounds. Animals and other heterotrophs must obtain carbon and energy from organic compounds already built by autotrophs.

2. Photosynthesis is the main biosynthetic pathway by which carbon and energy enter the web of life. It consists of two sets of reactions:

a. The light-dependent reactions take place at the thylakoid membrane system of chloroplasts. These reactions produce ATP and NADPH (or ATP alone).

b. The light-independent reactions take place in the stroma around the membrane system. They produce sugar phosphates that are used in building sucrose, starch, and other end products of photosynthesis.

3. These are the key points about the light-dependent reactions:

a. Photosystems are clusters of photosynthetic pigments that are bound to proteins in the thylakoid membrane. Light absorption causes the transfer of electrons from a chlorophyll to an acceptor molecule, which donates them to a transport system in the membrane.

b. Operation of electron transport systems causes $H^+$ to accumulate inside the thylakoid membrane system. This produces concentration and electric gradients that drive the formation of ATP.

c. In the cyclic pathway, electrons travel in a circle, back to the photosystem that originally gave them up (photosystem I). This pathway yields ATP only.

d. The noncyclic pathway also yields ATP, but the electrons end up in NADPH. In this pathway, electrons derived from the splitting of water molecules travel from photosystem II to a transport system, then to photosystem I and another transport system.

4. These are the key points about the light-independent reactions:

a. The ATP produced in the light-dependent reactions provides energy and NADPH provides hydrogen atoms and electrons for the "synthesis" part of photosynthesis.

b. Sugar phosphates form by operation of the Calvin-Benson cycle. The cycle begins when carbon dioxide from the air is affixed to RuBP, making an unstable intermediate that splits into two PGA.

c. Each PGA receives a phosphate group from ATP. The resulting molecule receives $H^+$ and electrons from NADPH to form PGAL. Two of every twelve PGAL are used to produce a six-carbon sugar phosphate. The rest of the PGAL are rearranged to regenerate RuBP for the cycle.

5. Some plants living in hot climates have an additional carbon-fixing system, the C4 pathway. This pathway helps circumvent photorespiration, in which oxygen instead of carbon dioxide becomes affixed to RuBP. (Otherwise, photorespiration undoes much of what photosynthesis accomplishes.)

## Review Questions

1. Define the difference between autotrophs and heterotrophs, and give examples of each. In what category do photosynthesizers fall? *105*

2. Summarize the photosynthesis reactions in words, then as an equation. Distinguish between the light-dependent and the light-independent stages of these reactions. *105–106*

3. A thylakoid compartment is a reservoir for which of the following substances: glucose, photosynthetic pigments, hydrogen ions, fatty acids? *107*

4. Sketch the reaction steps of noncyclic photophosphorylation, showing where the excited electrons eventually end up. Do the same for the cyclic pathway. *110–111*

5. Which of the following substances are *not* required for the light-independent reactions: ATP, NADPH, RuBP, chlorophyll, carotenoids, free oxygen, carbon dioxide, enzymes? *113*

6. Suppose a plant carrying out photosynthesis were exposed to carbon dioxide molecules that contain radioactively labeled carbon atoms ($^{14}CO_2$). In which of the following compounds will the labeled carbon first appear: NADPH, PGAL, pyruvate, PGA? *113*

7. How many $CO_2$ molecules must enter the Calvin-Benson cycle to produce one sugar phosphate molecule? Why? *114*

## Self-Quiz *(Answers in Appendix IV)*

1. Molecules with backbones of _____ serve as the main building blocks of all organisms.

2. Photosynthetic autotrophs use _____ from the air as their carbon source and _____ as their energy source.

3. In plant cells, light-*dependent* reactions occur _____.
   a. in the cytoplasm
   b. at the plasma membrane
   c. in the stroma
   d. in the thylakoid membrane

4. In plant cells, light-*independent* reactions occur _____.
   a. in the cytoplasm
   b. at the plasma membrane
   c. in the stroma
   d. in the grana

5. In the light-dependent reactions, _____.
   a. carbon dioxide is incorporated into carbohydrates
   b. ATP and NADPH are formed
   c. carbon dioxide accepts electrons
   d. sugar phosphates are formed

6. When light is absorbed by a photosystem, _____.
   a. sugar phosphates are produced
   b. electrons are transferred to an acceptor molecule
   c. RuBP accepts electrons
   d. the light-dependent reactions are initiated
   e. both b and d are correct

7. The Calvin-Benson cycle begins when _____.
   a. light is available
   b. light is not available
   c. carbon dioxide is attached to RuBP
   d. electrons leave a photosystem

8. In the light-independent reactions, ATP furnishes phosphate groups to _____.
   a. RuBP
   b. NADP$^+$
   c. PGA
   d. PGAL

9. Match each event in photosynthesis with its correct description.
   ___ uses RuBP; produces PGA
   ___ uses ATP and NADPH
   ___ forms NADPH
   ___ produces ATP and NADPH
   ___ produces ATP only

   a. cyclic pathway
   b. noncyclic pathway
   c. carbon dioxide fixation
   d. formation of PGAL
   e. Transfer of $H^+$ and electrons to NADP$^+$

## Selected Key Terms

autotroph *105*
C4 plant *115*
Calvin-Benson cycle *113*
carbon dioxide fixation *113*
chemiosmotic theory *112*
chlorophyll *108*
electron transport system *110*
granum *107*
heterotroph *105*
light-dependent reaction *107*
light-independent reaction *112*
PGA *113*
PGAL *113*
photolysis *111*
photophosphorylation *110*
photosynthesis *105*
photosystem *110*
RuBP *113*
stroma *107*
sugar phosphate *113*
thylakoid membrane *107*

## Readings

Alberts, B. 1989. "Chloroplasts and Photosynthesis." In *Molecular Biology of the Cell.* Second edition. New York: Garland.

Moore, P. 1981. "The Varied Ways Plants Tap the Sun." *New Scientist* 12:394–397. Clear, simple introduction to the C4 plants.

Youvan, D., and B. Marrs. 1987. "Molecular Mechanisms of Photosynthesis." *Scientific American* 256:42–50.

# 8 ENERGY-RELEASING PATHWAYS

## The Killers Are Coming!

"Killer" bees from South America buzzed across the border between Mexico and the United States in 1990 and are now working their way north. Alarming newspaper headlines accompanied their arrival, and for good reason. The bees are descended from aggressive African queen bees, and they can be terrifying.

In the 1950s, researchers had some queen bees shipped from Africa to Brazil for breeding experiments. Honeybees, you understand, are big business. Not only do they produce honey, they also are rented out to pollinate commercial orchards. (Put a screened cage around an orchard tree in bloom and less than 1 percent of the flowers will set fruit. Put a hive of honeybees in the same cage and 40 percent of the flowers will set fruit.) The honeybees in Brazil seemed sluggish, compared to their aggressive African relatives, so the researchers hoped to cross-breed the two types and come up with a mild-mannered but zippier pollinator.

One of the researchers accidentally released twenty-six of the African queens. This was bad enough. Then beekeepers got wind of the preliminary experimental results. After learning that the first few generations of offspring were jazzed-up but still nice honeybees, they imported hundreds of *additional* African queens and released them to mate with locals.

The genetic backlash came later. For some inexplicable reason, genes of the African bees became dominant in subsequent generations. Now there are "Africanized" honeybees with nasty tempers. When disturbed, they become astonishingly agitated. Whereas a mild-mannered honeybee might chase an animal fifty yards or so, a squadron of Africanized bees will chase it for a quarter of a mile. If they catch it, they can sting it to death.

Africanized bees do everything other bees do, but they do more of it faster. The developmental stages leading from egg to adult proceed more quickly. Adults fly more rapidly. (They beat their wings more than 200 times per second; by contrast, a butterfly wing flaps

about 4 times per second.) Adults produce more offspring in the same span of time. They die sooner. But while they are alive, they forage more vigorously than other bees do in the competition for nectar.

For the Africanized honeybee, doing things faster means having a nonstop supply of energy that can keep the metabolic fires burning. This bee's stomach can hold 30 milligrams of sugar-rich nectar—enough to keep the bee flying for 60 kilometers without running out of fuel. The cells of its flight muscles are packed with exceptionally large mitochondria—and mitochondria are truly efficient organelles in which energy originally stored in the sugar becomes converted to the energy of ATP. Only with mitochondria can the Africanized bee generate enough energy to sustain prolonged effort, whether this be day-long nectar gathering or chasing down intruders.

As you might deduce from all of this, Africanized bees probably will not survive in regions where winters are harsh and plants stop blooming for months at a time. Thus, although residents of California, Florida, and similar sunbelt states may have to keep an eye on the sky, it is not likely that Alaskans will be bothered by huge populations of killer bees.

In their reliance on a great deal of fuel and efficient metabolic furnaces, Africanized bees are not that different from you or any other highly active or rapidly growing organism. In their reliance on energy stored in sugars and other organic compounds, they are like *all* other organisms. Even though the energy-releasing pathways differ in some details from one organism to the next, each requires characteristic starting materials, then yields predictable products and by-products. And *all* of the energy-releasing pathways yield ATP. In fact, throughout the biosphere, there is startling similarity in the uses to which energy and raw materials are put. *At the biochemical level, there is undeniable unity among all forms of life.* We will return to this idea in the *Commentary* at the chapter's end.

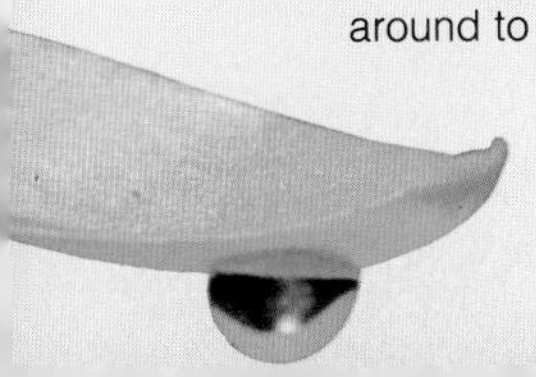

**Figure 8.1** A mild-mannered honeybee buzzing in for a landing on a flower, wings beating with energy provided by ATP. If this were one of its Africanized relatives, possibly you would not stay around to watch the landing.

KEY CONCEPTS

**1.** Plants produce and use ATP during photosynthesis. Plants and all other organisms also produce ATP by degrading (breaking apart) organic compounds such as glucose. Aerobic respiration is the degradative pathway with the greatest energy yield.

**2.** Aerobic respiration proceeds through three stages. First, glucose is partially broken down to pyruvate with a net energy yield of two ATP. Second, the pyruvate is completely broken down to carbon dioxide and water. Electrons liberated during the reactions are delivered to a transport system. Third, energy is released as electrons are transferred through the system, and it drives the formation of thirty-six or more ATP for every glucose molecule. At the end of the system, free oxygen combines with the electrons and with hydrogen ions to form water.

**3.** Over evolutionary time, photosynthesis and aerobic respiration have become linked on a global scale. The oxygen by-products of photosynthesis serve as final electron acceptors for the aerobic pathway. And the carbon dioxide and water released in aerobic respiration are raw materials used in building organic compounds during photosynthesis:

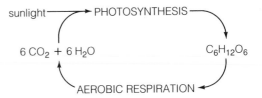

## ATP-PRODUCING PATHWAYS

No organism stays alive without taking in energy. Plants get energy from the sun; animals get energy secondhand, thirdhand, and so on, by eating plants and one another. But no matter what the *source* of energy may be, organisms must convert it to a form of chemical energy that can drive metabolic reactions. In all cells, the "coins" most commonly minted and then spent on metabolic reactions are molecules of adenosine triphosphate, or **ATP**. Figure 6.16 shows the structure of ATP, which has three phosphate groups attached to the rest of the molecule.

Plants produce ATP during photosynthesis, as described in the preceding chapter. But plants and all other organisms also can produce ATP through degradative pathways that release chemical energy stored in carbohydrates, lipids, or proteins.

The main degradative pathway is **aerobic respiration**. The "aerobic" part of the name means that oxygen is the final acceptor of electrons stripped from glucose or some other molecule during energy-releasing reactions. With every breath you take, you are replenishing oxygen supplies for your busily respiring cells. That is why astronauts who have ventured to the moon wouldn't have lasted more than a little while if their spacesuits had not been plugged into oxygen tanks as they clambered about on the moon's surface. (Unlike the earth, the moon has no oxygen-rich atmosphere.) When you breathe out, you are ridding your body of the carbon dioxide and water leftovers of aerobic respiration (Figure 8.2).

Other degradative pathways are "anaerobic," in that something besides oxygen serves as the final electron acceptor for energy-releasing reactions. The most common anaerobic pathways are called **fermentation** and **anaerobic electron transport**. Cells in your body rely mostly on aerobic respiration, but they can use a fermentation pathway for short periods when oxygen supplies are low. Many microbes rely exclusively on anaerobic pathways. Some are indifferent to the presence or absence of oxygen; they include bacteria that are used in the production of yogurt. Other microbes are "strict anaerobes" that die if exposed to oxygen. Among them are the bacteria that cause botulism, tetanus, and some other serious diseases.

All energy-releasing pathways proceed through an orderly sequence of steps. As we examine the pathways in more detail, keep in mind that they do not proceed all by themselves. Enzymes catalyze each step, and the intermediate produced at a given step serves as a substrate for the next enzyme in the pathway.

## AEROBIC RESPIRATION

### Overview of the Reactions

Of all degradative pathways, aerobic respiration produces the most ATP for each glucose molecule being dismantled. Whereas fermentation has a net yield of two ATP, the aerobic route yields *thirty-six* ATP or more. If you were a microscopic bacterium, you would not require much ATP. Being large, complex, and highly active, you depend absolutely on the high ATP yield of aerobic respiration.

a

Using glucose as the starting material, the aerobic route is often summarized this way:

$$C_6H_{12}O_6 + 6O_2 \longrightarrow 6CO_2 + 6H_2O$$

GLUCOSE
CARBON DIOXIDE

The summary equation only tells us what the substances are at the start and the finish of the aerobic route. In between are *three stages* of reactions.

In the first stage, *glycolysis*, glucose is partially degraded (oxidized) to pyruvate. By the end of the second stage, which includes the *Krebs cycle*, glucose has been completely degraded to carbon dioxide and water. Neither stage produces much ATP. However, protons and electrons are stripped from intermediates during both stages, and these are delivered to a transport system. That system is used in the third stage of reactions, *electron transport phosphorylation*, which yields many ATP. Oxygen accepts the "spent" electrons from the transport system (Figure 8.3).

b

c

**Figure 8.2** (**a**) Astronauts about to land in an anaerobic world—the Apollo 12 voyagers to the moon. (**b**) The view homeward, with the sun emerging from behind the earth—the only planet known to be enveloped in an oxygen-rich atmosphere. (**c**) A few humans and a number of trees. Like most kinds of organisms on earth, the trees as well as the women are using oxygen in metabolic reactions that sustain their activities.

**Figure 8.3** Overview of aerobic respiration, the main energy-releasing pathway, with glucose as the starting material. In glycolysis, glucose is partially broken down to pyruvate. (Glycolysis is also the initial set of reactions in other pathways, including fermentation.) In the Krebs cycle, pyruvate is completely broken down to carbon dioxide. Coenzymes (NADH and $FADH_2$) accept protons and electrons being stripped from intermediates of the reactions and deliver them to an electron transport system. Oxygen accepts the electrons from the transport system. From start (glycolysis) to finish, the aerobic pathway typically has a net energy yield of thirty-six ATP.

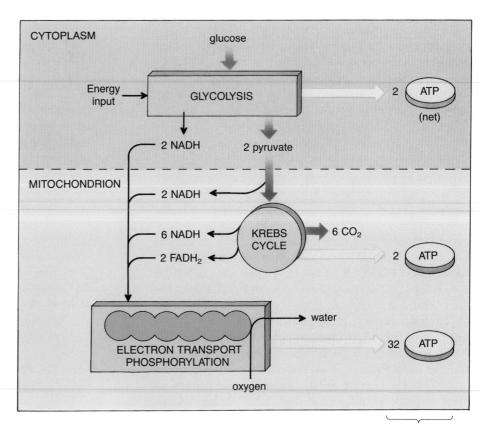

**Figure 8.4** Glycolysis, showing glucose as the starting material. These breakdown reactions take place in the cytoplasm of all cells. They end with the formation of two pyruvate molecules and have a net energy yield of two ATP.

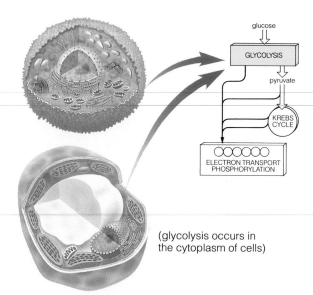

(glycolysis occurs in the cytoplasm of cells)

## Glycolysis

**Glycolysis** is the partial breakdown of glucose or some other organic compound into two molecules of pyruvate. The breakdown occurs by way of a controlled sequence of reactions that take place in the cytoplasm of all cells. Enzymes lower the activation energy for the formation of each intermediate along the route, and energy is released when some of the intermediates are converted to different molecular forms.

Each glucose molecule has a backbone of six carbon atoms to which hydrogen atoms are attached (Figure 3.5). Here we show the backbone in simplified fashion:

The first steps of glycolysis are *energy-requiring*; they do not proceed without an energy input. The energy becomes available when two ATP molecules are subjected to hydrolysis (page 37). Through enzyme action, each ATP transfers a phosphate group to the six-carbon backbone. The backbone then splits apart to form two molecules of **PGAL** (phosphoglyceraldehyde), each with a three-carbon backbone. Formation of PGAL marks the start of the *energy-releasing* steps of glycolysis (Figure 8.4).

Through enzyme action, each PGAL now gives up one proton and two electrons to NAD⁺, forming NADH. Recall that $NAD^+$ is a "reusable" coenzyme that functions at the active site of an enzyme. It accepts protons and electrons stripped from a substrate, then transfers them elsewhere and so becomes $NAD^+$ again. Meanwhile, the intermediate remaining at the site combines with inorganic phosphate ($P_i$). The result is a rather unstable molecule that readily gives up a phosphate group to ADP. In this manner, two ATP molecules form, one for each PGAL.

As you can see, *ATP has been formed by the direct, enzyme-mediated transfer of a phosphate group from a substrate to ADP.* The mechanism is called **substrate-level phosphorylation**. (This is different from "electron transport" phosphorylation, which requires oxygen and a transport system.)

Substrate-level phosphorylation occurs again during glycolysis. At a certain step, enzymes strip a proton and a hydroxide ion ($OH^-$) from each of two intermediate molecules. This leaves two molecules of PEP (phosphoenol pyruvate). PEP is rather unstable, and when it breaks down, the energy released is enough to drive the transfer of one of its phosphate groups to ADP. Because the phosphorylation occurs twice (once for each PGAL), two more ATP are formed. At this point, glucose has been broken down to two pyruvate molecules, each with a three-carbon backbone (Figure 8.4). Glycolysis is over.

In sum, the initial breakdown of glucose produces two NADH, four ATP (by substrate-level phosphorylation), and two pyruvate molecules. But remember, two ATP were invested at the start of glycolysis, so the *net* yield is only two ATP.

**Glycolysis produces two NADH, two ATP (net), and two pyruvate molecules for each glucose molecule entering the reactions.**

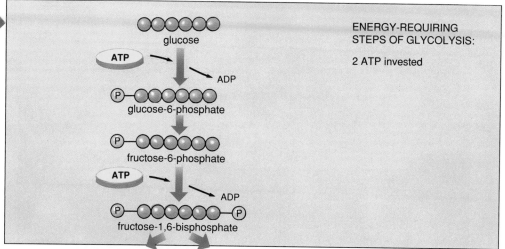

ENERGY-REQUIRING
STEPS OF GLYCOLYSIS:

2 ATP invested

**1.** Glucose, a six-carbon molecule, gets an ATP-derived phosphate group attached to it at one end, then gets rearranged into fructose-6-phosphate.

**2.** Another ATP-derived phosphate group gets attached at the other end, forming fructose-1,6-bisphosphate.

**3.** This intermediate is split into PGAL and DHAP. Because each of those molecules is easily converted to the other, we can say that two PGAL have formed.

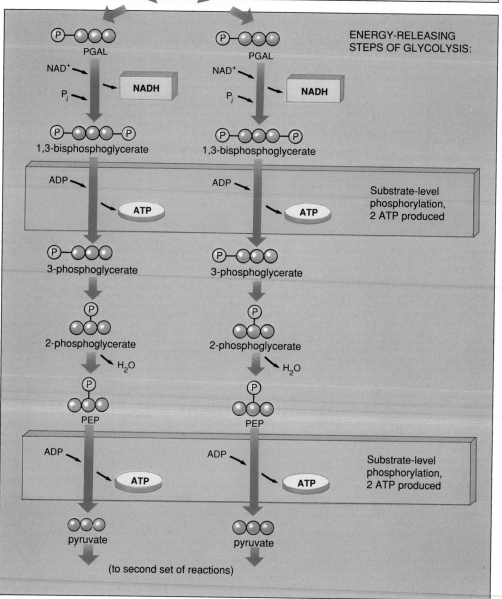

ENERGY-RELEASING
STEPS OF GLYCOLYSIS:

**4.** Each PGAL gives up two electrons and one proton ($H^+$) to $NAD^+$, forming NADH. Each PGAL also combines with inorganic phosphate ($P_i$). Each of the two resulting intermediates donates a phosphate group to ADP, forming ATP.

**5.** With this formation of two ATP, the original energy investment of two ATP is paid off.

**6.** In the next two conversions, the intermediate gives up one proton and one hydroxide ion ($OH^-$) which combine to form water. The resulting intermediate is phosphoenol pyruvate (PEP).

**7.** The rather unstable PEP molecule readily gives up a phosphate group to ADP, forming ATP. Thus the net energy yield from glycolysis is two ATP for each glucose molecule entering the reactions.

**8.** Two molecules of pyruvate (the ionized form of pyruvic acid) are the end products of glycolysis.

NET ENERGY YIELD: 2 ATP

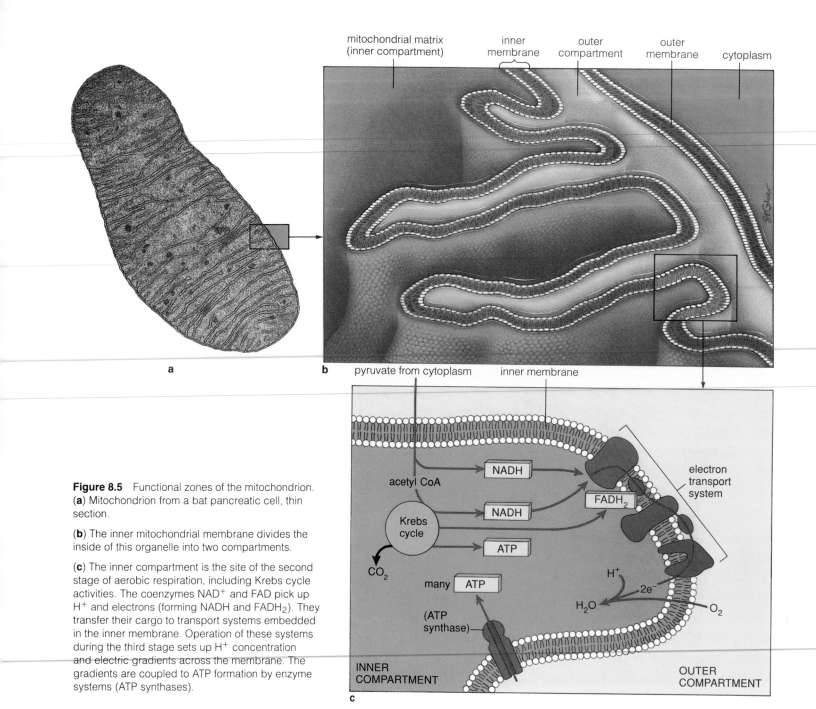

**Figure 8.5** Functional zones of the mitochondrion.
(**a**) Mitochondrion from a bat pancreatic cell, thin section.

(**b**) The inner mitochondrial membrane divides the inside of this organelle into two compartments.

(**c**) The inner compartment is the site of the second stage of aerobic respiration, including Krebs cycle activities. The coenzymes $NAD^+$ and FAD pick up $H^+$ and electrons (forming NADH and $FADH_2$). They transfer their cargo to transport systems embedded in the inner membrane. Operation of these systems during the third stage sets up $H^+$ concentration and electric gradients across the membrane. The gradients are coupled to ATP formation by enzyme systems (ATP synthases).

## Krebs Cycle

At the end of glycolysis, pyruvate molecules are still in the cytoplasm. They now may enter a *mitochondrion*, the only organelle where the second and third stages of the aerobic pathway can proceed. Figure 8.5 shows the structure of a representative mitochondrion.

The next series of reactions takes place in the inner compartment of the mitochondrion. There, the pyruvate molecules donate their carbon atoms to reactions that produce carbon dioxide molecules as by-products. Take a look at Figure 8.6. It shows that the second stage begins with conversion of each pyruvate to acetyl-CoA. The con-

version product becomes attached to oxaloacetate, the point of entry into a cyclic set of reactions called the **Krebs cycle**. (The cycle was named after Hans Krebs, who began working out its details in the 1930s.) Notice there is some symmetry here: Three carbon atoms enter the reactions (as the backbone of each pyruvate), and three carbon atoms leave (in three molecules of carbon dioxide).

The Krebs cycle itself serves three functions. First, many reaction steps are concerned with juggling the intermediate molecules back into the form of oxaloacetate. Cells have only so much oxaloacetate, and the cycle could not proceed over and over again if oxaloacetate

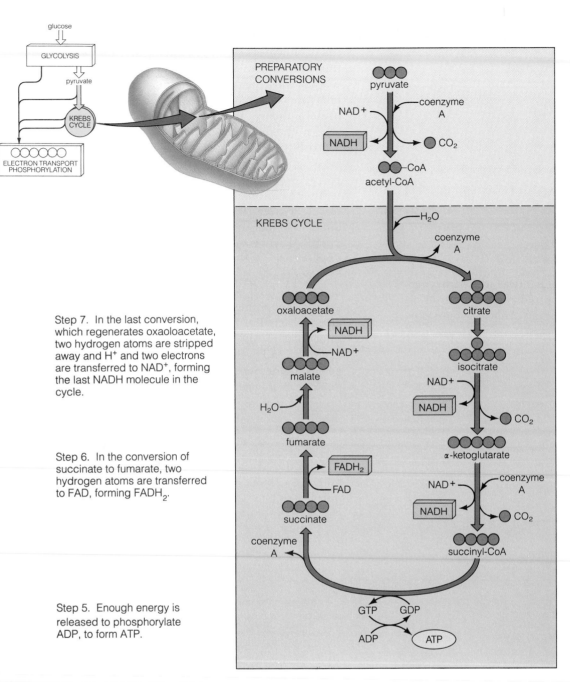

glucose

GLYCOLYSIS

pyruvate

KREBS CYCLE

ELECTRON TRANSPORT PHOSPHORYLATION

**PREPARATORY CONVERSIONS**

pyruvate

NAD+ → coenzyme A

NADH

$CO_2$

CoA
acetyl-CoA

**KREBS CYCLE**

$H_2O$

coenzyme A

oxaloacetate

citrate

NADH
NAD+

malate

isocitrate

NAD+
NADH
$CO_2$

$H_2O$

fumarate

α-ketoglutarate

FADH₂
FAD

NAD+ → coenzyme A
NADH
$CO_2$

succinate

succinyl-CoA

coenzyme A

GTP   GDP

ADP   ATP

**Step 1.** As three-carbon pyruvate enters the mitochondrion, enzymes split away its $COO^-$ group, which departs as $CO_2$. Enzymes also transfer a proton ($H^+$) and two electrons to $NAD^+$, forming NADH. The two-carbon molecule remaining is linked to coenzyme A, forming the intermediate acetyl-CoA.

**Step 2.** The two-carbon molecule becomes attached to oxaloacetate, the point of entry into the Krebs cycle, to form the six-carbon citrate.

**Step 3.** Citrate is rearranged into the six-carbon isocitrate, which is stripped of two hydrogen atoms, with all but one $H^+$ of those atoms being transferred to $NAD^+$ to form NADH. It is also stripped of a $COO^-$ group, this being the second $CO_2$ to depart.

**Step 4.** The resulting intermediate also gives up two hydrogen atoms and NADH forms. And it gives up a $COO^-$ group, which is the last $CO_2$ to depart.

The rest of the reactions work to convert the intermediate remaining (succinyl-CoA) back to the oxaloacetate on which the Krebs cycle turns.

**Step 7.** In the last conversion, which regenerates oxaloacetate, two hydrogen atoms are stripped away and $H^+$ and two electrons are transferred to $NAD^+$, forming the last NADH molecule in the cycle.

**Step 6.** In the conversion of succinate to fumarate, two hydrogen atoms are transferred to FAD, forming FADH₂.

**Step 5.** Enough energy is released to phosphorylate ADP, to form ATP.

were not regenerated. Second, one reaction step serves to produce ATP (by substrate-level phosphorylation). Third, some reaction steps liberate protons and electrons for transfer to the coenzymes $NAD^+$ and FAD, which will in turn transfer them to other reaction sites. When carrying their cargo, the coenzymes are abbreviated NADH and FADH₂.

It takes two preparatory reaction sequences and two turns of the Krebs cycle to use up the two pyruvate molecules. The two turns of the cycle add only two more ATP to the small yield from glycolysis. *But they also load many more coenzymes with protons and electrons that can be used in the third stage of the aerobic pathway.* The ATP

**Figure 8.6** The Krebs cycle and the preparatory reactions preceding it. For *each* pyruvate molecule, 3 $CO_2$, 1 ATP, 4 NADH, and 1 FADH₂ are formed. But remember the steps shown occur *twice* for each glucose molecule broken down.

payoff from that final stage will be substantial indeed, as you might well induce from this chart:

| Glycolysis: | | 2 NADH |
|---|---|---|
| Pyruvate conversion preceding Krebs cycle: | | 2 NADH |
| Krebs cycle: | 2 FADH$_2$ | 6 NADH |
| Total electron carriers sent to third stage of aerobic pathway: | 2 FADH$_2$ + | 10 NADH |

## Electron Transport Phosphorylation

In the third stage of the aerobic pathway, the coenzymes NADH and FADH$_2$ give up protons and electrons to highly organized transport systems. Those systems consist of enzymes and other proteins that are located in the inner membrane of the mitochondrion. Together, they work in ways that set up concentration and electric gradients across the membrane. And according to the **chemiosmotic theory**, those gradients drive the formation of ATP, in the following manner.

As Figures 8.5 and 8.7 show, the components of each transport system are arranged in series. Electrons are passed down the line, so to speak, and with each transfer they lose some energy. The energy released at certain transfers drives the pumping of unbound protons (H$^+$) out of the inner compartment of the mitochondrion. The pumping action produces H$^+$ concentration and electric gradients across the inner membrane. H$^+$ can flow across the membrane, down its combined electrochemical gradient. The flow occurs through ATP synthases, which are channel proteins that span the membrane and show enzyme activity. Here, enzyme action joins inorganic phosphate to ADP, forming ATP.

At the end of the electron transport system, free oxygen combines with the electrons and with hydrogen ions to form water. By removing electrons from the transport system, oxygen allows new electrons to flow through the system.

Operation of the transport system commonly leads to the formation of thirty-two ATP. When added to the ATP produced during the first and second stages, this brings the total net yield of the aerobic pathway to thirty-six ATP.

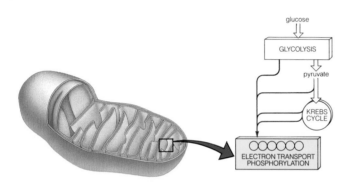

glucose

GLYCOLYSIS

pyruvate

KREBS CYCLE

ELECTRON TRANSPORT PHOSPHORYLATION

**Figure 8.7** Electron transport phosphorylation. The reactions occur at transport systems and enzyme systems (ATP synthases) embedded in the inner mitochondrial membrane. The transport system consists of enzymes and other proteins (including cytochrome molecules) that operate one after the other, as shown in Figure 8.5.

The membrane itself creates two compartments. The reactions begin in the inner compartment, when NADH and FADH$_2$ give up H$^+$ and electrons to the transport system. Electrons are accepted and passed through the system, but the H$^+$ is left behind—in the outer compartment:

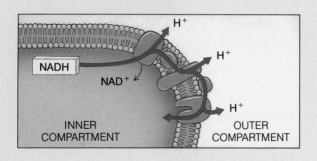

NADH

NAD$^+$

H$^+$

H$^+$

H$^+$

INNER COMPARTMENT

OUTER COMPARTMENT

Soon there is a higher concentration of H$^+$ ions in the outer compartment than in the inner one. In other words, concentration and electric gradients now exist across the membrane. The H$^+$ ions follow the gradients and move back into the inner compartment. They do this by flowing through the ATP synthases that span the membrane. Energy associated with the flow drives the coupling of ADP and inorganic phosphate into ATP:

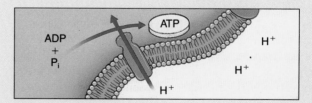

ADP + P$_i$

ATP

H$^+$

H$^+$

H$^+$

Do these events sound familiar? They should: ATP forms in much the same way in chloroplasts (Figure 7.7). The idea that concentration and electric gradients across a membrane drive ATP formation is called the chemiosmotic theory.

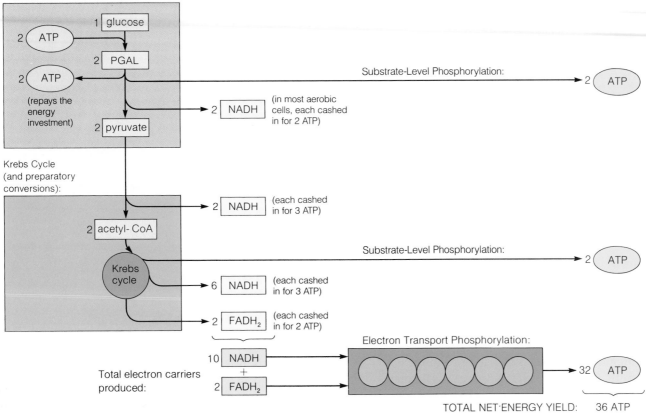

**Figure 8.8** Summary of the energy harvest from one glucose molecule sent through the aerobic respiration pathway. Actual ATP yields vary, depending on cellular conditions and on the mechanism used to transfer energy from cytoplasmic NADH into the mitochondrion.

**In aerobic respiration, glucose is completely broken down to carbon dioxide and water.**

**$NAD^+$ and FAD accept unbound protons ($H^+$) and electrons stripped from substrates of the reactions, then they deliver them to an electron transport system. Oxygen is the final acceptor of those electrons.**

**From glycolysis (in the cytoplasm) to the final reactions (in the mitochondrion), this pathway commonly yields thirty-six ATP for every glucose molecule.**

### Glucose Energy Yield

Figure 8.8 summarizes energy-yielding steps of aerobic respiration. Notice that the net yield from the complete breakdown of a glucose molecule commonly is thirty-six or thirty-eight ATP. The total varies, depending on what happens to the electrons being transferred by NADH molecules that have been produced in the cytoplasm (during glycolysis).

To understand what goes on, compare what happens to the electrons carried by an NADH or $FADH_2$ molecule formed *inside* the mitochondrion during the second-stage reactions. The NADH makes its delivery to the highest possible point of entry into a transport system. The delivery allows the pumping of enough $H^+$ to produce three ATP molecules. The $FADH_2$ makes its delivery at a lower point of entry into the transport system. As a result, fewer $H^+$ are pumped, and only two ATP are produced (Figure 8.5c).

It so happens that NADH formed in the cytoplasm makes its delivery *to* the mitochondrion, not *into* it. Once inside that organelle, its electrons might be transferred to an $NAD^+$ *or* FAD molecule.

For example, consider what happens in liver, heart, and kidney cells. A shuttle built into the outer membrane accepts the electrons, then gives them up to $NAD^+$ inside. Because NADH turns over electrons at the top of a transport system, three ATP form. Thus, in liver, heart, and kidney cells, the overall energy harvest is thirty-eight ATP.

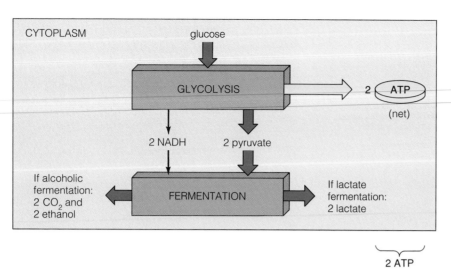

**Figure 8.9** Overview of fermentation, a type of degradative pathway in which an intermediate or product of the reactions themselves serves as the final electron acceptor. The photograph shows the dustlike coating on grapes that contains yeasts, single-celled organisms that use a fermentation pathway.

More commonly, as in skeletal muscle and brain cells, a different shuttle accepts deliveries from cytoplasmic NADH. That shuttle donates electrons to FAD inside the mitochondrion. Because the electrons are delivered to a lower point of entry into the transport system, only two ATP can form. In such cells, the overall energy harvest from the complete breakdown of each glucose molecule is thirty-six ATP.

Remember that glucose is an unstable compound; its covalent bonds are much weaker than the covalent bonds of carbon dioxide and water. In fact, when glucose is broken down completely to carbon dioxide and water, about 686 kilocalories of energy are released. Of that, about 7.5 kilocalories become conserved for further use in each of the thirty-six ATP molecules typically formed. Thus the energy-conserving efficiency of this pathway is $(36)(7.5)/(686)$, or 39 percent.

## ANAEROBIC ROUTES

Our planet has its share of anaerobic environments. Among them are marshes, bogs, and the sediments and mud of lakes, rivers, and seas. Other anaerobic settings include the animal gut, canned foods, and wastewater treatment facilities (Chapter 48). Many bacterial species thrive in totally anaerobic environments. Other species thrive in environments that are well oxygenated some of the time but not all of the time. The aerobic pathway predominates under such conditions, then other pathways kick in when the oxygen concentration drops.

Let's look briefly at three common anaerobic pathways of glucose breakdown. Two are fermentation pathways (Figure 8.9); the third is called anaerobic electron transport. Unlike aerobic respiration, the fermentation pathways do not require an "outside" electron acceptor such as oxygen. The other pathway uses some molecule other than oxygen as an electron acceptor.

**Alcoholic Fermentation**

Like the aerobic pathway, the fermentation pathways begin with glycolysis: glucose is broken down to two pyruvate molecules, and two NADH molecules are formed. In **alcoholic fermentation**, the pyruvate molecules from glycolysis undergo rearrangements but are not used up completely. The NADH that forms during the reactions donates electrons to an intermediate, acetaldehyde, which thereupon becomes ethanol. Figure 8.10 shows the reaction sequence.

Yeasts, which are single-celled fungi, can produce ATP through alcoholic fermentation. This metabolic activity has widespread commercial applications. For example, bakers use cells of the yeast *Saccharomyces cerevisiae* to make bread dough rise (Figure 8.11). They mix the yeast with a small amount of sugar and blend the mixture into the dough. As the yeast converts the sugar to ethanol and carbon dioxide, the gaseous $CO_2$ expands and causes the dough to rise. Oven heat causes the gas to escape from the dough, leaving behind a yeasty-tasting, porous (and so light-textured) product. Baking powder also produces carbon dioxide, but it cannot impart the signature yeasty taste to baked goods.

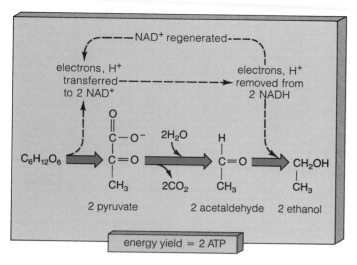

Figure 8.10 Alcoholic fermentation. The net ATP yield is from glycolysis, the first stage of the route.

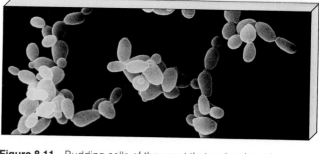

Figure 8.11 Budding cells of the yeast that makes bread dough rise.

Large-scale fermentation by yeasts also is the main-stay of the manufacture of beer and wine. Fruits are a natural home to wild yeasts, and the alcoholic results of their metabolic activities have been recognized since ancient times. Today, vintners still rely on the wild yeasts that are present on grapes (Figure 8.9), but they also add cultivated strains of *S. ellipsoideus* to the juice in fermentation vats. Unlike wild yeasts, which can tolerate an alcohol concentration of only 4 percent, cultivated strains can keep breaking down sugar molecules even when the alcohol concentration reaches 12 to 14 percent.

The fermentation products of wild yeasts still pack a punch. Robins get drunk in droves on naturally fermenting berries of *Pyracantha* shrubs. Where those shrubs are planted densely alongside freeways, the robins typically collide with cars and trees in a manner analogous to drunk drivers. Similarly, wild turkeys get a buzz on when they gobble up naturally fermenting apples in untended orchards.

## Lactate Fermentation

In **lactate fermentation**, pyruvate from glycolysis accepts protons and electrons from NADH. The result is a three-carbon product, lactate. Figure 8.12 shows the reaction sequence. Sometimes lactate is called "lactic acid." However, it is more accurate to refer to the ionized form of the compound (lactate), which is far more common in cells.

One group of bacteria produces lactate exclusively as the fermentation product; milk or cream turned sour is a sign of their activity. Muscle cells may or may not use the lactate fermentation pathway, depending on the

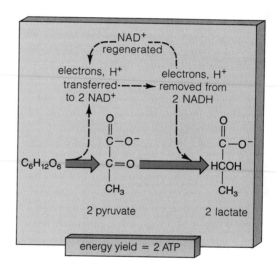

Figure 8.12 Lactate fermentation. The net ATP yield is from glycolysis, the first stage of the route.

demands being placed on the muscle to which they belong (page 633). Aerobic respiration is absolutely necessary for moderate, prolonged muscle action. But when demands are intense but brief—say, during a short race—lactate fermentation produces ATP quickly. This fermentation route cannot be used for long. It has such a low energy yield that the muscle cells would quickly exhaust their glycogen reserves. (Glycogen is the main storage form for glucose in animals.) Once such reserves are depleted, muscles fatigue quickly and lose their ability to contract.

Keep in mind that glucose is not completely degraded by either fermentation pathway, so considerable energy still remains in the products. No more ATP is produced, beyond the two molecules from glycolysis. *The final steps serve only to regenerate NAD$^+$.* The low energy yield is quite enough for some single-celled anaerobic organisms. It can even help carry some otherwise "aerobic" cells through times of stress. But it is not enough to sustain the activities of large, multicelled organisms (this being one of the reasons you never will come across an anaerobic elephant).

**Fermentation pathways have a net yield of two ATP (from glycolysis). The NAD$^+$ necessary for glycolysis is regenerated during the reactions.**

### Anaerobic Electron Transport

Aerobic respiration and fermentation proceed in our own cells, and fermentation is of great commercial interest to us. But a variety of energy-releasing pathways exists in the natural world, especially within the bacterial kingdom. These pathways are vitally important to the cycling of nitrogen, sulfur, and other elements that are necessary components of biological molecules.

One such pathway is called **anaerobic electron transport**. Here, electrons stripped from various substrates are donated to transport systems bound in the bacterial plasma membrane. For example, sulfate-reducing bacteria produce ATP by stripping electrons from a variety of organic compounds and sending them through membrane transport systems. In some cases, the inorganic compound sulfate ($SO_4^{--}$) serves as the final electron acceptor and so is converted into sulfide ($H_2S$). Sulfate-reducing bacteria are often abundant in waterlogged soils and aquatic habitats that are rich in decomposed organic material. They also thrive on dissolved substances that move through the animal gut, an anaerobic setting.

Other kinds of bacteria in soils can produce ATP by stripping electrons from nitrate ($NO_3^-$), leaving nitrite ($NO_2^-$) as the end product. In Chapter 46, we will consider the role of these bacteria in the global cycling of nitrogen.

Anaerobic electron transport also is a significant energy-releasing pathway used by sulfate-reducing bacteria, nitrifying bacteria, and other species that live deep in the ocean, around hydrothermal vents. As you will see in Chapter 47, they form the food production base for unique communities.

## ALTERNATIVE ENERGY SOURCES IN THE HUMAN BODY

Our cells require a steady supply of carbohydrates, lipids, and proteins for energy and raw materials. When we eat more carbohydrates than our cells are calling for, the excess can be stored as glycogen, most notably in liver and muscle cells. Excess fats are tucked away as glistening droplets in the cells of adipose tissue. Between meals, when demands for raw materials and energy exceed dietary intake, the body can draw upon a variety of stored organic compounds.

Of the foods we eat, carbohydrates are the main source of energy This is true of mammals generally. The carbohydrates are broken down into glucose and other monosaccharides, which are absorbed across the gut lining and then circulated to cells by the bloodstream. Our cells use free glucose for as long as it is available. Our brain cells essentially use nothing else, for they have virtually no stores of lipids for energy metabolism and only a limited amount of glycogen. During normal conditions, a mammal taps into its glycogen stores with great precision when the blood glucose level decreases slightly. A starving mammal will break into its fat reserves, thus "sparing" the glucose that is still available for the all-important brain. It will then start breaking down the body's proteins. However, after the blood glucose level falls precipitously, brain function may be so disrupted that death is likely to follow.

Figure 8.13 shows the points of entry into the aerobic pathway for complex carbohydrates, fats, and proteins. Chapter 37 describes the digestion of these compounds into their simpler components. Here we can simply out-

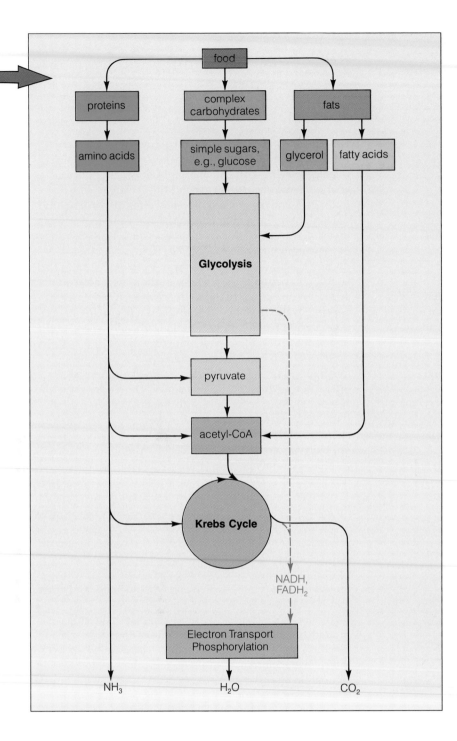

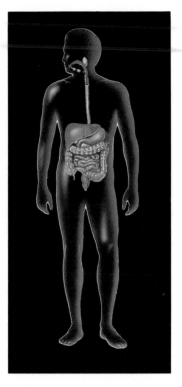

human digestive system, which breaks down food molecules into smaller molecules that can be taken up by individual cells

**Figure 8.13** Points of entry into the aerobic pathway for complex carbohydrates, fats, and proteins after they have been reduced to their simpler components in the human digestive system.

line how fats and proteins can be used as alternative energy sources.

A fat molecule, recall, consists of a glycerol head and one, two, or three fatty acid tails. Once a fat molecule has been digested into its component parts, the glycerol can be converted to PGAL and inserted into the glycolytic reactions. The long carbon backbone of each fatty acid can be split many times into two-carbon fragments, these being easily converted into acetyl-CoA and picked up for the Krebs cycle. Whereas each glucose molecule has six carbon atoms, a fatty acid has many more—and its degradation produces much more ATP.

Following protein digestion, the amino acid subunits can have their amino group ($NH_3$) removed. The remnants can enter the Krebs cycle, where the hydrogens can be removed from the remaining carbon atoms and transferred to coenzymes. The amino groups are converted into ammonia, a waste product.

## Perspective on Life

In this unit, you have read about pathways by which cells trap, store, and then release energy to drive their activities. Over evolutionary time, the main pathways—photosynthesis and aerobic respiration—have become interconnected on a grand scale.

When life began, little (if any) free oxygen was present in the earth's atmosphere. Most likely, early single-celled organisms produced ATP by pathways similar to glycolysis. And given the anaerobic conditions, fermentation routes must have predominated.

Photosynthetic organisms emerged more than 3 billion years ago, and they turned out to be a profound force in evolution. Oxygen, a by-product of their activities, began to accumulate in the atmosphere. Some photosynthesizers were opportunistic about the increasing oxygen levels. Perhaps through mutations in their metabolic machinery, they gained the capacity to use oxygen as an electron acceptor for degradative reactions—and in time some cells abandoned photosynthesis entirely. Among those cells were the forerunners of animals and other organisms able to survive with aerobic machinery alone.

With aerobic respiration, life became self-sustaining, for its final products—carbon dioxide and water—are precisely the materials used to build organic compounds in photosynthesis! Thus the flow of carbon, hydrogen, and oxygen through the metabolic pathways of living organisms came full circle:

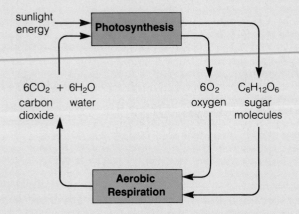

Perhaps one of the most difficult connections you are asked to perceive is the link between yourself—a living, intelligent being—and such remote-sounding things as energy, metabolic pathways, and the cycling of carbon, hydrogen, and oxygen. Is this really the stuff of humanity?

Think back, for a moment, to the description of a water molecule. A pair of hydrogen atoms competing with an oxygen atom for a share of the electron joining them doesn't exactly seem close to our daily lives. But from that simple competition, the polarity of the water molecule arises. As a result of the polarity, hydrogen bonds form between water molecules. And that is a beginning for the organization of lifeless matter that leads, ultimately, to the organization of matter in all living things.

For now you can imagine other kinds of molecules interspersed in water. Many are nonpolar and resist interaction with the water molecules. Others are polar and respond by dissolving in it. And the lipids among them (with water-soluble *and* water-insoluble regions) spontaneously assemble into a two-layered film. Such lipid bilayers are the basis for all cell membranes, hence all cells. The cell has been, from the beginning, the fundamental *living* unit.

With the boundary afforded by a cell membrane, chemical reactions can be contained and controlled. The essence of life *is* chemical control. This "control" is not some mysterious force. It is a chemical responsiveness to energy changes and to the kinds of molecules present in the environment. It operates by "telling" a class of protein molecules—enzymes—when and what to build, and when and what to tear down.

And it is not some mysterious force that creates the proteins themselves. DNA, the slender double strand of heredity, has the chemical structure—*the chemical message*—that allows molecule faithfully to reproduce molecule, one generation after the next. Those DNA strands tell many billions of cells in your body how countless molecules must be built and torn apart for their stored energy.

So yes, carbon, hydrogen, oxygen, and other organic molecules represent the stuff of you, and us, and all of life. But it takes more than molecules to complete the picture. You are alive because of the way molecules are organized and maintained by a constant flow of energy. It takes outside energy from sources such as the sun to drive their formation. Once molecules are assembled into cells, it takes outside energy derived from food, water, and air to sustain their organization. Plants, animals, fungi, protistans, and bacteria are part of a web of energy use and materials cycling that ties together all levels of biological organization. Should energy fail to reach any part of any level, life there will dwindle and cease.

For energy flows through time in only one direction—from forms rich in usable energy to forms having less usable stores of it. Only as long as sunlight flows into the web of life—and only as long as there are molecules to recombine, rearrange, and recycle—does life have the potential to continue in all its rich diversity.

In short, life is no more *and no less* than a marvelously complex system of prolonging order. Sustained by energy transfusions, it continues because of a capacity for self-reproduction—the handing down of hereditary instructions. With those instructions, energy and materials are organized, generation after generation. Even with the death of the individual, life is prolonged. With death, molecules are released and can be recycled once more, providing raw materials for new generations. In this flow of energy and cycling of material through time, each birth is affirmation of our ongoing capacity for organization, each death a renewal.

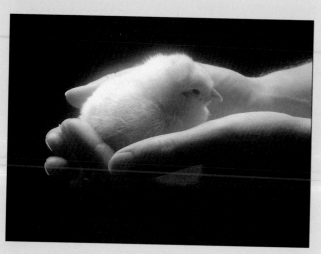

## SUMMARY

1. ATP, the main energy carrier in cells, can be produced by photosynthesis. It also can be produced by aerobic respiration or fermentation, these being degradative pathways by which chemical energy is released from glucose (or some other organic compound).

2. Glycolysis, the partial breakdown of a glucose molecule, is the first stage of all the main degradative pathways. It takes place in the cytoplasm and requires an initial energy input from ATP. During glycolysis, two ATP (net), two NADH, and two pyruvate molecules are produced for each glucose molecule.

3. In aerobic respiration, oxygen is the final acceptor of electrons stripped from glucose. The pathway proceeds from glycolysis, through the Krebs cycle, and through electron transport phosphorylation. Its net energy yield is commonly thirty-six ATP.

4. In the second stage of aerobic respiration, pyruvate from glycolysis is converted to a form that can enter the Krebs cycle (a cyclic metabolic pathway). The conversion reactions and the cycle itself produce eight NADH, two $FADH_2$, and two ATP. In this stage, the glucose molecule is degraded completely to carbon dioxide and water.

5. The coenzymes NADH and $FADH_2$ (formed during glycolysis and the Krebs cycle) deliver electrons to a transport system embedded in the inner membrane of mitochondria. The third stage of the aerobic pathway (electron transport phosphorylation) depends on electron transfers through the system *and* on phosphorylation reactions that take place at channel proteins spanning the membrane.

6. Operation of the transport system sets up $H^+$ concentration and electric gradients across the membrane. $H^+$ moves down the gradients, through channel proteins. Energy associated with the flow of hydrogen ions drives the coupling of ADP and inorganic phosphate to form ATP. Oxygen accepts the "spent" electrons and combines with $H^+$ to form water.

7. Many microbes are not metabolically equipped to use oxygen as a final electron acceptor. They rely instead on alcoholic fermentation, lactate fermentation, or anaerobic electron transport. Also, some cells that normally use the aerobic pathway (such as skeletal muscle cells) rely on lactate fermentation when the demand for muscle action is brief but intense.

8. Compared with aerobic respiration, the anaerobic pathways have a small net yield (two ATP, from glycolysis), because glucose is not completely degraded. Following glycolysis, the remaining reactions serve to regenerate $NAD^+$.

## Review Questions

1. ATP can be produced when carbohydrates are degraded. Define three types of energy-releasing pathways by which this occurs. Which yields the most ATP? *120*

2. Which energy-releasing pathways occur in the cytoplasm? In the mitochondrion of eukaryotes? *121, 122, 124*

3. Is the following statement true? Your muscle cells cannot function at all unless they are supplied with oxygen. *129*

4. Glycolysis is the first stage of all the main pathways by which glucose is degraded. Can you define those pathways in terms of the final electron acceptor for their reactions? If you include the two ATP molecules formed during glycolysis, what is the *net* energy yield from one glucose molecule for each pathway? *126–130*

5. In anaerobic routes of glucose breakdown, further conversions of pyruvate do not yield any more usable energy. What, then, is the advantage of the conversions? *130*

6. Describe the functions of the Krebs cycle. Describe the functions of electron transport phosphorylation. *120, 125, 126*

---

### Self-Quiz *(Answers in Appendix IV)*

1. Plants as well as bacteria, protistans, fungi, and animals can produce ATP by degrading _____.

2. Glucose can be degraded by way of two anaerobic pathways, called _____ and _____, as well as by aerobic respiration.

3. In the first stage of aerobic respiration, glucose is partially broken down to _____, which in the second stage is broken down completely to _____ and _____.

4. ATP is best described as _____.
   a. a high-energy phosphate compound
   b. a primary source of chemical energy
   c. being produced by plants, animals, bacteria, protistans, and fungi
   d. all of the above

5. Which of the following is *not* a product of glycolysis?
   a. two NADH
   b. two pyruvate
   c. two $H_2O$
   d. two ATP

6. Glycolysis occurs in which part of the cell?
   a. nucleus
   b. mitochondrion
   c. plasma membrane
   d. cytoplasm

7. The final acceptor of the electrons stripped from glucose during aerobic respiration is _____.
   a. water
   b. hydrogen
   c. oxygen
   d. NADH

8. Electron transport systems for the aerobic reactions are located in the _____.
   a. cytoplasm
   b. inner mitochondrial membrane
   c. outer mitochondrial membrane
   d. stroma

9. The flow of _____ through channel proteins in the inner mitochondrial membrane provides the energy to couple ADP and inorganic phosphate to form ATP.
   a. electrons
   b. hydrogen ions
   c. NADH
   d. $FADH_2$

10. Match each type of metabolic reaction with its function:
    _____ glycolysis
    _____ fermentation
    _____ Krebs cycle
    _____ electron transport phosphorylation

    a. produces ATP, NADH, and $CO_2$
    b. degrades glucose into two pyruvate
    c. regenerates $NAD^+$
    d. flow of $H^+$ through channel proteins that drives ATP formation

---

### Selected Key Terms

aerobic respiration *120*
alcoholic fermentation *128*
anaerobic electron transport *130*
anaerobic pathway *120*
ATP *120*
chemiosmotic theory *126*
electron transport phosphorylation *126*
$FADH_2$ *125*
glycolysis *122*
Krebs cycle *124*
lactate fermentation *129*
mitochondrion *124*
NADH *125*
substrate-level phosphorylation *122*

---

### Readings

Becker, W. 1986. *The World of the Cell*. Menlo Park, California: Benjamin/Cummings. Chapters 7 and 8 are a good place to start for further readings on anaerobic and aerobic metabolism.

Brock, T., B. Smith, and M. Madigan. 1988. *Biology of Microorganisms*. Fifth edition. Englewood Cliffs, New Jersey: Prentice-Hall. Clear descriptions of the energy-releasing pathways of microbes.

Lehninger, A. 1982. *Principles of Biochemistry*. New York: Worth. Clear, accessible introduction to metabolic pathways.

FACING PAGE: *Human sperm, one of which will penetrate this mature egg and so set the stage for the development of a new individual in the image of its parents.*

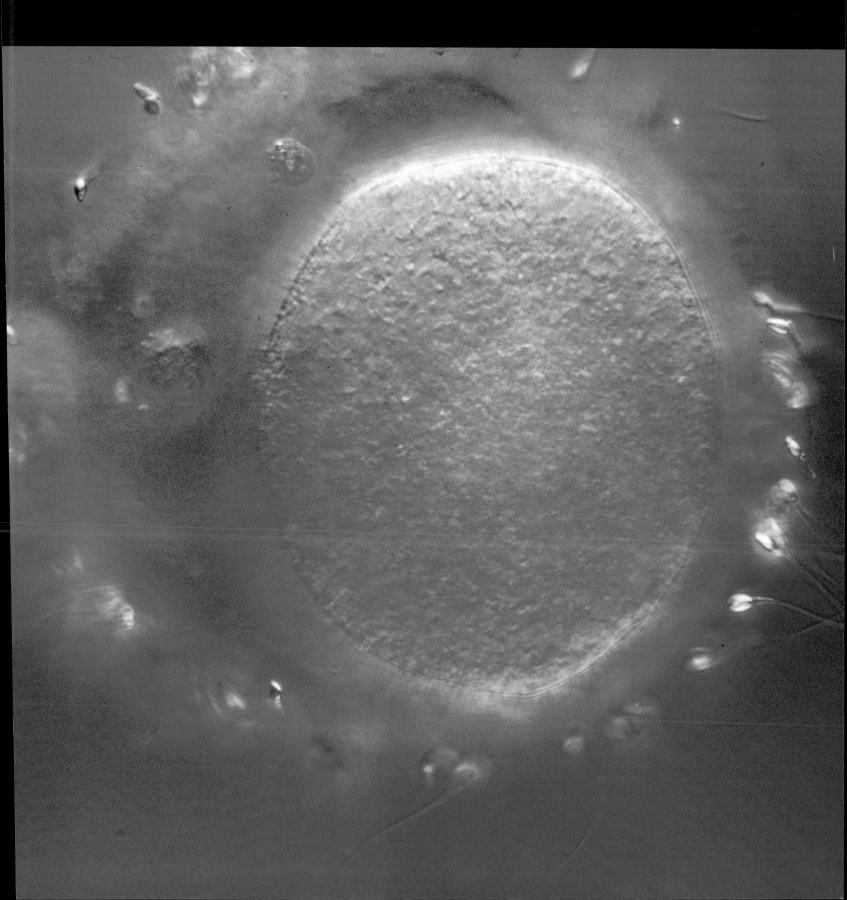

# CELL DIVISION AND MITOSIS

## *Silver in the Stream of Time*

Five o'clock, and the first rays of the sun dance over the wild Alagnak River of the Alaskan tundra. This September morning, as every morning for several weeks, life is both ending and beginning in the clear, frigid waters. By the thousands, mature silver salmon have returned from the open ocean to spawn in their shallow native home. A female salmon rests briefly in a quiet eddy, then continues upstream a bit more (Figure 9.1). She is tinged with red, the color of spawners, and she is dying.

On this morning the female salmon pauses, then quickly hollows out a shallow "nest" in the gravel riverbed. Now scores of translucent pink eggs emerge from her body. Within moments a male salmon appears and sheds a cloud of sperm near the eggs. Trout and other voracious predators of the Alagnak will consume most of the eggs. But a few will survive, and following successful fertilization, they will give rise to a new generation.

The female lingers on for a few hours, but depleted of eggs and with vital organs failing, she soon dies. On

the riverbank, a bald eagle loses no time in consuming her carcass. Yet her remains speak of a remarkable journey. That female had herself started life as a pea-sized egg that had been fertilized in the Alagnak's gravel bed. Within three years, she had become a streamlined vertebrate, fashioned from billions of cells. And she had reached reproductive maturity. Early in her development, some of her cells had been set aside for reproduction, and in time they had given rise to eggs. On this morning, together with the sperm that eventually fertilized them, those eggs became part of an ongoing story of birth, cell divisions and growth, death, and rebirth for the silver salmon.

For you, as for the silver salmon and all other organisms, reproduction depends on the capacity of cells to divide. Starting with the fertilized egg in your mother's body, a single cell divided in two, then the two into four, and so on until billions of cells were growing, developing in specialized ways, and dividing at different times to produce your genetically prescribed body parts. Today your body is composed of approximately 65

trillion cells. Cell divisions are still proceeding in many parts of it. Every five days, for example, cell divisions manage to replace the lining of your small intestine. Even now, in males who are reading this page, divisions are probably proceeding that will give rise to mature sperm cells—part of the reproductive bridge to the next generation.

Understanding the nature of cell division—and, ultimately, how new individuals are put together in the image of their parents—begins with answers to three questions. *First*, what structures and substances are necessary for inheritance? *Second*, how are they divided and distributed into daughter cells? *Third*, what are the division mechanisms themselves? We will require more than one chapter to consider the answers (and best guesses) about cell reproduction and other mechanisms of inheritance. However, the points made in the first half of this chapter can help you keep the overall picture in focus.

**Figure 9.1** The last of one generation and the first of the next in the Alagnak River of Alaska.

**1.** Each cell of a new generation will not grow or function properly unless it receives the necessary hereditary information, in the form of parental DNA, and a portion of the cytoplasm from the parental cell. In eukaryotes, DNA is parceled out to daughter cells by mitosis or meiosis, both of which are nuclear division mechanisms.

**2.** A chromosome is a DNA molecule with certain proteins attached to it. The cells of each species have a characteristic number of chromosomes. "Diploid" cells contain two of each type of chromosome characteristic of the species.

**3.** Mitosis maintains the number of chromosomes from one cell generation to the next. Thus each daughter cell formed by the division of the nucleus and cytoplasm of a diploid parental cell will be diploid also.

**4.** Mitosis proceeds through four continuous stages: prophase, metaphase, anaphase, and telophase. Actual cytoplasmic division (cytokinesis) occurs toward the end of the nuclear division or at some point afterward.

**5.** Mitosis is the basis of asexual reproduction of single-celled eukaryotes as well as the growth of multicelled eukaryotes. Meiosis is the basis of gamete formation, hence of sexual reproduction.

## DIVIDING CELLS: THE BRIDGE BETWEEN GENERATIONS

### Overview of Division Mechanisms

In biology, **reproduction** means producing a new generation of cells or multicelled individuals. Reproduction is part of a *life cycle*, a recurring frame of events in which individuals grow, develop, maintain themselves, and reproduce according to instructions encoded in DNA, which they inherit from their parents. Reproduction begins with the division of single cells. And the ground rule for cell division is this: *Each cell of a new generation must receive hereditary information (encoded in parental DNA) and enough cytoplasmic machinery to start up its own operation.*

Recall that DNA contains instructions for making proteins. Some proteins serve as structural materials; many serve as enzymes with roles in the synthesis of carbohydrates, lipids, and other building blocks of the cell. Thus, unless new cells receive the necessary instructions for making proteins, they will not grow or function properly.

Also, the cytoplasm of the parental cell already has operating machinery—enzymes, organelles, and so on. When a daughter cell inherits what looks merely like a blob of cytoplasm, it really is getting "start-up" machinery for its operation, until it has time to use its inherited DNA for growing and developing on its own.

In multicelled plants and animals, cell division typically begins with mitosis or meiosis and ends with cytokinesis. **Mitosis** and **meiosis** are *nuclear* division mechanisms—the means by which DNA instructions are sorted out and distributed into new nuclei for the forthcoming daughter cells. The actual splitting of a parental cell into two daughter cells occurs by way of **cytokinesis**, or *cytoplasmic* division.

In multicelled organisms, mitosis is the basis for growth through repeated divisions of the body's cells, which are called *somatic* cells. Mitosis also is the basis for asexual reproduction in many plants, animals, and other organisms.

In contrast, meiosis occurs only in germ cells, a cell lineage set aside for sexual reproduction. By definition, *sexual reproduction* is a process that begins with meiosis, proceeds through the formation of gametes (sperm and eggs), and ends at fertilization. At fertilization, a sperm nucleus and egg nucleus fuse together in the zygote, the first cell of the new individual.

This chapter focuses on mitosis, and the chapter to follow, on meiosis. As you will see, the two division mechanisms have much in common but they differ in their end result. Both are limited to eukaryotes. The prokaryotes (bacteria) use a different division mechanism, as listed in Table 9.1 and described on page 355.

## Some Key Points About Chromosome Structure

Before you track the distribution of DNA into daughter cells, reflect for a moment on its structural organization. Many proteins are attached to eukaryotic DNA, and they generally are equal in mass to the DNA itself. Together, the DNA and proteins form a structure called the **chromosome**.

Between divisions, a chromosome is stretched out in threadlike form. It is still threadlike when it is duplicated prior to cell division, and the two threads remain attached for a while as **sister chromatids** of the chromosome. Each "thread," of course, is a DNA double helix with its associated proteins, which we show here as a simplified version of a current model:

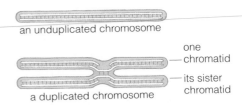

an unduplicated chromosome

a duplicated chromosome — one chromatid / its sister chromatid

Notice the small region of the chromosome where the DNA appears to be constricted. This is the **centromere**, a region having attachment sites for microtubules that will help move the chromosome during nuclear division:

centromere (constricted region)

Keep in mind that the location of the centromere varies from one type of chromosome to the next. Also keep in mind that this model is highly simplified. The DNA double helix in each chromatid is actually two molecular

---

| Table 9.1 Cell Division Mechanisms | |
|---|---|
| Mechanisms | Used By |
| Mitosis, cytokinesis | *Single-celled eukaryotes (for asexual reproduction)* |
| | *Multicelled eukaryotes (for bodily growth; also for asexual reproduction in some species)* |
| Meiosis, cytokinesis | *Eukaryotes (basis of gamete formation and sexual reproduction)* |
| Prokaryotic fission | *Bacterial cells* |

**Figure 9.2** (**a**) Photograph of the 46 chromosomes from a diploid cell of a human male. All are in the duplicated state. (**b**) By cutting apart and arranging the chromosomes according to length and centromere location, we see that there are two sets of 23 chromosomes, with all the chromosomes in one set having a partner, or homologue, in the other. The partners don't pair at all during mitosis, but they pair up with each other during meiosis.

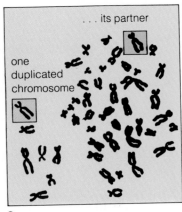

. . . its partner

one duplicated chromosome

a

strands twisted together repeatedly like a spiral staircase (that's what the "double helix" means), and it is much longer than can be shown here.

## Mitosis, Meiosis, and the Chromosome Number

All individuals of the same species have the same number of chromosomes in their somatic cells. For example, there are 46 chromosomes in your somatic cells, 48 in a gorilla's, and 14 in a garden pea's (Table 9.2). When a gorilla or pea plant grows, mitosis assures that all the new cells added to its body will end up with the parental number of chromosomes. *With mitosis, the chromosome number is maintained, division after division:*

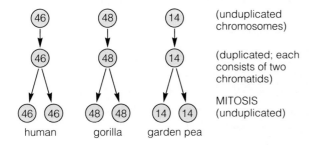

(unduplicated chromosomes)

(duplicated; each consists of two chromatids)

MITOSIS (unduplicated)

human    gorilla    garden pea

To get an initial sense of the difference between mitosis and meiosis, take a look at Figure 9.2, which shows all the chromosomes in a human somatic cell. Notice how the chromosomes can be lined up as pairs in two rows. (One row is coded pink and the other, green.) Except for sex chromosomes, designated X and Y, the members of each pair have the same length and centromere location, and their hereditary instructions deal with the same traits. The two members of each pair are **homologous chromosomes**.

Homologues don't interact at all during mitosis, but they pair with each other during meiosis. Although the X and Y chromosomes differ in most respects, they pair during meiosis so we still call them homologues.

*During meiosis, the chromosome number is reduced by half for forthcoming gametes.* And not just any half—each gamete ends up with *one of each pair* of homologous chromosomes. (It doesn't matter which of the two it gets.) Reducing the chromosome number requires two nuclear divisions:

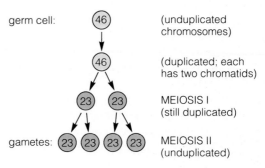

germ cell:    46    (unduplicated chromosomes)

46    (duplicated; each has two chromatids)

23    23    MEIOSIS I (still duplicated)

gametes: 23  23  23  23    MEIOSIS II (unduplicated)

Gametes end up with a **haploid** number of chromosomes, which we can take to mean as "half" of the parental number. When a sperm nucleus and an egg nucleus fuse at fertilization, the **diploid** number is restored. "Diploid" means having two chromosomes of each type in the somatic cells of sexually reproducing species.

**Mitosis is a type of nuclear division that *maintains* the parental number of chromosomes for forthcoming cells. It is the basis for bodily growth and, in some cases, asexual reproduction of eukaryotes.**

**Meiosis is a type of nuclear division that *reduces* the parental chromosome number by half—to the haploid number. It occurs only in sexual reproduction.**

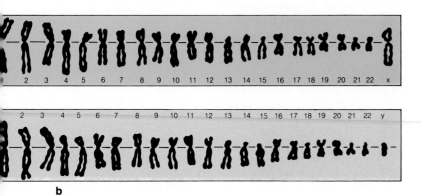

b

| Table 9.2 | Number of Chromosomes in the Somatic Cells of Some Eukaryotes* |
|---|---|
| Mosquito, *Culex pipiens* | 6 |
| Fruit fly, *Drosophila melanogaster* | 8 |
| Garden pea, *Pisum sativum* | 14 |
| Corn, *Zea mays* | 20 |
| Lily, *Lilium* | 24 |
| Yellow pine, *Pinus ponderosa* | 24 |
| Frog, *Rana pipiens* | 26 |
| Earthworm, *Lumbricus terrestris* | 36 |
| Rhesus monkey, *Macaca mulatta* | 42 |
| Human, *Homo sapiens* | 46 |
| Chimpanzee, *Pan troglodytes* | 48 |
| Gorilla, *Gorilla gorilla* | 48 |
| Potato, *Solanum tuberosum* | 48 |
| Amoeba, *Amoeba* | 50 |
| Horse, *Equus caballus* | 64 |
| Horsetail, *Equisetum* | 216 |
| Adder's tongue fern, *Ophioglossum reticulatum* | 1,000+ |

*These examples are a sampling only. Chromosome number for most species falls between 10 and 50.

# MITOSIS AND THE CELL CYCLE

Mitosis is a very small part of the **cell cycle**, which includes events that extend from the time a cell forms until its own division is completed. Normally, a cell destined to enter mitosis spends about 95 percent of the cell cycle in **interphase**, when it increases its mass, approximately doubles the number of its cytoplasmic components, and finally duplicates its DNA.

As Figure 9.3 and the following list indicate, "interphase" actually consists of three phases of activities:

| Mitosis | M | *nuclear division, commonly followed by cytokinesis* |
|---|---|---|
| Interphase | $G_1$ | *a "gap" (interval) before DNA replication* |
| | S | *"synthesis" (replication) of DNA and associated proteins* |
| | $G_2$ | *a second "gap" after DNA replication, before mitosis* |

The "gaps" were so named before biologists knew what was going on—specifically, that new cell components are synthesized during $G_1$ and assembled for distribution to daughter nuclei during $G_2$.

To get an idea of the challenges facing a cell that is destined to divide, imagine that one of your liver cells is blown up as large as a basketball. If you stretched out its diploid number of chromosomes end to end, the lineup would be about 20 miles long. Before that cell divides, another 20 miles' worth of chromosomes must be copied.

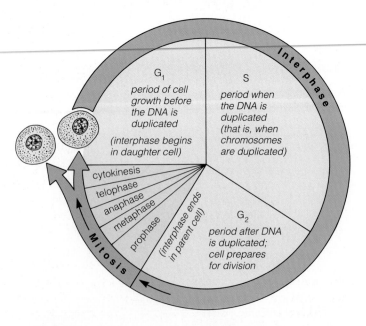

**Figure 9.3** Eukaryotic cell cycle. This drawing has been generalized; the length of different stages varies greatly from one type of cell to the next.

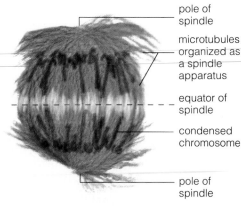

**Figure 9.4** Mitosis in a plant cell (*Haemanthus*). The chromosomes are stained blue, and the microtubules that move them about are stained red.

pole of spindle

microtubules organized as a spindle apparatus

equator of spindle

condensed chromosome

pole of spindle

Before its nucleus finally divides, the 40 miles must be sorted into two subsets—each with one of each type of chromosome. Before its cytoplasm divides, the subsets must be packaged so that each daughter cell ends up with 20 miles of the proper chromosomes. All of this goes on in a space the size of a basketball. And all in a few hours.

The duration of the cell cycle varies widely, but it is fairly consistent for all cells of a given type. For example, all of your brain cells are arrested at interphase and never will divide again. Cells of a newly forming sea urchin may double in number every two hours. To be sure, adverse environmental conditions may disrupt the cycle, as when cells atypically become arrested in the $G_1$ phase. (For example, this happens among amoebas and other protistans when they are deprived of a vital nutrient.) Even so, if a cell progresses past a certain point in $G_1$, the cycle normally will be completed regardless of outside conditions. Because the cycle is so predictable for each cell type in each kind of organism, you might suspect that cells have built-in controls over the cycle's duration. As you will read on page 232, this is indeed the case.

Good health depends on the successful completion of cell cycles. Every error in DNA duplication or misshipment of chromosomes may lead to problems. Even the timing and regulation of cell division must be precisely controlled. If the cell cycle stops in growing tissues, the tissues die. If the cycle proceeds uncontrollably in mature tissues, cancer follows. A landmark case of unchecked cell divisions is described in the *Commentary* at this chapter's end.

## STAGES OF MITOSIS

When a cell makes the transition from interphase to mitosis, it stops constructing new cell parts. Profound changes now proceed one after the other, through four stages. The sequential stages of mitosis are called *prophase, metaphase, anaphase,* and *telophase*.

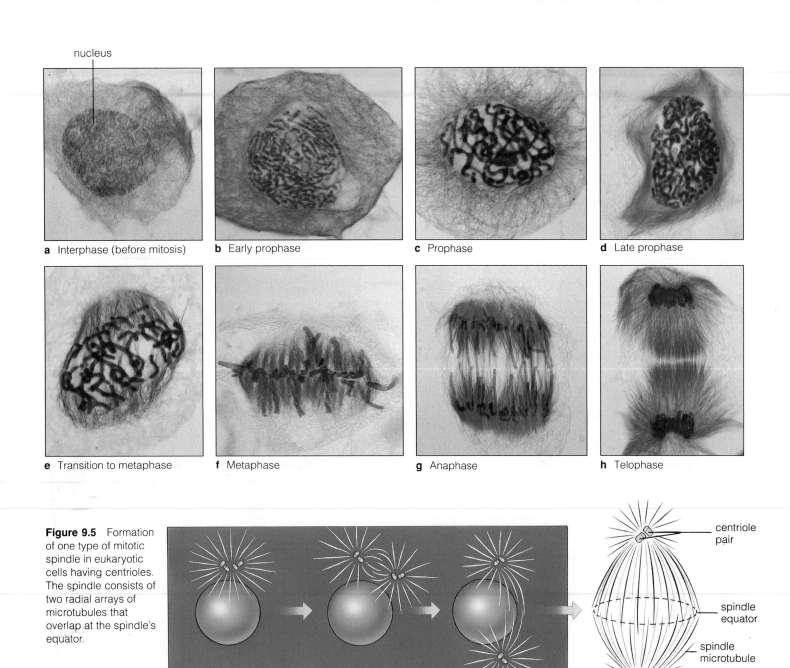

nucleus

**a** Interphase (before mitosis)    **b** Early prophase    **c** Prophase    **d** Late prophase

**e** Transition to metaphase    **f** Metaphase    **g** Anaphase    **h** Telophase

**Figure 9.5** Formation of one type of mitotic spindle in eukaryotic cells having centrioles. The spindle consists of two radial arrays of microtubules that overlap at the spindle's equator.

centriole pair

spindle equator

spindle microtubule

centriole pair

## The Microtubular Spindle

Before considering the details of mitosis, take a look at Figure 9.4, which will give you an idea of the extent of chromosome movements through the different stages. The chromosomes do not move about on their own. They are moved by a **spindle apparatus**, a bipolar structure composed of organized arrays of microtubules. During nuclear division, the chromosomes will move toward the two spindle poles.

Remember that microtubules are components of the cell's internal framework, the cytoskeleton. When a nucleus is about to divide, nearly all of the cell's existing microtubules disassemble into their protein subunits.

Then the subunits reassemble into new microtubules, just outside the nucleus. These are the microtubules that become arranged as a spindle.

In many cells, a microtubule organizing center dictates *where* the new microtubules will arise. As mentioned on page 69, the center often includes a pair of centrioles. The center (including its centriole pair) is duplicated during interphase. Then, while the spindle is forming, the two centers separate and become positioned at opposite spindle poles (Figure 9.5). Their positioning and orientation seem to influence the organization of the cytoskeleton that will form in each daughter cell. And proper cell functioning depends on that organization.

**Figure 9.6** Mitosis: the nuclear division mechanism that maintains the parental chromosome number in daughter cells. Shown here, a diploid animal cell (with pairs of homologous chromosomes, derived from two parents).

For the sake of clarity, only two pairs of homologous chromosomes are shown in the diagram and the spindle apparatus is simplified. With rare exceptions, the picture is more involved than this, as indicated by the micrographs of mitosis in a whitefish cell.

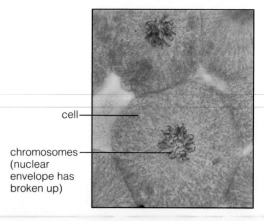

cell

chromosomes (nuclear envelope has broken up)

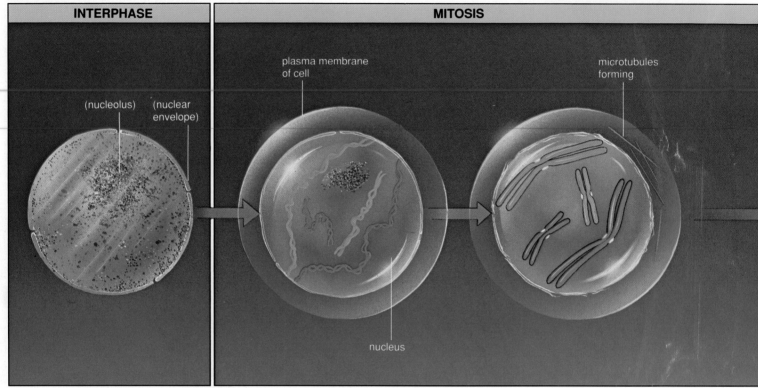

| INTERPHASE | MITOSIS |
|---|---|

(nucleolus) (nuclear envelope)

plasma membrane of cell

microtubules forming

nucleus

**Nucleus at Interphase**
The DNA is duplicated, then the cell prepares for division.

**Early Prophase**
The DNA and associated proteins start condensing into the threadlike chromosome form. (Chromosomes are already duplicated.) Two chromosomes derived from the male parent are in green; their homologues from the female are pink.

**Late Prophase**
Chromosomes continue to condense. Microtubules start to assemble outside the nucleus; they will form the spindle. Centrioles (if present) are moved by the microtubules toward opposite poles. The nuclear envelope starts breaking up during the transition to metaphase.

### Prophase: Mitosis Begins

Each molecule of eukaryotic DNA has many proteins attached to it like beads on a string. **Prophase**, the first stage of mitosis, is evident when the "beaded strings" begin to fold and twist into condensed chromosome structures. At this stage the chromosomes are visible in the light microscope as threadlike forms. ("Mitosis" comes from the Greek *mitos*, meaning thread.)

Each chromosome was duplicated earlier, during interphase. In other words, it already consists of two sister chromatids joined at the centromere. By late prophase, the chromatids of each chromosome have condensed into thicker, rodlike forms (Figure 9.6).

Microtubules of the cytoskeleton disassemble toward the end of prophase. Then new microtubules begin to reassemble, and these will eventually form the spindle. At this time, however, the nuclear envelope prevents

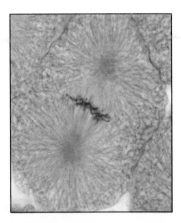
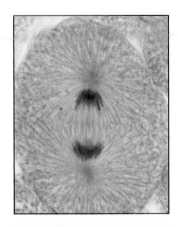
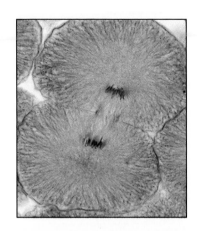

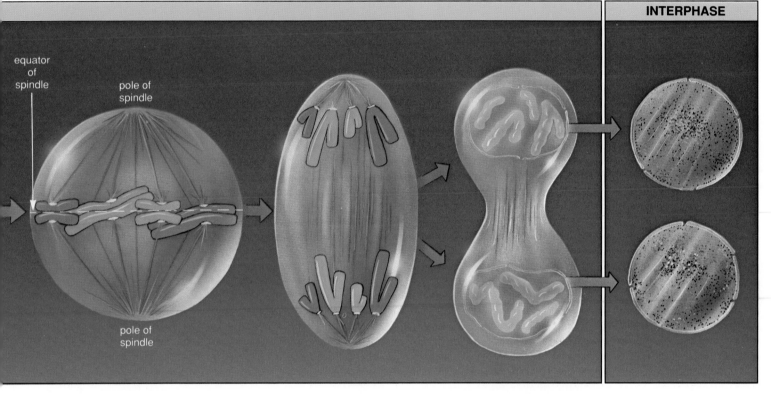

equator
of
spindle

pole of
spindle

pole of
spindle

**Metaphase**

Sister chromatids of each chromosome are attached to the spindle. All chromosomes are now lined up at the spindle equator.

**Anaphase**

Sister chromatids of each chromosome will now be separated from each other and moved to opposite poles.

**Telophase**

Chromosomes decondense. New nuclear membranes start forming. Most often, cytokinesis occurs before the end of telophase.

**Interphase**

Two daughter nuclei are formed, each with a diploid number of chromosomes (the same as the parental nucleus).

the microtubules from interacting with the chromosomes inside the nucleus.

## Metaphase

The nuclear envelope abruptly breaks up during the transition from prophase to metaphase. Now the chromosomes are free to interact with microtubules that had been assembled earlier, outside the nucleus. The chromosome-microtubule interactions culminate in the formation of a bipolar spindle. These events occur early in **metaphase**, the second stage of mitosis. (Many researchers now refer to early metaphase as "prometaphase.") While these interactions are proceeding, fragments of the nuclear envelope become small membranous vesicles.

To gain insight into the interaction between the chromosomes and microtubules, take a look at the meta-

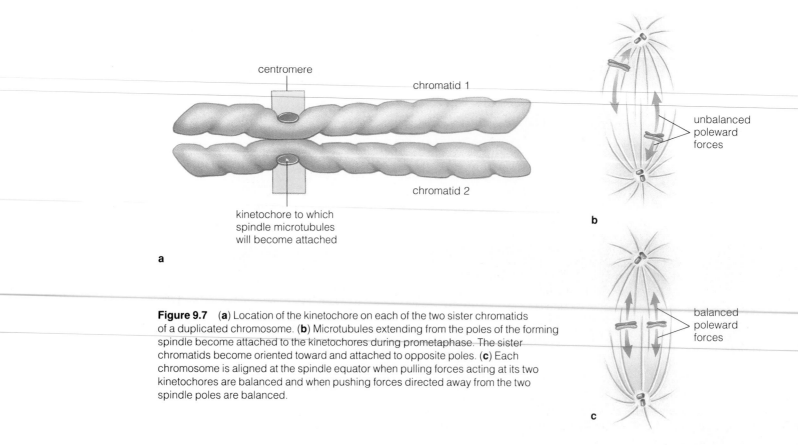

centromere

chromatid 1

unbalanced
poleward
forces

chromatid 2

kinetochore to which
spindle microtubules
will become attached

a

b

**Figure 9.7** (**a**) Location of the kinetochore on each of the two sister chromatids of a duplicated chromosome. (**b**) Microtubules extending from the poles of the forming spindle become attached to the kinetochores during prometaphase. The sister chromatids become oriented toward and attached to opposite poles. (**c**) Each chromosome is aligned at the spindle equator when pulling forces acting at its two kinetochores are balanced and when pushing forces directed away from the two spindle poles are balanced.

balanced
poleward
forces

c

phase chromosomes shown in Figures 9.6 and 9.7. Notice the centromere, the constricted region where the two sister chromatids of each chromosome are held together. At this region, each chromatid has a **kinetochore**, a specialized grouping of proteins and DNA to which several spindle microtubules become attached.

Early in metaphase, the chromosomes seem to go into a frenzy. This happens when kinetochores randomly start harnessing certain microtubules that extend toward them from the spindle poles. With the first random contacts, the chromosomes become attached to the forming spindle. Each chromosome is yanked back and forth until its sister chromatids become firmly oriented toward and attached to opposite poles. Now, forces work to pull the kinetochores of sister chromatids toward opposite poles. Other forces push the poles away from each other (Figure 9.7). The opposing forces are balanced by late metaphase. At that point, all the chromosomes lie halfway between the poles, at the spindle equator.

In sum, three events dominate metaphase. *First*, the spindle becomes fully formed. *Second*, as the spindle forms, the sister chromatids of each chromosome become oriented toward and attached to opposite spindle poles. *Third*, all chromosomes become aligned halfway between the poles. This alignment is crucial for the chromosomal movements that follow.

### Anaphase

During **anaphase**, sister chromatids of each chromosome are separated from each other, and those former partners are moved toward opposite poles. Once they do separate, the partners are no longer referred to as chromatids. Each is now an independent chromosome:

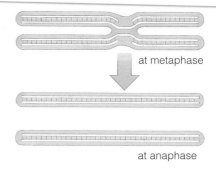

at metaphase

at anaphase

A chemical signal prods the sister chromatids into separating. Two mechanisms apparently account for the subsequent anaphase movements. First, microtubules attached to the kinetochores shorten as the chromosomes approach the spindle poles. Second, the spindle elongates when forces push the two spindle poles far-

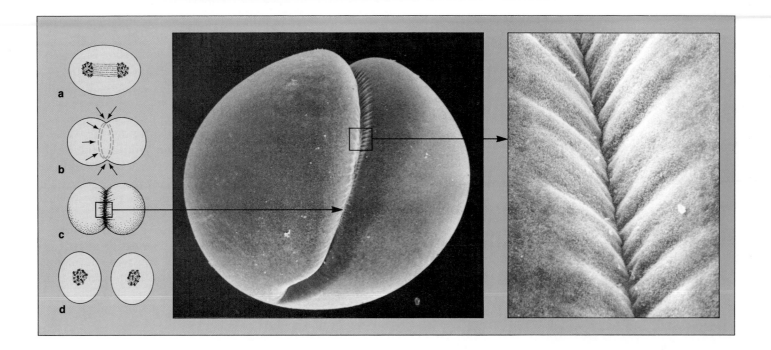

**Figure 9.8** Cytokinesis in an animal cell. The scanning electron micrographs show the furrowing of the plasma membrane caused by the contraction of a microfilament ring just beneath it. (**a**) Nuclear division is complete; the spindle is disassembling. (**b**) Microfilament rings at the former spindle equator contract, like a purse string closing. (**c**) Contractions cause furrowing at the cell surface. (**d**) The cytoplasm is pinched in two.

ther apart. Although metaphase may last a long time, anaphase begins abruptly and commonly ends within a few minutes. All the duplicated chromosomes split apart at about the same time and move to opposite poles at the same rate.

### Telophase

Telophase begins once the separated chromosomes arrive at opposite spindle poles. Now the kinetochores no longer have microtubules harnessed to them, and the chromosomes decondense into the threadlike form. The vesicles of the old nuclear envelope fuse together to form patches of membrane around the chromosomes. Patch joins with patch, and eventually a new, continuous nuclear envelope separates the hereditary material from the cytoplasm. Once the nucleus is completed, telophase is completed—and so is mitosis.

## CYTOKINESIS

Cytoplasmic division, or cytokinesis, usually coincides with the period from late anaphase through telophase. For most animal cells, deposits accumulate and form a layer around microtubules at the cell midsection. A shallow, ringlike depression appears above the layer, at the cell surface (Figure 9.8). At this depression, the **cleavage furrow**, contractile microfilaments pull the plasma membrane inward and cut the cell in two. The contractile force is so strong, it can bend a fine glass needle that is inserted into a dividing cell.

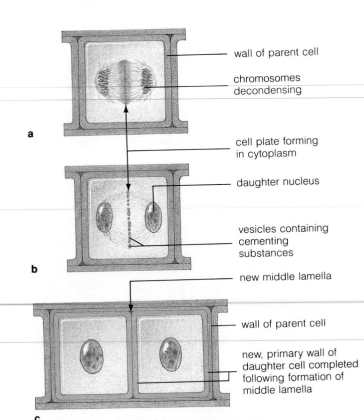

wall of parent cell

chromosomes decondensing

**a**

cell plate forming in cytoplasm

daughter nucleus

vesicles containing cementing substances

**b**

new middle lamella

wall of parent cell

new, primary wall of daughter cell completed following formation of middle lamella

**c**

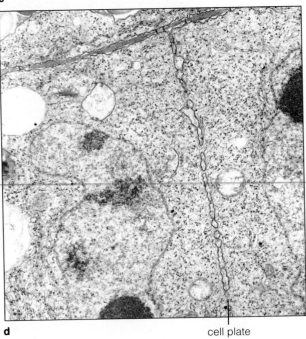

**d**

cell plate

**Figure 9.9** Cytokinesis following mitosis in a plant cell. (**a-c**) Vesicles form at the spindle equator and gradually fuse to form a cell plate. The cell plate grows outward until it reaches the parent cell wall. The vesicles contain substances that will form the middle lamella, which will cement together the primary walls of the daughter cells. (**d**) The membrane of the vesicles is used in forming the plasma membrane on both sides of the cell plate.

A different form of cytokinesis, **cell plate formation**, occurs in most land plants. (Plant cells typically have fairly rigid walls that preclude the formation of cleavage furrows.) Vesicles filled with wall-building material fuse with remnants from the spindle, forming a disklike structure (the "cell plate"). Here, cellulose deposits form a crosswall between the two daughter cells (Figure 9.9).

This concludes our introduction to mitosis. What is the take-home lesson? Simply this: You see the results of mitotic cell divisions whenever you look at *any* eukaryotic organism. Just look at your hands—and think of all the cells that are arranged in precise ways to form your palms, your thumbs, your fingers. Look at a tissue sample through a microscope and you will see individual cells, just as a special microscope might have revealed their forerunners when you were developing early on, inside your mother (Figure 9.10). And be grateful for the absolutely astonishing precision that led to their formation—because the alternatives can be terrible indeed (see the *Commentary*).

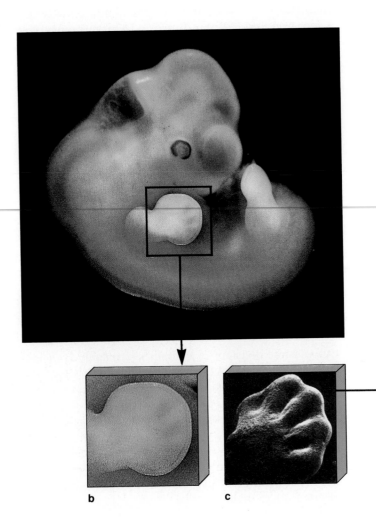

**b**          **c**

## Commentary

## Henrietta's Immortal Cells

Each human starts out as a single cell, a zygote, carrying 23 chromosomes from its mother and 23 from its father. At birth a human body has about a trillion cells, each carrying 46 copies of the chromosomes that were present in the zygote. Between conception and birth, then, a phenomenal number of cell divisions must have occurred.

Even in an adult, trillions of cells are still dividing. In the lining of the stomach, for example, cells divide every day. In the liver, cells usually do not divide—but if part of the liver is lost to injury or disease, cells start dividing and continue to do so until the damaged part is replaced. Only then does division stop.

In 1951, George and Margaret Gey of Johns Hopkins University were trying to develop a way to keep human cells dividing outside the body. Researchers could use those cells to gain insight into basic life processes. They also could use them to study diseases, including cancer, without having to experiment directly on humans.

The Geys obtained both normal and diseased human cells from local physicians, who had taken the cells from patients during normal medical procedures. But the Geys just couldn't stop the cell lines from dying out within a few weeks. Mary Kubicek, one of their assistants, was about to give up after dozens of failed attempts. When she received yet another sample of cancer cells on February 9, 1951, she expected to fail again.

Still, she took the sample and prepared it for culture. The sample was code-named HeLa, for the first two letters of the patient's first and last names.

The cells in this new sample began to divide. And divide. And divide again. By the fourth day there were so many cells that they had to be subdivided into more tubes. As months passed, the culture continued to thrive.

Unfortunately for the patient, the tumor cells inside her body were just as vigorous. Six months after she was first diagnosed as having cancer, tumor cells had spread through her body. Two months later, *Henrietta Lacks*, a young woman from Baltimore, was dead.

Although Henrietta was gone, some of her cells continued to live in the Geys' laboratory. As the first successful human cell culture, HeLa cells were soon being shipped to researchers. Recipients passed cells onto others, and soon HeLa cells were growing in laboratories all over the world. Some even traveled into space aboard the *Discoverer XVII* satellite. Every year, researchers publish hundreds of scientific papers based on work with HeLa cells.

Henrietta was only thirty-one years old when runaway cell divisions killed her in 1951. Now, more than forty years later, her legacy is still benefiting humans everywhere, in cells that are still alive and dividing, day after day after day.

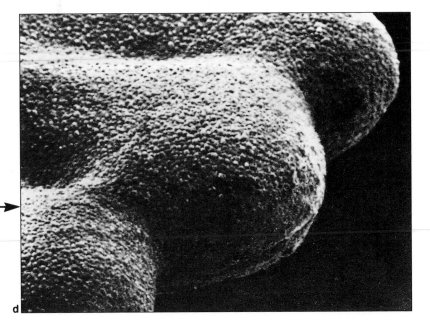

d

e

**Figure 9.10** Development of the human hand by way of cell divisions and other processes. Individual cells resulting from the mitotic cell divisions are clearly visible in (**d**). The hand is turned palm upward in (**e**).

## SUMMARY

1. Each cell of a new generation must receive the required hereditary instructions (in the form of parental DNA) and enough cytoplasmic machinery to start up its own operation.

2. Cells of multicelled eukaryotes commonly have a diploid number of chromosomes. That is, they have two of each type of chromosome characteristic of the species. The pairs of "homologous" chromosomes generally are alike in length, centromere location, and which heritable traits they deal with.

3. Eukaryotic chromosomes are duplicated during interphase (between cell divisions). Whereas each was one DNA molecule (and associated proteins), now it consists of two, which temporarily stay attached as sister chromatids.

4. Eukaryotes employ different division mechanisms that serve different functions:

   a. Mitosis is a type of nuclear division that maintains the parental number of chromosomes in each of two daughter nuclei. Thus if the parental cell's nucleus is diploid, the nucleus in each daughter cell also will be diploid.

   b. Mitosis is the basis of bodily growth for multicelled eukaryotes. It also is the basis of asexual reproduction for some eukaryotes.

   c. Meiosis is a type of nuclear division that reduces the parental chromosome number by half (to the haploid number) in each of four daughter nuclei.

   d. Meiosis is the basis of sexual reproduction. It is a necessary precursor to the formation of gametes.

   e. For most organisms, actual cytoplasmic division, or cytokinesis, occurs toward the end of nuclear division or at some point afterward.

5. A cell destined to divide by mitosis spends about 95 percent of the cell cycle in interphase, a period between mitotic divisions. During interphase, the cell increases in mass, doubles its number of cytoplasmic components, and duplicates its chromosomes.

6. Mitosis proceeds through five continuous stages:

   a. Prophase. Duplicated, threadlike chromosomes start to condense. New microtubules start to assemble in organized arrays near the nucleus; they will form a spindle apparatus.

   b. Metaphase. The nuclear envelope breaks up during the transition to metaphase. The kinetochores of chromosomes are free to capture microtubules of the forming spindle. Sister chromatids of each chromosome become oriented toward and attached to opposite spindle poles. All chromosomes are moved to and become aligned at the equator of the fully formed spindle.

   c. Anaphase. The two sister chromatids of each chromosome separate. Both are now independent chromosomes and they move to opposite poles.

   d. Telophase. Chromosomes decondense to the threadlike form. A new nuclear envelope forms around the two parcels of chromosomes. Mitosis is completed.

### Review Questions

1. Define the two types of nuclear division mechanisms that occur in eukaryotes. What is cytokinesis? *138–139*

2. Define somatic cell and germ cell. Which type of cell can undergo meiosis? *138*

3. What is a chromosome? What is a chromosome called in its unduplicated state? In its duplicated state (that is, with two sister chromatids)? *138*

4. Define homologous chromosomes. Do homologous chromosomes pair during mitosis, meiosis, or both? *139*

5. Describe the spindle apparatus and its general function in nuclear division processes. *141*

6. Name the four main stages of mitosis, and describe the main features of each stage. *140–145*

### Self-Quiz *(Answers in Appendix IV)*

1. Eukaryotic DNA is distributed to daughter cells by _____ or _____ , both of which are nuclear division mechanisms.

2. Each kind of organism contains a characteristic number of _____ in each cell; each of those structures is composed of a _____ molecule with its associated proteins.

3. A pair of chromosomes that are similar in length, shape, and the traits they govern are called _____ .
   a. diploid chromosomes    c. homologous chromosomes
   b. mitotic chromosomes    d. germ chromosomes

4. Somatic cells of multicelled eukaryotic organisms usually have a _____ number of chromosomes, whereas gametes have a _____ number.
   a. haploid; haploid    c. diploid; diploid
   b. haploid; diploid    d. diploid; haploid

5. Interphase is the stage when _____.
   a. nothing occurs
   b. a germ cell forms its spindle apparatus
   c. a cell grows and duplicates its DNA
   d. cytokinesis occurs

6. Following mitosis, a daughter cell will end up with genetic instructions that are _____ and with a chromosome number that is _____ the parent cell.
   a. identical to the parent cell's; the same as
   b. identical to the parent cell's; one-half
   c. rearranged; the same as
   d. rearranged; one-half

7. Cytokinesis is a term that describes _____.
   a. doubling the chromosome number
   b. nuclear division
   c. cytoplasmic division
   d. reducing the chromosome number

8. During interphase, a cell _____.
   a. grows
   b. doubles the number of cytoplasmic components
   c. duplicates its chromosomes
   d. all of the above

9. All of the following are stages of mitosis except _____.
   a. prophase
   b. interphase
   c. metaphase
   d. anaphase

10. A duplicated chromosome has _____.
    a. one chromatid
    b. two chromatids
    c. three chromatids
    d. four chromatids

11. Match each stage of mitosis with the following key events.
    _____ metaphase
    _____ prophase
    _____ telophase
    _____ anaphase
    a. sister chromatids of each chromosome separate and move to opposite poles
    b. threadlike chromosomes start to condense
    c. chromosomes decondense, daughter nuclei re-form
    d. spindle is fully formed and all chromosomes are aligned at its equator

## Selected Key Terms

anaphase *144*
cell cycle *140*
cell plate formation *146*
centromere *138*
chromosome *138*
cleavage furrow *145*
cytokinesis *138*
diploid *139*
haploid *139*
homologous chromosome *139*
interphase *140*
kinetochore *144*
life cycle *137*
meiosis *138*
metaphase *143*
mitosis *138*
prophase *142*
reproduction *137*
sister chromatid *138*
spindle apparatus *141*
telophase *145*

## Readings

Alberts, B., et al. 1989. *Molecular Biology of the Cell.* Second edition. New York: Garland Publishing.

John, B., and K. Lewis. 1980. *Somatic Cell Division.* Burlington, North Carolina: Carolina Biological Supply.

Prescott, D. 1988. *Cells: Principles of Molecular Structure and Function.* Boston: Jones and Bartlett. Chapter 7.

Smith-Klein, C., and V. Kish. 1988. *Principles of Cell Biology.* New York: Harper & Row.

## *Octopus Sex and Other Stories*

The couple clearly are interested in each other. He caresses her first with one tentacle, then another—then another, another, and another. She reciprocates. This goes on for hours; a tentacled hug here, an enveloping squeeze there. Finally the male reaches deftly under his mantle and removes a packet of sperm, which he inserts into the cavity under the female's mantle. For every one of his sperm that successfully performs its function, a fertilized egg can develop into a new octopus.

For the octopus, sex is an occasional event, preceded by a courtship ritual involving intermingled tentacles. Sex for the slipper limpet is a lifelong group activity. Slipper limpets are marine animals, relatives of land snails. Before a slipper limpet becomes a sexually mature adult, it passes through a free-living larval stage. When the time comes for the larva to become transformed into the adult form, it settles onto a substrate. If it settles down all by itself, the slipper limpet will develop into a female. If another larva settles down on the female, it will develop into a male. If another larva settles down on that male it, too, will become a male. Adult slipper limpets almost always live in such piles, with the one on the bottom invariably being female (Figure 10.1a). All of the males continually contribute sperm to the task of sexual reproduction. When the one female finally dies, the male at the bottom of the pile becomes transformed into a female, and so it goes.

Strawberry plants, too, engage in sexual reproduction, but a strawberry plant also can do something you could not even begin to do except in your wildest imagination. It can reproduce all by itself. Through mitosis, aboveground stems called runners grow outward from the plant—and brand new plants sprout up along the runners. Similarly, the entire body of a flatworm can split into two roughly equivalent parts—then, through mitosis, each part can grow into a whole flatworm.

Or consider the aphid, an insect common to gardens. All summer long, nearly every aphid is female, and nearly every one is reproducing asexually. Inside her reproductive organs, unfertilized egg cells are chemically stimulated to develop into female embryos, which develop into young aphids—which are born alive (Figure 10.1b). All of this occurs so rapidly that young aphids are born pregnant—they carry embryos inside! As autumn approaches, males are produced and they do their part in sexual reproduction. Still, females that survive the winter can do without males, and come summer they begin another round of producing offspring all by themselves.

**Figure 10.1** (**a**) Limpets busily perpetuating the species by way of sexual reproduction. (**b**) Live birth of an aphid, a type of insect that shows variations on the sexual theme.

a

Regardless of how—or how often—it takes place among plants and animals, reproduction involves predictable events. For asexually reproducing organisms, chromosome duplications and mitotic cell divisions provide each new cell with all required genetic instructions. For sexually reproducing organisms also, chromosome duplications precede cell divisions by which gametes (sex cells) are produced. In this case, however, *two* gametes fuse at fertilization, bearing genetic instructions from *two* parental cells. Mitosis won't work here. Only through *meiosis* will each gamete end up with half the parental number of chromosomes. Only then will a fertilized egg end up with the proper chromosome number, no more, no less. *Meiosis and fertilization, then, are the unifying theme of sexual reproduction, regardless of the species.* Intermingled tentacles and communal sex and pregnant newborns are simply variations on that theme.

b

## KEY CONCEPTS

**1.** Sexual reproduction of multicelled plants and animals depends on these events: meiosis, gamete formation, and fertilization. In plants, other reproductive events, including the formation of spores, may occur between meiosis and gamete formation.

**2.** The somatic cells of most animals and many plants have a diploid number of chromosomes (*two* of each type characteristic of the species), half of which are from one parent and half from another parent organism. In germ cells, which are a subpopulation of cells set aside for sexual reproduction, every two chromosomes that are alike (homologues) will pair with each other during meiosis.

**3.** Meiosis, a nuclear division process in germ cells only, reduces the diploid number of chromosomes by half for the forthcoming gametes. Each gamete produced is haploid; it has only *one* of each type of chromosome that was present in the germ cell. The union of two gametes at fertilization restores the diploid number in the new individual.

**4.** During meiosis, homologues exchange segments (through crossing over and recombination), then shufflings occur that will give rise to different mixes of chromosomes from two parent organisms in gametes.

## ON ASEXUAL AND SEXUAL REPRODUCTION

Strawberry plants, aphids, and transversely dividing flatworms give us a sense of how different organisms might rely, one way or another, on **asexual reproduction**. By this process, *one* parent always passes on a duplicate of all of its genes to offspring. "Genes" are specific portions of a DNA molecule, and they contain the inherited instructions for producing or influencing a trait in offspring. This means that, rare mutations aside, asexually produced offspring can only be genetically identical copies, or clones, of the parent.

Inheritance is much more interesting with sexual reproduction. Commonly, **sexual reproduction** involves two parents, each with two genes for nearly every trait. Both parents pass on one of each gene to offspring by way of meiosis, gamete formation, and fertilization. Thus the first cell of a new individual inherits two genes for every trait—one from each parent.

If the instructions in every pair of genes were identical down to the last detail, then sexual reproduction would produce clones, also. Just imagine—you, everyone you know, every member of the entire human population would be clones and might all end up looking exactly alike.

But it happens that the molecular structure of a gene can change; this is what we mean by "mutation." As a result of past mutations, different individuals of a species might be carrying different molecular forms of a gene that "say" slightly different things about how a trait will be expressed in offspring. Whenever there are different molecular forms of the same gene, each form is called an **allele**. Admittedly, this is not a word that is

easy to warm up to. It may help to know that it is short for an even worse word—allelomorph, after the Greek *allos* (meaning other) and *morphē* (meaning form).

As Figure 10.2 indicates, a gene that affects how the human chin develops can vary this way. One molecular form of that gene says "put a dimple in it" and another says "no dimple." Different genes govern thousands of different traits. And this brings us to a key reason why individuals of any sexually reproducing species don't all look alike. *Sexual reproduction puts together new combinations of alleles in offspring.*

This chapter provides us with a closer look at meiosis, the foundation for sexual reproduction. More importantly, it starts us thinking about some far-reaching consequences of the gene shufflings possible with sexual reproduction. New gene combinations among offspring lead to variations in their physical and behavioral traits. *Such variation is acted upon by agents of natural selection—and so it is a basis of evolutionary change.*

**Figure 10.2** (**a**) The chin fissure, a heritable trait arising from a rather uncommon form of a gene. Actor Kirk Douglas received a gene that influences this trait from each of his parents. One gene called for a chin fissure and the other didn't, but one is all it takes in this case. (**b**) This photograph shows what Mr. Douglas' chin might have looked like if he had inherited two ordinary forms of the gene instead.

Through meiosis and fertilization, old gene combinations are broken up and new ones are put together. The immediate consequence is variation in the physical and behavioral traits of offspring. The long-term consequence can be evolutionary change.

a                                                                 b

**Figure 10.3** Examples of the location of germ cells that give rise to sperm and eggs.

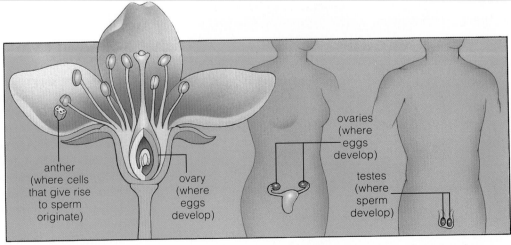

anther (where cells that give rise to sperm originate)

ovary (where eggs develop)

ovaries (where eggs develop)

testes (where sperm develop)

flowering plant                    human female                    human male

# OVERVIEW OF MEIOSIS

## Think "Homologues"

The preceding chapter made a few points about meiosis, and now we can put those points together in the following picture. Meiosis is a nuclear division mechanism. It sorts out the DNA present in a cell nucleus into parcels that can be distributed to forthcoming daughter cells. Meiosis happens only in *germ cells,* the cell lineage destined to give rise to the gametes (sperm or eggs) used in sexual reproduction. Germ cells develop in a variety of reproductive structures and organs (Figure 10.3). Like somatic cells, germ cells commonly have a diploid number of *chromosomes,* these being structures composed of DNA and proteins.

"Diploid" means there are two chromosomes of each type in a cell. The two are **homologous chromosomes**. Generally, homologues have the same length, the same centromere location, and the same genes—and they line up with each other during meiosis. Only the sex chromosomes, designated X and Y in humans and many other organisms, differ in form and in which genes they carry—but they still function as homologues during meiosis.

Meiosis reduces the diploid number (2*n*) by half, to the "haploid" number (*n*). And not just any half: *Each gamete ends up with one member of each pair of homologous chromosomes.* To give an example, the diploid number for humans is 46—that is, 23 + 23 homologues. A human gamete ends up with 23 chromosomes, one of each type.

## Overview of the Two Divisions

Meiosis resembles mitosis in some respects, even though the outcome is different. While a germ cell is still in interphase, each chromosome is duplicated by a process called DNA replication. The chromosome hangs onto its duplicate at a small, constricted region (the centromere), and as long as the two remain attached, they are called **sister chromatids**:

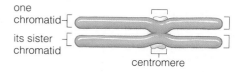

one
chromatid—[
its sister —[
chromatid
centromere

As in mitosis, microtubules of a spindle apparatus harness each chromosome and take part in its movement during nuclear division. Unlike mitosis, however, there are *two divisions,* which ultimately lead to the formation of four nuclei. The two different divisions are called meiosis I and II:

| DNA duplication during interphase | MEIOSIS I | No DNA duplication between divisions | MEIOSIS II |
|---|---|---|---|
| | Prophase I | | Prophase II |
| | Metaphase I | | Metaphase II |
| | Anaphase I | | Anaphase II |
| | Telophase I | | Telophase II |

During meiosis I, each duplicated chromosome lines up with its partner, *homologue to homologue,* then the partners are separated from each other. Here we show just one pair of homologous chromosomes, but the same thing happens to all pairs in the nucleus:

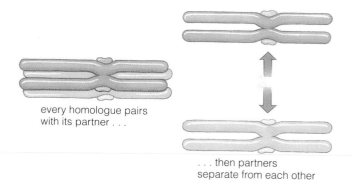

every homologue pairs
with its partner . . .

. . . then partners
separate from each other

Cytokinesis typically follows. At this point each daughter nucleus contains only one of each type of chromosome. But each chromosome is still in the duplicated state, consisting of two sister chromatids.

During meiosis II, *the sister chromatids of each chromosome are separated from each other:*

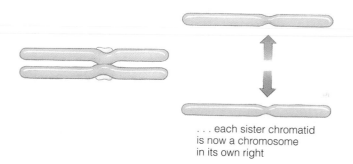

. . . each sister chromatid
is now a chromosome
in its own right

Cytokinesis typically follows this separation, also. Putting the preceding example in perspective, every chromosome in a diploid germ cell was duplicated before the onset of meiosis. Two nuclear divisions and two cytoplasmic divisions later, the outcome was four cells—that is, each with a haploid number of unduplicated chromosomes.

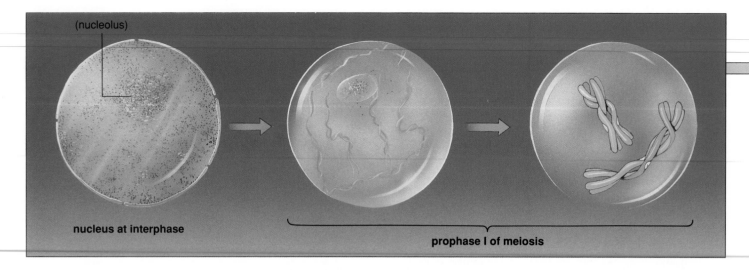

(nucleolus)

nucleus at interphase

prophase I of meiosis

Early in Prophase I, chromosomes are like thin threads, with each "thread" being two sister chromatids in close alignment. Shown here, two chromosomes from the male parent (green) and two from the female parent (pink).

Before prophase I ends, each chromosome pairs with its homologue; and *nonsister* chromatids undergo crossing over.

**Figure 10.4** Prophase I of meiosis, when crossing over occurs between homologous chromosomes.

## STAGES OF MEIOSIS

### Prophase I Activities

The first stage of meiosis, **prophase I**, is a time of major gene shufflings between homologous chromosomes. Homologues begin to pair at the onset of prophase I. At this stage a duplicated chromosome looks like a very long, thin thread, so closely are its two sister chromatids aligned with each other.

Each threadlike chromosome and its homologue are drawn together during a process called *synapsis*. It is as if they become stitched point by point along their entire length, with little space between them. (The X and Y chromosomes pair at one end only.)

The intimate parallel arrangement between homologues favors **crossing over**. By this mechanism, *nonsister* chromatids break at one or more sites along their length and exchange corresponding segments—that is, genes—at the breakage points.

Figure 10.4 shows only a single crossover. On the average, between two and three crossovers are thought to occur between each pair of homologues in human germ cells undergoing meiosis.

Gene-swapping would be rather pointless if each type of gene never varied from one chromosome to the next. But remember a gene can come in alternative forms—alleles. You can safely bet that all the genes running down the length of one chromosome will not be an identical match with those on the homologue. With each crossover, then, there is a chance that homologues may be swapping *different* instructions for some traits.

We will look at the mechanism of crossing over in later chapters. For now it is enough to know that crossing over leads to **genetic recombination**, which in turn leads to variation in the traits of offspring.

**Crossing over is an event by which old combinations of alleles in a chromosome are broken up and new ones put together during meiosis.**

After segments have been exchanged, all four chromatids thicken and can spread apart somewhat—but they remain joined at a few places where nonsister chromatids extend across each other (Figure 10.5). Each crosslike, temporary attachment between two nonsister chromatids is a *chiasma* (plural, chiasmata). Such attachments will play a role in aligning the homologues during metaphase I.

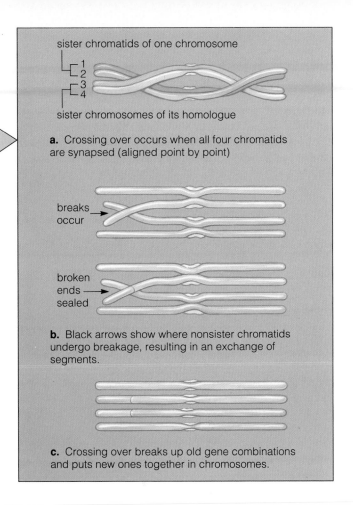

**a.** Crossing over occurs when all four chromatids are synapsed (aligned point by point)

breaks occur →

broken ends → sealed

**b.** Black arrows show where nonsister chromatids undergo breakage, resulting in an exchange of segments.

**c.** Crossing over breaks up old gene combinations and puts new ones together in chromosomes.

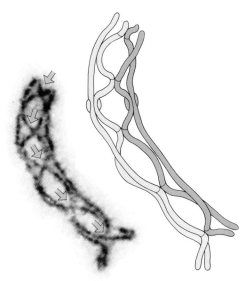

**Figure 10.5** Two homologous chromosomes held together by chiasmata (gold arrows). A chiasma is not the same thing as a crossover. Although it indicates that crossing over occurred earlier, it is not always located at the site of breakage and exchange. The reason is that chiasmata are not stationary attachments; they are forced toward the chromatid tips as prophase I draws to a close.

## Separating the Homologues

**Metaphase I**, the second stage of meiosis I, is a time of major shufflings of whole chromosomes, before their distribution into daughter nuclei. Suppose the shufflings are proceeding right now in one of your germ cells. We can call that cell's homologous chromosomes "maternal" and "paternal." (This is a short way of saying that one of each type was inherited from your mother and their homologues were inherited from your father.) Figure 10.6 shows how many maternal and paternal chromosomes are present in that cell, so you can get an idea of how complicated it would be to track their movements. To keep things simple, let's imagine that we are tracking only *two* pairs of homologous chromosomes.

During prophase I, microtubules had started to assemble outside the nucleus of the germ cell. As in mitosis, they ultimately will form a spindle apparatus (page 141). During the transition to metaphase, the nuclear envelope broke apart, so now the microtubules are free to interact with the chromosomes. The chromosomes become attached to microtubules extending from both spindle poles, then they are moved into position at the spindle equator. At **metaphase I**, the spindle is fully formed and all chromosomes are aligned at its equator. Then, during **anaphase I**, each homologue moves away

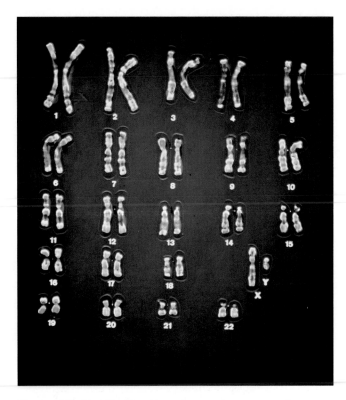

**Figure 10.6** The 23 pairs of homologous chromosomes from a human male.

from its partner and the two head toward opposite spindle poles, as shown in Figure 10.7.

Are all the maternal chromosomes destined to move to one pole and the paternal chromosomes to the other? Maybe, but probably not. The positioning of each pair of homologues at the spindle equator and their subsequent direction of movement are random events. *It doesn't matter which partner moves to which pole.*

Consider Figure 10.8, which shows how just three pairs of homologues can be shuffled into any one of four possible positions at metaphase I. In this case, $2^3$ or 8 combinations of maternal and paternal chromosomes are possible for the forthcoming gametes. A human germ cell has 23 pairs of homologous chromosomes, not just three. So $2^{23}$ or *8,388,608 combinations* of maternal and paternal chromosomes are possible every time a germ cell gives rise to sperm or eggs! (Are you beginning to get an idea of why such splendid mixes of inherited traits show up even in the same family?)

Typically, anaphase I proceeds to telophase I and **interkinesis**, which often are fleeting stages before the final nuclear division. There is no DNA duplication between the two meiotic divisions. But remember, each chromosome was duplicated earlier (during interphase). And it is still in the duplicated form when meiosis II gets under way.

**One or the other member of each pair of homologous chromosomes may end up at a given spindle pole during meiosis I. It doesn't matter which one arrives at which pole.**

**This means each pair of homologues is assorted into gametes *independently* of the other pairs present in the cell.**

**As a result of independent assortment, gametes end up with different mixes of maternal and paternal chromosomes.**

**Figure 10.7** Meiosis: the nuclear division mechanism by which the parental number of chromosomes is reduced by half (to the haploid number) for forthcoming gametes. Only two pairs of homologous chromosomes are shown. The green ones are derived from one parent, and the pink ones are their homologues from the other parent.

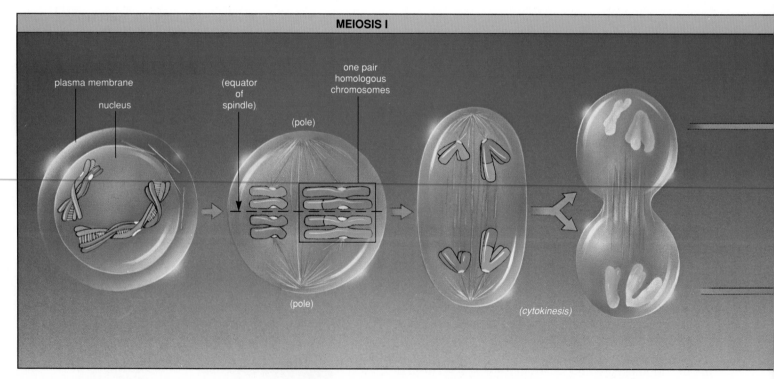

**MEIOSIS I**

plasma membrane
nucleus
(equator of spindle)
one pair homologous chromosomes
(pole)
(pole)
(cytokinesis)

**Prophase I**
Each chromosome condenses, then pairs with its homologue. Crossing over and recombination occur.

**Metaphase I**
Spindle apparatus forms, nuclear envelope (not shown) breaks down during transition to metaphase I. Homologous pairs align randomly at spindle equator.

**Anaphase I**
Each homologue is separated from its partner, and the two are moved to opposite poles.

**Telophase I**
A haploid number of chromosomes (still duplicated) ends up at each pole.

## Separating the Sister Chromatids

Meiosis II has one overriding function: separation of the two sister chromatids of each chromosome. Here again, microtubules of a spindle apparatus have roles in the separation.

At metaphase II, the chromosomes attach to the microtubules of the forming spindle and are moved to its equator. Each duplicated chromosome becomes aligned at the equator. At anaphase II, each is split, and its (formerly) sister chromatids are now unduplicated chromosomes in their own right.

Each spindle pole is the destination of half the parental number of chromosomes. But that haploid number includes one of each type of chromosome characteristic of the species. During telophase II, new nuclear membranes form around the chromosomes after they have become clustered at the two poles. Two nuclei are thus formed. Meiosis is completed.

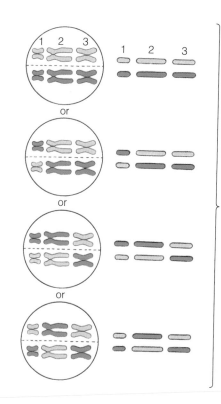

eight different haploid combinations of the maternal and paternal chromosomes are possible in the forthcoming gametes

**Figure 10.8** (*Right*) Possible outcomes of the random alignment of three pairs of homologous chromosomes at metaphase I of meiosis. The three types of chromosomes are labeled 1, 2, and 3. Maternal chromosomes are pink; paternal ones are green.

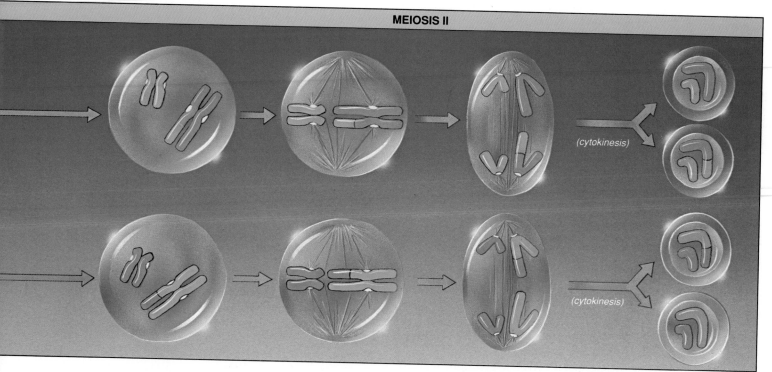

**MEIOSIS II**

**Prophase II**
There is no DNA replication between divisions.
Sister chromatids of each chromosome are still attached at the centromere.

**Metaphase II**
Each chromosome is aligned at the spindle equator.

**Anaphase II**
Each chromosome splits; what were once sister chromatids are now chromosomes in their own right and are moved to opposite poles.

**Telophase II**
Four daughter nuclei form. Following cytokinesis, each gamete has a haploid number of chromosomes, all in the unduplicated state.

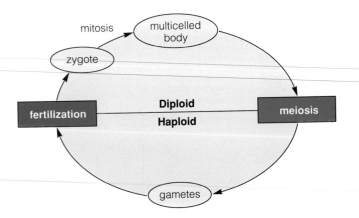

**Figure 10.9** Generalized life cycle for animals.

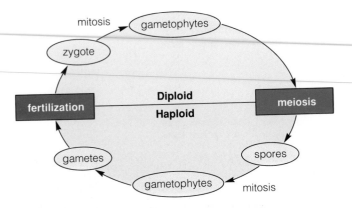

**Figure 10.10** Generalized life cycle for most plants. During the diploid stage of the life cycle, a *sporophyte* ("spore-producing plant body") forms. During the haploid stage, a *gametophyte* ("gamete-producing plant body") forms. A pine tree is an example of a sporophyte. Gametophytes develop in its pine cones (page 398).

## MEIOSIS AND THE LIFE CYCLES

Meiosis precedes the formation of mature gametes during the life cycle of all sexually reproducing organisms, as Figures 10.9 and 10.10 suggest. In later chapters, we will consider the details of some of those cycles, including the one for humans. Here we simply will make a few points that can help you keep the details in perspective.

### Gamete Formation

Although all sexually reproducing organisms produce gametes, keep in mind that the gametes do not all look alike. For example, human sperm have one tail, opossum sperm have two, and roundworm sperm have none. Crayfish sperm look like pinwheels. Most eggs are microscopic, yet ostrich eggs (tucked inside a protective shell) are as large as a softball. In outward appearance, the gametes of plants are not recognizably like the gametes of mammals.

**Gamete Formation in Animals.** Animal life cycles typically proceed from meiosis to gamete formation, fertilization, then growth by way of mitosis. Formation of the male gametes, called spermatogenesis, is shown in Figure 10.9. Inside the reproductive tract of *male* animals, a diploid germ cell increases in size. The resulting large, immature cell (a primary spermatocyte) undergoes meiosis. Following cytokinesis, the four resulting cells eventually develop into four haploid spermatids (Figure 10.11). The spermatids change in form, develop a tail, and become **sperm**, the mature male gametes.

Inside the reproductive tract of *female* animals, meiosis and gamete formation is called oogenesis, and it differs from what goes on in males in two important features. Compared to a primary spermatocyte, many more cytoplasmic components accumulate in the female germ cell, the primary oocyte. Also, the cells formed after meiosis differ in size and function (Figure 10.12).

Following meiosis I, one cell (the secondary oocyte) receives nearly all the cytoplasm. The other, much smaller cell is the first "polar body." Both cells may undergo meiosis II, and the outcome is one large cell and three extremely small polar bodies.

The large cell develops into the mature female gamete, or **ovum.** This also is known rather loosely as "the egg." The polar bodies do not function as gametes. In effect, they serve as dumping grounds for three sets of parental chromosomes, so that the egg ends up with the necessary haploid number. Besides this, polar bodies do not receive much cytoplasm, so they do not get much in the way of nutrients and metabolic machinery. Thus they are destined to degenerate.

**Gamete Formation in Plants.** For pine trees, roses, and other familiar plants, the life cycle proceeds in the same general way. As Figure 10.10 indicates, however, some additional events occur between meiosis and gamete formation. Among other things, plants form **spores**, which are haploid cells that are resistant to dry periods or other adverse environmental conditions. When favorable conditions return, spores germinate and develop into some kind of haploid body or structure that will go on to produce the gametes. In short, the life cycles of flowering plants include both spore-producing bodies and gamete-producing bodies.

### More Gene Shufflings at Fertilization

The diploid number of chromosomes is restored at **fertilization,** the fusion of nuclei of two gametes in the zygote (the first cell of a new multicelled individual). Here we see why meiosis is so necessary. If gametes ended up diploid, fertilization would double the number of chromosomes in cells of every new generation.

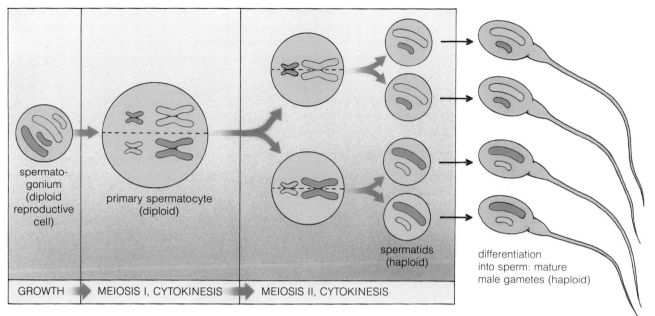

**Figure 10.11** Generalized picture of spermatogenesis in male animals. (For the sake of clarity, the nuclear envelopes are not shown in Figures 10.11 through 10.14.)

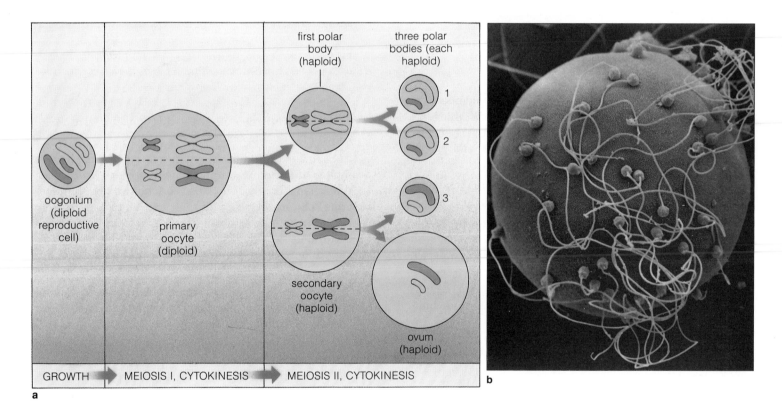

a

b

Like meiosis, fertilization also contributes to variation in the traits of offspring. Think about the possibilities for humans. First, genes are shuffled during prophase I, when each chromosome takes part in two or three crossovers, on the average. Second, random alignments at metaphase I lead to one of 8,388,608 possible combinations of maternal and paternal chromosomes in

**Figure 10.12** Generalized picture of oogenesis in female animals. This sketch is not drawn to the same scale as Figure 10.11. A primary oocyte is *much* larger than a primary spermatocyte, as suggested by the scanning electron micrograph of the egg and sperm of a clam to the right. Also, the polar bodies are extremely small compared to an ovum, as shown in Figure 43.11.

each gamete. Third, of all the genetically diverse male and female gametes that are produced, *which* two will get together is a matter of chance. As you can see, the sheer number of new combinations brought together at fertilization is staggering!

## MEIOSIS COMPARED WITH MITOSIS

In this unit, our main focus has been on mitosis and meiosis—two nuclear division mechanisms used in the reproduction of eukaryotic cells. Mitosis underlies asexual reproduction of single-celled eukaryotes as well as growth of multicelled eukaryotes. Meiosis occurs only in germ cells, which give rise to the haploid gametes used in sexual reproduction.

The major difference between them is this: Mitotic cell division produces clones—genetically identical copies of the parent cell. Meiosis and fertilization give rise to novel combinations of alleles in offspring which, as a consequence, vary from the parents and one another in the details of their traits. As we have seen, three events are responsible for the variation:

**1.** Crossing over and genetic recombination occur during prophase I of meiosis.

**2.** During anaphase I of meiosis, the two members of each pair of homologous chromosomes assort independently of the other pairs. Thus the forthcoming gametes can end up with different mixes of maternal and paternal chromosomes.

**3.** Fertilization is a chance mix of different combinations of alleles from two different gametes (Figure 10.13).

In later chapters, we will see how the variation in traits made possible by meiosis and fertilization is a testing ground for agents of natural selection. As such, both contribute to the evolution of sexually reproducing populations.

## SUMMARY

1. The life cycle of multicelled plants and animals generally includes meiosis and gamete formation, fertilization, and growth by way of cell divisions.

a. In animals, meiosis produces haploid gametes (sperm in males, eggs in females). Fusion of a sperm nucleus and an egg nucleus at fertilization produces a diploid cell (zygote), which develops into the multicelled individual by way of mitosis and cytokinesis.

b. In plants, meiosis gives rise to spores, which

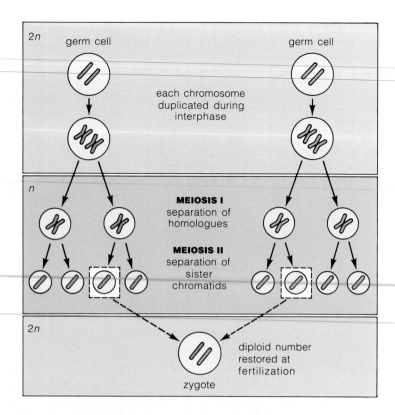

**Figure 10.13** How fertilization restores a diploid (2n) number of chromosomes that has been reduced by half during meiosis.

eventually undergo mitotic cell divisions that give rise to gamete-producing bodies.

2. Sexually reproducing organisms commonly have a diploid number of chromosomes (2n), or two of each type characteristic of the species. The two are "homologous" chromosomes, with the same length, centromere location, and gene sequence. Homologues pair with each other during meiosis. The X and Y chromosomes vary in length, shape, and which genes they carry, but they still pair as homologues.

3. Before meiosis, all chromosomes in a germ cell are duplicated. The duplicates remain attached (as sister chromatids) at the centromere.

4. Meiosis consists of two consecutive divisions that sort out the chromosomes in a germ cell. In meiosis I, each chromosome pairs with and then separates from its homologue. In meiosis II, the sister chromatids of each chromosome separate from each other. In both cases, microtubules of a spindle apparatus help move the chromosomes.

5. The following key events occur during meiosis I:

a. At prophase I, each chromosome comes into point-by-point alignment with its homologue. Crossing over occurs between nonsister chromatids. Crossing over breaks up old combinations of alleles and puts together new ones in the chromosomes. This genetic recombination leads to variation in traits among offspring.

b. Late in metaphase I, all pairs of homologous chromosomes are aligned at the spindle equator.

c. At anaphase I, each maternal chromosome is separated from its paternal homologue and moved to the opposite spindle pole.

6. The following key events occur during meiosis II:

a. At metaphase II, all chromosomes are moved to the spindle equator.

b. At anaphase II, the sister chromatids of each chromosome are separated for movement to opposite poles. Once separated, they are chromosomes in their own right. And each is in the unduplicated state.

7. Following cytokinesis (cytoplasmic division), there are four haploid cells, one or all of which may function as gametes (or give rise to gametes, in the case of spore-producing plants).

Figure 10.14 summarizes the similarities and differences between mitosis and meiosis.

## Review Questions

1. Define sexual reproduction. How does it differ from asexual reproduction? 151

2. Refer to Table 9.2, which gives the diploid number of chromosomes in the body cells of a few organisms. What would be the *haploid* number for the gametes of humans? For the garden pea? 153

3. Suppose the diploid cells of an organism have four pairs of homologous chromosomes, designated AA, BB, CC, and DD. How would its haploid set of chromosomes be designated? 153

4. When, and in which type of cells, does meiosis occur? 153

5. Define meiosis and characterize its main stages. In what respects is meiosis like mitosis? In what respects is it unique? 153–157, 160

6. Does crossing over occur during mitosis, meiosis, or both? At what stage of nuclear division does it occur, and what is its significance? 154

7. Outline the steps involved in spermatogenesis and oogenesis. 158

## Self-Quiz (Answers in Appendix IV)

1. The somatic (body) cells of sexually reproducing organisms commonly have a _____ number of chromosomes, or _____ of each type characteristic of that species.

2. Two homologous chromosomes generally contain the same _____.
   a. genes in reverse order
   b. alleles in reverse order
   c. alleles in the same order
   d. none of the above

3. Prior to meiosis, all the chromosomes in a diploid germ cell are _____.
   a. paired
   b. randomly mixed
   c. duplicated
   d. separated

4. Crossing over _____.
   a. alters the chromosome alignments at metaphase
   b. occurs between sperm DNA and egg DNA at fertilization
   c. leads to genetic recombination
   d. occurs only rarely

5. Because of the _____ alignment of homologous chromosomes at metaphase, gametes can end up with _____ mixes of maternal and paternal chromosomes.
   a. unvarying; different
   b. unvarying; duplicate
   c. random; duplicate
   d. random; different

6. Variation in the traits of offspring is increased by the mix of _____ allele combinations from two _____ gametes at fertilization.
   a. similar; similar
   b. different; similar
   c. different; different
   d. similar; different

7. Prior to the meiotic divisions, duplicated chromosomes remain attached as sister _____ at the area of the chromosome called the _____.
   a. chromosomes; centromere
   b. chromatids; centriole
   c. chromosomes; centriole
   d. chromatids; centromere

8. Following meiosis and cytokinesis, there are _____ haploid cells, one or all of which may function as _____.
   a. two; body cells
   b. two; gametes
   c. four; gametes
   d. four; body cells

9. The net result of meiosis is that the _____ chromosome number is _____.
   a. diploid; doubled
   b. diploid; halved
   c. haploid; doubled
   d. haploid; halved

## Selected Key Terms

allele 152
anaphase I 155
anaphase II 157
asexual
   reproduction 151
crossing over 154
fertilization 158
gene 151
genetic
   recombination 154

germ cell 153
homologous
   chromosome 153
interkinesis 156
metaphase I 155
oogenesis 158
ovum 158
prophase I 154

sexual
   reproduction 151
sister chromatid 153
sperm 158
spermatogenesis 158
telophase I 156
telophase II 157

## Readings

Cummings, M. 1988. *Human Heredity: Principles and Issues*. New York: West.

Strickberger, M. 1985. *Genetics*. Third edition. New York: Macmillan. Contains excellent introduction to chromosomes and meiosis.

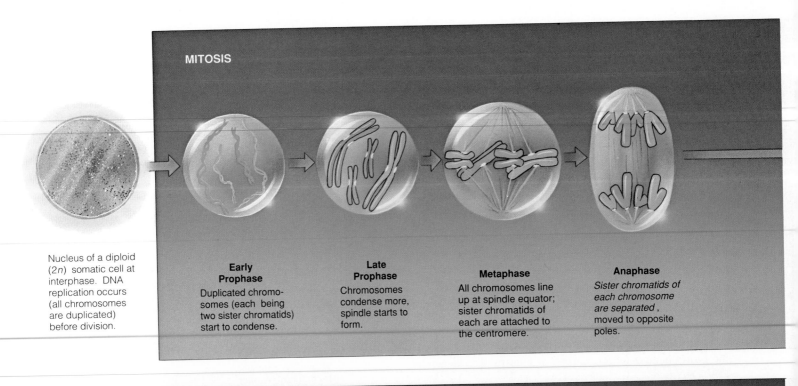

**MITOSIS**

Nucleus of a diploid (2n) somatic cell at interphase. DNA replication occurs (all chromosomes are duplicated) before division.

**Early Prophase**

Duplicated chromosomes (each being two sister chromatids) start to condense.

**Late Prophase**

Chromosomes condense more, spindle starts to form.

**Metaphase**

All chromosomes line up at spindle equator; sister chromatids of each are attached to the centromere.

**Anaphase**

*Sister chromatids of each chromosome are separated*, moved to opposite poles.

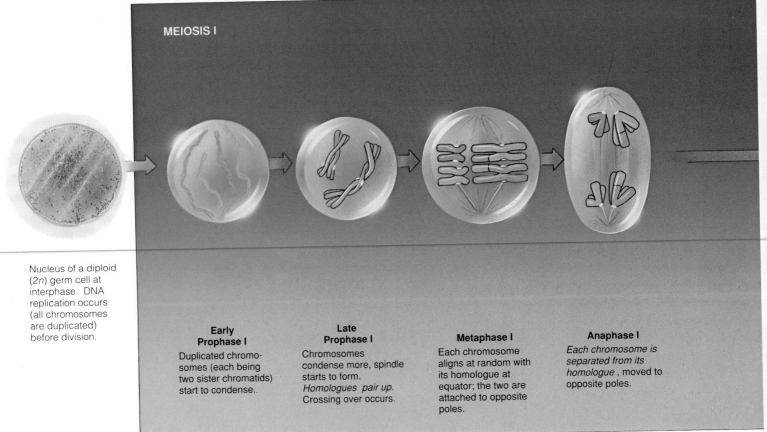

**MEIOSIS I**

Nucleus of a diploid (2n) germ cell at interphase. DNA replication occurs (all chromosomes are duplicated) before division.

**Early Prophase I**

Duplicated chromosomes (each being two sister chromatids) start to condense..

**Late Prophase I**

Chromosomes condense more, spindle starts to form. *Homologues pair up.* Crossing over occurs.

**Metaphase I**

Each chromosome aligns at random with its homologue at equator; the two are attached to opposite poles.

**Anaphase I**

*Each chromosome is separated from its homologue*, moved to opposite poles.

**Figure 10.14** Summary of mitosis and meiosis, using a diploid (2n) animal cell as the example. The diagram is arranged to help you compare the similarities and differences between the two division mechanisms. (Chromosomes derived from the male parent are green; their homologues from the female parent are pink.)

**Telophase**
Each daughter nucleus is *diploid* (2*n*), with pairs of homologous chromosomes. Each chromosome is in the unduplicated state.

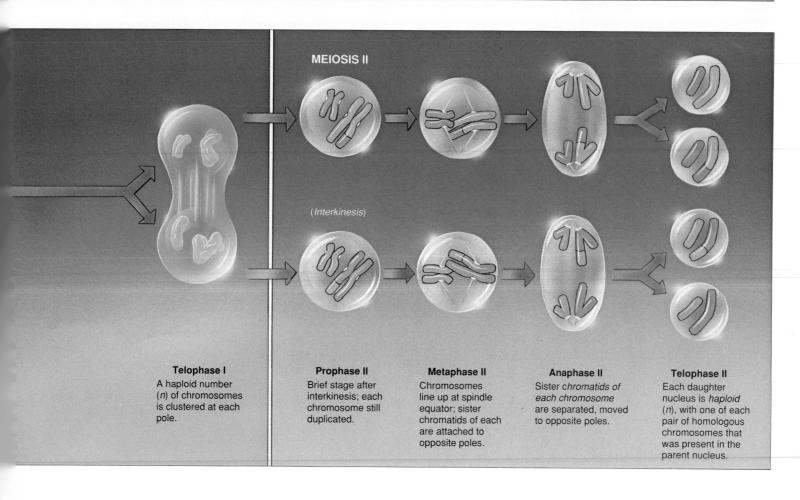

**MEIOSIS II**

*(Interkinesis)*

**Telophase I**
A haploid number (*n*) of chromosomes is clustered at each pole.

**Prophase II**
Brief stage after interkinesis; each chromosome still duplicated.

**Metaphase II**
Chromosomes line up at spindle equator; sister chromatids of each are attached to opposite poles.

**Anaphase II**
Sister *chromatids of each chromosome* are separated, moved to opposite poles.

**Telophase II**
Each daughter nucleus is *haploid* (*n*), with one of each pair of homologous chromosomes that was present in the parent nucleus.

## Sickled Cells and Garden Peas

Parties and champagne were the last things on Ernest Irons' mind. It was New Year's Eve, 1904, and Irons, a medical intern, was doodling sketches of peculiarly elongated red blood cells. He and his supervisor, James Herricks, had never seen anything like them. The cells were reminiscent of sickles, a type of short-handled farm tool having a crescent-shaped blade. They were present in a blood sample from a new patient.

The patient had complained of weakness, dizziness, and painful skin sores. He had been gasping for air during bouts of wheezing. Already his father and two sisters had died from mysterious ailments that had damaged their lungs or kidneys. Did those deceased family members also have sickled cells in their blood? Was there a connection between the abnormal cells and the ailments? How did the cells become sickled in the first place?

The medical problems that baffled Irons and Herricks killed their patient when he was only thirty-two. The symptoms themselves are characteristic of a genetic disorder now called *sickle-cell anemia*. The disorder arises when a person receives two copies of a mutated gene (one from each parent) that codes for the protein hemoglobin. Hemoglobin is an oxygen-transporting protein in red blood cells—and the oxygen is vital for aerobic respiration in the body's living cells. The skewed genetic instructions result in skewed hemoglobin molecules.

Consider what happens when blood pumped from the heart reaches capillaries, the blood vessels having the smallest diameter and the thinnest wall. There, red blood cells move along single file. And there they give up oxygen, which diffuses across the capillary wall, through the surrounding tissue fluid, and into individual cells. The blood concentration of oxygen is lower in capillaries than in the rest of the body's blood vessels. Low oxygen levels cause the abnormal hemoglobin molecules to stick together and form long, rodlike structures. Whereas normal red blood cells are shaped like a doughnut without the hole, those containing clumped-together hemoglobin molecules become distorted into a rigid, crescent shape (Figures 11.1a and 11.1b).

Rigid, sickled cells cannot move through the blood capillaries. They clog the tiny passages, even rupture them. The blood cells themselves are easily ruptured. Thus the tissues served by the capillaries become starved for oxygen and saturated with waste products of metabolism. The resulting symptoms range from shortness of breath and fatigue to irreparably damaged organs (Figure 11.16).

You will be reading more about sickle-cell anemia in this chapter and others. It is a disorder that has been studied in great detail at both the molecular and the ecological level. You may find it curious, however, that

**Figure 11.1** (**a**) Normal red blood cells. (**b**) Sickled cells, trademark of a well-known human genetic disorder. (**c**) A mere fifty years before such distorted cells were observed, Gregor Mendel identified rules that turned out to be the starting point for modern genetic analysis. Through such analysis, we have gained insight into many aspects of inheritance—including the chromosomal and molecular events that give rise to sickle-cell anemia.

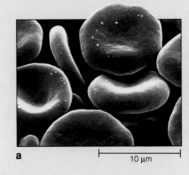

a          10 µm

b          10 µm

our understanding of sickle-cell anemia—and so many other special cases of heritable traits—actually began with studies of thousands of pea plants in a monastery garden.

Fifty years before Ernest Irons doodled red blood cells, a scholarly monk named Gregor Mendel started using peas to study *patterns* of inheritance among sexually reproducing organisms. To test his hypotheses about inheritance, which were novel at the time, Mendel bred generation after generation of pea plants. And he garnered indirect but *observable* evidence of how parents transmit discrete units of information about traits—genes—to offspring.

Mendel's findings seem simple enough today. The cells of pea plants carry two genes for each trait. Following meiosis, the two genes end up in different gametes. When two gametes combine at fertilization, each new plant again has two genes for each trait. At the time, however, no one—not even Mendel himself—knew that he had discovered some near-universal rules governing inheritance. As you will see, his insights still have the power to explain many of the puzzling and sometimes devastating aspects of inheritance that occupy our attention today.

c

KEY CONCEPTS

1. Genes, the units of instructions for producing heritable traits in offspring, have specific locations on chromosomes. The molecular form of the gene at a given location may be slightly different from one individual to the next. All of the different molecular forms of a gene are called alleles.

2. Diploid cells, which have pairs of homologous chromosomes, have pairs of genes. Gregor Mendel's monohybrid crosses of pea plants provided indirect evidence that the two genes of each pair segregate from each other during meiosis and end up in different gametes.

3. Mendel's dihybrid crosses of pea plants provided indirect evidence that a gene pair tends to assort into gametes independently of gene pairs that are located on other (nonhomologous) chromosomes.

4. Traits may be influenced by dominance relations. Here, the two genes of a pair are not identical, and one exerts more pronounced effects compared to its partner. Traits also can be influenced by interactions among different gene pairs, by single genes that affect more than one structure or function in the body, and by environmental conditions.

## MENDEL'S INSIGHTS INTO THE PATTERNS OF INHERITANCE

When Charles Darwin first proposed his theory of evolution by natural selection, he offered the world a new way of looking at life's diversity. In Darwin's view, members of a population vary in heritable traits. Variations that improve chances of surviving and reproducing show up more often in each generation. Those that don't become less frequent. In time the population changes—it evolves.

Not everyone accepted the theory, partly because it did not fit with a prevailing view of inheritance. It was common knowledge that instructions for heritable traits reside in sperm and eggs—but how were the

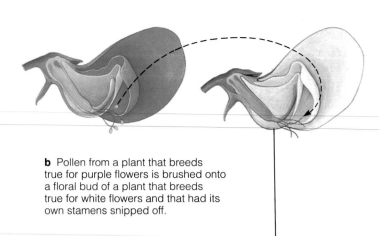

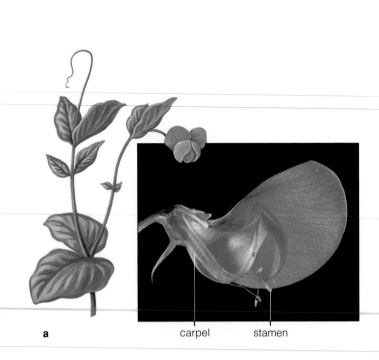

**b** Pollen from a plant that breeds true for purple flowers is brushed onto a floral bud of a plant that breeds true for white flowers and that had its own stamens snipped off.

**c** The cross-fertilized plant produces seeds, each of which is allowed to grow into a new plant.

**a**

carpel     stamen

**Figure 11.2** The garden pea plant (*Pisum sativum*), the focus of Mendel's experiments. The photograph shows a flower that has been sectioned to reveal the location of the stamens (where pollen grains develop) and the carpel. (Eggs develop, fertilization takes place, and seeds mature inside the carpel.)

**d** Flower color of new plants can be used as evidence of patterns in how hereditary material is transmitted from each parent.

instructions combined at fertilization? Many thought they blended together, like cream into coffee.

Yet if blending were true, why aren't distinctive traits diluted out of a population? Why do children with freckles keep turning up among nonfreckled generations? If it were true, why aren't all the descendants of a herd of white stallions and black mares uniformly gray? Blending scarcely explained what people could see with their own eyes, but it was considered a rule anyway. According to the blending theory, populations "had to be" uniform—and without variation for selective agents to act upon, evolution simply could not occur.

Even before Darwin presented his theory, however, someone was gathering evidence that eventually would support its premise about variation in heritable traits. In a monastery garden in Brünn, now in Czechoslovakia, a scholarly monk named Gregor Mendel was beginning to identify the rules governing inheritance.

The monastery of St. Thomas was somewhat removed from the European capitals, which were then the centers of scientific inquiry. Yet Mendel was not a man of narrow interests who simply stumbled by chance onto principles of great import. Having been raised on a farm, he was well aware of agricultural principles and their application. He kept abreast of breeding experiments and developments described in the available journals. Mendel was a founder of the regional agricultural society. He won several awards for developing improved varieties of fruits and vegetables. After entering the monastery, he spent two years studying mathematics at the University of Vienna.

Shortly after his university training, Mendel began experiments on the nature of plant diversity. Through his combined talents in plant breeding and mathematics, he perceived patterns in the emergence of traits from one generation to the next.

**Mendel's Experimental Approach**

Mendel experimented with the garden pea plant, *Pisum sativum* (Figure 11.2). This plant can fertilize itself. Its flowers produce male *and* female gametes, and fertilization can occur in the same flower. (To keep things simple, we will call those gametes "sperm" and "eggs," even though they bear little obvious resemblance to the sperm and eggs of animals.) Some pea plants are **true-**

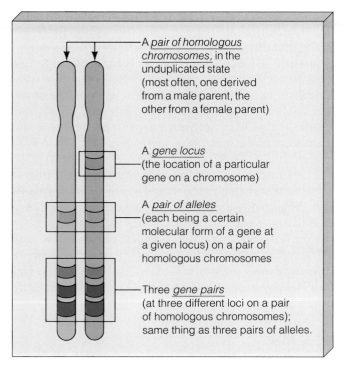

**Figure 11.3** A few genetic terms illustrated. In Mendel's time, no one knew about meiosis or chromosomes, but it was clear that offspring received hereditary material from parents by way of sperm and eggs. As we now know, the hereditary material (genes) is packaged in homologous chromosomes (one from the male, one from the female parent). Thus at each gene locus along the chromosomes, one allele has come from the male parent and its partner has come from the female parent.

**breeding**. Successive generations are exactly like the parents in one or more traits. For example, all offspring may "breed true" for white flowers. Of course, when left to their own devices, true-breeding pea plants show a rather monotonous and uninformative pattern of inheritance.

However, pea plants also lend themselves to artificial **cross-fertilization**, in which sperm from one plant are used to fertilize eggs from another. In some experiments, Mendel stopped plants from self-fertilizing by opening their flower buds and removing the stamens. (Stamens bear pollen grains, in which sperm develop.) Then he promoted cross-fertilization by brushing pollen from another plant on the "castrated" floral bud.

Why did Mendel tinker with plants this way? He wanted cross-fertilization to occur between two true-breeding plants that exhibited different forms of the same trait. For example, he crossed a white-flowered with a purple-flowered plant. If their offspring bore white *or* purple flowers, he could identify one plant or the other as the source of the hereditary material for that trait. If there *were* patterns in the way hereditary material is transmitted from parents to offspring, the use of variations in traits might be a way to identify them.

It will be useful to retrace a few of Mendel's experiments. The conclusions he drew from them have turned out to apply, with some modification, to all sexually reproducing organisms.

## Some Terms Used in Genetics

Having read the chapter on meiosis, you already have insight into the mechanisms of sexual reproduction—which is more than Mendel had. He did not know about chromosomes and so could not have known that the parental chromosome number is reduced by half in gametes, then restored at fertilization. Yet Mendel had some hunches about what was going on. As we follow his thinking, let's simplify things by substituting a few modern terms used in studies of inheritance (see also Figure 11.3):

1. **Genes** are units of instructions for producing or influencing a specific trait in offspring. Each gene has its own particular location (*locus*) on a chromosome.

2. Diploid cells have inherited a pair of genes for each trait, one on each of two homologous chromosomes.

3. Although both genes of a pair deal with the same trait, they may vary in their information about it. This happens when they have slight molecular differences, as when one gene for flower color specifies "red" and another specifies "white." All of the different molecular forms of a gene that exist are called **alleles**.

4. Gene shufflings during meiosis and fertilization can put together different mixes of alleles in offspring. If it turns out that the two alleles of a pair are the same, this is a *homozygous* condition. If different, this is a *heterozygous* condition.

5. Often one allele of a pair is "dominant," meaning its effect on a trait masks the effect of its "recessive" partner. We use capital letters for dominant alleles and lowercase letters for recessive ones (for example, alleles *A* and *a*).

6. Putting this together, we say a **homozygous dominant** individual has two dominant alleles (*AA*) for the trait being studied. A **homozygous recessive** individual has two recessive alleles (*aa*). A **heterozygous** individual has two different alleles (*Aa*).

7. To keep the distinction clear between genes and the traits they specify, we can use **genotype** when referring to the genes present in an individual, and **phenotype** when referring to an individual's observable traits.

## The Concept of Segregation

Mendel's first crosses were *monohybrid*. This means two parents that bred true for contrasting forms of a single trait were crossed in order to produce heterozygous offspring. Mendel tracked the trait through two generations of offspring, which can be designated as follows:

P    parental generation

$F_1$    first-generation offspring

$F_2$    second-generation offspring

In one case, Mendel crossed a purple-flowered plant with a white-flowered one. *All* the $F_1$ offspring from that cross had purple flowers. Then he allowed $F_1$ plants to self-fertilize and produce seeds—and some of the $F_2$ offspring had white flowers!

Mendel interpreted the results this way. Each plant inherits two "units" of instruction (two genes) for flower color, and those units retain their identity from one generation to the next. Each parent contributes only one unit to each of its offspring. Assume both units were purple in one parent and both were white in the other. Each $F_1$ plant would then inherit one purple and one white. If that were so, then purple must be the dominant form of the trait, because it masked the white in $F_1$ plants.

Let's rephrase Mendel's interpretation in terms of what we know about meiosis and fertilization. Pea plants are diploid, with pairs of homologous chromosomes. Assume one parent is homozygous dominant (*AA*) for flower color and the other, homozygous recessive (*aa*). After meiotic cell division, each sperm or egg will carry a gene for flower color on one of its chromosomes. As Figure 11.4 shows, when a sperm and egg combine at fertilization, only one outcome is possible: *A* + *a* = *Aa*.

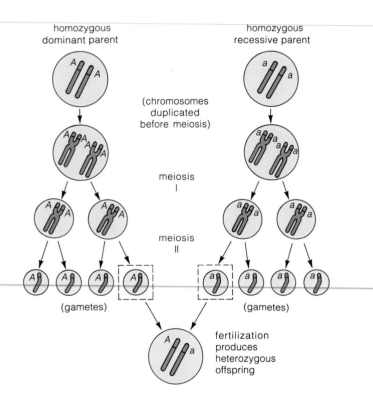

**Figure 11.4** Segregation of alleles in a monohybrid cross. Two parents that are true-breeding for contrasting forms of a single trait can give rise only to heterozygous offspring.

**Figure 11.5** (*Right*) Results from Mendel's monohybrid cross experiments with the garden pea. The numbers are his counts of $F_2$ plants showing the dominant or recessive form of the trait. On the average, the dominant-to-recessive ratio was 3:1.

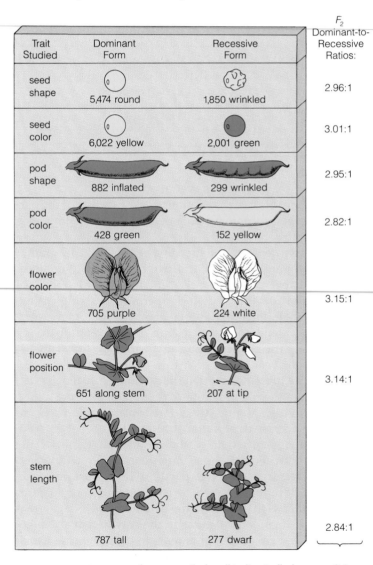

| Trait Studied | Dominant Form | Recessive Form | $F_2$ Dominant-to-Recessive Ratios: |
|---|---|---|---|
| seed shape | 5,474 round | 1,850 wrinkled | 2.96:1 |
| seed color | 6,022 yellow | 2,001 green | 3.01:1 |
| pod shape | 882 inflated | 299 wrinkled | 2.95:1 |
| pod color | 428 green | 152 yellow | 2.82:1 |
| flower color | 705 purple | 224 white | 3.15:1 |
| flower position | 651 along stem | 207 at tip | 3.14:1 |
| stem length | 787 tall | 277 dwarf | 2.84:1 |

Average ratio for all traits studied:    3:1

But what about the $F_2$ results? To understand this aspect of Mendel's monohybrid crosses, you have to know he crossed hundreds of plants and kept track of thousands of first- and second-generation offspring. He also *counted* and *recorded* the number of dominant and recessive offspring for each cross. An intriguing ratio emerged from his records. On the average, of every four $F_2$ plants, three showed the dominant form of the trait and one showed the recessive (Figure 11.5).

Mendel used his knowledge of mathematics to explain the 3:1 phenotypic ratio. He began by assuming each particular sperm is not precommitted to combining with one particular egg; fertilization has to be a chance event. This meant the monohybrid crosses could be interpreted according to rules of **probability**, which apply to chance events. "Probability" simply means the number of times a particular outcome will occur, divided by the total number of all possible outcomes.

A simple way to predict the probable outcome of a cross between two $F_1$ plants is the **Punnett-square method**, shown in Figure 11.6. Assume each $F_1$ plant produced two kinds of sperm or eggs in equal proportions: half were $A$, and half were $a$. If any sperm is equally likely to fertilize any egg, there are four possibilities for each encounter:

| Possible Event: | Probable Outcome: |
|---|---|
| sperm $A$ meets egg $A$ | 1/4 $AA$ offspring |
| sperm $A$ meets egg $a$ | 1/4 $Aa$ |
| sperm $a$ meets egg $A$ | 1/4 $Aa$ $\Big\}$ or 1/2 $Aa$ |
| sperm $a$ meets egg $a$ | 1/4 $aa$ |

Thus, as Figure 11.7 shows, a randomly selected $F_2$ plant had three chances in four of carrying at least one dominant allele and developing purple flowers. It had only one chance in four of carrying two recessive alleles and developing white flowers. As a result, the probable phenotypic ratio was 3 purple to 1 white, or 3:1.

Results from his monohybrid crosses led Mendel to formulate a principle. Stated in modern terms,

***Mendelian principle of segregation.*** **Diploid organisms inherit a pair of genes for each trait (on a pair of homologous chromosomes). The two genes segregate from each other at meiosis, so each gamete formed after meiosis has an equal chance of receiving one or the other gene, but not both.**

Keep in mind that Mendel's observed ratios weren't *exactly* 3:1 (see, for example, the numerical results in Figure 11.5). To understand why, flip a coin a few times.

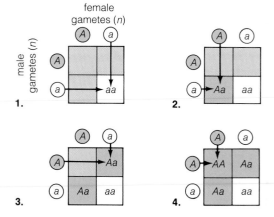

**Figure 11.6** Punnett-square method of predicting the probable ratio of traits that will show up in offspring of self-fertilizing individuals known to be heterozygous ($Aa$) for a trait. The circles represent gametes. The letters inside gametes represent the dominant or recessive form of the gene being tracked. Each square depicts the genotype of one kind of offspring.

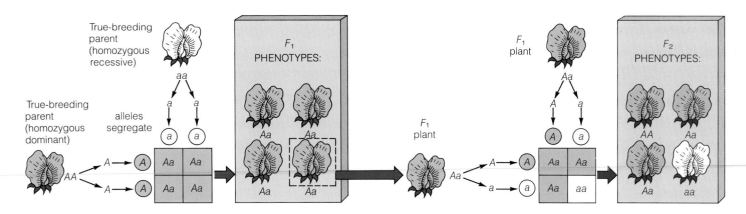

**Figure 11.7** Results from one of Mendel's monohybrid crosses to produce heterozygous offspring. Notice that the dominant-to-recessive ratio is 3:1 for the second-generation ($F_2$) plants.

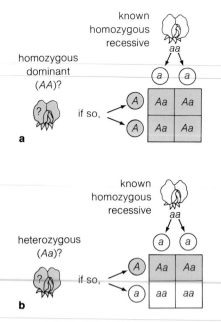

**a**

**b**

**Figure 11.8** Punnett-square method of predicting the outcomes of a testcross between an individual known to be homozygous recessive for a trait (here, white flower color) and an individual showing the dominant form of the trait. (**a**) If the individual of unknown genotype is homozygous dominant, all offspring will show the dominant form of the trait. (**b**) If the individual is heterozygous, about half the offspring will show the recessive form.

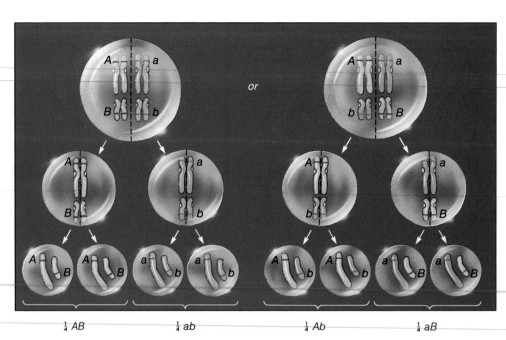

$\frac{1}{4}\ AB$      $\frac{1}{4}\ ab$      $\frac{1}{4}\ Ab$      $\frac{1}{4}\ aB$

**Figure 11.9** Example of independent assortment, showing just two pairs of homologous chromosomes. The different combinations of alleles possible in gametes arise in part through the random alignment of homologues during metaphase I of meiosis (page 156).

We all know that a coin is just as likely to end up heads as tails. But often it ends up heads, or tails, several times in a row. When you flip the coin only a few times, the observed ratio may differ greatly from the predicted ratio of 1:1. Only when you flip the coin many times will you come close to the predicted ratio. Almost certainly, Mendel's reliance on a large number of crosses and his understanding of probability kept him from being confused by minor deviations from the predicted results.

## Testcrosses

Mendel gained support for his concept of segregation through the **testcross**. In this type of cross, first-generation hybrids are crossed to an individual known to be true-breeding for the same recessive trait carried by the recessive parent. (In other words, that individual is homozygous recessive.)

For example, Mendel crossed purple-flowered $F_1$ plants with true-breeding, white-flowered plants. If his concept were correct, there would be about as many recessive as dominant plants in the offspring from the testcross. That is exactly what happened (Figure 11.8). As predicted, about half the testcross offspring were purple-flowered ($Aa$) and half were white-flowered ($aa$).

## The Concept of Independent Assortment

In another series of experiments, Mendel crossed true-breeding pea plants having contrasting forms of two traits. In such *dihybrid crosses*, the $F_1$ offspring inherit two gene pairs, neither of which consists of identical alleles. In the following example of a dihybrid cross, $A$ and $B$ stand for dominance in flower color and height, respectively; $a$ and $b$ stand for their recessive counterparts:

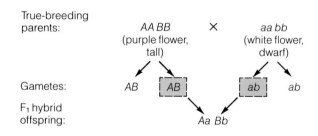

Mendel anticipated (correctly) that all the $F_1$ offspring of the cross would be purple-flowering and tall. But what would happen when those $Aa\,Bb$ offspring formed sperm and eggs of their own? Would a gene for flower color and a gene for height *travel together or independently of each other* into the gametes?

**Figure 11.10** Results from Mendel's dihybrid cross between true-breeding parent plants differing in two traits (flower color and height).

Here, *A* and *a* represent the dominant and recessive alleles for flower color. *B* and *b* represent the dominant and recessive alleles for height.

On the average, the phenotypic combinations in the $F_2$ generation occur in a 9:3:3:1 ratio. Keep in mind that working a Punnett square is really a way to show *probabilities* of certain combinations occurring as a result of allele shufflings during meiosis and fertilization.

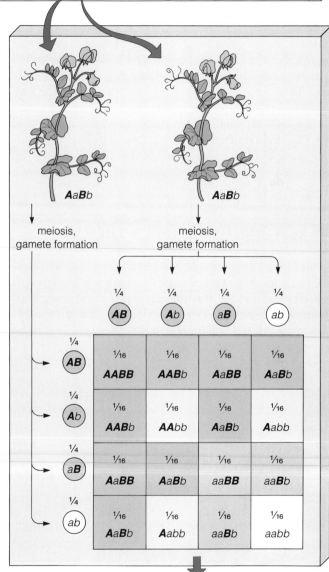

$F_1$ OUTCOME: All $F_1$ plants purple-flowered, tall (***AaBb*** heterozygotes)

As we now know, a particular gene pair tends to segregate into gametes independently of other gene pairs—when the others are located on *non*homologous chromosomes. In this case, four combinations of alleles are possible:

1/4 *AB*     1/4 *Ab*     1/4 *aB*     1/4 *ab*

These possibilities arise through the random alignment of maternal and paternal chromosomes during metaphase I of meiosis (page 156 and Figure 11.9).

Now think about the way alleles can be mixed at fertilization. Simple multiplication (four kinds of sperm times four kinds of eggs) tells us that *sixteen* gametic combinations are possible in dihybrid $F_2$ plants. Figure 11.10 lays out the possibilities, using the Punnett-square method. The sixteen gametic combinations result in nine different genotypes and four different phenotypes. What accounts for this numerical difference? As you can see from Figure 11.10, more than one combination of gametes can give rise to the same genotype, and some genotypes give rise to the same phenotype.

When we total the possible outcomes by phenotype, we get 9/16 tall purple-flowered, 3/16 dwarf purple-flowered, 3/16 tall white-flowered, and 1/16 dwarf white-flowered plants. That is a probable phenotypic ratio of 9:3:3:1.

Results from all of Mendel's dihybrid crosses were close to a 9:3:3:1 ratio, and they led to the formulation of another principle. In modern terms, here was evidence that the gene pairs located on two homologous chromosomes tend to travel into gametes independently of the gene pairs that are located on other homologous chromosomes.

---

*Mendelian principle of independent assortment.* **Each gene pair tends to assort into gametes independently of other gene pairs that are located on nonhomologous chromosomes.**

---

ADDING UP THE $F_2$ COMBINATIONS POSSIBLE:

☐ 9/16 or 9 purple-flowered, tall

☐ 3/16 or 3 purple-flowered, dwarf

☐ 3/16 or 3 white-flowered, tall

☐ 1/16 or 1 white-flowered, dwarf

The variety resulting from independent assortment and hybrid crossing is staggering. In a monohybrid cross (involving a single gene pair), only three genotypes are possible (*AA*, *Aa*, and *aa*). We can represent this as $3^n$, where *n* is the number of gene pairs. When more gene pairs are involved, the number of possible combinations increases dramatically. Even if parents differ in only ten gene pairs, almost 60,000 different genotypes are possible among their offspring. If they differ in twenty gene pairs, the number is close to 3.5 billion!

On the basis of his experimental results, Mendel was convinced that the hereditary material comes in units that retain their physical identity from one generation to the next (despite being segregated and assorted during meiosis). In 1865 he reported this idea before the Brünn Society for the Study of Natural Science. His report made no impact whatsoever. The following year his paper was published. Apparently it was read by few and its near-universal applicability was understood by no one.

In 1871 Mendel became an abbot of the monastery, and his experiments gave way to administrative tasks. He died in 1884, never to know that his work was the starting point for the development of modern genetics.

## VARIATIONS ON MENDEL'S THEMES

It was Mendel's genius or good fortune to limit his studies to genes that were expressed in completely dominant or recessive ways. But the phenotypic expression of many other genes is not as straightforward, as the following examples will demonstrate.

## Dominance Relations

Different degrees of dominance may exist between a pair of alleles. One or both may be fully dominant, or one may be incompletely dominant over the other.

In **incomplete dominance**, the phenotype of a heterozygote is intermediate between that of the homozygous dominant or recessive types. Suppose you cross true-breeding red-flowered and white-flowered snapdragons. All of the $F_1$ offspring will have *pink* flowers (Figure 11.11). This might appear to be an outcome of "blending"—until you perform a cross between two $F_1$ plants. You will discover that the "red" allele was not blended away, because the $F_2$ offspring of those first-generation plants will have red, pink, or white flowers in a predictable 1:2:1 ratio. The dominant allele calls for a red pigment, and it takes two alleles to produce enough pigment to give the flowers a red color. With their single dominant allele, heterozygotes can only produce enough pigment to give the flowers a pinkish cast.

In **codominance**, two alleles of a pair are not identical, yet the expression of *both* can be discerned in heterozygotes. Each gives rise to a different phenotype. For example, you probably have heard of **ABO blood typing**. It refers to a method of characterizing blood according to which forms of a certain protein occur at the surface of a person's red blood cells. (The protein is like a molecular fingerprint; it identifies the cell as being of a certain type.) Three alleles influence the protein's structure. Two, $I^A$ and $I^B$, are codominant when paired with each other. A third allele, *i*, is recessive. When paired with either $I^A$ or $I^B$, its expression is masked.

Both $I^A$ and $I^B$ code for enzymes that attach a sugar group to the protein after it has been synthesized, but

**Figure 11.11** Incomplete dominance at one gene locus. Red-flowering and white-flowering homozygous snapdragons produce pink-flowering plants in the first generation. The red allele ($R^1$) is only partially dominant over the white allele ($R^2$) in the heterozygous state.

the enzymes make the attachment in different ways. Two forms of the completed protein, called A and B, result from expression of alleles $I^A$ and $I^B$.

Whenever more than two forms of alleles exist for a given gene locus, we call them a **multiple allele system**. In this case, four blood types are possible, depending on which two of the codominant and recessive alleles of the system are present in a person's cells. As Figure 11.12 indicates, the four types are A, B, AB, and O. Both $I^A I^A$ and $I^A i$ individuals are type A. Both $I^B I^B$ and $I^B i$ individuals are type B. However, $I^A I^B$ individuals are type AB. Red blood cells of homozygous recessive people are neither A nor B; that is what the "O" means.

## Interactions Between Different Gene Pairs

On thinking about the traits described so far, you might conclude that each trait arises from the expression of only one pair of genes. However, one gene pair often influences the expression of other gene pairs.

**Comb Shape in Poultry.** In some interactions, two gene pairs *cooperate* to produce a phenotype that neither can produce alone. W. Bateson and R. Punnett identified two gene pairs that cooperate to produce comb shape in chickens. The allelic combination *rr* at one locus together with *pp* at the other gives rise to the most common phenotype, the single comb. Other phenotypes occur when the two dominant alleles, *R* and *P*, are present. Depending on the allelic combinations, the two gene pairs can produce rose, pea, or walnut combs as well as the single comb (Figure 11.13).

possible alleles in
gametes from mother:

$I^A$ or $I^B$ or $i$

resulting blood type:

$\blacksquare$ = A

$\blacksquare$ = AB

$\blacksquare$ = B

$\square$ = O

**Figure 11.12** Possible combinations of alleles associated with ABO blood typing.

**Figure 11.13** (*Below*) Interaction between two genes affecting the same trait in domestic breeds of chickens. The initial cross is between a Wyandotte (with a rose comb on the crest of its head) and a brahma (with a pea comb). With complete dominance at the gene locus for pea comb and at the gene locus for rose comb, the products of these two nonallelic genes interact and give a walnut comb (**a**). With complete recessiveness at both loci, the products interact and give rise to a single comb (**d**).

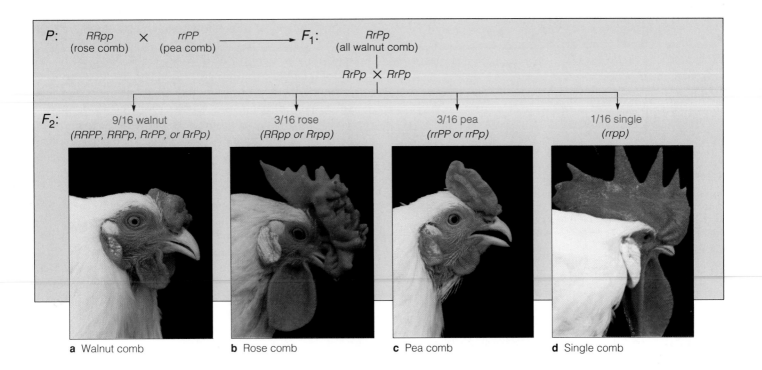

P: RRpp × rrPP ——→ $F_1$: RrPp
(rose comb)  (pea comb)  (all walnut comb)

RrPp × RrPp

$F_2$:
9/16 walnut
(RRPP, RRPp, RrPP, or RrPp)

3/16 rose
(RRpp or Rrpp)

3/16 pea
(rrPP or rrPp)

1/16 single
(rrpp)

**a** Walnut comb  **b** Rose comb  **c** Pea comb  **d** Single comb

**a** Black Labrador　　**b** Yellow Labrador

**c** Chocolate Labrador

P homozygotes: *BB EE* × *bb ee*

↓

All *F₁*:　　　　*Bb Ee*

↓

*F₂* combinations possible:

|  | BE | Be | bE | be |
|---|---|---|---|---|
| **BE** | BB EE | BB Ee | Bb EE | Bb Ee |
| **Be** | BB Ee | BB ee | Bb Ee | Bb ee |
| **bE** | Bb EE | Bb Ee | bb EE | bb Ee |
| **be** | Bb Ee | Bb ee | bb Ee | bb ee |

RESULTING PHENOTYPES:

- ▢ 9/16 or 9 black
- ▢ 3/16 or 3 brown
- ▢ 4/16 or 4 yellow

**Figure 11.14** Interaction among alleles of two gene pairs affecting the coat color of Labrador retrievers. At one gene locus concerned with melanin production, allele *B* (black) is dominant to allele *b* (brown). At a different gene locus, allele *E* allows melanin pigment to be deposited in individual hairs. But two recessive alleles at that locus (*ee*) prevent deposition and a yellow coat results (this is a case of recessive epistasis).

The *F₁* offspring of a dihybrid cross would produce *F₂* offspring in a 9:3:4 ratio, as shown above. The yellow Labrador in (**b**) probably has genotype *BB ee*. It has the capacity to produce melanin but not to deposit the pigment in hairs. (Looking at that photograph, can you say why?)

**Hair Color in Mammals.** In other interactions, two alleles of a gene *mask* the expression of alleles of another gene, and some expected phenotypes do not appear at all. Such interactions, called **epistasis**, are common among the gene pairs that affect the color of fur or skin in mammals.

Figure 11.14 shows a case of epistasis among Labrador retrievers, which may have black, brown, or yellow hair. The different colors result from variations in the amount and distribution of a brownish-black pigment called melanin. Many gene pairs and their alleles function in the metabolic processes leading to melanin production and its deposition in certain body regions. Alleles at one gene locus affect how much melanin will be produced. (They code for one of the enzymes involved in melanin production.) At this locus, a *B* allele (for black) is dominant to *b* (for brown). However, alleles at a different gene locus control whether pigment molecules will be deposited at all in a retriever's hairs. A dominant allele (*E*) at this locus permits pigment deposition in hairs. But a pair of recessive alleles (*ee*) will block deposition, resulting in a yellow coat color. Interaction between the *B* gene and two recessive alleles of the *E* gene is a case of recessive epistasis.

Epistasis involving still another gene locus (*C*) determines whether an individual will be an *albino*, a phenotype resulting from the complete absence of melanin. Alleles at this gene locus code for tyrosinase, the first enzyme required in the series of reactions by which melanin is produced. The *C* allele is dominant to *c*. Melanin can be produced in a *CC* or *Cc* individual. But its production is blocked in a *cc* individual, which will end up with the phenotype characteristic of albinism. Figure 11.15 shows an example.

**Figure 11.15** A rare albino rattlesnake, showing the pink eyes and white coloration of animals that are unable to produce the pigment melanin. (Eyes are pink because the absence of melanin allows red light to be reflected from blood vessels in the snake's eyes.)

Surface coloration in birds and mammals is due almost entirely to the color of feathers or fur. In fishes, amphibians, and reptiles, it is due to color-bearing cells in the skin. Some of the cells contain melanin (a brownish-black pigment) or red to yellow pigments. Others contain crystals that reflect light and alter the effect of other pigments present.

The mutation affecting melanin production in the snake shown here had no effect on the production of yellow-to-red pigments and light-reflecting crystals. So the snake's skin appears iridescent yellow as well as white.

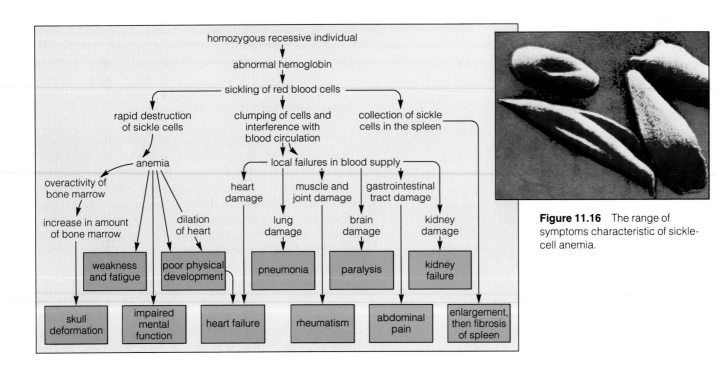

**Figure 11.16** The range of symptoms characteristic of sickle-cell anemia.

## Multiple Effects of Single Genes

A single gene can exert effects on seemingly unrelated aspects of an individual's phenotype. This form of gene expression is called **pleiotropy**.

For example, sickle-cell anemia, described at the start of this chapter, arises through a mutation at a single gene locus. The mutated allele gives rise to a defective form of hemoglobin, the oxygen-transporting pigment in red blood cells. When the defective pigment molecules give up oxygen, they stack together—and the cells become rigid and distorted. Blood flow through capillaries becomes impaired, the delivery of oxygen to cells is hampered, and so is the removal of carbon dioxide and other metabolic products from them.

As Figure 11.16 shows, the metabolic disruption leads to severe damage of many tissues and organs. The *symptoms* of this underlying genetic disorder are what we call sickle-cell anemia. Heterozygotes still have one functional allele and show few symptoms of the disorder. But homozygous recessives show severe phenotypic consequences.

**Figure 11.17** An environmental effect on phenotype. A Himalayan rabbit normally has black hair only on its long ears, nose, tail, and lower limbs of its legs. In one experiment, a patch of a rabbit's white fur was plucked clean, then an icepack was secured over the hairless patch. For as long as the artificially cold conditions were maintained, the hairs that grew back were black.

Himalayan rabbits are homozygous for the $c^h$ allele of a gene that codes for tyrosinase, one of the enzymes necessary to synthesize the pigment melanin. This particular allele produces a version of tyrosinase that is heat sensitive. It functions only when temperatures are below about 33°C. When the rabbit hair grows under warmer conditions, no melanin is produced, so the hair is light. Light fur normally covers the body regions that are warmer than the ears and other slender extremities (which tend to lose metabolically generated heat more rapidly).

Siamese cats that are homozygous for the same heat-sensitive allele also display this environmental effect on phenotype (Figure 6.12).

## Environmental Effects on Phenotype

The external environment has profound effect on gene expression. For example, fur growing on the ears, tail, and other extremities of a Himalayan rabbit (or Siamese cat) darkens at cooler temperatures. When a patch of dark fur is removed and the rabbit is kept in a warm place, light fur grows back in the bare patch. Similarly, a patch stripped of light hair will grow back dark if the rabbit is kept in a cold place (Figures 6.12 and 11.17). An allele of the *C* gene affects this phenotype. As we have seen, this gene codes for tyrosinase, an enzyme necessary for one of the metabolic steps leading to melanin production. The "Himalayan" allele ($c^h$) of this gene produces a version of tyrosinase that can catalyze the necessary step. But it is a heat-sensitive form of the enzyme—it is less active at warm temperatures. Thus, for homozygotes ($c^h c^h$), fur color depends on what the temperature is at the time hair is growing.

The water buttercup (*Ranunculus aquatilis*) provides another example of environmental effects on phenotype. This plant grows in shallow ponds, and some of its leaves develop underwater. The submerged leaves are finely divided, compared with the leaves growing in air. When a leaf-bearing stem is half in and half out of the water, its leaves display both phenotypes (Figure 11.18). The genes responsible for leaf shape produce very different phenotypes under different environmental conditions.

The internal environment (that is, conditions within the tissues surrounding the body's cells) also influences gene expression. At puberty, for example, the male body steps up its production of testosterone, a sex hormone. Among other things, increased levels of testosterone affect genes that govern the development of cartilage in the larynx. The voice deepens as a result of those changes. Without the hormonal increase, the voice will remain higher pitched later in life.

## Variable Gene Expression in a Population

**Penetrance and Expressivity.** Reflect, for a moment, on some of the concepts described so far. Many genes, such as the one responsible for purple flowers in pea plants, are expressed in a regular, consistent pattern. Others are expressed to varying degrees among the individuals of a population. Two terms, penetrance and expressivity, are used in describing variable gene expression in a population.

*Penetrance* is the frequency with which a dominant or homozygous recessive gene is actually expressed in the phenotype of the carriers of that gene. It's an all-or-none measure; either the gene is expressed in an individual or it is not. The dominant gene responsible for purple flowers is completely penetrant—if plants have the gene, the phenotype is fully expressed. By contrast, a dominant gene responsible for *campodactyly* is incompletely penetrant. This human genetic disorder is characterized by immobile, bent fingers. It results when muscles are improperly attached to bone when the little fingers are forming in an embryo. Because the gene is dominant, you might expect it to be expressed in all homozygotes and heterozygotes, but this is not the case. In one study, only eight of nine people who were known to carry the dominant gene exhibited the trait. Thus penetrance was 8/9, or 88 percent.

*Expressivity* is the *degree* to which a characteristic phenotype is exhibited among individuals of a population. In the case of campodactyly, the trait didn't show up at all in one individual. It showed up on both hands of some individuals but on only one hand of others.

Variable gene expression is not necessarily a property of the gene itself. Rather, it arises through gene interactions and through environmental influences. Keep in mind that *most* observable traits are the end products of a series of small, distinct metabolic steps. Some of those steps begin when certain genes first become active in a developing embryo. Other steps are taken later in life. As we have seen, a gene that is required for a step leading to a particular phenotype may be influenced by interactions with other genes. Some number of those genes occur in different allelic combinations in different indi-

**Figure 11.18** Variable expressivity resulting from variation in the external environment. Leaves of the water buttercup (*Ranunculus aquatilis*) show dramatic phenotypic variation, depending on whether they grow underwater or above it. This variation occurs even in the same leaf if it develops half in and half out of water. Compare Figure 20.5.

**Figure 11.19** A sampling from the range of continuous variation in human eye color. Many different genes are required to produce and deposit melanin in the iris of the eye, and the effects of those genes are cumulative. Slight differences in those genes and their gene products lead to small differences in the eye color trait, so that the frequency distribution for eye color appears to be continuous over the range from black to light blue.

**Figure 11.20** (*Right*) Continuous variation in traits.

In a given population, most traits show continuous variation. For example, humans are not all tall or short. They grow to various heights, ranging from very short to very tall, with average heights being more common than either extreme.

The following process is used to describe a population showing continuous variation in some trait. First, the full range of phenotypic variation is divided into measurable categories. Then individuals falling within a given category are counted. This process shows how the relative frequency of different categories is distributed across the possible range of measurable values for the phenotype.

Suppose you wish to determine the frequency distribution for height in a specific group of men, such as those shown in (**a**). First you decide on the degree to which you should divide up the range of possible heights, then you measure each individual and assign him to the appropriate category. Once this is done, you can divide the number in each category by the total number of all individuals in all categories.

(**b**) Often a bar graph is used to illustrate continuous variation. Here, the proportion of individuals falling into each category is plotted against the range of measured phenotypes.

viduals, and their products (proteins) may affect a given gene in different ways, at different times. Besides this, nutrition and other environmental factors may affect the activity of a given gene, and those factors may not be the same from one individual to the next.

**Continuous Variation.** Think about an observable trait such as eye color and hair color. These traits result from the cumulative expression of many different genes that are involved in the stepwise production and distribution of melanin. For the reasons given above, those genes are expressed to different degrees in different individuals in a population. Black eyes have abundant melanin deposits in the iris. Dark-brown eyes have less melanin, and light-brown or hazel eyes have still less (Figure 11.19). Green, gray, and blue eyes don't have green, gray, or blue pigments. They have so little mela-

nin that we readily see blue wavelengths of light being reflected from the iris.

For eye color and many other traits, a population of humans and other organisms shows **continuous variation**. The individuals within the population exhibit a range of smaller, less pronounced differences in some trait. Continuous variation is especially evident in traits that are easily measurable, such as height (Figure 11.20). As the number of gene pairs affecting a given phenotype increases, the expected phenotypic distribution appears to be more and more continuous.

Continuous variation is consistent with Mendel's views about inheritance. "Continuous" does not mean "blending," a term that suggests loss of the original identity of the genes governing heritable traits. The genes are still intact, regardless of the phenotype produced by their combined positive and negative effects.

(number of individuals)

(number of individuals)

1 0 0 1 5 7 7 22 25 26 27 17 11 17 4 4 1

(height, inches)

58 59 60 61 62 63 64 65 66 67 68 69 70 71 72 73 74

**a** Height distribution in a group of 175 U.S. Army recruits about the turn of the century.

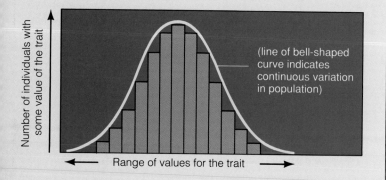

Number of individuals with some value of the trait

(line of bell-shaped curve indicates continuous variation in population)

Range of values for the trait

**b** Generalized bell-shaped curve typical of populations showing continuous variation in some trait.

The phenotypic expression of a given gene may vary, by different degrees, from one individual to the next in a population.

The degree to which the same gene is expressed in different individuals depends on gene interactions. It depends also on the physical and chemical environment in which that gene or its products must function.

## SUMMARY

1. Mendel's hybridization studies with garden pea plants demonstrated that diploid organisms have two units of hereditary material (genes) for each trait, and that genes retain their identity when passed on to offspring.

2. Mendel conducted monohybrid crosses, which are crosses between two true-breeding individuals having contrasting forms of a single trait. Interpreting the results in modern terms, we can say that the crosses provided indirect evidence that a gene at a given location on a chromosome can exist in slightly different molecular forms (alleles), some of which are dominant over other, recessive forms.

3. Homozygous dominant individuals have two dominant alleles ($AA$) for the trait being studied. Homozygous recessives have two recessive alleles ($aa$). Heterozygotes have two different alleles ($Aa$).

4. In Mendel's monohybrid crosses ($AA \times aa$), all $F_1$ offspring were $Aa$. Crosses between plants produced these combinations of alleles in $F_2$ offspring:

|  | $A$ | $a$ |
|---|---|---|
| $A$ | $AA$ | $Aa$ |
| $a$ | $Aa$ | $aa$ |

$AA$ (dominant)
$Aa$ (dominant)
$Aa$ (dominant)
$aa$ (recessive)

This produced an expected phenotypic ratio of 3:1.

5. Results from such monohybrid crosses supported the Mendelian principle of segregation: Diploid organisms have a pair of genes for each trait, on a pair of homologous chromosomes. The two genes segregate from each other during meiosis, such that each gamete formed will end up with one or the other, but not both.

6. Mendel also performed dihybrid crosses, which are crosses between two true-breeding individuals having contrasting forms of two different traits. Results from many experiments were close to a 9:3:3:1 phenotypic ratio:

9 dominant for both traits
3 dominant for $A$, recessive for $b$
3 dominant for $B$, recessive for $a$
1 recessive for both traits

7. On the basis of such dihybrid crosses, the Mendelian principle of independent assortment was formulated: Each gene pair (located on a pair of homologous chromosomes) tends to assort into gametes independently of other gene pairs that are located on other pairs of homologous chromosomes.

8. Since Mendel's time, we have learned that (1) degrees of dominance may exist between some gene pairs, (2) gene pairs can interact to produce some positive or negative effect on phenotype, (3) a single gene can have effects on many seemingly unrelated traits, and (4) the internal and external environments influence gene expression.

Chapter 11   Observable Patterns of Inheritance   **179**

## Review Questions

1. State the Mendelian principle of segregation. Does segregation occur during mitosis or meiosis? *169*

2. Distinguish between the following terms: (a) gene and allele, (b) dominant trait and recessive trait, (c) homozygote and heterozygote, (d) genotype and phenotype. *167*

3. Give an example of a self-fertilizing organism. What is cross-fertilization? *166–167*

4. Distinguish between monohybrid and dihybrid crosses. What is a testcross, and why is it valuable in genetic analysis? *168, 170*

5. State the Mendelian principle of independent assortment. Does independent assortment occur during mitosis or meiosis? *171*

## Self-Quiz *(Answers in Appendix IV)*

1. Alleles are _____ .
   a. alternative molecular forms of a gene
   b. alternative molecular forms of a chromosome
   c. self-fertilizing, true-breeding homozygotes
   d. self-fertilizing, true-breeding heterozygotes

2. A heterozygote is _____ .
   a. one of at least two forms of a gene
   b. a condition in which both alleles of a pair are the same
   c. a condition in which both alleles of a pair are different
   d. a haploid condition in genetic terms

3. The observable traits of an organism are called its _____ .
   a. phenotype      c. genotype
   b. sociobiology   d. pedigree

4. In the monohybrid cross $AA \times aa$, the $F_1$ offspring are
   _____ .
   a. all $AA$       d. 1/2 $AA$ and 1/2 $aa$
   b. all $aa$       e. none of the above
   c. all $Aa$

5. The second generation of offspring from a genetic cross is called the _____ .
   a. $F_1$ generation          c. hybrid generation
   b. $F_2$ generation          d. none of the above

6. In the genetic cross $Aa \times Aa$ where $A$ is completely dominant over $a$, the next generation will show a phenotypic ratio of
   _____ .
   a. 1:2:1          d. 9:3:3:1
   b. 1:1:1:1        e. none of the above
   c. 3:1

7. Which of the following statements most accurately explains Mendel's principle of segregation?
   a. particular units of heredity are transmitted to offspring
   b. two genes on a pair of homologous chromosomes segregate from each other in meiosis
   c. members of a population become segregated

8. Assuming simple dominance and independent assortment, dihybrid crosses between two true-breeding organisms with con-

trasting forms of two traits (as in $Aa\ Bb \times Aa\ Bb$) produce offspring phenotypic ratios close to _____ .
   a. 1:2:1
   b. 1:1:1:1
   c. 3:1
   d. 9:3:3:1

9. "Each pair of genes tends to assort into gametes independently of other gene pairs located on nonhomologous chromosomes" is a statement of Mendel's _____ .
   a. principle of dominance
   b. principle of segregation
   c. principle of independent assortment
   d. none of the above

10. Match each genetic term appropriately.
   _____ dihybrid cross            a. $AA \times aa$
   _____ monohybrid cross          b. $Aa$
   _____ homozygous condition      c. $Aa\ Bb \times Aa\ Bb$
   _____ heterozygous condition    d. $Aa \times Aa$
   _____ true-breeding parents     e. $aa$

## Genetics Problems *(Answers in Appendix III)*

1. One gene has alleles $A$ and $a$; another gene has alleles $B$ and $b$. For each of the following genotypes, what type(s) of gametes will be produced? (Independent assortment is expected.)
   a. $AA\ BB$          c. $Aa\ bb$
   b. $Aa\ BB$          d. $Aa\ Bb$

2. Still referring to the preceding problem, what genotypes will be present in the offspring from the following matings? (Indicate the frequencies of each genotype among the offspring.)
   a. $AA\ BB \times aa\ BB$      c. $Aa\ Bb \times aa\ bb$
   b. $Aa\ BB \times AA\ Bb$      d. $Aa\ Bb \times Aa\ Bb$

3. In one experiment, Mendel crossed a true-breeding pea plant having green pods with a true-breeding pea plant having yellow pods. All of the $F_1$ plants had green pods. Which trait (green or yellow pods) is recessive? Can you explain how you arrived at your conclusion?

4. In addition to the two genes mentioned in Problem 1, assume you now study a third independently assorting gene having alleles $C$ and $c$. For each of the following genotypes, indicate what type (or types) of gametes will be produced:
   a. $AA\ BB\ CC$      c. $Aa\ BB\ Cc$
   b. $Aa\ BB\ cc$      d. $Aa\ Bb\ Cc$

5. A man is homozygous dominant for ten different genes, which assort independently. How many genotypically different types of sperm could he produce? A woman is homozygous recessive for eight of these ten genes, and she is heterozygous for the other two. How many genotypically different types of eggs could she produce? What can you conclude regarding the relationship between the number of different gametes possible and the number of heterozygous and homozygous genes that are present?

6. Recall that Mendel crossed a true-breeding tall, purple-flowered pea plant with a true-breeding dwarf, white-flowered plant. All the $F_1$ plants were tall and purple-flowered. If an $F_1$ plant is now self-pollinated, what is the probability of obtaining an $F_2$ plant heterozygous for the genes controlling height and flower color?

7. Being able to curl up the sides of your tongue into a U-shape is under the control of a dominant allele at one gene locus. (When there is a recessive allele at this locus, the tongue cannot be rolled.) Having free earlobes is a trait controlled by a dominant allele at a different gene locus. (When there is a recessive allele at this locus, earlobes are attached at the jawline.) The two genes controlling tongue-rolling and free earlobes assort independently. Suppose a woman who has free earlobes and who can roll her tongue marries someone who has attached earlobes and who cannot roll his tongue. Their first child has attached earlobes and cannot roll the tongue.

a. What are the genotypes of the mother, the father, and the child?

b. If this same couple has a second child, what is the probability that it will have free earlobes and be unable to roll the tongue?

8. Assume that a gene was recently identified in canaries. One allele at this gene locus produces a yellow feather color. A second allele produces brown feathers. Suppose you are asked to determine the dominance relationship between these two alleles. (Is it one of simple dominance, incomplete dominance, or codominance?) What types of crosses would you make to find the answer? On what types of observations would you base your conclusions?

9. The ABO blood system has often been employed to settle cases of disputed paternity. Suppose, as an expert in genetics, you are called to testify in a case where the mother has type A blood, the child has type O blood, and the alleged father has type B blood. How would you respond to the following statements of the attorneys:

a. "Since the mother has type A blood, the type O blood of the child must have come from the father, and since my client has type B blood, he obviously could not have fathered this child." (*Made by the attorney of the alleged father*)

b. "Further tests revealed that this man is heterozygous and therefore he must be the father." (*Made by the mother's attorney*)

10. In mice, at one gene locus, the dominant allele (*B*) produces a dark-brown pigment; and the recessive allele (*b*) produces a light-brown, or tan, pigment. An independently assorting gene locus has a dominant allele (*C*) that permits the production of all pigments. Its recessive allele (*c*) makes it impossible to produce any pigment at all. The pigmentless condition is called "albino."

a. A homozygous *bb cc* albino mouse mates with a homozygous *BB CC* brown mouse. Assuming independent assortment, in what ratios would the phenotypes and genotypes be expected in the $F_1$ and $F_2$ generations?

b. If an $F_1$ mouse from part (a) above were backcrossed to its albino parent, what phenotypic and genotypic ratios would be expected?

11. Certain dominant alleles are so important for normal development that the mutant recessive alleles, when homozygous, lead to the death of the organism. However, such recessive alleles can be perpetuated as heterozygotes (*Ll*), which in many cases are not phenotypically different from homozygous (*LL*) normals. (In some cases, individuals carrying a recessive lethal allele do have a mutant phenotype.) Consider the mating of two such heterozygotes, *Ll* × *Ll*. Among their *surviving* progeny, what is the probability that any individual will be heterozygous?

12. In corn, a series of three independent pairs of gene loci (*A*, *C*, and *R*) affect the production of pigment that leads to kernel color. If any one of the three pairs is in the homozygous recessive state, then no pigment will form in the kernels. However, if at least one dominant allele of each locus is present, then pigment can form in the kernel. Two corn plants with the following genotypes were crossed:

$$Aa\ cc\ Rr \times aa\ Cc\ Rr$$

What fraction of the progeny kernels will be pigmented? (Note: each kernel represents a separate (potential) individual; it will exhibit the pigment phenotype of the plant that can be grown from it.)

## Selected Key Terms

| | |
|---|---|
| ABO blood typing 172 | homozygous recessiveness 167 |
| allele 167 | incomplete dominance 172 |
| codominance 172 | independent assortment 171 |
| continuous variation 178 | monohybrid cross 168 |
| dihybrid cross 170 | multiple allele system, 173 |
| epistasis 174 | penetrance, 177 |
| expressivity 177 | phenotype 167 |
| gene locus 167 | pleiotropy 175 |
| gene 167 | probability, 169 |
| genotype 167 | Punnett-square method 169 |
| heterozygous condition 167 | segregation 169 |
| homozygous condition 167 | testcross 170 |
| homozygous dominance 167 | true-breeding organism 166 |

## Readings

Dunn, L. 1965. *A Short History of Genetics.* New York: McGraw-Hill.

Mendel, G. 1959. "Experiments in Plant Hybridization." Translation in J. Peters (editor), *Classic Papers in Genetics.* Englewood Cliffs, New Jersey: Prentice-Hall.

Suzuki, D., et al. 1989. *An Introduction to Genetic Analysis.* Fourth edition. New York: Freeman. Chapters 2 and 4 are clear introductions to Mendelian analysis.

## Too Young To Be Old

Imagine being ten years old, trapped in a body that each day becomes a bit more shriveled, a bit more frail, *old*. You are just tall enough to peer over the top of the kitchen counter, and you weigh less than thirty-five pounds. Already you are bald, and your nose has become crinkled and beaklike.

Most likely, you have only a few more years to live. Yet in spite of this cruel twist of nature, you still have not lost your courage or your childlike curiosity about life. Like Mickey Hayes and Fransie Geringer (Figure 12.1), you play, laugh, celebrate birthdays, and hug your friends.

Of every 8 million humans born, one is destined to grow old far too soon, compared to the normal timetable for our species. Something has gone wrong with one gene on just one of the forty-six chromosomes brought together by chance at conception. From that moment on, the mistake is perpetuated each time cells of the embryo—then of the child—duplicate their chromosomes and divide. The outcome of that rare mistake will be an acceleration of the aging process and a greatly reduced life expectancy. This is the *Hutchinson-Gilford progeria syndrome*, once called the "leprechaun's disease." There is no cure.

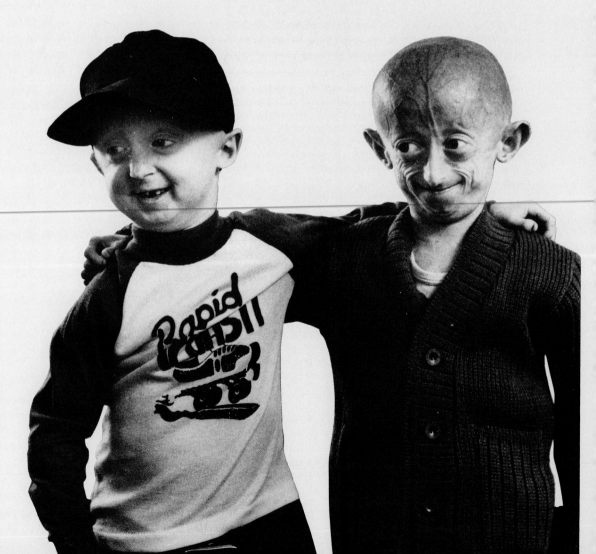

**Figure 12.1** Two boys, both less than ten years old, who met at Disneyland, California, during a gathering of progeriacs. Progeria is a genetic disorder characterized by accelerated aging and extremely reduced life expectancy.

Usually, symptoms begin to appear before affected individuals are two years old. In some unknown way, the interaction of the altered gene with other genes has absolutely devastating effects on normal cell division, growth, and development. The rate of growth quickly declines to abnormally low levels. Skin becomes thinner and muscles become flaccid. Limb bones that otherwise should lengthen and become stronger start to soften. Most of the time, all hair is lost and the individuals become bald. Progeriacs never reach puberty, the onset of sexual maturation. Most die in their early teens from a stroke or heart attack brought on by an earlier hardening of the arteries, a condition typical of advanced age.

Outwardly, none of the chromosomes of affected individuals appears to be defective when viewed with a microscope. Simple analysis tells us that the condition probably arises through a dominant gene mutation that strikes arbitrarily. The mutated gene is always expressed. It does not occur on a sex chromosome, because the resulting disorder can occur in either boys or girls. There are no documented cases of progeria running in families. As is the case for some other dominant gene mutations, there is a "paternal age effect," in that the father is four or five years older than the mother at the time of conception, on the average.

We began this unit of the book by looking at the mechanisms of cell division, the starting point of inheritance. Then we started thinking about the genes on chromosomes, and we began to analyze some of the phenotypic consequences of gene shufflings at meiosis and at fertilization. With this chapter we delve more deeply into the inheritance of chromosomes and the genes they carry, with emphasis on the patterns of inheritance in humans. At times the methods of analysis might seem abstract. But keep in mind that we are talking about normal and abnormal gene expression in yourself and in other human beings. At this writing, Mickey Hayes is close to eighteen years old and is the oldest living progeriac. Fransie was seventeen when he died.

1. Genes are arranged linearly along chromosomes. Although genes on the same chromosome tend to stay together during meiosis, crossing over disrupts such linkages. The farther apart two genes are along a chromosome, the greater will be the frequency of crossing over and recombination between them.

2. Crossing over during meiosis leads to genetic variation—hence to variation in traits among offspring. So does independent assortment of homologous chromosomes during meiosis, the outcome of which is the random mix of maternal and paternal chromosomes in gametes.

3. Abnormal changes in the structure or number of chromosomes occur on rare occasions. Such events often cause genetic disorders. Together with crossing over and independent assortment, they also may influence the course of evolution. They lead to different combinations of traits in offspring, and the differences can be acted upon by agents of selection.

## CHROMOSOMAL THEORY OF INHERITANCE

### Return of the Pea Plant

The year was 1884. Mendel's paper on hybridization of pea plants had been gathering dust in a hundred libraries for nearly two decades, and Mendel himself had just passed away. Ironically, the experiments described in that forgotten paper were about to be devised all over again, as a way to test ideas emerging from another line of research. Cytology, the study of cell structure and function, was about to converge with genetic analysis.

Improvements in microscopy had rekindled efforts to locate the cell's hereditary material, and researchers were zeroing in on the nucleus. By 1882, Walther Flemming had observed threadlike bodies—chromosomes—in the nuclei of dividing cells. By 1884, a question was taking shape: Could those threadlike chromosomes be the hereditary material?

Then researchers realized each gamete has half as many chromosomes as a fertilized egg. In 1887, August Weismann proposed that a special division process must reduce the chromosome number by half before gametes form. Sure enough, in that same year meiosis was discovered. Weismann now began to promote his theory of heredity. In essence, he argued that the chromosome number is halved during meiosis, then restored when sperm and egg combine at fertilization. Thus half the hereditary material in offspring is paternal in origin, and half is maternal. His views were hotly debated, and the debates drove researchers into testing the theory. Throughout Europe there was a flurry of experimental crosses—just like the ones Mendel had carried out.

Finally, in 1900, researchers came across Mendel's paper while checking for literature related to their own hybridization studies. To their chagrin, their results merely confirmed what Mendel already had said. *Diploid cells have two units of instruction (genes) for each heritable trait, and the units segregate prior to gamete formation.*

This chapter covers some observations and experiments that unfolded in the decades after the rediscovery of Mendel's work. They lend impressive support to what is now called the chromosomal theory of inheritance. From preceding chapters, we are already familiar with many of the points of this theory:

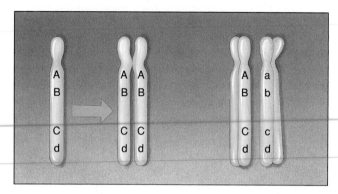

**a** Genes occur in linear sequence on a chromosome. Before a cell divides, each chromosome is copied; the copy has the same gene sequence. The capital and lowercase letters denote alleles—alternative forms of a gene at a particular location (locus) on the chromosome.

**b** Each pair of homologous chromosomes carries identical or different alleles at corresponding gene loci. Green represents paternal chromosomes; pink represents maternal ones.

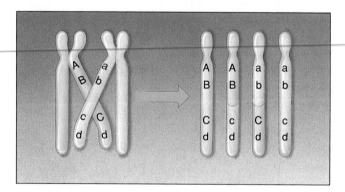

**c** Crossing over and genetic recombination typically occur during meiosis I. Each chromosome that is affected by a crossover event will carry both maternal and paternal alleles at different loci along its length.

**Figure 12.2** Crossing over and other key events described by the chromosomal theory of inheritance.

1. Genes, the units of instruction for heritable traits, are arranged one after the other along chromosomes.

2. From meiosis through fertilization, sexual reproduction assures that each new individual will have the same chromosome number as its parents.

3. Diploid (2*n*) cells have pairs of homologous chromosomes. Except for sex chromosomes, the homologues have the same length, shape, and gene sequence. All of them, including sex chromosomes, line up with their partner at meiosis.

4. Genes located on different chromosomes are inherited independently of each other, in accordance with the Mendelian principle of independent assortment (page 171).

5. Genes on the same chromosome tend to stay together when chromosomes move about during meiosis. But crossing over (breakage and exchange of corresponding segments of homologues) can disrupt such linkages and lead to genetic recombination (Figure 12.2).

6. Chromosomal abnormalities sometimes occur. A chromosome segment may be deleted, duplicated, inverted, or moved to a new location. Chromosomes also may not separate properly at meiosis, and so gametes end up with an abnormal chromosome number.

7. Independent assortment, crossing over, and chromosomal abnormalities play roles in evolution. By changing the genotype (genetic makeup), they lead to variations in phenotype (observable traits) upon which selective agents can act.

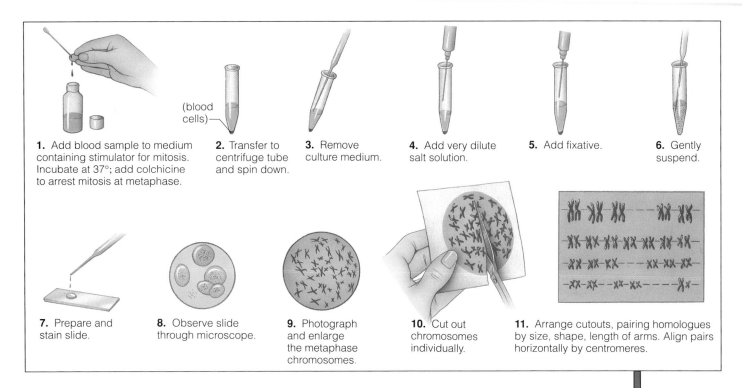

**1.** Add blood sample to medium containing stimulator for mitosis. Incubate at 37°; add colchicine to arrest mitosis at metaphase.

**2.** Transfer to centrifuge tube and spin down.

**3.** Remove culture medium.

**4.** Add very dilute salt solution.

**5.** Add fixative.

**6.** Gently suspend.

**7.** Prepare and stain slide.

**8.** Observe slide through microscope.

**9.** Photograph and enlarge the metaphase chromosomes.

**10.** Cut out chromosomes individually.

**11.** Arrange cutouts, pairing homologues by size, shape, length of arms. Align pairs horizontally by centromeres.

**Figure 12.3** Simplified picture of karyotype preparation, in this case a karyotype of a human male. Human cells have a diploid chromosome number of 46. The nucleus contains 22 pairs of autosomes and 1 pair of sex chromosomes (X and Y). Each chromosome of a given type has already undergone DNA replication.

## Autosomes and Sex Chromosomes

In the early 1900s, microscopists discovered a chromosomal difference between the sexes. Most of the chromosomes *are* the same number and the same type in both sexes; these were named **autosomes.** Depending on the species, however, one or two chromosomes *are not* the same in males and females; these were named sex chromosomes. **Sex chromosomes** carry hereditary instructions about *gender*—that is, whether a new individual will be male or female. For example, human females have two sex chromosomes designated "X." Males have one X chromosome and another, physically different chromosome designated "Y." (That is the most familiar pattern, but notable exceptions include birds, bees, moths, butterflies, turtles, and crocodiles.)

Today, human autosomes and sex chromosomes can be precisely characterized at metaphase, when they are in their most condensed form. Then, each type has a certain length, banding pattern, and so on. (The bands appear because some regions condense more than others and absorb stain differently. Figure 10.6 is an example.) These features are used to create a **karyotype:** a visual representation in which the chromosomes of a cell at metaphase are arranged in order, from largest to smallest (Figure 12.3).

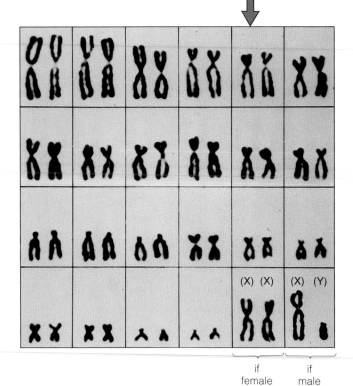

(X) (X)    (X) (Y)

if female     if male

Through microscopic studies, we know that each normal gamete produced by a female (XX) carries an X chromosome. Half the gametes produced by a male (XY) carry an X and half carry a Y chromosome. When a sperm and an egg both carry an X chromosome and combine at fertilization, the new individual will develop into a female. Conversely, when the sperm carries a Y chromosome, the new individual will develop into a male (Figure 12.4).

The Y chromosome of humans carries very few genes, but among them is a "male-determining gene." The presence of its product causes a new individual to develop testes. In the absence of that gene product, ovaries will develop automatically, as shown in the *Commentary*. The hormones secreted from testes and ovaries trigger the development of male *or* female characteristics.

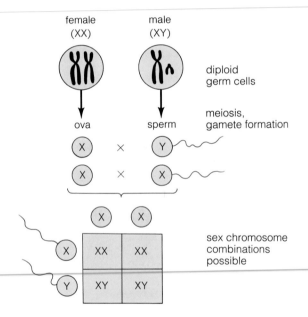

**Figure 12.4** Sex determination in humans. This same pattern occurs in many animal species. Only the sex chromosomes, not the autosomes, are shown. Males transmit their Y chromosome to their sons, but not to their daughters. Males receive their X chromosome only from their mother.

# *Commentary*

## Girls, Boys, and the Y Chromosome

The formation of male or female parts in a human embryo depends on gene interactions and on physical and chemical conditions in the mother's uterus. From the time of conception, the embryo that develops from the zygote is normally XX or XY. For the first month or so of development, the embryo is neither male nor female. However, ducts and other structures start forming that can go either way, so to speak (Figure *a*).

As the next four to six weeks of development unfold, testes—the primary male reproductive organs—begin to

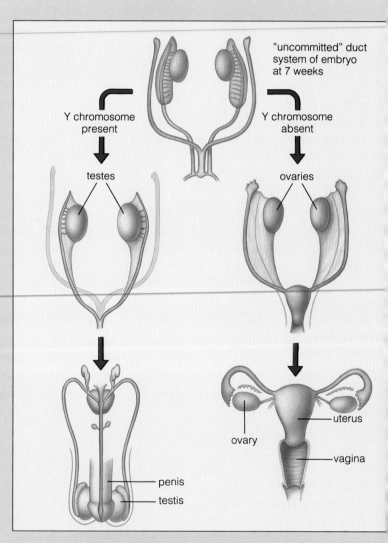

**a** Duct system in the early embryo that develops into a male *or* female reproductive system. Compare Figures 43.2 and 43.7.

develop in XY embryos. Apparently, a gene region on the Y chromosome governs the fork in the developmental road that can lead either to maleness or femaleness. If the embryo is XX, it has no such region, and it develops ovaries—the primary female reproductive organs. It does so automatically, in the absence of the Y chromosome.

Once their development is under way, the testes start producing sex hormones, including testosterone. These hormones influence the development of assorted parts that make up the male reproductive system, as described in Chapter 43. The newly forming ovaries also produce sex hormones, and these influence the development of the female reproductive system.

A newly identified region of the Y chromosome appears to be the master gene for sex determination. It has been called SRY (for sex-determining region of the Y chromosome). So far, the same gene has been identified in the DNA of human males and in male chimpanzees, rabbits, pigs, horses, cattle, and tigers. In all females tested, the gene was absent. Other tests with mice indicate that the gene region becomes active about the time that testes start developing.

In molecular structure, the SRY gene resembles the gene regions that are known to produce regulatory proteins. Such proteins can bind to certain parts of DNA and so turn genes on and off (page 234). Do the protein products of the SRY gene turn off the genes required for female development? Researchers are attempting to find out. For example, they are inserting the SRY gene into female mouse embryos. If those embryos go on to develop male characteristics, this will be evidence of a master regulatory role for the SRY gene.

**b** External appearance of developing reproductive organs.

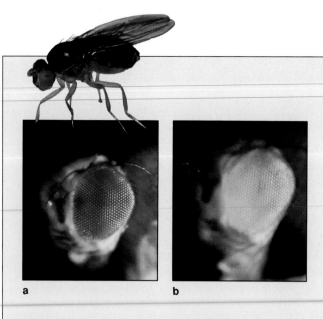

a      b

**Figure 12.5** X-linked genes: clues to patterns of inheritance.

In the early 1900s, the embryologist Thomas Hunt Morgan began work to explain the apparent connection between gender and certain nonsexual traits. For example, blood clotting in humans occurs in both males and females. Yet for centuries, it was known that hemophilia (a blood-clotting disorder) shows up most often in the males, not females, of a family lineage. This phenotypic outcome was not like anything Mendel saw in his hybrid crosses between pea plants. In those crosses, either one parent plant or the other could carry a recessive allele. It made no difference *which* parent carried it; the phenotypic outcome was the same.

Morgan studied eye color and other nonsexual traits in the fruit fly, *Drosophila melanogaster*. These small flies can be grown in bottles on bits of cornmeal, molasses, and agar. A female lays hundreds of eggs in a few days, and her offspring can reproduce in less than two weeks. Morgan could track hereditary traits through nearly thirty generations of thousands of flies in a year's time.

At first, all the flies were wild-type for eye color; they had brick-red eyes, as in (**a**). ("Wild-type" simply means the normal or most common form of a trait in a population.) Then, through an apparent mutation in a gene controlling eye color, a *white-eyed* male appeared (**b**).

Morgan established true-breeding strains of white-eyed males and females. Then he did a series of *reciprocal crosses*. (These are pairs of crosses. In the first, one parent displays the trait in question; in the second, the other parent displays the trait.)

White-eyed males were mated with true-breeding (homozygous) red-eyed females. All the $F_1$ offspring of the cross had red eyes—but of the $F_2$ offspring, only some of the *males* had white eyes. Then white-eyed females were mated with true-breeding red-eyed males. Of the $F_1$ offspring of that second cross, half were red-eyed females and half were white-eyed males. Of the $F_2$ offspring, 1/4 were red-eyed females, 1/4 were white-eyed females, 1/4 red-eyed males, and 1/4 white-eyed males!

The seemingly odd results indicated the gene for eye color was related to gender. Probably it was located on one of the sex chromosomes. But which one? Since females (XX) could be white-eyed, the recessive allele would have to be on one of their X chromosomes. Suppose white-eyed males (XY) also carry the recessive allele on their X chromosome—*and*

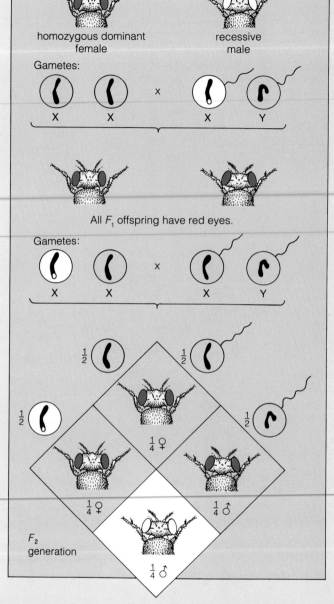

**c** Correlation between sex and eye color in *Drosophila*. Given the genetic makeup of the second generation, the recessive allele (depicted by the white dot) must be carried on the X chromosome only.

*suppose there is no corresponding eye-color allele on their Y chromosome.* Those males would have white eyes because the recessive allele would be the only eye-color gene they had!

In (**c**) are the results we can expect when the idea of an X-linked gene is combined with Mendel's concept of segregation. By proposing that a specific gene occurs on the X but not the Y chromosome, Morgan was able to explain the outcome of his reciprocal crosses. The results of the experiments matched the predicted outcomes.

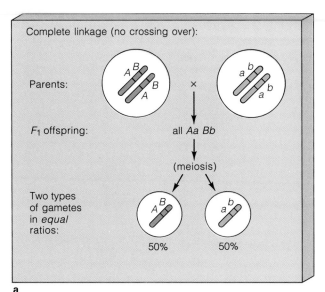

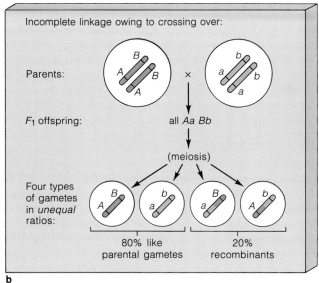

Complete linkage (no crossing over):

Incomplete linkage owing to crossing over:

**Figure 12.6** How crossing over can affect gene linkage, using two gene loci as the example.

The human X chromosome carries at least 100 and possibly more than 200 genes. Like all the other chromosomes, it carries some genes associated with sexual traits (such as the distribution of body fat). But most of the genes on the X chromosome are concerned with *nonsexual* traits, such as eye color.

Any gene on the X or Y chromosome may be called a "sex-linked gene." However, researchers now use the more precise designations, **X-linked** or **Y-linked genes**.

Some time ago, Thomas Hunt Morgan and his coworkers performed a series of hybridization experiments with the fruit fly, *Drosophila melanogaster*. As described in Figure 12.5, their work led to the discovery of X-linked genes, and it reinforced a major concept: *Each gene is located on a specific chromosome.*

### Linkage and Crossing Over

Through their studies of the fruit fly, researchers came to realize that many traits were being inherited *as a group* from one parent or the other. They identified four groups that apparently corresponded to the haploid number of chromosomes (four) in the fruit fly's gametes. They suspected that the genes on any one of those chromosomes were probably staying together during meiosis and gamete formation.

The term **linkage** is now used to describe the tendency of genes located on the same chromosome to end up together in the same gamete. Linkage is not inevitable, however. It can be disrupted by **crossing over**—the breakage and exchange of segments between homologous chromosomes (Figure 12.2).

Think about the location of any two genes on the same chromosome. The probability of a crossover occurring between those two genes is proportional to the distance separating them along the chromosome. Suppose two genes *A* and *B* are twice as far apart as two other genes, *C* and *D*:

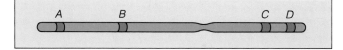

We would expect crossing over to disrupt the linkages between *A* and *B* much more frequently than those between *C* and *D*.

Two genes located physically close together on a chromosome nearly always end up in the same gamete; they are very closely linked (Figure 12.6a). Two genes relatively far apart are more vulnerable to crossing over and recombination, in comparison with closely linked genes (Figure 12.6b). Two genes very far apart on the same chromosome are affected by crossing over so often that they may appear to assort independently.

**The farther apart two genes are on a chromosome, the greater will be the frequency of crossing over and recombination between them.**

As an example of this, think about a watermelon with a green rind and one with a striped rind. The gene responsible for this trait happens to be closely linked to the gene for melon shape (round or oblong). Green and round are the dominant phenotypes. In one experiment, a homozygous dominant plant (true-breeding for green, round melons) was crossed with a homozygous recessive plant (true-breeding for striped, oblong melons). All plants of the $F_1$ generation produced green, round melons. In a testcross, $F_1$ plants were hybridized with the homozygous recessive parent. Knowing that the genes governing the two traits are closely linked, you might expect that the phenotypic ratio of the $F_2$ generation was 1:1, as shown here:

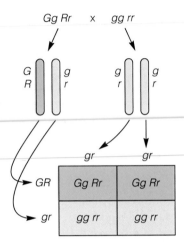

However, these were the observed $F_2$ phenotypes resulting from the cross:

| | |
|---|---|
| 46 green, round | 4 green, oblong |
| 47 striped, oblong | 3 striped, round |

We can explain the results by recognizing that 7 percent of the progeny (the 4 green, oblong and the 3 striped,

round) inherited a chromosome that had undergone a crossover between the gene locus for rind color and the gene locus for rind shape:

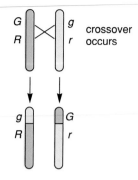

The patterns in which these and other genes are distributed into gametes tell us something about their organization on chromosomes. The patterns are so regular that they can be used to determine the *positions* of genes relative to one another. Plotting the positions of genes on a given chromosome is called linkage mapping. Figure 12.7 gives an example of a linkage map.

Of the several thousand known genes in the four chromosomes of *Drosophila* gametes, the positions of about a thousand have been mapped. There are many more genes in human chromosomes. As you will read in a later chapter, the locations of all human genes may be identified through an ambitious, long-term research project that is now under way. We have a long way to go in mapping the physical basis of heredity. But it is clear that genes are carried linearly, one after another, in human chromosomes—just as they are in fruit flies and all other organisms.

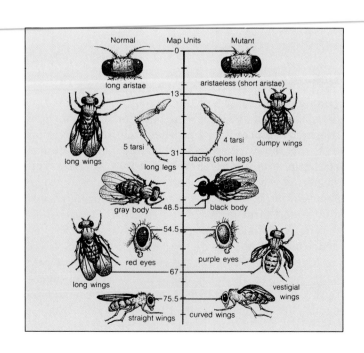

**Figure 12.7** Genetic mapping of genes on a segment of chromosome 2 in *D. melanogaster.* Such maps don't show actual physical distances between genes. Rather they show relative distance between gene locations that undergo crossing over and other chromosomal rearrangements. Only if the probability of crossing over were equal along the chromosome's length (which it is not) would it be possible to calculate physical distance exactly.

Here, distances between genes are measured in map units, based on the frequency of recombination between the genes. (One genetic map unit = 1 percent recombination.) Thus, if the frequency turns out to be 10 percent, the genes are said to be separated by 10 map units. The amount of recombination to be expected between "vestigial wings" and "curved wings," for instance, would be 8.5 percent (75.5−67).

# CHROMOSOME VARIATIONS IN HUMANS

Pea plants and fruit flies lend themselves to genetic analysis. They grow and reproduce rapidly in small spaces, under controlled conditions. Because flies or pea plants have a much shorter life span than the geneticist who studies them, their traits can be tracked through many generations in relatively little time.

Humans are another story. We humans live under variable conditions in diverse environments. Typically we find a mate by chance and reproduce if and when we want to. Human subjects live just as long as the geneticists who study them, so tracking traits through generations is rather tedious. And human families are generally so small, there aren't enough numbers for easy statistical inferences about inheritance patterns.

Even so, human genetics is a rapidly growing field. Researchers use standardized methods for constructing **pedigrees**, or charts of genetic relationships of individuals (Figure 12.8). With pedigrees, they can identify inheritance patterns and track genetic abnormalities through several generations. Studying the same trait in many families increases the numerical base for analysis.

Keep in mind that "abnormality" and "disorder" are not necessarily the same thing. *Abnormal* means deviation from the average. An abnormality is a rare or less common occurrence, as when a person is born with six toes on each foot instead of five. Whether such a trait is viewed as disfiguring or merely interesting is subjective—there is nothing inherently life-threatening or even ugly about it. Other abnormalities cause mild to severe medical problems, and genetic *disorder* is the more appropriate word here.

**Figure 12.8** (**a**) Some symbols used in constructing pedigree diagrams. (**b**) Example of a pedigree for *polydactyly*, a condition in which an individual has extra fingers, extra toes, or both. The number of fingers on each hand is shown in black numerals, and the number of toes on each foot is shown in blue. The phenotype of female 1 is uncertain.

Polydactyly is an example of how gene expression can vary. As a human embryo develops, a dominant allele *D* controls how many sets of bones will form within the body regions destined to become hands and feet. The *Dd* genotype varies in how it is expressed. The pedigree shown here indicates the kind of variation possible.

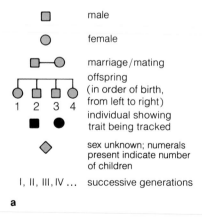

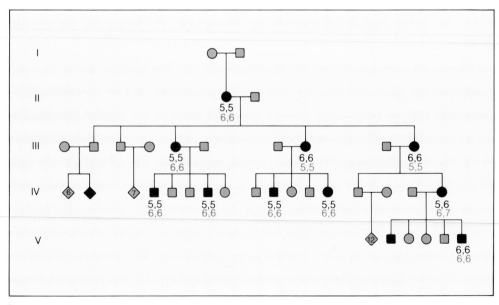

a

b

The next section of this chapter gives examples of the genetic disorders we deal with as individuals and as members of society. Table 12.1 lists some of them. The examples serve as a framework for considering some practical and ethical aspects of screening, counseling, and treatment programs.

**Autosomal Recessive Inheritance**

Sometimes a mutation produces a recessive allele on an autosome and gives rise to **autosomal recessive inheritance**. This condition has the following characteristics:

1. Males or females can carry the recessive allele on an autosome (not on a sex chromosome).

2. Heterozygotes generally are symptom-free. Homozygotes are affected.

3. When both parents are heterozygous, there is a 50 percent chance each child born to them will be heterozygous and a 25 percent chance it will be homozygous recessive (Figure 12.9). When both parents are homozygous, all of their children will be affected.

*Galactosemia*, an autosomal recessive condition, affects about 1 in 100,000 newborns. The condition arises when a breakdown product of lactose, or milk sugar, cannot be metabolized. Normally, lactose is first broken down to glucose and galactose, then ultimately to glucose-1-phosphate, which can be degraded by glycolysis. But galactosemics cannot produce molecules of an enzyme required for one of the conversion steps. They have two recessive alleles, coding for a defective form of the enzyme:

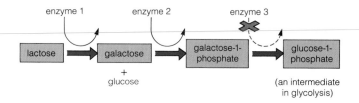

Galactose builds up in the blood, and in large concentrations it can damage the eyes, liver, and brain. Early symptoms include malnutrition, diarrhea, and vomiting. Without treatment, galactosemics usually die in childhood. However, unusually high concentrations can be detected in urine samples of homozygous recessive infants. If the disorder is detected early enough, infants can be put on a diet that includes milk substitutes and can grow up symptom-free.

**Autosomal Dominant Inheritance**

Recessive alleles that cause genetic disorders can persist at low but constant frequencies, because heterozygotes may still survive and reproduce. (Their one normal allele may yield enough of the required gene product.) But what if a *dominant* allele causes the disorder? In **autosomal dominant inheritance**, a dominant allele is usually expressed to some extent. If its expression reduces the chance of surviving and reproducing, its frequency among individuals in the population will decrease.

Even so, a few dominant alleles that cause pronounced disorders do remain in populations. How? Mutations can replenish the supply of defective alleles. Also, *some dominant alleles do not affect reproduction or they are only expressed after reproductive age*. Figure 12.10 shows an inheritance pattern for an autosomal dominant condition.

One autosomal dominant allele causes *achondroplasia*, a type of dwarfism, in about 1 in 10,000 individuals. When limb bones develop in affected children, cartilage forms in ways that lead to disproportionately short arms

| Table 12.1 | Examples of Human Genetic Disorders | |
|---|---|
| Disorder (or Abnormality)* | Main Consequences |
| **Autosomal Recessive Inheritance:** | |
| Albinism (174) | Absence of pigmentation (melanin) |
| Sickle-cell anemia (164, 175, 284) | Severe tissue, organ damage |
| Galactosemia (192) | Brain, liver, eye damage |
| Phenylketonuria (199) | Mental retardation |
| **Autosomal Dominant Inheritance:** | |
| Achondroplasia (193) | A type of dwarfism |
| Compodactyly (177) | Rigid, bent little fingers |
| Huntington's disorder (193) | Progressive, irreversible degeneration of nervous system |
| Polydactyly (191) | Extra digits |
| Progeria (192) | Premature aging |
| **X-Linked Inheritance:** | |
| Hemophilia A (194) | Deficient blood-clotting |
| Testicular feminizing syndrome (595) | Absence of male organs, sterility |
| **Changes in Chromosome Structure:** | |
| Cri-du-chat (195) | Mental retardation, skewed larynx |
| **Changes in Chromosome Number:** | |
| Down syndrome (198) | Mental retardation, heart defects |
| Turner syndrome (201) | Sterility, abnormal development of ovaries and sexual traits |
| Klinefelter syndrome (201) | Sterility, mental retardation |
| XYY condition (201) | Mild mental retardation in some cases; no symptoms in others |

*Number in parentheses indicates the page(s) on which the disorder is described.

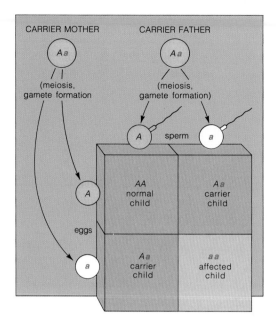

**Figure 12.9** Possible phenotypic outcomes for autosomal recessive inheritance when both parents are heterozygous carriers of the recessive allele (shaded red here).

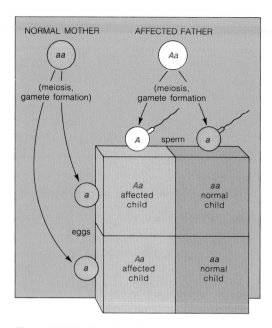

**Figure 12.10** Possible phenotypic outcomes for autosomal dominant inheritance, assuming the dominant allele is fully expressed in the carriers. (The dominant allele is shaded red.)

and legs. Affected persons are less than 4 feet, 4 inches tall. The dominant allele often has no other phenotypic effects in heterozygotes, which normally are fertile and so may reproduce. Homozygous dominant fetuses usually are stillborn.

*Huntington's disorder*, a rare form of autosomal dominant inheritance, causes progressive degeneration of the nervous system. In about half the cases, symptoms emerge from age forty onward—after most people have already had children. In time, movements become convulsive, brain function deteriorates rapidly, and death follows.

*Progeria*, or early aging is an example of autosomal dominant inheritance. Children affected by this rare disorder start to show signs of advanced aging when they are only five or six years old (figure 12.1). Their skin wrinkles, their hair thins, they start to suffer arthritis, and their blood vessels show arteriosclerosis. Frenquently, affected indivduals die of heart disease before they are ten years old.

### X-Linked Recessive Inheritance

Some genetic disorders fall in the category of **X-linked recessive inheritance**, which has these characteristics:

1. The mutated gene occurs on the X (not the Y) chromosome (Figure 12.11).

2. Heterozygous females are phenotypically normal; the nonmutated allele on their other X chromosome covers

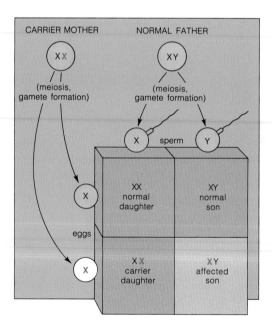

**Figure 12.11** Possible phenotypic outcomes for X-linked inheritance when the mother carries a recessive allele on one of her X chromosomes (shaded red here).

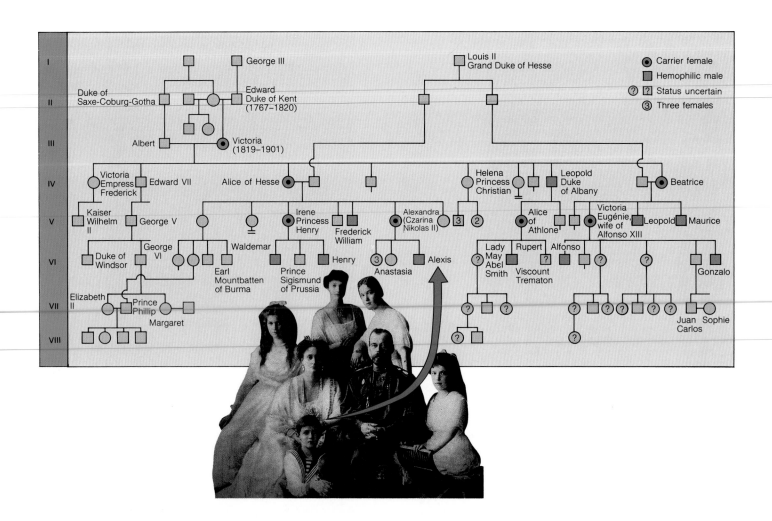

**Figure 12.12** Descendants of Queen Victoria, showing carriers and affected males that possessed the X-linked gene conferring the disorder hemophilia A. The photograph shows the Russian royal family members. The mother was a carrier of the mutated gene; Crown Prince Alexis was afflicted with the disorder. Many individuals of later generations are not included in the family pedigree.

the required function. Males typically are affected. They have only one allele for the trait on the X chromosome, and it is recessive.

3. When the male is normal but the female is heterozygous, there is a 50 percent chance each daughter born to them will be a carrier, and a 50 percent chance each son will be affected (Figure 12.11). When the female is homozygous recessive and the male is normal, all daughters will be carriers and all sons will be affected.

*Hemophilia A* is an example of X-linked recessive inheritance. Normally, a blood-clotting mechanism quickly stops bleeding from minor injuries. Some people bleed for an unusually long time because the mechanism is defective. The reactions leading to clot formation depend on the products of several genes. If any of the genes is

mutated, its defective product can cause one of several bleeding disorders.

In hemophilia A, the gene for a protein called clotting factor VIII is mutated. Males with a recessive allele on their X chromosome are always affected. They run the risk of dying from untreated bruises, cuts, or internal bleeding. Blood-clotting time is more or less normal in heterozygous females. The nonmutated gene on their other X chromosome produces enough factor VIII to cover the required function.

Hemophilia A affects only about 1 in 7,000 human males. The frequency of the recessive allele was unusually high among the royal families of nineteenth-century Europe, whose members often intermarried. Queen Victoria of England was a carrier, as were two of her daughters (Figure 12.12). At one time, eighteen of her sixty-nine descendants were affected males or female carriers.

Crown Prince Alexis of Russia was one of Victoria's hemophilic descendants. His affliction drew together an explosive cast of characters—Czar Nicholas II, Czarina Alexandra (a granddaughter of Victoria and a carrier), and the power-hungry monk Rasputin, who manipulated the aggrieved family to his political advantage. Indirectly, this hemophilic child helped catalyze events that brought an end to dynastic rule in the Western world.

## Changes in Chromosome Structure

On rare occasions, chromosome structure becomes abnormally rearranged. Deletions, duplications, inversions, and translocations are examples of such rearrangements.

A **deletion** is a loss of a chromosome segment. The loss occurs when an end segment of a chromosome breaks off or when viral attack, irradiation, or chemical action causes breaks in a chromosome region. The loss almost always means problems, for genes influencing one or more traits may be missing. For example, one deletion from human chromosome 5 leads to mental retardation and a malformed larynx. When affected infants cry, the sounds produced are more like meowing—hence the name of the disorder, *cri-du-chat* (meaning cat-cry). Figure 12.13 shows an infant affected by the disorder.

A **duplication** is a gene sequence in excess of its normal amount in a chromosome. This happens, for example, when a deletion from one chromosome is inserted into its homologue (Figure 12.14a). Duplications probably have been important in evolution. Cells require specific gene products, so mutations that alter the function of most genes probably would be selected against. But *duplicates* of a gene could change through mutation (the normal gene would still provide the required product). In time, they could yield products with related or even new functions. This apparently happened in chromosome regions coding for the polypeptide chains of the hemoglobin molecule. In humans and other primates, those regions have multiple copies of strikingly similar gene sequences and they produce whole families of slightly different chains.

An **inversion** is a chromosome segment that separated from the chromosome and then was inserted at the same place—but in reverse. The reversal alters the position and relative order of the genes on the chromosome (Figure 12.14b).

Most often, a **translocation** is the transfer of part of one chromosome to a *non*homologous chromosome. In some types of human cancer, for example, a segment of chromosome 8 has been transferred to chromosome 14. Genes on that segment had been precisely regulated at their normal chromosomal location, but controls over their expression apparently are lost at the new location.

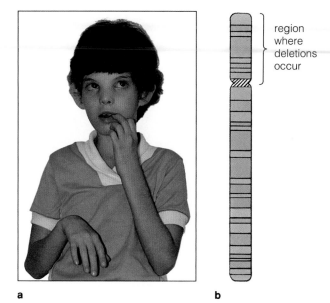

**a**        **b**

**Figure 12.13** Cri-du-chat syndrome. Affected infants have an improperly developed larynx, and their cries sound like a cat in distress. They show severe mental retardation. Outward symptoms include a rounded face, small cranium, and misshapen ears (**a**). The deletions occur in the short arm of chromosome 5, which must have a fragile region because breaks occur more frequently than at any other part of the chromosome (**b**).

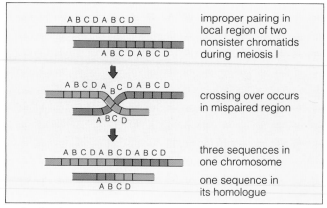

**a** Duplication

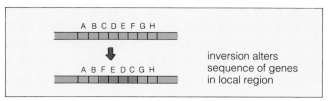

**b** Inversion

**Figure 12.14** Two kinds of changes in chromosome structure: a duplication (**a**) and an inversion (**b**).

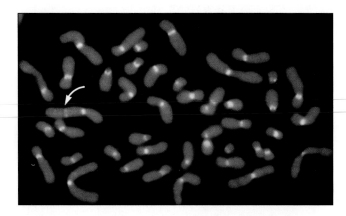

**Figure 12.15** Translocation. These metaphase chromosomes were stained with substances that react preferentially with kinetochore proteins, which show up as fluorescent yellow in micrographs. The arrow points to the inactivated kinetochores of chromosome 9—which fused with chromosome 11. (The normal kinetochores of chromosome 11 are visible to the right of the fusion point.)

Once in a while, translocation occurs in such a way that almost an entire chromosome becomes fused to a nonhomologous chromosome—and the two continue to function. This structural rearrangement changes the chromosome number. Figure 12.15 shows the result of such a fusion in a cell from a human female. Her diploid chromosome number is 45 instead of the normal 46.

Chromosomes have undergone structural changes during the evolution of humans and their closest primate relatives. Eighteen of the twenty-three pairs of human chromosomes are nearly identical to their counterparts in chimpanzees and gorillas. However, karyotype analysis of banding patterns shows that inversions and translocations occurred in the others.

### Changes in Chromosome Number

Sometimes new individuals end up with the wrong chromosome number. The phenotypic effects range from minor physical changes to devastating or lethal disruption of organ systems. More often, the affected individuals are miscarried—that is, spontaneously aborted before birth. There are two general categories of change in chromosome number. One is aneuploidy, the other is polyploidy.

**Categories of Change.** Gametes or cells of a new individual may end up with one extra or one less than the

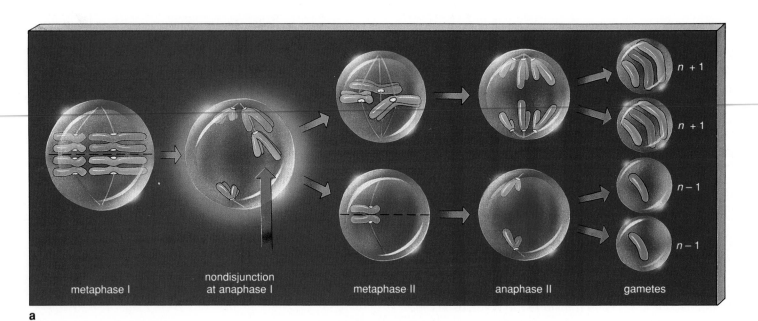

metaphase I     nondisjunction at anaphase I     metaphase II     anaphase II     gametes

a

**Figure 12.16** Two examples of nondisjunction, an event that can change the chromosome number in gametes (hence in offspring).

parental number of chromosomes. This abnormal condition is called **aneuploidy**. It is a major cause of reproductive failure in humans. For reasons not fully understood, its rate of occurrence among humans is higher (by as much as ten times) than it is in other mammals. On the average, aneuploidy probably affects one of every two newly fertilized human eggs. Of the human embryos that have been miscarried and autopsied each year, about 70 percent were aneuploids.

Gametes or cells of a new individual also may end up with three or more of each type of chromosome characteristic of the parental stock. This condition is called **polyploidy**. As described in Chapter 18, polyploidy is common in the plant kingdom—in fact, it has given rise to about half of all flowering plant species. Polyploidy is lethal for humans. All but 1 percent of human polyploids die before birth, and the rare few who are born die within a month. Most likely, the resulting imbalance between the genetic instructions of autosomes and sex chromosomes disrupts key steps in the long, complex pathways of development and reproduction.

**Mechanisms of Change.** An abnormal chromosome number may arise during mitotic or meiotic cell divisions or during the fertilization process.

For example, even before meiosis occurs, mitotic cell divisions produce the germ cells that give rise to sperm and eggs. Suppose DNA replication and mitosis itself proceed as usual—but something prevents cytoplasmic

division. The result will be a "tetraploid" germ cell with *four* of each type of chromosome instead of two. If meiotic cell division follows, the resulting gametes will be diploid instead of haploid—and so on down the line for an unfortunate embryo.

As another example, one or more pairs of chromosomes may fail to separate during mitosis or meiosis, an event called **nondisjunction**. Perhaps a chromosome does not separate from its homologue at anaphase I of meiosis. Or perhaps sister chromatids of a chromosome do not separate at anaphase II. As Figure 12.16 shows, some or all of the gametes can end up either with one extra chromosome or with one less than the parental number.

Suppose a human gamete has one *extra* chromosome ($n + 1$). If it combines with a normal gamete at fertilization, the diploid cells of the new individual will have three of one type of chromosome ($2n + 1$). Such a condition is called *trisomy*. If the gamete is *missing* a chromosome, the new individual will have a chromosome number of $2n - 1$. One chromosome in its diploid cells will not have a homologue—a condition called *monosomy*.

After asking whether their new baby is a girl or a boy, most parents apprehensively ask, "Is our baby normal?" They naturally want their baby to be free of genetic disorders, and most of the time it is. Chapter 43 describes the story of human reproduction and development when all goes well. Here, let's briefly consider some of the rare cases when it does not.

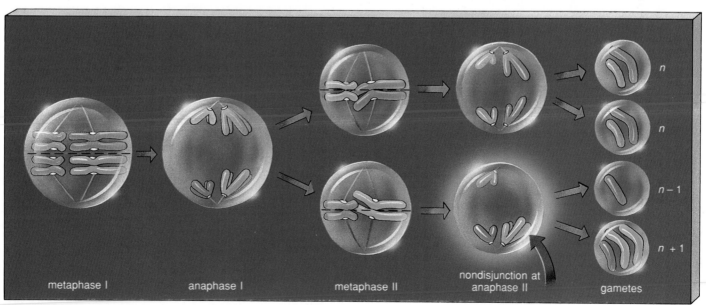

metaphase I          anaphase I          metaphase II          nondisjunction at anaphase II          gametes

$n$

$n$

$n - 1$

$n + 1$

b

**Down Syndrome.** Sometimes the cells of an individual have three copies of chromosome 21, which is one of the smallest chromosomes in human cells. That condition, called trisomy 21, leads to Down syndrome. ("Syndrome" simply means a set of symptoms characterizing a particular disorder; typically the symptoms occur together.) Figure 12.17 shows a karyotype of a girl with Down syndrome.

Trisomic 21 embryos are often miscarried, but about 1 of every 1,000 liveborns in North America alone will develop the disorder. Most affected children show moderate to severe mental retardation, and about 40 percent have heart defects. Skeletal development is slower than normal and muscles are rather slack. Older children are shorter than normal and have distinguishing facial features, including a small skin fold over the inner corner of the eyelid. With special training, affected individuals often participate in normal activities, and they enjoy life to the fullest extent allowed by their condition (Figure 12.18). Down syndrome is one of the genetic disorders that can be detected by prenatal diagnosis (see the *Commentary*).

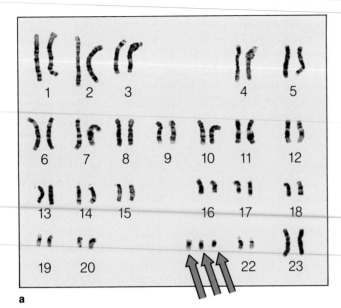

a

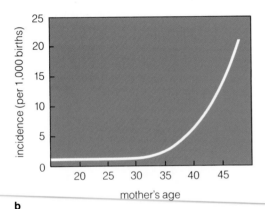

b

**Figure 12.17** (**a**) Karyotype of a girl with Down syndrome; red arrows identify the trisomy of chromosome 21. (**b**) Relationship between the frequency of Down syndrome and the mother's age. Results are from a study of 1,119 children with the disorder who were born in Victoria, Australia, between 1942 and 1957.

**Figure 12.18** Children with Down syndrome.

## Prospects and Problems in Human Genetics

Chances are, you know of someone who has a genetic disorder. Of all newborns, possibly 1 percent will have pronounced problems arising from a chromosomal aberration. Between 1 and 3 percent more will have problems because of mutant genes that produce defective proteins or none at all. Of all patients in children's hospitals, 10 to 25 percent are treated for problems arising from genetic disorders.

Human geneticists work to diagnose and treat heritable disorders. However, we apparently cannot approach the disorders in the same way we approach infectious diseases (such as influenza, measles, and polio). Infectious agents are enemies from the environment, so to speak. We have had no qualms about mounting counterattacks with immunizations and antibiotics that can eliminate the agents or bring them under control. With genetic disorders, the problem is inherent in the chromosomes of individual human beings.

How do we attack an "enemy" within? Do we institute regional, national, or global programs to identify affected persons? Do we tell them they are "defective" and run a risk of bestowing their disorder on their children? Who decides which alleles are "harmful"? Should society bear the cost of treating disorders such as Down syndrome? If so, should society also have some say in whether affected fetuses will be born at all, or aborted? These questions are only the tip of an ethical iceberg, and answers have not been worked out in universally acceptable ways.

### Phenotypic Treatments

Genetic disorders cannot be permanently cured, but sometimes we can get around their phenotypic consequences. Treatments include diet modifications, environmental adjustments, surgery, and chemical modification of gene products.

Controlling the diet can suppress or minimize the outward symptoms of several disorders. Galactosemia is controlled this way (page 192). So is *phenylketonuria*, or PKU. Normally, a gene product (an enzyme) converts one amino acid to another (phenylalanine to tyrosine). However, the first amino acid builds up in people who are homozygous recessive for a mutated form of the gene. If the excess is diverted into other metabolic pathways, phenylpyruvic acid and other compounds may be produced. At high levels, phenylpyruvic acid can lead to mental retardation. A diet that provides the minimum required amount of phenylalanine will alleviate the symptoms of PKU. When the body is not called upon to dispose of excess amounts, affected persons can lead normal lives.

Diet soft drinks and many other products often are artificially sweetened with aspartame, which contains phenylalanine. Such products carry warning labels directed at phenylketonurics.

Environmental adjustments can alleviate the outward symptoms of other genetic disorders. True albinos, for example, can avoid direct sunlight. Individuals affected by sickle-cell anemia can avoid strenuous activity when oxygen levels are low, as at high altitudes.

Surgical reconstructions can correct or minimize many phenotypic defects. One type of *cleft lip* is a genetic abnormality of the upper lip. A vertical fissure cuts through the lip midsection and often extends into the roof of the mouth. Surgery can usually correct the lip's appearance and function.

Phenotypic treatments also include chemical modification of gene products. *Wilson's disorder* arises from an inability to use copper. The body requires trace amounts of copper, which serves as a cofactor for several enzymes. Excess copper can damage the brain and liver, leading to convulsions and death. One drug binds with the copper, and the excess is eliminated by the urinary system.

### Genetic Screening

In some cases, genetic disorders can be detected early enough to start preventive measures *before* symptoms can develop. In other cases, carriers who show no outward symptoms can be identified before giving birth to affected children. "Genetic screening" refers to large-scale programs to detect affected persons or carriers in a population. Most hospitals in the United States routinely screen all newborns for PKU, for example, so it is becoming less common to see people with symptoms of the disorder.

### Genetic Counseling

Sometimes prospective parents suspect they are very likely to produce a severely afflicted child. Either their first child or a close relative shows an abnormality and they now wonder if future children will be affected the same way. In such cases, clinical psychologists, geneticists, social workers, and other consultants may be brought in to give emotional support to parents at risk.

Counseling begins with accurate diagnosis of parental genotypes; this may reveal the potential for a specific disorder. Biochemical tests can be used to detect many metabolic disorders. Detailed family pedigrees can be constructed to aid the diagnosis.

For disorders showing simple Mendelian inheritance patterns, it is possible to predict the chances of having an affected child—but not all disorders follow Mendelian patterns. Even ones that do can be influenced by other factors, some identifiable, others not. Even when the extent of risk has been determined with some confidence, prospective parents must know the risk is the same for *each* pregnancy. For example, if a pregnancy has one chance in four of producing a child with a genetic disorder, the same odds apply to every subsequent pregnancy, also.

## Prenatal Diagnosis

What happens when a woman is already pregnant? Suppose a woman forty-five years old wants to know if the child she is bearing will develop Down syndrome. Through prenatal diagnosis, this and more than a hundred other genetic disorders can be detected before birth.

One detection procedure is based on *amniocentesis:* sampling the contents of the fluid-filled sac (amnion) that contains the fetus in the mother's uterus (Figure *a*). During the fourteenth to sixteenth week of pregnancy, the thin needle of a syringe is inserted through the mother's abdominal wall and into the amnion. Epidermal cells shed from the fetus float about in the amniotic fluid. The syringe withdraws some fluid—along with its sample of fetal cells. The cells are cultured and allowed to undergo mitosis. Abnormalities can be diagnosed by karyotype analysis and other tests that can be completed within weeks. Cells obtained by amniocentesis also can be tested for many biochemical defects, such as the one causing sickle-cell anemia.

Amniocentesis carries a risk: Care must be taken not to puncture the fetus or cause infection.

*Chorionic villi sampling* (CVS), a newer procedure, uses cells drawn from the chorion (a membranous sac surrounding the amnion). This procedure can be used earlier in pregnancy (by the eighth week), and results often are available in one or two days. However, a greater risk is associated with CVS, compared with amniocentesis.

What happens if an embryo is diagnosed as having a severe disorder? Unfortunately, there are no known cures for changes in chromosome number or structure. Prospective parents might decide on an *induced abortion* (an induced expulsion of the embryo from the uterus). Such decisions are bound by ethical considerations. We can expect the medical community to provide prospective parents with the information they need to make their own choice. That choice must be consistent with their own values, within the broader constraints imposed by society.

Removal of about 20 ml of amniotic fluid containing suspended cells sloughed off from the fetus

Centrifugation

Amniotic fluid: a few biochemical analyses possible

Fetal cells

Quick determination of fetal sex and analysis of purified DNA

Growth for weeks in culture medium

Biochemical analysis for the presence of about 40 metabolic disorders

Karyotype analysis for chromosomal aberration

**Turner Syndrome**. About 1 in every 5,000 newborns is destined to have Turner syndrome. Through a nondisjunction, affected individuals have a chromosome number of 45 instead of 46 (Figure 12.19). They are missing a sex chromosome (XO), this being a type of sex chromosome abnormality.

Turner syndrome occurs less often than other sex chromosome abnormalities, probably because most XO embryos are miscarried early in pregnancy. Affected persons have a distorted female phenotype. Their ovaries are nonfunctional, they are sterile, and secondary sexual traits fail to develop at puberty. Often they age prematurely and have shortened life expectancies.

**Klinefelter Syndrome**. Nondisjunction can give rise to XXY males who show Klinefelter syndrome (Figure 12.19). This sex chromosome abnormality occurs at a frequency of about 1 in 1,000 liveborn males. Symptoms do not develop until after the onset of puberty. XXY males show low fertility and some degree of mental retardation. Their testes are much smaller than normal, body hair is sparse, and there may be some breast enlargement. Injections of the hormone testosterone can reverse the feminized phenotype but not the sterility or mental retardation.

A number of Klinefelter males are "mosaics," with both XY and XXY cell lineages. This condition commonly is related to nondisjunction during gamete formation in the individual's mother, especially in cases where pregnancy occurred late in life. Nondisjunction also has produced XXYY, XXXY, and XXXXY forms of Klinefelter syndrome. Symptoms are severe in these individuals, and mental retardation is pronounced.

**XYY Condition**. About 1 in every 1,000 males has one X and two Y chromosomes, this being called an XYY condition. It is probably inappropriate to apply the term "syndrome" to this sex chromosome abnormality. XYY males tend to be taller than average and some may show mild mental retardation, but most are phenotypically normal.

At one time, XYY males were thought to be genetically predisposed to become criminals. But a comprehensive study in Denmark showed that the number who do end up in prison is no more notable than the percentage of other tall men. Compared to normal (XY) males, their rate of conviction was indeed greater: 41.7 percent compared to 9.3 percent. But this is not necessarily proof of a predisposition to crime. With their moderately impaired mental ability, XYY males simply may be easier to catch.

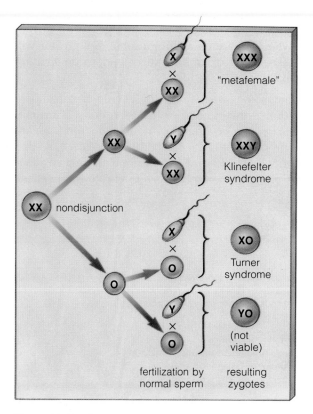

**Figure 12.19** Genetic disorders that result from nondisjunction of X chromosomes followed by fertilization involving normal sperm.

## SUMMARY

1. Genes, the units of instruction for heritable traits, are arranged linearly on a chromosome.

2. Sexual reproduction begins with meiosis and gamete formation and ends at fertilization. By this process, offspring receive the same number and type of chromosomes found in the body cells of their parents.

3. Human diploid cells have two chromosomes of each type (one from the mother, one from the father). The two are homologues; they pair at meiosis. The type called a sex chromosome is physically different in females (XX) and males (XY). All other pairs of chromosomes are autosomes; they are the same in males and females.

4. Homologues assort independently during meiosis.

5. Genes on one chromosome segregate from their partners on the homologous chromosome and end up in separate gametes.

6. Genes on the *same* chromosome tend to stay together during meiosis and end up in the same gamete. But

crossing over can disrupt such linkages. The farther apart two genes are on a chromosome, the greater will be the frequency of crossing over and recombination between them.

7. Chromosome *structure* can be altered by deletions, duplications, inversions, or translocations. Chromosome *number* can be altered by nondisjunction (the failure of chromosomes to separate during meiosis), so that gametes end up with one extra or one missing chromosome. Gene mutations, changes in chromosome structure, and changes in chromosome number cause many genetic disorders.

8. Variation in a population arises not only through mutation, but also through independent assortment, crossing over, and changes in the structure or number of chromosomes. These events can play roles in evolution. They change the genotypes (genetic makeup of individuals) and so lead to differences in phenotype (physical and behavioral traits) upon which selective agents can act.

## Self-Quiz *(Answers in Appendix IV)*

1. _____ segregate during _____ .
   a. Homologues; mitosis
   b. Genes on one chromosome; meiosis
   c. Homologues; meiosis
   d. Genes on one chromosome; mitosis

2. Two genes of a pair on homologous chromosomes end up in separate _____ .
   a. body cells
   b. gametes
   c. nonhomologous chromosomes
   d. offspring
   e. both b and d are possible

3. Genes on the same chromosome tend to remain together during _____ and end up in the same _____ .
   a. mitosis; body cell
   b. mitosis; gamete
   c. meiosis; body cell
   d. meiosis; gamete
   e. both a and d

4. The probability of a crossover occurring between two genes on the same chromosome is _____ .
   a. unrelated to the distance between them
   b. increased if they are closer together on the chromosome
   c. increased if they are farther apart on the chromosome
   d. impossible

5. Chromosome structure can be altered by _____ .
   a. deletions
   b. duplications
   c. inversions
   d. translocations
   e. all of the above

6. Nondisjunction can be caused by _____ .
   a. crossing over in meiosis
   b. segregation in meiosis
   c. failure of chromosomes to separate during meiosis
   d. multiple independent assortments

7. A gamete affected by nondisjunction would have _____ .
   a. a change from the normal chromosome number
   b. one extra or one missing chromosome
   c. the potential for a genetic disorder
   d. all of the above

8. Genetic disorders can be caused by _____ .
   a. gene mutations
   b. changes in chromosome structure
   c. changes in chromosome number
   d. all of the above

9. Which of the following contributes to variation in a population?
   a. independent assortment
   b. crossing over
   c. changes in chromosome structure and number
   d. all of the above

10. Match the chromosome terms appropriately.
    _____ crossing over
    _____ deletion
    _____ nondisjunction
    _____ translocation
    _____ gene mutation

    a. a change in DNA which may affect genotype and phenotype
    b. movement of a chromosome segment to a nonhomologous chromosome
    c. disrupts gene linkages during meiosis
    d. causes gametes to have abnormal chromosome numbers
    e. loss of a chromosome segment

## Genetics Problems *(Answers in Appendix III)*

1. Recall that human sex chromosomes are XX for females and XY for males.
   a. Does a male child inherit his X chromosome from his mother or father?
   b. With respect to an X-linked gene, how many different types of gametes can a male produce?
   c. If a female is homozygous for an X-linked gene, how many different types of gametes can she produce with respect to this gene?
   d. If a female is heterozygous for an X-linked gene, how many different types of gametes can she produce with respect to this gene?

2. One human gene, which may be Y-linked, controls the length of hair on men's ears. One allele at this gene locus produces nonhairy ears; another allele produces rather long hairs (hairy pinnae).

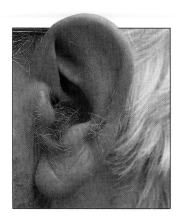

a. Why would you *not* expect females to have hairy pinnae?
b. If a man with hairy pinnae has sons, all of them will have hairy pinnae; if he has daughters, none of them will. Explain this statement.

3. Suppose that you have two linked genes with alleles *A,a* and *B,b* respectively. An individual is heterozygous for both genes, as in the following:

```
        A              B
  ——————————————————————————
        A              B

        a              b
  ——————————————————————————
        a              b
```

If the crossover frequency between these two genes is 0 percent, what genotypes would be expected among gametes from this individual, and with what frequencies?

4. In *D. melanogaster*, a gene influencing eye color has red (dominant) and purple (recessive) alleles. Linked to this gene is another that determines wing length. A dominant allele at this second gene locus produces long wings; a recessive allele produces vestigial (short) wings. Suppose a completely homozygous dominant female having red eyes and long wings mates with a male having purple eyes and vestigial wings. First-generation females are then crossed with purple-eyed, vestigial-winged males. From this second cross, offspring with the following characteristics are obtained:

252 red eyes, long wings
276 purple eyes, vestigial wings
 42 red eyes, vestigial wings
 30 purple eyes, long wings
———————————————————————
600 offspring total

Based on these data, how many map units separate the two genes?

5. Suppose you cross a homozygous dominant long-winged fruit fly with a homozygous recessive vestigial-winged fly. Shortly after mating, the fertilized eggs are exposed to a level of x-rays known to cause mutation and chromosomal deletions. When these fertilized eggs subsequently develop into adults, most of the flies are long-winged and heterozygous. However, a few are vestigial-winged.

Provide possible explanations for the unexpected appearance of these vestigial-winged adults.

6. Individuals affected by Down syndrome typically have an extra chromosome 21, so their cells have a total of 47 chromosomes. However, in a few cases of Down syndrome, 46 chromosomes are present. Included in this total are two normal-appearing chromosomes 21 and a longer-than-normal chromosome 14. Interpret this observation and indicate how these few individuals can have a normal chromosome number.

7. The mugwump, a type of tree-dwelling mammal, has a reversed sex-chromosome condition. The male is XX and the female is XY. However, perfectly good sex-linked genes are found to have the same effect as in humans. For example, a recessive, X-linked allele *c* produces red-green color blindness. If a normal female mugwump mates with a phenotypically normal male mugwump whose mother was color blind, what is the probability that a son from that mating will be color blind? A daughter?

8. One type of childhood *muscular dystrophy* is a recessive, X-linked trait in humans. A slowly progressing loss of muscle function leads to death, usually by age twenty or so. Unlike color blindness, this disorder is restricted to males, not ever having been found in a female. Suggest why.

---

## Selected Key Terms

aneuploidy *197*
autosomal dominant inheritance *192*
autosomal recessive inheritance *192*
autosome *185*
chromosomal deletion *195*
chromosomal duplication *195*
chromosomal inversion *195*
chromosomal translocation *195*
crossing over *189*
family pedigree *191*

karyotype *185*
linkage *189*
monosomy *197*
nondisjunction *197*
polyploidy *197*
sex chromosome *185*
trisomy *197*
X-linked gene *189*
X-linked recessive
 inheritance *193*

---

## Readings

Cummings, M. 1991. *Human Heredity: Principles and Issues.* St. Paul, Minnesota: West.

Edlin, G. 1988. *Genetic Principles: Human and Social Consequences.* Second edition. Portola Valley, California: Jones & Bartlett.

Fuhrmann, W., and F. Vogel. 1986. *Genetic Counseling.* Third edition. New York: Springer-Verlag.

Holden, C. 1987. "The Genetics of Personality." *Science* 237: 598–601. For students interested in human behavioral genetics.

Patterson, D. August 1987. "The Causes of Down Syndrome." *Scientific American* 257(2):52–60.

Weiss, R. November 1989. "Genetic Testing Possible Before Conception." *Science News* 136(21):326.

# 13 DNA STRUCTURE AND FUNCTION

## Cardboard Atoms and Bent-Wire Bonds

Linus Pauling in 1951 did something no one had done before. He figured out the three-dimensional shape of one of the main biological molecules. Through solid training in biochemistry, a talent for model building, and a few lucky hunches, he discovered the structure of the protein collagen. At the time, proteins were known to be composed of amino acids, strung together in polypeptide chains. Pauling perceived that each polypeptide chain of collagen twists helically, like a spiral staircase, and that hydrogen bonds hold the chain in its helical shape.

His discovery electrified the scientific community. If the structural secrets of proteins could be unraveled, why not other biological molecules? Further, wouldn't structural details about those molecules provide clues to how they function? And which would turn out to be the biggest prize of all—*the molecule that serves as a book of genetic information in every living cell*? Scientists all around the world started scrambling to be the first to find out.

Proteins seemed like good candidates. After all, heritable traits are spectacularly diverse. The molecules containing the information for those traits surely were structurally diverse also. With their potentially limitless combinations of amino acid subunits, proteins almost certainly could function as the sentences (genes) in each cell's book of inheritance.

Yet there was something about another substance—DNA—that tugged at more than a few good minds.

Certainly Pauling thought the possibility was there. Just when researchers in England were hot on the DNA trail, he wrote to Maurice Wilkins, one of the trailblazers. He requested copies of intriguing x-ray pictures that had been developed in Wilkins' lab. Wilkins' response was lukewarm, perhaps a delaying tactic. Why should he help a formidable competitor win the race to glory?

As it turned out, neither won the race. James Watson, a young postdoctoral student from Indiana University, had teamed up with Francis Crick, an energetic Cambridge University researcher. Watson and Crick spent long hours arguing over everything they had read about the size, shape, and bonding requirements of the known components of DNA. They fiddled with cardboard cutouts of those components. They badgered chemists to identify bonds they might have overlooked. They assembled models of bits of metal, held together with wire "bonds" bent at chemically correct angles.

In 1953, they finally put together a model that fit all the pertinent biochemical rules and all the facts about DNA that had been gleaned from other sources (Figure 13.1). Watson and Crick had discovered the structure of DNA. More than this, the breathtaking simplicity of that structure enabled them to solve a long-standing riddle about life—how it can show unity at the molecular level and yet give rise to so much diversity at the level of whole organisms.

With this chapter, we turn to some investigations that led to our current understanding of DNA structure and function, for they are revealing of how ideas are generated in science. On the one hand, having a shot at enduring fame and fortune quickens the pulse of competitive men and women in any profession, and scientists are no exception. On the other hand, science proceeds as a community effort, with individuals sharing not only what they can explain but also what they do not understand. Thus, even if an experiment "fails," it may turn up information that others can use or lead to questions that others can answer. Unexpected results, too, might be clues to something important about the natural world.

**Figure 13.1** James Watson and Francis Crick posing in 1953 by their newly unveiled model of DNA structure. Behind this photograph is a recent computer-generated model. It is more sophisticated in appearance, yet basically the same as the prototype that was built nearly four decades before.

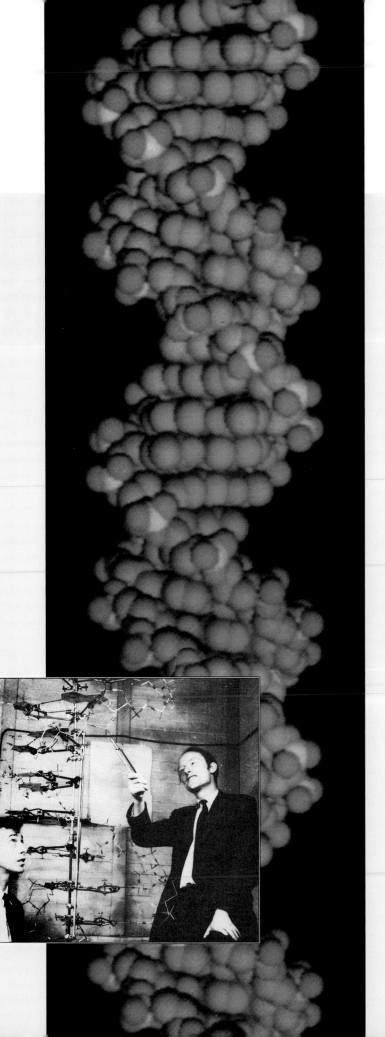

KEY CONCEPTS

**1.** Hereditary instructions in living cells are encoded in the linear sequence of nucleotides that make up DNA molecules. The four kinds of nucleotides in DNA differ in which nitrogen-containing base they contain. The bases are adenine, guanine, thymine, or cytosine.

**2.** In each DNA molecule, two strands of nucleotides are twisted together like a spiral staircase; they form a double helix. Hydrogen bonds occur between the bases of the two strands. As a rule, adenine pairs (hydrogen-bonds) only with thymine, and guanine pairs only with cytosine.

**3.** Before a cell divides, its DNA is replicated with the help of enzymes and other proteins. Each double-stranded DNA molecule starts unwinding, and a new, complementary strand is assembled on the exposed bases of each parent strand according to base-pairing rules.

**4.** There is only one DNA molecule (one double helix) in a chromosome. Except for bacterial cells, the chromosome also consists of many histones and other proteins that have roles in the structural organization of DNA.

## DISCOVERY OF DNA FUNCTION

One might have wondered, in the spring of 1868, why Johann Friedrich Miescher was collecting cells from the pus of open wounds and, later, from the sperm of a fish. Miescher wanted to identify the chemical composition of the nucleus, and he was interested in those cells because they are composed mostly of nuclear material, with very little cytoplasm. He succeeded in isolating an acidic substance, one with a notable amount of phosphorus. Miescher called it "nuclein." He had discovered what came to be known as deoxyribonucleic acid, or **DNA**.

The discovery caused scarcely a ripple through the scientific community. At the time, no one knew much about the physical basis of inheritance—that is, *which* chemical substance in cells actually encodes the instructions for reproducing parental traits in offspring. Only a

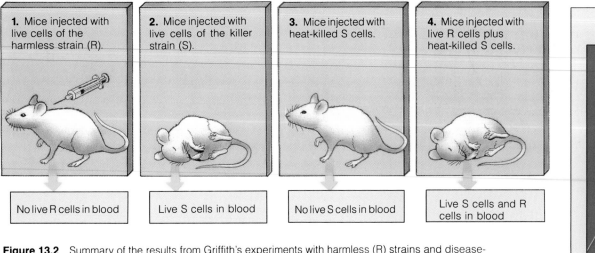

**1.** Mice injected with live cells of the harmless strain (R).

No live R cells in blood

**2.** Mice injected with live cells of the killer strain (S).

Live S cells in blood

**3.** Mice injected with heat-killed S cells.

No live S cells in blood

**4.** Mice injected with live R cells plus heat-killed S cells.

Live S cells and R cells in blood

DNA

protein coat

sheath

baseplate

tail fiber

a

**Figure 13.2** Summary of the results from Griffith's experiments with harmless (R) strains and disease-causing (S) strains of *Streptococcus pneumoniae*, as described in the text. You may be wondering why the S form is deadly and the R form harmless. The disease-causing strain produces a thick external capsule that protects the bacterial cells from attack by the host's immune system. Cells of the R strain form no such capsule. The host's defense system has the chance to destroy those cells before they can cause disease.

few researchers suspected that the nucleus might hold the answer. In fact, seventy-five years passed before DNA was recognized as having profound biological importance.

## A Puzzling Transformation

In 1928 an army medical officer, Fred Griffith, attempted to develop a vaccine against the bacterium *Streptococcus pneumoniae*, which causes the lung disease pneumonia. (Many vaccines are preparations of killed or weakened bacterial cells which, when introduced into the body, can mobilize the body's defenses against a later attack.) He never did create a vaccine, but his experiments unexpectedly opened a door to the molecular world of heredity.

Griffith isolated two strains of the bacterium, which he designated *S* and *R*. (When grown in culture, bacterial colonies of one strain have a *S*mooth surface appearance and colonies of the other have a *R*ough surface.) He used the strains in four experiments and came up with the results shown in Figure 13.2 and listed below:

1. Laboratory mice were injected with live R cells. They did not develop pneumonia; the R strain was harmless.

2. Mice were injected with live S cells. The mice died, and blood samples from them teemed with live S cells. The S strain was pathogenic, or disease-causing.

3. S cells were killed by exposure to high temperature. Mice injected with the heat-killed cells did not die.

4. Live R cells were mixed with heat-killed S cells and injected i to mice. Oddly, the mice died and blood samples fr om them teemed with live S cells!

What was going on in the fourth experiment? Maybe the heat-killed pathogens in the mixture were not really dead. But if that were true, the group of mice injected with heat-killed pathogenic cells alone would have contracted the disease, too. Maybe the harmless R cells in the mixture had mutated into the killer S form. But if that were true, the group of mice injected with R cells alone would have died.

The simplest explanation was this: Although heat did kill the pathogenic cells, it did not damage the chemical substance containing their hereditary information—including the part that specified "how to cause infection." Somehow the substance had been liberated from those dead cells, and it entered living cells of the harmless strain—where its instructions were expressed.

Further experiments made it clear that the harmless cells had indeed picked up instructions for causing infection and had been permanently transformed into pathogens because of it. Hundreds of generations of bacteria descended from the transformed cells also caused infections!

A few years later, researchers found that extracts of the killed pathogenic bacteria also could cause hereditary transformation. The microbiologist Oswald Avery and his colleagues began work to purify and experiment with the chemical substances of those extracts. This was the time when most biochemists believed that hereditary instructions were encoded in proteins, not DNA. That prevailing belief was challenged in 1944, when Avery's group reported that DNA probably was the substance of heredity.

Avery's key experiments could not be explained away. He could *block* transformation of harmless bacteria by adding an enzyme, pancreatic deoxyribonuclease, to

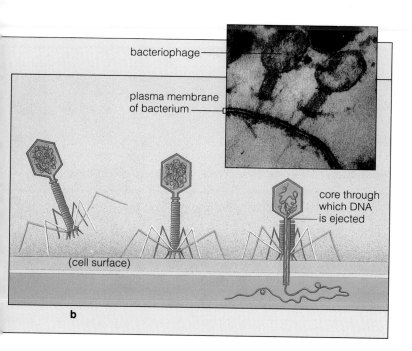

**Figure 13.3** (**a**) Components of a T4 bacteriophage. (**b**) When a T4 bacteriophage makes contact with the cell surface of *Escherichia coli*, proteins in its tail fibers chemically recognize molecules at the bacterial cell surface. The sheath contracts, and the DNA contained within the protein coat is injected into the cell.

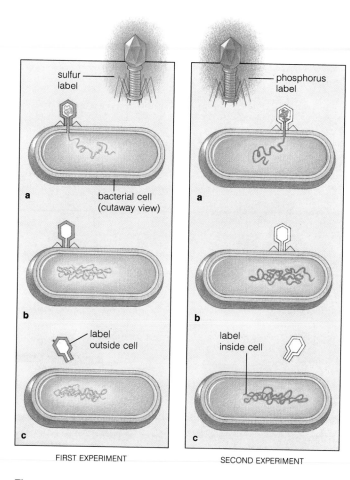

FIRST EXPERIMENT      SECOND EXPERIMENT

**Figure 13.4** Hershey-Chase bacteriophage studies pointing to DNA as the substance of heredity.

Bacteriophages, a class of viruses that infect bacteria, consist only of protein and DNA (Figure 13.3). When bacteriophages attach to a host cell, they inject their DNA into it. Soon the cell starts making viral nucleic acids and proteins (including enzymes) necessary to build new bacteriophages. Then viral enzymes degrade the bacterial cell wall. The cell bursts, releasing the new infectious generation.

(**a**) Bacteriophage proteins contain sulfur (S) but no phosphorus (P)—and the DNA contains phosphorus but no sulfur. In one experiment, bacteriophages were labeled with a radioisotope ($^{35}$S) to tag their proteins. In a second experiment, they were labeled with a radioisotope ($^{32}$P) to tag their DNA.

(**b**) Labeled bacteriophages were allowed to infect unlabeled cells suspended in fluid. Hershey and Chase whirred the fluid in a kitchen blender to remove the bacteriophage bodies from the cells. (**c**) Labeled protein remained in the fluid; it was associated with the bacteriophage bodies. Labeled DNA remained with the bacterial cells—it had to contain the hereditary instructions for producing new bacteriophages.

extracts of the pathogenic strain. The enzyme degrades DNA molecules but has no effect on proteins. In contrast, protein-degrading enzymes had no effect at all on the transforming activity.

Yet how were Avery's impressive findings received? Many (if not most) biochemists refused to give up on the proteins. His experimental results, they said, probably applied only to bacteria.

**Bacteriophage Studies**

While work was going on in Avery's laboratory, Max Delbrück, Alfred Hershey, and Salvador Luria were studying a class of viruses called **bacteriophages**, which infect bacterial cells. The infectious cycle starts when bacteriophages latch onto a target host cell. Within sixty seconds, an infected cell starts making the nucleic acids and proteins, including enzymes, necessary to build new bacteriophages. Then *lysis* occurs; the cell comes under chemical attack and dies. In this case, viral enzymes degrade the bacterial cell wall and the cell membrane becomes grossly leaky, thereby liberating a new infectious generation.

By 1952, researchers knew that some bacteriophages contain only DNA and protein. Electron micrographs had revealed that the main part of a virus particle remains at the surface of an infected host cell (Figure 13.3).

Clearly, genetic information was being injected *into* the cell body. Now Hershey and his colleague Martha Chase asked the question: Is the injected material DNA, protein, or both? They devised a way to track both substances through the infectious cycle (Figure 13.4).

Bacteriophage proteins contain sulfur but no phosphorus, and DNA contains phosphorus but no sulfur. Both elements have radioisotopes ($^{35}$S and $^{32}$P) that can

*All chromosomes contain DNA. What does DNA contain? Only four kinds of nucleotides. Each nucleotide has a five-carbon sugar (shaded red). That sugar has a phosphate group attached to the fifth carbon atom of its ring structure. It also has one of four kinds of nitrogen-containing bases (shaded blue) attached to its carbon atom. The nucleotides differ only in which base is attached to that atom:*

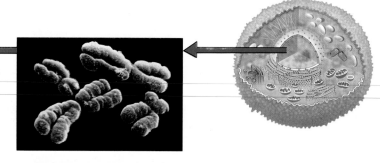

**Figure 13.5** The nucleotide subunits of DNA. The small numerals on the structural formulas identify the carbon atoms to which other parts of the molecule are attached.

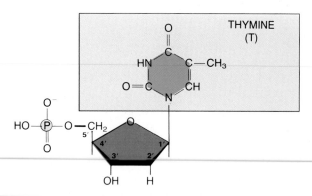

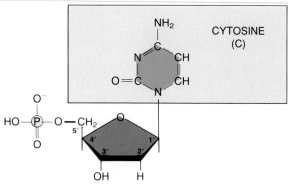

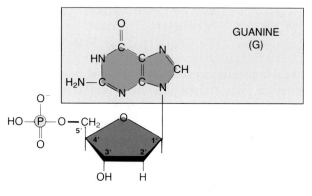

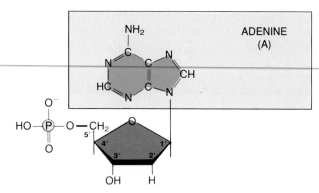

be used as tracers (page 22). Bacterial cells were grown on a culture medium containing $^{35}S$. When they synthesized proteins, the isotope became incorporated into the proteins and so made them radioactively labeled. When the cells were later infected, the new generation of bacteriophages assembled inside them also contained labeled protein. Similarly, cells grown on a culture medium containing $^{32}P$ ended up with labeled DNA.

As Figure 13.4 shows, labeled bacteriophages were now allowed to infect unlabeled cells. After the cycle of infection was under way, Hershey and Chase determined that the new bacteriophage generation incorporated labeled DNA—not the labeled protein.

Many bacteriophage studies confirmed Avery's conclusion that DNA is the hereditary substance. In fact, many different experiments have since been performed on cells of a variety of species from all five kingdoms. All confirm the following statement:

**Information for producing the heritable traits of single-celled and multicelled organisms are encoded in DNA.**

## DNA STRUCTURE

### Components of DNA

Long before the studies just described were proceeding, biochemists had shown that DNA contains only four kinds of nucleotides, the building blocks of nucleic acids. A **nucleotide** consists of a five-carbon sugar called deoxyribose, a phosphate group, and one of the following nitrogen-containing **bases**:

| adenine | guanine | thymine | cytosine |
|---------|---------|---------|----------|
| **(A)** | **(G)** | **(T)** | **(C)** |

Each type of nucleotide in DNA has its component parts joined together in much the same way as the others. But notice in Figure 13.5 that T and C are smaller, single-ring structures (called *pyrimidines*). A and G are larger, double-ring structures (called *purines*).

By 1949, the biochemist Erwin Chargaff had added these crucial insights about DNA structure. *First*, the four kinds of nucleotide bases making up a DNA molecule differ in relative amounts from species to species. *Second*, the amount of adenine always equals the amount of thymine (A = T), and the amount of guanine equals the amount of cytosine (G = C).

Now, here was something to think about! Could it be that the arrangement of the four kinds of bases in a DNA molecule represented the hereditary instructions? Maurice Wilkins, Rosalind Franklin, and others thought they might be able to identify that arrangement through *x-ray diffraction methods*. The atoms of any crystallized substance will disperse an x-ray beam. If the atoms in a crystal occur in a regular order, they will disperse the beam in a regular pattern. Such patterns show up as dots and streaks on a piece of film placed behind the crystal. By itself, the pattern on the exposed film does *not* reveal molecular structure. But it can be used to calculate the position of groups of atoms relative to one another in the crystal.

The researchers knew that DNA molecules are too large to be crystallized. But a suspension of DNA could be spun rapidly, spooled onto a rod, and gently pulled into gossamer fibers, like cotton candy. DNA molecules would be oriented in a regular pattern in such fibers, and those could be subjected to x-ray diffraction analysis.

Franklin obtained the best x-ray diffraction images, which provided convincing evidence that DNA had the following features. First, DNA had to be long and thin, with a uniform 2-nanometer diameter. Second, its structure had to be highly repetitive: Some part of the molecule was repeated every 0.34 nanometer, and a different part was repeated every 3.4 nanometers. Third, DNA might be helical, with a shape like a circular stairway.

## Patterns of Base Pairing

While work was proceeding in Wilkins' laboratory, Watson and Crick joined the search for the structure of DNA. According to Chargaff's data, the amount of adenine in DNA always equals the amount of thymine, and the amount of guanine always equals that of cytosine. According to Franklin's data, DNA has a uniform diameter. The sugar components of the different nucleotides were probably covalently bonded one after another along the length of the molecule. Thus DNA probably had some sort of sugar-phosphate backbone.

Watson and Crick reasoned that the *double*-ringed A and G bases of the nucleotides were probably paired with the *single*-ringed T and C bases along the entire length of DNA. Otherwise, DNA would bulge where two double rings were linked and narrow down where two

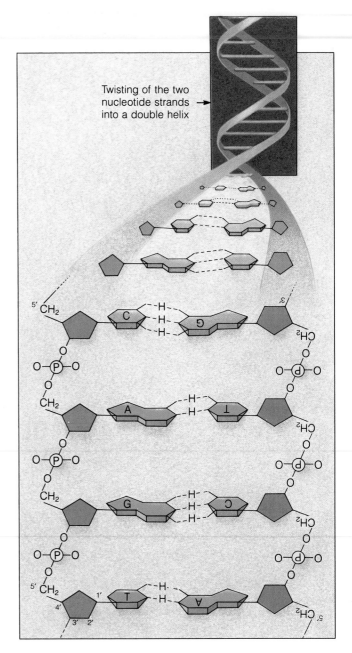

**Figure 13.6** Arrangement of bases (blue) in a DNA double helix.

single rings were linked. Watson and Crick shuffled and reshuffled paper cutouts of the nucleotides. They realized that in certain orientations, A and T could become linked by two hydrogen bonds, and G and C could become linked by three. Suppose there were *two strands* of nucleotides, with their bases facing each other. The hydrogen bonds could easily bridge the gap between them, like rungs of a ladder.

Scale models were constructed of how the "ladder" might look. The only model that fit all available data had A-T and G-C pairs. And those pairs formed the proper hydrogen bonds only when two sugar-phosphate backbones of two DNA strands ran in *opposing directions* and were twisted together to form a *double helix* (Figure 13.6).

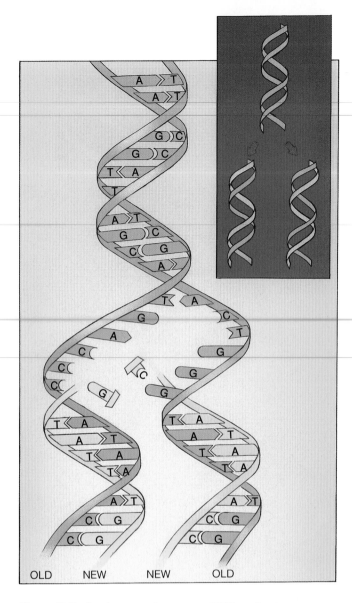

**Figure 13.7** Semiconservative nature of DNA replication. The original two-stranded DNA molecule is shown in blue. A new strand (yellow) is assembled on each of the two original strands.

OLD    NEW    NEW    OLD

As you can see, DNA molecules show both constancy and variation in their structure. This is the molecular foundation for the unity and diversity of life.

Base pairing between the two nucleotide strands in DNA is *constant* for all species (adenine to thymine, guanine to cytosine).

The base sequence (that is, which base follows the next in a nucleotide strand) is *different* from species to species.

## DNA REPLICATION

### Assembly of Nucleotide Strands

The discovery of DNA structure was a turning point in studies of inheritance. Until then, no one had any idea of how the hereditary material is replicated (that is, duplicated) prior to cell division. The Watson-Crick model suggested at once how this might be done.

Hydrogen bonds hold together the two nucleotide strands making up the DNA double helix, and those weak bonds are readily broken. Enzymes acting on a given region of the DNA molecule can cause one strand to unwind from the other, leaving bases exposed in the unwound region. Cells have stockpiles of free nucleotides, and these pair with exposed bases. Thus, each parent strand remains intact and a companion strand is assembled on each one according to the base-pairing rule.

As replication proceeds, each parent strand is twisted into a double helix with its new, partner strand. Because the parent strand is conserved, each "new" DNA molecule is really half-old, half-new. That is why the process is called **semiconservative replication** (Figure 13.7).

Prior to cell division, the double-stranded DNA molecule unwinds and is replicated. Each parent strand remains intact—it is conserved—and a new, complementary strand is assembled on each one.

### A Closer Look at Replication

**Origin and Direction of Replication.** Where does replication of the DNA molecule actually begin? The two strands of the double helix start to unwind at one or more distinct sites, each being a short, specific base sequence called the "origin." A viral or bacterial DNA molecule usually has one origin; a eukaryotic DNA molecule has many. Unwinding usually proceeds simul-

In the Watson-Crick model, then, hydrogen bonds join the bases of one strand with bases of the other. For the entire length of a DNA molecule, adenine always pairs with thymine, and cytosine always pairs with guanine. However, the *order* of bases in a nucleotide strand can vary greatly from one species to the next. In even a tiny stretch of DNA from a rose, gorilla, human, or any other organism, the base sequence might be:

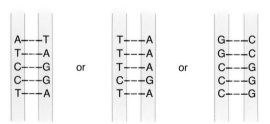

```
A----T          T----A          G----C
T----A          T----A          G----C
C----G   or     T----A   or     C----G
C----G          C----G          C----G
T----A          T----A          C----G
```

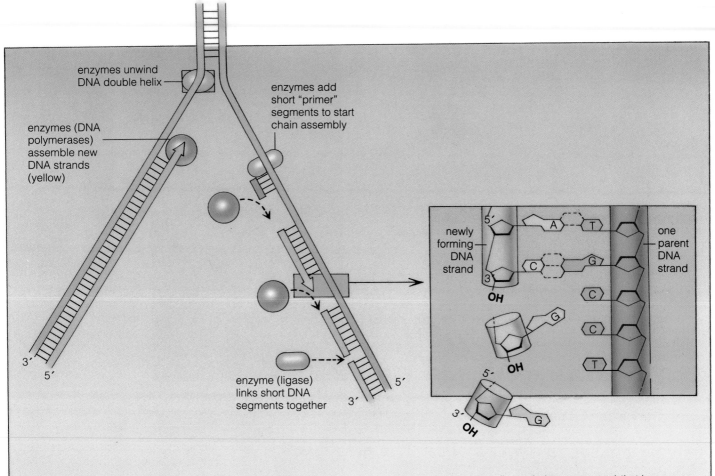

**Figure 13.8** A closer look at how a DNA molecule is replicated.

Assembly of new strands proceeds at replication forks. In these limited V-shaped regions, enzymes unwind the parent DNA double helix, and other enzymes assemble a new DNA strand on the exposed regions of each parent strand, which serves as a template. As discovered by Reiji Okazaki, DNA assembly is usually *continuous* on one parent template but *discontinuous* on the other. In discontinuous synthesis, short stretches of nucleotides are

assembled behind "start" tags (primer segments) that become positioned at intervals along a parent DNA strand. Then enzymes link the short stretches of DNA into a single chain.

As the boxed inset suggests, nucleotides can be added to a newly forming DNA chain in the 5′ → 3′ direction only. Bases projecting from a parent template dictate which kind of nucleotide can be added next. But an exposed —OH group must be present on the growing end of a DNA strand if enzymes are to catalyze the addition of more nucleotides to it.

taneously in both directions away from an origin. Strand assembly occurs behind each "fork" that continues to advance as the double helix is being unwound:

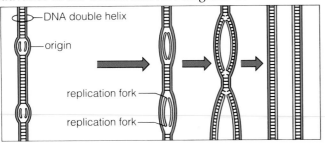

**Energy and Enzymes for Replication**. A DNA double helix does not unwind all by itself during replication. It takes a battery of enzymes and other proteins to unwind

the molecule, keep the two strands separate behind the replication forks, and assemble a new strand on each one. Even while one DNA region is being unwound, enzymes are winding up the replicated regions.

**DNA polymerases** are major replication enzymes. They govern nucleotide assembly on a parent strand (Figure 13.8). They also "proofread" the growing strands for mismatched base pairs, which are replaced with correct bases. The proofreading function is one reason DNA is replicated with such accuracy. On the average, for every 100 million nucleotides added to a growing strand, only *one* mistake slips through the proofreading net.

Where does the energy come from to drive replication? It happens that the free nucleotides brought up for strand assembly are not quite in the form shown in Figure 13.5. They are triphosphates, meaning they have

**Figure 13.9** (*Below*) Structural organization of DNA in mammalian metaphase chromosomes.

(**a**) Chromosomes in two chicken cells at mitosis, made visible by fluorescence microscopy. Blue-stained regions indicate where a fluorescent dye bound with DNA. Pink indicates where the dye bound with a molecule that can attach specifically to a type of chromosomal protein. The protein, topoisomerase II, is an enzyme that cuts and reseals DNA during replication. Its action helps prevent tangles and counteracts torsional stress when the DNA molecule twists about.

Notice how DNA loops fan out from the protein scaffold. The cells were treated with a buffer that caused the metaphase chromosomes to loosen up. The same kind of looping can be seen in electron micrographs of metaphase chromosomes from a HeLa cell (page 147). The chromosome in (**b**, **c**) had its histones removed by treating the cell with a mild detergent. The shape of the metaphase scaffolding is still evident. Topoisomerase II is the major protein of this scaffold.

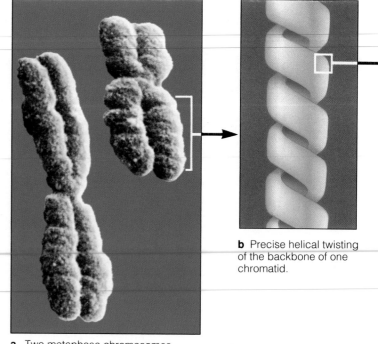

**b** Precise helical twisting of the backbone of one chromatid.

**a** Two metaphase chromosomes, each in the duplicated state.

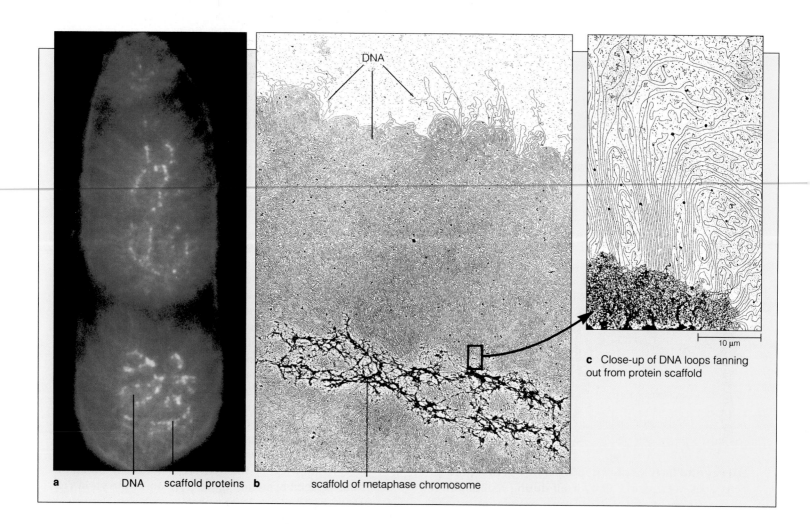

**a**      DNA    scaffold proteins

**b**      scaffold of metaphase chromosome

**c** Close-up of DNA loops fanning out from protein scaffold

10 µm

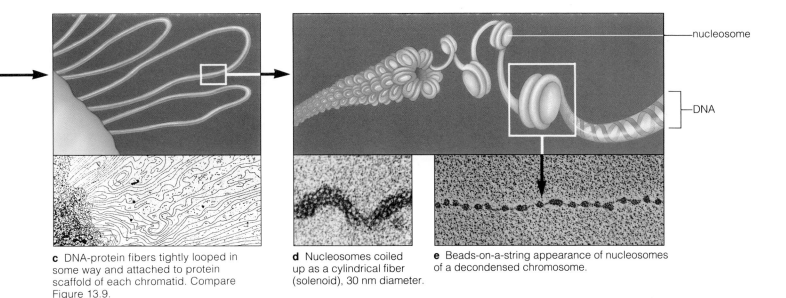

c DNA-protein fibers tightly looped in some way and attached to protein scaffold of each chromatid. Compare Figure 13.9.

d Nucleosomes coiled up as a cylindrical fiber (solenoid), 30 nm diameter.

e Beads-on-a-string appearance of nucleosomes of a decondensed chromosome.

**Figure 13.10** Levels of organization of DNA in a eukaryotic chromosome.

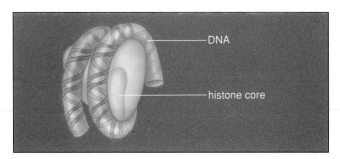

f Nucleosome (a double loop of DNA around a core of histone molecules).

three phosphate groups attached, not one. Triphosphates readily give up phosphate groups and so transfer energy to specific reactions. During replication, DNA polymerases use the energy released when two of the phosphate groups are split away. That energy drives the addition of nucleotides to a growing DNA strand. The unwinding process runs on energy provided by ATP.

## ORGANIZATION OF DNA IN CHROMOSOMES

There is one DNA molecule in each chromosome. If you put the DNA of all forty-six chromosomes of a human cell into a single line, end to end, it would extend over a meter. All that DNA might be a tangled mess if it were not for its precise organization (Figure 13.9).

The DNA of humans and all other eukaryotes is tightly bound with many proteins, including **histones.**

Some histones are like spools for winding up small stretches of DNA. Each histone-DNA spool is a **nucleosome** (Figure 13.10). Another histone (H1) stabilizes the arrangement and plays a role in higher levels of organization. The chromosome becomes coiled repeatedly through interactions between the histones and DNA; this process greatly increases its diameter.

Further folding results in a series of loops. Proteins other than histones may serve as a structural scaffold for the loops. Experiments suggest that at interphase the DNA and scaffold proteins remain in association. They also suggest that the scaffold regions occur *between* genes, not in regions that contain the information for building proteins. If this is true, it may be that the scaffold proteins organize the chromosome in functional "domains" that facilitate protein synthesis as well as DNA replication. The looped regions are known to vary in size. Does each one contain one or more gene sequences? This is the kind of question being asked by the new generation of molecular detectives.

## SUMMARY

1. Deoxyribonucleic acid, or DNA, is the master blueprint of hereditary instructions in cells. It is assembled from small organic molecules called nucleotides.

2. All nucleotides have a five-carbon sugar (deoxyribose) and a phosphate group. They also have one of four nitrogen-containing bases: adenine, thymine, guanine, or cytosine.

3. In a DNA molecule, two nucleotide strands are twisted together into a double helix. The bases of one strand pair with bases of the other strand (by hydrogen bonding).

4. There is constancy in base pairing in a DNA molecule. Adenine always pairs with thymine (A = T), and guanine always pairs with cytosine (G = C).

5. There is variation in the *sequence* of base pairs in the DNA molecules of different species.

6. DNA replication is semiconservative. The two strands of the DNA double helix unwind from each other, and a new strand is assembled on each one according to base-pairing rules. Two double-stranded molecules result, in which one strand is "old" (it is conserved) and the other is "new."

7. A variety of enzymes and other proteins take part in DNA replication. DNA polymerases are examples; they catalyze strand assembly and also perform proofreading functions.

8. A eukaryotic chromosome consists of one DNA molecule bound with many proteins. Interactions between the DNA and proteins give rise to a structural organization that is most pronounced at metaphase.

9. DNA replication usually is bidirectional. It proceeds simultaneously from each origin, which is a specific, short sequence of bases that functions as an initiation site for replication.

## Review Questions

1. How did Griffith's use of control groups help him deduce that the transformation of harmless *Streptococcus* strains into deadly ones involved a change in the hereditary material of the harmless forms? *206*

2. What is a bacteriophage? In the Hershey-Chase experiments, how did bacteriophages become labeled with radioactive sulfur and radioactive phosphorus? Why were these particular elements used instead of, say, carbon or nitrogen? *207–208*

3. DNA is composed of only four different kinds of nucleotides. Name the three molecular parts of a nucleotide. Name the four different kinds of nitrogen-containing bases that may occur in the nucleotides of DNA. *208–209*

4. What kind of bond holds two DNA chains together in a double helix? Which nucleotide base-pairs with adenine? Which pairs with guanine? Do the two DNA chains run in the same or opposite directions? *209–210*

5. The four bases in DNA may differ greatly in relative amounts from one species to the next—yet the relative amounts are always the *same* among all members of a single species. How does the concept of base pairing explain these twin properties—the unity and diversity—of DNA molecules? *210*

6. When regions of a double helix are unwound during DNA replication, do the two unwound strands join back together again after a new DNA molecule has formed? *210–211*

## Self-Quiz *(Answers in Appendix IV)*

1. _____ bonds hold the bases of one nucleotide strand to bases of the other nucleotide strand of a DNA double helix.

2. Which of the following is *not* a nitrogenous base of DNA?
   a. adenine
   b. thymine
   c. guanine
   d. cytosine
   e. uracil

3. Base pairing in the DNA molecule follows which configuration?
   a. A--G, T--C
   b. A--C, T--G
   c. A--U, C--G
   d. A--T, C--G

4. A single strand of DNA with the base sequence C–G–A–T–T–G would be complementary to the sequence _____.
   a. C–G–A–T–T–G
   b. G–C–T–A–A–G
   c. T–A–G–C–C–T
   d. G–C–T–A–A–C

5. The DNA of one species differs from others in its _____.
   a. sugars
   b. phosphate groups
   c. base-pair sequence
   d. all of the above

6. When DNA replication begins, _____.
   a. the two strands of the double helix unwind from each other
   b. the two strands condense tightly for base-pair transfers
   c. two DNA molecules bond
   d. old strands move to find new strands before bonding

7. DNA replication produces _____.
   a. two half-old, half-new double-stranded molecules
   b. two double-stranded molecules, one with the old strands and one with newly assembled strands
   c. three new double-stranded molecules, one with both strands completely new and two that are discarded
   d. none of the above

8. The process of DNA replication requires _____.
   a. a supply of new nucleotides
   b. forming of new hydrogen bonds
   c. many enzymes and other proteins
   d. all of the above

9. DNA polymerase has _____ functions.
   a. strand assembly
   b. phosphate attachment
   c. proofreading
   d. a and c are both correct

10. Match these DNA concepts appropriately.
    _____ base pair        a. two nucleotide strands twisted
    sequences                      together
    _____ metaphase        b. A = T, G = C
    chromosome                  c. one strand old (conserved), the
    _____ constancy in        other new
    base pairing                d. accounts for differences
    _____ replication         among species
    _____ double helix     e. structure results from
                                   interactions between DNA and
                                   proteins

## Selected Key Terms

adenine (A) *208*
bacteriophage *207*
base *208*
cytosine (C) *208*
DNA *205*
DNA polymerase *211*
guanine (G) *208*
histone *213*

nucleosome *213*
nucleotide *208*
purine *209*
pyrimidine *209*
semiconservative replication *210*
thymine (T) *208*
x-ray diffraction *209*

## Readings

Cairns, J., G. Stent, and J. Watson (editors). 1966. *Phage and the Origins of Molecular Biology*. Cold Spring Harbor, New York: Cold Spring Harbor Laboratories. Collection of essays by the founders of and converts to molecular genetics. Gives a sense of history in the making—the emergence of insights, the wit, the humility, the personalities of the individuals involved.

Darnell, J., et al. 1990. *Molecular Cell Biology*. Second edition. New York: Scientific American Books.

Felsenfeld, G. October 1985. "DNA." *Scientific American* 253(4): 58–67. Describes how the DNA double helix may change its shape during interactions with regulatory proteins.

Radman, M., and R. Wagner. August 1988. "The High Fidelity of DNA Duplication." *Scientific American* 259(2).

Taylor, J. (editor). 1965. *Selected Papers on Molecular Genetics*. New York: Academic Press.

Watson, J. 1978. *The Double Helix*. New York: Atheneum. Highly personal view of scientists and their methods, interwoven into an account of how DNA structure was discovered.

## Beyond Byssus

For the unattached mussel, creeping across a wave-scoured rock, time is of the essence. At any moment the pounding waves can whack it loose, hurl it repeatedly against the rock with shell-shattering force, and so offer up one more gooey lunch for gulls. That marine mussel —soft of body, nearly brainless—literally must hold on for dear life.

The mussel is fortunate. By chance, its muscular, probing foot comes across a suitable anchoring site—a small crevice in the rock. Now the mussel moves its foot, broomlike, and sweeps the site clean. Next it presses the foot down, like a rubber plunger, and expels water trapped beneath it. Then the flattened foot arches upward, creating a vacuum-sealed chamber (Figure 14.1).

A fluid flows into the chamber from ducts in the mussel's body. The fluid, which mussels stockpile in a special gland, contains enzymes, the protein keratin, and a resinous protein. Inside the chamber, the fluid bubbles into a sticky foam that gives the mussel a pre-liminary foothold on the rock. Now the foot flattens out, then curves into a deep groove. Like a spider spinning threads for a web, the mussel pumps the foam through the groove, converting it into a fine thread. Such threads, about as wide as a human whisker, become varnished with another protein. Together, they form an adhesive material called byssus.

Byssus is the world's premier underwater adhesive. Nothing that humans have manufactured comes close. (Sooner or later, water chemically degrades or deforms all synthetic adhesives.) Byssus fascinates biochemists, adhesive manufacturers, dentists, and surgeons looking for better ways to do tissue grafts and to rejoin severed nerves. Even now, genetic engineers are inserting a bit of mussel DNA into yeast cells, which reproduce in large numbers and serve as "factories" for translating the genes of mussels into useful amounts of proteins. This exciting work, like the mussel's own byssus-building, starts with one of life's universals. *Every protein is synthesized according to instructions that have been copied from DNA.*

You are about to trace the steps of protein synthesis, beginning with a linear sequence of code words in a strand of DNA. You will see how this becomes translated into a linear sequence of amino acids in a polypeptide chain. One or more of those newly crafted chains goes on to become a protein. Many enzymes and other proteins are players as well as products in this story, as are molecules of RNA. As you will see, it takes the same kinds of steps to produce all of the world's proteins, from mussel-inspired adhesives to the keratin of your own fingernails to the insect-digesting enzymes of a Venus flytrap.

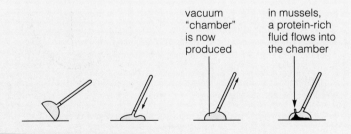

vacuum "chamber" is now produced

in mussels, a protein-rich fluid flows into the chamber

**Figure 14.1** Mussels (*Mytilus edulis*) busily demonstrating the importance of proteins for survival. When mussels come across a suitable anchoring site on a rocky shoreline, they use their foot like a plumber's plunger to create a vacuum chamber. In this chamber they manufacture the world's best underwater adhesive from a wonderful mix of proteins synthesized by some of their specialized cells.

KEY CONCEPTS

**1.** Life cannot exist without enzymes and other proteins. Genes, which are specific regions of DNA, contain the information required to build proteins. The "code words" of genes are sequences of nucleotide bases, read three at a time.

**2.** The path from genes to proteins has two steps. In transcription, an RNA strand is assembled on exposed bases of an unwound gene region. In translation, the code-word sequence of the RNA that was transcribed from DNA is converted into the amino acid sequence of a polypeptide chain. A protein molecule consists of one or more such chains.

**3.** Thus, DNA is used to build RNA, then RNA is used to build proteins—some of which take part in building DNA and RNA. This flow of information is the "central dogma" of molecular biology.

**4.** Replication and repair enzymes work to preserve genes. On rare occasions, however, one to several bases in a gene sequence may be deleted, added, or replaced. Such gene mutations are the original source of genetic variation in populations.

## PROTEIN SYNTHESIS

### The Central Dogma

DNA is like a book of instructions in each cell. As we have seen, the alphabet used to create the book is simple enough: A, T, G, and C. But how is the alphabet arranged into the sentences (genes) that become expressed as proteins? How does a cell skip through the book, reading only those genes that will provide certain proteins at certain times? Answers to these questions begin with the structure of DNA.

Each DNA molecule consists of two long strands, twisted together into a double helix (Figure 13.6). The four kinds of nucleotide subunits making up the strands differ only in their nitrogen-containing base (adenine, thymine, guanine, or cytosine). Which base follows the

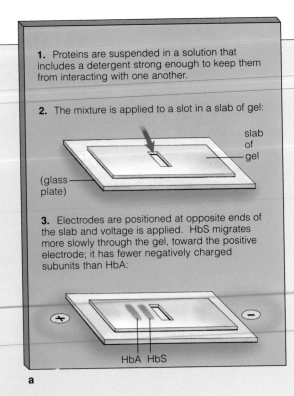

1. Proteins are suspended in a solution that includes a detergent strong enough to keep them from interacting with one another.

2. The mixture is applied to a slot in a slab of gel:

slab of gel

(glass plate)

3. Electrodes are positioned at opposite ends of the slab and voltage is applied. HbS migrates more slowly through the gel, toward the positive electrode; it has fewer negatively charged subunits than HbA:

HbA  HbS

**a**

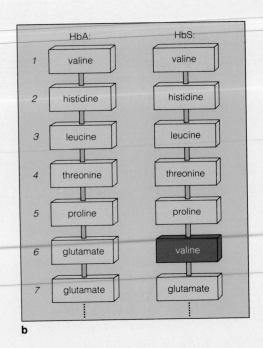

**b**

**Figure 14.2** The connection between genes and proteins.

In the early 1900s a physician, Archibald Garrod, was tracking metabolic disorders that seemed to be heritable (they kept recurring in the same families). Blood or urine samples from affected persons contained abnormally high levels of a substance known to be produced at a certain step in a metabolic pathway. Most likely, the enzyme at the *next* step in the pathway was defective and could not use that substance. Because the pathway was blocked from that step onward, unused molecules of the substance accumulated in the body:

$$A \longrightarrow B \xrightarrow{\begin{array}{c} C\ C\ C \\ C \end{array}} C \xrightarrow{\phantom{X}} D$$

*pathway is blocked*

Only one thing distinguished affected persons from normal ones. They had inherited one metabolic defect. Thus, Garrod concluded, specific "units" of inheritance (genes) function through the synthesis of specific enzymes.

Thirty-three years later, George Beadle and Edward Tatum were using the bread mold *Neurospora crassa* to study gene function. *N. crassa* will grow on a medium containing only sucrose, mineral salts, and biotin, one of the B vitamins. It synthesizes all other nutrients it requires, including other vitamins, and the steps of those synthesis pathways were known.

Suppose an enzyme of a synthesis pathway is defective as a result of a gene mutation. Beadle and Tatum suspected this had happened in some *N. crassa* strains. One strain grew only when supplied with vitamin $B_6$, another with vitamin $B_1$, and so on. Chemical analysis of cell extracts revealed a different defective enzyme in each mutant strain. *Each*

*inherited mutation corresponded to a defective enzyme.* Here was evidence favoring Garrod's "one gene, one enzyme" hypothesis.

The hypothesis was refined through studies of sickle-cell anemia (page 175). This heritable disorder arises from the presence of abnormal hemoglobin in red blood cells. The abnormal molecule is designated HbS instead of HbA. In 1949 Linus Pauling and Harvey Itano subjected HbS and HbA molecules to *electrophoresis*. This is a way to measure how fast and in what direction an organic molecule will move in response to an electric field.

As shown in (**a**), suppose you place a mixture of different proteins in a slab of gel. Each type will move toward one end of the slab or the other when voltage is applied to it. The rate and direction of movement depend partly on a molecule's net surface charge.

Electrophoresis studies showed that HbS and HbA molecules move toward the positive pole of the field—but HbS does so more slowly. HbS, it seemed, has fewer negatively charged amino acids.

Later, Vernon Ingram pinpointed the difference. Hemoglobin, recall, consists of four polypeptide chains (page 45). Two are designated alpha and the other two, beta. As (**b**) shows, in each beta chain of HbS, one amino acid (valine) has replaced another (glutamate). Glutamate carries a negative charge; valine has no net charge. Thus HbS behaved differently in the electrophoresis studies.

More importantly, this discovery suggested that *two* genes code for hemoglobin—one for each kind of polypeptide chain—and that genes code for proteins in general, not just for enzymes.

And so a more precise hypothesis emerged: *One gene codes for the amino acid sequence of one polypeptide chain—the structural unit of proteins.*

next in a strand—that is, the **base sequence**—differs to some degree from one species of organism to the next.

Before a cell divides, its DNA is replicated, so that each of its daughter cells will end up with a full complement of the required genetic information. At this time, the two strands of a DNA molecule unwind from each other and the exposed base sequence of each serves as a structural pattern, or **template**, upon which a new strand is built. At other times in a cell's life, however, *DNA regions are unwound so that the cell gains access to specific genes.*

For now, think of a **gene** as a region of DNA that calls for the assembly of specific amino acids into a polypeptide chain. Such chains are the basic structural units of proteins. Evidence for the connection between genes and proteins comes from many studies. Some of the classic research leading to this understanding is described in Figure 14.2.

The path from genes to proteins has two steps, called transcription and translation. Here our main focus will be on how the two steps occur in eukaryotic cells. In **transcription**, single-stranded molecules of ribonucleic acid, or **RNA**, are assembled on DNA templates in the nucleus. In **translation**, the RNA molecules are shipped from the nucleus into the cytoplasm, where they are used as templates for assembling polypeptide chains. (In bacterial cells, which have no nucleus, RNA molecules start getting translated while they are still peeling off the DNA.) Following translation, one or more chains become folded into the three-dimensional shape of protein molecules.

A circular relationship exists between DNA, RNA, and proteins. DNA is used in the synthesis of RNA, then RNA directs the synthesis of proteins. Those proteins have structural and functional roles in cells. They also have roles in transmitting genetic information from one cell generation to the next. Among them are all the enzymes and other proteins that take part in DNA replication, RNA synthesis, and protein synthesis. This flow of information in cells is the **central dogma** of molecular biology:

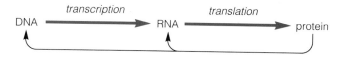

With this simple picture in mind, we are ready to expand our picture of the gene—and our description of RNA.

## Overview of the RNAs

Genes are transcribed into three different types of RNA molecules which are composed of nucleotide subunits of the sort shown in 14.3. Only *one* of the three types even-

tually becomes translated into a protein product. By contrast, the other two types of RNA molecules have specific roles during the process of translation.

| | |
|---|---|
| **ribosomal RNA** (rRNA) | a type of molecule that combines with certain proteins to form the *ribosome,* the structural "workbench" on which a polypeptide chain is assembled |
| **messenger RNA** (mRNA) | the "blueprint" (a linear sequence of nucleotides) delivered to the ribosome for translation into a polypeptide chain |
| **transfer RNA** (tRNA) | an adaptor molecule; it can pick up a specific amino acid *and* pair with an mRNA code word for that amino acid |

Therefore, three types of RNA are transcribed from DNA. All are shipped from the nucleus into the cytoplasm, and all take part in translation, the second stage of protein synthesis. *But only the mRNA molecules carry protein-building instructions out of the nucleus.*

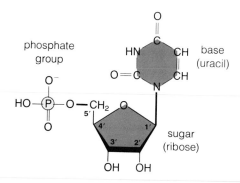

**Figure 14.3** Structure of one of the four nucleotides of RNA. The other three differ only in their component base (adenine, guanine, or cytosine instead of the uracil shown here). Compare Figure 13.5, which shows the nucleotides of DNA.

## TRANSCRIPTION OF DNA INTO RNA

### How RNA Is Assembled

Let's now consider how RNA is transcribed from a gene that codes for a specific polypeptide chain. A strand of RNA is almost, but not quite, like a strand of DNA. Its nucleotides consist of a sugar (ribose), a phosphate group, and a nitrogen-containing base. The bases in RNA are adenine, cytosine, guanine, and **uracil** (Figure 14.3). Like the thymine in DNA, uracil can base-pair with adenine. Thus a new RNA strand can be assembled on a DNA template according to base-pairing rules, in a manner similar to DNA replication:

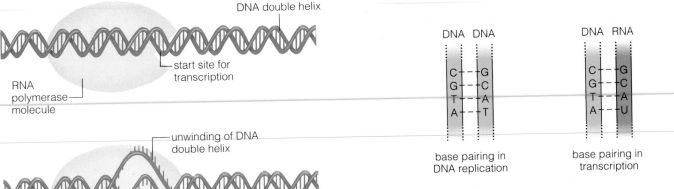

base pairing in DNA replication

base pairing in transcription

Transcription is similar to DNA replication in another respect. The nucleotides are added to a growing RNA

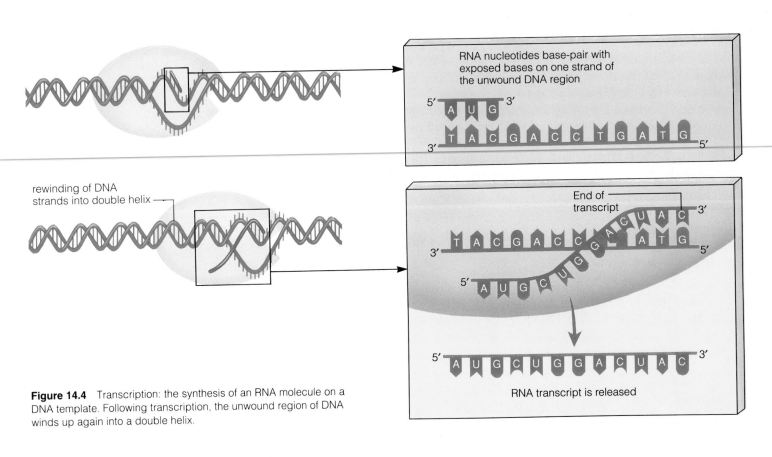

**Figure 14.4** Transcription: the synthesis of an RNA molecule on a DNA template. Following transcription, the unwound region of DNA winds up again into a double helix.

strand one at a time, in the 5' → 3' direction. (Here you may wish to refer to Figure 13.8.)

Transcription *differs* from DNA replication in three key respects. First, only one region of one DNA strand—not the whole strand—serves as the template. Typically, the genes coding for polypeptide chains are confined to a stretch of DNA that is between 70 and 10,000 nucleotides long. Second, transcription requires different enzymes. Three types of **RNA polymerases** put together the strands of rRNA, mRNA, and tRNA. Third, unlike DNA replication, the result is a single-stranded molecule, not a double-stranded one.

Transcription starts at a **promoter**, a specific sequence of bases on one of the two DNA strands that signals the start of a gene. An RNA polymerase does not find the location of a promoter all by itself. Rather, it recognizes and binds with one or more small proteins that have become positioned on the DNA strand a few nucleotides "up the road" from a promoter. After this happens, the enzyme can bind with the promoter and open up a local region of the DNA double helix.

As Figure 14.4 shows, the enzyme moves stepwise along the exposed nucleotides of one DNA strand, unwinding a bit more of the double helix as it goes. The RNA strand continues to grow until the enzyme encounters a base sequence that serves as a termination signal. Then the RNA is released from the template as a free, single-stranded transcript.

**Messenger-RNA Transcripts**

Again, of the three classes of RNA, only mRNA carries protein-building instructions from the nucleus into the cytoplasm. However, a newly formed mRNA transcript cannot even depart from the nucleus without first undergoing modification. Just as a dressmaker might snip off some threads or add bows on a dress before it leaves the shop, so does a eukaryotic cell tailor its mRNA.

For one thing, the first end of the mRNA to be synthesized (the 5' end) quickly gets capped. The cap is simply a nucleotide covalently bonded to a methyl group and phosphate groups. It seems to be recognized by other molecules (initiation factors) as the "start" signal for translation. Besides having a cap, most mature mRNA transcripts acquire a tail at the opposite end. This "poly-A tail" consists of about 100 to 200 adenine-containing molecules and seems to help prevent mRNA from being degraded in the cytoplasm.

For another thing, newly transcribed mRNA contains more than the code for a string of amino acids. The actual coding portions, which will become translated into proteins, are called **exons**. But new mRNA also contains **introns**. These are "noncoding" portions; they have no information about the amino acid sequence. Before the mRNA leaves the nucleus, its introns are snipped out and the exons spliced together, in the manner shown in Figure 14.5.

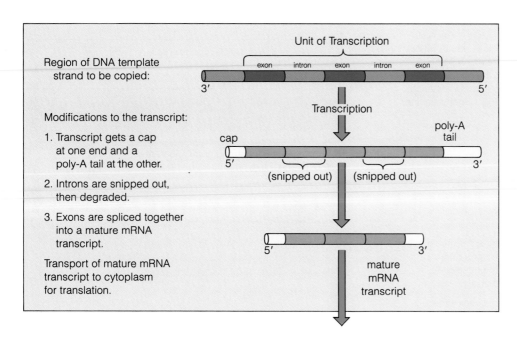

**Figure 14.5** Transcription and modification of newly formed mRNA in the nucleus of eukaryotic cells.

| First Letter | Second Letter | | | | Third Letter |
|---|---|---|---|---|---|
| | U | C | A | G | |
| U | phenylalanine | serine | tyrosine | cysteine | U |
| | phenylalanine | serine | tyrosine | cysteine | C |
| | leucine | serine | stop | stop | A |
| | leucine | serine | stop | tryptophan | G |
| C | leucine | proline | histidine | arginine | U |
| | leucine | proline | histidine | arginine | C |
| | leucine | proline | glutamine | arginine | A |
| | leucine | proline | glutamine | arginine | G |
| A | isoleucine | threonine | asparagine | serine | U |
| | isoleucine | threonine | asparagine | serine | C |
| | isoleucine | threonine | lysine | arginine | A |
| | (start) methionine | threonine | lysine | arginine | G |
| G | valine | alanine | aspartate | glycine | U |
| | valine | alanine | aspartate | glycine | C |
| | valine | alanine | glutamate | glycine | A |
| | valine | alanine | glutamate | glycine | G |

**Figure 14.6** The genetic code. The codons in an mRNA molecule are nucleotide bases, read in blocks of three. Each of those base triplets will call for a specific amino acid during mRNA translation. In this diagram, the first nucleotide of any triplet is given in the left column. The second is given in the middle columns; the third, in the right column. Thus we find (for instance) that trytophan is coded for by the triplet $\boxed{U}\boxed{G}\boxed{G}$. Phenylalanine is coded for by both $\boxed{U}\boxed{U}\boxed{U}$ and $\boxed{U}\boxed{U}\boxed{C}$.

**Figure 14.7** (*Right*) Genetic code in action. Notice the green-shaded blocks extending down through all three parts of this diagram (**a-c**). They show the relation between the nucleotide sequence of DNA and the amino acid sequence of proteins.

During transcription, the region of the DNA double helix shown here was unwound, and the exposed bases on one strand served as a template for assembling the mRNA strand. The bases of every three nucleotides in an mRNA strand equal one codon. Here, each codon called for one of the amino acids in this polypeptide chain. Using Figure 14.6 as a guide, can you fill in the blank codon for threonine in the chain?

# TRANSLATION

## The Genetic Code

Like a DNA strand, an mRNA transcript is a linear sequence of nucleotides. So we are still left with a central question: What are the protein-building "words" encoded in that sequence?

Francis Crick, Sidney Brenner, and others came up with the answer. They deduced the nature of the **genetic code**—that is, how the nucleotide sequence of DNA and then mRNA corresponds to the amino acid sequence of a polypeptide chain. The bases of the nucleotides in mRNA are read three at a time, and each of these *base triplets* calls for an amino acid. GGU, for example, calls for glycine (Figures 14.6 and 14.7). A start signal built into the mRNA strand establishes the correct "reading frame" for selecting three bases at a time in the sequence. Each base triplet in mRNA is now called a **codon**.

**a** A gene region in the DNA double helix.

**b** Part of an mRNA strand, transcribed from one of the two DNA strands of the double helix.

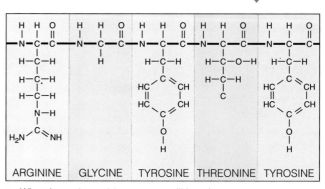

**c** What the amino acid sequence will be when the mRNA is translated into a polypeptide chain.

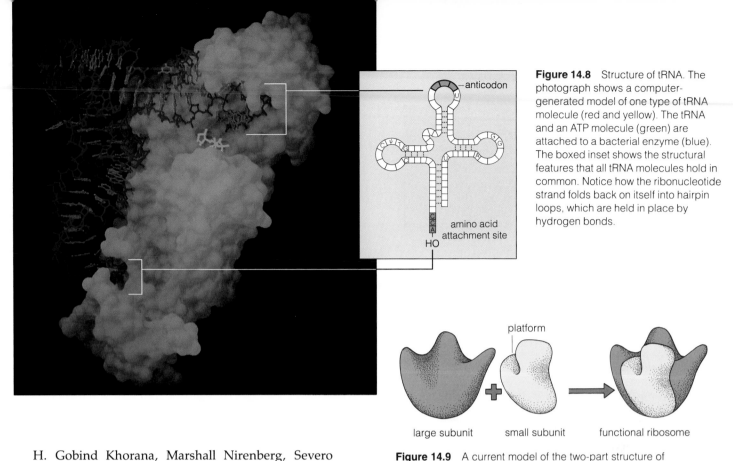

**Figure 14.8** Structure of tRNA. The photograph shows a computer-generated model of one type of tRNA molecule (red and yellow). The tRNA and an ATP molecule (green) are attached to a bacterial enzyme (blue). The boxed inset shows the structural features that all tRNA molecules hold in common. Notice how the ribonucleotide strand folds back on itself into hairpin loops, which are held in place by hydrogen bonds.

**Figure 14.9** A current model of the two-part structure of eukaryotic ribosomes.

H. Gobind Khorana, Marshall Nirenberg, Severo Ochoa, Robert Holley—these and so many others did the meticulous work to decipher the genetic code. Because there are 4 kinds of nucleotides in RNA and 3 bases in each codon, the researchers suspected that each mRNA strand must be assembled from a selection of $4^3$ or 64 different codons. They discovered that 61 codons actually specify amino acids. The other three (UAA, UAG, UGA) are *stop codons* that act like stop signs in an mRNA strand. Their presence "tells" enzymes that the end of the gene region has been reached and no more amino acids are to be added to the growing polypeptide chain.

As Figure 14.6 shows, the codon AUG specifies methionine. An mRNA strand typically contains a number of AUG codons. However, starting at the capped end of mRNA, the first suitable AUG that occurs in the strand also serves as the *start codon* for translation.

One final point should be made here. As described on page 362, mitochondria and chloroplasts have their own DNA, and their genetic code is almost but not quite like the one just described. But these are organelles, not organisms. The genetic code shown in Figure 14.6 is the language of protein synthesis *for nearly all organisms.* Codons calling for certain amino acids in bacteria call for the same amino acids in protistans, fungi, plants, and animals.

### Codon-Anticodon Interactions

Let's now turn to the fate of mRNA, with its string of codons, after it arrives in the cytoplasm. Sooner or later, it will interact with its molecular relatives, the tRNAs and rRNAs.

Thirty-one kinds of tRNA molecules are pooled together in the cytoplasm of eukaryotic cells. Each tRNA has an **anticodon**, a sequence of three nucleotide bases that can base-pair with a specific mRNA codon. Each tRNA also has a molecular "hook," an attachment site for a particular amino acid (Figure 14.8).

If an anticodon is to interact with any codon, its first two bases must not violate the base-pairing rules. (Adenine must always pair with uracil, and cytosine must always pair with guanine.) The rules loosen up with respect to the third base. For example, notice in Figure 14.6 that CCU, CCC, CCA, and CCG all specify proline. Such freedom in codon-anticodon pairing at the third base is called the *wobble effect.* Through the wobble effect, *sixty-one* kinds of codons that may be present in an mRNA molecule can call up amino acids, using as few as *thirty-one* kinds of tRNAs.

### Ribosome Structure

Polypeptide chains are assembled as a result of codon-anticodon interactions. And those interactions take place at specific binding sites on the surface of ribosomes. As Figure 14.9 shows, a **ribosome** has two subunits, each composed of rRNA and a number of proteins. The two subunits perform their function only during translation.

**Figure 14.10** Simplified picture of protein synthesis.

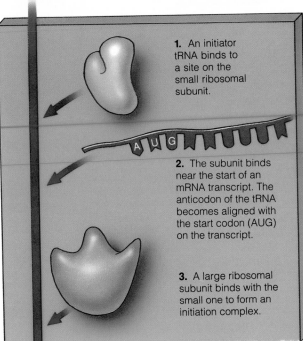

In the cytoplasmic regions of protein synthesis concentrated pools of amino acids, tRNAs, and ribosomal subunits exist.

**1.** An initiator tRNA binds to a site on the small ribosomal subunit.

**2.** The subunit binds near the start of an mRNA transcript. The anticodon of the tRNA becomes aligned with the start codon (AUG) on the transcript.

**3.** A large ribosomal subunit binds with the small one to form an initiation complex.

**a** Initiation

Transfer RNA molecules deliver amino acids to two binding sites, called the P and A sites, which are very close together on the smaller of the two ribosomal subunits. That same subunit also has a binding site for mRNA.

Ten thousand ribosomes may be present in the cytoplasm of a single bacterial cell. A eukaryotic cell may contain many tens of thousands. The bacterial ribosome is only 25 nanometers wide. That's about a millionth of an inch. Although the components of a eukaryotic ribosome are larger and more numerous, both kinds of ribosomes have nearly the same shape and function.

### Stages of Translation

Now that we have finally arrived at the ribosome, we are ready to consider how genes are actually translated into proteins. To keep things simple, we can portray the codon-anticodon interactions in this fashion:

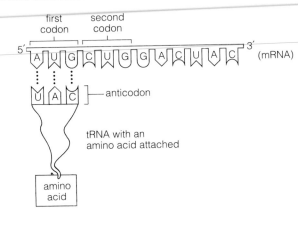

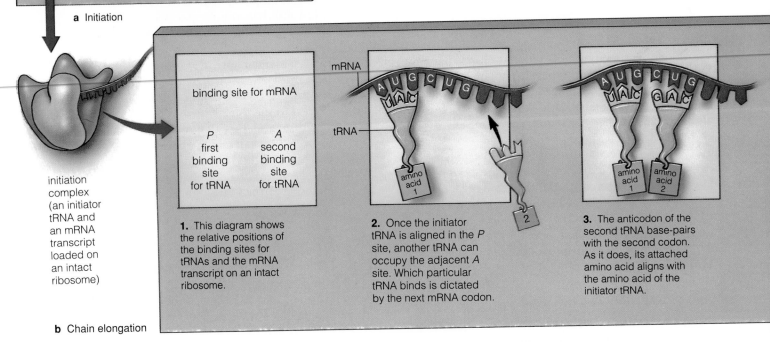

initiation complex (an initiator tRNA and an mRNA transcript loaded on an intact ribosome)

binding site for mRNA

*P* first binding site for tRNA

*A* second binding site for tRNA

mRNA

tRNA

amino acid 1

amino acid 1

amino acid 2

**1.** This diagram shows the relative positions of the binding sites for tRNAs and the mRNA transcript on an intact ribosome.

**2.** Once the initiator tRNA is aligned in the *P* site, another tRNA can occupy the adjacent *A* site. Which particular tRNA binds is dictated by the next mRNA codon.

**3.** The anticodon of the second tRNA base-pairs with the second codon. As it does, its attached amino acid aligns with the amino acid of the initiator tRNA.

**b** Chain elongation

Translation proceeds through three stages, called initiation, chain elongation, and chain termination. Figure 14.10 is a step-by-step picture of the stages of translation in eukaryotic cells.

In *initiation*, both a tRNA that can start transcription and an mRNA transcript become loaded onto an intact ribosome. First, the "initiator" tRNA binds with the small ribosomal subunit. Next, the cap on the mRNA molecule binds with the small ribosomal subunit in such a way that the proper start codon (AUG) becomes positioned in front of the initiator tRNA. Finally, a large ribosomal subunit joins with the small one. Chain elongation can begin.

In *chain elongation*, amino acids are strung together in the sequence dictated by the codons of mRNA. The mRNA's start codon defines the reading frame for the sequence. The mRNA strand passes between the two ribosomal subunits, like a thread being moved through the eye of a needle.

Figure 14.10b shows the initiator tRNA in position at the ribosome's P site. Another tRNA binds to the adjacent A site. Its anticodon is able to base-pair with the next codon "down the road" from the start codon of the mRNA molecule. Now the two tRNAs are positioned in such a way that a peptide bond can readily form between their attached amino acids. (Here you may wish to refer to Figure 3.16.) The initiator tRNA gives up its amino acid entirely and is removed from the ribosome. This leaves two amino acids attached to the second tRNA—which is shifted over to the P site. The mRNA molecule

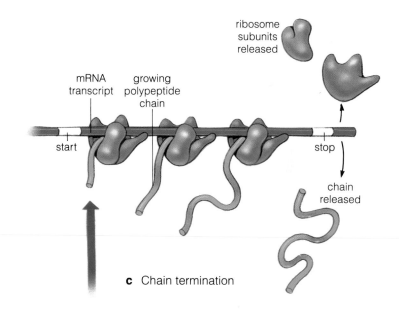

c  Chain termination

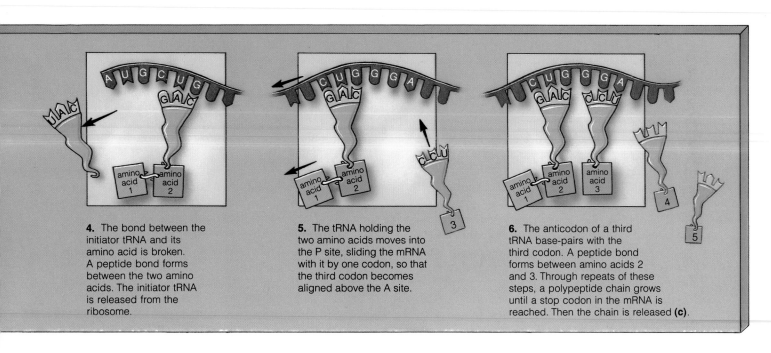

**4.** The bond between the initiator tRNA and its amino acid is broken. A peptide bond forms between the two amino acids. The initiator tRNA is released from the ribosome.

**5.** The tRNA holding the two amino acids moves into the P site, sliding the mRNA with it by one codon, so that the third codon becomes aligned above the A site.

**6.** The anticodon of a third tRNA base-pairs with the third codon. A peptide bond forms between amino acids 2 and 3. Through repeats of these steps, a polypeptide chain grows until a stop codon in the mRNA is reached. Then the chain is released **(c)**.

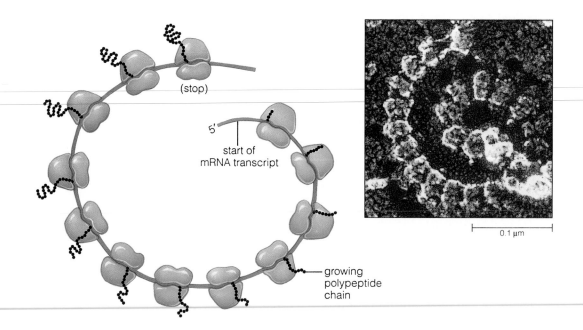

(stop)

5′

start of
mRNA transcript

growing
polypeptide
chain

0.1 µm

**Figure 14.11**  Micrograph and sketch of many ribosomes simultaneously translating the same mRNA molecule in a eukaryotic cell.

slides along with it. At this point *another* tRNA moves into the now-empty A site, and the steps are repeated. Enzymes built into the ribosome catalyze peptide bonds between every two amino acids delivered to the ribosome. In this way a polypeptide chain grows.

In *chain termination*, a stop codon in the mRNA signals that no more amino acids can be added to the polypeptide chain. Now, with the help of specific proteins called release factors, the ribosome and polypeptide chain are detached from the mRNA. The detached chain joins the pool of free proteins in the cytoplasm or enters the cytomembrane system for further processing. Page 63 outlines the final destinations of the completed proteins.

The steps just described can be repeated many times on the *same* mRNA transcript, with several copies of the polypeptide chain forming at the same time. What happens is this: A new ribosome hops onto the mRNA almost as soon as the preceding ribosome has translated enough of it to get out of the way. The word **polysome** refers to several ribosomes that are spaced closely together on the same mRNA. Figure 14.11 shows an example of a polysome. Such "assembly lines" for protein synthesis are a common feature of cells.

Figure 14.12 summarizes the flow of information along the path leading from genes to proteins.

## MUTATION AND PROTEIN SYNTHESIS

In general, the base sequence in DNA must be preserved from one generation to the next, otherwise offspring might not be able to synthesize all the proteins required for their own survival and reproduction. Yet changes do occur in the DNA, and they may affect one or more traits of the individual. Said another way, they give rise to variations in phenotype.

For instance, as we saw in the preceding chapters, phenotypic variation arises through crossing over and recombination, which put new mixes of alleles in chromosomes. It also arises through changes in the structure and number of chromosomes, as brought about by nondisjunction.

Another kind of change is called the **gene mutation**. This is a deletion, addition, or substitution of one to several bases in the nucleotide sequence of a gene. Gene mutations are rare events. On the average, the mutation rate for a gene is only one in a million replications.

Some gene mutations are induced by **mutagens**, environmental agents that can attack a DNA molecule and modify its structure. Viruses, ultraviolet radiation, and certain chemicals are examples of mutagens. Other gene mutations are spontaneous; they are not induced

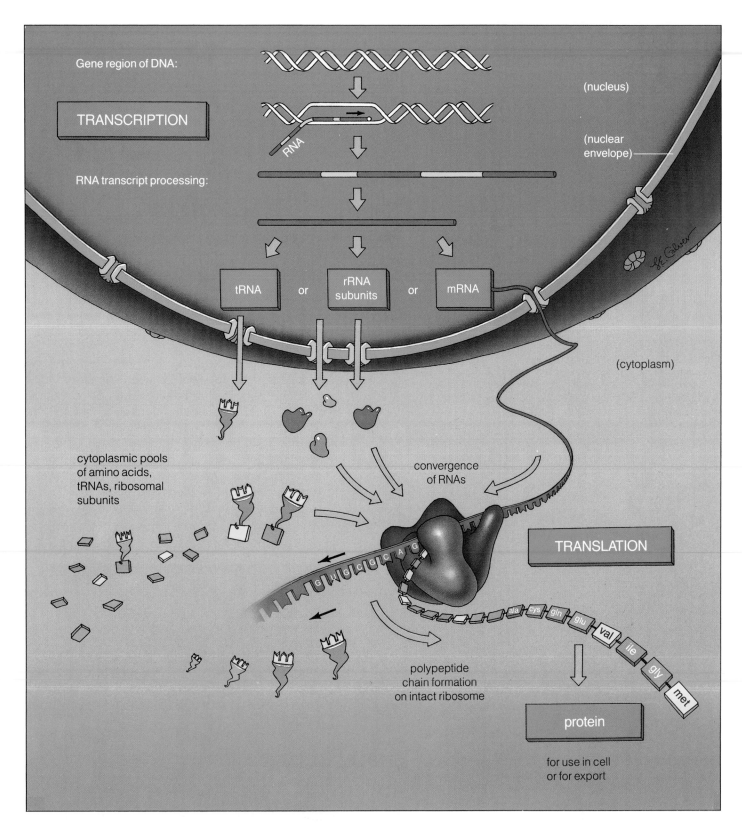

**Figure 14.12** Summary of the flow of genetic information in protein synthesis in eukaryotic cells.

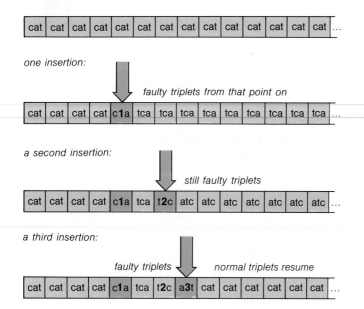

one insertion:

faulty triplets from that point on

| cat | cat | cat | cat | c1a | tca | tca | tca | tca | tca | tca | tca | tca | tca | ...

a second insertion:

still faulty triplets

| cat | cat | cat | cat | c1a | tca | t2c | atc | atc | atc | atc | atc | atc | atc | ...

a third insertion:

faulty triplets      normal triplets resume

| cat | cat | cat | cat | c1a | tca | t2c | a3t | cat | cat | cat | cat | cat | cat | ...

**Figure 14.13** Simplified picture of frameshift mutation. Studies of such mutations in bacteriophage led to the discovery of the genetic code. During translation, the base sequence of mRNA is read in blocks of three (that is, as base triplets) according to the genetic code.

The word *cat* is used here to represent every base triplet. Suppose an extra nucleotide ("1") is inserted into a gene. When mRNA is transcribed off that gene, the insertion will put the reading frame out of phase and the wrong amino acids will be called up during translation. A second insertion ("2") would not improve matters. A third insertion ("3") will restore the reading frame, so only part of the resulting protein will be defective. Insertions of extra bases into a DNA molecule often give rise to mutant phenotypes.

by agents outside the cell. During DNA replication, for example, an A might become paired with C instead of T. (Even enzymes make mistakes.) Proofreading enzymes might detect the mistake—but will it "fix" the A or the C? If the enzymes remove the wrong base from a mismatched pair, the result is a type of spontaneous mutation called a *base-pair substitution.*

You might be thinking that a change in a single base pair is insignificant. But as Figure 14.2 made clear, sickle-cell anemia has been traced to a single mutation in the DNA strand coding for the beta chain of hemoglobin. Only one amino acid is substituted for another in the resulting chain—yet the substitution can have severe consequences.

Another type of spontaneous change is the *frameshift mutation.* Here, the insertion or deletion of one to several base pairs in a DNA molecule puts the nucleotide sequence out of phase, so that the reading frame shifts during protein synthesis (Figure 14.13). Because genetic instructions are not read correctly, an abnormal protein is synthesized.

Barbara McClintock discovered still another type of spontaneous gene mutation. Through studies of Indian corn (maize), she realized that certain DNA regions frequently "jump" to new locations in the same DNA molecule or in a different one. These *transposable elements* often inactivate the genes into which they become inserted and give rise to observable changes in phenotype (Figure 14.14).

What is the point to remember about these mutations and others? The *Commentary* provides us with an answer.

**Figure 14.14** Color variations in kernels of Indian corn. All the cells in a kernel have pigment-coding genes, so you might expect the whole kernel to be the same color. Some are indeed fully colored, but others are spotted or entirely colorless. Early in plant growth, transposable elements (movable DNA regions) were present in a gene necessary to produce the kernel pigment. The transposable elements caused a mutation that gave rise to colorless kernels. When the transposable element moved out of the gene, the gene's activity was restored. Thus, all cells that were descendants of the cells in which the transposition occurred were able to produce pigment. The outcome was pigmented spots on the kernel.

*Commentary*

## Gene Mutation and Evolution

Each gene is said to have a characteristic *mutation rate*, which simply is the probability of its mutating between or during DNA replications. On the average, the mutation rate for a gene is one in a million ($10^6$) replications.

Genes mutate independently of one another. To determine the probability of any two mutations occurring in a given cell, we would have to multiply the individual mutation rates for two of its genes. For example, if the rate for the first gene is one in a million per cell generation and the rate for the second is one in a billion ($10^9$), then there is only one chance in a million billion ($10^{15}$) that both genes will mutate in the same cell.

In the natural world, gene mutations are rare, chance events. It is impossible to predict exactly when, and in which organism, they will appear. They also are inevitable. They arise not only through mistakes in DNA replication. They arise also through the action of mutagens, including ultraviolet light, ionizing radiation, and various chemicals in our environment.

A mutation may turn out to be beneficial, neutral, or harmful. The outcome depends on how the protein specified by the mutated DNA region functions in the cells that require the protein, and on how the protein affects the coordinated workings of the entire individual (page 277). Because of this, most mutations do not bode well for the individual. No matter what the species, each organism generally inherits a combination of many genes that already are fine-tuned for a given range of operating conditions in the body. A mutant gene is likely to code for a protein that is less functional, not more so, under those conditions.

Over evolutionary time, agents of natural selection undoubtedly perpetuated those packages of genes having a history of survival value. They must have favored DNA polymerases that showed little tolerance of mismatched base pairs. They also must have favored enzymes that could chemically recognize, remove, and replace mismatched base pairs. What is clear is that replication enzymes have protected the overall stability of the vulnerable molecules of inheritance that have been replicated through billions of years.

Yet every so often, some mutations provided their bearers with advantages. We saw this with the example of the peppered moth, described briefly in Chapter 1. Selection processes worked to perpetuate mutations having adaptive value. Besides this, other mutations produced DNA regions with no known function—but they did their bearers no harm and they, too, have been perpetuated. After more than three billion years, molecular descendants of the first strands of DNA are replete with mutations.

What does all this mean? It means that every living thing on earth shares the same chemical heritage with all others. Your DNA has the same kinds of substances, and follows the same base-pairing rules, as the DNA of earthworms in Missouri and grasses on the Mongolian steppes. Your DNA is replicated in much the same way as theirs, and the same genetic code is followed in translating its messages into proteins. In the evolutionary view, the reason you don't look like an earthworm or a grass plant is largely a result of selection of different mutations that originated in different lineages of organisms. Thus the sequence of base pairs along the DNA molecule has come to be different in you, the plant, and the worm.

And so we have three concepts of profound importance. *First, DNA is the source of the unity of life. Second, mutations and other changes in the structure and number of DNA molecules are the source of life's diversity. Finally, the changing environment is the testing ground for the success or failure of the proteins specified by each novel DNA sequence and assortment that appears on the evolutionary scene.*

# SUMMARY

1. Protein-building instructions are encoded in genes, each of which is a specific, linear sequence of nucleotides in one strand of a DNA double helix. The path leading from genes to proteins has two steps, called transcription and translation.

   a. In transcription, an exposed region of a DNA strand serves as the template for assembling an RNA strand.

   b. In translation, RNA molecules interact to convert the gene's message into a linear sequence of amino acids—that is, a polypeptide chain. Such chains are the structural units of proteins.

2. There are three classes of RNA:

   a. Molecules of rRNA are components of the ribosome on which polypeptide chains are assembled.

   b. An mRNA strand is the blueprint of genetic information for building a specific chain.

   c. Many different tRNA molecules deliver amino acids to the ribosome in a sequence dictated by their interaction with the mRNA.

3. Thus DNA is used to synthesize RNA, the RNA is used to synthesize proteins (some of which will take part in DNA and RNA synthesis). This flow of information is the central dogma of molecular biology:

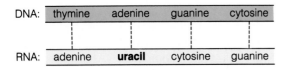

4. The genetic code is the relation between a linear sequence of nucleotides in DNA (then mRNA) and a linear sequence of amino acids in a polypeptide chain. The code words are a sequence of nucleotide bases that are read in blocks of three (base triplets). Each base triplet in mRNA is a codon; a complementary base triplet in tRNA is an anticodon.

5. Transcription follows essentially the same base-pairing rules that apply to DNA replication. However, uracil—not thymine—pairs with the adenine present in the DNA template strand:

6. In eukaryotes, a new mRNA transcript becomes modified in the nucleus before moving into the cytoplasm for translation. A cap and commonly a tail are added to it. Its introns (nucleotide sequences that do not code for parts of the polypeptide chain) are excised and its exons (coding sequences) are spliced together.

7. Translation proceeds through three stages:

   a. In initiation, a small ribosomal subunit binds with an initiator tRNA, then with an mRNA transcript. The small subunit then binds with a large ribosomal subunit to form the initiation complex.

   b. In chain elongation, tRNAs deliver amino acids to the ribosome. The tRNA anticodons pair appropriately with mRNA codons. Then peptide bonds form between their amino acids, forming the chain.

   c. In chain termination, a stop codon triggers events that cause the polypeptide chain to detach from the ribosome.

8. There is an underlying chemical unity among all organisms. Regardless of the species, DNA is composed of the same substances, follows the same base-pairing rules, and is replicated in much the same way. The genetic code by which its instructions are translated into proteins is nearly universal.

9. Overall, the protein-building instructions encoded in DNA are preserved through the generations. But crossing over and recombination, changes in chromosome structure or number, and gene mutations can change parts of those instructions. Such changes lead to phenotypic variation among individuals, and so provide grist for the evolutionary mill.

---

## Review Questions

1. Are the proteins specified by eukaryotic DNA assembled *on* the DNA molecule? If so, state how. If not, tell where they are assembled, and on which molecules. *219*

2. Figure 14.12 shows the steps by which hereditary instructions are transcribed from DNA into RNA, which is then translated into proteins. Study this figure and then, on your own, write a description of this sequence, taking care to define the terms *transcription* and *translation*. *219, 220–226*

3. Define *genetic code*. Is the same basic genetic code used for protein synthesis in all living organisms? *222–223*

4. Define the three types of RNA. What is a codon? *219, 222, 223*

5. If sixty-one codons in mRNA actually specify amino acids, and if there are only twenty common amino acids, then more than one

codon combination must specify some of the amino acids. How do triplets that code for the same thing usually differ? *223*

6. Define intron and exon. What happens to introns before an mRNA transcript is shipped to the cytoplasm? *221*

7. If genetic information were transmitted precisely from generation to generation, organisms would never change. What are some mutations that give rise to phenotypic diversity? *226, 228*

## Self-Quiz (Answers in Appendix IV)

1. Nucleotide bases, read _____ at a time, serve as the "code words" of genes.

2. Genetic information in DNA is transferred to RNA strands during _____, the first step in protein synthesis.
   a. replication
   b. duplication
   c. multiplication
   d. transcription

3. The RNA molecule is _____.
   a. a double helix
   b. usually single-stranded
   c. always double-stranded
   d. usually double-stranded

4. During transcription, base-pairing is similar to that of DNA replication except _____.
   a. cytosine in DNA pairs with guanine in RNA
   b. adenine in DNA pairs with uracil in RNA
   c. thymine in DNA pairs with adenine in RNA
   d. guanine in DNA pairs with cytosine in RNA

5. _____ starts when two ribosomal subunits, an initiator tRNA, and an mRNA transcript come together.
   a. Transcription
   b. Replication
   c. Subduction
   d. Translation

6. The coded genetic instructions for forming polypeptide chains are carried to the ribosome by _____.
   a. DNA
   b. rRNA
   c. mRNA
   d. tRNA

7. The function of tRNA is to _____.
   a. deliver amino acids to the ribosome
   b. pick up genetic messages from rRNA
   c. synthesize mRNA
   d. all of the above

8. How many amino acids are coded for in this mRNA sequence: CGUUUACACCGUCAC?
   a. three
   b. five
   c. six
   d. seven
   e. more than seven

9. An anticodon pairs with the nitrogen-containing bases of _____.
   a. mRNA codon
   b. DNA codons
   c. tRNA anticodon
   d. amino acids

10. Using the genetic code shown in Figure 14.6, translate the mRNA sequence UAUCGCACCUCAGGAUGAGAU. Which of the following polypeptide chains does this sequence specify?
   a. tyr-arg-thr-ser-gly-stop-asp...
   b. tyr-arg-thr-ser-gly...
   c. tyr-arg-tyr-ser-gly-stop-asp...
   d. none is correct

## Selected Key Terms

anticodon *223*
base-pair substitution *228*
base sequence *219*
base triplet *222*
central dogma *219*
codon *222*
electrophoresis *218*
exon *221*
frameshift mutation *228*
gene *219*
gene mutation *226*
genetic code *222*
intron *221*
messenger RNA (mRNA) *219*
polysome *226*
promoter *221*
ribosomal RNA (rRNA) *219*
ribosome *223*
RNA *219*
RNA polymerase *221*
start codon *223*
template *219*
transcription *219*
transfer RNA (tRNA) *219*
translation *219*
transposable element *228*
uracil *220*

## Readings

Alberts, B., et al. 1989. *Molecular Biology of the Cell.* Second edition. New York: Garland.

Amato, I. January 1991. "Stuck on Mussels." *Science News* 139: 8–15.

Darnell, J. October 1985. "RNA." *Scientific American* 253(4):68–78.

Nomura, M. January 1984. "The Control of Ribosome Synthesis." *Scientific American* 250(1):102–114.

Prescott, D. 1988. *Cells.* Boston: Jones and Bartlett. Chapter 8 of this textbook contains an excellent introduction to protein synthesis.

# 15 CONTROL OF GENE EXPRESSION

## A Cascade of Proteins and Cancer

Every second of the day, millions of cells in your skin, gut lining, liver, and other body regions divide and replace their worn-out, dead, and dying predecessors. They do not divide willy-nilly. They cannot divide at all unless they first synthesize and stockpile molecules of cyclin, a protein. Inside the cell that makes it, cyclin binds with another protein, called cdc2, and sets the division machinery in motion.

The cdc2 is the first of a series of enzymes that catalyze the transfer of phosphate from ATP to the next enzyme in line. The enzyme molecules act more than once. Each molecule activates many others, which activate many others, and so on in a growing cascade of reactions that ripple through the cell.

The first enzymes go to work while the cell is still in interphase; they replicate its DNA. Later, other enzymes act directly on proteins that are part of the nuclear envelope and so trigger its breakdown. Others help assemble and operate a spindle of microtubules. That spindle harnesses and moves the cell's chromo-

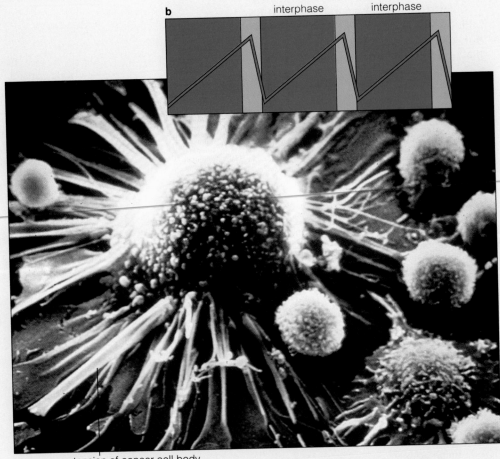

**Figure 15.1** (a) Scanning electron micrograph of a cancer cell, surrounded by some of the body's white blood cells that may or may not be able to destroy it. Cancer cells are skewed in structure and function. Worse, they have lost the controls over the genes and gene products that can suppress cell division. Because cancer cells cannot stop dividing, they form abnormal masses of cells that can destroy surrounding tissues. They are graphic examples of why gene expression must be regulated with precision. (b) From one study, a chart of the controlled changes in the intracellular levels of cyclin (brown line) in normal cells. Cyclin is the protein that guides cells into mitosis (light-blue bands) during the cell cycle.

interphase          interphase

**b**

**a**          extension of cancer cell body

somes into two parcels before the cytoplasm divides in two. Still other enzymes orchestrate the split. Following division, cyclin-degrading enzymes are activated and destroy the cell's entire batch of cyclin. Without cyclin, the division machinery is put to rest. Now the daughter cells, starting life at interphase, start stockpiling cyclin. If they, too, go on to divide, cyclin will again be destroyed.

Of all the cell's proteins, only cyclin accumulates at a constant rate, disappears abruptly, then accumulates again during cell cycles (Figure 15.1). Through its interaction with cdc enzymes, cyclin guides cells into the division process.

How does a cell "know" when to start and stop building cyclin? Just as it takes a turn from an ignition key to start and stop the engine of your car, so does it take signals from regulator molecules, such as hormones, to control the cyclin-driven engine. *Such signals lift the controls that otherwise suppress cell division, then put on the brakes by reinstating controls when division is completed.*

Researchers are only now unraveling the mystery of how those signals work, and their sleuthing is of more than passing interest to us. Why? Sometimes the controls over cell division are lost. It is not that cells start dividing at a berserk pace. Rather, it is that the cell division cycle cannot stop. This is what has happened in the body of a family member, a friend, or an acquaintance who has been stricken with cancer.

Unless cancer cells are eradicated, their chronically dividing descendants will kill the individual. In fact, cancer is a leading cause of human death. In the United States, it is second only to heart disease. Cancer is not just a human affliction. It has been observed in most animal species that have been studied to date. Comparable abnormalities have even been observed in many plants. At the chapter's end, we will consider the nature of cancerous transformations. To gain insight into what is going wrong in affected individuals, however, we must start with how cells use and control their genes when things are going right.

## KEY CONCEPTS

**1.** All cells precisely control when, how, and to what extent their various genes are expressed.

**2.** Control is exerted through regulatory proteins and other molecules that interact with DNA, with RNA transcribed off the DNA, or with the resulting polypeptide chains. Especially among vertebrates, hormones have major roles in controlling gene expression.

**3.** In all cells, controls come into play during transcription, translation, and post-translation (when new polypeptide chains become modified, as by having simple sugars attached to them). In eukaryotes, controls also govern the processing of new RNA transcripts and their shipment out of the nucleus for translation in the cytoplasm.

**4.** In multicelled eukaryotes, all cells have the same genes, but they activate or suppress many of those genes in different ways. This selective gene expression leads to cell differentiation—to pronounced structural and functional variations among the cells that make up different tissues.

## THE NATURE OF GENE CONTROL

All the different cells of your body carry the same genes, and they use most of them to synthesize proteins that are basic to any cell's structure and functions. That is why the protein subunits of microtubules are the same from one cell to the next, as are many of the enzymes used in metabolism.

Yet each type of cell also uses a small fraction of genes in highly specialized ways. Even though they all carry the genes for hemoglobin, only red blood cells activate those genes. Even though they all carry the genes for antibodies, which are protein "weapons" against specific agents of disease, only certain white blood cells activate them. *These and all other living cells control which genes are active and which gene products appear, at what times, and in what amounts.*

## When Controls Come Into Play

In any organism, gene controls operate in response to chemical changes within the cell or its surroundings. In terms of your own cells, chemical conditions change when you vary your diet or level of activity. Conditions also have been changing in inevitable ways ever since you were a tiny mass of cells growing in your mother's body. Within each responding cell, gene activity has been changing appropriately, either to keep the cell itself alive or to contribute to your overall growth and development through time.

*The elements of control operate in response to changing chemical conditions within a cell or its surroundings.*

### Control Agents and Where They Operate

Gene controls are exerted through the action of regulatory proteins and other molecules that interact with DNA, with mRNA transcribed from DNA, and with gene products resulting from mRNA translation. (Here you may wish to refer to page 219.) Transcriptional controls are the most common. They depend on two types of regulatory proteins, called repressors and activators, that can change the rates at which particular genes are transcribed:

**repressor protein**    *prevents the enzymes of transcription (RNA polymerases) from binding to the DNA; affords negative control of gene activity*

**activator protein**    *enhances the binding of RNA polymerases to the DNA; affords positive control of gene activity*

Many control agents bind to sites in the DNA, such as promoters. A **promoter**, recall, is a specific base sequence that signals the start of a gene in a DNA strand. Before enzymes can even assemble RNA on the DNA, they must first bind with a promoter. As you will see, some control agents bind to **operators**, which are short base sequences between a promoter and the start of a gene. Typically, a control agent does not stay permanently attached to a binding site. The binding may be reversed, for example, when the conditions that called for the synthesis of a particular protein have changed.

Among eukaryotes, **hormones** are major control agents. They are signaling molecules secreted from specific types of cells. They travel the bloodstream and affect gene expression in target cells somewhere in the body. (Any cell is a "target" if it has receptors to which a specific signaling molecule can bind.)

*Gene expression is controlled through regulatory proteins, hormones, and other molecules that interact with DNA, RNA, and the protein products of genes.*

## GENE CONTROL IN PROKARYOTES

Studies of *Escherichia coli*, a type of bacterial cell, yielded the first insights into gene controls. As is the case for most prokaryotes, its gene activity depends largely on negative and positive controls over the rate of transcription.

### Negative Control of Lactose Metabolism

*E. coli* makes its home in the intestines of mammals, and it survives on sugars and other nutrients of its host's diet. Some controls allow it to produce enzymes *only when needed* to degrade lactose, a sugar in milk.

After you drink a glass of milk, *E. coli* rapidly transcribes three adjacent genes that code for the lactose-degrading enzymes. A promoter and an operator precede those genes in *E. coli* DNA and have roles in the transcription of all three. This type of arrangement, in which the same promoter-operator sequence services more than a single gene, is called an **operon:**

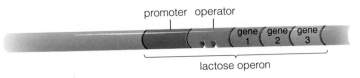

A gene at a different location in the DNA codes for a type of repressor protein:

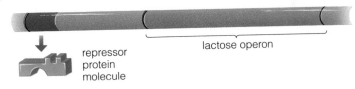

Depending on cellular conditions, this repressor can lock onto the operator *or* a lactose molecule. The repressor is part of a negative control mechanism, for it prevents the genes of the lactose operon from being transcribed. It binds with the operator when concentrations of lactose are low—as when you have not been drinking any milk. Being a rather large molecule, the repressor overlaps the promoter and so blocks RNA polymerase's access to the genes. Figure 15.2b illustrates this effect. Thus lactose-degrading enzymes are not produced when they are not required.

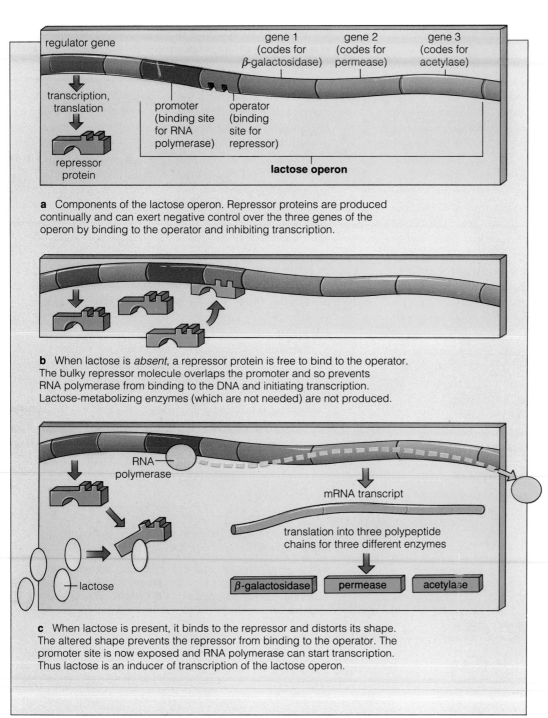

**a** Components of the lactose operon. Repressor proteins are produced continually and can exert negative control over the three genes of the operon by binding to the operator and inhibiting transcription.

**b** When lactose is *absent*, a repressor protein is free to bind to the operator. The bulky repressor molecule overlaps the promoter and so prevents RNA polymerase from binding to the DNA and initiating transcription. Lactose-metabolizing enzymes (which are not needed) are not produced.

**c** When lactose is present, it binds to the repressor and distorts its shape. The altered shape prevents the repressor from binding to the operator. The promoter site is now exposed and RNA polymerase can start transcription. Thus lactose is an inducer of transcription of the lactose operon.

**Figure 15.2** Negative control of the lactose operon. The first gene of the operon codes for an enzyme that splits lactose into two subunits (glucose and galactose). The second one codes for an enzyme that transports lactose molecules across the plasma membrane and into the cytoplasm. The third plays a complex role in lactose metabolism.

When you have been drinking milk and lactose has entered your small intestines, the repressor does not block transcription. What happens is this: A lactose molecule binds with and alters the shape of the repressor. In its altered shape, the repressor cannot bind to the operator. As a result, RNA polymerase has access to the genes, which are transcribed and translated into proteins (Figure 15.2c). Thus lactose-degrading enzymes are produced when they are required.

**Positive Control of Nitrogen Metabolism**

The positive control mechanisms afforded by activator proteins resemble the negative control just described—but with opposite results. Here, the promoter sequence is such that RNA polymerase binds very inefficiently to it without assistance. The promoter becomes functional when an activator protein is bound to it. Then, RNA polymerase binds and initiates transcription with greater efficiency. The transcription rate slows when the activator protein is removed from the promoter.

Consider what happens when you have not eaten much protein. The food material moving through your gut is therefore low in nitrogen, a vital nutrient for *E. coli* as well as for yourself. At such times, the bacterial cell steps up its synthesis of glutamine synthetase and other enzymes of the pathways by which nitrogen can be obtained from the surroundings. The genes coding for these enzymes are called the **nitrogen-related operon**. The more mRNA molecules that can be transcribed from those genes, the more enzyme molecules will be produced. The more enzymes the cell produces, the better will be its chance of assimilating what little nitrogen *is* available at that time.

The drop in nitrogen triggers a cascade of reactions, similar to the one described at the start of this chapter. One type of enzyme activates many molecules of another type. Each of these activates many of another type, and so on. At the end of the cascade, a large number of activator molecules have been called into service by having a phosphate group transferred to them. Only when they are phosphorylated can these activators turn on the gene coding for glutamine synthetase.

These reactions permit *E. coli* to make a very big metabolic response to a dilute amount of nitrogen in its surroundings. When nitrogen becomes plentiful, the same enzyme that transferred phosphate to the activator protein removes it, so the response can now be reversed.

In this example, a regulatory protein is controlled by its conversions between active (phosphorylated) and inactive forms. Prokaryotes do not rely heavily on such interconversions as a means of gene control. Eukaryotes do. The cascade of reactions that culminates in cell division is but one example.

## GENE CONTROL IN EUKARYOTES

### Selective Gene Expression

Compared to prokaryotes, less is known about gene controls in multicelled eukaryotes. The main reason is that patterns of gene expression vary within and between different body tissues. Consider that all the cells in your body are descendants of the same fertilized egg and so have the same genes. But cells of your brain, liver, and other tissues are **differentiated**: they have become specialized in composition, structure, and function.

Differentiation arises through *selective* gene expression in different cells. Depending on the cell type and the control agents acting on it, some genes might be turned on only at one particular stage of the life cycle. Others might be left on all the time or never activated at all. Still other genes might be switched on and off throughout an individual's life.

Think about hormones and other signaling molecules, which play crucial roles in selective gene expression. Hormones are secreted from glands or glandular cells, picked up by the bloodstream, and distributed throughout the body. Some have widespread effects on gene activity in many cell types. In vertebrates, for example, the pituitary gland secretes somatotropin (also called growth hormone). This hormone helps control the synthesis of proteins required for cell division and, ultimately, the body's growth. Most of the body's cells have receptors for somatotropin.

Other hormones affect gene expression only in certain cells at certain times. Prolactin is like this. Its target cells are in mammary glands. Beginning a few days after a mammalian female gives birth, prolactin activates genes in those cells alone—genes that have exclusive responsibility for milk production. Liver cells and heart cells also have those genes, but they have no means of responding to signals from prolactin and they never will have any role in milk production.

Explaining hormonal control of gene activity is like explaining a full symphony orchestra to someone who has never seen one or heard it perform. Many separate parts must be defined before their intricate interactions can be understood! We will be returning to this topic, starting with Chapter 34 on the endocrine system. As you will see in Chapters 42 and 43, some of the most elegant examples of hormonal controls are drawn from studies of animal reproduction and development.

**Cell differentiation occurs in multicelled eukaryotes as a result of** *selective gene expression.*

**Although all the cells in the body inherit the same genes, they activate or suppress some fraction of those genes in different ways to produce pronounced differences in their structure or function.**

### Levels of Control in Eukaryotes

Let's now consider a few examples of the levels of control of gene expression in eukaryotes. As Figure 15.3 and the following list indicate, controls are exerted at different levels:

1. *Transcriptional controls* influence when and to what degree a particular gene will be transcribed (if at all).

2. *Transcript-processing controls* govern modification of the initial mRNA transcripts in the nucleus.

3. *Transport controls* dictate which mature mRNA transcripts will be shipped out of the nucleus and into the cytoplasm for translation.

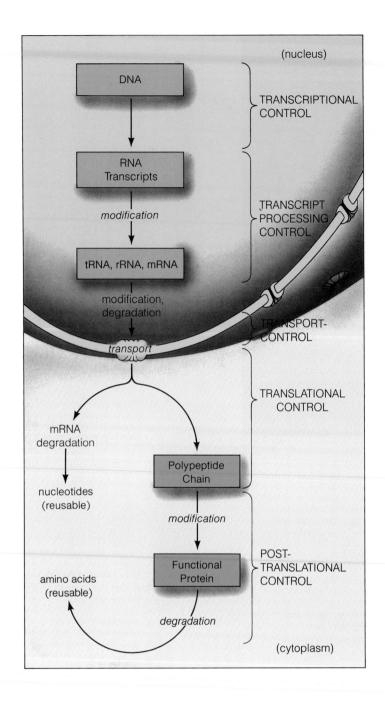

**Figure 15.3** (*Left*) Control of eukaryotic gene expression: levels at which regulatory mechanisms can be brought into play. (Here, the steps are superimposed on a sketch of nuclear and cytoplasmic regions of a eukaryotic cell.)

4. *Translational controls* govern the rates at which mRNA transcripts that reach the cytoplasm will be translated into polypeptide chains at the ribosomes.

5. *Post-translational controls* govern how the polypeptide chains become modified into functional enzymes and other proteins. (For example, some chains have specific sugar or phosphate groups attached to them.) They also govern the enzymes themselves by enhancing or repressing their action.

We know the most about controls that operate during transcription and transcript processing. Among these are regulatory proteins, especially activators that are turned on and off (by the addition and removal of phosphate). Transcription also is controlled through the way eukaryotic DNA is packed up with proteins in chromosomes. Figure 15.4 shows the packing at the most basic level of chromosome structure—the nucleosome, which consists of DNA looped around a core of histone proteins. Besides histones, other chromosomal proteins may

**Figure 15.4** (*Below*) One model of how DNA loosens up during transcription. As indicated earlier in Figure 13.10, nucleosomes are the basic packing unit of the eukaryotic chromosome. Each consists of a portion of the DNA double helix, looped twice around a core of histone proteins. The diagrams show a nucleosome from *Physarum polycephalum*. At one point in the life cycle of this slime mold, only the genes coding for rRNA are transcribed. The histones remain associated with the DNA in the gene regions being transcribed. But the tight packing becomes more relaxed.

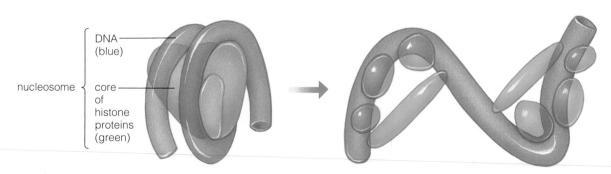

**a** Appearance of nucleosome when genes are inactive

**b** Appearance of nucleosome when genes are being actively transcribed

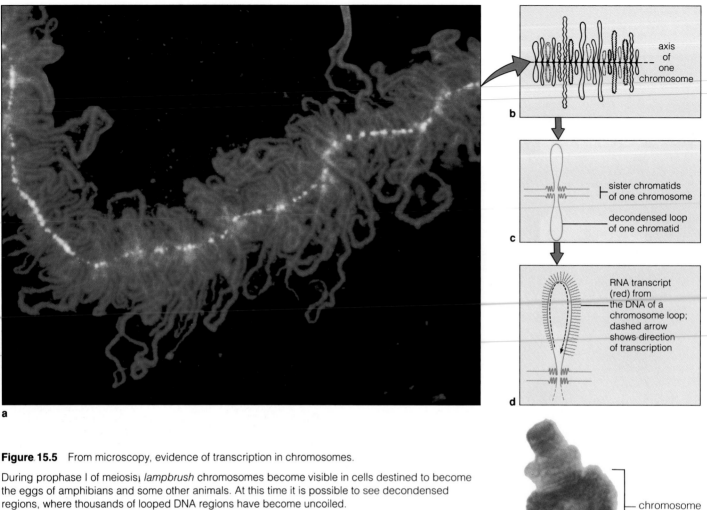

a

b axis
of
one
chromosome

c sister chromatids
of one chromosome

decondensed loop
of one chromatid

d RNA transcript
(red) from
the DNA of a
chromosome loop;
dashed arrow
shows direction
of transcription

e chromosome
puff

**Figure 15.5** From microscopy, evidence of transcription in chromosomes.

During prophase I of meiosis, *lampbrush* chromosomes become visible in cells destined to become the eggs of amphibians and some other animals. At this time it is possible to see decondensed regions, where thousands of looped DNA regions have become uncoiled.

(**a–d**) This is part of a lampbrush chromosome from a newt, *Notophthalmus viridiscens*. A dye that binds with DNA caused the axis of the chromosome to fluoresce white. A dye that binds with a ribonucleoprotein caused the loops to fluoresce red. Ribonucleoproteins are a clear sign of transcriptional activity.

Intense transcription also can be observed in the *polytene* chromosomes of certain fly larvae. Years ago, staining techniques revealed their faint banding patterns. Today we know that DNA replication has occurred repeatedly in such chromosomes, with the duplicated strands remaining packed in parallel. A hormone (ecdysone) prods a regulatory protein into binding with the DNA, and binding promotes transcription. In regions being transcribed, the chromosome has loosened up, forming *chromosome puffs*. The degree of transcription correlates with how large and diffuse the puffs become. (**e**) Puffing in a polytene chromosome from a midge. The red-violet stain indicates transcriptionally active regions.

organize the chromosome into functional domains, which can be kept locked up or loosened up in a given cell at a given time. Certainly packing variations are known to occur during the life cycle of many organisms, and these affect the accessibility and activity of different genes. In an example to be described shortly, the genes on an entire chromosome are never allowed to be transcribed.

## Evidence of Control Mechanisms

**Chromosome Loops and Puffs.** Evidence of controls over transcription can be observed in amphibian eggs and the larvae of certain insects, both of which grow very rapidly. It takes a great deal of mRNA and protein synthesis to sustain that rapid growth. Electron micrographs of amphibian eggs reveal visible changes in chromosome structure that can be correlated directly with transcription of the genes necessary for growth. During prophase I of meiosis, the chromosomes decondense into thousands of looped domains. The DNA has been selectively loosened, making specific gene regions accessible for transcription. The chromosomes look so bristly at this time, they are said to have a "lampbrush" configuration (Figure 15.5a).

As another example, many insect larvae are like feeding machines that munch incessantly on plant or animal tissues. It takes quite a bit of saliva to prepare chunks of food for digestion. Cells of their salivary glands contain rather unusual chromosomes in which the DNA molecule has been replicated repeatedly. Thus the cells have multiple copies of the same genes that can be transcribed at the same time to yield many copies of the protein components of saliva. These so-called "polytene" chromosomes puff out when the genes are being transcribed, as Figure 15.5e shows.

**X Chromosome Inactivation.** Each cell in a female mammal has two X chromosomes, one maternal and one paternal in origin. Only the genes of one are available for transcription. Most genes of the other chromosome are completely inaccessible. This condition is normal. Her cells function perfectly well with the gene products of only one X chromosome. It may even be that a double dose of gene products from two X chromosomes would prove lethal.

When each female is developing as an embryo in her mother's body, one X chromosome in each cell becomes condensed and transcription of most of its genes is permanently suppressed. Which of the two becomes condensed is a matter of chance. As Figure 15.6 shows, the condensed X chromosome is quite distinct in the interphase nucleus. It is called a "Barr body" after its discoverer, Murray Barr.

The embryo grows through cell divisions, and each daughter cell inherits the same pattern of X chromosome inactivation that occurred in its parent cell. *Either* the maternal *or* paternal X chromosome (never both) can be randomly inactivated in the cell lineage that gives rise to a given tissue region. Thus every adult female is a "mosaic" of X-linked traits. She has patches of tissue in which an allele on the maternal *or* paternal X chromosome is being expressed.

The mosaic tissue effect arising from random X chromosome inactivation is called *Lyonization* (after its discoverer, Mary Lyon). It is especially visible in female calico cats, which are heterozygous for black and yellow coat-color alleles on the X chromosome. Coat color in a given body region depends on which of the two X chromosomes is functioning and on which of the particular alleles is available for transcription (Figure 15.7).

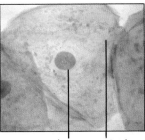

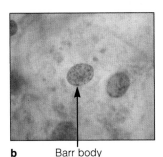

a   nucleus   cytoplasm

b   Barr body

**Figure 15.6** (**a**) Nucleus from a male cell, which has no Barr body.

(**b**) Nucleus from a female cell, showing a Barr body.

**Figure 15.7** Why is this female calico cat "calico"? Each of her cells contains two X chromosomes. One chromosome carries an allele coding for black coat color and the other carries an allele coding for yellow. When she was an embryo developing in her mother's body, one of the two X chromosomes was inactivated at random in each of the cells that had formed by that time. In all the cellular descendants of each cell, the same chromosome remains inactivated, leaving only one functional allele for the coat-color trait.

And so we see different patches of color, depending on which allele was inactivated in cells making up the tissue in a given region. (The white patches result from interaction with another gene locus—the so-called spotting gene—that determines whether any color appears at all.)

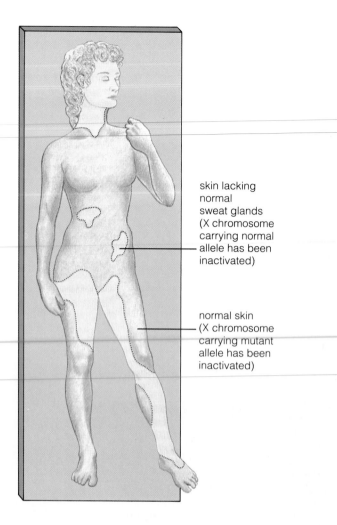

The mosaic effect also is evident in human females affected by *anhidrotic ectodermal dysplasia*. A mutant allele on one X chromosome gives rise to this skin disorder, which is characterized by an absence of sweat glands. In patches of defective skin, the X chromosome bearing the normal allele has been inactivated and the X chromosome bearing the mutant allele is functional (Figure 15.8).

**Transcript-Processing Controls.** So far, we have considered evidence of transcriptional control. Keep in mind that other levels of gene control may be equally important for normal cell activities; we just don't know as much about them.

For example, we still have a long way to go in identifying the controls over transcript processing—that is, over how newly formed mRNA transcripts get modified. Even so, researchers are giving us some interesting things to think about. Experiments show that the primary mRNA transcript from a single gene is sometimes processed in alternative ways. For instance, transcripts from a gene that codes for a contractile protein (troponin-T) are processed differently in different cell types! The outcome is two or more different mRNAs, each coding for a distinct kind of protein. Although the proteins are very similar, each is unique in a certain region of its amino acid sequence (Figure 15.9). All of the resulting proteins still function in contraction. But they do so in different ways—which may account for subtle variations in the way different types of muscles in your body function.

skin lacking
normal
sweat glands
(X chromosome
carrying normal
allele has been
inactivated)

normal skin
(X chromosome
carrying mutant
allele has been
inactivated)

**Figure 15.8** (*Above*) Pattern of gene expression in a woman affected by anhidrotic ectodermal dysplasia, a disorder in which patches of skin do not have normal sweat glands. The pattern arises through random inactivation of the X chromosome during embryonic development. The mutant allele responsible for the disorder is on one X chromosome, and its corresponding allele on the other chromosome is normal. Depending on which of the two chromosomes is inactivated in an embryonic cell, all of the clonal descendants of that cell will display the same pattern of gene expression.

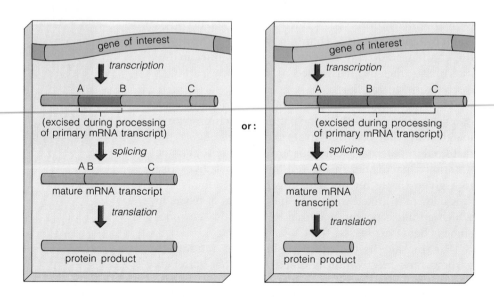

**Figure 15.9** Alternative processing of a primary mRNA transcript from a single gene. The primary transcript can be processed in different ways to produce distinct mRNA molecules that code for similar but distinct proteins.

*Commentary*

## Characteristics of Cancer

On rare occasions, certain cells in the body that should not be doing so divide again and again, until their offspring begin to crowd surrounding cells and interfere with tissue functions. The parent cell that started it all had become mutated, and it spawned a tumor. By definition, a **tumor** is any tissue mass composed of cells that are not responding to normal controls over cell growth and division.

The problem is not that tumor cells divide at a horrendous rate. Normal cells divide much faster when they replace a surgically removed portion of the liver. Rather, tumor cells have lost the controls telling them when to stop. They will not stop as long as conditions for growth remain favorable.

When a tumor is *benign*, it may continue to grow more rapidly than normal but it remains in the same place in the body. Surgical removal of the tissue mass removes its threat to health. When a tumor is *malignant*, its cells can migrate and then grow and divide in other organs.

Normally, recognition proteins at the plasma membrane allow cells to bind together in tissues and organs. When genes for those proteins are altered or suppressed, the cell can leave its proper place and travel (in blood or lymph) to other tissues, where it can form a new growth. This process of invasion is called **metastasis**.

There are many types of malignant tumors, but all are grouped into the general category of **cancer**. At the minimum, all cancer cells have these characteristics:

1. *Profound changes in the plasma membrane and cytoplasm.* Membrane permeability is amplified. Some membrane proteins are lost or altered, and new ones appear. The cytoskeleton shrinks, becomes disorganized, or both. Enzyme activity shifts (as in an amplified reliance on glycolysis).

2. *Abnormal growth and division.* Inhibitors of overcrowding in tissues are lost. Cell populations increase to unusually high densities. New proteins trigger an abnormal increase in small blood vessels that service the growing cell mass.

3. *Weakened capacity for adhesion.* Cells cannot become properly anchored in the parent tissue.

4. *Lethality.* Unless cancer cells are eradicated, they will kill the individual.

Any gene having the potential to induce cancerous transformations is called an **oncogene**. When introduced into a normal cell, an oncogene may transform it into a tumor cell. Such genes were first identified in retroviruses, a type of infectious agent (page 350). But nearly identical gene sequences *also* occur in the normal DNA of many species and rarely trigger cancer! These sequences are called **proto-oncogenes**.

Proto-oncogenes code for proteins necessary in normal cell function. They may become cancer-causing genes only on rare occasions, when specific mutations alter their structure or their expression. In other words, the *normal* expression of proto-oncogenes is vital, even though their abnormal expression may be lethal.

Insertion of viral DNA into the DNA of a host cell can skew transcription of a proto-oncogene. **Carcinogens** may do the same thing. Carcinogens bind to DNA and can cause a mutation. They include many natural and synthetic compounds (such as asbestos and components of cigarette smoke), x-rays, gamma rays, and ultraviolet radiation.

Yet cancer seems to be a multistep process, with mutations in more than one proto-oncogene required to bring it about. Look again at the listed characteristics of cancer cells. Now think about some of the products of proto-oncogenes. *They include growth factors (signals sent by one cell to trigger growth in other cells), regulatory proteins involved in cell adhesion, and the protein signals for cell division.*

### When the Controls Break Down

This chapter has barely touched on the controls over gene expression in eukaryotes. How is it possible to leave you with a strong impression of their absolutely crucial importance? Perhaps with a close look at what happens when controls over some basic cell functions are disrupted.

As you saw at the start of this chapter, we are beginning to understand which genes govern cell growth and division. On rare occasions, controls over cell division are lost and cells become cancerous, a transformation that is described in the *Commentary*. Possibly more than any other example, this transformation brings home the critical extent to which you and all other organisms depend on controls over gene expression.

# SUMMARY

1. In both prokaryotes and eukaryotes, shifts in chemical conditions inside and outside cells can trigger changes in gene activity. In multicelled eukaryotes, for example, conditions vary as a result of changes in diet and levels of activity. They vary inevitably during growth and development.

2. Gene expression is controlled by many interacting elements, including control sites built into DNA molecules, regulatory proteins, enzymes, and hormones. Their interactions govern which gene products appear, at what times, and in what amounts.

3. The best understood gene controls are transcriptional. In prokaryotes, operon controls influence transcription rates. In eukaryotes, the timing and rate of transcription are influenced by chromosome organization as well as by control factors that bind to the DNA.

4. In prokaryotes especially, negative control of transcription is afforded by repressor proteins that can bind to control sites near specific genes. A bound repressor protein inhibits transcription by blocking the access of RNA polymerase to those genes.

5. Positive control of transcription is afforded by activator proteins that also bind to control sites near specific genes. Bound activator proteins promote transcription by helping RNA polymerase bind and start transcription. Positive controls may be the more common type in eukaryotes.

6. In complex eukaryotes, cells differentiate: They become different in appearance, composition, function, and often position. Differentiation arises through "selective gene expression." This term means that different types of cells activate and suppress some fraction of their genes in a variety of ways that lead to pronounced differences in cell structure and function.

## Review Questions

1. Define these terms: promoter, operator, repressor protein, and activator protein. 234

2. Cells depend on controls over which gene products are synthesized, at what times, at what rates, and in what amounts. Describe one type of control over transcription in *E. coli*, a type of prokaryote. Then list five general kinds of controls involved in eukaryotes. 234–236

3. Define cell differentiation. How does it arise? 236

4. A plant, fungus, or animal is composed of diverse cell types. How might this diversity arise, given that all of the body cells in each organism inherit the *same* set of genetic instructions? 236

5. Somatic cells of human females have two X chromosomes. During what developmental stage are genes on *both* chromosomes active? Explain what happens to each of those chromosomes after that stage. 239

6. What are the characteristics of cancer cells? Explain the difference between a benign tumor and one that is malignant. 241

## Self-Quiz *(Answers in Appendix IV)*

1. In all cells, gene activity may be altered in response to changes in _____ conditions or environmental conditions.

2. Selective gene expression results in cell _____, or changes in the cell's appearance, structure, and function.

3. The best-understood gene control mechanisms are the ones concerned with _____.
   a. translation
   b. replication
   c. post-translation
   d. transcription
   e. mRNA transport

4. Gene expression is controlled by _____.
   a. control sites built into DNA
   b. regulatory proteins
   c. enzymes
   d. hormones
   e. all of the above may have control functions

5. _____ is an aspect of protein synthesis that occurs *only* in eukaryotic cells.
   a. transcription in the cytoplasm
   b. translation in the nucleus
   c. post-translation in the nucleus
   d. processing mRNA transcripts

6. In bacteria (prokaryotes), _____ have the most important roles in the negative control of transcription.
   a. activator proteins
   b. repressor proteins
   c. RNA polymerase
   d. DNA polymerase
   e. hormones

7. Positive control of transcription appears to be more common in _____; this control involves the action of _____.
   a. eukaryotes; repressor proteins
   b. prokaryotes; repressor proteins
   c. eukaryotes; activator proteins
   d. prokaryotes, activator proteins

8. Activator proteins promote RNA transcription by _____.
   a. binding to repressor proteins
   b. binding near control sites to assist release of RNA polymerase

c. inhibiting the action of repressor proteins

d. assisting the binding of RNA polymerase to begin transcription

e. directly initiating transcription

9. Cell differentiation in complex eukaryotes arises through _____.

   a. selective gene expression
   b. activating and suppressing the same genes in all cells
   c. mostly negative transcriptional controls
   d. operon controls

10. Match the gene control concepts.

   _____ bound repressor protein
   _____ cell differentiation
   _____ bound activator protein
   _____ transcript processing controls
   _____ post-translational controls

   a. governs modification of protein into functional form
   b. blocks access of RNA polymerase to the promoter
   c. assists binding of RNA polymerase to the promoter
   d. changes sequence of mRNA to specify different proteins
   e. occurs through selective gene expression

## Readings

Feldman, M., and L. Eisenbach. November 1988. "What Makes a Tumor Cell Metastatic?" *Scientific American* 259(5):60–85.

Kupchella, C. 1987. *Dimensions of Cancer*. Belmont, California: Wadsworth.

Murray, A., and M. Kirschner. March 1991. "What Controls the Cell Cycle." *Scientific American* 264(3):56–63.

Ptashne, M. January 1989. "How Gene Activators Work." *Scientific American* 260(1):41–47.

Weintraub, H. January 1990. "Antisense RNA and DNA." *Scientific American* 262(1):40–46.

## Selected Key Terms

activator protein *234*
Barr body *239*
cancer *241*
carcinogen *241*
chromosome puff *238*
differentiation *236*
hormone *234*
Lyonization *239*
metastasis *241*
oncogene *241*
operator *234*
operon *234*
post-translational control *237*
promoter *234*
proto-oncogene *241*
repressor protein *234*
selective gene expression *236*
transcriptional control *236*
transcript-processing control *236*
translational control *237*
transport control *236*
X chromosome inactivation *239*

# 16 RECOMBINANT DNA AND GENETIC ENGINEERING

## Life and Death on the Threshold of a New Technology

For much of human history, we have been dealing with a world that is often harsh. In a bad year, an influenza virus might strike hundreds of thousands of us. In any year, wheat rusts, viruses, and other agents of disease might destroy millions of acres of crops and contribute to starvation on a global scale. Every year, some of our kind are born with crippling genetic disorders.

And yet, through research that began only a few decades ago, we soon may have more control over our individual and combined destinies. A new technology is giving us the means to alter, to our advantage, the DNA molecules of viruses, crop plants and any other kind of organism—including human beings.

In September 1990, for example, a four-year-old girl received what may be a historic genetic reprieve. Her problem is chillingly simple. Due to a single gene mutation, she was born without defenses against viruses, bacteria, and other agents of disease. She has

no immune system. Of the forty-six chromosomes she inherited from her parents, one copy of chromosome 20 bore a defective gene. That gene codes for an enzyme, adenosine deaminase (ADA).

Normally, the enzyme functions in a pathway by which excess adenosine monophosphate (AMP) is stripped of its phosphate group and degraded to uric acid, which the body excretes. In ADA-deficient individuals, a related nucleotide phosphate accumulates. For reasons not fully understood, the excess nucleotide phosphate is toxic to lymphoblasts, which are a type of stem cell in bone marrow. Lymphoblasts divide again and again. Their progeny later differentiate into new recruits and replacements for the immune system's army, the infection-fighting white blood cells.

The ADA-deficient girl is a victim of *severe combined immune deficiency* (SCID). This is a set of disorders

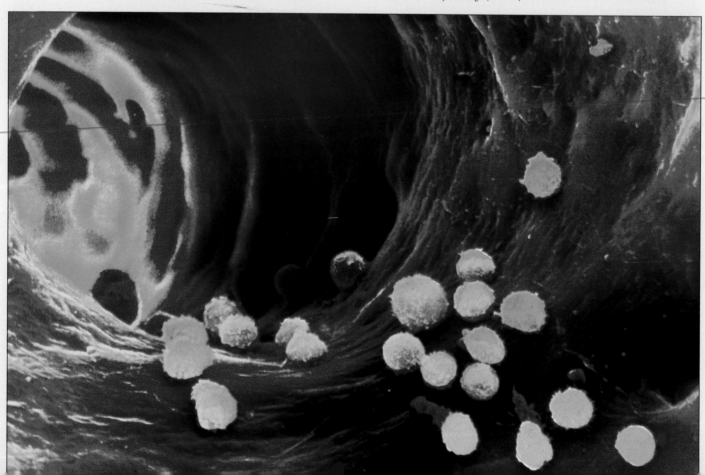

brought on by a drastic reduction or complete absence of two subpopulations of white blood cells—the B and T cells. In their absence, the girl risks death even from minor infections, even from bacteria that otherwise live harmlessly in the body.

Some individuals with SCID benefit from bone marrow transplants. If the transplants take hold, the donated stem cells may produce functional B and T cells. Similarly, ADA injections help some children. Both therapeutic approaches buy time, but neither solves the underlying problem of ADA deficiency. And neither apparently ends the threat of a killer infection.

Given the options, the parents of the four-year-old girl allowed her to participate in the first federally approved gene therapy test for humans. Using recombinant DNA methods of the sort described in this chapter, medical researchers introduced copies of the ADA gene into some of her T cells, which were then encouraged to divide repeatedly. Later, they placed about a billion copies of the genetically engineered cells in a saline solution and delivered them, through a plastic tube, into her bloodstream. Each month, the girl will receive another T cell infusion—and another chance at a longer life. In time, her doctors hope to extract lymphoblasts from her bone marrow, introduce functional ADA genes into them, and put them back in place. Only then might she be assured of a constant, lifelong supply of the crucial enzyme—and functional disease fighters.

As this example suggests, recombinant DNA technology has staggering potential for medicine. It has equally staggering potential for agriculture and industry. It does not come without risks. With this chapter, we consider some basic aspects of the new technology, and we address ecological, social, and ethical questions related to its application.

**Figure 16.1** A few white blood cells that were on patrol inside a blood vessel. Individuals with a severely compromised immune system have drastically reduced numbers of these infection-fighting cells—or none at all. Such individuals are prime candidates for gene therapy, one of the beneficial applications of recombinant DNA research.

KEY CONCEPTS

**1.** Genetic experiments have been occurring in nature for billions of years as a result of gene mutations, crossing over and recombination, and other events. Humans are now engineering genetic changes by way of recombinant DNA technology.

**2.** With recombinant DNA technology, DNA molecules are cut into fragments, and the fragments of interest are inserted into cloning tools such as plasmids. Then they are amplified rapidly into quantities suitable for research and practical applications.

**3.** The new technology raises social, legal, ecological, and ethical questions regarding its benefits and risks.

For more than 3 billion years, nature has been conducting genetic experiments through mutation, chromosomal crossing over, and other events. Genetic messages have changed countless times; this is the source of life's diversity.

We humans have been changing the genetic character of species for thousands of years. Through artificial selection, we coaxed modern crop plants and new breeds of cattle, birds, dogs, and cats from wild ancestral stocks. We developed meatier turkeys, sweeter oranges, seedless watermelons, and flamboyant ornamental plants. We produced the tangelo (tangerine × grapefruit) and the mule (donkey × horse).

Today we are analyzing and even engineering genetic changes through **recombinant DNA technology**. With this technology, DNA from different species can be cut, spliced together, then inserted into bacteria or other types of rapidly dividing cells—which multiply in quantity the recombinant DNA molecules. Genes can be isolated, modified, and reinserted into the organism (or transplanted into a different one). In many cases, the engineered genes produce functional proteins. Before looking at how this work is done, let's start out by considering a few recombination mechanisms in nature that actually paved the way for the new technology.

a

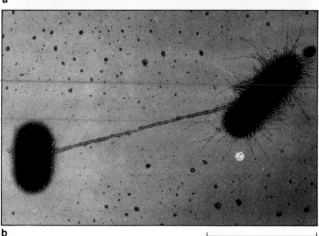

b

|——————————| 1 μm

**Figure 16.2** (**a**) A ruptured bacterial cell (*Escherichia coli*). Notice the larger bacterial chromosome and the several small, circular plasmids (blue arrows). (**b**) Early stage of conjugation between a recipient (F⁻) cell, at left, and a donor (F⁺) cell of *E. coli*. The long appendage joining the two bacteria is an F pilus; it will bring the two participants into close contact so that DNA can be transferred.

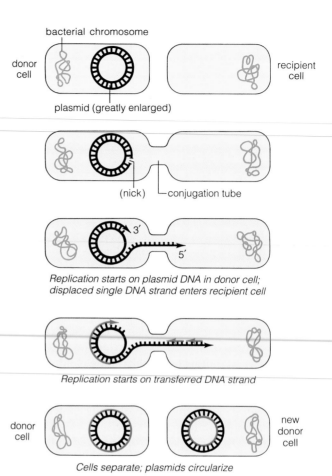

bacterial chromosome

donor cell

recipient cell

plasmid (greatly enlarged)

(nick)   conjugation tube

*Replication starts on plasmid DNA in donor cell; displaced single DNA strand enters recipient cell*

*Replication starts on transferred DNA strand*

donor cell

new donor cell

*Cells separate; plasmids circularize*

**Figure 16.3** Transfer of a plasmid between two bacterial cells during conjugation. For clarity, the bacterial chromosomes are not drawn full size. (Each bacterial chromosome actually contains about forty times more genetic information than the largest plasmids do.)

## RECOMBINATION IN NATURE: SOME EXAMPLES

### Transfer of Plasmid Genes

Bacteria have played a central role in the development of recombinant DNA technology. In nature, as in the laboratory, they have characteristics that suit them to the task. All bacteria have a single chromosome, and many bacterial species also have plasmids. The bacterial chromosome, a circular DNA molecule, contains all the genes necessary for normal growth and development. A **plasmid**, a small, circular molecule of "extra" DNA or RNA, carries only a few genes and is self-replicating (Figure 16.2a). In other words, a plasmid can produce more copies of itself regardless of whether the bacterial chromosome is undergoing replication.

Plasmid genes are transmitted through successive generations of bacterial cells. They are even transferred to bacterial cells of different species. What might be the selective advantage of spreading them around? Although not essential for normal growth and development, the products of plasmid genes do serve useful functions under some circumstances, as we know from studies of the **F plasmids** (the F stands for *Fertility*).

Genes carried on an F plasmid permit *bacterial conjugation*, a process by which one bacterial cell transfers DNA to another. Only a cell having an F plasmid can be a donor; it is designated F⁺. Only a cell lacking an F plasmid (designated F⁻) can be a recipient. Sometimes the transfer is said to be a form of sexual reproduction between "male" and "female" bacterial cells, although comparing this to sex among the eukaryotes requires a rather breathtaking leap of the imagination.

Some F plasmid genes code for the proteins necessary to construct an F pilus, a long appendage that can latch onto a recipient cell and draw it right up against the donor (Figure 16.2b). The attachment apparently activates an enzyme, which cuts one strand of the plasmid DNA. The cut strand starts to unwind from the other strand and enters an F⁻ cell. As Figure 16.3 shows, DNA replication proceeds on the exposed bases of both strands in both cells. Once the transfer and replication are completed, the cells separate. Each is now an F⁺ cell.

Once in a great while, a donated plasmid becomes integrated into the recipient's chromosome through natural recombination mechanisms. Like all cells, bacteria have enzymes capable of cutting DNA strands during normal repair operations. One such enzyme recognizes a short nucleotide sequence that happens to occur on both the plasmid and the chromosome. It cuts both molecules at the sequence, then splices their cut ends together (Figure 16.4).

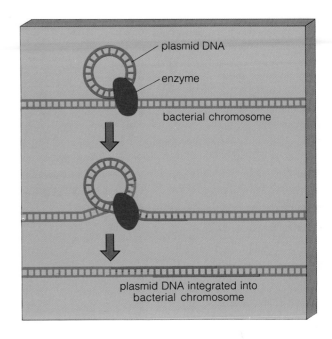

**Figure 16.4** Integration of a plasmid into a bacterial chromosome. Only a small stretch of the circular bacterial chromosome is shown.

### Transfer of Viral Genes

Among the viruses called bacteriophages are types that can integrate their DNA into the chromosome of bacterial host cells. Lambda bacteriophage is one of them. It can spread through a population of *Escherichia coli* through repeated cycles of infection. On rare occasions, a viral enzyme cuts the bacterial chromosome at a specific site, then the viral DNA is inserted between the cuts and sealed in place. The modified bacterial chromosome is replicated and the viral DNA is passed on in latent form to succeeding generations of *E. coli* (page 349). Later, the viral DNA may move out of the chromosome, and an infectious cycle begins again.

To date, transfer of genes by recombination has been discovered in many organisms, including a variety of bacteria, bacteriophages, yeasts, fruit flies, and mammals. It appears that gene transfer may be common to most, if not all, organisms. As we turn now to the kind of recombination techniques going on in the laboratory, keep in mind this basic point:

**Gene transfer and recombination is a common occurrence in nature, made possible by specific enzymes that can make cuts in DNA molecules.**

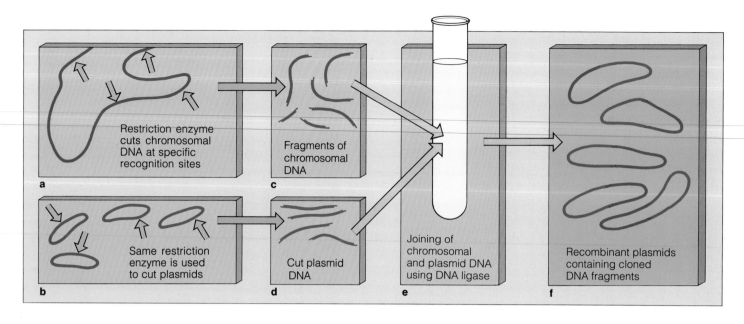

**Figure 16.5** Formation of a DNA library.

## RECOMBINANT DNA TECHNOLOGY

Recombinant DNA technology grew out of experiments with *E. coli* and the bacteriophages that infect it. In the late 1960s and early 1970s, researchers learned how to use different bacterial enzymes to cut DNA into fragments and "package" the fragments into plasmids for insertion into host cells. They developed ways to pinpoint DNA fragments of interest in dividing cells. They also started to identify nucleotide sequences of individual genes. And they used that information to map the positions of various genes in the DNA of different species.

### Producing Restriction Fragments

Different bacterial species produce **restriction enzymes**, the sole function of which is to cut apart and destroy foreign DNA molecules that often are injected into the cell by viruses. Several hundred restriction enzymes have been identified. Each makes its cut only at sites having a short, specific nucleotide sequence. This feature enables researchers to select those enzymes that will produce a DNA fragment containing a particular gene.

Restriction enzymes are also useful in another respect. Many produce staggered cuts, so both ends of a DNA fragment end up with short, single-stranded tails:

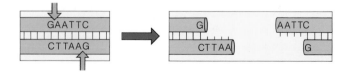

These "sticky ends" of the fragment have a terrific property. They can base-pair with any other DNA molecule cut by the same restriction enzyme.

Suppose we use the same restriction enzyme to cut plasmids and a chromosome. When the chromosomal fragments and cut plasmids are mixed together, they base-pair at the cut sites (Figure 16.5). The base-pairing can be made permanent by using another enzyme, **DNA ligase**. As shown earlier in Figure 13.8, DNA ligase is a replication enzyme that joins the short fragments of DNA formed by discontinuous synthesis.

We now have a **DNA library**—a collection of DNA fragments produced by restriction enzymes and incorporated into plasmids. The amount of DNA in a library is almost vanishingly small. To do anything useful with it, we must first amplify it—that is, allow the fragments to be copied again and again to produce huge numbers of them.

### DNA Amplification

**By Cloning.** A DNA library of recombinant plasmids can be inserted into many host cells for propagation. (Bacteria and yeast cells are commonly used for such a purpose. Both can be grown easily in the laboratory, and their rapid reproduction rates are quite impressive.) When put to such use, a plasmid or any other self-replicating genetic element is called a "cloning vector." After repeated replications and divisions of the host cells, we end up with **cloned DNA**—multiple, identical copies of DNA fragments contained within their cloning vectors.

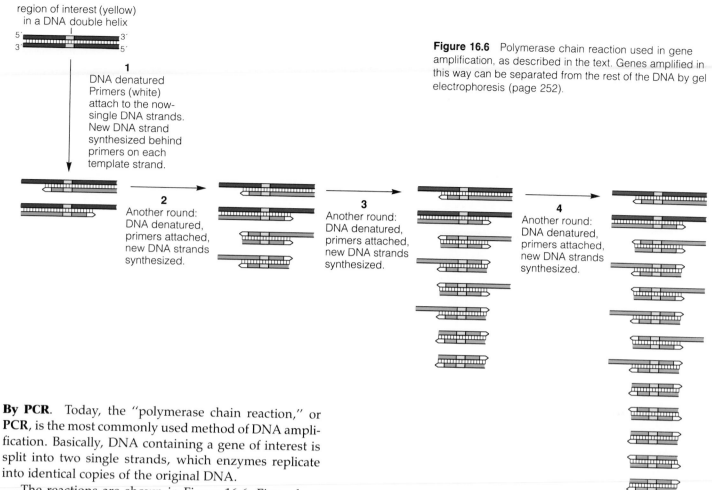

region of interest (yellow)
in a DNA double helix

5′ 3′
3′ 5′

**1**

DNA denatured
Primers (white)
attach to the now-
single DNA strands.
New DNA strand
synthesized behind
primers on each
template strand.

**2**
Another round:
DNA denatured,
primers attached,
new DNA strands
synthesized.

**3**
Another round:
DNA denatured,
primers attached,
new DNA strands
synthesized.

**4**
Another round:
DNA denatured,
primers attached,
new DNA strands
synthesized.

**Figure 16.6** Polymerase chain reaction used in gene amplification, as described in the text. Genes amplified in this way can be separated from the rest of the DNA by gel electrophoresis (page 252).

**5**
Continued rounds of amplification quickly yield great numbers of identical fragments that have primers at their 5′ ends—and those fragments contain only the region of interest.

**By PCR**. Today, the "polymerase chain reaction," or **PCR**, is the most commonly used method of DNA amplification. Basically, DNA containing a gene of interest is split into two single strands, which enzymes replicate into identical copies of the original DNA.

The reactions are shown in Figure 16.6. First, short nucleotide sequences complementary to the ones that flank the desired stretch of DNA are synthesized. They are added to the DNA, along with DNA polymerase, free nucleotides, and a buffer. Because they will base-pair with the flanking regions, those short nucleotide sequences can serve as "primers" for DNA polymerase. That enzyme, recall, cannot synthesize a new DNA strand unless a primer is already positioned on the existing one (Figure 13.8).

In this case, researchers use a DNA polymerase from *Thermus aquaticus*, a type of bacterium that thrives in hot springs, even in hot water heaters. The polymerase functions at high temperatures (for example, 72°C), so it remains stable at the elevated temperatures required to denature DNA.

The mixture of DNA, primers, and enzymes is subjected to multiple rounds of three temperatures. In each round, the DNA first is denatured by exposure to near-boiling temperatures (about 94°C). Then the temperature is lowered to between 37°C and 60°C, which is the range in which the primers base-pair most readily with the DNA. Then the temperature is raised to 72°C—and the DNA polymerases go to work. As the number of cycles increases, the newly synthesized DNA strands themselves become templates for primer base-pairing and copying.

After several rounds of PCR synthesis, nearly all of the DNA fragments will be the same length, corresponding to the distance between the ends of the two primers. This amplified DNA greatly outnumbers the original DNA template. It can be easily detected and analyzed by gel electrophoresis, as described in the *Doing Science* essay on page 252.

The advantage of PCR is that extremely small samples of DNA can be increased to high concentrations that can be studied easily, cloned, or both. A single DNA molecule can be quickly amplified to many *billions* of molecules. Thus it can be used to amplify samples with too little DNA, as might be found in a single hair left at the scene of a crime. Besides this, PCR by its very nature can be used to repair fragmented DNA. That is why it is being used in studies of samples that are too old to contain intact DNA, like those that might be found in an Egyptian mummy. In the near future, PCR may be used to identify genetic disorders in very early human embryos that consist only of eight cells!

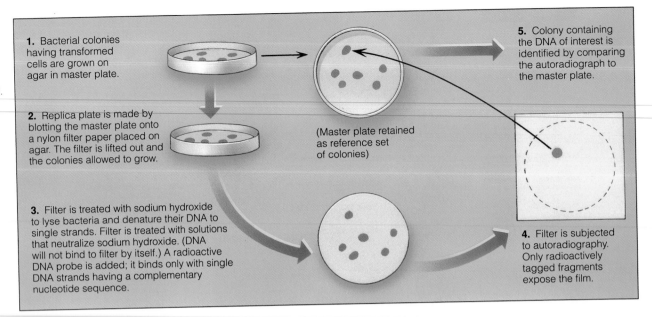

**Figure 16.7** Use of a DNA probe to identify the colony of transformed bacterial cells that have taken up the DNA of interest.

1. Bacterial colonies having transformed cells are grown on agar in master plate.

2. Replica plate is made by blotting the master plate onto a nylon filter paper placed on agar. The filter is lifted out and the colonies allowed to grow.

3. Filter is treated with sodium hydroxide to lyse bacteria and denature their DNA to single strands. Filter is treated with solutions that neutralize sodium hydroxide. (DNA will not bind to filter by itself.) A radioactive DNA probe is added; it binds only with single DNA strands having a complementary nucleotide sequence.

(Master plate retained as reference set of colonies)

4. Filter is subjected to autoradiography. Only radioactively tagged fragments expose the film.

5. Colony containing the DNA of interest is identified by comparing the autoradiograph to the master plate.

## Identifying Modified Host Cells

Suppose we mix recombinant plasmids with potential host cells on a culture medium in a petri dish. Not all of the cells will take up a plasmid, but they all may grow and reproduce to form large colonies on the culture medium. The following procedure can be used to identify the colonies that do harbor the DNA of interest. This procedure is an example of **nucleic acid hybridization**.

First, a plasmid with antibiotic-resistance genes is selected as a cloning vector. If the culture medium has been supplemented with antibiotics, all the colonies *except* the ones with transformed cells will be destroyed or their growth inhibited. Only cells that take up and successfully express the antibiotic-resistance genes will survive.

Second, a specific probe is prepared so that the colonies can be analyzed by nucleic acid hybridization techniques, using DNA probes. A **DNA probe** is a short DNA sequence that is assembled from radioactively labeled nucleotides. Part of the sequence must be complementary to that of the desired gene.

Nucleic acid hybridization requires that a replica plate be made of the colonies growing in the petri dish (the master plate). The colonies are blotted onto a nylon filter and the transferred cells are allowed to grow into new colonies. Their locations correspond to the locations of the original colonies on the master plate (Figure 16.7). Then the filter is treated with solutions that cause the cells to rupture and that allow the released DNA to become permanently affixed to the nylon filter. Nucleic acid hybridization follows when the radioactive DNA probe is added to such filters. The probe will hybridize only with the DNA having the complementary base sequence and will thereby tag the location of the transformed bacterial colony or colonies.

## Expressing the Gene of Interest

Even when a host cell has taken up a cloned gene, it may not be able to transcribe and translate it into functional protein. For example, human genes contain noncoding regions (introns) as well as coding regions (exons). The genes cannot be translated unless the introns are spliced out and the exons spliced together into a mature mRNA transcript (page 221). As it happens, bacterial host cells don't have enzymes that recognize the splice signals on eukaryotic genes. That is why cDNA typically is the choice of researchers who study the products of human genes.

The term **cDNA** refers to any DNA molecule that has been "copied" from a mature mRNA transcript by a process called *reverse transcription*. The single-stranded mRNA molecule serves as a template for assembling a DNA strand that is identical in sequence to the gene of interest. This seemingly "backwards" process (by which mRNA is transcribed into DNA) requires a special viral enzyme, *reverse transcriptase*. After the "hybrid" DNA/RNA molecule is assembled, the RNA strand is degraded. Other enzymes convert the remaining strand of DNA to double-stranded form (Figure 16.8). Once cDNA has been obtained, it typically is cloned into a vector.

Despite the obstacles, several human gene products already are being mass-produced and many more are being developed. Table 16.1 lists some of them. The availability of these gene products already has had major impact on genetic research.

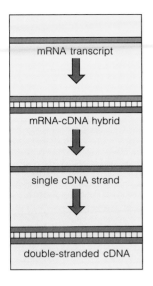

**1**

mRNA transcript of desired gene is used as template for the assembly of a DNA strand by enzyme action (reverse transcriptase). The result is an RNA-DNA hybrid molecule.

**2**

RNA polymerase nicks only the RNA in the hybrid molecule. DNA polymerase in the same sample uses the nicked RNA fragments as "primers" to start synthesis of DNA on the template strand. As it proceeds, the DNA polymerase displaces other sections of RNA from the template strand. DNA ligase in the same sample joins together the new lengths of DNA in the template strand, forming double-stranded cDNA.

**Figure 16.8** Formation of cDNA from an mRNA transcript.

| Table 16.1 Examples of Cloned Human Gene Products Approved for Use or Under Development | |
|---|---|
| Protein | Used in Treating |
| Insulin | Diabetes |
| Somatotropin (growth hormone) | Pituitary dwarfism |
| Erythropoetin | Anemia |
| Factor VIII | Hemophilia |
| Factor IX | Hemophilia |
| Interleukin-2 | Cancer |
| Tumor necrosis factor | Cancer |
| Interferons | Some cancers, viral infections |
| Atrial natriuretic factor | High blood pressure |
| Tissue plasminogen factor | Heart attack, stroke |

## RISKS AND PROSPECTS OF THE NEW TECHNOLOGY

### Uses in Basic Biological Research

Centuries ago a new invention—the microscope— revealed the existence of a world of diverse organisms that live in, on, and around us. That invention eventually gave us great insight into the nature of life itself— first by allowing us to explore the secrets of cells, then of chromosomes, then finally of DNA. Now, through implementation of DNA technology, that long road of basic research is forking almost daily in new directions.

Think about Gregor Mendel growing pea plants, selecting and planting seeds, growing new plants, and forming hypotheses about what *might* be giving rise to their patterns of inheritance, season after season. Now think about how radioactive DNA probes can be used to study genes *directly* in cells, in chromosomes, in DNA. Think about those who are destined to develop a life-threatening disease such as cancer. Without the new technology, we never would have known about proto-oncogenes, oncogenes, and skewed controls over cell division.

Or think about Darwin and others who have puzzled over the meaning of life's diversity. Through DNA technology, we can study a selected nucleotide sequence in the DNA of a given species. Then we can compare the extent to which the equivalent sequence in the DNA of another species differs from it. Chapter 19 will describe how the degrees of divergence in DNA sequences are an important source of evidence of evolutionary relationships among organisms.

As one final example, consider the **human genome project**. In a truly ambitious undertaking, researchers are working to sequence the estimated 3 billion nucleotides present in human chromosomes. They also are sequencing the chromosomes of *E. coli*, yeasts, mice, and other organisms that are commonly used in genetic research.

The potential benefits from the project are enormous. Imagine being able to pinpoint the gene or genes responsible for any specific genetic disorder. Such information surely will enhance efforts to diagnose, prevent, or treat the disorder, as the parents of the ADA-deficient girl already know. Imagine how advantageous such information will be to researchers investigating the controls over gene expression or studying the evolution of life.

By current calculations, it will take researchers working in about 1,000 laboratories ten years to complete the project. Even when the sequence of every chromosome is deciphered, it may take *many* decades to decipher what the sequences mean. Even so, it is likely that the sequencing technology itself will advance rapidly. In the meantime, many laboratories are collaborating in the mapping attempt.

### Applications of RFLPs

One recombinant DNA technique is being used to good advantage in the human genome project and other basic research efforts. It also is being put to use in some rather startling ways.

Earlier, we saw how restriction enzymes cut the DNA of an individual into fragments of specific lengths. The fragments separate from one another into distinct bands when subjected to gel electrophoresis (page 252). As it happens, certain DNA fragments show slight variations in the banding patterns from one person to the next.

**Figure 16.14** Ten-week-old mouse littermates, the one on the left weighing 29 grams, and the one on the right, 44 grams. The larger mouse grew from a fertilized egg into which the gene for human somatotropin (growth hormone) had been inserted.

crop plants are monocots. In some cases, genetic engineers use chemicals or electric shocks to deliver DNA directly into protoplasts (plant cells stripped of their walls). For some species, however, regenerating whole plants from protoplast cultures is not yet possible. Recently, someone came up with the idea to deliver genes into cultured plant cells by shooting them with a pistol. Instead of bullets, blanks are used to drive DNA-coated, microscopic tungsten particles into the cells. Although this might seem analogous to using a battleship cannon to light a match, the shooter is reporting some success.

**Genetic Modification of Animals**. In 1982, Ralph Brinster and Richard Palmiter introduced the rat gene for somatotropin (growth hormone) into fertilized mouse eggs. When the mice grew, it became clear that the rat gene had become integrated into the mouse DNA and was being expressed. The mice grew much larger than their normal littermates. Their cells had up to thirty-five copies of the gene, and blood concentrations of the hormone were several hundred times higher than normal values. More recently, the gene for human somatotropin was successfully introduced and expressed in mice, giving rise to the "super rodent" shown in Figure 16.14.

Similar experiments with large, domesticated animals have not been successful. For example, when the somatotropin gene was inserted into pigs, it was expressed—but the pigs developed arthritis-like symptoms and other disorders.

As these examples suggest, gene modification in animals is extremely unpredictable. Not only must new or modified genes be inserted in the body cells of an animal or in gametes, they also must end up in specific locations in chromosomes so that their expression will be properly regulated. And they must not disrupt the function of other genes.

For instance, genes can be inserted into a sperm nucleus just after it penetrates an egg. But the eggs are so sensitive to being poked that the procedure has a high rate of failure. Even when gene delivery is successful, researchers still cannot control *where* in the DNA the inserted gene will end up. In its new location, will the inserted gene activate an oncogene, with its potential to cause cancer? Or cause a mutation? Or alter expression of related genes?

Despite such obstacles, research is advancing on many fronts. The genes responsible for a few disorders have been cloned recently. DNA probes are being produced for use in prenatal diagnosis of heritable disorders. In combination with other technical advances, such as PCR, such probes are having dramatic impact on the incidence of many diseases, including sickle-cell anemia.

## Human Gene Therapy

Inserting one or more normal genes into the body cells of an organism to correct a genetic defect is called **gene therapy**. The idea of doing this to offer relief from severe genetic disorders seems to be socially acceptable at present, even though the technology by which gene therapy might be accomplished has yet to be developed and refined.

In contrast, inserting genes into a normal human (or sperm or egg) in order to modify or enhance a particular trait is called many things, including **eugenic engineering**. Yet who decides which traits are "desirable"? What if prospective parents could pick the sex of their children by way of genetic engineering? (Three-fourths of one recently surveyed group said they would choose a boy. What would be the long-term social implications of a drastic shortage of girls?) If it is okay to engineer taller or blue-eyed individuals, would it be okay to engineer "superhuman" offspring with exceptional strength or intelligence? Actually, intelligence and most other traits arise through complex interactions among many genes, so this will put them outside the reach of genetic manipulation for some time.

We have only touched on some of the social and ethical issues raised by recombinant DNA technology and genetic engineering. Some individuals say that no matter what the species of organism, DNA should never be altered. But as an earlier discussion in the chapter made clear, nature itself alters DNA much of the time. The real argument, of course, is whether we have the wisdom to bring about beneficial changes without causing harm to ourselves or the environment.

When it comes to manipulating human genes, one is reminded of our very human tendency to leap before we look. When it comes to restricting genetic modifications of any sort, one also is reminded of an old saying: "If God had wanted us to fly, he would have given us wings." And yet, something about the human experience gave us the *capacity* to imagine wings of our own making—and that capacity carried us to the frontiers of space.

Where are we going from here with recombinant DNA technology, this new product of our imagination? To gain perspective on the question, spend some time reading the history of our species. It is a history of survival in the face of all manner of threats, expansions, bumblings, and sometimes disasters on a grand scale. It is also a story of increasingly intertwined interactions with the environment and with one another. The questions confronting you today are these: Should we be more cautious, believing that one day the risk takers may go too far? And what do we as a species stand to lose if the risks are *not* taken?

## SUMMARY

1. Genetic "experiments" have been occurring in nature for billions of years. Gene mutation, crossing over and recombination at meiosis, and other natural events have all contributed to the current diversity among organisms.

2. Humans have been manipulating the genetic character of different species for thousands of years. The emergence of recombinant DNA technology in the past few decades has enormously expanded our capacity to modify organisms genetically.

3. Recombinant DNA technology is founded on procedures by which DNA molecules can be cut into fragments, inserted into plasmids or some other cloning tool, then propagated in a population of rapidly dividing cells.

4. Restriction enzymes and DNA ligase are used to insert DNA from a single species of interest into cloning vectors. A cloning vector is a plasmid, virus, or any other self-replicating genetic element that can be inserted into a host cell for propagation.

5. A collection of DNA fragments produced by restriction enzymes and incorporated into cloning vectors is called a DNA library. A DNA clone is any DNA sequence that has been amplified in dividing cells. DNA sequences also can be amplified in test tubes by the polymerase chain reaction.

6. Recombinant DNA technology and genetic engineering have enormous potential for research and applications in medicine, agriculture, and home and industry. As with any new technology, potential benefits must be weighed against potential risks, including ecological and social disruptions.

7. Although the new technology has not developed to the extent that human genes can be modified, it seems appropriate that the social, legal, ecological, and ethical questions should be explored in detail before such an application is possible.

1. What is a plasmid? What is a restriction enzyme? Do such enzymes occur naturally in organisms? *246, 248*

2. Recombinant DNA technology involves the following:
   a. Producing DNA restriction fragments. *248*
   b. Amplifying the DNA. *248–249*
   c. Identifying modified host cells. *250*

Briefly describe one of the methods used in each of these categories.

3. Having read about examples of genetic engineering in this chapter, can you think of some additional potential benefits of this technology? Can you envision other potential problems? *251–257*

1. Gene mutations, crossing over and recombination during meiosis, and other natural events are the basis of the _____ observed in present-day organisms.

2. Causing genetic change by deliberately manipulating DNA is known as _____.

3. _____ are small circles of bacterial DNA that are separate from the bacterial chromosome.

4. Genetic researchers use plasmids as _____.

5. Rejoining cut DNA fragments from any organism is best known as _____.
   a. cloning genes
   b. mapping genes
   c. recombinant DNA technology
   d. conjugating DNA

6. Using the metabolic machinery of a bacterial cell to produce multiple copies of genes carried on hybrid plasmids is _____.
   a. a way to create a DNA library
   b. bacterial conjugation
   c. mapping a genome
   d. DNA amplification

7. Any DNA sequence that has been amplified in dividing cells is a _____.
   a. DNA clone
   b. DNA library
   c. chunk of foreign DNA
   d. gene map

8. The polymerase chain reaction _____.
   a. is a natural reaction in bacterial DNA
   b. cuts DNA into fragments
   c. amplifies DNA sequences in test tubes
   d. inserts foreign DNA into bacterial DNA

9. Which may benefit from recombinant DNA technology?
   a. households
   b. industry
   c. medicine
   d. agriculture
   e. all of the above

10. Match the recombinant DNA information appropriately.
    _____ DNA clone
    _____ bacterial plasmid
    _____ natural genetic "experiments"
    _____ polymerase chain reaction
    _____ modification of human genes

    a. mutation, recombination
    b. raises social, legal, and ethical questions
    c. a method of test tube gene amplification
    d. any cellular amplification of DNA sequences
    e. a cloning tool

bacterial conjugation *246*
cDNA *250*
cloned DNA *248*
DNA library *248*
DNA ligase *248*
eugenic engineering *257*
gene therapy *257*
genetic fingerprint *253*
nucleic acid hybridization *250*
plasmid *246*
polymerase chain reaction *249*
recombinant DNA technology *245*
restriction enzyme *248*
reverse transcriptase *250*
RFLP *253*

Alberts, B., et al. 1989. *Molecular Biology of the Cell*. Second edition. New York: Garland. Chapter 5 has excellent coverage of DNA recombination and genetic engineering.

Anderson, W. F. 1985. "Human Gene Therapy: Scientific and Ethical Considerations." *Journal of Medicine and Philosophy* 10:274–291.

Brill, W. 1985. "Safety Concerns and Genetic Engineering in Agriculture." *Science* 227:381–384.

Guyer, R., and D. Koshland, Jr. December 1989. "The Molecule of the Year." *Science* 246(4937):1543–1546.

Palmiter, R., et al. 1983. "Metallothionein-Human GH Fusion Genes Stimulate Growth of Mice." *Science* 222:809–814. Report on landmark experiments in mammalian gene transfers.

White, R., and J. Lalouel. February 1988. "Chromosome Mapping with DNA Markers." *Scientific American* 258(2):40–48.

FACING PAGE: *Millions of years ago, a bony fish died, and sediments gradually buried it. Today its fossilized remains are studied as one more piece of the evolutionary puzzle.*

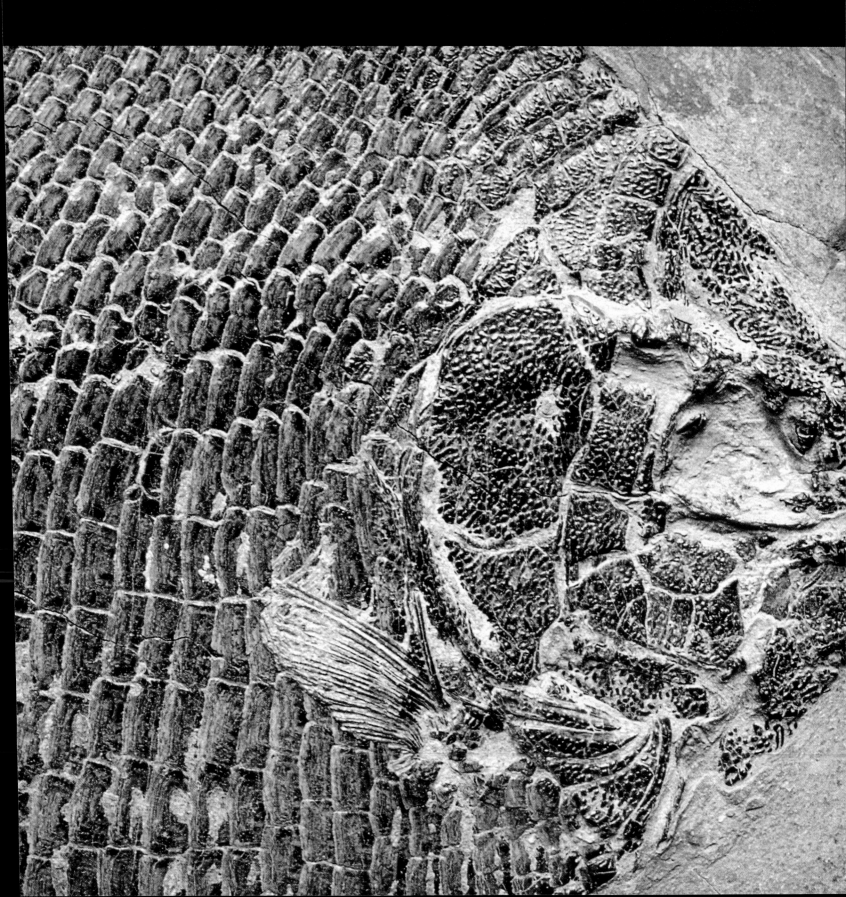

## Fire, Brimstone, and Human History

Lying in the Mediterranean waters between Greece and Crete is a crescent-shaped island, Thera, that wraps around a submerged crater 400 meters deep. It is the remnant of a volcano that blew apart around 3,500 years ago. The eruption was so violent, it generated seismic sea waves that were probably 100 meters high. Within twenty minutes, those giant waves would have reached the island of Crete to the south (Figure 17.1). Most likely, the catastrophic event brought about the abrupt collapse of the Minoan civilization on Crete, the earliest civilization in the history of Europe. Knossos, the Minoan capital, was virtually leveled at about the same time as the eruption. Prevailing winds carrying huge volumes of volcanic ash across the island would have darkened the skies for days, further terrifying those who survived the deluge.

Between the Red Sea to the south and the Dead Sea to the north is a long depression in the earth's crust, the Jordan Valley. Apparently the notorious cities of Sodom and Gomorrah flourished at the south end of the Dead Sea in this valley around 4,000 years ago. By biblical account, "brimstone and fire rained upon the cities . . . and the smoke of the land went up like the smoke of a furnace." To the everlasting terror of their residents, both cities were destroyed.

Today, geologists look at satellite images of the straight walls of the Jordan Valley. They see evidence of hot springs, past and present. They see evidence of ancient, great lava flows and violent earth tremors. Taken together, the straight valley walls, hot springs, lava flows, and earthquakes are signs of deep cracks (faults) in the earth's crust. Geologists postulate that severe earthquakes must have tilted part of the crust along the fault at the southern end of the Dead Sea. In what surely must have been one of the all-time nightmares, the earth heaved, incandescent lava poured out from the depths, hot springs spewed forth a sulfurous brew—and the violently displaced waters of the sea flooded Sodom and Gomorrah.

You can understand how catastrophic events of this sort could have impressed the people of the early Mediterranean civilizations. They still impress us. Even though our seismographs, satellite images, and numerous other technological advances give us a better picture of what causes the earth to tremble beneath our feet and skies to darken or giant waves to sweep across oceans when island volcanoes erupt, it doesn't make us any less uneasy.

Imagine yourself living 2,000 years ago, ignorant of the geologic forces responsible for earthquakes, giant waves, and other events. How would you have interpreted what was going on? Most likely then, as now, your interpretation would have been shaped by the prevailing beliefs of your society. Floods obviously occurred, and one sudden, devastating flood in particular was interpreted as punishment for bad human behavior. Bones and shells were observed in deep layers of the earth, and they were interpreted as

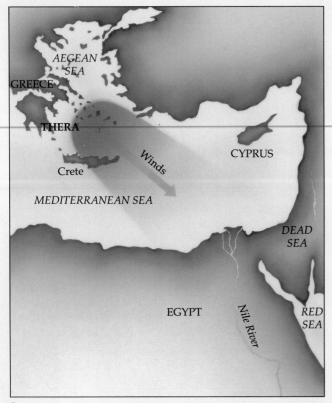

AEGEAN SEA

GREECE

THERA

Crete

Winds

CYPRUS

MEDITERRANEAN SEA

Nile River

EGYPT

DEAD SEA

RED SEA

a

**b**

**Figure 17.1** (**a**) Location of Thera, a volcanic island in the eastern Mediterranean Sea, and the area most devastated by its violent eruption in prehistoric times. (**b**) Immense lava flow from a modern-day volcanic eruption.

evidence of how nearly all creatures were swept away and buried during the Great Deluge.

Biological science emerged within the framework of such prevailing beliefs. And so did awareness of change not only in the earth, but also in its creatures. Understand this fact, and perhaps you will see why acceptance of the very idea of biological evolution was so long in coming.

KEY CONCEPTS

**1.** Evidence that species evolve comes from three lines of investigation that began nearly two centuries ago. First, relationships were discerned among major groups of animals, based on comparisons of body structure and patterning. Second, explorers discovered differences in the world distribution of species that could not be explained unless those species had evolved in different places. Third, geologists discovered apparent sequences of changing fossils in distinct layers of the earth.

**2.** Charles Darwin and Alfred Wallace observed that the individuals of populations vary in traits that might influence the ability to survive and reproduce. If those traits have a heritable basis, each generation will have more individuals bearing those traits. A population changes (evolves) when some forms of traits increase in frequency and others decrease or disappear over time. These points are central to the theory of evolution by natural selection.

## GROWING AWARENESS OF CHANGE

### The Great Chain of Being

More than 2,000 years ago, the seeds of biological inquiry were taking hold among the ancient Greeks. This was a time when popular belief held that supernatural beings intervened directly in human affairs. For example, the gods were said to cause a common ailment known as the sacred disease. Yet from a physician of the school of Hippocrates, these thoughts come down to us:

*It seems to me that the disease called sacred . . . has a natural cause, just as other diseases have. Men think it divine merely because they do not understand it. But if they called everything divine that they did not understand, there would be no end of divine things! . . . If you watch these fellows treating the disease, you see them use all kinds of incantations and magic— but they are also very careful in regulating diet. Now if food makes the disease better or worse, how can they say it is the gods who do this? . . . It does not really matter whether you*

**Figure 17.2** Examples of three different species that are native to three geographically separate parts of the world. (**a**) The ostrich of Africa, (**b**) rhea of South America, and (**c**) emu of Australia. All three species of birds have many features in common, most notably an inability to fly.

*call such things divine or not. In Nature, all things are alike in this, in that they can be traced to preceding causes.*

—On the Sacred Disease (400 B.C.)

Such passages reflect the start of a commitment to finding natural explanations for observable events.

Aristotle was foremost among the early naturalists, and he described the world around him in excellent detail. He had no reference books or instruments to guide him, for biological science in the Western world began with the great thinkers of this age. Yet here was a man who was no mere collector of random bits of information. In his descriptions is evidence of a mind perceiving connections between observations and attempting to explain the order of things.

When Aristotle began his studies, he believed (as did others) that each kind of organism was distinct from all the rest. Later he wondered about bizarre forms that could not be readily classified. In structure or function, they so resembled other forms that their place in nature seemed blurred. Some sponges looked like plants to him, for example, but in their feeding habits they were animals. Aristotle came to view nature as gradual levels of organization, from lifeless matter through complex forms of plant and animal life.

By the fourteenth century, this line of thought had become transformed into a rigid view of life. A great Chain of Being was seen to extend from the lowest forms to humans and on to spiritual beings. Each kind of being, or *species*, as it was called, was seen to have a fixed, separate place in the divine order of things. Each had remained unchanged since the time of creation, a permanent link in the chain. Scholars thought that once they had discovered, named, and described all the links, the meaning of life would be revealed.

## Questions from Biogeography

As long as naturalists believed that the world of living things did not extend much beyond Europe, the task of discovering and describing all species seemed manageable. With the global explorations of the sixteenth century, however, "the world" expanded enormously. Naturalists were soon overwhelmed by descriptions of thousands of plants and animals discovered in Asia, Africa, the Pacific islands, and the New World. Some kinds appeared to be quite similar to common European forms, but some were unique to different lands (Figure 17.2). The picture emerging from the studies of the world distribution of plants and animals (now called *biogeography*) raised an interesting question. Where did all those species "fit" in the great chain?

The naturalist Thomas Moufet, in attempting to sort through the bewildering array, simply gave up and recorded such gems as this description of grasshoppers and locusts: "Some are green, some black, some blue. Some fly with one pair of wings; others with more; those that have no wings they leap; those that cannot fly or

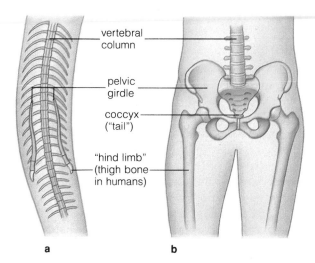

Figure 17.3 Bony parts of a python (**a**) that correspond to the pelvic girdle of other vertebrates, including humans (**b**). Small "hind limbs" protrude through the skin on the underside of the snake.

c

leap they walk; some have long shanks, some shorter. Some there are that sing, others are silent. . . ." This passage comes from his *Theater of Insects*, written about 1590. It was not exactly a time of subtle distinctions.

The emerging biogeographical picture raised another interesting question. If all species had been created at the same time in the same place, as most scholars then believed, *why were certain species found in only some parts of the world but not others?*

## Questions from Comparative Anatomy

During the eighteenth century, anatomists were asking questions of their own. Some puzzling similarities were becoming obvious during their comparisons of the body structure and patterning of different mammals (such as humans, whales, and bats) and other major animal groups. (Such studies are called *comparative anatomy*.) For example, the arms of humans, flippers of whales, and wings of bats differ in size, shape, and function. Yet why are those different structures made of the same materials? Why do they have similar locations in the body? And why do they form and develop in similar ways in the animal embryo?

According to one explanation, some body plans were so perfect there was no need to come up with new ones for each organism at the time of creation. Yet if that were true, how could there be body parts with no apparent function at all? For example, some snakes have bones corresponding to a pelvic girdle, a set of bones to which *hind limbs* attach (Figure 17.3). What were the bones doing there if snakes had been created in a state of limbless perfection? Similarly, humans have bony parts

exactly like the bones in a tail. What were parts of a tail doing in a perfectly designed human body?

## Questions About Fossils

Geologists were adding to the confusion. By the mid-eighteenth century, they had already started to map the *stratification*, or horizontal layering, of sedimentary rocks beneath the earth's surface. (Figure 19.3 shows an example of stratification.) Most of them agreed that such layers had been deposited slowly, one above the other, over time. Intriguingly, different layers held different kinds of **fossils**, the recognizable remains or body impressions of organisms that lived in the past. For example, fossils of simple marine organisms are restricted to deep layers in many parts of the world. Fossils in layers above them are similar but more complex in structure. And fossils in the uppermost layers closely resemble living marine organisms.

Many naturalists of this time tried to reconcile observations about biogeography, comparative anatomy, and the unmistakable layering of fossils with the traditional view of creation. George-Louis Leclerc de Buffon, for example, suggested that if all species were created at the same time and place, they would not now be found throughout the world; mountain barriers or oceans would have stopped their dispersal. *Perhaps species originated in more than one place.* Also, the "imperfections" in body plans and the fossil sequences could be interpreted to mean that species were not unalterably perfect. *Perhaps species became modified over time.* Awareness of **evolution**—of changes in lines of descent over time—was in the wind.

## ATTEMPTS TO RECONCILE THE EVIDENCE WITH PREVAILING BELIEFS

### Cuvier's Theory of Catastrophism

Naturalists now tried to reconcile the evidence of change with a traditional conceptual framework that did not allow for change. The nineteenth-century anatomist Georges Cuvier had spent more than twenty-five years comparing fossils with living organisms. He perceived that the fossil record changed abruptly in certain rock layers. The layers were so distinct from ones that had been deposited before them that they seemed to be boundaries for dramatic change in ancient environments. Cuvier actually made some astute inferences about past episodes of catastrophic change (Chapter 19). His attempt to explain them came to be known as *catastrophism*.

There was only one time of creation, said Cuvier, which had populated the world with all species. Many species had been destroyed in a global catastrophe. The few survivors repopulated the world. It was not that the survivors were new species. Naturalists simply hadn't got around to discovering earlier fossils of them, fossils that would date to the time of creation. Another catastrophe wiped out more species and led to repopulation by the survivors, and so on through various catastrophes.

Investigations never have turned up the fossils needed to support Cuvier's theory. Rather, they have turned up considerable fossil evidence against it, as you will see in the next chapter. The concept is illuminating, however, *for it shows how prevailing beliefs may influence explanations of what is being observed.*

### Lamarck's Theory of Desired Evolution

One of Cuvier's contemporaries viewed the fossil record differently. Jean-Baptiste Lamarck believed that life had been created long ago in a simple state. He believed further that it gradually improved and changed into complex levels of organization. The force for change was a built-in drive for perfection, up the Chain of Being. The drive was centered in nerve fibers, which directed "fluida" (vaguely defined substances) to body parts in need of change (in a manner unspecified).

For instance, suppose the ancestor of the modern giraffe was a short-necked animal. Pressed by the need to find food, this animal constantly stretched its neck to browse on leaves beyond the reach of other animals. Stretching directed fluida to its neck, making the neck permanently longer. The slightly stretched neck was bestowed on offspring, which stretched their necks also. Thus generations of animals desiring to reach higher leaves led to the modern giraffe.

a

Such was the Lamarckian *theory of inheritance of acquired characteristics*—the notion that changes acquired during an individual's life are brought about by environmental pressure and internal "desires," and that offspring inherit the desired changes.

Lamarck's contemporaries considered the theory a wretched piece of science, largely because Lamarck habitually made sweeping assertions but saw no need to support them with observations and tests. In retrospect, perhaps we can find kinder words for the man. His work in zoology was respected. And he did indeed piece together a foundation for an evolutionary theory: Species change over time, and the environment is a factor in that change. It was his misfortune that he made some crucial observations but came up with a theory that has never been supportable by tests.

## DARWIN'S JOURNEY

In 1831, in the midst of this intellectual ferment, Charles Darwin was twenty-two years old and wondering what to do with his life. He had some hard acts to follow. His grandfather, a physician and naturalist, was one of the first to propose that all organisms are related by descent. Darwin's father was a successful physician. Being from a wealthy family, Darwin had the means to indulge his interests while he was growing up. When he was eight years old, he was an enthusiastic but haphazard collector of shells. At ten, he focused on the habits of insects and birds. At fifteen, he found schoolwork boring compared to the pursuit of hunting, fishing, and observing the natural world.

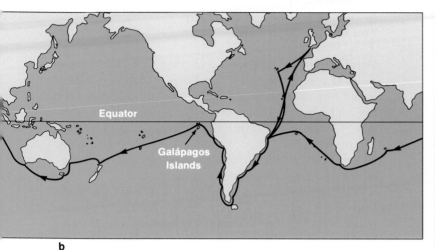

b

c

**Figure 17.4** (**a**) Charles Darwin a few years after he returned from his five-year voyage aboard H.M.S. *Beagle* (**b**). A replica of the *Beagle* is shown in (**c**) sailing off the coast of South America.

Darwin attempted to study medicine in college. He abandoned the study after realizing he never could practice surgery on his fellow humans, given the crude and painful procedures available. For a while he followed his own inclinations toward natural history. Then his father suggested that a career as a clergyman might be more to his liking and more respectable than diddling with nature, so Darwin packed for Cambridge. His grades were good enough to earn him a degree. But he spent most of his time among faculty members with leanings toward natural history.

It was the botanist John Henslow who perceived and respected Darwin's real interests. Henslow arranged for him to take part in a training expedition led by an eminent geologist. At the pivotal moment when Darwin had to decide on a career, Henslow arranged that he be offered the position of ship's naturalist aboard H.M.S. *Beagle*.

**Voyage of the *Beagle***

The *Beagle* was about to sail for South America to complete earlier work on mapping the coastline (Figure 17.4). Prolonged stops at islands, near mountain ranges, and along rivers would give Darwin a chance to study many diverse forms of life. Almost from the start of the voyage, the young man who had hated work suddenly began to work with enthusiasm, despite lack of formal training. Throughout the journey to South America, he collected and examined marine life. And he read Henslow's parting gift, the first volume of Charles Lyell's *Principles of Geology*.

Amplifying earlier ideas of the geologist James Hutton, Lyell argued that gradual processes now molding the earth's surface had also been at work in the past. These processes include the almost imperceptibly slow uplifting of mountain ranges, their weathering by wind and water, and the filling of valleys and the sea floor with sediments. Earthquakes and other catastrophic events were more like abrupt punctuation points in this gradual history of geologic change. This concept, which came to be known as *uniformitarianism*, called into question prevailing ideas about the age of the earth. (For example, Jewish calendar-years were based on the concept that the earth was less than 6,000 years old.) Given the rates at which known geologic processes proceed, Lyell had reckoned that it would have taken not a few thousand years but millions of years to mold the land into its current configurations.

The implications of this concept were staggering. One of the reasons that evolution initially seemed so implausible was that it was hard to imagine all of life's exuberant diversity evolving in the space of a few millennia. If Lyell were interpreting the geologic record correctly, then there had indeed been time enough for evolution.

## Evolution by Natural Selection: The Theory Takes Form

Darwin returned to England in 1836, after nearly five years at sea. In the years to follow, his writings established him as a respected figure in natural history. All the while, his consuming interest was the "species problem." What could explain the remarkable diversity among organisms? As it turned out, field observations he had made during his voyage enabled him later to recognize two clues that pointed to the answer.

First, while the Argentine coast was being mapped, Darwin repeatedly got off the ship. (He was prone to seasickness.) During his many exploratory trips inland, he made detailed field observations and collected fossils. He saw for the first time many unusual species, including an armadillo (Figure 17.5). Among the fossils were remains of the now-extinct glyptodonts. Glyptodonts were very large animals that bore suspicious resemblance to living armadillos. If both kinds of animals had been created at the same time, if they lived in the same part of the world, and if they were so much alike, then why were armadillos still lumbering about but the glyptodonts long gone and buried? Nothing else in the world resembled either animal. Although neither Darwin nor anybody else had ever seen one species evolve into another, he later wondered whether armadillos were descended from the glyptodonts.

Second, Darwin had observed that populations of similar kinds of organisms that were confined to different geographic regions often showed pronounced differences in some of their traits. For example, the Galápagos Islands are almost 1,000 kilometers off the coast of Ecuador (Figure 17.6). Every island or island cluster is home to diverse species, including birds called finches. Although Darwin didn't think much about it during his voyage, later discussions with colleagues back in London made him realize that the islands were home to more than a dozen closely related species. Perhaps all those species were descended from the same ancestral form and had become modified slightly after they became isolated on different islands.

a

b

**Figure 17.5** (a) View of the Andes Mountains of South America. Darwin explored parts of the Andes, where he observed fossils of marine organisms in rock layers that were 12,000 feet (3.6 kilometers) above sea level.

(b) Photograph of an armadillo and, above it, a reconstruction of an extinct animal, the glyptodont. Resemblances between these unusual, scaly animals and their similarly restricted geographic distribution provided Darwin with a clue that helped him develop his theory of how evolution occurs.

**Figure 17.6** (a) The Galápagos Islands, isolated bits of land in the ocean, far to the west of Ecuador. About 5 million years ago, the islands arose through volcanic action. (The photograph above the map shows clear signs of their volcanic origins.) Thus living organisms could not have originated there; they had to have been carried to the Galápagos by winds or ocean currents. The islands were named after the giant tortoises found there (the Spanish *galápa* means tortoise). Figure 26.19 shows one of the tortoises.

Scenes from the Galápagos. (b) Marine iguanas. (c) Tropical plants of the highlands of Santa Cruz Island. (d) A blue-footed male booby, engaged in a courtship display.

Darwin

Wolf

Pinta

Marchena          Genovesa

Santiago                              *EQUATOR*
              Bartolomé
                       Seymour
          Rábida      Baltra
     Pinzón
Fernandina                  Santa Cruz

              Santa Fe
                          San Cristóbal
     Isabela  Tortuga
                       Española
          Floreana

a

b

c

d

**Figure 17.7** Examples of variation in beak shape among different finch species of the Galápagos Islands. These and eight other species apparently are descended from a common ancestor, a seed-eating ground finch.

(**a**) *Certhidea olivacea*, a tiny tree-dwelling finch that resembles a warbler in song and behavior. It uses its slender beak to probe for insects.

(**b**) *Geospiza scandens* and (**c**) *G. conirostris*, two species with a beak adapted for eating cactus flowers and fruits. (The former is shown in the midst of dinner in **e**.) Other species in the same genus have thick, strong beaks adapted for crushing cactus seeds.

(**d**) *Camarhynchus pallidus*, a finch that feeds on wood-boring insects such as termites. It swings its small body like a woodpecker does, to hammer at bark. It does not have the woodpecker's long, probing tongue, but it has learned to break cactus spines to appropriate lengths, then hold the "tools" in its beak and use them as probes.

But *how* could such modifications occur? A clue came from an essay by Thomas Malthus, a clergyman and economist. In Malthus' view, any population tends to outgrow its resources, and its members must compete for what is available. Darwin thought about all the populations he had observed during his voyage. He thought about how the *individual members* of those populations had varied in body size, form, coloring, and other traits. It dawned on him that some traits could lead to differences in the ability to secure resources.

Today, for example, we can observe that each species of Galápagos finch has a distinct type of beak suitable for crushing seeds, spearing insects, or securing some other foodstuff (Figure 17.7). We might well conclude that variations in beak size and shape make some birds better equipped than others to obtain certain foods available on certain islands.

If there were struggles for existence (competition) within a population, then individuals born with a stronger seed-crushing beak or some other favorable trait might have an edge in surviving and reproducing. *Nature would select individuals with advantageous traits and eliminate others—and so a population could change.* Favored individuals would pass on the useful traits to offspring, their offspring would do the same, and so on. In time, descendants of the favored individuals would make up most of the population, and less favored individuals might have no descendants at all.

In similar fashion, Darwin correlated his observations of inheritance with certain features of populations

**Figure 17.8** Alfred Wallace. Although Darwin and Wallace had worked independently, they both arrived at the same concept of natural selection. Darwin tried to insist that Wallace be credited as originator of the theory, being the first to circulate a report of his work. Wallace refused; he would not ignore the decades of work Darwin had invested in accumulating his supporting evidence.

and the environment. And he saw that "natural selection" among variant individuals could be a mechanism of evolution. Here we outline the key correlations of the **theory of natural selection**:

### First Correlation

*Observation:* All natural populations have the reproductive capacity to exceed the resources required to sustain them. (This is true even of slow reproducers such as elephants, the females of which bear no more than four offspring, one at a time, over a course of fifty years.)

*Observation:* Despite their enormous reproductive potential, populations do not keep increasing indefinitely in size over time. (To give a simple example, a single sea star can release 2,500,000 eggs every year, but the oceans obviously do not fill with sea stars.)

*Observation:* In natural environments, food supplies and other resources do not increase explosively. In fact, they remain much the same over time.

*Observation:* The limited availability of resources puts limits on population growth. (There is only so much water, nutrients, and growing space for a plant population; only so many plants to feed an elephant population; and so on.)

*Inference:* When a population outstrips the supplies of necessary resources, there must be competition among its members for the resources that *are* available. Because

of this competition, not all of the individuals who were born will themselves survive and reproduce.

### Second Correlation

*Observation:* The members of a natural population show great variation in their traits, and much of the variation is passed on through generations (it has a heritable, or genetic, basis).

*Inference:* Some heritable traits are more adaptive than others. They give the individual a competitive edge in surviving and reproducing.

*Inference:* Over the generations, there is **natural selection**—a measurable difference in survival and reproduction among individuals that differ from one another in one or more traits.

*Inference:* Thus the character of the population changes over time—it evolves—as some forms of traits increase in frequency and others decrease or disappear.

Darwin did not formally announce his theory right away. He wanted first to sift through the evidence for flaws in his reasoning. Then, in 1858, he received a paper from the naturalist Alfred Wallace (Figure 17.8), who had arrived at the same conclusion! Darwin's colleagues prevailed upon him to formally present a paper along with Wallace's. The next year Darwin's detailed evidence in support of the theory was published in book form.

Darwin's theory still faced a crucial test. If evolution occurs, there should be evidence of one kind of organism changing into another. Yet, at the time, the known fossil record seemed to contain no transitional forms, the so-called "missing links" between major groups of organisms. Oddly, two years after Darwin's book was published such a fossil did turn up, but few paid much attention to it. *Archaeopteryx* resembled reptiles *and* birds (Figure 17.9). Like fossils of small two-legged reptiles, it had teeth and a long, bony tail. Like modern birds, its body was covered with feathers!

Other fossil evidence of evolution accumulated during Darwin's time. Even so, nearly seventy years passed before advances in genetics led to widespread acceptance of his theory of natural selection. In the meantime, his name was associated mostly with the idea that life evolves—something others had proposed before him.

## EVOLUTION—A STUDY OF PROCESS AND PATTERN

With this bit of history behind us, we are ready to consider some of the current views of evolution. In the chapter that follows, we will consider how modern genetics gives us a better understanding of the mechanisms by which evolution occurs. Later, in Chapter 19, we will examine the historical results of those processes—that is, the large-scale patterns of evolution through time and their relationship with earth history.

## SUMMARY

1. The idea that species evolve—change over time—emerged through the following:

   a. Comparisons of the body structure and patterning among major groups of animals (comparative anatomy).

   b. Questions about the world distribution of plants and animals (biogeography).

   c. Observations of fossils of different types (structurally simple to complex) buried in a series of distinctly stratified layers of the earth (most ancient layers to more recent layers).

2. Here are the key points of the theory of evolution by natural selection, as first proposed by Darwin and Wallace:

   a. Individuals in a population of a species vary in size, form, and other traits that might influence their ability to acquire resources—hence to survive and reproduce.

   b. If those traits have a heritable basis, then individuals with the adaptive traits will tend to make up more of the reproductive base in each new generation.

*Archaeopteryx*

A small, two-legged dinosaur

**Figure 17.9** One of six fossils of *Archaeopteryx* from limestone deposits that are more than 140 million years old. *Archaeopteryx* was a transitional form on (or very near) the evolutionary road leading from reptiles to birds. In fact, the fossils might have been classified as reptilian, had it not been for the clear imprints of feathers in the finely grained limestone.

*Archaeopteryx* had teeth and hind limbs like its reptilian ancestors, but in other features (such as its feathers and wishbone) it was already like modern birds. It did not have the strong muscle attachments and bony framework of modern birds, but it certainly could glide and perhaps it could fly feebly.

c. There can be natural selection—that is, a measurable difference in survival and reproduction that has occurred among individuals that differ from one another in one or more traits.

d. A population "evolves" when some forms of traits increase in frequency and others decrease or disappear over the generations.

3. One "test" of Darwin's theory would be evidence of one major kind of organism changing into another kind. *Archaeopteryx* provided early evidence. But not until many decades later would studies at many levels of biological organization provide a large body of substantiating evidence for his theory.

## Review Questions

1. Define biogeography and comparative anatomy. How did studies in both disciplines contradict the idea that species have remained unchanged since the time of creation? *262–263*

2. Cuvier and Lamarck interpreted the fossil record differently. Briefly state how their interpretations differed. *264*

3. Was Cuvier's perception of catastrophic events in earth history at odds with Lyell's view of uniformitarianism? *264, 265*

4. Darwin made two key correlations when he developed his theory of natural selection. Outline the observations and inferences of those correlations. *269*

5. Define evolution. Define evolution by natural selection. Can an individual evolve? *263, 268–269*

## Self-Quiz *(Answers in Appendix IV)*

1. Based on comparisons of _____ structure and patterning, we can discern relationships among major groups of animals.

2. Differences in the world distribution of species can be explained by the hypothesis that different species _____ in different parts of the world.

3. The observation that distinct layers of the earth reveal sequences of _____ showing gradual changes in structure supports the hypothesis that species evolve.

4. A/an _____ evolves when some forms of traits increase in frequency while others decrease or disappear over time.

5. Two keen observers of nature who outlined the theory of evolution by natural selection were _____.
   a. Cuvier and Lamarck    c. Buffon and Darwin
   b. Darwin and Wallace    d. Malthus and Lyell

6. Given that traits are inherited, individuals in a population who possess the _____ will tend to make up more of the reproductive base for the next generation.
   a. greatest variation in traits
   b. least variation in traits
   c. most adaptive traits
   d. most ancestors in the fossil record
   e. widest world distribution of their species

7. Natural selection may occur when there are _____.
   a. heritable traits
   b. variation in traits within a population
   c. adaptive traits

d. differences in survival and reproduction among members of a population
   e. all of the above

8. Most biologists strongly support the theory that species evolve because _____.
   a. it was first suggested by Darwin and Wallace
   b. the discovery of the *Archaeopteryx* fossil makes it certain
   c. studies at many levels of biological organization provide a large body of evidence that the theory is correct
   d. another, better explanation of the great body of observations and test results has yet to appear
   e. both c and d are correct

9. Match the following individuals and ideas or perceptions.
   _____ Cuvier       a. theory of natural selection
   _____ Lamarck      b. populations outgrow resources
   _____ Lyell        c. catastrophism
   _____ Malthus      d. inheritance of traits acquired
   _____ Darwin          through environmental pressures
         and Wallace      and internal desires for change
                       e. geologic evidence that the earth is
                          extremely ancient

## Selected Key Terms

*Archaeopteryx* 270
biogeography 262
comparative anatomy 263
Darwin's theory of
   natural selection 269
evolution 263
fossil 263

Great Chain of Being 261
species 262
stratification 263
theory of catastrophism 264
theory of inheritance of acquired
   characteristics 264
uniformitarianism 265

## Readings

Darwin, C. 1957. *Voyage of the Beagle*. New York: Dutton. In his own words, what Darwin saw and thought about during his global voyage.

Gould, S. 1982. "The Importance of Trifles." *Natural History* 91(4): 16–23. Gould discusses the principles of reasoning that are evident in one of Darwin's last books (on worms). He argues persuasively that Darwin was indeed one of the great thinkers and not "a great assembler of facts and a poor joiner of ideas," as a detractor once wrote.

Singer, C. 1962. *A History of Biology to About the Year 1900*. New York: Abelard-Schuman. Out of date, but contains absorbing portrayals of the men and women who led the way in developing basic biological concepts.

## Designer Dogs

We humans have tinkered rather ruthlessly with the modern-day descendants of a long and distinguished lineage. That lineage began some 40 million years ago with small, weasel-shaped, tree-dwelling carnivores, the forerunners of bears, raccoons, pandas, badgers— and dogs. About 10,000 years ago, we began our domestication of wild dogs.

It is probably safe to assume that we didn't do this out of self-indulgence. More likely, we brought dogs to our side and shared precious food with them because there were advantages to doing so. Times were tough in the olden days without supermarkets, forced-air heating, and police protection. Dogs could guard us and our possessions (most dogs have an inherent capacity to bark and bare teeth at perceived threats). They could dine on and so exterminate the rats and other vermin that took up residence in our shelters, uninvited.

It didn't take us long to develop different varieties, or breeds, through artificial selection. Individuals of larger or smaller body size and other traits that seemed useful were kept and encouraged to breed. Those without the desired traits were discarded. Over time we shaped sheepherding collies, badger-hunting dachshunds, wily retrievers, and snow-traversing, sled-pulling huskies. And at some point we began to delight in the rare individual, the odd, extraordinary dog. Imagine! In no time at all, evolutionarily speaking,

we picked our way through the pool of variant dog genes and came up with such extremes as the Great Dane and the chihuahua (Figure 18.1).

The chihuahua exhibits traits that have been stretched past the limits of biological common sense. In the wild, the mutations that gave rise to its tiny, spindly-legged, nearly hairless body would probably be selected against. Imagine how long it would last when confronted with a snowstorm or mountain lion. The chihuahua survives because we vaccinate it against disease, keep it protected from cold and large moving objects, and cater to its finicky eating habits.

Or consider what it now takes for an English bulldog to get through life. Long ago, breeders favored English bulldogs with a short snout and a compressed face, possibly because they expected the dogs to bite bulls on the nose. The dogs are no longer expected to do this—but they are left with a roof of the mouth that is too wide and often flabby. They have such trouble breathing that they snore, loudly. Sometimes they get so short of breath they lose consciousness.

We have been changing the genetic character of more than just dogs for thousands of years. Artificial selection practices also have given us modern crop plants and new breeds of cats, cattle, and birds. As you read in Chapter 16, we are now genetically designing plants for disease resistance and other desired traits. So when you hear someone wonder about whether "evolution" occurs, remind yourself that evolution simply means *change through time*—and that our selective breeding practices give us abundant and quite tangible evidence that changes do, indeed, occur. *How* those changes occur is the subject of this chapter.

**Figure 18.1**  About 10,000 years ago, humans began domesticating wild dogs. From that ancestral stock, artificial selection produced such diverse yet rather closely related breeds as the Great Dane (*legs, left*) and the chihuahua (*possibly fearful of being stepped on, right*).

**1.** All members of a population share the same traits (all rabbits have fur, for example). But the forms of a given trait may vary (the fur may be black or white). Variation exists when individuals carry different alleles for the same trait. Over time, each kind of allele may increase or decrease in frequency, or even disappear. *Microevolution* refers to such changes in allele frequencies in a population over time.

**2.** Allele frequencies change as a result of mutation, genetic drift, gene flow, and natural selection. *New* alleles arise only through mutation. *Existing* alleles are shuffled into, through, or out of populations by the other processes.

**3.** Natural selection is not an "agent," combing actively and purposefully through populations for the "best" individuals. It is simply a measure of the difference in survival and reproduction that has occurred among individuals who differ from one another in one or more traits.

**4.** For sexually reproducing organisms, speciation may occur when populations of a species become reproductively isolated from one another and differences build up in their pools of alleles. The divergence between populations can become pronounced, so that their members no longer interbreed and produce fertile offspring under natural conditions.

## MICROEVOLUTIONARY PROCESSES

### Variation in Populations

Evolution is a fact of nature, and the key to understanding the processes by which it occurs is this: Individuals don't evolve; *populations* do. By definition, a **population** is a group of individuals occupying a given area and belonging to the same species.

All individuals of the same population have a certain number of traits in common. They have the same overall form and appearance, as when all members of a bird

**Figure 18.2** Variation in shell color and banding patterns among populations of one species of snail found on islands of the Caribbean. For most traits, different members can carry different alleles. Alleles are alternative molecular forms of the same gene, and they code for alternative forms of the same trait. Members of a population vary in their traits because they can carry different combinations of alleles at most gene locations along their chromosomes.

population have blue feathers, a slender beak, a four-chambered heart, a gizzard, and so on. These are *morphological traits* (*morpho-* means form). The physical and chemical operation of cells and body parts proceeds in much the same way for all individuals during their growth, development, day-to-day housekeeping activities, and reproduction. These are *physiological traits* (they relate to body functioning). All individuals of a population make similar basic responses to stimuli, as when humans reflexively yank their hands away from a spider on a light switch. These are *behavioral traits*.

In most natural populations, the manifestations of different traits are not quite the same from one individual to the next. One member of a flock of mostly blue, straight-feathered birds might have white, frazzled feathers. A few members might be more sensitive to winter cold or more adept at attracting a mate with a dazzling courtship display. Humans normally have hair, but the hair differs in color, texture, amount, and distribution from one person to the next, as any bald person will tell you. Figure 18.2 illustrates the stunning variation in the color and banding patterns of snails. And these examples only hint at the immense variety among individuals of most populations, past and present.

**A population is a group of individuals occupying a given area and belonging to the same species.**

**Sources of Variation.** Variation in traits has a heritable basis. This is apparent simply by observing that children resemble their parents more than they resemble anyone else. Information about heritable traits resides in hundreds or thousands of genes, which are specific regions of DNA molecules. All individuals of the same species inherit the same number and kinds of genes. But some number of the genes in one individual have a slightly different molecular structure than their counterparts in another individual; this is what **genetic variation** means.

Genetic variation manifests itself as variation in *phenotype*—in some aspect of the individual's structure, function, or behavior. Whether your hair is black, brown, red, or white depends on *which* molecular forms of certain genes you happened to inherit from your mother and father.

The environment can mediate phenotypic variation by influencing how the genes governing different traits are expressed. Ivy plants grown from cuttings of the same parent all have the same genes, but the amount of sunlight affects the genes governing leaf growth. The plants (or even parts of a single plant) growing in full sun have smaller leaves than those growing in full shade. Figure 18.3 shows another example of environmental influences on variation. It is possible to distinguish between the genetic and environmental components of variation through artificial selection experiments and molecular studies.

**Figure 18.3** Example of environmental effects on phenotype. These floral clusters occur on plants of the same species (*Hydrangea macrophylla*). However, the flower color can range from reddish-blue to pink, depending on the acidity of the soil in which the plants are growing.

Genetic variation among individuals arises through several events. These events, which were introduced in earlier chapters, fall into five categories:

1. Gene mutation

2. Abnormal changes in chromosome structure or number

3. Crossing over and genetic recombination during meiosis

4. Independent assortment of chromosomes during meiosis

5. Fertilization between genetically different gametes

Of the five kinds of events listed, only mutation *creates* new molecular forms of a gene. The others simply shuffle *existing* genes into new combinations in new individuals.

But what a shuffle! By one estimate, more than $10^{600}$ combinations of genes are possible in human gametes, yet there are not even $10^{10}$ humans alive today. So unless you have an identical twin, it is extremely unlikely that another person with your exact genetic makeup has ever lived, or ever will.

**Far more genetic variation is possible than can ever be expressed in the individuals of any population.**

**Genetic Equilibrium**. Think about a population of snails or some other kind of sexually reproducing, diploid animal. Each snail has pairs of homologous chromosomes. This means it has two genes, not one, for every specific gene location (locus) on each type of chromosome. Recall that the different molecular forms of a gene are called *alleles* (page 152). Now, suppose one snail allele codes for yellow shells and another for white. A simple count shows there are many more yellow shells than white, so you might tentatively conclude that more snails carry the yellow allele. Later, through genetic analysis, you might be able to determine the actual **allele frequencies**: the abundance of each kind of allele in the whole population.

When a population is evolving, some of its *allele frequencies* are changing through successive generations. A formula called the "Hardy-Weinberg principle" is used to establish a theoretical reference point for measuring the rates of evolutionary change. At this point, called **genetic equilibrium**, allele frequencies for a trait remain stable through the generations; there is zero evolution.

## Hardy-Weinberg Principle:

The genotypic frequencies for a population in equilibrium will fit the formula

$$AA \quad Aa \quad aa$$
$$p^2 + 2pq + q^2$$

where $p$ = the frequency of allele $A$ and $q$ = the frequency of allele $a$.

The allele frequencies and the genotypic frequencies will be stable from generation to generation if the following assumptions are true:

There is no mutation, the population is infinitely large and is isolated from other populations, mating is random, and all genotypes are equally viable and fertile.

**Figure 18.4** Hardy-Weinberg equilibrium. To prove the validity of the Hardy-Weinberg rule stated above, let's follow the course of two alleles, $A$ and $a$, through succeeding generations.

For all members of the population, the gene locus must be occupied by either $A$ or $a$. In mathematical terms, the frequencies of $A$ and $a$ must add up to 1. For example, if $A$ occupies half of all the gene loci and $a$ occupies the other half, then $0.5 + 0.5 = 1$. If $A$ occupies 90 percent of all the gene loci, then $a$ must occupy the remaining 10 percent ($0.9 + 0.1 = 1$). No matter what the proportions of alleles $A$ and $a$,

$$p + q = 1$$

You know that during sexual reproduction of diploid organisms, the two alleles at a gene locus segregate and end up in separate gametes. Thus $p$ is also the proportion of gametes carrying the $A$ allele, and $q$ the proportion of gametes carrying the $a$ allele. To find the expected frequencies of the three possible genotypes ($AA$, $Aa$, and $aa$) in the next generation, we can construct a Punnett square. Here, the genotypes exhibiting $A$ phenotypes are shown as yellow squares:

|  | $p\,Ⓐ$ | $q\,ⓐ$ |
|---|---|---|
| $p\,Ⓐ$ | $AA$ $(p^2)$ | $Aa$ $(pq)$ |
| $q\,ⓐ$ | $Aa$ $(pq)$ | $aa$ $(q^2)$ |

Because the frequencies of genotypes must add up to 1,

$$p^2 + 2pq + q^2 = 1$$

To see how these calculations can be applied, let's follow the allele frequencies for a population of 1,000 diploid individuals made up of the following genotypes:

$$450 \ AA$$
$$500 \ Aa$$
$$\underline{50 \ aa}$$
$$1,000 \text{ individuals (or 2,000 alleles)}$$

Theoretically, of every 1,000 gametes produced, the frequency of $A$ will be $p^2 + \frac{1}{2}(2pq)$ or $450 + \frac{1}{2}(500) = 700$, or $p = 0.7$. The frequency of $a$ will be $\frac{1}{2}(2pq) + q^2$ or $\frac{1}{2}(500) + 50 = 300$, or $q = 0.3$. Notice that

$$p + q = 0.7 + 0.3 = 1$$

After one round of random mating, the frequencies of the three genotypes possible in the next generation will be as follows:

$$AA = p^2 = 0.7 \times 0.7 = 0.49$$
$$Aa = 2pq = 2 \times 0.7 \times 0.3 = 0.42$$
$$aa = q^2 = 0.3 \times 0.3 = 0.09$$

and

$$p^2 + 2pq + q^2 = 0.49 + 0.42 + 0.09 = 1$$

Actually, genetic equilibrium would be possible *only* if the following conditions were being met, as Figure 18.4 makes clear:

1. No mutations are occurring.

2. The population is very, very large.

3. It is isolated from other populations of the species.

4. All members survive and reproduce (no selection).

5. Mating is random.

A natural population rarely (if ever) meets all five conditions simultaneously. Mutation changes its allele frequencies over long spans of time. Three other processes—genetic drift, gene flow, and natural selection—also can drive the population away from genetic equilibrium, even in the space of a few generations. Changes in allele frequencies brought about by mutation, genetic drift, gene flow, and natural selection are called **microevolution**. These microevolutionary processes are listed in Table 18.1 and will now be described.

### Mutation

A **mutation** is a heritable change in DNA. In terms of individual genes in individual organisms, mutations are rare events (page 229). In evolutionary terms, their numbers are significant—they have been accumulating in different lineages of organisms for billions of years. They are the ultimate source of all heritable variation.

Mutations are random in terms of which gene will be affected and, most importantly, *whether they will be harmful, neutral, or beneficial to the individual*. As indicated earlier, on page 229, the effect of each mutated gene

Notice that the allele frequencies have not changed:

$$A = \frac{2 \times 490 + 420}{2,000 \text{ alleles}} = \frac{1,400}{2,000} = 0.7 = p$$

$$a = \frac{2 \times 90 + 420}{2,000 \text{ alleles}} = \frac{600}{2,000} = 0.3 = q$$

The genotypic frequencies have changed initially. However, given that the distribution of genotypes fits the equation $p^2 + 2pq + q^2$, the genotypic frequencies will be stable over succeeding generations. You can verify this by calculating the most probable allele frequencies for gametes produced by the second-generation individuals:

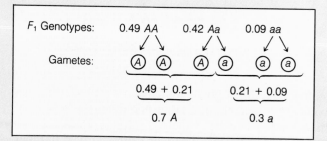

which is back where we started from. Because the allele frequencies are exactly the same as those of the original gametes, they will yield the same frequencies of genotypes as in the second generation.

You could go on with the calculations until you ran out of paper, or patience. As long as the population adheres to the conditions stated in the boxed inset for the Hardy-Weinberg principle, you would end up with the same results. When the frequencies of different alleles and different genotypes remain constant through successive generations, the population is in Hardy-Weinberg equilibrium: It is not evolving.

usually depends on its interactions with other genes and with the environment. No matter what the species, each individual inherits *many* genes that must work together in coordinated fashion. And the products of those genes must be able to perform their functions under a specific range of operating conditions.

Gene mutations are harmful when they alter an individual's structure, function, or behavior in ways that work against survival and reproduction. For example, a certain gene codes for a protein that is required for cartilage formation. Cartilage is a key component of many pathways in normal animal development. A mutation at this gene locus may lead to such deformities as blocked nostrils, narrowed tracheal passageways, thickened ribs, and the loss of elasticity in lung tissue. Because the mutated gene is expressed in the context of an intricate developmental program, it may have lethal effects. A **lethal**

| Table 18.1 | Major Microevolutionary Processes |
| --- | --- |
| **Mutation** | A heritable change in DNA |
| **Genetic drift** | Random fluctuation in allele frequencies over time, due to chance occurrences alone |
| **Gene flow** | Change in allele frequencies as individuals leave or enter a population |
| **Natural selection** | Change or stabilization of allele frequencies due to differences in survival and reproduction among variant members of a population |

**mutation** is one in which expression of the mutated gene always leads to the death of the individual.

**Neutral mutations** are neither harmful nor helpful to the individual. Some of these mutations do have an effect on phenotype. For example, a mutated gene for shell color might specify white instead of yellow. But this may have no effect whatsoever on survival and reproduction, so it would not be a factor in natural selection. Other neutral mutations have no effect on phenotype. Also, a mutated gene with unpleasant but not lethal effects might be closely linked to highly conserved genes on the same chromosome. (Such conserved genes include histones, which are essential for survival.) The close linkage would allow it to be passed along with vital genes during meiosis and remain in the population. The neutral mutation might even turn out to be beneficial if environmental conditions change.

Finally, every so often, a mutation provides its bearer with advantages. Increases or decreases in, say, body size would not have occurred without mutations in genes that influence patterns of growth and development. Think about the advantages bestowed on a tree that has a genetic capacity to grow taller, wider, and faster than its neighbors. Such a tree would get the best shot at incoming rays of sunlight and at soil water, with its limited supplies of dissolved nutrients. Think about the advantage that large predatory fishes have over small ones during a feeding frenzy. The small ones may end up on the menu themselves. Or think about the advantage of smaller size in a coral reef community, where being able to hide in nooks and crannies in the reef might mean the difference between life and death.

**Gene mutations, the source of all heritable variation, are chance events. Whether they prove harmful, neutral, or beneficial depends on the performance of the altered gene products under prevailing environmental conditions.**

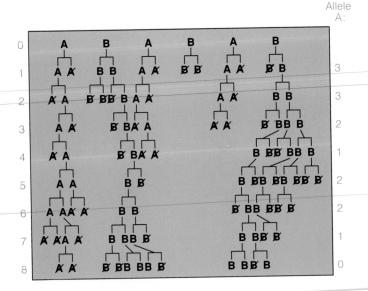
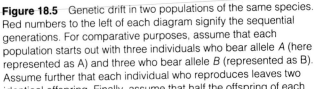
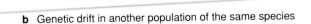

**a** Genetic drift in one population

**b** Genetic drift in another population of the same species

**Figure 18.5** Genetic drift in two populations of the same species. Red numbers to the left of each diagram signify the sequential generations. For comparative purposes, assume that each population starts out with three individuals who bear allele A (here represented as A) and three who bear allele B (represented as B). Assume further that each individual who reproduces leaves two identical offspring. Finally, assume that half the offspring of each generation die before they themselves are old enough to reproduce (population size remains constant). But *which* ones live or die is random—you simply toss a coin to determine the "survivors" of each generation.

In (**a**), the blue numbers to the right of the diagram signify the number of individuals bearing allele A who survived in each generation. Notice that the relative abundance of the two types of individuals changes until A disappears and B becomes established.

In (**b**), the blue numbers to the right of the diagram signify the number of individuals bearing allele B who survived in each generation. Notice that the relative abundance of the two alleles changes here, also—but in this case B nearly disappears and A dominates by the eighth generation.

### Genetic Drift

Allele frequencies can change randomly through the generations because of chance events alone. This process is called **genetic drift**. Genetic drift can be most rapid when population size is small. Although it can lead to a loss of variation within a population, it can lead to an increase in variation *between* populations.

Suppose only some members of a small population carry an allele A, and suppose none of them reproduces over several generations. Whether you call it chance or bad luck, some do not mate and others fall ill or die accidentally for reasons that have nothing to do with their genotype. As a result, the A allele disappears from the population (this is shown in Figure 18.5). It does so regardless of whether it is the "best" (most advanta-

geous) allele for a given trait. Such a run of bad luck would be less likely in a large population, where hundreds might carry the A allele.

Two extreme cases of genetic drift are called founder effects and bottlenecks. In both cases, a population originates or is rebuilt from very few individuals.

With the *founder effect*, a few individuals leave a population and manage to establish a new one. Simply by chance, allele frequencies for many traits probably will not be the same as they were in the original population—and a different range of phenotypes will become available for agents of selection.

The founder effect is important on newly formed volcanic islands. Seabirds, for example, can inadvertently bring in a few seeds from mainland species of plants (Figure 18.6). The alleles of those particular seeds will dictate the range of phenotypes possible for those plants. The founder effect also has consequences for programs to save endangered species. The allele frequencies of individuals that are captured and used to start a breeding population in captivity may differ greatly from those of the natural population.

With *bottlenecks*, disease, starvation, or some other stressful situation nearly wipes out a large population, and even though the population recovers, its relative abundances of alleles have been altered at random. Just before the turn of the century, hunters nearly wiped out the large population of northern elephant seals. Only about twenty seals survived. Since that time, the population has increased to more than 30,000. Interestingly, there is no allelic variation whatsoever at twenty-four gene loci that have been studied. The lack of variation is unique, compared to other seal species and populations that have not gone through comparable bottlenecks. It suggests that a number of alleles were lost during the bottleneck.

It may be that cheetahs underwent a similarly severe bottleneck. These sleek, swift cats are so genetically uniform that researchers can successfully graft a patch of skin from one cheetah onto another, even when they are not closely related. In other species of mammals, such grafts rarely take hold, even among littermates or siblings.

Conservationists worry about the extreme genetic uniformity in populations of northern elephant seals, cheetahs, and other species. Without genetic variation the whole population, no matter how large, is extremely susceptible to diseases or environmental changes. Such species could suddenly become extinct.

**Founder effects and bottlenecks occur when a population originates or is rebuilt from very few individuals. In both of these cases of genetic drift, the amount of genetic variation in the population may be severely limited.**

### Gene Flow

Allele frequencies change when individuals leave a population (emigration) or new individuals enter it (immigration). This microevolutionary process is called **gene flow**. The physical flow of alleles tends to make the genetic composition of adjacent populations more similar to each other. Over time, gene flow tends to decrease the effects of mutation, genetic drift, and natural selection.

**Gene flow is the physical movement of alleles into and out of populations. It tends to decrease the genetic variation between populations that arises through natural selection and other evolutionary processes.**

In this age of international travel, genes flow rather freely among human populations. They also flow rather freely between baboon troops in Africa. Male baboons periodically wander off or are driven out of one troop and sometimes join up with another some distance away. When they mate, they have a homogenizing effect on the genetic character of the population.

Bluejays promote gene flow among populations of oaks and other trees. Like many plant and animal species, most of the new generation of oaks (acorns) would end up sprouting close to the parent tree. Left to their own devices, each stand of oaks would establish a somewhat unique pool of alleles over time. But bluejays disperse acorns when they forage and build up their caches of nuts for the winter. Each fall they may make hun-

**Figure 18.6** One of the travel agents for the founder effect for certain plant populations. Seabirds that island-hop over long distances might have a few seeds intermingled with their feathers, as shown here. Simply by chance, the allele frequencies for some traits might not be the same in those few seeds as they were in the original plant population. If those seeds successfully germinate in a new environment, they will give rise to a new population with its own characteristic range of phenotypes.

**Figure 18.7** One of the travel agents for gene flow among different oak populations. Bluejays hoard supplies of acorns in their home territory, but they might do their shopping at nut-bearing trees up to a mile away. Some of the acorns do not get eaten. They germinate and grow into new trees, which contribute to the allele pool of an oak population some distance away.

dreds of round trips from nut-bearing trees to their home territories, which may be a mile away (Figure 18.7). This bird-nut connection has a homogenizing effect on oak populations, with the "immigrant acorns" keeping them genetically similar to one another.

## Natural Selection

Natural selection is the most influential microevolutionary process, along with genetic drift. As we saw in the preceding chapter, Charles Darwin gained insight into this process when he correlated his observations of inheritance with certain features of populations and the environment. The key points of the theory of natural selection were given in earlier chapters, but we repeat them here because of their importance:

**1.** Members of a population vary in form, function, and behavior, and much of this variation is heritable.

**2.** Some varieties of heritable traits are more adaptive than others; they improve chances of surviving and reproducing.

**3.** Because members with adaptive traits are more likely to reproduce, their offspring tend to make up more of the members of the next generation who actually do reproduce.

**4. Natural selection** is a measurable difference in survival and reproduction that has occurred among members of a population that differ in one or more traits. It is a microevolutionary process; it can change the character of a population over time.

Evolution by natural selection is now a well-documented phenomenon in nature. Hundreds of studies of plants, animals, and microorganisms have provided evidence in favor of the theory and how the process actually works. Here we will consider a few examples.

## EVIDENCE OF NATURAL SELECTION

Natural selection may have stabilizing, directional, or disruptive effects on the range of phenotypes in a population. Figure 18.8 is a generalized overview of these effects.

### Stabilizing Selection

In **stabilizing selection**, the most common phenotypes in a population are favored (Figure 18.8a). Over time, this mode of selection eliminates alleles that produce uncommon phenotypes. Thus stabilizing selection tends to counter the effects of mutation, genetic drift, and gene flow.

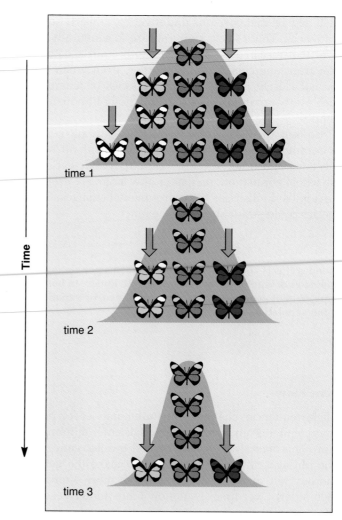

**a Stabilizing selection**

**Figure 18.8** Three modes of natural selection, using the phenotypic variation of a small population of butterflies as the example. The bell-shaped curve represents the range of continuous variation in wing color. The most common forms (powder blue) occur between extreme forms of the trait (white at one end of the curve, deep purple at the other). Orange arrows signify which forms are being selected against over time.

Examples of pronounced stabilizing selection can be drawn from the plant kingdom. Members of the sphenophyte lineage were abundant and diverse some 300 million years ago. Its only existing representatives, of the genus *Equisetum* (horsetails), retain the traits characteristic of its ancient relatives (Figure 18.9).

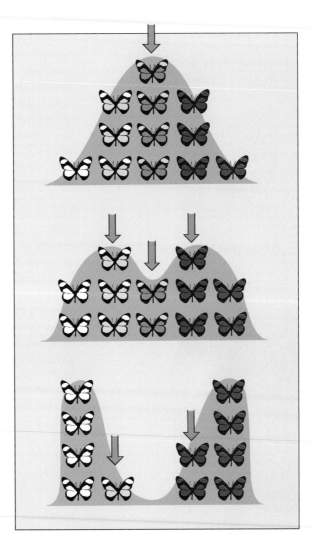

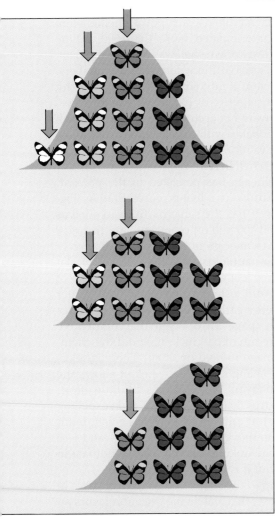

**b  Directional selection**

**c  Disruptive selection**

**a** Fossil

**b** Living representatives of lineage

**Figure 18.9**  A fossil (**a**) of the sphenophyte lineage, which extends back more than 380 million years. Horsetails (*Equisetum*) are the only living members (**b**).

cord of
vascularized
tissue

gas-filled
chamber

Or consider the nautiloids, a type of shelled, tentacled mollusk that originated more than 500 million years ago. The earliest known shell was only about 1 centimeter long, and it was partitioned into many chambers. Some shells of ancient nautiloids were rather straight; others were coiled (Figures 18.10a and b). In all cases, the chambers that were not occupied by the animal itself were gas-filled. (Probably with oxygen and carbon dioxide, supplied by a reinforced cord of blood vessels that ran through all the chambers.)

The nautiloid lineage has persisted to the present. The chambered nautilus is a living representative (Figures 18.10c and d). It has the same partitioned shell as nautiluses that lived 400 million years ago, and this may be viewed as an extreme case of stabilizing selection.

At one time the nautiloids were the premier predators of the seas. They lost ground with the emergence of fishes that could swim faster and maneuver better after prey; then they were nearly wiped out during episodes of mass extinction (page 310). Given the competitive pressures and their run of bad luck, how did the chambered nautilus manage to hold on at all? There is selective advantage to having a gas-filled shell; it allows the animal to remain buoyant in the water with little effort. Also, a nautilus moves by taking in water, then expelling it in a jet (page 424). It can't move as fast as a fish, but it takes less energy to move the same distance—and less energy to keep from sinking. The nautilus is not fast in the race for food, but it *is* energy efficient.

As a final example, the shape of the survival curve in Figure 18.11 suggests that stabilizing selection favors an average human birth weight of about 7 pounds. Data from widely different populations over the past few centuries show that newborns weighing significantly more or less than this tend not to survive.

**Figure 18.10**  (**a, b**) Diagrams of a type of tentacled mollusk called nautiloids. The one in (**a**) flourished about 450 million years ago; the one in (**b**), about 400 million years ago. The chambered shell of the only existing representative of the nautiloid lineage, shown in (**c**) and (**d**), is sliced in half lengthwise to show the intricate partitions. These partitions have persisted over time and may be viewed as an extreme case of stabilizing selection.

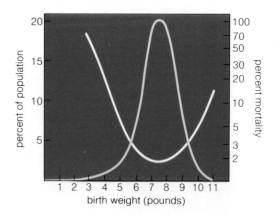

**Figure 18.11**  Weight distribution for 13,730 human newborns (yellow curve) and their survival rate (white curve). Here, stabilizing selection favors a birth weight between 7½ and 8 pounds and operates against newborns whose weight is significantly higher or lower than this.

## Directional Selection

In **directional selection**, allele frequencies shift in a steady, consistent direction in response to a new environment or a directional change in the old one. By this process, forms of traits at one end of the range of phenotypic variation become more common than the intermediate forms (Figure 18.8b).

**Peppered Moths.** Directional selection has been documented in populations of the peppered moth (*Biston betularia*) in England. At one time, a speckled light-gray form was common and a dark-gray form was extremely rare. Between 1848 and 1898, the dark form increased in frequency. For example, all but 2 percent of the moth population near one city had become dark gray. (Much later, researchers identified two genes coding for wing and body color. So the trait definitely has a heritable basis and is subject to agents of selection.)

Peppered moths are active at night. Although field observations are lacking, researchers believe that they rest during the day on the trunks of birches and other trees, where they are vulnerable to bird predators. Before the industrial revolution, tree trunks were cloaked with light-gray speckled lichens. Light-gray speckled moths resting on the lichens were camouflaged from the birds, but dark moths stood out like sore thumbs (Figure 18.12). Then soot and other pollutants from factories started killing the lichens and darkening the trunks. Now the rare dark moths blended with the changing background and the light moths did not. Apparently, dark moths survived and reproduced more in the altered environment, so allele frequencies in the populations changed.

In the 1950s, H. B. Kettlewell used a **mark-release-recapture method** to test that hypothesis. He bred both forms of the moth in captivity, then marked hundreds of them so that they could be identified. He released the moths in two areas, one near the heavily industrialized area around Birmingham and the other in the unpolluted area of Dorset. After a time he recaptured as many moths as he could.

Table 18.2 shows the results. More dark moths were recaptured in the polluted area—and more light moths in the pollution-free area. By stationing watchers in blinds near groups of moths tethered to tree trunks, Kettlewell also found by direct observations that birds captured more light moths around Birmingham and more dark moths around Dorset.

About a hundred moth species underwent directional selection in response to pollution in industrial regions throughout Great Britain. Strict pollution controls went into effect in 1952, however. Lichens have made comebacks in some areas, and tree trunks are now largely free of soot. As you might predict, the frequency of dark moths is now declining.

a

b

**Figure 18.12** An example of variation that is subject to directional selection in changing environments. (**a**) The light- and dark-colored forms of the peppered moth are resting on a lichen-covered tree trunk. (**b**) This is how they appear on a soot-covered tree trunk, which was darkened by industrial air pollution.

| Table 18.2 | Marked *Biston betularia* Moths Recaptured in a Polluted Area and a Nonpolluted Area | | | |
|---|---|---|---|---|
| | Light-Gray Moths | | Dark-Gray Moths | |
| Area | Number Released | Number Recaptured | Number Released | Number Recaptured |
| Near Birmingham (pollution high) | 64 | 16 (25%) | 154 | 82 (53%) |
| Near Dorset (pollution low) | 393 | 54 (13.7%) | 406 | 19 (4.7%) |

Data after H. B. Kettlewell.

## Sickle-Cell Anemia—Lesser of Two Evils?

Sickle-cell anemia, a genetic disorder, is caused by a mutant allele that codes for a defective form of hemoglobin (Figure 14.2b). In tropical and subtropical regions of Africa, this harmful allele (HbS) remains in the population along with the normal one (HbA). HbS/HbS homozygotes often die in their early teens or early twenties, but HbS/HbA *heterozygotes* make up nearly a third of the population.

The sickle-cell trait in humans is an example of "balanced polymorphism." The condition is an outcome of stabilizing selection in which two or more phenotypes are maintained in fairly stable proportions over the generations.

Why is the harmful allele maintained at such high frequency? Where the sickle-cell trait is most prevalent, so also is malaria. The parasite that causes malaria is transmitted to humans by a type of mosquito that is most common in the tropics and subtropics. Individuals who do not carry the mutant allele have a far greater chance of

surviving and reproducing than individuals who do—provided they don't get malaria. Specifically, the HbA/HbA homozygote has an 85 percent greater probability of surviving to reproductive age, but malarial infection dramatically alters that probability. The HbS/HbA heterozygote has greater resistance to malaria and is more likely to survive severe infections! (In one study, severe or fatal infections were found to be twice as high in HbA/HbA as in HbS/HbA individuals.) So the persistence of the harmful sickle-cell trait becomes a matter of relative evils.

In Central Africa, malaria has been an agent of selection for less than 2,000 years. In tropical and subtropical regions of the Middle East and Asia, it has been around for much longer. Even though the sickle-cell trait occurs at high frequencies in these regions also, the symptoms are not as pronounced as they are in Africa. Apparently, other alleles at other loci reduce the serious effects of the HbS allele.

**Insecticide Resistance.** Increasing resistance to insecticides is another example of directional selection. An initial insecticide application kills most of the targeted insects. However, a few insects may survive (some aspect of their structure, physiology, or behavior allows them to resist the pesticide's chemical effects). If the resistance has a genetic basis, there will be a greater proportion of resistant individuals in the next generation.

Often, farmers have resorted to heavier and more frequent insecticide applications to counter the increased numbers of resistant forms. In essence, the insecticides have become selective agents. They actually *favor* the resistant forms. Ironically, crop damage from insects is now greater than it was before the widespread use of insecticides (page 18).

### Disruptive Selection

In **disruptive selection**, forms at both ends of the phenotypic range are favored and intermediate forms are selected against (Figure 18.8c). Researchers observed the effect of disruptive selection on a small population of finches living on one of the Galápagos Islands. The finches differ from one another in many traits, including

beak size and shape (see, for example, Figure 17.7). At one end of the phenotypic range are birds with longer beaks, which are used to open cactus fruits and expose the seeds. At the other end are birds with deep, wide beaks. They can crack hard cactus seeds on the ground or strip away tree bark to get at insects. The finches were observed during a severe drought, when seeds and a few wood-boring insects were the only types of food available. At that time, the birds with extreme beak variations survived at a greater frequency than birds in between.

## SELECTION AND THE MAINTENANCE OF DIFFERENT PHENOTYPES

### Balanced Polymorphism

Earlier, we saw how stabilizing selection has maintained the most common phenotype among generation after generation of horsetail plants. A variation on the stabilizing theme results in a condition called *balanced polymorphism*. Here, two or more forms of a trait are maintained in fairly stable proportions over the generations. Often, the condition exists when heterozygotes for a

**Figure 18.13** Some results of sexual selection. (**a**) Male bird of paradise (*Paradisaea raggiana*) engaged in a spectacular courtship display that has caught the eye (and, perhaps, sexual interest) of the female. Males of this species compete fiercely for females, which serve as selective agents. (**b**) Northern seal with his harem of females. Males not able to secure a territory do not mate at all, so they contribute nothing to the allele pool of the next generation. As an outcome of this sexual selection, males weigh about twice as much as females. (**c**) Brilliantly hued male sugarbird with a subdued-hued female.

a

c

trait have a competitive edge over the homozygotes. In some environmental contexts, one allele is favored—but in other contexts, its nonidentical partner is favored. *The survival value of any allele must be weighed in the context of the environment in which it is being expressed.*

As an example of balanced polymorphism, take a look at the *Commentary.* It describes how this condition exists among members of the human population who carry alleles coding for HbS. As we have seen in earlier chapters, HbS is a mutated allele that can give rise to the genetic disorder called sickle-cell anemia.

### Sexual Dimorphism

Natural selection has been a major force in promoting and maintaining *sexual dimorphism,* or differences in appearance between males and females of a species. The differences are striking among certain birds and mam-

mals (Figure 18.13). As Darwin noted, male birds and mammals generally are larger, more splendid in color and patterning, and more aggressive than females. The females act as agents of selection. They choose their mates and so are directly associated with reproductive success. **Sexual selection** is based on any trait that gives the individual a competitive edge in mating and producing offspring.

Wonderful cases of sexual selection are described in Chapter 50, but a simple example will do here. Northern fur seals mate only on small islets and rocky beaches. Males that are large enough to command the rocks enjoy mating privileges with about ten to twenty females. The males that cannot secure a territory do not mate at all, hence they contribute nothing to the allele pool of the next generation. As an evolutionary outcome of sexual selection, males weigh as much as 1,000 kilograms, about twice the weight of the females (Figure 18.13).

# SPECIATION

## Defining the Species

As we have seen, changes in allele frequencies are not irreversible. Selection may be reversed if the environment changes in an appropriate direction. Even if an allele is lost from a population (through genetic drift or selection), it may be reinstated by mutation or gene flow. So what prevents a population from cycling the same alleles, over and over again?

An irreversible step, acting like a ratchet, causes genetically isolated populations to branch in different evolutionary directions. This step, "evolution's ratchet," is part of the process by which species originate—a process called **speciation.** By this process, a population of individuals that can interbreed successfully with one another becomes reproductively isolated from other such populations. Let's now consider the process, using the following definition as our point of reference for all sexually reproducing organisms:

A species is a unit of one or more populations of individuals that can interbreed under natural conditions and produce fertile offspring, and that are reproductively isolated from other such units.

## Divergence

No matter how diverse the individuals of a population become, they remain members of the same species as long as they continue to interbreed successfully and share a common pool of alleles. But sometimes, one or more parts of the population become isolated from other

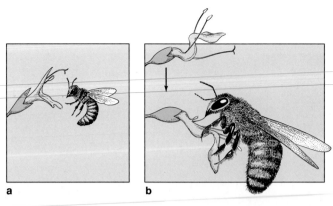

**a**         **b**

**Figure 18.15** Mechanical isolation between two species of sage (*Salvia mellifera* and *S. apiana*). The first species (**a**) has a small floral landing platform for its small or medium-size pollinators. The second species (**b**) has a large landing platform and long stamens, which extend some distance away from the nectary. Even though small bees can land on this larger platform, they can do so without brushing against the pollen-bearing stamens. It takes larger pollinators to do this. Hence the small pollinators of *S. mellifera* are mostly incapable of spreading pollen to flowers of *S. apiana*; and the large pollinators of *S. apiana* cannot land on and cross-pollinate *S. mellifera*. The plants and their pollinators are all drawn to the same scale.

parts. The isolation creates local breeding units. Then, two or more pools of alleles may exist where there had been only one before.

Suppose that, over time, there is no gene flow between the local units, and selection, genetic drift, and mutation operate independently in each one. The units are now reproductively isolated populations, and over time, differences in allele frequencies will accumulate between them. This process is called **divergence**. When divergence becomes great enough, members of the two populations will not be able to interbreed successfully under natural conditions; the populations will have become separate species (Figure 18.14).

## Reproductive Isolating Mechanisms

Any aspect of structure, function, or behavior that prevents interbreeding is a **reproductive isolating mechanism.** Some take effect before or during fertilization and so prevent the formation of hybrid zygotes. Some take effect as the embryo develops and lead to its early death, to sterility or, later on, to the failure of the hybrid animal to be recognized as an acceptable mate by members of either parental species. All reproductive isolating mechanisms prevent the movement of alleles (gene flow) between populations.

A reproductive isolating mechanism is any aspect of structure, function, or behavior that prevents successful interbreeding (hence gene flow) between populations.

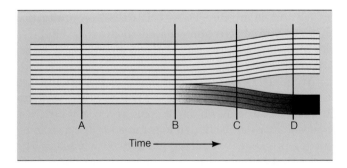

**Figure 18.14** Divergence leading to speciation. Because evolution is gradual here, we cannot say at any one point in time that there are now two species rather than one. Each horizontal line represents a different population. At time A, there is only one species. At D, there are two. At B and C, the divergence has begun but is far from complete.

**Figure 18.16** A mixed herd of horses and zebroids (hybrid animals resulting from crosses between horses and zebras). Zebroids generally are infertile, owing to chromosomal differences between the two parents.

**Table 18.3 Behavioral Isolation Between Three Closely Related *Drosophila* Species**

| Contact Limited to the Following Combinations: | | Number of Females | Number of Matings | Percent Matings |
|---|---|---|---|---|
| Females | Males | | | |
| *D. serrata* | *D. serrata* | 3,841 | 3,466 | 90.2 |
| *D. serrata* | *D. birchii* | 1,246 | 9 | 0.7 |
| *D. serrata* | *D. dominicana* | 395 | 5 | 1.3 |
| *D. birchii* | *D. birchii* | 2,458 | 1,891 | 76.9 |
| *D. birchii* | *D. serrata* | 699 | 7 | 1.0 |
| *D. birchii* | *D. dominicana* | 250 | 1 | 0.4 |
| *D. dominicana* | *D. dominicana* | 43 | 40 | 93.0 |
| *D. dominicana* | *D. serrata* | 163 | 0 | 0.0 |
| *D. dominicana* | *D. birchii* | 537 | 20 | 3.7 |

Data from F. Ayala. 1965. "Evolution of Fitness in Experimental Populations of *Drosophila serrata*." *Science* 150:903–905.

**Isolation of Gametes**. Incompatibilities between the sperm of one species and the egg (or the female reproductive system) of another may prevent fertilization. This is the case with certain marine animals that rely on external fertilization. Even when the eggs and sperm of two species of sea urchin are released at the same time, in the same place, the gametes rarely fuse. Probably there is no mechanism of molecular recognition between the sperm and eggs of different species. Similarly, incompatibilities between the pollen grains of one plant species and the pollen-receiving platform of female reproductive structures of another plant species can prevent fertilization.

**Mechanical Isolation**. Differences in reproductive structures or other body parts may prevent members of two populations from interbreeding. For example, two species of sage plants differ in the size and arrangement of their floral parts. Both reproduce with the aid of pollinators, which carry pollen from plant to plant. But each sage plant species has a "landing platform" that only one type of pollinator fits on, so cross-fertilization between the plant species is inhibited (Figure 18.15).

**Hybrid Inviability and Infertility**. Sometimes fertilization occurs between the gametes of different species. However, the resulting embryo usually dies because of physical or chemical incompatibilities with the mother. Hybrids that do live commonly are weak and their survival chances are not good. In a few cases, hybrid off-

spring are vigorous but sterile. A cross between a female horse and a male donkey produces a mule—a sturdy hybrid that is fully functional *except* in its capacity to reproduce. The zebroids shown in Figure 18.16 are another example of hybrid infertility.

**Behavioral Isolation**. Behavioral isolation is one of the strongest mechanisms that prevent interbreeding among related species in the same territory. A case in point is the complex courtship rituals that often precede mating. The song, head-bobbing, wing-spreading, and dancing of a male bird of one species may stimulate a female of his species—but the female of a related species probably would not even recognize his behavior as a sexual overture. Table 18.3 reveals the effectiveness of behavioral isolation among some closely related species of *Drosophila*.

**Isolation in Time**. Finally, differences in reproductive timing may serve as an isolating mechanism. For most animals and plants, mating or pollination is a seasonal event of relatively short duration, sometimes less than a day. Even closely related species may be reproductively isolated simply because their times of reproduction do not coincide. Two closely related species of cicadas provide an extreme example. In any one location, insects of one of these species emerge and mate every 13 years. Insects of the other species emerge every 17 years. The possibility of their meeting arises only once every 221 years!

## Speciation Routes

**Allopatric Speciation.** The populations of most species are not strung out continuously, with one merging into the others. Most often they are isolated geographically to some extent, with gene flow being more of an intermittent trickle than a steady stream. But sometimes barriers form and shut off even the trickles. This can happen rapidly, as when a major earthquake changed the course of the Mississippi River in the 1800s and isolated some populations of insects that could not swim or fly. Geographic isolation also happens very slowly, as when gradual shifts in climate caused the breakup of formerly continuous environments. Millions of years ago, for example, extensive forests in Africa gave way to grasslands with isolated stands of trees as a result of long-term shifts in rainfall.

Once geographic isolation is absolute, genetic drift or selection may lead to divergence, then to speciation. When new species form as a result of geographic isolation, the process is called **allopatric speciation** (from *allos*, meaning "different," and *patria*, "native land"). There are some excellent, well-documented cases of allopatric speciation, one of which is described in the *Doing Science* essay.

**Sympatric Speciation.** The word sympatric means "same native land." Speciation that follows after ecological, behavioral, or genetic barriers arise *within* the boundaries of a single population is called **sympatric speciation.**

Suppose a mutation leading to a shift in food preference or to a shift in the timing of reproduction begins to spread in an insect population. In such cases, individuals with that mutation will be in a position to breed only with one another. Similarly, various types of chromosome abnormalities can result in instantaneous speciation. These abnormalities include duplications of some or all of the chromosomes and changes in chromosome number, of the sort to be described next. The bearers of such chromosome abnormalities can successfully reproduce only with one another, if at all, and not with other members of the population.

**Polyploidy.** Sympatric speciation may occur by way of **polyploidy**. In this condition, offspring end up with three or more of each type of chromosome characteristic of the parental stock. Page 197 described how polyploidy can arise through complete nondisjunction at meiosis, followed by the formation of diploid instead of haploid gametes. It also can arise when germ cells duplicate their DNA but fail to divide.

## *Doing Science*

## Speciation and the Isthmus of Panama

The "solid earth" beneath our feet actually has more in common with the "moving sidewalks" you may have seen in large airport terminals. As described in the next chapter, it is cracked into huge fragments, or plates. The plates have been slowly drifting about on molten stuff, probably for as long as the earth has had a crust.

Roughly 3,100,000 years ago, two crustal plates collided beneath the waters of an ancient ocean. The advancing edge of one plate thrust under the other and gradually lifted it upward, forming the sinuous ridge of land we now call the Isthmus of Panama.

Biologists interested in exploring the mechanisms by which species originate could scarcely have asked for a better natural laboratory. Why? When the land rose high and dry enough, it divided the ocean basin in two. Gene flow between fish populations on either side of the newly forming coastlines declined to a trickle, then stopped. What is more, the new isthmus disrupted the prevailing ocean currents. Along both coasts, ocean currents moved in new directions, and their temperature and chemical composition changed. Thus, populations of fishes and other marine organisms had become divided into local breeding units in what were now different environments.

Today, biologists recognize a host of closely related "Pacific" and "Atlantic" fish species that presumably diverged from ancestral stocks after the Isthmus of Panama had formed. In many cases, related species look so much alike that only experts can tell them apart.

Now consider this. Some of the recent evidence of evolution comes from comparisons of gene products (proteins) from different species in the same major group of organisms. For example, comparisons might be made of the hemoglobin molecule in raccoons, sloths, polar bears, and other members of the order of mammals. Hemoglobin is one of the highly conserved proteins. Mutations cannot cause drastic changes in the molecular structure of the gene coding for it without dire consequences for the

About half of all flowering plants are polyploid species. The ability of many flowering plants to reproduce asexually or to undergo self-fertilization probably accounts for the widespread occurrence of this condition. (Because the polyploid individual need not wait for a sexual partner with the same chromosome number,

individual. However, *slight* changes might, in the long run, turn out to be neutral or adaptive. Such small-scale adaptations might be the evolutionary "foot in the door," an early sign that populations are diverging.

Suppose you were working with John Graves in the early 1980s, when he started comparing the enzymes in muscle cells of Isthmus fishes. You already knew that enzymes are proteins, each composed of one or more polypeptide chains. You knew also that each kind of enzyme functions best within a specific temperature range.

As Graves did, you think about two facts. First, the ocean water on the Pacific side of the Isthmus is generally cooler than the water on the Atlantic side by about 2°–3°C. Second, it shows more pronounced seasonal changes in temperature. The related fishes on both sides of the Isthmus are strong swimmers, and swimming depends on strong muscles. In turn, muscle action depends on efficient enzymes. Suppose gene mutations arose that slightly improved enzyme function under new conditions in the ocean environment. Could even a small variation in water temperature serve as an agent of natural selection for certain variant enzymes in fish muscle cells?

To test this hypothesis, you study the same type of enzyme from the muscle cells of four pairs of closely related fish species. You compare their functioning under a range of temperatures, and you subject them to gel electrophoresis (page 218). You discover that all of the "Pacific" enzymes function better at lower temperatures than the "Atlantic" enzymes do. In two of the four pairs of enzymes, you detect a slight difference in electric charge. In only one of those two pairs, electrophoresis results indicate that the enzymes must differ slightly in their amino acid sequence.

Graves actually discovered these differences and drew three tentative conclusions from them. First, even seemingly small differences in environmental conditions can favor molecular adaptation. Second, some gene

**a**

**b**

Two kinds of fishes, apparently related by descent from a common ancestral population that became divided when geologic forces created the Isthmus of Panama. (**a**) *Thalassoma bifasciatum*, a blue-headed wrasse from the Atlantic side of the Isthmus, and (**b**) *T. lucasanum*, a Cortez rainbow wrasse from the Pacific side. In this case, differences in body coloration and patterning do not provide much evolutionary insight. Why? Reef fishes commonly show stunning phenotypic variation among individuals of the *same* species. Information more revealing of evolutionary relationship comes from studies such as the one described in this essay.

mutations may have neutral effects. (Remember that all of the enzymes performed as expected, even though the electrophoresis studies revealed slight differences among them.) And finally, some changes in the amino acid sequence of a polypeptide chain may be too small to detect with electrophoresis, yet they can lead to detectable differences in enzyme function.

speciation is instantaneous. That individual can give rise to a whole population just like itself.) Polyploidy is less common in animals, probably because it upsets the balance between autosomes and sex chromosomes—a balance that is crucial for animal development and reproduction.

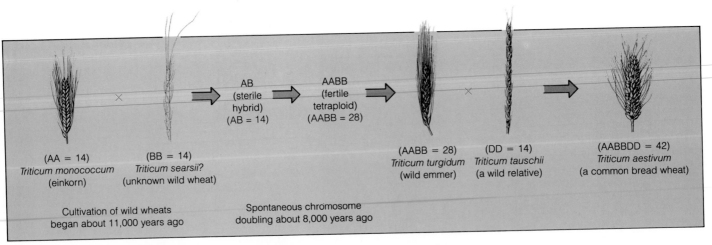

**Figure 18.17** Proposed speciation in wheat by way of polyploidy and hybridizations. Wheat grains dating from 11,000 B.C. have been found in the Near East. Several species of wild diploid wheat still grow there. They have 14 chromosomes (two sets of 7, designated AA). Also growing in the region is a wild grass with 14 chromosomes, designated BB. They differ from the A chromosomes, judging from their failure to pair with them at meiosis. One tetraploid wheat species has 28 chromosomes; analysis during meiosis shows that they are AABB. The A chromosomes pair with As, and the B chromosomes pair with Bs. A hexaploid wheat has 42 chromosomes (six sets of seven). Its chromosomes are AABBDD, the last set (DD) coming from *Triticum taushchii*, another wild grass.

Speciation also occurs when polyploidy is followed by successful hybridization. Most hybrids between two species are sterile because they have different numbers or types of chromosomes. This usually prevents homologous pairing at meiosis. But if polyploidy happens to occur in the hybrid's germ cells, then the "extra" set of chromosomes can pair with the original ones at meiosis, and viable gametes can form. Wheat is an example of a polyploid hybrid (Figure 18.17).

## SUMMARY

1. Individuals of a population show variation in their traits, corresponding to differences in the relative abundances of alleles for most of their genes.

2. Allele frequencies change as a result of mutation, genetic drift, gene flow, and natural selection. Mutation is the only source of new alleles. The other processes simply shuffle existing alleles into, through, or out of the population.

3. In terms of a given gene locus in individual organisms, mutations arise rarely and at random. In evolutionary terms, they have been accumulating to significant numbers in different lineages for billions of years and are the ultimate source of all heritable variation.

4. Most mutations are harmful or neutral. Individuals inherit a combination of many genes already fine-tuned for a given range of operating conditions, and a muta-

tion usually disrupts interdependencies among gene products. But if environmental conditions change, the product of a mutated gene may prove beneficial.

5. Genetic drift is the chance increase or decrease in the relative abundances of alleles in a population. Gene flow is a change in allele frequencies following the movement of individuals from one population to another and their subsequent reproduction.

6. Natural selection is a measurable difference in survival and reproduction among members of a population that vary in one or more traits. Agents of selection can have stabilizing, directional, or disruptive effects on the range of phenotypes.

7. A species is one or more populations whose members are able to interbreed under natural conditions and produce fertile offspring, and are reproductively isolated from other populations. (This definition applies only to sexually reproducing species.)

8. Sometimes two populations of a species (or local breeding units within a population) become reproductively isolated from one another, and divergence occurs. Divergence is a buildup of differences in allele frequencies between reproductively isolated populations. When divergence is great enough, successful interbreeding is no longer possible under natural conditions; speciation has occurred.

9. Geographic barriers as well as mechanical, physiological, or behavioral barriers can keep different populations reproductively isolated from each other.

## Review Questions

1. What is the Hardy-Weinberg baseline against which changes in allele frequencies may be measured? 275

2. Changes in allele frequencies may be brought about by mutation, genetic drift, gene flow, and selection pressure. Define these occurrences, then describe the way each one can send allele frequencies out of equilibrium. 276–280

3. What implications might the effect of genetic drift hold for an earlier concept of "survival of the fittest"? 278

4. Define stabilizing, directional, and disruptive forms of selection and give a brief example of each. 280–284

5. Give two examples of reproductive isolating mechanisms, and outline what they accomplish. 286–287

## Self-Quiz *(Answers in Appendix IV)*

1. A _____ is a group of individuals occupying a given area and belonging to the same species.

2. Variation in the individuals of a population corresponds to _____ in the relative abundances of alleles for most gene locations along the chromosomes.

3. Allele frequencies change as a result of _____.
   a. mutation          d. natural selection
   b. genetic drift     e. all of the above
   c. gene flow

4. The only source of new alleles is _____.
   a. mutation          d. natural selection
   b. genetic drift     e. all of the above
   c. gene flow

5. Existing alleles are shuffled into, through, or out of populations by _____.
   a. mutation          d. natural selection
   b. genetic drift     e. b, c, and d only
   c. gene flow

6. Which of the following statements about mutation is *not* true?
   a. mutations arise randomly
   b. most mutations are harmful or neutral
   c. mutations arise rather infrequently
   d. most mutations are beneficial

7. Speciation is _____.
   a. the extinction of a distinct population
   b. the accumulation of environmental factors that cause geographic isolation
   c. the process whereby different species originate
   d. a means of altering gene frequencies in a population

8. The sickle-cell trait evolved in tropical and subtropical regions of Asia, the Middle East, and Africa, then it appeared in the United States population with the influx of individuals who were forcibly brought over from Africa prior to the Civil War. In terms of microevolution, this is an example of _____.
   a. mutation          c. gene flow
   b. genetic drift     d. natural selection

9. Divergence is defined by _____.
   a. accumulated differences in allele frequencies between reproductively isolated populations
   b. a loss in the ability of two populations to interbreed
   c. eventual occurrence of speciation
   d. all of the above

10. Match the evolution concepts appropriately.
    _____ gene flow              a. the sole source of new alleles
    _____ sexually               b. chance change in relative allele
           reproducing               frequencies of a population
           species               c. one or more populations whose
    _____ natural                   members interbreed and produce
           selection                 fertile offspring
    _____ mutation               d. change in allele frequencies in a
    _____ genetic drift             population due to immigration,
                                     emigration
                                  e. differential survival and reproduction
                                     of variant members of population

## Selected Key Terms

allele frequency 275
allopatric speciation 288
balanced polymorphism 284
bottleneck 278
directional selection 283
disruptive selection 284
divergence 286
founder effect 278
gene flow 279
genetic drift 278
genetic equilibrium 275
genetic variation 274
Hardy-Weinberg principle 275
heritable trait 274

lethal mutation 277
microevolution 276
mutation 276
natural selection 280
neutral mutation 277
polyploidy 288
population 273
reproductive isolating
   mechanism 286
sexual dimorphism 285
sexual selection 285
speciation 286
stabilizing selection 280
sympatric speciation 288

## Readings

Ayala, F. J., and J. W. Valentine. 1979. *Evolving.* Menlo Park, California: Benjamin/Cummings. Short introduction to evolutionary theory.

Cook, L., G. Mani, and M. Varley. 1986. "Postindustrial Melanism in the Peppered Moth." *Science* 231:611–613.

Dobzhansky, T. 1973. "Nothing in Biology Makes Sense Except in the Light of Evolution." *The American Biology Teacher* 35(3):125–129. Personal views of a leading geneticist, who argues that the principle of evolution does not clash with religious faith.

Futuyma, D. 1987. *Evolutionary Biology.* Second edition. Sunderland, Massachusetts: Sinauer.

Grant, V. 1981. *Plant Speciation.* Second edition. New York: Columbia University Press. Good discussion of speciation in plants.

## On Floods and Fossils

Nearly 500 years ago, Michaelangelo was just finishing his magnificent painting of the Great Deluge on the ceiling of the Sistine Chapel in Rome (Figure 19.1). And another artist, Leonardo da Vinci, was brooding about wonderfully intact seashells that were embedded in rock layers in the high mountains of northern Italy. How did they get there? The traditional view was that they were strewn about during the Deluge. Yet how could even a great flood wash them inland and up the mountainsides, which were hundreds of kilometers away from the sea? Many of the shells were so thin and fragile, they would have disintegrated if turbulent floodwaters had swept them across such distances and then battered them against the rocks.

With his artist's eye, da Vinci also perceived that the rock layers were conspicuously stratified—that is, arranged like cake layers one above the other. Some layers contained distinct kinds of shells, while others had none. Then da Vinci thought about how large

rivers deposit quantities of silt when they flood each spring. Could it be that the shell-rich and shell-barren layers in Italy's mountains were silent testimony to a *succession* of floods and silt deposition? If so, then those distinct assemblages of shells might be the fossilized remains of ancient marine communities, buried gradually over time. Da Vinci did not run about discussing his novel idea, perhaps knowing it would have been met with deafening silence, imprisonment, or worse.

By the early 1700s, fossilized shells, bones, plant parts, and impressions of footprints, tracks, and burrows were generally acknowledged as evidence of life in the past. But the fossils were still being interpreted according to the traditional view, as when a Swiss naturalist excitedly unveiled the remains of a giant salamander and announced they were the skeleton of a man who had drowned in the Deluge.

It was not until the mid-eighteenth century that the traditional view was even seriously questioned.

**Figure 19.1** (*Left*) Michaelangelo's interpretation of the Great Deluge. Puzzling events of this magnitude have punctuated geologic time, although they have been explained in different ways throughout human history. (*Right*) A modern-day photographer captures the intricate structural pattern of fossilized ammonite shells. About 65 million years ago, all ammonites perished, along with many other groups of organisms. That mass extinction is but one piece of the macroevolutionary puzzle.

Extensive canal-building, quarrying, and mining operations were revealing regular patterns in the color and composition of stratified rock layers. Soon it became apparent that certain unique fossils occur in distinctive types of layers—and that some of them do so over considerable distances. For example, examination of the cliffs of England on one side of the English Channel and the cliffs of France on the other revealed the same fossil sequences.

Some scholars simply used the correlation between fossils and strata as a geologic mapping tool. Others, like Georges Cuvier, started using them as a way to analyze the possible connections between past geologic events and the history of life (page 264). Fossils have been analyzed in ever refined ways to the present day. As you will see in this chapter, they afford convincing evidence of change through time—*changes in the geologic stage, and changes in the organisms that have marched across it.*

KEY CONCEPTS

**1.** Speciation occurs through microevolution—changes in allele frequencies within reproductively isolated populations of a species. In contrast, macroevolution refers to the large-scale patterns, trends, and rates of change in groups of species over geologic time. The patterns include the retention of certain traits and the modification of others within major groups.

**2.** Two trends—mass extinction and adaptive radiation—have changed the course of biological evolution many times. The pace of evolution has been gradual in some lines of descent. In others it has proceeded with bursts of speciation events followed by long periods of little change.

**3.** Evidence of macroevolution comes from the fossil record in combination with the geologic record. It also comes from radioactive dating methods, comparative morphology, and comparative biochemistry.

**4.** The evolution of life has been linked, from the time of origin to the present, to the physical and chemical evolution of the earth.

## EVIDENCE OF MACROEVOLUTION

The history of life spans nearly 4 billion years. It is a story of how species originated, persisted or became extinct, and stayed put or radiated into new environments. **Macroevolution** refers to the large-scale patterns, trends, and rates of change among groups of species. Those groupings, the so-called *higher taxa*, include all the different genera, families, phyla (or divisions), and so on up to the most inclusive groups of species, the kingdoms.

Given our knowledge of natural selection and other microevolutionary processes, we can interpret the evidence of large-scale patterns and trends by starting with a simple fact: *Evolution proceeds by modifications of organisms that already exist.* As we saw in the preceding chapter, "new" species don't appear out of thin air. They

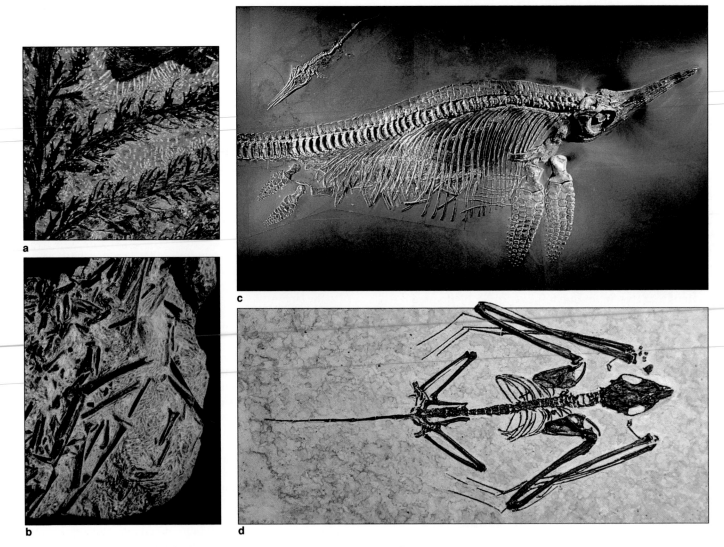

**Figure 19.2** Some representative fossils. (**a**) Fossilized leaves from *Archeopteris*, which probably was on the evolutionary road that led to gymnosperms (such as pine trees) and angiosperms (flowering plants). Some *Archeopteris* trees were more than three stories tall. (**b**) A good find: parts of the fossilized skeletons of several individuals of a ducklike bird. Many hours of careful preparation and study will be required to establish the identity of the organism and to piece together its morphological characteristics.

Two fossils of the sort that paleontologists dream of finding. (**c**) A complete fossil of a female ichthyosaur, about 200 million years old, that died while giving birth. The extinction of these dolphin-like marine reptiles coincided with the rise of modern sharks, which were much more efficient at feeding and swimming. (**d**) The complete skeleton of a bat that lived 50 million years ago. Fossils of this quality are extremely rare.

emerge as mutation, natural selection, and genetic drift change the allele frequencies in reproductively isolated populations of an existing species. Given this fact, there must be underlying threads of relatedness connecting all species since the origin of life. Today, attempts to reconstruct the past and to classify life's diversity take this relatedness into account.

**Because evolution proceeds by modifications within already existing species, there must be a continuity of relationship among all species that have ever appeared on earth.**

### The Fossil Record

"Fossil" comes from a Latin word for something that has been "dug up." As generally used, however, a **fossil** is recognizable, physical evidence of an organism that lived long ago. Physical evidence for the long history of life comes from fossilized skeletons, shells, leaves, seeds, tracks, and such. To be preserved as a fossil, body parts or impressions must be buried before they decompose, and the rock layers in which they are entombed must not be disturbed much over time.

The fossil record is uneven in terms of which organisms and environments are represented. For example, fossils of mollusks and other animals with hard shells or

**Figure 19.3** Stratification at the Grand Canyon of the American Southwest, a region that was once part of an ocean basin. The layers of sedimentary rocks were laid down gradually over hundreds of millions of years. Tectonic forces lifted them above sea level. Over time, rivers changed course and the erosive force of moving water formed ever deeper canyon walls. This great canyon is one of the truly splendid slices through geologic time.

skeletons are abundant, but fossils of jellyfishes and other soft-bodied animals are not. Sea floors, floodplains, swamps, and natural traps such as caves and tar pits favor fossilization; rapidly eroding hills do not.

The fossil record also is uneven in terms of the quality of specimens represented. Most fossils are broken, incomplete, and often crushed or deformed. A few, such as the plant, ichthyosaur, and bat shown in Figure 19.2, are spectacularly complete and well preserved. They tell us a great deal about the structure, function, and behavior of extinct organisms.

Some parts of the record are gone forever. Other parts have yet to be discovered. We have fossils of about 250,000 species, dating mostly from the past 600 million years. Although that is not much, compared to the estimated tens of millions of existing and extinct species, it's better than nothing at all. Besides, the fossil record for some lines of descent, or **lineages**, *is* quite extensive. This is especially true of former inhabitants of shallow seas in which sediments were steadily deposited. There, the number of species of shelled animals known from

fossils is closer to the number of such species now living in the same type of environment.

**The completeness of the fossil record varies as a function of the kinds of organisms represented, where they lived, and the stability of the region since the time of fossilization.**

### Dating Fossils

When geologists of the mid-1800s started mapping the layers of sedimentary rock beneath the earth's surface, they discovered that certain fossils occur in the same kinds of layers over vast areas, even on different continents. Sedimentary rocks form by a gradual "rain" of erosion products and skeletons of tiny marine organisms, so it was logical to assume the deepest layers formed first and the ones closest to the surface formed last (Figure 19.3). The layered pattern of fossils was used as the basis of a relative time scale for biological events—but it was a time scale with no absolute dates.

Earth history simply was divided into four great eras, based on four abrupt transitions in the fossil record. The oldest fossils came from the *Proterozoic*, followed by fossils from the *Paleozoic*, *Mesozoic*, and finally fossils from the "modern" era, the *Cenozoic*. The four divisions are still in use. We now know they mark the boundaries of four immense, mass extinctions in which major groups of organisms simultaneously perished. The actual time spans between boundaries were finally established by radioisotope dating methods (see the *Commentary* on page 21). The span that extended back in time from the Proterozoic-Paleozoic boundary turned out to be far more immense than people had even imagined. Thus the Proterozoic was subdivided, and its first half is now called the *Archean* (after the Greek *arché-*, meaning the beginning).

We are accustomed to thinking of a year as a long time. Few of us will last a century. Yet one million years is ten thousand centuries end to end. And that is still only $\frac{1}{3800}$ of the history of life.

### Comparative Morphology

Another source of evidence of macroevolution comes from detailed comparisons of body form and structural patterns in major taxa. Through **comparative morphology**, evolutionary history is reconstructed on the basis of information contained in the observed patterns of body form. This work involves detailed studies of embryos at different stages of development as well as the adult form. The premise is that an organism's body form (morphology) is an end product of evolution.

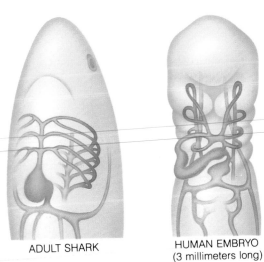

ADULT SHARK            HUMAN EMBRYO
                      (3 millimeters long)

**Figure 19.4** Evidence of evolution from comparative embryology. Aortic arches, a two-chambered heart, and other structures develop in a fish embryo. These structures, which persist in adult fishes, function in respiration in aquatic environments. Fishlike respiratory structures also develop in amphibian embryos and are used during aquatic larval stages. Some disappear during development and others become altered for adult life on land.

Similarly, the early embryos of reptiles, birds, and mammals also develop fishlike structures. Shown here, the aortic arches (*red*), a two-chambered heart (*orange*), and certain veins (*blue*) in an early human embryo. Notice how these structures resemble those of an adult shark. They provide evidence of the retention of basic developmental processes during the evolution of different groups of vertebrates.

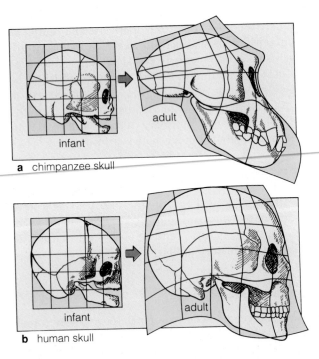

**a** chimpanzee skull

**b** human skull

**Figure 19.5** Comparison of the proportional changes in a chimpanzee skull and a human skull, both of which are remarkably similar in infants. (Imagine that these representations of the infant skulls are paintings on a blue rubber sheet divided into a grid. Stretching the sheet deforms the squares of the grid, and the resulting changes between the two adult skulls reflect differences in the growth patterns in each square of the grid.)

**Stages of Development**. At certain stages of development, the embryos of different organisms within a phylum or some other major group bear striking resemblance to one another. Often the similarities indicate evolutionary relationships. For example, certain structures of the sort shown in Figure 19.4 emerge in all vertebrate embryos at corresponding stages of development. Such structural similarities are one of the reasons why fishes, amphibians, reptiles, birds, and mammals are said to belong to the same subphylum (Vertebrata) despite the large variation among adult forms.

Why do similar embryonic stages persist in different vertebrates? Cells have specific positions and roles in the tiny embryo, and interactions among their hormones and other gene products are vital for normal growth and development. That is why most gene mutations affecting embryonic cells usually have devastating effects on later stages. Nearly all gene mutations affecting the early stages were probably selected against during vertebrate history. For most vertebrate embryos, complex interactions between cells, between developing body parts, and between the embryo and its "maternal environment" act as *developmental constraints*. Such constraints limit the range of forms possible in the adults.

Then why are there such variations in form among adult vertebrates? At least some variations probably arose through mutations in **regulatory genes** that control the *rate of growth* of different body parts.

Figure 19.5 shows how differences in growth rates between two very similar embryos can produce large differences between their adult forms. The skullbones of newborn chimpanzees and humans are proportionally alike. From infancy onward, changes in skull proportions are dramatic for chimps but only minor for humans. We cannot account for the difference by assuming chimps and humans have different sets of genes; their genes are very nearly identical. But suppose a regulatory gene underwent mutation in an isolated population of a species that was ancestral to both chimps and humans. Suppose the mutation blocked the rapid growth required for major changes in the skullbones. That microevolutionary event could have put the early ancestors of humans on a separate evolutionary road.

**Homologous Structures**. Different species may have "homologous" structures, meaning they resemble one another in body form or patterning due to descent from a common ancestor. In **morphological divergence**, one or more homologous structures have departed in appearance, function, or both from the ancestral form.

The vertebrate forelimb is an example of divergence (Figure 19.6). Most land-dwelling vertebrates have a five-toed limb. Such a limb was the point of departure for the evolution of wings in three groups of vertebrates (pterosaurs, birds, and bats). All three types of wings

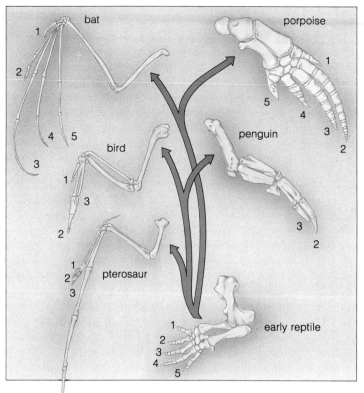

**Figure 19.6** Morphological divergence in the vertebrate forelimb, starting with the generalized form of ancestral early reptiles. Diverse forms have emerged even while many similarities in the number and position of bones have been preserved.

have the same component parts, as do the flippers of porpoises. Similarly, the five-toed limb was the forerunner of the long, one-toed limbs of modern horses, the stubby limbs of moles and other burrowing mammals, and the pillarlike limbs of elephants.

**Analogous Structures.** In **morphological convergence**, dissimilar and only distantly related species adopt a similar way of life, and body parts that take on similar functions end up resembling one another. Similar body parts used for similar functions in evolutionarily remote lineages are said to be "analogous" to one another.

Convergence is an outcome of natural selection. Often it results when the physical requirements of a particular life-style are quite strict. Sharks, penguins, and porpoises, for example, are among the most distantly related vertebrates. Yet as Figure 19.7 shows, they are similar to one another in terms of the proportion, position, and function of body parts. The similarities reflect similar lifestyles.

Sharks, penguins, and porpoises are all fast-swimming predators of the seas. Sharks have pectoral fins that stabilize the body in water; penguins and porpoises have similarly shaped flippers that do the same thing.

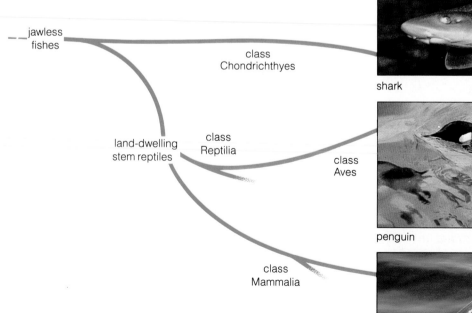

shark

penguin

porpoise

**Figure 19.7** Morphological convergence among sharks, penguins, and porpoises. Selection for adaptations that permitted rapid swimming resulted in superficially similar shapes among these three kinds of vertebrates, even though they are only remotely related.

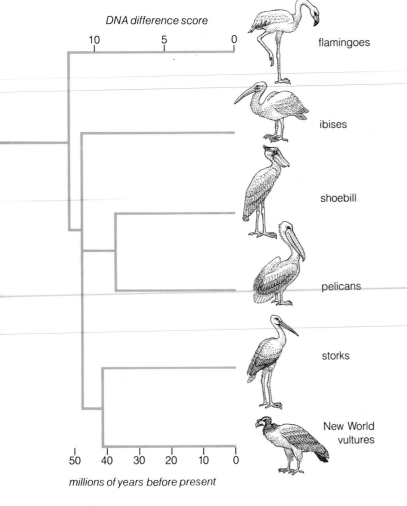

DNA difference score

10    5    0

flamingoes

ibises

shoebill

pelicans

storks

New World
vultures

50  40  30  20  10  0

*millions of years before present*

**Figure 19.8** Relationships among some New World vultures, storks, and other birds, as indicated by DNA hybridization studies.

The shark lineage never left the water, and the shark fin has not undergone much modification in structure or function. Penguins are descended from flying birds that became adapted to life in water instead of air. The penguin flipper is a modified wing. Porpoises are descended from land-dwelling, four-legged mammals that became adapted to life in the seas. The porpoise flipper is a modified front leg. In other words, the flippers of penguins and porpoises have converged on the pectoral fins of the shark. Similarly, all three vertebrates have a streamlined shape that reduces drag during rapid swimming.

Morphological convergence is not restricted to the animal kingdom. As you will see in Chapter 47, dramatic examples can be observed among plants, such as certain cactuses of North America and euphorbs of Africa. These two kinds of plants diverged from common ancestral stock in the very remote past, yet today they bear striking resemblances to each other.

## Comparative Biochemistry

The genes and gene products (proteins) of different species contain information about evolutionary relationships. Consider that some kinds of genes have been highly conserved over time and are present in most living organisms. For example, more than 90 percent of the known gene products in humans have counterparts in yeasts, even though it has been many hundreds of millions of years since yeasts and humans shared a common ancestor.

Mutations have been accumulating in those conserved genes probably since life began. It even appears that they have accumulated at regular rates, although the rates vary among classes of proteins and from one group of organisms to the next. Among them are *neutral mutations* that confer neither advantages nor disadvantages on the individual and so are not subject to agents of selection (page 277). In general, neutral mutations seem to be reliable enough to be used as a "molecular clock" for dating the divergence of two species from a common ancestor. In other words, when mapped against the known geologic time scale, they give us a rough idea of speciation events over evolutionary time. Certainly they have helped answer some thorny questions about evolutionary relationships.

Different methods can be used to identify the mutations. Some, such as protein comparisons, are useful for examining closely related organisms. Other methods, such as DNA hybridization studies, are more useful when the organisms are more distantly related.

**DNA Comparisons.** Various methods are being used to study the evolutionary secrets contained in the DNA of different species. With *DNA-DNA hybridization studies,* the extent to which the DNA from one species base-pairs with DNA from another is a rough measure of evolutionary distance between them. First the DNA is converted to single-stranded form, then single strands from both species are allowed to recombine into "hybrid" molecules. The hybrid DNA is heated to break the hydrogen bonds between the two strands. The heat energy required to pull the strands apart is a measure of the similarity between them. Hybridizations between the DNA of different groups of birds yielded the chart of relationships in Figure 19.8.

Today, precise information about evolutionary relationships is being obtained through DNA sequencing methods and the use of RFLPs, as described in Chapter 16.

**Protein Comparisons.** Figure 14.2 described how gel electrophoresis of proteins can be used to identify even a single amino acid substitution in the polypeptide chain of a protein. Genes dictate the amino acid sequence in

**Figure 19.9** Representation of the primordial earth, about 4 billion years ago. Within another 500 million years, living cells would be present on the surface. (During its formation, the moon presumably was closer to the earth. Here it looms on the horizon.)

those chains. If the sequence is the same (or nearly so) in comparable proteins from two different species, the genes coding for those proteins point to a shared ancestor. The closer the match, the more recently the two species shared that common ancestor. The more remote the match, the more distantly the two species are related.

Consider cytochrome $c$, a component of electron transport chains in organisms ranging from aerobic bacteria to corn plants to humans. This protein consists of 104 amino acids. The amino acid sequence in the cytochrome $c$ of humans precisely matches the sequence in chimpanzees. The sequence in rhesus monkeys differs by only one amino acid. Humans, chimpanzees, and rhesus monkeys have been placed in the same group (Primates) on the basis of other evidence, but protein comparisons provide objective confirmation of the grouping. By contrast, compared to human cytochrome $c$, the sequence differs by 18 amino acids in chickens, 19 in turtles, and 56 in yeasts (*Saccharomyces*). Such numbers tell us that the gene coding for cytochrome $c$ is highly conserved, with "workable" mutations occurring only about once every 21 million years.

## MACROEVOLUTION AND EARTH HISTORY

The fossil record, comparative morphology, and comparative biochemistry provide evidence of the evolution of life on a grand scale. When this diverse evidence is carefully pieced together, an important fact emerges. *The evolution of life has been linked, from its origin to the present, to the physical and chemical evolution of the earth.*

### Origin of Life

**Early Earth and Its Atmosphere.** Billions of years ago, explosions of dying stars ripped through our galaxy and left behind a dense cloud of dust and gas that extended trillions of kilometers in space. As the cloud cooled, countless bits of matter gravitated toward one another. By 4.6 billion years ago, the cloud had flattened out into a slowly rotating disk. Our sun was born at the extremely dense, hot center of the disk. There, thermonuclear reactions began that would perpetuate themselves for the next 10 billion years.

Farther out from the center, the earth was forming along with other planets. By 4 billion years ago, it was hurtling through space as a thin-crusted inferno (Figure 19.9). An early atmosphere developed as gases formed in the earth's molten interior or trapped below the crust were forced to the outside. This first atmosphere had very little free oxygen—a condition that favored the origin of life. Free oxygen disrupts the structure of amino acids, nucleotides, and other biological molecules exposed to it.

Early on, water vapor must have been released from the breakdown of rocks during volcanic eruptions, but it would have evaporated in the intense heat blanketing the crust. In time the crust cooled, and rains started stripping mineral salts from the parched rocks. Salt-laden waters collected in depressions in the crust and formed the early seas.

If the earth had become a smaller planet, it would not have had enough gravitational mass to hold onto an atmosphere. If it had settled into an orbit closer to the sun, its surface would have remained too hot for rain to form. If the orbit had been more distant, its surface would have become too cold, locking up any water as ice. *Because of its size and distance from the sun, the early earth retained liquid water on its surface.* Without liquid water, life as we know it never would have originated.

The first living cells emerged between 4 billion and 3.8 billion years ago. There is no record of the event. Most rocks from that period melted, solidified, and remelted many times because of large-scale movements in the earth's mantle and crust. Some rocks were buried so deeply that heat and compression altered any clues they might have held.

Even so, we can gain insight into the manner in which life originated by considering four questions:

1. What were physical and chemical conditions like at the time of origin?

2. Based on known physical, chemical, and evolutionary principles, could life have originated spontaneously under those conditions?

3. Can we postulate a sequence of events by which the first living systems developed?

4. Can we devise experiments to test whether that sequence could indeed have taken place?

**Synthesis of Biological Molecules.** All the components found in biological molecules were present on the early earth. We know this from rock samples of meteorites, the earth's moon, and Mars that were formed 4.5–4.6 billion years ago (Figure 19.10). We also know that lightning, hot volcanic ash, even shock waves have enough energy to drive the synthesis of biological molecules under abiotic conditions. ("Abiotic" means not involving or produced by organisms.) For example, Stanley Miller mixed hydrogen, methane, ammonia, and water in a reaction chamber (Figure 19.11). He recirculated the mixture and kept bombarding it with a spark discharge to simulate lightning. Within one week, many amino acids and other organic compounds had formed.

**Figure 19.10** Evidence of an ancient impact at the earth's surface—part of a glassy spherelike particle, no bigger than a sand grain. In 1990, Gary Byerly and Donald Lowe discovered sedimentary layers in South Africa that are peppered with the particles. Radioactive dating has indicated that the layers are 3.5 billion to 2.5 billion years old. Unlike rocks on earth, the particles contain high amounts of iridium, an element that is rare on earth but abundant in meteorites. The particles formed as a result of four separate impacts, possibly during the tail end of the same bombardments that cratered the moon. Their glassiness indicates they were molten at first, then cooled rapidly. The fact that they were buried in sedimentary layers indicates they were not volcanic in origin.

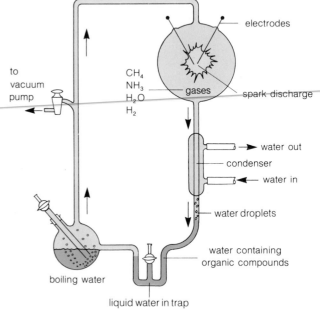

**Figure 19.11** Stanley Miller's apparatus used in studying the synthesis of organic compounds under conditions believed to have been present on the early earth. (The condenser cools circulating steam and causes water to condense into droplets.)

**Self-Replicating Systems**. During the 300 million years after the first rains began, organic compounds accumulated in the shallow waters of the earth. The first self-replicating systems must have emerged in this organic "soup." By *systems* we mean the following interacting molecules:

DNA  *protein-building instructions*

RNA  *transcribers and translators of DNA*

proteins *including enzymes required for DNA replication and protein synthesis*

Clay crystals at the bottom of tidal flats and estuaries may have been the first templates (structural patterns) for protein synthesis. Iron, zinc, and other metal ions often are embedded in such crystals. (If you have ever baked meat in a clay or metal pot, you know that bits of protein stick to both kinds of heated surfaces.) Although amino acids could have joined together through random collisions in the water, clay templates would have allowed longer chains to form in less time and to resist breaking apart.

Suppose amino acids became linked together on a clay template. Suppose the resulting protein's shape and chemical behavior allowed it to function as a weak enzyme in hastening the linkage between other amino acids. Clay templates promoting such linkages would have had selective advantage over other templates. There would have been *chemical competition* for available amino acids—and selection for the first large molecules characteristic of living cells.

Now suppose a clay template (or amino acids stuck to it) also attracted nucleotides. Intriguingly, most existing enzymes are assisted by small molecules—coenzymes—some of which are structurally identical to RNA nucleotides. Simple systems of enzymes, coenzymes, and RNA have been created in the laboratory. Experiments with these systems suggest that RNA could have replaced clay as a template for protein synthesis.

At this point we have no idea how DNA entered the picture. We do know this: The reactions leading to the first self-replicating systems could not have been purely random. *Physical and chemical conditions of the early earth made some reactions more probable than others.* For example, amino acids can exist in two forms, like mirror-images of two hands. Almost all living things have "lefthanded" amino acids. In contrast, meteorites contain disorganized arrays of lefthanded and righthanded forms. When the disorganized arrays are exposed to clay crystals, the clay attracts only lefthanded amino acids!

**The First Plasma Membranes**. The plasma membrane of all living cells is a lipid bilayer with its associated proteins. Metabolism cannot proceed—life cannot exist—without this vital barrier against random chemi-

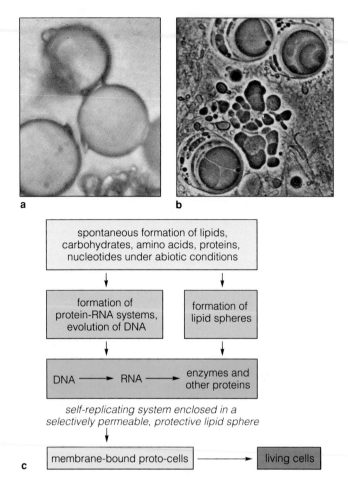

**Figure 19.12** Microscopic spheres of (**a**) proteins and (**b**) lipids that self-assembled under abiotic conditions. (**c**) One proposed sequence of events that led to the first self-replicating systems and, later, to the first living cells.

cal fluctuations in the outside world. How did plasma membranes originate?

Most likely, the first living cells were little more than membrane-bound sacs holding nucleic acids that served as templates for protein synthesis. Experiments show that such sacs can form spontaneously. For example, Sidney Fox heated amino acids under dry conditions to form protein chains, which he placed in hot water. After the chains were allowed to cool, they self-assembled into small, stable spheres (Figure 19.12a). The spheres were selectively permeable to some substances. They also tended to pick up lipids from the water and so formed a lipid-protein film at their surface. In other experiments, lipids alone self-assembled into small, water-filled sacs that displayed many properties characteristic of cell membranes (Figure 19.12b).

And so self-replicating systems and the membranes required to protect them from the environment could have arisen through spontaneous but inevitable chemical events. Figure 19.12c summarizes the sequence of events that could have led to the first cells.

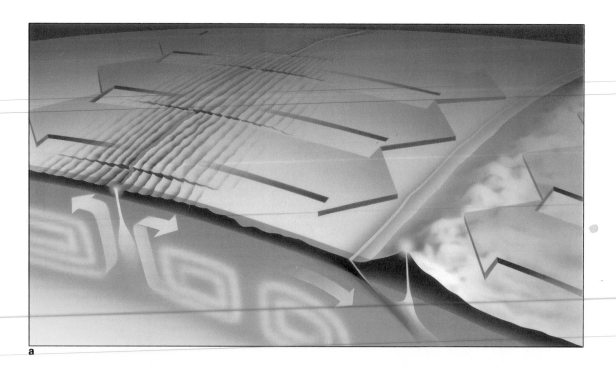

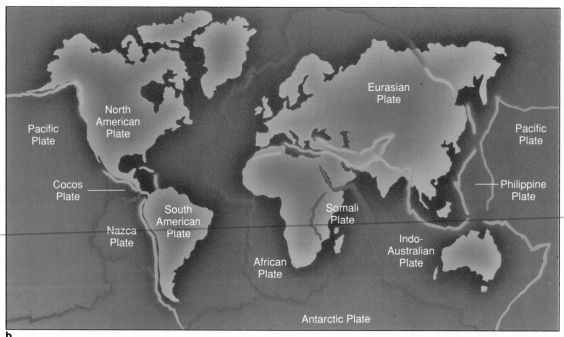

**Figure 19.13** Plate tectonics, the arrangement of the earth's crust in rigid plates that split apart, move about, and collide with one another.

(**a**) The seafloor is slowly spreading from mid-oceanic ridges, measurably displacing the continents away from those ridges. Hot, molten material deep inside the earth slowly wells up and spreads out laterally beneath the crust, much as hot air rises from a stove and spreads out at the ceiling. Oceanic ridges are places where the material has ruptured the crust. As cooled material moves away from the ridges, it acts like a conveyor belt, carrying older continental crust with it.

Similarly, heat builds up beneath the continents, causing deep rifts that eventually cause land masses to split apart. We see this happening today at the Great Rift Valley in eastern Africa.

The major plates of the earth's crust are shown in (**b**). As the plates push against a continental margin, they are often thrust beneath it. The thrusting causes the crumpling and upheavals that have created most major mountain ranges.

## Drifting Continents and Changing Seas

The first living cells did not emerge in a wonderfully stable environment, and things haven't settled down much for their single-celled and multicelled descendants. Even today, living things contend with more than volcanic eruptions, earthquakes, and other local upheavals. They contend with long-term consequences of drifting and colliding continents, with newly forming and disappearing ocean basins, and with bombardments from outer space!

According to the **plate tectonic theory**, the earth's outer layer (lithosphere) is fragmented into slablike plates that float on a hot, plastic layer of the underlying mantle. Driven by a process called seafloor spreading (Figure 19.13), the plates move a few centimeters a year, on the average. The positions of the continents and oceans change as a result of these movements.

For much of earth history, continents have drifted and collided to form supercontinents, which later split open at deep rifts that eventually became new ocean basins. During the Paleozoic, for example, an early continent (Gondwana) drifted southward from the tropics, across the south polar region, then northward. Later, Gondwana and other land masses crunched together to form Pangea. This single world continent extended from pole to pole, and an immense ocean spanned the rest of the globe. Pangea started breaking up during the Mesozoic, and the drifting and collisions among its fragments continue today.

*All such changes in land masses, shorelines, and oceans had profound effects on the evolution of life.* Early on, when land masses were widely dispersed, many populations of marine organisms lived in isolation from others. Speciation was favored, especially in warm coastal waters. When land masses collided, shorelines disappeared and volcanic activity was intense. Marine habitats were lost and species diversity tended to decline. Also, with each major shift in land masses, warm or cold ocean currents shifted directions and climates changed drastically. Immense glaciers formed many times and tied up great volumes of ocean water, so shallow seas disappeared. When the glaciers melted, continental margins were flooded. Once again there were repercussions for the evolution of life on land and in the seas.

**Figure 19.14** Two interpretations of how morphological diversity develops in a lineage. Each vertical line represents a single species. In a *gradual* model (**a**), changes occur more or less steadily but at different rates among species. In a *punctuational* model (**b**), rapid morphological change is associated with speciation, as indicated by the horizontal lines that signify branch points, and the morphologies of established species remain constant through time.

As if these planetary insults were not enough, asteroids or some other huge extraterrestrial objects have repeatedly struck the earth's surface. Some of the impacts irreversibly changed the course of biological evolution.

## Speciation and Rates of Change

So far, we have considered how life may have originated on a changing geologic stage. We turn now to the question of how its subsequent branchings occurred through time.

In the evolutionary view, all species are related by way of descent. The evolutionary relationships among groups of species can be depicted as a "family tree," in which each twig or branch is a single line of descent (a lineage). The branch points are speciation events. Microevolutionary processes, including mutation, genetic drift, and natural selection, were responsible for the branchings. But were those processes operating more or less steadily within lineages, or were their effects most pronounced during speciation events?

Take a look at Figure 19.14, which shows two ways to interpret *rates* of morphological change in an evolving lineage. According to the traditional view, termed the **gradualistic model,** the branchings on family trees diverged gradually, with each new species emerging through many small changes in form over long spans of time. Some cases of gradualism are well documented. One of the most compelling is the fossil sequence for

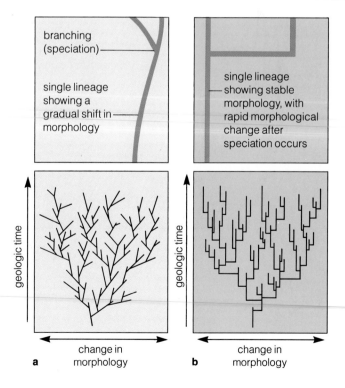

branching (speciation)

single lineage showing a gradual shift in morphology

single lineage showing stable morphology, with rapid morphological change after speciation occurs

geologic time

geologic time

**a** change in morphology

**b** change in morphology

foraminiferans, a type of single-celled eukaryote. Continuous layers of certain deep-sea sediments clearly show gradual changes in form in the foraminiferan lineage. Other examples, including the much-repeated "gradual" evolution of the modern horse, have not held up under close examination.

According to an alternative model, most morphological changes took place rapidly *during* speciation. Each species underwent a spurt of changes in form when it first branched from the parental lineage, then changed little for the rest of its duration on earth—for 2 million to 6 million years, on the average. Compared to that time span, the hundreds or thousands of years required for speciation would be a short period. Yet founder effects, bottlenecks, and strong directional selection could accomplish a great deal of genetic change in only a thousand years. Thereafter, stabilizing selection could maintain the traits of a well-adapted species within relatively narrow limits. This is the thrust of the **punctuational model**.

The punctuational model is consistent with the observation that there are few transitional forms between closely related species. It also is consistent with the observation that most existing species are clearly separate, with "morphological gaps" between them. Thus, for example, we see no intermediate forms between lions and tigers, ocelots and pumas, and other species of cats.

Probably neither model alone is enough to explain the whole history of life. Lineages vary in terms of when their member species originated, how rapidly those species changed in form, and how long they persisted on the evolutionary stage. Certain lineages dribbled along, producing a species here, losing a species there, for tens of millions of years. Others filled the environment with spectacular bursts of evolutionary activity, called adaptive radiations. And some entire lineages came abruptly to an end during episodes of mass extinctions. Let's take a look at the nature of these important macroevolutionary events.

| Table 19.1 | Examples of the Estimated Average Durations of Species | |
|---|---|---|
| | Group | Duration (millions of years) |
| Protistans: | | |
| | Foraminiferans | 20–30 |
| | Diatoms | 25 |
| Plants: | | |
| | Bryophytes | 20+ |
| | Higher plants | 8–20+ |
| Animals: | | |
| | Bivalves | 11–14 |
| | Gastropods | 10–13.5 |
| | Graptolites | 2–3 |
| | Ammonites | 1–2, 6–15 |
| | Trilobites | 1+ |
| | Beetles | 2+ |
| | Freshwater fishes | 3 |
| | Snakes | 2+ |
| | Mammals | 1–2+ |

From Stanley, 1985.

## Extinctions and Adaptive Radiations

**Extinctions**. Some number of species within a lineage inevitably disappear as local conditions change. The rather steady rate of their disappearance over time is called "background extinction." Table 19.1 lists some examples of this. In contrast, a **mass extinction** is an abrupt rise in extinction rates above the background level. It is a catastrophic, global event in which not just one species but major groups of species are wiped out simultaneously.

Major groups tend to survive episodes of mass extinction when their members are widely dispersed in different regions. The ones hit hardest generally are adapted for specialized ways of life in tropical regions. Most importantly, luck is the rule of the game during mass extinctions. For example, what *had* been the most adaptive traits may make no difference whatsoever if an asteroid the size of Vermont hits the earth.

**Adaptive Radiations**. In an **adaptive radiation**, a lineage has filled a wide range of habitats with new species through bursts of microevolutionary activity. Adaptive radiations are common during the first few million years following a mass extinction. For example, most orders of living mammals appeared during the 12 million years after the mass extinction that serves as the boundary between the Mesozoic and Cenozoic eras (Figure 19.15). However, radiations may occur whenever there are unfilled adaptive zones. *Adaptive zones* are most easily defined as ways of life, such as "burrowing in the sea floor" or "catching insects in the air at night."

Before a lineage can successfully occupy an adaptive zone, it must have physical, ecological, and evolutionary access to it. Consider first an example of *physical* access. All mammalian species are at least distantly related, owing to a common origin. Mammals originally were small insect eaters that were spread out through a rather uniform tropical environment on a single world continent (Pangea). Some mammals living in the tropics today have changed little from their forerunners. However, some populations became physically isolated from others on different land masses after Pangea split apart. The resulting changes in land formations, climate, and food resources, as well as variations in the composition of species that ended up by chance on each land mass, set the stage for independent radiations.

Next consider *evolutionary* access, which may result when a key innovation develops in a species. A **key innovation** is a modification in structure or function that permits individuals of a species to exploit the environment in an improved or novel way. For example, the modification of forelimbs into wings opened new adaptive zones to the ancestors of birds and bats. Finally, *ecological* access can occur if an adaptive zone is unoccupied or if the invading species can outcompete the resident species. We will have more to say about this in Chapter 45.

We turn now to a brief version of the history of life, beginning with the overview shown in Figure 19.16. This figure correlates milestones in the evolution of life with key events in the evolution of the earth.

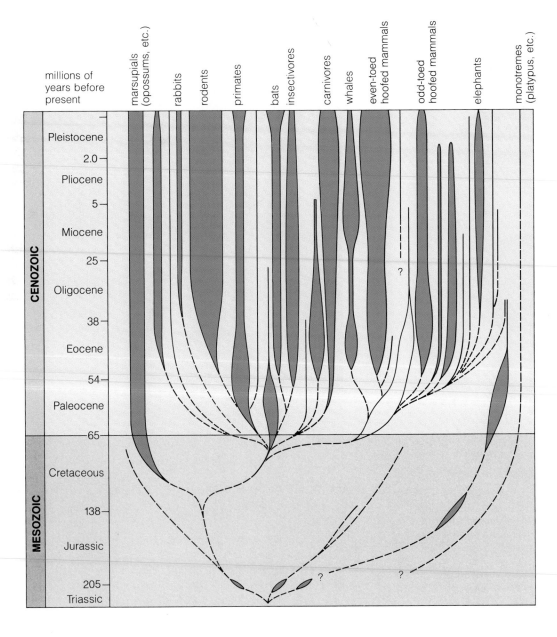

**Figure 19.15** The great adaptive radiation of mammals in the first 10–12 million years of the Cenozoic Era. This radiation is thought to have resulted from the invasion of adaptive zones vacated by dinosaurs at the close of the Mesozoic.

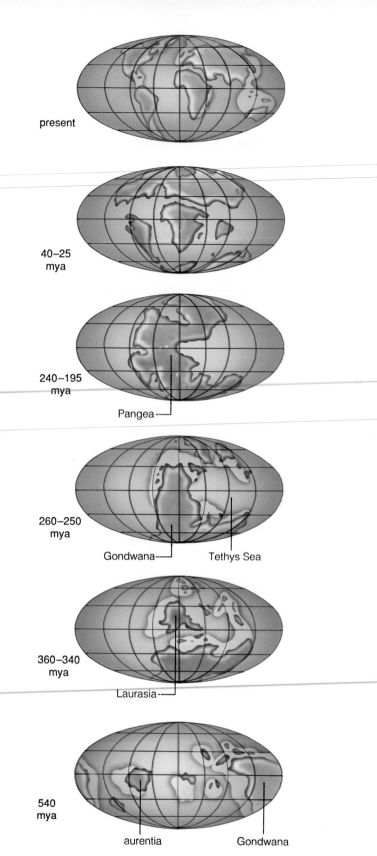

present

40–25 mya

240–195 mya

Pangea

260–250 mya

Gondwana — Tethys Sea

360–340 mya

Laurasia

540 mya

aurentia          Gondwana

**Figure 19.16**   Major events in the evolution of the earth and its organisms. The time spans of the different eras are not to scale; if they were, the chart would run off the page. Think of the spans as minutes on a clock, with life originating at midnight. The Paleozoic would not even begin until 10:04 a.m. The Mesozoic would begin at 11:09 a.m., the Cenozoic at 11:47 a.m., and the Recent epoch of the Cenozoic would begin during the last 0.1 *second* before noon.

| Era | Period | Epoch | Millions of Years Ago (mya) | |
|---|---|---|---|---|
| CENOZOIC | Quaternary | Recent | 0.01- | |
| | | Pleistocene | 1.65 | |
| | Tertiary | Pliocene | 5 | |
| | | Miocene | 25 | |
| | | Oligocene | 38 | |
| | | Eocene | 54 | |
| | | Paleocene | 65 | |
| MESOZOIC | Cretaceous | Late | 100 | |
| | | Early | 138 | |
| | Jurassic | | 205 | |
| | Triassic | | 240 | |
| PALEOZOIC | Permian | | 290 | |
| | Carboniferous | | 360 | |
| | Devonian | | 410 | |
| | Silurian | | 435 | |
| | Ordovician | | 505 | |
| | Cambrian | | 550 | |
| PROTEROZOIC | | | 2,500 | |
| ARCHEAN | | | | |

**Range of Global Diversity**
**(marine and terrestrial)**

**Times of Major Geologic and Biological Events**

1.65 mya to present. Major glaciations. Modern humans emerge and begin what may be greatest **mass extinction** of all time on land, starting with Ice Age hunters.

65-1.65 mya. Unprecedented mountain building as continents rupture, drift, collide. Major climatic shifts; vast grasslands emerge. Major **radiations** of flowering plants, insects, birds, mammals. Origin of earliest human forms.

65 mya. Asteroid impact? **Mass extinction** of all dinosaurs and many marine organisms.

135-65 mya. Pangea breakup continues, broad inland seas form. Major **radiations** of marine invertebrates, fishes, insects, dinosaurs. Origin of angiosperms (flowering plants).

181-135 mya. Pangea breakup begins. Rich marine communities. Major **radiations** of dinosaurs.

205 mya. Asteroid impact? Mass extinction of many organisms in seas, some on land; dinosaurs, mammals survive.

240-205 mya. Recovery, **radiations** of marine invertebrates, fishes, dinosaurs. Gymnosperms the dominant land plants. Origin of mammals.

240 mya. **Mass extinction.** Nearly all species in seas and on land perish.

280-240 mya. Pangea, worldwide ocean form; shallow seas squeezed out. Major **radiations** of reptiles, gymnosperms.

360-280 mya. Tethys Sea forms. Recurring glaciations. Major **radiations** of insects, amphibians. Spore-bearing plants dominate; gymnosperms present. Origin of reptiles.

370 mya. **Mass extinction** of many marine invertebrates, most fishes.

435-360 mya. Laurasia forms, Gondwana moves north. Vast swamplands, early vascular plants. **Radiations** of fishes continue. Origin of amphibians.

435 mya. Glaciations as Gondwana crosses South Pole. **Mass extinction** of many marine organisms.

500-435 mya. Gondwana moves south. Major **radiations** of marine invertebrates, early fishes.

550-500 mya. Land masses dispersed near equator. Simple marine communities. Origin of animals with hard parts.

700-550 mya. Supercontinent Laurentia breaks up; widespread glaciations.

2,500-570 mya. Oxygen present in atmosphere. Origin of aerobic metabolism. Origin of protistans, algae, fungi, animals.

3,800-2,500 mya. Origin of photosynthetic bacteria.

4,600-3,800 mya. Formation of earth's crust, early atmosphere, oceans. Chemical evolution leading to origin of life (anaerobic bacteria).

4,600 mya. Origin of earth.

As you study Figure 19.16, keep in mind that it is only a generalized picture of life's history. It shows the five greatest mass extinctions—but there were many others in between. It shows the shrinking and expanding range of species diversity for all the major groups combined. But within that overall pattern, each major group has its own distinctive history of extinctions and radiations (Figure 19.17). With these qualifications in mind, let's discuss briefly the macroevolutionary trends that dominated the five eras of geologic time—the Archean, Proterozoic, Paleozoic, Mesozoic, and Cenozoic eras. This discussion will set the stage for the next unit of the book, which describes the history and current range of diversity for all five kingdoms of organisms.

## The Archean and Proterozoic Eras

Until about 3.7 billion years ago, the earth's crust was highly unstable. Even after somewhat rigid land masses had formed, they were fringed with active volcanoes. Yet rocks 3.5 billion years old contain fossils of well-developed prokaryotic cells that probably lived in tidal mud flats. Those cells resembled the simple bacteria found today in mud flats, bogs, and pond mud where oxygen is absent. Fermentation, the most common anaerobic pathway, was probably the first to evolve. Before the close of the Archean era, the first photosynthetic bacteria had evolved.

Between 2.5 billion and 700 million years ago, photosynthetic bacteria dominated the shallow seas. Mats of photosynthetic populations and sediments that became trapped in them accumulated one atop the other, forming the curiously shaped structures called stromatolites (Figure 19.18). Oxygen, a by-product of photosynthesis, accumulated in the atmosphere during the Proterozoic. It had two irreversible effects on the course of evolution. First, the abundance of free oxygen prevented further abiotic synthesis of organic compounds; *the chemical origin of life would be a one-time event.* Second, the oxygen-

rich atmosphere was an adaptive zone of global dimensions. In some bacterial lineages, *metabolic machinery became modified in ways that permitted aerobic respiration.* This key innovation foreshadowed the evolution of multicelled plants, fungi, and animals. By 1.2 billion years ago, green algae had evolved. Rock formations in Australia, dated at 900 million years, have yielded fossils of at least fifty-six species of green and red algae, plant spores, and fungi.

About 800 million years ago, stromatolites began to decline dramatically. Tiny grazing animals had appeared on the evolutionary scene, and the stromatolites did not respond well to grazing. Small, soft-bodied animals also made their appearance about this time, at the margins of a supercontinent (Laurentia). They were not blessed with abundant nutrients, and their tracks and burrows suggest that they were scavengers of what little was available. Nevertheless, they were the likely point of departure for a burst of evolutionary activity that spanned the Proterozoic–Paleozoic boundary and produced the world's first predators.

## The Paleozoic Era

The Paleozoic is divided into the Cambrian, Ordovician, Silurian, Devonian, Carboniferous, and Permian periods (Figure 19.16). During the transition from Precambrian to Cambrian times, the supercontinent broke up. The newly separated land masses straddled the equator, and warm, shallow waters lapped at their margins. Geographical conditions did not favor pronounced seasonal changes in winds, sea surface temperatures, and ocean currents, so there could not have been much seasonal variation in the supplies of nutrients. Using existing tropical seas as our model, we can say that supplies were probably stable but limited.

Early Cambrian organisms had truly flattened bodies. *Dickensonia,* for example, was pancake-thin yet could grow to about the same diameter as a manhole cover (Figure 19.19a). Flattened body plans are highly adaptive when environmental concentrations of nutrients are low, for they have a favorable surface-to-volume ratio for nutrient uptake. Not surprisingly, most of the early Cambrian animals lived on or just beneath the seafloor, where organic debris and suspended particles were most readily available.

**Figure 19.17** A limited representation of the evolutionary histories of organisms. Notice the diverse patterns of radiation and extinction for the sample shown here. Notice also the spectacular current success of the insects. The widths of the lineages, all shown in blue, represent the approximate numbers of *families* in the case of plants and animals.

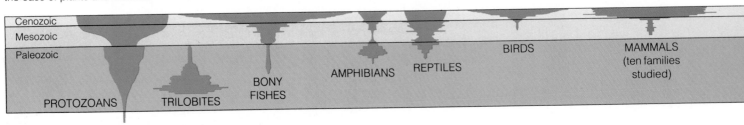

**Figure 19.18** (a) One of the oldest known fossils, dated at 3.5 billion years old. It is a filament formed of walled cells. (b) Stromatolites in Western Australia, which started to form between 2,000 and 1,000 years ago in shallow seawater. Calcium deposits preserved their structure. They are identical to stromatolites more than 3 billion years old.

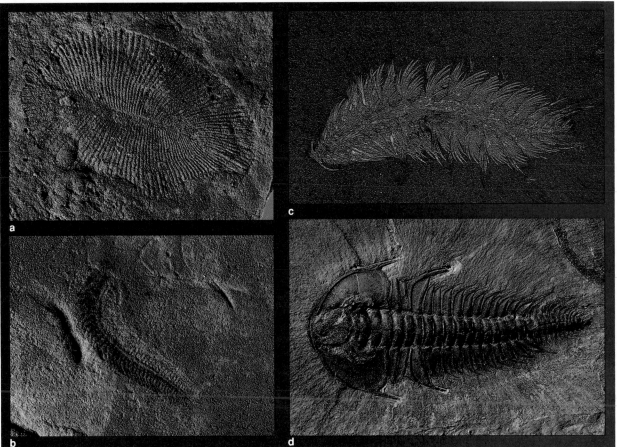

**Figure 19.19** (a, b) From Australia's Ediacara Hills, fossils of two early animals, about 600 million years old. (c) From the Burgess Shale of British Columbia, a remarkably well-preserved fossil of a marine worm of Cambrian times. Notice the bundles of fine bristles that may have functioned in locomotion. (d) One of the earliest Cambrian trilobites.

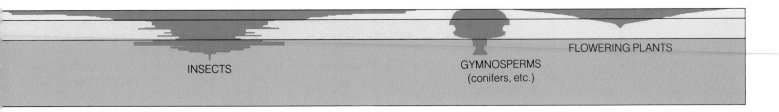

Diverse animals representing most of the major phyla evolved in rather short order in the Cambrian seas. James Valentine and Douglas Erwin argue that genetic flexibility gave rise to the abrupt increase in diversity. Diversity is an outcome of genetic variation, and that variation arises through mutations. Perhaps in the comparatively simple Cambrian animals, most mutations would not

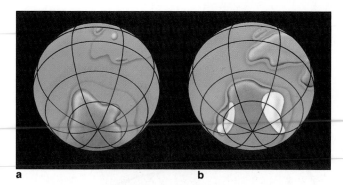

**a**               **b**

**Figure 19.20** Plate tectonics and global changes in climate. Gondwana drifted southward during the Ordovician and vast, warm, shallow seas inundated most of the land. This was a time of major adaptive radiations in marine environments. Late in the Ordovician, Gondwana straddled the pole (**a**), and within 5 million years, huge ice sheets had formed over it, locking up enough ocean water to cause a drop in sea level. (**b**) Colder waters moved into the tropics and large land areas were drained, with dire consequences for many families of marine organisms.

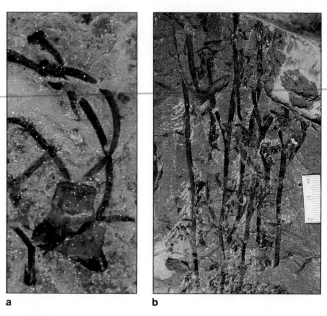

**a**               **b**

**Figure 19.21** Fossils of plants that pioneered the invasion of land during Silurian times. (**a**) Stems of the oldest known plant, *Cooksonia*, were less than 7 centimeters tall. (**b**) Fossils of a type of Devonian plant (*Psilophyton*) that may have been the earliest ancestors of seed-bearing plants.

have been as harmful as they usually are in existing animals, for the functions of genes governing developmental programs were not as intertwined as they are today.

Besides this, the extensive shorelines around the different equatorial land masses were separate and largely unoccupied adaptive zones. Communities developed, with different plants and animals becoming linked through increasingly different types of predators and prey. Besides fossils of obvious predators, we find fossils of animals with chunks and holes in the body—and the ones with wounds healed over tell us that the damage was not inflicted after the animal died. The relatively abrupt appearance of armorlike shells, protective spines, and nightmarish mouths and feeding appendages tells us that things were starting to get lively.

Late in the Cambrian, for as yet unknown reasons, the warm, oxygen-rich waters of the shallow seas started mixing abruptly with cold, oxygen-poor waters from the deep oceans. Until then, animals called trilobites were common (Figure 19.19d), but now they were nearly wiped out. Trilobite populations adapted to cold water at the fringes of continents survived—but those in the shallow seas were devastated.

During the Ordovician, one of the continents (Gondwana) drifted southward and seas flooded more of the land. Major adaptive radiations were favored as new and vast, shallow marine environments opened up. Reef organisms flourished, and fast-swimming predators—the nautiloids—dominated the evolutionary stage. By the late Ordovician, Gondwana straddled the South Pole. Immense glaciers formed on its surface and the shallow seas were drained (Figure 19.20). This first ice age triggered the first global mass extinction; reef life everywhere collapsed.

Gondwana drifted northward during the Silurian and on into the Devonian. Reef organisms recovered, and there was a major radiation of predatory fishes equipped with armor plates and massive jaws. The invasion of land was beginning rather inconspicuously. Small stalked plants became established along the muddy margins of the land (Figure 19.21). Lobe-finned fishes ancestral to amphibians were starting to move onto land. Like many existing fishes of stagnant waters, they were able to breathe air (Figure 19.22).

Another global mass extinction occurred at the Devonian–Carboniferous boundary, when sea levels swung dramatically. Nearshore species were washed out to sea and became entombed in deep ocean sediments, and huge boulders broke loose from cliffs along the coasts. The cause of the Devonian extinction is not known.

Major adaptive radiations of plants and insects occurred during the Carboniferous, when land masses were submerged and drained repeatedly. As described on page 34, the accumulating organic debris became

**Figure 19.22** (**a**) Reconstruction of lobe-finned fishes, venturing onto a Devonian shoreline. Those fast-swimming predators probably gulped air at the surface of water that had become shallow and stagnant. Over the course of several million years, the lobed fins in some lineages evolved into limbs. Such fishes were the ancestors of amphibians, reptiles, birds, and mammals. (**b**) A modern-day catfish making an excursion onto land. Catfish simultaneously thrash sideways and use their fins to pull the body forward.

compacted into the extensive coal deposits of Britain and of the Appalachians of North America.

Insects, amphibians, and early reptiles flourished in the vast swamp forests of the Permian era. The forests themselves included cycads, ginkgos, and the ancestors of the modern conifers. As the Permian drew to a close, nearly all known species on land and in the seas perished in the greatest of all mass extinctions. This happened at a time when all land masses collided to form a vast continent (Pangea) surrounded by a single world ocean. Changes in the distribution of water, land area, and land elevation had catastrophic effects on world temperature and climate. To the north, arid lowlands and humid uplands emerged. To the south, near the pole, glaciers advanced over the land. The reptilian ancestors of dinosaurs, birds, turtles, crocodiles, snakes, and lizards survived. Those predatory animals met the challenge of drastic swings in climate.

### The Mesozoic Era

The Mesozoic lasted about 175 million years. It is subdivided into Triassic, Jurassic, and Cretaceous periods. The most ecologically specialized lineages had been hit hardest during the Permian extinction, particularly tropical reef communities. And now recovery occurred through adaptive radiations of new lineages. This time the increase in diversity did not level off, as it did in Paleozoic times. It continued to increase to the present day.

The major radiation of reptiles had ended abruptly before the Mesozoic, but divergences in a few lineages now gave rise to mammals and some wonderful reptilian "monsters," including the dinosaurs.

Many marine organisms perished during a mass extinction of unknown causes at the Triassic–Jurassic boundary. On land, some dinosaurs and small mammals were among the survivors. The supercontinent Pangea began to break up during the Jurassic and early Cretaceous (Figure 19.16). This was the time when dinosaurs emerged as rulers of the evolutionary stage. Did they die out with a whimper or a bang at the close of the Mesozoic? This question is being hotly debated (see the *Commentary* on page 312).

Dinosaurs so dominate our thinking about the Mesozoic that we often overlook one of the most spectacular events of all. Early in the Cretaceous, flowering plants emerged and underwent a major radiation in the evolutionarily short space of 10 million years. In most habitats they overwhelmed the already declining gymnosperms (conifers and their relatives). The radiation of flowering plants continues today. It is being fueled by a speciation rate exceeding that of any other plant lineage.

## The Dinosaurs—A Tale of Global Impacts, Radiations, and Extinctions

By 200 million years ago, near the close of the Triassic, the first dinosaurs had evolved. They were not much larger than a wild turkey. Most had high metabolic rates—some researchers say they were warmblooded—and many ran swiftly on two legs. The Triassic dinosaurs were not the dominant lineage on land. Center stage belonged to *Lystrosaurus* and other plant-eating, mammal-like reptiles that were too large to be bothered by most predators. Then *Lystrosaurus* ran out of luck and adaptive zones opened up for the dinosaurs, perhaps when an extraterrestrial object struck the earth. There is a crater about the size of Rhode Island in central Quebec (Figure *a*), and even if it is not *the* impact crater, it certainly tells us that huge impacts have occurred.

The blast wave and resulting firestorm, earthquakes, and possibly lava flows from an impact that size would have been extremely destructive. The atmospheric distribution of rocks and water vaporized during the impact could have caused global darkening, and would have been followed by months or years of acid rain. The animals that survived this time of extinction and later ones tended to be small and metabolically active—and less vulnerable to long-term swings in climate, whatever their cause.

The surviving dinosaurs underwent a major adaptive radiation. Over the next 140 million years, their descendants were the ruling reptiles (Figure *b*). Some, including

**a** Manicougan crater, Quebec—an example of earth's past encounters with extraterrestrial objects.

**b** (*Right*) And how might the small, unobtrusive mammals of this time ever have ventured out from under the shrubbery if dinosaurs had not disappeared? Would *you* even be here today?

the "ultrasaurs," reached monstrous proportions. They weighed 70 tons and towered 50 feet (15 meters) above the ground. Many dinosaurs perished during a mass extinction at the end of the Jurassic, then during a pulse of extinctions during the Cretaceous. Each time, some lineages recovered and new forms replaced them on the evolutionary stage. By the late Cretaceous, there were perhaps a hundred different genera of dinosaurs. Duckbilled dinosaurs appeared in forests and swamps. Tanklike *Triceratops* and other plant eaters flourished in more open regions. They were prey for the agile, fearsomely toothed *Tyrannosaurus rex*.

The final blow came at the Cretaceous–Tertiary (K–T) boundary. Perhaps long-term changes in climate or other planetary events had been winnowing out some of the dinosaurs for a few million years before then. Or perhaps a shower of comets bombarded the earth during that same time span. Whatever the cause, there was a pulse of extinctions culminating with the disappearance of all (or nearly all) of the remaining dinosaurs.

The impact associated with the K–T boundary probably was stupendous (Figure *c*). A thin, worldwide layer of iridium-rich rock has been dated precisely at that boundary. Iridium is rare on the earth's surface but common in asteroids. A few asteroids still remain in earth-crossing orbits, and perhaps there was one more 63 million years ago than there is now. The thin, iridium-rich layer may have formed from dust settling after a collision.

So where is the crater? A leading candidate is a huge underground ring recently discovered in the northern Yucatan peninsula of Mexico. Based on gravity maps, iridium concentration in soil samples, and other evidence, the circular formation may be the impact site. It is about 10 kilometers in diameter. An asteroid of that size slamming against the earth would have been the equivalent of an explosion of a billion tons of TNT, certainly a cataclysmic force by any standards. And the ring has been dated to the K–T boundary.

**c**  An artist/astronomer's interpretation of what might have happened during the last few minutes of the Cretaceous.

Even more exotic forces may be at work in and around the solar system. There is some evidence that mass extinctions have occurred at regular intervals of 26–30 million years. Were dinosaurs and other major lineages the victims of a cosmic timetable? Consider that comets and other debris form a dense cloud at the fringes of our solar system. Does the orbit of an unknown star swing near that cloud to the extent that comets are hurled toward the earth?

Intriguingly, 2.3 million years ago, a huge object from space hit the Pacific Ocean. About the same time, vast ice sheets started forming abruptly in the Northern Hemisphere. Long-term shifts in climate may have been ushering in this most recent ice age, but a global impact might have accelerated the process. (Water vaporized during the impact would have formed a cloud cover that prevented sunlight from reaching the earth's surface.) By that time, the early ancestors of humans were on the evolutionary stage, and the extreme shift in climate surely put their adaptability to the test (Chapter 21). The formation of ice sheets following the global impact is one more bit of information that compels us to look skyward, also, in our attempts to piece together the evolutionary story of life.

### The Cenozoic Era

The world's land masses were already undergoing major reorganization at the dawn of the modern era, the Cenozoic. Lava poured through faults and fissures in the earth's crust. Coastlines fractured; intense volcanic activity and uplifting produced mountains along the margins of massive rifts and plate boundaries. The Alps, Andes, Himalayas, and Cascades were born through these upheavals (Figure 19.23).

As the continents assumed their current configurations, major shifts in climate affected the further evolution of life. For example, vast, semiarid, cooler grasslands emerged—new adaptive zones into which plant-eating mammals and their predators radiated (Figure 19.24).

Today the continents are dispersed to nearly their maximum separation, and they intersect the tropics and confine the polar regions. These conditions have favored a period of unparalleled richness in species diversity. The vast tropical forests of South America, Africa, and Southeast Asia may well be the richest ecosystems ever to appear on earth. And the marine ecosystems of the island chains of the tropical Pacific—remnants and omens of tectonic movements—are probably not far behind.

This is the geologic stage for what may turn out to be the greatest mass extinction of all time. Beginning about 50,000 years ago, early humans started following the migrating herds of wild animals around the Northern Hemisphere. Within a few thousand years, major groups

**Figure 19.23** The view northward across Crater Lake in Oregon, a collapsed volcanic cone aligned with Cascade volcanoes reaching into Washington—and all paralleling the Pacific Coast. These formations are testimony to the violent upheavals that began in Jurassic times.

**Figure 19.24** (**a**) Scene from the Eocene, showing small, four-toed horses (*Orohippus*) and a much larger herbivore (*Uintatherium*). (**b**) Scene from the Pleistocene, reconstructed on the basis of fossils recovered from a pitch pool at Rancho La Brea, California. Shown here, the saber-tooth cat (*Smilodon*), a large bird (*Teratornis*), and an extinct species of horse.

a

b

of large mammals, including the woolly mammoth, had disappeared. The pace of extinction has been accelerating ever since, as humans hunt animals for food, fur, feathers, or fun; as they destroy habitats to clear land for cattle or crops. The repercussions are global in scope. We will return to this topic in Chapter 48.

## SUMMARY

1. Because evolution proceeds by modifications of already existing species, there is a continuity of relationship among all species. The large-scale patterns, trends, and rates of change among groups of species (higher taxa) over time are called macroevolution.

2. Evidence of evolutionary relationships comes from the fossil record and earth history, radioactive dating methods, comparative morphology, and comparative biochemistry. The evidence reveals similarities and differences in form, function, behavior, and biochemistry.

3. The fossil record varies as a function of the kinds of organisms represented, where they lived, and the geologic stability of the region since the time of fossilization. Together with the geologic record, it shows that the evolution of life has been linked, since the time of its origin to the present, to the evolution of the earth.

4. Comparative morphology often reveals similarities in embryonic development stages that indicate evolutionary relationships. It shows evidence of divergences and convergences in body parts among certain major groups.

5. Comparative biochemistry is based on mutations that have accumulated in the DNA of different species. The mutations serve as a molecular clock for dating the time of divergence between two species from a common ancestor. Comparisons of proteins and DNA from different species reveal similarities and differences among them.

6. Over geologic time, tectonic movements of the earth's crust as well as bombardments by extraterrestrial objects have caused profound shifts in land masses, shorelines, and oceans. The course of biological evolution was redirected many times by these events.

7. Family trees may be constructed to show evolutionary relationships. The branches of such trees are lines of descent (lineages). The branch points are speciation events, brought about by mutation, natural selection, and other microevolutionary processes.
   a. In the gradualistic model, most morphological change occurred gradually within species.
   b. In the punctuational model, most morphological change occurred during the times of speciation, with little change thereafter.
   c. Probably neither model alone explains the history of life. Lineages vary in terms of when their member species originated, how rapidly morphological change occurred, and how long their species have persisted through time.

8. Two important macroevolutionary trends are called mass extinction (a catastrophic, global event in which major groups of species perish abruptly) and adaptive radiation (bursts of speciation events by which a lineage fills the environment with new species). Highly specialized organisms confined to tropical regions tend to be hit hardest by mass extinction, but luck is the most important factor here.

### Review Questions

1. What is the difference between microevolution and macroevolution? *293*

2. What factors have influenced the fossil record? *294–295*

3. Explain the difference between:
   a. homologous and analogous structures *296–297*
   b. morphological divergence and convergence *296–297*

4. How is an adaptive radiation defined, and what criteria must be met before a radiation can occur? *304–305*

5. Describe the chemical and physical characteristics of the earth 4 billion years ago. How do we know what it was like? *299–300*

6. Describe the experimental evidence for the spontaneous origin of large organic molecules, the self-assembly of proteins, and the formation of organic membranes and spheres, under conditions similar to those of the early earth. *300–301*

7. How does continental drift occur, and in what ways does this process influence changes in biological communities? *303*

8. When did plants and vertebrates invade the land? *307*

9. The Atlantic Ocean is widening, and the Pacific and Indian oceans are closing. Write a short essay on the possible biological consequences of the forthcoming formation of a second Pangea.

### Self-Quiz (Answers in Appendix IV)

1. Large-scale patterns, trends, and rates of changes in *groups of species* over time are called _____.

2. Two trends have changed the course of biological evolution repeatedly; they are _____ and _____.

3. The rate of evolution has been _____ in some lineages but in others it has been much more rapid with _____ of specia-

tion events followed by lengthy periods with little evidence of change.

4. The fossil record of evolution correlates with evidence from _____ .
   a. the geologic record
   b. radioactive dating
   c. comparative morphology
   d. comparative biochemistry
   e. all of the above

5. Comparative biochemistry _____ .
   a. is based mainly on the fossil record
   b. often reveals similarities in embryonic development stages that indicate evolutionary relationship
   c. is based on mutations that have accumulated in the DNA of different species
   d. compares the proteins and the DNA from different species to reveal relationships
   e. both c and d are correct

6. _____ reveals mutations that have accumulated in the DNA of different species.
   a. Fossil evidence
   b. Embryonic development
   c. DNA hybridization
   d. Comparative morphology

7. Comparative morphology _____ .
   a. is based mainly on the fossil record
   b. often reveals similarities in embryonic development stages that indicate evolutionary relationship
   c. shows evidence of divergences and convergences in body parts among certain major groups
   d. compares the proteins and the DNA from different species to reveal relationships
   e. both b and c are correct

8. Through study of the geologic record, we know that the evolution of life is linked to _____ .
   a. tectonic movements of the earth's crust
   b. bombardment of the earth by terrestrial objects
   c. profound shifts in land masses, shorelines, and oceans
   d. physical and chemical evolution of the earth
   e. all of the above are correct

9. Match the era with the events listed.
   _____ Archean
   _____ Proterozoic
   _____ Paleozoic
   _____ Mesozoic
   _____ Cenozoic

   a. dinosaurs, flowering plants, mammals
   b. formation of earth's crust, oceans, early atmosphere; chemical origin of life
   c. major radiations of flowering plants, insects, birds, mammals; emergence of human species
   d. origin of protistans, aquatic plants, fungi, animals
   e. origin of amphibians, spread of early vascular plants, formation of Pangea

## Selected Key Terms

| | |
|---|---|
| adaptive radiation *304* | mass extinction *304* |
| analogous structure *297* | Mesozoic Era *295* |
| Archean Era *295* | morphological convergence *297* |
| Cenozoic Era *295* | |
| comparative morphology *295* | morphological divergence *296* |
| fossil *294* | neutral mutation *298* |
| gradualistic model *303* | Paleozoic Era *295* |
| homologous structure *296* | plate tectonics *303* |
| lineage *295* | Proterozoic Era *295* |
| macroevolution *293* | punctuational model *304* |

## Readings

Bambach, R., C. Scotese, and A. Ziegler. 1980. "Before Pangea: The Geographies of the Paleozoic World." *American Scientist* 68(1): 26–38.

Cech, T. November 1986. "RNA as an Enzyme." *Scientific American* 255(5):64–75.

Futuyma, D. 1986. *Evolutionary Biology*. Second edition. Sunderland, Massachusetts: Sinauer.

Gore, R. June 1989. "The March Toward Extinction." *National Geographic* 175(6):662–699.

Grieve, R. April 1990. "Impact Cratering on the Earth." *Scientific American* 262(4):66–73.

Margulis, L. 1982. *Early Life*. Boston: Science Books International. Easy-to-read introduction to the origin and evolution of prokaryotes and eukaryotes. Paperback.

Simpson, G. 1983. *Fossils and the History of Life*. New York: Scientific American Books.

Stanley, S. 1987. *Extinction*. New York: Scientific American Books.

## Pandas in the Tree of Life

Probably for as long as they have been talking to each other, humans have had the urge to name living things and lump similar ones together. That urge nearly went out of control when Europeans took it upon themselves to start sailing around the world in the sixteenth century. Suddenly, descriptions of thousands upon thousands of newly discovered species started pouring in from the globetrotting explorers of Asia, Africa, the Pacific islands, and the New World. Today we might smile at the nearly useless cataloguing by Thomas Moufet and others who attempted to make sense of the bewildering news (page 262). Yet we still have a long way to go in deciding where all the known species fit in the tree of life.

Consider the giant panda, a plant-eating animal. Unlike antelope, deer, or cattle, the panda does a poor job of digesting the tough cellulose fibers in plants. As mentioned on page 640, antelope pummel plant material in their elaborately chambered gut, and they benefit from symbiotic gut-dwelling bacteria that produce cellulose-digesting enzymes for them. The panda does not have the gut of a plant eater. It has the same type of gut as a meat eater. Yet it eats only bamboo—and huge amounts of it—to get enough carbohydrates, proteins, and fats. Panda paws are rounded, with short toes. These are not much good at holding and stripping leaves from bamboo stalks. Yet the panda also has a thumblike bony digit on each front paw. When the panda uses a "thumb" in opposition to the five toes on one of its paws, a bamboo salad is a little easier to deal with.

With its peculiar assortment of traits, the giant panda is not easy to classify. Is it related to bears, which it resembles, or to raccoons? The giant panda shares physical and behavioral features with the red panda, although the red panda has no thumb. Both live in the same general area of China, and both eat bamboo. Yet the red panda *is* remarkably similar in appearance to raccoons (Figure 20.1).

Molecular studies shed light on this question of evolutionary relationship. For example, researchers studied the extent to which single-stranded DNA from the giant panda would hybridize with DNA from the red panda and from bears. The greater the number of mismatched nucleotide bases, the greater the evolutionary distance between two organisms (page 298). The DNA hybridization results reinforced an earlier view that the three kinds of animals are related by descent from a common ancestor that lived more than 40 million years ago. Apparently a divergence occurred among the descendants, with one branch eventually leading to modern raccoons and the red panda, and another to bears. About 15–25 million years ago, a split from the bear lineage put the ancestors of the giant panda on their unique evolutionary road. The giant panda is more closely related to bears than it is to the red panda.

Pandas and bears are only a tiny sampling of the many millions of kinds of organisms that have appeared and disappeared over the past 3.8 billion years. Very few of those alive today have been scrutinized with the new molecular technologies. Yet widely accepted systems of classifying different species have been around since the eighteenth century. At first the categories were based only on perceived similarities and differences in body features. They served to reinforce the prevailing view among scholars that each species was a unique, unchanging kind of organism, locked in place in a great Chain of Being (page 262). In time, the traditional systems became modified to reflect **phylogeny**—the evolutionary relationships among species, starting with the most ancestral forms and including all the branches leading to all of their descendants. Investigation of these evolutionary patterns is at the heart of the field of biological systematics, and the subject of this chapter.

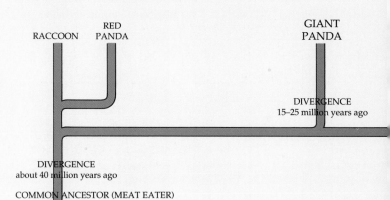

RACCOON  RED PANDA  GIANT PANDA

DIVERGENCE
15–25 million years ago

DIVERGENCE
about 40 million years ago

COMMON ANCESTOR (MEAT EATER)

brown
bear

giant
panda

red
panda

SPECTACLED   SLOTH        SUN      BLACK              POLAR    BROWN
BEAR         BEAR         BEAR     BEARS              BEAR     BEAR

DIVERGENCES
2 million years ago

**Figure 20.1** A long-standing question about relatedness answered—the giant panda, red panda, and their apparent evolutionary relationship to each other, to raccoons, and to bears.

KEY CONCEPTS

**1.** Biological systematics deals with the patterns of life's diversity on two fronts: How diverse organisms are distributed in the space of the environment, and how they are distributed over evolutionary time.

**2.** Systematics encompasses three lines of inquiry: taxonomy, phylogenetic reconstruction, and classification. Taxonomy involves identifying organisms and assigning names to them. Phylogenetic reconstruction involves defining evolutionary patterns that connect different organisms. Classification involves organizing phylogenetic information into retrieval systems in which organisms are grouped in a hierarchy of ever more inclusive levels.

**3.** Identification of species (the units of evolution) may be complicated by sexual, developmental, and environmental variation, as well as by morphological convergences that occurred in the past. Thorough understanding of the biology of organisms is necessary for the correct recognition of species.

**4.** Homologous features—those shared as a result of descent from a common ancestor—are a key to understanding and reconstructing evolutionary patterns.

## SYSTEMATICS

When considering the diversity of life, what sort of assessment will provide us with useful insights? We might simply count the total number of species, but such a tally would not be particularly informative. For example, knowing how many kinds of lizards there are does not tell us anything about the lizards themselves—how various types came to live where they do, and how their characteristics came into being. We might learn far more by employing methods that will allow us to assess the *patterns of diversity* among organisms in space and time. Thus we might hope to discover the relationships between lizard species in a Costa Rican rain forest and in the Gila desert, and between living species of lizards and those that have gone before—maybe all the way back to Jurassic times, when lizards originated. The

male phenotype
*Papilio dardanus*

**Inedible Models:**          **Mimics:**

*Danaus niavius dominicanus*          *P. dardanus* (female)

*Amauris jacksoni*          *P. dardanus* (female)

*Bematistes poggeii*          *P. dardanus* (female)

**Figure 20.2** Sexual dimorphism within populations of the African swallowtail butterfly (*Papilio dardanus*). The males in all populations of *P. dardanus* have yellow and black wings with "tails" at the tips. In most of tropical Africa, the females of *P. dardanus* are conspicuously different in appearance from the males—and from one another.

Variation among the females is associated with *mimicry*, in which one species is deceptively similar in color, form, or behavior to another species that has a selective advantage. One species is the mimic, the other is the model. For example, if a model has warning coloration that identifies it as inedible to potential predators, then a tasty species that mimics its coloration (as does *P. dardanus*) will have a better chance of being left alone.

The wing patterns and coloration of each form of the female mimic those of an inedible species present in the same region. In regions where models are absent, there are no mimicking females. Bird predators are apparently selective agents promoting variation in these populations. Unlike the females, males of *P. dardanus* are not mimics. In terms of individual reproductive success, being recognized by females probably has a greater advantage for the males than does the avoidance of predation by mimicry.

branch of biology called **systematics** deals with such patterns of diversity in an evolutionary context.

Recall, from Chapter 17, that evolution can be viewed as a study of both process and pattern. Microevolution deals primarily with the *process* of evolution—that is, with the origin and success of given genotypes and their associated phenotypes. Macroevolution deals with the large-scale *patterns* of diversity through time. For example, mutation, natural selection, and other microevolutionary processes put the forerunners of reptiles on the evolutionary stage. Then divergences and speciation events put the ancestors of dinosaurs, lizards, and other reptilian groups on separate roads through time.

In large part, reconstructing evolutionary history must be based on observations of living organisms and on the available fossil record. As a practical matter, this means that the study of macroevolution often must be carried out in the absence of specific information about (1) the genetic makeup of the organisms of interest and (2) the selective pressures acting to modify populations over time.

Systematics is a three-pronged field of inquiry by which patterns of diversity are assessed. First, it includes **taxonomy**, the identification of organisms and the assigning of names to them. Second, it relies heavily on **phylogenetic reconstruction**. The goal of this work is to identify the evolutionary patterns that unite different organisms—say, an iguana and a horned lizard or a red panda and a giant panda. Third, systematics uses **classification**—categorizing phylogenetic information into a "retrieval system" consisting of many hierarchical levels, or ranks. This work feeds back into taxonomy, since the names we apply to organisms (or to groups of them) should reflect their place in the hierarchy.

## TAXONOMY

### Recognizing Species

In any systematic study, the immediate task is to identify **taxa** (singular, taxon). These are groups of organisms that are evolving independently of other such groups. The "species" is the taxon studied most often. Systematists view the species as a group of populations that have maintained genetic contact and that are evolving independently of other such groups. All the species that are connected this way—by genetic contact in the space of the environment *and* through time—represent a **lineage**. The lizard lineage is an example.

**All species of the same lineage have part of their evolutionary history in common and are on a separate evolutionary road from species of other lineages.**

Few among us would have trouble recognizing a giant panda as a giant panda, but identifying species is not always that easy. Dramatic differences in color, body form, and behavior may reflect sex-specific or age-related variation within a species. Butterflies are a good example of this. The larval forms of all butterflies are markedly different from the adult forms, the adult females often differ markedly from the males—and sometimes adult females differ markedly from one another! Take a look at the female swallowtail butterflies (*Papilio dardanus*) in Figure 20.2. On the basis of appearance alone, would you recognize them as belonging to the same species? Similarly, would you even guess wildly on the basis of appearance alone that the two fishes shown in Figure 20.3 are female and male of the same species?

The potential for confusion doesn't stop there. Sometimes, variations exist between different populations of the same species. Would you know, for example, that the plants shown in Figure 20.4 are all specimens of *Potentilla glandulosa*, a perennial herb that is a close relative of the strawberry? Even within a population, closely related individuals of the same age and sex may differ considerably in appearance, perhaps in response to different environmental effects. The arrowhead plants shown in Figure 20.5 are an example.

Alternatively, organisms that outwardly appear to be almost identical may be genetically distinct and have separate evolutionary histories. Consider two species of gray treefrogs (*Hyla versicolor* and *H. chrysoscelis*), both of which live throughout the central and eastern United States. Both have warty skin and pronounced sticky pads on their toes. Both have a white belly, yellow or yellow-orange coloration on the inside of the hind legs,

coastal specimen     mid-Sierran specimen     alpine specimen

**Figure 20.4** (*Above*) Variation among plants from different populations of the same species. Three specimens of *Potentilla glandulosa* were grown in the same garden in Mather, California. The three plants had been collected years before from populations growing in three different locations—in coastal, mid-Sierran, and alpine habitats.

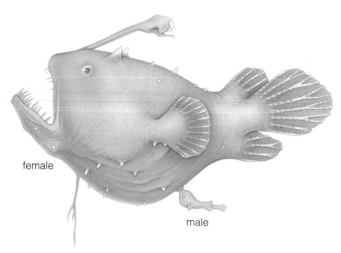

**Figure 20.3** Sexual dimorphism in a deep-sea angler fish (*Linophryne*), also known as the net devil. The female is about 10 centimeters long. Notice the minuscule male attached to her.

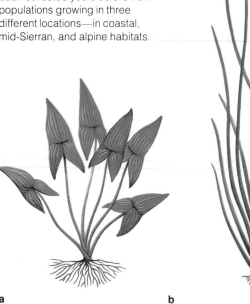

a             b

**Figure 20.5** Adaptive response by members of the same population to different environmental conditions. The shape of mature leaves of *Sagittaria sagittifolia* depends on whether the individual plant grows on land (**a**) or in water (**b**). It does not result from genetic differences.

a

b

c

**Figure 20.6** (**a**) Which frog is shown here—*Hyla versicolor* or *H. chrysoscelis*? You cannot tell the two species apart on the basis of external appearance. Identification depends on other clues, such as the female-attracting calls made by the male frogs during the breeding season. The sound spectrograms of (**b**) *H. versicolor* and (**c**) *H. chrysoscelis* reveal a behavioral difference between the two.

and a large white spot below each eye (Figure 20.6). Both live in the same type of forest habitats, eating the same kind of prey.

How were *H. versicolor* and *H. chrysoscelis* identified as separate species? During the breeding season, the males attract *only females of the same species* with a distinctive call. Besides this behavioral isolating mechanism, the two treefrogs show a chromosomal difference—one is diploid and the other is tetraploid. This example reinforces an important point. Correct identification of species often requires a knowledge of the anatomy, biochemistry, physiology, ecology, and behavior of the organisms of interest.

**Outward appearance alone is not enough to identify an organism as belonging to a particular species. Correct identification often requires knowledge of the anatomy, biochemistry, physiology, ecology, and behavior of organisms that appear to be related.**

### The Linnean Scheme

Whenever a new species is identified, it is given a scientific name. The modern practice of providing scientific names for organisms has its origins in the work of an eighteenth-century Swedish naturalist, Carl von Linné, now known by his latinized name, Linnaeus.

Linnaeus was a naturalist whose enthusiasm knew no bounds. He sent ill-prepared students around the world to gather specimens of plants and animals, and is said to have lost a third of his collectors to the rigors of their expeditions. Although perhaps not very commendable as a student adviser, Linnaeus did go on to develop a **binominal system** by which species are assigned a two-part Latin name. The first part was generic; it was descriptive of similar species that were thought to be the same type of organism and so were grouped together. Such a grouping is still called a **genus** (plural, genera). The second part was a specific name which, in combination with the generic name, referred to one kind of organism only.

For example, there is only one *Ursus maritimus*, or polar bear. Other bears are *Ursus arctos* (the brown bear) and *Ursus americanus* (the black bear). Notice that the first letter of the generic name is capitalized and the second names are uncapitalized. The specific name is never used without the full or abbreviated generic name preceding it, for it also can be the second name of a species in an entirely different group. As an example, *U. americanus* does indeed mean black bear, but *Homarus americanus* means Atlantic lobster, and *Bufo americanus* means American toad. (Hence one would not order *americanus* for dinner unless one is willing to take what one gets.)

Latin or scientific names and the conventions associated with them provide a universal way for people to discuss the same organism unambiguously, everywhere. Figure 20.7 shows the many different common names that are used in Latin America for the same venomous snake (*Bothrops asper*). Many different names are often used for the same organism even within a single country, and the same common name is sometimes used for more than one species. Only by using the standardized Latin name can we avoid confusion about which organism is being discussed and, possibly, be warned in time that we are about to step on a bad snake.

Shortly after it was devised, the binominal system became the heart of a classification scheme that was thought to mirror the patterns of links in the great Chain of Being (page 262). The scheme was based only on perceived similarities or differences in physical features—the number of legs, body size, wings or no wings, coloration, and so forth. In time, more inclusive groupings were invented to show broader relationships. A family was said to include all genera that resemble one another more than they do the genera of other families. An order was said to include all families that resemble one another, and so on up to a few great kingdoms of organisms.

The Linnean scheme proved to be quite successful, for it came at a time when classification was desperately needed. The names of its ever more inclusive groupings of species, the higher taxa, are still in use. And its hierarchical framework lends itself readily to an evolutionary approach.

**Figure 20.7** Common names applied to the same species of snake that is widely distributed through Latin America. To add to the confusion, some of the names, including vibora, are used for as many as fifty other species in the region.

**Scientific Name: *Bothrops asper***

**Sampling of the Local Names:**

YELLOW-JAW TOMMYGOFF (Belize)

BARBA AMARILLA (Colombian coast)

BOQUIDORÁ, BOQUIDORADA (Caribbean coast)

CUATRO NARICES, EQUIS, EQUIS NEGRA, EQUIS RABO DE CHUCHA (Isla Gorgona, Cauca)

GATA, MACABREL, MACAUREL, MAPANÁ, MAPANÁ DE UÑA, MAPANÁ EQUIS, MAPANÁ PRIETA, MAPANÁ RABO BLANCO, MAPANÁ TIGRE (Chocó)

EQUIS (Ecuador, coast)

BARBA AMARILLA, CANTIL BOCA DORADA (Guatemala, south coast)

BARBA AMARILLA (Honduras)

AHUEYACTLI, BARBA AMARILLA, COLA BLANCA, CUATRO NARICES, NAUHYACACÓATL, NAUYACA, NAUYACA COLA DE HUESO, NAUYACA REAL, NAUYAQUE, PALANCA, PALANCA LOCA, PALANCA LORA, PALANCACOATE, PALANCACÓATL, PALANCACUATE, RABO DE HUESO, TEPOCHO, TEPOTZO, VÍBORA SORDA, XOCHINAUYAQUE (Mexico)

BARBA AMARILLA, TERCIOPELO (Nicaragua)

FER-DE-LANCE, MAPEPIRE BALSAIN (also spelled: balsin, balcin, valsin, barcin), RABO FRITO (juveniles) (Trinidad)

MACAGUA, MACAUREL (Venezuela, central coast)

MAPANARÉ (Andes)

TERCIOPELO (western Andes)

CACHETE DE PUINCA, CHANGUANGA, CHIGDU, CUAIMA CARBÓN, CUATRONARIZ, DEROYA, DOROYA; JUBA-VITU (Indigenous tribal name), MACAO, MAPANARÉ TERCIOPELO, RABOAMARILLO, SAPAMANARE, TALLA EQUIS, TIGRA, TUKEKA (Venezuela)

# PHYLOGENY

## Three Interpretative Approaches

Having named organisms according to their species and genus, how do we go about grouping them? There are different schools of thought about this. At one time, an attempt was made to reconstruct the tree of life by looking in a subjective way at both differences and similarities between organisms. Fossils were particularly important in this respect, for one idea was to identify ancestors of living forms from the fossil record. This approach, known as "evolutionary systematics," produced many useful results. However, its methodology was imprecise, so any two researchers working with the same data might arrive at very different conclusions.

In a second approach to systematics, called "phenetics," organisms are grouped according to their similarities. A potential pitfall to this approach is that some similarities are not always the result of a shared evolutionary history. Think about the example of morphological convergence shown in Figure 19.7. By this evolutionary process, dissimilar species that were not at all closely related may have adopted a similar way of life, such that body parts took on similar functions and ended up resembling one another.

In still another approach to systematics, organisms are grouped according to similarities that are derived from a common ancestry. This approach is called "phylogenetic systematics," or **cladistics**. Independently evolving lineages that share a common evolutionary heritage are termed **monophyletic** (meaning one genealogy).

As Figure 20.8 suggests, the different approaches just described may translate into some very large differences in the ways that organisms end up being grouped. At present, cladistics is viewed as the least treacherous way to portray relationships, and it is the approach that will guide us through the remainder of this chapter.

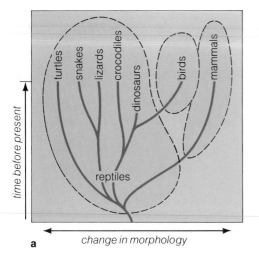

a  *change in morphology*

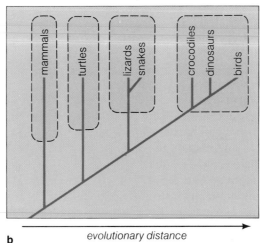

b  *evolutionary distance*

**Figure 20.8** How a few major groups of vertebrates are defined according to two different schools of systematics. In evolutionary systematics (**a**), the groups are defined by such traditionally accepted characteristics as scales, feathers, or fur. A cladistic, or phylogenetic, scheme (**b**) links together the organisms having shared ancestries.

## Portraying Relationships Among Organisms

A phylogeny is a pattern of evolutionary relationships. When we construct a phylogeny, we are making *hypotheses* about how organisms are related to one another. Phylogenetic reconstruction is a complicated process that requires a thorough understanding of the structure, evolution, and development of the organisms of interest. It depends on the identification of **homologous structures**—features that can be attributed to a common ancestry. Thus, as Figure 19.6 showed, bat wings and penguin flippers are both derived from the forelimb of a common ancestor and so are homologous.

By examining the patterns of distribution of many homologous structures, it is possible to construct a **cladogram**, a branching diagram that represents the patterns of relationships of the whole organism. Cladograms don't convey direct information about ancestors and descendants—"who came from whom." Rather, they portray *relative relationships* among organisms. Figure 20.8b is an example of this. Taxa that are closer together on this cladogram simply share a more recent common ancestor than those that are farther apart.

The *Doing Science* essay describes how cladograms are constructed and interpreted. Briefly, an organism is analyzed in terms of discrete morphological, physiological, and behavioral traits, or "characters," that vary among the taxa being studied. The analysis requires at least some understanding of likely relationships among organisms in the group under study. This in turn requires selection of an "outgroup," a different organism (but one not too distantly related) to serve as a point of departure for evaluating relative evolutionary distances within the "ingroup."

## CLASSIFICATION

### Classification as a Retrieval System

Much of biology is a comparative endeavor, with researchers arriving at generalizations after comparing similar attributes among a variety of organisms. Inferences about one organism may be based on observations of or experiments with another.

Knowledge of phylogeny provides a basis for making an informed decision about which organisms to study. For example, how might you go about deciding which type of habitat should be protected to help conserve a rare, poorly understood species? You might study the habitat requirements for species that are closely related to it—and better understood. How might you go about testing the effectiveness of a new antibiotic against a dangerous bacterium, such as the one responsible for Legionnaires' disease in humans? You might choose a close primate relative of humans, the assumption being

*Doing Science*

## Constructing a Cladogram

To see how phylogenetic reconstruction works in practice, consider a group of vertebrate animals—a lamprey, trout, lungfish, turtle, cat, gorilla, and human. For each of these animals, the six different traits, or characters, shown in Figure *a* have been recorded. To keep things simple, you can focus only on the presence (+) or absence (−) of the traits shown.

For various reasons that will become apparent in Chapter 26, the lamprey is known to be only distantly related to the other vertebrates listed. For example, it lacks jaws and paired appendages. This means you can use the lamprey as the outgroup. Compared to the other groups being considered, an *outgroup* is the one exhibiting the fewest derived characters. It allows us to make inferences about the common ancestor of the organism being considered. You may interpret any deviation from an outgroup characteristic as an evolutionary change, or *derived trait*. In Figure *b*, the outgroup condition is marked by a zero (0) and the derived condition by a one (1).

The next step is to look for the *shared* derived traits. For example, you see that the human and gorilla share five

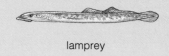

lamprey

turtle

cat

gorilla

lungfish

trout

human

| Taxon | Characters | | | | | |
|---|---|---|---|---|---|---|
| | Jaws | Limbs | Hair | Lungs | Tail | Shell |
| Lamprey | − | − | − | − | + | − |
| Turtle | + | + | − | + | + | + |
| Cat | + | + | + | + | + | − |
| Gorilla | + | + | + | + | − | − |
| Lungfish | + | − | − | + | + | − |
| Trout | + | − | − | − | + | − |
| Human | + | + | + | + | − | − |

a

| Taxon | Characters | | | | | |
|---|---|---|---|---|---|---|
| | Jaws | Limbs | Hair | Lungs | Tail | Shell |
| Lamprey | 0 | 0 | 0 | 0 | 0 | 0 |
| Turtle | 1 | 1 | 0 | 1 | 0 | 1 |
| Cat | 1 | 1 | 1 | 1 | 0 | 0 |
| Gorilla | 1 | 1 | 1 | 1 | 1 | 0 |
| Lungfish | 1 | 0 | 0 | 1 | 0 | 0 |
| Trout | 1 | 0 | 0 | 0 | 0 | 0 |
| Human | 1 | 1 | 1 | 1 | 1 | 0 |

b

derived traits, whereas the human and cat share only four. For such a small set of traits, you can simply sketch out a cladogram, a type of family tree that shows patterns of relationships of organisms, based on their derived traits. (In practice, systematists often resort to computers. They may use many, many traits to examine relationships between many taxa, and it takes a computer to find the pattern that is best supported by so much data.)

Figure c shows how you would sketch information based on the jaw trait. All of the vertebrates other than the outgroup have jaws. This is a derived trait. Figure d shows what happens when you add information based on lungs. All of the vertebrates except the lamprey and the trout have lungs. The lungfish, human, gorilla, turtle, and cat thus appear to be a *monophyletic group*, meaning they have a common evolutionary heritage. (They share two derived traits with respect to the lamprey and one with respect to the trout.)

Figures e through g show the resulting changes in the cladogram as information about other derived traits is added to it.

Notice that one character, the presence of a shell, was not used. Why not? Only the turtle has a shell. Whereas a shell is a derived trait relative to the outgroup, it is not shared with any of the other animals under consideration. Therefore, it is not useful in reconstructing this phylogeny.

Of course, every species has many such characters. The traits may be useful in identifying the species but not in determining relationships. In your case, none of the traits provides information that is contradictory to other traits. In reality, such contradictions often occur. Why?

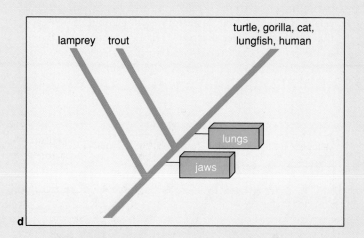

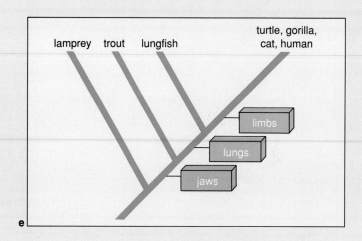

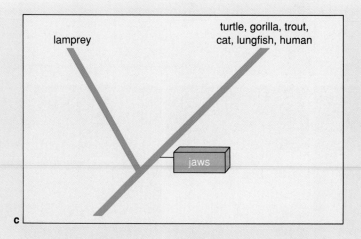

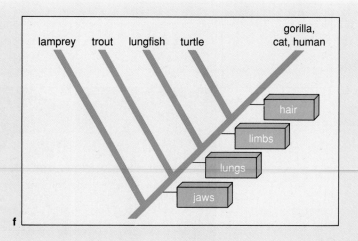

All organisms are a mixture of evolutionarily ancient and derived traits. Also, not all traits you initially regarded as homologous may actually be homologous. As you saw in Chapter 19, morphological convergences may have occurred.

How do you read the cladogram in Figure g? Keep in mind that cladograms give you information about relative relatedness, not about ancestors. Your cladogram tells you that the human is more closely related to the gorilla than to the cat. The cat, human, and gorilla as a group are more closely related to each other than to the turtle.

These examples are easy to follow. Perhaps less obviously, your cladogram also tells you that the lungfish is more closely related to the human than to the trout. Follow the lines or branches of the cladogram from the human and lungfish back to the intersection, or node, where they meet (Figure h, node 1). Next, trace back the branches of the lungfish and trout to their intersection (Figure h, node 2). Note that the trout-lungfish intersection is nearer to the base of the cladogram than the lungfish-human intersection. The higher up the cladogram, the more derived traits are shared by the taxa. The lower they are on the cladogram, the fewer are shared.

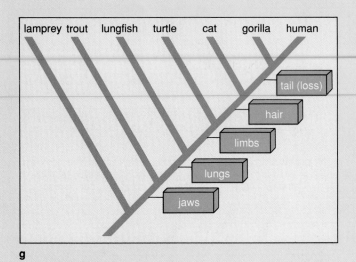

g

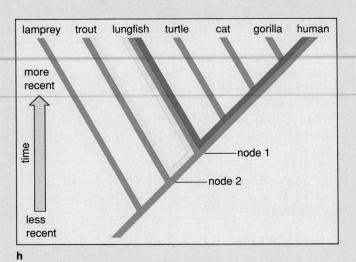

h

| Table 20.1 Classification of Four Organisms | | | | |
|---|---|---|---|---|
| Category (taxon) | Corn | Vanilla Orchid | Housefly | Human |
| Kingdom | Plantae | Plantae | Animalia | Animalia |
| Phylum (or Division, botanical schemes) | Anthophyta (flowering plants) | Anthophyta | Arthropoda | Chordata |
| Class | Monocotyledonae (monocots) | Monocotyledonae | Insecta | Mammalia |
| Order | Commelinales | Orchidales | Diptera | Primates |
| Family | Poaceae | Orchidaceae | Muscidae | Hominidae |
| Genus | *Zea* | *Vanilla* | *Musca* | *Homo* |
| Species | *mays* | *planifolia* | *domestica* | *sapiens* |

Another way to think about this is to regard the axis of the cladogram as a time bar, but one without absolute dates. It simply signifies this: The lower the position of a group on the cladogram, the more distant is the most recent common ancestor of the taxa being compared.

Finally, even though the axis of the cladogram can be viewed as a time line, keep in mind that the ordering of taxa from left to right is not important. The cladogram shows only the *pattern* of their relationships to one another. In Figure *i*, the order of the taxa across the top of the cladogram is very different from what it was Figure *h*—but the information portrayed is exactly the same. Following the branches back to the lungfish-human and lungfish-trout intersections gives the same result.

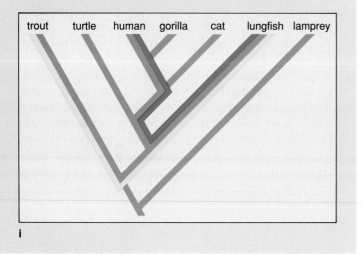

i

that its physiological responses to the infection and to the antibiotic will be similar, if not the same.

In this respect, a **classification** is a useful system for retrieving information about relationships among organisms. The classification systems in use today consist of a series of ever more inclusive taxa, beginning with the species. Consider the following taxa:

> kingdom
> > phylum (or division)
> > > class
> > > > order
> > > > > family
> > > > > > genus
> > > > > > > species

Now take a look at Table 20.1, which shows how they are used in the classification of four different organisms. As you can see, the ranking alone does not provide information about the exact pattern of relationships, but the groupings reflect the relative degree of relationship. A corn plant and a vanilla orchid are close relatives in the sense that they are both monocots, but they belong to separate orders. A housefly and a human are both animals, but they are only distantly related—they are not even in the same phylum. Similarly, Figure 20.9 shows that humans are more closely related to other members of the superfamily Hominoidea than to those in the Ceboidea, but they share a more recent common ancestry with the Ceboidea than with any member of the suborder Prosimii.

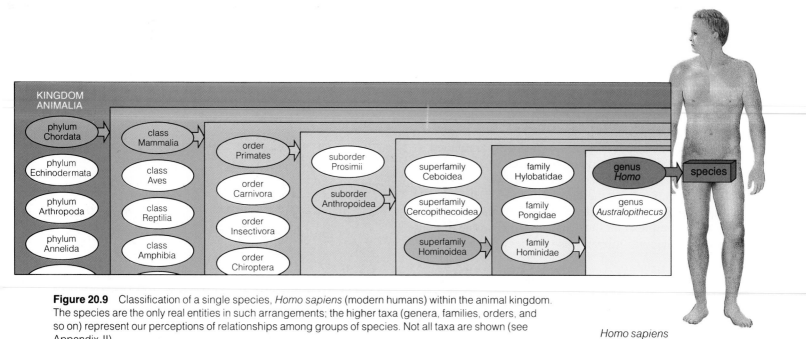

**Figure 20.9** Classification of a single species, *Homo sapiens* (modern humans) within the animal kingdom. The species are the only real entities in such arrangements; the higher taxa (genera, families, orders, and so on) represent our perceptions of relationships among groups of species. Not all taxa are shown (see Appendix II).

*Homo sapiens*
(only living species of this genus)

In the next unit of this book, you will encounter trees of descent that portray presumed evolutionary relationships for major groups of organisms. It is important to understand that only the *species* are real entities in those trees. The "twigs, branches, and limbs" are not real; they are *categories of relationship* that we superimpose on the species themselves. Obviously, then, we cannot talk about "the evolution" of genera, families, and so on except in terms of the developments that occurred during the history of all the species contained in those increasingly inclusive categories. As we saw in Chapter 19, these developments include the origins, persistence, radiations, and extinctions of different groups of species.

## Five-Kingdom Classification

Five kingdoms of organisms are recognized in this book, following the classification scheme established by Robert Whittaker:

| | |
|---|---|
| **Monera** | *monerans; single-celled prokaryotes, some autotrophs (photosynthetic or chemosynthetic), others heterotrophs* |
| **Protista** | *protistans; diverse single-celled eukaryotes, some photosynthetic autotrophs, many heterotrophs* |
| **Fungi** | *fungi; multicelled heterotrophs that feed by extracellular digestion and absorption* |
| **Plantae** | *plants; multicelled photosynthetic autotrophs* |
| **Animalia** | *animals; diverse multicelled heterotrophs, including predators and parasites* |

As you will perceive by the end of the next unit, the set of traits used to characterize each kingdom suggests that the existing members of each are the result of very long, independent lines of evolution. In many respects, the five kingdoms are so distinctive that it is difficult to find homologous features to compare. As suggested by Figure 20.10, a family tree can show only broad relationships among them, with ancient single-celled monerans being the evolutionary roots.

The classification used in this book, like any classification, is not definitive. Rather it is a stored, simplified version of a phylogenetic hypothesis, itself based on the analysis of incomplete data. If we have learned anything about systematics since the days of Linnaeus, it is that the diversity of life reflects evolutionary patterns so exquisitely complex that they challenge the simple categorizations that we humans use to define the world around us.

**Figure 20.10** Modified version of Robert Whittaker's five-kingdom system of classification.

## SUMMARY

1. Systematics is the branch of biology that deals with phylogeny, the evolutionary relationships among species, beginning with the most ancestral forms and including all the branches leading to all of their descendants.

2. Systematics takes three approaches to assess the patterns of diversity in the distribution of organisms in space and time.

   a. Taxonomy involves identifying organisms and assigning names to them.

   b. Phylogenetic reconstruction involves defining evolutionary patterns that give evidence of inherited relationships among organisms.

   c. Classification involves categorizing phylogenetic information into a useful retrieval system.

3. The binominal system, developed by Linnaeus, is still in use. Each type of organism is assigned a two-part Latin name. The first part, the genus, identifies a group of similar species that share a recent common ancestor. The second part is the name of the particular species. When used in combination with the generic part, the species designation refers only to one kind of organism.

4. Identifying a species is often a vexing task, given the possible variations in sex-specific, developmental, and environmental factors. Morphological convergences

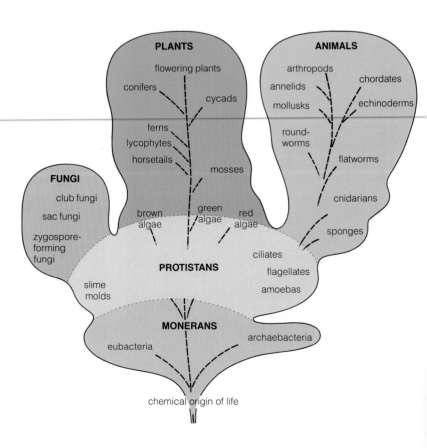

also may complicate the picture. Identification requires good understanding of the biology of the organisms of interest.

5. Currently, the most widely used approach to studying patterns of diversity is phylogenetic systematics, or cladistics. In this approach, organisms are grouped together according to shared similarities that are perceived to be derived from common ancestry. Phylogenetic reconstruction involves thorough understanding of the structure, evolution, and development of the organisms of interest.

6. The classification system that is most widely used today consists of a series of ever more inclusive taxa, from species to kingdoms. Although the classification does not reveal explicit phylogenetic patterns, it does reflect the general level of relatedness among organisms.

## Review Questions

1. Patterns of diversity among species are assessed by using taxonomy, phylogenetic reconstruction, and classification. Distinguish among these three investigative approaches. *320*

2. Give reasons why two organisms that seem identical in outward appearance may not belong to the same species. *321–322*

3. Give reasons why two organisms that are quite different in outward appearance may belong to the same species. *321–322*

4. True or false: The species name alone may be used to refer to an organism in a document, provided that its full name (genus and species) has been spelled out the first time it is used. *322*

## Self-Quiz *(Answers in Appendix IV)*

1. A taxon is _____ evolving independently of other such groups.
   a. a species        c. a phylum (or division)
   b. a genus         d. any group of organisms

2. A lineage consists of a group of populations that _____ .
   a. are living at the same time and maintaining genetic contract
   b. have maintained genetic contact in the space of the environment and through time
   c. have followed one another in succession through time
   d. are evolving independently of other such groups
   e. both b and d are correct

3. Defining evolutionary patterns that suggest inherited relationships among organisms is a concern of _____ .
   a. cladistics         c. phylogenetic reconstruction
   b. classification     d. taxonomy

4. The names of all genera and species are italicized. Besides this, the generic name is _____; the species name is _____ .
   a. capitalized; capitalized
   b. capitalized; uncapitalized and may be used alone
   c. capitalized; uncapitalized and is not used alone
   d. uncapitalized; uncapitalized

5. When used in combination with the genus name, the species name can refer to _____ .
   a. several kinds of organisms that belong to the same genus
   b. only one kind of organism
   c. only one kind of organism and the group ancestral to it
   d. the person who discovered the genus

6. Categorizing organisms according to phylogenetic information into a useful retrieval system is called _____ .
   a. classification       c. evolution
   b. taxonomy         d. the binominal system

7. Phylogenetic reconstruction typically involves thorough understanding of an organism's _____ .
   a. body structure      c. embryonic development
   b. evolution          d. all of the above

8. A branching diagram that represents the patterns of relationships of whole organisms is a _____ .
   a. taxon          c. binominal system
   b. cladogram       d. Whittaker scheme

9. Classify a corn plant by arranging the following taxa in sequence (*a* through *g*), with the most inclusive taxon first:
   _____ a. order Commelinales
   _____ b. species *mays*
   _____ c. division Anthophyta
   _____ d. family Poaceae
   _____ e. kingdom Plantae
   _____ f. genus *Zea*
   _____ g. class Monocotyledonae

## Selected Key Terms

| | | |
|---|---|---|
| binominal system *322* | genus *322* | phylogenetic reconstruction *320* |
| cladistics *323* | homologous structures *324* | phylogeny *318* |
| cladogram *324* | kingdom *327* | phylum *327* |
| class *327* | lineage *320* | systematics *320* |
| classification *320* | mimicry *320* | taxon *320* |
| derived character (trait) *324* | monophyletic group *325* | taxonomy *320* |
| family *327* | order *327* | |

## Readings

Brooks, D. R., and D. A. McLennan. 1991. *Phylogeny, Ecology, and Behavior*. Chicago: University of Chicago Press. An up-to-date examination of the approaches to studying the diversity of life.

Sneath, P., and R. Sokal. 1973. *Numerical Taxonomy*. San Francisco: Freeman. A leading text in the phenetic approach to systematics.

Wiley, E. 1981. *Phylogenetics: The Theory and Practice of Phylogenetic Systematics*. New York: Wiley.

# 21 HUMAN EVOLUTION: A CASE STUDY

## The Cave at Lascaux and the Hands of Gargas

Half a century ago, on a warm autumn day, four boys out for a romp stumbled into a cave near Lascaux, a town in the Perigord region of France. Inside the intricately tunneled cave, they made a discovery that stunned the world. On the cave walls were magnificent sketches, engravings, and paintings (Figure 21.1). The red, yellow, purple, and brown pigments were as vivid as if they had been painted only recently. Yet they turned out to be 17,000 to 20,000 years old. They had been created deep inside the cave, where sunlight could not fade them and winds and water could not wear them away. By the light of crude oil lamps, prehistoric artists had captured the graceful, dynamic lines of bison, stags, stallions, ibexes, lions, a rhinoceros, and a heifer, now known as the Great Black Cow.

Many caves in southern France, northern Spain, and Africa hold treasures from even earlier times. About 25,000 years ago, for example, more than 150 imprints and outlines of human hands were committed with great care to walls in the cave of Gargas, in the Pyrenees.

Who were the people who did this? What events led them to leave such evidence of their existence? We know that they were modern humans, anatomically like us. We can only sense that, through the refined conception and execution of their art, they were expressing uniquely human qualities that began to evolve in the remote past.

Humans did not materialize out of thin air. The story of human evolution has its beginnings 65 million

**Figure 21.1** Part of the human cultural heritage—prehistoric cave paintings, a unique outcome of a long history of biological evolution.

KEY CONCEPTS

**1.** The fossil record gives evidence of evolutionary trends among the primates.

**2.** Among those trends were skeletal modifications that began when four-legged primates were adapting to life in the trees. Those primates and their presumed descendants foreshadowed the evolution of humanlike forms that were capable of walking upright, on two legs.

**3.** Modifications in the hands led to increased dexterity and manipulative skills. There was less reliance on the sense of smell and more on daytime vision. The brain became larger and more complex; among humans and their predecessors, this trend was accompanied by refined technologies and cultural evolution.

**4.** All of the trends in primate evolution were the foundation for a remarkable characteristic of humans, their plasticity. The term refers to their capacity to remain flexible and adapt to a wide range of challenges imposed by unpredictable, complex environments.

years ago, with the origin of primates in long-since-vanished tropical forests. In turn, primates did not pop up out of nowhere. *Their* story began with the evolution of mammals more than 200 million years ago. The story extends back to the origin of animals between 2,500,000 and 750 million years ago—and so on back in time, to the origin of the first living cells.

As we poke about the roots of our family tree, keep this greater evolutionary story in mind. If there is one point that is central to understanding our evolution, it is this. *Our "uniquely" human traits emerged through modification of traits that had already evolved in ancestral forms.* At each branch in the family tree, mutations introduced *workable* changes in genes—and so introduced changes in phenotype that proved to be successful in the prevailing environment.

From this perspective, when you look at "ancient" cave paintings, you are looking at the legacy of individuals who departed only yesterday, so to speak. The artists of Lascaux and Gargas are not remote from us. They *are* us.

In the preceding chapters, we likened the history of life to a great tree. Each branch of the tree represents a line of descent, and the branch points represent divergences leading to new species. There are more than 40,000 existing species of fishes, amphibians, reptiles, birds, and mammals on the vertebrate branch of that tree. When we turn to the evolution of any one of those species—as we do here—it helps to keep a key point in mind. At each crossroad leading to a new species, complex traits were already in place and functioning—*and new traits emerged only through modification of what went before.* The evolution of the human species speaks eloquently of this characteristic of life.

## THE MAMMALIAN HERITAGE

Humans are members of the class Mammalia, one of the seven classes of vertebrates described in Chapter 26. Like other vertebrates, mammals have an internal skele-

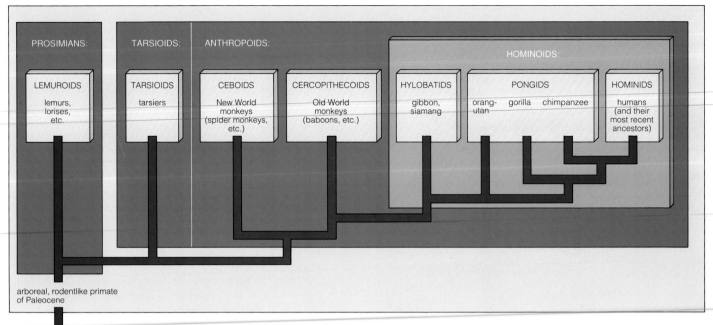

**Figure 21.2** Simplified family tree for primates. The two highest groupings (the prosimians and the tarsioids and anthropoids) are shown in green. Representative members of the major groups of living primates are indicated. All monkeys, apes, and humans are anthropoids.

ton with two key features: a nerve cord within a column of bones (backbone) and a skull that houses sense organs and a three-part brain (hindbrain, midbrain, and forebrain). Two other mammalian features are central to the story of human evolution. The first is their *dentition*—the type, number, and size of teeth. In this respect mammals differ from their reptilian ancestors, which had peglike upper teeth that didn't meet at the surface of their peglike lower ones (compare Figure 26.21b). Mammals have four types of upper and lower teeth that match up and work together to crush, grind, or cut food:

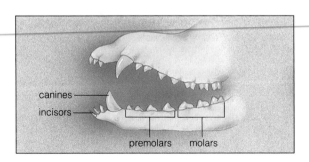

Teeth tell us quite a bit about what an animal eats, hence something of its life-style. *Incisors*, shaped like flat chisels or cones, are used to nip or cut food. They are pronounced in horses and other grazing animals that live in open grasslands. Pointed *canines* are used in biting and piercing. Meat-eating animals have long, sharp canines, useful for piercing their prey. Some monkeys

and apes also have long canines that can be used to split open bamboo stalks or other tough plant parts. *Premolars* and *molars* (cheek teeth) are a platform for food, and their surface bumps (cusps) are used in crushing, grinding, and shearing. Large, flat-surfaced cheek teeth are characteristic of animals that live (or whose ancestors lived) in environments having quantities of fibrous plant material.

Teeth also happen to fossilize very well. Many inferences about the diet and life-styles of human ancestors are drawn from fragments of jaws and teeth they left behind.

The second feature characterizing mammals is their extended period of *infant dependency and learning*. Young mammals depend on adults for nourishment and protection and as models for behavior. They rely on a limited set of social behaviors that can be learned and repeated. Some also show a capacity to add quickly to their behavioral repertoire. This behavioral flexibility has reached its fullest expression among the primates.

## THE PRIMATES

### Primate Classification

Figure 21.2 shows a family tree for primates. Its members include prosimians, tarsioids, and the anthropoids, which in turn include monkeys, apes, and humans. Figure 21.3 shows a few representatives. The smallest prosimians are the size of a mouse. The larger ones gener-

a

b

c

**Figure 21.3** Representative primates. Gibbons (**a**) have limbs and a body adapted for *brachiation* (swinging arm over arm through the trees). Monkeys are quadrupedal (four-legged) climbers, leapers, and runners, as the spider monkey in (**b**) demonstrates. Tarsiers (**c**) are vertical clingers and leapers.

ally are about the size of a miniature poodle. Most forage at night and have large, forward-directed eyes specialized for night vision. As their name suggests, prosimians are the oldest primate lineage (*pro*, before, *simian*, ape). For millions of years, before the first monkeys or apes appeared, the prosimians dominated the trees in North America, Europe, and Asia. In time, the more agile and larger brained monkeys and, later, the apes replaced them in all but a few restricted forests of Africa and Asia. For example, except for zoo specimens, lemurs are now restricted to Madagascar.

The so-called Old World monkeys make their home in Africa, Asia, and Gibraltar. The New World monkeys live in Central and South America. Even though both groups are called monkeys, they differ in the number of teeth and other traits. If you go through the primate section of a zoo, for example, you will see that only the New World monkeys can grasp objects with their tail. They also have a flat nose, with nostrils far apart and facing outward. Thus New World monkeys are said to be platyrrhine, a Greek term meaning flat-nosed. In contrast, Old World monkeys have close-together nostrils

that face forward and downward. They are catarrhine, after the Greek word for hook-nosed. Actually, in terms of nose shape, number of teeth, and many other traits, Old World monkeys are more like apes and humans.

Monkeys and apes are not the same. In many anatomical details and in biochemistry, the apes are much more similar to humans. Thus apes, humans, and their recent ancestors are classified together, as **hominoids**. Millions of years ago, the evolutionary road leading to modern humans branched off from an ape lineage. All species to appear on that road since the time of divergence are classified separately as the **hominids**.

### Trends in Primate Evolution

Most primates live in tropical or subtropical forests, woodlands, or savannas, which are open grasslands with a few stands of trees. Like their ancient forerunners, the vast majority of primates are tree dwellers. As a group, they are difficult to define, for no one feature sets them apart from other mammals. Perhaps the primates are best defined in terms of *key trends in their evolu-*

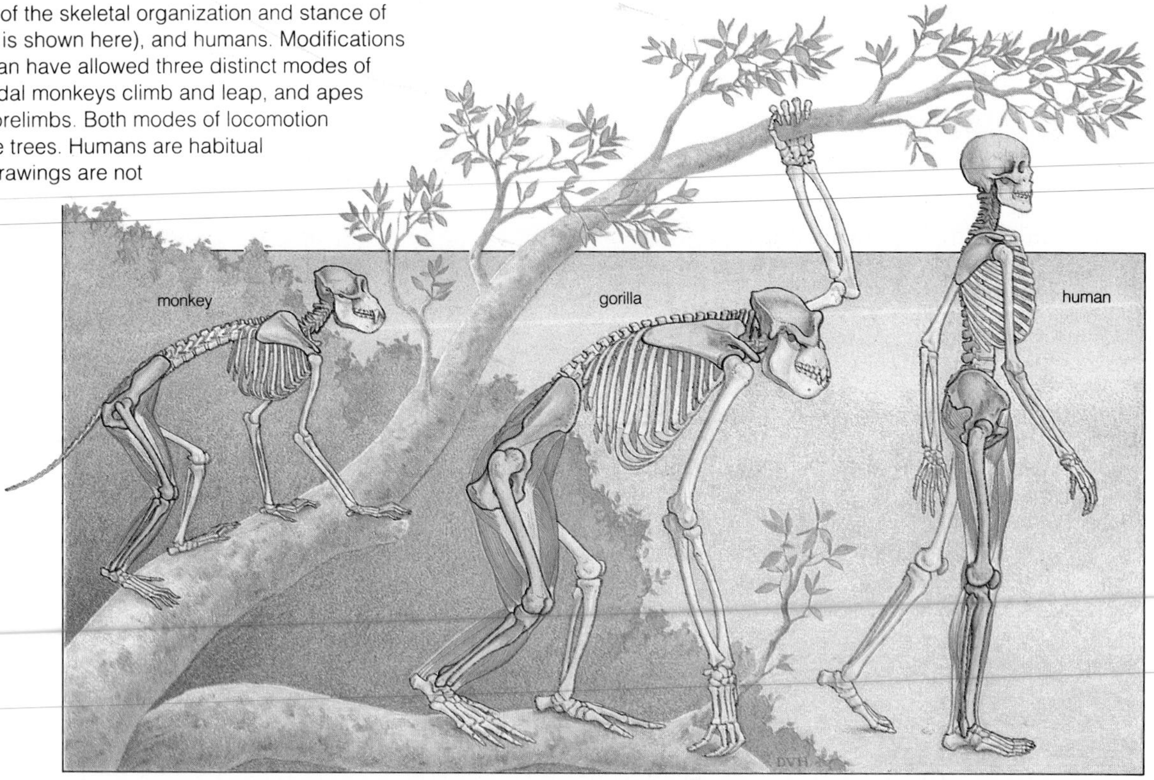

**Figure 21.4** Comparison of the skeletal organization and stance of monkeys, apes (the gorilla is shown here), and humans. Modifications of the basic mammalian plan have allowed three distinct modes of locomotion. The quadrupedal monkeys climb and leap, and apes climb and swing by their forelimbs. Both modes of locomotion are well suited for life in the trees. Humans are habitual two-legged walkers. The drawings are not to the same scale.

*tion.* Many of those trends began when ancestral forms were adapting to life in the trees:

1. Change in overall skeletal structure and mode of locomotion

2. Modification of the hands, leading to increased dexterity and manipulation

3. Less reliance on the sense of smell and more on daytime vision, including color and depth perception

4. Change in dentition, toward fewer and less specialized teeth

5. Brain expansion and elaboration

6. Behavioral evolution

The trends did not proceed at the same time or same pace in different lineages. And some trends proceeded in a few groups only. Existing prosimians have retained many "primitive" features, for example, including a snout. Taken as a whole, however, the list gives us a sense of the adaptive potential that is uniquely "primate."

**Locomotion**. Of all the primates, only humans can stride freely on two legs for long periods of time. Their habitual two-legged gait, called **bipedalism**, emerged through skeletal modifications in primates ancestral to humans.

A monkey skeleton is suitable for life in the trees. It allows rapid climbing, leaping, and running along branches (Figures 21.3 and 21.4). For example, the armbones and legbones are about the same length, so monkeys can run palms-down. Try doing this yourself and see what happens.

An ape skeleton is suitable for hanging on to overhead branches and using the arms to carry some of the body weight. Many large apes use their elongated arms to support body weight while on the ground. The shoulder blade positioning allows apes to swivel their arms freely above the head while their body is in an erect or semi-erect position. (Monkeys cannot do this.) The vertebral column in humans is shortened, S-shaped, and moderately flexible. Together with the position and shape of the shoulder blades and pelvic girdle, it allows a fully upright stance.

All of these skeletal and locomotor adaptations may be viewed as responses to conditions that prevailed when the ancestors of modern monkeys, apes, and humans were evolving.

**Modification of Hands**. The first mammals were four-legged, and they spread their toes apart to help support body weight as they walked or ran. Primates still can spread their toes or fingers apart. Many also make cupping motions, as when monkeys bring food to the mouth.

Two other hand movements developed among ancestral tree-dwelling primates. Modifications in the handbones allowed fingers to be wrapped around objects (*prehensile* movements) and the thumb and tip of each finger to touch each other (*opposable* movements).

Hands began to be freed from load-bearing functions among early tree-dwelling primates. Much later, on the evolutionary road leading to humans, refinements in hand movements led to the precision grip and power grip:

These hand positions gave early humans the capacity to make and use tools—which were the foundation for unique technologies and cultural development.

**Enhanced Daytime Vision.**   Early primates had an eye socket on each side of the skull. Later ones had forward-directed eyes, so they had overlapping visual fields, stereoscopic vision, and depth perception. Over time, they became quite good at discerning shape, movement, and variations in color and light intensity (dim to bright). These visual stimuli are typical of conditions up in the trees.

**Changes in Dentition.**   Monkeys have long canines and rather rectangular jaws. Human teeth are smaller and more uniform in length, and the jaw is bow-shaped. The jaws and teeth became less specialized during the evolution of forms leading to humans. Beginning with the earliest primates, there was a shift from eating insects, then fruit and leaves, and on to a mixed diet.

**Brain Expansion and Elaboration.**   During primate evolution, the brain increased in mass and complexity. The largest brain size in living apes is 600 cubic centimeters (for the gorilla); the average size in humans is 1,350. However, size alone is not the best indicator of capabilities. Rather, we must consider brain size relative to overall body size. Specifically, there has been expansion in brain regions that are responsible for our human way of life. Even ancient vertebrates were equipped with a forebrain, for example. But in primates—humans especially—parts of this expanded to become an intricate cerebral cortex (page 572). Neural connections in the human cerebral cortex are the foundation for thought processes, language, intricate hand movements, and so forth.

**Behavioral Evolution.**   By using existing primates as our model, we can identify a behavioral trend in primate evolution. The trend involves longer life spans, longer periods between pregnancies, single births rather than litters, and extended periods of infant dependency (Figure 21.5). The increased parental investment in fewer offspring requires strong bonds between parents and offspring, intense maternal care, and longer periods of learning.

Like mammals generally, primates rely on a set of social behaviors that can be learned and repeated by the young. And they show a capacity to add quickly to their behavioral repertoire. To give one example, when researchers in Japan introduced a group of macaque monkeys to the sweet potato, one adolescent female monkey carefully dipped it in water before eating it. Her peers imitated her, and after 3-1/2 years, potato-dipping had become common behavior among the new generation.

The capacity for learning expanded during human evolution. Brain modifications and behavioral complexity were interlocked, with new developments in one stimulating development of the other. Nowhere is this more evident than in the parallel evolution of human culture and the human brain.

Here we define **culture** as the sum total of behavior patterns of a social group, passed between generations by learning and by symbolic behavior, especially language. Culture is the culmination of a long evolutionary history.

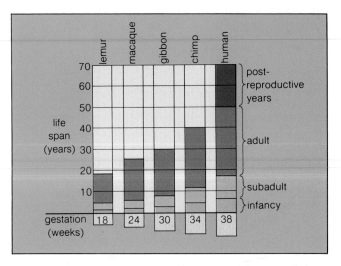

**Figure 21.5**  Trend toward longer life spans and longer periods of infant dependency among existing primates.

# PRIMATE ORIGINS

Primates evolved from ancestral mammals more than 60 million years ago, during the Paleocene. The first known primates resembled small rodents or tree shrews (Figures 21.6 and 21.7). Like tree shrews, they probably had huge appetites and foraged at night for insects, seeds, buds, and eggs on the forest floor. They had a long snout and a well-developed sense of smell, useful for detecting food or predators. They could claw their way upward through the shrubbery, although not with much speed or grace.

Between 54 and 38 million years ago (the Eocene), some descendants of the first primates were living in the trees. Fossils give evidence of increased brain size, a shorter snout, enhanced daytime vision, and refined grasping movements. How did these traits evolve?

Consider that the trees were a promising adaptive zone, with new food sources and safety from ground-dwelling predators. Yet the trees also were visually complex. Imagine the dappled sunlight, boughs swaying in the wind, colorful fruit tucked among the leaves, darting insects and other prey, perhaps predatory birds. A long snout would not have been of much use; air currents disperse odors. But individuals with a refined sense of color, shape, and movement would have a competitive edge. So would individuals skilled at running, swinging, and leaping (especially!) from branch to branch. There would have been uncompromising selection for a brain that could assess sensory input on distance, body weight, winds, and suitability of the destination—and that could compensate quickly for miscalculations.

By 35 million years ago, during Oligocene times, the tree-dwelling ancestors of anthropoids had emerged. Some forms lived above humid swamps lining the rivers of tropical rain forests. Given the altogether nightmarish predatory reptiles that inhabited the swamps, perhaps we can sense why the Oligocene primates rarely ventured to the ground—and why it became imperative to think fast and grip strongly. Slip-ups were always possible; a surprising number of primates still fall out of the trees.

Here the story moves on to an adaptive radiation of apelike forms—the first hominoids—that occurred between 23 million and 20 million years ago.

Then, the continents were beginning to assume their current positions, and climates became cooler and drier (Figure 19.16). Forests gave way to grasslands, and perhaps speciation was favored as subpopulations of apes became reproductively isolated within the shrinking stands of trees. This was the time of origin for the "dryopiths" and other forms that ranged through Africa and on into Europe and Eurasia (Figure 21.7). Eventually most of these forms became extinct. Between 10 million and 6 million years ago, however, some of their descendants may have given rise to the forerunners of modern gorillas, chimpanzees—and humans. The fossil record suggests that divergences marking the origin of those hominoid lineages occurred during this epoch. So do many comparative biochemical and immunological studies, of the sort that were described earlier in Chapters 16 and 19.

**Figure 21.6** One of the night-foraging tree shrews of Indonesia.

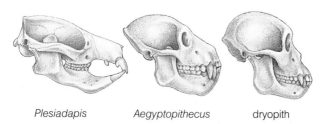

Plesiadapis     Aegyptopithecus     dryopith

**Figure 21.7** Comparison of skull shape and dentition of some extinct primates. *Plesiadapis* was a Paleocene primate with rodentlike teeth. *Aegyptopithecus*, an Oligocene anthropoid, probably predates the divergence leading to Old World monkeys and the apes. The dryopiths originated during an adaptive radiation that began in the Miocene. The drawings are not to the same scale.

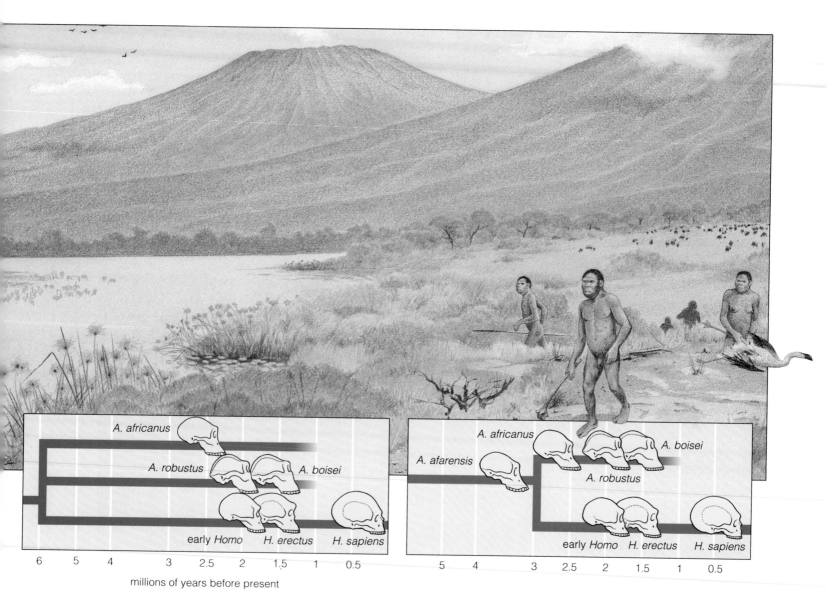

millions of years before present

# THE HOMINIDS

The first hominids probably emerged between 10 million and 5 million years ago, during the late Miocene. Fossils of humanlike forms that are 4 million years old have been discovered in Africa. There appear to have been many varieties of early hominids, but many had three features in common:

Bipedalism
Omnivorous feeding behavior
Further brain expansion and elaboration

These features probably coincided with the emergence of cooler, drier climates. When African rain forests started shrinking, some hominids made the transition to life in mixed woodlands and in the grasslands that were opening up (Figure 21.8).

**Figure 21.8** Reconstruction of one of the environments that prevailed during the emergence of the first humanlike forms, the hominids. There is general agreement that the hominids emerged during the late Miocene, between 10 and 5 million years ago. The *Homo* lineage is well documented; the australopith connections are not yet understood. Thus several phylogenetic trees for the hominids have been proposed; two are shown here.

Rather than speculate on how bipedal, omnivorous, and large-brained hominids evolved during the transition, think about the plasticity inherent in those three features. Here, **plasticity** means the ability to be flexible and to adapt to a wide range of demands. The first hominids were faced with a new, complex, and unpredictable world. Yet they had the use of hands freed from their load-bearing functions. They could survive on more than one type of food. And they had the brains to *learn* how to adapt to a changing environment.

Thus, the hominid lineage emerged through modifications of traits that can be observed among the other primates. *It was based on the primate heritage.*

### Australopiths

The earliest known hominids, the australopiths, can be grouped into two broad categories:

1. Gracile forms (slightly built), currently named *Australopithecus afarensis* and *A. africanus*

2. Robust forms (muscular, heavily built), including *A. boisei* and *A. robustus*

We really have little idea of how these early hominids were related. Apparently this was a "bushy" period of evolution—that is, a time of rapid divergences and radiations into new adaptive zones (page 304). In addition, inferences about those hominids must be drawn from a limited number of fossil fragments. Figure 21.8 shows two of several interpretations of the fossil record.

Legbone fragments 4 million years old show muscle attachment sites that are typical of two-legged walkers. The skeleton of one female, dubbed Lucy, is particularly revealing. The degree to which her thighs converged at her knees allowed body weight to be centered directly beneath the pelvis. This is a hallmark of bipedalism.

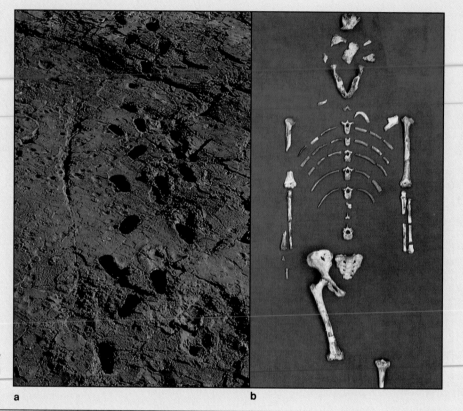

*Following the path produces, at least for me, a kind of poignant time wrench. At one point, and you need not be an expert tracker to discern this, the traveler stops, pauses, turns to the left to glance at some possible threat or irregularity, then continues to the north. This motion, so intensely human, transcends time. Three million seven hundred thousand years ago, a remote ancestor—just as you or I—experienced a moment of doubt.*

—Mary Leakey

**Figure 21.9** (**a**) Footprints made in soft, damp volcanic ash 3.7 million years ago at Laetoli, Tanzania, as discovered by Mary Leakey. The arch, big toe, and heel marks are those of upright early hominids. (**b**) Fossil remains of Lucy, one of the earliest known australopiths. Lucy was only 1.1 meters (3 feet, 8 inches) tall. The density of her limb bones is indicative of very strong muscles.

a

b

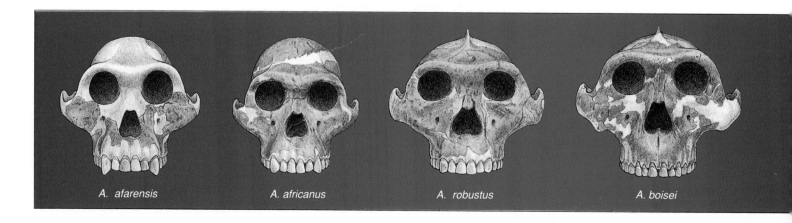

A. afarensis      A. africanus      A. robustus      A. boisei

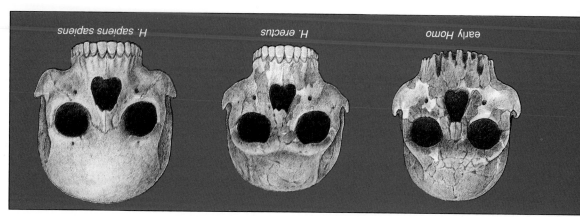

**Figure 21.11** Comparison of skull shapes of the early hominids relative to modern humans (*Homo sapiens*). In general, australopiths had a small brain and large face, compared with the larger brained, smaller faced forms of the genus *Homo*. The drawings are not all to the same scale. White areas of skulls are reconstructions.

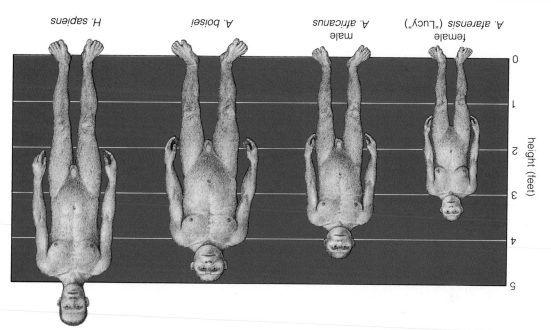

**Figure 21.10** Comparison of size and stature of australopiths with modern humans.

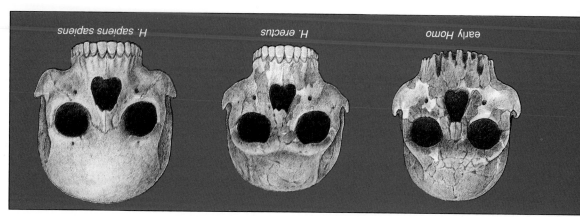

(Ape thigh bones splay out from the pelvis, and apes have a waddling, quadrupedal gait.) Besides Lucy's bones, the australopiths left unmistakable evidence of bipedalism—the footprints shown in Figure 21.9.

In other traits, the australopiths were transitional between the Miocene apes and later hominids. Bipedalism had freed their hands—and fossil handbones suggest they were as dexterous as we are. Like small apes, their cranial capacity was about 400 cubic centimeters. Unlike apes, their jaw was slightly bow-shaped. And their dentition suggests that some were omnivores and others, vegetarians.

For example, *A. africanus* was probably omnivorous. Its cheek teeth formed a platform that could grind plants, but its incisors were relatively large, as is the case for carnivores. *A. robustus* had strong jaw muscles and a large grinding platform typical of plant-eaters. *A. boisei* had very large, heavily cusped molars. This hominid may have specialized in chewing seeds, nuts, and other tough plant material.

### Stone Tools and Early *Homo*

By about 2.5 million years ago, hominids started making stone tools. (Before then, they could have used sticks and other perishable materials as tools, much like apes do today.) The first toolmaker may have been the hominid called "early *Homo*" in one scheme (and "late australopith" or *H. habilis* in others). Compared to australopiths, early *Homo* had a smaller face, more generalized teeth, and a larger brain (Figures 21.10 and 21.11). This hominid apparently was a scavenger and gatherer of plant material, small animals, and insects. And it may have been ancestral to modern humans.

At some point, early *Homo* began using rocks to crack open animal bones and expose the soft, edible marrow. This hominid also started using sharp-edged flakes, formed naturally (and not too often) as rocks tumbled through a river or down a hill. It used such flakes to scrape flesh from animal bones. At some point, it started to *shape* stone implements.

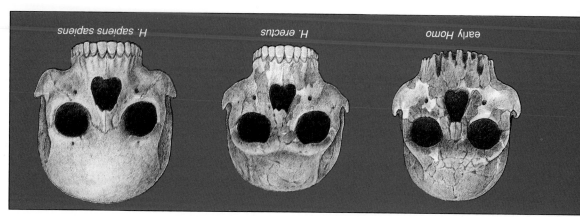

### Homo erectus

Between 1.5 million and 300,000 years ago, during the Pleistocene, extreme shifts in climate put the adaptive potential of the early humans to the test. A new, larger brained species, *Homo erectus*, emerged at this time.

Global temperatures declined during the Pleistocene. The polar ice caps grew and advanced outward until glaciers covered over one-fourth of the land, including vast areas of northern Europe, Asia, and North America. Some glaciers were 2 miles high! They advanced and retreated more than once at their southern limits. When they grew, sea levels fell. When they melted, sea levels rose. Concurrently, land bridges and coastal regions were alternately submerged and exposed. It was during the interglacials that *H. erectus* migrated out of Africa and into Southeast Asia, China, and Europe.

Compared with modern humans, *H. erectus* had a thick-walled, primitively shaped skull (Figure 21.12), but brain size was much the same. The large brain size correlates with the spectacular travels of *H. erectus*. No other hominids had ranged so far from their original environment. It correlates also with refinements in tool-making. And *H. erectus* seemed to have made controlled use of fire—and how advantageous campfires would have been—in the moves to colder regions.

The earliest known "manufactured" tools were crudely chipped pebbles, first discovered by Mary Leakey at Olduvai Gorge (Figure 21.12). This African gorge cuts through a sequence of sedimentary deposits, with the more recently deposited layers containing ever more sophisticated tools. The sequence gives insight into the increasingly refined ways in which hominids exploited a major food source—the abundant game of open grasslands.

Today, the stone tools of early *Homo* are barely recognizable as tools, and we still don't know exactly how they were put to use. It seems likely that a single tool served different functions, such as digging the ground for roots and tubers, smashing bones for marrow, and poking insects out from tree bark.

The skeletal remains of our early hominid ancestors do indicate that the males were notably bigger than the females. We can observe the same sort of sexual dimorphism among a number of primate species. Those primates live in multiple-male groups, with no long-term relationship between a given male and female. Thus, even though our early ancestors were adopting some behavioral traits that we would consider "human," their social groups probably were very different from what we consider to be the families of most modern Western societies.

**Figure 21.12** Representatives of the more than 37,000 stone tools recovered from Olduvai Gorge. *Upper row:* A crude chopper and more advanced forms having a joint, as well as a sharp edge. The stone ball may represent a transition from passive to aggressive tool use. It resembles the Argentine bolas, which are strung together on lengths of hide and thrown at animals to entangle the legs and bring them down. *Lower row:* A hand ax and a cleaver.

### Homo sapiens

Somewhere between 300,000 and 200,000 years ago, modern humans (*H. sapiens*) apparently arose from *H. erectus* stock. There was a transitional period, often with different forms coexisting in time, then *H. erectus* gradually disappeared. Early *H. sapiens* had a rounder, higher skull, with a face more delicately structured. The teeth and jaw were rather small. Many forms had a chin (their predecessors did not). It was not until about 40,000 years ago that anatomically modern humans, *H. sapiens sapiens*, emerged. Before then, a number of archaic groups lived in Europe, the Near East, and China. Among them were the Neandertals.

Neandertals had heavy facial bones and often large browridges. They had larger brains than we do—ranging between 1,300 and 1,750 cubic centimeters. And they were proficient hunters and gatherers in diverse environments. They lived in caves, rock shelters, and open-air camps, in varied climates. Yet they disappeared suddenly about 35,000 or 40,000 years ago.

When we think about the anatomically modern humans who created the cultural treasures at Lascaux and elsewhere, many questions come to mind. First and foremost, *where did they come from?* Think about the char-acteristic range of phenotypes in African, European, East Asian, and aboriginal Australian populations. The available evidence indicates that ancient hominid stocks evolved in each major geographic region, but whether they did so in isolation of one another is open to question. Did gene flow knit them together over the past 700,000 years? Or did anatomically modern humans evolve in Africa, then radiate during the past 100,000 years into different parts of the world? Some evidence suggests so (see the *Doing Science* essay). If so, did they displace native hominid populations entirely or were there instances of gene flow? At this writing we simply do not know the answers to any of these questions. We can look forward to new discoveries that may help clarify the picture.

From 40,000 years ago to the present, human evolution has been almost entirely cultural rather than bio-logical, and so we leave the story. From the biological perspective, however, we can make these concluding remarks: Humans have spread throughout the world by rapidly devising the cultural means to deal with a broad range of environmental conditions. Compared with their predecessors, modern humans have developed spec-tacularly rich and varied cultures, moving from "stone-age" technology to the age of "high tech." Yet hunters and gatherers persist in parts of the world, attesting to the great plasticity and depth of human adaptations.

---

*Doing Science*

## Mitochondria and Human Evolution

Mitochondria are organelles, about the size of a bacterium, that produce wonderful amounts of ATP by the aerobic pathway. They also have their own store of DNA, which they can replicate more or less independently of what goes on in the rest of the cell.

In 1988, geneticists made a startling pronouncement. After comparing samples of mitochondrial DNA (mtDNA) from human populations around the world, they reported that all living humans appear to be descended from a single female who lived in Africa about 200,000 years ago. News reporters quickly dubbed this female the "Mitochondrial Eve."

What prompted the researchers to study mitochondria in the first place? Consider that humans inherit pairs of genes, on pairs of maternal and paternal chromosomes. With mitochondria, the situation is different. All the mitochondria in a human body are inherited from the mother.

A human sperm delivers the paternal chromosomes into an egg at fertilization, but that is just about all it does. All of the cytoplasmic machinery that the first cell of the new individual will require—and this includes mitochondria—is already in place in the egg cytoplasm. Before any cell divided in the growing embryo, mitochondria had replicated their DNA and divided beforehand. Thus mtDNA cannot be "contaminated" by the random gene shufflings of sexual reproduction. The only things that can alter mtDNA are copying mistakes—mutations—that arise when it is being replicated. Such mutations have occurred much more frequently in mtDNA than in nuclear DNA.

And a number of those mutations are neutral. They have had no significant effect on phenotype, and so have not been culled from the mtDNA. Thus, to the geneticists who were studying human populations, the neutral mutations in mtDNA seemed like an excellent choice for

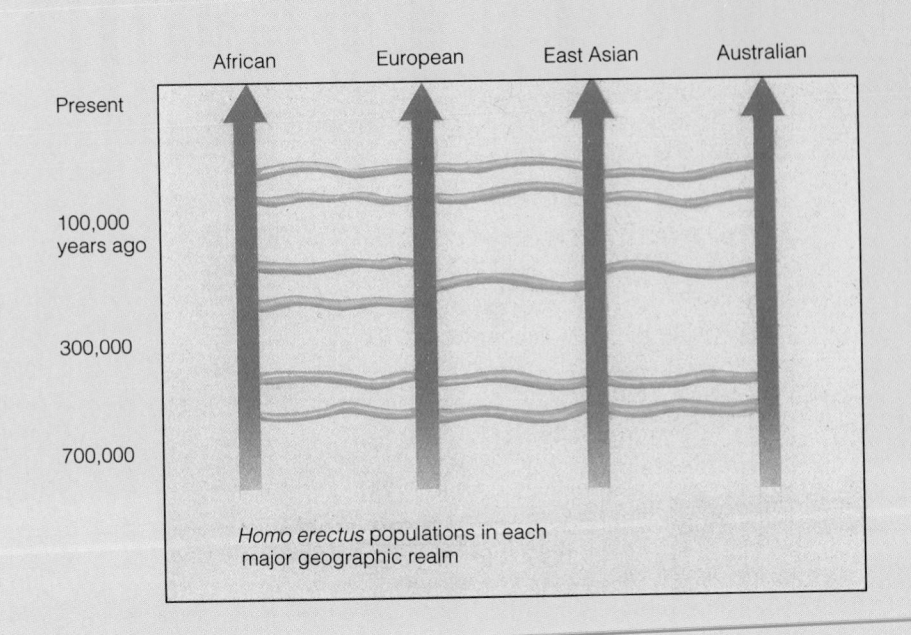

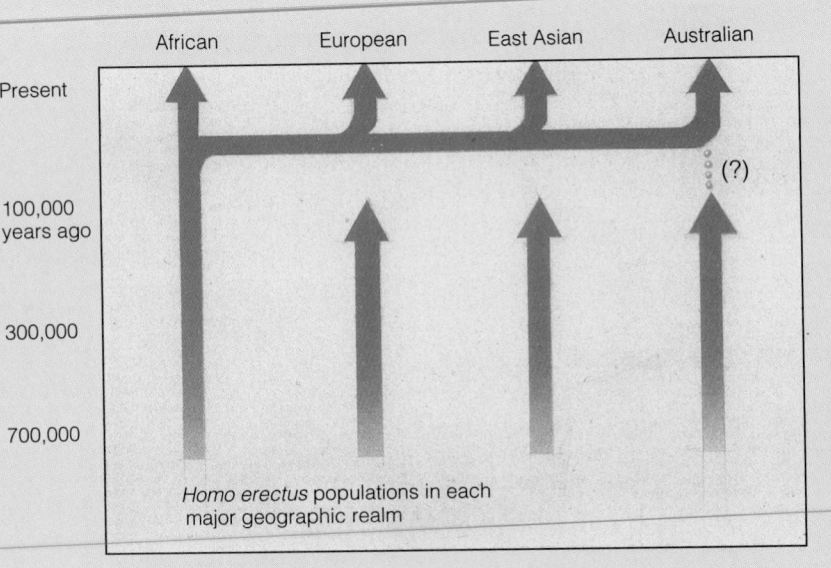

**a** Two models of evolutionary roads that led to the characteristic phenotypes of modern human populations. Green arrows indicate evolutionary lines, yellow bands indicate gene flow.

constructing a molecular clock. Such a clock could be used to assign a date to every branch point on an evolutionary tree.

Of course, the clock makes sense only when we assume that neutral mutations accumulate at a constant rate over evolutionary time, as described on page 298. Today there is disagreement over the average rate of change. Some researchers argue for a slow or inconsistent rate, which would place the common human ancestor at about 500,000 years ago. Others argue for a faster, more constant rate—which would place the common ancestor at about 150,000 years ago.

Data on mtDNA have now been correlated with new evaluations of the fossil record as well as with results from refined methods of dating fossils. When all the data are pooled, they suggest to some researchers that regional variations in modern human populations arose *after* modern humans had evolved in isolation in Africa. By this interpretation, Neandertalers and other long-since-vanished hominids may represent separate lineages that were displaced as the modern hominids radiated out of Africa. If gene flow did occur during this time, it may have been restricted to matings with the Australian populations that were already established when the modern hominids arrived.

The controversy goes on, however. Some researchers still present a strong case for the evolutionary model shown in the upper diagram of Figure *a*.

## SUMMARY

1. Primates include prosimians (lemurs and related forms) and anthropoids (including tarsioids, monkeys, apes, and humans). All are descended from small rodentlike mammals that evolved about 65 million years ago.

2. The anthropoids include the New World monkeys, Old World monkeys, apes, and humans. Apes and humans are classified further as hominoids. Only the early and modern human forms are further classified as hominids.

3. The following evolutionary trends occurred among the primates as a whole. Many of the trends are related to the tree-dwelling ancestry of the group:

   a. From a four-legged gait to specialized modes of locomotion, including bipedalism (habitual free-striding, two-legged gait). This trend involved changes in the shoulders, backbone, pelvic girdle, legs, and feet.

   b. Increased manipulative skills owing to modification of the hands, which began to be freed from their load-bearing function among tree-dwelling primates. Primates able to stand or sit upright had an advantage in reaching for fruit, holding onto infants, and not falling out of trees and becoming prey to ground-dwelling predators.

   c. Less reliance on the sense of smell, more reliance on enhanced daytime vision, including color vision and depth perception. This trend began in the visually complex environment of the trees.

   d. From specialized to omnivorous eating habits. This trend was pronounced during the Miocene, when tropical forests were giving way to mixed woodlands and savannas.

   e. Brain expansion and reorganization. This trend began among mammals generally but accelerated during hominid evolution. Larger, more complex brains are cor-

related with increasingly sophisticated technology (from simple to refined tools) and with social development.

   f. All of these trends were the foundation for the remarkable plasticity of the hominids (their capacity to remain flexible and to adapt to a wide range of demands imposed by an unpredictable, complex environment).

4. The oldest known primates date from the Paleocene (65 to 54 million years ago). The first prosimians date from the Eocene (54 to 38 million years ago). Anthropoids emerged by the Oligocene (38 to 25 million years ago).

5. An adaptive radiation among anthropoids started in the Miocene as environments became cooler and drier. Forests shrank, grasslands spread, and speciation was favored as subpopulations of anthropoids became reproductively isolated and adapted to local conditions. Eventually the great apes and hominids emerged.

6. The first hominids (australopiths) emerged between 10 and 5 million years ago. All were bipedal, with a larger brain than their predecessors. Some were slightly built and omnivorous. Others were heavily built, taller, and muscular vegetarians.

7. Fossils of early Homo, the first known representative of the human lineage, date from about 2.5 million years ago. Early Homo was omnivorous, larger brained than its predecessors, used simple tools, and showed some social development.

8. Homo erectus fossils associated with abundant cultural artifacts date from about 1.5 million to 300,000 years ago. It was larger brained than its predecessors, its cultural evolution was more pronounced, and it was adapted to a wide range of habitats.

9. Fully human forms (H. sapiens) emerged between 300,000 and 200,000 years ago, possibly from H. erectus stock. By 40,000 years ago, fully modern forms (H. sapiens sapiens) had evolved. From that point on, cultural evolution outstripped biological evolution of the human form.

## Review Questions

1. What are the general evolutionary trends that occurred among the primates as a group? What way of life apparently was the foundation for these trends? 333-335

2. What conditions seem to have been responsible for the great adaptive radiation of apelike forms during the Miocene? 336

3. What is the difference between "hominoid" and "hominid"? Are we hominoids, hominids, or both? 333

4. Describe the key characteristics of hominid evolution. How do they relate to the concept of plasticity? 337-338

## Self-Quiz (Answers in Appendix IV)

1. Primates include _____.
a. lemurs    c. apes    e. all of the above
b. monkeys    d. humans

2. _____ are hominoids; only _____ are hominids.
   a. lemurs; monkeys and their immediate ancestors
   b. apes and humans; humans and their recent ancestors
   c. monkeys; apes and their recent ancestors
   d. monkeys, apes, and humans; apes and their recent ancestors

3. The key trends in primate evolution began in the _____.
   a. savanna          c. trees
   b. water            d. forest floor

4. Which of the following did not occur on the evolutionary road leading to humans?
   a. bipedalism
   b. hand modification that increased manipulative skills
   c. shift from omnivorous to specialized eating habits
   d. less reliance on smell, more on vision
   e. brain expansion and reorganization
   f. all were evolutionary trends

5. Early hominids displayed great plasticity. This means that _____.
   a. they were adapted to a wide range of demands in complex environments
   b. they had flexible bones that cracked easily
   c. they were limber enough to swing through the trees
   d. they were adapted for a narrow range of demands in complex environments

6. The oldest known primates date from the _____.
   a. Miocene (25 million to 13 million years ago)
   b. Paleocene (65 million to 54 million years ago)
   c. Oligocene (38 million to 25 million years ago)
   d. Pliocene (13 million to 2 million years ago)

7. The first known hominids are generally classified as _____.
   a. *Homo*          c. cercopiths
   b. dryopiths       d. australopiths

8. Fossils of the early *Homo*, the first known representative of the human line, date from _____ million years ago.
   a. 8               c. 4
   b. 6               d. 2

9. Match the primates with their descriptions.
   _____ australopiths
   _____ *Homo erectus*
   _____ anthropoids
   _____ *Homo sapiens sapiens*
   _____ hominids
   
   a. extinct Pleistocene species of our genus
   b. only humans and their recent ancestors
   c. fully modern humans
   d. the first known hominids
   e. monkeys, apes, humans

## Selected Key Terms

anthropoid 332
australopith 338
bipedalism 334
brachiation 333
culture 335
dentition 332
early *Homo* 339
hominid 333
hominoid 333

*Homo erectus* 340
*Homo sapiens* 341
mitochondrial DNA 341
opposable movement 335
plasticity 337
prehensile movement 335
primate 332
prosimian 332

## Readings

Conroy, G. C. 1990. *Primate Evolution*. New York: Norton. A detailed look at the history and adaptations of primates.

Lewin, R. 1989. *Bones of Contention: Controversies in the Search for Human Origins*. New York: Simon & Schuster. A readable account of what is known, and what is still being debated, about human evolution.

Reader, J. 1981. *Missing Links*. Boston: Little, Brown. Readable, exquisitely illustrated historical account of discoveries concerning human evolution.

Rensberger, R. 1981. "Facing the Past." *Science 81* 2(8):41–50. Intriguing look at how artists' reconstructions can bias our perceptions of what the early hominids looked like.

Stringer, C. 1990. "The Emergence of Modern Humans." *Scientific American*. 263:98–104. Examines fossil and biochemical evidence regarding human origins.

Weiss, M., and A. Mann. 1990. *Human Biology and Behavior*. Fifth edition. New York: Harper Collins. Outstanding introduction to human evolution from a biological perspective.

FACING PAGE: *Patterns of diversity in nature, here represented by different species of plants and fungi.*

## The Unseen Multitudes

From *E. coli* to elephants, from clams to coast redwoods, all organisms are related to one another, even though the evolutionary distance between any two types might be stupendous. In the preceding unit, this concept guided us through some of the observations and methods being used to discern evolutionary relationships. We turn now to the existing spectrum of life, beginning with the two kingdoms of single-celled forms—the monerans (bacteria) and protistans. The different types of infectious particles called viruses also warrant our attention here. Viruses are *not* alive, but they do have impact on organisms in all five kingdoms.

Bacteria and most protistans are *microorganisms*, meaning that they are too small to be seen without a microscope. Exactly how small is "small"? The answer may stun you. Let's start with the size of a more familiar organism—ourselves. Have you ever heard someone say they are 1/1000 of a mile tall? Probably not. We rarely think of ourselves in units that are so much larger than we are. By contrast, microscopists measure bacteria and protistans in terms of micrometers. A micrometer is 1/1000 of a millimeter in length—and a millimeter is only about as wide as the dot of this "i." Yet a *thousand* cells of a common bacterial species could line up across the space of that dot. Also take a look at Figure 22.1, which shows the size of one type of bacterium compared to the tip of a pin. Viruses are even smaller than this. They are measured in nanometers. A nanometer is a mindboggling *one-millionth* of a millimeter!

Without microscopes, we never would have been able to observe anything smaller than about 100 micrometers—even though the inhabitants of the microscopic

100 μm

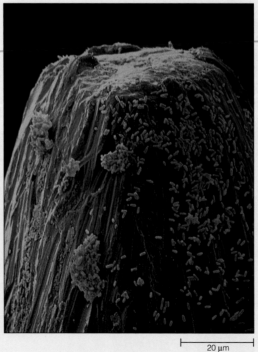

20 μm

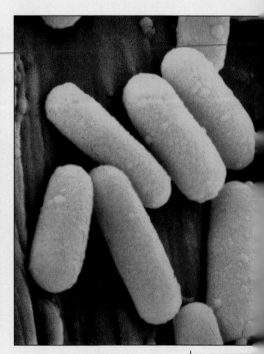

1 μm

world vastly outnumber all other organisms combined. There may be billions or trillions of bacteria in the gut of just one large mammal. There may be just as many protistans floating about in a few cubic meters of a woodland pond. Such numbers result from a staggering potential for reproduction, which bacteria accomplish by dividing in two. Under the best of all possible conditions—say, in a nutrient-stocked petri dish—some types divide about every twenty minutes. At that rate, a single bacterial cell might produce more than a billion descendants in less than half a day.

Fortunately, microorganisms do encounter limits to growth. Sooner or later, the by-products of their metabolic activities typically ruin the very conditions that initially favored their rapid reproduction. Besides, other organisms or viruses have ways to kill them off and so help stop them from taking over the world.

Actually, we have no reason to love or even be indifferent to a good number of bacteria, protistans, or viruses. Many of them are *pathogens*, or disease-causing agents. The pathogenic types infect other organisms, meaning that they can invade and multiply inside the cells or tissues of a host organism. Disease follows an infection when the pathogen's activities interfere with the host's normal body functions.

Even so, the pathogenic types should not give the members of the entire microscopic world a bad name. A great many microorganisms are food producers—photosynthetic autotrophs—for many communities. Given their vast numbers, their photosynthetic activities also play major roles in the global cycling of carbon (page 50). Many more microorganisms are decomposers and they, too, play major roles in the cycling of vital materials on a global scale.

**Figure 22.1** How small are bacteria? Shown here, *Bacillus* cells on the tip of a pin.

KEY CONCEPTS

1. All living organisms have the metabolic means to maintain themselves, grow, and reproduce. Viruses are infectious agents that are not alive. It takes the metabolic machinery of a living host cell to assemble the viral nucleic acids and proteins necessary to construct new virus particles.

2. Bacteria, the sole members of the kingdom Monera, are structurally simple. They are prokaryotic, meaning they do not have a membrane-bound nucleus and other specialized organelles that are characteristic of eukaryotic cells. However, bacteria as a group show far more metabolic diversity than any other kind of organism, and many make complex behavioral responses to their environment.

3. The kingdom Protista is dominated by eukaryotic organisms that are single-celled. Beyond this, different protistans are diverse in their modes of nutrition and other characteristics. Protistans may be living examples of the kinds of ancient single cells that gave rise to the kingdoms of multicelled eukaryotes—the fungi, plants, and animals.

## VIRUSES

### General Characteristics of Viruses

In ancient Rome, *virus* meant "poison" or "venomous secretion." In the late 1800s, this rather nasty word was bestowed on a newly discovered class of pathogens, smaller than the bacteria being studied by Louis Pasteur and others of that era. Many viruses deserve their name, for they can have devastating effects on humans, cats, cattle, insects, crop plants, fungi, and bacteria. Name any organism, and there probably are one or more kinds of viruses that can infect it.

Today we define a **virus** as a noncellular infectious agent having two characteristics. *First*, a virus consists of a nucleic acid core (its genetic material) within a pro-

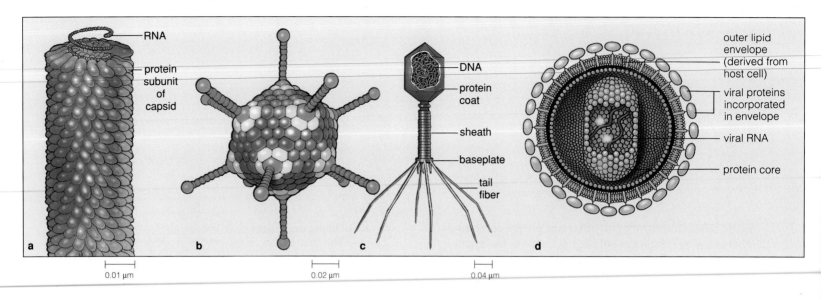

**Figure 22.2** Representative viruses, showing a few of the diverse kinds of capsids. (**a**) Rod-shaped capsid of the tobacco mosaic virus. (**b**) Polyhedral capsid of an adenovirus. (**c**) Component parts of a T4 bacteriophage. (**d**) Lipid envelope around the capsid of HIV, the causative agent of AIDS.

tective protein coat. *Second*, a virus can be replicated only after its genetic material has entered a host cell and subverted the cell's biosynthetic machinery. When outside of a host, a virus is no more alive than a chromosome is alive.

The genetic material of a given virus is DNA or RNA, but not both, and it may be a single- or double-stranded molecule. The viral coat, or **capsid**, comes in many different forms. Figure 22.2 shows four examples. The rod-shaped capsid from the tobacco mosaic virus has 2,200 identical protein subunits coiled helically around the nucleic acid core. Another capsid has protein subunits arranged as a many-sided structure (polyhedron). The third capsid is equipped with a sheath, tail fibers, and other components. The fourth capsid shown is surrounded by a lipid envelope. Such envelopes are remnants of membrane from the last cell that served as a host, although viral proteins have become incorporated into it.

### Viral Replication

Viruses replicate in a variety of ways. In all cases, however, the viral DNA or RNA is copied repeatedly, many viral proteins are synthesized, and many new virus particles are assembled inside a suitable host cell. By definition, a cell is a potential *host* if a virus can chemically recognize and lock onto specific molecular groups at its surface.

Regardless of the variations, viral replication always proceeds through the following stages:

**1.** The virus chemically recognizes and becomes attached to a host cell.

**2.** The whole virus or its genetic material alone (DNA or RNA) enters the cell's cytoplasm.

**3.** In an act of molecular piracy, information contained in the viral DNA or RNA directs the host cell into replicating viral nucleic acids and synthesizing viral enzymes, capsid proteins, and sometimes other viral proteins (which become incorporated into the host plasma membrane).

**4.** The viral nucleic acids, enzymes, and capsid proteins are assembled into new virus particles.

**5.** The newly formed virus particles are released from the infected cell.

Viruses usually replicate by lytic or temperate pathways. In a **lytic pathway**, stages 1 through 4 above proceed quickly, and virus particles are released as the cell undergoes lysis. In this context, lysis means the host cell ruptures and then dies after its contents are lost.

In **temperate pathways**, the virus does not kill the host cell outright. Instead, the infection enters a period of **latency**, in which viral genes remain inactive inside the host cell and any of its descendants. In some cases of latency, the viral genes actually become integrated into the host DNA, are replicated along with it, and so are passed along to all the daughter cells. In time, damage

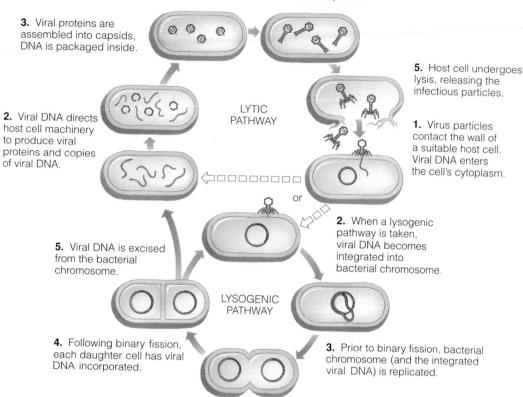

**4.** Tail fibers, other components are added to capsids. New viral particles are completed.

**3.** Viral proteins are assembled into capsids, DNA is packaged inside.

**2.** Viral DNA directs host cell machinery to produce viral proteins and copies of viral DNA.

LYTIC PATHWAY

**5.** Host cell undergoes lysis, releasing the infectious particles.

**1.** Virus particles contact the wall of a suitable host cell. Viral DNA enters the cell's cytoplasm.

or

**2.** When a lysogenic pathway is taken, viral DNA becomes integrated into bacterial chromosome.

**5.** Viral DNA is excised from the bacterial chromosome.

LYSOGENIC PATHWAY

**4.** Following binary fission, each daughter cell has viral DNA incorporated.

**3.** Prior to binary fission, bacterial chromosome (and the integrated viral DNA) is replicated.

**Figure 22.3** Replication of a bacteriophage. In a lytic pathway, a virulent bacteriophage replicates in a host bacterial cell, which then undergoes lysis to release new virus particles that can infect other host cells. Some of the so-called "temperate" bacteriophages can enter a lysogenic pathway. In this case, the viral DNA becomes integrated into the host chromosome and is replicated along with it through successive cell divisions. DNA damage or other events can induce the bacterium to enter the lytic pathway.

to the DNA or some other event may activate transcription of the viral genes. Ultimately, new virus particles can be produced and the infected cells destroyed.

## Major Classes of Viruses

**Bacteriophages.** The **bacteriophages** are a class of viruses that infect bacterial cells. Although bacteriophages may have adverse effects on a host cell, they have been put to good use in genetics research. Chapter 13, for example, describes how bacteriophages were used as research tools in early experiments that were designed to reveal whether DNA or proteins are the molecules of inheritance. Bacteriophages are now being used as research tools in genetic engineering.

For bacteriophages, replication can proceed by lytic or temperate pathways (Figure 22.3).

**Animal Viruses.** Table 22.1 lists a few types of animal viruses that are responsible for diseases as varied as warts, chickenpox, the common cold, and several forms of cancer. Among them are the influenza viruses responsible for recurring pandemics. (A pandemic is

**Table 22.1 Classification of Animal Viruses**

| DNA Viruses | Some Diseases |
|---|---|
| Adenoviruses | Respiratory infections |
| Herpesviruses: | |
| *H. simplex* type I | Oral herpes, cold sores |
| *H. simplex* type II | Genital herpes |
| Varicella-zoster | Chickenpox, shingles |
| Epstein-Barr | Infectious mononucleosis, implicated in some cancers |
| Papovaviruses | Benign and malignant warts |
| Parvoviruses | Roseola (fever, rash) in small children; aggravates sickle-cell anemia |
| Poxviruses | Smallpox, cowpox |

| RNA Viruses | Some Diseases |
|---|---|
| Picornaviruses: | |
| Enteroviruses | Polio, hemorrhagic eye disease; hepatitis A (infectious hepatitis) |
| Rhinoviruses | Common cold |
| Togaviruses | Encephalitis, yellow fever, dengue fever |
| Paramyxoviruses | Measles, mumps |
| Rhabdoviruses | Rabies |
| Coronaviruses | Respiratory infections |
| Orthomyxoviruses | Influenza |
| Arenaviruses | Hemorrhagic fevers |
| Reoviruses | Respiratory, intestinal infections |
| Retroviruses: | |
| HTLV I, II | Associated with cancer |
| HIV | AIDS, ARC |

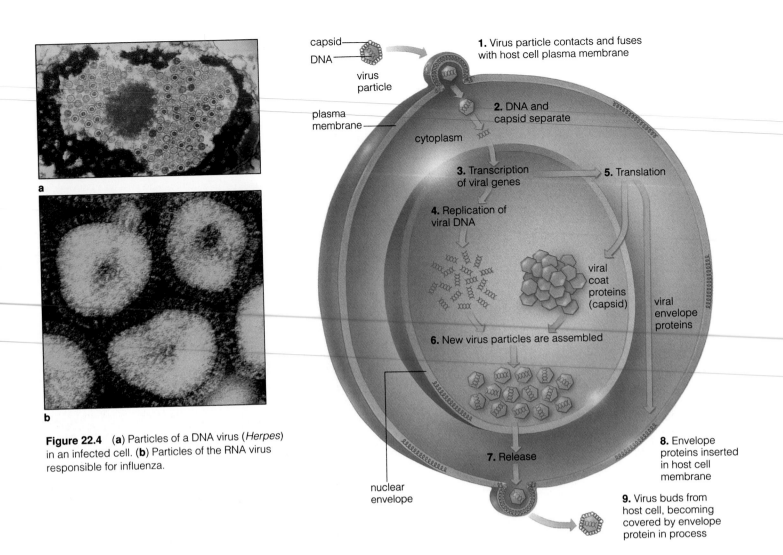

capsid
DNA
virus
particle

**1.** Virus particle contacts and fuses with host cell plasma membrane

**2.** DNA and capsid separate

plasma membrane

cytoplasm

**3.** Transcription of viral genes

**5.** Translation

**4.** Replication of viral DNA

viral coat proteins (capsid)

viral envelope proteins

**6.** New virus particles are assembled

nuclear envelope

**7.** Release

**8.** Envelope proteins inserted in host cell membrane

**9.** Virus buds from host cell, becoming covered by envelope protein in process

**a**

**b**

**Figure 22.4** (**a**) Particles of a DNA virus (*Herpes*) in an infected cell. (**b**) Particles of the RNA virus responsible for influenza.

**Figure 22.5** Replication of an enveloped animal virus. In this case, replication occurs in the nucleus; in some viruses, it takes place in the cytoplasm. Also, some other viruses acquire their envelope as they emerge from the nuclear envelope.

a worldwide blitz of human hosts, as described in the *Commentary* on page 352.) Figure 22.4 shows two representative DNA and RNA animal viruses.

Many animal viruses infect animal cells through endocytosis and depart from them either by exocytosis or by lysis. (The mechanisms of endocytosis and exocytosis are described on page 86.) Briefly, the plasma membrane of a host cell is stimulated into dimpling inward after virus particles have become attached to it (Figure 22.5). The resulting endocytic vesicle transports the virus particles into the cytoplasm. Later, new particles escape from the infected cell in vesicles that bud from the host cell's plasma membrane. You can see an example of this in the illustration on page 694. It shows particles of the HIV virus budding away from the surface of an infected cell.

Animals have defenses against the initial attack of different animal viruses. Even so, some virus particles may escape detection and enter latency, only to be reactivated at some future time. For example, the *Herpes*

*simplex* virus is latent in the vast majority of people—and it can cause cold sores each time it is reactivated. Similarly, humans who have had chickenpox might still harbor the virus—which may cause a skin disease (shingles) if it becomes reactivated later in life.

*Retroviruses* are RNA viruses that infect animal cells and that follow temperate pathways of replication. You may wonder how the genetic material of an *RNA* virus becomes integrated into host cell DNA. After the viral RNA molecule enters the cytoplasm, a viral enzyme (reverse transcriptase) uses it as a template and synthesizes a DNA "transcript" on it. The transcript, not the RNA itself, becomes integrated into the host DNA. This is what has happened in people infected by the human immune deficiency virus, or HIV—the causative agent of AIDS (page 693).

**Plant Viruses**. Table 22.2 lists some of the known plant viruses. In all cases, such viruses can cause an infectious disease only after they have successfully penetrated the protective wall of plant cells. Typically, aphids and other insects that feed on plants assist in the infection. Virus particles may be clinging to their piercing or sucking devices, and when those devices penetrate plant cells, the virus enters with them.

Viruses are known to cause more than a thousand diseases among plants. Most are RNA viruses. Viral diseases can greatly reduce the yields of a variety of crops, including potatoes, tomatoes, cauliflowers, cucumbers, turnips, and barley. Outward symptoms of infection include mottled and often blistered leaves, misshapen or unusually small fruit, tumors on roots, and color changes in flowers.

Figure 22.6 shows how certain viruses affect the outward appearance of tulips and some other plants. The tulip blossom has colorless streaks because some of its pigment-containing cells were attacked by the color-breaking virus. Because tulip fanciers sometimes admire the variegated blossoms, commercial growers keep virus-infected bulbs in their inventory—but they keep them isolated from noninfected bulbs.

### Viroids and Other Unconventional Agents

Viruses may not be the most stripped-down disease agents. *Viroids*, a category of plant pathogens, are naked strands or circles of RNA, with no protein coat. Viroids are mere snippets of genes, smaller than the smallest known viral DNA or RNA molecule. Yet they can have devastating effects on citrus, avocados, potatoes, and other crop plants. Apparently, enzymes already present in a host cell synthesize viroid RNA, then use it as templates for building new viroids.

As-yet unidentified infectious agents cause some rare, fatal diseases of the nervous system, including *scrapie* in sheep, and *kuru* and *Creutzfeldt-Jakob* disease in humans. Possibly the diseases are caused by infectious protein particles, tentatively dubbed *prions*. Prions might be synthesized according to information contained in mutated genes. Researchers studying scrapie have isolated the genes coding for altered forms of a protein in infected cells. However, such diseases might also be caused by a snippet of foreign nucleic acid.

In conclusion, viruses, viroids, and possibly other infectious agents operate at the molecular level of organization. And they operate only inside *living* cells. We now cross the threshold into the living world, to the simplest representatives of the cellular level of organization—the bacteria.

**Table 22.2  Some Common Plant Viruses**

| Type of Virus | Target |
|---|---|
| **RNA Viruses:** | |
| Closterovirus | Beets |
| Comovirus | Cowpeas |
| Cucumovirus | Cucumbers |
| Hordeivirus | Barley |
| Potaxvirus | Potatoes |
| Tobamovirus (tobacco mosaic virus) | Tobacco |
| **DNA Viruses:** | |
| Caulimovirus | Cauliflower |
| Geminivirus | Maize |

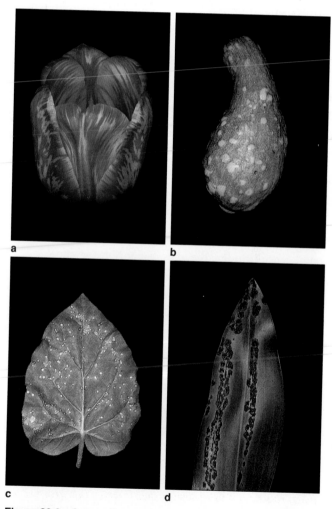

**Figure 22.6** Outward symptoms of some viral infections of plants. (**a**) Variegated blossom of a tulip (*Tulipa gesneriana*), resulting from a relatively benign viral infection that affects pigment formation in different tissue regions. (**b**) Mottling of a summer crookneck squash (*Curcurbita pepo*), caused by the watermelon mosaic virus. (**c**) Leaf of a tobacco plant (*Nicotiana glutinosa*) after infection by the tobacco mosaic virus. (**d**) Leaf of an orchid (*Grammatophyllum scriptum*) infected by a rhabdovirus.

# Infection and Human Disease

The human body is continually under siege. It is the preferred home for a variety of pathogens—including viruses, bacteria, fungi, protozoans, and parasitic worms. "Pathogen" means that these agents are capable of *infection*—of invading and then multiplying in the cells and tissues of a host organism. *Disease* may be the outcome of an infection. It results whenever the body cannot mobilize its defenses quickly enough to prevent the pathogen's activities from interfering with normal body functions.

## Modes of Transmission

The following modes of disease transmission are common:

1. Direct contact, as by touching open sores or other lesions on an infected person. (This is where "contagious disease" comes from; the Latin *contagio* means touch or contact.) Infected people also can transfer pathogens from such lesions to their own hands or mouth and so contaminate someone else through a handshake or a kiss. Gonorrhea, syphilis, and other sexually transmitted diseases are spread primarily by direct contact.

2. Indirect contact, as by touching doorknobs, food, diapers, hypodermic needles (as used by drug abusers), or other objects that were previously in contact with an infected person. Food that is moist, not refrigerated, and not too acidic can be contaminated by an assortment of pathogens, including the ones responsible for amoebic dysentery, bacterial dysentery (*Shigella*), and typhoid fever.

3. Inhaling pathogens that have been injected into the air, as by coughs and sneezes (Figure *a*).

**a** A full-blown sneeze.

4. Encounters with biological *vectors*. These include mosquitoes, flies, fleas, ticks, and other arthropods that transport pathogens from infected people or contaminated material to new hosts. Many pathogens only use vectors like taxis, so to speak. Many other pathogens depend on vectors as *intermediate hosts*. This means that a portion of the pathogen's life cycle must be carried out inside the vector. Mosquitoes, for instance, are intermediate hosts for the causative agent of malaria (page 368).

## Patterns of Occurrence

Infectious diseases often are described in terms of their patterns of occurrence.

As the name suggests, *sporadic* diseases break out irregularly and affect only a few people. Whooping cough is an example. *Endemic* diseases occur more or less continuously, but they are localized to a relatively small portion of the population. Leprosy and tuberculosis are examples. So is impetigo, a highly contagious bacterial infection that is endemic to many day-care centers.

During an *epidemic*, a disease abruptly spreads through large portions of the population for a limited period. When influenza breaks out along the east coast of North America, this is an epidemic. When epidemics break out in several countries around the world in a given time span, they collectively are called a *pandemic*. AIDS is now pandemic; as many as 300 million may be infected by the year 1992.

## Virulence

*Virulence* refers to the damage that a pathogen can do to a susceptible host, and this depends on how easily it can avoid detection and destruction in the tissues of a particular individual. Every pathogen has its own virulence factors.

For example, a certain virus may cause a mild, "24-hour flu," whereas toxins produced by the bacterium *Clostridium botulinum* may kill a person within a few hours. As far as we know, the virus responsible for AIDS is difficult to transmit, but it eventually kills its victims.

There can be great individual differences in the responses to many pathogens. Most important is the overall state of the body's immune system. As you will see in Chapter 39, immune systems can be weakened (as by chronic fatigue or stress). In some cases, they can be essentially shut down (as during AIDS).

## What Antibiotics Can and Cannot Do

When your grandparents were children, possibly one-fourth of all annual deaths in the United States resulted from tuberculosis, pneumonia, and influenza. Other common killers included scarlet fever, smallpox, dysentery, whooping cough, and diphtheria. Infections associated with childbirth killed or maimed thousands of women. Today, antibiotics are used to treat these and other infectious diseases.

An *antibiotic* is a normal metabolic by-product of certain microorganisms, including some of the actinomycetes. Researchers have identified more than 2,500 naturally occurring antibiotics. These include penicillins, tetra-cyclines, and streptomycins. Antibiotics afford protection from different pathogens by interfering with gene expression or other functions. Streptomycin antibiotics, for example, block protein synthesis. Penicillins disrupt the formation of bonds necessary to hold molecules together in the cell walls of eubacteria. At least one penicillin derivative causes the wall to elongate abnormally until it ruptures; the bacterium sort of stretches itself to death.

Antibiotics do not work against viruses, although some fairly new *antiviral drugs* show promise. Acyclovir, for example, is used to treat cold sores and genital herpes. But it does not effect a cure; it only eases the patient's symptoms between recurring outbreaks. AZT, another antiviral drug, at least temporarily helps some patients.

Drug treatments often do not come without a price. Many antibiotics can have potent side effects on the body's own cells. Penicillins, tetracyclines, and other drugs all inhibit or destroy normal intestinal bacteria as well as the targeted pathogen, so they effectively disrupt digestive functions. Women who take antibiotics often simultaneously use an antifungal drug to combat yeast infections; the antibiotics upset the normal balance of microorganisms in the vaginal tract.

Bacterial resistance to antibiotics is a more serious problem. Use of antibiotics increased dramatically in the early 1960s. In effect, they became powerful agents of natural selection. They killed off the members of bacterial populations that were most susceptible to antibiotic treatment—and actually favored the more resistant ones. Today, strains of bacteria have become established that are resistant to prescribed drugs. New drugs are continually being sought. With an ability to multiply many times over in the space of a few hours, however, the mutated "enemies within" keep the battles raging.

# MONERANS—THE BACTERIA

Bacteria, the sole members of the kingdom Monera, are the most abundant and far-flung microorganisms. There are two great lineages, the **eubacteria** and the **archaebacteria**. Different members of those lineages live in hot springs, snow, deserts, "pristine" lakes, and deep oceans. There may be tens of billions of them in a handful of rich soil—several times the number of people on earth. Many bacteria live in or on other organisms. The ones in your gut and on your skin outnumber the cells of your body. Fortunately, animal cells are much larger than bacteria, so you can think of yourself as being only a small percentage "bacterial" by weight.

## Characteristics of Bacteria

**Metabolic Diversity.** Compared with other kinds of organisms, bacteria as a group show the most metabolic diversity. Like plants, some are *photosynthetic autotrophs*. They produce their own energy-rich organic compounds from simple inorganic substances, using sunlight as an energy source. But the pathways of photosynthesis among bacteria are more varied than they are among plants. A few kinds of bacteria are *chemosynthetic autotrophs*. They produce their own organic compounds, using a rather astonishing variety of simple inorganic substances as the energy source. The vast majority of bacteria are *heterotrophs*. One type or another uses organic compounds that have already been synthesized by protistans, fungi, plants, and animals.

**Structural Features.** Bacteria differ from all other kinds of organisms in being prokaryotic. They have no nucleus or other membrane-bound organelles (Table 22.3). Their metabolic reactions take place at the plasma membrane—the bacterial equivalent of organelle membranes. Proteins are synthesized rapidly at numerous ribosomes, which are distributed through the cytoplasm or attached to the plasma membrane.

---

**Table 22.3  Characteristics of Bacterial Cells**

1. Bacterial cells are prokaryotic (they have no membrane-bound nucleus or other organelles in the cytoplasm).

2. Bacterial cells have a single chromosome (a circularized DNA molecule); many species also have plasmids (page 246).

3. Most bacteria have a cell wall composed of peptidoglycan.

4. Most bacteria reproduce by binary fission.

5. Collectively, bacteria show great diversity in their modes of metabolism.

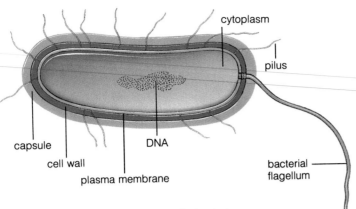

Figure 22.7 Generalized body plan of a bacterium.

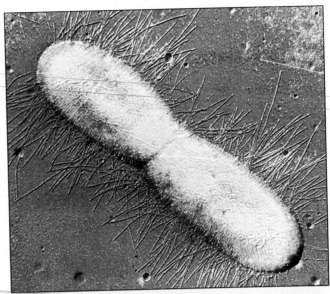

a

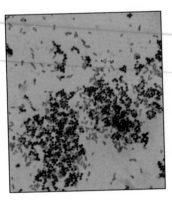

Figure 22.8 Example of Gram staining. Gram-positive *Staphylococcus aureus* remains purple when washed with organic solvents. Gram-negative *Escherichia coli* loses color easily. *E. coli* cells shown here appear red because they have been treated with a light-red dye (safranin) after being washed. Without this "counter-stain," they would be colorless.

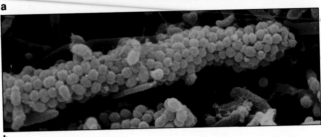

b

Figure 22.9 (a) Filamentous structures (pili) at the surface of a dividing *Escherichia coli* cell. (b) Bacteria attached by means of their glycocalyx to the surface of a human tooth.

Nearly all bacteria have a cell wall (Figure 22.7). In eubacteria, it is composed of a tough mesh of *peptidoglycan*. This is a type of molecule in which complex polysaccharide strands are cross-linked to one another by peptides. Peptidoglycan makes the wall strong and semi-rigid, and it helps maintain the bacterial cell in one of three shapes:

coccus
(plural, cocci);
*spherical*

rod or bacillus
(plural, bacilli);
*cylindrical*

spirillum
(plural, spirilla);
*helical*

Many bacterial species can be identified partly on the basis of the structure and composition of their cell wall. For example, in one type of staining reaction, the *Gram stain*, a purple dye is applied to bacterial cells. Then the cells are washed with alcohol and a pink dye is applied. Cell walls of "Gram-positive" bacteria retain the purple

stain when washed with alcohol. Those of "Gram-negative" bacteria lose the color when washed and become pink (Figure 22.8).

Many layers of peptidoglycan form the cell wall of Gram-positive bacteria. A thin layer of peptidoglycan forms the wall of Gram-negative bacteria, and it is surrounded by an outer membrane. One membrane component, endotoxin, causes some of the damage that the pathogenic types inflict on host organisms.

Exterior to the wall is a *glycocalyx*, a cover that is usually composed of simple polysaccharides. These molecules sometimes form a jellylike capsule that helps pathogenic types escape detection by infection-fighting cells of the host organism. The glycocalyx also helps bacterial cells attach to a surface. For example, it does this for certain bacteria that inhabit your mouth.

Among some bacterial species, filamentous structures may be anchored to the cell wall and plasma membrane. One is the *bacterial flagellum* (plural, flagella), a rather rigid protein filament that rotates like a propeller and so moves the cell through its fluid surroundings. The other kind of filament is a *pilus* (plural, pili) of the sort shown in Figure 22.9. Long pili help bacteria attach

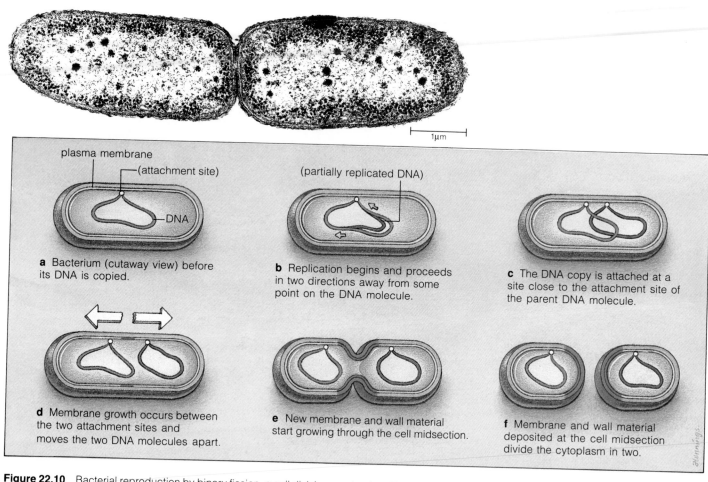

plasma membrane
(attachment site)

DNA

**a** Bacterium (cutaway view) before its DNA is copied.

(partially replicated DNA)

**b** Replication begins and proceeds in two directions away from some point on the DNA molecule.

**c** The DNA copy is attached at a site close to the attachment site of the parent DNA molecule.

**d** Membrane growth occurs between the two attachment sites and moves the two DNA molecules apart.

**e** New membrane and wall material start growing through the cell midsection.

**f** Membrane and wall material deposited at the cell midsection divide the cytoplasm in two.

**Figure 22.10** Bacterial reproduction by binary fission, a cell division mechanism. The micrograph shows the bacterium *Pseudomonas* after new membrane and wall material divided the cytoplasm of the parent cell.

to one another as a prelude to conjugation. As we saw in Chapter 16, bacterial conjugation is the transfer of genetic material from donor to recipient cells. Short pili help the cell adhere to a variety of surfaces. For example, the bacterium responsible for gonorrhea, a sexually transmitted disease, uses pili to attach to mucous membranes. Unless such attachment occurs, the bacterium cannot inflict damage.

## Bacterial Reproduction

Most bacterial cells reproduce by way of **binary fission**. By this cell division mechanism, a parent cell divides into two genetically identical daughter cells following DNA replication. The mechanism is simpler than mitosis and meiosis, the nuclear division mechanisms used by eukaryotes. For one thing, bacterial cells have only a single chromosome—a circular DNA molecule—to replicate and parcel out to the daughter cells.

Figure 22.10 is a step-by-step picture of binary fission. Daughter cells may stick together in pairs, clusters, or chains when they fail to separate completely after division.

## Plasmids Revisited

Besides the bacterial chromosome, the cells of many bacterial species have plasmids in their cytoplasm. As described on page 246, a **plasmid** is a small circle of "extra" DNA that carries only a few genes and that can be replicated independently of the bacterial chromosome.

For example, two kinds of plasmids, designated F and R, carry genes that permit bacterial conjugation (Figure 16.3). The R plasmids also carry genes that confer resistance to ampicillin or other antibiotics. An *antibiotic* is a substance that can kill or inhibit the growth of bacteria and other microorganisms (see the *Commentary*, page 352). R plasmids can be transferred among bacteria of different genera. This apparently happened with *Shigella*, a type of bacterium responsible for a serious form of dysentery. Because *Shigella* now has antibiotic-resistance genes, infected individuals do not respond well to treatment. Antibiotic resistance also has been conferred upon the microbes responsible for typhoid, meningitis, assorted intestinal tract disorders, gonorrhea, and the "staph" infections by *Staphylococcus aureus* that are so troublesome today in hospitals.

## Table 22.4 Some Major Groups of Bacteria

| Group | Main Habitats | Characteristics | Representatives |
|---|---|---|---|
| **Archaebacteria** | | | |
| Methanogens | Anaerobic sediments of lakes, swamps; also animal gut | Chemosynthetic; methane producers; used in sewage treatment facilities | *Methanobacterium* |
| Extreme halophiles | Brines (extremely salty water) | Heterotrophic; also have photosynthetic machinery of a unique sort | *Halobacterium* |
| Extreme thermophiles | Acidic soil, hot springs, hydrothermal vents on seafloor | Heterotrophic or chemosynthetic; use inorganic substances such as sulfur as a source of electrons for ATP formation | *Sulfolobus, Thermoplasma, Thermus* |
| **Photosynthetic eubacteria** | | | |
| Cyanobacteria | Mostly lakes, ponds; some marine, terrestrial | In photosynthesis, water is electron donor, oxygen a by-product; some fix nitrogen | *Anabaena, Nostoc* |
| Prochlorobacteria | Live in tissues of marine invertebrates | In photosynthesis, water is electron donor, oxygen a by-product | *Prochloron* |
| Purple or green bacteria | Generally anaerobic sediments of lakes, ponds | In photosynthesis, $H_2$, $H_2S$, or S is electron donor, oxygen *not* a by-product | *Rhodospirillum, Chlorobium* |
| **Chemosynthetic eubacteria** | | | |
| Nitrifying bacteria | Soil, freshwater, marine habitats | Major ecological role (nitrogen cycle) | *Nitrosomonas, Nitrobacter* |
| **Heterotrophic eubacteria** | | | |
| Spirochetes | Aquatic habitats; parasites of animals | Helically coiled, motile; free-living and parasitic species; some major pathogens | *Spirochaeta, Treponema* |
| Gram-negative, aerobic rods and cocci | Soil, aquatic habitats; parasites of animals, plants | Some major pathogens; some (e.g., *Rhizobium*) fix nitrogen | *Pseudomonas, Neisseria, Rhizobium, Agrobacterium* |
| Gram-negative, facultative anaerobic rods | Soil, plants, animal gut | Many are major pathogens; one (*Photobacterium*) is bioluminescent | *Salmonella, Shigella, Proteus, Escherichia, Photobacterium* |
| Rickettsias, chlamydias | Cells of insects, other animals | Intracellular parasites; many pathogens | *Rickettsia, Chlamydia* |
| Sulfur-, sulfate-reducing bacteria | Anaerobic muds, sediments (as in bogs, marshes) | Use sulfur or sulfur compounds as final electron acceptor in ATP formation | *Desulfovibrio* |
| Myxobacteria | Decaying plant, animal matter; bark of living trees | Gliding, rod-shaped; aggregate to form fruiting bodies with sporelike structures | *Myxococcus, Chondromyces* |
| Mycoplasmas | Cells of plants, animals | Intracellular parasites; many pathogens | *Mycoplasma* |
| Gram-positive cocci | Soil; skin and mucous membranes of animals | Some major pathogens | *Staphylococcus, Streptococcus* |
| Endospore-forming rods and cocci | Soil; animal gut | Some major pathogens | *Bacillus, Clostridium* |
| Gram-positive nonsporulating rods | Fermenting plant, animal material; human oral cavity, gut, vaginal tract | Some important in dairy industry, others serious contaminators of milk, cheese | *Lactobacillus, Listeria* |
| Actinomycetes | Soil; some aquatic habitats | Include anaerobes and strict aerobes; major producers of antibiotics | *Actinomyces, Streptomyces* |

**Figure 22.11** Great Salt Lake, Utah. Halophilic bacteria and algae growing in this vast, saline lake impart a pink cast to the water. The diagonal strip across the photograph is a raised bed for a railroad track.

## Classification of Bacteria

How are the many thousands of known bacterial species related to one another? Traditionally, they have been assigned to one group or another on the basis of Gram reactions, cell shape, mode of nutrition, metabolic patterns, and other characteristics. Table 22.4, for example, lists major bacterial groups according to their predominant mode of nutrition.

But characterizing a given bacterium is not the same thing as classifying it. Bacteria are not well represented in the fossil record. Until recently, they kept their evolutionary secrets to themselves. Today, however, comparative biochemistry studies by Carl Woese and others are revealing significant information about many species of archaebacteria and the eubacteria. In time, such studies may give us a system of bacterial classification based on evolutionary relationships.

## Archaebacteria

Three intriguing types of archaebacteria—the methanogens, extreme halophiles, and extreme thermophiles—are closely related to one another but not to any other bacteria. They all live in harsh settings reminiscent of the ancient environments in which life began. Hence the name, "archaebacteria" (after *arche-*, which means beginning). They are genetically and structurally distinct from all other organisms.

**Methanogens.** The methanogens, or "methanemakers," inhabit swamps. They also inhabit mud, sewage, and the animal gut. In fact, they are notably abundant in the gut of elephants, cows, and other ruminants (compare page 640).

Methanogens are strictly anaerobic. They make ATP by converting carbon dioxide and hydrogen gases to methane ($CH_4$). The pungent fumes hanging over stock-yards are testimony to their dedicated metabolic activity. So is the "marsh gas" of swamps and sewage treatment facilities. Collectively, methanogens play a role in the global cycling of carbon. They produce about *2 billion tons* of methane gas each year and so influence the concentration of carbon dioxide in the earth's atmosphere.

**Extreme Halophiles.** The extreme halophiles, or "salt-lovers," require or tolerate high salt concentrations in their surroundings. Different species of *Halobacterium*, for example, thrive in salt lakes, salt ponds, and brackish seas. They also can cause spoilage of salted fish and salted animal hides, and they can contaminate "sea salt" produced commercially from seawater evaporation. Large colonies of halobacteria appear pink, red, or red-orange, owing to carotenoid pigments in individual cells. At times, the colonies and certain algal populations turn the water pink in Great Salt Lake, Utah (Figure 22.11). The pigments seem to protect the cells against the strong sunlight characteristic of their habitats.

Most halophiles are heterotrophs that use aerobic pathways of ATP formation. When oxygen levels are low, however, some strains use sunlight for a unique mode of photosynthesis. Patches of a special pigment (bacteriorhodopsin) form in the plasma membrane. When the pigments absorb light energy, $H^+$ ions are pumped out of the cell. Then the ions reenter the cell through enzyme systems—and ATP forms by chemiosmosis (compare page 112).

**Extreme Thermophiles.** Hot springs, highly acidic soils, even sediments around volcanic, hydrothermal vents at the ocean floor and other seemingly inhospitable environments are home to thermophiles ("heat-lovers"). *Thermoplasma*, for instance, has only one known habitat—the waste piles of coal mines! Because this is a habitat of recent origin—coal mines have been around for only a few hundred years—*Thermoplasma* must have evolved in places we don't know about. Or consider the thermophiles at hydrothermal vents—fissures in the floor of deep oceans, where the water temperature can exceed 250°C. Hydrogen sulfide ($H_2S$) is spewed from such vents, and the thermophiles use it as a source of electrons for ATP formation. These enterprising bacteria are the basis of food webs that include clams, mussels, barnacles, and tube worms (Figure 47.32).

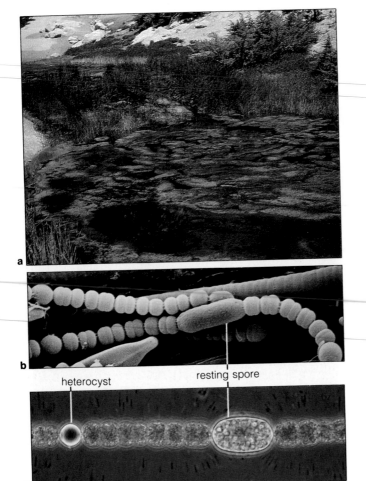

**Figure 22.12** (a) A population of cyanobacteria floating near the surface of a pond enriched with nutrients. The resting spores shown in the scanning electron micrograph (b) and phase-contrast micrograph (c) form when conditions do not favor growth. When favorable conditions return, a resting spore germinates and gives rise to a new chain of cells. A heterocyst, the type of differentiated cell that is specialized for nitrogen fixation, is shown in (c).

### Eubacteria

There are thousands of diverse eubacteria. Here we will consider only a few examples, using the mode of nutrition (photosynthetic, chemosynthetic, and heterotrophic) as a conceptual framework.

**Photosynthetic Eubacteria.** Like plants, some eubacteria use sunlight energy and oxygen for photosynthesis and ATP formation. Cyanobacteria (also called blúe-green algae) are an example. You can see them at the surface of freshwater ponds. Many species grow as chains of cells that surround themselves with a mucous sheath. Often the chains intertwine into dense, slimy mats that float near the water's surface. One group, the prochlorobacteria, was discovered as recently as 1975 and reminds us that we still have much to learn about life on our planet.

Some cyanobacteria, including *Anabaena*, engage in nitrogen fixation. They can convert nitrogen gas ($N_2$) to ammonia, which can be incorporated into organic compounds. When environmental supplies of nitrogen-containing compounds run low, some cells in the cyanobacterial chains develop into **heterocysts**. These are a special type of cell that can produce molecules of a nitrogen-fixing enzyme. Nitrogen compounds form and are shared with photosynthetic cells by moving through junctions that connect the cytoplasm of neighboring cells in the chain. Similarly, carbohydrates formed in the photosynthetic cells are shared with the heterocysts (Figure 22.12).

Unlike *Anabaena*, the green bacteria and purple bacteria do not use water as a source of electrons during photosynthesis. The green bacteria, for example, strip electrons from hydrogen sulfide ($H_2S$) or hydrogen gas ($H_2$). In this respect, they may resemble the ancient anaerobic bacteria in which photosynthesis first emerged. Oxygen is not a by-product of their activities.

**Chemosynthetic Eubacteria.** Chemosynthetic eubacteria use inorganic substances in the environment as an energy source. They are key players in the global cycling of nitrogen, sulfur, iron, phosphorus, and other nutrients. The nitrifying bacteria, for example, have roles in the nitrogen cycle. Without nitrogen—a component of all amino acids and proteins—there would be no life. Nitrifying bacteria attack ammonia or nitrite in soil, stripping electrons from them for use in reactions leading to ATP formation (page 840). In some species, the reactions are carried out across extensive infoldings of the plasma membrane (Figure 22.13).

**Heterotrophic Eubacteria.** Most of the world's eubacteria are heterotrophs. The majority are important decomposers, although many types also have impact, good and bad, on human affairs. *Azotobacter* and *Rhizobium*, for example, are among the few kinds of organisms that can use the nitrogen gas in the atmosphere. Their nitrogen-fixing activities, described on page 840, enhance soil fertility and plant growth—and plant growth is the food production base for communities on land. *E. coli* cells in the mammalian gut help newborns digest milk by breaking apart lactose (milk sugar) molecules. Humans enlist *Lactobacillus* species for the large-scale manufacture of cheese, sour cream, yogurt, and other fermented milk products. They also use actinomycetes in the manufacture of antibiotics.

Many pathogenic heterotrophs do damage to humans and their crops, farm animals, and pets. Ticks and other insects may transmit certain pathogens from one host to another. Suppose a tick sucks blood from a deer that is infected with the spirochete *Borrelia burgdorferi*. If that tick later bites and infects a person walking through the woods, the outcome will be *Lyme disease*. Symptoms include a quarter-shaped rash around the tick bite, then severe headaches, backaches, chills, and fatigue. Without prompt treatment, infected persons can develop conditions that resemble rheumatoid arthritis, heart disorders, and damage to the nervous system.

Some heterotrophic eubacteria can enter a dormant stage. They form a resistant body, an **endospore**, around their genetic material and a bit of cytoplasm. Endospores are extraordinarily resistant to heat, drying, boiling, radiation, and other environmental insults. They germinate and give rise to new bacterial cells when favorable conditions return. One endospore-former, *Clostridium tetani*, is shown in Figure 22.14. It causes the disease tetanus (page 557). Another of this same genus, *C. botulinum*, produces a deadly toxin. Cattle and birds die after eating fermented grains tainted with the toxin. *C. botulinum* also produces toxins that can taint food in improperly sterilized and sealed cans or jars. Eating the tainted food can lead to *botulism*, a form of poisoning that affects muscle activity. Death can follow as a result of respiratory failure.

*Escherichia coli*, another heterotrophic eubacterium, normally benefits its human hosts. *E. coli* inhabits the gut, where it produces vitamin K and compounds useful in fat digestion. Its activities help prevent many pathogens that are ingested with food from colonizing the gut. Even so, some *E. coli* strains produce a potent toxin and cause a serious type of diarrhea that is the leading cause of infant mortality in developing countries.

## About the "Simple" Bacteria

Bacteria are small. They are not as structurally complex as eukaryotic cells. *But bacteria are not simple.* We can reinforce this point by concluding with a look at bacterial behavior.

Like multicelled organisms, bacteria deal with the environment in sophisticated ways. Photosynthetic species sense the intensity of light and move toward its source (or away if it is too bright). Heterotrophic species sense and move toward higher concentrations of nutrients. Aerobes move toward oxygen, anaerobes move away from it. Many bacteria can detect and avoid toxins.

A bacterium detects the presence of a stimulus through membrane receptors. The receptors change shape when they absorb light or when chemical compounds bind to them. When the bacterium moves in different directions, variations are introduced in the activity of membrane receptors. The variations are the basis of a fleeting biochemical "memory," allowing the bacterial cell to compare present conditions against the immediate past.

Many bacteria reverse direction when they sense that conditions are becoming unfavorable. Others swim

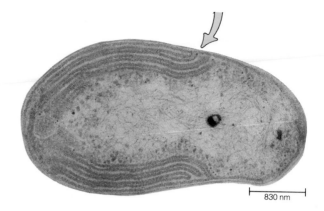

**Figure 22.13**  *Nitrobacter*, one of the nitrifying bacteria. The arrow points to a site where the plasma membrane folds into the cytoplasm. The membrane infoldings greatly increase the membrane surface area available for metabolic reactions.

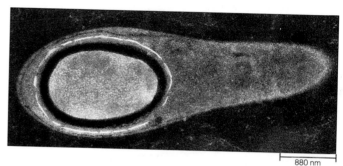

880 nm

**Figure 22.14**  Transmission electron micrograph of a developing endospore of the bacterium *Clostridium tetani*.

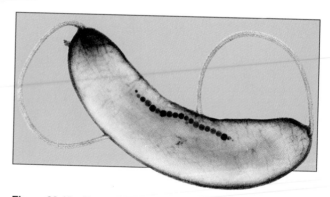

**Figure 22.15**  Transmission electron micrograph of a magnetotactic bacterium, showing the chain of magnetite particles that act like a tiny compass.

for a period of time, stop and tumble for a few seconds, then swim again. The tumbles orient them in random ways. But a bacterium rarely tumbles when it is moving toward favorable conditions and so tends to keep moving in that direction. A bacterium moving away from favorable conditions tumbles frequently and so keeps changing direction until it senses that it is going the "right" way.

Magnetotactic bacteria have in their cytoplasm a chain of magnetite particles that serves as a tiny compass (Figure 22.15). The compass helps them sense which way is north and also down. These bacteria swim toward the bottom of a body of water, where oxygen concentrations are lower and therefore more suitable for their growth.

Some bacterial species even show collective behavior. Even though the myxobacteria are not multicellular, they give a good imitation of it. Millions of cells of *Myxococcus xanthus* collect into "predatory" colonies that trap cyanobacteria and other microbes. Their enzyme secretions digest "prey" that becomes stuck to the colony, and the cells absorb the breakdown products. More than this, migrating *M. xanthus* cells change direction and *move as a single unit* toward what may be food, then move away if their target is not a source of nutrients.

Many myxobacterial species also form **fruiting bodies**, or spore-bearing structures (Figure 22.16). Under appropriate conditions, some cells in the colony differentiate and produce a slime stalk, other cells form branches, and still others form clusters of single-celled spores. When the clusters burst open, the spores are dispersed; each may form a new colony. As you will now see, such structures also form during the life cycles of many eukaryotes.

**Figure 22.16** Fruiting body of *Chondromyces crocatus*, a myxobacterium in which cells differentiate.

## THE RISE OF EUKARYOTIC CELLS

It is one of nature's greatest secrets. *Where did eukaryotic organisms come from?* The biochemical "records" built into living bacteria offer tantalizing hints about their origins. Comparisons of nucleotide sequences of the DNA and RNA from different species suggest that archaebacteria, eubacteria, and the forerunners of eukaryotes diverged from a common ancestor long before fossil evidence of eukaryotes started accumulating.

Single-celled protistans and multicelled plants, fungi, and animals are eukaryotes. They are like one another—and *unlike* bacteria—in having a nucleus and other membrane-bound organelles that separate different metabolic activities in the space of the cell cytoplasm. The simplest eukaryotes are the bloblike protistans called amoebas. Because amoebas have no hard parts, they vanish quickly after they die. If the ancestors of eukaryotes also lacked hard parts, fossils of them may never be found. At present, we have no direct evidence of how single-celled and then multicelled eukaryotes arose, although the kind of speculation described in the *Commentary* gives us some interesting things to think about.

## PROTISTANS

### Classification of Protistans

By definition, **protistans** are single-celled eukaryotic organisms, but the boundaries of their kingdom are poorly defined. Remember, the boundaries of higher taxa are artificial. *We simply impose boundaries on the continuous threads of descent that tie together all species, past and present.* Without doubt, the protistan lineages continue into the kingdoms of plants, fungi, and animals. Thus, what may be "a photosynthetic protistan" to one biologist may be "a plant" to another. In this book, we classify the slime molds, euglenoids, chrysophytes, dinoflagellates, amoebas, flagellated protozoans, sporozoans, and ciliates as protistans (Table 22.5 and Appendix II).

**Table 22.5  Classification of Protistans**

| Phylum (Division) | Common Name |
|---|---|
| Gymnomycota | Slime molds |
| Euglenophyta | Euglenoids |
| Chrysophyta | Yellow-green algae, golden algae, diatoms |
| Pyrrhophyta | Dinoflagellates |
| Mastigophora | Flagellated protozoans |
| Sarcodina | Amoeboid protozoans (amoebas, foraminiferans, radiolarians, heliozoans) |
| Ciliophora | Ciliated protozoans |
| | Sporozoans* |

*The name has no formal taxonomic status.

# Speculations on the Origin of Eukaryotes

Between 3.8 billion and 2.5 billion years ago, divergences from an ancestral prokaryote marked the beginning of three major lineages: archaebacteria, eubacteria, and the forerunner of eukaryotes. Figure *a* is a family tree showing the presumed relationships among these lineages. It is based on recent work in comparative biochemistry.

## Origin of Organelles

Archaebacteria and eubacteria are fundamentally different in cell wall characteristics and many other properties. And eukaryotes differ from both of those prokaryotic lineages in having membrane-bound organelles. How did organelles emerge in the cells ancestral to eukaryotes? The most conspicuous organelle, the nucleus, may have evolved through modification of infoldings of the plasma membrane. Such infoldings occur in many bacteria (Figure

22.13 shows an example). In some species the infoldings even form a saclike structure around the DNA after it has been replicated.

According to Lynn Margulis and many other biologists, some organelles had symbiotic origins. **Symbiosis** means "living together." It refers to interactions in which one species serves as host to another species (the guest, or symbiont). Such partnerships can be observed today between different species, as Figure *b* suggests. According to the theories of symbiotic origins of eukaryotes, mito-

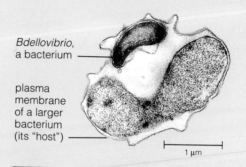

*Bdellovibrio,* a bacterium

plasma membrane of a larger bacterium (its "host")

1 µm

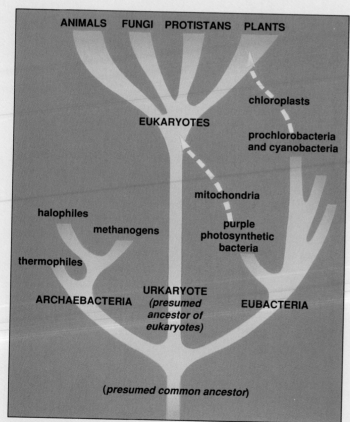

**a** Proposed evolutionary relationships between prokaryotes and eukaryotes. Dashed lines indicate symbiotic origins of mitochondria and chloroplasts.

(Figure a labels: ANIMALS, FUNGI, PROTISTANS, PLANTS; chloroplasts; EUKARYOTES; prochlorobacteria and cyanobacteria; mitochondria; halophiles; methanogens; purple photosynthetic bacteria; thermophiles; ARCHAEBACTERIA; URKARYOTE (*presumed ancestor of eukaryotes*); EUBACTERIA; (*presumed common ancestor*))

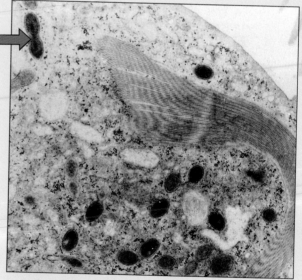

**b** Symbiosis between existing prokaryotic and eukaryotic cells. Top micrograph: *Bdellovibrio,* a bacterium (the dark eggplant-shaped cell) has ended up living in the space between the cell wall and plasma membrane of a larger bacterium. Lower micrograph: Bacteria (dark ovals) living inside *Personympha,* a protistan that dwells in the termite gut. The bacteria resemble mitochondria in size and distribution in the host cytoplasm. They probably use glucose breakdown products from the cell's metabolic activities, as mitochondria do. The arrow points to a bacterium dividing in the cytoplasm—as mitochondria do.

chondria, chloroplasts, flagella, and other eukaryotic features arose through inadvertent partnerships that formed between bacterial cells of different species.

Consider one line of evidence in favor of this theory. The basic genetic code is identical for all living species, from bacterial to human. The absence of variation among species suggests that the genetic code arose before the first cells on earth diverged into separate evolutionary lines. Once the code had been established, the rare mutations that did change one or more code words (codons) must have been selected against.

Yet the mitochondria of several species follow a slightly altered code. Mitochondria, recall, are organelles specialized for aerobic respiration. They are uncannily like bacteria in size and structure. In some respects, a mitochondrion operates somewhat independently of the cell in which it is housed. It has its own DNA, which is replicated independently of the nuclear DNA. Mitochondrial DNA governs the synthesis of some RNA and proteins that the organelle requires for its specialized tasks. Although mitochondrial genetic codes are almost the same as the one used by cells, a *few* codons have different meanings. (For instance, UGA is a termination codon in most genetic messages; for mitochondria, it specifies tryptophan.)

Suppose some early free-living cells, similar to existing amoebalike cells that weakly tolerate free oxygen, preyed on aerobic bacteria. Suppose some of those "prey," the forerunners of mitochondria, resisted digestion. We can imagine that the undigested guest found itself in a protected, nutrient-rich environment. And the host was supplied with "extra" ATP that could be channeled into growth, greater activity, and the assembly of more structures, such as hard body parts. In time, the symbiotic bacteria became modified as their increasingly complex host cells performed some functions for them. They became mitochondria, incapable of independent existence.

Similarly, chloroplasts are like photosynthetic bacteria in their metabolism and overall DNA sequences, and they replicate somewhat independently of the cell. Chloroplasts may have evolved a number of times, in a number of different lineages. Among existing eukaryotic species, they vary in shape and their light-absorbing pigments—just as different species of photosynthetic bacteria do.

## Slime Molds

In many respects, slime molds (phylum Gymnomycota) resemble organisms of other kingdoms. Like myxobacteria, the cells of some slime molds differentiate and form fruiting bodies—stalked structures bearing spores at their tips. Certain spores are thick-walled and are dispersed by air currents, just like the spores of many fungi. Slime molds also spend part of their life creeping about like animals and engulfing food, which for them include bacteria, spores, and organic remains.

Slime molds are classified in two groups, the **cellular slime molds** and **plasmodial slime molds**. Both have a phagocytic phase during the life cycle. The organism moves along decaying logs, twigs, and leaves, engulfing food as it goes. But the body of a *cellular* slime mold is an aggregation of distinct, amoebalike cells. The body of a *plasmodial* slime mold is a multinucleate blob of cytoplasm (Figure 22.17). When migrating, both types leave a slimy track (hence the name).

*Dictyostelium discoideum* is the best known of the cellular slime molds. Figure 22.18 shows details of its life cycle, including the formation of a fruiting body. Similar spore-bearing structures form during the life cycle of a plasmodial slime mold. Under suitable conditions, a plasmodium may grow until it spreads out over several square feet. Upon dehydration and exposure to light, the mass gives rise to a fruiting body.

a

b

**Figure 22.17** Plasmodial slime molds. (**a**) The multinucleate mass of cytoplasm of *Physarum*, migrating along a decaying log. Some plasmodia spread over a square meter. (**b**) Spore-bearing structures of *Stemonitis splendens*.

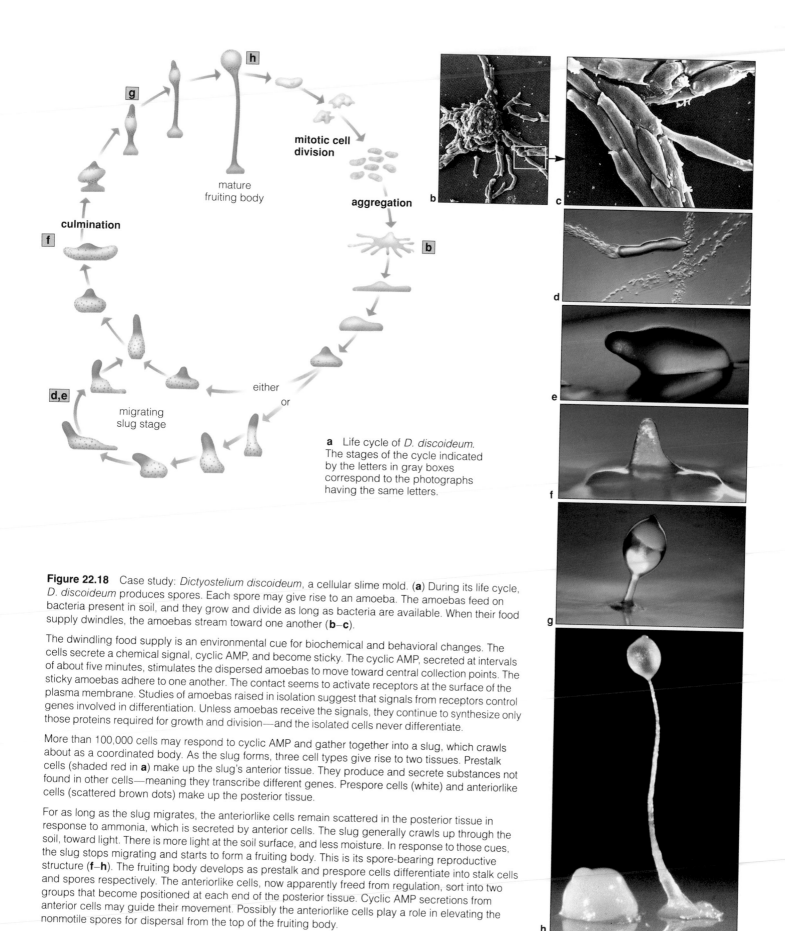

**a** Life cycle of *D. discoideum.* The stages of the cycle indicated by the letters in gray boxes correspond to the photographs having the same letters.

**Figure 22.18** Case study: *Dictyostelium discoideum*, a cellular slime mold. **(a)** During its life cycle, *D. discoideum* produces spores. Each spore may give rise to an amoeba. The amoebas feed on bacteria present in soil, and they grow and divide as long as bacteria are available. When their food supply dwindles, the amoebas stream toward one another (**b–c**).

The dwindling food supply is an environmental cue for biochemical and behavioral changes. The cells secrete a chemical signal, cyclic AMP, and become sticky. The cyclic AMP, secreted at intervals of about five minutes, stimulates the dispersed amoebas to move toward central collection points. The sticky amoebas adhere to one another. The contact seems to activate receptors at the surface of the plasma membrane. Studies of amoebas raised in isolation suggest that signals from receptors control genes involved in differentiation. Unless amoebas receive the signals, they continue to synthesize only those proteins required for growth and division—and the isolated cells never differentiate.

More than 100,000 cells may respond to cyclic AMP and gather together into a slug, which crawls about as a coordinated body. As the slug forms, three cell types give rise to two tissues. Prestalk cells (shaded red in **a**) make up the slug's anterior tissue. They produce and secrete substances not found in other cells—meaning they transcribe different genes. Prespore cells (white) and anteriorlike cells (scattered brown dots) make up the posterior tissue.

For as long as the slug migrates, the anteriorlike cells remain scattered in the posterior tissue in response to ammonia, which is secreted by anterior cells. The slug generally crawls up through the soil, toward light. There is more light at the soil surface, and less moisture. In response to those cues, the slug stops migrating and starts to form a fruiting body. This is its spore-bearing reproductive structure (**f–h**). The fruiting body develops as prestalk and prespore cells differentiate into stalk cells and spores respectively. The anteriorlike cells, now apparently freed from regulation, sort into two groups that become positioned at each end of the posterior tissue. Cyclic AMP secretions from anterior cells may guide their movement. Possibly the anteriorlike cells play a role in elevating the nonmotile spores for dispersal from the top of the fruiting body.

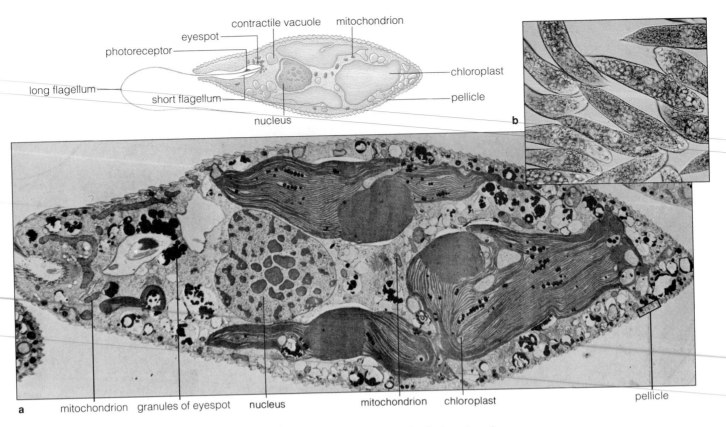

**Figure 22.19** Example of the complexity possible in a eukaryotic cell. (**a**) This longitudinal section of *Euglena* shows the profusion of internal organelles. (**b**) Light micrograph of living *Euglena* cells.

## Euglenoids

Ponds and lakes, especially stagnant ones, are home to most of the world's free-living **euglenoids** (phylum Euglenophyta), although some live in seawater. Most of the 800 species are photosynthetic, and most have two flagella. Like most other flagellated protistans, the euglenoids reproduce by **longitudinal fission**. The cell grows in circumference while all organelles are being duplicated, then the cell divides along its long axis.

Typically, euglenoids are only about 40 or 50 micrometers long. For their size, they show astonishing complexity. For example, spiral strips of a translucent, protein-rich material occur beneath the plasma membrane of *Euglena*. They are the main components of a *pellicle*, the cell's firm yet flexible outer layer (Figure 22.19). Granules of carotenoid pigments are massed into an "eyespot." They shield one side of a light-sensitive receptor, which is at the base of the long flagellum. By moving that flagellum, euglenoids keep the receptor exposed to light—and so stay positioned where light intensity is most suitable for their activities.

In sunlight, *Euglena* is photosynthetic, but not perfectly so. It cannot synthesize vitamin $B_{12}$ and depends on neighboring microorganisms as suppliers. When deprived of sunlight, *Euglena* actually becomes hetero-

trophic; it subsists on organic substances dissolved in the surroundings. It has been suggested that the photosynthetic euglenoids acquired their chloroplasts long ago, through endosymbiosis (see *Commentary*, page 362). Intriguingly, the chlorophyll pigments in their chloroplasts are identical to those in green algae and land plants.

## Chrysophytes

Chrysophytes (phylum Chrysophyta) live in freshwater and marine environments. Most are photosynthetic; some are heterotrophs. Chrysophytes include 450 species of "yellow-green algae," about 500 species of "golden algae," and more than 5,000 species of golden-brown "diatoms." The photosynthesizers contain xanthophylls and beta carotene. Those pigments mask the color of chlorophyll in golden algae and diatoms (Figure 22.20).

Most golden algae are flagellated and many have silica scales or "skeletons." Diatoms are not flagellated. Their walls are two thin, overlapping shells that fit together like a pillbox (Figure 22.20d). Many holes perforating each shell facilitate the movement of materials into and out of the cell, across the plasma membrane. In some cases, part of the cell pokes out through slots in the wall. Such extensions move the cell over rocks and other solid surfaces. Together with golden algae, diatoms are oxygen suppliers and food producers. They are hit hard by water pollution, as are other organisms that depend on their activities.

About 100 million years ago, diatom shells started building up in ocean sediments. They now form extensive deposits of a fine, crumbly material used in manufacturing abrasives, filtering materials, and insulating materials. More than 270,000 metric tons of it are quarried annually near Lompoc, California.

### Dinoflagellates

Most of the 1,200 species of **dinoflagellates** (phylum Pyrrhophyta) are photosynthetic members of marine plankton. **Plankton** is derived from the Greek *planktos*, meaning to wander. It refers to aquatic communities of protistans and animal larvae, mostly microscopic, that drift or swim weakly through water. Some dinoflagellates are marine heterotrophs and freshwater photosynthesizers. Some species have flagella that fit like ribbons in grooves between stiff cellulose plates at the body surface.

Dinoflagellates appear yellow-green, green, brown, blue, or red, depending on the main photosynthetic pigments. Every so often, red dinoflagellates undergo population explosions and color the seas red or brown. Because some forms produce a neurotoxin, the resulting **red tides** can have devastating effects. Hundreds of thousands of fish that feed on plankton may be poisoned and wash up along the coasts (Figure 22.21). The neurotoxin does not affect clams, oysters, and other mollusks, but it builds up in their tissues. Humans who eat the tainted mollusks may die.

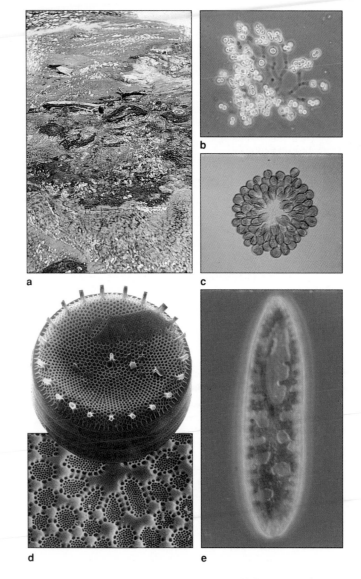

**Figure 22.20** Representative chrysophytes. Yellow-green algae: (**a**) *Vaucheria*, growing over red sandstone in Arizona and (**b**) *Mischococcus*, a colonial, planktonic type. (**c**) *Synura*, a colonial, planktonic golden alga known for its fishy odor. (**d**) Pillbox structure of a diatom shell and a closer look at how intricate the diatom shell perforations can be. Compare with the diatoms shown in Figure 4.1. (**e**) The large chloroplast inside the diatom *Surirella constricta*.

10 µm

**Figure 22.21** (*Left*) Partial view of a fish kill resulting from a dinoflagellate "bloom." (*Right*) Source of red tides along the Florida coast—the dinoflagellate *Ptychodiscus brevis*.

**Figure 22.22** Some flagellated protozoans. (**a**) A trypanosome, *Trypanosoma brucei*, responsible for African sleeping sickness. (**b**) *Giardia lamblia*, a cyst-forming parasite that causes intestinal disturbances. (**c**) *Trichomona vaginalis*, a parasite that causes trichomoniasis. This is a sexually transmitted disease of the sort described on page 780.

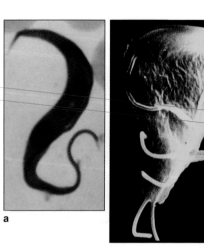

a

b

c

**Figure 22.23** (*Below*) Some amoeboid protozoans. (**a**) *Amoeba proteus*, crawling about in a water droplet. Shells of foraminiferans (**b**) and a radiolarian (**c**), no longer housing the amoeboid cells that produced them. (**d**) A living heliozoan, with needlelike pseudopods visible.

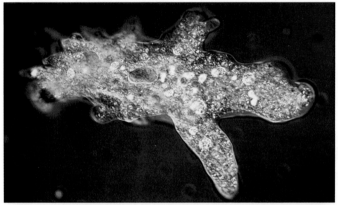

a

b

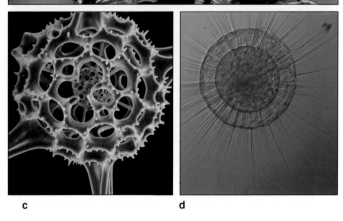

c

d

## Protozoans

Some protistans were once viewed as primitive relatives of animals because of their predatory or parasitic habits. They were named **protozoans** ("first animals"), although their evolutionary links with animals are not clear. There are more than 65,000 protozoan species. Fewer than two dozen cause diseases in humans, yet their influence is staggering. In any year, hundreds of millions of people suffer protozoan infections!

By one classification scheme, there are four major groups of protozoans. These are the flagellated forms, amoeboid forms, ciliated forms, and the sporozoans (Table 22.5).

**Flagellated Protozoans.** Among the ranks of flagellated protozoans (phylum Mastigophora) are free-living and parasitic species. The free-living types often are important in freshwater and marine communities. The parasitic types include the trypanosomes. *Trypanosoma brucei*, shown in Figure 22.22a, is responsible for a dangerous disease called *African sleeping sickness*. Bites of the tsetse fly transmit the disease from one human to another. Early symptoms include fever, headaches, rashes, and anemia. Later, the central nervous system becomes damaged. Without treatment, death follows. Another trypanosome (*Trypanosoma cruzi*) is prevalent in Mexico and South America, where it causes *Chagas disease*. Assorted bugs pick up the parasite when they feed on infected humans, armadillos, opossums, and other animals. The parasite multiplies in the insect gut, then is excreted onto human skin. Scratches or abrasions on the skin are open doors for infection. A terrible disease follows infection. The liver and spleen enlarge, the face and

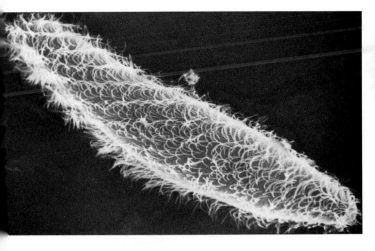

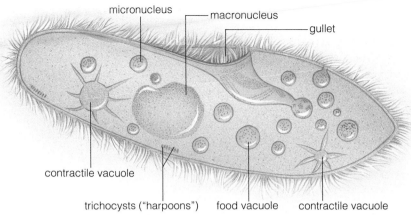

**Figure 22.24** *Paramecium*, a ciliated protozoan.

eyelids swell, then the brain and heart become severely damaged. There is no specific treatment or cure.

By some estimates, about 10 percent of the human population in the United States is infected with another flagellated protozoan, *Giardia lamblia* (Figure 22.22b). Infection leads to mild intestinal disturbances, including diarrhea, but it has severe and sometimes fatal consequences in a few susceptible people. This flagellated protozoan forms **cysts** (walled, resting structures) that leave the body in feces. It infects a new host who ingests food or water contaminated with the feces. Even the water of remote mountain streams may contain the cysts and should be boiled before drinking.

Flagellated protozoans also include trichomonads, of the sort shown in Figure 22.22c. *Trichomonas vaginalis*, a worldwide nuisance, is transferred to new human hosts during sexual intercourse. Without treatment, trichomonad infection damages membranes in the urinary and reproductive tracts.

**Amoeboid Protozoans.** Among the amoeboid protozoans (phylum Sarcodina) are the amoebas, foraminiferans, heliozoans, and radiolarians. Adult forms move or capture prey by sending out pseudopods ("false feet," which really are temporary extensions of the cell body). Most amoeboid protozoans feed on algae, bacteria, and other protozoans.

The **amoebas** live in freshwater, seawater, and soil. They include *Amoeba proteus* of biology laboratory fame (Figure 22.23a). A parasitic form, *Entamoeba histolytica*, causes amoebic dysentery, a severe intestinal disorder. It travels in cysts within feces, which may contaminate water and soil in regions with inadequate sewage treatment.

**Foraminiferans** live mostly in the seas. Their hardened shells are often peppered with hundreds of thousands of tiny holes through which sticky, threadlike pseudopods extend. Figure 22.23b shows some foramini-

feran shells. Often the shells bear spines, which in some species are long enough that the shell can be seen with the naked eye.

Among the structurally stunning protozoans are the **radiolarians**, which are found mostly in marine plankton. The cells basically resemble the heliozoans, but most also have a skeleton of silica (Figure 22.23c). Some radiolarian species form colonies in which many individual cells are cemented together. Accumulated shells of radiolarians and foraminiferans are key components of many ocean sediments, and are testimony to the abundance of these organisms in the past.

The **heliozoans**, or "sun animals," have fine, needlelike pseudopods that radiate from the body like sun rays (Figure 22.23d). These largely freshwater protozoans are generally floaters or bottom-dwellers. Part of the cytoplasm forms an outer sphere around a core composed of denser cytoplasm and the bases of microtubular rods.

**Ciliated Protozoans.** The **ciliated protozoans** are important members of aquatic communities. They feed on algae and in turn are food for larger animals. Some of the ciliated types enjoy symbiotic relationships with algae that reside in their cytoplasm.

Cilia at the cell surface are synchronized for swimming. In some species they number in the thousands. Often, hundreds of poison-charged, harpoonlike structures are arrayed at the cell surface. These structures, called trichocysts, are probably useful in defense against other predators and prey.

The 8,000 species of ciliated protozoans live in freshwater and marine habitats. *Paramecium* is one of them (Figure 22.24). Like most ciliated protozoans, *Paramecium* has a gullet, a cavity that opens to the outside at the cell surface. The synchronized beating of rows of specialized cilia cause water laden with bacteria and food particles to move into the gullet. Once inside, the particles become enclosed in enzyme-filled vesicles and are digested. Wastes are moved to an "anal pore" and eliminated. Like the amoeboid protozoans, *Paramecium* relies on contractile vacuoles to rid the cell of excess water (Figure 5.6).

Although some ciliates crawl about or stay put, *Paramecium* is built for speed. Between 10,000 and 14,000 cilia project like rows of tiny, flexible oars from the cell surface. So efficient is the coordinated beating of the rows that some species of *Paramecium* are propelled through their surroundings at a remarkable 1,000 micrometers per second. Even so, *Paramecium* is often outmaneuvered and eaten by another free-swimming ciliate, *Didinium* (Figure 22.25).

**Sporozoans.** The **sporozoans** are a diverse group of parasites. All have an infectious, sporelike stage (sporozoites) that often is transmitted to new hosts by insects. Probably the most notorious sporozoans are the species of *Plasmodium* that cause malaria (Figure 22.26). Mosquitoes transmit them to bird or human hosts. The infection persists for long periods. Each year about 200 million people are stricken with the disease, mostly in tropical and subtropical regions—and each year about 1 to 2 million die from it.

Work is under way to develop a genetically engineered vaccine against this organism. Until a vaccine

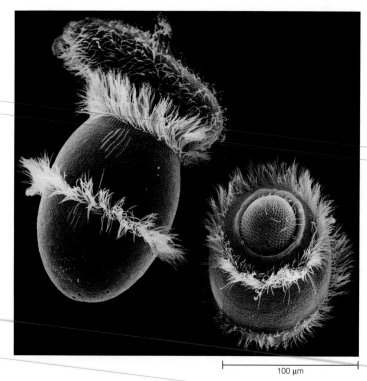

100 μm

**Figure 22.25** Mealtime for *Didinium*, a ciliated protozoan with a big mouth. Dinner in this case is *Paramecium*, poised at the mouth (*left*) and swallowed (*right*).

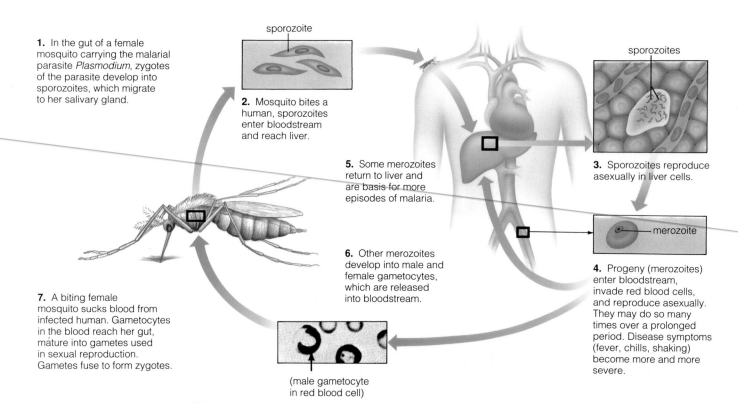

**1.** In the gut of a female mosquito carrying the malarial parasite *Plasmodium*, zygotes of the parasite develop into sporozoites, which migrate to her salivary gland.

sporozoite

**2.** Mosquito bites a human, sporozoites enter bloodstream and reach liver.

**5.** Some merozoites return to liver and are basis for more episodes of malaria.

**6.** Other merozoites develop into male and female gametocytes, which are released into bloodstream.

**7.** A biting female mosquito sucks blood from infected human. Gametocytes in the blood reach her gut, mature into gametes used in sexual reproduction. Gametes fuse to form zygotes.

(male gametocyte in red blood cell)

sporozoites

**3.** Sporozoites reproduce asexually in liver cells.

merozoite

**4.** Progeny (merozoites) enter bloodstream, invade red blood cells, and reproduce asexually. They may do so many times over a prolonged period. Disease symptoms (fever, chills, shaking) become more and more severe.

**Figure 22.26** Life cycle of the sporozoan *Plasmodium*, which causes the disease malaria. The life cycle unfolds in the human body and in an insect (the female *Anopheles* mosquito), which transfers the sporozoan to new hosts during bites.

## Commentary

# Beginnings of Multicellularity

*Multicellularity* signifies interdependence and division of labor among specialized cell types. Think about the advantages it bestows—say, on one type of organism that is tasty to another. Compared to single cells, a multicelled organism might be too large to eat. It may have muscles and other specialized tissues that provide the means for rapid response to danger. Each tissue or organ performs a single function very well, and it depends on other tissues or organs to help maintain favorable operating conditions for the organism as a whole.

Among the simple eukaryotes are forms that may be reminiscent of nature's early experiments in multicellularity. Think about a small mollusk that eats green algae, then stockpiles their chloroplasts in one of its body tissues (Figure *a*). The chloroplasts continue functioning and provide their "host" with oxygen!

Think about the division of labor in *Volvox*, a photosynthetic, colonial organism. Some *Volvox* colonies are single-layered, hollow spheres of many tens of thousands of flagellated cells. The sphere can be moved forward when the flagella beat in unison. Reproduction is assigned to a few cells; they divide and give rise to daughter colonies that develop inside the parent sphere (Figure 6.2). The daughter colonies are released after they produce and secrete certain enzymes that dissolve the jellylike secretions holding the parent colony together. Some species show sexual differentiation (certain cells produce eggs or sperm).

Think about the green algae that are merely chains or sheets of cells. Tiny multicelled plants that first invaded the land were not much more complex than this. Or think about how a tiny marine animal (*Trichoplax*) is no more than a flattened ball of ciliated cells, half a millimeter across (Figure *b*). It has no right side, left side, front, or back. It simply moves in any direction. *Trichoplax* may resemble some of the first multicelled animals.

Finally, think about just one of your own multicelled systems—the calcium-containing bones of your skeleton. Given the central roles of calcium in cell division and cell movements, the earliest eukaryotes must have had proteins that served as calcium storage centers. Such proteins could latch onto calcium entering the cell, much like calcium-binding proteins do in existing eukaryotes. In time, proteins of this sort must have become incorporated into shells and internal structures. Certainly by 750 million years ago, simple multicelled animals with hard parts exploded onto the evolutionary stage (page 308). In the increasingly elaborate shells and internal skeletons of the early animals, we have hints of the origins of bones—of our own skeletal system, and our own calcium reservoirs.

**a** One of nature's experiments—*Plakobranchus*, a marine mollusk that feeds on algae.

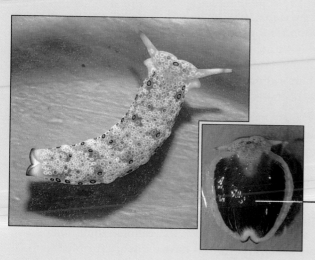

body folds pushed back to show chloroplasts

**b** Another experiment—*Trichoplax adhaerens*, one of the simplest multicelled animals. Little more than a flattened ball of tiny ciliated cells, this *Trichoplax* was discovered crawling about in a seawater aquarium.

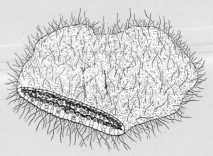

becomes available, individuals traveling through countries with high rates of malaria are advised to use antimalarial drugs such as chloroquine.

This concludes our survey of the kingdoms of single-celled organisms. When we think about redwoods, whales, and other complex organisms, it seems almost incomprehensible that forms similar to existing protistans gave rise to species in the kingdoms of multicelled eukaryotes. Yet as the examples in the above *Commentary* suggest, it may be that we are forgetting to consider what must have been an immense evolutionary parade of intermediate forms.

**Table 22.6 Comparison of Prokaryotes With Eukaryotes**

|  | Prokaryotes | Eukaryotes |
|---|---|---|
| Organisms represented | Bacteria only | Protistans, fungi, plants, and animals |
| Level of organization | Single-celled | Mostly multicelled; division of labor among differentiated cells |
| Typical cell size | Small (1–10 micrometers) | Large (10–100 micrometers) |
| Cell wall | Mostly distinctive sugars and peptides | Cellulose or chitin; none in animal cells |
| Membrane-bound organelles | Absent | Present |
| Modes of metabolism | Both anaerobic and aerobic prevalent | Aerobic modes predominate |
| Genetic material | Single bacterial chromosome (and sometimes plasmids) | Complex chromosomes (DNA and many associated proteins) within a nucleus |
| Mode of cell division | Binary fission | Mitosis, meiosis, or both |

The features that distinguish prokaryotes from eukaryotes have been touched on in many parts of this book. Table 22.6 brings together some of the key features as a simple way of comparing the differences before we turn to the multicelled forms.

## SUMMARY

1. In the microscopic world are two kingdoms of single-celled organisms: the monerans (bacteria) and protistans. Viruses also are microscopic, but they are not alive.

2. A virus is a noncellular infectious agent with two characteristics. It consists only of a nucleic acid core and a protein coat that sometimes is enclosed in a lipid envelope. And it can replicate only after its genetic material enters a host cell and directs the cellular machinery into synthesizing the materials necessary to produce new virus particles.

3. All bacteria are prokaryotic; they have no membrane-bound nucleus or other organelles. All have one chromosome (a circular DNA molecule). Many also have extra DNA in the form of plasmids. Plasmids are replicated independently of the bacterial chromosome.

4. Most bacteria reproduce by binary fission. Bacteria are metabolically diverse, with many different types of photosynthesizers, chemosynthesizers, and heterotrophs. They also show varied behavior.

5. Comparative biochemistry studies indicate that three prokaryotic lineages diverged from a common ancestor during the Archean Era: archaebacteria, eubacteria, and the forerunner of eukaryotes.

6. Archaebacteria include the methanogens, extreme halophiles, and extreme thermophiles. They thrive in harsh settings, much like the environments in which life probably originated. Eubacteria (all other existing bacteria) are fundamentally different from the other lineage in wall characteristics and other properties.

7. Protistans are single-celled eukaryotes (having membrane-bound organelles) with diverse modes of nutrition. The boundaries of their kingdom are poorly defined, with some lineages extending into the kingdoms of multicelled eukaryotes (plants, fungi, and animals).

8. In this book, the following organisms are classified as protistans:

a. Slime molds. Both the cellular slime molds and plasmodial slime molds have a phagocytic phase.

b. Euglenoids, chrysophytes, and dinoflagellates. Most members of these groups are photosynthetic. The chrysophytes include yellow-green algae, golden algae, and diatoms.

c. Flagellated, amoeboid, and ciliated protozoans. These are motile predators or parasites. The amoeboid protozoans are further classified as amoebas, foraminiferans, heliozoans, and radiolarians.

d. Sporozoans. These are parasites.

## Review Questions

1. Define a virus. Why is a virus considered to be no more alive than a chromosome? *347–348*

2. Outline the replication cycle of a virus that enters the lytic pathway. *348–349*

3. Describe the key characteristics of a bacterium. *353*

4. Name a few photosynthetic, chemosynthetic, and heterotrophic eubacteria. Describe some that are likely to give you the most trouble recreationally (if you enjoy water sports), medically, and ecologically (if you are worried about the greenhouse effect, for example; see page *838*). *358–359*

5. What is an endospore? Are all endospore-forming bacteria dangerous? *359*

6. Name the main categories of protistans. Think about where most of them live. Can you draw a few correlations between biotic and abiotic conditions in their environment and their structural characteristics? *360, 362–369*

## Self-Quiz *(Answers in Appendix IV)*

1. Viruses cannot reproduce without the metabolic machinery of _____.

2. Viruses are _____.
   a. the simplest living organisms   d. a and b above
   b. agents of infection             e. b and c above
   c. nonliving

3. Viruses infect _____.
   a. bacteria        c. animals
   b. plants          d. all of the above

4. The two main structural features of viruses are a _____ and a _____.
   a. DNA core; protein coat
   b. nucleic acid core; plasma membrane
   c. DNA-containing nucleus; lipid envelope
   d. nucleic acid core; protein coat

5. The two kingdoms of single-celled organisms are the _____.
   a. animals and plants       d. viruses and monerans
   b. fungi and plants         e. fungi and protistans
   c. monerans and protistans

6. All bacteria have _____ circular chromosome(s) and may have smaller, extra circles of _____ known as plasmids.
   a. one; RNA        c. one; DNA
   b. two; RNA        d. two; DNA
   c. one; DNA

7. Among the diverse bacteria are _____.
   a. photosynthetic autotrophs
   b. chemosynthetic autotrophs
   c. heterotrophs
   d. a and b
   e. a, b, and c above

8. Archaebacteria are thought to be like ancient prokaryotes because they _____.
   a. have RNA rather than DNA as their hereditary material
   b. live in places reminiscent of conditions on early earth
   c. photosynthesize by mechanisms similar to that of their ancestors
   d. all of the above

9. Bacteria reproduce by _____.
   a. mitosis          c. binary fission
   b. meiosis          d. use of elaborate sexual systems

10. Match the groups to their descriptions.
   _____ monerans and       a. cannot reproduce without pirating
         protistans             metabolic machinery of living cells
   _____ archaebacteria     b. all bacteria except archaebacteria
   _____ protistans         c. all single-celled organisms
   _____ viruses            d. single-celled eukaryotes
   _____ eubacteria         e. methanogens, extreme halophiles, and thermophiles

## Selected Key Terms

archaebacterium *353*
bacterial flagellum *354*
bacteriophage *349*
binary fission *355*
capsid *348*
chemosynthetic autotroph *353*
chrysophyte *364*
cyst *367*
endospore *359*
eubacterium *353*
euglenoid *364*
fruiting body *360*
Gram stain *354*
heterocyst *358*
heterotroph *353*
latency *348*

longitudinal fission *364*
lytic pathway *348*
microorganism *346*
multicellularity *369*
pathogen *347*
photosynthetic autotroph *353*
plasmid *355*
protistan *360*
protozoan *366*
retrovirus *350*
slime mold *362*
sporozoan *368*
symbiosis *361*
temperate pathway *348*
thermophile *357*
virus *347*

## Readings

Brock, T., and M. Madigan. 1988. *Biology of Microorganisms.* Fifth edition. Englewood Cliffs, New Jersey: Prentice-Hall.

Frankel-Conrat, H., P. Kimball, and J. Levy. 1988. *Virology.* Second edition. Englewood Cliffs, New Jersey: Prentice-Hall.

Frazier, W., and D. Westoff. 1988. *Food Microbiology.* Fourth edition. New York: McGraw-Hill. Good reference on the microbes that have major effects on our food supplies.

Margulis, L., and K. Schwartz. 1988. *Five Kingdoms.* New York: Freeman. An illustrated guide to the diversity of life. Paperback.

Stanier, R., et al. 1986. *The Microbial World.* Fifth edition. Englewood Cliffs, New Jersey: Prentice-Hall.

Woese, C. 1981. "Archaebacteria." *Scientific American* 244(6): 98–125.

# 23 FUNGI

**a** Sulfur shelf fungus (*Polyporus*) on a living tree

**Figure 23.1** Fungal species from southeastern Virginia. This small sampling merely hints at the rich diversity within the boundaries of the kingdom Fungi.

## Dragon Run

When the first winter storm blasts through southeastern Virginia, Dragon Run—part swamp, part marsh, part old-growth woodland—pays tribute to the wind. Oaks, maples, gums, beeches, and other trees release dead leaves by the millions and these shower to earth, where they pile up as crisp, ankle-deep mounds. Branches snap off; sometimes whole trees come crashing down. Sheaves of dead grasses sink into the mud along the margins of the marsh. Other plants crumple into the shallow, murky waters of the bottomlands. The storm kills uncounted numbers of insects, some birds, a few squirrels. By late December, Dragon Run is partially buried in organic debris.

This is not the first time death reigns at Dragon Run, nor will it be the last. Dig down through the mounds of debris and you will discover the accumulated litter of many past seasons—moist cushions of decayed leaves, remnant hard parts of spiders and insects, mouldering branches, perhaps the carcass of a small mammal. Each year, an astounding amount of organic matter is produced, cast off to the surroundings, then broken down, with its component nutrients cycled back to the producers.

A new growing season begins with the warm rains of February. Buds expand and develop on shrubs and trees, including a massive silver beech that has been budding each spring for three hundred years. Tiny woodland violets are the first to sprout in the thawing soil. By April, the resurgence of plant growth has imparted such a sharp, green freshness to Dragon Run that you might overlook signs of the very organisms on which the resurgence depends. On logs, under leaves, on the banks and in the waters of the marsh, members of the kingdom Fungi are commandeering resources and engaging in the business of vegetative growth.

The cottony molds, puffballs, mushrooms, and other fungi at Dragon Run are living representatives of a distinguished lineage (Figure 23.1). Fungi have been around for at least 570 million years. Through all that time, their peculiar metabolic activities have established them as nature's premier decomposers. Like other heterotrophs, fungi feast on organic compounds that have already been synthesized by other organisms.

But very few organisms besides fungi digest their dinner out on the table, so to speak. As fungi grow in or on organic matter, they secrete digestive enzymes that break it down into small bits that their individual cells can absorb. This "extracellular digestion" of organic matter liberates carbon and other nutrients *that also can be absorbed by plants*—the primary producers of Dragon Run and nearly all other ecosystems on earth.

Keep this global perspective in mind as you consider the range of fungal diversity, which is the focus of this chapter. As you will see, the metabolic activities of some fungi also happen to cause diseases in humans, pets and farm animals, ornamental plants, and important crop plants. Some species are notorious spoilers of food supplies. Others help us produce substances ranging from antibiotics to excellent cheeses. We tend to assign "value" to fungi and other organisms in terms of their direct effect on our lives. There is nothing wrong with battling dangerous species and appreciating beneficial ones—as long as we do not lose sight of the greater roles of fungi or any other kind of organism in nature.

b Scarlet hood (*Hygrophorus*)

c Frost's bolete (*Boletus*)

d Yellow coral fungus (*Clavaria*)

e Trumpet chanterelle (*Craterellus*)

f Big laughing mushroom (*Gymnophilus*)

g Rubber cup fungus (*Sarcosoma*)    h Purple coral fungus (*Clavaria*)

KEY CONCEPTS

**1.** Fungi are heterotrophs. They secrete enzymes that digest food outside their body and fungal cells absorb the breakdown products. Their activities are essential in the decomposition of organic material and cycling of nutrients in nature.

**2.** Some fungi, termed saprobes, obtain nutrients from nonliving organic matter. Other fungi are parasites; they obtain nutrients from the tissues of living host organisms.

**3.** The great majority of fungi are multicelled and filamentous. The mycelium, the food-absorbing part of the fungal body, is a mesh of tiny, elongated filaments called hyphae. Commonly, modified hyphae form a reproductive structure in or upon which spores develop. A "mushroom" is such a structure. Each spore dispersed from it may germinate and grow into a new mycelium.

**4.** Many fungi interact symbiotically with the young roots of conifers and many other plants. In this mutually beneficial relationship, called a mycorrhiza, fungal hyphae provide the plant with nutrients, and the plant provides the fungus with carbohydrates.

## GENERAL CHARACTERISTICS OF FUNGI

### Mode of Nutrition

Fungi are heterotrophs that survive by decomposing living or nonliving organic matter. Different kinds can break down just about anything organic—nature's garbage (including dead plants and animal wastes), your groceries, clothing, paper, photographic film, leather, and paint. Some fungi even grow in jet fuel and cause trouble by clogging the fuel lines.

Most fungi are **saprobes**, meaning they get nutrients from nonliving organic matter. Some are **parasites**; they get nutrients directly from the tissues of a living host. In all cases, however, they rely on **extracellular digestion and absorption**. Enzymes secreted by their cells break down large organic molecules in their surroundings into smaller components, which the cells then absorb.

373

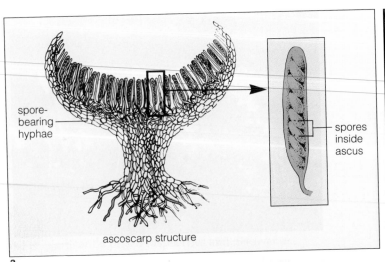

**Figure 23.9** A few of the sac fungi. (**a**) A cup-shaped ascocarp, composed of tightly interwoven hyphae. The spore-producing structures (asci) occur inside the cup. (**b**) Scarlet cup fungus (*Sarcoscypha*). (**c**) True morel (*Morchella*), which is edible. There is a "false morel" that can be poisonous. (**d**) Apple scab, a disease caused by the sac fungus *Venturia inaequalis*.

In figure a: spore-bearing hyphae, spores inside ascus, ascoscarp structure

## Sac Fungi

There are more than 30,000 species of sac fungi (Ascomycetes). All produce haploid sexual spores in pouchlike cells called **asci** (singular, ascus). Some of the yeasts, which usually are single celled, are the simplest members of the group. Yeasts occur naturally in the nectar of flowers and on fruits and leaves. One commercially important yeast, *Saccharomyces cerevisiae*, produces the carbon dioxide that leavens bread and the ethanol in wine, beer, and other alcoholic beverages (Figure 8.11). Most often, yeasts reproduce asexually, but many also can reproduce sexually following the fusion of two cells.

Filamentous sac fungi are far more numerous than the yeasts. They usually produce **ascocarps**, which are complex reproductive structures that bear or contain the asci. Some ascocarps look like globes, others like flasks or open dishes (Figure 23.9). Their mycelia grow through soil, wood, and other substrates.

Many sac fungi are plant pathogens. (A pathogen is a disease-causing agent.) One species of *Neurospora* is a nuisance in bakeries, but another (*N. crassa*) is an important organism in genetic research. Also included in this class of fungi are the edible truffles and morels. Trained dogs are used to snuffle out truffles, which grow underground. Truffles are now cultivated commercially on the roots of inoculated trees in France. Even so, these fungi remain one of the most expensive luxury foods; they are priced by the gram, not by the pound.

## Club Fungi

You probably are familiar with some of the 25,000 or so species of club fungi (Basidiomycetes). Members of this group include the mushrooms, shelf fungi, coral fungi, bird's nest fungi, stinkhorns, and puffballs. Some club fungi are saprobes that are important decomposers of plant debris. As you will see, many are important symbionts that live in association with the roots of forest trees, conifers especially. Others, including the rust and smut fungi, cause serious plant diseases. Some species are edible; in fact, cultivation of the common mushroom

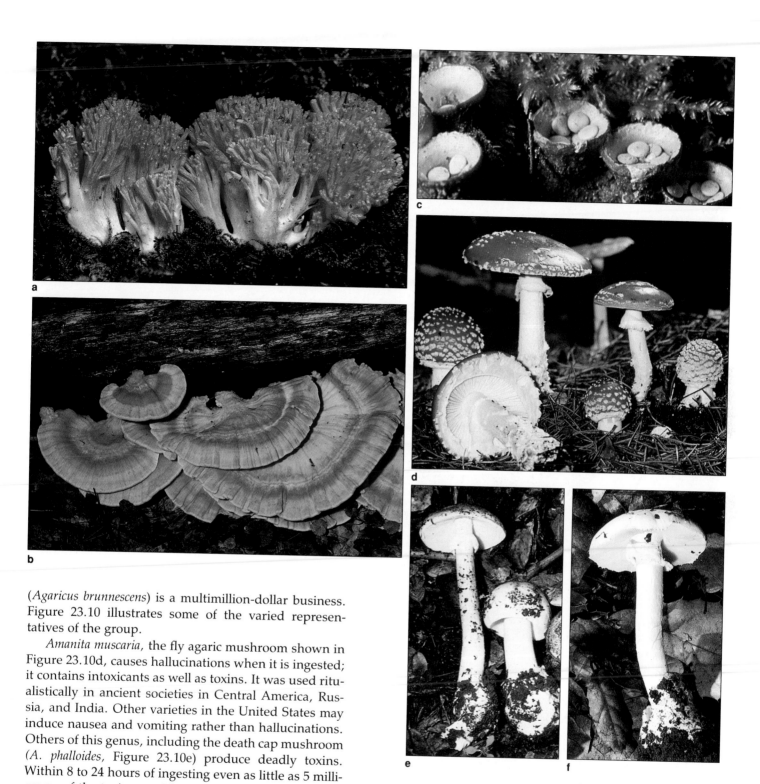

(*Agaricus brunnescens*) is a multimillion-dollar business. Figure 23.10 illustrates some of the varied representatives of the group.

*Amanita muscaria*, the fly agaric mushroom shown in Figure 23.10d, causes hallucinations when it is ingested; it contains intoxicants as well as toxins. It was used ritualistically in ancient societies in Central America, Russia, and India. Other varieties in the United States may induce nausea and vomiting rather than hallucinations. Others of this genus, including the death cap mushroom (*A. phalloides*, Figure 23.10e) produce deadly toxins. Within 8 to 24 hours of ingesting even as little as 5 milligrams of the toxin, vomiting and diarrhea begin. Later, kidney and liver cells start to degenerate; death can follow within a few days.

There is no general rule that will enable you to distinguish harmless from deadly mushrooms. *No one should eat mushrooms gathered in the wild unless they have been first accurately identified as edible.* Said another way, there are old mushroom hunters and bold mushroom hunters, but no old, bold mushroom hunters.

**Figure 23.10** Representative club fungi. (**a**) Light-red coral fungus (*Ramaria*). (**b**) Shelf fungus (*Polyporus*). (**c**) Bird's nest fungi. Figure 1.7c shows another splendid member of this group of mushrooms, the stinkhorn fungus.

Species of *Amanita* with hallucinogenic or deadly effects: (**d**) *A. muscaria*, the fly agaric mushroom, contains intoxicants as well as toxins. (**e**) *A. phalloides*, the death cap mushroom, is usually fatal when ingested. (**f**) In California, *A. ocreata* also has caused fatalities.

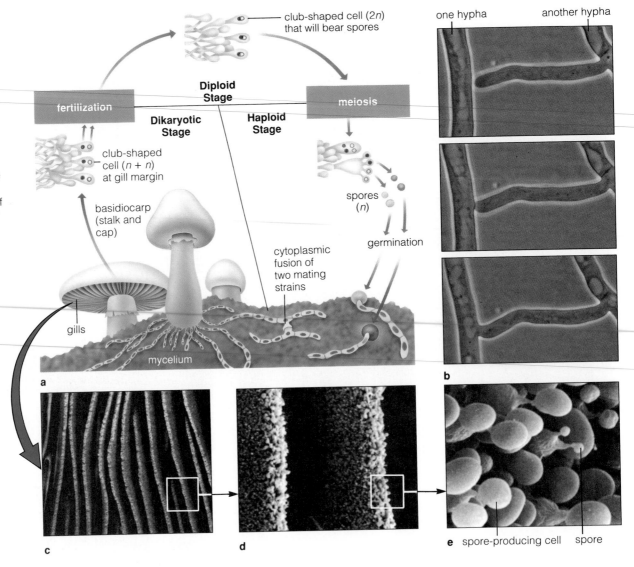

**Figure 23.11** (**a**) Generalized life cycle for mushroom-forming club fungi. During the dikaryotic stage, cells contain two distinct nuclei. (**b**) This series of micrographs shows one hypha fusing with another hypha of a genetically compatible mycelium. Such hyphae may be part of the same mycelium or from a different one. (**c–e**) Closer views of the gills of a basidiocarp from a common mushroom.

Labels in figure: club-shaped cell (2n) that will bear spores; fertilization; **Diploid Stage**; meiosis; **Dikaryotic Stage**; **Haploid Stage**; club-shaped cell (n + n) at gill margin; basidiocarp (stalk and cap); spores (n); germination; cytoplasmic fusion of two mating strains; gills; mycelium; **a**; **b**; one hypha; another hypha; **c**; **d**; **e** spore-producing cell; spore.

The spore-producing cells of club fungi are called basidia (singular, basidium). They usually are club-shaped, and they always bear the sexual spores on their outer surface (Figure 23.11). The club-shaped cells typically develop on a short-lived reproductive structure, the **basidiocarp**. The part of the fungus that is visible above the ground or on the surface of a log is the basidiocarp. The living mycelium is buried in the soil or decaying wood.

When most people hear the word fungus, they think of a mushroom. About 10,000 species of club fungi produce mushrooms, but they are the *only* fungi to do so. Each mushroom is a short-lived basidiocarp consisting of a stalk and a cap. Its spore-producing cells occur on the sides of gills, which are sheets of tissue in the cap. The rest of the fungus is an extensive mycelium growing through soil or decaying wood.

When a spore dispersed from a mushroom lands on a suitable site, it germinates and gives rise to a haploid mycelium. When hyphae of two compatible mating strains grow next to each other, cytoplasmic fusion may occur between them. Figure 23.11b shows an example of this. Nuclear fusion does not follow at once, so the resulting mycelium is dikaryotic. In other words, each of its cells contains one nucleus of each mating type. After an extensive mycelium develops and when conditions are favorable, mushrooms form. At first, each spore-producing cell of a mushroom is dikaryotic, but then its two nuclei fuse to form a zygote. The zygote's existence is brief, however. It soon undergoes meiosis and haploid spores (basidiospores) are produced. Later, the spores are dispersed by air currents.

### "Imperfect" Fungi

Fungi are classified mainly on the basis of their reproductive structures. When a sexual phase is absent (or undetected), the fungal species is said to be "imperfect." When researchers do discover a sexual phase for one of

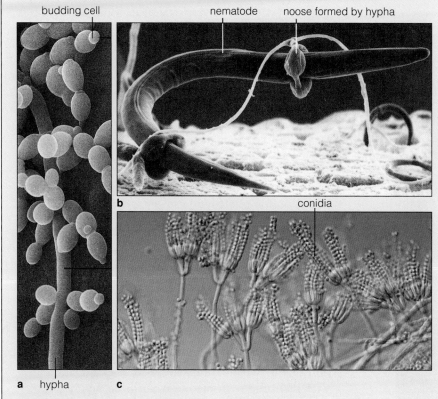

budding cell | nematode | noose formed by hypha

conidia

a hypha c

**Figure 23.12** Some imperfect fungi. (**a**) *Candida albicans*, cause of "yeast infections" of the mouth and vagina. (**b**) Hyphae of *Arthrobotrys dactyloides*, a predatory fungus, form nooselike rings that swell rapidly with incoming water when stimulated. The increased turgor pressure shrinks the "hole" in the noose and captures this nematode. Hyphae grow into the victim and release digestive enzymes. (**c**) Rows of conidia (asexual spores) and the structures that produce them in *Penicillium*. Spores of this type are common among imperfect fungi.

**Figure 23.13** Lichens. Upper photograph, *Usnea*, commonly called old man's beard. Lower photograph, *Cladonia rangiferina*, sometimes called reindeer "moss."

the species, they assign it to a recognized group—the sac fungi or club fungi, most often.

The yeast *Candida albicans* and other imperfect fungi cause many human diseases (Figure 23.12 and the *Commentary*). Some saprobic types grow in damp grain, and their toxic metabolic wastes can cause cancer in humans who eat the poisoned grain over an extended period. As Figure 23.12b demonstrates, some imperfect fungi even show predatory behavior by ensnaring tiny worms.

Some imperfect fungi produce a type of asexual spore on a hypha called a **conidium** (plural, conidia). Among them are commercially important species. *Aspergillus* is used to produce the citric acid that imparts a lemon flavor to candies and soft drinks. It also is used in the manufacture of soy sauce (the fungus ferments the soybeans used to make the sauce). Certain species of *Penicillium* produce the aroma and distinctive flavors of Camembert and Roquefort cheeses. The antibiotic penicillin, which is effective against many pathogenic bacteria, is derived from other *Penicillium* species.

## Lichens

Recall that a *symbiotic relationship* is a mutually beneficial interaction in which one or both species comes to depend permanently on the other. Thousands of sac fungi and club fungi are symbionts with cyanobacteria or green algae. Associations between a fungus and its photosynthetic partner are **lichens** (Figure 23.13).

Lichens often live in inhospitable places, including bare rock and wind-whipped tree trunks. Their secretions help break down rock and convert it to soil that can support larger plants. In the arctic tundra, where large plants are scarce, reindeer and musk oxen can survive on lichens. Air pollution is often monitored by observing lichens, which cannot grow in heavily polluted air.

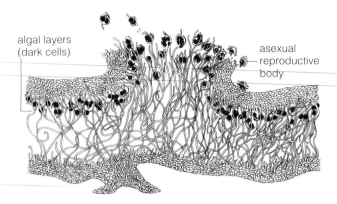

**Figure 23.14** Cross-section through a complex lichen, *Lobaria verrucosa*. Fungal hyphae form the upper, protective layer. Just below is a layer of algal cells, which are functionally connected with loosely interwoven, thin-walled hyphae. Notice the fragments containing both hyphae and algal cells; lichens reproduce asexually by such fragmentation.

algal layers
(dark cells)

asexual
reproductive
body

The relationship often begins when a fungal mycelium contacts a free-living cyanobacterium, an algal cell, or both. The fungus parasitizes a photosynthetic cell, sometimes killing it in the process. If the photosynthetic cell can survive, however, it multiplies in association with the fungal hyphae (Figure 23.14). The fungus probes into cells of its partner with short, specialized hyphae. The fungus depends entirely on its captive algal cells for food.

Often, the fungal member of a lichen absorbs all but 20 percent of the carbohydrates produced by its partner. It is difficult to see what the partner gets out of being enslaved. After all, the drain on nutrients has adverse effects on the photosynthesizer's growth and reproductive capacity. Only a few green algae truly benefit from the relationship. They grow very slowly and cannot compete well with other plants on their own, but they thrive in the shelter provided by a lichen.

## MYCORRHIZAE

A **mycorrhiza**, or "fungus-root," is a symbiotic relationship in which fungal hyphae associate intimately with plant roots. In many mycorrhizae, hyphae thread densely around the young roots of forest trees and shrubs (Figure 23.15). The fungus gets carbohydrates from the plant. The plant in turn absorbs some of the ions from the fungus during growth. When viewed collectively, the fungal hyphae have a tremendous surface area for absorbing mineral ions from a large volume of the surrounding soil. The hyphae conserve dissolved mineral ions when they are plentiful and release them to the plant when they are scarce in the soil. In the absence of mycorrhizae, many plants cannot readily absorb mineral ions, particularly phosphorus (Figure 23.16).

Mycorrhizae happen to be highly susceptible to damage from acid rain. Their susceptibility is having repercussions in the world's forests, a topic to which we will be returning in Chapter 48.

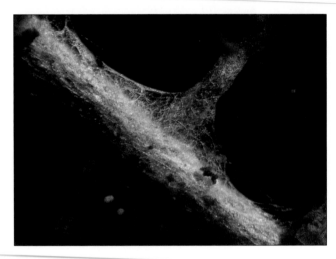

**Figure 23.15** Mycorrhiza (fungus-root) of a hemlock tree. White threads are fungal strands.

**Figure 23.16** Effect of mycorrhizal fungi on plant growth. The six-month-old juniper seedlings on the left were grown in sterilized low-phosphorous soil inoculated with a mycorrhizal fungus. The seedlings on the right were grown without the fungus.

## SUMMARY

1. Fungi are heterotrophs that are important decomposers for the world of life. Many are symbiotic; others are serious pathogens of plants and animals.

2. Saprobic fungi feed on nonliving organic matter. Parasitic types obtain nutrients from tissues of living organisms. The cells of all types secrete digestive enzymes that break down food into small molecules which are absorbed across the cell's plasma membrane.

3. Most fungi are multicelled and filamentous, with haploid and diploid stages alternating in the life cycle.

The multicelled body (mycelium) is composed of microscopic filaments (hyphae).

4. Fungi can reproduce asexually by spore formation, budding, or fragmentation of the parent body. Many sexually reproducing fungi have a dikaryotic stage that intervenes between cytoplasmic fusion and nuclear fusion of two gametes.

5. The main groups of fungi are the chytrids, water molds, zygosporangium-forming fungi, sac fungi, and club fungi. They are categorized mainly on the basis of their reproductive structures. When a sexual phase cannot be detected or is absent from the life cycle of a given specimen, that specimen is assigned to an informal category—the "imperfect" fungi.

6. Many sac fungi and some club fungi enter into symbiotic relationships with cyanobacteria or green algae. The term *lichen* refers to the intimate association between such a fungus and its photosynthetic partner.

7. Many fungi are symbiotic with young roots of shrubs and trees. In this association (a mycorrhiza), fungal hyphae provide the plant with nutrients; the plant provides the fungus with carbohydrates.

## Review Questions

1. Distinguish among saprobic and parasitic fungi. What is meant by extracellular digestion and absorption? *373*

2. How is a mycelium constructed and what is its function? *374*

3. Describe one of the main groups of fungi. *375–380*

4. Some fungi trap living nematodes as food. Do these fungi feed as parasites or saprobes? Why? *381*

5. Define sporangium, gametangium, and basidiocarp. *375, 380*

6. What is the difference between a mycorrhiza and a lichen? *381–382*

## Self-Quiz (Answers in Appendix IV)

1. Heterotrophic fungi are important _____ , breaking down organic materials and so assisting in the cycling of nutrients through communities.

2. A multicelled fungal body is called a _____ and it is composed of microscopic filaments called _____ .

3. Parasitic fungi obtain nutrients from _____ .
   a. tissues of living host organisms
   b. nonliving organic matter
   c. only living plants
   d. only living animals
   e. none of the above is correct

4. The main groups of fungi are categorized mainly on the basis of their _____ .
   a. habitat
   b. color and size
   c. reproductive structures
   d. digestive enzymes

5. A mycorrhiza is a _____ .
   a. fungal disease of the foot
   b. fungus-plant relationship
   c. parasitic water mold
   d. fungus endemic to barnyards

6. A "mushroom" is _____ .
   a. the food-absorbing part of a fungal body
   b. part of the fungal body not constructed of hyphae
   c. a reproductive structure
   d. a nonessential part of the fungus

7. New mycelia form following the germination of _____ .
   a. hyphae          c. mycelia
   b. spores          d. mushrooms

8. A lichen is an intimate symbiotic association between a fungus and a _____ .
   a. mycorrhiza          d. tree trunk
   b. green alga          e. cyanobacteria
   c. parasitic fungus    f. b and e are correct

9. Fungi can reproduce asexually by _____ .
   a. forming spores          c. fragmenting of parent bodies
   b. budding                 d. all of the above are correct

10. Match each concept with its description.
    _____ saprobic fungi
    _____ fungal mode of nutrition
    _____ imperfect fungi
    _____ dikaryotic stage
    _____ mycelium

    a. mass of hyphae; the multicelled fungal body
    b. for some fungal life cycles, stage between cytoplasmic fusion and fusion of nuclei of two gametes
    c. feed on nonliving organic matter
    d. no detectable or detected sexual phase
    e. extracellular digestion and absorption

## Selected Key Terms

ascus *378*
ascocarp *378*
basidiocarp *380*
conidium *381*
dikaryotic stage *374*
extracellular digestion and absorption *373*
fungal spore *374*
gametangium *375*
hypha *374*
lichen *381*
mycelium *374*
mycorrhiza *382*
parasite *373*
rhizoid *375*
saprobe *373*
sporangium *375*

## Readings

Bold, H., and J. LaClaire. 1987. *The Plant Kingdom*. Fifth edition. Englewood Cliffs, New Jersey: Prentice-Hall. Paperback.

Kendrick, B. 1991. *The Fifth Kingdom*. Second edition. Waterloo: Mycologue Publications.

Moore-Landecker, E. 1990. *Fundamentals of the Fungi*. Third edition. Englewood Cliffs, New Jersey: Prentice-Hall. Well-written introduction to the kingdom of fungi.

Among existing gymnosperms and angiosperms, the developing male gametophytes—*pollen grains*—are released from the parent plant and travel by air currents or by insects, birds, or other animals to the structures that contain the female gametophytes. The evolution of pollen grains that carry sperm without requiring liquid water allowed these plants to radiate into diverse land environments. Their female gametophytes are surrounded by protective tissues, some of which develop into a seed coat, and they mature while still attached to the parent plant. Seeds are ideal packages for surviving hostile conditions. It is probably no coincidence that the dominant seed plants arose during the extreme climates of Permian times (page 311).

---

**During the evolution of complex land plants, vascular tissues and other structural adaptations to dry conditions emerged.**

**Haploid (gametophyte) dominance gave way to diploid (sporophyte) dominance. Sporophytes with well-developed roots and shoots had the means to nourish and protect spores and gametophytes through unfavorable conditions.**

**Homospory gave way to heterospory. This permitted the evolution of male gametes specialized for dispersal without liquid water, and of seeds—the packaging of embryos with protective and nutritive tissues.**

---

With these evolutionary trends in mind, we turn to the spectrum of plant diversity, beginning with the algae.

## THE "ALGAE"

Algae is a term that originally came into use to define simple aquatic "plants." It no longer has any formal significance in classification schemes, because the organisms once lumped together under the term are now assigned to different kingdoms. The problem is that taxonomists do not always agree on the assignments. Botanists traditionally have viewed the red, brown, and green algae as plants, largely because all three groups include multicelled species that show complex growth patterns as well as reliance on photosynthesis. But these groups also include single-celled and filamentous forms that are decidedly different from leafy plants. For these and other reasons, some botanists argue that the red, brown, and green algae should be assigned to the kingdom Protista.

Remember that *we* impose the boundaries of kingdoms across unbroken evolutionary lines. Until there is greater consensus that the boundaries should be shifted once more, this book will take the traditional approach and keep the red, brown, and green algae in the plant kingdom:

| Kingdom | Division |
|---------|----------|
| Plantae: | Rhodophyta (red algae) |
| | Phaeophyta (brown algae) |
| | Chlorophyta (green algae) |
| Protista: | Chrysophyta (golden algae, diatoms) |
| | Euglenophyta (euglenoids) |
| | Pyrrhophyta (dinoflagellates) |
| Monera: | Cyanobacteria (blue-green algae) |

### Red Algae

If you have ever tried to raise fruit flies or some other small organism in the laboratory, you may be familiar with *agar*. It is a rather inert, gelatinous substance that helps solidify culture media. *Gelidium* and other types of red algae secrete agar as part of their cell walls. Agar is used as a moisture-retaining agent in baked goods and as a setting agent in gelatin desserts.

Nearly 4,000 known species of algae have been assigned to the division Rhodophyta (after the Greek *rhodon*, meaning rose; and *phyton*, meaning plant). None produces motile cells. Different species appear green, red, purple, or greenish-black, depending on the types and abundance of different photosynthetic pigments. These pigments include chlorophyll *a* and the phycobilins, which can trap blue-green light in deep water. (Chlorophylls are more efficient at absorbing red and blue wavelengths that may not penetrate far below the water's surface.)

Figure 24.3 shows representatives of this group. Several types of red algae make their home in freshwater lakes, streams, and springs. Most, however, live in marine habitats. They abound in tropical seas, often at surprising depths (more than 200 meters) when the water is clear. Many marine species have stonelike cell walls in which calcium carbonate has been deposited; they are among the builders of coral reefs (page 868).

Red algae differ considerably in their mode of sexual reproduction. In most species, the sporophytes and gametophytes look alike, although they can be distinguished from each other by their reproductive structures. Spore-producing structures develop on the sporophyte. Meiosis occurs in these structures, which are called **sporangia** (singular, sporangium). The haploid spores formed by way of meiosis give rise to the gametophytes. Later, the gametophytes produce nonmotile sperm and egglike structures by way of mitosis.

### Brown Algae

If you consume commercial ice cream, pudding, salad dressing, canned and frozen foods, jellybeans, or beer, if you use cough syrup, toothpaste, a variety of cosmetics, paper, textiles, ceramics, or floor polish, thank the brown algae. Some species produce *algin* in their cell

**Figure 24.3** (**a**) A red alga showing the most common growth pattern (filamentous and branched). This alga reproduces asexually when the hooklike branchlets break off and grow into new plants after they become caught on branches of other algae. (**b**) From a tropical reef, an alga with sheetlike growth.

**Figure 24.4** One of the brown algae, *Postelsia palmaeformis*, commonly called the sea palm. It grows in intertidal zones, where it is alternately submerged and exposed to air.

blade

float (bladder)

stipe

holdfast

**Figure 24.5** (**a**) A kelp that usually is submerged even at low tide. Only the uppermost parts are visible at the water's surface; long stipes are anchored to rocks below. (**b**) Leaflike and stemlike parts of many brown algae. Gas-filled bladders (floats) occur along the stipes of several species.

walls, and this substance can be used as thickening, emulsifying, or suspension agents.

About 1,500 olive-green, golden, and dark-brown species are classified as brown algae (Phaeophyta, after the Greek *phaios*, meaning dusky). An abundance of xanthophylls, together with chlorophylls and other pigments, gives these plants their distinctive coloration.

Structurally, the simplest brown algae have a branching, filamentous body plan. Many others are quite complex, with leaflike, stemlike, and rootlike parts. The sea palms shown in Figure 24.4 have leaflike *blades* growing from a *stipe*, a stout, fleshy, stemlike structure. The giant kelps (*Macrocystis*) have hollow, gas-filled bladders, or *floats*, at the base of each leafy blade. These help keep the blades near the surface, where sunlight available for photosynthesis is greatest. The kelps also have specialized conducting cells that are similar to those in phloem of vascular plants. Kelps and other brown algae living offshore or in the intertidal zone attach to submerged rocks and sediments by *holdfasts*, which are rootlike structures at the base of the plant (Figure 24.5). The

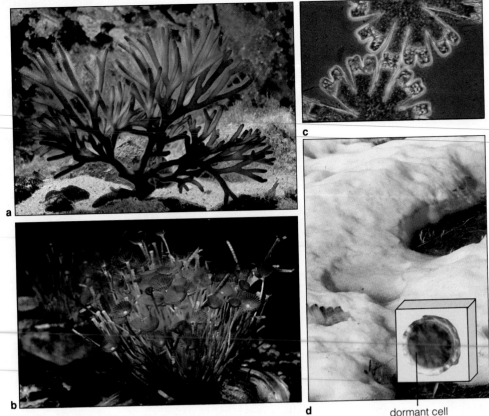

**Figure 24.6** Green algae from marine habitats. (**a**) A species of *Codium*, with multinucleate, branched cells. *Codium* from Puget Sound, Washington, was accidentally introduced into the mouth of the Connecticut River in 1956. There are no native species of *Codium* in that part of the Atlantic—and no predators that evolved with and are adapted to eating them. The introduced alga spread from the coasts of Maine to North Carolina in a little over two decades. (**b**) A group of mermaid's wineglasses (*Acetabularia*). Each plant, a multinucleate cell mass, has a rootlike structure, a stalk, and a cap in which gametes form. (**c**) *Micrasterias*, a freshwater green alga, undergoing cell division. (**d**) "Red snow" above the summer timberline in Utah. It results from the presence of dormant cells of *Chlamydomonas nivalis*, a green alga with red accessory pigments that protect its chlorophyll.

dormant cell

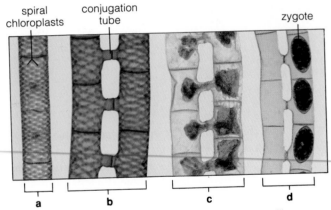

spiral chloroplasts    conjugation tube    zygote

a        b        c        d

**Figure 24.7** One mode of sexual reproduction in watersilk (*Spirogyra*), a filamentous green alga with ribbonlike, spiral chloroplasts (**a**). A conjugation tube forms between cells of adjacent, haploid filaments of different mating strains (**b**). The cellular contents of one strain pass through the tubes into cells of the other strain, where zygotes form (**c,d**). The zygotes will develop thick walls, and meiosis will occur in them when they germinate to form new haploid filaments.

brown alga *Sargassum* is unusual. It floats as extensive tangled masses through the vast Sargasso Sea, which lies between the Azores and the Bahamas.

The giant kelps sometimes grow more than 50 meters long, forming underwater forests that sway with the currents. These "kelp forests" are among the most productive ecosystems, the home and shelter for diverse marine organisms. Where kelps have been removed for commercial purposes—for example, for algin production or to simplify abalone harvesting for the California abalone industry—productivity in the region plummets.

For many brown algae, the life cycle includes a multicelled haploid stage as well as a multicelled diploid stage. This also is true of many red and green algal species. Among simpler, filamentous species, the gametophyte and sporophyte may be outwardly similar. In more complex species, the gametophyte is extremely small and a large sporophyte dominates the life cycle. The kelps shown in Figure 24.5 are sporophytes; their gametophytes consist of microscopic filaments.

### Green Algae

Think about a crew of astronauts spending years in a space station. Or, closer to home, a crew on a nuclear submarine with orders not to surface for a couple of years. Where would they get their oxygen? Green algae (*Chlorella*) are being looked at as a possible supplier. Algae would not take up much space and they would require only light, carbon dioxide, and some minerals to grow. More to the point, algae would give off oxygen as a by-product of photosynthesis while taking up the carbon dioxide exhaled by the aerobically respiring crew.

Of all the aquatic plants, the green algae bear the greatest resemblance to land plants. They store excess carbohydrates as starch inside chloroplasts, just as land

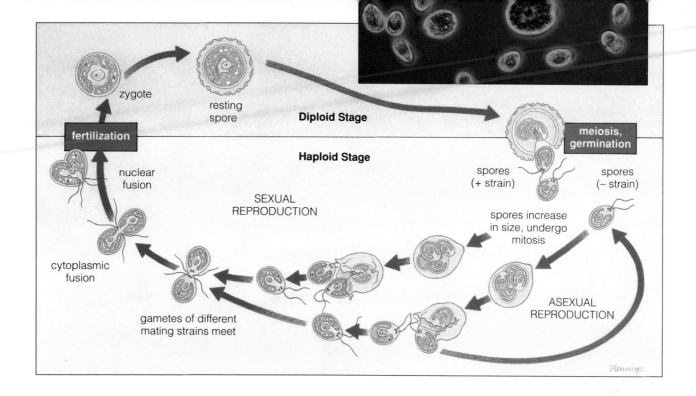

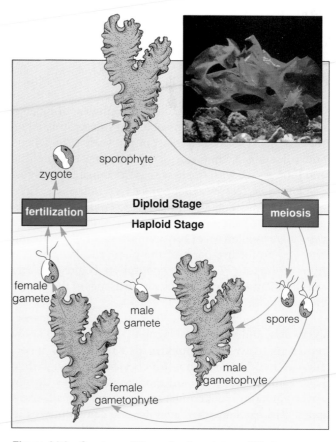

plants do. They, too, have a distinct array of photosynthetic pigments: chlorophylls *a* and *b*, carotenoids, and xanthophylls. And the majority have cellulose in their cell walls, as land plants do. These and other similarities suggest that land plants evolved from ancestral forms of green algae.

More than 7,000 species of green algae are known. Their dominant color comes from chlorophyll pigments. Hence the division name, Chlorophyta (after the Greek *chloros*, meaning green). Green algae grow all around the world—in freshwater, on the surface of the open oceans, just below the surface of soil and marine sediments, on rocks, on treebark as well as on other organisms, even on snow (Figure 24.6).

Green algae show diverse patterns of reproduction. Figure 24.7 shows one of the reproductive modes for *Spirogyra*, a filamentous form. For some species of *Chlamydomonas*, haploid cells of different mating types function as gametes, although they are identical in size and structure. There is no multicelled stage, only an alternation between haploid and diploid cells. For other species of *Chlamydomonas*, gametes differ in size and motility. Figure 24.8 shows how spores of the leaflike *Ulva* give rise to male and female gametophytes that look alike. The gametophytes also look just like the sporophyte.

**Most red, brown, and green algae are multicelled aquatic plants, with the haploid (gametophyte) phase dominating their life cycle.**

**Figure 24.8** Life cycle of *Chlamydomonas*, a single-celled green alga. Asexual reproduction is most common, but sexual reproduction also can occur between gametes of different mating strains.

**Figure 24.9** One type of life cycle of sea lettuce (*Ulva*).

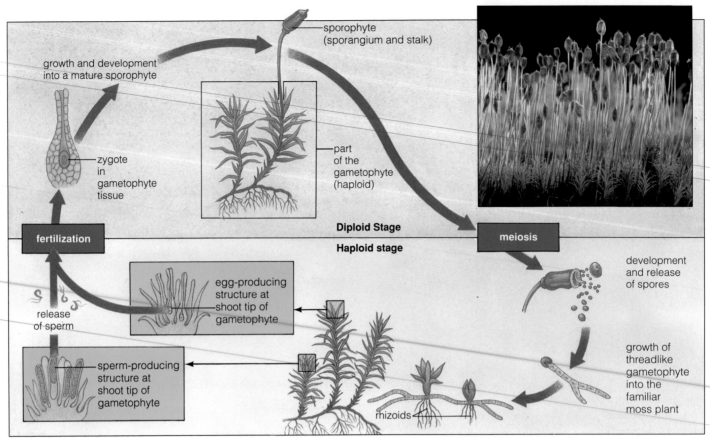

**Figure 24.10** Life cycle of a moss, a representative bryophyte. Notice how the sporophyte is attached to (and dependent upon) the gametophyte.

## Bryophytes

The nonvascular plants called bryophytes originated more than 350 million years ago. They grow in habitats that are moist for at least part of the year. Like lichens and some algae, they can dry out during dry months and then revive with the new season's rains. Bryophytes display three features that surely were adaptive when ancient plants made the transition to land. *First*, aboveground parts have a water-conserving cuticle. *Second*, a protective cellular jacket surrounds the sperm-producing and egg-producing parts of the plant and keeps them from drying out. *Third*, the sporophyte begins its early development as an embryo *inside* the tissues of the female gametophyte.

Bryophytes include 16,000 or so species of mosses, liverworts, and hornworts. Figures 24.10 and 24.11 show representatives. All bryophytes are small plants, generally less than 20 centimeters (8 inches) tall. Although they have leaflike, stemlike, and rootlike parts, they do not have xylem or phloem. Most species have *rhizoids*, elongated cells or filaments that attach the gametophytes to the soil. Rhizoids also absorb water and dissolved mineral ions.

Mosses are the most common bryophytes. Their spores give rise to threadlike gametophytes that grow into the familiar moss plants. Generally, eggs and sperm develop in saclike, jacketed structures at the shoot tips (Figure 24.9). Each structure is a "gamete vessel," or **gametangium** (plural, gametangia). Sperm reach the eggs by swimming through a film of water on plant parts.

After fertilization, the zygotes develop into mature sporophytes, each consisting of a stalk and a sporangium. The sporangium is a saclike, jacketed structure in which spores develop. The sporophytes eventually may nourish themselves photosynthetically, but initially they depend on the parent gametophytes for food and continue to depend on them for nutrients and water. The bryophytes are distinct from all other land plants in having independent gametophytes and dependent sporophytes.

Bryophytes are nonvascular land plants that require the presence of liquid water for fertilization. Their sporophytes are attached to and dependent upon the gametophytes for nutrients and water.

male gametophyte    female gametophyte    gemmae

a    b    c

**Figure 24.11** The bryophyte *Marchantia*. It is a type of liverwort, and the only one to produce male and female reproductive structures on separate plants (**a**,**b**). *Marchantia* also reproduces asexually by way of gemmae, multicelled vegetative bodies that develop in tiny cups on the plant body (**c**). Gemmae can grow into new plants when splashing raindrops transport them to suitable sites.

## SEEDLESS VASCULAR PLANTS

The fossil record tells us that seedless vascular plants were established by the late Silurian, some 420 million years ago, and they flourished for about 60 million years before most kinds became extinct. *Cooksonia* was one of the earliest ancestors of these plants (Figure 19.21a). Like other members of the division Rhyniophyta, *Cooksonia* was leafless, but its internal structure was similar to that of many existing vascular plants. Individuals of at least one species grew as tall as 0.5 meter. *Psilophyton*, of the division Trimerophyta, grew taller than this and was structurally more complex (Figure 19.21b). Members of another division, the Progymnospermophyta, may have been ancestral to the seed-bearing plants. Figure 19.2a shows the fossilized leaves of one tree-sized specimen.

Some lineages of seedless vascular plants survived to the present. They are commonly called whisk ferns, lycophytes, horsetails, and ferns. Like their ancestors, the existing members of these four divisions differ from the bryophytes in some important respects. *First*, their sporophytes develop independently of the gametophytes. *Second*, their sporophytes have well-developed vascular tissues. *Third*, the sporophyte is the larger, longer lived stage of the life cycle; the gametophytes are very small and some lack chlorophyll. In addition, members of all four divisions differ from the gymnosperms and angiosperms in this respect: none produces seeds.

Although their sporophytes are adapted for life on land, the seedless vascular plants are confined largely to wet, humid regions. Why? Their gametophytes have no vascular tissues for water transport. Moreover, the male gametes must have water to reach the eggs. Thus the whisk ferns, lycophytes, horsetails, and ferns are the

**Figure 24.12** Sporophytes of a whisk fern (*Psilotum*), a seedless vascular plant. Pumpkin-shaped, spore-producing structures occur at the ends of stubby branchlets.

"amphibians" of the plant kingdom: *For part of the life cycle they are, in essence, aquatic.* The few species that are adapted for survival in extreme environments such as deserts can reproduce sexually only when adequate water is available.

**Psilophytes, lycophytes, horsetails, and ferns require liquid water for fertilization. Their sporophytes dominate the gametophyte stage of the life cycle.**

### Psilophytes

Whisk ferns (Psilophyta) look more like whisk brooms than true ferns. You will find them growing in moist tropical and subtropical regions, including Hawaii, Florida, Louisiana, Texas, and Puerto Rico. You also will find them in florist shops; they are popular ornamental plants. The *Psilotum* sporophytes shown in Figure 24.12 have no roots or leaves. Aboveground, they have a system of branches with scalelike projections. Belowground, they have a system of short, branching outgrowths. These are *rhizomes* (underground, mainly horizontal stems). These are absorptive structures, and they benefit from their association with mycorrhizal fungi (page 382). Photosynthesis occurs in the outer cells of the stem, which has a vascular cylinder of xylem and phloem.

**Figure 24.13** *Lycopodium* sporophyte, showing the conelike strobili in which spores are produced.

**Figure 24.14** Strobili borne on stems of horsetails (*Equisetum*). The close-up shows clusters of sporangia on a strobilus.

Like the early vascular plants, the whisk ferns lack roots and leaves. Evolutionarily speaking, it is doubtful that their ancestors ever would have evolved without mycorrhizal interactions. Ancient "fungus-roots" surely performed the absorptive functions of roots for them.

### Lycophytes

The lycophytes (Lycophyta) were highly diverse 350 million years ago, when some forms were tree-sized members of swamp forests. We can thank ancient lycophytes and other plants of that time for our existing coal deposits (page 34).

The most familiar of the existing lycophytes are tiny club mosses, members of the genera *Lycopodium* and *Selaginella*, that grow on forest floors. The sporophyte of a club moss has true roots, stems, and small leaves with a single strand of vascular tissue. Often it also has cone-shaped clusters of leaves bearing the spore sacs. Figure 24.13 shows an example. Each cluster is a *strobilus* (plural, strobili). Spores are dispersed from the spore sacs and germinate to form small, free-living gametophytes. Although species of *Lycopodium* are homosporous, those of other genera are heterosporous. All require ample water in which sperm can swim to the eggs.

### Horsetails

The ancient relatives of modern-day horsetails (Sphenophyta) included treelike forms taller than a two-story building. A single genus, *Equisetum*, has survived to the present. Horsetails grow in moist soil along streams and in disturbed habitats, such as vacant lots, roadsides, and beds of railroad tracks. Their sporophytes typically have underground stems (rhizomes). The scalelike leaves are arranged in whorls about an aboveground, photosynthetic stem that is hollow inside. Pioneers of the American West used horsetails to scrub their cooking pots. The walls of stem cells contain silica, giving the stems the gritty quality of sandpaper.

Figure 18.9 shows the vegetative, photosynthetic stems of one horsetail species. Figure 24.14 shows the fertile stems. As for all horsetail species, the spores form inside cone-shaped clusters of tiny branches at the shoot tips. Air currents disperse them. They must germinate within a few days to produce gametophytes, which are free-living plants about the size of a small pea.

### Ferns

Among the 12,000 or so species of ferns (Pterophyta) are some of the most popular house plants. Most ferns are native to tropical and temperate regions (Figure 24.15). The size range for members of this division is stunning. Some floating species are less than 1 centimeter (0.4 inch) across—while some tropical tree ferns are 25 meters (82 feet) tall. Except for tropical tree ferns, the stems are mostly underground. Fern leaves, or *fronds*, are usually featherlike, with the blades finely divided into segments, but extraordinary diversity exists on the basic plan.

You may have noticed rust-colored patches on the lower surface of many fern fronds. Each patch is a *sorus* (plural, sori), which is a cluster of sporangia. Each sporangium looks rather like a baby rattle. At dispersal time, the sporangium snaps open and causes the spores to catapult through the air. A germinating spore develops into a small gametophyte, such as the green, heart-shaped type shown in Figure 24.15.

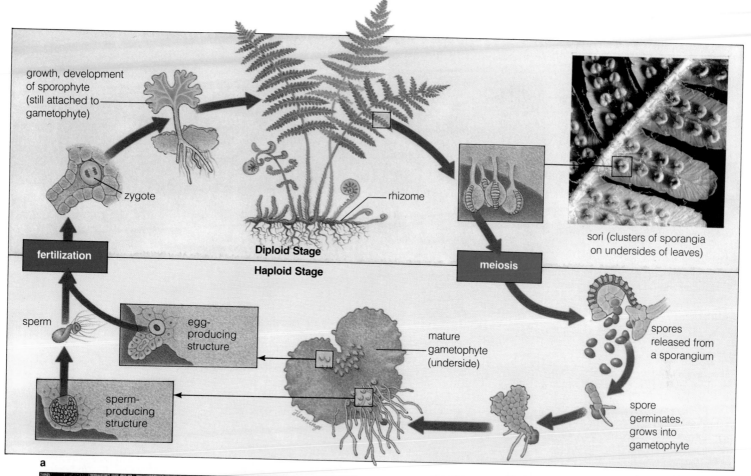

growth, development of sporophyte (still attached to gametophyte)

zygote

fertilization

Diploid Stage

Haploid Stage

meiosis

sori (clusters of sporangia on undersides of leaves)

rhizome

sperm

egg-producing structure

mature gametophyte (underside)

sperm-producing structure

spores released from a sporangium

spore germinates, grows into gametophyte

a

b

c

**Figure 24.15** (**a**) Life cycle of a fern. (**b**) Ferns in a moist habitat in Indiana. (**c**) Tree ferns in a temperate rain forest of Tasmania.

**Table 24.2 Summary of Major Groups of Plants and Their Characteristics**

| Division | Number of Species | Some General Characteristics |
|---|---|---|
| **Simple, Mostly Aquatic Plants** | | |
| **Red algae** | 4,000 | Single-celled to multicelled, with branching filaments or fan shapes. Most in warm marine habitats, some deep-water species, some reef builders. Abundant phycobilin pigments. |
| **Brown algae** | 1,500 | Multicelled, some with structures that resemble roots, stems, leaves. Almost all marine, many in colder seas. Abundant xanthophyll pigments. Diploid dominance in many species. |
| **Green algae** | 7,000 | Single-celled to multicelled, with filamentous, colonial, or sheetlike forms. Common in wet soil, freshwater, many in warm seas. Pigments like those of land plants. |
| **Nonvascular Land Plants. Free water required for fertilization.** | | |
| **Bryophytes** | 16,000 | Mosses, liverworts, hornworts. Underground "stems," rootlike and leaflike structures. Many in moist, humid habitats. Diploid sporophyte depends on dominant (haploid) gametophyte. Some heterosporous. |
| **Seedless Vascular Plants. Free water required for fertilization.** **Diploid dominance. Cuticle, stomata present.** | | |
| **Psilophytes** | 10 | Simple sporophytes—no roots (rhizomes as absorptive system), photosynthetic branching stem. Homosporous. |
| **Lycophytes** | 1,000 | Club mosses, etc. Simple leaves. Mostly wet or shady habitats. Some heterosporous. |
| **Horsetails** | 15 | One existing genus. Hollow photosynthetic stem with spore-bearing, scalelike leaves. Swamps, moist woodlands, lake edges, railroad beds. Homosporous. |
| **Ferns** | 12,000 | Mostly tropical, temperate plants. Finely divided leaves typical, rhizomes; tree ferns have woody stems. Wet, humid habitats. Mostly homosporous. |
| **Vascular Plants with "Naked" Seeds (Gymnosperms). Diploid dominance.** | | |
| **Conifers** | 550 | Mostly evergreen, woody trees and shrubs with needlelike or scalelike leaves, pollen- and seed-bearing cones. Widespread in Northern and Southern hemispheres. |
| **Cycads** | 100 | Shrubby or treelike with palmlike leaves, pithy stems, pollen- and seed-bearing cones on different plants. Very slow growth. Tropics, subtropics. |
| **Ginkgos** | 1 | Woody-stemmed tree, deciduous fan-shaped leaves. Pollen, ovules on different plants. Fleshy seeds. |
| **Gnetophytes** | 70 | Some woody branching shrubs or woody vines, one has strappy leaves. Pollen- and seed-bearing cones on different plants. |
| **Vascular Plants with Flowers, Protected Seeds (Angiosperms). Diploid dominance.** | | |
| **Flowering plants:** | | Woody and nonwoody (herbaceous) plants. Depend on wind pollination or pollinators. Almost every land habitat, some aquatic. |
| Monocots | 65,000 | Grasses, palms, lilies, orchids, onions, etc. Floral parts often arranged in threes or multiples of three; one cotyledon (seed leaf), parallel-veined leaves common. |
| Dicots | 170,000+ | Most fruit trees, roses, cabbages, melons, beans, potatoes, etc. Floral parts often arranged in fours, fives, or multiples of these; two cotyledons, net-veined leaves common. |

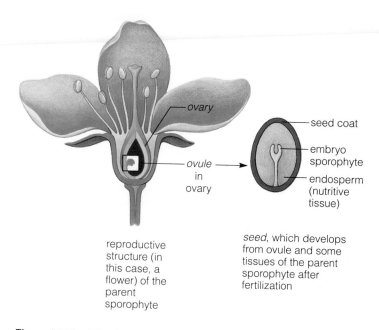

Figure 24.16 A few botanical terms illustrated.

ovary

ovule
in
ovary

reproductive
structure (in
this case, a
flower) of the
parent
sporophyte

seed coat

embryo
sporophyte

endosperm
(nutritive
tissue)

seed, which develops
from ovule and some
tissues of the parent
sporophyte after
fertilization

Figure 24.17 One of the lesser known gymnosperms, a cycad. Notice its cones.

We now leave the seedless vascular plants. Take a moment to review Table 24.2, which compares their characteristics with those of the remaining major plant groups.

## EXISTING SEED PLANTS

In terms of sheer numbers and distribution, the most successful vascular plants are the seed-bearing species. Each seed starts out as an ovule. An **ovule** is a structure containing a female gametophyte, complete with egg cell, that is surrounded by tissues and a jacket of protective cell layers. A fertilized egg develops into an embryo sporophyte and the outer tissue layer develops into a seed coat. In short, a **seed** is a structure formed by the maturation of an ovule following fertilization (Figure 24.16).

The ancestors of seed plants first appeared during Devonian times. Fossil evidence of one species, a progymnosperm, was shown earlier in Figure 19.2. Plants of this type were on the evolutionary road leading to both gymnosperms and angiosperms.

The word **gymnosperm** is derived from the Greek *gymnos* (meaning naked), and *sperma* (which is taken to mean seed). As the name implies, gymnosperm seeds are rather unprotected; they are perched at the surface of reproductive parts. The word **angiosperm** is derived from the Greek *angeion* (meaning a case or vessel) and *sperma*. The "vessel" part of the name refers to the tissues that surround and protect the ovules and later, during development, the seeds.

### Gymnosperms

Gymnosperms were abundant in the past. Today there are about 650 species. The sporophytes of nearly all species are conspicuous trees or shrubs; a few are woody vines. The most widespread and familiar gymnosperms are the conifers (Coniferophyta). Others are the cycads (Cycadophyta), ginkgo (Ginkgophyta), and the gnetophytes (Gnetophyta). Members of the last three divisions listed are restricted in their native distribution and numbers of species. In all species, the small female gametophytes are not free-living but rather are enclosed by sporophyte tissues. The small male gametophytes develop in pollen grains.

**Cycads.** During the Mesozoic era, the cycads flourished along with the dinosaurs. About 100 species have survived to the present, but they are confined to the tropics and subtropics. At first glance you might mistake a cycad for a small palm tree (Figure 24.17). Despite having similar leaves and stems, the palms and cycads are not closely related (palms are angiosperms). Cycads have massive, cone-shaped strobili that bear either pollen or ovules. Pollen from the "male" plants is transferred by air currents or crawling insects to the developing seeds on "female" plants. In tropical Asia, cycad seeds and a starchy flour made from cycad trunks are used as food sources—but only after their poisonous alkaloids have been washed out.

**Figure 24.18** Two more representatives of the lesser known gymnosperms. Gingko trees (**a**) and their fleshy seeds (**b**). The gnetophyte *Welwitschia* (**c**) and its female cones (**d**).

**Ginkgos**. The ginkgos are even more restricted in native distribution than the cycads. Only a single species has survived to the present, despite the diversity of the group during the Mesozoic. Several thousand years ago, ginkgo trees were planted extensively in cultivated grounds around Buddhist temples in China. Since then, the natural populations from which the domesticated trees were derived must have disappeared. The near-extinction of this living fossil from the age of dinosaurs is puzzling, for ginkgos seem to be hardier than many other trees. It may be that as human populations expanded in size, ginkgos were chopped down for fuel. It is known that ginkgo wood also is highly prized by wood carvers. Male ginkgo trees are now being planted in cities because of their attractive, fan-shaped leaves (Figure 24.18a) and their resistance to insects, disease, and air pollutants. (The fruits of female trees produce an awful stench when squished on sidewalks and pavements.)

**Gnetophytes**. There are about seventy species of gnetophytes, which are divided into three genera: *Gnetum*, *Ephedra*, and *Welwitschia*. Moist, tropical regions are home to about thirty species of *Gnetum*, which includes both trees and leathery leafed vines. About thirty-five species of *Ephedra* grow in desert regions of the world. Of all the gymnosperms, *Welwitschia* is the most bizarre. This seed-producing plant grows in the hot des-

erts of south and west Africa. The bulk of the plant is a deep-reaching taproot. The only exposed part is a woody disk-shaped stem that bears cone-shaped strobili and leaves. The plant never produces more than two strap-shaped leaves, which split lengthwise repeatedly as the plant grows older, the outcome being a rather scraggly pile (Figure 24.18c).

**Conifers**. Conifers generally are woody trees and shrubs with needlelike or scalelike leaves. Familiar examples are the pines, spruces, firs, hemlocks, junipers, cypresses, and redwoods. Most are evergreen species; although they shed old leaves throughout the year, they retain enough leaves to distinguish them from deciduous species. Conifers have **cones**. Nearly all of these are cone-shaped clusters of modified leaves bearing the sporangia (hence the name conifer, which means cone-bearing). Seeds develop on the shelflike scales of female cones.

We can use the pine as a general example of conifer life cycles (Figure 24.19). A pine tree produces two kinds of spores in two kinds of cones. The scales of male cones bear sporangia in which spore mother cells undergo meiosis and give rise to haploid **microspores**. The spores develop into pollen grains, each containing a male gametophyte. The scales of female cones bear ovules. Inside each ovule, a mother cell undergoes meiosis. Only one of the resulting haploid spores, the **megaspore**, survives and develops into a female gametophyte.

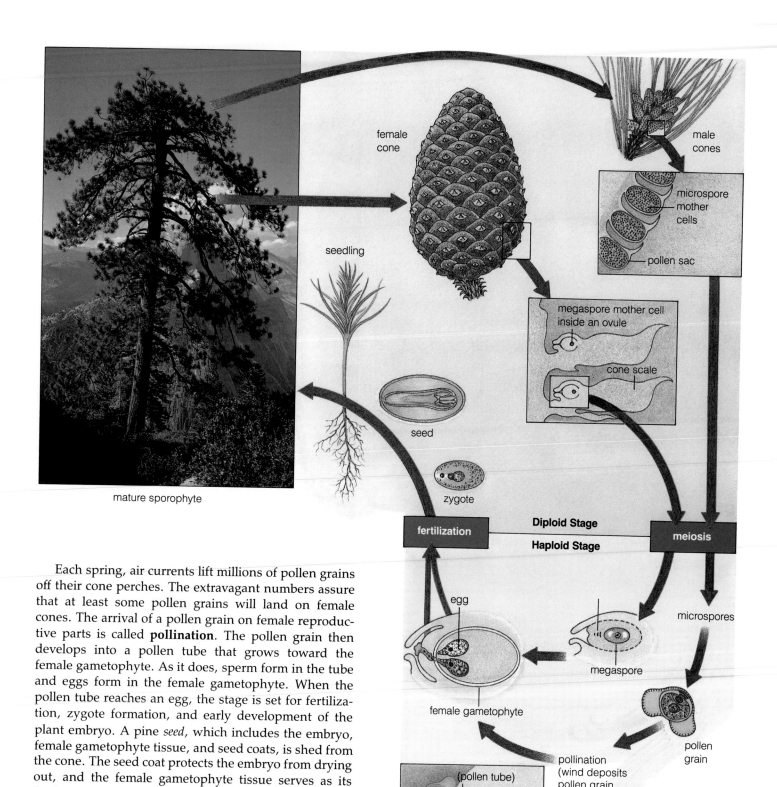

female cone

seedling

seed

zygote

male cones

microspore mother cells

pollen sac

megaspore mother cell inside an ovule

cone scale

mature sporophyte

**Diploid Stage**

fertilization

**Haploid Stage**

meiosis

egg

microspores

megaspore

female gametophyte

pollination (wind deposits pollen grain near ovule)

pollen grain

(pollen tube)

(sperm nuclei)

This cell will form two sperm nuclei as pollen tube grows toward the egg.

male gametophyte

Each spring, air currents lift millions of pollen grains off their cone perches. The extravagant numbers assure that at least some pollen grains will land on female cones. The arrival of a pollen grain on female reproductive parts is called **pollination**. The pollen grain then develops into a pollen tube that grows toward the female gametophyte. As it does, sperm form in the tube and eggs form in the female gametophyte. When the pollen tube reaches an egg, the stage is set for fertilization, zygote formation, and early development of the plant embryo. A pine *seed*, which includes the embryo, female gametophyte tissue, and seed coats, is shed from the cone. The seed coat protects the embryo from drying out, and the female gametophyte tissue serves as its food reserve.

With their mechanisms of producing, protecting, and dispersing seeds, the conifers radiated into many land environments during the Mesozoic. Today there are about 550 species, some of which are major sources of lumber, paper, and other products. They are still dominant in many regions, especially to the north and at high altitudes (page 860). Many conifers also grow in the Southern Hemisphere.

**Figure 24.19** Life cycle of a representative conifer, a ponderosa pine.

a

b

c

d

## Angiosperms—The Flowering Plants

Of all the divisions of plants, the angiosperms (flowering plants) are the most successful—they have dominated the land for more than 100 million years. Today there are about 235,000 known species (Figures 24.20 and 24.21 show a few examples), and new ones are discovered almost daily in previously unexplored regions of the tropics.

Angiosperms are also the most diverse types of plants. You can find them growing on dry land and in wetlands, freshwater, and seawater. Angiosperms range in size from duckweeds about a millimeter long to *Eucalyptus* trees more than 100 meters tall. Most angiosperms are free-living and photosynthetic. A few are saprobes that feed on nonliving organic matter. The Indian pipes that grow on the floor of eastern deciduous forests are an example of a saprobe (Figure 24.20f). Having no chlorophyll of their own, they obtain food by associating

**Figure 24.20** Flowering plants. Diverse photosynthetic species are adapted to nearly all environments, ranging from deserts (**a**) to snowlines of high mountains (**b**). Orchids (**c**), water lilies (**d**), and sugarcane plants (**e**) are successful in numbers and distribution, being planted, tended, and enjoyed by humans. (**f**) A saprobic flowering plant—Indian pipe (*Monotropa uniflora*). (**g**) A parasitic species—dwarf mistletoe (*Arceuthobium*). Its activities seriously limit the growth and productivity of forest trees in the western United States.

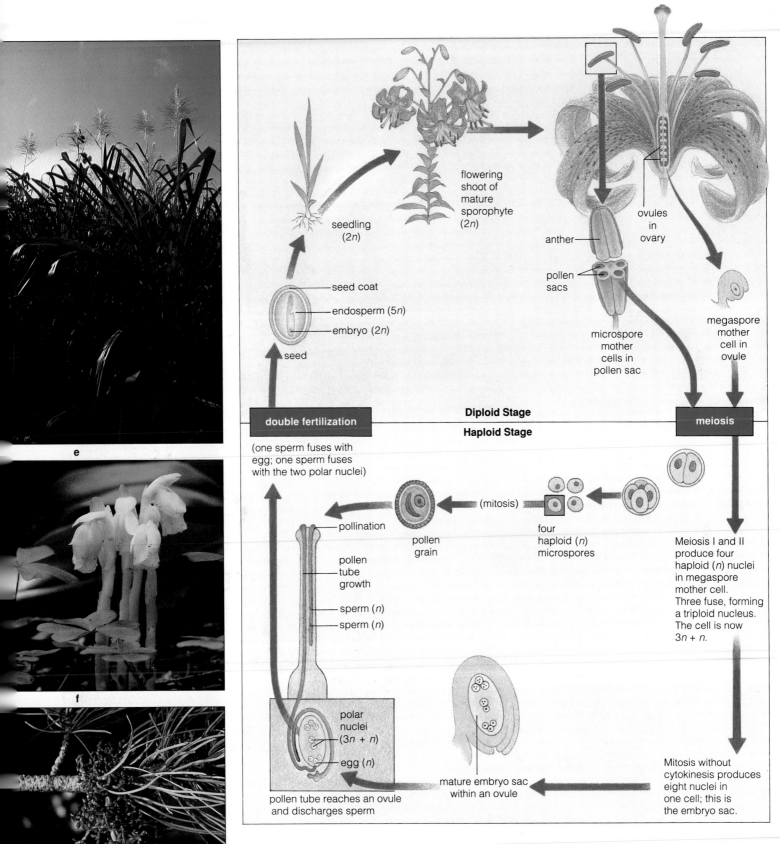

e

f

g

flowering
shoot of
mature
sporophyte
(2n)

seedling
(2n)

seed coat

endosperm (5n)

embryo (2n)

seed

ovules
in
ovary

anther

pollen
sacs

microspore
mother
cells in
pollen sac

megaspore
mother
cell in
ovule

**Diploid Stage**

**Haploid Stage**

**double fertilization**

**meiosis**

(one sperm fuses with
egg; one sperm fuses
with the two polar nuclei)

(mitosis)

four
haploid (n)
microspores

pollen
grain

pollination

Meiosis I and II
produce four
haploid (n) nuclei
in megaspore
mother cell.
Three fuse, forming
a triploid nucleus.
The cell is now
3n + n.

pollen
tube
growth

sperm (n)
sperm (n)

polar
nuclei
(3n + n)

egg (n)

pollen tube reaches an ovule
and discharges sperm

mature embryo sac
within an ovule

Mitosis without
cytokinesis produces
eight nuclei in
one cell; this is
the embryo sac.

**Figure 24.21**  Life cycle of a flowering plant, the monocot *Lilium*. Although about 70 percent
of known flowering plant species follow the pattern shown in Figure 29.8, microscope slides
of lilies abound, and you are likely to encounter such slides in the classroom.

with mycorrhizae, which are in turn associated with the roots of photosynthetic plants. Still other angiosperms are parasites that feed on living host organisms. Mistletoe plants are parasitic (Figure 24.20g). So is *Rafflesia*, which obtains food by tapping into the roots of members of the grape family (Figure 48.1c).

There are two classes of angiosperms, the **monocots** and **dicots**. The formal names are Monocotyledonae and Dicotyledonae. The monocots include grasses, palms, lilies, and orchids. The main crop plants (wheat, corn, rice, rye, sugarcane, and barley) are domesticated grasses—all monocots. Although there are 20,000 species each of grasses and orchids, dicots are more diverse; there are nearly 200,000 species. They include most of the familiar shrubs and trees other than conifers, most nonwoody (herbaceous) plants, the cacti, and the water lilies. For comparative purposes, we show the life cycle of a monocot in Figure 24.21. The life cycle of a typical dicot is described in detail in the next unit of the book, which focuses on the structure and function of flowering plants.

Many factors contributed to the adaptive success of angiosperms. As with other seed plants, the diploid sporophyte dominates the life cycle. The sporophyte of the land-dwelling types has root and shoot systems, and other features that allow it to take up and conserve water and dissolved minerals. In all types, the sporophyte retains and nourishes the gametophyte. The embryos are nourished by a unique tissue (endosperm) within the seed. Also, the seeds are packaged in fruits, which help protect and disperse them. Above all, angiosperms have unique reproductive structures called **flowers**.

As you will see in the next unit of the book, many diverse floral structures have coevolved with animal pollinators—insects, bats, birds, and rodents. This innovation probably figured in the rise of angiosperms and the gradual decline of gymnosperms in so many regions over the past 100 million years.

The evolution of pollen grains freed gymnosperms and angiosperms from dependence on free water for fertilization. It provided for wider dispersal of gametes, more opportunities for genetic combinations, and more evolutionary possibilities.

## SUMMARY

1. The plant kingdom includes aquatic and terrestrial species, nearly all of which are photosynthetic autotrophs. Land plants are believed to have evolved from multicelled green algae over 400 million years ago.

2. The following evolutionary trends among plants have been identified:

    a. Structural adaptations to dry conditions, especially the development of the vascular tissues called xylem and phloem.

    b. A shift from haploid (gametophyte) dominance to diploid (sporophyte) dominance.

    c. A shift from spores of the same type to spores of two different types—that is, from homospory to heterospory. This paved the way for the development of male gametophytes that became specialized for dispersal without liquid water, and female gametophytes that became specialized for holding onto, protecting, and nourishing the embryo sporophytes (in seeds).

    d. Female gametophytes serve this protective and nutritive function for all vascular plants except the angiosperms. In this case, the parent sporophyte holds and protects the new "generation" and provides nutritive endosperm for it.

3. The red, brown, and green algae are mostly multi-celled aquatic plants.

4. Existing nonvascular land plants (those without well-developed xylem and phloem) include the bryophytes—the mosses, liverworts, and hornworts.

5. Existing seedless vascular land plants include the psilophytes (whisk ferns), lycophytes, horsetails, and ferns. Existing seed-bearing vascular plants include the gymnosperms and angiosperms (flowering plants).

6. Vascular land plants typically have a cuticle and stomata that help control water loss. They have sporangia (protective tissue layers around their spores). They have ovules or similar structures with protective and nutritive tissue layers around their gametophytes. The embryo sporophyte begins its development *within* gametophyte tissues.

7. Bryophytes are small, nonvascular, and restricted to moist habitats. Like seedless vascular plants, they require liquid water for fertilization.

8. The most complex vascular plants (gymnosperms and angiosperms) have escaped from dependency on liquid water for reproduction. Their female gametophytes are attached to and protected by the sporophyte. With its root and shoot systems, the sporophyte is well adapted to conditions on dry land.

9. The evolution of pollen and pollen tubes in gymnosperms and angiosperms freed those plants from dependence on free water for fertilization. It also paved the way for much wider dispersal of pollen, more opportunities for new genetic combinations, and consequent adaptive and evolutionary possibilities.

10. The seeds of gymnosperms and angiosperms are efficient means of dispersing the new generation and helping it through hostile conditions.

## Review Questions

1. Describe the evolutionary trends among plants that figured in the invasion of land. *386–388*

2. What are some differences between bryophytes and the vascular plants? *393*

3. How does the life cycle of a gymnosperm (such as pine) differ from that of an angiosperm (such as a lily)? *397, 402*

## Self-Quiz *(Answers in Appendix IV)*

1. Most plants are multicelled, photosynthetic _____ that directly or indirectly nourish nearly all other forms of life.

2. Land plants are believed to have evolved from multicelled _____ over 400 million years ago.

3. Which of the following was *not* a major trend in the evolution of complex land plants?
   a. from nonvascular to vascular body plants
   b. from sporophyte to gametophyte dominance
   c. from one spore type to two specialized types
   d. away from reliance on free-standing water for fertilization
   e. among some lineages, reliance on pollinators, seed formation, and seed dispersal mechanisms
   f. all were major trends

4. Which of the following statements is *not* true?
   a. Monocots and dicots are two classes of angiosperms.
   b. Bryophyptes are nonvascular plants.
   c. Lycophytes, horsetails, ferns, gymnosperms, and angiosperms are vascular plants.
   d. Horsetails and gymnosperms are the simplest vascular plants.

5. Red, brown, and green algae are mostly _____.
   a. multicelled aquatic plants
   b. nonvascular land plants
   c. seedless vascular plants
   d. seed-bearing vascular plants

6. Of all land plants, bryophytes alone have independent _____ and attached, dependent _____.
   a. sporophytes; gametophytes
   b. gametophytes; sporophytes
   c. rhizoids; zygotes
   d. rhizoids; sporangium with stalk

7. Psilophytes, lycophytes, horsetails, and ferns are _____.
   a. multicelled aquatic plants
   b. nonvascular land plants
   c. seedless vascular plants
   d. seed-bearing vascular plants

8. Gymnosperms and angiosperms are _____.
   a. multicelled aquatic plants
   b. nonvascular seed plants
   c. seedless vascular plants
   d. seed-bearing vascular plants

9. Which of the following is *not* characteristic of gymnosperms and angiosperms?
   a. vascular tissues
   b. diploid dominance
   c. single spore type
   d. all are correct

10. Match each structure and its function.
   _____ cuticle, stomata
   _____ seed
   _____ sporophyte
   _____ ovule
   _____ gametophyte
   _____ flower

   a. attract pollinator
   b. produce gametes
   c. produce spores
   d. control water loss
   e. protect, disperse embryo sporophyte
   f. female gametophyte with egg cell, plus surrounding tissue and protective layers

## Selected Key Terms

alga *388*
angiosperm *397*
bryophyte *392*
cone *398*
conifer *398*
cycad *397*
dicot *402*
endosperm *402*
flower *402*
gametangium *392*
gametophyte *387*
ginkgo *398*
gnetophyte *398*
gymnosperm *397*
heterospory *387*
homospory *387*

megaspore *398*
microspore *398*
monocot *402*
nonvascular plant *386*
ovule *397*
phloem *386*
pollination *399*
rhizoid *392*
rhizome *393*
root system *386*
seed *397*
shoot system *386*
sporangium *388*
sporophyte *387*
vascular plant *386*
xylem *386*

## Readings

Bold, H., and J. LaClaire. 1987. *The Plant Kingdom*. Fifth edition. Englewood Cliffs, New Jersey: Prentice-Hall. Paperback.

Gensel, P., and H. Andrews. 1987. "The Evolution of Early Land Plants." *American Scientist* 75:478–489.

Raven, P., R. Evert, and S. Eichhorn. 1986. *Biology of Plants*. Fourth edition. New York: Worth. Lavishly illustrated.

Stern, K. 1991. *Introductory Plant Biology*. Fifth edition. Dubuque, Iowa: Boston. Paperback.

## Animals of the Burgess Shale

It happened some 530 million years ago, in the deep and silent waters alongside a massive reef that paralleled the coast of an early continent. Protected from ocean currents, great banks of sediments and mud had formed in the basin behind the reef. No animals lived in the still, anaerobic mud at the bottom of the basin. At a depth of about 500 feet, however, the dimly lit waters were oxygenated and clear. There, an astonishing array of marine animals flourished in, on, and above the muddy sediments piled against the steep flank of the reef.

Like castles built from wet sand along a seashore, their home was unstable. And 530 million years ago, during the Cambrian era, part of the bank slumped suddenly and buried an entire community of organisms. Although catastrophic for the community, the underwater avalanche also assured that scavengers would not be able to reach and obliterate all traces of the dead animals.

Over time, muddy silt rained down on the natural tomb, and gradually it all became compacted into finely stratified shale. Soft parts of the by-now flattened carcasses were transformed into shimmering films of calcium aluminosilicate. All the while, imperceptibly, the continents were on the move. By the dawn of the Cenozoic, the Pacific plate of the earth's crust was plowing under and lifting the western edge of the North American plate (page 314). During this period of geologic upheaval, the magnificent mountain ranges of western Canada were born.

By the year 1909, the buried fossils were high up in the mountains that run along the eastern border of British Columbia. In that year a fossil hunter, Charles Doolittle Wolcott, tripped over a chunk of shale that had fallen onto a mountain path. Among the fine layers of shale— which split apart easily—Wolcott made the discovery of a lifetime. Some animals entombed and fossilized in the Burgess Shale had come to light.

Soon afterward, Wolcott and others recovered fossils of more than 120 animal species. Some bore obvious resemblance to existing species. Others were bizarre evolutionary experiments that apparently led nowhere. *Opabinia* was one of these. About as long as a tube of lipstick, *Opabinia* had five eyes perched on its head and a grasping organ that possibly was used in prey capture (Figure 25.1). The smaller but rather nightmarish *Hallucigenia* moved about on seven pairs of stiltlike spines. Seven tentacles running along its back may have been used when the animal scavenged on dead worms and other organic tidbits.

The Burgess Shale fossils are testimony to a great adaptive radiation of early animals. That radiation almost certainly was triggered by the breakup of a supercontinent just before the dawn of the Cambrian. Think of the adaptive zones that must have opened up in the waters along the vast new coastlines! Yet where are signs of the *first* animals—those that predate the Burgess Shale species? Were they no more complex than *Trichoplax adhaerens*? That existing soft-bodied animal is little more than a miniature, two-layered pancake, yet it

**Figure 25.1** (**a**) From the Burgess Shale, fossils of two kinds of animals—the five-eyed *Opabinia*, with a grasping organ at its head end, and *Hallucigenia*, decked out with multiple spines and tentacles. (**b**) A sudden, underwater slide off the coast of Baja California, reminiscent of the catastrophic slumping that entombed the organisms of the Burgess Shale.

b

is capable of sexual reproduction (page 369). Were they like existing ball-shaped colonies of flagellated cells, as in *Volvox*, or like the tiny ciliated larvae of marine animals? Here the fossil record is a closed book.

One thing is clear. Every major animal phylum was established before the time of the Burgess Shale animals. By then, representatives were already in place on the evolutionary roads that would lead, ultimately, to modern sponges, cnidarians, flatworms, roundworms, mollusks, annelids, arthropods, echinoderms, and other invertebrates. Even before then, there were animals with muscles arrayed in a zigzag pattern and a stiffened rod running partway down their back. These may have been on or near the evolutionary road leading to the vertebrates—to humans and all other animals with backbones. These are the roads traced in this chapter and the next.

KEY CONCEPTS

**1.** There are about 1.5 million known species of animals—multicellular, motile heterotrophs that pass through a period of embryonic development during their life cycle.

**2.** By comparing the body plans of existing animal groups and integrating this information with the fossil record, we can identify major trends that occurred in animal evolution. The most revealing aspects of the animal body plan are the type of symmetry, gut, and cavity (if any) between the gut and body wall; whether it has a distinct head end; and whether it is divided into a series of segments.

**3.** Sponges show no body symmetry and no tissue organization. Cnidarians show radial symmetry and some tissue organization. Nearly all other invertebrates show bilateral symmetry and well-developed tissues, organs, and organ systems.

**4.** In terms of numbers and distribution, insects and other arthropods are the most successful invertebrates. Their hardened exoskeleton, specialized segments, jointed appendages, specialized respiratory and nervous systems, and (among insects especially) a division of labor in the life cycle have contributed to their success.

Mammals, birds, reptiles, amphibians, fishes—these are all **vertebrates**, the only animals with a "backbone" and the ones most familiar to us. Yet, of the approximately 1.5 million known species of animals, fewer than 50,000 are vertebrates! Among the **invertebrates** (animals without a backbone) are lesser known but diverse species. The photographs in this chapter only hint at how spectacularly diverse they are.

Suppose we take representatives of each animal phylum and arrange them in sequence, from the structurally simplest to the most complex. Sponges would be at one end of the spectrum and vertebrates at the other. If we accept that new species arise only from preexisting species, then all the phyla in our arrangement are related by descent—some closely, others distantly. Thus, comparing their similarities and differences may allow us to identify major trends in animal evolution.

# OVERVIEW OF THE ANIMAL KINGDOM

## General Characteristics of Animals

There are more than thirty phyla of animals, but we can follow several evolutionary trends without considering every single one. In this chapter and the next, our sampling will include sponges, cnidarians, comb jellies, flatworms, roundworms, ribbon worms, rotifers, mollusks, annelids, arthropods, echinoderms, and chordates (Table 25.1). Like other animals, they are defined by the following characteristics:

1. Animals are multicellular, and except for sponges, their cells form tissues. The tissues usually are arranged into organs and organ systems.

2. Animals are heterotrophs; they cannot produce their own food. They eat other organisms or absorb nutrients produced by them.

3. Animals are diploid organisms that reproduce sexually and, in many cases, asexually.

4. Animal life cycles include a period of embryonic development. In brief, cell divisions transform the zygote into a ball-shaped multicelled embryo. Then cells become arranged into germ tissue layers (*ectoderm, endoderm* and, in most species, *mesoderm*), and these give rise to all tissues and organs of the adult (Chapter 42).

5. Most animals are motile, at least during part of the life cycle.

## Body Plans

We can use five body features to track the increasing complexity among different animal groups. These are *body symmetry, cephalization, type of gut, type of body cavity,* and *segmentation.*

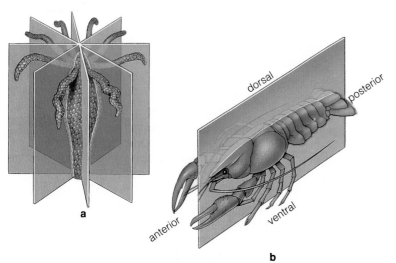

a

anterior

dorsal

posterior

ventral

b

**Body Symmetry and Cephalization.** Nearly all animals are radial or bilateral. In animals with **radial symmetry**, body parts are arranged regularly around a central axis, like spokes of a bike wheel (Figure 25.2a). A slice down the center of a hydra divides it into equal halves; another slice at right angles to the first divides it into equal quarters. You cannot get the same result with a crayfish, which has a bilateral body plan. In animals with **bilateral symmetry**, the right and left halves of the animal are mirror images of each other (Figure 25.2b). Even if some body parts are positioned to one side of the midline, as is often the case, this does not violate the basic bilateral plan.

Certain directional terms are used when describing the location of body parts of bilateral animals. Such animals generally have an *anterior* end (head), *posterior* end, *ventral* surface (the surface closest to a substrate), and *dorsal* surface (the upper surface, or back).

Radial animals are aquatic, and their body plan allows them to respond to food or danger floating past in any direction. Their ancestors predated the first bilateral animals, which crawled on the seafloor. The anterior end of the crawlers was the first to encounter food and other stimuli, so there surely were selective advantages in **cephalization**—in having sensory structures and nerve cells concentrated up front, in a head. We can indeed identify trends toward bilateral symmetry and cephalization. They involved the development of paired muscles, paired sensory structures, paired nerves, and paired brain regions (Chapter 33).

**Type of Gut.** The part of the body where food is actually digested and absorbed is informally called the **gut**. The gut of some animals is saclike, and food moves into it and residues are expelled from it by way of the same opening. In most animals, however, the gut is part of a "complete" digestive system. Basically, the system is a tube with an opening (mouth) at one end and another opening (anus) at the opposite end. Different regions of the gut have separate, specialized functions, such as mechanically breaking down food, absorbing nutrients and water, or compacting wastes. It is an efficient arrangement. In certain animal lineages, the emergence of complete digestive systems helped pave the way for increases in body size and activity.

**Body Cavities.** A body cavity lies between the gut and the body wall of most animals. One type of cavity, the **coelom**, has a cellular lining called the *peritoneum* (Figure 25.3). The lining also covers organs inside the coelom and helps hold them in place. You are one of the

**Figure 25.2** Planes of radial symmetry in a hydra (**a**) and of bilateral symmetry in a crayfish (**b**).

coelomate animals. In your case, a muscular partition divides the coelom into an upper *thoracic* cavity and a lower *abdominal* cavity. Your heart and lungs occupy the thoracic cavity. Your stomach, intestines, liver, bladder, and other organs occupy the abdominal cavity.

Some invertebrates, including flatworms, have no body cavity. In such *acoelomate* animals, the space between the gut and body wall is packed with tissues. Other invertebrates, including roundworms, have a body cavity but no peritoneum. They are *pseudocoelomate* animals. (The term means they have a "false coelom.")

A true coelom apparently was a prerequisite for the evolution of large, complex animals. Being cushioned and protected inside a body cavity, the organs of early coelomate animals had the potential for increases in size and activity.

**Segmentation.** "Segmented" animals consist of a series of body units that may or may not be similar to one another. In earthworms, for example, most segments look alike on the outside, and some also show similarities on the inside. In evolutionary terms, segmentation proved to have great potential for the development of increasingly specialized head parts, legs, wings, and other appendages. A spider, for example, has two very different body parts, each composed of groupings of specialized segments. Similarly, each of the three parts of an insect body (the head, thorax, and abdomen) is composed of such groupings.

| Table 25.1 | Animal Phyla Described in This Book | |
|---|---|---|
| Phylum | Some Representatives | Number of Known Species |
| **Placozoa** | *Trichoplax adhaerens* | 1 |
| **Porifera** (sponges) | Tubular, cuplike, vaselike, sprawling animals | 8,000 |
| **Cnidaria** (cnidarians) | *Hydra*, jellyfishes, corals, sea anemones | 11,000 |
| **Ctenophora** (comb jellies) | *Pleurobrachia* | 100 |
| **Platyhelminthes** (flatworms) | Turbellarians, flukes, tapeworms | 15,000 |
| **Nemertea** (ribbon worms) | Soft-bodied animals shaped like ribbons, rubber bands, shoelaces | 800 |
| **Nematoda** (roundworms) | Pinworms, hookworms | 20,000 |
| **Rotifera** (rotifers) | Small animals with a crown of cilia | 1,800 |
| **Mollusca** (mollusks) | Snails, slugs, clams, squids, octopuses | 110,000 |
| **Annelida** (annelids) | Earthworms, leeches, polychaetes | 15,000 |
| **Arthropoda** (arthropods) | Crabs, lobsters, spiders, insects | 1,000,000+ |
| **Echinodermata** (echinoderms) | Sea stars, sea urchins, sea cucumbers | 6,000 |
| **Chordata** (chordates) | Invertebrate chordates: Tunicates, lancelets | 2,100 |
| | Vertebrates: Fishes | 21,000 |
| | Amphibians | 3,900 |
| | Reptiles | 7,000 |
| | Birds | 8,600 |
| | Mammals | 4,500 |

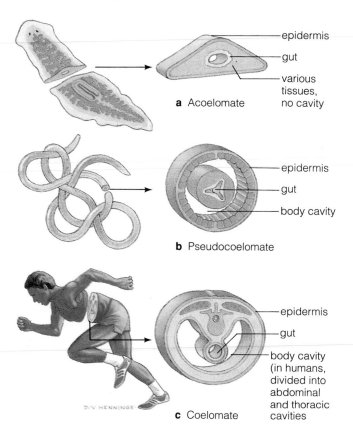

**a** Acoelomate

epidermis
gut
various tissues, no cavity

epidermis
gut
body cavity

**b** Pseudocoelomate

epidermis
gut
body cavity (in humans, divided into abdominal and thoracic cavities)

**c** Coelomate

**Figure 25.3** Body plans of bilateral animals. (**a**) Acoelomate, or without a body cavity between the gut and body wall, as in flatworms.

(**b**) Pseudocoelomate, with a body cavity that has no continuous peritoneal lining. Various organs, especially of the reproductive system, occupy the pseudocoel.

(**c**) Coelomate, a plan typical of vertebrates and several invertebrate groups (including annelids and echinoderms).

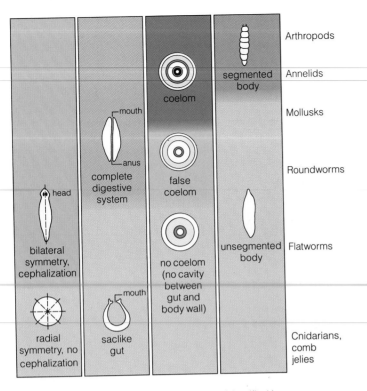

**Figure 25.4** Key trends in animal evolution, identified by comparing body plans of major groups of existing animals.

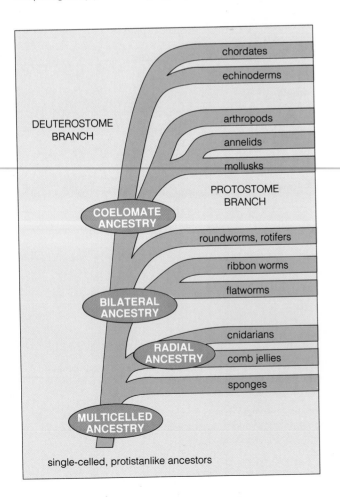

**Figure 25.5** A family tree showing broad evolutionary relationships among major groups of animals. All groups shown were established by 570 million years ago. Take a moment to study this diagram. It will be used repeatedly as our road map through discussions of each animal group.

Not all of the body features just described appeared in every lineage, as Figure 25.4 shows. Their absence does not mean an animal is "primitive" or evolutionarily stunted. Even the simplest animals are exquisitely adapted to their environment. Taken together, however, the similarities and differences among animals help us perceive broad evolutionary relationships (Figure 25.5). Let us turn now to examples from each major phylum.

## SPONGES

Simple as they are, sponges (Porifera) are one of nature's success stories. Ever since Cambrian times, they have been among the most abundant marine animals. Today they are common in coastal waters, tropical reefs, and many other marine habitats. Only about 100 of the 8,000 or so species live in freshwater. Many animals, including small fishes, shrimps, barnacles, and a variety of worms, make their home in or on sponges.

Some sponges are as small as a fingernail and a few are big enough to sit in. Many are flat and sprawling. Others are compact, lobed, tubular, cuplike, or vaselike (Figures 25.6 and 25.7). Regardless of the shape, the sponge body has no symmetry and no organs. Flattened cells do line the outer surface and parts of the internal cavities, but these are not the same as the specialized tissues that cover the body surface and line the internal cavities of other animals. Between the linings is a semifluid matrix with skeletal elements and amoebalike cells. The skeletal elements called spicules are needles of silica or calcium carbonate; others are fibers of a flexible protein. Many spicules project outward, forming a protective barrier of sharp points, but most function to support the body. The fibrous skeleton of natural bath sponges gives them a soft, squeezable texture.

A large volume of water moves through an irregular array of microscopic pores and chambers that perforate the sponge body wall. More specifically, it moves past thousands or millions of flagellated *collar cells* that are components of the sponge's interior linings. Collar cells help keep the water moving by the collective beating of their flagella (Figure 25.7). Bacteria and particles of food become trapped in the "collars" of microvilli. The collar cells may transfer some of the food to amoebalike cells inside the matrix for further digestion, storage, and distribution. The water flowing continually through the sponge body moves out through one or more outcurrent openings called oscula (singular, osculum).

Sponges reproduce sexually, with eggs and sperm being derived from collar cells and perhaps from amoeboid cells. Sperm are released into the surrounding water (Figure 25.6). After fertilization, the eggs may be released or they may spend some time in the canals or chambers of the parent body. Young sponges pass through a microscopic, swimming larval stage. A **larva** (plural, larvae) is a sexually immature form that precedes the juvenile and adult stage of the life cycle of many animals.

Some sponges reproduce asexually by *fragmentation*. Small fragments break away from the parent body, settle on a substrate, and grow and develop into new sponges. Most freshwater sponges can produce *gemmules*. These are aggregations of sponge cells, some of which form a hard covering around the others. The living cells in a gemmule resist extreme cold or drying out. When favorable conditions return, a gemmule germinates and a new colony is established. The new sponge is a "chip off the old block," so to speak, given that sexual reproduction was not involved in its formation.

a        b

**Figure 25.6** (**a**) Two types of sponges in their marine habitats. (**b**) A basket sponge of the Caribbean, releasing sperm.

**Figure 25.7** (*Below*) Body plan of a simple sponge (**a**) and a more complex type (**b**). Purple shows locations of collar cells, which trap food particles on their "collar" of microvilli. Black arrows show the direction of water flow into the sponge, caused by the collective beating of collar cell flagella. (**c**) Section through a sponge body wall. (**d**) One type of collar cell.

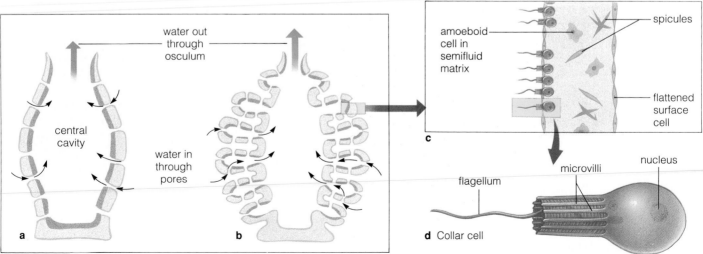

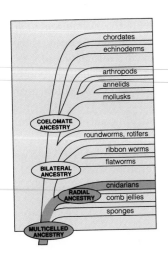

# CNIDARIANS

The phylum Cnidaria includes approximately 11,000 species of aquatic animals. Among them are scyphozoans (jellyfishes), anthozoans (including sea anemones and corals), and hydrozoans (including *Hydra* and *Obelia*). Most are marine animals. Fewer than fifty species of sponges are adapted to freshwater habitats. Figures 25.8 through 25.12 show some representatives of the group.

Two body forms are common among cnidarians. Both show radial symmetry and are wonderfully suited for capturing prey floating by from any direction. One, the **medusa** (plural, medusae), floats in the water like a tentacle-fringed bell or upside-down saucer. Centered under the bell is a mouth, sometimes with extensions called "oral arms" that assist in prey capture and feeding. The other body form, the **polyp**, is somewhat tubular. Its mouth end has a fringe of tentacles, and the opposite end usually is attached at one end to a rock or some other firm substrate:

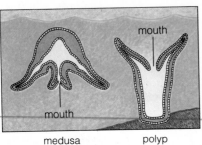

Cnidarians have a saclike gut, with one opening. They also have definite tissues, of the type called **epithelia** (singular, epithelium). One epithelium, the epidermis, covers the external body surface. Another, the gastrodermis, lines the gut. Both tissues have nerve cells threading through them. **Nerve cells** coordinate an animal's responses to stimuli. In cnidarians they form a *nerve net*, a type of nervous system that works with contractile cells in the epithelia to bring about movement, shape changes, and feeding behavior (Figure 25.8a). Some jellyfishes also have **sensory cells**, which detect specific environmental stimuli. Certain sensory cells are organized into structures that detect changes in orientation.

Between the epidermis and gastrodermis is a layer of secreted material, the *mesoglea* ("middle jelly"). Jellyfishes have an abundance of mesoglea—enough to

impart buoyancy and serve as a firm yet deformable skeleton against which the contractile cells can act. This is important for swimming. Each contraction that narrows the bell squirts out a jet of water from the underside, then the bell recovers quickly to its original shape.

Polyps generally have little mesoglea. They are more likely to use the water in their fluid-filled gut as a hydrostatic skeleton. A *hydrostatic skeleton* is a fluid-filled cavity or cell mass, the volume of which remains the same when its shape is changed by muscles surrounding it. A dramatic example of this is the response of certain sea anemones to predators (Figure 25.9e).

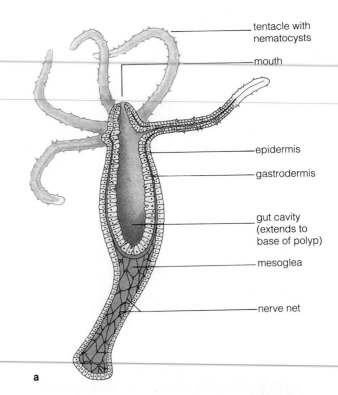

a

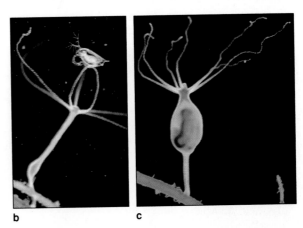

b                                    c

**Figure 25.8** (**a**) Body plan of a hydrozoan (*Hydra*), showing its tissue organization. (**b,c**) Polyp of *Hydra*, capturing and then digesting a small crustacean.

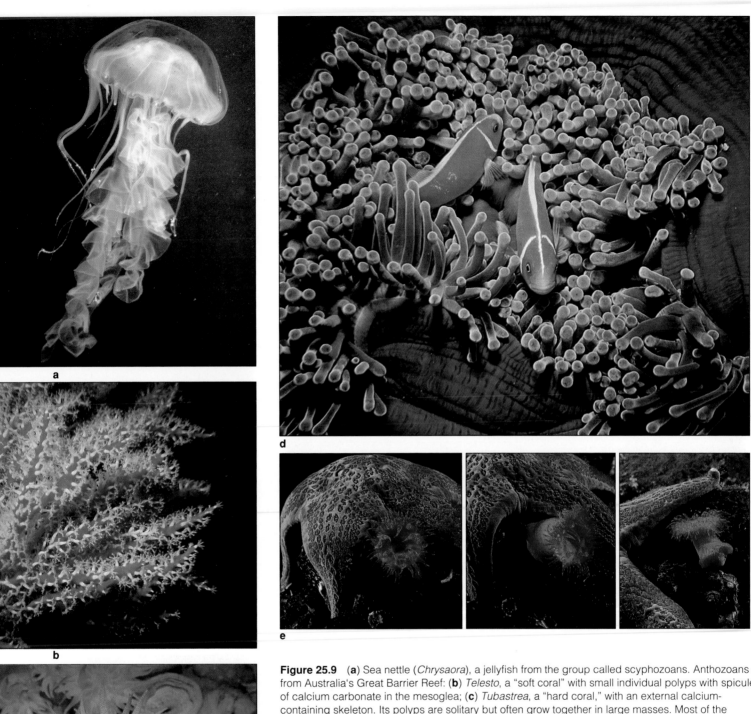

**Figure 25.9** (**a**) Sea nettle (*Chrysaora*), a jellyfish from the group called scyphozoans. Anthozoans from Australia's Great Barrier Reef: (**b**) *Telesto*, a "soft coral" with small individual polyps with spicules of calcium carbonate in the mesoglea; (**c**) *Tubastrea*, a "hard coral," with an external calcium-containing skeleton. Its polyps are solitary but often grow together in large masses. Most of the massive, reef-forming hard corals are colonial (composed of many interconnected polyps). Compare Figure 47.31.

(**d**) Tentacles fringing the mouth of a sea anemone. Sea anemones eat many fishes, but not the clownfishes shown here. Clownfishes swim away, capture food, then return to the anemone's tentacles—which protect them from predators. The anemone eats food scraps falling from the fishes' mouths. Both animals mutually benefit from their association.

(**e**) A bright-orange sea anemone escaping from a sea star. It closes its mouth, and water inside its gut serves as a hydrostatic skeleton that its muscles can act against. Muscle contractions allow the sea anemone to detach from rocks and thrash about, away from the predator's mouth. Only a few kinds of sea anemones can do this.

c   mouth of one polyp    another polyp

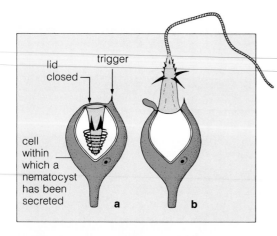

**Figure 25.10** One type of nematocyst, a capsule with an inverted tubular thread inside. This kind has a bristlelike trigger (**a**). When prey (or predators) touch the trigger, the capsule becomes more "leaky" to water. As water diffuses inward, pressure inside the capsule increases and the thread is forced to turn inside out (**b**). The thread's tip may penetrate the prey, releasing a toxin as it does.

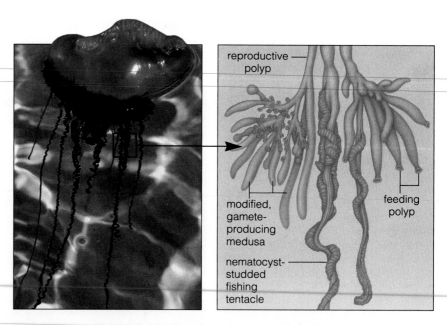

**Figure 25.11** A floating colonial hydrozoan, the Portuguese man-of-war (*Physalia*). The colony has many modified polyps and medusae.

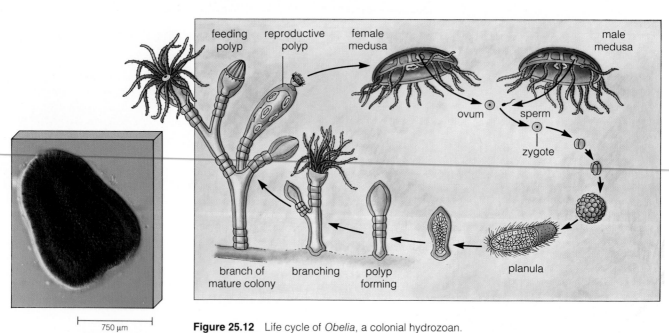

750 μm

**Figure 25.12** Life cycle of *Obelia*, a colonial hydrozoan.

A reproductive polyp in the colony produces the medusa stage, which is free-swimming. Fusion of gametes from male and female medusae produces a zygote. The zygote develops into a swimming or crawling larval stage, a planula. The planula develops into a polyp, which starts a new colony. New polyps form as the colony continues growing.

The boxed inset shows an example of a planula, this from the jellyfish *Aurelia*.

Of all the animals, only cnidarians produce *nematocysts*, capsules that discharge threads with roles in capturing prey or fending off predators (Figure 25.10). Both medusae and polyps are equipped with them, and the toxin-laden types produced by some species can give humans a painful sting. (Hence the phylum name Cnidaria, after the Greek word for "nettle.") A single cnidarian tentacle may have hundreds or thousands of nematocysts embedded in the epidermis.

Besides the body features just described, most of the corals have calcium-reinforced external skeletons. These skeletons happen to serve as the main building material for reefs. The most extravagant coral accomplishment, the Great Barrier Reef, parallels the eastern Australian coast for about 1,600 kilometers. All reef-forming corals require water that is clear and relatively warm (at least 20°C). The northernmost reefs are around Bermuda, where the waters are warmed by the Gulf Stream.

Reef-forming hard corals often are **colonial organisms**, meaning they are composed of many interconnected individuals. Another remarkable example of colonial organization is the hydrozoan *Physalia*, the Portuguese man-of-war (Figure 25.11). Several kinds of polyps and medusae are joined together in *Physalia*. They are specialized for different functions, such as feeding, defense, and reproduction.

Many cnidarians have only a polyp stage or a medusa stage in the life cycle. Others such as *Obelia* have both, with the medusa being the sexual form (Figure 25.12). Simple reproductive organs, called **gonads**, are associated with the epidermis or gastrodermis, and they just rupture and release the gametes. The zygote formed at fertilization nearly always develops into a swimming or creeping larva called a *planula* (Figure 25.12). With few exceptions, planulas have ciliated epidermal cells. Eventually a mouth opens up at one end, the larva is transformed into a polyp or medusa, and so the cycle begins anew.

## COMB JELLIES

The comb jellies belong to the phylum Ctenophora (which means "comb-bearing"). These predatory marine animals have eight rows of comblike structures. The combs consist of thick, fused cilia. When all the combs in each row beat in waves, they propel the animal forward, usually mouth first.

Comb jellies show modified radial symmetry. You can slice them lengthwise into two equal halves, and even into quarters—but two of the quarters will be mirror-images of the other two. Some comb jellies have an efficient feeding net. It consists of two long, muscular tentacles with many branches that are equipped with sticky cells (Figure 25.13). Others have sticky lips for capturing prey, which may include other comb jellies or jellyfishes.

Many comb jellies are luminescent; they glow in the dark. The *Commentary* on page 101 describes the basis of bioluminescence, which involves special enzymes and electron transfers. Together with some jellyfishes, dinoflagellates, and other organisms, comb jellies make the water glow at night when a boat or a swimmer cuts through it.

Comb jellies do not produce nematocysts, although sometimes they opportunistically save and use the ones from jellyfish they have eaten. Intriguingly, they have cells with multiple cilia—a trait characteristic of many of the more complex animals. And they have embryonic tissue comparable to the mesoderm of flatworms and more complex animals. This tissue gives rise to most of their muscles.

chordates
echinoderms
arthropods
annelids
mollusks
COELOMATE ANCESTRY
roundworms, rotifers
ribbon worms
flatworms
BILATERAL ANCESTRY
cnidarians
RADIAL ANCESTRY  comb jellies
sponges
MULTICELLED ANCESTRY

**Figure 25.13** *Pleurobrachia*, a comb jelly with long, sticky tentacles.

## FLATWORMS

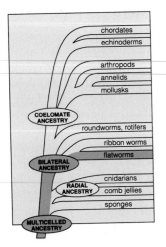

About 15,000 species of flatworms (phylum Platyhelminthes) have been identified. They include turbellarians, flukes, and tapeworms. Most have more or less flattened bodies, hence the name. Unlike the sponges or cnidarians, the flatworms show bilateral symmetry and cephalization. They also are equipped with organ systems. All have a saclike gut, with food usually entering through a pharynx (a muscular tube).

More importantly for our evolutionary story, three germ layers form in the flatworm embryo. The midlayer, mesoderm, gives rise to muscles and reproductive structures, and it was pivotal in the evolution of complex animals. *Mesoderm allowed contractile cells to evolve independently of the other layers, and it became the embryonic source of blood, bones, and many other complex tissues and organs.*

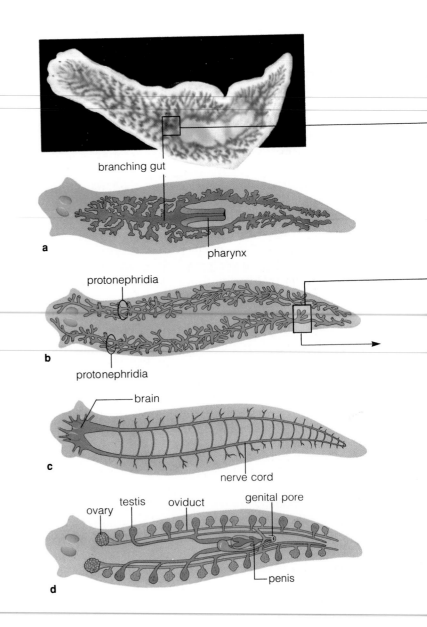

### Turbellarians

Although they are not what you usually think of when you hear the word "carnivores," many turbellarians are right up there with lions and other meat-eaters. Typically they feed on whole small animals or suck tissues from dead or wounded ones. The herbivorous types generally prefer diatoms. Most turbellarians live in the seas. A few, including the planarians, live in freshwater habitats.

Planarians have an organ system for regulating the volume and composition of body fluids. The system consists of one or more small, branched tubes, called *protonephridia* (singular, protonephridium). These extend from one or more pores at the body surface to many bulb-shaped *flame cells* in the body tissues (Figure 25.14). Flame cells get their name from a tuft of cilia that "flickers" inside the hollow interior of the bulb. When excess water moves into the bulb, the flickering drives it down the system of tubes to the outside. Some soluble substances are probably reabsorbed from the tubes. Nitrogenous wastes may be carried away with the water, but they also are eliminated across the epidermis and the gut lining.

Planarians commonly reproduce asexually, by dividing in half, with each half then regenerating the missing portion. Some planarians rely almost exclusively on such asexual modes of reproduction. They do not even have a functional system for reproducing sexually.

However, most flatworms do reproduce sexually. They are **hermaphrodites**, meaning each individual has both female and male gonads. Sexual reproduction usually involves the mutual transfer of sperm. In each worm, the reproductive system includes a sperm-delivery structure (a penis), a structure for storing incoming sperm, and glands that help produce a protective capsule around one or more fertilized eggs.

### Flukes

Adult flukes (trematodes) are parasites of vertebrates. A **parasite** is a type of heterotroph, a rather unwelcome guest that lives on or in another living organism (its host). A parasite obtains nutrients from its host's tissues and may or may not end up killing it.

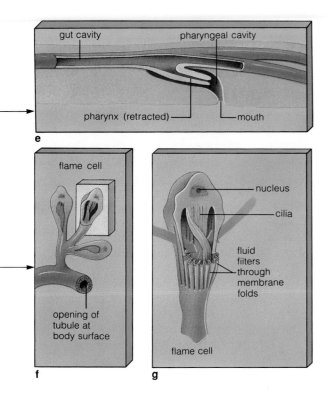

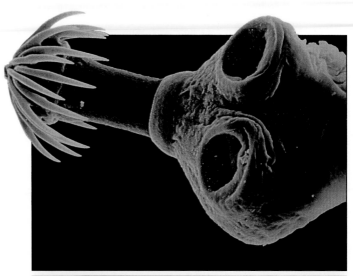

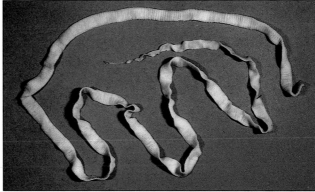

**Figure 25.14** Organ systems in a planarian, a type of flatworm. (**a**) Its branching, saclike gut has a pharynx that opens to the outside. Between feedings, the pharynx is retracted into a narrow chamber in the body; it extends out past the body surface while the worm feeds. (**b**) The flatworm system of controlling the volume and composition of body fluids. Two networks of branching tubules (protonephridia) use ciliated flame cells to drive excess water from the body. (**c**) The nervous system, with two nerve cords and a rudimentary brain. (**d**) The reproductive system, with male and female parts.

**Figure 25.15** (*Above*) Scanning electron micrograph of the scolex of a tapeworm that parasitizes shorebirds. (*Below*) A sheep tapeworm.

Some flukes live on the body surface of fishes and amphibians, sucking in cells, tissue juices, or blood. The vast majority, however, are internal parasites of the gut, liver, lungs, bladder, or blood vessels of vertebrates.

Flukes generally have complex life cycles that include both sexual and asexual phases. Like many other parasites, they require at least two kinds of organisms to complete the cycle. They reach sexual maturity in the *primary* host. Their larval stages develop or become encysted in an *intermediate* host. Humans are the primary host of many flukes, which return the favor by causing some awful diseases. The *Commentary* on page 416 describes one of the most notorious species.

**Tapeworms**

Tapeworms (cestodes) are intestinal parasites of vertebrates. The ancestors of tapeworms almost certainly had a gut, but presumably it disappeared during their evolution in an environment chockful of digested food. Present-day tapeworms simply absorb soluble nutrients from their hosts.

A tapeworm attaches to the intestinal wall by a *scolex*, a structure with suckers, hooks, or both (Figure 25.15). Just behind the scolex is a region where new tapeworm body units, the *proglottids*, form by budding. Figure *b* of the *Commentary* shows one of these units. Each proglottid is almost like an individual. It has a hermaphroditic reproductive system. Proglottids mate and transfer sperm, and fertilized eggs accumulate in older proglottids (the ones farthest from the scolex). Sooner or later, the oldest ones leave the body by way of feces and so carry the eggs to the outside. There they become available to infect an intermediate host. Eventually they may reach the host in which they mature.

In some respects, the simplest turbellarians and the larval stages of flukes and tapeworms resemble the planulas of cnidarian life cycles. The resemblance has inspired speculation that ancient bilateral animals evolved from ancestors that were much like planulas. Such evolution could have occurred through increased cephalization and the emergence of tissues derived from mesoderm. In this way, planula-like ancestors may have given rise to most groups of complex animals.

# Commentary

## A Rogues' Gallery of Parasitic Worms

Many parasitic flatworms and roundworms call the human body home, much to our enormous discomfort. In any given year, about 200 million people house blood flukes responsible for *schistosomiasis*. Figure *a* shows how the life cycle of a Southeast Asian blood fluke (*Schistosoma japonicum*) requires a human host, a watery medium for swimming larval stages, and an intermediate host (an aquatic snail). Fluke eggs hatch into ciliated, swimming larvae (1). These burrow into a snail and multiply asexually. In time they produce many fork-tailed larval forms (2,3). These larvae leave the snail and actively swim about until they encounter human skin (4). They bore inward and migrate to thin-walled veins in the intestines where they mature and where sexual reproduction (5) occurs. The eggs leave the body by way of feces, and the cycle begins anew.

Infected humans typically mount an immune response to the masses of fluke eggs being produced in their body. White blood cells and other immune fighters infiltrate the infected areas, and grainy masses form in tissues. In time, the liver, spleen, bladder, and kidneys deteriorate.

Tapeworms also do damage to humans. One kind uses pigs as intermediate hosts; another uses cattle. Humans become infected when they eat insufficiently cooked pork or beef (Figure *b*). Or consider how the larvae of one tapeworm are eaten by copepods (tiny crustaceans). The larvae avoid digestion, then they develop further in fishes that eat the copepods. Humans who eat infected fishes that are raw, improperly pickled, or insufficiently cooked can become hosts for the adult tapeworms.

Then there are the roundworms called pinworms and hookworms. The pinworm *Enterobius vermicularis* parasitizes humans (especially children) in temperate regions. It lives in the large intestine. At night, the centimeter-long female pinworms migrate to the anal region to lay eggs.

Their presence at the body surface causes itching, and scratchings made in response will transfer the eggs to other objects. Newly laid eggs contain embryos, but within a few hours they are juveniles and ready to hatch if the eggs are inadvertently ingested by another human.

Hookworms can be a serious problem for humans in the tropics or subtropics. Adult hookworms live in the small intestine, where they feed on blood and other tissues. Teeth or sharp ridges bordering their mouth cut into the intestinal wall. Adult females, about a centimeter long, can release a thousand eggs daily. These leave the body by way of feces, then hatch into juveniles. A juvenile may penetrate the skin of a barefoot person. Inside a host, the parasite travels the bloodstream to the lungs, where it works its way into the air spaces. After moving up the windpipe, it is swallowed. Soon it is in the small intestine, where it may mature and live for several years.

Another roundworm, *Trichinella spiralis*, causes painful and sometimes fatal symptoms. The adults live in

**a** Life cycle of a dangerous blood fluke, *Schistosoma japonicum*. The micrograph shows an adult male fluke.

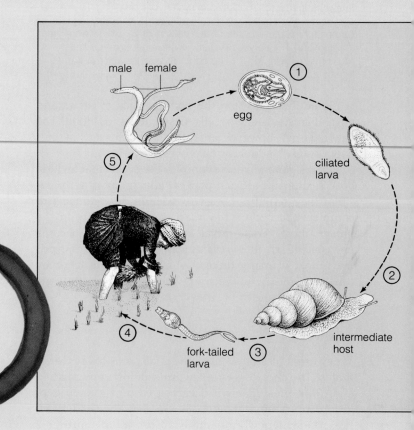

the lining of the small intestine. Female worms release juveniles (Figure c), and these work their way into blood vessels and travel to muscles. There they become *encysted* (they produce a covering around themselves and enter a resting stage). Humans usually become infected by eating insufficiently cooked meat from pigs or certain game animals. The presence of encysted juveniles cannot easily be detected when fresh meat is examined, even in a slaughterhouse.

Finally, Figure *d* shows the results of prolonged and repeated infections by the roundworm *Wuchereria bancrofti*. Adult worms live in the lymph nodes, where they can obstruct the flow of lymph that normally is returned to the bloodstream. The obstruction causes fluid to accumulate in the legs and other body regions, which undergo grotesque enlargement, a condition called *elephantiasis*. A mosquito is the intermediate host. The female roundworms produce active young that travel the bloodstream at night. If a mosquito sucks blood from a human, the juveniles may

enter the insect's tissues. After some growth, they move near the insect's sucking device, where they are ready to enter another human host when the mosquito draws blood again.

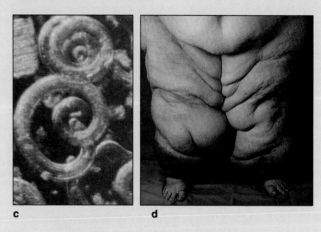

**c**  **d**

(**c**) Juveniles of a roundworm, *Trichinella spiralis*, inside muscle tissue. (**d**) Elephantiasis in a woman, caused by the roundworm *Wuchereria bancrofti*.

**b** Life cycle of a beef tapeworm, *Taena saginata*.

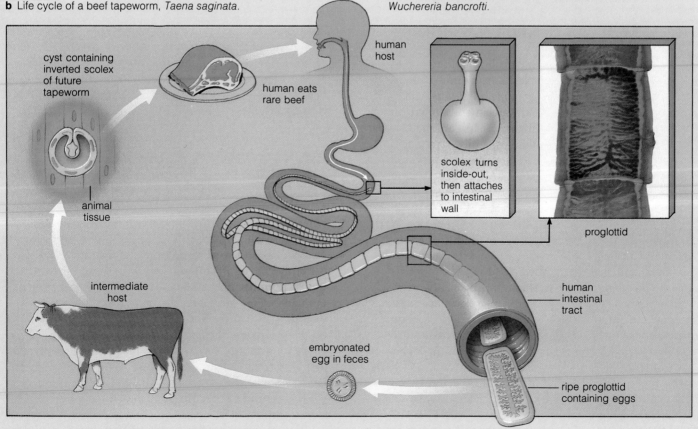

cyst containing inverted scolex of future tapeworm

animal tissue

human eats rare beef

human host

scolex turns inside-out, then attaches to intestinal wall

proglottid

intermediate host

human intestinal tract

embryonated egg in feces

ripe proglottid containing eggs

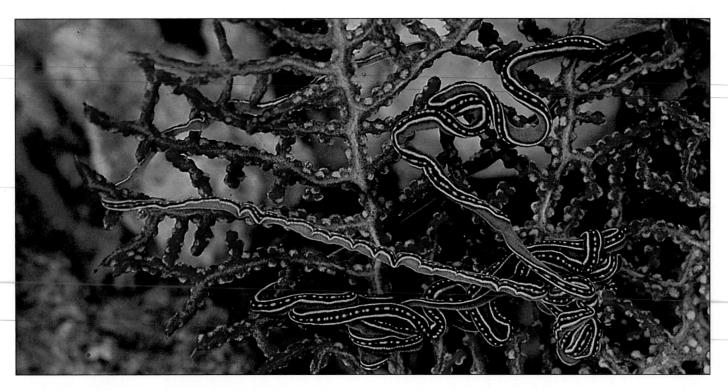

**Figure 25.16** A ribbon worm, one of the lesser known invertebrates.

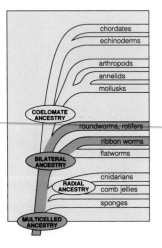

## RIBBON WORMS

Practically all of the ribbon worms (phylum Nemertea), including the splendid specimen shown in Figure 25.16, are marine animals. Only a few species have been discovered in freshwater habitats or in moist habitats on land.

It seems likely that ribbon worms are closely related to the turbellarians just described. They, too, are ciliated externally and their tissue organization is similar to that of flatworms in general. Unlike flatworms, however, they have a complete gut and a circulatory system, and the sexes are nearly always separate. Ribbon worms also have a *proboscis*, a tubular device that they use to capture prey. All of these features are significant departures from the flatworm body plan.

The proboscis lies within a fluid-filled cavity. When circular muscles around the cavity contract, hydrostatic pressure inside is raised, such that the proboscis is forced to turn inside-out and emerge from the mouth or from a separate opening at the anterior end. Glands associated with the proboscis secrete a paralytic venom. Often the proboscis is equipped with a stylet, a penetrating device that is plunged into prey—another kind of worm, perhaps, or a small mollusk or crustacean. Ribbon worms swallow their prey or suck out some of their tissues or juices.

## ROUNDWORMS

Roundworms (nematodes) are truly abundant, yet most people have never seen one. They live just about everywhere, from snowfields to deserts and hot springs. A cupful of rich soil has thousands of them, and a dead earthworm or a fruit rotting on the ground will almost certainly have an interesting variety of scavenging types. Roundworms also parasitize plants and animals. Humans alone are infected by about thirty species, including pinworms and hookworms. (The *Commentary* on page 416 describes a few of them.) One of the larger types causes thin, serpentlike ridges when it is present just under the skin. For thousands of years, healers have removed the "serpents" by winding them out very slowly, around a stick. The symbol of the medical profession continues to be a serpent wound around a staff.

Roundworms are bilateral, with a cylindrical body that is usually tapered at both ends. They have a com-

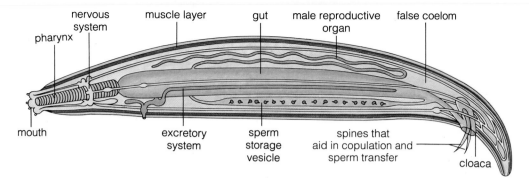

**Figure 25.17** Body plan of a roundworm, which is specialized for feeding and reproducing, and little else. (In males, gametes as well as digestive residues leave the body by way of the cloaca, the last part of the gut. In females, there is a separate genital pore, used in both copulation and egg laying.)

Labels in figure:
nervous system, muscle layer, gut, male reproductive organ, false coelom, pharynx, mouth, excretory system, sperm storage vesicle, spines that aid in copulation and sperm transfer, cloaca

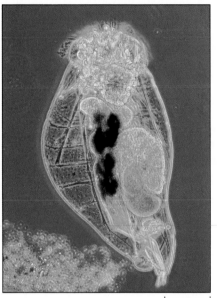

50 μm

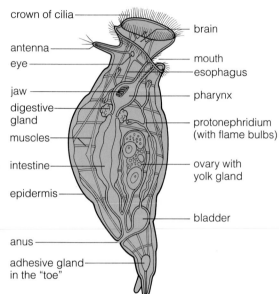

Labels in figure:
crown of cilia, brain, antenna, eye, mouth, esophagus, jaw, pharynx, digestive gland, protonephridium (with flame bulbs), muscles, intestine, ovary with yolk gland, epidermis, bladder, anus, adhesive gland in the "toe"

**Figure 25.18** Lateral view of a rotifer, busily laying eggs. Males are unknown in many species; the females produce diploid eggs that develop into diploid females. Females of other species do the same, but they also can produce haploid eggs that develop into haploid males. If a haploid egg happens to be fertilized by a male, it develops into a female. The male rotifers appear only occasionally, and they are dwarfed and short-lived.

plete digestive system (Figure 25.17). Between the gut and body wall is a pseudocoel that often is nearly filled with reproductive organs. Fluid in this body cavity circulates nutrients through the body. Roundworms also have a **cuticle**, which in animals is a tough, flexible body covering. Even though different roundworms live in diverse habitats, their body plan remains much the same. It is as if the plan worked well under many different conditions, so it just stayed with them.

## ROTIFERS

Of the rotifers, we can say this: Seldom has so much been packed in so little space. Most of the 1,800 or so species of rotifers are less than a millimeter long. Some are members of **plankton**, the communities of small aquatic organisms that swim weakly or drift with currents. Others crawl over aquatic plants, wet moss, and other substrates. They feed mainly on bacteria and single-celled algae, and other animals in turn feed on them. They are often abundant and probably important components of food webs.

Early microscopists called these animals rotifers (meaning "wheel-bearers"). Most rotifers have a crown of cilia, and the ciliary motion reminded the microscopists of a turning wheel (Figure 25.18). The cilia function in swimming and in creating currents that direct food to the mouth. Behind the crown, rotifers have salivary glands, a pharynx with jaws, an esophagus, digestive glands, a stomach from which most nutrients are absorbed, and usually an intestine and anus. Protonephridia similar to those of flatworms remove excess water from the rotifer's body.

Nerve cells are clustered at the head end, and there often are masses of light-absorbing pigments called eye—which, when stimulated, help orient the rotifer in directions favorable for its activities.

Many rotifers have a pair of flexible "toes." Sticky substances exuding from the toe enable rotifers to attach to a substrate while feeding.

Rotifers are probably one of the side roads of animal evolution, although they do show the structural complexity that is possible in animals with a false coelom. However, it is only when we turn to the coelomate animals that we see complexity on a spectacular scale.

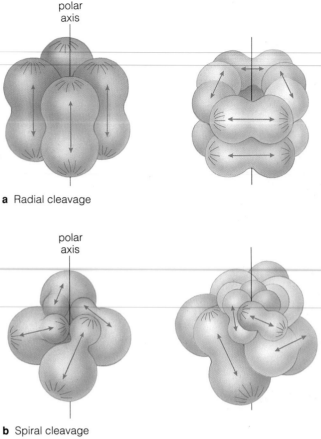

**a** Radial cleavage

**b** Spiral cleavage

**Figure 25.19** Two major patterns of cleavage, or early cell divisions, in animal embryos. (**a**) Radial cleavage, characteristic of deuterostomes. (**b**) Spiral cleavage, characteristic of protostomes.

# TWO MAIN EVOLUTIONARY ROADS

Bilateral animals not much more complex than flatworms emerged during Cambrian times. A major divergence occurred not long afterward, and it led to two distinctly different evolutionary lines of animals, the **protostomes** and **deuterostomes**. As shown in Figure 25.5, mollusks, annelids, and arthropods are protostomes. Echinoderms and chordates are deuterostomes.

Animals of both lineages usually have a coelom and a complete digestive system. But their embryos develop in different ways. For example, the cell divisions in the early embryo do not occur in the same way (Figure 25.19). In deuterostomes, they divide the fertilized egg as you might divide an apple, by cutting it into four wedges from top to bottom. Each wedge is then cut in half at its midsection, then each piece is cut at *its* midsection. Such cell divisions, running parallel and perpendicular to the polar axis of the early embryo, are called **radial cleavage**. In protostomes, the "wedges" are cut at oblique angles to the original polar axis, a pattern called **spiral cleavage**.

Another difference is in the way the mouth and anus form with respect to indentations that develop in the early ball-shaped embryo. In deuterostomes, the opening of the first indentation becomes the anus, and a second one becomes the mouth. In protostomes, the opening of the first indentation becomes the mouth; the anus forms elsewhere. As a final example, in deuterostomes the coelom begins as outpouchings of the gut wall. In protostomes it arises within solid masses of tissue at the sides of the gut. This pattern is shown in Figure 25.20.

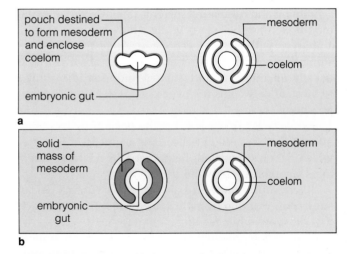

**Figure 25.20** Differences in the way the coelom forms in the embryos of deuterostomes (**a**) and protostomes (**b**). The "circles" are diagrams of what you would see if you sliced embryos through their midsection at two successive stages of development.

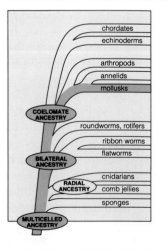

# MOLLUSKS

Today there are probably more than 100,000 species in the phylum Mollusca. The name means "soft-bodied animals." Snails, clams, and octopuses are familiar examples. Mollusks have a head, a foot, and a body region called the visceral mass. They have a *mantle*, a tissue fold that hangs down like a skirt around some or all of the body (Figure 25.21). Mollusks in general have a shell consisting largely of calcium carbonate. And the mollusks that have a well-developed head have distinctive eyes and tentacles.

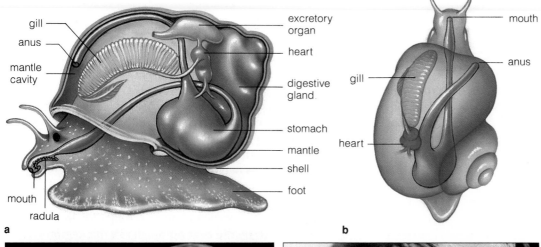

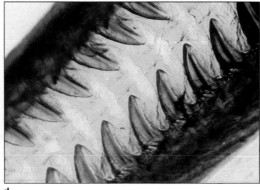

a

b

c

d

**Figure 25.21** (**a**,**b**) Body plan of an aquatic snail that has a single gill, of the type called a ctenidium. (**c**) A land snail, a familiar mollusk. Generally, the radula (**d**) is protracted and retracted in a somewhat rhythmic way. Food is rasped off and drawn to the gut on the retraction stroke.

Most of the visceral mass is a gut, along with a heart, kidneys, and reproductive organs. In most mollusks other than clams and their relatives, food entering the gut is preshredded, having been processed by a *radula*, a tonguelike organ with hard teeth.

Mollusks commonly have a pair of *ctenidia* (singular, ctenidium), which are specialized respiratory structures that fall in the general category of **gills**. The ctenidia are outgrowths of the mantle wall, and they are positioned within the mantle cavity (Figure 25.21a).

## Chitons

The marine mollusks called chitons are rather elongated animals. Most of them use their radula to graze on algae, hydrozoans, and other low-growing organisms. A large, broad foot enables them to creep over and cling tenaciously to rocks and other hard surfaces.

When it is dislodged, a chiton can roll up into a ball and protect itself until it can safely unroll and become reattached elsewhere. This maneuver is possible because its dorsal shell is divided into a series of eight plates (Figure 25.22). When a chiton is disturbed or exposed by a receding tide, the muscles in its foot can pull the animal down tightly. The edge of the mantle, which more or less covers the shell plates, now functions like the rim of a suction cup. The animal is then difficult to detach.

**Figure 25.22** Chitons—one variation on the basic molluscan plan. Walk along a rocky shore and you will probably see chitons. The ones shown here live in the intertidal zone of Monterey Bay, California. Members of this class of mollusks show beautiful variations in the color and patterns of their elongated shells.

## Gastropods

Snails and slugs make up the largest group of mollusks, the gastropods ("belly foots"). They are so named because the foot spreads out under the animal while it is crawling. Many snails have a spirally coiled or cone-shaped shell (Figures 18.2 and 25.21). Coiling is a way of compacting the organs into a mass that can be balanced above the rest of the body, much as you would balance a backpack full of books.

As most gastropods grow and develop, some body parts undergo a strange internal realignment. In a process called *torsion*, certain muscles contract and different body parts grow at different rates. The outcome? The posterior mantle cavity twists around to the right and then forward, until it is above the head (Figure 25.23). The twisting produces a cavity into which the head can withdraw in times of danger. But it also brings the gills, anus, and kidney openings above the head. This could create a sanitation problem, given that wastes are dumped near the respiratory structures and the mouth. In most species, however, the beating of cilia in this region creates currents that sweep wastes away.

During the course of evolution, nudibranchs (sea slugs) and some other gastropods apparently underwent detorsion. The larval body still twists to some extent, but then muscle contractions and asymmetrical growth causes the body to untwist. Besides this, the mantle cavity largely disappeared. Nudibranchs also lost the ctenidia, but most have other outgrowths that function as secondary gills. Many nudibranchs are striking in their coloration and their array of outgrowths (Figures 25.24a and b).

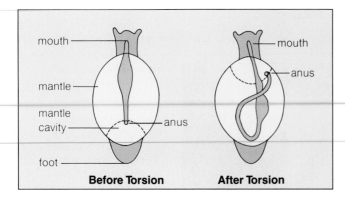

**Figure 25.23** Internal realignment of body parts that occurs during the development of most gastropods.

**Figure 25.24** (**a,b**) Two of the soft-bodied gastropods of the type called nudibranchs. The pair of "Mexican dancers" in (**b**) are engaged in mating behavior. (**c**) A sea hare (*Aplysia*), widely used in experimental studies of behavior and physiology. The two flaps over its dorsal surface are extensions of the foot. They undulate and help ventilate the underlying mantle cavity. Like many other gastropods, *Aplysia* is hermaphroditic. It may act as a male, a female, or both during encounters with others like itself. Neurobiologists are much taken with *Aplysia* and have tracked several of its nerve pathways.

## Bivalves

Clams, scallops, oysters, and mussels are well-known bivalves (animals with a "two-valved shell"). Humans have been eating one type or another since prehistoric times. The shell of many bivalves, including a few pearl-producing types, has an inner lining of iridescent mother-of-pearl.

Some bivalves are only 1 or 2 millimeters across. A few giant clams of the South Pacific are more than a meter across and weigh 225 kilograms (close to 500 pounds). The bivalve head is not much to speak of, but the foot is usually large and specialized for burrowing. In nearly all bivalves, gills function in collecting food and in respiration. As cilia in the gills move water through the mantle cavity, mucus on the gills traps tiny bits of food (Figure 25.25). Tracts of cilia move the mucus and food to palps, where final sorting takes place before acceptable bits are driven to the mouth.

Bivalves hunkered in mud or sand have a pair of *siphons*, which are extensions of the mantle edges, fused into tubes. Water is drawn into the mantle cavity through one siphon and leaves through the other, carrying wastes from the anus and kidneys. Siphons of the giant geoduck (pronounced "gooey duck") of the Pacific Northwest may be more than a meter long.

Scallops and a few other bivalves can swim by clapping their valves together, thus producing a localized jet of water that propels the animal for some distance. This fast, jerky locomotion often helps scallops to escape from a sea star, octopus, or some other predator (Figure 25.26).

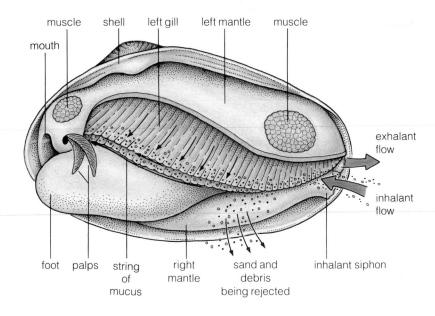

**Figure 25.25** Anatomy of a clam. The left shell has been lifted off for this diagram. Food trapped in mucus on the gills is sorted by the palps, and suitable particles are swept by cilia to the mouth.

**Figure 25.26** A scallop swimming away from a predator (a sea star). Scallops and a few other bivalves can clap their two valves together rapidly enough to force a strong jet of water from the mantle cavity.

**Figure 25.27** An octopus, one of the cephalopods.

## Cephalopods

The cephalopods include squids, octopuses, nautiluses, and cuttlefish, all fast-swimming predators of the oceans. Figures 25.27 and 25.28 show representative species. Some cephalopods are only a few centimeters long. The giant squid can grow to 18 meters (about 60 feet); it is the largest invertebrate known.

Instead of a foot, cephalopods have tentacles, usually equipped with suction pads as illustrated in Figure 25.27. They use the tentacles to capture prey, a radula to draw it into the mouth, and a beaklike pair of jaws to bite or crush it. Venomous secretions often speed the captive's death.

Cephalopods move rapidly by a type of *jet propulsion* in which a stream of water is forced out of the mantle cavity through a funnel-shaped siphon. When muscles in the mantle relax, water is drawn into the mantle cavity. When they contract, a jet of water is squeezed out. If, at the same time, the free edge of the mantle is closed down on the animal's head and siphon, water shoots out in a jet through the siphon. By manipulating the siphon, the animal partly controls the direction of its own movement.

**Figure 25.28** (**a**) A female and a male cuttlefish (*Sepia*) mating, head end to head end. (**b**) General anatomy of a squid or a cuttlefish. The tentacles are similar to the arms, but they are more slender and specialized for capturing prey. The shell of a cuttlefish is calcified and layered, as shown; squids have only a vestigial shell and it is not calcified.

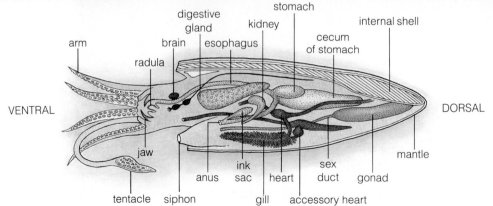

Being highly active, cephalopods have great demands for oxygen, and they alone among the mollusks have a closed circulatory system (page 660). Blood is pumped from the main heart to the two gills. Each gill has an accessory, "booster" heart at its base that speeds the flow and therefore the uptake of oxygen and elimination of carbon dioxide.

Even when a cephalopod is resting, it rhythmically draws water into its mantle cavity and then expels it. The movements enhance gas exchange and also help to eliminate wastes that the excretory organs and anus discharge into the mantle cavity.

The nervous system of cephalopods is well developed, and the brain is larger, relative to body size, than in other mollusks. Giant nerve fibers connect the brain with muscles used in jet propulsion, making it possible for a cephalopod to respond quickly to food or danger. Cephalopod eyes bear some resemblance to yours, although they are formed in a different way. Finally, cephalopods can learn. Give an octopus a mild electric shock after showing it an object with a distinctive shape, for example, and it will thereafter avoid the object. In terms of memory and learning, the cephalopods are the world's most complex invertebrates.

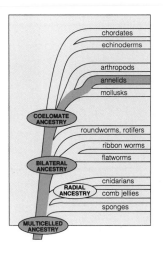

# ANNELIDS

About 15,000 species of annelids, or segmented worms, have been catalogued. The best-known members of this phylum (Annelida) are the earthworms and leeches. Less familiar are the polychaetes.

## Earthworms

Earthworms are the usual textbook examples of annelids (Figure 25.29). They belong to the group of worms called oligochaetes, which has both land-dwelling and aquatic members. Earthworms are scavengers on land. They burrow in moist soil or mud, where they feed on decomposing plant material and other organic matter. Every twenty-four hours, an earthworm can ingest its own weight in soil. Earthworms aerate soil, to the benefit of many plants. They also make nutrients available to other organisms by carrying subsoil to the surface.

## Leeches

Leeches live in freshwater, the seas, and, in tropical regions, moist habitats on land. They use the suckers at both ends of their muscular body to inch about. Most leeches swallow small animals or kill them and suck out their juices. The ones most people have heard about feed on vertebrate blood (Figure 25.30).

Blood-sucking leeches have sharp jaws. They plunge a sucking apparatus into incisions they make in prey. They secrete a substance that prevents blood from coagulating while they eat. The leech gut has many side branches for storing food taken in during a big meal, this being handy for an animal that may have long waits between meals.

**Figure 25.29** A well-known annelid, an earthworm.

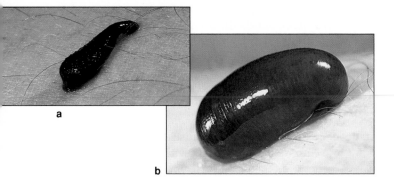

**Figure 25.30** A leech before feeding (**a**) and afterward, when it is engorged with blood from a human host (**b**). Medical practitioners have used *Hirudo medicinalis*, a freshwater leech, as a bloodletting tool for at least 2,000 years. Especially during the nineteenth century, doctors used leeches to "cure" disorders ranging from nosebleeds to obesity. By one estimate, leeches were drawing more than 300,000 liters of blood annually from hapless patients in France alone. Nowadays, leeches are used more selectively, as in relieving congested skin grafts. They also are used for removing blood from a severed ear, lip, or fingertip that has been surgically reattached to the body. The leeches take up blood that otherwise cannot move away until blood circulation routes are reestablished.

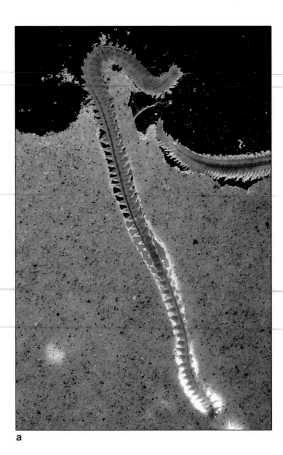

a

b

**Figure 25.31** Polychaetes. (**a**) A marine polychaete inside its burrow. Notice the setae (chitin-reinforced bristles) along the length of the body.

(**b**) A tube-dwelling polychaete, with featherlike structures coated with mucus. The mucus traps small food particles suspended in the water.

### Polychaetes

For the most part, the annelid worms called polychaetes live in marine habitats. Different species burrow in sand or soft mud, swim freely, or become attached to substrates. Some types excavate vertical or U-shaped burrows. Others construct tubes from calcium carbonate secretions or from sand grains or bits of shells.

Many polychaetes feed on small invertebrates, some feed on algae, and others scavenge organic matter from sediments. The predatory types have a muscular pharynx equipped with jaws that can be everted ("popped out") to capture prey. Herbivorous types use their jaws to grasp and tear algae. Most of the tube dwellers have ciliated, mucus-coated tentacles or featherlike structures near the mouth (Figure 25.31). Bacteria and other bits of food become trapped in the mucus, then the coordinated beating of cilia sweeps the mixture to the mouth.

### Annelid Adaptations

As a group, the annelids are characterized partly by their notable segmentation and also by the chitin-reinforced bristles, called **setae**, on nearly all of the segments. You can see some of these bristles on the polychaete shown in Figure 25.31a. Leeches are among the few annelids that have none. Setae, which occur in pairs or clusters at the

body surface, can be protracted or retracted by special muscles. They provide the traction required for crawling and burrowing through soil, and in swimming species they often are broadened into paddles. Although earthworms have only a few setae per body segment, the marine polychaete worms typically have many of them, often concentrated on fleshy lobes. Those lobes, called **parapodia**, have roles in respiration and locomotion.

Annelid body segments typically have a cuticle, a layer of secreted material at the outer surface (Figure 25.32). The cuticle is thin and flexible enough to permit bending as well as gas exchange. Inside, partitions usually separate the segments from one another, so there is a series of coelomic chambers. The gut extends continuously through all the chambers, from mouth to anus. A brain at the head end integrates sensory input for the whole worm. Leading away from the brain is a double nerve cord that extends the length of the body. A **nerve cord** is a bundle of slender extensions of nerve cell bodies. In each segment, the nerve cord broadens into a **ganglion** (plural, ganglia), a cluster of nerve cell bodies that controls local activity. You will read more about these cells in Chapters 32 and 33.

The coelomic chambers of annelids serve as a hydrostatic skeleton against which muscles can operate. Circular muscles are located mostly within the body wall of each segment, and longitudinal ones span several seg-

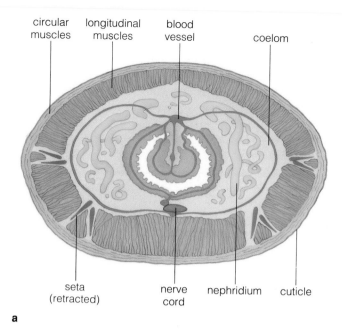

circular muscles · longitudinal muscles · blood vessel · coelom

seta (retracted) · nerve cord · nephridium · cuticle

a

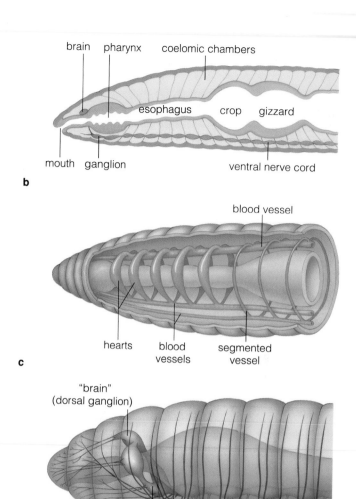

brain · pharynx · coelomic chambers

esophagus · crop · gizzard

mouth · ganglion · ventral nerve cord

b

blood vessel

hearts · blood vessels · segmented vessel

c

"brain" (dorsal ganglion)

ganglion · nerve

d

**Figure 25.32** Earthworm anatomy. (**a**) Cross-section through an earthworm. Anterior end of (**b**) the digestive system, (**c**) nervous system, and (**d**) circulatory system. (**e**) A nephridium, the functional unit of the earthworm's system of maintaining the volume and composition of body fluids.

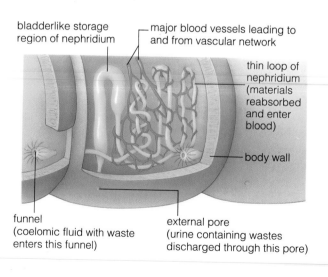

bladderlike storage region of nephridium · major blood vessels leading to and from vascular network

thin loop of nephridium (materials reabsorbed and enter blood)

body wall

funnel (coelomic fluid with waste enters this funnel) · external pore (urine containing wastes discharged through this pore)

e

ments. When the longitudinal muscles contract and the circular ones relax, segments shorten and fatten. They lengthen when the pattern reverses. An earthworm moves forward when its first few segments elongate and the segments just behind them hold their position (by protruding their setae). Then the first segments contract and protrude their setae—and the ones behind retract their setae and are pulled forward. Alternating contractions and elongations proceed along the length of the body and move the worm forward.

Annelids generally have a well-developed circulatory system. Contractions of muscularized blood vessels keep blood circulating in one direction. Smaller blood vessels lead to and from the gut, nerve cord, and body wall. Functionally linked with the circulatory system is a system of **nephridia** (singular, nephridium), which regulate the volume and composition of body fluids. Often the units of this system have cells similar to flame cells. Thus they resemble protonephridia, implying an evolutionary link between flatworms and annelids. More commonly, the beginning of each nephridium is a funnel-like structure that collects fluid from a coelomic chamber. The funnel leads into a tubular portion of the nephridium that carries the fluid to a pore at the body surface (Figure 25.32e). The funnel is located in one coelomic chamber but its terminal pore is located in the body wall of the next chamber in line.

All of these features suggest the annelids were an early offshoot of the protostome branch, and they remind us of the developments that led to increased size and more complex internal organs.

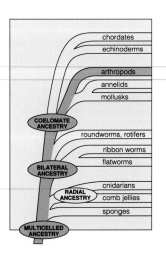

# ARTHROPODS

Of all the invertebrate animals, the arthropods show the most diversity. More than a million species have already been catalogued and bewildering numbers of new species are being discovered each week, especially in many previously unexplored regions of the tropics.

Ancestral annelids or animals like them apparently gave rise to four different arthropod groups. Figure 25.33 shows an existing animal that bears striking resemblances to both annelids and arthropods. The groups of arthropods are so distinct that each is now considered to be a separate subphylum:

| Subphylum | Representatives |
|---|---|
| Trilobites | *Now extinct (Figure 19.19d)* |
| Chelicerates | *Horseshoe crabs, spiders, scorpions, ticks, mites* |
| Crustaceans | *Copepods, crabs, lobsters, shrimps, barnacles* |
| Uniramians | *Centipedes, millipedes, insects* |

## Arthropod Adaptations

Arthropods are spectacularly abundant. They are distributed through more habitats than any other animals. They eat many kinds of food (and lots of it), and they are very good at defending themselves against predators. Six adaptations in particular have contributed to their success as a group:

1. Hardened exoskeleton

2. Specialization of segments

3. Jointed appendages

4. Specialized respiratory systems

5. Efficient nervous system and sensory organs

6. Among insects especially, a division of labor in the life cycle

**The Arthropod Exoskeleton.** Think back on that thin cuticle of annelid worms. The thinness has a drawback, for predators can pierce right through it. When arthropods were evolving, the pressures of predation must have favored thicker, hardened cuticles. The cuticle of existing arthropods is actually an external skeleton, or **exoskeleton**. It contains protein and chitin, a combination that is flexible, lightweight, and protective. Some

**Figure 25.33** A velvety "walking worm" of the small phylum Onychophora, the members of which resemble both the annelids and the arthropods. These worms live mostly in humid forests, especially of the tropics. They hide by day under rotting logs and in leaf litter and hunt for prey at night. The largest is about 15 centimeters (nearly 6 inches) long. Depending on the species, there are between fourteen and forty-three pairs of stumpy, unjointed legs with claws at the tip (onychophoran means "claw bearer").

Like annelids, these worms have ciliated nephridia. Like arthropods, they molt and they have an open circulatory system, with blood entering spaces in body tissues and reentering the heart through pores. Moreover, they have tracheal tubes similar to those of millipedes, centipedes, and insects.

Such similarities do not mean that onychophorans are "missing links" between annelids and arthropods. They are probably one of many invertebrates that evolved from annelid ancestors and have survived to the present.

cuticles even have calcium deposits. They are less flexible, but they are like armor plates.

Although exoskeletons probably evolved as a defense against predators, they took on other functions when aquatic arthropods began invading the land. *An arthropod exoskeleton is a superb barrier to evaporative water loss, and it can support a body deprived of water's buoyancy.*

Exoskeletons do restrict increases in size. But arthropods grow in spurts by **molting**. At different stages in the life cycle, they shed their exoskeleton and grow a new one (Figure 36.3). Aquatic arthropods swell with water and so enlarge before their new exoskeleton hardens. Land-dwelling arthropods swell up with air. Just before molting, some arthropods go into hiding to protect themselves until their new armor hardens.

**Specialization of Segments.** As arthropods evolved, body segments became more specialized, reduced in number, and grouped or fused together in a variety of ways. In some lineages, for example, several segments became organized to form the head, others a thorax and an abdomen. In other lineages, a few segments formed the forebody, and the rest, a hindbody.

**Jointed Appendages.** Arthropods have jointed appendages (the word arthropod means "jointed foot"). These are specialized for diverse tasks. For example, the arthropod head has pairs of appendages used in feeding and sensing information about the surroundings. Other appendages perform such tasks as walking, swimming, flying, sperm transfer, and spinning silk. The origins of jointed appendages are obscure. But it seems likely that they were derived from fleshy, segmental outgrowths similar to the parapodia of some annelids.

**Respiratory Systems.** Arthropods have a variety of respiratory structures, but the ones called *tracheas* contributed most to their diversity, especially among the insects. Tracheas begin at pores on the body surface and branch to supply oxygen directly to body tissues (see, for example, Figure 40.5). Tracheas support high metabolic rates in small-bodied insects. And they were the foundation for such energy-consuming activities as insect flight.

**Specialized Sensory Structures.** The arthropod eye and other sensory organs contributed immensely to the success of these animals. Many arthropods have a wide angle of vision, and they can process visual information from many directions. That is partly why dragonflies are so good at capturing other insects in midair; their eyes contain more than *30,000* photoreceptor units. We will consider these and other sensory structures in Chapter 35.

**Division of Labor in the Life Cycle.** By far, insects are the most successful animals on earth. For many of them, a division of labor in the life cycle contributes to their success. To be sure, there are types of insects that become sexually mature adults simply by growing in size and molting. But many other types first develop as sexually immature, free-living larvae that molt and *change* as they grow. Then the larvae are transformed into adults by way of a "resting" pupal stage, which requires massive tissue reorganization and remodeling. We saw an example of this in Figure 1.4, which showed the emergence of an adult moth. Growth and transformation of a larva into the adult form is called **metamorphosis**.

Moths, butterflies, beetles, and flies are examples of metamorphosing insects. Their larval stages specialize in *feeding and growth*, whereas the adult is concerned mostly with *dispersal and reproduction*. With this division of labor, these insects are highly adapted to seasonal changes in food sources and environmental conditions.

With this overview of arthropod adaptations in mind, we turn now to the specializations among the living members of this phylum: the chelicerates, crustaceans, and insects and their kin.

## Chelicerates

Chelicerates originated in the shallow seas of the early Paleozoic. Except for some mites, the only truly marine survivors are the so-called horseshoe crabs (Figure 25.34) and sea spiders. Of the familiar existing chelicerates—the spiders, scorpions, ticks, and mites—we might say this: Never have so many been loved by so few.

a

b

c

**Figure 25.34** (**a**) Body imprint of a horseshoe crab made about 250 million years ago—imprints that could well be made by a modern-day horseshoe crab (*Limulus polyphemus*), as suggested by **b** and **c**.

a

b

c

**Figure 25.35** Representative spiders. (**a**) The brown recluse, the bite of which can be severe to fatal for humans. This North American spider lives under bark and rocks, and in and around buildings. It can be identified by the violin-shaped mark on its forebody. (**b**) A female black widow, the bite of which can be painful and sometimes dangerous. (**c**) Wolf spider. Like most spiders, it is harmless to humans and plays a major role in keeping insect populations in check.

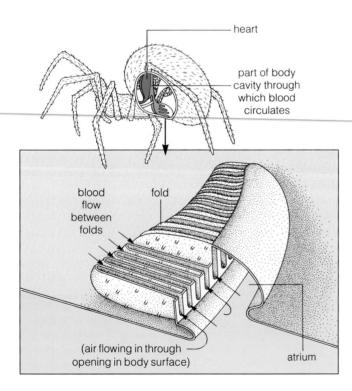

**Figure 25.36** A book lung, a type of respiratory organ that occurs among spiders and scorpions.

Like scorpions, spiders are predators, and many help keep populations of insect pests in check. Even so, the venomous types that occasionally bite humans have managed to give the whole group a bad reputation. Ticks are blood-sucking parasites that often serve as vectors for pathogenic microorganisms. Thus they have roles in spreading infectious diseases, including Rocky Mountain spotted fever and Lyme disease (page 359). Most mites are free-living scavengers, but some parasitic forms live in such interesting places as the vertebrate ear.

A spider or scorpion has a forebody and hindbody (Figure 25.35). Six pairs of appendages project from the forebody. Four pairs are walking legs. The other two pairs, chelicerae and pedipalps, function in subduing prey and handling food. The chelicerae inflict wounds and discharge venom. The spider or scorpion pumps digestive enzymes from its gut into the wounds, and then it sucks up the liquefied remains of its prey. Male spiders also use the pedipalps to transfer sperm to the female, and scorpions use their pincerlike pedipalps to grasp their partner in the mating ritual. The forebody has several eyes, used to detect moving prey.

Appendages on the hindbody, if any, are extremely modified. Spiders use small posterior appendages to spin out threads of silk for webs and egg cases. The silk

a

b

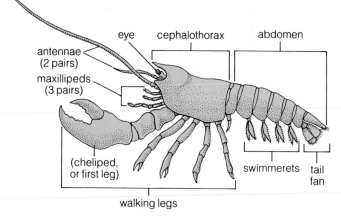

eye    cephalothorax    abdomen

antennae
(2 pairs)

maxillipeds
(3 pairs)

(cheliped,
or first leg)

swimmerets    tail
fan

walking legs

**Figure 25.37**  Paired appendages of a crab (**a**) and lobster (**b**). The body is organized into a cephalothorax (head plus thorax) and an abdomen. The first appendage on the lobster abdomen shown in the diagram transfers sperm to the female.

threads of most spider webs are nets for capturing prey. They also serve as lines of vibration that inform the spider of a disturbance in the web, as when a prey animal attempts to free itself. One spider spins a vertical thread from which a ball of sticky material is suspended. Then it uses one of its legs to swing the ball at passing insects!

Spiders and scorpions have respiratory organs called *book lungs*, these being deep pockets on the underside of the body (Figure 25.36). The pockets have extensively folded walls that resemble the pages of a book, and they greatly increase the surface area available for gas exchange with blood. The folds are moist, and blood circulating between them picks up oxygen that diffuses in from the surrounding air.

## Crustaceans

Shrimps, crayfishes, lobsters, crabs, copepods, and pill bugs belong to this group. All are important components of food webs, and many edible varieties are commercially important. Most crustaceans live in the seas, but there are freshwater and land-dwelling species.

The simplest crustaceans have rather uniform, unspecialized body segments and appendages. They may resemble the ancient arthropods that gave rise to the group. Most crustaceans, however, have many segments that are elaborated upon in diverse ways. For example, crabs have strong claws for shredding seaweed, collecting organic debris, or intimidating or attacking another animal; barnacles use featherlike appendages for combing microscopic food from the water.

Crustaceans commonly have between sixteen and twenty segments, although some have more than sixty. The segmentation is not readily apparent in crabs,

lobsters, and some other types because the dorsal cuticle of the head extends backward, covering some or all segments with a *carapace* (a shieldlike cover). The head appendages are two pairs of antennae, a pair of mandibles, and two pairs of maxillae. The *antennae* are mostly sensory, although they are sometimes used for swimming or other functions. *Mandibles* are jawlike appendages; *maxillae* are food-handling and food-sorting appendages. Figure 25.37 shows how some of these body parts are arranged in a crab and a lobster.

The shrimps, crayfish, lobsters, crabs, and their relatives are called decapods because they have ten legs projecting from the thorax (*deca*- means ten). Many decapods are commercially important. Most live in the seas, although there are a few freshwater crabs and shrimps. A few crabs burrow on land, but these must return to the seas to reproduce. Lobsters walk about on the sea-

**Figure 25.38** *Pollicipes polymerus*, a type of barnacle that lives in the intertidal zone of the Pacific Northwest. This aggregation is from the coast of Vancouver Island, British Columbia. Compare them with the sea-going barnacles (*Lepas*) shown in Figure 5.1.

**Figure 25.39** A marine copepod. The copepods of greatest ecological importance are free-swimming herbivores of aquatic communities. They form an important part of the diet of fishes and other carnivores.

floor, preying on snails, clams, crabs, and small fishes. Although crabs are rather broad and flat, their weight is distributed so evenly over their legs that they are remarkably agile walkers.

Barnacles are unusual among arthropods in having a strong, calcified shell (actually a modified exoskeleton) that protects them from predators, drying out, and the force of waves or currents (Figure 25.38). Some barnacles attach to rocks and other surfaces by an elongated stalk. Others are stalkless forms that attach directly to rocks, wharf pilings, ship hulls, and similar surfaces. A few species attach only to the skin of whales!

The 7,500 species of copepods are among the most abundant aquatic animals. The single eye in the middle of their head is a distinctive feature (Figure 25.39). Most copepods are weakly swimming, planktonic forms less than 2 millimeters long. They dine on microscopic algae and other small particles and in turn are eaten by larger aquatic animals. About 1,500 species are parasitic on fishes and various invertebrates.

### Insects and Their Relatives

**Millipedes and Centipedes**. Millipedes and centipedes are close relatives of the insects. Both are notable for their numerous legs along the trunk, this being a long, segmented region behind the head. Millipedes have a cylindrical or slightly flattened trunk (Figure 25.40a). They range from about 2 millimeters to nearly 30 centimeters in length. All millipedes are slow-moving, nonaggressive scavengers of decaying plant material.

Centipedes have a flattened trunk. They are fastmoving, aggressive carnivores, complete with fangs and venom glands. Centipedes generally live in damp places under rocks, logs, or forest litter. They prey mostly on insects, earthworms, and snails, although some tropical species can subdue small lizards and toads (Figure 25.40b). One long-legged, insect-eating centipede (*Scutigera coleoptrata*) is harmless to humans, and its unexpected appearance in a house occasionally startles people. Nevertheless, species in various parts of the world can inflict painful bites.

**Insects**. Insects are the most varied group of animals on earth. More than 800,000 species have already been catalogued! Despite the diversity, insects as a group share the adaptations listed on page 428. Here we add a few more details to that list.

a                                                    b

**Figure 25.40** (**a**) One of the mild-mannered millipedes, which scavenge on decaying plant parts. (**b**) One of the centipedes of Southeast Asia, an aggressive predator that can bring down small frogs and lizards.

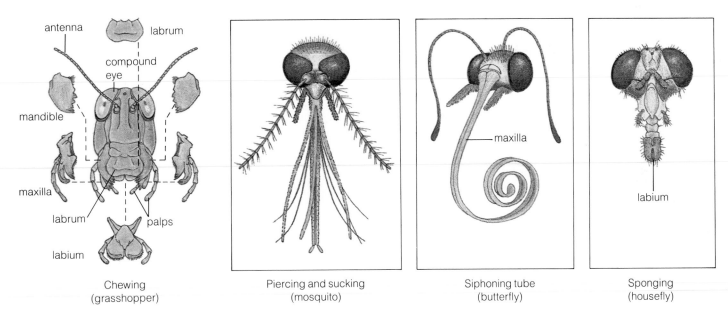

Chewing
(grasshopper)

Piercing and sucking
(mosquito)

Siphoning tube
(butterfly)

Sponging
(housefly)

**Figure 25.41** Examples of the specialized appendages of arthropods. Shown here, insect headparts, adapted for feeding in specialized ways.

The insect body consists of a head, thorax, and abdomen. The head has a pair of sensory antennae and paired mouthparts that are modified for biting, chewing, puncturing, or sucking (Figure 25.41). The thorax has three pairs of legs, used in walking and clinging, and usually two pairs of wings. For most insects, the only abdominal appendages are reproductive structures, such as egg-laying devices.

The insect gut has three specialized regions, the foregut, midgut, and hindgut. Digestion proceeds mostly in the midgut. Water is reabsorbed in the hindgut. Insects rid themselves of waste material through small tubes, the *Malpighian tubules*, which connect with the latter part of the midgut. Nitrogen-containing wastes (from protein breakdown) diffuse from the blood into the tubules, where they are converted into harmless crystals of uric acid. The crystals enter the midgut and are eliminated with feces. This system allows land-dwelling insects to get rid of potentially toxic wastes without losing precious water.

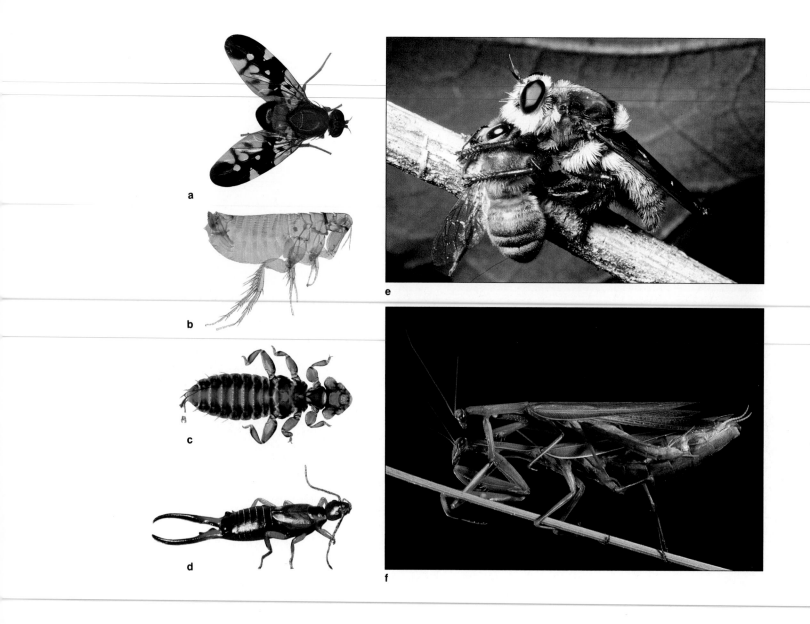

**Figure 25.42** Representatives of some orders of insects.

(**a**) Mediterranean fruit fly (order Diptera). The larval forms destroy citrus fruit and other valuable crops. (**b**) Flea (order Siphonaptera), with big strong legs, excellent for jumping onto and off animal hosts. (**c**) Duck louse (order Mallophaga), which feeds on particles of feathers and bits of skin. (**d**) European earwig (order Dermaptera), a common household pest. (**e**) The robberfly (order Diptera) looks beelike but is a true fly that preys on honeybees (order Hymenoptera) and other insects. Its relatives include mosquitoes, fruit flies, tsetse flies, and other insects we often would rather do without.

(**f**) A male praying mantid (order Mantodea) mating with a larger female. He may be eaten during or immediately after mating with her. Mantids are not useful as a biological control of insect pests. They cannot make a dent in large pest populations and do not discriminate between "good" and "bad" insects. They themselves are pests around honeybee hives.

(**g**) Stinkbugs (order Hemiptera), newly hatched. (**h**) Ladybird beetles (order Coleoptera) swarming. These beetles are raised commercially and released in great numbers as biological controls of aphids and other pests. Also in this order, the scarab beetle (**i**). With more than 300,000 named species, Coleoptera is the largest order in the animal kingdom. (**j**) Luna moth (order Lepidoptera), a flying insect of North America. Like most other moths and butterflies, its wings and body are covered with microscopic scales. (**k**) Dragonfly (order Odonata). This remarkable aerialist captures and eats other insects in midflight.

Figure 25.42 shows members of major orders of insects. Generally speaking, the small size of insects, their widespread reliance on metamorphosis, and their wings contribute to their success. Small insects grow and reproduce in great numbers on a single plant that might be only an appetizer for another animal. The reproductive capacity often is staggering. (By one estimate, if all the progeny of a single female fly were to survive and reproduce through only six more generations, that fly would have more than 5 trillion descendants!) Metamorphosing insects use different resources at dif-

ferent times. And wings can carry insects to widely scattered food sources, even to different regions with the changing seasons. (Unlike a bird wing, the insect wing develops as a lateral fold of the exoskeleton.)

The factors that contribute to their success also make insects our most aggressive competitors. They destroy vegetable crops, stored food, wool, paper, and timber. They draw blood from us and our pets, and transmit diseases. On the bright side, some insects pollinate certain crop plants, and many "good" insects attack or parasitize the ones we would rather do without.

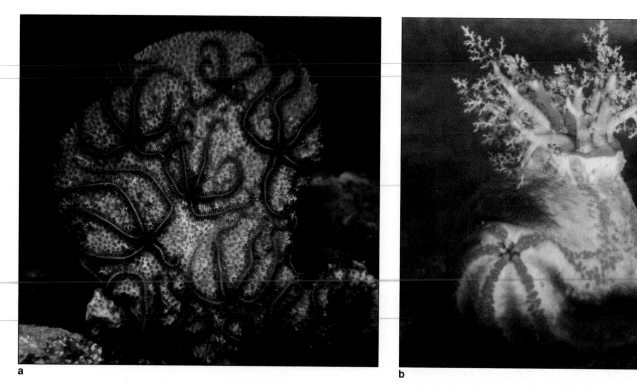

a
b

**Figure 25.43** Representative echinoderms. (**a**) Brittle stars, which move by rapid, snakelike action of their rays (arms). (**b**) Sea cucumber, with an elongated body and lengthwise rows of tube feet. The branching tentacles at the head end are modified tube feet, used in collecting food. (**c**) Sea urchin, which uses its spines and tube feet for moving about. Spines also afford protection from predators, as do the venom glands of some species. If you step on a sea urchin and one of the spines punctures and breaks off under your skin, it can inflame tissues. (**d**) A feather star, of the group called crinoids.

## ECHINODERMS

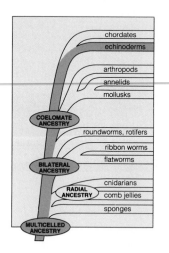

We turn now to the deuterostome lineage. The invertebrate members of this lineage include the echinoderms (Echinodermata). The name means "spiny-skinned" and alludes to the calcified spines, spicules, or plates in the body wall of these animals.

Among the members of this phylum are the sea stars, sea urchins, brittle stars, and sea cucumbers, a few of which are shown in Figures 25.43 and 25.44.

Adult echinoderms have a curious body plan. They are radially symmetrical but have some bilateral features. Some of the early echinoderms were bilaterally symmetrical, others appear to have been asymmetrical. When the life cycle proceeds through a larval stage, the larva is bilateral, suggesting that some group of ancient bilateral invertebrates gave rise to the ancestral echi-

noderms. Today, the radial arrangement in adults is pronounced. The nervous system, for example, is decentralized; there is no brain. This arrangement is probably advantageous for radial animals, for it enables them to respond to food and danger coming from different directions. Any arm of a sea star can become the leader, for instance, with the rest of the body following in that direction.

If you turn a sea star over, you will see *tube feet*. These are fluid-filled, muscular structures with sucker-like adhesive disks (Figure 25.44). Tube feet are used for walking, burrowing, clinging to a rock, or gripping a clam or snail about to become a meal. They are part of a *water-vascular system* unique to echinoderms. In sea stars, the system includes a main canal in each arm. Short side canals extend from them and deliver water to the tube feet. Each tube foot has an ampulla, a fluid-filled, muscular structure something like the rubber bulb on a medicine dropper. When the ampulla contracts, it forces fluid into the foot, which thereby lengthens.

Tube feet change shape constantly as muscle action redistributes fluid through the water-vascular system. Hundreds of tube feet may move at a time. After being released, each one swings forward, reattaches to the

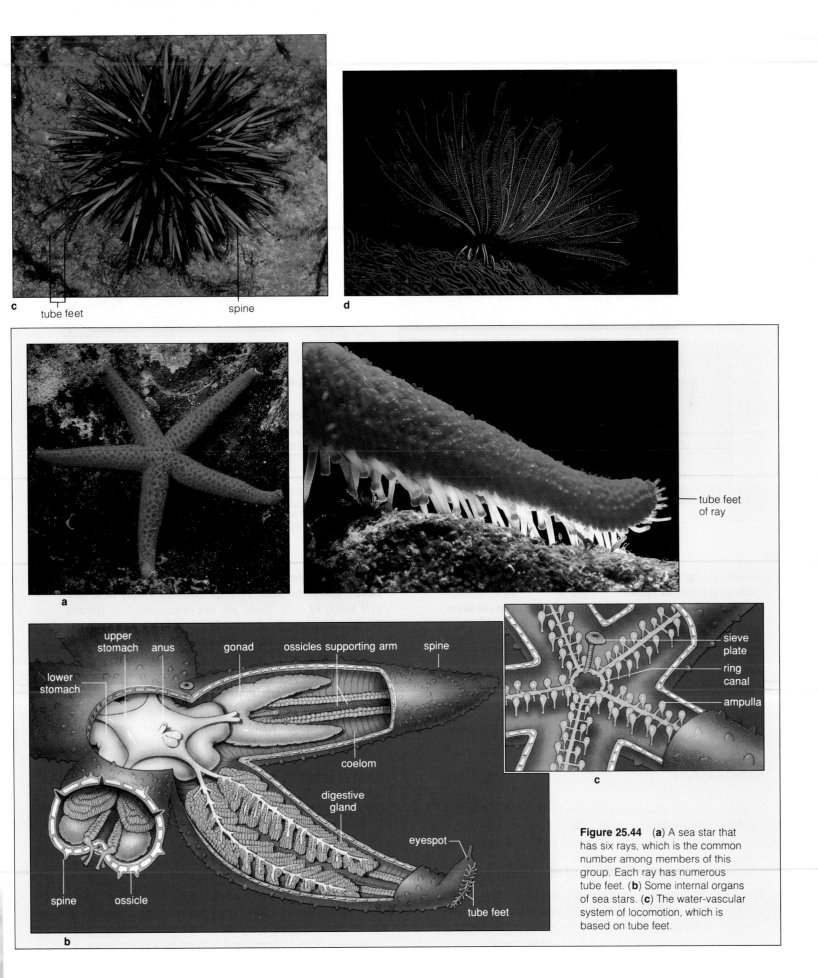

**c** tube feet

spine

**d**

**a**

tube feet
of ray

**b**

upper
stomach    anus         gonad      ossicles supporting arm      spine

lower
stomach

coelom

digestive
gland

eyespot

spine    ossicle                                              tube feet

sieve
plate

ring
canal

ampulla

**c**

**Figure 25.44** (**a**) A sea star that
has six rays, which is the common
number among members of this
group. Each ray has numerous
tube feet. (**b**) Some internal organs
of sea stars. (**c**) The water-vascular
system of locomotion, which is
based on tube feet.

**Table 25.2 Summary of Characteristics of the Major Animal Phyla**

| Phylum (and some representatives) | Typical Environment | Typical Life Style of Adult Form | Nervous System | Support and Movement |
|---|---|---|---|---|
| **Porifera** (8,000)* sponges | Most marine, some freshwater | Attached filter feeders | No nervous system (cell-to-cell transmission) | Support by spicules, protein fibers, or both; contractile cells change openings at body surfaces |
| **Cnidaria** (11,000) hydras, sea anemones, jellyfishes, corals | Most marine, some freshwater | Attached, creeping, swimming, or floating carnivores | Nerve net, cell-to-cell transmission | Hydrostatic support (by fluid in gut), by secreted jellylike mesoglea or by skeletal elements; contractile fibers in epithelial cells |
| **Ctenophora** (100) comb jellies | Marine | Mostly planktonic; few attached or creeping | Nerve net | Support by mesoglea; muscles separate from epithelia; locomotion by cilia in eight rows of comblike structures; sometimes by muscular activity |
| **Platyhelminthes** (15,000) flatworms, flukes, tapeworms | Marine, freshwater, some terrestrial in moist places; many parasitic in or on other animals | Herbivores, carnivores, scavengers, parasites | Brain, nerve cords | Hydrostatic support (no secreted skeleton); well-developed muscle tissue |
| **Nemertea** (800) ribbon worms | Most marine; few freshwater, terrestrial | Mostly carnivores | Brain, nerve cords | Hydrostatic support; well-developed muscle tissue |
| **Nematoda** (20,000) roundworms | Marine, freshwater, terrestrial; many parasitic | Scavengers, carnivores, parasites | Nerve ring, nerve cords | Hydrostatic support (by pseudocoel); tough cuticle; longitudinal muscle in body wall |
| **Rotifera** (1,800) rotifers | Marine, freshwater, moisture on mosses | Mostly filter feeders, capturing bacteria, unicellular algae | Brain, nerve cords | Locomotion by cilia; muscles for shape changes |
| **Mollusca** (110,000) snails, slugs, clams; squids, octopuses | Marine, freshwater, terrestrial | Herbivores, carnivores, scavengers, detritus or filter feeders; mostly free-moving, some attached | Brain, nerve cords, major ganglia other than brain | Hydrostatic skeleton in most; well-developed musculature in foot, mantle, other structures |
| **Annelida** (15,000) earthworms, leeches, polychaetes | Marine, freshwater, terrestrial in moist places | Herbivores, carnivores, scavengers, detritus or filter feeders; mostly free-moving | Brain, double ventral nerve cord | Hydrostatic skeleton (using coelom); well-developed musculature in body wall |
| **Arthropoda** (1,000,000+) crustaceans, spiders, insects | Marine, freshwater, terrestrial | Herbivores, carnivores, scavengers, detritus or filter feeders, parasites; mostly free-moving | Brain, double ventral nerve cord | Exoskeleton (of cuticle); jointed appendages; muscles mostly in bundles |
| **Echinodermata** (6,000) sea stars, brittle stars, sea urchins, sea lilies, sea cucumbers | Strictly marine | Mostly carnivores, detritus feeders, few herbivores; most free-moving, some attached | Radially arranged nervous system | Endoskeleton (of spines, etc.); muscles for body movement, tube feet often used in locomotion |
| **Chordata** (47,100) tunicates, lancelets, jawless fishes, jawed fishes, amphibians, reptiles, birds, mammals | Marine, freshwater, terrestrial | Herbivores, carnivores, scavengers, filter feeders; generally free-moving (most tunicates attached as adults) | Well-developed brain, dorsal and tubular nerve cord in most | Notochord or a bony or cartilaginous endoskeleton; well-developed musculature in most |

*Number in parentheses indicates approximate number of known species.

| Digestive System | Respiratory System | Circulatory System | Mode of Reproduction |
|---|---|---|---|
| No gut; microscopic food particles secured by individual cells | None; respiration by individual cells | None | Sexual (certain cells become or produce gametes); some asexual budding; production of resistant bodies (gemmules) |
| Saclike gut (may be branched) | None; respiration by individual cells; gut may distribute oxygen | None, other than via gut | Sexual (usually separate sexes; gonads discharge gametes into gut or to exterior); asexual |
| Saclike, but branched | None; respiration by individual cells; gut may distribute oxygen | None, other than via gut | Sexual (usually hermaphroditic); gonads closely associated with gut |
| Saclike gut (may be branched) | None; gas exchange across body surface | None | Sexual (usually hermaphroditic, with complex reproductive system); some asexual |
| Complete gut | None; gas exchange across body surface | Closed system | Sexual (sexes usually separate; reproductive system simple) |
| Complete gut | None; gas exchange across body surface | Pseudocoel | Sexual (sexes separate; reproductive system fairly complex) |
| Usually complete gut (sometimes saclike) | None; gas exchange across body surface | Pseudocoel | Mostly parthenogenetic (eggs develop without fertilization); males appear only occasionally |
| Complete gut | Ctenidia, other gills; mantle can be modified as lung; gas exchange across body surface | Usually open (closed system in cephalopods) | Sexual (hermaphroditic or separate sexes; reproductive system usually complex) |
| Complete gut | Gas exchange across body surface; varied outgrowths of surface in many | Usually closed; coelom also may function in distribution | Sexual (hermaphroditic or sexes separate; reproductive system simple or complex); asexual also in many |
| Complete gut | Gills; tracheal tubes; book lungs; general body surface in some | Open system | Sexual (usually separate sexes; reproductive system fairly complex) |
| Usually complete gut (sometimes saclike) | Gas exchange across general body surface or surface of outgrowths of it (such as tube feet) | Coelom around viscera, also water-vascular coelom | Sexual (reproductive system simple; gonads usually discharge gametes directly to exterior); some asexual |
| Complete gut | Lungs in most vertebrates other than fishes; perforated pharynx; gills; gas exchange across body surface | Closed system in most (open system in most tunicates); lymphatic system in many vertebrates | Sexual (sexes usually separate, except in most tunicates); asexual in some tunicates |

## Making Do—Rather Well—With What You've Got

It has taken the platypus nearly two centuries to earn a little respect. In 1798, skeptical naturalists at the British Museum poked and probed a specimen with a scalpel, looking for signs that a prankster had stitched the bill of an oversized duck onto the pelt of a small furry mammal. And no wonder. At first blush the platypus looks like evolution's practical joke (Figure 26.1).

The platypus is a mosaic of reptilian, avian, and mammalian traits. A wary animal, it hides in underground burrows by day and forages in streams and lagoons by night. Like other mammals, the platypus has fur, it maintains a fairly constant body temperature (even when submerged in near-freezing water), and it nourishes its young with milk produced by impressively large mammary glands. And yet its pectoral girdle, the bones to which forelimbs attach, would be equally at home in a lizard. Like reptiles and birds, it has a cloaca, a single external opening that functions both in reproduction and in excretion. More jarringly, the platypus lays shelled eggs, as do most reptiles and all birds.

And about that "duckbill"! A hungry platypus uses its flattened bill to scoop up freshwater snails, mussels, insect larvae, worms, and shrimps. When submerged, its nostrils close, a fleshy groove around the eyes and ears snaps shut, and the bill takes over sensory functions. In fact, the platypus has two separate sensory systems— one that operates on land and the other, underwater!

The bill's velvety surface conceals more than 800,000 sensory receptors. Many receptors detect mechanical pressure—waves produced by prey organi ing in the water. Other receptors detect we currents—even the tiny electric field create of a shrimp tail will do the trick. Once the p has scooped up shrimps and other edible b ground up on the horny pads of the platyp adult platypus has no teeth).

a

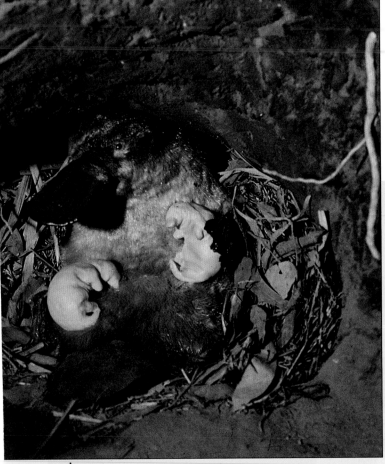

**b**

**Figure 26.1** Body plan (**a**) and burrow (**b**) of one of evolution's successful experiments, the platypus.

The platypus is about half the size of a housecat. Its flat, furry, fat-filled tail serves as a rudder in water, an energy storehouse, and a blanket that helps keep the body—and eggs—warm inside burrows. In water, its oversized, webbed front feet become paddles when the webbing flares out. On land, its strong, clawed hind feet are great for digging. When a combative rival comes around, the males can wield venom-dispensing spurs on their hind limbs. The venom is potent enough to kill a wild dog, and it can make a human miserable with partial paralysis and severe pain for weeks.

We know now that the platypus belongs to a wonderfully ancient lineage. Fossils representative of that lineage, discovered in Australian sedimentary rocks, may be 100 million years old. The forerunners of the modern platypus were present when flowering plants first took hold; they coexisted with the dinosaurs. Those ancestral forms were residents of Gondwanaland, and when that supercontinent broke up, they booked passage on the huge fragment that eventually would become Australia. There, perhaps largely by virtue of secretive behavior, the lineage has endured in splendid isolation. With that track record, who are we to chuckle at the curious collection of platypus traits? As is the case for every other animal around today, those traits meet nature's most important test. They work.

### KEY CONCEPTS

**1.** The invertebrate and vertebrate chordates are bilateral animals. They have a notochord (a supporting rod for the body), dorsal nerve cord, pharynx, and gill slits in the pharynx wall. These features appear in chordate embryos and often in adults. Vertebrate chordates additionally have a cartilaginous or bony backbone and a protective chamber for the brain.

**2.** Living invertebrate chordates include tunicates and lancelets. Living vertebrates include jawless fishes, cartilaginous fishes, bony fishes, amphibians, reptiles, birds, and mammals. Another group, the placoderms, became extinct early in vertebrate history.

**3.** Five key trends occurred during vertebrate evolution. Structural support and locomotor functions came to depend less on the notochord and more on a vertebral column. Jaws evolved. The nerve cord expanded into a spinal cord and brain. Among the aquatic ancestors of land-dwelling vertebrates, gas exchange came to depend less on gills and more on lungs and an efficient circulatory system. Also among those ancestors, fleshy fins with skeletal supports evolved into legs, which are variously modified among amphibians, reptiles, birds, and mammals.

## ON THE ROAD TO VERTEBRATES

The preceding chapter concluded with a look at the echinoderms, some of our distant relatives on the deuterostome branch of the animal family tree. When we poke about among the more recent branches of that tree, we come across a kind of animal whose portrait we probably would not put up on the wall with Grandma's. These are the sea squirts, small, baglike organisms that live their adult lives attached to the seafloor, straining out edible stuff from the water.

Surprisingly, animals something like the larval forms of sea squirts may have been the forerunners of vertebrates, beginning with the first fishlike forms. From some fish lineages, amphibians arose. Then certain amphibians gave rise to the reptiles; and out of different reptilian lineages came the birds and mammals.

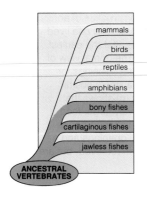

**Jawless Fishes**. Living jawless fishes are limited to about seventy-five species of lampreys and hagfishes. They have cylindrical, eel-like bodies, a notochord, and a cartilaginous skeleton. Neither has paired fins, and their vertebrae consist only of tiny chips of cartilage.

Lampreys are parasites. They have a suckerlike oral disk with horny, toothlike plates that rasp flesh from prey (Figure 26.9). Some species give the whole group a poor reputation; they latch onto salmon, trout, and other commercially valuable fishes, then suck out body juices and tissues. Just before the turn of the century, lampreys began invading the Great Lakes of North America. Populations of lake trout and other large fishes collapsed. The populations are recovering since the development of a chemical that poisons lamprey larvae, but the battle goes on.

Hagfishes are scavengers. They look like large worms with "feelers" around the mouth. They are not the favorite of fishermen. Not only do they burrow into fish trapped by setlines or nets, the ones that end up on deck secrete copious amounts of sticky, slimy mucus.

**Figure 26.9** A lamprey, here pressing its toothed oral disk to the glass wall of an aquarium. Like other living members of the class Agnatha, it has a slender, rounded, eel-like body with skin that is soft and lacks scales. The skeleton is cartilaginous.

**Figure 26.10** (*Right*) Representative cartilaginous fishes, of the class Chondrichthyes (**a**–**c**). Representative ray-finned fishes, of the class Osteichthyes: (**d**) A long-nose gar and (**e**–**g**) three of the diverse teleosts.

**Cartilaginous Fishes**. Sharks, skates, sawfishes, rays, and chimaeras belong to this group, the Chondrichthyes (Figure 26.10a through c). Most of the 850 species are specialized marine predators with a streamlined body. All have cartilage endoskeletons and five to seven gill slits on both sides of the pharynx. Their fins enhance maneuverability and stability.

Most members of this group have scales of one sort or another. Shark scales are shaped like irregular cones and give the skin a texture like sandpaper. Skates and rays typically have a few rows of scales on the back that sometimes are modified into spines. Chimaeras have almost no scales at all.

At 15 meters from head to tail, some sharks are among the largest living vertebrates. That is longer than two pickup trucks parked end to end! With their formidable jaws and unnerving capacity to detect even traces of blood in water, sharks are reputed to be monsters of the deep. Surprisingly, perhaps, most of the larger types feed on small invertebrates. Those that do eat large fishes and marine mammals use sharp, triangular teeth to capture prey and rip off chunks of flesh. Their teeth (modified scales) are shed and replaced continually.

Skates and rays are mostly bottom dwellers with flattened teeth for crushing hard-shelled invertebrates. When their mouth is buried in sediments or sand, water

is drawn into the gill chamber through their spiracle (Figure 26.8). Skates and rays both have distinctive, enlarged fins extending onto the side of the head. They vary in size, with the largest being the manta ray. Some specimens measure over 6 meters from the tip of one fin to the other. Certain rays have electric organs in the tail or fins. They can stun prey fishes with up to 200 volts of electricity. The stingray tail is equipped with a spine (a modified scale) with a venom gland at its base. Stingrays eat invertebrates, mostly, so the spine is probably used in defense against predators.

The chimaeras are commonly called ratfishes. With their bulky body and long, slender tail, they do look rather like a rat (Figure 26.10c). All of the thirty or so species live in moderately deep water. Most use their highly modified jaws to crush hard-shelled mollusks. Chimaeras have a venom gland associated with a spine in front of the dorsal fin.

**Bony Fishes**. The bony fishes (Osteichthyes) may date from the Silurian. Before that period drew to a close, they had diverged into two major lineages—the ray-finned and the lobe-finned fishes. As the sampling in Figure 26.10 suggests, the *ray-finned fishes* are spectacularly diverse. The number of species possibly exceeds the number of all species of land vertebrates combined. All

equippe
trils tha
sible for
was clos

One
has not
19.22). T
poor wa
African l
dry seaso
fish burr
When mi
protects t
season.

**a** Blue-spotted reef ray

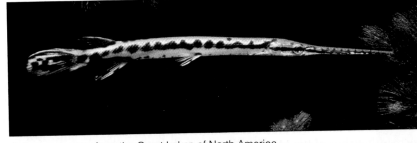

**d** Long-nose gar from the Great Lakes of North America

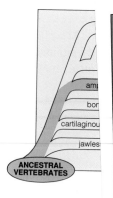

**b** Shark

an **amphibi**
aquatic, half
reptiles in its

For the fi
danger—and
there might b
available. Th
does not fluct
themselves ba
and buoys up
to support its
At the same t
and modificat
tory organs co
increase the er

**e** Goatfish exhibiting schooling behavior

**c** Chimaera (ratfish)

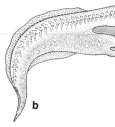

**b**

**f** Sea horse

**g** Deep-sea angler fish

membe

that ori

Most h

scales,

Their w

of oxyg

shows tl

One

larities t

the pad

group, t

23,000 n

rockfish,

teleosts

caudal fin

a        p

**Figure 26.11**

**Figure 26.12**
resembles the e

a

b

**Figure 26.19** Representative turtles. (**a**) A heavily shelled Galápagos tortoise, one of the reptiles that made a lasting impression on Charles Darwin during his five-year voyage around the world. (**b**) A marine turtle, with its streamlined shell.

a

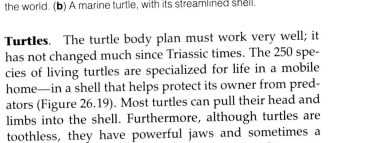

b

**Figure 26.20** (**a**) A frilled lizard, flaring a ruff of skin on the neck in a defensive display. (**b**) A chameleon capturing an insect with its long tongue. (**c**) A rattlesnake of the American Southwest. (**d**) J-shaped tracks of "sidewinding" rattlesnakes. (**e**) Green python. (**f**) Coral snake, one of the most venomous types.

**Turtles**. The turtle body plan must work very well; it has not changed much since Triassic times. The 250 species of living turtles are specialized for life in a mobile home—in a shell that helps protect its owner from predators (Figure 26.19). Most turtles can pull their head and limbs into the shell. Furthermore, although turtles are toothless, they have powerful jaws and sometimes a fierce disposition that helps keep predators at bay.

The shell also helps land-dwelling turtles conserve water. Besides this, it conserves body heat, which may be beneficial on cool days but less helpful on hot days. Only among sea turtles and other notably mobile types is the shell reduced in size and strength.

Even marine species of turtles return to land to lay their eggs. Although many eggs are laid, only a few

turtle offspring actually survive to maturity. In a few species, the survivors may live for a century.

Marine turtles especially are vulnerable when they come ashore to lay eggs; humans commonly kill them for their shell and meat. Most are endangered species as a result of these and other human activities, but steps are now being taken to protect them (page 712).

**Lizards and Snakes**. More than 90 percent of the living reptiles are lizards and snakes. They are distantly related to dinosaurs, including some giants. Existing species are generally quite small. Lizards typically weigh no more than about 20 grams. A notable exception, the Komodo monitor lizard, may weigh 75 kilograms. It's large enough to hunt deer and water buffalo.

Lizards survive in many parts of the world, but they are most numerous in deserts and tropical forests. Most of the 3,750 known species are insect eaters. Their small, peglike teeth serve as prey-grabbing devices. The chameleons zap insects with their long tongue (Figure 26.20). Being small themselves, most lizards are potential dinner for many other animals. When grabbed by predators, many give up their tail rather than fight. The disconnected tail usually wriggles about for a bit, so it sometimes distracts a predator long enough to give the lizard time for a getaway. Some lizards also buy time by startling a predator with a sudden flaring of their throat fan (Figure 26.20). More commonly, such flares serve as communication signals between lizards of the same species.

Many burrowing lizards and some lizards of the open grasslands have shorter legs and a longer body than most. Long ago, one such group of lizards gave rise to the elongated and limbless snakes. In fact, some modern snakes retain bony remnants of the hindlimbs of their ancestors.

More than 2,300 species of snakes have been identified. Some are shorter than the length of your hand (about 11 centimeters). The longest would stretch out across 10 yards (about 9 meters) of a football field. They all are limited in the way they can move their long body. Most move in S-shaped waves, much as fishes and salamanders do. The "sidewinders" maneuver themselves with J-shaped movements across the unstable sands of hot deserts (Figure 26.20).

Most snakes eat other vertebrates, although some eat insects, slugs, or bird eggs. As Figure 26.20 shows, their jaws are wonderfully movable; some types can swallow animals that are larger in diameter than they are. Pythons and boas coil their body around a prey animal, then hold it tightly until it suffocates. Rattlesnakes, coral snakes, and other types inject venom into prey and so disrupt neural function, blood circulation, or some other vital process. Although snakes generally are not aggressive toward humans, each year the venomous ones cause as many as 40,000 deaths worldwide.

a

b

c

**Figure 26.21** (**a**) Tuatara, a "living fossil." Its body plan has not changed much since the age of dinosaurs, about 150 million years ago. Two crocodilians: (**b**) A spectacled caiman from Brazil. Notice how its upper peglike teeth do not match its bottom teeth, the way yours do. (**c**) An African crocodile sunning itself.

**Tuataras.** Despite its appearance, the tuatara (*Sphenodon*) shown in Figure 26.21a is not a lizard. It has a more evolutionarily ancient skull than lizards, and it does not have male copulatory organs. Today there are only two recognized species, although the group was somewhat diverse during Mesozoic times. Existing tuataras are restricted to the small offshore islands of New Zealand, where they are metabolically adapted to the cold climate. Tuataras may be 20 years old by the time they reach sexual maturity. Like turtles, they may live well beyond 60 years.

**Crocodilians.** This group includes twenty-two living species of crocodiles and alligators. All live in or near water. All have similar body plans, including a long snout (Figure 26.21). Variations in the shape of the snout are correlated with specialized diets.

Like other living reptiles, the crocodilians are ectotherms. They show complex behavioral and physiological responses for regulating body temperature (page 730). They also show complex social behavior, as when male and female parents assist hatchlings in their move out of the egg and into the water. Young alligators of the Gulf Coast swamps remain with the mother for two years after hatching.

All crocodilians are large, by reptilian standards, and some are huge. Until recently, saltwater crocodiles grew up to 7 meters long. Unfortunately for them, their belly skin is highly prized as leather. They have been hunted with such a vengeance over the past two centuries that few live long enough to grow to impressive size.

Although crocodilians have a lizardlike body, detailed studies of their anatomy and physiology reveal that they are in fact relatives of another group of vertebrates, the birds. For example, unlike most reptiles, which have conserved the amphibian system of circulation, crocodilians have a ventricle that is divided into right and left chambers. Blood from the right ventricle goes to the lungs; that from the left ventricle goes to the rest of the body. (They still retain the ability to shunt blood between the great vessels leaving the heart. This allows them to adjust blood pressure, as when they are diving.) Their overall pattern of blood circulation is more nearly comparable to that of birds and mammals.

## BIRDS

There are nearly 9,000 known species of birds, including the types shown in Figure 26.22. Birds apparently descended from reptiles that ran around on two legs during Jurassic times, some 160 million years ago. The oldest known bird (*Archaeopteryx*) resembled those reptiles in its

mammals
birds
reptiles
amphibians
bony fishes
cartilaginous fishes
jawless fishes
ANCESTRAL VERTEBRATES

a

b

c

d

limb bones and other features, as shown in Figure 17.10. Birds still resemble reptiles in many of their internal structures, their horny beaks and scaly legs, and their habit of laying eggs. So close is the resemblance that Thomas Huxley called birds "glorified reptiles" and placed both groups of vertebrates in the same class. Today birds are indeed viewed as a branch of the reptilian lineage.

Most birds have a similar body plan. And unlike their reptilian ancestors, they all have **feathers**—light-

**Figure 26.22** Some characteristics of birds. (**a**) Bahama woodstar, sipping nectar from a hibiscus blossom. Like other hummingbirds, it can forage for nectar in midflight. Its long, narrow bill coevolved with specific floral structures (Chapter 29). (**b**) A male Himalayan moral, a pheasant with flamboyant plumage (an outcome of sexual selection). This native of the mountains of India is an endangered species. As is the case for many birds, its jewel-colored feathers end up adorning humans—in this case, on the caps of native tribespeople. (**c**) Speckled eggs of a magpie. All birds lay hard-shelled eggs. (**d**) Canada geese, one of the types of birds that show migratory behavior, at their wintering grounds in New Mexico.

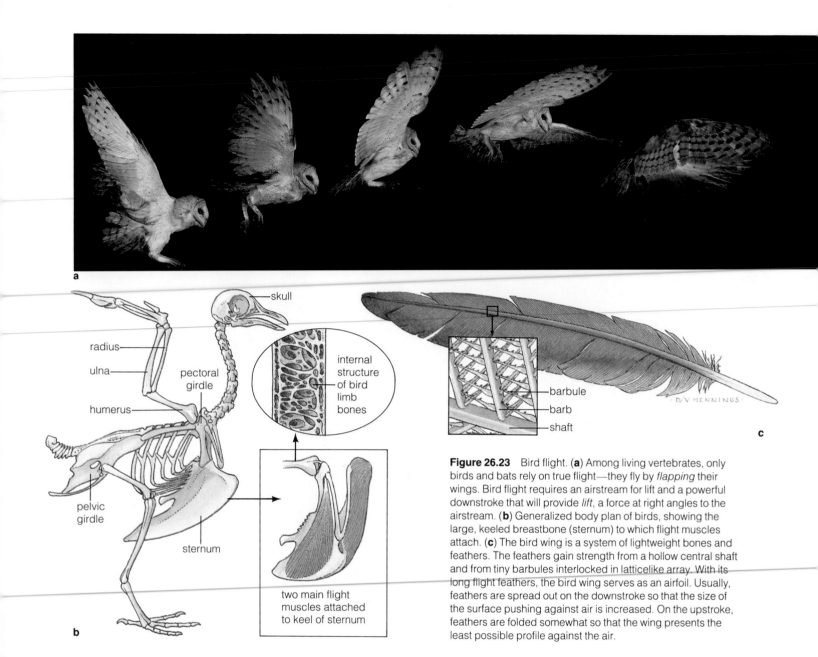

**Figure 26.23** Bird flight. (**a**) Among living vertebrates, only birds and bats rely on true flight—they fly by *flapping* their wings. Bird flight requires an airstream for lift and a powerful downstroke that will provide *lift*, a force at right angles to the airstream. (**b**) Generalized body plan of birds, showing the large, keeled breastbone (sternum) to which flight muscles attach. (**c**) The bird wing is a system of lightweight bones and feathers. The feathers gain strength from a hollow central shaft and from tiny barbules interlocked in latticelike array. With its long flight feathers, the bird wing serves as an airfoil. Usually, feathers are spread out on the downstroke so that the size of the surface pushing against air is increased. On the upstroke, feathers are folded somewhat so that the wing presents the least possible profile against the air.

weight structures that have roles in flight, insulation, or both. Birds do vary tremendously in size, body proportions, coloration, and capacity for flight. A small hummingbird barely tips the scales at 2.25 grams (0.08 ounce). The largest bird, the ostrich, weighs about 150 kilograms (330 pounds). As Figure 17.2 illustrates, ostriches cannot fly, but they are impressively long-legged sprinters. Especially among the warblers and other perching birds, differences in coloration and territorial songs help us distinguish even closely related species from one another. Bird songs and other social behaviors are remarkably complex, as you will see in later chapters.

The body plan of nearly all birds meets the two key requirements for flight: low weight and high power. The **bird wing** is a forelimb constructed of feathers, power-

ful muscles, and lightweight bones. Generally, birds have a greatly enlarged sternum (breastbone) to which flight muscles are attached (Figure 26.23). Those muscles also attach to the upper limb bone (humerus) on both sides of the breastbone. When those muscles contract, they produce the powerful downstroke required for flight.

Bird bones are strong and yet weigh very little because of air cavities in the bony tissue. The endoskeleton of a frigate bird, which has a 7-foot wingspan, weighs only 4 ounces. That is less than the feathers weigh! All birds have a large, strong, four-chambered heart. They also have an efficient system for oxygen uptake in which air flows *through* sacs, not into and out of them, for gas exchange in the lungs (Figure 40.7).

## MAMMALS

The characteristics and evolutionary trends among mammals were described earlier, in Chapter 21. In brief, the ancestors of mammals originated during the Carboniferous, from synapsid reptiles that diverged from the lineages that led, ultimately, to dinosaurs and to modern reptiles and birds (Figure 26.18). Mammal-like reptiles—therapsids—appeared on that separate evolutionary road. They gave rise to diverse mammals that flourished throughout the late Mesozoic.

Unlike reptiles, female mammals have milk-secreting glands that provide nourishment for their young. Unlike reptiles, mammals have hair that covers at least part of their body (although this trait has been lost in most whales). And unlike reptiles, which normally swallow their prey whole, most mammals have several specialized kinds of teeth by which food is killed, cut, and chewed before being swallowed. Mammalian teeth include molars that serve as platforms for grinding food (page 332). They help break down food faster—and so help fuel the high metabolic rates that sustain the active life-style of most mammals.

Mammals have built-in mechanisms for maintaining their body temperature despite fluctuations in environmental temperatures (page 731). Their hair or thick skin helps retain metabolic heat; some of their neural and hormonal controls deal with hot and cold conditions. Their excellent temperature-regulating capacity is one of the reasons why mammals have radiated throughout the world's continents and oceans.

Another point to remember is this: It was not until the great mammalian radiation at the dawn of the Cenozoic

**Figure 26.24** An egg-laying mammal of arid habitats of Australia: the short-nosed anteater (*Tachyglossus aculeatus*). The platypus in Figure 26.1 is another egg-laying mammal.

that the brain began to reveal its true potential. Especially among the primates, it began expanding to include larger, interconnected masses of information-encoding and information-processing cells. This development was the foundation for our own remarkable capacity for memory, learning, and conscious thought.

Today there are three major groups of mammals: the egg-laying, pouched, and placental mammals.

### Egg-Laying Mammals

Two major types of egg-laying mammals (subclass Prototheria), have survived to the present day. They are the platypus, described at the start of this chapter, and several species of spiny anteaters, which are burrowing animals of Australia and New Guinea (Figure 26.24). Both types lost most or all of their teeth during the course of evolution. The loss is correlated with their specialized diets. Whereas the platypus feeds mostly on small aquatic invertebrates, the spiny anteaters feed on termites and ants, which they capture with their long, sticky tongue.

Several rather archaic traits have persisted in both types of egg-laying mammals. Besides being egg layers, these animals do not show the same kind of skeletal developments of other mammals, and their metabolic rates are lower. Yet their young are suckled after hatching. Like other mammals, the platypus is covered with hair, and the spiny anteaters are covered with spines derived from hair. Despite their low metabolic rates, both mammals maintain a relatively constant body temperature that is well above that of their surroundings.

## Pouched Mammals

There are about 260 species of pouched mammals, or marsupials (subclass Metatheria). Nearly all are native to Australia and nearby islands; a few live in North and South America. Marsupial young are born live but not quite "finished"—they are tiny, blind, and hairless. However, the newborns have an excellent sense of smell and strong forelimbs, which they use to locate and reach the mother's pouch, which is some distance away from the birth canal. They are suckled in that pouch, and there they complete their early development.

The kangaroos, probably the most familiar marsupials, are also the largest. Some types weigh 90 kilograms. Even larger species once lived in Australia and may have been hunted to extinction by early human populations. At least one marsupial has managed to coexist successfully with humans, however. During the past century, the Virginia opossum has greatly expanded its range in the United States. It now thrives in urban areas as well as in its native forest habitats (Figure 26.25a).

For at least 50 million years, Australian marsupials evolved in relative isolation from placental mammals. Their ancestors apparently were able to cross a narrow sea that had opened up between two land masses, one of which became Australia. Placentals, inexplicably, did not make the crossing. Thus the marsupials radiated through their island continent, filling adaptive zones that might not have been available to them under more competitive circumstances. For example, whereas other continents are home to deer, antelope, and other herbivorous mammals, Australia has its kangaroos and wallabies. North American and Russian forests have their wolves; Australia has (or had) its Tasmanian "wolf," a marsupial which is now probably extinct. Other marsupials glide like flying squirrels, and some climb like monkeys. In recent times, human populations introduced cattle, sheep, horses, and other placental mammals to Australia, and many of the native marsupials are being threatened with displacement.

## Placental Mammals

More than 4,500 placental mammals (subclass Eutheria) are known. One or more members of this group live in virtually every kind of aquatic and terrestrial environment (Figure 25.25b through f).

As you will see in Chapter 43, a *placenta* is a spongy tissue that develops in the mother's uterus and has functional connections with the embryo. This tissue, composed of maternal tissue and embryonic membranes, is the means by which the embryo receives nutrients and oxygen and gets rid of metabolic wastes. Placental mammals grow faster in the uterus than marsupials do in a pouch. At birth, many are fully developed and can

a

b

move about almost immediately. Even then, they all remain with the mother for some time.

Take a quick look at Appendix II, and you will see that placental mammals include a great variety of species—many familiar to us, and others obscure. In terms of diversity, the rodents are most successful. These include rats, mice, squirrels, and prairie dogs. Bats are next in terms of diversity. Other familiar placental mammals are the carnivorous dogs, bears, cats, walruses, and dolphins, as well as the herbivorous horses, camels, deer, and elephants. Exotic types, including manatees and anteaters, also hold membership in this group. The physiology, behavior, and ecology of many of these animals will occupy our attention in chapters to follow.

## SUMMARY

1. The chordates are animals with four distinguishing characteristics. Nearly all chordate embryos, and commonly the adults, have a notochord, a dorsal hollow nerve cord, pharynx, and gill slits (or hints of these) in the pharynx wall.

2. The invertebrate chordates include the tunicates (sea squirts) and the lancelets. The vertebrates include fishes, amphibians, reptiles, birds, and mammals.

3. In the course of the early evolution of vertebrates, the brain and dorsal nerve cord became protected by an internal skeleton of cartilage or bone. Other skeletal ele-

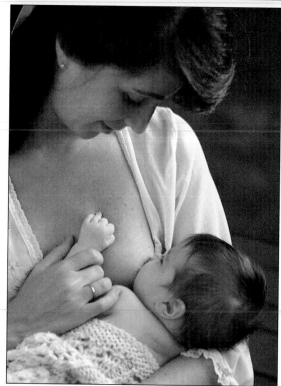

**Figure 26.25** (**a**) A pouched mammal: a female opossum with her young.

Among the placental mammals are species that feed on submerged seaweed (the manatee in **b**), traverse deserts with impunity (the camels in **c**), climb trees (the juvenile raccoons in **d**), and swim in frigid waters and sunbathe on ice (the walruses in **e**).

The human baby in (**f**) also belongs to the placental group. It is busily demonstrating a trait characteristic of all three mammalian subclasses—it derives nourishment from mammary glands.

ments, such as jaws, ribs, and supports for the limbs, also developed. The vertebrae around the spinal cord allow for considerable flexibility, and nearly all skeletal structures are sites to which muscles concerned with chewing, locomotion, and other functions are attached.

4. Lungs and fleshy fins with skeletal elements evolved in the fishes ancestral to amphibians. Reptiles evolved from certain amphibians. Reptiles have protective scaly skins and physiological adaptations to life on land. They can reproduce without going back to water, which most amphibians must do.

5. Birds and mammals evolved from separate groups of reptiles. Their success is related in part to their constant and warm body temperature and to the insulation provided by feathers or hair which help birds and mammals to live in many situations. Good vision, well-developed brains, appendages for flight, swimming, running, grasping, and other purposes are a few of the other reasons birds and mammals are so successful.

## Review Questions

1. List and describe the features that distinguish chordates from other animals. *444*

2. Name and describe the characteristics of an organism from each of the two groups of invertebrate chordates. *445–447*

3. List four trends that occurred during vertebrate evolution. *447*

4. Which evolutionary modifications in fishes set the stage for the emergence of amphibians? *452–453*

5. List some of the identifying characteristics of reptiles, birds, and mammals. *456–464*

## Self-Quiz (Answers in Appendix IV)

1. All chordate animals show _____ symmetry.

2. Besides the tunicates, the invertebrate chordates include:
   a. acorn worms
   b. lampreys
   c. lancelets
   d. all of the above

3. Tunicates and lancelets are _____, but tunicates probably _____ on the evolutionary branch leading to vertebrates.
   a. not chordates; were
   b. invertebrate chordates; were not
   c. placoderms; were not
   d. both urochordates; were

4. Vertebrate chordates also have a cartilaginous or bony chamber for the _____.

5. Which of the following are *not* living vertebrates?
   a. jawless fishes and cartilaginous fishes
   b. bony fishes
   c. placoderms
   d. amphibians and reptiles
   e. birds and mammals

6. Lungs and fleshy fins with skeletal elements were adaptations that foreshadowed the evolution of _____.
   a. lancelets
   b. lampreys
   c. amphibians
   d. all of the above

7. Key events in the evolution of reptiles were _____.
   a. matched molars for grinding food
   b. internal fertilization, shelled amniote egg
   c. bony, scaly exoskeleton
   d. all of the above were key events

8. Protective scaly skin, physiological adaptations that reduce water loss, and an ability to reproduce without returning to water are characteristics of _____.
   a. fishes
   b. amphibians
   c. reptiles
   d. birds
   e. mammals

9. The evolutionary success of birds and mammals is due to _____.
   a. their constant and warm body temperature
   b. insulation provided by feathers or hair
   c. good vision and well-developed brain
   d. appendages specialized for flight, swimming, running, grasping, and other functions
   e. all of the above

## Selected Key Terms

| | | |
|---|---|---|
| agnathan *448* | feather *461* | notochord *444* |
| amniote egg *456* | fin *449* | ostracoderm *448* |
| amphibian *453* | gill slit *444* | pharynx *444* |
| bird *460* | internal fertilization | placenta *464* |
| bony fish *450* | *456* | placoderm *448* |
| cartilaginous fish *450* | invertebrate | reptile *456* |
| cephalochordate *446* | chordate *444* | scale *449* |
| closed circulatory | jawless fish *450* | swim bladder *449* |
| system *455* | lobe-finned fish *452* | urochordate *445* |
| dorsal nerve cord *444* | mammal *463* | vertebrate *444* |
| endoskeleton *447* | nerve cord *444* | |

## Readings

Carroll, R. L. 1988. *Vertebrate Paleontology and Evolution.* New York: Freeman.

Hoffman, E. 1990. "Paradox of the Platypus." *International Wildlife* 20(1):18–21.

Romer, A. S., and T. S. Parsons. 1986. *The Vertebrate Body.* Sixth edition. Philadelphia: Saunders.

Welty, J., and L. Baptista. 1988. *The Life of Birds.* Fourth edition. New York: Saunders.

FACING PAGE: *A flowering plant (Prunus) busily doing what it does best: producing flowers for the fine art of reproduction.*

# 27 PLANT TISSUES

## The Greening of the Volcano

In the spring of 1980, Mount St. Helens in southwestern Washington exploded, and 540 million tons of ash blew skyward. Within minutes, hundreds of thousands of mature trees near the volcano's northern flank were incinerated or blown down like matchsticks by the shock waves. Rivers of ash and fire surged down the slopes faster than a hundred miles an hour. Twenty billion gallons of water, released when snow and glacial ice melted under the intense heat, turned those nightmarish rivers into torrents of cementlike mud. In one brief moment in time, nearly 100,000 acres of magnificent forests had become a scarred, barren sweep of land.

Yet in less than a year, fireweed, blackberry, and other plants were sprouting around the grayed trunks of fallen trees. Within a decade, willows and alders were flourishing along the rivers and shrubs were cloaking much of the land, making conditions more hospitable for hemlocks and Douglas fir seedlings. Figure 27.1b shows the view from a ridge overlooking Spirit Lake not long after the eruption; Figure 27.1c shows the same view nine years later. Within fifty years, a young forest will rise above the low, green shrubs. Within a century, the forest will be as it once was.

Events of this magnitude dramatize what the world would be like without plants. It is almost mind-numbing to think about how the forests we take for granted could be turned so capriciously into rock-strewn desolation. And it is somehow reassuring to know that grasses, shrubs, and trees have the tenacious capacity to start all over again.

With this chapter we open a unit dedicated to the flowering plants. In terms of sheer diversity and distribution, they are the most successful plants on earth. What characterizes the flowering plants, which directly or indirectly nourish other organisms and make the land habitable? Can we identify patterns of structural organization among them? Do they, like animals, have internal systems for transporting and distributing materials among their component cells? How do they reproduce? What governs their growth and development? These are questions addressed in this unit.

a

**1.** Most of the 275,000 plant species are flowering plants. Despite their diversity, flowering plants are composed of only three kinds of tissues, called dermal, ground, and vascular tissues.

**2.** Dermal tissues form a protective cover at the surface of all plant parts. Ground tissues make up the bulk of the plant body. Vascular tissues are like pipelines through the ground tissues. They distribute water, dissolved minerals, and products of photosynthesis through roots, stems, and leaves.

**3.** Plants grow at the tips of their roots and shoots. Each tip has a region of undifferentiated cells (apical meristem). These cells divide repeatedly, elongate, and develop into the specialized types that make up the dermal, ground, and vascular tissues. The growth originating at root and shoot tips produces the "primary" tissues of the plant body.

**4.** Many plants also show secondary growth, which increases the diameter of roots and stems. The regions responsible for secondary growth (lateral meristems) do not occur at the tips of plant parts. Rather, a fully developed lateral meristem is like a long, thin cylinder that produces new cells along the length of an older root or stem. Extensive secondary growth results in the tissues called wood.

**Figure 27.1** (**a**) Eruption of Mount St. Helens in 1980. (**b**) Nothing remained of the forest that had surrounded the volcano's flanks. (**c**) Less than a decade later, plants were making a comeback.

## THE PLANT BODY: AN OVERVIEW

As we saw in the preceding unit, there are more than 275,000 species of plants, and no one species can be used as a "typical example" of their body plans. The most familiar ones are the angiosperms and gymnosperms.

**Angiosperms** are the flowering plants, such as roses, apple trees, and corn. Besides producing flowers (reproductive structures), they produce seeds that are completely enclosed in protective tissue layers. **Gymnosperms** are chiefly conifers, such as pine trees and junipers. They produce "naked" seeds that are borne on surfaces of reproductive structures rather than being surrounded by tissues. The majority of vascular plants are angiosperms, however, and they will be our focus here.

## Shoot and Root Systems

Flowering plants typically have well-developed shoot and root systems. Both have internal pipelines that conduct water, minerals, and organic substances throughout the plant body (Figure 27.2). The **shoot system** typically consists of stems, leaves, and reproductive structures. Stems serve as frameworks for upright growth. With upright growth, photosynthetic tissues in leaves are favorably exposed to light, and flowers are favorably situated for pollination. Some parts of the shoot system store food. The **root system**, which usually grows below ground, absorbs water and dissolved minerals from soil. The root system also stores food, anchors the plant, and sometimes structurally supports it.

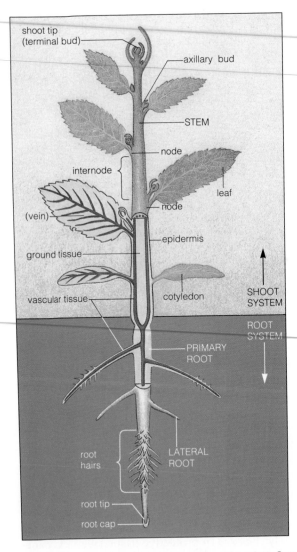

**Figure 27.2** Body plan for one type of flowering plant. Some vascular tissues by which water and nutrients move through the plant body are shown in red. Notice how they thread through the ground tissue (shaded yellow), which makes up the bulk of the plant body. Dermal tissue (epidermis) forms a covering for the root and shoot systems shown here.

Three kinds of tissues predominate in root and shoot systems. The bulk of the plant body consists of different types of *ground* tissues, which have *vascular* tissues dispersed through them (Figure 27.2). *Dermal* tissues serve as a protective covering for the plant body.

## Ground Tissues

Most of what you eat when you chew on a celery stalk is ground tissue. As Figure 27.3 shows, there are three types of ground tissue—**parenchyma**, **collenchyma**, and **sclerenchyma**. They differ primarily in the wall structure of the majority of cells each type contains.

Here it will help to remember that all new plant cells develop a *primary* cell wall, composed of strands of cellulose bundled together. Later, many types of plant cells deposit more cellulose and other materials inside the primary wall, forming a *secondary* cell wall (Figure 4.26). The cellulose in these walls is a polysaccharide in which parallel arrays of threadlike molecules are hydrogen-bonded to one another. Pectin, another type of wall material, is a polysaccharide that incorporates calcium and magnesium salts. It is abundant in the *middle lamella*, the cementing layer between the primary walls of neighboring cells, and helps bind adjacent cells together.

**Parenchyma.** Cells of the most abundant ground tissue, parenchyma, generally have thin primary walls. This is the soft, moist stuff in the celery stalk. In stems, roots, leaves, flowers, and the flesh of fruits, parenchyma cells are massed together with ample air spaces between them. Different types of parenchyma cells take part in photosynthesis, storage, secretion, and other tasks. Parenchyma cells are still alive at maturity, and they retain the capacity for cell division. When you see scars on a plant, parenchyma cells have been at work, healing the wounds and often regenerating missing parts.

**Collenchyma.** The ground tissue called collenchyma helps strengthen the plant body. Commonly it is arranged as strands or cylinders just beneath the dermal tissue of stem and leaf stalks. The pliable "strings" in the celery stalk are an example. The cells of collenchyma are alive at maturity. The primary cell walls become thickened with cellulose and pectin, often at their corners. Bonding interactions between the two substances make collenchyma quite pliable. When stretched during growth, the cells can easily retain thir new shape.

**Sclerenchyma.** Mature plant parts gain mechanical support and protection from sclerenchyma. In this ground tissue, cells have thick secondary walls that are rigid *or* pliable. Commonly the secondary walls are impregnated with *lignin*, a rather inert substance that has yet to be defined chemically.

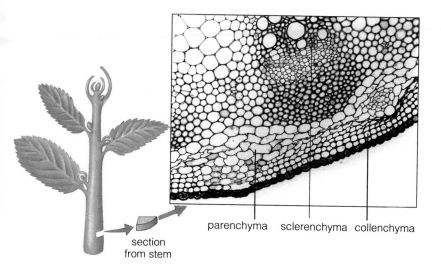

section
from stem

parenchyma   sclerenchyma  collenchyma

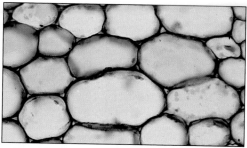

**a** Parenchyma

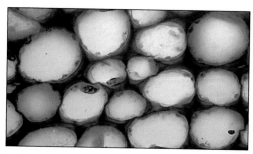

**b** Collenchyma

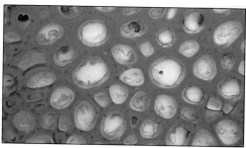

**c** Sclerenchyma

**Figure 27.3** Examples of ground tissues, which make up the bulk of the plant body. All three are cross-sections from the stem of a sunflower plant (*Helianthus*).

Lignin contains different sugar alcohols in amounts that vary, depending on the species. In a process called **lignification**, this substance is deposited first at the corners of cells, then in middle lamellae. Its presence has three effects. First, it anchors the cellulose in the walls and imparts more strength and rigidity to them. Second, it forms a stable coating around the other wall components and protects them from physical or chemical damage. Third, it forms a waterproof barrier around the cellulose strands. Where lignin deposits occur, water cannot keep the wall hydrated and soft. Indeed, many biologists believe that vascular plants arose when cells developed the capacity for lignification.

Sclerenchyma cells, called *sclereids* and *fibers*, are concerned with strengthening and protecting plant parts. Flax plants have long, tapered fibers massed together in parallel. The fibers, which flex and twist without stretching, are manufactured into rope, paper, and thread. Sheetlike arrays of sclereids form a strong, protective coat around seeds (try cracking a coconut shell or a peach pit). Sclereids scattered through the flesh of a pear give that fruit its gritty texture (Figure 27.4).

Of course, if any cell encased itself completely in a lignified wall, it would end up killing itself. There would be no way to exchange gases, nutrients, and other materials with its environment. Where there are lignified walls, there also are pits and other openings that serve as channels between cells and to the outside world. As you will see, the openings vary in size and number, depending on the cell's role.

**Ground tissues make up the bulk of the vascular plant body. Parenchyma, the most abundant type, has many functions, including photosynthesis. Collenchyma and sclerenchyma offer structural support. Sclerenchyma also offers protection.**

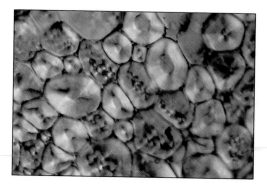

**Figure 27.4** From the flesh of a pear (*Pyrus*), some thick, lignin-impregnated stone cells, a type of sclereid.

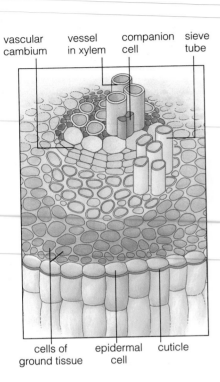

vascular cambium | vessel in xylem | companion cell | sieve tube

cells of ground tissue | epidermal cell | cuticle

**Figure 27.5** Example of how the cells of xylem and phloem are arranged in bundles in the ground tissue of a stem.

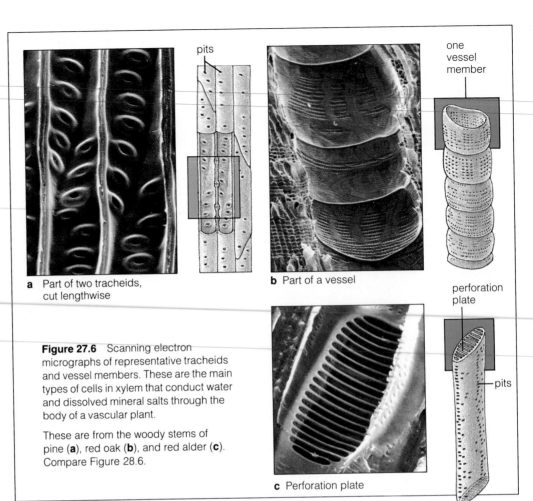

pits

**a** Part of two tracheids, cut lengthwise

**b** Part of a vessel

one vessel member

perforation plate

pits

**c** Perforation plate

**Figure 27.6** Scanning electron micrographs of representative tracheids and vessel members. These are the main types of cells in xylem that conduct water and dissolved mineral salts through the body of a vascular plant.

These are from the woody stems of pine (**a**), red oak (**b**), and red alder (**c**). Compare Figure 28.6.

## Vascular Tissues

Flowering plants have two kinds of vascular tissues, called **xylem** and **phloem**. Both have specialized conducting cells, some fibers, and some parenchyma cells clustered together in distinctive bundles, such as the one diagrammed in Figure 27.5.

**Xylem**. Xylem conducts water and dissolved minerals absorbed from the soil; it also mechanically supports the plant. Its water-conducting cells overlap or are joined at their ends and form continuous pipelines through roots, stems, and leaves. The main types of cells in xylem are *tracheids* and *vessel members*. Both types are dead at maturity, with all or part of their walls lignified. Tracheids are long cells with tapered, overlapping ends. Water flows from cell to cell through thin wall regions called pits. Vessel members are shorter cells, joined end to end to form a *vessel*, a tube through which water flows freely (Figure 27.6). Vessel members have pits and perforation plates, which are openings in their end walls. Some vessel members have only one large opening.

Others have ladderlike bars extending across the open end or a cluster of small, round perforations.

**Phloem**. Phloem is the vascular tissue by which sugars and other solutes are transported through the plant body. Its main conducting cells, the *sieve tube members*, are alive at maturity. Their walls have clusters of pores through which the cytoplasmic contents of adjacent cells are connected. In many plants, "companion cells" in the phloem help the sieve tube members load sugars produced by photosynthetic cells in leaves and unload them in other plant regions engaged in food storage or growth. The structure and function of these living cells are main topics in the chapter to follow.

**Xylem, a vascular tissue, passively conducts water and dissolved minerals absorbed from soil through the plant and helps support it structurally. Phloem, another vascular tissue, rapidly transports sugars and other solutes through the plant.**

## Dermal Tissues

A continuous layer of tightly packed cells, the **epidermis**, covers the primary plant body. The surface of epidermal cell walls is coated with waxes that have become embedded in cutin, a fatty substance. This surface coating, a **cuticle**, restricts water loss and offers some resistance to microbial attack. A protective cover, the **periderm**, replaces epidermis when roots and stems increase in diameter and become woody.

For the most part, water vapor as well as carbon dioxide and oxygen move across the epidermis at openings between pairs of guard cells. These are specialized epidermal cells. Each opening between them is called a *stoma* (plural, stomata); Figure 27.7 shows examples. Stomatal closure is a highly regulated activity, as you will read in the next chapter.

Dermal tissues help restrict water loss from the plant, provide some resistance to microbial attack and, in woody plants, physically protect the underlying tissues.

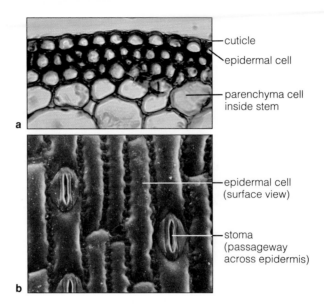

**Figure 27.7** (**a**) Cross-section through the stem of a corn plant, showing part of the epidermis. (**b**) Surface view of leaf epidermis from a corn plant (*Zea mays*).

## How Plant Tissues Arise: The Meristems

New plants grow and older plant parts lengthen through cell division and enlargement at the tips of roots and shoots. Growth that is initiated at root and shoot tips is known as *primary* growth. Each tip has a dome-shaped mass of cells, an **apical meristem**. Descendants of some of those cells develop into the specialized tissues of the lengthening root or stem. One cell lineage, *protoderm*, gives rise to epidermis; another, *ground meristem*, to ground tissues. The third meristematic lineage, *procambium*, gives rise to primary xylem and phloem. Other cells in the mass divide and so perpetuate the apical meristem.

Many plants, including corn, die after only one season of primary growth. Plants with a woody body show *secondary* growth at regions other than root and shoot tips. Secondary growth originates at self-perpetuating tissue masses called **lateral meristems**, and it increases the diameter of older roots and stems. Each spring, for example, a maple tree undergoes primary growth at its root and shoot tips, and secondary growth adds to its woody parts. Figure 27.8 shows the general location of meristems in one type of flowering plant stem.

New plants grow and older plant parts lengthen through cell divisions at apical meristems present at root and shoot tips. Older roots and stems of woody plants increase in diameter through cell divisions at lateral meristems.

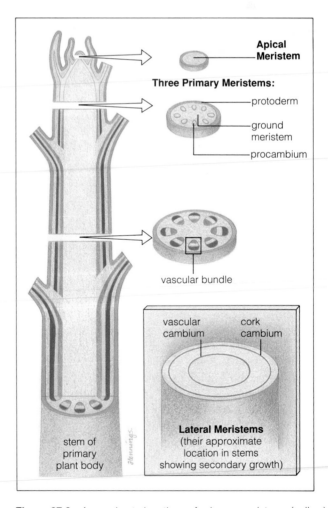

**Figure 27.8** Approximate locations of primary meristems (yellow) and lateral meristems (red) in plants that show both primary and secondary growth.

**a** Typical monocot flower

**b** Typical dicot flower

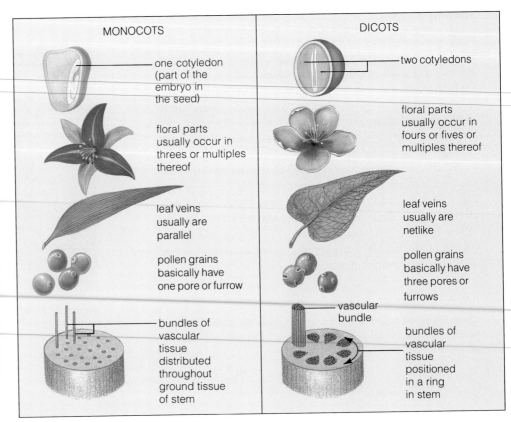

| MONOCOTS | DICOTS |
|---|---|
| one cotyledon (part of the embryo in the seed) | two cotyledons |
| floral parts usually occur in threes or multiples thereof | floral parts usually occur in fours or fives or multiples thereof |
| leaf veins usually are parallel | leaf veins usually are netlike |
| pollen grains basically have one pore or furrow | pollen grains basically have three pores or furrows |
| bundles of vascular tissue distributed throughout ground tissue of stem | vascular bundle / bundles of vascular tissue positioned in a ring in stem |

**c**

**Figure 27.9** A comparison of the main features by which the two classes of flowering plants, the monocots and dicots, are distinguished from each other. The photographs show the flower of a wild iris (*Iris*, a monocot) and of St. John's wort (*Hypericum*, a dicot).

## Monocots and Dicots Compared

As we have seen, there are two classes of flowering plants, the Monocotyledonae and Dicotyledonae (page 402). Fortunately for us, they also go by the names **monocots** and **dicots**. Grasses, lilies, orchids, irises, cattails, and palms are examples of monocots. Nearly all familiar trees and shrubs, other than the gymnosperms, are dicots.

Monocots and dicots are similiar in structure and function, but they differ in some distinctive ways. For example, monocot seeds have one cotyledon and dicot seeds have two. A "cotyledon" is a leaflike structure originating in the seed, as part of the plant embryo. After the seed germinates, the cotyledons may unfurl somewhere along the length of the tiny seedling. Figure 27.9 shows other notable differences between monocots and dicots.

The tissue organization of stems, leaves, and roots will be described next. When looking at the photographs accompanying the text, keep in mind the following terms, which identify the way a given tissue specimen was cut from the plant:

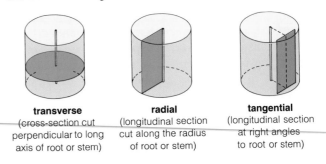

| transverse | radial | tangential |
|---|---|---|
| (cross-section cut perpendicular to long axis of root or stem) | (longitudinal section cut along the radius of root or stem) | (longitudinal section at right angles to root or stem) |

## SHOOT SYSTEM

### Arrangement of Vascular Bundles

The primary tissues of monocot and dicot stems usually are organized in one of two patterns, based on the distribution of vascular bundles. A **vascular bundle** is a multistranded cord of primary xylem and phloem that threads lengthwise through the stem ground tissue.

The stems of most monocots and some dicots have the vascular bundles distributed throughout the ground tissue. Figure 27.10 shows an example of this arrangement. The stems of most dicots and conifers have the vascular bundles arranged as a ring that divides the ground tissue into two zones, an outer **cortex** and inner **pith** (Figure 27.11).

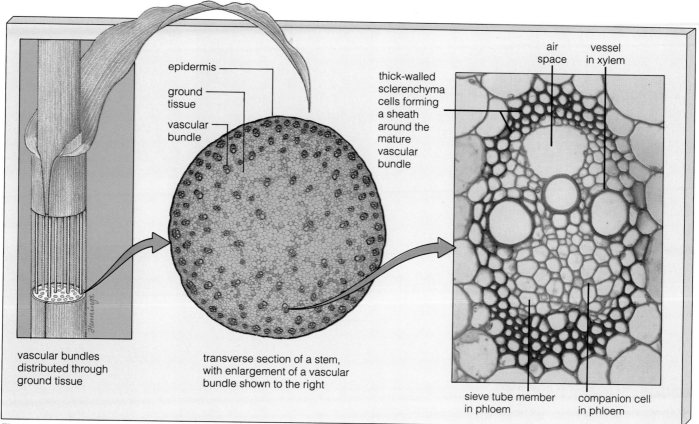

epidermis

ground
tissue

vascular
bundle

thick-walled
sclerenchyma
cells forming
a sheath
around the
mature
vascular
bundle

air
space

vessel
in xylem

sieve tube member
in phloem

companion cell
in phloem

vascular bundles
distributed through
ground tissue

transverse section of a stem,
with enlargement of a vascular
bundle shown to the right

**Figure 27.10**   Stem structure of corn, a monocot.

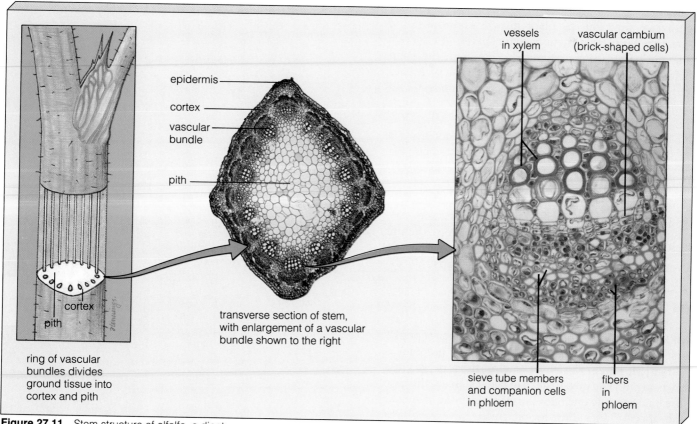

epidermis

cortex

vascular
bundle

pith

vessels
in xylem

vascular cambium
(brick-shaped cells)

cortex

pith

ring of vascular
bundles divides
ground tissue into
cortex and pith

transverse section of stem,
with enlargement of a vascular
bundle shown to the right

sieve tube members
and companion cells
in phloem

fibers
in
phloem

**Figure 27.11**   Stem structure of alfalfa, a dicot.

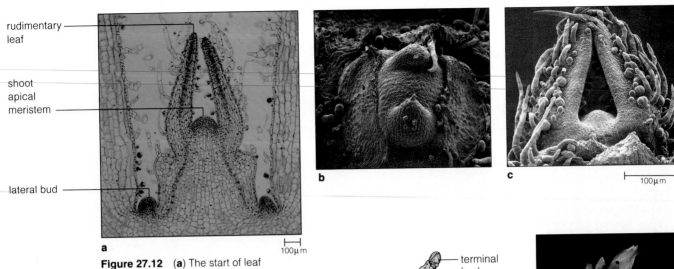

rudimentary leaf

shoot apical meristem

lateral bud

a
100μm

b

c
100μm

**Figure 27.12** (**a**) The start of leaf development at the shoot tip of *Coleus*, as seen in thin section. (**b**, **c**) Scanning electron micrographs of the same type of shoot tip.

## Arrangement of Leaves and Buds

For most vascular plants, **leaves** are the main sites of photosynthesis. Leaves develop on the flanks of the tip of a main stem or its branches. Each starts out as a slight bulge from apical meristem and enlarges into a thin, rudimentary leaf (Figure 27.12). The bulges are close together at first, but as plant growth continues, the leaves that form from them become spaced at intervals along the length of the stem. The point on the stem where one or more leaves are attached is a *node*, and each stem region between two successive nodes is an *internode* (Figure 27.13).

Look at a twig of a walnut tree in winter, when it is devoid of leaves (Figure 27.13a). At the shoot tip is a **bud**, an undeveloped shoot of mostly meristematic tissue, often protected by a covering of modified leaves. In addition to this "terminal bud," other buds occur at regular intervals along the stem. These "lateral buds" form in the upper angle where a leaf is attached to the stem.

Buds give rise to leaves, flowers, or both (Figure 27.13b through d). Depending on the species, each node has one, two, or three or more leaves and buds.

## Leaf Structure

**Leaf Shapes**. Many dicot leaves, such as those of poplar trees, have a broad blade attached by a stalk (petiole) to the stem. Most monocot leaves, such as those of rye grass or corn, are not like this. Instead of being stalked, the base of the blade simply encircles the stem, forming a sheath. Some species, including locust trees, have *compound leaves*. This means the leaf blade is deeply lobed or

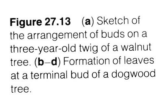

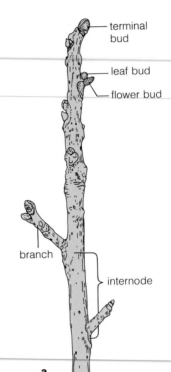

terminal bud

leaf bud

flower bud

branch

internode

a

b

c

d

**Figure 27.13** (**a**) Sketch of the arrangement of buds on a three-year-old twig of a walnut tree. (**b**–**d**) Formation of leaves at a terminal bud of a dogwood tree.

# Commentary

## Uses and Abuses of Leaves

Imagine a meal without leaves. Our plates would not be graced with lettuces, parsley, spinach, or chard; our senses would not be tweaked by splendid spices and seasonings from the leaves of parsley, sage, rosemary, and thyme. There would be no more of the teas brewed from a variety of leaves.

But leaves have uses beyond the table. Oils from the leaves of orange trees and lavender find their way into perfumes and scented soaps. Oils of eucalyptus, camphor, and other leaves find their way into medicine chests. Doctors prescribe drugs isolated from belladonna, an extract of leaves of nightshade plants. They use digitalis, isolated from foxglove leaves, to help regulate heartbeat and blood circulation. They treat skin that has been damaged from the sun or other forms of radiation with juices from the leaves of *Aloe vera*.

Landscape architects and imaginative homeowners use the leaves of different trees and shrubs much as an artist uses different colors and textures. Ecologically aware people plant shrubs next to buildings and so help keep them from heating up in summer and losing heat in winter. We get twine and rope from leaves of century plants (*Agave*), cords and textiles from leaf fibers of Manila hemp, hats from leaves of Panamanian palms, and thatched roofs from the leaves of palms and grasses.

We kill cockroaches, fleas, lice, and flies with insecticides derived from Mexican cockroach plants. We kill over a hundred kinds of insects, mites, and nematodes with extracts of Neem tree leaves—without killing off the natural predators of all those far-too-common pests.

Some people also smoke, chew, or tuck into their mouth the leaves of tobacco plants—and so became candidates for the hundreds of thousands of deaths each year from lung, mouth, and throat cancers. Cocaine, derived from leaves of the coca plant, is used medicinally and abused by increasing numbers of individuals—with devastating social and economic effects, as described on page 562 and in the *Commentary* on page 576.

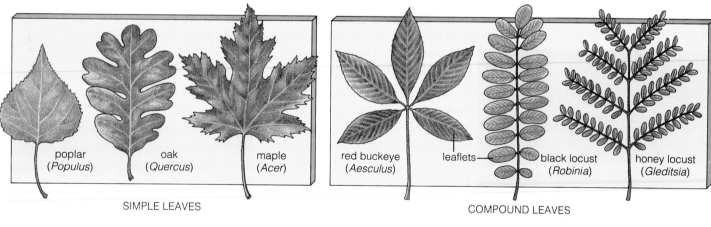

SIMPLE LEAVES          COMPOUND LEAVES

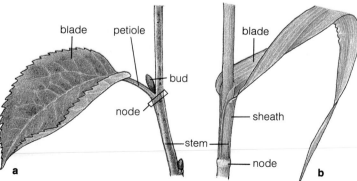

**Figure 27.14** Common forms of (**a**) dicot and (**b**) monocot leaves. The boxes show a few simple and compound leaves.

divided into smaller leaflets, and each leaflet may have its own tiny stalk (Figure 27.14). By contrast, *simple leaves* are not divided in such ways. There are many variations on these basic leaf plans. For example, some leaves have hairs and scales, others have hooks that impale predators.

In temperate regions, most leaves are short-lived. In "deciduous" species such as birches, the leaves drop away from the stem as winter approaches. Species such as camellias also drop leaves, but they appear "evergreen" because the leaves do not all drop at the same time.

**Root Cap.**  A root tip has a dome-shaped cell mass, the *root cap*. Root apical meristem produces the cell mass and in turn is protected by it. The root cap is pushed forward as the root grows, and some of its cells are torn loose. The slippery remnants lubricate the cap.

**Root Epidermis.**  Behind the root cap, the epidermis, ground tissue, and vascular tissues form. *Root epidermis* is the absorptive interface with the environment. Some of the epidermal cells send out long extensions called **root hairs** (Figures 27.18 and 28.2). Root hairs greatly increase the surface available for taking up water and solutes. That is why gardeners learn never to yank a plant out of the ground when transplanting it. Too much of the fragile absorptive surface would be torn off.

**Vascular Cylinder.**  Most often, the vascular tissues of a root are arranged as a central column, or **vascular cylinder**. Ground tissues called the root cortex surround the cylinder (Figure 27.19). In corn and some other species, the vascular tissues are arranged as a ring that divides the ground tissue into cortex and pith (Figure 27.20). Abundant air spaces in the ground tissue allow

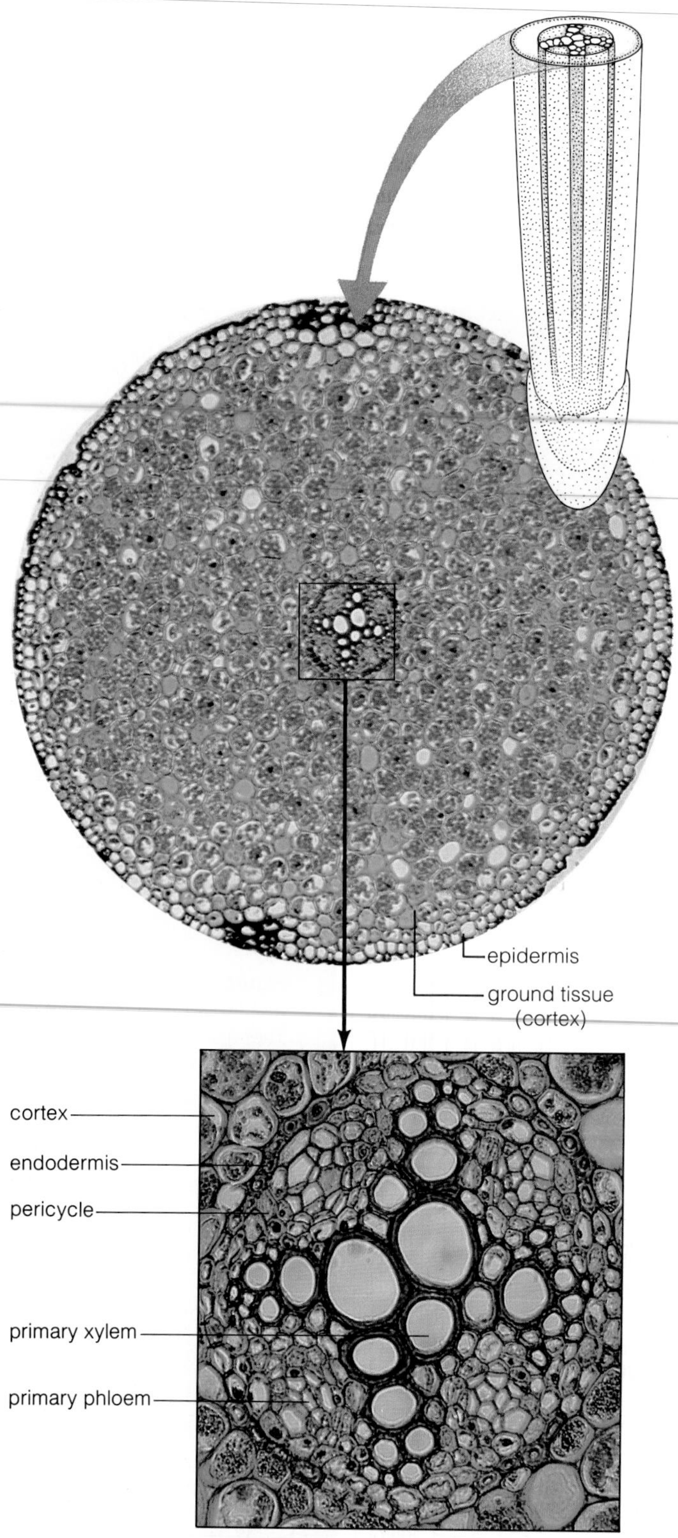

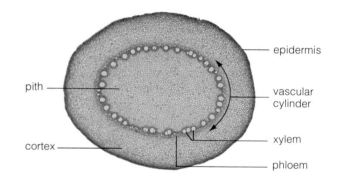

**Figure 27.20**  Transverse section through the root of a corn plant (*Zea mays*). Notice the division of the ground tissue into cortex and pith.

epidermis

ground tissue (cortex)

cortex
endodermis
pericycle
primary xylem
primary phloem

**Figure 27.19**  Section through a young root from a buttercup (*Ranunculus*). The closeup shows details of its vascular cylinder.

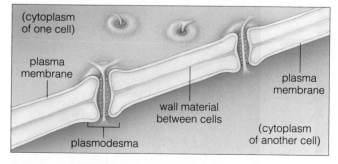

**Figure 27.21**  Plasmodesmata—channels that cross the wall material between plant cells. Such channels provide direct links between the cytoplasm of adjacent cells.

oxygen to reach the living root cells, which depend on oxygen for aerobic respiration. Also, many cell junctions connect the cytoplasm of adjacent cells of the cortex. Figure 27.21 shows an example of this type of junction, which is called a *plasmodesma* (plural, *plasmodesmata*).

Water entering the root moves from cell to cell until it reaches the *endodermis*, the innermost part of the root cortex. The endodermis is a sheetlike layer, one cell thick, around the vascular cylinder. As you will see in the next chapter, the abutting walls of endodermal cells are cemented together and help control the movement of water and dissolved minerals into the vascular cylinder.

Just inside the endodermis is the *pericycle*. This part of the vascular column has one or more layers of cells that give rise to **lateral roots**, which grow out through the cortex and epidermis (Figure 27.22).

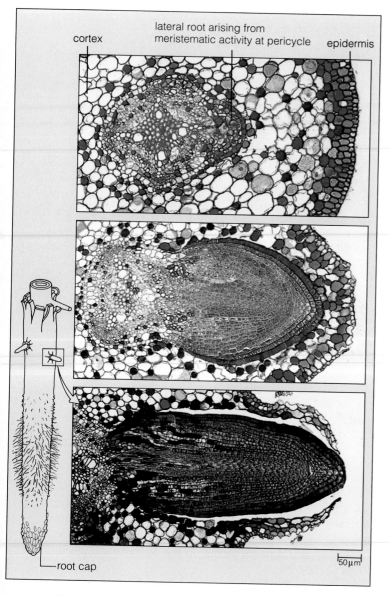

cortex

lateral root arising from meristematic activity at pericycle

epidermis

root cap

50 μm

**Figure 27.22**  Lateral root formation in a willow (*Salix*).

# WOODY PLANTS

## Herbaceous and Woody Plants Compared

The life cycle of flowering plants extends from seed germination to seed formation, then eventual death. Most monocots and some dicots show little or no secondary growth during their life cycle. They are nonwoody, or "herbaceous," plants. In contrast, many dicots and all gymnosperms show secondary growth during two or more growing seasons; they are "woody" plants. Herbaceous and woody plants are characterized as follows:

annuals
: *Life cycle completed in one growing season; little (if any) secondary growth. Examples: snap beans, corn, marigolds.*

biennials
: *Life cycle completed in two growing seasons (root, stem, leaf formation the first season; flowering, seed formation, death the second). Example: carrots.*

perennials
: *Vegetative growth and seed formation continue year after year. Some have secondary tissues, others do not. Examples: woody shrubs (roses), vines (ivy, grape), and trees (apples, elms, magnolias).*

Figure 27.23 shows the structure of a tree trunk—an older woody stem that has undergone extensive secondary growth. Notice how the living phloem is confined to a thin zone just beneath the corky surface. Stripping off a band of phloem all the way around a tree's circumference, an activity called "girdling," will kill the tree. When the vertical phloem pipelines are broken, there is no way to transport photosynthetically derived food to the roots, which eventually die.

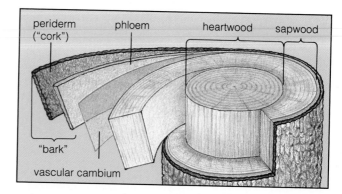

periderm ("cork")     phloem     heartwood     sapwood

"bark"

vascular cambium

**Figure 27.23**  Structure of a woody stem showing extensive secondary growth. Heartwood, the core of the mature tree, is devoid of living cells. Sapwood is the cylindrical zone of xylem between the heartwood and vascular cambium; it contains some living parenchyma cells among the nonliving vessels and tracheids. Everything outside the vascular cambium is often called "bark." Everything inside it is called wood.

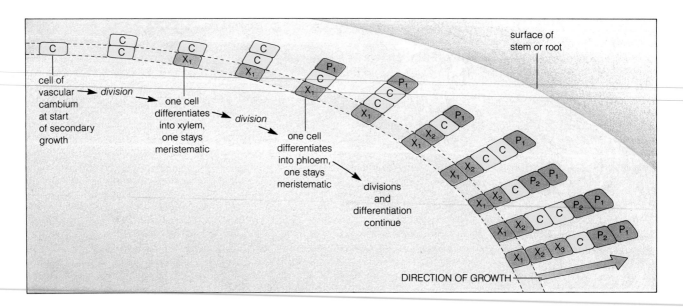

**Figure 27.24** Relationship between the vascular cambium and its derivative cells (secondary xylem and phloem). This is a composite drawing of growth in a stem through successive seasons. Notice how the ongoing divisions displace the cambial cells, moving them steadily outward even as the core of xylem increases the stem or root thickness.

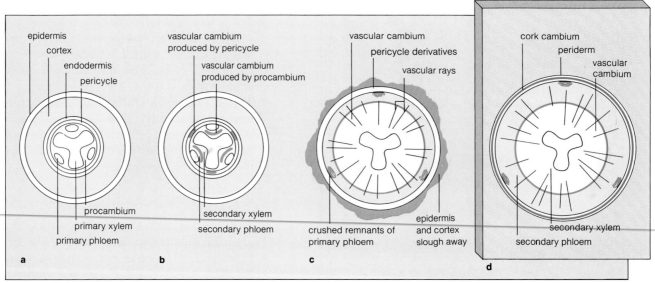

**Figure 27.25** Secondary growth in a dicot root, transverse section. (**a**) Arrangement of tissues at the end of primary growth. (**b**, **c**) Formation of a complete ring of vascular cambium. The vascular cambium gives rise to secondary xylem and phloem. The root increases in diameter as cell divisions proceed parallel to the vascular cambium. The cortex ruptures as the tissue mass increases. (**d**) Epidermis is replaced by periderm, which arises from cork cambium.

## Tissue Formation During Secondary Growth

How do older stems and roots of plants become more massive and woody? They do so through the activity of two types of lateral meristems. These meristems are **vascular cambium** and **cork cambium**. When fully developed, vascular cambium is like a cylinder, one or a few cells thick (Figure 27.8). Its meristematic cells give rise to secondary xylem and phloem tissues that conduct water up, down, and horizontally through the enlarging stem or root. Xylem forms on the inner face of the vascular cambium, and phloem forms on the outer face (Figures 27.24 and 27.25).

The mass of xylem increases season after season, and it usually crushes the thin-walled phloem cells from the

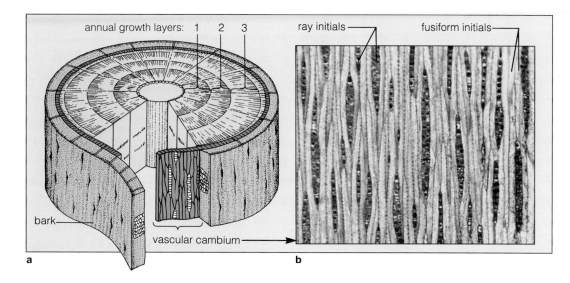

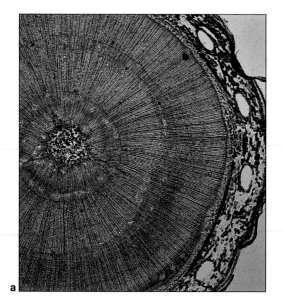

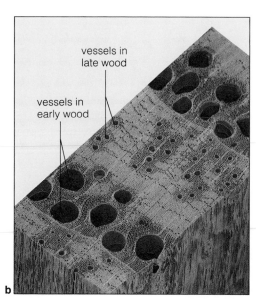

Figure 27.26 (a) Location of vascular cambium in an older stem showing secondary growth. Some of the meristematic cells of the vascular cambium (the fusiform initials) produce the secondary xylem and phloem that conduct water and food vertically through the stem. Other meristematic cells (ray initials) produce parenchyma and other cells that serve as lateral channels for water and food, and as food storage centers. (b) Vascular cambium from the trunk of an apple tree (Malus).

Figure 27.27 (a) Annual growth layers, or "tree rings," in the trunk of a pine (Pinus), a conifer. (b) Scanning electron micrograph of the pattern of annual growth of red oak (Quercus rubra), a woody dicot.

preceding growth period. New phloem cells are added each year, outside the growing inner core of xylem.

In time, the mass of new tissue inside a stem or root causes the cortex and outer phloem to rupture. Parts of the cortex split away and carry epidermis with them. But cork cambium is produced by meristematic cells. In turn, the cork cambium produces the periderm, a corky replacement for the lost epidermis. "Cork" is not the same as "bark," which refers to all living and nonliving tissues between the vascular cambium and the stem or root surface (Figure 27.26).

### Early and Late Wood

In regions having prolonged dry spells or cool winters, the vascular cambium of a woody plant's stems and roots becomes inactive during parts of the year. The first xylem cells produced at the start of the growing season tend to have large diameters and thin walls; they represent **early wood** (Figure 27.27). As the season progresses, the cell diameters become smaller and the walls thicker. These cells represent the **late wood**.

The last-formed, small-diameter cells of late wood end up next to the first-formed, large-diameter cells of the next season's growth. When you look at a full-diameter slice from an old tree trunk, you won't see the individual cells. But early and late wood reflect light differently, and it is possible to identify them as alternating light and dark bands. The alternating bands represent annual growth layers, or "tree rings" (Figure 27.27).

In some wet tropical regions, the growing season is continuous. Conditions in other tropical regions allow several spurts of growth during one year. As a result, the growth layers of many tropical plants are faint, nonexistent, or do not correspond to a single year's growth.

**Figure 27.28** Summary of primary and secondary growth during the development of a stem from a vascular plant.

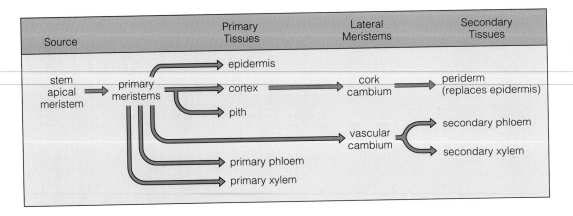

**Figure 27.29** Summary of primary and secondary growth during the development of a root from a vascular plant.

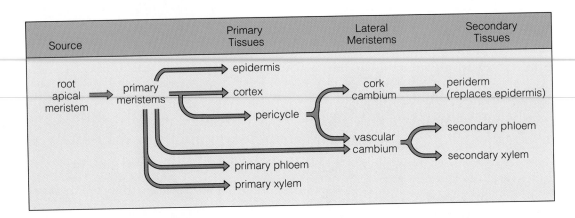

## SUMMARY

1. There are two classes of flowering plants, or angiosperms, that are informally called monocots and dicots. The root systems and shoot systems (stems, leaves) of monocots and dicots are composed of three kinds of tissues (dermal, ground, and vascular tissues).

2. Dermal tissues include epidermis, which covers and protects the surfaces of primary plant parts; and periderm, which replaces epidermis on plants showing secondary growth.

3. There are three types of ground tissues. One (parenchyma) makes up the bulk of fleshy plant parts; its generally thin-walled, living cells function in photosynthesis, storage, and other tasks. The other two types (collenchyma and sclerenchyma) have cells with thickened walls, and both provide mechanical support for growing plant parts.

4. Vascular tissues include xylem and phloem. Xylem contains cells (actually the water-permeable walls of cells dead at maturity) that interconnect to form pipelines for conducting water and dissolved minerals. Phloem, the food-conducting tissue, contains living cells joined end to end to form tubes. The cytoplasm of all adjacent living cells is interconnected at cell junctions (plasmodesmata).

5. Plant growth originates at meristems, as summarized in Figures 27.28 and 27.29.

6. Primary growth, in which roots and shoots elongate, originates at apical meristems, the undifferentiated tissue of self-perpetuating cells at root and shoot tips. Descendants of some cells of the apical meristem produce the primary tissues. Lateral meristems (vascular cambium and cork cambium) increase the diameter of stems and roots of plants showing secondary growth.

7. Stems give photosynthetic tissues of leaves favorable exposure to light and display the flowers. Their vascular tissues distribute substances to and from roots, leaves, and other plant parts. Monocot stems have vascular bundles distributed throughout the ground tissue; dicot stems have vascular bundles arrayed as a cylinder that separates the ground tissue into cortex and pith.

8. Photosynthetic parenchyma cells are organized between the upper and lower epidermis of leaves, with abundant air spaces around them. Numerous openings (stomata) in the lower epidermis allow water vapor and gases to move across the waxy cuticle.

9. Roots absorb water and dissolved minerals from the surroundings and conduct them to aerial plant parts; they anchor and sometimes support the plant and often have food storage regions.

## Review Questions

1. List some of the functions of the root system and the shoot systems of a flowering plant. *470*

2. What are the three main types of tissues in flowering plants, and what are the functions of each? *470–473*

3. What are some of the differences between monocots and dicots? Which of the following stem sections is typical of most monocots? Of most dicots? Label the main tissue regions of each: *474*

 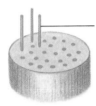

4. Where are apical meristems and lateral meristems located, and what kinds of tissues originate from them? *473, 482–484*

5. How are annual growth layers formed in woody stems? If you were to strip away a band of phloem all the way around a tree's circumference, what would happen to the tree? Why? *481, 483*

## Self-Quiz *(Answers in Appendix IV)*

1. In vascular plants, _____ tissues cover and protect the root and shoot systems; _____ tissues make up the greater portion of any plant body; and strands of _____ tissues conduct food, water, and minerals.

2. The three basic kinds of plant tissues arise from cell division and elongation in regions called _____ meristems at the tips of roots and shoots. Increases in root or stem diameter occur at _____ meristems.

3. The two classes of flowering plants are _____.
   a. angiosperms and gymnosperms    c. shrubs and trees
   b. monocots and dicots    d. herbs and shrubs

4. One type of ground tissue, _____, makes up the bulk of fleshy plant parts.
   a. parenchyma    c. collenchyma
   b. sclerenchyma    d. epidermis

5. _____ cells are thin-walled, alive at maturity, and function in photosynthesis, storage, and other tasks.
   a. Parenchyma    c. Collenchyma
   b. Sclerenchyma    d. Epidermis

6. Cells of the ground tissues _____ and _____ have thick walls and help mechanically to support the plant parts.
   a. parenchyma; collenchyma
   b. collenchyma; sclerenchyma
   c. parenchyma; sclerenchyma
   d. parenchyma; epidermis

7. _____ is a vascular tissue that conducts water and minerals; _____ is a vascular tissue that conducts food.
   a. Phloem; xylem    c. Xylem; phloem
   b. Vascular cambium; phloem    d. Xylem; vascular cambium

8. Herbaceous annual plants show little or no _____ growth; perennial plants show considerable _____ growth.
   a. secondary; secondary    c. secondary, primary
   b. primary; secondary    d. primary; primary

9. _____ stems have vascular bundles scattered throughout the ground tissue; _____ stems have vascular bundles arranged as a cylinder that separates ground tissue into cortex and pith.
   a. Dicot; dicot    c. Monocot; monocot
   b. Dicot; monocot    d. Monocot; dicot

10. Match these plant parts with their structure or function.

| | |
|---|---|
| ___ roots | a. expose photosynthetic tissues to light, display flowers; distribute materials to and from roots, leaves |
| ___ leaves | |
| ___ dermal tissues | |
| ___ stems | |
| ___ ground tissues | b. contain xylem and phloem |
| ___ vascular tissues | c. absorb, conduct water and minerals, anchor the plant, store food |
| | d. have photosynthetic cells between upper, lower epidermis |
| | e. contain parenchyma, collenchyma, sclerenchyma |
| | f. contain epidermis, periderm |

## Selected Key Terms

| | | |
|---|---|---|
| adventitious root *479* | lateral meristem *473* | root system *470* |
| angiosperm *469* | lateral root *481* | sclerenchyma *470* |
| apical meristem *473* | leaf *476* | shoot system *470* |
| bud *476* | lignification *471* | sieve tube member *472* |
| collenchyma *470* | monocot *474* | spongy mesophyll *478* |
| cork cambium *482* | node *476* | stoma *473* |
| cortex *474* | palisade mesophyll *478* | taproot system *479* |
| cuticle *473* | parenchyma *470* | vascular bundle *474* |
| dicot *474* | pericycle *481* | vascular cambium *482* |
| endodermis *481* | periderm *473* | |
| epidermis *473* | phloem *472* | vascular cylinder *480* |
| fibrous root system *479* | pith *474* | |
| | plasmodesma *481* | vein *478* |
| gymnosperm *469* | root cap *480* | xylem *472* |
| internode *476* | root hair *480* | |

## Readings

Bold, H., C. Alexopoulos, and T. Delevoryas. 1987. *Morphology of Plants and Fungi*. Fifth edition. New York: Harper & Row.

Raven, P., R. Evert, and S. Eichhorn. 1986. *Biology of Plants*. Fourth edition. New York: Worth. Exquisite color micrographs and illustrations of plant cells and tissues.

Rost, T., et al. 1984. *Botany: An Introduction to Plant Biology*. Second edition. New York: Wiley.

Stern, K. 1988. *Introductory Plant Biology*. Fourth edition. Dubuque, Iowa: Brown. Beautifully illustrated, accessible introduction to plant structure and function. Paperback.

# 28 PLANT NUTRITION AND TRANSPORT

## Flies for Dinner

It took you eighteen years or so to grow to your present height. A corn plant grows more than that in three months! Yet how often do we stop to think that plants actually do anything impressive? Being endowed with great mobility, intelligence, and rich emotions, we tend to be fascinated more with ourselves than with immobile, expressionless plants.

Plants don't just stand around soaking up sunlight. Like all other organisms, they have their own intricate adaptations to particular environments—adaptations that allow them to secure energy and raw materials, then juggle both in ways that assure their growth and development.

Consider a flytrap that evolved long before we ever thought about constructing one. Soggy soils in North and South Carolina are home to the Venus flytrap, a plant bearing an unsettling resemblance to the devices humans have used to trap animals. As Figure 28.1a shows, its two-lobed leaves look almost like they are hinged at the blade. They open and close much like a steel trap. And stiff spines fringing the leaf margins intermesh when the lobes come together, forming a cagelike structure.

Unless Venus flytraps obtain nitrogen and other minerals, they can't grow. Nitrogen is a component of all amino acids, the building blocks of proteins—and nothing much happens in *any* cell without enzymes and other proteins. Minerals are leached out of the soggy soils where these plants grow and so are scarce. But flying insects are abundant. For the Venus flytrap they are, in a manner of speaking, movable feasts—ready sources of nitrogen.

Sticky, sugary secretions ooze out of some glands and glisten on the leaf. This stuff is the bait—and part of the snare. Insects landing on a leaf might start to feed, but also present on the leaf are tiny hairlike projections that serve as triggers for the trap. If the insect touches any two hairs at the same time, even if it touches the same hair twice in rapid succession, the leaf snaps shut. The movement occurs when epidermal cells on the underside swell rapidly (it takes fully a third of their ATP stores to make this happen). The leaf will open only when epidermal cells on the inner face of the leaf do the same thing.

And now a solution that contains digestive enzymes starts pouring out of pincushion-shaped cells at the leaf

**Figure 28.1** Two "carnivorous" plants that turn the dinner table on animals. **(a)** A Venus flytrap (*Diondea muscipula*) and **(b)** a bladderwort (*Utricularia*).

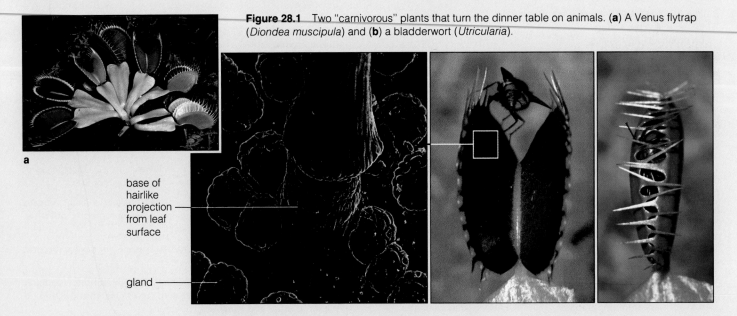

base of
hairlike
projection
from leaf
surface

gland

a

**b**

surface (Figure 28.1b). The solution forms a pool around the insect, the insect's minerals dissolve in the pool—and plant cells absorb the mineral-enriched liquid.

The Venus flytrap is one of several kinds of plants that are called carnivorous ("meat-eating"), although it takes a flying leap of the imagination to put their extracellular digestion and absorption in the same category as the chompings of lions, dogs, and similar meat eaters. Bladderworts, which grow in shallow waters of lakes and streams, are also carnivorous plants (Figure 28.1b). They are decked out with hollow, pear-shaped bladders, each with an opening guarded by a bristle-fringed door. When a small aquatic insect brushes against the bristles, the door springs open and water rushes into the bladder—carrying the insect with it. When the door snaps shut, cells in the bladder wall secrete digestive enzymes, as do bacteria that have taken up residence in the traps.

With this chapter we turn to adaptations by which land plants function in the environment. You know from previous chapters that plants generally are photosynthetic autotrophs; they require only sunlight, water, carbon dioxide, and some minerals to nourish themselves. Yet plants, like people, do not have unlimited supplies of all necessary resources. The air is only 1 part carbon dioxide to 350 million parts of everything else. Unlike the wet homes of Venus flytraps, most soils are frequently dry. And nowhere except in overfertilized gardens does soil water hold lavish amounts of dissolved minerals. As you will see, *many aspects of plant structure and function are responses to low concentrations of environmental resources.*

### KEY CONCEPTS

**1.** Many aspects of plant structure and function are adaptive responses to low concentrations of water, minerals, and other environmental resources.

**2.** Leaves have passageways (stomata) across the epidermis. Most plants conserve water by closing their stomata at night. They lose water during the day, when stomata remain open and so allow carbon dioxide (used in photosynthesis) to move into the leaves.

**3.** Plants have specialized mechanisms for water transport. Water is pulled "uphill" from roots to aerial plant parts. It moves through xylem as a result of evaporation from plant parts (transpiration), combined with hydrogen bonding between water molecules.

**4.** Plants transport sucrose and other organic compounds by a specialized mechanism called translocation. Transport occurs inside a system of interconnecting sieve tubes that extends throughout the plant. Translocation results from differences in sieve-tube turgor pressure where compounds are loaded into the system and where they are taken out.

## NUTRITIONAL REQUIREMENTS

No plant, including the Venus flytrap, can grow normally when it is deprived of any element that is essential for its metabolism. Generally speaking, plants require sixteen essential elements. Three of them—oxygen, carbon, and hydrogen—are used as the main building blocks in the synthesis of carbohydrates, lipids, proteins, and nucleic acids. Plants obtain those three elements from water ($H_2O$), and from gaseous oxygen ($O_2$) and carbon dioxide ($CO_2$) in the air.

The other essential elements become available to plants as dissolved salts; they are "mineral ions." Several of the mineral ions are *macronutrients*, meaning a significant fraction of each becomes incorporated in plant tissues. The rest are *micronutrients*; only small traces occur in plant tissues. Micronutrients as well as macronutrients play vital roles in photosynthesis and

## Table 28.1 Essential Elements for Most of the Complex Land Plants

| Element | Symbol | Form Available to Plants | Percent by Weight in Dry Tissue | |
|---|---|---|---|---|
| Carbon | C | $CO_2$ | 45 | |
| Oxygen | O | $O_2$, $H_2O$, $CO_2$ | 45 | 96% of total dry weight |
| Hydrogen | H | $H_2O$ | 6 | |
| Nitrogen | N | $NO_3^-$, $NH_4^+$ | 1.5 | |
| Potassium | K | $K^+$ | 1.0 | |
| Calcium | Ca | $Ca^{++}$ | 0.5 | |
| Magnesium | Mg | $Mg^{++}$ | 0.2 | |
| Phosphorus | P | $H_2PO_4^-$, $HPO_4^{--}$ | 0.2 | |
| Sulfur | S | $SO_4^{--}$ | 0.1 | |
| Chlorine | Cl | $Cl^-$ | 0.010 | |
| Iron | Fe | $Fe^{++}$, $Fe^{+++}$ | 0.010 | |
| Copper | Cu | $Cu^+$, $Cu^{++}$ | 0.006 | |
| Boron | B | $H_3BO_3$ | 0.002 | |
| Manganese | Mn | $Mn^{++}$ | 0.0050 | |
| Zinc | Zn | $Zn^{++}$ | 0.0020 | |
| Molybdenum | Mo | $MoO_4^-$ | 0.00001 | |

other metabolic events. Both categories of mineral elements contribute to solute concentration gradients, which are necessary for moving substances into and out of cells.

Table 28.1 lists the sixteen elements that plants require. Table 28.2 lists which mineral ions are macronutrients and micronutrients, some of their known functions, and some observable symptoms of mineral deficiencies.

## UPTAKE OF WATER AND NUTRIENTS

The availability of water and dissolved mineral ions profoundly affects root development, and root development affects the growth of the entire plant. The first roots branch out through some areas of the surrounding soil. Then, as soil conditions change, new roots that branch out into different areas replace them. It is not that the roots "explore" soil for resources. Rather, those areas having greater concentrations of water and mineral ions provide greater stimulation for outward root growth.

## Table 28.2 Role of Mineral Elements in Plant Function

| Macronutrient | Some Known Functions | Some Deficiency Symptoms | Micronutrient | Some Known Functions | Some Deficiency Symptoms |
|---|---|---|---|---|---|
| Nitrogen | Component of proteins, nucleic acids, coenzymes, chlorophyll | Stunted growth; light green older leaves; older leaves yellow and die (chlorosis) | Chlorine | Role in root, shoot growth; role in photolysis | Wilting; chlorosis; some leaves die |
| Potassium | Activation of enzymes, role in maintaining water-solute balance* and thus affecting osmosis | Reduced growth; curled, mottled, or spotted older leaves; burned leaf margins; weakened roots and stems | Iron | Roles in chlorophyll synthesis, electron transport | Chlorosis; yellow and green striping in grasses |
| Calcium | Roles in cementing cell walls, regulation of many cell functions | Leaves deformed; terminal buds die; reduced root growth | Boron | Roles in flowering, germination, fruiting, cell division, nitrogen metabolism | Terminal buds, lateral branches die; leaves thicken, curl, become brittle |
| Magnesium | Component of chlorophyll; activation of enzymes | Chlorosis; drooped leaves | Manganese | Role in chlorophyll synthesis; coenzyme activity | Light green leaves with green major veins; leaves whiten and fall off |
| Phosphorus | Component of nucleic acids, phospholipids, ATP | Purplish veins in older leaves; fewer seeds and fruits; stunted growth | Zinc | Role in formation of auxin, chloroplasts, and starch; enzyme component | Chlorosis; mottled or bronzed leaves; abnormal roots |
| Sulfur | Component of most proteins, two vitamins | Light green or yellow leaves; reduced growth | Copper | Component of several enzymes | Chlorosis; dead spots in leaves; stunted growth; terminal buds die |
| | | | Molybdenum | Component of enzyme used in nitrogen metabolism | Possible nitrogen deficiency; pale green, rolled or cupped leaves |

*All mineral elements contribute to the water-solute balance, but potassium is notable because there is so much of it.

## Root Nodules

Many flowering plants take up nutrients with the help of other organisms, which get something from the plants in return. Such associations are examples of *mutualism*. The two species interact on a permanent basis in a mutually beneficial way (page 806).

Consider the type of mutualistic relationship that helps legumes secure nitrogen. Legumes include string beans, soybeans, peas, alfalfa, clover, and other economically important plants. In many agricultural regions, harvests suffer from nitrogen scarcity. There actually is plenty of gaseous nitrogen (N≡N) in the air, but plants do not have the metabolic means to break apart the three covalent bonds in each molecule. High yields of commercial crop plants depend on applications of nitrogen-rich fertilizers or on the activity of "nitrogen-fixing" bacteria in the soil. As described in more detail in Chapter 46, those bacteria convert nitrogen to forms that they—and the plants—can use.

Legumes have an advantage in this respect. Nitrogen-fixing bacteria actually reside in localized swellings, or **root nodules**, on the plant roots (Figure 28.2). The bacteria feed on some of the plant's photosynthetically produced organic molecules. But they also provide the plant with usable nitrogen.

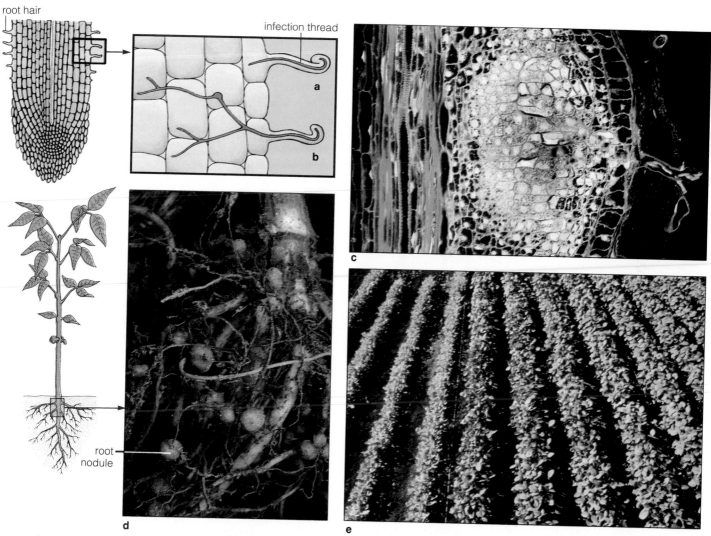

**Figure 28.2** Nutrient uptake at root nodules of legumes, which interact mutualistically with nitrogen-fixing bacteria (*Rhizobium* and *Bradyrhizobium*). (**a**) When bacterial cells infect slender extensions of root epidermal cells (root hairs), the cells are induced to form an "infection thread" of cellulose deposits. (**b**) The bacteria use the thread like a highway to invade cells in the root cortex, some of which are tetraploid. (**c**) When infected, those cells and the bacteria within them divide repeatedly, forming a swollen mass that is destined to become a root nodule. The bacteria start fixing nitrogen when plant cell membranes surround them. The plant takes up some of the nitrogen, and the bacteria take up some compounds produced by the plant. (**d**) Root nodules on a soybean plant. (**e**) Soybean plants growing in nitrogen-poor soil. The plants on the right were inoculated with *Rhizobium* cells and developed root nodules.

## Mycorrhizae

Other structures besides root nodules promote the uptake of water and dissolved mineral ions in many species of flowering plants. **Mycorrhizae** (singular, mycorrhiza) are a prime example. The name means "fungus-root," and it refers to a mutually beneficial association between a fungus and a young root. As shown earlier in Figure 23.15, the fungus often grows as a mat of thin filaments around roots. Collectively, the filaments have a tremendous surface area for absorbing mineral ions from a large volume of soil. The fungus uses some of the plant's sugars and nitrogen-containing compounds. As it grows, the root uses some of the minerals—scarce resources—that the fungus has secured.

In other types of mycorrhizae, the fungus lives *inside* cells of the root cortex. Orchids are among the plants that depend on this mutualistic association.

## Root Hairs

As a final example of specialized absorptive structures, think about **root hairs**, the slender extensions of specialized epidermal cells. As mentioned in the preceding chapter, root hairs greatly increase the surface area available for absorbing water and mineral ions from the soil. A single root system might develop millions or billions of root hairs. Figure 28.3 illustrates their structure.

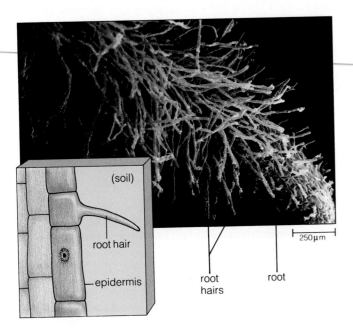

**Figure 28.3** Scanning electron micrograph of root hairs.

250 μm

(soil)

root hair

epidermis

root hairs

root

Root nodules, mycorrhizae, and root hairs are examples of specialized structures that enhance a plant's capacity to absorb water and dissolved mineral ions.

### Controls Over Nutrient Uptake

Refer to Figure 27.19, which shows the internal structure of a typical root. Once water has been absorbed from the surrounding soil, it moves through the root cortex until it reaches a sheetlike layer of single cells wrapped around the vascular cylinder. This cell layer is the **endodermis**. A waxy band, the **Casparian strip**, acts as an impermeable barrier between the walls of abutting endodermal cells. Water cannot cross the strip. As Figure 28.4 shows, water can move into the vascular cylinder only by crossing the plasma membrane of endodermal cells, diffusing through their cytoplasm, then crossing the plasma membrane on the other side. Plasma membranes, recall, selectively permit the movement of some substances but not others across the lipid bilayer. *Membrane transport mechanisms help control the types of absorbed solutes that will become distributed through the plant.*

Recent work has revealed that the roots of most flowering plants also have an **exodermis**, a layer of cells just inside the epidermis. This layer also has a Casparian strip that functions like the one inside the endodermis.

Once nutrients have entered the vascular cylinder, their distribution and uptake in different tissue regions is coordinated in ways that affect growth (Figure 28.5). Living cells throughout the plant take up nutrients by active transport mechanisms at the plasma membrane. Energy from ATP drives the membrane "pumps" by which solutes are moved into cells. The pumps are transport proteins embedded in the plasma membrane (page 85).

In photosynthetic cells, the ATP necessary for the membrane pump operation is formed during both photosynthesis and aerobic respiration. What about nonphotosynthetic cells, such as parenchyma cells in roots? How do they get all the ATP necessary for active transport? In those cells, ATP is formed almost entirely through aerobic respiration.

Control over the uptake of nutrients is exerted at a cell layer (exodermis) near the root surface, then at the endodermis next to the root vascular cylinder, and finally at the plasma membranes of living cells throughout the plant body.

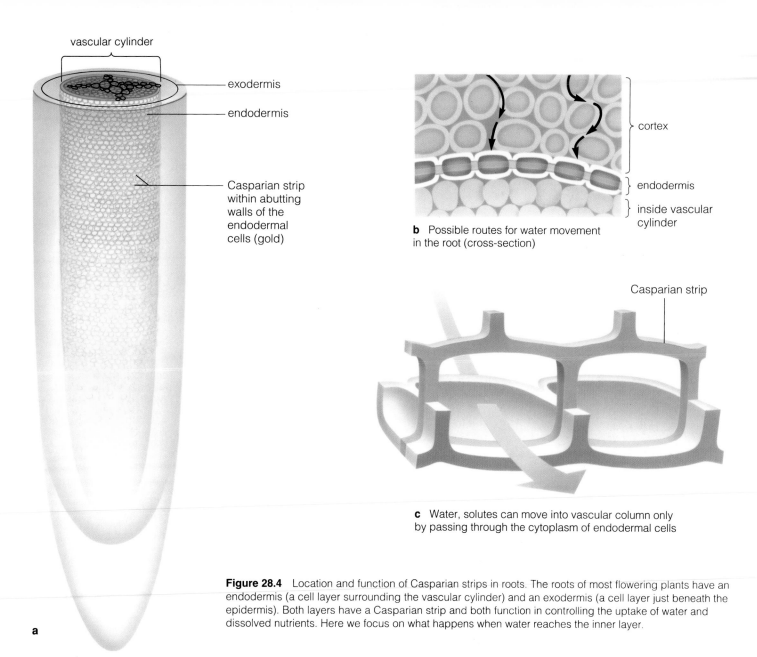

vascular cylinder

exodermis

endodermis

Casparian strip within abutting walls of the endodermal cells (gold)

cortex

endodermis

inside vascular cylinder

**b** Possible routes for water movement in the root (cross-section)

Casparian strip

**c** Water, solutes can move into vascular column only by passing through the cytoplasm of endodermal cells

**Figure 28.4** Location and function of Casparian strips in roots. The roots of most flowering plants have an endodermis (a cell layer surrounding the vascular cylinder) and an exodermis (a cell layer just beneath the epidermis). Both layers have a Casparian strip and both function in controlling the uptake of water and dissolved nutrients. Here we focus on what happens when water reaches the inner layer.

a

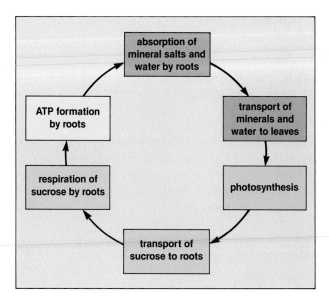

absorption of mineral salts and water by roots

transport of minerals and water to leaves

photosynthesis

transport of sucrose to roots

respiration of sucrose by roots

ATP formation by roots

**Figure 28.5** Interrelated processes that influence the coordinated growth of roots, stems, and leaves. When one process is rapid, the others also speed up. Any environmental factor limiting one process eventually slows growth of all plant parts.

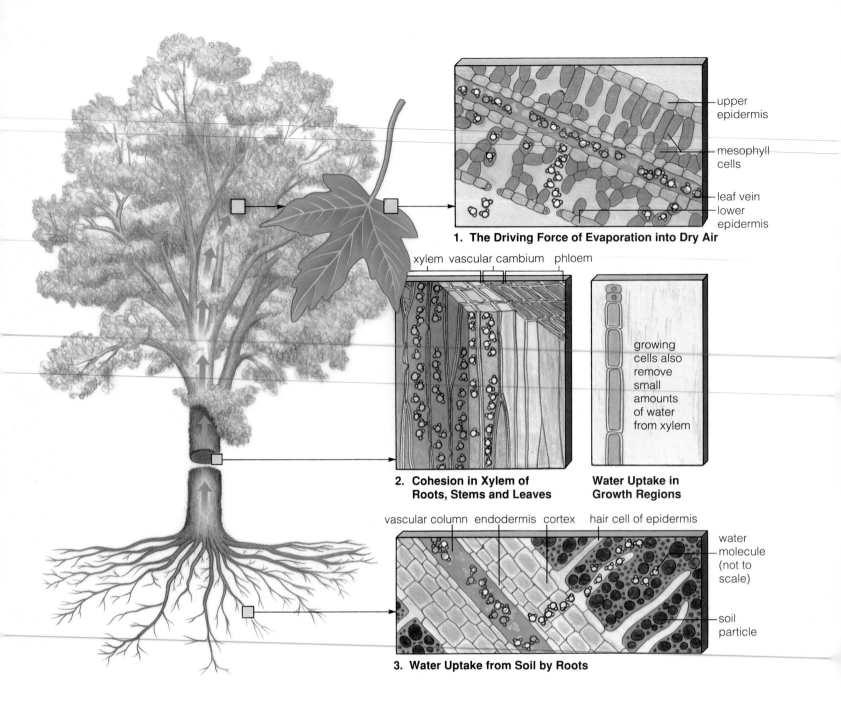

**1. The Driving Force of Evaporation into Dry Air**

upper epidermis

mesophyll cells

leaf vein
lower epidermis

xylem  vascular cambium  phloem

**2. Cohesion in Xylem of Roots, Stems and Leaves**

growing cells also remove small amounts of water from xylem

**Water Uptake in Growth Regions**

vascular column  endodermis  cortex  hair cell of epidermis

water molecule (not to scale)

soil particle

**3. Water Uptake from Soil by Roots**

**Figure 28.6** Cohesion-tension theory of water transport. Tensions in water in the xylem extend from leaf to root. These tensions are caused mostly by transpiration (the evaporation of water from plant parts). As a result of the tensions, columns of water molecules that are hydrogen-bonded to one another are pulled upward. The hydrogen bonding underlies water's cohesive properties (page 28).

## WATER TRANSPORT AND CONSERVATION

### Transpiration

Let's turn now to the actual mechanisms by which water (and the nutrients dissolved in it) moves from roots to stems, then into leaves. A fraction of the water is used in growth and metabolism, but most evaporates into the air. Water evaporation from stems, leaves, and other plant parts is called **transpiration**.

How does water get all the way to the top of plants, including trees that are 50, 80, even 100 meters tall? As shown in Figure 28.6, water moves through conducting cells of the vascular tissue called xylem. Actually, those cells are dead at maturity and only their walls remain, so the cells themselves are not pulling water "uphill." Rather, *water is pulled up by the drying power of air, which creates continuous negative pressures (tensions) that extend downward from the leaves to the roots.*

Transpiration involves a continuum of events. *First,* water evaporates from the walls of photosynthetic cells inside leaves. As water molecules escape, they are replaced by others from the cell cytoplasm. Then water from xylem in the leaf veins replaces the water being lost from the cells. *Second,* when water molecules move out of veins, replacements are pulled in from xylem in the stem. The pulling action puts water inside the xylem in a state of tension. *Third,* replacement water moves into the roots—and more soil water is drawn into the plant, following its osmotic gradient (from higher to lower concentration). This inward movement continues until the soil becomes so dry that an osmotic gradient no longer exists.

When water moves as a continuous, fluid column through xylem pipelines, why doesn't the "stretching" cause the molecules to snap away from each other? Some time ago the botanist Henry Dixon came up with a good explanation, which has since been named the **cohesion-tension theory of water transport:**

**1.** The drying power of air causes *transpiration*, the evaporation of water from leaves and other plant parts exposed to air.

**2.** Transpiration puts the water in xylem in a state of tension that is continuous from leaves, down through the stems, to roots.

**3.** As long as water molecules continue to escape from the plant, the continuous tension in xylem permits more molecules to be pulled up to replace them.

**4.** Columns of water are pulled up by the collective strength of hydrogen bonds between water molecules, which are confined in the narrow, tubular xylem cells.

**5.** Hydrogen bonds are enough to hold water molecules together in the xylem, but they are not strong enough to prevent them from breaking away from each other during transpiration and escaping from the leaves.

## Control of Water Loss

Of the water moving into a leaf, more than 90 percent is usually lost through transpiration. About 2 percent of the water that is conserved in the leaf is used in photosynthesis, membrane functions, and other activities. However, when water loss by transpiration exceeds water uptake by roots, the resulting dehydration of plant tissues will interfere with these water-requiring activities.

Even under mild conditions, plants would rapidly wilt and die if it were not for the *cuticle,* the waxy covering that reduces the rate of water loss from aboveground plant parts (page 473). The cuticle does conserve water, but it also limits the rate of diffusion of carbon dioxide into the leaf.

Transpiration occurs mostly at **stomata** (singular, stoma), the small passageways across the cuticle-covered epidermis of leaves and stems. So does the inward diffusion of carbon dioxide. Figure 28.7 shows how two *guard cells* flank each opening.

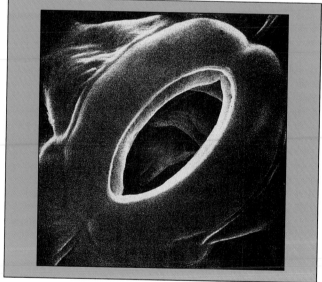

a

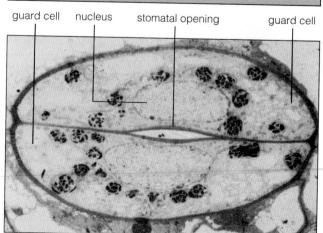

guard cell    nucleus    stomatal opening    guard cell

b

**Figure 28.7** (**a**) From the lower epidermis of a dicot leaf, a stoma, defined by a pair of guard cells. In this scanning electron micrograph, mesophyll cells inside the leaf are visible through the stomatal gap. (**b**) Transmission electron micrograph of the structure of guard cells from the stem of a beavertail cactus (*Opuntia*), thinsection.

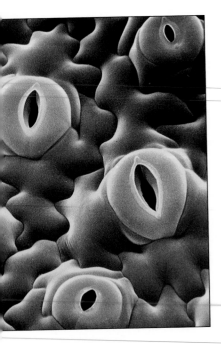

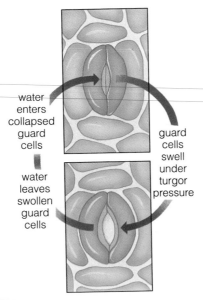

water enters collapsed guard cells

water leaves swollen guard cells

guard cells swell under turgor pressure

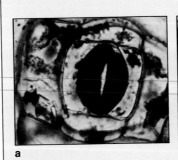

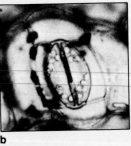

a b

**Figure 28.9** Evidence for potassium accumulation in stomatal guard cells undergoing expansion. Strips from the leaf epidermis of a dayflower (*Commelina communis*) were immersed in solutions containing dark-staining substances that bind preferentially with potassium ions. (**a**) In leaf samples having opened stomata, most of the potassium was concentrated in the guard cells. (**b**) In leaf samples having closed stomata, very little potassium was in guard cells; most was present in normal epidermal cells.

**Figure 28.8** Sketch of stomatal action. The micrograph shows stomata from a tobacco plant (*Nicotiana tabacum*).

Take a look now at Figure 28.8. When two paired guard cells are swollen with water, turgor pressure distorts their shape in such a way that they move apart from each other. Their separation produces a gap between them—the actual stoma. When the water content of the guard cells dwindles, turgor pressure drops and the stoma closes once more.

When stomata are open, carbon dioxide can be absorbed for photosynthesis. But when they are open, and unless the relative humidity is 100 percent, water always moves out! *Stomata must open and close at different times to control water loss and carbon dioxide uptake.*

A stoma opens and closes according to how much water and carbon dioxide are present in the two guard cells flanking it. When the sun comes up, carbon dioxide is used in photosynthesis. Eventually, carbon dioxide levels drop in cells, including the guard cells. In addition, blue wavelengths in the sun's rays act on the guard cells directly. The effect of blue light as well as the decrease in carbon dioxide concentration triggers the active transport of potassium ions into guard cells (Figure 28.9). This is followed by an inward movement of water by osmosis. As water pressure builds up inside, the guard cells become swollen and move apart, producing the stomatal opening. Thus, water vapor is lost and carbon dioxide moves into the leaf during the day.

Photosynthesis stops when the sun goes down, but carbon dioxide accumulates in cells as a by-product of aerobic respiration. Potassium in the guard cells moves

out now, followed by water. The guard cells collapse against each other, closing the gap between them. Thus, transpiration is reduced and water is conserved during the night.

**In most plants, stomata remain open during daylight, when photosynthesis occurs. Water is lost, but carbon dioxide can enter the leaves.**

**Stomata remain closed during the night, when carbon dioxide accumulates through aerobic respiration. Then, water is conserved.**

As long as soil is moist, the stomata of plants growing in it can remain open during daylight. When the soil is dry and the air is also dry and hot, the stomata close or do not open as much, so little water is absorbed and transpired. Although photosynthesis and growth slow as a consequence, the plants still can survive short drought periods. They can do so repeatedly. Briefly, such stressful conditions trigger the production of a plant hormone called **abscisic acid** in roots, and this ends up in leaves. The hormone is synthesized faster when a leaf is water stressed. When abscisic acid accumulates in a leaf, it causes guard cells to give up potassium ions, hence water, so the stomata close.

We see a splendid variation on the water-conservation theme in **CAM plants**, which include cacti and most other succulents. CAM plants open stomata at *night*, when they fix carbon dioxide by way of a special C4 metabolic pathway (compare page 115). The fixed carbon dioxide is used in photosynthesis the next day—when stomata are closed.

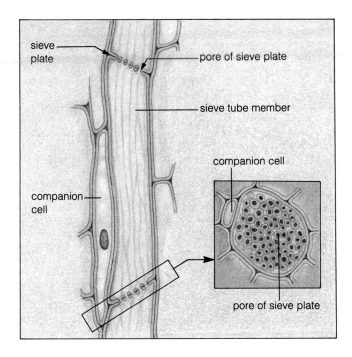

**Figure 28.10** The main cell type in phloem, sliced lengthwise. The boxed inset shows how the wall of this cell, called a sieve tube member, is perforated at plates between adjacent cells. Companion cells metabolically support the sieve tube members and may direct their activities.

**Figure 28.11** Honeydew droplet exuding from the tail end of an aphid feeding on the sugars present in the phloem of a plant.

## TRANSPORT OF ORGANIC SUBSTANCES

### Storage and Transport Forms of Organic Compounds

Sucrose and other organic compounds resulting from photosynthesis are used throughout the plant. Leaf cells use some of the compounds and the rest are transported to roots, stems, buds, flowers, and fruits. Carbohydrates become stored as starch in most plant cells. Quantities of fats become stored in some fruits, including the rich flesh of avocados. Proteins and fats become stored in many seeds.

Starch molecules are too large to cross cell membranes, so they cannot leave the cells in which they are formed. They also are too insoluble to be transported to other regions of the plant body. Fats are largely insoluble in water, and they cannot be transported out of their storage sites. Storage proteins do not lend themselves to transport, either.

Storage forms of organic compounds are converted to transportable forms through specific reactions, including hydrolysis (page 37). For example, hydrolysis of starch liberates glucose units, which combine with fructose. The resulting molecule, sucrose, is the main form in which sugars are transported through the roots, stems, and leaves of most plants.

Storage starch, fats, and proteins are converted to smaller subunits that are soluble and transportable through the plant body.

### Translocation

Sucrose and other organic compounds are distributed through the plant by **translocation**, a process that occurs in the vascular tissue called phloem. As we have seen, the phloem of flowering plants contains interconnecting tubes that are formed by living cells. Figure 28.10 shows one of these cells, which are called **sieve tube members**. The conducting tubes lie end to end within vascular bundles, and they extend through all parts of the plant. Water and organic compounds can flow rapidly through large pores on their end walls. **Companion cells** are nonconducting cells that are adjacent to the sieve tube members. As you will see, companion cells have central roles in translocation.

The familiar small insects called aphids have helped demonstrate that organic compounds flow under pressure in the phloem. An aphid feeds on leaves and stems. It forces its stylet (a mouthpart) into sieve tubes and feeds on the dissolved sugars inside. The contents of the tubes are under high pressure, often five times as much as in an automobile tire. This pressure forces the fluid through the aphid gut and out the other end as "honeydew" (Figure 28.11). Park your car under trees being attacked by aphids and it might get a spattering of sticky honeydew droplets, thanks to high fluid pressure in the phloem.

In some experiments, feeding aphids were anesthetized by exposure to high concentrations of carbon dioxide. Then their bodies were severed from their stylets, which were left embedded in sieve tubes that the aphids had been feeding upon. Analysis of the fluid being forced out of the tubes verified that in most plant species, sucrose is the main carbohydrate being translocated under pressure through the phloem.

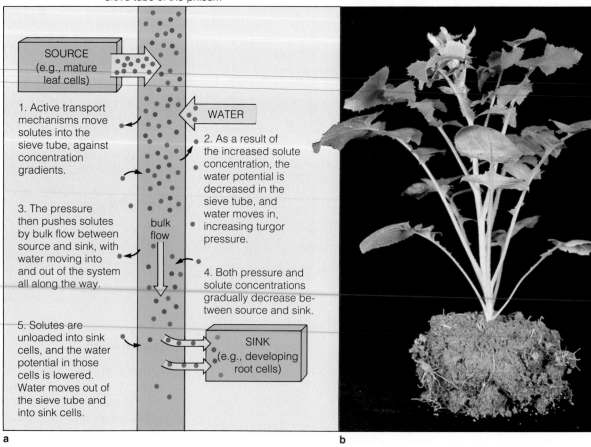

**Figure 28.12** (**a**) Proposed mechanism of pressure flow in the phloem of flowering plants. (**b-e**) Translocation in *Sonchus*, commonly called sow thistle. Compare these illustrations with Figure 7.2 to see how translocation is linked with the process of photosynthesis.

Sieve tube of the phloem

SOURCE (e.g., mature leaf cells)

1. Active transport mechanisms move solutes into the sieve tube, against concentration gradients.

WATER

2. As a result of the increased solute concentration, the water potential is decreased in the sieve tube, and water moves in, increasing turgor pressure.

3. The pressure then pushes solutes by bulk flow between source and sink, with water moving into and out of the system all along the way.

bulk flow

4. Both pressure and solute concentrations gradually decrease between source and sink.

5. Solutes are unloaded into sink cells, and the water potential in those cells is lowered. Water moves out of the sieve tube and into sink cells.

SINK (e.g., developing root cells)

a

b

## Pressure Flow Theory

Within flowering plants, sugars and other organic compounds flow from a "source" to a "sink" along gradients of decreasing solute concentration and pressure. A **source** is any region of the plant where organic compounds are being loaded into the sieve tube system. A **sink** is any region where organic compounds are being unloaded from the sieve tube system and used or stored.

The site of photosynthesis in mature leaves is an example of a source. Another example is a tulip bulb at springtime, when stored food is being mobilized for transport to growing plant parts. By contrast, young and growing flowers are sink regions. So are the apples, peaches, pears, and other fruits you see growing on trees. Actually, young leaves, roots, and many other plant regions start out as sinks, only to mature into sources later on.

What makes sugars and other organic compounds flow from a source to a sink? According to the **pressure flow theory**, pressure builds up at the source end of a sieve tube system and *pushes* those solutes toward a sink, where they are removed.

Consider how sucrose flows from a source to a sink. In leaves, the major sources, sucrose formed inside photosynthetic mesophyll cells is transported to the phloem of the smallest veins. There, it is loaded into small sieve tubes (Figure 28.12). The sieve tube members are alive at maturity but play only a passive role in translocation. The loading process depends on energy expenditures by the companion cells adjacent to them.

As loading proceeds, the sucrose concentration increases inside the sieve tubes. The water potential decreases as a result. Recall that turgor pressure—the internal fluid pressure exerted against a cell wall—increases when water moves into a cell by osmosis. (Here you may wish to review pages 82 and 83 on osmosis and water potential.) In this case, the lower water potential causes water to rush into the sieve tubes. As turgor pressure increases, the sucrose-laden fluid moves by bulk flow into the increasingly larger sieve tubes of larger veins. Eventually the fluid is pushed out of the leaf, into the stem, and on to the sink. There it is unloaded from the sieve tube system (Figure 28.12e).

Because sucrose is added at the source end of the sieve tubes and removed at the sink end, its concentration gradually decreases along the route, so there is a simultaneous decrease in pressure. Most of the sucrose loaded into the small sieve tubes still reaches the sink. By contrast, water enters and leaves the sieve tube system all along the route. Thus few (if any) of the water molecules entering the sieve tubes at the source make it all the way

**LOADING**
at the source

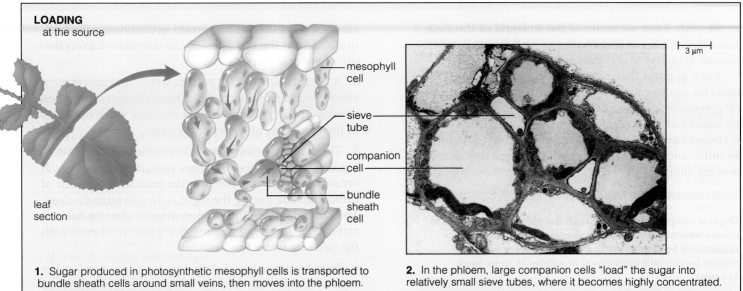

mesophyll cell

sieve tube

companion cell

bundle sheath cell

leaf section

**1.** Sugar produced in photosynthetic mesophyll cells is transported to bundle sheath cells around small veins, then moves into the phloem.

**2.** In the phloem, large companion cells "load" the sugar into relatively small sieve tubes, where it becomes highly concentrated.

c

**TRANSLOCATION**
along the path

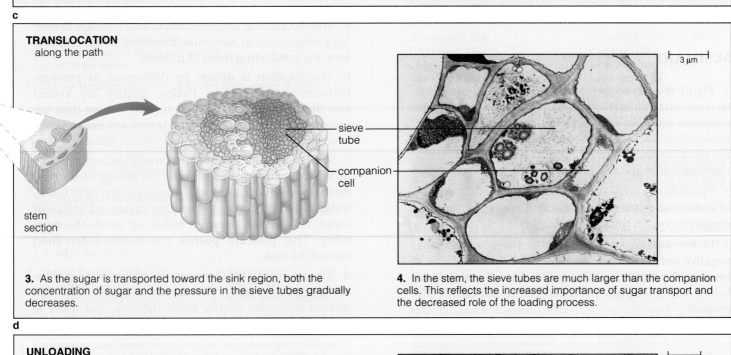

sieve tube

companion cell

stem section

**3.** As the sugar is transported toward the sink region, both the concentration of sugar and the pressure in the sieve tubes gradually decreases.

**4.** In the stem, the sieve tubes are much larger than the companion cells. This reflects the increased importance of sugar transport and the decreased role of the loading process.

d

**UNLOADING**
at the sink

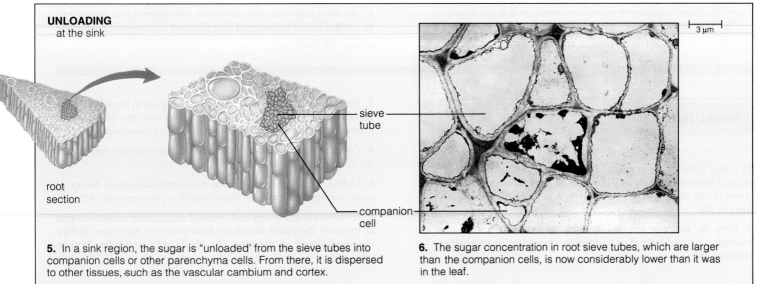

sieve tube

companion cell

root section

**5.** In a sink region, the sugar is "unloaded' from the sieve tubes into companion cells or other parenchyma cells. From there, it is dispersed to other tissues, such as the vascular cambium and cortex.

**6.** The sugar concentration in root sieve tubes, which are larger than the companion cells, is now considerably lower than it was in the leaf.

e

# 29 PLANT REPRODUCTION

## *Chocolate from the Tree's Point of View*

When the tax collector came around in ancient Mexico, he liked to be paid in beans. And not just any beans. They had to come from the colorful pods of the cacao tree (*Theobroma cacao*). The tax payment went straight to the king, Montezuma, and from him to the royal kitchen. Fermented, roasted, and ground to a buttery paste, the beans served as the base for chocolatl, a very rich brown drink. Chocolatl was heady stuff. Cinnamon, hot chili peppers, extracts from vanilla orchids, and a pinch of anise added zip to the cacao bean paste. Montezuma drank chocolatl from gold goblets after meals, during religious rites, and reportedly before visits to his harem. After each goblet had the privilege of touching the king's lips, it was ceremonially tossed into the lake outside the palace.

In 1519 Hernando Cortez dethroned Montezuma, drained the lake, and recovered the goblets. These he shipped to Spain, along with three large chests filled with the prized beans—and so began the worldwide fascination with chocolate.

**Figure 29.1** (**a**) Fruits growing from the trunk of a cacao tree (*Theobroma cacao*), shown here in its native habitat. (**b**) Chocolates—one of the products derived from its seeds and an indirect contributor to this flowering plant's reproductive success.

a

**b**

*Theobroma cacao* evolved in the undergrowth of tropical rain forests in Central America, and there it was first domesticated by the Mayas and Aztecs. If numbers and distribution are any yardstick, the species has become an evolutionary success story. Today cacao trees flourish on vast plantations in the tropical lowlands of Central America, the West Indies—and West Africa. Often rubber trees are planted alongside the treasured cacao trees to serve as their umbrellas. In this way, plantation owners provide *T. cacao* with the filtered shade characteristic of its native habitat.

Like many other successful plants, *T. cacao* produces flowers. Unlike most, its flowers grow directly from buds on the tree trunk instead of on separate floral shoots. About six months after the flowers are pollinated and the eggs fertilized, large, heavy fruits have developed from them (Figure 29.1a). Each fruit, or "pod," contains as many as forty seeds—the cacao "beans."

Chocolate manufacturers process the beans into creamy cocoa butter and chocolate essence (liquor). These become ingredients for the 8 to 10 pounds of chocolate that the average American consumes each year (Figure 29.1b). The average Swiss consumes a whopping 22 pounds of chocolate each year. Cocoa researchers have yet to figure out how theobromine, methylxanthin, and other cocoa compounds execute their compelling (some say addictive) effects on the human brain. They have yet to come up with a quality synthetic chocolate, although they recently discovered that the tasty properties of the real thing result from interactions among as many as 1,000 different chemical compounds.

In the meantime, *T. cacao* reigns supreme. Like orange trees, corn plants, and every other major crop plant that comes to mind, it has entered into a relationship with humans that helps assure its reproductive success and proliferation as a species. As you will see in this chapter, most flowering plants similarly benefit from beetles, bees, birds, bats, and other animals that contribute to their reproductive success. And in evolutionary terms, reproductive success is what it is all about.

## KEY CONCEPTS

**1.** For flowering plants, sexual reproduction requires the production of spores as well as gametes. The spores form in specialized reproductive structures called flowers. In many species, the flowers coevolved with insects, birds, and other animals that assist in pollination and seed dispersal.

**2.** Male floral structures produce haploid microspores that develop into immature male gametophytes (pollen grains). Sperm form inside pollen grains. Female floral structures produce haploid megaspores that give rise to female gametophytes. Eggs form inside female gametophytes.

**3.** Pollen grains are released from the parent plant and are adapted for traveling to the eggs. Female gametophytes remain attached to the parent plant and are nourished by it.

**4.** Following fertilization, seeds develop. Each seed consists of an embryo sporophyte and tissues that function in its nutrition, protection, and dispersal.

## REPRODUCTIVE MODES

Although it probably is not something you think about very often, flowering plants engage in sex. As in humans, they have elaborate reproductive systems that produce, protect, and nourish sperm and eggs. As in humans, female structures of flowering plants house the embryo during its early development. Flowers serve as invitations to third parties—pollinators—that function in getting sperm and egg together. Long before humans ever thought of it, flowering plants were using tantalizing colors and titillating fragrances as ways of improving the odds for sexual success.

Many plants also do something humans cannot do (at least not yet). They can reproduce asexually. Recall that *sexual* reproduction requires the formation of gametes, followed by fertilization. This means two sets of genetic instructions, from two gametes, are present in the fertilized egg. *Asexual* reproduction occurs by way of mitosis, so individuals of the new generation are clones—that is, genetically identical to the parent plant.

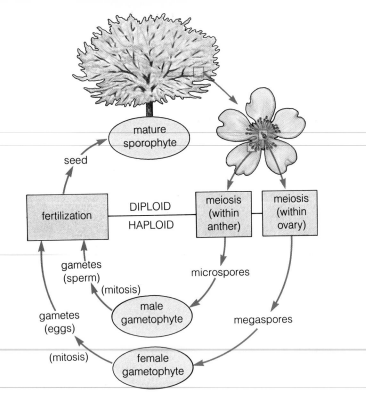

**Figure 29.2** Generalized life cycle for flowering plants. During the haploid phase, gametophytes produce gametes. The diploid phase begins when two gametes fuse to form the zygote, which develops into a sporophyte. In the sporophyte, meiotic divisions result in the formation of spores, which give rise to new gametophytes. Such a reproductive cycle, in which haploid gametophytes alternate with a diploid sporophyte, is sometimes called "alternation of generations."

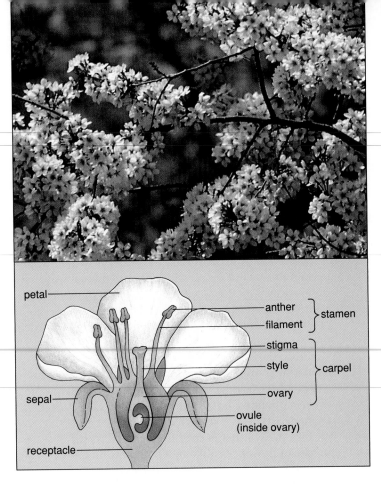

**Figure 29.3** Common arrangement of floral appendages. Shown here, a cherry (*Prunus*) flower, with a single carpel that will form the cherry.

What we usually think of as "the plant" is the **sporophyte**, a vegetative body that develops after a fertilized egg—the zygote—embarks on a developmental course of mitotic cell divisions and cell differentiation. Radish plants, cactus plants, and the cherry tree depicted in Figure 29.2 are examples of sporophytes. At some point during one or more seasons of growth and development, the sporophyte produces the reproductive shoots called **flowers**. Some cells in the flowers divide by meiosis and give rise to **gametophytes**, which will produce haploid sex cells—gametes. Male gametophytes produce sperm; female ones produce eggs.

The female gametophytes are usually tiny multicelled bodies embedded within floral tissues. Immature male gametophytes are released from flowers as small pollen grains; they are like shipping crates for sperm-producing cells until they actually land on a female flower part. Chemical and molecular cues guide the tubelike growth of male gametophytes down through the female floral tissues, toward the egg chamber and sexual destiny.

Sporophytes may also reproduce themselves asexually by several means. Strawberry plants send out horizontal aboveground stems, and new roots and shoots develop at every other node along the stems. Short underground stems of onions or lilies put out buds that grow into new plants. In fall, when you think it is doing nothing except being dormant, Bermuda grass is giving birth to new little plants at nodes along its underground horizontal stems. Asexual reproduction also occurs with help from humans. Whole orchards of pear trees, for example, have been grown from cuttings or buds of a parent tree.

Sexual reproduction dominates the life cycle of flowering plants, however, and it will be our focus here.

## GAMETE FORMATION IN FLOWERS

### Floral Structure

A flower develops at a shoot tip (Figure 29.3). During its development, cells differentiate and tissues become arranged into nonfertile parts (sepals and petals) and fertile parts (stamens and carpels). These specialized parts grow out from the modified end of the floral shoot, the receptacle.

The outermost whorl of parts, the leaflike *sepals*, are the flower's "calyx." Commonly the calyx encloses all the other parts, as it does in roses and lilacs before a bud opens (Figure 29.4). Next inside are the *petals*, also leaf-like, which ring the male and female parts. Collectively, petals are the flower's "corolla." Corollas are distinctive in their coloration, patterning, and shape, as Figure 1.6e suggests. These features often function in attracting bees and other pollinators.

Inside the corolla are **stamens**, the male reproductive parts. In nearly all living species, a stamen consists of a slender stalk capped by an anther. *Anthers* are two-lobed structures that contain four **pollen sacs**, the chambers in which pollen grains develop.

The innermost, central part of a flower consists of one or more **carpels**, the female reproductive parts. A carpel is a vessel with no opening, a house with no door. The lower part of the carpel is the **ovary**. The ovary is where eggs develop, fertilization takes place, and seeds mature. ("Angiosperm" refers to the carpel; the name is derived from the Greek *angeion*, meaning a case or vessel, and *sperma*, meaning seed.) Many flowers have a group of carpels fused together in such a way that they form a common ovary.

Commonly, the upper part of a carpel narrows into a slender column (style) that terminates at the *stigma*, a landing platform for pollen. Fused carpels may share a common stigma and style, or each may retain separate ones.

So-called "perfect" flowers have both male and female parts. "Imperfect" flowers have male *or* female parts, but not both. Some species, including oaks, have male and female flowers on the same plant. Other species, including willows, have them on separate plants.

corolla (all petals combined)

calyx (all sepals combined)

receptacle

**Figure 29.4** Location of a few floral parts in one of the most prized of all cultivated plants, the rose (*Rosa*).

## Microspores to Pollen Grains

Let's now turn to pollen grain formation. A **pollen grain** is a two-celled, immature male gametophyte. Figure 29.5 shows a few examples. In nearly all cases, a given family of plants can be distinguished on the basis of the size and wall sculpturing of its pollen grains, as well as the number of pores in the wall. Those walls are tough

**Figure 29.5** Scanning electron micrographs of pollen grains from (**a**) rose, (**b**) grass, and (**c**) ragweed plants.

c

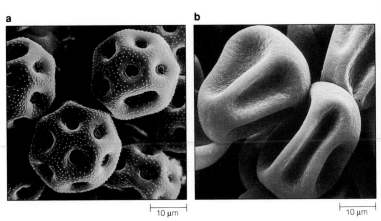

a

b

10 µm

10 µm

10 µm

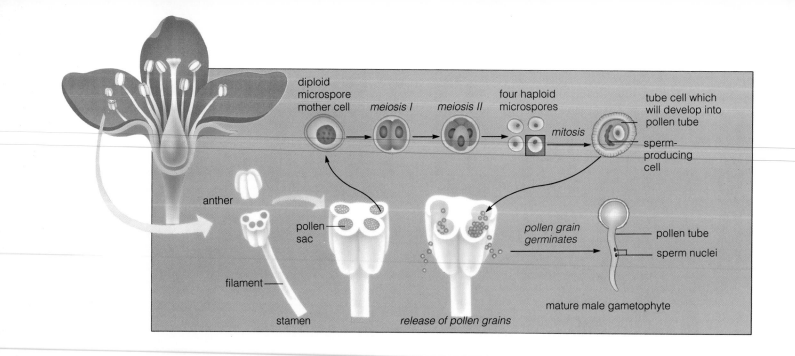

**Figure 29.6** Stages in the development of a male gametophyte.

enough to protect the sperm cells during the somewhat chancy journey from the anther to a stigma. They also are strong enough to withstand decomposition. This is one reason why pollen grains fossilize well and provide such good clues about the evolution of flowering plants.

Pollen grains start to form while an anther is growing inside a flower bud. As Figure 29.6 indicates, each microspore in the anther undergoes mitotic cell division to produce four masses of "mother" cells. Several layers of cells form a walled chamber around each cell mass. This is the "sac" in which the pollen will develop. Each cell inside a pollen sac undergoes meiosis, and eventually four haploid cells form. The haploid cells resulting from meiosis are *not* gametes; they are a type of spore. The ones produced in pollen sacs are designated **microspores**.

Each microspore undergoes mitotic cell division, the result being a two-celled haploid body. This is the pollen grain. Later on, after a pollen grain lands on a stigma, one cell will give rise to sperm cells. The other will develop into a *pollen tube*, which will grow through tissues of a carpel and so transport the sperm to the ovary.

### Megaspores to Eggs

Meanwhile, in the carpel of a flower, one or more dome-shaped cell masses have been developing on the inner wall of the ovary. Each mass is the start of an ovule, which, if all goes well, will become a seed. Only one

dome-shaped mass forms in the carpel of a cherry flower. Hundreds or thousands may form in the carpels of other flowers. Cut open a ripe papaya and the abundance of seeds inside will give you an idea of how warty the ovarian wall of the papaya flower must have looked at one time.

As the cell mass grows, some of its cells form a stalk. As Figure 29.7 shows, the rest develop into an inner tissue (nucellus), and one or two protective layers (the integuments) form around it. Only a tiny part of the nucellus does not become covered with integuments. Most commonly this tiny gap, the micropyle, is where a pollen tube will penetrate the ovule. Within the cell mass, a diploid mother cell divides by meiosis.

In the ovule, the four haploid cells that form after meiosis are called **megaspores**. Typically, three of the four disintegrate. The one remaining undergoes mitosis three times *without* cytoplasmic division, so at first it is a single cell with eight nuclei (Figure 29.8). The cytoplasm divides only after each nucleus migrates to a specific location in the cell. The result is a seven-celled **embryo sac**, the female gametophyte. One cell is the egg. Another cell, the "endosperm mother cell," has two nuclei. It will help form **endosperm**, a nutritive tissue around the forthcoming embryo.

---

An *ovule* is a stalked structure that develops on the ovary wall of a carpel. It consists of an egg cell within an embryo sac (female gametophyte), a surrounding tissue (nucellus), and one or two protective layers (integuments).

When mature, an ovule becomes a seed.

---

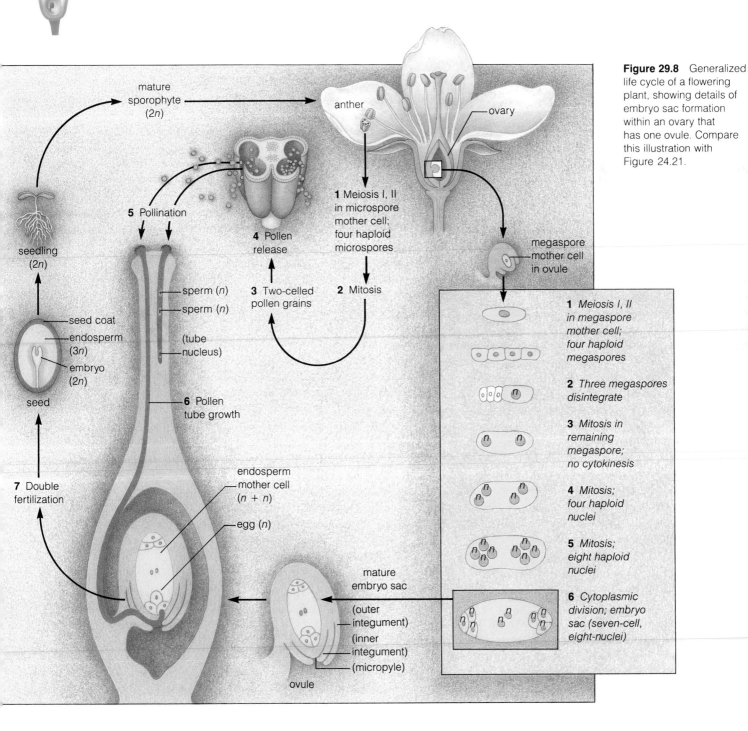

**Figure 29.7** Formation of a megaspore mother cell (gold) in a developing ovule.

an ovule

integument
nucellus
megaspore
mother cell

micropyle

ovary
wall

stalk

# POLLINATION AND FERTILIZATION

## Pollination

The transfer of pollen grains to a stigma is called **pollination**. Wind, insects, birds, or other agents make the transfer. The relationship between flowering plants and their pollinators is one of the most intriguing of all evolutionary stories. It is the topic of the *Commentary* on the next page.

**Figure 29.8** Generalized life cycle of a flowering plant, showing details of embryo sac formation within an ovary that has one ovule. Compare this illustration with Figure 24.21.

mature
sporophyte
(2n)

anther

ovary

**5** Pollination

**4** Pollen
release

**1** Meiosis I, II
in microspore
mother cell;
four haploid
microspores

megaspore
mother cell
in ovule

seedling
(2n)

sperm (n)
sperm (n)

**3** Two-celled
pollen grains

**2** Mitosis

seed coat
endosperm
(3n)
embryo
(2n)

(tube
nucleus)

**1** *Meiosis I, II
in megaspore
mother cell;
four haploid
megaspores*

seed

**6** Pollen
tube growth

**2** *Three megaspores
disintegrate*

**3** *Mitosis in
remaining
megaspore;
no cytokinesis*

**7** Double
fertilization

endosperm
mother cell
(n + n)

egg (n)

**4** *Mitosis;
four haploid
nuclei*

**5** *Mitosis;
eight haploid
nuclei*

mature
embryo sac

(outer
integument)

(inner
integument)

(micropyle)

**6** *Cytoplasmic
division; embryo
sac (seven-cell,
eight-nuclei)*

ovule

# Coevolution of Flowering Plants and Their Pollinators

An astonishing variety of flowering plants can be found almost everywhere, from snow-covered flanks of mountains to low deserts, from freshwater ponds to the surface of open oceans. How did this distribution and diversity come about, given that plants (unlike animals) cannot just pick up and move to new places?

For the answer, we must go back 430 million years or so, to the time when plants began invading the land. Insects that could scavenge on moist, decaying plant parts and tiny spores were probably right behind them. Fossils provide us with evidence of the evolution of stronger stems and taller plants. We also have fossils of scavengers that had become adapted to withstanding exposure to air. The absence of competition for edible but aerial plant parts seems to have favored a variety of feeding adaptations (such as sucking, piercing, and chewing mouthparts). It also favored the development of insect wings.

The first *seed*-bearing plants made their entrance in humid coastal forests, some 395 million years ago. They were ancestral to existing gymnosperms and flowering plants. Often their ovules and pollen sacs were located in conelike formations of modified leaves, and it seems the pollen grains simply drifted on air currents to the ovules.

Pollen grains happen to be rich sources of protein. Suppose some insects came to associate "cone" with "food source." *They would have begun serving as pollinating agents.* Some of the dustlike pollen would be eaten, but some would cling to the insect body and be transported to ovules. Insects clambering about reproductive cones would not be precision pollinators—but they would be more effective than air currents alone (which are not much to speak of, in dense forests). Pollen would be delivered right to the door, so to speak. The tastier the pollen, the more home deliveries, and the more seeds formed. *And the greater the number of seeds formed, the greater the reproductive success.*

What we are describing here is a case of **coevolution**. The word means the joint evolution of two (or more) species interacting in close ecological fashion. When one species evolves, the change affects selection pressures operating between the two species and so the other also evolves. In our evolutionary story, there was natural selection of variant plants able to attract beneficial insects. At the same time, there was selection of pollinator insects. Because of their ability to recognize a particular food and locate it quickly, the pollinators were able to outcompete other foraging insects.

Another, perhaps related change should be mentioned. Existing pollen-eating beetles have strong mouthparts,

(a) Above, a sunbird visiting a flower with red components. To the right (b), a hummingbird visiting a red-tinted passion flower.

(c) Nectar guide of the marsh marigold. The petals of this flower appear solid yellow to us. But its distinctive markings become apparent when the flower is exposed to ultraviolet light, which the bee eye can detect (our eyes cannot do this).

b

pollen

**(d)** Coevolutionary match between a flower and its pollinator. The color of Scotch broom attracts bees. Some petals serve as a landing platform that corresponds to the size and shape of the bee body. The weight of a bee on the platform forces petals apart. The pollen-laden stamens, positioned to strike against the bee, are released and pollen is dusted onto dense hairs on the bee. This orange mass of pollen was packed into a "basket" of stiff hairs when the bee groomed itself before returning to the hive.

and many chew on the ovules they pollinate. *Chewing behavior may have been a selective force in the evolution of floral structure.* At one time, ovules were naked and vulnerable on cone scales. The ovules of today's flowering plants are protected inside closed carpels—which afford some protection against hungry insects (see also page 386).

Many aspects of floral structure can be correlated with specific pollinators. Figures *a* and *b* show examples. Petals of these flowers have red components, and they form a tube for nectar. The colors do not visually attract beetles or honeybees, neither of which can detect red wavelengths. (Besides, in flowers with large, deep tubes, insects also face the distinct possibility of drowning in the nectar.) Birds, however, have a keen sense of vision and can detect flowers with red as well as yellow components. And certain birds have a beak as long as the floral tube, as these two figures suggest. The flower's anthers and stigma are located

where the bird will brush against them. Because birds visit the nectar cups of many plants of the same species, they promote cross-pollination.

Flowers visited by birds typically do not have strong floral odors. Birds have a poor sense of smell, so attractive fragrances would be wasted on them. By contrast, some flowers pollinated by beetles (and flies) have strong odors that smell like decaying meat or moist dung. Perhaps such odors originally resembled the smells of decaying matter in forest litter, where beetles first evolved.

Flowers that are commonly pollinated by bees have strong, sweet odors and bright yellow, blue, purple, or ultraviolet components (Figures *c* and *d*). The plumelike hairs on most bees retain pollen from the flowers being visited. The landing platforms of some bee-pollinated flowers favorably position the bee body so that it brushes against pollen-laden anthers (Figure *d*).

Butterflies forage by day and often are attracted to sweet-smelling, red, and upright flowers having a more or less horizontal landing platform. Most moths forage by night. They pollinate flowers with strong, sweet odors and white or pale-colored petals, which are more visible in the dark (Figure *e*). Butterflies and moths have long, narrow mouthparts, corresponding to narrow floral tubes or spurs. When uncoiled, the mouthpart of the Madagascar hawkmoth is a record 22 centimeters long—the same length as the floral tube of an orchid (*Angraecum sesquipedale*)! Hawkmoths do not require a landing platform; they hover in front of a floral tube.

(e) Stephanotis, like other night-flowering plants, has no distinctive color pattern. Its white petals and strong scent attract moth pollinators (white or pale colors reflect more light and so are more visible at night).

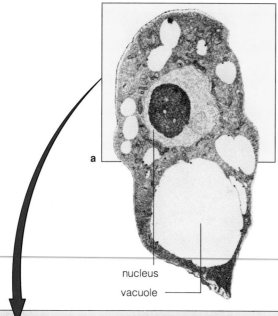

nucleus

vacuole

**Figure 29.9** Some stages in the development of shepherd's purse (*Capsella bursa-pastoris*), a dicot. The micrographs are not to the same scale. (**a**) Internal organization of the zygote. The organization is heritable and contributes to the shaping and cell differentiations of the early embryo, identified by the yellow boxes in (**b**–**d**). The embryo in (**e**) is well differentiated; the one in (**f**) is mature.

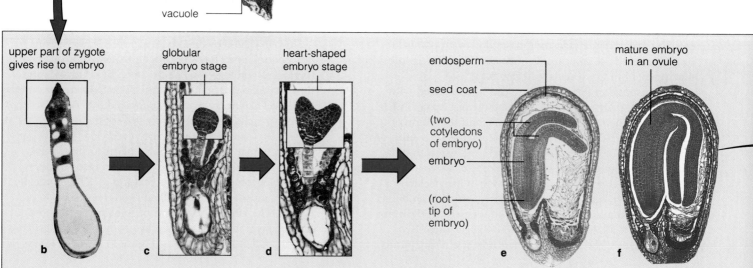

upper part of zygote gives rise to embryo

globular embryo stage

heart-shaped embryo stage

endosperm

seed coat

(two cotyledons of embryo)

embryo

(root tip of embryo)

mature embryo in an ovule

# SEED AND FRUIT FORMATION

## Fertilization and Endosperm Formation

Once a pollen grain lands on the proper stigma, it resumes growth (germinates). A pollen tube develops from one of its cells and starts the journey to an ovule. Before or during the tube's growth through the stigma and style, the pollen grain's sperm-producing cell undergoes mitosis, forming two sperm nuclei. What happens when the pollen tube reaches the ovary? It grows toward an ovule. When it penetrates the embryo sac, its tip ruptures and the two sperm are released (Figure 29.8).

"Fertilization" generally means the fusion of a sperm nucleus and an egg nucleus. **Double fertilization** occurs in flowering plants. One sperm nucleus fuses with that of the egg, forming a *diploid* (2*n*) zygote. Meanwhile, the other sperm nucleus and both nuclei of the endosperm mother cell all fuse together, forming a cell with a *triploid* (3*n*) nucleus. Tissues derived from that 3*n* cell are called endosperm. They will nourish the embryo and the seedling (the young sporophyte) until its leaves form and photosynthesis is under way.

Endosperm forms only in flowering plants. Its evolution coincided with a reduction in the female gametophyte—the source of nourishment for the embryo sporophytes of other land plants.

## Seed Formation

When the zygote first forms, the ovule containing it is still attached to the parent plant (by the stalk that developed out of the ovarian wall). It undergoes some development even before the mitotic divisions begin that will lead to the mature embryo. For example, notice in Figure 29.9 how most organelles, including the nucleus, reside in the top half of a *Capsella* zygote. A vacuole takes up most of the lower half. Once divisions begin, some of the daughter cells give rise only to a simple row of cells (the suspensor) that transfer nutrients from the parent plant to the embryo. Other daughter cells give rise to the mature, multicelled embryo.

**Cotyledons**, or "seed leaves," develop as part of the embryo (Figures 29.9 and 29.10). Many plants have large cotyledons that absorb the endosperm and function in food storage. Other plants have thin cotyledons that may produce enzymes for digesting and transferring stored food from the endosperm to the germinating seedling.

As the embryo continues to develop, the endosperm expands and integuments of the ovule harden and thicken. When the ovule is mature—that is, a seed—its integuments have become the seed coat.

In double fertilization, one sperm nucleus fuses with the egg nucleus, and a diploid zygote results. The other sperm nucleus fuses with the two nuclei of the endosperm mother cell, which gives rise to triploid (3*n*) nutritive tissue.

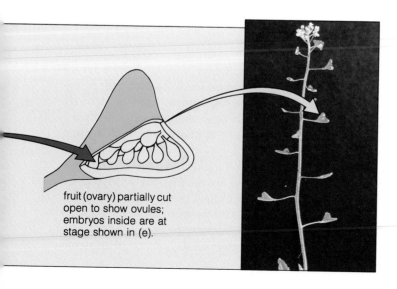

fruit (ovary) partially cut open to show ovules; embryos inside are at stage shown in (e).

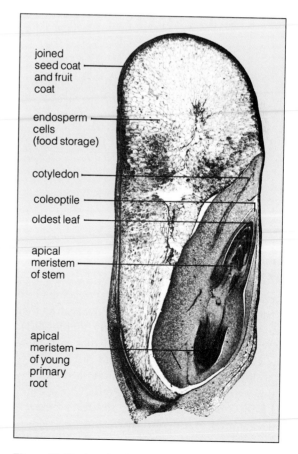

joined seed coat and fruit coat

endosperm cells (food storage)

cotyledon

coleoptile

oldest leaf

apical meristem of stem

apical meristem of young primary root

**Figure 29.10** Longitudinal section through a grain of corn (*Zea mays*).

**Table 29.1  Kinds of Fruits of Some Flowering Plants**

| Type | Characteristics | Some Examples |
|---|---|---|
| **Simple** (formed from single carpel, or two or more united carpels of one flower) | 1. Fruit wall *dry; split* at maturity | Pea, magnolia, tulip, mustard |
| | 2. Fruit wall *dry; intact* at maturity | Sunflower, wheat, rice, maple |
| | 3. Fruit wall *fleshy*, sometimes with leathery skin | Grape, banana, lemon, cherry |
| **Aggregate** (formed from numerous but separate carpels of single flower) | *Aggregate* (cluster) of matured ovaries (fruits), all attached to receptacle (modified stem end) | Blackberry, raspberry |
| **Multiple** (formed from carpels of several associated flowers) | *Multiple* matured ovaries, grown together into a mass; may include accessory structures (such as receptacle, sepal, and petal bases) | Pineapple, fig, mulberry |
| **Accessory** (formed from one or more ovaries *plus* receptacle tissue that becomes fleshy) | 1. *Simple:* a single ovary surrounded by receptacle tissue | Apple, pear |
| | 2. *Aggregate:* swollen, fleshy receptacle with dry fruits on its surface | Strawberry |

**a** Apple blossoms

**b** Petals fallen away

**c** Enlarging ovaries

**Figure 29.11** Fruit formation on an apple (*Malus*) tree. Petals dropping from the flower usually signify successful fertilization. After this, the ovary and receptacle expand (**b**). Sepals and stamens are still visible on the immature fruit (**c**).

## Fruit Formation and Seed Dispersal

An ovary containing the ovule (or ovules) develops into a **fruit**. A fruit is any matured or ripened ovary, together with other associated parts of the flower (Table 29.1). Many fruits, including apples and tomatoes, are juicy and fleshy. Others, including grains and nuts, are dry. The fruit wall of a bean is dry and intact at maturity. Multiple flowers remain clustered together in a pineapple and form a multiple fruit. Figures 29.11 through 29.13 show examples of different fruits.

*Fruits function in seed protection and dispersal in specific environments.* For example, maple fruits have winglike extensions. When the fruit drops, the wings cause it to spin sideways. With such spinnings, seeds can be dispersed to new locations, where they will not have to compete with the parent plant for soil water and minerals. Many fruits have hooks, spines, hairs, and sticky surfaces. They are taxied to new locations when they adhere to feathers or fur of animals that brush against them.

Fleshy fruits such as blueberries and cherries are tasty to many animals and are adapted for surviving the digestive enzymes in the animal gut (see the *Doing Science* essay, page 512). The enzymes remove just enough of the hard seed coats to increase the chance of successful germination when the seeds are expelled from the body.

seed (in carpel)

wing

floral remnants

ovary

**a**

**b**

receptacle

ovary

**c**

**Figure 29.12** (**a**) Pineapple, a multiple fruit; (**b**) maple, a dry fruit; and (**c**) strawberry, an aggregate fruit.

**a**

**b**

**Figure 29.13** Two examples of economically important crop plants. (**a**) Sunflower (*Helianthus*), valued for its seeds. (**b**) Nutmeg (*Myristica fragrans*), the seeds of which are ground to produce one of the most popular spices in the world. Another spice, mace, comes from the fleshy seed coat, which appears as red bands in this photograph.

## Why So Many Flowers?

A giant saguaro cactus growing in the Sonoran desert of Arizona (Figure *a*) may produce as many as a hundred flowers at the tips of its huge, spiny arms. The large, showy white flowers, shown in Figure *b*, bloom for only 24 hours. During that time, insects visit the flowers by day and bats visit them by night. Both types of animals are attracted to the plant's nectar—and they happen to transport pollen grains that stick to their body from one cactus plant to another.

Flowers that are pollinated courtesy of insects or bats may then begin the task of producing seeds and fruits. Petals wilt, then wither as the many egg cells inside the ovary become fertilized. Each egg-containing ovule now expands and matures into a seed. At the same time, the ovary itself develops into a fruit (Figures *c* and *d*). After a number of weeks, the plum-sized fruit splits open, exposing its brilliant red interior dotted with blackish seeds.

White-winged doves feast on the ripe fruits. When they fly away from the cactus they might carry hundreds of seeds in their gut. A few of the seeds may pass through the gut undigested and end up deposited on the ground. There, each will have a tiny chance of growing into a new giant saguaro.

One of the very puzzling aspects of this annual event is the frequency with which saguaro flowers *fail* to give rise to fruit (see Figure *c*, for example). Would you expect a cactus to invest energy in constructing a hundred flowers but set fruit in only thirty? Imagine comparing one of these "inefficient" plants with one that produced a hundred flowers, *all* of which developed into fruits. Which do you think would have more descendants?

Common sense might lead you to conclude, "The more fruits, the better." Yet saguaros *and many other plants* commonly produce far more flowers than mature fruits. Such plants seem to be producing "excess" flowers and giving up chances to make seeds and leave descendants—which is not a reproductive adaptation that you would expect, based on Darwinian evolutionary theory.

a

b

Well, suppose some of the plants simply never received enough pollen to have all their egg cells fertilized. After all, unfertilized flowers cannot produce seeds and fruits. That hypothesis has been experimentally tested for some plant species. Researchers have carefully brushed pollen on every single flower. In many instances, the plants still fail to produce fruit for every flower!

Alternatively, suppose the so-called "excess" flowers are formed strictly to produce pollen for export to other plants. Pollen grains are small. In energy terms, they are inexpensive to produce, compared to the large, calorie-rich, seed-containing fruits. Thus, for a fairly small investment, a plant might reap a large reward in offspring that carry its genes.

The idea that some plants actually set aside flowers exclusively for pollen export has yet to be tested for saguaros. Perhaps you might like to design and carry out experiments that will help clear up the mystery of the "excess" flowers of saguaros and similar plants.

fruit sets

fruit fails to set

**c**

**d**      maturing fruit

| Table 29.2 | Asexual Reproductive Modes of Flowering Plants | |
|---|---|---|
| Mechanism | Representative Species | Characteristics |
| **Reproduction on modified stems:** | | |
| 1. Runner | Strawberry | New plants arise at nodes on an aboveground horizontal stem |
| 2. Rhizome | Bermuda grass | New plants arise at nodes of underground horizontal stem |
| 3. Corm | Gladiolus | New plant arises from axillary bud on short, thick, vertical underground stem |
| 4. Tuber | Potato | New shoots arise from axillary buds on tubers (enlarged tips of slender underground rhizomes) |
| 5. Bulb | Onion, lily | New bulb arises from axillary bud on short underground stem |
| Parthenogenesis | Orange, rose | Embryo develops without nuclear or cellular fusion (e.g., from unfertilized haploid egg; or develops adventitiously, from tissue surrounding embryo sac) |
| Vegetative propagation | Jade plant, African violet | New plant develops from tissue or organ (e.g., a leaf) that drops or is separated from plant |
| Tissue culture propagation | Carrot, corn, wheat, rice | New plant arises from cell in parent plant that is not irreversibly differentiated; laboratory technique only |

## ASEXUAL REPRODUCTION OF FLOWERING PLANTS

The sexual reproductive modes just described are the most prevalent among flowering plants. As noted earlier, however, plants also can reproduce asexually by several means, as listed in Table 29.2.

A strawberry plant sends out horizontal aboveground stems, known as *runners*. Along such runners, new roots and shoots develop at every other node. Oranges reproduce every so often by *parthenogenesis:* the development of an embryo from an unfertilized egg. In some plants,

## Foolish Seedlings and Gorgeous Grapes

A few years before the American stock market grew feverishly and then collapsed catastrophically in 1929, a researcher in Japan came across a substance that caused runaway growth and then subsequent collapse of rice plants. Ewiti Kurosawa was studying what the Japanese called *bakane*—the "foolish seedling" effect on rice plants. Stems of rice seedlings that had become infected with a fungus (*Gibberella fujikuroi*) elongated twice as much as uninfected plants. The long stems were spindly and weak, and eventually they collapsed and the plants died. Kurosawa discovered that applying extracts of the fungus also could trigger the disease. Many years later, other researchers purified the disease-causing substance from fungal extracts. The substance was given the name gibberellin.

Today we know that gibberellin is one of the premier plant hormones. More than eighty different kinds have been isolated from the seeds of flowering plants as well as from fungi, and possibly they exist in other plants as well. Gibberellins and other plant hormones are signaling molecules. They are produced by some cells and transported to different cells elsewhere in the plant body, where they trigger changes in metabolic activities. Through their action, the gibberellins make stems grow longer (Figure 30.1a). They help seeds and buds break dormancy and resume growth in spring. There is evidence that they help induce flowering.

A radish plant exposed to the proper concentration of a gibberellin may grow as large as a beach ball. A cabbage plant similarly stimulated may grow 6 feet tall. Gibberellin applications can make celery longer and crispier; they can prevent the skin of navel oranges from growing old too soon in orchards.

And gibberellins have had dramatic effect on the market appeal of seedless grapes (Figure 30.1b). Grapes grow in bunches along stems. By lengthening internodes along the stems, gibberellins provide better air circulation around the fruits and give them more room to grow larger. The enhanced air circulation also makes it harder for pathogenic fungi to become established on the fruit.

At present, we do not know much about how plant hormones work. We know that at least five different types have major, predictable effects on flowering plants, beginning with the germination of each new seed. As with humans and other animals, plants grow and develop according to genetic information in their DNA. That information governs the synthesis of

a

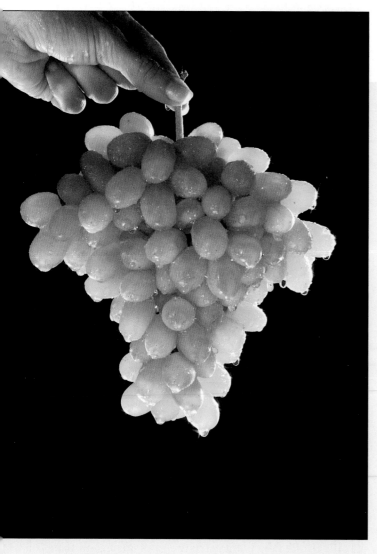

**Figure 30.1** (a) Effects of one hormone, gibberellin, on stem growth of a popular flowering plant, the California poppy. (b) A come-hither bunch of seedless grapes, radiating market appeal that results from gibberellin applications.

enzymes and other proteins necessary for metabolism and other aspects of cell function. How those enzymes function depends partly on hormonal action.

As you will see in this chapter, plant hormones operate through interactions with one another. They also operate in response to cues afforded by the rhythmic changing of the seasons, as when days become longer and warmer in spring, after the short, cold nights of winter. That grape, cabbage leaf, or celery stalk you pop into your mouth is the culmination of a beautifully regulated program of plant growth and development.

KEY CONCEPTS

**1.** Plant growth and development depend on the action of different hormones. These substances are produced by cells in certain plant regions, then are often transported to cells in other regions where they trigger changes in metabolic activities. Those metabolic changes have predictable effects, as when they cause a stem to elongate.

**2.** The known plant hormones are auxins, gibberellins, cytokinins, abscisic acid, and ethylene. The existence of another hormone, tentatively named florigen, is suspected.

**3.** Plant hormones interact with one another to bring about specific effects on growth and development. They help adjust patterns of growth in response to environmental rhythms, including seasonal changes in daylength and temperature. They also help adjust those patterns in response to the environmental circumstances in which a plant finds itself—that is, the amount of sunlight or shade, moisture, and so on at a given site.

In the preceding chapter, we traced the events by which a zygote of a flowering plant becomes transformed into an embryo, housed inside a protective seed coat. At some point following its dispersal from the parent plant, the mature embryo grows into a seedling, which in turn develops into a mature sporophyte. Flowers, fruits, and new seeds form. Often, old leaves drop away from the plant in autumn. Let's begin with the kinds of internal mechanisms that govern these aspects of plant growth and development. Then we will look at environmental signals that set them in motion.

## SEED GERMINATION

Before or after seed dispersal, embryonic growth idles. Then, at **germination**, the embryo absorbs water, resumes growth, and breaks through the seed coat. The amount of moisture and oxygen, the temperature, the number of daylight hours, and other aspects of the environment influence germination. Mature seeds do not contain enough water for cell expansion or metabolism.

**517**

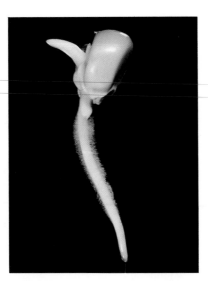

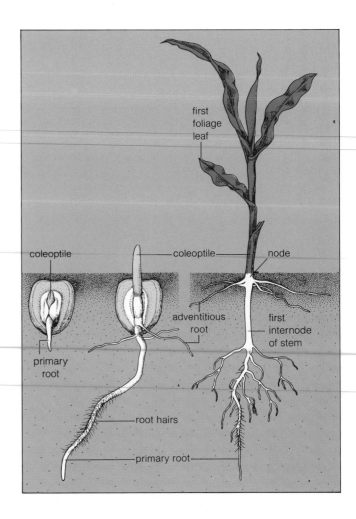

**Figure 30.2** Some stages in the development of a corn plant, a monocot. The shoot's young leaves are enclosed in a coleoptile, which protects them during upward growth through the soil. The first node develops at the coleoptile's base, and adventitious roots develop here. When a corn seed is planted deeply, the first internode of the stem elongates as shown in the sketch. When a seed is planted close to the soil surface, light inhibits elongation of the first internode. In this case, the adventitious roots and the first root look as if they originate in the same region (just below the soil surface).

Usually, water availability is seasonal and germination coincides with the return of spring rains. In a process called *imbibition*, water molecules move into the seed, being especially attracted to hydrophilic groups of the stored proteins. As more and more water molecules move inside, the seed swells and the coat finally ruptures.

Once the seed coat splits, oxygen moves more easily from the surrounding air into the seed. Cells of the embryo engage in aerobic respiration, and metabolism moves into high gear. Cells divide and elongate continuously to produce the seedling.

Inside a germinating seed, root cells of the embryo generally are the first to grow. As they divide and elongate, they give rise to the first, or *primary*, root of the new plant. When the primary root pokes out from the seed coat, germination is completed.

## PATTERNS OF EARLY GROWTH

Figure 30.2 shows the pattern of growth and development for a corn plant, which is a monocot. Figure 30.3 shows the comparable pattern for a soybean plant, a dicot. How do those two kinds of plants actually grow from a small seed? The answer begins with cell divisions.

Often the cell divisions in a growing plant are unequal, with one daughter cell ending up with more cytoplasm than the other. The unequal distribution of cytoplasm means that such cells differ in their composition and structure, and the differences affect how they interact with their neighbors during growth. Obviously, all cells in the new plant carry the same genes (all were derived from the zygote). But now their cytoplasmic differences and interactions with one another trigger selective gene expression, as described on page 236. Certain genes governing, say, the synthesis of growth-stimulating hormones are activated in some cells and remain silent in others. Such events seal the developmental fate of various cell lineages. Their descendant cells divide in prescribed planes and expand in specific directions. The eventual outcome will be plant parts of diverse shapes and functions.

Plant growth and development are under genetic controls.

Selective gene expression, which begins in the cells of a plant embryo, influences the direction in which cells divide and how they will differentiate.

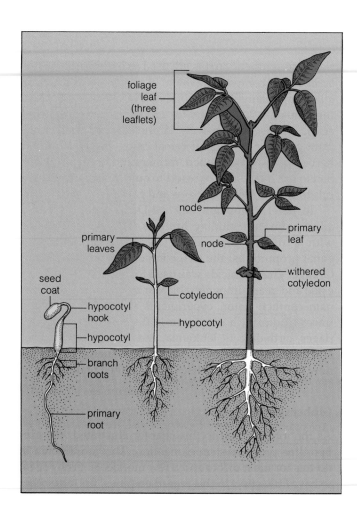

**Figure 30.3** Some stages in the development of a soybean plant, a dicot. Food-storing cotyledons are lifted above the soil surface when cells of the hypocotyl (the stem *below* the cotyledon) elongate. The hypocotyl becomes hook-shaped and forces a channel through the soil as it grows. The cotyledons can be pulled up through the channel without being torn apart. At the soil surface, light causes the hook to straighten. For several days, cells of the cotyledons function in photosynthesis; then the cotyledons wither and fall off. Photosynthesis is taken over by the first leaves that develop along the stem and later by foliage leaves, each divided into three leaflets. Flowers develop in buds at the nodes shown here.

Keep in mind that cell division in itself is only one aspect of growth. On the average, about half of the daughter cells being produced also undergo *enlargement*—often by twenty times! The other half of the daughter cells grow no larger than the parent cell, but they retain the capacity to divide again.

Water uptake drives cell enlargement. When water moves into new cells, fluid pressure increases inside them and forces the primary cell wall to expand. The expansion resembles a balloon being blown up with air. Balloons with soft walls are easy to inflate; cells with soft walls grow rapidly under little pressure. However, a balloon wall gets thinner as it "grows," but a cell wall does not. New polysaccharides are added to the wall during growth, and more cytoplasm forms between the wall and the central vacuole (Figure 30.4).

Primary walls of young cells are *plastic* and tend to retain much of their stretched shape (as does a blown-up bubble of bubble gum when the air has been let out). This expansive property encourages plant growth, as defined by increases in cell volume and mass.

**Figure 30.4** (Right) The nature and direction of plant growth.

nucleus
vacuole

**a** Through divisions, meristems provide new cells for growth. At the tip of a young root, small meristematic cells double in size, then divide.

**b** The lower cell remains meristematic. The upper cell grows into a mature parenchyma cell of pith or cortex, for example. Tiny vacuoles in young cells absorb water, fuse, and form the central vacuole of mature parenchyma cells, such as the one shown to the left.

nucleus

central vacuole

primary wall

## From Embryo to Mature Oak

Where the ocean breaks along the central California coast, the land rolls inward as steep and rounded hills. These sandstone hills started forming 65 million years ago, when violent movements in the earth's crust caused parts of the submerged ocean floor to crumple upward into a jagged new coastal range. Since then, rains and winds played across the land, softening the stark contours and sending mineral-laden sediments into the canyons. In time, grasses cloaked the inland hills, and their organic remains accumulated and enriched the soil. On these hillsides, in these canyons, the coast live oak (*Quercus agrifolia*) began to evolve more than 10 million years ago.

*Quercus agrifolia* is a long-lived giant; some trees are known to be 300 years old. They can grow more than 100 feet tall, and their evergreen branches can spread even wider. In early spring, clusters of male flowers develop and form golden catkins among the leaves. Wind carries pollen grains from the catkins to female flowers near the branch tips of the same or neighboring trees. After pollination, a sperm-bearing pollen tube grows through the style and into the ovary, where fertilization takes place. The newly formed zygote undergoes repeated cell divisions, which give rise to root and shoot tips and to cotyledons. Integuments of the ovule form the seed coat, and ovary walls develop into a shell. By early fall, the seed reaches maturity and is shed from the tree as a hard-shelled acorn.

Three centuries ago, long before Gaspar de Portola sent landing parties ashore to found colonies throughout upper California, oaks were shedding the seeds of a new generation. Suppose it was then that a scrubjay, foraging at the foot of a hillside, came across a worm-free acorn. In storing away food for leaner days, the bird used its beak to scrape a small crater in the soil, then dropped in the acorn and covered it with decaying leaves. Although a scrubjay remembers its hiding places most of the time, this particular acorn lay forgotten. The next spring, it germinated.

The oak seed embarked on a journey of continued growth. Cells divided repeatedly, grew longer, and increased in diameter. Water pressure forced the enlargement—water taken up osmotically as ions accumulated in the seedling's first root. A root cap formed and protected the root as it grew downward through the soil. Cell differentiations gave rise to the cortex, epidermis, and a vascular column through which water and ions would flow. Lateral roots emerged, probably under the influence of hormones. As new roots grew longer, their absorptive surfaces increased. When the first shoot began its upward

oak seed (acorn)

surge, separate vascular bundles began forming; eventually they would form a continuous cylinder of secondary xylem and phloem.

In parenchyma cells of the new roots, stems, and leaves, central vacuoles grew and pressed the cytoplasm outward until it became only a thin zone against the cell wall. The cell's surface area increased relative to the volume of cytoplasm, and this enhanced absorption. Water, mineral ions, and carbon dioxide could be harvested rapidly in spite of their dilute concentrations in soil and air. Mycorrhizae, the symbiotic association of roots and fungi, further enhanced absorption. Stomata developed in leaves and began to control water and carbon dioxide movements.

As the oak seedling grew, vascular pipelines served as functional links among all plant parts. Through xylem, water and minerals moved from roots to stems and leaves. Through phloem, organic compounds were shuttled from one region of the oak to another.

At the whim of a scrubjay, the seed had sprouted in a well-drained, sunlit basin at the foot of a canyon. Rainwater accumulated each winter, keeping the soil moist enough to encourage luxuriant growth through spring and the dry summers. Out in the open, red wavelengths of sunlight activated phytochrome in the seedling, triggering hormonal events that encouraged stem branching and leaf expansion. All the while, delicate hormone-mediated responses were being made to the winds, the sun, the tug of gravity, the changing seasons.

And so the oak increased in size. Every season, year after year, century after century, roots continued to develop and snake through a tremendous volume of the moist soil. Branches continued to spread beneath the sun. Leaves proliferated—leaves where the oak put together its own

food with sunlight energy, water, carbon dioxide, and the few simple minerals it mined from the soil.

On their way to the gold fields, prospectors of the California gold rush rested in the shade of the oak's immense canopy. The great earthquake of 1906 scarcely disturbed the giant, anchored as it was by a root system extending 80 feet through the soil. By chance, the brush fires that periodically sweep through California's coastal canyons did not seriously damage the tree. Fungi that could have rotted its roots never took hold; the soil was too well drained and the water table too deep. Leaf-chewing insects were kept in check by protective chemicals in the leaves and by predatory birds living in the canyon.

During the 1960s, human population growth surged in California. The land outside cities began to show the effects of population overflow as native plants gave way to suburban housing. A developer turned his tractors into the canyon but was so impressed with the giant oak that the tree was not felled. Death came later.

The new homeowners were not aware of the ancient, delicate relationships between the giant tree and the land that sustained it. Soil was graded between the trunk and the drip line of the overhanging canopy. Flower beds were mounded against the trunk. Lawns were planted beneath the branches and sprinklers installed. Overwatering in summer created standing water next to the great trunk— and the oak root fungus (*Armillaria*) that had been so successfully resisted until then became established. With its roots rotting away, the oak began to suffer the effects of massive disruption to the feedback relationships among its roots, stems, and leaves. Eventually it had to be cut down. In their fifth winter, in their red brick fireplace, the homeowners began burning three centuries of firewood.

## SUMMARY

1. Following dispersal from the parent, the seeds of flowering plants germinate: the embryo inside the seed absorbs water, resumes growth, and its primary root breaks through the seed coat.

2. Following germination, the plant increases in volume and mass. Tissues and organs of the seedling develop; later, flowers, fruits, and new seeds form, then older leaves drop away from the plant. Plant hormones govern these developmental events.

3. Five hormones have been identified in most flowering plants:
   a. Auxins promote elongation of coleoptile and stem cells.
   b. Gibberellins promote stem elongation; they also may help seeds and buds break dormancy.
   c. Cytokinins stimulate cell division, promote leaf expansion, and retard leaf aging.
   d. Abscisic acid promotes stomatal closure (retards water loss) and may trigger seed and bud dormancy.
   e. Ethylene promotes fruit ripening, abscission, and other processes.

4. Plants adjust their patterns of growth in response to environmental rhythms and to unique environmental circumstances. Among these responses are tropisms: differences in the rate and direction of growth on two sides of an organ such as a stem or root. Phototropism and gravitropism are examples.

5. Plants have biological clocks, or internal time-measuring mechanisms that have a biochemical basis. They can "reset" the clocks and so make seasonal adjustments in their patterns of growth, development, and reproduction.

6. In photoperiodism, plants respond to a change in the relative length of daylight and darkness in a 24-hour period. A switching mechanism involving phytochrome (a blue-green pigment) promotes or inhibits germination, stem elongation, leaf expansion, stem branching, and formation of flowers, fruits, and seeds.

7. Long-day plants flower in spring or summer, when daylength is long relative to night; short-day plants flower when daylength is relatively short. Flowering of day-neutral plants is not regulated by light. The flowering response of all may interact with phytochrome.

8. Senescence is the sum total of processes leading to the death of a plant or plant structure.

9. Dormancy is a state in which a perennial or biennial stops growing even though conditions appear to be suitable for continued growth. A decrease in Pfr levels may trigger dormancy.

## Review Questions

1. List the five known types of plant hormones and describe the main functions of each. *520–521*

2. What is phytochrome, and what is its role in flowering or some other process? *524–526*

3. Define plant tropism and give a specific example. *522*

4. Distinguish among vernalization, senescence, and dormancy. *526–527*

## Self-Quiz *(Answers in Appendix IV)*

1. Seed germination is over when the _____ .
   a. embryo absorbs water
   b. embryo resumes growth
   c. primary root pokes out of seed coat
   d. cotyledons unfurl

2. Which of the following statements is false?
   a. Auxins and gibberellins promote stem elongation.
   b. Cytokinins promote cell division and leaf expansion but retard leaf aging.
   c. Abscisic acid promotes water loss, promotes dormancy.
   d. Ethylene promotes fruit ripening and abscission.

3. Plant hormones _____ .
   a. interact with one another
   b. are influenced by environmental cues
   c. are active in plant embryos within seeds
   d. are active in adult plants
   e. all of the above are correct

4. Plant growth depends on _____ .
   a. cell division
   b. cell enlargement
   c. hormones
   d. all of the above are correct

5. Light of _____ wavelengths is the main stimulus for phototropism.
   a. red               c. green
   b. far-red           d. blue

6. Light of _____ wavelengths causes phytochrome to switch from inactive to active form; light of _____ wavelengths has the opposite effect.
   a. red; far-red      c. far-red; red
   b. red; blue         d. far-red; blue

7. The flowering process is a _____ response.
   a. phototropic       c. photoperiodic
   b. gravitropic       d. thigmotropic

8. Abscission occurs during _____ .
   a. seed germination  c. senescence
   b. flowering         d. dormancy

9. Match the plant reproduction and development terms.
   _____ gibberellin        a. promotes stem elongation
   _____ senescence         b. unequal growth following contact
   _____ phytochrome           with solid objects
   _____ phototropism       c. switching mechanism for
   _____ thigmotropism         photoperiodism
                            d. response to blue light, mainly
                            e. all processes leading to death of
                               plant or plant part

## Selected Key Terms

| | | |
|---|---|---|
| abscisic acid (ABA) *520* | coleoptile *520* | imbibition *518* |
| abscission *521* | cytokinin *520* | photoperiodism *524* |
| apical dominance *522* | dormancy *527* | phototropism *522* |
| auxin *520* | ethylene *521* | phytochrome *524* |
| biological clock *524* | flavoprotein *522* | plant hormone *520* |
| circadian rhythm *524* | florigen *522* | senescence *527* |
| | germination *517* | target cell *520* |
| | gibberellin *520* | thigmotropism *523* |
| | gravitropism *522* | tropism *522* |
| | herbicide *520* | vernalization *526* |

## Readings

Bowley, J. D., and M. Black. 1985. *Seeds: Physiology of Development and Germination*. New York: Plenum Press.

Nickell, L. 1982. *Plant Growth Regulators: Agricultural Uses*. New York: Springer-Verlag. Concise explanations of agricultural practices that include use of growth regulators.

Salisbury, F., and C. Ross. 1991. *Plant Physiology*. Fourth edition. Belmont, California: Wadsworth.

Villiers, T. 1975. *Dormancy and the Survival of Plants*. London: Edward Arnold. A short book summarizing major dormancy mechanisms.

Whatley, F. R., and J. M. Whatley. 1980. *Light and Plant Life*. London: Edward Arnold. A short book summarizing effects of light on germination and development with ecological descriptions, too.

FACING PAGE: *How many and what kinds of body parts does it take to function as a lizard in a tropical forest? Make a list of what comes to mind as you start reading Unit VI, then see how resplendent the list can become at the unit's end.*

## Meerkats, Humans, It's All the Same

After a bitingly cold night in the Kalahari Desert of Africa, animals small enough to fit in the pocket of an overcoat emerge from their underground burrows. These animals, called meerkats, stand on their hind legs and face eastward, exposing a large surface area of their chilled bodies to the warm rays of the morning sun (Figure 31.1).

The meerkats don't know it, but they are working to ensure good operating conditions for their enzymes. The animal body runs on enzymes, and if its temperature exceeds or falls below a tolerable range, the rate of enzyme activity drops sharply and metabolism suffers. Like many other animals, meerkats rely on behavioral adjustments to help maintain "inside" temperatures even though outside temperatures change.

After gathering warmth from the sun, the meerkats venture into the open and begin foraging. The farther they venture from their burrow, the more vulnerable they become to jackals and other predators—but they don't have much choice. They are hungry. The day before, nutritious little insects and the occasional lizard provided them with glucose and other required molecules. They used the glucose as a quick energy fix

and to keep their brain supplied with the stuff; glucose is *the* major nutrient that brain cells can actually use.

The meerkats that were lucky enough to take in many more food molecules than their cells could use stored away the excess, mostly by converting them to fats and to "animal starch" (glycogen). After returning to the safety of their burrow and hunkering in for the night, the animals relied on built-in controls in their body to trigger a shift in the type of molecules used to support their cells. Fats that had been stored away earlier in adipose tissue were broken down, transported by the bloodstream to the liver, and there converted to glucose. Glycogen that had been stored in liver cells was rapidly broken down to glucose, which was released into the bloodstream.

Each of those small furry bodies clearly depended on more than a digestive system. A circulatory system picked up nutrients that had been absorbed from the gut and transported them to cells throughout the body. A respiratory system helped cells use the nutrients by supplying them with oxygen (for aerobic respiration) and relieving them of carbon dioxide wastes. It did not matter that one meerkat had feasted on a tasty assort-

ment of insects, and another on a scrawny lizard. Even with variations in diet, their urinary system helped maintain the volume and composition of the vital fluids bathing their cells. As the meerkats slept, as they awakened and faced another day of hunger, thirst, and possibly heart-thumping flights from predators, they were never aware that their nervous system and endocrine system were working together in response to the challenges of the new day. The functioning of meerkats, humans, and all other complex animals depends on how well those two systems maintain operating conditions for all living cells in the body.

With no more than a cursory look at the meerkats, we have started thinking about the central topics of this unit. These topics are the structure of the animal body (its *anatomy*) and the mechanisms by which the body functions in the environment (its *physiology*). This chapter provides us with an overview of the kinds of tissues and organ systems we will be considering. It provides us also with an overview of the kinds of *homeostatic mechanisms* by which cells, tissues, organs, and organ systems work together in ways that maintain a stable environment *inside* the body.

**Figure 31.1** In the vast Kalahari Desert of Africa, meerkats line up and face the warming rays of the morning sun, just as they do every morning.

This simple behavior helps meerkats maintain their internal body temperature even when the outside temperature changes. How animals function in their environment is the subject of this unit.

KEY CONCEPTS

**1.** The cells of most animals interact at three levels of organization—in *tissues*, many of which are combined in *organs*, which are components of *organ systems*.

**2.** Most animals are constructed of only four types of tissues: epithelial, connective, nervous, and muscle tissues.

**3.** Each cell engages in basic metabolic activities that assure its own survival. At the same time, cells of a tissue or organ perform activities that contribute to the survival of the animal as a whole.

**4.** The combined contributions of cells, tissues, organs, and organ systems help maintain a stable "internal environment" that is required for individual cell survival. This concept is central to understanding the functions of any organ system, regardless of its complexity.

## ANIMAL STRUCTURE AND FUNCTION: AN OVERVIEW

Regardless of whether you are talking about a flatworm or salmon, a meerkat or human being, each animal is structurally and physiologically adapted to perform the following tasks:

**1.** Maintain internal "operating conditions" within some tolerable range even though external conditions change.

**2.** Locate and take in nutrients and other raw materials, distribute them through the body, and dispose of wastes.

**3.** Protect itself against injury or attack from viruses, bacteria, and other foreign agents.

**4.** Reproduce, and often help nourish and protect the new individuals during their early development.

Even the most complex animal is constructed of only four basic types of tissues, called epithelial, connective, muscle, and nervous tissues. A **tissue** is a group of cells and intercellular substances that function together in

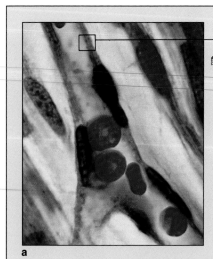

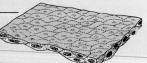

TYPE: Simple squamous

DESCRIPTION: Single layer flattened cells

COMMON LOCATION: Blood vessel walls, air sacs of lungs

FUNCTION: Diffusion

a

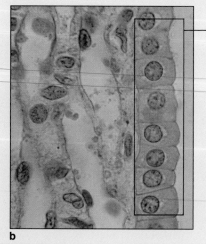

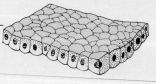

TYPE: Simple cuboidal

DESCRIPTION: Single layer cubelike cells; may have microvilli at its free surface

COMMON LOCATIONS: Part of gut lining, part of respiratory tract lining

FUNCTION: Secretion, absorption

b

one or more of the specialized tasks listed above. Complex animals consist of millions, even trillions of cells organized into tissues. Tissues split up the work, so to speak, in ways that contribute to the survival of the animal as a whole. This is sometimes called "a division of labor."

Different tissues become organized in specific proportions and patterns to form **organs**, such as a stomach. In **organ systems**, two or more organs interact chemically, physically, or both in performing a common task. For example, different organs of your digestive system ingest and prepare food for absorption by cells, then eliminate food residues.

Not all of the 2 million or so known species of animals have elaborate organ systems. Sponges and *Trichoplax adhaerens*, the organless, pancake-shaped marine animal shown on page 369, do not even have organs. The giant squid, one of the most complex invertebrates, has organ systems that are as sophisticated as yours. We mention these animals only to emphasize that there is no such thing as a "typical" animal. We humans tend to be interested in vertebrates (ourselves especially), and so they will be our focus in this unit. But comparative examples also will be drawn from the invertebrates to keep the structural and functional diversity of the animal kingdom in perspective.

## ANIMAL TISSUES

### Tissue Formation

Animal tissues are formed by cell divisions and cell differentiation that begin when an animal embryo embarks on its course of development. Chapter 42 sketches out the developmental mechanisms involved. Here it is enough to say that, at some point in the life cycle, meiotic cell division occurs in *germ cells*, which are immature reproductive cells that develop into gametes. In male animals, the gametes are sperm; in females, they are ova, or eggs. All other cells in the body are said to be *somatic* (from the Greek *soma*, meaning body). Fusion of a sperm and egg results in the formation of a zygote, which undergoes mitotic cell divisions that produce the early embryo.

In the case of vertebrates, cells in the early embryo soon become arranged into three "primary" tissues—ectoderm, mesoderm, and endoderm. These are the embryonic forerunners of all the specialized tissues of the body. *Ectoderm* gives rise to the outer layer of skin and to tissues of the nervous system. *Mesoderm* gives rise to muscle; the organs of circulation, reproduction, and excretion; most of the internal skeleton; and connective tissue layers of the gut and body covering. *Endoderm* gives rise to the lining of the gut and the major organs derived from it.

With this overview in mind, let's now consider some of the features that allow us to identify the four types of specialized tissues present in the adult body.

### Epithelial Tissue

Figure 31.2 shows a few examples of epithelial tissues. An epithelial tissue also is called an **epithelium** (plural, epithelia), and different types serve different functions. The epidermis of skin, for example, is a protective covering for the body's exterior surface. Epithelia that line parts of the gut and other internal cavities function in secreting or absorbing substances.

**General Features**. The cells of epithelium adhere to one another closely, with little space or extracellular material between them, and they are organized as one or more layers. Epithelium always has one free surface,

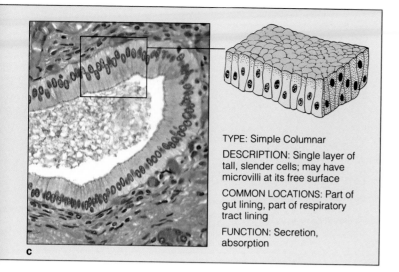

c

TYPE: Simple Columnar

DESCRIPTION: Single layer of tall, slender cells; may have microvilli at its free surface

COMMON LOCATIONS: Part of gut lining, part of respiratory tract lining

FUNCTION: Secretion, absorption

**Figure 31.2** Examples of simple epithelium, showing the three basic cell shapes in this type of tissue.

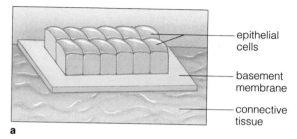

epithelial cells

basement membrane

connective tissue

a

which means that no other cells adhere to it. In many cases, cells at the free surface are crowned with cilia or microvilli (singular, microvillus), which are fingerlike protrusions of cytoplasm. These surface modifications have special roles, as you will see in later chapters. The opposite surface of epithelium adheres to a *basement membrane*. This is a noncellular layer, rich with proteins and polysaccharides, that lies between the epithelium and an underlying connective tissue (Figure 31.3a).

*Simple epithelium* consists of a single layer of cells, which typically have roles in the diffusion, secretion, absorption, or filtering of substances across the layer. For example, oxygen and carbon dioxide diffuse readily across the simple epithelium making up the wall of fine blood vessels, as shown in Figure 31.2a. *Stratified epithelium* consists of two or more layers of cells, as shown in Figure 31.3b, and it typically functions in protection.

**Glandular Epithelium.** It takes digestive enzymes, mucus, and many other glandular secretions to keep your body functioning properly. *Glands* are single cells or multicelled secretory structures that are derived from and composed of epithelia. Mucus-secreting, goblet-shaped cells, for instance, are embedded in epithelia that line some of the tubes leading to your lungs. The entire epithelium of your stomach is a sheet of glandular cells that secrete mucus and enzymes.

We classify glands according to how their products are distributed. **Exocrine glands** secrete products onto a free epithelial surface, usually through ducts or tubes. Exocrine products include cell-produced substances such as mucus, saliva, earwax, oil, milk, and digestive enzymes. **Endocrine glands** are ductless. As described in Chapter 34, their products (hormones) are secreted directly into the fluid bathing cells and are picked up for distribution by the bloodstream.

free surface

b

**Figure 31.3** Characteristics of epithelium. (**a**) All epithelia have a free surface, and a basement membrane is interposed between the opposite surface and an underlying connective tissue. (**b**) In stratified epithelium, which consists of two or more layers, cells at the free surface typically are flattened as shown here.

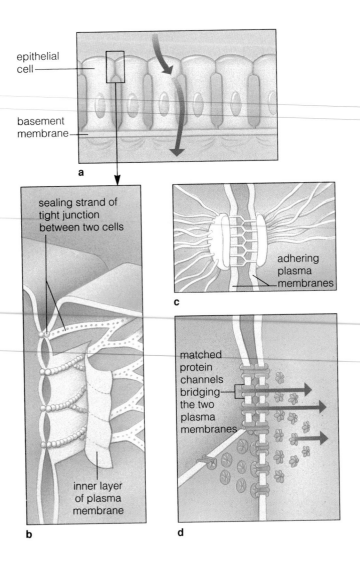

epithelial cell

basement membrane

a

sealing strand of tight junction between two cells

inner layer of plasma membrane

b

adhering plasma membranes

c

matched protein channels bridging the two plasma membranes

d

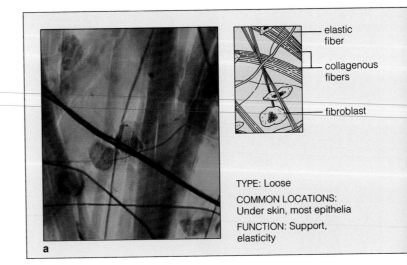

elastic fiber

collagenous fibers

fibroblast

a

TYPE: Loose

COMMON LOCATIONS: Under skin, most epithelia

FUNCTION: Support, elasticity

**Figure 31.4** Examples of cell-to-cell contacts in animal tissues.

**(a)** A complex of cell junctions serves as a barrier to the leakage of substances *between* the cells of some epithelia, such as the one lining the small intestine. Here, *tight junctions* ring each cell and seal it to its neighbors near their free surface. Protein strands are the sealing devices; they are embedded in the lipid bilayer of the plasma membranes **(b)**. To move across the lining, any substance must be absorbed by the cells themselves.

**(c)** Below the tight junction are *desmosomes*, which are something like spot welds that mechanically link cells and permit them to function as a structural unit. Each has cytoskeletal filaments attached to dense, "cementing" material on the cytoplasmic side of the plasma membrane, and these extend into both of the adjoining cells. Desmosomes occur within all types of animal tissues. They are especially abundant in the surface layer of skin and other kinds of epithelia that are subjected to abrasion and other mechanical insults.

**(d)** *Gap junctions* also occur in epithelia. Here, clustered protein channels across the plasma membrane of one cell match up with clustered channels of an adjoining cell. Gap junctions provide a low-resistance path for the diffusion of ions and small molecules from cell to cell. They are abundant in certain tissues of the heart, liver, and other organs that depend on rapid coordination of chemical activities among their cells.

**Cell-to-Cell Contacts in Epithelium.** With few exceptions, epithelial cells adhere strongly to one another by means of specialized attachment sites. These sites of cell-to-cell contact are especially thick when substances must not leak from one compartment to another in the body.

If the potently acidic fluid in your stomach were to leak through its epithelial lining, for instance, it would start digesting your body's own proteins instead of the ones brought in with your meals. (Actually, this is an end result of a peptic ulcer, as described on page 646.) This epithelium and others have intricate cell-to-cell contacts that make up a leakproof barrier between cells near their free surface. As Figure 31.4 shows, epithelia also include junctions that serve as spot welds and as open channels between cells.

## Connective Tissue

Connective tissues are diverse, and they serve very diverse functions. They range from the ones called connective tissue proper to the specialized types, which include cartilage, bone, adipose tissue, and blood (Table 31.1).

Spaces intervene between the cells of connective tissue. In this respect, connective tissue is notably different from epithelia. Except for blood, some of the cells produce fibers that serve as structural elements. The fibers contain collagen (which makes them strong) or elastin (which makes them elastic). The cells of connective tissue also secrete a *ground substance* which, together with the fibers, make up the extracellular matrix (page 70).

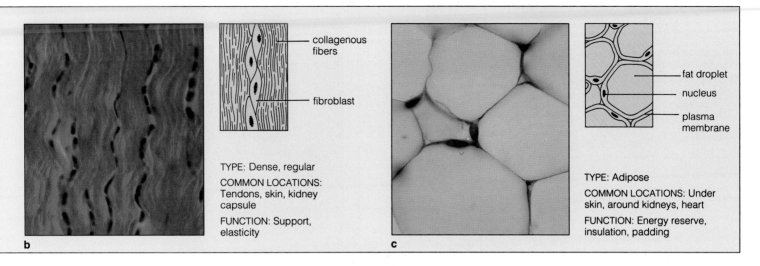

**Figure 31.5** Three examples drawn from the diverse kinds of tissues collectively called connective tissues. (**a**) Loose connective tissue, (**b**) dense connective tissue, and (**c**) adipose tissue.

**Connective Tissue Proper**. The components of both types of connective tissue in this category (loose or dense) are irregularly arranged. **Loose connective tissue** supports epithelia and many organs, and it surrounds blood vessels and nerves. As the name implies, it contains loosely arranged collagenous and elastic fibers as well as fibroblasts, a type of cell that produces the fibers and ground substance (Figure 31.5a). It also contains macrophages and other cells that migrate through tissues or take up residence in them. These cells perform housekeeping tasks and help protect the body against disease. In fact, the ones in connective tissue underlying the skin serve as a first line of defense when bacteria and other agents enter the body through cuts, abrasions, and other wounds.

Compared to its loose counterpart, **dense, irregular connective tissue** has thicker fibers and more of them, but far fewer cells. Its fibers interweave with one another but not in a regular orientation. If this suggests to you that dense connective tissues occur in organs that are not subjected to continual stretching, you would be correct. For example, such tissues occur in the protective capsule around the testis, the primary reproductive organ in males.

**Specialized Connective Tissue**. Unlike the tissues just described, **dense, regular connective tissue** has its collagen fibers oriented in parallel (Figure 31.5b). With this organized orientation, the fibers strongly resist being pulled apart under tension when the tissue is stretched. Often, parallel bundles of fibers have rows of fibroblasts between them. This is the arrangement you see in ligaments (which attach bone to bone) and tendons (which attach muscle to bone).

**Table 31.1  Types of Connective Tissue**

**Connective tissue proper:**

Loose connective tissue
Dense, irregular connective tissue

**Specialized connective tissue:**

Dense, regular connective tissue (ligaments, tendons)
Cartilage
Bone
Adipose tissue
Blood

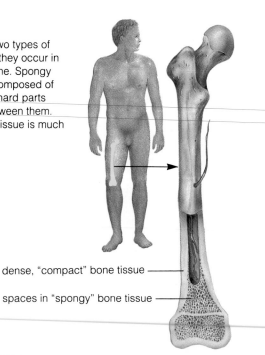

**Figure 31.6** Two types of bone tissue, as they occur in a human leg bone. Spongy bone tissue is composed of tiny, needlelike hard parts with spaces between them. Compact bone tissue is much more dense.

dense, "compact" bone tissue

spaces in "spongy" bone tissue

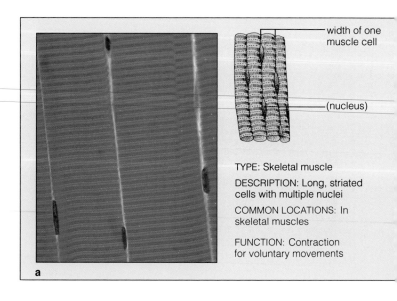

width of one muscle cell

(nucleus)

TYPE: Skeletal muscle

DESCRIPTION: Long, striated cells with multiple nuclei

COMMON LOCATIONS: In skeletal muscles

FUNCTION: Contraction for voluntary movements

a

**Cartilage**, another specialized connective tissue, cushions and helps maintain the shape of many body parts. It has a solid yet somewhat pliable matrix through which substances diffuse from nearby blood vessels. This pliability allows cartilage to resist compression and stay resilient, like a piece of solid rubber. As you will see, cartilage is an important tissue in growing bones. It also is present at the ends of many mature bones as well as parts of the nose, ear, and backbone.

**Bone** is unlike all other connective tissues in being mineralized and hardened; its collagen fibers and ground substance are loaded with calcium salts. Bone tissue is not completely solid, however. A variety of spaces occur within the ground substance, some of which harbor living bone cells (Figure 36.13). This specialized connective tissue becomes organized into bony structures that support and protect softer tissues and organs. Figure 31.6 shows such a bone in the human body. The body's arm and leg bones act with muscles; they form a leverlike system for movement. Some bones also function in red blood cell production.

The connective tissue called **adipose tissue** has large, densely clustered cells that are specialized for fat storage (Figure 31.5c). The animal body can store only so much carbohydrate and protein, and the excess is converted to storage fats that are tucked away in these cells. The tissue has a rich supply of blood, which serves as an immediately accessible "highway" for the movement of fats to and from individual adipose cells.

The specialized connective tissue called **blood** transports oxygen to cells and wastes away from them; it also transports hormones and enzymes. Some of its compo-

nents protect against blood loss (through clotting mechanisms), and others defend against disease-causing agents. Chapter 38 describes this tissue and its complex functions.

**Muscle Tissue**

All types of **muscle tissue** contain specialized cells that contract (shorten) in response to stimulation, then passively lengthen and so return to their resting state. Muscle tissue helps move the whole body as well as its individual parts. There are three categories, called skeletal, smooth, and cardiac muscle tissues.

*Skeletal* muscle tissue contains many long, cylindrical cells (Figure 31.7a). Typically, a number of skeletal muscle cells are bundled together, then several bundles are enclosed in a tough connective tissue sheath to form "a muscle," such as a biceps. Chapter 36 describes how this type of tissue functions.

*Smooth* muscle tissue consists of spindle-shaped cells, which are tapered at both ends (Figure 31.7b). Connective tissue holds the cells together. Smooth muscle tissue occurs in walls of blood vessels, the stomach, and other internal organs. In vertebrates, smooth muscle is said to be "involuntary," because the individual usually cannot directly control its contraction.

*Cardiac* muscle tissue is the contractile tissue of the heart (Figure 31.7c). The plasma membranes of adjacent cardiac muscle cells are fused together. Cell junctions at these fusion points allow the cells to contract as a unit. When one muscle cell receives a signal to contract, its neighbors are also stimulated into contracting.

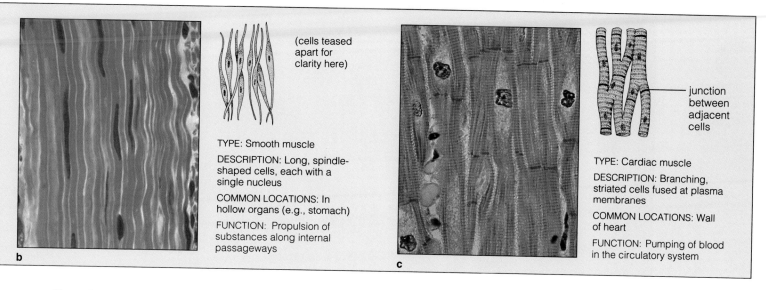

(cells teased apart for clarity here)

TYPE: Smooth muscle

DESCRIPTION: Long, spindle-shaped cells, each with a single nucleus

COMMON LOCATIONS: In hollow organs (e.g., stomach)

FUNCTION: Propulsion of substances along internal passageways

junction between adjacent cells

TYPE: Cardiac muscle

DESCRIPTION: Branching, striated cells fused at plasma membranes

COMMON LOCATIONS: Wall of heart

FUNCTION: Pumping of blood in the circulatory system

b

c

**Figure 31.7** Examples of skeletal, smooth, and cardiac muscle tissues.

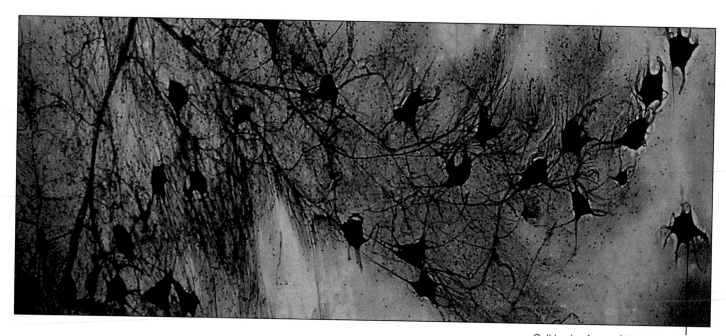

Cell body of one of the motor neurons in this nervous tissue sample

## Nervous Tissue

In **nervous tissue**, cells called neurons are organized as lines of communication that extend throughout the body. Some types of neurons detect specific changes in environmental conditions. Others coordinate the body's immediate and long-term responses to change. Still others relay signals to muscles and glands that can carry out those responses. Examples of this last type are shown in Figure 31.8, and their functioning will be a key topic in Chapter 33.

**Figure 31.8** A sampling of the millions of neurons that form communication lines within and between different regions of the human body. Shown here, motor neurons, which relay signals from the brain or spinal cord to muscles and glands. Collectively, these and other neurons sense environmental change, integrate a great number and variety of signals about those changes, and initiate appropriate responses.

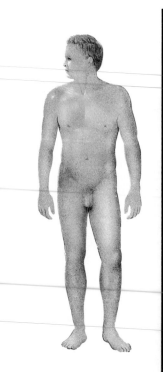

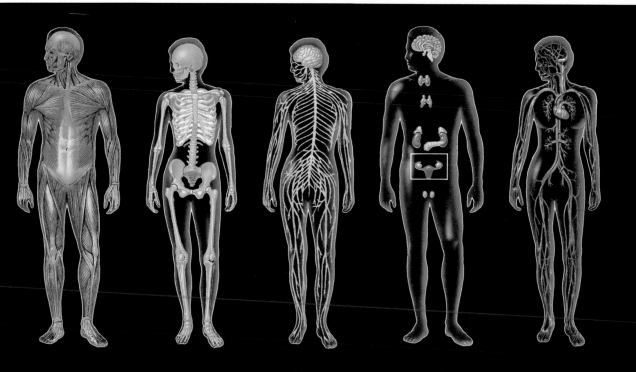

| INTEGUMENTARY SYSTEM | MUSCULAR SYSTEM | SKELETAL SYSTEM | NERVOUS SYSTEM | ENDOCRINE SYSTEM | CIRCULATORY SYSTEM |
|---|---|---|---|---|---|
| Protection from injury and dehydration; body temperature control; excretion of some wastes; reception of external stimuli; defense against microbes. | Movement of internal body parts; movement of whole body; maintenance of posture; heat production. | Support, protection of body parts; sites for muscle attachment, blood cell production, and calcium and phosphate storage. | Detection of external and internal stimuli; control and coordination of responses to stimuli; integration of activities of all organ systems. | Hormonal control of body functioning; works with nervous system in integrative tasks. | Rapid internal transport of many materials to and from cells; helps stabilize internal temperature and pH. |

**Figure 31.9** Organ systems of the human body. All vertebrates have the same types of systems, serving similar functions.

## MAJOR ORGAN SYSTEMS

During embryonic development, recall, epithelial tissues, connective tissues, nervous tissue, and muscle tissue begin their formation in the animal embryo. They eventually become organized into organs and then into organ systems that are much the same in all vertebrates. Figure 31.9 shows the general arrangement of these systems in humans. Each type of organ system will be described in subsequent chapters.

You might think we are stretching things a bit when we say that each one of those organ systems contributes to the survival of all living cells in the body. After all, what could the body's skeleton and musculature have to do with the life of a tiny cell? Yet interactions between

the skeletal and muscular systems provide a means of moving about—toward sources of nutrients and water, for example. Some of their components help circulate blood through the body, as when contractions of certain leg muscles help move blood in veins back to the heart. Through blood circulation, nutrients and other substances are transported to individual cells, and wastes are carried away from them.

Throughout this unit, we will be using some standard terms for describing the location of organs and organ systems in the vertebrate body. Take a moment to study Figure 31.10a, which shows the location of some major body cavities in which organs occur. Also study Figure 31.10b, which defines some anatomical terms that apply to most animals.

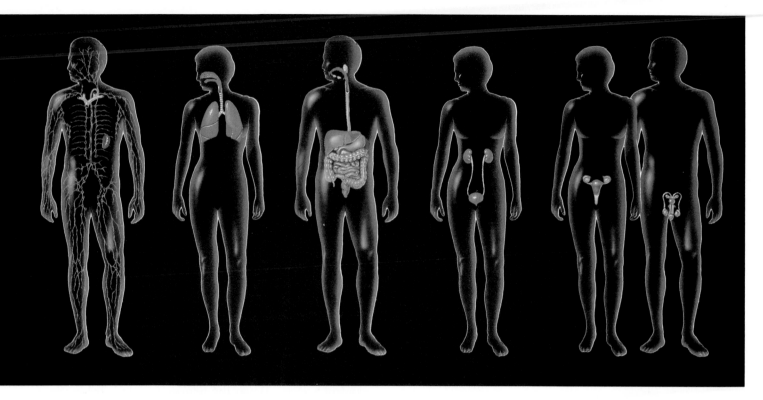

| LYMPHATIC SYSTEM | RESPIRATORY SYSTEM | DIGESTIVE SYSTEM | URINARY SYSTEM | REPRODUCTIVE SYSTEM |
|---|---|---|---|---|
| Return of some tissue fluid to blood; roles in immunity (defense against specific invaders of the body). | Provisioning of cells with oxygen; removal of carbon dioxide wastes produced by cells; pH regulation. | Ingestion of food, water; preparation of food molecules for absorption; elimination of food residues from the body. | Maintenance of the volume and composition of extracellular fluid. Excretion of blood-borne wastes. | Male: production and transfer of sperm to the female. Female: production of eggs; provision of a protected, nutritive environment for developing embryo and fetus. Both systems have hormonal influences on other organ systems. |

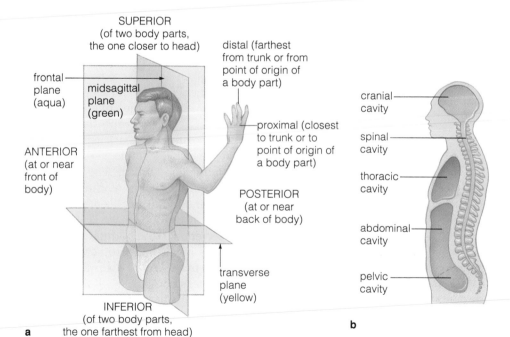

**Figure 31.10** (**a**) Directional terms and planes of symmetry for the human body. The midsagittal plane divides the body into right and left halves. The frontal plane divides it into anterior (front) and posterior (back) parts. For humans, the main body axis is perpendicular to the earth. For rabbits and other animals that move with the main body axis parallel to the earth, *ventral* corresponds to anterior, and *dorsal* corresponds to posterior. (Compare Figure 25.2.)

(**b**) Major cavities in the human body.

# HOMEOSTASIS AND SYSTEMS CONTROL

## The Internal Environment

To stay alive, your body's cells must be continually bathed in fluid that supplies them with nutrients and carries away metabolic wastes. In this they are no different from an amoeba or any other free-living, single-celled organism. However, many *trillions* of cells are crowded together in your body—and they all must draw nutrients from and dump wastes into the same 15 liters of fluid. That is less than 16 quarts.

The fluid *not* inside cells is called **extracellular fluid**. Much of it is *interstitial*, meaning it occupies the spaces between cells and tissues. The rest is *plasma*, the fluid portion of blood. Interstitial fluid exchanges substances with blood and with the cells it bathes.

In functional terms, the extracellular fluid is continuous with the fluid inside cells. That is why drastic changes in its composition and volume have drastic effects on cell activities. Its concentrations of hydrogen, potassium, calcium, and other ions are especially important in this regard. Those concentrations must be maintained at levels that are compatible with the survival of the body's individual cells. Otherwise, the animal itself cannot survive.

It makes no difference whether the animal is simple or complex. *The component parts of any animal work together to maintain the stable fluid environment required by its living cells.* This concept is absolutely central to our understanding of the structure and function of animals, and it may be summarized this way:

1. Each cell of the animal body engages in basic metabolic activities that ensure its own survival.

2. Concurrently, the cells of a given tissue typically perform one or more activities that contribute to the survival of the whole organism.

3. The combined contributions of individual cells, organs, and organ systems help maintain the stable internal environment—that is, the extracellular fluid—required for individual cell survival.

## Mechanisms of Homeostasis

The word **homeostasis** refers to stable operating conditions in the internal environment. This state is maintained through homeostatic controls, which operate in coordinated ways to keep physical and chemical aspects of the body within tolerable ranges. Many of the controls work through feedback mechanisms.

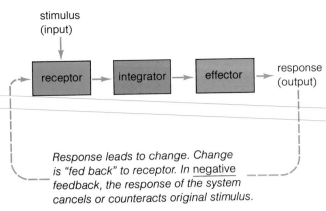

*Response leads to change. Change is "fed back" to receptor. In negative feedback, the response of the system cancels or counteracts original stimulus.*

**Figure 31.11** Components necessary for negative feedback at the organ level.

In a **negative feedback mechanism**, an activity changes some condition in the internal environment, and this in turn triggers a response that reverses the changed condition. Think about a furnace with a thermostat. The thermostat senses the air temperature and "compares" it to a preset point on a thermometer built into the furnace control system. When the temperature falls below the preset point, the thermostat signals a switching mechanism that turns on the heating unit. When the air becomes heated enough to match the prescribed level, the thermostat signals the switching mechanism, which shuts off the heating unit.

Similarly, meerkats and many other animals rely on feedback mechanisms to raise or lower body temperature so that it is maintained near 37°C (98.6°F) even during extremely hot or cold weather. When the body senses that its skin is getting too hot outside in the summer sun, for example, mechanisms are set in motion that slow down metabolic activity and overall body activity. Thus internal controls work to counteract the possibility of an intolerably high body temperature by slowing down the body's heat-generating activities.

Under some circumstances, **positive feedback mechanisms** set in motion a chain of events that *intensify* the original condition. Positive feedback is associated with instability in a system. For example, sexual arousal leads to increased stimulation, which leads to more stimulation, and so on until an explosive, climax level is reached (page 766). As another example, during childbirth, pressure of the fetus on the uterine walls stimulates production and secretion of the hormone oxytocin. Oxytocin causes muscles in the walls to contract, this increases pressure on the fetus, and so on until the fetus is expelled from the mother's body.

Homeostatic control mechanisms require three components: sensory receptors, integrators, and effectors (Figures 31.11 and 31.12). *Sensory receptors* are cells or parts of cells that can detect a specific change in the

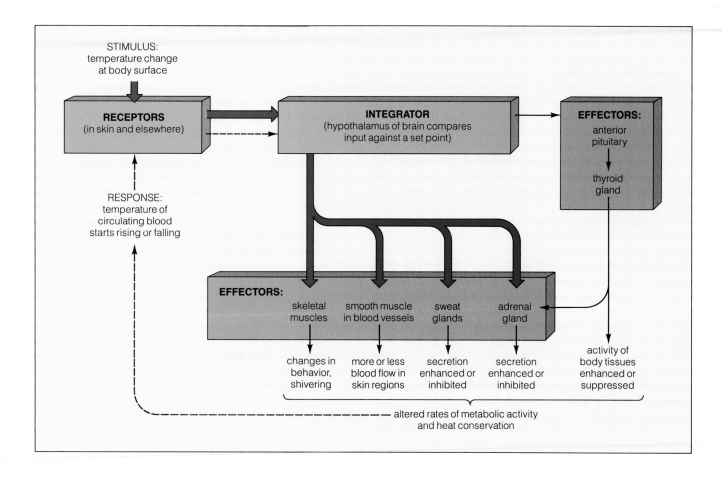

**Figure 31.12** Homeostatic controls over the internal temperature of the human body. The dashed line shows how the feedback loop is completed. The blue arrows indicate the main control pathways.

environment. For example, when someone kisses you, there is a change in pressure on your lips. Receptors in the skin of your lips translate the stimulus energy into a signal, which can be sent to the brain. Your brain is an *integrator*, a control point where different bits of information are pulled together in the selection of a response. It can send signals to your muscles or glands (or both), which are *effectors* that carry out the response. In this case, the response might include flushing with pleasure and kissing the person back. Or it might include flushing with rage and shoving the person away from your face.

Thus your brain continually receives information about how things *are* operating—that is, the information from receptors. It also receives information about how things *should be* operating—that is, information from "set points." When conditions deviate significantly from a set point, the brain functions to bring them back to the most effective operating range. It does this by way of signals that cause specific muscles and glands to increase or decrease their activity.

---

**Homeostatic control mechanisms maintain physical and chemical aspects of the internal environment within ranges that are most favorable for cell activities.**

---

What we have been describing here is a general pattern of monitoring and responding to a constant flow of information about the external and internal environments. During this activity, organ systems operate together in coordinated fashion. Throughout this unit, we will be asking the following questions about their operation:

1. What physical or chemical aspect of the internal environment are organ systems working to maintain as conditions change?

2. By what means are organ systems kept informed of change?

3. By what means do they process incoming information?

4. What mechanisms are set in motion in response?

As you will see in the chapters that follow, the operation of all organ systems is under neural and endocrine control.

## SUMMARY

1. A tissue is an aggregation of cells and intercellular substances united in the performance of a specialized activity. An organ is a structural unit in which tissues are combined in definite proportions and patterns that allow them to perform a common task. An organ system has two or more organs interacting chemically, physically, or both in ways that contribute to the survival of the animal as a whole.

2. Epithelial tissues are one or more layers of adhering cells with one free surface and the opposite surface resting on a basement membrane that intervenes between it and an underlying connective tissue. Epithelia cover external body surfaces and line internal cavities and tubes.

3. Connective tissues bind together other tissues or offer them mechanical or metabolic support. They include connective tissue proper and specialized connective tissues (such as cartilage, bone, and blood).

4. Muscle tissue, which is specialized for contraction, functions in the movements of body parts and movement through the environment. Nervous tissue detects and coordinates information about change in the internal and external environments, and it controls responses to those changes.

5. Organ systems work largely through homeostatic control mechanisms that help maintain a stable internal environment. In negative feedback (the most common mechanism), the response to a disturbance in a system decreases the original disturbance. In positive feedback, a response intensifies the original disturbance.

6. Tissues, organs, and organ systems work in ways that help maintain the stable internal environment (the extracellular fluid) required for individual cell survival.

7. Control of the internal environment depends on the body's receptors, integrators, and effectors. Receptors detect stimuli, which are specific changes in the environment. Integrating centers (such as the brain) receive information from receptors about how some aspect of the body's receptors, integrators, and effectors. Receptors detect stimuli, which are specific changes in the environment. Integrating centers (such as the brain) receive both), which carry out the appropriate response.

### Review Questions

1. Label the following tissues appropriately: *534–539*

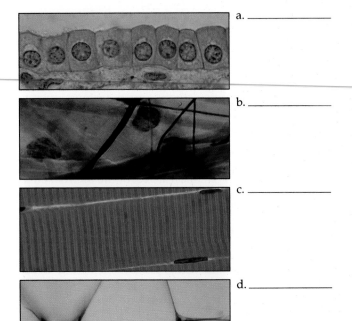

a. _____

b. _____

c. _____

d. _____

2. What is an animal tissue? An organ? An organ system? List the major organ systems of the human body, along with their functions. *533–534, 540*

3. Define extracellular fluid and interstitial fluid. *542*

4. Epithelial tissue and connective tissue differ from each other in overall structure and function. Describe how. *534–538*

5. State the overall functions of (a) muscle tissue and (b) nervous tissue. *538–539*

6. A major important concept in animal physiology relates the functioning of cells, organs, and organ systems to the internal environment. Can you state the three main points of this concept? *542*

7. Define homeostasis. What are the three components necessary for homeostatic control over the internal environment? *542–543*

8. What are the differences between negative feedback and positive feedback mechanisms? *542*

### Self-Quiz *(Answers in Appendix IV)*

1. The four main types of tissues in most animals are _____, _____, _____, and _____.

2. Animals are structurally and functionally adapted for these tasks:
   a. maintenance of the internal environment
   b. nutrient acquisition, processing, distribution, disposal
   c. self-protection against injury or attack
   d. reproduction
   e. all of the above

3. _____ tissues cover external body surfaces, line internal cavities and tubes, and some form the secretory portions of glands.
   a. Muscle
   b. Nervous
   c. Connective
   d. Epithelial

4. Most _____ tissues bind or mechanically support other tissues; but one type functions in physiological support of other tissues.
   a. muscle
   b. nervous
   c. connective
   d. epithelial

5. _____ tissues detect and coordinate information about environmental changes and control responses to those changes.
   a. Muscle
   b. Nervous
   c. Connective
   d. Epithelial

6. _____ tissues contract and make possible internal body movements as well as movements through the external environment.
   a. Muscle
   b. Nervous
   c. Connective
   d. Epithelial

7. Cells in the animal body _____.
   a. engage in metabolic activities that ensure their survival
   b. perform activities that contribute to the survival of the animal
   c. contribute to maintaining the extracellular fluid
   d. all of the above

8. In a state of _____, physical and chemical aspects of the body are being kept within tolerable ranges by controlling mechanisms.
   a. positive feedback
   b. negative feedback
   c. homeostatis
   d. metastasis

9. In negative feedback mechanisms, _____.
   a. a detected change brings about a response that tends to return internal operating conditions to the original state
   b. a detected change suppresses internal operating conditions to levels below the set point
   c. a detected change raises internal operating conditions to levels above the set point
   d. fewer solutes are fed back to the affected cells

10. _____ detect specific environmental changes, an _____ pulls different bits of information together in the selection of a response, and _____ carry out the response.

11. Match the concepts.
    _____ muscles and glands
    _____ positive feedback
    _____ body receptors
    _____ negative feedback
    _____ brain

    a. integrating center
    b. the most common homeostatic mechanism
    c. eyes and ears
    d. effectors
    e. chain of events intensifies the original condition

## Selected Key Terms

adipose tissue *538*
blood *538*
bone *538*
cardiac muscle *538*
cartilage *538*
dense, irregular connective tissue *537*
dense, regular connective tissue *537*
endocrine gland *535*
effector *543*
epithelium *534*
exocrine gland *535*
extracellular fluid *542*
germ cell *534*
homeostasis *542*

integrator *543*
loose connective tissue *537*
muscle tissue *538*
negative feedback mechanism *542*
nervous tissue *539*
organ *534*
organ system *534*
positive feedback mechanism *542*
sensory receptor *542*
skeletal muscle *538*
smooth muscle *538*
somatic cell *534*
tissue *533*

## Readings

Bloom, W., and D. W. Fawcett. 1986. *A Textbook of Histology.* Eleventh edition. Philadelphia: Saunders. Outstanding reference text.

Leeson, C. R., T. Leeson, and A. Paparo. 1985. *Textbook of Histology.* Philadelphia: Saunders.

Ross, M., and E. Reith. 1985. *Histology: A Text and Atlas.* New York: Harper & Row.

Vander, A., J. Sherman, and D. Luciano. 1990. "Homeostatic Mechanisms and Cellular Communication" in *Human Physiology.* Fifth edition. New York: McGraw-Hill.

# 32 INFORMATION FLOW AND THE NEURON

## Tornado!

It is spring in the American Midwest, and you are fully engrossed in photographing the magnificent wildflowers all around you in an expanse of shortgrass prairie. So intent are you on capturing all the different species on film that you fail to notice the rapidly darkening sky. By the time you finally look up, the sky is darkening ominously. What's that rumbling you hear in the distance? It sounds like a freight train. Yet how can that be, when there is no track, anywhere, in this part of the prairie? You turn to identify the source of the sound. And you see it—but you don't want to believe your eyes. A funnel cloud is advancing across the prairie and heading right for you! *TORNADO!* The terrifying image rivets your attention as nothing else has ever done before.

You sense instantly that you cannot remain where you are and survive. Suddenly you remember that hiding in a low area is better than standing out in the open. Commands flash from your brain to your limbs: *GET MOVING OUT OF HERE!* With heart thumping, you start to run along a path, looking frantically for safety. You're in luck! Just ahead is a steep-banked creek. With a tremendous burst of speed you reach the creek in less than a minute, then scramble down the muddy bank. There you find a small ledge hanging over the water. You wedge yourself under it, hoping wildly to be inconspicuous, to be overlooked by a force of nature on the rampage.

It takes a few minutes before you realize that the tornado has roared past the creek. You remain motionless, heart still thumping, fingers still clutching mud. Finally, cautiously, you sit up and look around. Some distance away, you see a swath of twisted prairie grass that marks the tornado's path. Your heart no longer feels like it is slamming against your chest wall, although your legs feel like rubber when you stand up.

You can thank your nervous system for every perception, every memory, and nearly every action that helped you escape the tornado. You can thank the coordinated signals that traveled rapidly through its communication lines, which are composed of cells called neurons.

Were all of those lines silent until they received signals from the outside, much as telephone lines wait to carry calls from all over the country? Not even

**Figure 32.1** A truly terrifying view across the Kansas prairie—a tornado about to touch down. Imagine yourself alone in the prairie when you first see the tornado, knowing you have but minutes to remove yourself from harm's way. What thoughts flash through your mind? How rapidly do you accept or reject possible plans of action?

remotely. Even before you were born, your newly developing neurons became organized in vast gridworks and started chattering among themselves. All through your life, in moments of danger or reflection, excitement or sleep, their chatter has never ceased. Constant communication among neurons, such as the ones controlling the muscles concerned with breathing, keep you alive. It keeps your body primed for rapid response to change—even to the totally unexpected appearance of a tornado.

This chapter begins with the structure and function of individual neurons. Then it starts us thinking about how neurons interact. The next two chapters deal with the nervous system and endocrine system, which are inseparable in their integrative and control functions. Chapter 35 provides a closer look at sensory structures and sensory organs, including eyes, by which you detect tornadoes and other events that may have bearing on whether you survive from one day to the next.

KEY CONCEPTS

**1.** A nervous system senses, interprets, and issues commands for responses to specific aspects of the environment. Its communication lines are highly organized gridworks of cells called neurons.

**2.** A steady difference in electric charge exists across a neuron's plasma membrane. Sudden, brief reversals in that difference are the basis of messages sent through the nervous system. The reversals, called action potentials, occur when a neuron is adequately stimulated. They occur from the point of stimulation to the neuron's signal output zone, where it forms a junction with another cell.

**3.** At junctions called chemical synapses, action potentials trigger the release of a chemical substance that stimulates or inhibits the activities of the next cell in line.

## CELLS OF THE NERVOUS SYSTEM

In all nervous systems, the basic unit of communication is the nerve cell, or **neuron**. Neurons do not act alone. They *collectively* sense changing conditions, integrate sensory inputs, then activate different body parts that can carry out responses. These tasks involve signals among different classes of nerve cells, called sensory neurons, interneurons, and motor neurons.

We can define each class in terms of its role in a control scheme, described in Chapter 31, by which the nervous system monitors and responds to change:

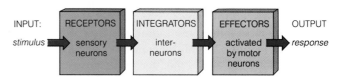

INPUT: RECEPTORS → INTEGRATORS → EFFECTORS OUTPUT
*stimulus* → sensory neurons → inter-neurons → activated by motor neurons → *response*

**Sensory neurons** have receptor regions that can detect specific stimuli, such as light energy. These neurons relay signals about the stimulus *to* the brain and spinal cord (the integrators in our control scheme). In the brain and spinal cord are **interneurons**, which integrate infor-

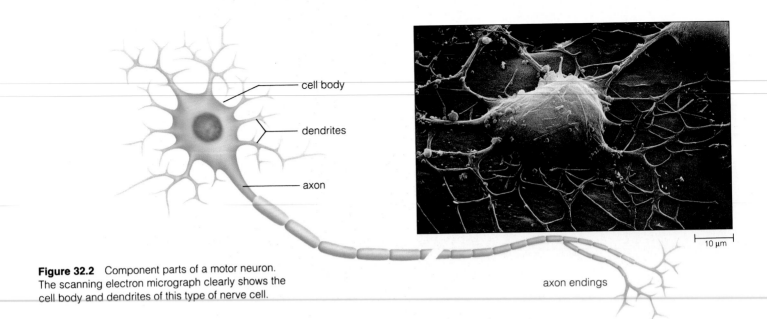

cell body

dendrites

axon

axon endings

10 µm

**Figure 32.2** Component parts of a motor neuron. The scanning electron micrograph clearly shows the cell body and dendrites of this type of nerve cell.

mation arriving on sensory lines and then influence other neurons in turn. **Motor neurons** relay information *away* from the integrator to muscle cells or gland cells. Muscles and glands are the body's effectors, which carry out responses.

Keep in mind that the three classes of neurons just defined make up only about half the volume of vertebrate nervous systems. A variety of **neuroglial cells** also are present in great numbers. These specialized cells physically support and protect the neurons and help them carry out their tasks. Some impart structure to the brain, much as connective tissues do for other body regions. Some segregate groups of neurons. Others wrap like a series of jellyrolls around parts of sensory and motor neurons, and they affect how fast a signal travels along the neuron.

## FUNCTIONAL ZONES OF THE NEURON

All neurons have a cell body, which contains the nucleus and the metabolic machinery for protein synthesis. Most neurons have slender cytoplasmic extensions of the cell body, although these differ enormously in number and length. So great are the structural differences that there really is no such thing as a "typical" neuron. The ones described most often are motor neurons of the sort shown in Figure 32.2. Motor neurons have many **dendrites**, which are short, slender extensions of the cell body. They also have one long, cylindrical extension called an **axon**.

An axon of motor neurons has finely branched endings that terminate on muscle cells. As Figure 32.3 suggests, dendrites and axons also occur on many sensory neurons and interneurons.

Generally speaking, we can think of dendrites and the cell body as "input zones," where signals about changing conditions are received. Axon endings are "output zones," where messages are sent to other cells.

Each neuron has input zones for receiving signals as well as output zones, where signals are sent on to other cells.

## NEURAL MESSAGES

### Membrane Excitability

Different kinds of signals, some electrical, some chemical, constitute the messages that travel through a nervous system. Let's start with the signals that travel along the plasma membrane of individual neurons.

A neuron "at rest," not doing anything special, shows a steady difference in electric charge—a *voltage difference*—across its plasma membrane. In this case, the fluid just inside the membrane is more negatively charged than the fluid outside. That steady voltage difference is the **resting membrane potential**. Think of it as a "potential for activity" across the membrane.

When a neuron receives signals at its input zone, the resting membrane potential there may change temporarily. For example, imagine you accidentally trip over a cat that is snoozing on the kitchen floor. In response to the unfortunate stimulation, a veritable tidal wave of signals washes swiftly through the cat's nervous system, and those that reach, say, motor neurons with axons in the cat's tail will no doubt disturb the resting membrane potential of those neurons.

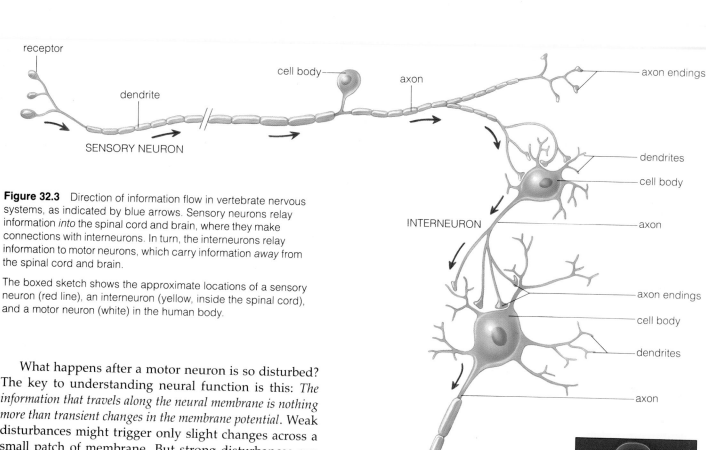

receptor

cell body

axon

axon endings

dendrite

dendrites

cell body

**SENSORY NEURON**

axon

INTERNEURON

axon endings

cell body

dendrites

axon

MOTOR NEURON

**Figure 32.3** Direction of information flow in vertebrate nervous systems, as indicated by blue arrows. Sensory neurons relay information *into* the spinal cord and brain, where they make connections with interneurons. In turn, the interneurons relay information to motor neurons, which carry information *away* from the spinal cord and brain.

The boxed sketch shows the approximate locations of a sensory neuron (red line), an interneuron (yellow, inside the spinal cord), and a motor neuron (white) in the human body.

What happens after a motor neuron is so disturbed? The key to understanding neural function is this: *The information that travels along the neural membrane is nothing more than transient changes in the membrane potential*. Weak disturbances might trigger only slight changes across a small patch of membrane. But strong disturbances can trigger an **action potential**: an abrupt, short-lived reversal in the polarity of charge across the plasma membrane. For a fraction of a second, the cytoplasmic side of a patch of membrane becomes positive with respect to the outside. This transient change in membrane potential moves rapidly along the entire length of the axon.

Any cell that can respond to stimulation by producing action potentials is said to show *membrane excitability*.

Keep in mind that action potentials usually begin and end on the same neuron; they are not transferred to neighboring cells. When action potentials reach the axon endings of the cat's motor neurons, for example, they trigger the release of molecules that serve as chemical signals to adjacent muscle cells. Muscles contract in response to the signals and the cat's tail shoots straight up. The cat simultaneously makes other, violent responses to the stimulation, but these need not concern us here.

A neuron at rest shows a steady voltage difference across its plasma membrane, with the inside more negative than the outside. That difference is the resting membrane potential.

A neuron shows membrane excitability. It can produce action potentials in response to strong stimulation.

An action potential is an abrupt, transient reversal in the voltage difference across the membrane: The inside becomes more positive with respect to the outside.

## Neurons "At Rest"

For information to flow along a neuron, the neuron must first be in the "resting" state. Only then can it undergo the changes that result in an action potential. The question becomes this: What establishes the resting membrane potential and maintains it between action potentials? The answer starts with three factors.

*First*, the concentrations of potassium ions ($K^+$), sodium ions ($Na^+$), and other charged substances are not the same on the two sides of the plasma membrane. *Second*, channel proteins that span the membrane control the diffusion of specific types of ions across it. *Third*, transport proteins that span the membrane actively pump sodium and potassium ions across it.

The unequal concentration of ions across the membrane can be depicted this way:

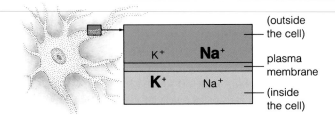

In the above sketch, the large letters denote which side of the membrane has the greater concentration of the two ions. How great is "greater"? Think about a motor neuron in the tail of that cat on the kitchen floor. For every 150 potassium ions on the cytoplasmic side of the plasma membrane, there are only 5 in the same volume of fluid outside. For every 15 sodium ions inside, there are about 150 on the outside.

If $K^+$, $Na^+$, and other charged substances could move freely across the neural membrane, they would each dif-

**Figure 32.4** (*Below*) Pathways for ions across the plasma membrane of a neuron. These pathways are provided by proteins embedded in the lipid bilayer. Compare Figure 32.6 to this model of membrane structure.

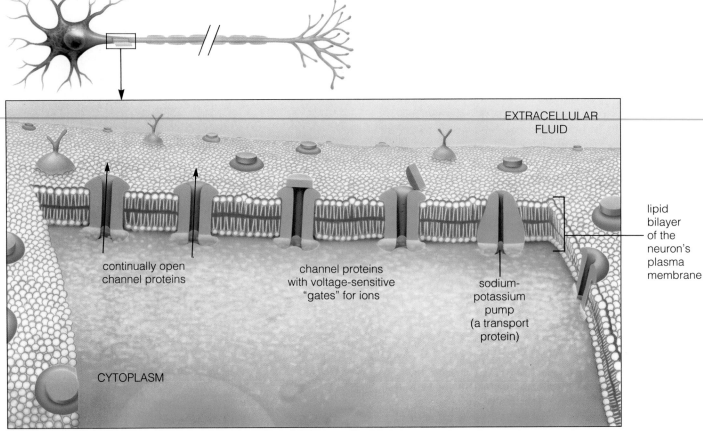

fuse down their respective concentration gradients—and those gradients eventually would disappear. However, the ions can move only through the interior of proteins that span the membrane (Figure 32.4). Their passage through the type called channel proteins is called facilitated diffusion. (As described on page 80, a channel protein passively allows the ions to move through their interior in the direction that a concentration gradient would take them.) Some channel proteins permit ions to "leak" (diffuse) through them all the time. Others have "gates" that open only when stimulated.

Imagine now that you are small enough to stand on an input zone of a motor neuron in the cat's tail. A profusion of surface bumps—the tops of membrane proteins—spread out before you. The sleeping cat has not yet been tripped over, so the neuron is at rest. More precisely, most of its channels for sodium are shut. And some channels for potassium are open, so potassium is leaking out through channel proteins, following its concentration gradient. This makes the interior of the neuron more negative than the extracellular fluid, and some potassium ions (which are positively charged) are attracted back inside.

When the inward pull of opposite charge balances the outward force of diffusion, there is no more *net* movement of potassium across the membrane. There is a steady voltage difference across the membrane, which is the amount of energy inherent in the concentration and electric gradients between the two differently charged regions. For many neurons, this amount—the resting membrane potential—is about −70 millivolts.

Neurons cannot maintain a resting potential indefinitely without expending energy. It happens that a small amount of sodium does leak into the neuron through a few open sodium channels. Unless the inward leakage of positive ions is countered, the resting membrane potential eventually will disappear. Transport proteins called **sodium-potassium pumps** do the countering. Using energy from ATP, they actively transport potassium into the neuron, and they pump sodium ions out at the same time. This is an example of the active transport mechanism shown earlier in Figure 5.10.

Figure 32.5 summarizes the balancing effect of the pumping and leaking mechanisms that maintain membrane conditions between action potentials.

Passive transport mechanisms establish the concentration and electric gradients across the plasma membrane of a neuron.

In a resting neuron, an active transport mechanism maintains those gradients by pumping potassium ions into the neuron and sodium ions out of it.

## Local Disturbances in Membrane Potential

In all neurons, not just motor neurons, stimulation at an input zone produces localized signals that do not spread very far. When you tripped over the cat, for example, you disturbed many sensory neurons that had their receptors (input zones) embedded in the connective tissue beneath the cat's skin. At each receptor site, the stimulus—in this case, mechanical pressure—affected ion movements across a small patch of plasma membrane. The voltage difference changed slightly at this patch, producing a type of graded, local signal.

"*Graded*" means the signals can vary in magnitude—they can be small or large—depending on the intensity and duration of the stimulus. In the cat's case, the stronger the pressure on its skin, the greater the disturbance to the sensory receptors.

"*Local*" means the signal does not spread far—half a millimeter or less, most often. Input zones simply do not have the type of ion channels needed to propagate a signal farther than this. However, when stimulation is intense or prolonged, graded signals can spread into an adjacent **trigger zone** of the membrane—the site where action potentials can be initiated.

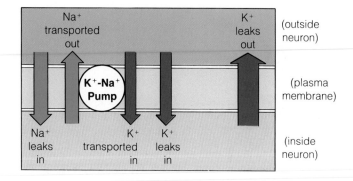

**Figure 32.5** Balance between pumping and leaking processes that maintain the distribution of sodium and potassium ions across the plasma membrane of a neuron at rest. The relative widths of the arrows indicate the magnitude of the movements. The total inward movement counteracts the total outward movement for each kind of ion; hence the ion distributions are maintained.

# A CLOSER LOOK AT ACTION POTENTIALS

## Mechanism of Excitation

The action potential is analogous to a pulse of electrical activity (hence its original name, nerve impulse). Measurements of the voltage difference across a membrane before, during, and after an action potential reveal this pattern:

1. The inside of a neuron at rest is more negative with respect to the outside (its membrane is *polarized*).

2. During an action potential, the inside is more positive than the outside (the membrane is *depolarized*).

3. Following an action potential, resting conditions are restored (the membrane is *repolarized*).

An action potential is triggered when a disturbance causes the voltage difference to change by a certain minimum amount, a *threshold* level. The change occurs when voltage-sensitive gated channels for sodium ions open in an accelerating way (Figure 32.6). The inward flow of these positively charged sodium ions makes the fluid on the cytoplasmic side of the membrane less negative. This causes more gates to open, more sodium to enter, and so on until the charge difference reverses. The accelerating flow of sodium is an example of positive feed-

back, whereby an event intensifies as a result of its own occurrence:

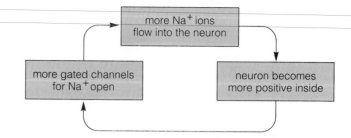

Once threshold is reached, the opening of more sodium gates no longer depends on the strength of the stimulus. It proceeds automatically because the positive-feedback cycle has started. That is why all action potentials in a given neuron "spike" to the same level above threshold as an *all-or-nothing event*. In other words, if threshold is reached, nothing can stop the full spiking. If threshold is not reached, the membrane disturbance will subside when the stimulus is removed.

The squid *Loligo* helped provide researchers with evidence of the spiking that occurs during an action potential. Certain squid axons have such large diameters that a fine electrode can be inserted easily into one of them, then another electrode can be positioned outside the axon membrane. Electrodes are devices used to measure voltage changes. When the two electrodes are connected to a voltage source and an oscilloscope, voltage changes show up as deflections of a beam traveling across the scope's fluorescent screen. Figure 32.7 shows examples of this.

**Figure 32.6** Propagation of action potentials along the axon of a neuron. The plasma membrane is shown in yellow.

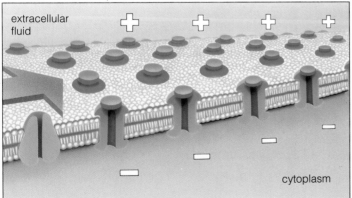

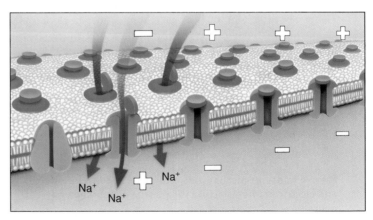

**a** Membrane at rest (inside negative with respect to the outside). An electrical disturbance (red arrow) spreads from an input zone to an adjacent trigger region of the membrane, which has many gated sodium channels.

**b** A strong disturbance initiates an action potential. Sodium gates open, the inflow decreases the negativity inside; this causes more gates to open, and so on, until threshold is reached and the voltage difference across the membrane reverses.

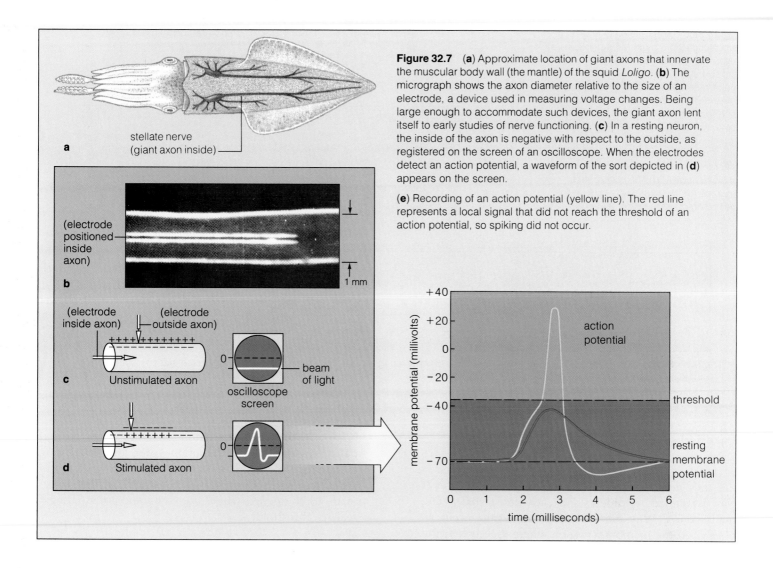

**Figure 32.7** (**a**) Approximate location of giant axons that innervate the muscular body wall (the mantle) of the squid *Loligo*. (**b**) The micrograph shows the axon diameter relative to the size of an electrode, a device used in measuring voltage changes. Being large enough to accommodate such devices, the giant axon lent itself to early studies of nerve functioning. (**c**) In a resting neuron, the inside of the axon is negative with respect to the outside, as registered on the screen of an oscilloscope. When the electrodes detect an action potential, a waveform of the sort depicted in (**d**) appears on the screen.

(**e**) Recording of an action potential (yellow line). The red line represents a local signal that did not reach the threshold of an action potential, so spiking did not occur.

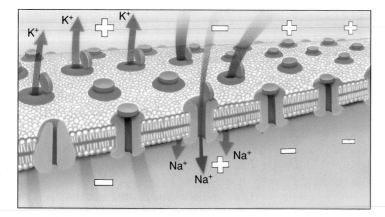

**c** The reversal causes sodium gates to shut and potassium gates to open (at purple arrows). Potassium follows its gradient (out of the neuron). Voltage is restored. The disturbance produced by the action potential triggers another action potential at the adjacent membrane site, and so on, away from the point of stimulation.

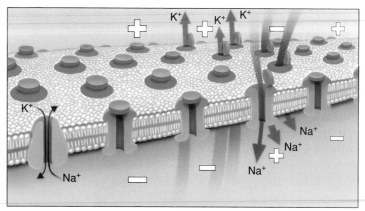

**d** The inside of the membrane becomes negative again following each action potential, but the sodium and potassium concentration gradients are not yet fully restored. Active transport at sodium-potassium pumps restores the gradients.

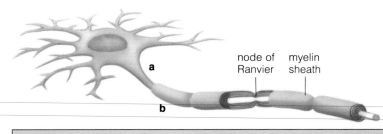

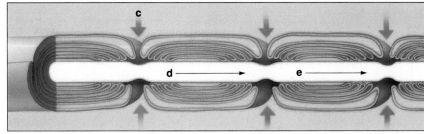

node of    myelin
Ranvier    sheath

**Figure 32.8** Propagation of an action potential along a motor neuron having a myelin sheath. (**a**) An action potential is initiated at a trigger zone in the axon membrane. (**b**) The sheath hinders ion movements across the membrane, so the disturbance spreads rapidly down the axon. (**c**) The small nodes are not sheathed, and they have very dense arrays of gated sodium channels. The voltage difference across the membrane reverses at these nodes. (**d**) The disturbance spreads rapidly to the next node in line, and so on down the axon (**e**).

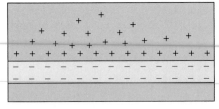

Charge density along an unsheathed axon

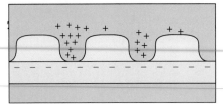

Charge density along a sheathed axon

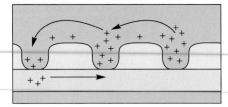

Propagation of action potential along a sheathed axon

## Duration of Action Potentials

Several hundred new action potentials were triggered in the cat's nervous system in the single second after it was awakened so rudely, and each one lasted a few milliseconds. Why did each action potential occur so briefly? At the membrane region where it occurred, the gates of some channel proteins closed and shut off the flow of sodium ions across the membrane. Then, about halfway through the action potential, the gates of other channel proteins opened and potassium ions were free to diffuse out of the neuron. The increased outward flow of many more potassium ions restored the original voltage difference across the membrane. And the sodium-potassium pumps restored the gradients.

## Propagation of Action Potentials

**Refractory Period.** After an action potential occurs in a trigger zone, it is propagated along the membrane without becoming diminished in magnitude. In brief, the disturbance spreads to adjacent patches of membrane, where the opening of gated channels is repeated. The new disturbance causes channels to open in the next patch of membrane, and so on away from the stimulation site. Notice, in Figure 32.6, how action potentials travel *away* from the stimulation site. A *refractory period* following each one helps prevent backflow. This is the period when the sodium gates at a given patch of mem-

brane are shut and potassium gates are open, so that patch is insensitive to stimulation. Later, after the resting membrane potential has been restored, most potassium gates close and sodium gates return to their initial state, ready to be opened when the membrane potential next reaches threshold.

**Sheathed Axons.** A *myelin sheath* wraps around the axons of many sensory and motor neurons that serve as cord. The sheath consists of the plasma membranes of specialized neuroglial cells called Schwann cells. These cells grow around and around the axon, rather like a jellyroll, and they form many layers of insulation (Figure 32.8).

Each Schwann cell is separated from adjacent ones by a small, exposed gap, or node, where the axon membrane is loaded with voltage-sensitive gated sodium channels. In a manner of speaking, the action potentials jump from node to node. The sheathed regions hinder the flow of ions across the membrane, and this forces the ions to flow along the length of the axon until they can exit at a node and generate a new action potential there.

The node-to-node hopping in myelinated neurons is called *saltatory conduction*, after the Latin word meaning "to jump." Saltatory conduction affords the most rapid signal propagation with the least metabolic effort by the cell. In the largest myelinated axon, signals travel 120 meters per second. That's 270 miles per hour.

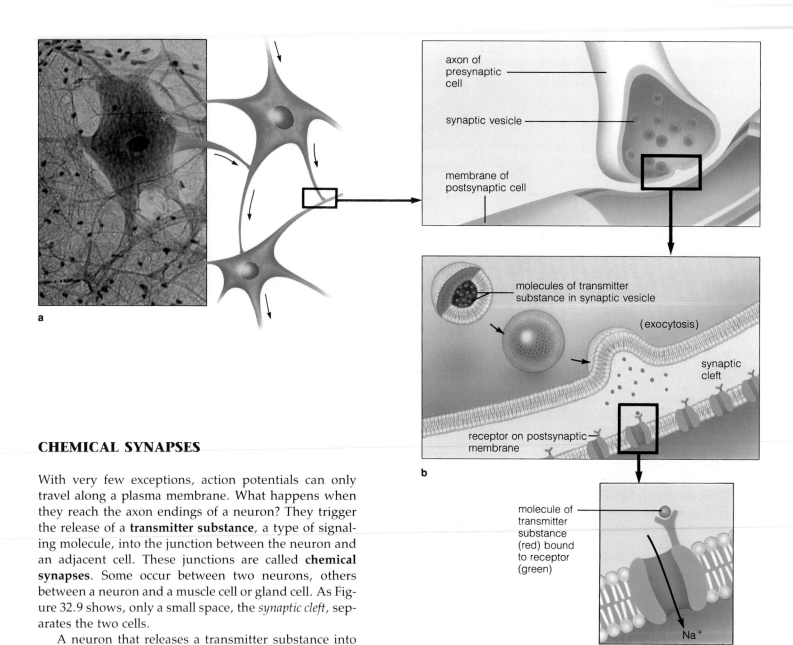

## CHEMICAL SYNAPSES

With very few exceptions, action potentials can only travel along a plasma membrane. What happens when they reach the axon endings of a neuron? They trigger the release of a **transmitter substance**, a type of signaling molecule, into the junction between the neuron and an adjacent cell. These junctions are called **chemical synapses**. Some occur between two neurons, others between a neuron and a muscle cell or gland cell. As Figure 32.9 shows, only a small space, the *synaptic cleft*, separates the two cells.

A neuron that releases a transmitter substance into the cleft is called the "*pre*synaptic cell." The one whose behavior is affected by the transmitter substance is the "*post*synaptic cell." The presynaptic cell contains numerous vesicles filled with transmitter substances (Figure 32.9).

When an action potential arrives at the presynaptic cell membrane facing the cleft, it causes voltage-sensitive gated channels for *calcium ions* to open. Calcium ions are more concentrated outside the cell, and when they move inside (down their gradient), they cause synaptic vesicles to fuse with the plasma membrane. Thus the contents of the vesicles are released into the cleft, and diffusion carries them to the postsynaptic cell. There the molecules of transmitter substance bind briefly to membrane receptor molecules, and after they exert their effects, they are picked up again by the presynaptic cell or inactivated by enzymes.

**Figure 32.9** Chemical synapses. Typically, action potentials spread along axons, away from the neuron cell body (**a**). In (**b**), the axon terminates next to another neuron, this being an example of a chemical synapse. Information flows from the presynaptic cell to the postsynaptic cell by way of a transmitter substance (**c**).

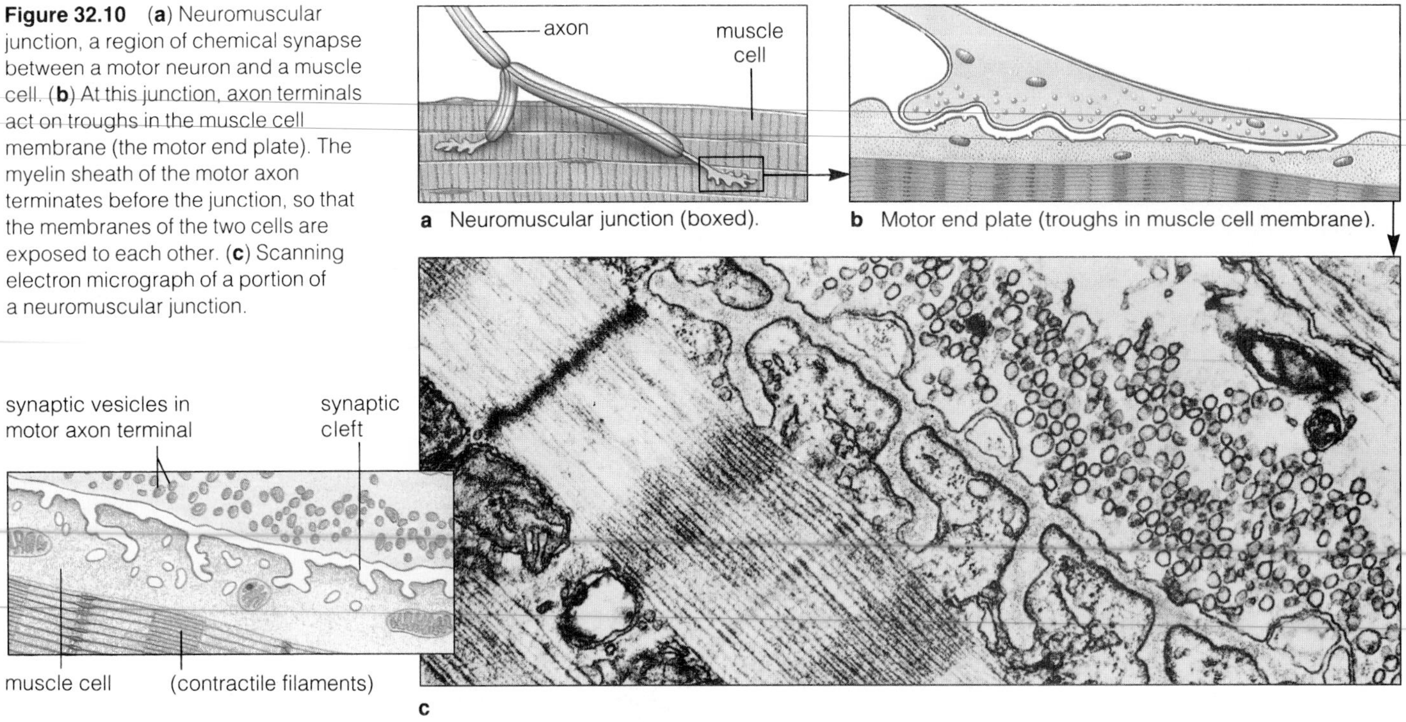

**Figure 32.10** (**a**) Neuromuscular junction, a region of chemical synapse between a motor neuron and a muscle cell. (**b**) At this junction, axon terminals act on troughs in the muscle cell membrane (the motor end plate). The myelin sheath of the motor axon terminates before the junction, so that the membranes of the two cells are exposed to each other. (**c**) Scanning electron micrograph of a portion of a neuromuscular junction.

**a** Neuromuscular junction (boxed).

**b** Motor end plate (troughs in muscle cell membrane).

synaptic vesicles in motor axon terminal

synaptic cleft

muscle cell    (contractile filaments)

**c**

## Effects of Transmitter Substances

A transmitter substance affects a postsynaptic cell in one or two ways. If it has an *excitatory* effect, it helps drive the cell's membrane toward the threshold of an action potential. If it has an *inhibitory* effect, it helps drive the membrane away from threshold. A given transmitter substance can have either excitatory or inhibitory effects, depending on which type protein channels it opens up in the postsynaptic membrane.

*Acetylcholine* (ACh) is a transmitter substance with excitatory and inhibitory effects on the cells of muscles and glands throughout the body. It also acts on certain cells in the brain and spinal cord. Think about what ACh does at **neuromuscular junctions**, which are synapses between a motor neuron and muscle cells. At this type of junction, the branched axon endings of the motor neuron are positioned on the muscle cell membranes (Figure 32.10). An action potential traveling down the motor neuron spreads through all the endings and causes the release of ACh into each synaptic cleft. When ACh binds to receptors on the muscle cell membrane, it has an excitatory effect. It may trigger action potentials, which in turn initiate events in the muscle cells that lead to contraction. Those events are a topic of Chapter 35.

*Serotonin*, another transmitter substance, acts on brain cells that govern sleeping, sensory perception, temperature regulation, and emotional states. *Norepinephrine* affects brain regions that apparently are concerned with emotional states as well as dreaming and

awaking. Other transmitter substances that act on different parts of the brain are *dopamine* and *GABA* (gamma aminobutyric acid).

## Neuromodulators

The effects of transmitter substances are often influenced by signaling molecules known generally as **neuromodulators**. These molecules enhance or reduce membrane responses in target neurons. Among them are the *endorphins*, which function in inhibiting perceptions of pain. These naturally occurring peptide molecules are much more potent than morphine, a painkiller derived from opium poppies. Endorphins also may have roles in memory and learning, temperature regulation, and sexual behavior as well as emotional depression and other mental disorders.

## Synaptic Integration

How will a given postsynaptic cell respond to incoming signals? That depends on which of its membrane receptors are called into play and on the nature of the signals reaching those receptors. *At any given moment, excitatory and inhibitory signals are competing for control of the membrane.*

In a process called **synaptic integration**, the competing signals at an input zone of a neuron are summed up. Those signals are two kinds of graded potentials that often are abbreviated EPSP and IPSP. An *excitatory post-*

## Deadly Imbalances at Chemical Synapses

An anaerobic bacterium, *Clostridium tetani*, lives in the gut of horses and other grazing animals and its endospores can survive in soil, especially soil enriched with manure. This bacterium can enter a human body through a deep puncture or cut, and it can multiply if tissues around the wound die off and become anaerobic. A product of this bacterium's metabolic activities is a neurotoxin. It interferes with the effect of acetylcholine (ACh) on motor neurons. The result is a severe disorder called *tetanus*, the symptoms of which are prolonged, spastic paralysis that can lead to death.

During normal muscle contraction, action potentials from the brain travel through the spinal cord. They excite motor neurons that trigger the release of acetylcholine at neuromuscular junctions, and the muscle cells are stimulated into contracting. Many of the body's muscles occur as paired sets, such as those shown in Figure *a*. When one set contracts, an opposing set is stretched. Bend your arm at the elbow and you can feel two such sets (biceps and triceps) in your upper arm. When the biceps contracts, inhibitory signals are sent to the triceps and it

relaxes. The tetanus toxin blocks the release of inhibitory signals—so *both* sets of muscles contract! Within four to ten days, paired muscles attempt to work in opposition

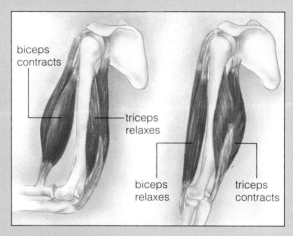

**a** Antagonistic muscle pair.

---

*synaptic potential* (EPSP) brings the membrane closer to threshold; it has a *depolarizing* effect. An *inhibitory postsynaptic potential* (IPSP) drives the membrane away from threshold—it has a *hyper*polarizing effect—or maintains the membrane at its resting level.

Take a look at Figure 32.11. The yellow line in this diagram shows how an EPSP of a given magnitude would register on an oscilloscope screen *if it were occurring alone*. The purple line shows the same thing for an IPSP. The red line shows what happens when the two occur simultaneously. In this case, synaptic integration pulls the membrane potential away from threshold.

The balancing acts that go on at many chemical synapses are essential for survival. We know this because foreign substances that interfere with synaptic integration can have deadly consequences, as the *Commentary* suggests.

---

**Synaptic integration is the moment-by-moment combining of excitatory and inhibitory signals acting on adjacent membrane regions of a neuron.**

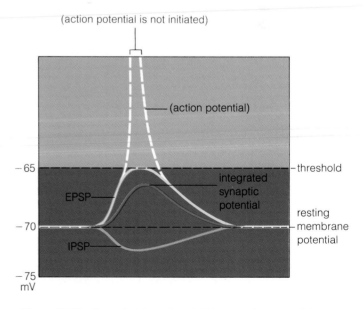

**Figure 32.11** Synaptic integration. In this example, an excitatory synapse and an inhibitory synapse nearby are activated at the same time. The IPSP reduces the magnitude of the EPSP from what it could have been, pulling it away from threshold. The red line represents the integration of these two synaptic potentials. Threshold is not reached in this case; hence an action potential cannot be initiated.

to each other. This is the start of spastic paralysis—the muscles simply cannot be released from contraction. The increase in muscle tension (spasms) can become violent enough to break bones in the body. Fists and jaws may undergo prolonged clenching (hence the name lockjaw, which is sometimes used for the disorder). The back may become paralyzed in a permanent arch. Muscles of the respiratory system and heart also may undergo spastic paralysis, in which case the affected individual nearly always dies.

Since the development of effective vaccines, tetanus occurs only rarely in the United States. But the disease was terrifying to the soldiers of early wars, when battlefields were littered with manure from calvary horses and with corpses of the horses themselves. At that time, *C. tetani* endospores were like profuse biological landmines. Battle wounds commonly became contaminated with those endospores. We sense the agony of one young victim of such contamination; the dramatic painting in Figure *b* was made as he lay dying in a military hospital.

*C. botulinum*, a relative of the bacterial agent of tetanus, has a different effect on synaptic integration. It produces a toxin that can block the release of ACh from motor neurons. In this case, muscle contraction cannot occur, so the body shows the symptoms of the disease *botulism*. It shows flaccid paralysis, meaning its muscles remain relaxed. Without prompt treatment, victims simply will stop breathing.

**b** A soldier in 1809 dying of tetanus.

## PATHS OF INFORMATION FLOW

Through synaptic integration, signals arriving at any given neuron in the body can be reinforced or dampened, sent on or suppressed. What determines the direction in which a given signal will travel? That depends on the organization of neurons into circuits or pathways.

The brain has many "local" circuits in which the chattering of neurons is confined to a single region. In contrast, signals between the brain or spinal cord and other body regions travel by cordlike communication lines called **nerves** (Figure 32.12). Axons of sensory neurons, motor neurons, or both are bundled together in a nerve. Within the brain and spinal cord, such bundles are called nerve "tracts."

The sensory and motor neurons of many nerves take part in reflexes, which are simple, stereotyped movements made in response to sensory stimulation. In the simplest **reflex arc**, sensory neurons directly synapse on motor neurons. The *stretch reflex* is an example; it works to contract a muscle when that muscle has been stretched.

Think about how you can hold out a large glass and keep it stationary when someone pours lemonade into it. As the lemonade adds weight to the glass and your hand starts to drop, a muscle in your arm (the biceps) is stretched. The stretching activates certain receptors in the muscle. The stretch-sensitive receptors are part of **muscle spindles**—sensory organs in which small, specialized cells are enclosed in a sheath that runs parallel with the muscle itself. The receptors are the input zone of sensory neurons that synapse with motor neurons in the spinal cord (Figure 32.13). Axons of the motor neurons lead right back to the stretched muscle, and action potentials that reach the axon endings trigger the release of ACh, which initiates contraction. Continued receptor activity excites the motor neurons further, allowing them to maintain your hand's position.

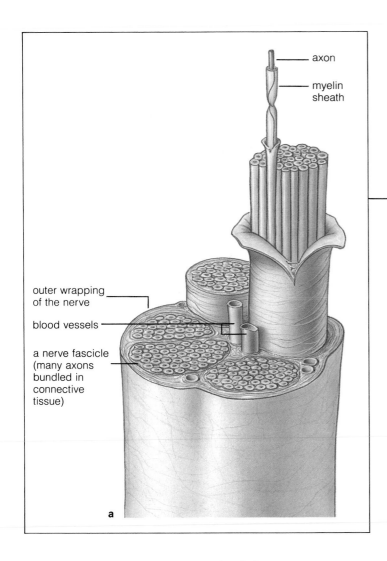

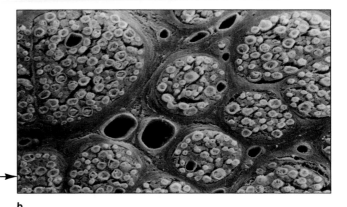

b

**Figure 32.12** Structure of a nerve. The sketch (**a**) and the scanning electron micrograph (**b**) show bundles of axons in cylindrical wrappings of connective tissue inside the nerve.

In figure (a):
- axon
- myelin sheath
- outer wrapping of the nerve
- blood vessels
- a nerve fascicle (many axons bundled in connective tissue)
- a

**Figure 32.13** Simple reflex arc governing the stretch reflex. A sensory axon is shown in purple, a motor axon in red. Stretch-sensitive receptors of the sensory neuron are located in muscle spindles within a skeletal muscle. Stretching the muscle disturbs the receptors, and action potentials are generated in the sensory neuron. They travel to the axon endings that synapse with motor neurons—which have axons leading right back to the stretched muscle. Signals from the motor neuron can stimulate the muscle cell membrane and initiate contraction (page 634).

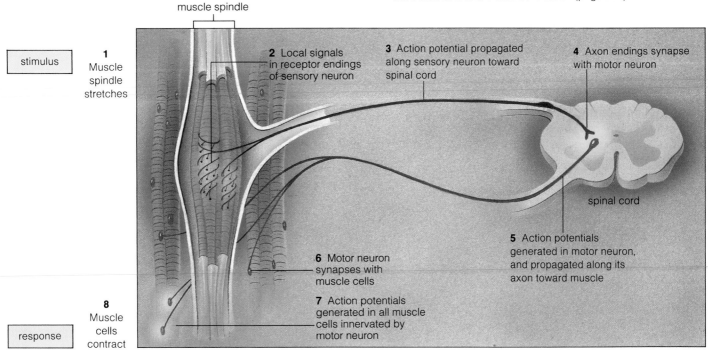

muscle spindle

stimulus

**1** Muscle spindle stretches

**2** Local signals in receptor endings of sensory neuron

**3** Action potential propagated along sensory neuron toward spinal cord

**4** Axon endings synapse with motor neuron

spinal cord

**5** Action potentials generated in motor neuron, and propagated along its axon toward muscle

**6** Motor neuron synapses with muscle cells

**7** Action potentials generated in all muscle cells innervated by motor neuron

**8** Muscle cells contract

response

In the vast majority of reflexes, sensory neurons make connections with a number of interneurons, which then activate or suppress the motor neurons necessary for a coordinated response. An example is the *withdrawal reflex*, a rapid pulling away from an unpleasant or harmful stimulus. If you have ever accidentally touched a hot stove, you know this reflex action can be completed even before you are conscious it has occurred.

## SUMMARY

1. The neuron, or nerve cell, is the basic unit of communication in all nervous systems. Neurons *collectively* sense environmental change, swiftly integrate sensory inputs, and then activate muscle cells and gland cells (effectors) that can carry out responses.

2. Vertebrate nervous systems contain sensory neurons, interneurons, and motor neurons. They also contain neuroglial cells, which support and protect the neurons.

3. Like all cells, the neuron at rest shows a steady voltage difference across its plasma membrane, with the inside more negative than the outside. This results from differences in the concentrations of potassium ions, sodium ions, and other charged substances present in cytoplasm and extracellular fluid. The amount of energy inherent in the concentration and electric gradients across the membrane is the resting membrane potential.

4. Most neurons and some other cells (including muscle cells) show membrane excitability. In response to stimulation, the voltage difference across the membrane can undergo brief but sudden reversals, called action potentials—the inside becomes positive with respect to the outside.

5. Excitability depends on these membrane features:
   a. The lipid bilayer is impermeable to ions.
   b. Certain ions can diffuse passively across the membrane through channel proteins. Some channel proteins are always open, others have gates that open only under stimulation.

   c. Transport proteins (sodium-potassium pumps) in the membrane restore and maintain concentration gradients between action potentials.

6. The neuron receives and integrates signals at *input* zones, usually dendrites and the cell body. The disturbance may produce local, graded potentials that may spread to a *trigger* zone, where action potentials can be generated. Action potentials travel to *output* zones (axon terminals), where other kinds of signals are sent to target cells.

7. Action potentials occur when the voltage difference across the membrane changes dramatically, past a certain minimum amount called the threshold level. Then, gates on channel proteins open in an accelerating way and suddenly reverse the voltage difference, which registers as a spike on recording devices.

8. The overall pattern before, during, and after an action potential is this: polarization (inside negative with respect to outside), depolarization (inside becomes positive), and repolarization (inside becomes negative again).

9. Chemical synapses are junctions between two neurons, or between a neuron and a muscle cell or gland cell. Here, the *pre*synaptic cell releases a transmitter substance into a cleft between it and the *post*synaptic cell. A transmitter substance may have an excitatory or inhibitory effect, depending on which type of ion channels it opens up in the postsynaptic cell membrane. Integration is the moment-by-moment combining of all signals—excitatory and inhibitory—acting at all the different synapses on a neuron.

10. The direction of information flow through the body depends on the organization of neurons into circuits and pathways. Local circuits are sets of interacting neurons confined to a single region in the brain or spinal cord. Nerve pathways extend from neurons in one body region to neurons in different regions.

11. Reflexes (simple, stereotyped movements made in response to sensory stimuli) are examples of how signals are sent through nervous systems. In simple reflexes, sensory neurons directly signal motor neurons that act on muscle cells. In more complex reflexes, interneurons coordinate and refine the responses.

## Review Questions

1. Define sensory neuron, interneuron, and motor neuron. *547–548*

2. What is the difference between a neuron and a nerve? *547, 558*

3. Two major concentration gradients exist across a neural membrane. What substances are involved, and how are the gradients maintained? *550–551*

4. An electric gradient also exists across a neural membrane. Explain what the electric and concentration gradients together represent. *548–549, 550–551*

5. Label the functional zones of a motor neuron on the following diagram: *548*

6. Distinguish between an action potential and a graded potential. What is meant by "all-or-nothing" messages? *549, 551, 552*

7. What is a synapse? Explain the difference between an excitatory and an inhibitory synapse. Define neural integration. *555–557*

8. What is a reflex? Describe the sequence of events in a stretch reflex. *558–560*

## Self-Quiz *(Answers in Appendix IV)*

1. The communication lines of vertebrate nervous systems are organized networks of cells called _____.

2. In vertebrate nervous systems, _____ are receptors of specific stimuli, _____ integrate information and send commands for responses to muscles and glands by way of _____.

3. In a neuron at rest, there is a steady voltage difference across the plasma membrane, with the _____ being more _____ than the _____.
   a. inside; negative; outside
   b. outside; negative; inside
   c. inside; positive; outside
   d. outside; positive; inside

4. The "resting membrane potential" is _____ across the neuron's plasma membrane.
   a. an action potential
   b. a graded potential
   c. a steady voltage difference
   d. both a and c are correct

5. Action potentials occur _____.
   a. when a neuron is adequately stimulated
   b. when potassium gates open in an accelerating way
   c. when sodium-potassium pumps kick into action
   d. both a and b are correct

6. An action potential lasts only briefly because _____ at the membrane region where it occurred.
   a. gates for sodium open, gates for potassium close
   b. gates for sodium close, gates for potassium open

c. sodium-potassium pumps restore gradients
   d. both b and c are correct

7. All transmitter substances diffuse across a _____.
   a. neuromuscular junction
   b. synaptic cleft
   c. myelin sheath
   d. both a and b are correct

8. _____ is an example of a transmitter substance; _____ is an example of a neuromodulator.
   a. Serotonin; an endorphin
   b. Serotonin; GABA
   c. Ach; an endorphin
   d. both a and c are correct

9. A nerve may consist of bundled-together axons of _____.
   a. sensory neurons
   b. motor neurons
   c. sensory and motor neurons
   d. all of the above are correct

10. Match the following concepts and descriptions.
   _____ simple reflex
   _____ local circuit
   _____ transmitter substances
   _____ neuron
   _____ complex reflex

   a. have excitatory or inhibitory effects on postsynaptic cells
   b. basic unit of communication in all nervous systems
   c. sensory neuron directly signals motor neuron
   d. set of interacting neurons in region of brain or spinal cord
   e. interneurons coordinate and refine responses

## Selected Key Terms

acetylcholine (ACh) *556*
action potential *549*
axon *548*
chemical synapse *555*
dendrite *548*
excitatory postsynaptic potential (EPSP) *557*
graded potential *551*
inhibitory postsynaptic potential (IPSP) *557*
interneuron *547*
membrane excitability *549*
motor neuron *548*
muscle spindle *558*
myelin sheath *554*
nerve *558*
neuroglial cell *548*

neuromodulator *556*
neuromuscular junction *556*
neuron *547*
reflex arc *558*
refractory period *554*
resting membrane potential *548*
saltatory conduction *554*
Schwann cell *554*
sensory neuron *547*
sodium-potassium pump *551*
stretch reflex *558*
synaptic integration *556*
threshold *552*
transmitter substance *555*
trigger zone *551*

## Readings

Berne, R., and M. Levy (editors). 1988. *Physiology*. Second edition. St. Louis: Mosby. Section II is an authoritative introduction to neural functioning.

Dunant, Y., and M. Israel. April 1985. "The Release of Acetylcholine." *Scientific American* 252(4):58–83. Experiments showing how this major neurotransmitter functions.

Lent, C., and M. Dickenson. June 1988. "The Neurobiology of Feeding Behavior in Leeches." *Scientific American* 258(6):98–103. Describes the relationship between serotonin (a transmitter substance) and feeding behavior in an invertebrate.

## Why Crack the System?

James Kalat, a professor at North Carolina State University, sometimes asks students to volunteer for an experiment. He tells them he would like to implant a device inside their brain that will make them feel really good. There are risks. The device compromises health, reduces life expectancy by a decade or so, and possibly causes permanent brain damage. The behavior of the volunteers will change for the worse, so they might have trouble completing their education, getting a job, keeping a job, or holding a family together. Volunteers can quit the experiment at any time but the longer the device is in their brain, the harder it will be to get out. They will not be paid. Rather, they must pay the experimenter—at bargain rates at first, then a little more each week. The device is illegal, so if volunteers get caught using it, they as well as the experimenter will probably go to jail.

Very, very few students volunteer for the experiment (which of course is hypothetical). Yet when Kalat changes "brain device" to *drug* and "experimenter" to *drug dealer*, an amazing number come forward. Like 30 million people in the United States alone, those students appear more than willing to engage in the self-destructive use of the so-called psychoactive drugs, which alter emotional and behavioral states.

The consequences show up in unexpected places. Each year, about 300,000 women addicted to crack give birth—and their newborns are already addicts. *Crack* is a cheap, potent form of cocaine, and it disrupts basic functions of the nervous system. Remember those chemical synapses described in the preceding chapter? Crack disrupts synapses between the neurons of a "pleasure center" within the brain. It stimulates the release of transmitter substances from presynaptic cells into the synaptic cleft—then blocks their reabsorption. The transmitter substances accumulate next to the postsynaptic cells and relentlessly stimulate them. Normal impulses to eat and sleep are suppressed. Blood

pressure rises. Feelings of euphoria and sexual desire become intense. All the while, the brain demands constant stimulation—but the molecules of transmitter substances in the synaptic cleft gradually break down, and the presynaptic cells cannot keep up with the incessant demand to provide more. Crack users become frantic, then profoundly depressed. Only crack makes them "feel good" again.

Addicted newborns cannot know all of this, of course. They can only quiver with "the shakes." Overstimulation of their brain neurons makes them chronically irritable. Their body is abnormally small. As they were developing inside their mother, their body tissues were not provided with enough oxygen and nutrients—crack causes blood vessels to constrict. Paradoxically, even though crack babies are abnormally fussy, they do not respond to rocking and other kinds of normal stimulation. It may be a year or more before they recognize their mother. Without treatment they are likely to grow up as emotionally unstable children, prone to aggressive outbursts and stony silences.

Each of us possesses a body of great complexity. Its architecture, its functioning are legacies of millions of years of evolution. It is unique in the living world because of its highly developed nervous system—a system that is capable of processing far more than the experience of the individual. One of its most astonishing products is language, the encoding of shared experiences of groups of individuals in time and space. Through the evolution of our nervous system, the sense of history was born, and the sense of destiny. Through this system we can ask how we have come to be what we are, and where we are headed from here. Perhaps the sorriest consequence of drug abuse is its implicit denial of this legacy—the denial of self when we cease to ask, and cease to care.

**Figure 33.1** Owners of an evolutionary treasure—a complex brain that is the foundation for our memory and reasoning, and our future.

## KEY CONCEPTS

**1.** Nervous systems provide a means of sensing specific information about external and internal conditions, integrating that information, and issuing commands for response from the body's effectors (muscles and glands).

**2.** The simplest nervous systems are the nerve nets of sea anemones and other cnidarians. The vertebrate nervous system shows pronounced cephalization and bilateral symmetry. It includes a brain, spinal cord, and many paired nerves. The brain is complex, with centers for receiving, integrating, processing, and responding to information.

**3.** The oldest parts of the vertebrate brain provide reflex control over breathing, blood circulation, and other essential functions. During the evolution of certain vertebrates, the brain expanded in complexity and its newer regions appropriated more and more control over the ancient reflex functions.

Neurologically speaking, sponges are just about the simplest members of the animal kingdom. Prick one with a pin and it will contract slowly, in a diffuse sort of way, and its response will never extend more than a few millimeters beyond the point of stimulation. Even a Venus flytrap makes a showier response to stimulation, as when its spiny trap slams shut around insect prey. However, animals as a group are unexcelled in their means of detecting specific stimuli and responding swiftly to them.

Almost all animals reach out or lunge after food; they pull back, crawl, swim, run, or fly when they are about to become food themselves. And think about what they have to do to find a mate and slow it down or hold its attention. (Think about all the things you have to do.) *The more complex the life-style, the more elaborate are the animal modes of receiving, integrating, and responding to information about the external and internal world.* Nervous systems provide for those three functions. As we saw in the preceding chapter, the communication lines of such systems are composed of nerve cells, or neurons. Messages traveling along those lines are coordinated with one another to produce complex patterns of behavior.

There are more than a million known species of animals, each with special features in its neural wiring, so the examples used in this chapter are necessarily limited. Even so, the following list is a useful starting point for understanding the organization of nearly all nervous systems:

1. Reflexes provide the basic operating machinery of nervous systems. *Reflexes* are simple, stereotyped movements made in response to specific types of stimulation. In the simplest reflexes, a sensory neuron signals a motor neuron, which acts on muscle cells that help carry out responses to the stimulation (page 558).

2. Nervous systems evolved as more nervous tissue became layered over ancient reflex pathways. In bilateral animals, the layerings became most pronounced at the head end of the body, with many neurons becoming concentrated into a brain. This evolutionary process is called *cephalization*.

3. In existing vertebrates, the oldest parts of the brain still deal with reflex coordination of vital functions, such as blood circulation and breathing. Other parts deal with storing and comparing information about experiences—and using that information to initiate novel actions. The neural connections within the most recent layerings are the basis of memory, learning, and reasoning.

4. Nervous systems coevolved with complex sensory organs, such as eyes, and motor structures, such as legs and wings. The coevolution of nervous, sensory, and motor systems was central to the development of more intricate behavior.

## INVERTEBRATE NERVOUS SYSTEMS

### Nerve Nets

Sea anemones, hydras, jellyfishes, and other cnidarians are aquatic animals, and they have the simplest nervous systems. As described in Chapter 25, these animals show radial symmetry. Their body parts are arranged about a central axis, much like the spokes of a bike wheel. Their nervous system extends through all the spokes, so to speak, and allows the animal to respond to food and danger coming from any direction.

The cnidarian nervous system, a **nerve net**, has sensory cells, nerve cells, and contractile epithelial cells (Figure 33.2). The three types of cells interact in simple reflex pathways. In the pathway concerned with feeding behavior, for example, nerve cells extend from sensory receptors in tentacles to contractile cells around the mouth. In jellyfishes, reflex pathways permit slow swimming movements and keep the body right-side up.

### Bilateral, Cephalized Nervous Systems

Flatworms are the simplest animals with bilateral symmetry, meaning their body has equivalent parts on the left and right side of its midsagittal plane. (Imagine yourself sliding down a staircase banister that turns into a razor and you probably never will forget where the midsagittal plane is.) Muscles that move the body forward are arranged the same way on both sides of the body, not just one. Nerves controlling the muscles are arranged the same way on both sides, and so on.

Bilateral nervous systems may have evolved from arrangements as simple as nerve nets. Some cnidarians pass through a self-feeding larval stage during the life cycle. Like flatworms, the larva (a planula) has a somewhat flattened body, and it uses cilia to swim or crawl about before it develops into an adult:

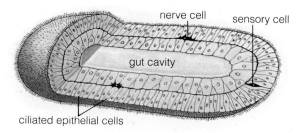

Imagine a few such planulas crawling about on the seafloor in Cambrian times. Suppose mutations of regulatory genes blocked their transformation into the adult form but allowed their reproductive organs to mature. As indicated on page 446, this actually happens among certain animals. The planulas would keep on crawling, they would reproduce—and so pass on the mutated genes.

The forward end of a crawling, aquatic animal is the first to encounter the presence or odor of food in the water. We can speculate that selective agents favored planulalike animals in which sensory cells became concentrated at the forward end, for that arrangement would permit more rapid and effective responses to important stimuli.

All flatworms have a ladderlike nervous system (Figure 33.3). Intriguingly, some also have **ganglia** (singular, ganglion). These are small, regional clusterings of the cell bodies of neurons. They also have **nerves** and two **nerve cords**, these being bundled-together axons of neu-

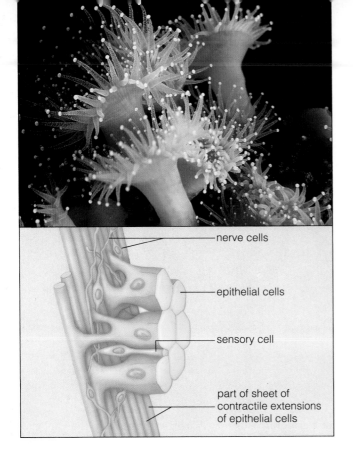

nerve cells

epithelial cells

sensory cell

part of sheet of
contractile extensions
of epithelial cells

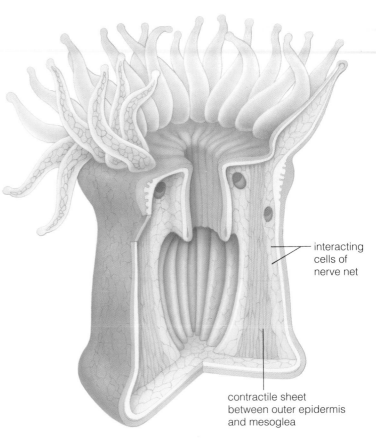

interacting
cells of
nerve net

contractile sheet
between outer epidermis
and mesoglea

**Figure 33.2** (*Above*) Nerve net of a sea anemone, one of the cnidarians.

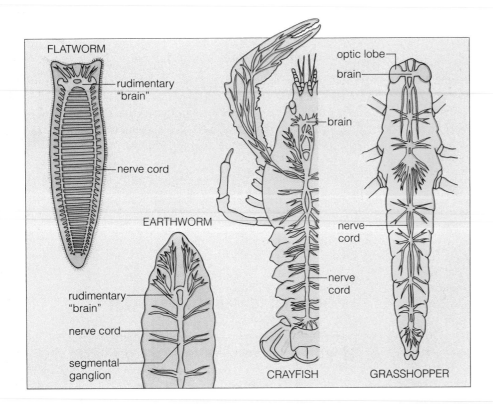

FLATWORM

rudimentary "brain"

nerve cord

EARTHWORM

rudimentary "brain"

nerve cord

segmental ganglion

optic lobe

brain

brain

nerve cord

nerve cord

CRAYFISH

GRASSHOPPER

**Figure 33.3** Bilateral symmetry evident in the nervous systems of a few invertebrates. The sketches are not to scale relative to one another.

rons. Some flatworm ganglia are arranged as a two-part brainlike structure at the head end. They coordinate signals from paired sensory organs, including two eyespots, and provide some control over the nerve cords.

As you will now see, *the patterns of bilateral symmetry and cephalization have echoes in the paired nerves and muscles, paired sensory structures, and paired brain center of yourself and all other vertebrates.*

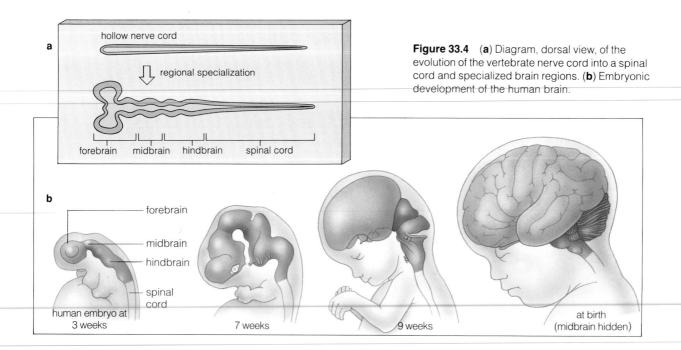

hollow nerve cord

regional specialization

forebrain    midbrain    hindbrain    spinal cord

**Figure 33.4** (**a**) Diagram, dorsal view, of the evolution of the vertebrate nerve cord into a spinal cord and specialized brain regions. (**b**) Embryonic development of the human brain.

b

forebrain

midbrain

hindbrain

spinal cord

human embryo at 3 weeks

7 weeks

9 weeks

at birth (midbrain hidden)

## VERTEBRATE NERVOUS SYSTEMS

### How the System Developed

The vertebrate nervous system shows more than cephalization and bilateral symmetry. It shows features that were shaped by a shift from reliance on a notochord to reliance on a vertebral column and nerve cord.

The *notochord*, a long rod of stiffened tissue, helps support the body. All vertebrate embryos have one, but it is greatly reduced or absent in adults. Most often, a backbone takes over the function of the notochord. The backbone contains hard, bony segments (vertebrae) arranged one after the other in a *vertebral column*. As described in Chapter 26, the vertebral column evolved many millions of years ago, and it was pivotal in the evolution of fast-moving, jawed predators.

The first vertebrates also had a *nerve cord*, a hollow, tubular structure running dorsally above the notochord. As Figure 33.4 suggests, this nerve cord was the forerunner of the spinal cord and brain. In the changing world of fast-moving vertebrates, predators and prey that were better equipped to sense and respond to one another's presence had the competitive edge. Their senses of smell, hearing, and balance became keener. The brain itself became variably thickened with nervous tissue that could integrate the rich sensory information and issue commands for complex, coordinated responses. In time, the thickening brain tissues became divided into three functionally specialized parts, called the forebrain, midbrain, and hindbrain.

Today, a nerve cord still develops in all vertebrate embryos. We call it the "neural tube." The neural tube undergoes expansion and regional modification into the brain and spinal cord, and it becomes enclosed within the vertebral column. Adjacent tissues in the embryo give rise to nerves that thread through all body regions and connect with the spinal cord and brain. Figure 33.5 will give you a sense of how intricate the communication lines become in the human nervous system.

### Functional Divisions of the System

For descriptive purposes, we can divide the vertebrate nervous system into central and peripheral regions. The **central nervous system** includes the brain and spinal cord. The **peripheral nervous system** includes all the nerves carrying signals to and from the brain and spinal cord. Both divisions also have neuroglial cells, which protect or assist neurons. The Schwann cells described in the preceding chapter are an example. They wrap around the axons of neurons that are bundled together in nerves of the peripheral nervous system (see, for example, page 554).

## PERIPHERAL NERVOUS SYSTEM

The peripheral nervous system of humans has thirty-one pairs of spinal nerves, which connect with the spinal cord. It also has twelve pairs of cranial nerves, which connect directly with the brain. Some nerves of the peripheral system carry only sensory information. The optic nerves, which carry visual signals from the eyes, are like this. Other nerves contain both sensory and motor axons. For example, the vagus nerves have sensory axons leading into the brain as well as motor axons leading out to the lungs, gut, and heart.

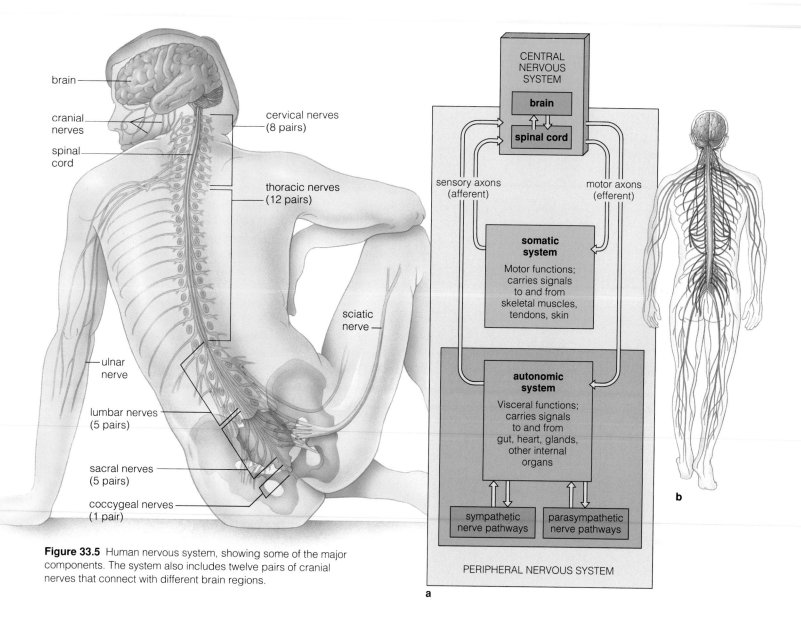

**Figure 33.5** Human nervous system, showing some of the major components. The system also includes twelve pairs of cranial nerves that connect with different brain regions.

brain

cranial nerves

spinal cord

cervical nerves (8 pairs)

thoracic nerves (12 pairs)

sciatic nerve

ulnar nerve

lumbar nerves (5 pairs)

sacral nerves (5 pairs)

coccygeal nerves (1 pair)

CENTRAL NERVOUS SYSTEM

**brain**

**spinal cord**

sensory axons (afferent)

motor axons (efferent)

**somatic system**

Motor functions; carries signals to and from skeletal muscles, tendons, skin

**autonomic system**

Visceral functions; carries signals to and from gut, heart, glands, other internal organs

sympathetic nerve pathways

parasympathetic nerve pathways

PERIPHERAL NERVOUS SYSTEM

a

b

**Figure 33.6** (**a**) Functional divisions of the vertebrate nervous system. In (**b**), the central nervous system is color-coded blue, the somatic nerves green, and the autonomic nerves, red. Sometimes the nerves carrying sensory input to the central nervous system are said to be *afferent* (a word meaning "to bring to"). The ones carrying motor output away from the central nervous system to muscles and glands are *efferent* ("to carry outward").

## Somatic and Autonomic Subdivisions

The peripheral nervous system has two subdivisions, called somatic and autonomic (Figure 33.6). The **somatic system** deals with movements of the body's head, trunk, and limbs. Its sensory axons carry signals inward from receptors in the skin, skeletal muscles, and tendons; and its motor axons carry signals out to the body's skeletal muscles. The **autonomic system** deals with the "visceral" portion of the body—that is, the internal organs and structures. Its sensory and motor axons carry signals from and to smooth muscle, cardiac (heart) muscle, and glands in different regions inside the body.

The reflex pathway shown earlier in Figure 32.14 is a simple example of how somatic nerves work. Autonomic nerves are more intricate than this, for they play off one another during the body's overall functioning in ways that will now be described.

## The Sympathetic and Parasympathetic Nerves

There are two subdivisions of autonomic nerves, called parasympathetic and sympathetic. Excitatory and inhibitory signals from **parasympathetic nerves** tend to slow down the body overall and divert energy to basic "housekeeping" tasks, such as digestion. This nerve action dominates when the body is not receiving much outside stimulation. Signals from its **sympathetic nerves** tend to slow down housekeeping tasks and increase overall body

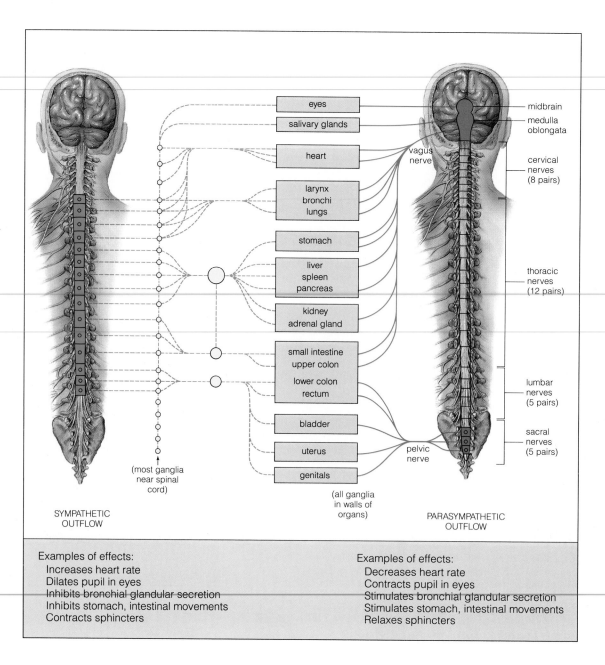

**Figure 33.7** Autonomic nervous system. Shown are the main sympathetic and parasympathetic pathways leading out from the central nervous system to some major organs. Keep in mind that both systems have *paired* nerves leading out from the central nervous system. Ganglia (singular, ganglion) are simply clusters of cell bodies of the neurons that are bundled together in nerves.

eyes
salivary glands
heart
larynx
bronchi
lungs
stomach
liver
spleen
pancreas
kidney
adrenal gland
small intestine
upper colon
lower colon
rectum
bladder
uterus
genitals

vagus nerve

midbrain
medulla oblongata
cervical nerves (8 pairs)
thoracic nerves (12 pairs)
lumbar nerves (5 pairs)
sacral nerves (5 pairs)

pelvic nerve

(most ganglia near spinal cord)

(all ganglia in walls of organs)

SYMPATHETIC OUTFLOW

PARASYMPATHETIC OUTFLOW

Examples of effects:
  Increases heart rate
  Dilates pupil in eyes
  Inhibits bronchial glandular secretion
  Inhibits stomach, intestinal movements
  Contracts sphincters

Examples of effects:
  Decreases heart rate
  Contracts pupil in eyes
  Stimulates bronchial glandular secretion
  Stimulates stomach, intestinal movements
  Relaxes sphincters

activities during times of heightened awareness, excitement, or danger. Sympathetic nerves prepare the animal to fight or flee when threatened or to frolic intensely, as in play and sexual behavior.

*Both* kinds of autonomic nerves are *continually* carrying signals to and from the central nervous system, and so help bring about minor adjustments in the activity of internal organs. Even while low levels of sympathetic signals are causing your heart to beat a little faster, low levels of parasympathetic signals are opposing this effect. At any moment, your heart rate is the net outcome of opposing signals (Figure 33.7). However, the parasympathetic input is reduced and the sympathetic system dominates in times of emergency or intense

excitement. Then, it calls a hormone (epinephrine) into action, the heart rate and breathing rate increase, and if the individual is capable of sweating, it sweats. In this state of intense arousal, the individual is primed to fight (or play) hard or get away fast. Hence the name, the *fight-flight response*.

Once the stimulus that triggered a fight-flight response is removed, sympathetic activity may decrease abruptly and parasympathetic activity may rise suddenly. Evidence of this "rebound effect" might be observed after a person has become instantly mobilized to rush onto a highway to save a child from an oncoming car. The person may well faint as soon as the child has been swept out of danger.

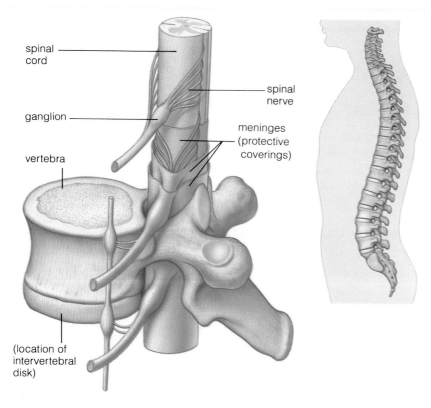

**Figure 33.8** Organization of the spinal cord and its relation to the vertebral column.

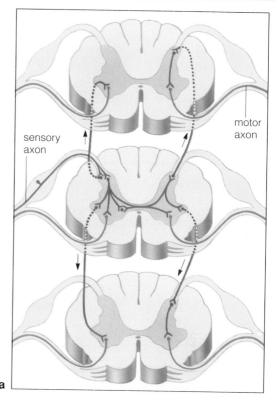

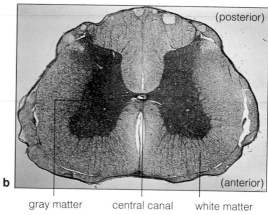

**Figure 33.9** (**a**) Examples of vertical and lateral connections among interneurons (color-coded green) in the spinal cord. (**b**) Photograph showing the arrangement of gray matter and white matter of the spinal cord.

# CENTRAL NERVOUS SYSTEM

## The Spinal Cord

The **spinal cord** is a vital expressway for signals between the peripheral nervous system and the brain. It also is a center for controlling some reflex activities. Here, in the cord, the sensory and motor neurons governing the movements of the body's limbs make direct reflex connections. The stretch reflex, described in the preceding chapter, arcs through the spinal cord in this manner.

The spinal cord threads through a canal formed by the stacked bones of the vertebral column (Figure 33.8). The bones and ligaments attached to them protect the cord. So do the *meninges*—three tough, tubelike coverings around the spinal cord and brain.

Propagation of signals up and down the spinal cord occurs in major **nerve tracts**, which are bundles of sheathed axons. The glistening myelin sheaths of these axons give the tracts the name *white matter*. The *gray matter* of the spinal cord consists of dendrites, cell bodies of neurons, as well as neuroglial cells. In cross-section, the gray matter of the cord looks vaguely like a butterfly (Figure 33.9). This part of the spinal cord deals mainly with reflex connections for limb movements (such as

walking) and internal organ activity (such as bladder emptying).

Maybe you have been on a farm when a chicken is destined for the stewpot. Even though the chicken has its head cut off, it still runs around for awhile. Chicken legs, you might correctly deduce, are governed to a great extent by stereotyped reflex pathways in the spinal cord. Experiments with frogs provide evidence of the importance of such pathways. Ascending nerve tracts between the frog's spinal cord and brain contain neu-

| Divisions | Main Components | Some Functions |
|---|---|---|
| FOREBRAIN | Cerebrum | Two cerebral hemispheres. Centers for coordinating sensory and motor functions, for memory, and for abstract thought. Most complex coordinating center; intersensory association, memory circuits |
| | Olfactory lobes | Relaying of sensory input from the nose to olfactory structures of cerebrum |
| | Limbic system | Scattered brain centers. With hypothalamus, coordination of skeletal muscle and internal organ activity underlying emotional expression |
| | Thalamus | Major coordinating center for sensory signals; relay station for sensory impulses to cerebrum |
| | Hypothalamus | Neural-endocrine coordination of visceral activities (e.g., solute-water balance, temperature control, carbohydrate metabolism) |
| | Pituitary gland | "Master" endocrine gland (controlled by hypothalamus). Control of growth, metabolism, etc. |
| | Pineal gland | Control of some circadian rhythms; role in mammalian reproductive physiology |
| MIDBRAIN | Tectum | Largely reflex coordination of visual, tactile, auditory input; contains nerve tracts ascending to thalamus, descending from cerebrum |
| HINDBRAIN | Pons | "Bridge" of transverse nerve tracts from cerebrum to both sides of cerebellum. Also contains longitudinal tracts connecting forebrain and spinal cord |
| | Cerebellum | Coordination of motor activity underlying limb movements, maintaining posture, spatial orientation |
| | Medulla oblongata | Contains tracts extending between pons and spinal cord; reflex centers involved in respiration, cardiovascular function, gastric secretion, etc. |

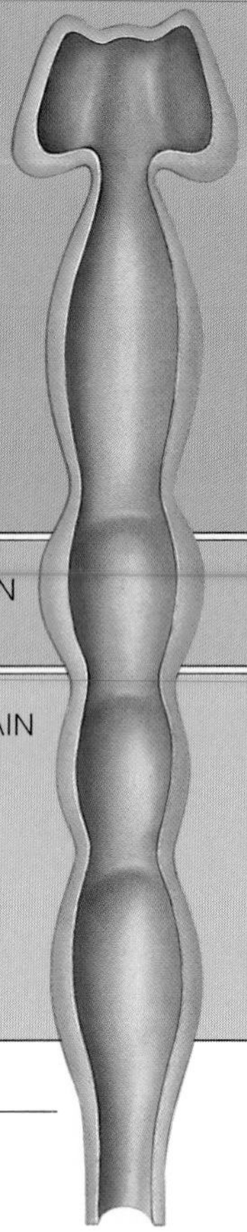

anterior end of spinal cord

**Figure 33.10** Summary of the three parts of the vertebrate brain and their main subdivisions. The drawing is highly simplified and flattened. While the vertebrate embryo is developing, the brain bends forward, and complex folds form in its wall regions, as shown in Figure 33.4. The midbrain, pons, and medulla oblongata also are called the *brain stem*. A network of interneurons, the reticular formation, extends the length of the brain stem and helps govern the activity of the nervous system as a whole.

rons that deal with straightening the legs after they have been bent. If those neurons are severed at the base of a frog brain, the legs become paralyzed—but only for about a minute. The so-called extensor reflex pathways in its spinal cord recover quickly and have the frog hopping around in no time. By comparison, it takes a few days for such pathways to recover in cats, days or weeks in monkeys, and many months in humans (who show the greatest cephalization).

## Divisions of the Brain

The **brain** is the body's master control panel. It receives, integrates, stores, and retrieves information, and it coordinates appropriate responses by intricately stimulating and inhibiting the activities of different body parts. The brain starts out as a continuation of the anterior end of the spinal cord. Like the spinal cord, it is protected by bones (of the cranial cavity) and meninges.

Figure 33.10 summarizes the functions of the three divisions of the brain—the hindbrain, midbrain, and forebrain. As we have seen, these regions develop from a hollow tube of nervous tissue in the embryo.

**Hindbrain.** The **hindbrain** consists of the medulla oblongata, cerebellum, and pons. The *medulla oblongata* has reflex centers for respiration, blood circulation, and other vital tasks. Here also, motor responses and complex reflexes (such as coughing) are coordinated. Its centers influence other brain centers that help you sleep or wake up.

The *cerebellum* has reflex centers for maintaining posture and refining limb movements. It integrates signals from the eyes, muscle spindles, skin, and elsewhere. It keeps other parts of your brain informed about how your trunk and limbs are positioned, how much different muscles are contracted or relaxed, and in which direction the body or limbs happen to be moving.

The *pons* (meaning bridge) is a major traffic center for nerve tracts passing between brain centers. The name refers to prominent bands of axons that extend into each side of the cerebellum.

**Midbrain.** The **midbrain** evolved as a coordinating center for reflex responses to visual and auditory input. Its roof of gray matter, the *tectum*, is important in fishes, amphibians, reptiles, and birds for integrating signals from the eyes and ears. (You can surgically remove a frog's cerebrum, its highest integrative center, and the frog can still do just about everything it normally does.) In mammals, sensory input still converges on the tectum, but it is rapidly sent on to higher centers.

The midbrain, pons, and medulla oblongata together represent the brain's "stem." Within the core of the brain stem and extending through its entire length is a major network of interneurons, the **reticular formation**. The reticular formation has extensive connections with the forebrain and helps govern the activity of the nervous system as a whole.

**Forebrain.** The **forebrain** has the most recently evolved layers of nerve tissues. A pair of olfactory lobes, which deal with the sense of smell, dominated early vertebrate forebrains. A brain center, the *cerebrum*, integrated input about odors and selected motor responses to it. Sensory signals were relayed and coordinated at the *thalamus*, a center below the cerebrum (some motor pathways also converged here). Another brain center, the *hypothalamus*, monitored internal organs and influenced forms of behavior related to their activities, such as thirst, hunger, and sex. In time, a thin layer of gray matter developed over each half of the cerebrum. In mammals, this *cerebral cortex* expanded into information-encoding and information-processing centers. It has become most highly developed in the human brain.

### Cerebrospinal Fluid

To get an idea of how soft nervous tissue is, try holding a jiggling blob of Jell-O in your hands. The brain and spinal cord are fragile! Besides being protected by bones and meninges, both actually float in **cerebrospinal fluid**. This clear extracellular fluid cushions the brain and spinal cord from abrupt, jarring movements. The fluid also fills four ventricles, which are interconnected cavities

within the brain that connect with one another and with the central canal of the spinal cord (Figure 33.11).

The bloodstream exchanges substances with the cerebrospinal fluid, which in turn exchanges substances with neurons. Mechanisms called the **blood-brain barrier** help control *which* blood-borne substances are allowed to reach the neurons. The mechanisms are built into more than 99 percent of the blood capillaries servicing the brain. Endothelial cells making up the walls of those special capillaries are fused together by continuous tight junctions (Figure 31.0). This means that substances cannot reach the brain without passing *through* the cells. Transport proteins embedded in the plasma membrane of those cells selectively transport glucose and other water-soluble substances across the barrier. Lipid-soluble substances quickly diffuse through the lipid bilayer of the plasma membrane. This is why caffeine, nicotine, alcohol, barbiturates, heroin, and other lipid-soluble drugs have such rapid effects on brain function.

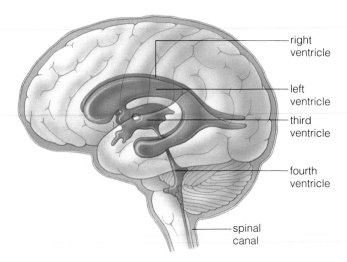

**Figure 33.11** Location of the cerebrospinal fluid in the human brain. This extracellular fluid surrounds and cushions the brain and spinal cord. It also fills the four interconnected cavities (cerebral ventricles) within the brain and the central canal of the spinal cord.

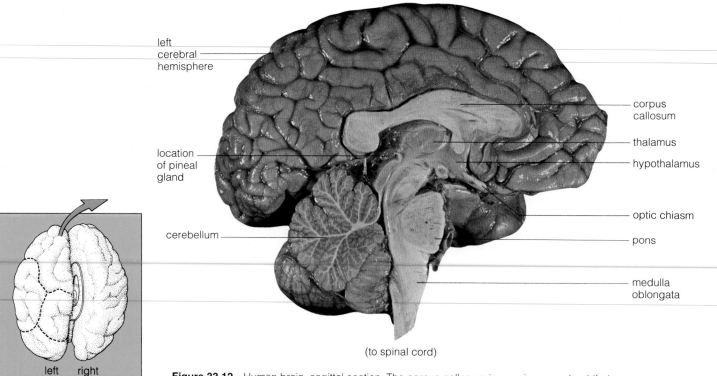

left cerebral hemisphere

location of pineal gland

cerebellum

corpus callosum

thalamus

hypothalamus

optic chiasm

pons

medulla oblongata

left cerebral hemisphere    right cerebral hemisphere

(to spinal cord)

**Figure 33.12** Human brain, sagittal section. The corpus callosum is a major nerve tract that runs transversely, connecting the two cerebral hemispheres. The boxed inset shows the two hemispheres pulled slightly apart; normally they are pressed together, with only a longitudinal fissure separating them.

## THE HUMAN BRAIN

Although there may be occasional argument about how wisely we use it, we humans have an impressively large, intricate brain. On the average, the human brain weighs about 1,300 grams (3 pounds). It contains about a hundred billion neurons.

### The Cerebrum

The human cerebrum vaguely resembles the much-folded nut inside a walnut shell. The folding suggests that expansion of the mammalian cerebrum outpaced the enlargement of the hard skullbones housing it. A deep fissure divides the human cerebrum into two parts, the left and right *cerebral hemispheres*. Other fissures and folds in each hemisphere follow certain patterns and divide it into the lobes named in Figure 33.12.

Much of the gray matter of the cerebral hemispheres is arranged as a thin surface layer, the **cerebral cortex**. The cerebral cortex weighs about a pound, and if you were to stretch it flat, it would cover a surface area of 2-1/2 square feet. The white matter consists of major nerve tracts that keep the hemispheres in communication with each other and with the rest of the body. Some

of the tracts originate in the brainstem, where they are rather jammed together, then fan out extensively in the cerebral hemispheres. Each hemisphere has its own set of tracts that serve as communication lines among its different regions. A prominent band of 200 million axons, the *corpus callosum*, keeps the two hemispheres in communication with each other. We know this through experiments of the sort described in the *Doing Science* essay.

The functioning of the cerebral hemispheres has been the focus of many experiments. Taken together, the results have revealed the following information:

1. Each cerebral hemisphere can function separately. However, the left cerebral hemisphere responds primarily to signals from the right side of the body. The opposite is true for the right cerebral hemisphere. Signals that travel by way of the corpus callosum coordinate the functioning of both hemispheres.

2. The main regions responsible for spoken language skills generally reside in the left hemisphere.

3. The main regions responsible for nonverbal skills (music, mathematics, and other abstract abilities) generally reside in the right hemisphere.

# *Doing Science*

## Sperry's Split-Brain Experiments

Experiments performed by Roger Sperry and his coworkers demonstrated some intriguing differences in perception between the two halves of the cerebrum. The subjects of the experiments were epileptics. Persons with severe epilepsy are wracked with seizures, sometimes as often as every half hour of their lives. The seizures have a neurological basis, analogous to an electrical storm in the brain.

What would happen if the corpus callosum of epileptics were cut? Would the electrical storm be confined to one cerebral hemisphere, leaving at least the other to function normally? Earlier studies of animals and of humans whose corpus callosum had been damaged suggested this might be so.

The surgery was performed. The electrical storms subsided in frequency and intensity. Cutting the neural bridge between the two hemispheres put an end to what must have been positive feedback of ever intensified electrical disturbances between them. Beyond this, the "split-brain" individuals were able to lead what seemed, on the surface, entirely normal lives.

But then Sperry devised some elegant experiments to find out whether the conscious experience of those individuals was indeed "normal." After all, the corpus callosum contains no less than 200 million axons; surely something was different. Something was. "The surgery," Sperry later reported, "left these people with two separate minds, that is, two spheres of consciousness. What is experienced in the right hemisphere seems to be entirely outside the realm of awareness of the left."

In Sperry's experiments, the left and right hemispheres of split-brain individuals were presented with different stimuli. It was known at the time that visual connections to and from one hemisphere are mainly concerned with the opposite visual field (Figure *a*).

Sperry projected words—say, COWBOY—onto a screen. He did this in such a way that COW fell only on the left visual field, and BOY fell on the right (Figure *b*). The subject reported seeing the word BOY. The left hemisphere, which controls language, received only the letters BOY. However, when asked to write the perceived word with the left hand—a hand that was deliberately blocked from the subject's view—the subject wrote COW, as shown in Figure *b*.

The right hemisphere, which "knew" the other half of the word (COW) had directed the left hand's motor response. (That hemisphere controls muscles on the left side of the body, and vice versa.) But it couldn't tell the

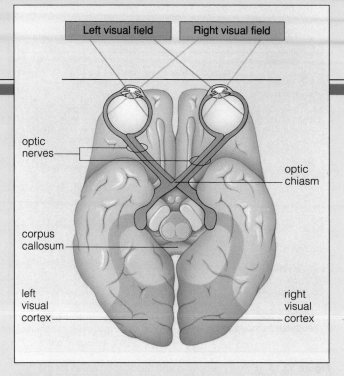

**a** In the human eye, visual information is gathered at the retina, a layer of densely packed light receptors. Light from the *left* half of the viewing field strikes receptors on the right side of both retinas. Parts of the two optic nerves carry signals from those receptors to the right cerebral hemisphere. Light from the *right* half of the viewing field strikes receptors on the left side of both retinas. Parts of the two optic nerves carry signals from those receptors to the left hemisphere.

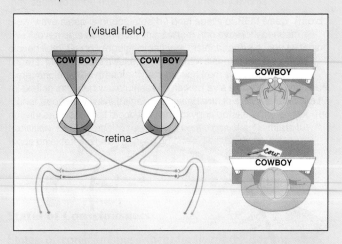

**b** Responses of split-brain individual to different visual stimuli.

left hemisphere what was going on because of the severed corpus callosum. The subject knew that a word was being written, but could not say what it was!

Thus, the two cerebral hemispheres were functioning separately, and each was responding to visual signals from the opposite side of the body. Sperry's work showed that the corpus callosum is necessary to coordinate their functioning.

# Drug Action on Integration and Control

Broadly speaking, a drug is any substance introduced into the body to provoke a specific physiological response. Some drugs help a person cope with discomforts of an illness or stress. Others act on the *pleasure center* in the hypothalamus and artificially fan the sense of pleasure that we associate with eating, sexual activity, and other self-gratifying behaviors.

Many drugs are habit-forming. Even if the body functions well enough without them, they continue to be used for the real or imagined relief they afford. Often the body develops *tolerance* of such drugs, meaning it takes larger or more frequent doses to produce the same effect. Habituation and tolerance are signs of **addiction**, or chemical dependence, on a drug. *The drug has taken on an "essential" biochemical role in the body.* Abruptly deprive an addict of the drug, and he or she will suffer agonizing physical pain as well as mental anguish. Such withdrawal symptoms are manifestations of major biochemical upheavals throughout the body.

Here we consider four classes of psychoactive drugs, which act on brain regions governing states of consciousness and behavior. They are the depressants and hypnotics, stimulants, narcotic analgesics, and hallucinogens and psychedelics.

## Stimulants

Stimulants include caffeine, nicotine, amphetamines, and cocaine. First they increase alertness and body activity, then they lead to depression.

Coffee, tea, chocolate, and many soft drinks contain caffeine, one of the most widely used stimulants. Low doses of caffeine stimulate the cerebral cortex first, and cause increased alertness and restlessness. Higher doses act at the medulla oblongata to disrupt motor coordination and intellectual coherence.

Nicotine, a component of tobacco, has powerful effects on the central and peripheral nervous systems. It mimics acetylcholine and can directly stimulate a number of sensory receptors. Its short-term effects include water retention, irritability, increased heart rate and blood pressure, and gastric upsets. Its long-term effects can be devastating.

Like dopamine and norepinephrine (which they resemble), the amphetamines (including "speed") stimulate the pleasure center. In time, the brain produces less and less of its own signaling molecules and comes to depend on artificial stimulation.

Cocaine stimulates the pleasure center in a different way. It produces a rush of pleasure by blocking the reabsorption of dopamine, norepinephrine, serotonin, and other signaling molecules that are normally released at synapses. Receptor cells are incessantly stimulated over an extended period. Heart rate and blood pressure rise; sexual appetite increases. But then the effects change. The signaling molecules that have accumulated in synaptic clefts diffuse away—but the cells that produce them cannot make up for the extraordinary loss. The sense of pleasure evaporates as the receptor cells (which are now hypersensitive to stimulation) demand stimulation. The cocaine user becomes anxious and depressed. After prolonged, heavy use of cocaine, "pleasure" is impossible to experience. The addict loses weight and cannot sleep properly. The immune system becomes compromised, and heart abnormalities set in.

Granular cocaine, which is inhaled (snorted), has been around for some time. Crack cocaine, a cheaper but more potent form, is burned and the smoke inhaled. As suggested at the start of this chapter, crack is incredibly addictive; its highs are higher, but the crashes are more devastating.

In the brain stem, a branch of the reticular formation controls the changing levels of consciousness. It sends signals to the spinal cord, cerebellum, and cerebrum as well as back to itself. The flow of signals along these circuits—and the inhibitory or excitatory chemical changes accompanying them—affects whether you stay awake or fall asleep. Damage to parts of the circuits can lead to unconsciousness and coma.

Interneurons of one of the "sleep centers" of the reticular formation release serotonin. This transmitter substance inhibits other neurons that arouse the brain and maintain wakefulness. Thus, high serotonin levels are linked to drowsiness and sleep. Substances released from another brain center counteract serotonin's effects and bring about wakefulness.

### Emotional States

Our emotions are governed by the cerebral cortex and by different brain regions collectively called the **limbic**

## Depressants, Hypnotics

These drugs lower the activity in nerves and parts of the brain, so they reduce activity throughout the body. Some act at synapses in the reticular formation system and in the thalamus.

Depending on the dosage, most of these drugs can produce responses ranging from emotional relief, sedation, sleep, anesthesia and coma, to death. At low doses, inhibitory synapses are often suppressed slightly more than excitatory synapses, so the person feels excited or euphoric at first. Increased doses also suppress excitatory synapses, leading to depression. Depressants and hypnotics have additive effects; one amplifies another. For example, combining alcohol with barbiturates amplifies behavioral depression.

Alcohol (ethyl alcohol) differs from the drugs just described because it acts directly on the plasma membrane to alter cell function. Some persons mistakenly think of it as a harmless stimulant (it produces an initial "high"). But alcohol is one of the most powerful psychoactive drugs and a major cause of death. Small doses even over the short term can produce disorientation, uncoordinated motor functions, and diminished judgment. Long-term addiction destroys nerve cells and causes permanent brain damage; it can permanently damage the liver (cirrhosis).

## Analgesics

When stress leads to physical or emotional pain, the brain produces its own analgesics, or natural pain relievers. Endorphins and enkephalins are examples. These substances seem to inhibit activity in many parts of the nervous system, including brain centers concerned with emotions and perception of pain.

| Classes of Psychoactive Drugs | |
|---|---|
| Class | Examples |
| Depressants, hypnotics | Barbiturates (e.g., Nembutal, Quaalude) Antianxiety drugs (e.g., Valium, alcohol) |
| Stimulants | Caffeine Nicotine Amphetamines (e.g., Dexedrine) Cocaine |
| Narcotic analgesics | Codeine Opium Heroin |
| Psychedelics, hallucinogens | Lysergic acid diethylamide (LSD) *Cannabis* (marijuana) |

The narcotic analgesics, including codeine and heroin, sedate the body and relieve pain. They are extremely addictive. Deprivation following massive doses of heroin leads to fever, chills, hyperactivity and anxiety, violent vomiting, cramping, and diarrhea.

## Psychedelics, Hallucinogens

These drugs, which alter sensory perception, have been described as "mind-expanding." Some skew acetylcholine or norepinephrine activity. Others, such as LSD (lysergic acid diethylamide), affect serotonin activity. Even in small doses, LSD dramatically warps perceptions.

Marijuana is another hallucinogen. The name refers to the drug made from crushed leaves, flowers, and stems of the plant *Cannabis*. In low doses marijuana is like a depressant. It slows down but does not impair motor activity; it relaxes the body and elicits mild euphoria. However, it can produce disorientation, increased anxiety bordering on panic, delusions (including paranoia), and hallucinations.

Like alcohol, marijuana can affect an individual's ability to perform complex tasks, such as driving a car. In one study, commercial pilots showed a marked deterioration in instrument-flying ability for more than two hours after smoking marijuana. Recent studies point to a link between marijuana smoking and suppression of the immune system.

---

**system** (Figure 33.15). The limbic system of humans is only distantly related to the sense of smell that figured so prominently in vertebrate evolution. Even so, interconnections have been maintained that play some role in memory function and cognition.

The connections between the sense of smell and the limbic system are the reason why you may "smell" a cologne all over again when you have a pleasant memory of the person who wore it; or why you smell a bad odor when you remember a confrontation with a skunk.

The hypothalamus is the gatekeeper of the limbic system. Many connections from the cerebral cortex and lower brain centers pass through it. Through these connections, the reasoning possible in the cerebral cortex can dampen rage, hatred, and other so-called "gut reactions."

The hypothalamus also monitors internal organs in addition to emotional states. This is what keeps your heart and stomach on fire when you are sick with passion (or indigestion).

# SUMMARY

1. Nervous systems provide specialized means of detecting stimuli and responding to them swiftly.

2. The simplest nervous systems are nerve nets, such as those of sea anemones, hydra, and jellyfishes. They are based on reflex connections between nerve cells and contractile cells of the epithelium.

3. The nervous system of vertebrates shows pronounced cephalization and bilateral symmetry, as evident in its complex brain centers and its many paired nerves.

4. Parts of the vertebrate brain deal with reflex coordination of sensory inputs and motor outputs beyond that afforded by the spinal cord alone. Its most recent layerings also deal with storing, comparing, and using experiences to initiate novel, nonstereotyped action. These regions are the basis of memory, learning, and reasoning.

5. The brain and spinal cord represent the central nervous system. Many pairs of nerves carry signals between the central nervous system and the body's various organs and structures. These nerves are the basis of the peripheral nervous system.

6. The peripheral nervous system has a somatic subdivision, which deals with skeletal muscles concerned with voluntary body movements. It also has an autonomic subdivision, which deals with the functions of the heart, lungs, glands, and other internal organs.

7. The spinal cord has nerve tracts that carry signals between the brain and the peripheral nervous system. It also is a center for some direct reflex connections that underlie limb movements and internal organ activity.

8. The brain has three regional divisions (hindbrain, midbrain, and forebrain).

   a. The hindbrain includes the medulla oblongata, pons, and cerebellum and contains reflex centers for vital functions and muscle coordination.

   b. The midbrain functions in coordinating and relaying visual and auditory information.

   c. The medulla oblongata, pons, and midbrain constitute the brain stem. The reticular formation, an extensive network of interneurons, extends the length of the brain stem and helps govern activities of the nervous system as a whole.

   d. The forebrain includes the cerebrum, thalamus, hypothalamus, and limbic system. The thalamus relays sensory information and helps coordinate motor responses. The hypothalamus monitors internal organs and influences thirst, hunger, sexual activity, and other behaviors related to their functioning. It is gatekeeper to the limbic system, which has roles in learning, memory, and emotional behavior.

9. The cerebral cortex has regions devoted to specific functions, such as receiving information from the various sense organs, integrating this information with memories of past events, and coordinating motor responses.

10. Memory apparently occurs in two stages: a short-term formative period and long-term storage, which depends on chemical or structural changes in the brain.

11. States of consciousness vary between total alertness and deep coma. The levels are governed by the reticular activating system. They are subject to the influence of psychoactive drugs.

## Review Questions

1. What are some of the organizational features that nearly all nervous systems have in common? *564*

2. What constitutes the central nervous system? The peripheral nervous system? *566*

3. Can you distinguish among the following:
   a. ganglia and nerves *564*
   b. spinal nerves and cranial nerves *566*
   c. somatic system and autonomic system *567*
   d. parasympathetic and sympathetic nerves *567–568*

4. Review Figure 33.10. Then, on your own, describe the components of the three main subdivisions of the vertebrate brain. *570*

5. What is a psychoactive drug? Can you describe the effects of one such drug on the central nervous system? *576–577*

6. Label the parts of the human brain: *572*

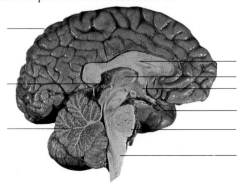

## Self-Quiz *(Answers in Appendix IV)*

1. Sea anemones, hydras, and jellyfishes have simple nervous systems called _____.

2. Structurally, the vertebrate nervous system shows pronounced _____ and _____ symmetry.

3. The oldest parts of the vertebrate brain provide _____.
   a. reflex control of breathing, blood circulation, and other basic activities
   b. coordinating and relaying visual and auditory information
   c. storing, comparing, and using experiences to initiate novel, nonstereotyped action
   d. both a and c are correct

4. The central nervous system includes _____. The peripheral nervous system includes _____.
   a. nerves and ganglia; brain and spinal cord
   b. brain and spinal cord; nerves and ganglia
   c. spinal cord; brain
   d. nerves and interneurons; brain and spinal cord

5. Overall, _____ nerves slow down the body and divert energy to digestion and other housekeeping tasks; _____ nerves slow down housekeeping tasks and increase overall activity during times of heightened awareness, excitement, or danger.
   a. autonomic; somatic
   b. sympathetic; parasympathetic
   c. parasympathetic; sympathetic
   d. peripheral; central

6. Parasympathetic and sympathetic nerves carry signals to and from the nervous system _____ to bring about minor adjustments in internal organ activity.
   a. continually
   b. in separate, alternating fashion
   c. only during a fight-flight response
   d. none of the above is correct

7. The _____ of the spinal cord consists of nerve tracts; the _____ consists of dendrites, neuron cell bodies, and neuroglial cells.
   a. gray matter; white matter
   b. white matter; gray matter

8. The hindbrain (medulla oblongata, cerebellum, and pons) contains _____.
   a. the reticular formation
   b. major nerve tracts between brain centers
   c. reflex centers for limb movements, respiration, breathing, and other vital tasks
   d. both b and c are correct

9. Extending through the entire length of the brain stem (midbrain, pons, and medulla oblongata) is the _____.
   a. reticular formation
   b. blood-brain barrier
   c. olfactory lobe
   d. tectum

10. The most highly developed part of the human brain (the forebrain) includes the _____.
    a. medulla oblongata, pons, and cerebellum
    b. cerebrum, thalamus, hypothalamus, and limbic system
    c. medulla oblongata, pons, and cerebral cortex
    d. cerebellum, medulla oblongata, pons, and limbic system
    e. hypothalamus, limbic system, pons, and cerebral cortex

11. Match the central nervous system region with some of its functions.
    _____ spinal cord
    _____ medulla oblongata
    _____ hypothalamus
    _____ limbic system
    _____ cerebral cortex

    a. receives sensory input, integrates it with stored information
    b. monitors internal organs and related behavior (e.g., hunger)
    c. with cerebral cortex, governs emotions
    d. reflex control of respiration, blood circulation, other basic activities
    e. expressway for signals between brain and peripheral nervous system

## Selected Key Terms

addiction *576*
autonomic system *567*
blood-brain barrier *571*
brain *570*
brain stem *570*
central nervous system *566*
cephalization *564*
cerebellum *571*
cerebral cortex *571*
cerebrospinal fluid *571*
cerebrum *571*
corpus callosum *572*
forebrain *571*
ganglion *564*
gray matter *569*
hindbrain *570*
hippocampus *574*
hypothalamus *571*
limbic system *576*
medulla oblongata *570*
memory *574*
meninges *569*
midbrain *571*
nerve *564*
nerve cord *564*
nerve net *564*
nerve tract *569*
notochord *566*
parasympathetic nerve *567*
peripheral nervous system *566*
pons *571*
reflex *564*
reticular formation *571*
somatic system *567*
spinal cord *569*
sympathetic nerve *567*
tectum *571*
thalamus *571*
white matter *569*

## Readings

Barlow, R., Jr. April 1990. "What the Brain Tells the Eye." *Scientific American* 262(4):90–95.

Bloom, F., and A. Lazerson. 1988. *Brain, Mind, and Behavior*. Second edition. New York: Freeman.

Churchland, P., and P. Churchland. January 1990. "Could a Machine Think?" *Scientific American* 262(1):32–37.

Julien, R. 1985. *A Primer of Drug Action*. Fourth edition. New York: Freeman. Effectively fills the gap between popularized (and often superficial or misleading) accounts of drug action and the upper-division books in pharmacology. Paperback.

Kalil, R. December 1989. "Synapse Formation in the Developing Brain." *Scientific American* 261(6):76–85.

Romer, A., and T. Parsons. 1986. *The Vertebrate Body*. Sixth edition. Philadelphia: Saunders. Insights into the evolution of vertebrate nervous systems.

Shepherd, G. 1988. *Neurobiology*. Second edition. New York: Oxford. Paperback.

Springer, S., and Deutsch, G. 1985. *Left Brain, Right Brain*. Revised edition. New York: Freeman.

## *Hormone Jamboree*

In the early 1960s, at her camp in the forests along the shores of Lake Tanganyika in Africa, the primatologist Jane Goodall let it be known that bananas were available. Among the first chimpanzees attracted to the delicious new food was a female—Flo, as she came to be called. Flo brought along her offspring, an infant female and a juvenile male, and tended carefully to them. Three years later, Flo's preoccupation with being a mother gave way to a preoccupation with sex. Male chimpanzees followed Flo to the camp, and stayed for more than the bananas.

Sex, Goodall discovered, is the premier force in the social life of chimpanzees. These primates do not mate for life, as eagles do, or wolves. Before the rainy season begins, the mature females that are undergoing a fertile cycle become sexually active. In response to changing blood concentrations of sex hormones, their external genitalia become enormously swollen and pink, a visual signal that is magnetic to males. Their swellings are the flags of sexual jamborees, of great gatherings of highly

stimulated chimps in which any males present may copulate in sequence with the same female.

Although a swollen bottom makes it rather difficult to sit and is vulnerable to being torn, it has its advantages. The gathering of many flag-waving females in the same place draws together individuals that otherwise would spend most of the year foraging alone or in small family groups. Now they spend time together, reestablishing bonds that hold them together in their rather fluid community. Infants and juveniles interact with one another and with the adults. Future dominance hierarchies have their foundations in the playful and aggressive jostlings. Consider Flo, a high-ranking member of the social hierarchy. Through her sexual attractiveness and direct solicitations, she built alliances with many male chimps. Through her high

**Figure 34.1** (a) Primatologist Jane Goodall in Gombe National Park, near the shores of Lake Tanganyika, scouting for chimpanzees. (b) The socially dominant Flo, center, and three of her offspring.

a

**b**

status and aggressive behavior, she helped her male offspring win confrontations with other young male chimps.

Intriguingly, the sexual swelling lasts somewhere between ten and sixteen days—yet the female actually is fertile for only one to five days. Furthermore, swellings may occur during nonfertile periods and at irregular times even after a female has become pregnant. It is not difficult to imagine why prolonged swelling has been favored during the course of chimpanzee evolution. The males groom a sexually attractive female more often and give her a larger share of their food. She is allowed to travel with males to new, peripheral food sources and is protected by them. The greater her acceptance by males, the higher up she goes in the social hierarchy—and the more her offspring benefit.

Through their effects, hormones help orchestrate the reproductive cycle of chimpanzees. They do the same for nearly all animals, from invertebrate worms to humans. Besides this, hormones help orchestrate growth and development. They help control minute-to-minute and day-to-day metabolic functions. Through their interplays with one another and with the nervous system, hormones have major influence over the physical appearance of individuals, their well being, and how they will behave. How individuals behave has major influence on whether they survive, either on their own or as members of social groups.

This chapter is about hormones and other signaling molecules—their sources, targets, and interactions. It is about their effects and the molecular mechanisms governing their secretion and action. If the details start to seem remote, remember that this is the stuff of life. Hormones underwrote Flo's appearance, behavior, and rise through the chimpanzee's social hierarchy—and just imagine what they have been doing for you.

KEY CONCEPTS

**1.** Hormones and other signaling molecules help integrate cell activities in ways that benefit the whole body. Some hormones help the body adjust to short-term changes in diet and levels of activity. Other hormones have roles in long-term adjustments underlying growth, development, and reproduction.

**2.** The hypothalamus and pituitary gland interact in ways that coordinate the stimulation and inhibition of many endocrine glands. Together they control many of the body's functions.

**3.** Neural signals, hormonal signals, chemical changes in the blood, and environmental cues trigger hormone secretions. Only cells with receptors for a specific hormone are its targets. Steroid hormones trigger the activation of genes and protein synthesis in target cells. Protein hormones alter the activity of existing enzymes in target cells.

## "THE ENDOCRINE SYSTEM"

The word "hormone" dates back to the early 1900s, when W. Bayliss and E. Starling were trying to figure out what triggers the secretion of pancreatic juices that act on food traveling through the canine gut. At the time, it was known that acids mix with food in the stomach, and that the pancreas secretes an alkaline solution when the acidic mixture has passed into the small intestine. Was the nervous system or something else stimulating the pancreatic response?

To find the answer, Bayliss and Starling cut off the nerve supply to the upper small intestine but left its blood vessels intact. When an acid was introduced into the intestinal region, the pancreas still was able to make the secretory response. More telling, extracts of cells taken from the epithelial lining of the small intestine also induced the response. Glandular cells in the lining had to be the source of a substance that stimulated the pancreas into action.

The substance itself came to be called secretin. Demonstration of its existence and its mode of action was the

| Table 34.1 | Effect of Releasing and Inhibiting Hormones on Anterior Pituitary | |
|---|---|---|
| Hormone | Influences Secretion of: | Effect* |
| Corticotropin-releasing hormone (CRH) | Corticotropin (ACTH) | + |
| Thyrotropin-releasing hormone (TRH) | Thyrotropin (TSH) | + |
| Gonadotropin-releasing hormone (GnRH) | Follicle-stimulating hormone (FSH) | + |
|  | Luteinizing hormone (LH) | + |
| STH-releasing hormone (STHRH) | Somatotropin (STH); also called growth hormone (GH) | + |
| Somatostatin | Somatotropin, TSH | − |
| Dopamine | Prolactin (PRL), LH, FSH | − |

*Stimulatory (+) or inhibitory (−).

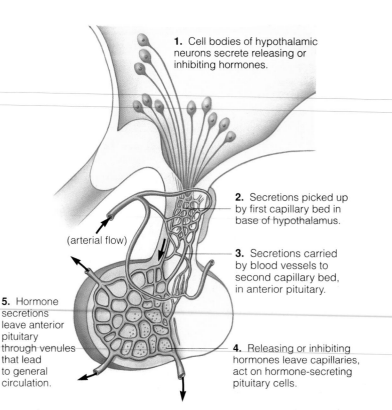

1. Cell bodies of hypothalamic neurons secrete releasing or inhibiting hormones.

2. Secretions picked up by first capillary bed in base of hypothalamus.

(arterial flow)

3. Secretions carried by blood vessels to second capillary bed, in anterior pituitary.

5. Hormone secretions leave anterior pituitary through venules that lead to general circulation.

4. Releasing or inhibiting hormones leave capillaries, act on hormone-secreting pituitary cells.

**Figure 34.6** Functional links between the hypothalamus and the anterior lobe of the pituitary.

different regions of the body. It causes arterioles in some tissues to constrict and so helps divert blood flow to other tissues, where metabolic demands are greater. Oxytocin has roles in reproduction. For example, it triggers muscle contractions in the uterus during labor and causes milk to be released when the young are being nursed.

### Anterior Lobe Secretions

**Role of the Hypothalamus**. The hypothalamus produces and secretes **releasing hormones** and **inhibiting hormones**. Both are signaling molecules that act on specific cells of the anterior lobe of the pituitary. Most stimulate target cells to secrete other hormones; others slow down the secretions (Table 34.1). As Figure 34.6 shows, the releasing or inhibiting hormones travel through two capillary beds before leaving the blood and binding to receptors on target cells in the anterior lobe.

Releasing hormones were identified by monumental research efforts that began in 1955, most notably by Roger Guillemin and Andrew Schally. Over one four-year period, Guillemin's team purchased 500 tons of sheep brains from meat processing plants and extracted 7 tons of hypothalamic tissue from them. They eventually ended up with a single milligram of a substance that stimulates the release of "TSH," a hormone that controls the functions of the thyroid gland.

**Anterior Pituitary Hormones**. In response to commands from the hypothalamus, different cells of the anterior pituitary secrete the following hormones of their own:

| | |
|---|---|
| Corticotropin | ACTH |
| Thyrotropin | TSH |
| Follicle-stimulating hormone | FSH |
| Luteinizing hormone | LH |
| Prolactin | PRL |
| Somatotropin (or growth hormone) | STH (or GH) |

The first four hormones listed act on endocrine glands, which in turn produce other hormones. The effect of ACTH on adrenal glands and TSH on the thyroid will be described shortly. FSH and LH play elegant roles in reproductive function, a topic that will occupy our attention in Chapter 43.

The last two hormones listed, prolactin and somatotropin, have effects on body tissues in general (Figure 34.7 and Table 34.2). Prolactin influences a variety of activities among vertebrate species ranging from primitive fishes to humans. One of its functions is to stimulate and sustain milk production in mammary glands during lactation (page 775). Prolactin does this only when the tissues in those glands have been primed by other hormones. Prolactin also affects the production of hormones by ovaries.

**Table 34.2   Hormones Released from the Mammalian Pituitary Gland**

| Pituitary Lobe | Secretions | Abbreviation | Main Targets | Primary Actions |
|---|---|---|---|---|
| **Posterior** | | | | |
| Nervous tissue (extension of hypothalamus) | Antidiuretic hormone | ADH | Kidneys | Induces water conservation required in control of extracellular fluid volume (and, indirectly, solute concentrations) |
| | Oxytocin | | Mammary glands Uterus | Induces milk movement into secretory ducts Induces uterine contractions |
| **Anterior** | | | | |
| Mostly glandular tissue | Corticotropin | ACTH | Adrenal cortex | Stimulates release of adrenal steroid hormones |
| | Thyrotropin | TSH | Thyroid gland | Stimulates release of thyroid hormones |
| | Gonadotropins: Follicle-stimulating hormone | FSH | Ovaries, testes | In females, stimulates follicle growth, helps stimulate estrogen secretion, ovulation; in males, promotes spermatogenesis |
| | Luteinizing hormone | LH | Ovaries, testes | In females, stimulates ovulation, corpus luteum formation; in males, promotes testosterone secretion, sperm release |
| | Prolactin | PRL | Mammary glands | Stimulates and sustains milk production |
| | Somatotropin (also called growth hormone) | STH (GH) | Most cells | Has growth-promoting effects in young; induces protein synthesis, cell division; has role in glucose, protein metabolism in adults |
| **Intermediate*** | | | | |
| Mostly glandular tissue | Melanocyte-stimulating hormone | MSH | Pigmented cells in skin, other surface coverings | Induces color changes in response to external stimuli; affects behavior |

*Present in most vertebrates (not adult humans). MSH is associated with the anterior lobe in humans.

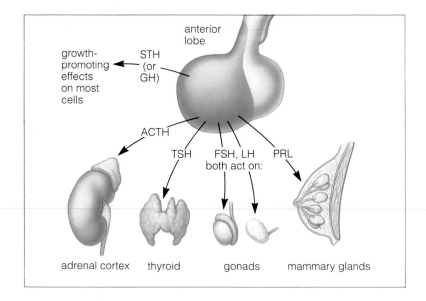

**Figure 34.7**   Secretions of the anterior lobe of the pituitary and some of their targets.

**Figure 34.8** (**a**) Manute Bol, an NBA center, is 7 feet 6-3/4 inches tall owing to excessive STH production during childhood.

(**b**) Effect of somatotropin (STH) on overall body growth. The person at the center is affected by gigantism, which resulted from excessive STH production during childhood. The person at right displays pituitary dwarfism, which resulted from underproduction of STH during childhood. The person at the left is average in size.

a

b

**Figure 34.9** Acromegaly, which resulted from excessive production of STH during adulthood. Before this female reached maturity, she was symptom-free.

age nine

sixteen

thirty-three

fifty-two

Somatotropin, or growth hormone, stimulates protein synthesis and cell division in target cells. It profoundly influences overall growth, especially of cartilage and bone. Figure 34.8 shows what can happen with too little or too much somatotropin. *Pituitary dwarfism* results when not enough somatotropin was produced during childhood. The adult is similar in proportion to a normal person but much smaller. *Gigantism* results when excessive amounts of somatotropin were produced during childhood. The adult is similar in proportion to a normal person but is much larger.

*Acromegaly* results from excessive secretion of somatotropin during adulthood, when long bones no longer can lengthen. Cartilage, bone, and other connective tissues of the hands, feet, and jaws thicken, as do epithelial tissues of the skin, nose, eyelids, lips, and tongue (Figure 34.9). Skin thickening is pronounced on the forehead and soles of the feet.

## SELECTED EXAMPLES OF HORMONAL ACTION

Table 34.3 lists hormones from endocrine glands other than the pituitary. Here we will focus on a few examples of endocrine activity to show how hormonal controls work. These examples will lay the groundwork for understanding how hormones affect digestion, circulation, reproduction, and other activities described in chapters to come.

**Table 34.3  Hormone Sources Other Than the Mammalian Hypothalamus and Pituitary**

| Source | Its Secretion(s) | Main Targets | Primary Actions |
|---|---|---|---|
| **Adrenal cortex** | Glucocorticoids (including cortisol) | Most cells | Promote protein breakdown and conversion to glucose |
| | Mineralocorticoids (including aldosterone) | Kidney | Promote sodium reabsorption; control salt, water balance |
| **Adrenal medulla** | Epinephrine (adrenalin) | Liver, muscle, adipose tissue | Raises blood level of sugar, fatty acids; increases heart rate, force of contraction |
| | Norepinephrine | Smooth muscle of blood vessels | Promotes constriction or dilation of blood vessel diameter |
| **Thyroid** | Triiodothyronine, thyroxine | Most cells | Regulates metabolism; has roles in growth, development |
| | Calcitonin | Bone | Lowers calcium levels in blood |
| **Parathyroids** | Parathyroid hormone | Bone, kidney | Elevates calcium levels in blood |
| **Gonads:** | | | |
| Testis (in males) | Androgens (including testosterone) | General | Required in sperm formation, development of genitals, maintenance of sexual traits; influences growth, development |
| Ovary (in females) | Estrogens | General | Required in egg maturation and release; prepares uterine lining for pregnancy; other actions same as above |
| | Progesterone | Uterus, breast | Prepares, maintains uterine lining for pregnancy; stimulates breast development |
| **Pancreatic islets** | Insulin | Muscle, adipose tissue | Lowers blood sugar level |
| | Glucagon | Liver | Raises blood sugar level |
| | Somatostatin | Insulin-secreting cells of pancreas | Influences carbohydrate metabolism |
| **Endocrine cells of stomach, gut** | Gastrin, secretin, etc. | Stomach, pancreas, gallbladder | Stimulates activity of stomach, pancreas, liver, gallbladder |
| **Liver** | Somatomedins | Most cells | Stimulates overall growth, development |
| **Kidney** | Erythropoietin* | Bone marrow | Stimulates red blood cell production |
| | Angiotensin* | Adrenal cortex, arterioles | Helps control blood pressure, aldosterone secretion |
| | Vitamin $D_3$* | Bone, gut | Enhances calcium resorption and uptake |
| **Heart** | Atrial natriuretic hormone | Kidney, blood vessels | Increases sodium excretion; lowers blood pressure |
| **Thymus** | Thymosin, etc. | Lymphocytes | Has roles in immune responses |
| **Pineal** | Melatonin | Gonads (indirectly) | Influences daily biorhythms, seasonal sexual activity |

*These hormones are not produced in the kidneys but are formed when *enzymes* produced in kidneys activate specific substances in the blood.

There are three points to keep in mind about hormonal action. *First*, the response of target cells depends on the number of receptors they have and on the concentration of hormone molecules at a given time. *Second*, many hormones are linked to the neuroendocrine control center by **homeostatic feedback loops**. In such loops, the hypothalamus, pituitary, or both detect a change in the concentration of a hormone in some body region, then respond by inhibiting or stimulating the gland that secretes the hormone. *Third*, hormones interact to produce some effect on body functions.

As you will see, three kinds of hormonal interactions are common:

**1.** *Antagonistic interaction.* The effect of one hormone may oppose the effect of another. Insulin, for example, promotes a decrease in the glucose level in the blood and glucagon promotes an increase.

**2.** *Synergistic interaction.* The sum total of the action of two or more hormones is necessary to produce the required effect on target cells. Mammals, for example, cannot produce and secrete milk without the synergistic interaction of the hormones prolactin, oxytocin, estrogen, and progesterone.

**3.** *Permissive interaction.* One hormone exerts its effect only when a target cell has become "primed" to respond in an enhanced way to that hormone. The priming is accomplished by previous exposure to another hormone. Getting pregnant, for example, depends on the lining of the uterus being exposed first to estrogens, then to progesterone.

### Adrenal Glands

**Adrenal Cortex**. Humans have a pair of adrenal glands, one above each kidney (Figure 34.10). We call the outer portion of each gland the **adrenal cortex**. Glucocorticoids are among the hormones secreted by the adrenal cortex. And homeostatic feedback loops to the neuroendocrine control center govern those secretions.

Glucocorticoids influence metabolic reactions that help maintain the glucose level in the blood. They also suppress inflammatory responses to tissue injury or infection. Cortisol is an example. Cortisol blocks the uptake and use of glucose by muscle cells, stimulates liver cells to store glucose (as glycogen), and makes more glucose in blood available to the brain. It also stimulates liver cells to form glucose from amino acids when the blood glucose level falls.

When the blood level of glucose falls below a set point, a condition called *hypoglycemia*, the hypothala-

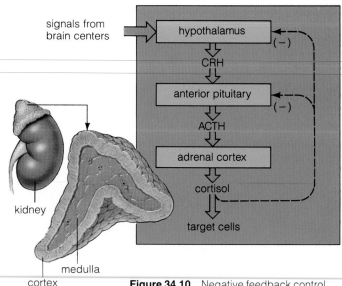

**Figure 34.10** Negative feedback control of cortisol secretion.

mus is called into action. The hypothalamus initiates a *stress response* by secreting a releasing hormone (CRH). As Figure 34.10 suggests, the CRH prods the anterior pituitary into secreting corticotropin (ACTH). The ACTH stimulates the adrenal cortex into secreting cortisol—which works to prevent muscle cells from withdrawing glucose from the blood. As you can see, control of cortisol secretion is based on *negative* feedback mechanisms.

The feedback control of cortisol secretion is overridden by the nervous system when the body is abnormally stressed. A painful injury or severe illness may trigger shock, tissue inflammation, or both. Then, increased secretion of cortisol and other signaling molecules is essential to recovery. That is why cortisol-like drugs, such as cortisone, are administered to persons suffering from asthma and serious inflammatory disorders. Massive doses of such drugs also block certain immune reactions and so help prevent the body from rejecting surgically transplanted organs.

**Adrenal Medulla**. Hormone-secreting neurons are located in the **adrenal medulla**, the inner region of the adrenal gland (Figure 34.10). Their secretions, epinephrine and norepinephrine, help regulate blood circulation and carbohydrate metabolism. The hypothalamus and other brain centers govern their secretion by issuing commands along sympathetic nerves that service the adrenal glands.

The adrenal medulla helps mobilize the body's defenses during times of excitement or stress. For example, in response to epinephrine and norepinephrine, the heart beats faster and harder, blood flow is diverted to

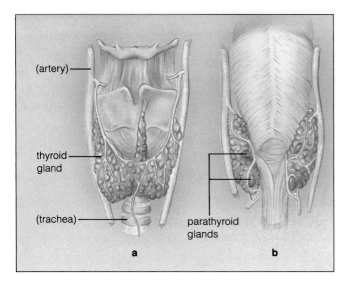

**Figure 34.11** (**a**) Anterior view of the human thyroid gland. (**b**) Posterior view, showing the location of the four parathyroid glands adjacent to it.

**Figure 34.12** A mild case of goiter, as displayed by Maria de Medici in 1625. During the late Renaissance, a rounded neck was considered a sign of beauty; it occurred regularly in parts of the world where iodine supplies were insufficient for normal thyroid function.

heart and muscle cells from other body regions, airways in the lungs dilate, and more oxygen is delivered to cells throughout the body. These are features of the "fight-flight" response described on page 568.

## Thyroid Gland

Earlier we saw how excessive or insufficient output from the pituitary gland affects body functioning. Abnormal secretions from other endocrine glands also can have profound effects on the body. Consider the human **thyroid gland**. Figure 34.11 shows its location at the base of the neck, in front of the trachea (windpipe). The main secretions of this gland, thyroxine and triiodothyronine, influence overall metabolic rates, growth, and development.

Thyroid hormones contain iodine and cannot be produced without it. In the absence of iodine, thyroid hormone levels in the blood decrease. The anterior pituitary responds by secreting thyroid-stimulating hormone (TSH). Excess TSH overstimulates the thyroid gland and causes it to enlarge. The resulting tissue enlargement is a form of *goiter* (Figure 34.12). Goiter caused by iodine deficiency is no longer common in countries where iodized table salt is used. Elsewhere, hundreds of thousands of people still suffer from the disorder, which is easily preventable.

Insufficient thyroid output is called *hypothyroidism*. Hypothyroid adults are sluggish, dry-skinned, and intolerant of cold. If the disorder is present at birth and is not detected early in infants, *cretinism* may result. Cretinism may arise from a genetic disorder that affects the thyroid gland in the fetus. Affected children show mental retardation, and if they do not receive treatment, their growth will be stunted. When cretinism is not identified in time, the mental retardation cannot be reversed.

Excessive thyroid output can lead to *hyperthyroidism*. The most common hyperthyroid disorder is called *Graves' disease*. Affected people show increases in metabolic rates, heart rate, and blood flow, and they lose weight even when they take in normal or increased amounts of food. They are excessively nervous, agitated, and typically have trouble sleeping. They are intolerant of heat and sweat profusely. Apparently, Graves' disease is an *autoimmune disorder* (in which the immune system mounts an attack against the body's own cells). In this case, cells of the immune system attack thyroid cells in much the same way that TSH stimulates them. Often individuals have a genetic predisposition to the disorder, which may be triggered by some environmental event. Treatment involves the surgical removal of some (or all) of the thyroid gland or treatment with drugs that suppress the synthesis of thyroid hormones.

## Parathyroid Glands

Some glands are not stimulated directly by hormones or nerves. Rather, they respond homeostatically to a chemical change in their immediate surroundings. The **parathyroid glands** are like this. (Figure 34.11 shows how four of these glands are adjacent to the back of the human thyroid.) In response to a drop in extracellular levels of calcium ions, the glands secrete parathyroid hormone (PTH), which helps restore blood calcium levels. By its

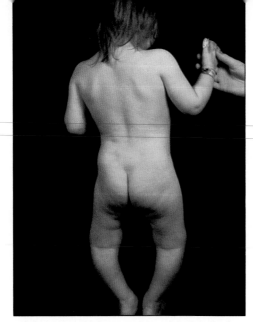

**Figure 34.13** A child with rickets. Notice the bowed legs characteristic of the disorder.

action, then, the parathyroid glands influence the availability of calcium ions for enzyme activation, muscle contraction, blood clotting, and many other tasks.

PTH stimulates calcium and phosphate removal from bone and its movement into extracellular fluid. It stimulates the kidneys to conserve calcium. It also helps activate vitamin D. The activated form, which is a type of hormone, enhances calcium absorption from food moving through the gut. Vitamin D deficiency leads to *rickets*, a disorder arising from insufficient calcium and phosphorus for proper bone development (Figure 34.13).

### Gonads

Homeostatic feedback loops also govern the function of gonads—the primary reproductive organs. Male gonads are called testes (singular, testis) and female gonads, ovaries. Gonads produce gametes. They also secrete estrogens, progesterone, and androgens (including testosterone). These are sex hormones that control reproductive function and the development of secondary sexual traits, as they did for Flo and others in the chimpanzee community described at the start of this chapter. How they do this is an elegant but intricate story, and for that reason we postpone our discussion of them until Chapter 43.

### Pancreatic Islets

In one respect, the pancreas is an exocrine gland associated with the digestive system; it secretes digestive enzymes. (Exocrine gland products, recall, are not picked up by the bloodstream; they are secreted onto a free epithelial surface.) But the pancreas also has small clusters of endocrine cells scattered through it. There are about

2 million of these clusters, which are called the **pancreatic islets**. The following are three types of hormone-secreting cells in the islets:

**1.** *Alpha cells* secrete the hormone **glucagon**. Between meals, cells use the glucose delivered to them by the bloodstream. The blood glucose level decreases, at which time glucagon secretions cause glycogen (a storage polysaccharide) and amino acids to be converted to glucose in the liver. In such ways, *glucagon raises the glucose level in the blood.*

**2.** *Beta cells* secrete the hormone **insulin**. After meals, when the blood glucose level is high, insulin stimulates uptake of glucose by liver, muscle, and adipose cells especially. It also promotes synthesis of proteins and fats, and inhibits protein conversion to glucose. Thus *insulin lowers the glucose level in the blood.*

**3.** *Delta cells* secrete **somatostatin**, a hormone with regulatory functions in the digestive system. Somatostatin also can block the secretion of insulin and glucagon.

Figure 34.14 shows how interplays among the pancreatic hormones help keep blood glucose levels fairly constant despite great variation in when—and how much—we eat. The importance of this function is clear if we consider what happens when the body cannot produce enough insulin or when insulin's target cells cannot respond to it. Disorders in carbohydrate, protein, and fat metabolism occur.

Insulin deficiency can lead to *diabetes mellitus*. Blood glucose levels rise, and glucose accumulates in the urine. This promotes urination to the extent that the body's water-solute balance is disrupted. Affected persons become dehydrated and excessively thirsty. Their insulin-deprived (glucose-starved) cells start degrading proteins and fats for energy, and this leads to weight loss. Ketones, which are normal acidic products of fat breakdown, accumulate in the blood and urine, and they promote excess water loss. The imbalance disrupts brain function and, in extreme cases, death may follow.

In "type 1 diabetes," the body mounts an autoimmune response against its own insulin-secreting beta cells and destroys them. Genetic susceptibility and environmental triggers combine to produce the disorder, which is the less common but more immediately dangerous of the two types of diabetes. Symptoms usually appear during childhood and adolescence, hence the disorder also is called "juvenile-onset diabetes." Type 1 diabetic patients survive with insulin injections.

In "type 2 diabetes," insulin levels are close to or above normal—but target cells cannot respond to the hormone. As affected persons grow older, their beta

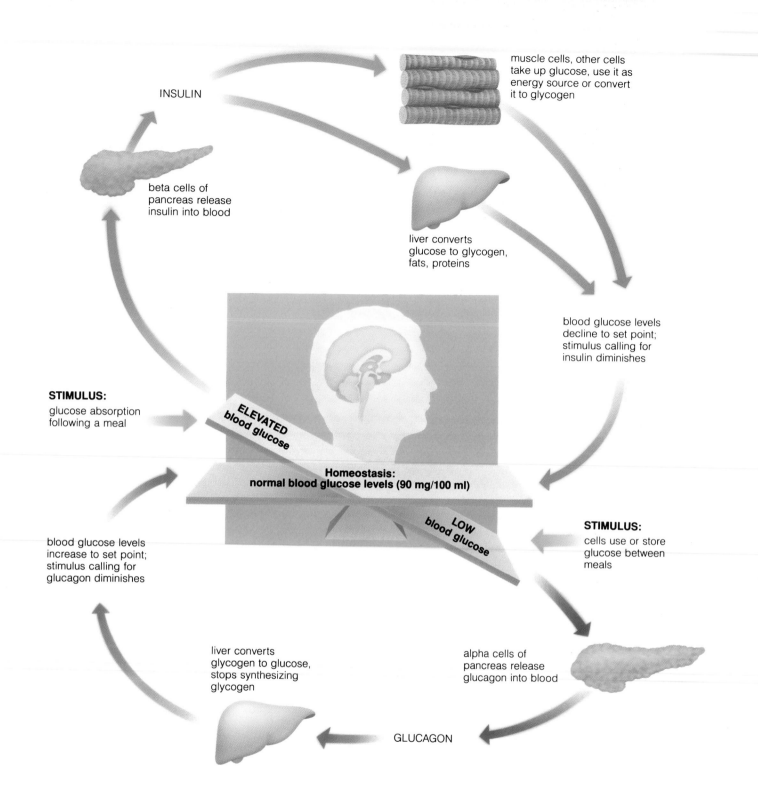

INSULIN

muscle cells, other cells take up glucose, use it as energy source or convert it to glycogen

beta cells of pancreas release insulin into blood

liver converts glucose to glycogen, fats, proteins

blood glucose levels decline to set point; stimulus calling for insulin diminishes

**STIMULUS:**
glucose absorption following a meal

**ELEVATED blood glucose**

**Homeostasis: normal blood glucose levels (90 mg/100 ml)**

**LOW blood glucose**

blood glucose levels increase to set point; stimulus calling for glucagon diminishes

**STIMULUS:**
cells use or store glucose between meals

liver converts glycogen to glucose, stops synthesizing glycogen

alpha cells of pancreas release glucagon into blood

GLUCAGON

**Figure 34.14** Some of the homeostatic controls over glucose metabolism. *Following* a meal, glucose enters the bloodstream faster than cells can use it. Blood glucose levels rise. Pancreatic beta cells are stimulated to secrete insulin. The hormonal targets (mainly liver, fat, and muscle cells) use glucose or store it as glycogen.

*Between* meals, blood glucose levels drop. Pancreatic alpha cells are stimulated to secrete glucagon. The hormonal targets convert glycogen back to glucose, which enters the blood. Also, the hypothalamus prods the adrenal medulla into secreting other hormones that slow down the conversion of glucose to glycogen in liver, fat, and muscle cells.

## Rhythms and Blues

The pineal gland, a bump of tissue in your brain, secretes the hormone melatonin. When the brain receives sensory signals from your eyes about the waning light at sunset, the pineal gland steps up its melatonin secretion, which is picked up by the bloodstream, which transports it to target cells—in this case, certain brain neurons. Those neurons are involved in sleep behavior, a lowering of body temperature, and possibly other physiological events. At sunrise, when the eye detects the light of a new day, melatonin production slows down. Your body temperature increases, and you wake up and become active.

The cycle of sleep and arousal is evidence of an internal biological clock that seems to tick in synchrony with daylength. Think about what happens when circumstances disturb the clock. There is the night worker who tries to sleep in the morning but ends up staring groggily at sunbeams on the ceiling. There is the traveler from the United States to Paris who starts off a vacation with four days of disoriented "jet lag." Two hours past midnight he is sitting upright in bed, wondering where the coffee and croissants are. Two hours past noon he is ready for bed. His body will gradually shift to a new routine when melatonin's signals begin arriving at their target neurons on Paris time.

Maybe you have heard of people affected by severe *winter blues*—depression, carbohydrate binges, and an overwhelming desire to sleep. Their discomfort may result from a biological clock that is out of synchrony with the changes in daylength in winter (days are shorter and nights longer). Their symptoms worsen when they are given doses of melatonin. And they improve dramatically when they are exposed to intense light—which shuts down pineal activity.

Melatonin is a long-time player in the course of vertebrate evolution. It is known to affect physiological rhythms in trout, alligators, sea turtles, sparrows, armadillos, hamsters, and rats. Researchers have identified its action in sexual behavior and other aspects of the reproductive cycle. They have identified its role in the seasonal deposition of body fat. Melatonin's roles may be varied, but they are part and parcel of life's tempos.

---

cells deteriorate and they produce less and less insulin. Type 2 diabetes usually occurs in middle age and is less dramatically dangerous than the other type. Affected persons can lead a normal life by controlling their diet, controlling their weight, and sometimes taking drugs that enhance insulin action or secretion.

### Thymus Gland

The lobed **thymus gland** is located behind the breastbone and between the lungs. Certain lymphocytes (white blood cells) multiply, differentiate, and mature in this gland, which secretes a group of hormones collectively called thymosins. These hormones affect the functioning of lymphocytes that defend the body against disease (Chapter 39).

### Pineal Gland

So far, we have seen how endocrine glands and endocrine cells respond to other hormones, to signals from the nervous system, and to chemical changes in their surroundings. Now we can start thinking about a larger picture, in which reproduction and development of the body are controlled by hormonal responses to environmental cues.

Until about 240 million years ago, it seems, vertebrates commonly had a third eye, on top of the head. Lampreys still have one, beneath the skin. A modified form of this photosensitive organ persists in nearly all vertebrates. We call it the **pineal gland** (Figure 34.2). The pineal gland secretes melatonin, a hormone that functions in the development of gonads and in reproductive cycles.

As described in the *Commentary*, melatonin is secreted in the absence of light. This means melatonin levels vary from day to night. The levels also change with the seasons, as when winter days are shorter than summer days. The hormonal effects can be observed during hamster reproductive cycles. High melatonin levels in winter suppress sexual activity, and in summer, when melatonin levels are low, sexual activity peaks. In humans, decreased melatonin secretion might help trigger the onset of *puberty*, the age at which reproductive organs and structures start to mature. If disease destroys the pineal gland, puberty may begin prematurely.

## LOCAL SIGNALING MOLECULES

In mammals, cells with mediating functions detect changes in the surrounding chemical environment and alter their activity, often in ways that either counteract or amplify the change. The cells secrete local signaling molecules, the action of which is confined to the immediate vicinity of change. Most of the signaling molecules are taken up so rapidly that not many are left to enter the general circulation. Prostaglandins and growth factors are examples of such secretions.

### Prostaglandins

More than sixteen different prostaglandins have been identified in tissues throughout the body. They are released continually, but the rate of synthesis often increases in response to local chemical changes. The stepped-up secretion can influence neighboring cells as well as the prostaglandin-releasing cells themselves.

At least two prostaglandins help adjust blood flow through local tissues. When their secretion is stimulated by epinephrine and norepinephrine, they cause smooth muscle in the walls of blood vessels to constrict or dilate. Prostaglandins have similar effects on smooth muscle of airways in the lungs. Allergic responses to airborne dust and pollen may be aggravated by prostaglandins (page 693).

Prostaglandins have major effects on some mammalian reproductive events. When women menstruate, many of them experience painful cramping and excessive bleeding—both of which have been traced to prostaglandin action. (Aspirin and other anti-prostaglandin drugs block synthesis of this local signaling molecule and alleviate the discomfort.) Prostaglandins also influence the corpus luteum, a glandular structure that develops from cells that earlier surrounded a developing ovum in the ovary. When pregnancy does not follow ovulation (that is, the release of an ovum from the ovary), a corpus luteum self-destructs. It produces copious amounts of prostaglandins that interfere with its own function. Prostaglandins also have roles in stimulating uterine contractions during labor.

### Growth Factors

Signaling molecules called **growth factors** influence growth by regulating the rate at which certain cells divide. Epidermal growth factor (EGF), discovered by Stanley Cohen, influences the growth of many cell types. Nerve growth factor (NGF) is another example. NGF, discovered by Rita Levi-Montalcini, promotes survival and growth of neurons in the developing embryo. One experiment demonstrated that certain immature neurons survive indefinitely in tissue culture when NGF is present but die within a few days if it is not. NGF also may define the direction of growth for these embryonic neurons, laying down a chemical path that leads the elongating processes to target cells.

## SIGNALING MECHANISMS

Hormones and other signaling molecules induce diverse responses in target cells. They can trigger the entry of substances into cells. They can alter the rate of protein synthesis, cause modification in proteins already present in the cell, and induce changes in the cell's shape and internal structure. What dictates the nature of the target cell's response?

That depends largely on two things. First, different signals activate different cellular mechanisms. Second, not all cells *can* respond to all types of signals. Many types of cells have receptors for cortisol and some other hormones; that is why those hormones have such widespread effects. But only a few cell types have receptors for more specific hormones, which have highly directed effects.

Let's think about some responses to just two of the main categories of hormones, as listed below and in Table 34.4:

**1.** *Steroid hormones*, which are synthesized from cholesterol. Steroid hormones are lipid-soluble and readily cross plasma membranes.

**2.** *Nonsteroid hormones*, which are synthesized from amino acids. They include protein and peptide hormones, and catecholamines. All are water-soluble and cannot cross the lipid bilayer of the plasma membrane.

**Table 34.4  Two Main Categories of Hormones**

| Type of Hormone | Examples |
|---|---|
| Steroid | Estrogens, testosterone, aldosterone, cortisol |
| Nonsteroid: | |
| Amines | Norepinephrine, epinephrine |
| Peptides | ADH, oxytocin, TRH |
| Proteins | Insulin, somatotropin, prolactin |
| Glycoproteins | FSH, LH, TSH |

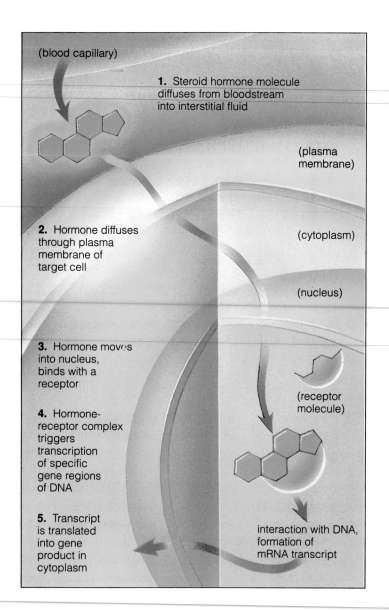

**1.** Steroid hormone molecule diffuses from bloodstream into interstitial fluid

(blood capillary)

(plasma membrane)

**2.** Hormone diffuses through plasma membrane of target cell

(cytoplasm)

(nucleus)

**3.** Hormone moves into nucleus, binds with a receptor

(receptor molecule)

**4.** Hormone-receptor complex triggers transcription of specific gene regions of DNA

**5.** Transcript is translated into gene product in cytoplasm

interaction with DNA, formation of mRNA transcript

**Figure 34.15** Proposed mechanism of steroid hormone action on a target cell. This same type of mechanism is also thought to occur for thyroid hormones, with one qualification. Recent studies suggest that membrane proteins facilitate the movement of thyroid hormones across the plasma membrane.

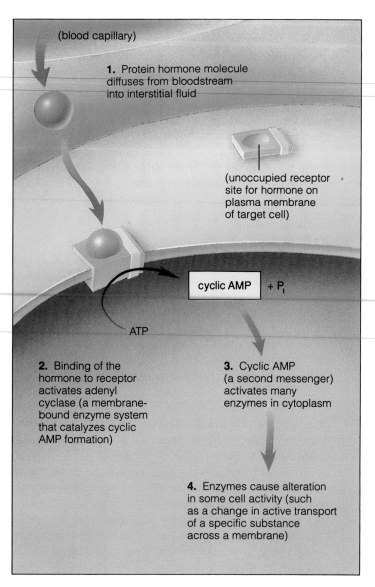

(blood capillary)

**1.** Protein hormone molecule diffuses from bloodstream into interstitial fluid

(unoccupied receptor site for hormone on plasma membrane of target cell)

cyclic AMP $+ P_i$

ATP

**2.** Binding of the hormone to receptor activates adenyl cyclase (a membrane-bound enzyme system that catalyzes cyclic AMP formation)

**3.** Cyclic AMP (a second messenger) activates many enzymes in cytoplasm

**4.** Enzymes cause alteration in some cell activity (such as a change in active transport of a specific substance across a membrane)

**Figure 34.16** Proposed mechanism of protein hormone action on a target cell. The response is mediated by a second messenger inside the cell—in this case, cyclic AMP. Other chemical messengers may be involved, depending on the particular hormone and its particular target cell.

## Steroid Hormone Action

Steroid hormones stimulate or inhibit protein synthesis by switching certain genes on or off. They do not alter the activity of already existing proteins. Being lipid-soluble, steroid hormones diffuse easily across the plasma membrane of target cells (Figure 34.15). Once inside, they move into the nucleus, where they bind with receptors for them.

The three-dimensional shape of the hormone-receptor complex allows it to interact with the DNA. The complex triggers the transcription of certain gene regions. Translation of the transcripts into specific proteins follows.

Testosterone is an example of a steroid hormone. It influences the development of male sexual traits. In *testicular feminization syndrome*, the receptor to which testosterone binds is defective. The affected individuals are males (XY), and they have functional testes that secrete testosterone. But none of the target cells can respond to the hormone, so the secondary sexual traits that do develop are like those of females.

## Nonsteroid Hormone Action

Protein hormones and other water-soluble signaling molecules cannot cross the plasma membrane of target cells without assistance. First they bind to receptors at the plasma membrane. In some cases, the hormone-receptor complex moves into the cytoplasm by way of endocytosis, then further action occurs inside the cell. Figure 5.13 shows photomicrographs of this type of endocytosis. Some hormones bind to receptors that recruit transport proteins into action or trigger the opening of channel proteins across the membrane. Certain ions or other substances move inward, and their cytoplasmic concentration changes in ways that affect cell activities.

Most peptide and protein hormones, including glucagon, activate **second messengers**. These are molecules inside the cell that mediate the response to a hormone. An example is *cyclic AMP*. (The full name is cyclic adenosine monophosphate.) First a hormone binds to a membrane receptor on a target cell. Binding alters the activity of a membrane-bound enzyme system (Figure 34.16). An enzyme, adenylate cyclase, is prodded into action. This enzyme speeds the conversion of ATP to cyclic AMP.

The hormone-receptor complex activates many molecules of the enzyme, not just one. Each enzyme molecule increases the rate at which many ATP molecules are converted to cyclic AMP. Each cyclic AMP molecule so formed then activates many enzyme molecules. Each of the enzyme molecules so activated can convert a very large number of substrate molecules into activated enzymes, and so on. Soon the number of molecules representing the final cellular response to the initial signal is enormous. Thus, second messengers *amplify* the response to a signaling molecule.

## SUMMARY

This chapter concludes our survey of controls over the integration of body activities in multicelled animals. Throughout the remainder of this unit, we will be looking at specific examples of these controls, so keep the following key concepts in mind:

1. For metabolic activity to proceed smoothly, the chemical environment of a cell must be maintained within fairly narrow limits.

2. In complex animals, thousands to billions of cells continually take up some substances from the extracellular fluid and secrete other substances into it. The nature and amount of the substances can change with the diet or level of activity; they inevitably change during the course of development.

3. It follows that the myriad withdrawals and secretions must be integrated in ways that ensure cell survival through the whole body.

4. Integration is accomplished by signaling molecules: chemical secretions by one cell that adjust the behavior of other, target cells. A target is one having receptors to which specific signaling molecules can bind and elicit a cellular response. It may or may not be adjacent to the signaling cell.

5. Signaling molecules include hormones, transmitter substances, local signaling molecules, and pheromones.

6. A neuroendocrine control center integrates many activities for the vertebrate body. This center consists of the hypothalamus and pituitary gland.

7. Two hypothalamic hormones—ADH and oxytocin—are stored in and released from the posterior lobe of the pituitary. ADH influences extracellular fluid volume. Oxytocin influences contraction of the uterus and milk release from mammary glands.

8. Six other hypothalamic hormones, called releasing or inhibiting hormones, control the secretions by cells of the anterior lobe of the pituitary.

9. The anterior lobe of the pituitary produces and secretes six hormones. Two (prolactin and somatotropin, or growth hormone) have general effects on body tissues. The remainder (ACTH, TSH, FSH, and LH) act on specific endocrine glands.

10. Hormone secretion is influenced by neural signals, hormonal interactions, local chemical changes in the surrounding tissues, and changes in the external environment.

11. Antagonistic, synergistic, and permissive interactions occur among hormones. The secretion of many hormones is controlled by homeostatic feedback of the neuroendocrine control center.

12. Fast-acting hormones such as parathyroid hormone (PTH) or insulin generally come into play when the extracellular concentration of a substance must be homeostatically controlled. Slow-acting hormones such as somatotropin have more prolonged, gradual, and often irreversible effects, such as those on development.

13. Cells respond to specific hormones or other signaling molecules only if they have receptors for them. Steroid hormones have receptors in the nucleus of target cells. Nonsteroid hormones (the amines, peptides, proteins, and glycoproteins) have receptors on the plasma membrane of target cells; responses to them are often mediated by a second messenger (such as cyclic AMP) inside the cell.

14. Steroid hormones trigger gene activation and protein synthesis. Most nonsteroid hormones alter the activity of proteins already present in target cells. These cellular responses contribute in some way to maintaining the internal environment or to the developmental or reproductive program.

## Review Questions

1. Which secretions of the posterior and anterior lobes of the pituitary glands have the targets indicated? (Fill in the blanks; *see pages 583 and 585.*)

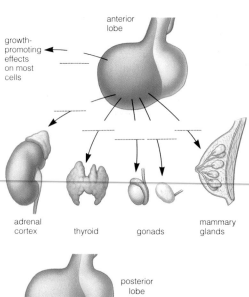

anterior lobe

growth-promoting effects on most cells

adrenal cortex    thyroid    gonads    mammary glands

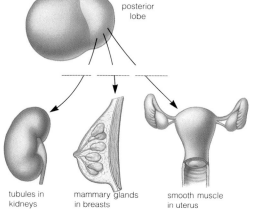

posterior lobe

tubules in kidneys    mammary glands in breasts    smooth muscle in uterus

2. Name the main endocrine glands and state where each is located in the human body. *582*

3. Define hormone. What functions do hormones serve? How do these functions differ from those of transmitter substances? 582

4. How do steroid and polypeptide hormones act on a target cell? *593–595*

5. The hypothalamus and pituitary are considered to be a neuroendocrine control center. Can you describe some of the functional links between these two organs? *583*

6. How does the hypothalamus control secretions of the posterior lobe of the pituitary? The anterior lobe? *583, 584*

7. Name three endocrine glands and a substance that each one secretes. What are the main consequences of their secretion? *585, 587*

8. Which hormone secreted by the anterior pituitary has an effect on most body cells rather than on a specific cell type? What are the clinical consequences of too little or too much secretion of this hormone? *584, 586*

## Self-Quiz (Answers in Appendix IV)

1. _____, _____, _____, and _____ are types of signaling molecules that help integrate cell activities in ways that benefit the whole body.

2. The _____ and _____ gland interact as a neuroendocrine control center to integrate endocrine gland secretions.

3. Stimulation or inhibition of hormone secretions often involves _____ loops between the neuroendocrine control center and glands.

4. A target cell for a specific hormone or some other signaling molecule has _____ for that molecule.

5. The hypothalamus produces two hormones that are released from the posterior lobe of the pituitary gland. One hormone,

_____ , affects kidney function; the other, _____ , affects some reproductive events.

    a. ADH; oxytocin
    b. prolactin; ADH
    c. oxytocin; ADH
    d. ADH; prolactin

6. The anterior lobe of the pituitary gland produces two hormones, _____ and _____ , that have general effects on body tissues in general.

    a. ACTH; somatotropin
    b. prolactin; FSH
    c. ACTH; FSH
    d. prolactin; somatotropin

7. Which of the following does *not* stimulate hormone secretion?

    a. neural signals
    b. local chemical changes
    c. hormonal signals
    d. environmental cues
    e. all of the above can stimulate hormone secretion

8. Insulin is an example of a _____ hormone which must work to accomplish homeostatic control of an extracellular substance; somatotropin is an example of a _____ hormone which operates during body development.

    a. growth; metabolic
    b. fast-acting; slow-acting
    c. metabolic; growth
    d. slow-acting; fast-acting

9. Which of the following statements is true?

    a. Steroid hormones have receptors on the plasma membrane of target cells.
    b. Protein hormones have receptors in target cell nuclei.
    c. Most protein hormones alter the activity of genes.
    d. Steroid hormones activate genes and protein synthesis.

10. Match the endocrine control concepts

    _____ oxytocin
    _____ ADH
    _____ steroid hormone
    _____ somatotropin (GH)
    _____ hypothalamus/ pituitary

    a. neuroendocrine control center
    b. affects kidney function
    c. has general effects on growth
    d. affects reproductive events
    e. triggers protein synthesis

## Selected Key Terms

adrenal cortex *588*
adrenal medulla *588*
anterior lobe of pituitary *583*
cyclic AMP *595*
endocrine system *581*
glucagon *590*
homeostatic feedback loop *588*
hormone *581*
inhibiting hormone *584*
insulin *590*
local signaling molecule *582*

neuroendocrine control center *583*
nonsteroid hormone *593*
pancreatic islet *590*
parathyroid gland *589*
pheromone *582*
pineal gland *592*
posterior lobe of pituitary *583*
releasing hormone *584*
second messenger *595*
steroid hormone *593*
thyroid gland *589*

## Readings

Cantin, M., and J. Genest. February 1986. "The Heart As an Endocrine Gland." *Scientific American* 254(2):76–81.

Fellman, B. May 1985. "A Clockwork Gland." *Science* 85 6(4): 76–81. Describes some of the known functions of the pineal gland.

Goodall, J. 1986. *The Chimpanzees of Gombe.* Cambridge, Massachusetts: Belknap Press of Harvard University Press.

Hadley, M. 1988. *Endocrinology.* Second edition. Englewood Cliffs, New Jersey: Prentice-Hall.

Sapolsky, R. January 1990. "Stress in the Wild." *Scientific American* 262(1):116–123. Study of hormonal effects on stress responses in baboons.

Snyder, S. October 1985. "The Molecular Basis of Communication Between Cells." *Scientific American* 253(4):132–141.

## Nobody Calls the Bat Man's Best Friend

How many of your friends have pet bats? None, most likely. Something about bats makes them unloved, the models for gargoyles and other imagined monsters. Nearly all bats sleep by day and spread their webbed wings at dusk, when different species take to the air in search of nectar, fruit, frogs, or insects. Probably it is the few blood-sucking species among them that have given the entire order of bats a bad name. Yet bats are rather close to us on the family tree for animals. They are every bit as mammalian as a dog or cat, and even though you might be reluctant to scratch one behind the ears, you have to give them credit for being relatives with some distinguishing sensory traits.

Consider that many of the sensory receptors in the eyes, nose, ears, mouth, and skin of bats are not that different from your own. When male frogs attempt to attract females by croaking as they float about in a pond, sensory receptors in your ears allow you to hear the relatively low-pitched sounds and use them to locate the frog's general whereabouts. The frog-eating bat shown in Figure 35.1 can do the same thing—with greater accuracy. Such bats routinely zero in on vocal but unlucky male frogs.

Other kinds of bats hunt in different ways, with a sense of hearing that we cannot begin to match. Think about the species of bats with tiny eyes, nearly blind, yet able to navigate with ease through forests and capture flying insects with great precision, even in the dark! Such bats are masters of *echolocation*. They emit calls, and when the sound waves of those calls bounce off objects in the environment—insects, tree branches—the echoes are assessed by the bat brain and used to judge the precise location and shape of those objects.

As an echolocating bat flies, it emits a steady stream of about ten clicking sounds every second—sounds *you* cannot hear at all. The clicks are intense. But they are "ultrasounds," meaning they are above the range of sound waves that sensory receptors in human ears can detect. Bats can hear even the extremely faint echoes of ultrasounds that are returning from distant objects.

**Figure 35.1** (**a**) Having used its well-developed sense of hearing to track down a singing tropical frog, this frog-eating bat is about to scoop dinner from the water. (**b**) Other bats, including this one, detect prey by listening to echoes of their own high-frequency cries. The echoes bounce back from objects in the environment, and the bat brain deciphers them in ways that provide the bat with a "sound map." With this map, the bat easily captures mosquitoes and moths in midair, without even seeing them.

When an echolocating bat hears a pattern of distant echoes from, say, an airborne mosquito or moth, it increases the rate of ultrasonic clicks to as many as 200 per second—which is faster than a machine gun fires bullets. There are only a few milliseconds of silence between clicks—but in that blip of silence the bat's receptors detect the echoes. Sensory nerves carry signals to the bat's brain, where the signals are decoded and processed. The brain somehow constructs a "map" of sounds that the bat uses in its maneuvers through the night world. With this map, the bat can swoop in on its insect snack without ever having caught a glimpse of it.

With this chapter, we turn to the means by which animals receive signals from the external and internal environments—and then decode those signals in ways that give rise to awareness of sounds, sights, odors, pain, and other sensations. Sensory neurons, nerve pathways, and brain regions are required for these tasks. Together, they represent the portions of the nervous system that are called *sensory systems*.

## SENSORY SYSTEMS: AN OVERVIEW

**Sensory systems** are the front doors of the nervous system, the parts that let in information about changing conditions and allow the brain to become aware of pertinent changes that are going on outside and inside the body. Sponges are among the very few animals that get along without sensory systems—but then, not much changes from one day to the next in the life of a sponge. You as well as most other animals would never even survive without elaborate sensory systems, given the sheer volume of information that you deal with every day. (Think about what it takes simply to cross a busy street during rush hour.)

Each sensory system has three component parts. These are (1) sensory receptors, (2) nerve pathways

**599**

**Table 35.1　Receptors Associated with the Major Senses**

| Category of Receptor | Examples | Stimulus |
|---|---|---|
| **Chemoreceptors:** | | |
| Internal chemical senses | Carotid bodies in blood vessel wall | Substances ($CO_2$, etc.) dissolved in extracellular fluid |
| Taste | Taste receptors of tongue | Substances dissolved in saliva, etc. |
| Smell | Olfactory receptors of nose | Odors in air, water |
| **Mechanoreceptors:** | | |
| Touch, pressure | Pacinian corpuscles in skin | Mechanical pressure against body surface |
| Stretch | Muscle spindle in skeletal muscle | Stretching |
| Auditory | Hair cells within ear | Vibrations (sound or ultrasound waves) |
| Balance | Hair cells within ear | Fluid movement |
| **Photoreceptors:** | | |
| Visual | Rods, cones of eye | Wavelengths of light |
| **Thermoreceptors** | Cold or warm receptors in skin; central thermoreceptors in hypothalamus, etc. | Presence of or change in radiant energy (heat) |
| **Nociceptors** | Free nerve endings in skin | Any stimulus that causes tissue damage and leads to sensation of pain |

leading from the receptors to the brain, and (3) brain regions where sensory information is processed.

**Sensory receptors** are finely branched endings of sensory neurons or specialized cells adjacent to them, and they detect specific stimuli. Think of a *stimulus* as light, heat, mechanical pressure, or some other form of energy that is capable of eliciting a response from a sensory receptor. Once detected, the stimulus energy is converted to action potentials—the form in which information travels along nerve pathways to the brain. Finally, specific brain regions translate information about the stimulus into a sensation.

What we call a **sensation** is a conscious awareness of a stimulus. It is not the same thing as **perception**, or understanding what the sensation means. Plunge your finger accidentally into boiling water and you are conscious of the stimulus—but beyond this, you understand acutely that the intense heat is hurting you. Plunge a live lobster into a cookpot and its reactions tell us that the stimulus (boiling water) has indeed registered. But lobsters and other invertebrates probably do not have the neural means to understand the meaning of their predicament.

**A sensory system consists of sensory receptors for specific stimuli, nerve pathways that conduct information from those receptors to the brain, and brain regions where the information is processed.**

**Information flowing through a sensory system may or may not lead to conscious awareness of the stimulus.**

## Types of Sensory Receptors

Different sensory receptors are specialized to detect different kinds of stimuli. Many sensory receptors are positioned individually, like sentinels, in the skin and other body tissues. Others are part of sensory organs, such as eyes, that amplify or focus the stimulus energy.

By using the different types of stimulus energy as a guide, we can define five major types of sensory receptors, as listed below and in Table 35.1:

**1. Chemoreceptors** detect the chemical energy of specific substances dissolved in the fluid surrounding them.

**2. Mechanoreceptors** detect mechanical energy associated with changes in pressure, changes in position, or acceleration.

**3. Photoreceptors** detect visible and ultraviolet light.

**4. Thermoreceptors** detect infrared energy (heat).

**5. Nociceptors** (pain receptors) detect tissue damage.

Keep in mind that different animals do not have the same kinds or numbers of sensory receptors, so they sample the environment in different ways and differ in their awareness of it. You don't have photoreceptors for ultraviolet light, as bees do—so you do not "see" many flowers the way they do (page 506). Unlike the bat in Figure 35.1b, you do not have mechanoreceptors for ultrasound. Unlike the python in Figure 35.2, you do not have thermoreceptors that detect warm-blooded prey in the dark.

**Figure 35.2** A python of southern Asia, equipped with thermoreceptors inside the pits, shown here, above and below its mouth. The python eats small, night-foraging mammals. Its thermoreceptors are sensitive to body heat (infrared energy) of its prey, which are much warmer than the night air. They notify the snake brain, which has a program for assessing signals about the location of objects. The program works very well—the snake's strike may be only a few degrees off-center. Yet the same snake might slither past a motionless, edible frog. Frog skin is cool and blends with background colors. The snake does not have receptors for detecting it or a neural program for responding to it.

## Sensory Pathways

Each sensory pathway starts at receptors of sensory neurons that are sensitive to the same type of stimulus, such as light, cold, or pressure. As Figure 35.3 shows, the axons of sensory neurons lead into the central nervous system. There they might converge on a single interneuron or branch out and connect with several to many interneurons.

Sensory nerve pathways from different receptors lead to different parts of the cerebral cortex, the outermost layer of the brain. For example, signals from receptors in the skin and joints travel to the *somatic sensory cortex*. Cells of this cortical region are laid out like a map corresponding to the body surface. Some map regions are larger than others; the body regions they represent are functionally more important and have more

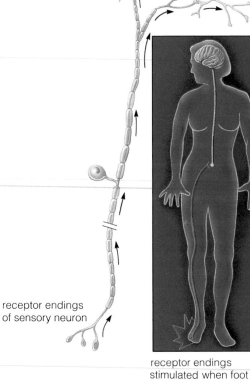

message sent on to interneurons in brain

interneuron in spinal cord

receptor endings of sensory neuron

receptor endings stimulated when foot lands on a tack

**Figure 35.3** Example of a sensory nerve pathway leading from a sensory receptor to the brain. The sensory neuron is coded red; interneurons are coded yellow.

# Disorders of the Eye

Two-thirds of all the sensory receptors your body requires are located in your eyes. Those photoreceptors do more than detect light. They also allow you to see the world in a rainbow of colors. Your eyes are the single most important source of information about the outside world.

Injuries, disease, inherited abnormalities, and advancing age can disrupt functions of the eyes. The consequences range from relatively harmless conditions, such as near-sightedness, to total blindness. Each year, many millions of people must deal with such consequences.

## Color Blindness

Consider a common heritable abnormality, *red-green color blindness*. It is an X-linked, recessive trait that shows up most often in males. The retina lacks some or all of the cone cells with pigments that normally respond to light of red or green wavelengths. Most of the time, color-blind persons merely have trouble distinguishing red from green in dim light. However, some cannot distinguish between the two even in bright light. The rare few who are totally color blind have only one of three kinds of pigments that selectively respond to red, green, or blue wavelengths. They see the world only in shades of gray.

## Focusing Problems

Other heritable abnormalities arise from misshapen features of the eye that affect the focusing of light. *Astigmatism*, for example, results from corneas with an uneven curvature; they cannot bend incoming light rays to the same focal point.

*Nearsightedness* (myopia) commonly occurs when the vertical axis of the eyeball is longer than the horizontal axis. It also occurs when the ciliary muscle responsible for adjustments in the lens contracts too strongly. The outcome is that images of distant objects are focused in front of the retina instead of on it (Figure *a*). *Farsightedness* (hyperopia) is the opposite problem. The horizontal axis of the eyeball is longer than the vertical axis (or the lens is "lazy"), so close images are focused behind the retina (Figure *b*).

Even a normal lens loses some of its natural flexibility as a person grows older. That is why people over forty years old often start wearing eyeglasses.

## Eye Diseases

The structure of the eye and its functions are vulnerable to infection and disease. Especially in the southeastern United States, for example, a fungal infection of the lungs (histo-

plasmosis) can lead to retinal damage. This complication can cause partial or total loss of vision. As another example, *Herpes simplex*, a virus that causes skin sores, also can infect the cornea and cause it to ulcerate.

*Trachoma* is a highly contagious disease that has blinded millions, mostly in North Africa and the Middle East. The culprit is a bacterium that also is responsible for the sexually transmitted disease chlamydia (page 782). The eyeball and the lining of the eyelids (conjunctiva) become damaged. The damaged tissues are entry points for bacteria that can cause secondary infections. In time the cornea can become so scarred that blindness follows.

## Age-Related Problems

*Cataracts*, a gradual clouding of the lens, is a problem associated with aging, although it also may arise through injury or diabetes. Possibly the condition arises when the transparent proteins making up the lens undergo structural changes. The clouding may skew the trajectory of incoming light rays. If the lens becomes totally opaque, light cannot enter the eye at all.

*Glaucoma* results when excess aqueous humor accumulates inside the eyeball. Blood vessels that service the retina collapse under the increased fluid pressure. Vision deteriorates as neurons of the retina and optic nerve die off. Although chronic glaucoma often is associated with advanced age, the problem actually starts in middle age. If detected early, the fluid pressure can be relieved by drugs or surgery before the damage becomes severe.

## Eye Injuries

*Retinal detachment* is the eye injury read about most often. It may follow a physical blow to the head or an illness that tears the retina. As the semifluid vitreous body oozes through the torn region, the retina becomes lifted from the underlying choroid. In time it may peel away entirely, leaving its blood supply behind. Early symptoms of injury include blurred vision, flashes of light that occur in the absence of outside stimulation, and loss of peripheral vision. Without medical intervention, the injured person may become totally blind in the injured eye.

## New Technologies

Today a variety of tools are used to correct some eye disorders. In *corneal transplant surgery*, the defective cornea is removed, then an artificial cornea (made of clear plastic) or

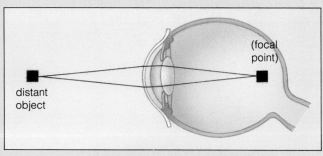

**a** Focal point in nearsighted vision. The example shows flamingos in Tanzania, East Africa.

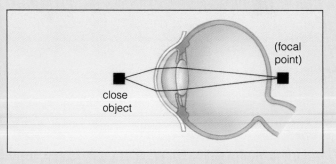

**b** Focal point in farsighted vision.

a natural cornea from a donor is stitched in place. Within a year, the patient is fitted with eyeglasses or contact lenses. Similarly, cataracts can be surgically corrected by removing the lens and replacing it with an artificial one, although the operation is not always successful.

Severely nearsighted people may opt for *radial keratotomy*, a still-controversial surgical procedure in which tiny, spokelike incisions are made around the edge of the cornea

to flatten it more. When all goes well, the adjustment eliminates the need for corrective lenses. Sometimes, however, the result is overcorrected or undercorrected vision.

Retinal detachment may be treatable with *laser coagulation*, a painless technique in which a laser beam seals off leaky blood vessels and "spot welds" the retina to the underlying choroid.

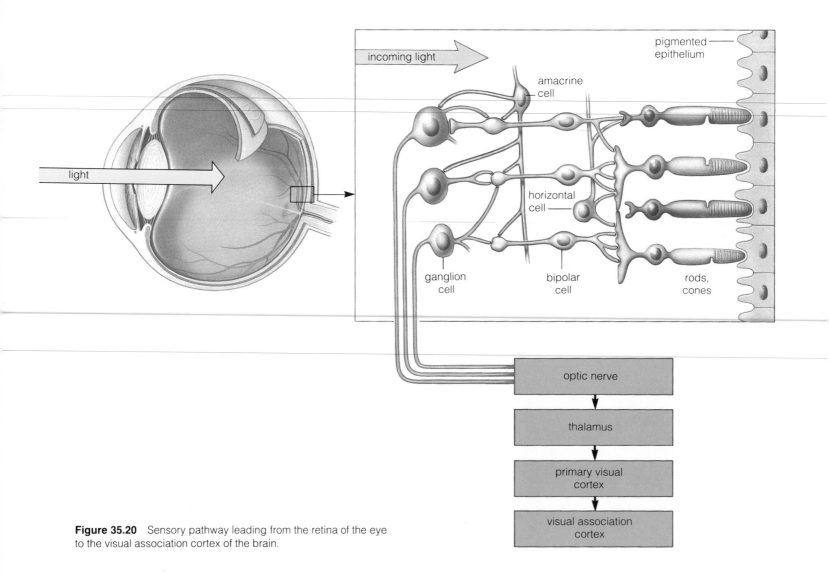

**Figure 35.20** Sensory pathway leading from the retina of the eye to the visual association cortex of the brain.

Rods contain molecules of rhodopsin in their membranes. Each molecule consists of a protein (opsin) to which a side group (*cis*-retinal) is attached. The retinal is derived from vitamin A. When the side group absorbs light energy, it is temporarily converted to a slightly different form (*trans*-retinal):

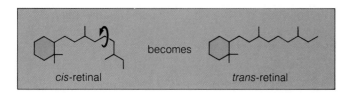

In this altered form, the side group initiates a series of chemical reactions within the photoreceptor. The reactions lead to a voltage change across the photoreceptor's plasma membrane. This local, graded potential affects the release of a transmitter substance from the photoreceptor. That substance acts on neighboring neurons, in ways that will now be described.

**Processing Visual Information.** How does the information from photoreceptors become translated into the sense of vision? Part of the answer is that the information moves in increasingly organized ways through *levels* of synapsing neurons—first in the retina, then in different parts of the brain. That information includes signals about form, movement, depth, color, and texture. Each type of signal seems to be processed along a separate communication channel. And many different channels run in parallel to the brain.

Only the rods and cones respond directly to light. When stimulated, they pass graded signals horizontally

STIMULI:                    RESPONSES:

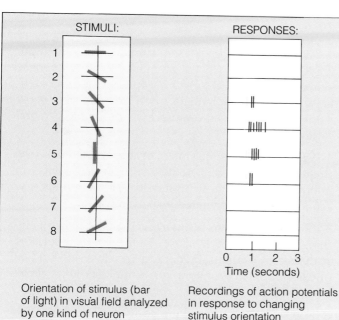

Time (seconds)

Orientation of stimulus (bar of light) in visual field analyzed by one kind of neuron

Recordings of action potentials in response to changing stimulus orientation

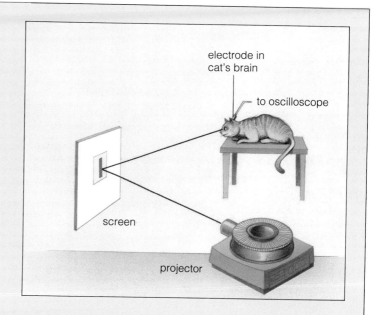

electrode in cat's brain

to oscilloscope

screen

projector

**Figure 35.21** From signaling to visual perception. Neurons in the visual cortex are stacked in columns at right angles to the brain's surface. Connections run between neurons in each column and between different columns. Each column apparently deals with only one kind of stimulus, received from only one location. Visual perception seems to be based on the organization and synaptic connections between neurons in these columns. The neurons fall into a few categories. In each category, they seem to be tripped into action the same way. For instance, excitatory signals traveling up through the cortex activate certain neurons, which then send out inhibitory signals to other neurons. The excitatory and inhibitory signals between neurons form narrow bands of electrical activity.

Experiments show that the pattern of excitation through specific columns of neurons is highly focused. For example, David

Hubel and Torsten Wiesel implanted electrodes in individual neurons in the brain of an anesthetized cat. Then they positioned the cat in front of a small screen. They projected images of different shapes (including a bar) onto the screen. When the bar was tilted at different angles, changes in electrical activity that corresponded to the different angles were recorded.

The strongest activity was recorded for one type of neuron when the bar image was vertical (numbered 5 in the sketch). When the bar image was tilted slightly, the signals were less frequent. When the image was tilted past a certain angle, the signals stopped. In other experiments, a certain neuron fired only when an image of a block was moved from left to right across the screen; another fired when the image was moved from right to left.

to one another as well as to adjacent neurons, including the bipolar cells shown in Figure 35.20. Action potentials start with functional groups of bipolar cells, which differ from one another in their sensitivity to contrast and color. Those neurons pass signals to ganglion cells, the axons of which converge to form the optic nerve leading to the brain. The optic nerves from both eyes converge at the base of the brain. A portion of the axons of each nerve cross over here before continuing onward. (This partial crossover, the optic chiasm shown on page 573, ensures that information from both eyes will reach both cerebral hemispheres.) Most axons of the optic nerves lead into the thalamus, which passes on information to the visual cortex.

The visual cortex has several subdivisions, but each has the whole visual field mapped onto it. The *visual field* is the portion of the outside world that is being detected by photoreceptors at any given time. A particular portion of each subdivision thus receives input from a particular portion of the visual field. Even within such portions, some neurons are sensitive to stimuli in the visual field that are oriented in one direction only—a line in the outside world that is horizontal, vertical, tilted left, or tilted right. (Figure 35.21 shows an example of this.) And nearby neurons may respond to stimuli oriented in a different direction.

Now different bits of information that have reached the visual cortex are sent to different parts of the cerebral cortex. Some parts analyze what the stimulus might be, another part analyzes where it is located in the visual field, and so on. All the information is processed rapidly, at the same time, in different cortical regions. Finally, signals are integrated to produce the organized electrical activity that gives rise to the sensation of sight.

## SUMMARY

1. A stimulus is a specific form of energy that the body is able to detect by means of sensory receptors. A sensation is an awareness that stimulation has occurred. Perception is understanding what the sensation means.

2. Sensory receptors are endings of sensory neurons or specialized cells adjacent to them. They respond to specific stimuli, such as light and particular forms of mechanical energy. Animals can respond to specific events only if they have receptors sensitive to the energy of the stimulus.

    a. Chemoreceptors, such as taste receptors, detect chemical substances dissolved in the body fluids that are bathing them.

    b. Mechanoreceptors, such as free nerve endings, detect mechanical energy associated with changes in pressure, changes in position, or acceleration.

    c. Photoreceptors, such as rods and cones of the retina, detect light.

    d. Thermoreceptors detect the presence of or changes in radiant energy from heat sources.

3. At receptor endings, the stimulus triggers local, graded signals. When the stimulus is strong enough, summation of graded signals may produce action potentials. The action potentials travel on particular nerve pathways from the receptors to parts of the cerebral cortex.

4. Variations in stimulus intensity are encoded in (1) the frequency of action potentials propagated along an information-carrying neuron and (2) the number of action potentials generated in a given tissue.

5. Somatic sensations include touch, pressure, temperature, pain, and muscle sense. The receptors associated with these sensations are not localized in a single organ or tissue. Stretch receptors, for example, occur in skeletal muscles throughout the body.

6. The special senses include taste, smell, hearing, balance, and vision. The receptors associated with these senses typically reside in sensory organs or some other particular region.

---

### Review Questions

1. Label the component parts of the human eye: *610*

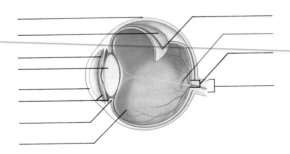

2. What is a stimulus? Receptor cells detect specific kinds of stimuli. When they do, what happens to the stimulus energy? *600*

3. Give some examples of chemoreceptors and mechanoreceptors. *600, 603–607*

4. What is sound? How are amplitude and frequency related to sound? Give some examples of animals that apparently perceive sounds. *606–607*

5. What is pain? Can you name one of the receptors associated with pain? *604*

6. How does vision differ from photoreception? What sensory apparatus does vision require? *608*

7. How does the vertebrate eye focus the light rays of an image? What is meant by *nearsighted* and *farsighted*? *610–611, 612*

---

### Self-Quiz *(Answers in Appendix IV)*

1. A _____ is a specific form of energy that is capable of eliciting a response from a sensory receptor.

2. Conscious awareness of a stimulus is called a _____.

3. _____ is understanding what particular sensations mean.

4. Each sensory system is composed of _____.
    a. nerve pathways from specific receptors to the brain
    b. sensory receptors
    c. brain regions that deal with sensory information
    d. all of the above are components of sensory systems

5. _____ detect mechanical energy associated with changes in pressure, in position, or acceleration.
    a. Chemoreceptors    c. Photoreceptors
    b. Mechanoreceptors    d. Thermoreceptors

6. Detecting chemical substances present in the body fluids that bathe them is the function of _____.
    a. thermoreceptors    c. mechanoreceptors
    b. photoreceptors    d. chemoreceptors

7. Which of the special senses is based on the following events: Membrane vibrations cause fluid movements, which bend mechanoreceptors and so trigger action potentials.
   - a. taste
   - b. smell
   - c. hearing
   - d. vision

8. The outer layer of the human eye includes the _____ .
   - a. lens and choroid
   - b. sclera and cornea
   - c. retina
   - d. both a and c are correct

9. The middle layer of the human eye includes the _____ .
   - a. lens and choroid
   - b. sclera and cornea
   - c. retina
   - d. start of optic nerve

10. Match each term with the appropriate description.
    - _____ somatic senses
    - _____ stimulus
    - _____ special senses
    - _____ variations in stimulus intensity
    - _____ action potential
    - _____ sensory receptor

    - a. produced by strong stimulation and summation of graded signals
    - b. endings of sensory neurons or specialized cells next to them
    - c. taste, smell, hearing, balance, and vision
    - d. a specific form of energy that can elicit a response from a sensory receptor
    - e. frequency and number of action potentials
    - f. touch, pressure, temperature, pain, and muscle sense

## Selected Key Terms

adaptation *603*
chemoreceptors *600*
compound eye *609*
cone cell *611*
ear *607*
echolocation *599*
eye *608*
eyespot *608*
hair cell *607*
iris *609*
lens *608*
mechanoreceptor *600*
mosaic theory *609*
nociceptor *604*
olfactory receptor *605*
pain *604*
perception *600*
pheromone *605*
photoreceptor *600*
rod cell *611*
sensation *600*
sensory receptor *600*
sensory system *599*
somatic sensation *603*
somatic sensory cortex *601*
taste receptor *604*
thermoreceptor *600*
vision *608*
visual cortex *602*
visual field *615*

## Readings

Eckert, R., D. Randall, and G. Augustine. 1988. *Animal Physiology: Mechanisms and Adaptations*. Third edition. New York: Freeman.

Hubel, D. H., and T. N. Wiesel. September 1979. "Brain Mechanisms of Vision." *Scientific American* 241(3):150–162. Describes studies on information processing in the primary visual cortex.

Hudspeth, A. January 1983. "The Hair Cells of the Inner Ear." *Scientific American* 248(1):54–66.

Jacobs, G. 1983. "Colour Vision in Animals." *Endeavour* 7(3): 137–140.

Kandel, E., and J. Schwartz. 1985. *Principles of Neural Science*. Second edition. New York: Elsevier. Advanced reading, but good coverage of sensory perception.

Newman, E. A., and P. H. Hartline. March 1982. "The Infrared 'Vision' of Snakes." *Scientific American* 246(3):116–127.

Parker, D. November 1980. "The Vestibular Apparatus." *Scientific American* 243(5):118–130.

Stryer, L. July 1987. "The Molecules of Visual Excitation." *Scientific American* 257(1)42–50. Well-written description of the cascade reactions that give rise to nerve signals in the retina.

Vander, A., J. Sherman, and D. Luciano. 1990. *Human Physiology*. Fifth edition. New York: McGraw-Hill. Chapter 9 is a clear introduction to sensory systems.

Wu, C. H. November–December 1984. "Electric Fish and the Discovery of Animal Electricity." *American Scientist* 72(6):598–607.

Young, J. 1978. *Programs of the Brain*. New York: Oxford University Press. An extraordinary book, beautifully written.

## The Challenge of the Iditarod

*"All right—GO!"*

Once again Susan Butcher is mushing out. On command, her trained Alaskan huskies leap forward, gathering momentum for the long haul across 1,157 miles of the Iditarod Trail. They will be towing a 200-pound sled in a race that will take them on a frozen, isolated route between Anchorage and Nome, Alaska. At the minimum, they will face eleven days and nights of snow, ice, and treacherous river crossings. Butcher is confident; many consider her to be the finest long-distance sled-dog racer of all time.

When it comes to speed, stamina, and built-in protection against the elements, we humans are not the superstars of the animal kingdom. Long before the Iditarod race began, Butcher began following a marathoner's regimen of diet and exercise to put her arm and leg muscles in peak condition. Through physical workouts, she increased the capacity of her skeletal-muscular system to support and help move her body. Lacking the fur coat of mammals that are native to the Far North, Butcher selected clothing that would insulate and protect her while still allowing freedom for strenuous movements.

Protection, support, movement—these aspects of animal anatomy and physiology are the topics of this chapter. Some basic rules apply for all of the examples we will be using. For example, the body must have a system of structural support—some type of skeleton. The skeleton must have fairly rigid parts that muscles can work against and so transmute force into body movement. Those skeletal parts must be lightweight as well as strong, thus minimizing the amount of energy required for movement. The muscle cells and tissues must be organized to work with one another as well as with the skeleton. Only then will the animal be able to execute the movements and positional changes required for a particular life-style in a particular environment.

Consider Granite, the lead dog of Butcher's team when she won the Iditarod in 1986, then in 1987, and again in 1988. Through artificial selection practices, huskies have been bred for remarkable strength and endurance. They can haul a light load at moderate speed over great distances, and most of the time in bitter cold. The long bones of Granite's legs are large in diameter and quite sturdy, yet much of the structural material inside them is lightweight. Granite's chest is

deep but not too broad. Breeders of huskies favor dogs with a rib cage that is a bit flattened on the sides, so the forelegs will have full freedom of movement. Although lean overall, Granite has well-developed muscles that ripple across the upper bones of his hind legs. These are not the muscles of a greyhound, cheetah, or some other sprinter. They are the muscles of a load-pulling, long-distance runner. The pads on Granite's feet are notably thick and tough. Packed with the protein collagen, they are built-in cushions against sharp ice and frozen rock. Granite also has a superior fur coat. Dense, fine hair serves as an insulative layer next to the skin. Above it is a slightly oily layer of tougher, longer hairs that take the first brunt of biting winds and near-freezing moisture.

Huskies, humans, and other animals show considerable variation in their systems of support, movement, and protection. When thinking about those variations, keep in mind that they are evolutionary responses to life in a particular environment. From this standpoint, the Iditarod is a testing ground—not only for variations that enhance speed and endurance, but for the functioning of all the interacting systems.

**1.** The body of nearly all animals has an outer covering (integument), muscle cells, and some form of skeleton. The contractile force of the muscle cells acts against a fluid medium, rigid structures (such as bones), or both.

**2.** The vertebrate integumentary system (skin and its component structures) protects the body from dehydration, ultraviolet radiation, abrasion, bacterial attack, and other environmental insults. It contributes to overall body functioning, as when it helps control moisture loss and when some of its cells synthesize the vitamin D required for calcium metabolism.

**3.** Smooth, cardiac, and skeletal muscle tissues occur in vertebrates. When adequately stimulated, the cells of all three types of muscle tissue can contract (shorten), then return to the resting position.

**4.** In vertebrates, the force of skeletal muscle contraction acts against an internal skeleton of bone and cartilage. Together, the skeletal and muscular systems change the positions of body parts and move the body through the environment. The bones function not only in movement but also in protection and support for soft organs, in mineral storage, and in blood cell formation.

**Figure 36.1** To the left, Susan Butcher and her Alaskan huskies, superbly illustrating their systems of support, movement, and protection along the Iditarod Trail.

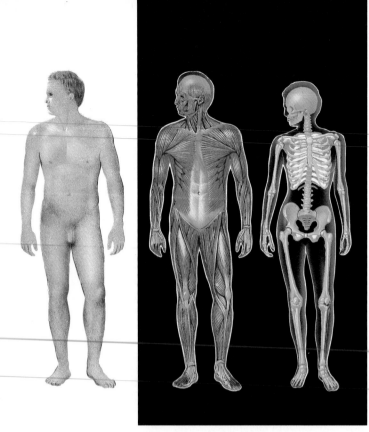

**Figure 36.2** Starting point for a tour of three types of organ systems, using the human body as an example. Traveling from the outside in, these diagrams show the integumentary system (skin and its derivatives), muscle system, and skeletal system.

I n the image you hold of yourself, you are tall or short, pale or dark, sparsely or profusely haired, taut-skinned or flabby, slow-moving or always on the move (or somewhere in between). Like most other animals, you have three organ systems to thank for your body's shape, superficial features, and capacity for movement. These are the integumentary, skeletal, and muscular systems. Figure 36.2 gives a general picture of what the systems look like for humans.

## INTEGUMENTARY SYSTEM

Animals ranging from worms to humans have an outer cover for the body. It is called an **integument** (after the Latin *integere*, meaning "to cover"). In most cases, the integument is tough yet pliable, a barrier against a great variety of environmental insults. For insects, crabs, and other arthropods, the integument is a hardened covering called a *cuticle*. Figure 36.3 shows examples. Arthropod cuticles consist of chitin, protein, and sometimes lipid secretions. For vertebrates, the integument is skin and the structures derived from epidermal cells of its outer layers of tissue.

Integuments vary among vertebrates. Depending on the vertebrate group, the skin may be decked out with scales, feathers, hair, beaks, hooves, horns, claws, nails, quills, and other structures (Figure 36.4). Variation exists within groups as well. Thus, as you read in Chapter 26,

**Figure 36.3** Off with the old, on with the new. (**a**) A green cicada (*Tibicen superbus*) and (**b**) a centipede (*Lithobius*), each shedding its outgrown cuticle during a molting cycle. The new cuticle is pale and soft, but soon will harden and darken.

a

b

some fishes have hard scales, others have bare skin coated with slimy mucus. Here our focus will be on the properties of human skin.

## Functions of Skin

No garment ever made approaches the qualities of the one covering your body—your skin. What besides skin maintains its shape in spite of repeated stretchings and washings, kills many bacteria on contact, screens out harmful rays from the sun, is waterproof, repairs small cuts and burns on its own, and with a little care, will last as long as you do?

Skin does more than protect the rest of the body from dehydration, abrasion, and bacterial attack. It helps control the body's internal temperature. It has so many small blood vessels that it serves as one of the reservoirs for blood. The reservoirs can be tapped and shunted to metabolically active regions, such as leg muscles during strenuous pushes along the Iditarod Trail. Skin produces vitamin D, which is required for calcium metabolism. And signals from sensory receptors in skin help the brain assess what is happening in the outside world.

## Structure of Skin

Assuming you are an average-sized adult, your skin weighs about 9 pounds. Stretched out, it would have a surface area of 15 to 20 square feet. For the most part, your skin is as thin as a paper towel. It thickens only on

the soles of your feet and in other regions subjected to pounding or abrasion.

As is the case for other vertebrates, skin has two distinct regions. As Figure 36.5 shows, the outermost region is the **epidermis** and the underlying region, the **dermis**. Beneath this is the *hypodermis*, a tissue that anchors the skin and yet allows it some freedom of movement. Fat stored in the hypodermis insulates the body against cold. Strictly speaking, the hypodermis is not part of skin.

**Figure 36.4** (*Above*) Feathers—one of the diverse kinds of structures arising from cell differentiations in the epidermis.

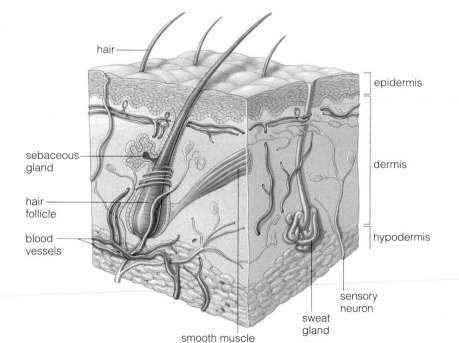

**Figure 36.5** The two-layered structure of human skin. The hypodermis is a subcutaneous layer, not part of skin.

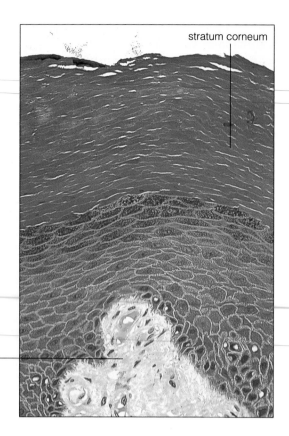

stratum corneum

dermis

**Figure 36.6** Section through human skin, showing the uppermost layer of epidermis (stratum corneum), the deeper epidermal layers, and the underlying dermis.

**Epidermis.** The epidermis consists mostly of stratified epithelium, a tissue described earlier on page 535. An abundance of cell junctions of the sort shown in Figure 31.6 knits the epithelial cells together. The cells themselves arise within the epidermis but are pushed toward its free surface, as rapid and ongoing mitotic divisions produce new cells beneath them.

Most cells of the epidermis are keratinocytes. Each is a tiny factory for manufacturing *keratin*, a tough, water-insoluble protein. Those cells start producing keratin when they are in mid-epidermal regions. By the time they reach the skin's free surface, they are dead and flattened. All that remain are fibers of keratin, packed inside plasma membranes. This is the composition of the outermost layer of skin—the tough, waterproof "stratum corneum" (Figure 36.6). Millions of the flattened keratin packages are worn off daily, but cell divisions continually push up replacements. The rapid divisions also contribute to skin's capacity to mend itself quickly after cuts or burns.

In the deepest epidermal layer, cells called melanocytes produce *melanin*, a brownish-black pigment. This pigment is transferred to keratin-producing cells and accumulates inside them, forming a shield against ultraviolet radiation. Melanin also contributes to skin color.

Humans generally have the same number of melanocytes. Variations in skin color arise through differences in melanocyte distribution and activity. Albinos, for example, lack melanin; their melanocytes cannot produce all of the enzymes required for its production (page 174). Melanocyte activity is increased in suntanned skin (Figure 36.7).

Skin color also is influenced by *hemoglobin* (the oxygen-carrying pigment of red blood cells) and *carotene* (a yellow-orange pigment). Pale skin, for example, has a pinkish cast. It does not have much melanin, so the presence of hemoglobin is not masked. Hemoglobin's red color shows through thin-walled blood vessels and the epidermis, both of which are transparent.

**With its multiple layers of keratinized, melanin-shielded epidermal cells, skin helps the body conserve water, avoid damage from ultraviolet radiation, and resist mechanical stress.**

**Rapid, continuous cell divisions in deep epidermal layers underlie skin's capacity to heal itself after being abraded, burned, or cut.**

**Dermis.** Dense connective tissue makes up most of the dermis, and it fends off damage from everyday stretching and other mechanical insults. There are limits to this protection. For example, the dermis tears when skin over the abdomen is stretched too much during pregnancy, leaving white scars ("stretch marks"). With persistent abrasion, the epidermis separates from the dermis and you get a "blister."

Blood vessels, lymph vessels, and the receptor endings of sensory nerves thread through the dermis. Nutrients from the bloodstream reach epidermal cells by diffusing through the dermal tissue. Sweat glands, oil glands, and the husklike cavities called hair follicles reside mostly in the dermis, even though they are derived from epidermal tissue.

The fluid secreted from *sweat glands* is 99 percent water, along with dissolved salts, traces of ammonia and other metabolic wastes, vitamin C, and other substances. You have about 2.5 million sweat glands, which are controlled by sympathetic nerves. One type abounds in the palms of the hands, soles of the feet, forehead, and armpits. They function mainly in temperature regulation (page 733). They also function in "cold sweats," one of the responses you make when you are frightened, nervous, or merely embarrassed. Another type of sweat gland prevails in skin around the sex organs. Their secretion steps up during stress, pain, sexual foreplay, and estrus. Do they have functions similar to those of scent glands in other animals? No one knows.

**Figure 36.7** Sunlight and the skin. Melanin-producing cells of the epidermis are stimulated by exposure to ultraviolet radiation. With prolonged sun exposure, melanin levels increase and light-skinned people become tanned (visibly darkened). Tanning provides some protection against ultraviolet radiation, but prolonged exposure can damage the skin. Over the years, tanning causes elastin fibers of the dermis to clump together. The skin loses its resiliency and begins to look like old leather.

Prolonged exposure to ultraviolet radiation also suppresses the immune system. Certain phagocytes and other specialized cells in the epidermis defend the body against specific viruses and bacteria. Sunburns interfere with the functioning of these cells. This may be why sunburns can trigger the small, painful blisters called "cold sores." The blisters are a symptom of a viral infection. Nearly everyone harbors this virus (*Herpes simplex*); usually it becomes localized in a nerve ending near the skin surface, where it remains dormant. Stress factors—including sunburn—can activate the virus and trigger the skin eruptions.

Ultraviolet radiation from sunlight or from the lamps of tanning salons also can activate proto-oncogenes in skin cells (page 241). Epidermal skin cancers start out as scaly, reddened bumps. They grow rapidly and can spread to adjacent lymph nodes unless they are surgically removed. Basal cell carcinomas start out as small, shiny bumps and slowly grow into ulcers with beaded margins. Their threat to the individual ceases when they are surgically removed, provided they are removed in time.

*Oil glands* (also called sebaceous glands) are everywhere except on the palms and soles. They function to soften and lubricate both the hair and the skin—and to kill surface bacteria. *Acne* is a skin condition in which the ducts of oil glands have become infected by bacteria, followed by inflammation of the glands.

*Hairs* are flexible structures, composed mostly of keratinized cells. Each has a root embedded in skin and a shaft that projects above the skin's surface. As living cells divide near the base of the root, older cells are pushed upward, then flatten and die. The outermost layer of the shaft consists of flattened cells that overlap one another like roof shingles (Figure 36.8). The most abused of these cells tend to frizz out near the end of the hair shaft; we call these "split ends."

The average scalp has about 100,000 hairs, but a person's genes, nutrition, and hormones influence hair growth and density. Protein deficiency causes hair to thin, for hair cannot grow without the amino acids required for keratin synthesis. Severe fever, emotional stress, and excessive vitamin A intake also cause hair thinning. Excessive hairiness (hirsutism) may result when the body produces abnormal amounts of testosterone. This hormone influences patterns of hair growth and other secondary sexual traits.

As we age, epidermal cells divide less often, and our skin becomes thinner and more susceptible to injury. Glandular secretions that kept the skin soft and moistened start dwindling. Collagen and elastin fibers in the

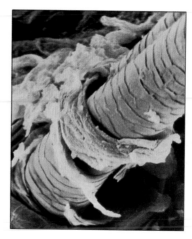

**Figure 36.8** Close look at a hair. This scanning electron micrograph shows overlapping cells of the outer layers of a hair shaft, here emerging from the epidermal surface of skin. Compare Figure 3.20, which shows a hair's molecular structure.

dermis break down and become sparser, so the skin loses its elasticity and wrinkles deepen. Excessive tanning, prolonged exposure to drying winds, and tobacco smoke accelerate the skin aging processes.

**Hairs, oil glands, sweat glands, and other structures associated with skin are derived from epidermal cells, but they are largely embedded in skin's underlying region, the dermis.**

**Blood and lymph vessels as well as receptor endings of sensory neurons also reside in the dermis.**

**Figure 36.9** Effects of muscle contractions on the hydrostatic skeleton of sea anemones. Compare this photograph with Figure 33.2. In (**a**), radial muscles that ring the gut cavity are relaxed, and longitudinal muscles running parallel with the body axis are contracted. Anemones typically look like this at low tide, when currents are not bringing in morsels of food. In (**b**), the radial muscle cells are contracted and the longitudinal muscles are stretched, so that the body is stretched into an upright position. This is the way sea anemones look when they attempt to gather food.

**a** Relaxed position    **b** Feeding position

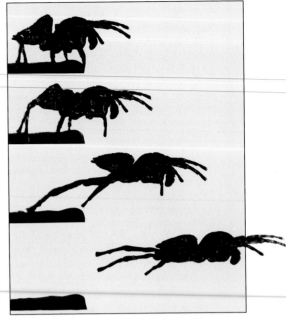

**Figure 36.11** Leap of the jumping spider *Sitticus pubescens*, based on the hydraulic extension of the hind legs when blood surges into them under high pressure.

**Figure 36.10** Example of a hinged region of an insect exoskeleton. This hinge is composed of layers of chitin and a highly elastic protein that withstands rapid, sustained movement characteristic of insect flight.

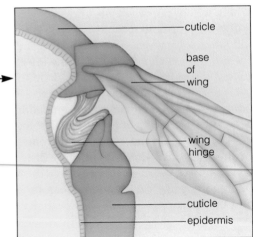

cuticle

base of wing

wing hinge

cuticle

epidermis

## SKELETAL SYSTEMS

So far in this unit, we have considered how the nervous system samples the external and internal environments with great precision and keeps informed of change. Many responses to those changes require movements—either of the whole body or some parts of it. Those movements occur by the activation, contraction, and relaxation of muscle cells. But muscle cells alone cannot produce movement. They require the presence of some medium or structural element against which the force of contraction can be applied. An internal or external skeleton fulfills this requirement. Three kinds of skeletal systems predominate in the animal world:

| | |
|---|---|
| **hydrostatic skeleton** | *Force of contraction applied against internal body fluids, which transmit the force* |
| **exoskeleton** | *Force of contraction applied against rigid external body parts, such as shells or armor plates* |
| **endoskeleton** | *Force of contraction applied against rigid internal body parts (cartilage and bone)* |

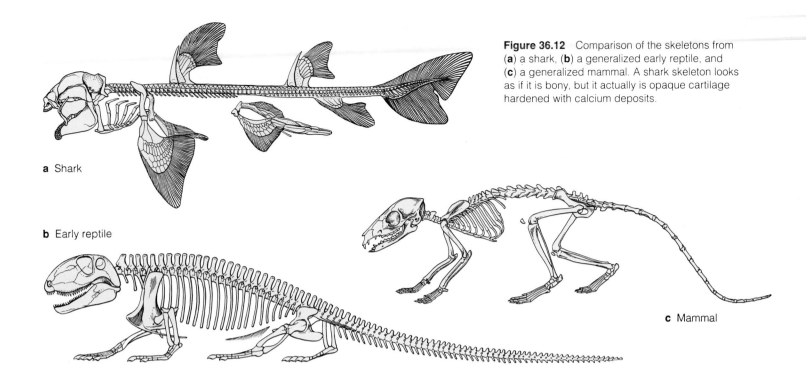

**Figure 36.12** Comparison of the skeletons from (**a**) a shark, (**b**) a generalized early reptile, and (**c**) a generalized mammal. A shark skeleton looks as if it is bony, but it actually is opaque cartilage hardened with calcium deposits.

**a** Shark

**b** Early reptile

**c** Mammal

## Invertebrate Skeletons

As we saw in Chapter 25, many soft-bodied invertebrates have hydrostatic skeletons. In all hydrostatic skeletons, some type of fluid is confined to a limited space. Like a fully filled waterbed, the hydrostatic skeleton resists compression, thereby serving as the medium against which muscles can work.

Sea anemones, for example, apply the force of contraction against their fluid-filled gut cavity. Between meals, longitudinal muscles are contracted, rings of muscles in the body wall are relaxed, and the animal looks rather like a flattened blob (Figure 36.9). When the radial muscles contract, fluid in the gut cavity is forced out, and longitudinal muscles in the body wall are stretched. Thus the body lengthens into its upright feeding position.

Animals with hydrostatic skeletons do not move with Olympian precision, although invertebrates with segmented bodies show more complex body movements than sea anemones do. Recall that earthworms have a series of coelomic compartments, each with its own muscles and nerves (page 426). Each segment has its own nerve supply, as well as longitudinal and radial muscles in the body wall. This means that the body can lengthen and shorten a few segments at a time in controlled ways. By coordinating muscle contractions on one side or the other of different segments, the earthworm can thrash from side to side as well as move forward and back.

Invertebrates with rigid exoskeletons lack the flexibility of soft bodies, but they benefit in other ways. Hard external parts work like armor against predators. They also provide support for increased body size, especially on land, where animals are deprived of water's buoyancy. Because hard parts can be moved like levers by the sets of muscles attached to them, they afford more precise and often more rapid movements. With an exoskeleton, large movements of body parts (such as wings) can result from small muscle contractions. Think about the cuticle of flying insects. It extends over all body segments *and* over the gaps between segments (Figure 36.10). The cuticle remains pliable at these gaps and acts like a hinge when muscles raise and lower either the wing or the body parts to which wings are attached.

Jumping spiders have hinged exoskeletons, but they also use body fluids to transmit force when they leap at prey. Muscle contractions cause blood inside the spider's body tissues to surge rapidly into the hind leg spines. It's something like giving a water-filled rubber glove a quick, hard squeeze, so that the glove's skinny fingers become rigidly erect. Figure 36.11 shows the outcome of this hydraulic pressure ("hydraulic" meaning fluid pressure inside tubes).

## Vertebrate Skeletons

Humans and other vertebrates have an endoskeleton of bone and cartilage (or cartilage alone). Some fishes have a flexible skeleton of an elastic, translucent form of cartilage that almost looks like glass. Sharks have a skeleton of an opaque form of cartilage, hardened with calcium deposits (Figure 36.12). However, most vertebrate skeletons are constructed primarily of bone. Let's turn now to the functions and characteristics of bones, using the human skeletal system as our example.

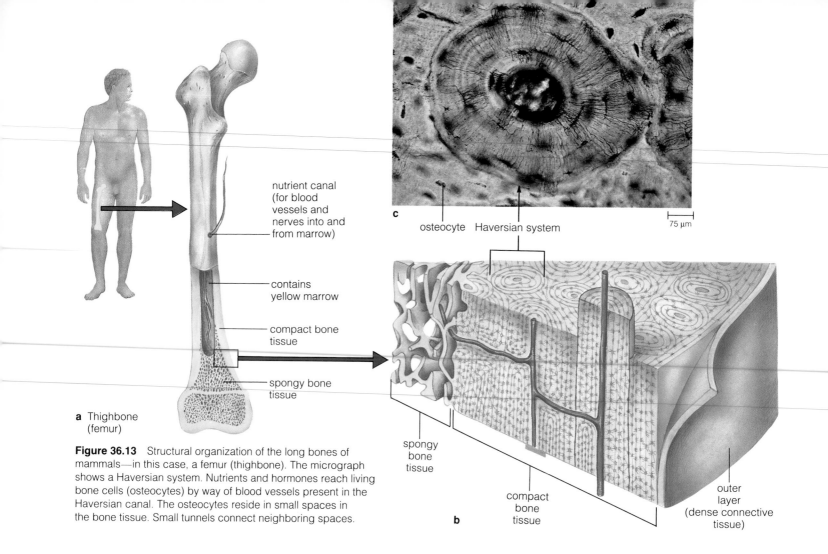

**a** Thighbone (femur)

nutrient canal (for blood vessels and nerves into and from marrow)

contains yellow marrow

compact bone tissue

spongy bone tissue

osteocyte   Haversian system

75 µm

**c**

spongy bone tissue

compact bone tissue

outer layer (dense connective tissue)

**b**

**Figure 36.13**  Structural organization of the long bones of mammals—in this case, a femur (thighbone). The micrograph shows a Haversian system. Nutrients and hormones reach living bone cells (osteocytes) by way of blood vessels present in the Haversian canal. The osteocytes reside in small spaces in the bone tissue. Small tunnels connect neighboring spaces.

## Functions of Bone

Just as skin is more than a baglike covering, so is the skeletal system more than a frame to hang muscles on. Its major component parts, called **bones**, are complex organs composed of a number of tissues. Those organs function in movement, protection, support, mineral storage, and blood cell formation:

1. *Movement:* Through interactions with skeletal muscle, bones maintain or change the position of body parts.

2. *Protection:* Bones are hard compartments that enclose and protect the brain, lungs, and other vital organs.

3. *Support:* Bones of the skeletal system support and anchor muscles and soft organs.

4. *Mineral Storage:* Bone tissue serves as a "bank" for calcium, phosphorus, and other mineral ions. The body makes deposits and withdrawals of these reserves, depending on metabolic needs.

5. *Blood Cell Formation:* Parts of some mature bones (such as the breastbone) are sites of blood cell production.

## Characteristics of Bone

In size and shape, human bones range from tiny ear-bones to pea-size wrist bones to strong, clublike thigh-bones. Bones are classified as long, short (or cubelike), flat, and irregular. Here we will focus mainly on long bones that occur in the body's limbs.

**Bone Structure.**  Like other organs, bones are made of tissues, including epithelium and various connective tissues, but they alone incorporate *bone tissue*. Bone tissue consists of living cells and collagen fibers distributed through a ground substance. Both the fibers and the ground substance are hardened by deposits of calcium salts.

Take a look at Figure 36.13, which shows the internal organization of a thighbone. The tissue that forms the bone's shaft and outer portion of its two ends is dense, or *compact bone tissue*. Such tissue forms the shaft of all long bones and allows them to withstand mechanical shocks. Notice, in Figure 36.13b, how the tissue is organized as thin, concentric layers around small canals. These "Haversian canals" are interconnected channels for blood vessels and nerves that service the living bone cells that reside in compact bone tissue.

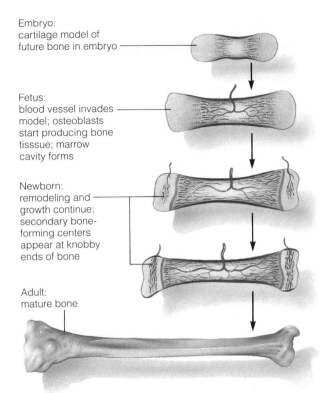

Embryo:
cartilage model of
future bone in embryo

Fetus:
blood vessel invades
model; osteoblasts
start producing bone
tisssue; marrow
cavity forms

Newborn:
remodeling and
growth continue;
secondary bone-
forming centers
appear at knobby
ends of bone

Adult:
mature bone

**Figure 36.14** Long bone formation, starting with osteoblast activity in a cartilage model (here, already formed in the animal embryo). Bone-forming cells are active in the shaft region first. Their activities are repeated in the knobby bone ends until only cartilage is left in the joints at both ends of the shaft.

The bone tissue *inside* the shaft and the ends is less packed; it has a spongelike appearance. Tiny, flattened parts make up this *spongy bone tissue*, which actually is quite firm and strong. In many bones, **red marrow** fills the spaces in the spongy tissue, which serves as a major site of blood cell formation. Most mature bones have **yellow marrow** in interior cavities. Yellow marrow is mostly fat. It converts to red marrow and produces red blood cells if blood loss from the body is severe.

---

Bones are complex organs composed of living cells and various tissues, the most notable of which is bone tissue.

A distinguishing feature of bone tissue is its extracellular matrix of collagen fibers and ground substance, both of which are mineralized.

---

**How Bones Develop**. Long bones form in cartilage models that develop in the embryo. Bone-forming cells (osteoblasts) secrete material inside the shaft and on the surface of the cartilage model. Gradually, the cartilage breaks down in the shaft region and the marrow cavity forms (Figure 36.14). The bone-forming cells continue to secrete bone tissue and eventually become trapped by

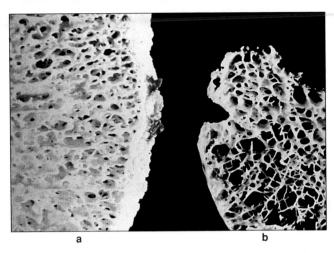

a                b

**Figure 36.15** Effect of osteoporosis on bone tissue. In normal tissue (**a**), mineral deposits continually replenish mineral withdrawals, so the tissue is maintained. (**b**) After the onset of osteoporosis, mineral replacements cannot keep pace with the withdrawals, and the tissue gradually erodes. Bones become progressively hollow and brittle.

their own secretions. Then, they are called **osteocytes** (living bone cells). Figure 36.13 shows some of these cells, which are responsible for maintaining mature bones.

**Bone Tissue Turnover**.  Bone tissue is like a bank from which minerals are constantly deposited and withdrawn. The turnover occurs when adult bone is subjected to exercise, which generally increases bone density. It occurs also after stress or injury. In bone remodeling programs that occur when young individuals are growing, this turnover is especially important. For example, the diameter of the thighbone increases as certain bone cells deposit minerals at the surface of the shaft. At the same time, other bone cells destroy a small amount of bone tissue inside the shaft. Thus the thighbone becomes thicker and stronger—but not too heavy.

Bone turnover also helps maintain calcium levels for the body as a whole. Consider how the body resorbs calcium. First, bone cells secrete enzymes that break down bone tissue. As the component minerals dissolve, the released calcium enters interstitial fluid; from there, it is taken up by the blood. This resorption activity is central to the hormonal control of calcium balance, as described on page 589.

Bone turnover can deteriorate with increasing age, especially among older women. The bone mass decreases in the backbone, hips, and elsewhere (Figure 36.15). The backbone can collapse and curve so much that the ribcage position is lowered, leading to complications in internal organs. The syndrome is called *osteoporosis*. Decreasing osteoblast activity, calcium and sex hormone deficiencies, excessive protein intake, and decreased physical activity are suspected of contributing to osteoporosis.

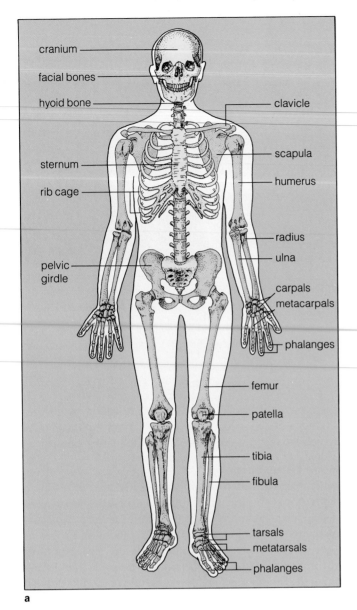

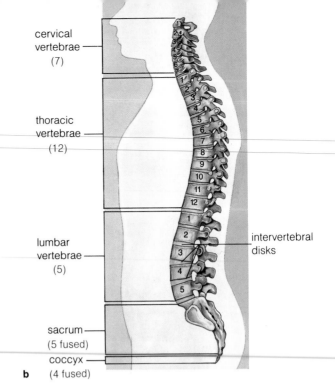

**Figure 36.16** (**a**) Human skeleton, with the axial portion color-coded in yellow and the appendicular portion, in tan. Can you identify similar structures in the endoskeleton of the mammal diagramed in Figure 36.12? (**b**) Side view of the vertebral column, or backbone. The cranium balances on the column's uppermost vertebra. Compare Figure 33.8.

## HUMAN SKELETAL SYSTEM

### Skeletal Structure

Humans started walking on their hind legs about 3 million years ago and haven't stopped since. As Figure 21.3 suggests, the upright posture puts the backbone into an S-shaped curve, which is not an ideal arrangement. (The older we get, the longer we have resisted the pull of gravity in an imperfect way, and the more lower back pain we have.) But evolution occurs through modifications of preexisting structures, and the skeleton of ancestral four-legged vertebrates was one of them.

Your body has 206 bones, in the two skeletal regions shown in Figure 36.15. The **axial skeleton** includes the skull, vertebral column (backbone), ribs, and sternum (the breastbone). The **appendicular skeleton** includes bones of the pectoral girdles (at the shoulders), arms, hands, pelvic girdle (at the hips), legs, and feet. Straps of dense, regular connective tissue called **ligaments** connect the bones at joints. Cords or straps of dense, regular connective tissue called **tendons** attach muscles to the bones (or to other muscles).

The flexible, curved backbone extends from the base of the skull to the pelvic girdle, where it transmits the weight of your torso to the lower limbs. The delicate spinal cord threads through a cavity formed by bony parts of the vertebrae, which are arranged one above the other (Figure 33.8).

**Intervertebral disks**, which contain cartilage, occur between the vertebrae (Figure 36.16b). They serve as shock absorbers and flex points. However, severe or rapid shocks may cause a disk to herniate. A *herniated disk* has slipped out of place and possibly may rupture. The protruding disk may press against neighboring nerves or the spinal cord and cause excruciating pain.

Each pectoral girdle has a large, flat shoulder blade and a long, slender collarbone that connects to the breastbone. It is not a sturdy arrangement. Fall on an outstretched arm and you might end up with a fractured clavicle or dislocated shoulder. Of all bones, the collarbone is the one most frequently broken.

### Joints

"Joints" are areas of contact or near-contact between bones. The most familiar type, the **synovial joint**, is

## On Runner's Knee

When you run, one foot and then the other is pounding hard against the ground. Each time a foot hits the ground, the knee joint above it must absorb the full force of your body weight. The knee joint allows us to do many things. It allows the leg bones beneath it to swing and, to some degree, to bend and twist. And the joint can absorb a force nearly seven times the body's weight—but there is no guarantee that it can do so repeatedly. Nearly 5 million of the 15 million joggers and runners in the United States alone suffer from "runner's knee," which refers generally to various disruptions of the bone, cartilage, muscle, tendons, and ligaments at the knee joint.

Like most joints, the knee joint permits considerable movement. The two long bones joined here (the femur and tibia) are actually separated by a cavity. They are held together by ligaments, tendons, and a few fibers that form a capsule around the joint. A membrane that lines the capsule produces a fluid that lubricates the joint, and where the bone ends meet, they are capped with a cushioning layer of cartilage.

Between the femur and tibia are wedges of cartilage that add stability and act like shock absorbers for the weight placed on the joint. Here also are thirteen fluid-filled sacs (bursae) that help cut down friction.

When the knee joint is hit hard or twisted too much, its cartilage can be torn. Once cartilage is torn, the body often cannot repair the damage. Orthopedic surgeons usually recommend removing most or all of the torn tissue; otherwise it can cause arthritis. Each year, more than 50,000 pieces of torn cartilage are surgically removed from the knees of football players alone. Football players,

tennis players, basketball players, weekend joggers—all are helping to support the burgeoning field of "sports medicine."

The seven ligaments that strap the femur and tibia together are also vulnerable to injury. A ligament is not meant to be stretched too far, and blows to the knee during collision sports (such as football) can tear it apart. A ligament is composed of many connective tissue fibers. If only some of the fibers are torn, it may heal itself. If the ligament is severed, however, it must be surgically repaired. (Edward Percy likens the surgery to sewing two hairbrushes together.) Severed ligaments must be repaired within ten days. The fluid that lubricates the knee joint happens to contain phagocytic cells that remove the debris resulting from day-to-day wear and tear in the joint. The cells will also go to work indiscriminately on torn ligaments and turn the tissue to mush.

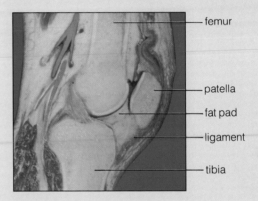

Longitudinal section through the knee joint.

freely movable. Such joints are stabilized in part by straplike ligaments that are capable of stretching. A flexible capsule of dense connective tissue surrounds the bones of a synovial joint. Cells of a membrane that lines the interior of the capsule secrete a lubricating fluid into the joint.

Unfortunately, freely movable joints sometimes move too freely and their structural organization is disrupted (see *Commentary*).

As a person ages, the cartilage covering the bone ends of freely movable joints may simply wear away, a condition called *osteoarthritis*. In contrast, *rheumatoid arthritis* is a degenerative disorder with a genetic basis. The synovial membrane becomes inflamed and thick-

ened, cartilage degenerates, and bone becomes deposited in the joint.

At **cartilaginous joints**, cartilage fills the space between bones and permits only slight movement. Such joints occur between vertebrae and between the breastbone and ribs. At **fibrous joints**, fibrous tissue unites the bones and no cavity is present. Fibrous joints loosely connect the flat skull bones of a fetus. During childbirth, the loose connections allow the bones to slide over each other and so prevent skull fractures. The skull of a newborn still has fibrous joints and membranous areas that are known as "soft spots" (fontanels). But the fibrous tissue hardens completely during childhood, so the skull bones become fused into a single unit.

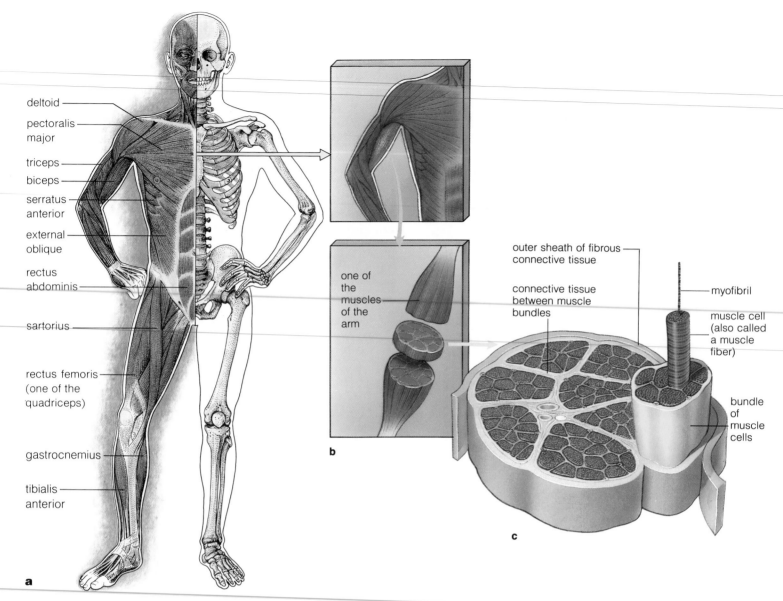

**Figure 36.17** (**a**) Some of the major skeletal muscles of the human skeletal-muscular system. (**b**) Closer view of the fine structure of an individual skeletal muscle. (**c**) Location of myofibrils, the threadlike structures inside each muscle cell. As shown in Figure 36.18, each myofibril has units of contraction (sarcomeres) arranged one after another along its length.

Labels in figure:
- deltoid
- pectoralis major
- triceps
- biceps
- serratus anterior
- external oblique
- rectus abdominis
- sartorius
- rectus femoris (one of the quadriceps)
- gastrocnemius
- tibialis anterior
- one of the muscles of the arm
- outer sheath of fibrous connective tissue
- connective tissue between muscle bundles
- myofibril
- muscle cell (also called a muscle fiber)
- bundle of muscle cells

## MUSCULAR SYSTEM

### Comparison of Muscle Tissues

We turn now to the skeletal muscle—the functional partner of bone. Recall that there are three types of muscle tissues: skeletal, cardiac, and smooth (page 538). Although skeletal and cardiac muscle are quite different in appearance from smooth muscle, they are all alike in three respects.

First, muscle cells show *excitability*. All cells have a voltage difference across the plasma membrane (the outside is more positively charged than the inside). In excit-

able cells, the voltage reverses suddenly and briefly in response to adequate stimulation. This sudden reversal in charge, called an action potential, was described in earlier chapters. Second, muscle cells can *contract* (shorten) in response to action potentials. Third, muscle cells are *elastic*; after contracting, they return to their original, relaxed position.

*Skeletal muscle is the only type of muscle tissue that interacts with the skeleton to bring positional changes of body parts and locomotion.* In vertebrates, smooth muscle occurs mostly in the wall of internal organs. (For example, smooth muscle in the stomach and intestinal walls helps

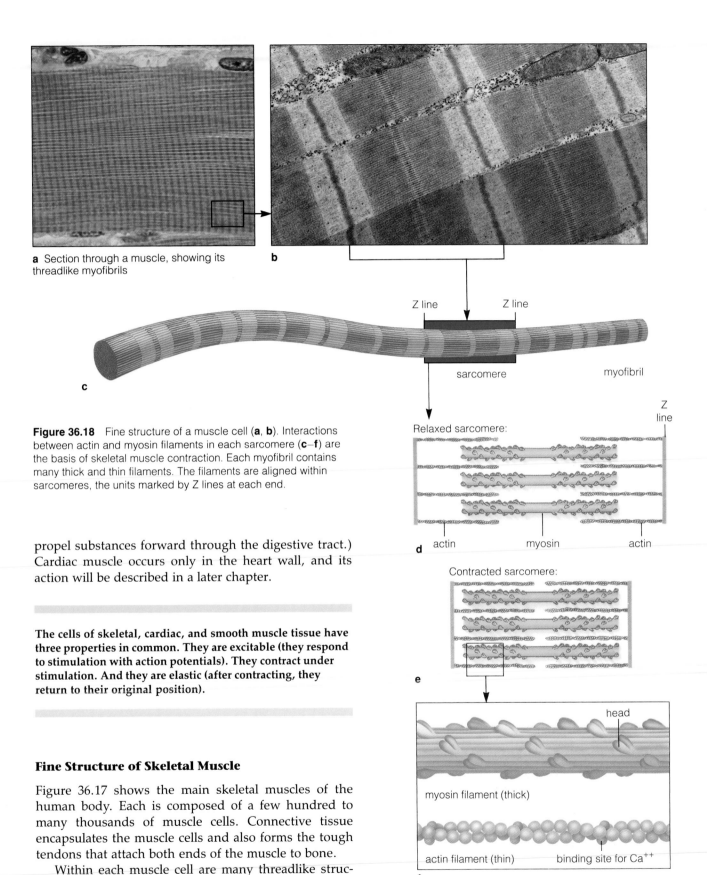

**a** Section through a muscle, showing its threadlike myofibrils

**b**

Z line          Z line

sarcomere          myofibril

**c**

Z line

Relaxed sarcomere:

**d** actin          myosin          actin

Contracted sarcomere:

**e**

head

myosin filament (thick)

actin filament (thin)          binding site for Ca$^{++}$

**f**

**Figure 36.18** Fine structure of a muscle cell (**a**, **b**). Interactions between actin and myosin filaments in each sarcomere (**c**–**f**) are the basis of skeletal muscle contraction. Each myofibril contains many thick and thin filaments. The filaments are aligned within sarcomeres, the units marked by Z lines at each end.

propel substances forward through the digestive tract.) Cardiac muscle occurs only in the heart wall, and its action will be described in a later chapter.

The cells of skeletal, cardiac, and smooth muscle tissue have three properties in common. They are excitable (they respond to stimulation with action potentials). They contract under stimulation. And they are elastic (after contracting, they return to their original position).

## Fine Structure of Skeletal Muscle

Figure 36.17 shows the main skeletal muscles of the human body. Each is composed of a few hundred to many thousands of muscle cells. Connective tissue encapsulates the muscle cells and also forms the tough tendons that attach both ends of the muscle to bone.

Within each muscle cell are many threadlike structures called **myofibrils**. You can see these in Figures 36.17 and 36.18. In turn, within each myofibril are many thin and thick filaments, side by side in parallel array. Close

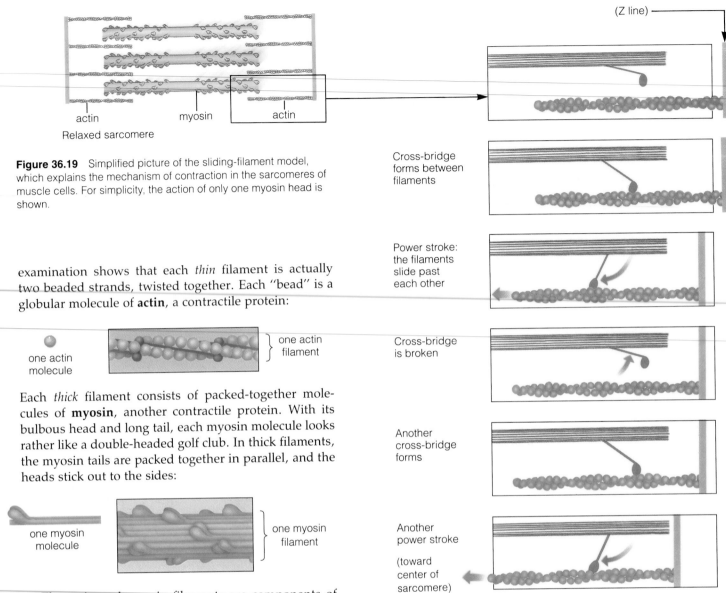

actin          myosin          actin

Relaxed sarcomere

(Z line)

Cross-bridge forms between filaments

Power stroke: the filaments slide past each other

Cross-bridge is broken

Another cross-bridge forms

Another power stroke

(toward center of sarcomere)

**Figure 36.19** Simplified picture of the sliding-filament model, which explains the mechanism of contraction in the sarcomeres of muscle cells. For simplicity, the action of only one myosin head is shown.

examination shows that each *thin* filament is actually two beaded strands, twisted together. Each "bead" is a globular molecule of **actin**, a contractile protein:

one actin molecule          one actin filament

Each *thick* filament consists of packed-together molecules of **myosin**, another contractile protein. With its bulbous head and long tail, each myosin molecule looks rather like a double-headed golf club. In thick filaments, the myosin tails are packed together in parallel, and the heads stick out to the sides:

one myosin molecule          one myosin filament

The actin and myosin filaments are components of **sarcomeres**, the basic units of muscle contraction. The organization of actin and myosin filaments in sarcomeres is so highly ordered, it gives skeletal and cardiac muscles a striped appearance (Figure 36.18c).

### Mechanism of Skeletal Muscle Contraction

The only way that skeletal muscles can move the body parts to which they are attached is to shorten. When a skeletal muscle shortens, its cells are shortening. And when a muscle cell shortens, its component sarcomeres are shortening. *The combined decreases in length of the individual sarcomeres account for contraction of the whole muscle.*

How does a sarcomere contract? According to the **sliding-filament model**, myosin filaments physically slide along and pull the actin filaments toward the center of a sarcomere during contraction.

The sliding movement depends on the formation of cross-bridges between adjacent actin and myosin filaments. A cross-bridge forms when the "head" of a myosin molecule attaches to binding sites on actin (Figure 36.19). An ATP molecule is associated with each myosin head. When some of its energy is released, the myosin head tilts in a short power stroke, toward the center of the sarcomere. As actin filaments become attached to the myosin heads, they also move toward the center. Now another energy input (from ATP) causes each myosin head to detach, reattach at the next actin binding site in line, and move the actin filaments a bit more. A single contraction takes a whole series of power strokes by myosin heads in each sarcomere.

In the absence of ATP, the cross-bridges never do detach. Following death, for instance, ATP production stops along with other metabolic activities. Cross-

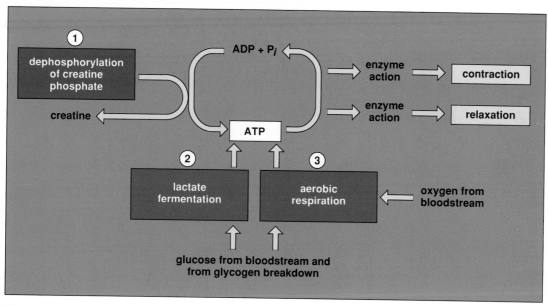

**Figure 36.20** Three possible metabolic pathways for producing ATP in muscles.

bridges remain locked in place and all skeletal muscles in the body become rigid. This condition, *rigor mortis*, lasts up to sixty hours after death.

In skeletal muscle, contraction occurs in sarcomeres, which are *contractile units* organized one after another in the myofibrils of muscle cells. Each sarcomere contains parallel arrays of actin filaments and myosin filaments.

Each sarcomere *shortens* when its actin and myosin filaments slide past each other, propelled by cross-bridge formation.

The combined decrease in length of the individual sarcomeres accounts for contraction of the muscle.

### Energy Metabolism in Muscles

How are muscle cells assured of getting enough ATP? As Figure 36.20 and the following list indicate, three metabolic supply routes are available to them:

1. Creatine phosphate metabolism

2. Lactate fermentation

3. Aerobic respiration

*Which* metabolic pathway dominates at a given time depends on the demands being placed on the muscle. During the sudden onset of contraction, many ATP molecules are stripped of a phosphate group (during the formation and detachment of cross-bridges). Creatine phosphate instantaneously restores the ATP by donat-

ing a phosphate group to ADP. Supplies of creatine phosphate are so limited, however, that they would dwindle within a few seconds unless synthesis reactions met the demands of the contracting cells.

When the demand for muscle action is *intense but brief* (say, a 100-meter race), muscles use an anaerobic pathway. As described in Chapter 8, lactate fermentation is an anaerobic route by which glucose is broken down to lactate, with a net yield of two ATP molecules. (Lactate is the ionized form of lactic acid.) Muscle cells get the glucose from blood and by tapping into their "storage form" of glucose—glycogen molecules. This pathway also produces energy very quickly, but lactate builds up in the muscles.

If the demand for muscle action is *moderate*, it also can be *prolonged*, as during the Iditarod sled-dog race. At such times, most of the required ATP forms by electron transport phosphorylation, the final stage of the aerobic pathway. As described on page 126, this event proceeds within mitochondria. And it has a net energy yield of 36 ATP per glucose molecule.

In the muscles of humans and other complex animals, the rate of ATP production by the aerobic pathway is linked to the rate at which the circulatory system is delivering oxygen to mitochondria. It is linked also to the number of mitochondria in muscle cells. When world-class marathoners engage in rigorous training, one of the physiological results is an increase in the number of mitochondria in their muscle cells. (Remember that mitochondria can divide independently of the cell in which they are located.) This is one outcome of the training that Butcher and her king-of-the-sled, Granite, undergo before a race.

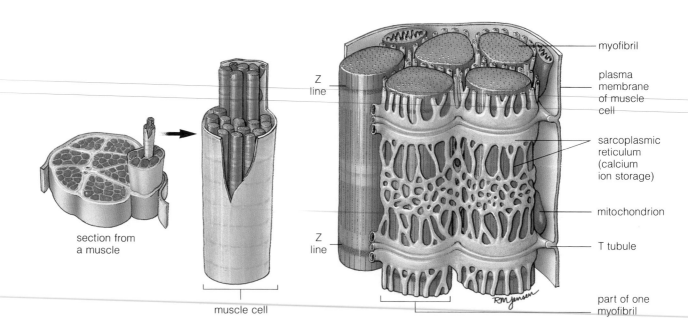

Labels for the top figure:
- myofibril
- plasma membrane of muscle cell
- sarcoplasmic reticulum (calcium ion storage)
- mitochondrion
- T tubule
- part of one myofibril
- Z line
- Z line
- section from a muscle
- muscle cell

**Figure 36.21** Membrane systems of a muscle cell. The plasma membrane surrounds the myofibrils. The plasma membrane is continuous with membranous tubes (T tubules) that thread inward. The tubes are located very close to a calcium-storing system (sarcoplasmic reticulum). Signals travel along the plasma membrane and the T tubules, and then trigger calcium release from the sarcoplasmic reticulum. Without calcium ions, actin and myosin filaments in the myofibrils cannot interact to bring about contraction.

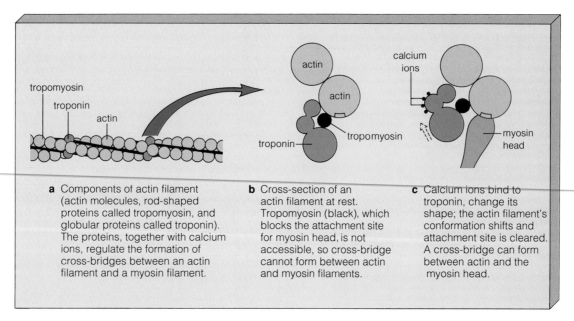

**a** Components of actin filament (actin molecules, rod-shaped proteins called tropomyosin, and globular proteins called troponin). The proteins, together with calcium ions, regulate the formation of cross-bridges between an actin filament and a myosin filament.

**b** Cross-section of an actin filament at rest. Tropomyosin (black), which blocks the attachment site for myosin head, is not accessible, so cross-bridge cannot form between actin and myosin filaments.

**c** Calcium ions bind to troponin, change its shape; the actin filament's conformation shifts and attachment site is cleared. A cross-bridge can form between actin and the myosin head.

**Figure 36.22** Role of calcium in the formation of cross-bridges between actin and myosin.

## Control of Contraction

Skeletal muscle contracts under commands from motor neurons. Appropriate stimulation from those neurons can trigger action potentials that travel along the plasma membrane of a muscle cell. They continue along infoldings of the plasma membrane that form many small tubes (the transverse tubule system). Those tubes reach the **sarcoplasmic reticulum**, a continuous system of membranous chambers around the myofibrils (Figure 36.21). The chambers actually are a compartment inside the cell that serves as a storehouse for calcium ions. The arrival of action potentials causes the compartment's membrane to become more permeable to calcium ions. After they

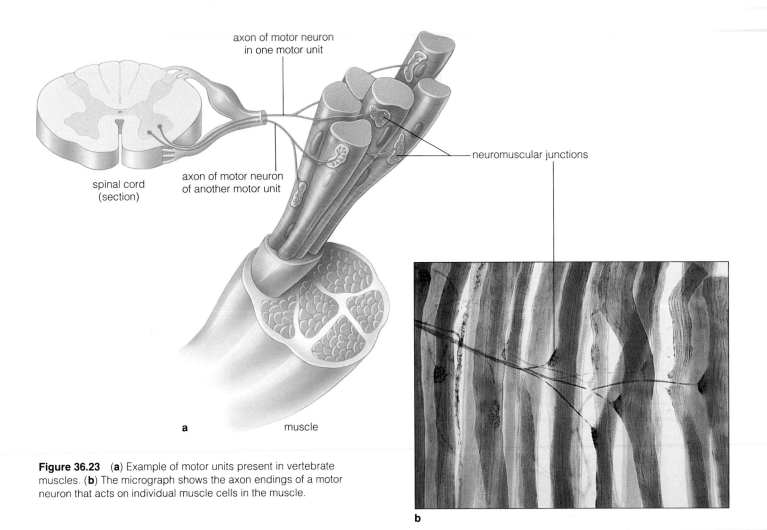

axon of motor neuron
in one motor unit

neuromuscular junctions

spinal cord
(section)

axon of motor neuron
of another motor unit

a

muscle

b

**Figure 36.23** (**a**) Example of motor units present in vertebrate muscles. (**b**) The micrograph shows the axon endings of a motor neuron that acts on individual muscle cells in the muscle.

are released, the ions diffuse through the cytoplasm and attach to a protein component (troponin) of actin filaments.

When calcium is attached to troponin, the protein changes shape, as Figure 36.22 shows. The change causes a shift in the position of an adjacent protein (tropomyosin) and a resulting conformational change in the actin filament. Thus, *calcium ions clear the myosin binding sites on actin, allowing cross-bridges to form.*

**Muscle contracts when calcium ions are released from the sarcoplasmic reticulum.**

**Muscle relaxes when calcium ions are actively taken up after contraction and stored in a membrane system around myofibrils (the sarcoplasmic reticulum).**

**By controlling the action potentials that reach the sarcoplasmic reticulum in the first place, the nervous system controls calcium ion levels in muscle tissue—and so exerts control over muscle contraction.**

## CONTRACTION OF A SKELETAL MUSCLE

Arnold Schwarzenegger, with his bulging muscles, is one of Hollywood's icons of great physical strength. Ultimately, his strength depends on how forcefully his muscles can contract. That, in turn, depends on the size of a given muscle, how many of its cells are contracting, and the frequency of stimulation by the nervous system. The larger a muscle is in diameter, the more potential it has for strength. By exercising regularly, you can prod your muscle cells to increase in size. (The *Commentary* on the page that follows describes what can happen when this line of reasoning is carried to an extreme.)

A skeletal muscle contains a large number of cells, and these may contract with different degrees of force for different periods of time. To get a sense of what is going on in that muscle, think back on the neuromuscular junction, where a motor neuron makes functional connections with one or more muscle cells (page 556). Together, a motor neuron and the muscle cells under its control are called a **motor unit**. Figure 36.23 shows an example of this.

Physiologists artificially stimulate a motor neuron with electrical impulses and make recordings of the changes in muscle contraction. When a single, brief stimulus activates a motor unit, the muscle contracts briefly, then relaxes. This muscle response is a **muscle twitch** (Figure 36.24). When a motor unit is stimulated again before a twitch response is completed, it twitches again. The strength of the contraction depends on how far the twitch response has proceeded by the time the second signal arrives. A motor unit stimulated repeatedly does not have time to relax (there is not enough time for all of the calcium ions to be transported back into the sarcoplasmic reticulum). Instead, the motor unit is maintained in a state of contraction called **tetanus**. (Recall that in the disease tetanus, described on page 558, muscles cannot be released from the state of contraction.)

In a weak contraction, the nervous system activates only a small number of motor units. In a stronger one, a larger number are activated, at a high frequency.

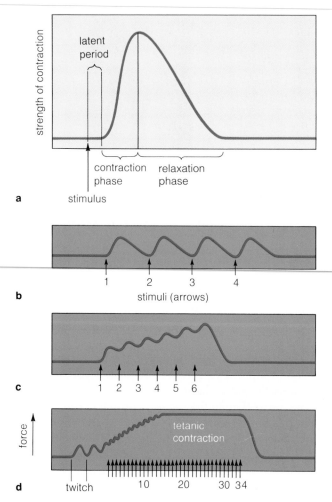

**Figure 36.24** Recording of a muscle twitch (**a**). Recordings of a series of muscle twitches caused by about two stimulations per second (**b**); recordings of a summation of twitches resulting from about six stimulations per second (**c**); and a tetanic contraction resulting from about twenty stimulations per second (**d**).

# Muscle Mania

Some call it the will-to-win gone bonkers, a consuming desire for muscles that are larger than life. Others say it is a modern requirement for excellence in athletic competition. Either way, they are talking about athletes who illegally use performance-enhancing drugs, mainly anabolic steroids.

Ten athletes were disqualified from the 1988 Olympics—one was even stripped of a gold medal—for using banned drugs, including the anabolic steroid stanazolol. Other competitors dropped out when they heard of the stringent new drug tests.

Cream-of-the-crop amateur athletes account for only a fraction of the surreptitious users of anabolic steroids. By one estimate, 85 percent of all professional football players use or have used the drugs. Adolescent boys as well as their parents sometimes look to the drugs to gain a winning edge in wrestling, football, and weight-lifting tournaments. Each year in the United States alone, perhaps as many as 1 million athletes use anabolic steroids.

## What Anabolic Steroids Are

Anabolic steroids are synthetic hormones. They were developed in the 1930s as therapeutic drugs that could mimic the effects of a sex hormone, testosterone. Secondary sexual traits, among other things, depend on testosterone. Under its influence, boys turning into men get a deeper voice; more hair on the skin of their face, underarms, and pubic region; more secretions from sweat glands; and increased muscle mass in the arms, legs, shoulders, and elsewhere. Testosterone also seems to stimulate the more aggressive behavior often associated with maleness. Anabolic steroids can also do these things—but at a significant physical and psychological price.

## What Anabolic Steroids Do

The "steroid" part of *anabolic steroid* tells you that molecules of this drug have a backbone of four carbon rings (page 42). The "anabolic" part echoes a name that chemists have for the synthesis of organic compounds (anabolism). The roughly twenty varieties of anabolic steroids stimulate the synthesis of protein molecules—including muscle proteins.

Supposedly, using anabolic steroids while engaged in a weight-training exercise program can lead to rapid gains in lean muscle mass and strength. The claim is disputed, because the results of most studies are based on too few subjects. Even so, testimonials pour in from weight lifters, football players, and athletes who specialize in shotput, discus, hammer throw, and other "brute power" events. Users commonly "stack" their steroid intake, combining daily oral doses with a single hefty injection each month. Much of their self-medication is on the sly, since the nonprescription use of anabolic steroids is illegal.

What, if anything, is bad about anabolic steroids? Physicians, researchers, and athletes themselves report a long list of minor and major side effects.

In men, acne, baldness, shrinking testes, and infertility are the first signs of toxicity. These symptoms are attributable to the fact that high blood levels of anabolic steroids cause the normal production of testosterone to drop precipitously. The drugs may be linked to an early onset of a cardiovascular disease, atherosclerosis. Even brief or occasional use may contribute to kidney damage and to cancer of the liver, testes, and prostate gland.

In women, anabolic steroids trigger the development of a deep voice and pronounced facial hair. Menstrual periods become irregular. Breasts may shrink and the clitoris may become grossly enlarged.

## Roid Rage

Not all steroid users have developed severe physical side effects. In fact, studies suggest that severe mental difficulties are more common. Called everything from 'roid rage to body-builders' psychosis, the symptoms range from annoying to frightening. Some men experience irritability and increased aggressiveness. Many competitive athletes look upon the added aggressiveness as a plus. Other men, however, experience uncontrollable aggression, delusions, and wildly manic behavior. In 1988, one steroid-using athlete traveling at 35 miles per hour deliberately drove his car into a tree.

With all of the suspected dangers associated with anabolic steroids, some may wonder why anyone would place his or her body and future in such jeopardy. Possibly not everyone is convinced that the drugs do enough damage to outweigh the "edge" they give in competition. What should a competitor do in a world that accords winning athletes wealth and the status of hero, while relegating others to the pile of also-rans? What would *you* do?

## Sorry, Have to Eat and Run

For the pronghorn antelope (*Antilocapra americanus*), home is where the food is. Populations of this medium-sized mammal range from central Canada, down through the American Southwest, and on into northern Mexico. Late winter and fall, you may find herds on windblown mountain ridges, where wild sage grows. In spring you may find them moving to open grasslands and deserts, wherever low grasses and tasty shrubs are sprouting.

Young antelope are vulnerable and tasty to coyotes, bobcats, and golden eagles, so while the herds are browsing, they keep a constant eye out for danger. They can do this even while their head is bent low in the grasses, given how far back their eye sockets are positioned in the skull (Figure 37.1a). And can those animals eat and run! If danger appears imminent, they leave the table with bursts of speed that have been clocked at 95 kilometers per hour.

Just as you can do for other mammals, you can look at a pronghorn antelope's teeth and gain insight into its life-style. Think about your own cheek teeth, for instance, with their flattened crown that serves as a grinding platform. The crown of the antelope's cheek teeth dwarfs them (Figure 37.1b). You probably do not have your mouth pressed close to the ground while eating, but the antelope does. Because the antelope ends up with abrasive soil particles as well as tough plant material in its mouth, its teeth wear down far more rapidly than yours do. For them, natural selection apparently has favored more crown to wear down.

Like other ruminants, pronghorn antelopes spend nearly all their lives alternately eating and then bedding down to chew their cud. Pound for pound, it takes far more plant tissue to provide energy than the same amount of animal tissue provides for predators. To be sure, predators face energy shortages if they have to

a

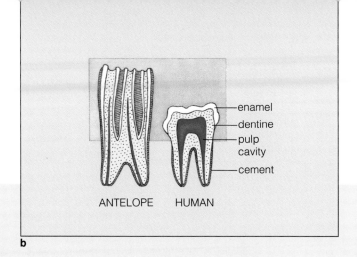

**b**

**Figure 37.1** (**a**) Pronghorn antelopes (*Antilocapra americanus*) busy at work, taking in nutrients. (**b**) A comparison of the general structure of cheek teeth of herbivorous mammals, including antelopes, and humans. Each tooth's crown is positioned over the green background; its root is beneath this.

wait a long time between meals, and the energy deficit increases when they have to run down dinner. But ruminants have to spend far more time *digesting* meals. The plants they eat consist largely of cellulose, the structure of which was shown on page 39. Cellulose digestion requires specific enzymes and a rather long processing time. Not surprisingly, pronghorn antelopes have one of the world's most elaborate stomachs. Not one stomach sac for them—their stomach is partitioned into four interconnected sacs!

The first two sacs of the stomach house vast microbial populations. Among these are symbiotic bacteria. The bacteria synthesize digestive enzymes that act specifically on cellulose. By degrading cellulose, they make its component nutrients available to their host as well as to themselves. While bacterial enzymes are attacking cellulose, the antelope is regurgitating the contents of the first two stomach sacs and rechewing the stuff before swallowing again. (This is what "chewing the cud" means.) Thus, plant material is mixed and pummeled more than once—so more of the cellulose fibers are exposed to agents of digestion before continuing on through the antelope gut.

We humans don't chew cud, but our nutrient-acquiring strategies are just as amazing. An Eskimo might eat only raw whale blubber in a given day, a Nepalese might eat only rice, and an American might partake of pepperoni pizza, chocolate, kiwi fruit, couscous, snake meat, or dandelion wine. Yet through its metabolic magic, the human body converts these and a dizzying variety of other substances into usable energy and tissues of its own.

With this chapter we start our tour of nutrition, which will take us through processes by which food is ingested, digested, absorbed, and later converted to the body's own carbohydrates, lipids, and proteins.

KEY CONCEPTS

**1.** Interactions among the digestive, circulatory, respiratory, and urinary systems supply the body's cells with raw materials, dispose of wastes, and maintain the volume and composition of extracellular fluid.

**2.** Most digestive systems have specialized regions for food transport, processing, and storage. Different regions are concerned with mechanical and chemical breakdown of food, absorption of the breakdown products, and elimination of unabsorbed residues.

**3.** To maintain an acceptable body weight and overall health, energy intake must balance energy output (by way of metabolic activity, physical exertion, and so on). Complex carbohydrates are the main energy source.

**4.** Nutrition requires the intake of vitamins, minerals, and certain amino acids and fatty acids that the body cannot produce itself.

## TYPES OF DIGESTIVE SYSTEMS AND THEIR FUNCTIONS

Generally speaking, a **digestive system** is some form of body cavity or tube in which food is first reduced to particles, then to small molecules. A layer of cells lines the body cavity or tube, and nutrients cross this lining and so enter the internal environment.

Chapter 25 described some invertebrates that are equipped with a saclike gut. Such animals are said to have an **incomplete digestive system** because the gut has only one opening. What goes in but cannot be digested goes out the same way. Recall that planarians, a type of flatworm, have this system. A muscular pharynx opens into a highly branched cavity that serves both digestive and circulatory functions. Food is partly digested and transported to cells even as residues are being sent back out through the pharynx. During the course of flatworm evolution, the two-way traffic must have worked against modification of the gut into specialized regions for food transport, processing, and storage.

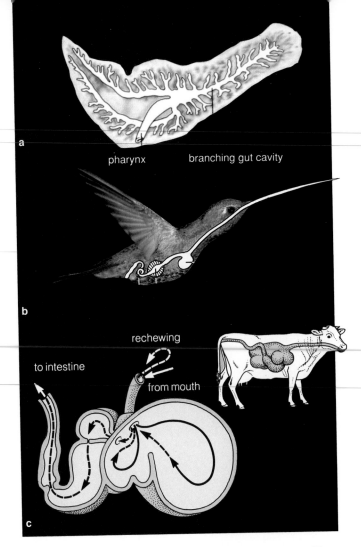

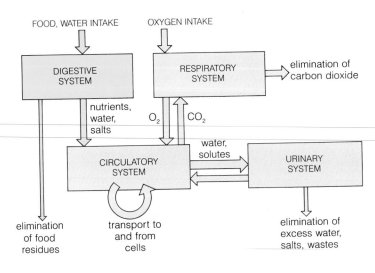

**Figure 37.3** Links between the digestive, respiratory, circulatory, and urinary systems. These organ systems work together to supply the body's cells with raw materials and eliminate wastes. This chapter focuses on the digestive system; subsequent chapters will address the other systems shown here.

**Figure 37.2** (**a**) Incomplete digestive system of a flatworm, with two-way traffic of food and undigested material through one opening (the pharynx). The branched gut cavity serves both digestive and circulatory functions. (**b**) Complete digestive system of a bird—basically a tube with regional specializations and an opening at each end. (**c**) Complete digestive system of a cow. Cattle, antelopes, and other ruminants have a stomach with multiple chambers in which cellulose is digested before being sent on to the small intestine, where nutrients are absorbed.

Specialized regions of complete digestive systems can be correlated with feeding behavior. Predators and scavengers (such as vultures) have *discontinuous* feeding habits. They generally gorge themselves with food when it is available, then may go for long periods without eating at all. Parts of their digestive system store food that is taken in faster than it can be digested and absorbed. Other, accessory parts help maintain an adequate distribution of nutrients between meals. By contrast, antelope, deer, goats, cattle, and other ruminants eat almost continuously when they are not bedding down. (A *ruminant* is any hoofed mammal having multiple stomach chambers in which cellulose can be digested.) Figure 37.2c is a generalized diagram of their digestive system.

We can summarize the overall functions of complete digestive systems in the following way:

**1. Motility** Muscular movement of the gut wall, leading to the mechanical breakdown, mixing, and passage of ingested nutrients, then elimination of undigested and unabsorbed residues.

**2. Secretion** Release into the lumen of enzymes, fluids, and other substances required for the functions of the digestive tract.

**3. Digestion** Breakdown of nutrients into particles, then into molecules small enough to be absorbed.

**4. Absorption** Passage of digested nutrients, fluid, and ions across the tube wall and into the blood or lymph, which will distribute them through the body.

Recall also from Chapter 25 that chordates, including antelopes, have a **complete digestive system**. So do annelids, mollusks, arthropods, and echinoderms. All of these animals have an internal tube with an opening at one end for taking in food and an opening at the other end for eliminating unabsorbed residues (Figure 37.2). Between the two openings, food generally moves in one direction through the lumen. (*Lumen* refers to the space inside a tube.) The tube itself is subdivided into specialized regions for food transport, processing, and storage. For instance, one part of the digestive tube of birds is modified into a crop, a food storage organ. Another part is modified into a gizzard, a muscular organ that grinds food into smaller bits.

| Table 37.1 | Components of the Human Digestive System |
|---|---|
| Organ | Main Functions |
| Mouth | Mechanically break down food, mix it with saliva |
| Salivary glands | Moisten food; start polysaccharide breakdown; buffer acidic foods in mouth |
| Stomach | Store, mix, dissolve food; kill many microorganisms; start protein breakdown; empty contents in a controlled way |
| Small intestine | Digest and absorb most nutrients |
| Pancreas | Enzymatically break down all major food molecules; buffer hydrochloric acid from stomach |
| Liver | Secrete bile for fat absorption; secrete bicarbonate, which buffers hydrochloric acid from stomach |
| Gallbladder | Store, concentrate bile from liver |
| Large intestine | Store, concentrate undigested matter by absorbing water and salts (mineral ions) |
| Rectum | Control over elimination of undigested and unabsorbed residues |

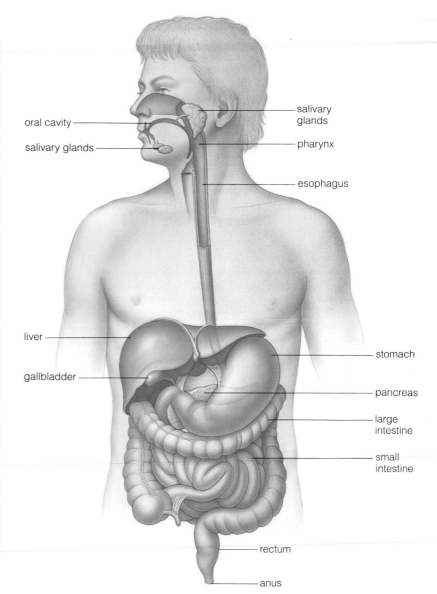

**Figure 37.4** Simplified picture of the human digestive system.

The remainder of this chapter will focus on the digestive system and nutritional requirements of humans. As Figure 37.3 suggests, this system does not act alone to meet the body's metabolic needs. Digested nutrients enter the internal environment—that is, the body's extracellular fluid. A circulatory system typically distributes the nutrients to cells throughout the body. A respiratory system helps cells use the nutrients by supplying them with oxygen (for aerobic respiration) and relieving them of carbon dioxide wastes. And even though the kinds and amounts of nutrients being absorbed can vary, depending on the diet, a urinary system helps maintain the volume and composition of extracellular fluid.

Keep these vital interactions in mind as we proceed with our nutritional tour in this chapter and the ones to follow.

## HUMAN DIGESTIVE SYSTEM: AN OVERVIEW

### Components

Figure 37.4 shows the human digestive system, and Table 37.1 lists the functions of its components. Humans have discontinuous feeding habits, and they ingest a variety of foods. From this you might deduce, correctly, that the human digestive system is a tube with many regional specializations. Stretched out, the tube would be 6.5 to 9 meters (21 to 30 feet) long in adults. Its specialized regions are the mouth (oral cavity), pharynx, esophagus, and the *gut*, or gastrointestinal tract. The gut itself is subdivided into a stomach, small intestine, large intestine (colon), rectum, and anus. Enzymes and other substances from the salivary glands, liver, gallbladder, pancreas, and the gut wall are secreted into different parts of the tube. The secretions assist in digestion and absorption.

## Gut Structure and Motility

Figure 37.5 shows the structure of the gut wall. The *mucosa* (an epithelium and underlying layer of connective tissue) faces the gut lumen. It is surrounded by the *submucosa*, a connective tissue layer with blood and lymph vessels and nerve plexuses (local networks of neurons). Next is *smooth muscle*—usually two sublayers, one circular and the other longitudinal in orientation. The outer layer of connective tissue (*serosa*) is almost as thin as Saran wrap. *Sphincters* (rings of muscle in the wall) occur at the beginning and end of the stomach and other specialized regions. They help control the forward movement of food and prevent backflow.

The muscle layers engage in mixing and wavelike contractions (Figure 37.6). During *peristalsis*, rings of circular muscles contract behind food and relax in front of it. The food distends the tube wall, peristaltic movement forces the food onward and expands the next wall region, and so on. During *segmentation*, rings of smooth muscle in the gut wall repeatedly contract and relax, creating an oscillating (back-and-forth) movement in the same place. This movement constantly mixes and forces the contents of the lumen against the absorptive surface of the intestinal wall.

## Control of the Digestive System

Recall that homeostatic control mechanisms operate when physical and chemical conditions change in the *internal* environment. By contrast, controls over the stomach and intestines respond to the volume and composition of material in the gut lumen. The nervous system, local nerve plexuses in the gut wall, and the endocrine system interact to exert control.

After a meal, for instance, food distends the gut wall. Signals from mechanoreceptors travel on short reflex pathways that are confined to nerve plexuses. They also may travel on long reflex pathways to the central nervous system. Signals along one or both types of pathways can lead to muscle contractions in the gut wall or secretion of enzymes and other substances into the gut lumen. The hypothalamus and other parts of the brain monitor such activities and coordinate them with other events (Figure 37.7).

Four gastrointestinal hormones are known. *Gastrin* is secreted by endocrine cells in the stomach's lining when amino acids and peptides are in the stomach. It mainly stimulates the secretion of acid into the stomach. Endocrine cells in the lining of the small intestine secrete the other three hormones. *Secretin*, a peptide hormone, stimulates the pancreas to secrete bicarbonate (page 581). *CCK* (cholecystokinin) enhances the actions of secretin and stimulates gallbladder contractions. *GIP*

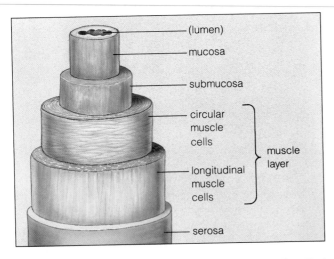

**Figure 37.5** Generalized sketch of the wall of the gastrointestinal tract. (The layers are not drawn to scale.)

**Figure 37.6** (**a**) Peristaltic wave down the stomach, produced by alternating contraction and relaxation of muscles in the stomach wall. (**b**) Segmentation, or oscillating movement, in the intestines.

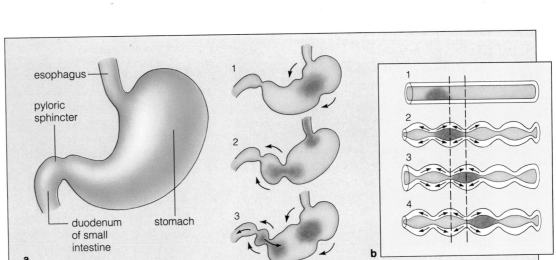

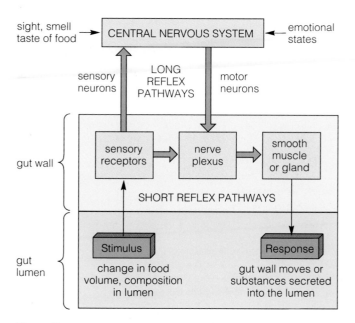

**Figure 37.7** Local and long-distance reflex pathways called into action when food is in the digestive tract.

(glucose insulinotropic peptide) is released in response to the presence of glucose and fat in the small intestine. It stimulates insulin secretion.

**The nervous system, endocrine system, and local nerve plexuses control conditions in the digestive system in response to the volume and composition of material in the gut lumen.**

**In this respect, controls over the digestive system differ from homeostatic control mechanisms (which maintain conditions in the internal environment).**

## INTO THE MOUTH, DOWN THE TUBE

### Mouth and Salivary Glands

Food starts getting pummeled and polysaccharide digestion begins in the mouth (oral cavity). Only humans and other mammals *chew* food. Adult humans normally have thirty-two teeth to do this. Each tooth is an engineering marvel, able to withstand many years of chemical insults and mechanical stress. It has an enamel coat (hardened calcium deposits), dentine (a thick bonelike layer), and an inner pulp (with nerves and blood vessels). The chisel-shaped incisors bite off chunks of food, the cone-shaped cuspids tear it, and the flat-topped molars grind it (page 332 and Figure 37.1).

Food in the mouth becomes mixed with saliva, a fluid secreted from **salivary glands**. Saliva includes a starch-degrading enzyme (salivary amylase), bicarbonate ($HCO_3^-$), and mucins. The buffering action of bicarbonate ions keeps the pH of your mouth between 6.5

and 7.5, even when you eat tomatoes and other acidic foods. Mucins (modified proteins) bind bits of food into a softened, lubricated ball called a bolus.

Muscle contractions of the tongue force the softened ball of food into the **pharynx**. This muscular tube connects with the **esophagus**, which leads to the stomach. Swallowing is initiated when an individual voluntarily pushes a bolus into the pharynx. Sensory receptors in the wall of the pharynx are stimulated, and they trigger contractions (an involuntary response). The pharynx and esophagus do not have roles in digestion; contractions in their walls simply propel food into the stomach.

The pharynx also connects with the trachea, which leads to the lungs. Swallowing opens a sphincter at the start of the esophagus. You normally don't choke on food because a flaplike valve, the epiglottis, closes off the opening into the respiratory tract and keeps you from breathing while food is moving into the esophagus.

### The Stomach

The **stomach**, a muscular, stretchable sac, has three main functions. First, it stores and mixes food. Second, its secretions help dissolve and degrade food. Third, it helps control the movement of food into the small intestine.

**Stomach Acidity.** Each day, cells in the stomach lining secrete about 2 liters of substances, including hydrochloric acid (HCl), pepsinogens, and mucus. The substances make up the fluid in the stomach, the so-called *gastric fluid*. The HCl separates into $H^+$ and $Cl^-$, and the increase in acidity helps dissolve bits of food to form a solution called chyme. It also kills most of the microorganisms hitching rides into the body in food.

Stomach secretion begins when your brain responds to the sight, aroma, and taste of food (even to hungry thoughts about it) and fires off signals to the acid-secreting and endocrine cells in the stomach lining. But most of the secretions occur in response to food in the stomach. When food stretches the stomach, it activates receptors in the stomach wall and gives rise to neural signals that call for stepped-up secretions. Also, secretory cells are stimulated directly by certain substances, including partially dismantled proteins as well as the caffeine in coffee, tea, chocolate, and cola drinks.

Protein digestion begins in the stomach. High stomach acidity structurally changes proteins and exposes the peptide bonds. It also converts pepsinogens to active forms (pepsins) that break down proteins. Protein fragments directly stimulate the secretion of gastrin, a hormone that acts on HCl-secreting cells. The more protein you eat, the more gastrin and HCl are released.

What protects the stomach lining itself from HCl and pepsin? Control mechanisms assure that enough mucus

**Table 37.2   Major Enzymes of Digestion**

| Enzyme | Source | Where Active | Substrate | Main Breakdown Products* |
|---|---|---|---|---|
| **Carbohydrate Digestion:** | | | | |
| Salivary amylase | Salivary glands | Mouth | Polysaccharides | Disaccharides |
| Pancreatic amylase | Pancreas | Small intestine | Polysaccharides | Disaccharides |
| Disaccharidases | Small intestine | Small intestine | Disaccharides | Monosaccharides (e.g., glucose) |
| **Protein Digestion:** | | | | |
| Pepsins | Stomach mucosa | Stomach | Proteins | Peptide fragments |
| Trypsin and chymotrypsin | Pancreas | Small intestine | Proteins, polypeptides | Peptide fragments |
| Carboxypeptidase | Pancreas | Small intestine | Peptide fragments | Amino acids |
| Aminopeptidase | Intestinal mucosa | Small intestine | Peptide fragments | Amino acids |
| **Fat Digestion:** | | | | |
| Lipase | Pancreas | Small intestine | Triglycerides | Free fatty acids, monoglycerides |
| **Nucleic Acid Digestion:** | | | | |
| Pancreatic nucleases | Pancreas | Small intestine | DNA, RNA | Nucleotides |
| Intestinal nucleases | Intestinal mucosa | Small intestine | Nucleotides | Nucleotide bases, monosaccharides |

*Yellow parts of table identify breakdown products that can be absorbed into the internal environment.

and buffering molecules (bicarbonate ions especially) are secreted to protect the lining from their destructive effects. Sometimes, however, normal controls are blocked. When the surface of the stomach breaks down, $H^+$ diffuses into the lining and triggers the release of a chemical called histamine from tissue cells. Histamine acts on local blood vessels—and it stimulates more HCl secretion. A positive-feedback loop is set up, tissues become further damaged, and bleeding into the stomach and possibly into the abdomen may occur. This outcome is called a *peptic ulcer.*

**Stomach Emptying**.   In the stomach, peristaltic waves mix the chyme and build up force as they approach the sphincter between the stomach and small intestine (Figure 37.6). The arrival of a strong contraction closes the sphincter, so most of the chyme is squeezed back. But a small amount moves into the small intestine.

The volume and composition of chyme affect how fast the stomach empties. For example, large meals activate more receptors in the stomach wall, these call for increases in the force of contraction, and the stomach empties faster. As another example, receptors in the small intestine sense increases in acidity, fat content, and so on, and they call for the release of hormones that slow stomach emptying. Through such slowdowns, food is not moved along faster than it can be processed. Fear, depression, and other emotional upsets also can slow stomach motility.

**The Small Intestine**

Table 37.2 summarizes the regions where carbohydrates, fats, proteins, and nucleic acids are digested and absorbed. Digestion is completed and most nutrients are absorbed in the **small intestine**, which has three regions: the duodenum, jejunum, and ileum (Figure 37.4). Secretions from the **pancreas** and **liver** enter a common duct that empties into the small intestine. Each day, about 9 liters of fluid travel from the stomach, liver, and pancreas into this part of the gut. All but 5 percent is absorbed across the intestinal lining.

**Digestion Processes**.   Enzymes secreted from the pancreas act on carbohydrates, fats, proteins, and nucleic acids. For example, like pepsin in the stomach, the pancreatic enzymes trypsin and chymotrypsin digest proteins into peptide fragments. The fragments are then degraded to free amino acids by carboxypeptidase (from the pancreas) and by aminopeptidase (present on the surface of the intestinal mucosa). The pancreas also secretes bicarbonate. The buffering action of bicarbonate helps neutralize the HCl arriving from the stomach. Two hormones secreted from the pancreas (insulin and glucagon) do not function in digestion, but they still have roles in nutrition, as described on page 590.

*Bile*, a secretion from the liver, has a key role in digestion. This secretion contains bile salts, bile pigments, cholesterol, and lecithin (a phospholipid). The

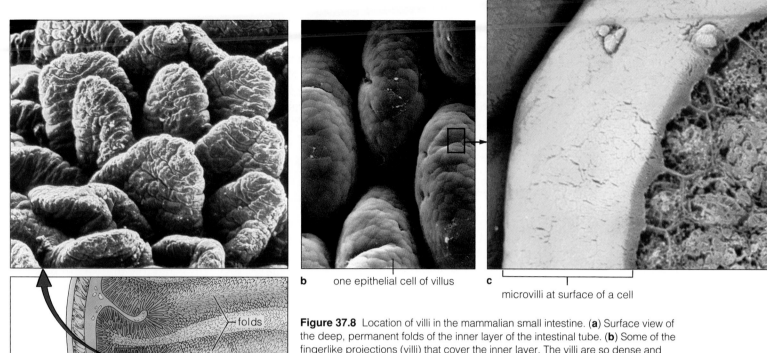

**b** one epithelial cell of villus   **c**

microvilli at surface of a cell

**Figure 37.8** Location of villi in the mammalian small intestine. (**a**) Surface view of the deep, permanent folds of the inner layer of the intestinal tube. (**b**) Some of the fingerlike projections (villi) that cover the inner layer. The villi are so dense and numerous, they give the surface a velvety appearance. Individual epithelial cells are visible. (**c**) The dense crown of microvilli at the surface of a single cell.

**a**   location of intestinal villi

folds

bile salts assist in fat breakdown and absorption. Most fats in our diet are triglycerides, clumped into large fat globules. The globules are mechanically broken apart into smaller droplets in the small intestine. Bile salts keep the droplets from clumping back together into globules, a process called *emulsification*. Through the emulsifying effect of bile salts, fat-degrading enzymes have greater access to more triglycerides, so fat digestion is enhanced.

Between meals, bile is stored and concentrated in the **gallbladder** (Figure 37.4).

**Absorption Processes.** By the time proteins, lipids, and carbohydrates are halfway through the small intestine, mechanical action and enzymes have broken down most of them to smaller molecules. The breakdown products include glucose and other monosaccharides, amino acids, fatty acids, and monoglycerides, all of which can move across the intestinal lining. The lining is densely folded into **villi** (singular, villus). Villi are absorptive structures that increase the surface area available for interactions with chyme. Epithelial cells at their surface have a crown of **microvilli**. Each microvillus is a threadlike projection of the plasma membrane. Collectively, microvilli greatly increase the surface area available for absorption (Figure 37.8).

At each villus, glucose and most amino acids cross the gut lining by active transport mechanisms, which move them across the plasma membranes of epithelial

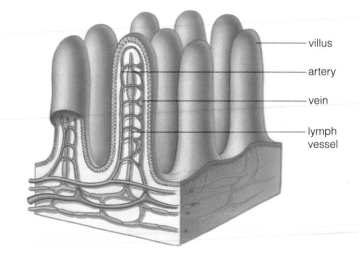

villus

artery

vein

lymph vessel

**Figure 37.9** Location of blood vessels and lymph vessels in intestinal villi. Monosaccharides and most amino acids moving across the intestinal lining enter the blood vessels; fats enter the lymph vessels, which drain into the general circulation.

cells. Then they diffuse through extracellular fluid and enter small blood vessels inside the villus (Figure 37.9).

The fatty acids and monoglycerides diffuse across the lipid bilayer of the membranes. (Bile salts help maintain the required concentration gradients for those free lipids. They combine with a number of the lipid molecules, forming small aggregates called *micelles*. The micelles themselves don't diffuse across the membrane. They are more like holding stations *at* the membrane. The bound lipids readily depart from micelles when their concentrations decrease in the lumen as a result of absorption.) Inside the epithelial cells, the fatty acids and monoglycerides recombine into fats. The fats cluster together as small droplets. Then the droplets leave the cells by exocytosis and enter lymph vessels, which drain into the blood circulation system. Water and mineral ions also are absorbed at the intestinal villi.

### The Large Intestine

Material not absorbed in the small intestine moves into the **large intestine**, or **colon**. The colon stores and concentrates *feces*, a mixture of undigested and unabsorbed material, water, and bacteria. The concentrating mechanism involves the active transport of sodium ions across the lining of the colon; water follows passively as a result.

The colon is about 1.2 meters long and starts out as a blind pouch (Figure 37.10). The *appendix*, a narrow projection from the pouch, has no known digestive functions. (It may have roles in defense against infectious agents.) The colon ascends on the right side of the abdominal cavity, continues across to the other side, then descends and connects with a small tube, the **rectum** (Figure 37.4). Distension of the rectal wall triggers the expulsion of fecal matter from the body. Expulsion is controlled by the nervous system, which can stimulate or inhibit contractions of a muscle sphincter at the **anus**, the terminal opening of the gut.

The average American diet does not include enough bulk. *Bulk* is the volume of fiber (mainly cellulose) and other undigested food material that cannot be decreased by absorption in the colon. Because of the insufficient volume, it takes longer for feces to move through the colon, with irritating and perhaps carcinogenic effects.

Most people of rural Africa and India cannot afford to eat much more than whole grains—which are high in fiber content—and they rarely suffer colon cancer or appendicitis. (In *appendicitis*, the appendix becomes infected and may rupture; bacteria normally living in the colon may spread into the abdominal cavity and cause serious infection.) When those people move to wealthier nations and leave behind their fiber-rich diet, they are more likely to suffer colon cancer and appendicitis. And this example leads us into a closer look at what constitutes good and bad nutrition.

## HUMAN NUTRITIONAL REQUIREMENTS

The earliest human ancestors, it seems, dined on fresh fruits and other fibrous plant material. From this nutritional beginning, humans in many parts of the world have moved to diets rich in saturated fats, cholesterol, refined sugars, and salts—and low in fiber. To be sure, our life span is much longer than that of our earliest ancestors. But many of us are probably suffering more than they did from colon cancer, kidney stones, breast cancer, and circulatory disorders—all of which may be correlated with the long-term shift in diet.

### Energy Needs and Body Weight

The body grows and maintains itself when kept supplied with energy and materials from foods of certain types, in certain amounts. Nutritionists measure energy in units called "calories," which unfortunately is supposed to mean the same thing as "kilocalories." A *kilocalorie* is 1,000 calories of energy, and that is the term we will use here.

To maintain an acceptable weight and keep the body functioning normally, caloric intake must be balanced with energy output. The output varies from one person to the next because of differences in physical activity, basic rate of metabolism, age, sex, hormone activity, and emotional state. Some of these factors are influenced by a person's social environment. But others have a genetic basis. For example, long-term studies were made of many identical twins (with identical genes) who were separated at birth and raised apart, in different households. At adulthood, the body weights of the separated twins were remarkably similar.

In most adults, energy input balances the output, so body weight remains much the same over long periods. As any dieter knows, it is as if the body has a set point for what that weight is going to be and works to counteract deviations from its set point.

How many kilocalories should you take in each day to maintain what you consider to be an acceptable body weight? One way to answer this question is to estimate your body's energy requirements. First, multiply the desired weight (in pounds) by 10 if you are not very active physically, by 15 if you are moderately active, and by 20 if you are quite active. Then, depending on your

**Figure 37.10**
Location of the appendix, a narrow projection from the cecum (a cup-shaped pouch at the start of the large intestine).

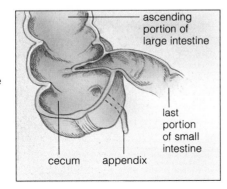

ascending portion of large intestine

last portion of small intestine

cecum    appendix

| Man's Height | Size of Frame | | |
|---|---|---|---|
| | Small | Medium | Large |
| 5' 2" | 128–134 | 131–141 | 138–150 |
| 5' 3" | 130–136 | 133–143 | 140–153 |
| 5' 4" | 132–138 | 135–145 | 142–156 |
| 5' 5" | 134–140 | 137–148 | 144–160 |
| 5' 6" | 136–142 | 139–151 | 146–164 |
| 5' 7" | 138–145 | 142–154 | 149–168 |
| 5' 8" | 140–148 | 145–157 | 152–172 |
| 5' 9" | 142–151 | 148–160 | 155–176 |
| 5'10" | 144–154 | 151–163 | 158–180 |
| 5'11" | 146–157 | 154–166 | 161–184 |
| 6' 0" | 149–160 | 157–170 | 164–188 |
| 6' 1" | 152–164 | 160–174 | 168–192 |
| 6' 2" | 155–168 | 164–178 | 172–197 |
| 6' 3" | 158–172 | 167–182 | 176–202 |
| 6' 4" | 162–176 | 171–187 | 181–207 |

a

| Woman's Height | Size of Frame | | |
|---|---|---|---|
| | Small | Medium | Large |
| 4'10" | 102–111 | 109–121 | 118–131 |
| 4'11" | 103–113 | 111–123 | 120–134 |
| 5' 0" | 104–115 | 113–126 | 122–137 |
| 5' 1" | 106–118 | 115–129 | 125–140 |
| 5' 2" | 108–121 | 118–132 | 128–143 |
| 5' 3" | 111–124 | 121–135 | 131–147 |
| 5' 4" | 114–127 | 124–138 | 134–151 |
| 5' 5" | 117–130 | 127–141 | 137–155 |
| 5' 6" | 120–133 | 130–144 | 140–159 |
| 5' 7" | 123–136 | 133–147 | 143–163 |
| 5' 8" | 126–139 | 136–150 | 146–167 |
| 5' 9" | 129–142 | 139–153 | 149–170 |
| 5'10" | 132–145 | 142–158 | 152–173 |
| 5'11" | 135–148 | 145–159 | 155–176 |
| 6' 0" | 138–151 | 148–162 | 158–179 |

b

**Figure 37.11** (**a**) "Ideal" weights for adults according to one insurance company in 1983. Values shown are for people 25 to 59 years old wearing shoes with 1-inch heels and 3 pounds of clothing (for women) or 5 pounds (for men). (**b**) Extreme obesity puts severe strain on the circulatory system. The body produces many more capillaries to service the increased tissue masses, so blood volume drops. The heart becomes more stressed. It must pump harder to keep blood circulating.

age, subtract the following amount from the value obtained from the first step:

| Age | Subtract |
|---|---|
| 25–34 | 0 |
| 35–44 | 100 |
| 45–54 | 200 |
| 55–64 | 300 |
| Over 65 | 400 |

For example, if you want to weigh 120 pounds and are highly active, 120 × 20 = 2,400 kilocalories. If you are thirty-five years old, then you should take in a total of (2,400 − 100), or 2,300 kilocalories a day. Such calculations provide a rough estimate, but other factors, such as height, must be considered also. (An active person who is 5 feet, 2 inches tall doesn't need as many kilocalories as an active person who weighs the same but is 6 feet tall.)

By definition, **obesity** is an excess of fat in the body's adipose tissues, caused by imbalances between caloric intake and energy output. Yet, what is too fat or too thin? What is a person's "ideal weight"? Many charts have been developed (Figure 37.11), mostly by insurance companies that want to identify overweight people who are considered to be insurance risks. Such charts factor in height. People who are 25 percent heavier than the "ideal" are viewed as obese.

Some researchers who study causes of death suspect that the "ideal" actually may be 10 to 15 pounds heavier than the charts indicate. Some nutritionists are convinced the chart values should be less. Whatever the ideal range may be, serious disorders do arise with extremes at either end of that range (see the *Commentary* on page 650).

**To maintain an acceptable body weight, energy input (caloric intake) must be balanced with energy output (as through metabolic activity and exercise).**

### Carbohydrates

Complex carbohydrates are the body's main sources of energy. As we have seen, they can be readily broken down into glucose units (Chapter 6). Glucose is the primary energy source for your brain, muscles, and other body tissues. According to many nutritionists, the fleshy fruits, cereal grains, legumes (including beans and peas), and other fibrous carbohydrates should make up at least 50 to 60 percent of the daily caloric intake.

Each year, the average American eats as much as 128 pounds of refined sugar, or sucrose. That's more than 2 pounds a week! You may think this a far-fetched statement, but take a look at the ingredients listed on your packages of cereal, frozen dinners, soft drinks, and other prepared foods. A common ingredient is sucrose. This simple sugar adds calories to the diet but does so without the fiber of complex carbohydrates.

### Proteins

Proteins should make up about 12 percent of the total diet. When proteins are digested and absorbed, their amino acids become available for the body's own protein-building programs. Of the twenty common amino acids, eight are **essential amino acids**. Our cells cannot build these amino acids; they must obtain them indirectly, from food. The eight are cysteine (or methionine), isoleucine, leucine, lysine, phenylalanine (or tyrosine), threonine, tryptophan, and valine.

# Human Nutrition Gone Awry

## Eating Disorders

**Anorexia Nervosa.** Millions of Americans are dieting in any given day. Unfortunately, in a growing number of cases, obsessive dieting leads to a potentially fatal eating disorder called *anorexia nervosa* (Figure *a*). The disorder occurs primarily in women in their teens and early twenties.

Individuals with anorexia nervosa have a skewed perception of their body weight. They have an overwhelming fear of being fat and being hungry. They embark on a course of self-induced starvation and, frequently, overexercising. Emotional factors contribute to the disorder. Some individuals fear growing up in general and maturing sexually in particular; others have irrational expectations of what they can accomplish. Severe cases require psychiatric treatment.

**Bulimia.** Another eating disorder on the rise is *bulimia* ("an oxlike appetite"). At least 20 percent of college-age women are now suffering to varying degrees from this disorder. Those afflicted may look outwardly healthy, but their food intake is out of control. During an hour-long eating binge, a bulimic may take in more than 50,000 kilocalories. This is followed by vomiting or purging the body with laxatives, sometimes in doses of 200 tablets or more. The binge–purge routine may occur once a month; it may occur several times a day.

Some women start doing this because it seems like a simple way to lose weight. Others have emotional problems. Often they are well-educated, accomplished individuals, but they strive for perfection and may have problems with control exerted by other family members. According to one view, eating may actually be an unpleasant event for them, but the purging (which they themselves control) relieves them of anger and frustration.

Repeated purgings, however, can damage the gastrointestinal tract. Repeated vomiting, which brings stomach acids into the mouth, can erode teeth to stubs. At its most extreme, bulimia can lead to death through heart failure, stomach rupturing, or kidney failure. Psychiatric treatment and hospitalization may be required in severe cases.

## Digestive Disorders

Generally, Americans are among the best-fed people in the world—yet at the same time, they suffer a high incidence of digestive disorders. Along with affluence, it appears that other bad eating habits besides anorexia nervosa and bulimia are rampant. Americans skip meals, eat too much

and too fast when they do sit down at the table, and generally give their gut erratic workouts. Worse yet, their diet tends to be rich in sugar, cholesterol, and salt—and low in bulk. The problem with too little bulk in the diet comes from the longer transit time of feces through the colon. As indicated in the text, this material has irritating and even potentially carcinogenic effects. The longer the material is in contact with the colon walls, the more damage it can do. (Thus, the more steadily the contents of the colon are cleared out by natural processes, the better. Increased bulk produces increased pressure on the colon walls, which stimulates expulsion of the material from the body.)

In addition, diet affects the distribution and diversity of bacterial populations living in the gut. Do changes in those populations contribute to digestive disorders? That is not known.

The emotional stress of living in complex societies seems to compound the nutritional problem. Urban populations seem to be more susceptible to the irritable colon syndrome (colitis). Its symptoms include abdominal pain, diarrhea (excretion of watery feces), and constipation. Diarrhea can be brought on by emotional stress. Although the development of ulcers may have a genetic basis, emotional stress also is a contributing factor.

Learning to handle stress is one way to ease up on the digestive tract, and learning how to eat properly is another.

Yet what is "eating properly"? As long ago as 1979, the United States Surgeon General released a report representing a medical consensus on how to promote health and avoid such afflictions as high blood pressure, heart disorders, cancer of the colon, and bad teeth. The report advised Americans to eat "less saturated fat and cholesterol; less salt; less sugar; relatively more complex carbohydrates such as whole grains, cereals, fruits, and vegetables; and relatively more fish, poultry, legumes (for example, peas, beans, and peanuts); and less red meat."

The controversies over what constitutes proper nutrition are still raging today. In the meantime, it might not be a bad idea to think about your own eating habits and how moderation in some things might help you hedge your bets. Put the question to yourself: Do you look upon a bowl of bran cereal with the same passion as you look upon, say, french fries and ice cream, prime rib, and chocolate mousse? Now put the same question to your colon.

| No Limiting Amino Acid | Low in Lysine | Low in Methionine, Other Sulfur-Containing Amino Acids | Low in Tryptophan |
|---|---|---|---|
| legumes:<br> soybean<br> tofu<br> soy milk<br><br>cereal grains:<br> wheat germ<br><br>nuts:<br><br><br><br>milk<br>cheeses (except<br> cream cheese)<br>yogurt<br>eggs<br>meats | legumes:<br> peanuts<br><br>cereal grains:<br> barley<br> buckwheat<br> corn meal<br> oats<br> rice<br> rye<br> wheat<br><br>nuts, seeds:<br> almonds<br> cashews<br> coconut<br> English walnuts<br> hazelnuts<br> pecans<br> pumpkin seeds<br> sunflower seeds | legumes:<br> beans (dried)<br> black-eyed peas<br> garbanzos<br> lentils<br> lima beans<br> mung beans<br> peanuts<br><br>nuts:<br> hazelnuts<br><br>fresh vegetables:<br> asparagus<br> broccoli<br><br><br>green peas<br>mushrooms<br>parsley<br>potatoes<br>soybeans<br>Swiss chard | legumes:<br> beans (dried)<br> garbanzos<br> lima beans<br> mung beans<br> peanuts<br><br>cereal grains:<br> corn meal<br><br>nuts:<br> almonds<br> English walnuts<br><br>fresh vegetables:<br> corn<br> green peas<br> mushrooms<br> Swiss chard |

**a**

**b**

Most animal proteins are "complete," meaning they contain high amounts of all essential amino acids. Plant proteins are "incomplete." They have all the essential amino acids, but not in the proportions required for proper human nutrition. To get enough protein, vegetarians must eat certain combinations of different plants (Figure 37.12). Nutritionists use a measure called *net protein utilization*, or NPU, to compare proteins from different sources (Table 37.3). NPU values range from 100 (all essential amino acids present in ideal proportions) to 0 (one or more absent; the protein is useless when eaten alone).

You know that enzymes and other proteins are vital for the body's structure and function, so it should be readily apparent that protein-deficient diets are no joking matter. Protein deficiency is most damaging among the young, for rapid brain growth and development occur early in life. Unless enough protein is taken in just before and just after birth, irreversible mental retardation occurs. Even mild protein starvation can retard growth and affect mental and physical performance.

## Lipids

Fats and other lipids have important roles in the body. For example, phospholipids (such as lecithin) and cholesterol are components of animal cell membranes. Besides being used as energy reserves, fat deposits serve as cushions for many organs, including the eyes and kidneys, and they provide insulation beneath the skin. Fats from the diet also help the body absorb fat-soluble vitamins.

**Figure 37.12** (a) Essential amino acids—a small portion of the total protein intake. All must be available at the same time, in certain amounts, if cells are to build their own proteins. Milk and eggs have high amounts of all eight essential amino acids in required proportions; they are among the complete proteins.

Nearly all plant proteins are incomplete, so vegetarians should plan their meals carefully to avoid protein deficiency. For example, they can combine *different* foods from any two of the columns shown in (b). Also, vegetarians who avoid dairy products and eggs should take vitamin $B_{12}$ and $B_2$ (riboflavin) supplements. Animal protein is a luxury in most traditional societies, yet good combinations of plant proteins are worked into their cuisines—including rice/beans, chili/cornbread, tofu/rice, lentils/wheat bread, and macaroni/cheese.

**Table 37.3 Efficiency of Some Single Protein Sources in Meeting Minimum Daily Requirements**

| Source | Protein Content (%) | Net Protein Utilization (NPU) | Amount Needed to Satisfy Minimum Daily Requirement (grams) | (ounces) |
|---|---|---|---|---|
| Eggs | 11 | 97 | 403 | 14.1 |
| Milk | 4 | 82 | 1,311*** | 45.9*** |
| Fish* | 22 | 80 | 244 | 8.5 |
| Cheese* | 27 | 70 | 227 | 7.2 |
| Red meat* | 25 | 68 | 253 | 8.8 |
| Soybeans | 34 | 60 | 210** | 7.3** |
| Kidney beans | 23 | 40 | 468** | 16.4** |
| Corn | 10 | 50 | 860** | 30.0** |

*Average values.
**Dry weight values.
***Equivalent of 6 cups. The figure is somewhat misleading, for most of the volume of milk is water. Milk is actually a rich source of high-quality protein.

## Table 37.4   Vitamins Required for Normal Cell Functioning

| Vitamin | RDA* (milligrams) Females | Males | Common Sources | Some Known Functions |
|---|---|---|---|---|
| **Water-Soluble Vitamins:** | | | | |
| B₁ (Thiamin) | 1.1 | 1.5 | Lean meats, liver, eggs, whole grains, green leafy vegetables, legumes | Connective tissue formation; iron, folic acid utilization |
| B₂ (Riboflavin) | 1.3 | 1.7 | Milk, egg white, yeast, whole grains, poultry, fish, meat | Coenzyme action (FAD, FMN) |
| Niacin | 15 | 19 | Meat, poultry, fish; also peanuts, potatoes, green leafy vegetables, liver | Coenzyme action (NAD⁺, NADP⁺) |
| B₆ | 1.6 | 2 | Meats, potatoes, tomatoes, spinach | Coenzyme role in amino acid metabolism |
| Pantothenic acid | ** | ** | In many foods, especially meat, yeast, egg yolk | Coenzyme role in glucose metabolism; fatty acid and steroid synthesis |
| Folic acid (folate) | 0.18 | 0.2 | Dark green vegetables, eggs, liver, yeast, lean meat, whole grains; produced by bacteria in gut | Coenzyme role in nucleic acid and amino acid metabolism |
| B₁₂ | 0.002 | 0.002 | Meat, poultry, fish, eggs, dairy foods (not butter) | Coenzyme role in nucleic acid metabolism |
| Biotin | ** | ** | Legumes, nuts, liver, egg yolk; some produced by bacteria in gut | Coenzyme action in fat and glycogen formation; amino acid metabolism |
| Choline | ** | ** | Whole grains, legumes, egg yolk, liver | Component of phospholipids, acetylcholine |
| C (Ascorbic acid) | 60 | 60 | Citrus, papaya, cantaloupe, berries, tomatoes, potatoes, green leafy vegetables | Structural role in bone, cartilage, teeth; roles in collagen formation, carbohydrate metabolism |
| **Fat-Soluble Vitamins:** | | | | |
| A (Retinol) | 0.8 | 1 | Formed from carotene in deep-yellow, deep-green leafy vegetables; already present in fish liver oil, liver, egg yolk, fortified milk | Role in synthesis of visual pigments; required for bone, tooth development; maintains epithelial tissues |
| D | 0.01/0.005 | 0.01/0.005 | Vitamin D₃ formed in skin cells (also in fish liver oils, egg yolk, fortified milk); converted to active form in other body regions | Promotes bone growth, mineralization; increases calcium absorption |
| E | 8 | 10 | Vegetable oils, margarine, whole grains, dark-green vegetables | Prevents breakdown of vitamins A, C in gut; helps maintain cell membranes |
| K | 0.06/0.065 | 0.07/0.08 | Most formed by bacteria in colon; also in green leafy vegetables, cauliflower, cabbage | Role in clot formation; electron transport role in ATP formation |

*1989 recommended daily allowance for two age groups: 19–24/25–50.
**Not established.

Today, lipids make up 40 percent of the average diet in the United States. Most of the medical community agrees it should be less than 30 percent. The body can synthesize most of its own fats, including cholesterol, from protein and carbohydrates. (That is exactly what it does when you eat too much protein and carbohydrates.) You only need to take in *one tablespoon a day* of a polyunsaturated fat, such as corn oil or olive oil. These oils contain linoleic acid. It is one of the **essential fatty acids**. This means the body cannot produce this fatty acid; it must be provided by the diet.

Butter and other animal fats are saturated fats, which tend to raise the level of cholesterol in the blood. Cho-lesterol is necessary in the synthesis of bile acids and steroid hormones. However, too much cholesterol may have devastating effects on the circulatory system (page 671).

### Vitamins and Minerals

Normal metabolic activity depends on small amounts of more than a dozen organic substances called **vitamins**. Most plant cells synthesize all of these substances. In general, animal cells have lost the ability to do so, so animals must obtain vitamins from food. Human cells need at least thirteen different vitamins, each with specific metabolic roles. Many metabolic reactions depend on

**Table 37.5  Minerals Required for Normal Cell Functioning**

| Mineral | RDA* (milligrams) Females | Males | Common Sources | Some Known Functions |
|---|---|---|---|---|
| Calcium | 1200/800 | 1200/800 | Dairy products, dark-green vegetables, dried legumes | Bone, tooth formation; clotting, neural signals |
| Chlorine | ** | ** | Table salt; usually too much in diet | HCl formation by stomach; helps maintain body pH |
| Copper | ** | ** | Meats, legumes, drinking water | Used in synthesis of hemoglobin, melanin, transport chain components |
| Fluorine | ** | ** | Fluoridated water, seafood | Bone, tooth maintenance |
| Iodine | 0.15 | 0.15 | Marine fish, shellfish; dairy products | Thyroid hormone formation |
| Iron | 10 | 15 | Liver, lean meats, yolk, shellfish, nuts, molasses, legumes, dried fruit | Hemoglobin, cytochrome formation |
| Magnesium | 280 | 350 | Dairy products, nuts, whole grains, legumes | Coenzyme role in ATP–ADP cycle; role in muscle, nerve function |
| Phosphorus | 1200/800 | 1200/800 | Dairy products, red meat, poultry, whole grains | Component of bone, teeth, nucleic acids, proteins, ATP, phospholipids |
| Potassium | ** | ** | Diet provides ample amounts | Muscle, nerve function; role in protein synthesis; acid-base balance |
| Sodium | ** | ** | Table salt; diet provides adequate to excess amounts | Key salt in solute-water balance, muscle and nerve function |
| Sulfur | ** | ** | Dietary proteins | Component of body proteins |
| Zinc | 12 | 15 | Seafood, meat, cereals, legumes, nuts, yeast | Component of digestive enzymes; roles in normal growth, wound healing, taste and smell, sperm formation |

*1981 recommended daily allowance for two age groups: 18–24/25–50.
**Not established.

*several* vitamins, and the absence of one vitamin can affect the functions of others (Table 37.4).

Metabolic activity also depends on inorganic substances called **minerals** (Table 37.5). For example, most cells use calcium and magnesium in many different reactions. All cells use potassium during muscle activity and nerve function. All cells require iron for cytochrome molecules, these being components of electron transport chains. Red blood cells contain iron in hemoglobin, the oxygen-carrying pigment in blood.

The sensible way to supply cells with essential vitamins and minerals is to eat a well-balanced selection of carbohydrates, lipids, and proteins. About 250–500 grams of carbohydrates, 66–83 grams of lipids, and 32–42 grams of protein should do the trick. Some people claim the body will benefit from massive doses of certain vitamins and minerals. To date, no clear evidence exists that vitamin intake above recommended daily amounts leads to better health. To the contrary, excessive vitamin doses are often merely wasted and may cause chemical imbalances.

For example, the body simply will not hold more vitamin C than it needs for normal functioning. Vitamin C is not fat-soluble, and the excess is eliminated in urine. In fact, any amount above the recommended daily allowance ends up in the urine almost immediately after it is absorbed from the gut! Abnormal intake of at least two other vitamins (A and D) can cause serious disorders. Like all fat-soluble vitamins, vitamins A and D can accumulate in tissues and interfere with normal metabolic function.

Similarly, sodium is present in plant and animal tissues, and it is a component of table salt. Sodium has roles in the body's salt–water balance, muscle activity, and nerve function. Yet, prolonged, excessive intake of sodium is thought to be a cause of high blood pressure.

**Severe shortage or massive excess of vitamins and minerals can disturb the delicate balances that promote health.**

**Figure 37.13** Summary of major pathways of organic metabolism. Urea formation occurs primarily in the liver. Carbohydrates, fats, and proteins are continually being broken down and resynthesized.

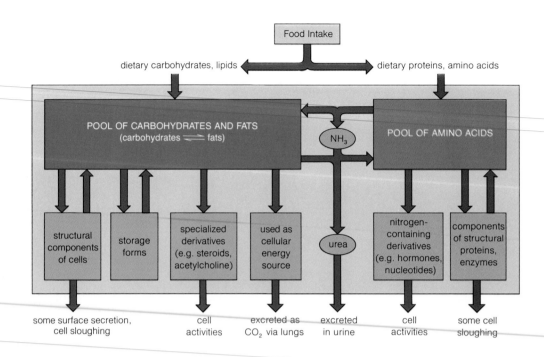

## Table 37.6 Some Activities That Depend on Liver Functioning

1. Carbohydrate metabolism

2. Control over some aspects of plasma protein synthesis

3. Assembly and disassembly of certain proteins

4. Urea formation from nitrogen-containing wastes

5. Assembly and storage of some fats

6. Fat digestion (bile is formed by the liver)

7. Inactivation of many chemicals (such as hormones and some drugs)

8. Detoxification of many poisons

9. Degradation of worn-out red blood cells

10. Immune response (removal of some foreign particles)

11. Red blood cell formation (liver absorbs, stores factors needed for red blood cell maturation)

## NUTRITION AND METABOLISM

### Storage and Interconversion of Nutrients

Figure 37.13 summarizes the main routes by which nutrient molecules are shuffled and reshuffled once they have been absorbed into the body. With few exceptions (such as DNA), most carbohydrates, lipids, and proteins are broken down continually, with their component parts picked up and used again in new molecules. At the molecular level, your body undergoes massive and sometimes rapid turnovers.

When you eat, the body builds up its pools of organic molecules. Excess carbohydrates and other dietary molecules are transformed mostly into fats, which are stored in adipose tissue. Some are also converted to glycogen in the liver and in muscle tissue. Most cells use glucose as their main energy source at this time; there is no net breakdown of protein in muscle or other tissues.

Between meals, there is a notable shift in the type of food molecules used to support cell activities. A key factor in this shift is the need to provide brain cells with glucose, the major nutrient they use for energy.

When glucose is being absorbed, its blood levels are readily maintained. How does the body maintain blood glucose concentrations between meals? First, glycogen, stored mainly in the liver, is rapidly broken down to glucose, which is released into blood. Second, body proteins are broken down to amino acids, which are sent to the liver for conversion to glucose that is released into the blood.

Most cells use fats as the main energy source between meals. Fats stored in adipose tissue are broken down into glycerol and fatty acids, which are released into blood. The glycerol can be converted to glucose in the liver; the circulating fatty acids can be used in ATP production.

**During a meal, glucose moves into cells, where it can be used for energy and where the excess can be stored.**

**Between meals, most cells use fat as the main energy source. Stored fats are mobilized, and brain cells are kept supplied with glucose (their major energy source).**

# Commentary

## Case Study: Feasting, Fasting, and Systems Integration

With the possible exception of gastroenterologists, most of us probably would not use the control of digestion as a riveting topic of conversation at a party. Yet every day of your life, you depend absolutely on the integrated functions of those controls.

Suppose, this morning, you are vacationing in the mountains and decide on impulse to follow a forested trail. You fail to notice a wooden trail marker that bears the intriguing name, "Fat Man's Misery." As you walk down the tree-lined corridor, you are enjoying one of the benefits of discontinuous feeding. Having eaten a large breakfast, you have assured your cells of ongoing nourishment; you do not have to forage constantly amongst the ferns as, say, a roundworm must do. Food partly digested in the stomach has already entered the small intestine. Right now, amino acids, simple sugars, and fatty acids are moving across the intestinal wall, then into the bloodstream.

With the surge of nutrients, glucose molecules are entering the bloodstream faster than your cells can use them. The level of blood glucose begins to rise slightly. This is no problem, for your body has a homeostatic program to convert glucose into storage form when it is flooding in, then to release some of the stores when glucose is scarce.

With the rise in blood glucose, pancreatic beta cells are called upon to secrete insulin. Blood concentrations of

insulin rise—and the hormonal targets (liver, fat, and muscle cells) quickly begin using or storing the glucose molecules (Figure 34.14). At the same time, alpha cells are prevented from secreting glucagon—which slows the liver's conversion of stored glycogen into glucose.

As you can see, the liver is central to the storage and interconversion of absorbed carbohydrates, lipids, and proteins. Keep in mind that it has other vital roles (Table 37.6). For example, it helps maintain the concentrations of blood's organic substances and removes many toxic substances from it. The liver inactivates most hormone molecules and sends them on to the kidneys for excretion (in urine). Also, ammonia ($NH_3$) is produced when cells break down amino acids, and it can be toxic to cells. The circulatory system carries ammonia to the liver, and there it is converted to urea. This is a much less toxic-waste product, and it leaves the body by way of the kidneys, in urine.

### Controls Over Metabolism

Both endocrine and neural controls govern metabolism during and between meals. As we saw on page 590, the most important control agents are hormones secreted by the pancreas. One of the hormones, **glucagon**, is secreted in response to a drop in the glucose level in blood between meals. (Between meals, cells use the glucose circulated to them by the bloodstream.) Among other things, glucagon stimulates the conversion of glycogen into glucose and so *increases* the blood glucose level. Another hormone is **insulin**, which is secreted in response to a rise in the blood glucose level after a meal. Among other things, insulin stimulates glucose uptake by cells—and so *decreases* the blood glucose level.

Figure 34.14 illustrates some of the interplays among hormones that help keep the blood glucose level relatively constant, despite great variation in when (and how much) we eat. A discussion of the control mechanisms themselves would be beyond the scope of this book. However, the *Commentary* above will give you a sense of the splendid nature of their interactions.

What is the outcome? High levels of glucose that have entered the circulation from your gut move out of the blood and into cells, where it can be burned as fuel or stored for later use.

Even though you are no longer feeding your body, your brain cells have not lessened their high demands for glucose. Neither have your muscle cells, which are getting a strenuous workout. Little by little, blood glucose levels drop. Now endocrine activities shift in the pancreas. With less glucose binding to them, beta cells decrease their insulin output. With less glucose to inhibit them, alpha cells increase their glucagon output. When glucagon reaches your liver, it causes the conversion of glycogen back to glucose—which is returned to your blood. This prevents blood glucose from falling below levels required to maintain brain function.

Yet the best-laid balance of internal conditions can go astray when external conditions change. In your case, the "miserable" part of the trail has begun. You find yourself scrambling higher and higher on steep inclines. Suddenly you stop, surprised, in pain. You forgot to reckon with the lower oxygen pressure of mountain air, and your leg muscles cramped. Your body has already detected its deficiency of oxygen-carrying red blood cells at this altitude, but it will take days before enough additional red blood cells are available.

In the meantime, your muscle cells are not being supplied with enough oxygen for the prolonged climb. They are relying on an anaerobic pathway in which lactate is the end product (page 129).

Again, systems interact to return conditions to a homeostatic state. Your body detects the reduced oxygen pressure and an accompanying increase in hydrogen ion concentrations in cerebrospinal fluid. Nerve impulses course toward the respiratory center in the medulla. The result: The diaphragm and other muscles associated with inflating and deflating your lungs contract more rapidly. You breathe faster now, and more deeply. In the liver, lactate is converted to glucose—which is returned to the blood.

On checking the sun's position, you see it is well past noon. And guess what: you forgot about lunch. When you start the long walk back, the drop in blood glucose levels triggers new homeostatic control mechanisms.

Under hypothalamic commands, your adrenal medulla begins secreting epinephrine and norepinephrine. Its main targets: the liver, adipose tissue, and muscles. In the liver, glycogen synthesis stops. In body tissues generally, glucose uptake is blocked. In fat cells, fats are converted to fatty acids, which are routed to the liver, muscles, and other tissues as alternative energy sources. For every fatty acid molecule sent down metabolic pathways in those tissues, several glucose molecules are held in reserve for the brain.

You do get back to the start of the trail by sundown. However, your body had enough stored energy to sustain you for many more days, so the situation was never really desperate. The balance of blood sugar and fat is constantly monitored by the liver and controlled by hormones. Glucose levels only drop beyond the set point to stimulate glycogen conversion and fat conversion, and vice versa. It takes several days of fasting before blood sugar levels are markedly reduced.

Even after several days of fasting, your energy supplies would not have run out. Another command from the hypothalamus would have prodded your anterior pituitary into secreting ACTH (page 584). The ACTH would have signaled adrenal cortex cells to secrete glucocorticoid hormones, which have a potent effect on the synthesis of carbohydrates from proteins and on the further breakdown of fat. Slowly, in muscles and other tissues, your body's proteins would have been disassembled. Amino acids from these structural tissues would have been used in the liver to build new glucose molecules—and once more your brain would have been kept active.

As extreme as this last pathway might be, it would be a small price to pay for keeping your brain functional enough to figure out how to take in more nutrients and bring you back to a homeostatic state.

## SUMMARY

1. Nutrition includes all the processes by which the body takes in, digests, absorbs, and uses food.

2. A digestive system breaks down food molecules by mechanical, enzymatic, and hormone-assisted means. It also enhances absorption of the breakdown products into the internal environment, and it eliminates the unabsorbed residues.

3. The human digestive system includes the mouth, pharynx, esophagus, stomach, small intestine, large intestine (colon), rectum, and anus. Glands associated with digestion are the salivary glands, liver, gallbladder, and pancreas.

4. Controls over the digestive system operate in response to the volume and composition of food passing through. The response can be a change in muscle activity, the secretion rate of hormones or enzymes, or both.

5. Starch digestion begins in the mouth; protein digestion begins in the stomach. Digestion is completed and most nutrients are absorbed in the small intestine. Following absorption, monosaccharides (including glucose) and most amino acids are sent directly to the liver. Fatty acids and monoglycerides enter lymph vessels and then enter the general circulation.

6. To maintain acceptable weight and overall health, caloric intake must balance energy output. Complex carbohydrates are the body's main energy source. The body produces fats as storage forms of carbohydrates and proteins. Eight essential amino acids, a few essential fatty acids, vitamins, and minerals must be provided by the diet.

## Review Questions

1. Study Figure 37.3. Then, on your own, diagram the connections between metabolism and the digestive, circulatory, and respiratory systems. *642*

2. What are the main functions of the stomach? The small intestine? The large intestine? *645, 646, 648*

3. Name four kinds of breakdown products that are actually small enough to be absorbed across the intestinal lining and into the internal environment. *646–648*

4. A glass of milk contains lactose, protein, butterfat, vitamins, and minerals. Explain what happens to each component when it passes through your digestive tract. *645–648, 653*

5. Describe some of the reasons why each of the following is nutritionally important: carbohydrates, proteins, fats, vitamins, and minerals. *648–653*

## Self-Quiz *(Answers in Appendix IV)*

1. The _____, _____, _____, and _____, interact in supplying body cells with raw materials, disposing of wastes, and maintaining the volume and composition of extracellular fluid.

2. Different specialized regions of the digestive system function in _____ and _____ food and in _____ unabsorbed food residues.

3. Maintaining good health and normal body weight requires that _____ intake be balanced by _____ output.

4. The main energy sources for the body are complex _____.

5. The human body cannot produce its own vitamins or minerals, and it also cannot produce certain _____ and _____.

6. Which glands are *not* associated with digestion?
   a. salivary glands
   b. thymus gland
   c. liver
   d. gallbladder
   e. pancreas

7. Digestion is completed and breakdown products are absorbed in the _____.
   a. mouth
   b. stomach
   c. small intestine
   d. large intestine

8. After absorption, fatty acids and monoglycerides move into the _____.
   a. bloodstream
   b. intestinal cells
   c. liver
   d. lymph vessels

9. _____ are storage forms of excess carbohydrates and proteins.
   a. Amino acids
   b. Starches
   c. Fats
   d. Monosaccharides

10. Match each digestive system component with its description.
   _____ liver
   _____ small intestine
   _____ human digestive system
   _____ nutrition
   _____ digestive system controls

   a. begins at mouth, ends at anus
   b. operate in response to food volume and composition
   c. functions are digestion, absorption, use of food
   d. where most digestion is completed
   e. receives monosaccharides and amino acids

## Selected Key Terms

absorption *642*
bile *646*
bulk *648*
digestion *642*
digestive system *641*
emulsification *647*
essential amino acid *649*
essential fatty acid *652*
gallbladder *647*
gastrointestinal tract (gut), *643*
large intestine (colon) *648*
liver *655*
micelle *647*
microvillus *647*

mineral *653*
motility *642*
net protein utilization *651*
pancreas *646*
peristalsis *644*
pharynx *645*
ruminant *642*
salivary gland *645*
secretion *642*
segmentation *644*
small intestine *646*
stomach *645*
villus *647*
vitamin *652*

## Readings

Campbell-Platt, G. May 1988. "The Food We Eat." *New Scientist*, 19:1–4.

Cohen, L. 1987. "Diet and Cancer." *Scientific American* 257(5): 42–68.

Hamilton, W. 1985. *Nutrition: Concepts and Controversies*. Third edition. Menlo Park, California: West.

Krause, M., and L. Mahan. 1984. *Food, Nutrition, and Diet Therapy*. Seventh edition. Philadelphia: Saunders.

# 38 CIRCULATION

## Heartworks

For Augustus Waller, Jimmie the bulldog was no ordinary pooch. Connected to wires and soaked to his ankles in buckets of salty water, Jimmie was a four-footed window into the workings of the heart.

Feel the pulse of blood that is coursing through the artery at your wrists or the repeated thumpings of your heart at the chest wall. Those same rhythms fascinated Waller and other physiologists of the nineteenth century. They suspected that every heartbeat might produce a characteristic pattern of electrical currents—a pattern that could be recorded painlessly at the body surface. That is where Jimmie and the buckets came in. Saltwater is an efficient conductor of electricity—so efficient that it carried faint signals from Jimmie's beating heart, through the skin of his legs, to a crude monitoring device. With this device, Waller made one of the world's first recordings of a beating heart—an electrocardiogram (Figure 38.1).

Look at the series of peaks of that simple graph. Taken together, they resemble the pattern that shows up in recordings of the electrical activity of your own heart (or of any other vertebrate). In fact, your heart started beating this way within weeks after you began to develop from a fertilized egg. Inside your tiny embryonic body, patches of newly formed cardiac muscle started contracting until one patch took the lead. Ever since that moment, it has been the pacemaker for your heart's activity—and normally it will remain so until the day you die.

**Figure 38.1** History in the making—Dr. Augustus Waller's pet bulldog, Jimmie, taking part in a painless experiment (**a**) that yielded one of the world's first electrocardiograms (**b**).

It is the pacemaker that faithfully sets, adjusts, and resets the rate at which blood is pumped from your heart, through a vast network of blood vessels, then back to the heart. Do nothing more than stretch out in a meadow and stare mindlessly at the sky and your heart rate will be moderate, somewhere around 70 beats a minute. Find yourself suddenly confronted by a mean-tempered bull in the meadow, and the demands by your muscles for blood-borne oxygen and glucose will escalate. Then, your heart may start pounding 150 times a minute to deliver sufficient blood to them.

Sometimes injury or disease interrupts the heart's programmed cadence. The heart may race wildly, slow ominously, or alternate sporadically between the two extremes. Blood flow to the body's tissues becomes skewed, so nutrients cannot be delivered and metabolic wastes cannot be removed properly. Each year in the

BUCKET

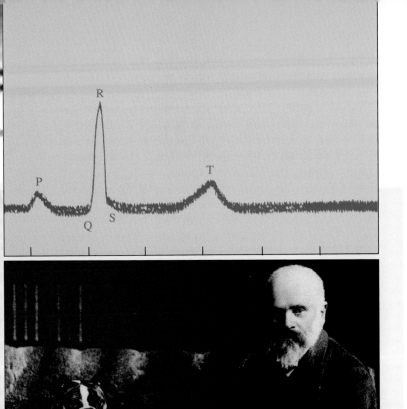

b    JIMMIE          DR. WALLER

United States alone, more than 400,000 individuals face the consequences of disrupted heart function.

We have come a long way from Jimmie and the buckets in our understanding of the patterns and changes in the heart's tempo. Sophisticated external sensors can pick up the faintest of signals. Minute variations that might signal an impending heart attack can now be pinpointed. Hospital computers analyze a patient's beating heart and instantaneously build color images of it on a video screen. Heart surgeons implant artificial pacemakers—small, battery-powered heart regulators that pinch-hit for a malfunctioning one. Even physical fitness buffs attach sensors to their wrists, earlobes, or fingertips and monitor their heart rate as they exercise.

With this chapter, we turn to the circulatory system, the means by which substances are rapidly moved to and from all living cells in animals ranging from worms to bulldogs and other mammals. As you will see, the system is absolutely central to the body's ability to maintain stable operating conditions in the internal environment—a state we call homeostasis.

**1.** Cells survive by exchanging substances with their surroundings. In complex animals, a closed circulatory system allows rapid movement of substances to and from all living cells. Most circulatory systems consist of a heart, blood, and blood vessels, which are supplemented by a lymphatic system.

**2.** The human body has two circuits for blood flow. In the pulmonary circuit, the heart pumps oxygen-poor blood to the lungs (where it picks up oxygen), then the oxygen-enriched blood flows back to the heart. In the systemic circuit, the heart pumps the oxygen-enriched blood to all body regions. After giving up oxygen in those regions, the blood flows back to the heart. Carbon dioxide, plasma proteins, vitamins, hormones, lipids, and other solutes also make the circuits.

**3.** Arteries and veins are large-diameter transport tubes. Capillaries and venules are fine-diameter tubes for diffusion. Arterioles, with adjustable diameters, serve as control points for the distribution of different volumes of blood flow to different regions of the body. Metabolically active regions get more of the total volume at a given time.

## CIRCULATORY SYSTEM: AN OVERVIEW

Imagine what would happen if an earthquake or flood closed off the highways around your neighborhood. Grocery trucks couldn't enter and waste-disposal trucks couldn't leave—so food supplies would dwindle and garbage would pile up. Every living cell in your body would face similar predicaments if your body's highways were disrupted. Those highways are part of the **circulatory system**, which functions in the rapid internal transport of substances to and from cells. Together with other important organ systems in the body, the circulatory system helps maintain favorable neighborhood conditions, so to speak.

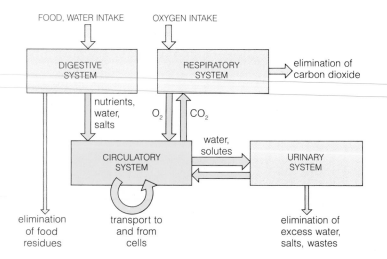

**Figure 38.2** The central role of the circulatory system in transporting substances to and from the body's living cells. Together with the other systems shown, it helps maintain favorable operating conditions in the internal environment.

Throughout this unit of the book, we've seen that differentiated cells abound in complex animals. Differentiated cells perform specific tasks with great efficiency, but not without cost. Such cells cannot really fend for themselves. They have few mechanisms for adjusting to drastic changes in the composition, volume, and temperature of the tissue fluid surrounding them, the *interstitial fluid*. To stay alive, they depend on circulating blood to maintain stable operating conditions in their internal environment. Figure 38.2 diagrams the central role of the circulatory system in this task.

### Components of the System

In most invertebrates and all vertebrates, the circulatory system has these components:

| | |
|---|---|
| **blood** | *a fluid connective tissue composed of water, solutes, and formed elements (blood cells and platelets)* |
| **heart** | *a muscular pump that generates the pressure required to help keep blood flowing through the body* |
| **blood vessels** | *tubes of different diameters through which blood is transported (for example, arteries and fine capillaries)* |

Circulatory systems can be open or closed. Arthropods and most mollusks have an open system, with blood (or bloodlike fluid) pumped from the heart into tubes that dump it into a space or cavity in body tissues. There, blood mingles with tissue fluids. The fluid has nowhere to go except through open-ended tubes

leading back to the heart, which pumps it out again (Figure 38.3a).

Most animals have a closed system, in which the walls of the heart and blood vessels are continuously connected. Before considering the components of this system, think about its overall "design." The heart constantly pumps the blood it receives, so the volume of flow through the entire system is equal to the blood returned to the heart. *Yet the rate and volume of flow through individual blood vessels must be adjusted along the route.* Blood flows rapidly through the large-diameter vessels, but it must be slowed somewhere in the system to allow enough time for substances to diffuse to and from cells. As you will see, blood flow is rather leisurely in *capillary beds*, where it is funneled through vast numbers of small-diameter tubes. As a result, there is sufficient time for diffusion (Figure 38.3b).

### Functional Links with the Lymphatic System

Whether open or closed, a circulatory system unfortunately is an ideal highway by which bacteria and other agents of disease can spread through the body. Fortunately, the body can detour the invaders through supplementary highways that are part of the **lymphatic system**. This system consists of a network of tubes (lymph vessels) as well as structures and organs that house vast numbers of infection-fighting cells. The fluid within the system is called *lymph*.

The lymphatic system picks up excess fluid, reclaimable solutes, *and disease agents* from interstitial fluid—and runs them past its armies of infection-fighting cells. The system then returns the cleansed fluid to the circulatory system.

The lymphatic system works to prevent the spread of disease agents through the circulatory system. It also reclaims water and solutes from interstitial fluid.

## CHARACTERISTICS OF BLOOD

### Functions of Blood

Blood is classified as a connective tissue, and it serves multiple functions. It carries oxygen as well as nutrients to cells, and it carries away secretions (including hormones) and metabolic wastes. Phagocytic cells travel the blood highways as mobile scavengers and infection fighters. Blood helps stabilize internal pH. In birds and mammals, blood also helps equalize body temperature

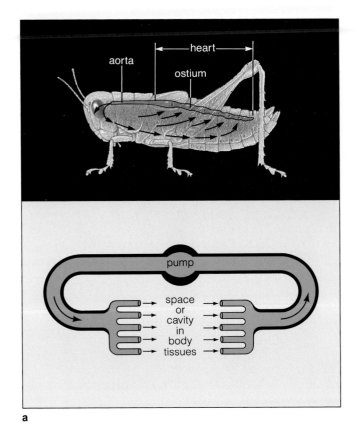

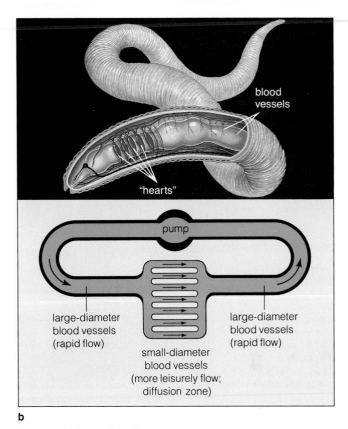

**a**

**b**

**Figure 38.3** (**a**) Fluid flow through an open circulatory system. A grasshopper, for example, has a "heart" that pumps blood through a vessel (aorta), which dumps the blood into body tissues. Blood diffuses through the tissues, then back into the heart through openings (ostia). (**b**) Fluid flow through a closed circulatory system. An example is the earthworm, with blood vessels leading away from and back to several muscular "hearts" near its head end. The walls of the hearts and blood vessels interconnect.

by carrying excess heat from regions of high metabolic activity (such as skeletal muscles) to the skin, where the heat can be dissipated.

---

In most animals, blood is a circulating fluid that carries raw materials to cells, carries products and wastes from them, and helps maintain an internal environment that is favorable for cell activities.

---

### Blood Volume and Composition

The volume of blood in a human individual depends on body size and on the concentrations of water and solutes. For average-sized adults, blood volume generally is about 6 to 8 percent of the body weight. That amounts to about 4 to 5 quarts.

Blood is a rather sticky, viscous fluid, thicker than water and more slow-flowing. As for all vertebrates, human blood consists of plasma, red blood cells, white blood cells, and platelets. When you place a sample of blood in a test tube and prevent it from clotting, it separates into a layer of straw-colored liquid (the plasma) that floats over the red-colored cellular portion of blood. Normally, the plasma portion accounts for 50 to 60 percent of the total blood volume.

**Plasma.** The portion of blood called plasma is mostly water. Besides serving as a transport medium for the cellular components of blood, the water functions as a solvent for various molecules. Among these molecules are hundreds of different *plasma proteins*, including albumins, globulins, and fibrinogen. The concentration of plasma proteins influences the distribution of water between blood and interstitial fluid. Therefore, it also influences blood's fluid volume. Albumin is important in this water-balancing act, for it represents as much as 60 percent of the total amount of plasma proteins. Some alpha and beta globulins transport lipids and fat-soluble vitamins. As you will see later in this chapter and the next, gamma globulins function in immune responses, and fibrinogen serves in blood clotting.

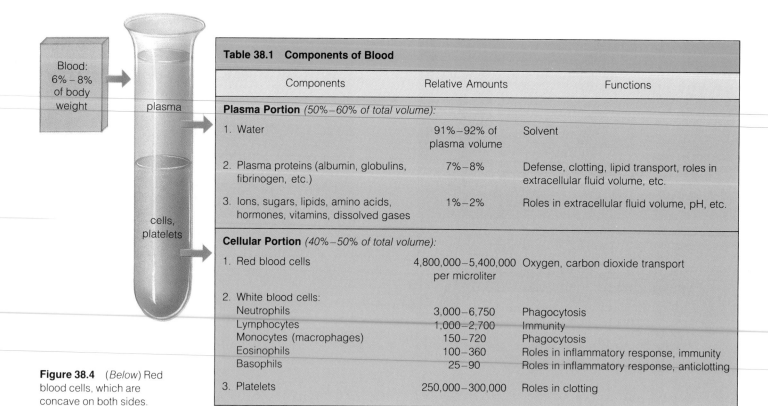

**Blood: 6%–8% of body weight**

plasma

cells, platelets

**Table 38.1 Components of Blood**

| Components | Relative Amounts | Functions |
|---|---|---|
| **Plasma Portion** *(50%–60% of total volume):* | | |
| 1. Water | 91%–92% of plasma volume | Solvent |
| 2. Plasma proteins (albumin, globulins, fibrinogen, etc.) | 7%–8% | Defense, clotting, lipid transport, roles in extracellular fluid volume, etc. |
| 3. Ions, sugars, lipids, amino acids, hormones, vitamins, dissolved gases | 1%–2% | Roles in extracellular fluid volume, pH, etc. |
| **Cellular Portion** *(40%–50% of total volume):* | | |
| 1. Red blood cells | 4,800,000–5,400,000 per microliter | Oxygen, carbon dioxide transport |
| 2. White blood cells: | | |
| Neutrophils | 3,000–6,750 | Phagocytosis |
| Lymphocytes | 1,000–2,700 | Immunity |
| Monocytes (macrophages) | 150–720 | Phagocytosis |
| Eosinophils | 100–360 | Roles in inflammatory response, immunity |
| Basophils | 25–90 | Roles in inflammatory response, anticlotting |
| 3. Platelets | 250,000–300,000 | Roles in clotting |

**Figure 38.4** (*Below*) Red blood cells, which are concave on both sides.

8 μm average diameter

a

red blood cell

capillary

b

10 μm

Plasma also contains ions, glucose and other simple sugars, lipids, amino acids, vitamins, hormones, and dissolved gases—mostly oxygen, carbon dioxide, and nitrogen (Table 38.1). The ions help maintain extracellular pH and fluid volume. The lipids include fats, phospholipids, and cholesterol. Lipids being transported from the liver to different regions generally are bound with proteins, forming lipoproteins.

**Red Blood Cells**. Erythrocytes, or **red blood cells**, transport oxygen to cells. Your own red blood cells are shaped like doughnuts without the hole (Figure 38.4). Their red color comes from hemoglobin, the iron-containing protein described on page 44. When oxygen from your lungs diffuses into the bloodstream, it binds with hemoglobin. Oxygenated blood is bright red. Blood somewhat depleted of oxygen is darker red but appears blue when observed through blood vessel walls. (That is why veins close to the body surface, as near our wrists, look "blue.") Hemoglobin also transports some of the carbon dioxide wastes of aerobic metabolism.

As Figure 38.5 shows, red blood cells form in red bone marrow. Like white blood cells and platelets, they are derived from stem cells. (By definition, *stem cells* retain the capacity to divide and give rise to different populations of cells.) Mature red blood cells no longer have their nucleus, but they also no longer require its protein-synthesizing instructions. They have enough enzymes and other proteins to remain functional for about 120 days.

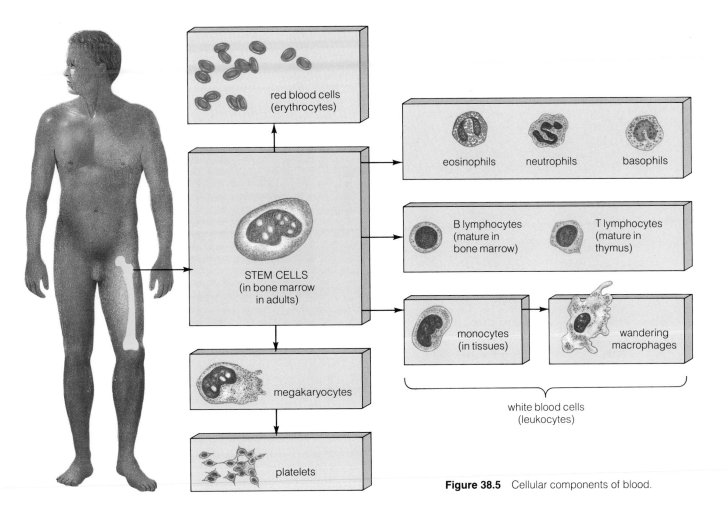

**Figure 38.5** Cellular components of blood.

Phagocytic cells continually remove the oldest red blood cells from the bloodstream, but ongoing replacements keep the red blood cell count fairly stable. A *cell count* is the number of cells of a given type in a microliter of blood. The red blood cell count is 5.4 million in males and 4.8 million in females, on the average.

Feedback mechanisms help stabilize the red blood cell count. Suppose you have just started vacationing in the Swiss Alps. Because air is "thinner" at high altitudes, your body must work harder to obtain the required oxygen. First your kidneys secrete an enzyme that converts a plasma protein into a hormone (erythropoietin). Then the hormone stimulates an increase in red blood cell production in red bone marrow. New oxygen-carrying red blood cells enter the bloodstream, and within a few days there is a rise in the oxygen level in your tissues. Information about the increase is fed back to the kidneys, production of the hormone dwindles, and the production of red blood cells drops accordingly.

**White Blood Cells.** Leukocytes, or **white blood cells**, function in day-to-day housekeeping and defense. These are the cells that scavenge dead or worn-out cells in the body and that respond to tissue damage or invasion by bacteria, viruses, and other foreign agents. All white blood cells arise from stem cells in bone marrow (Figure 38.5). They travel the circulation highways, but they perform most housekeeping and defense functions after they squeeze out of blood capillaries and enter tissues.

There are five types of white blood cells, based on differences in size, nuclear shape, and staining traits. They are lymphocytes, monocytes, neutrophils, eosinophils, and basophils (Table 38.1 and Figure 38.5). Two major classes of lymphocytes, the "B cells" and "T cells," are central to immune responses, which are described in the next chapter.

Neutrophils and mature monocytes are "search-and-destroy" cells of the body's immune system. Monocytes follow chemical trails to inflamed tissues. There, they differentiate into wandering macrophages ("big eaters") that engulf invaders and cellular debris.

White blood cell counts vary, depending on whether the body is highly active, healthy, or under siege. Table 38.1 shows the general range of counts for each type of white blood cell. Their vast numbers and continual replacement testify to the fact that the human body is an inviting environment for a great variety of bacteria, viruses, fungi, and protozoans. At any moment, some of those foreign agents can penetrate the body by way of the mouth, nose, skin pores, or injuries. A phago-

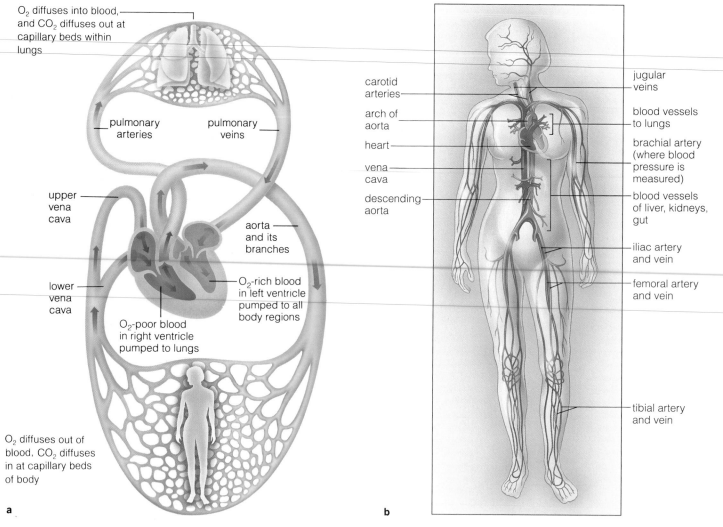

O₂ diffuses into blood, and CO₂ diffuses out at capillary beds within lungs

pulmonary arteries

pulmonary veins

upper vena cava

aorta and its branches

lower vena cava

O₂-rich blood in left ventricle pumped to all body regions

O₂-poor blood in right ventricle pumped to lungs

O₂ diffuses out of blood, CO₂ diffuses in at capillary beds of body

**a**

carotid arteries

arch of aorta

heart

vena cava

descending aorta

jugular veins

blood vessels to lungs

brachial artery (where blood pressure is measured)

blood vessels of liver, kidneys, gut

iliac artery and vein

femoral artery and vein

tibial artery and vein

**b**

**Figure 38.6** (**a**) Diagram of the pulmonary and systemic circuits. The right half of the heart pumps blood through the pulmonary circuit; the left half, through the systemic circuit. (**b**) Human circulatory system, showing the locations of the heart and some major blood vessels.

cytic white blood cell may engulf so much foreign material that its own metabolism becomes disrupted. Life for most white blood cells is challenging and short, typically measured in days or, during a major battle, a few hours.

**Platelets**. In bone marrow, some stem cells develop into "giant" cells (megakaryocytes). Those cells shed fragments of cytoplasm, which become enclosed in a bit of their plasma membrane. The fragments, called **platelets**, are oval or rounded disks about 2 to 4 micrometers across. They have no nucleus.

Each platelet lasts only five to nine days, but hundreds of thousands are always circulating in blood. They

are essential for preventing blood loss from slightly damaged blood vessels. Substances released from platelets initiate a chain of reactions leading to blood clotting, as will be described shortly.

## CARDIOVASCULAR SYSTEM OF VERTEBRATES

"Cardiovascular" comes from the Greek *kardia*, meaning heart, and the Latin *vasculum*, meaning vessel. In the cardiovascular system of all vertebrates, a heart pumps blood into large-diameter arteries. From there the blood flows into small, muscular arterioles, which branch into tiny capillaries. Blood flows from capillaries into small venules, then into large veins. The veins return blood to the heart. Figure 26.00 is a general picture of the system in fishes, amphibians, birds, and mammals. Figure 38.6 shows where some of its major components are located in the human body.

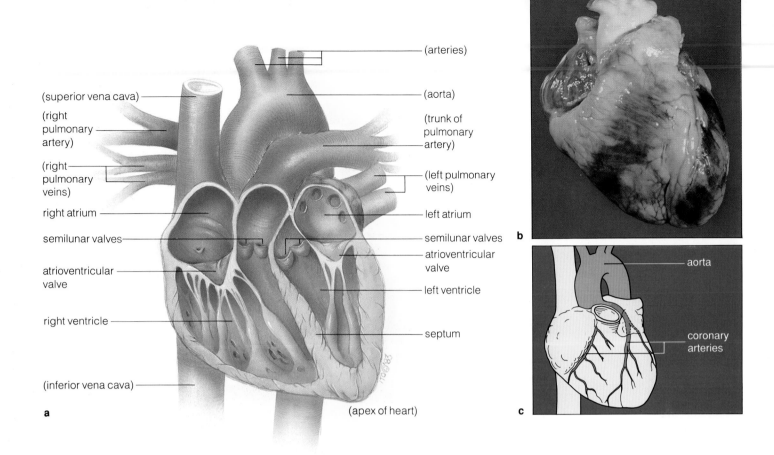

(arteries)

(superior vena cava)

(aorta)

(right pulmonary artery)

(trunk of pulmonary artery)

(right pulmonary veins)

(left pulmonary veins)

right atrium

left atrium

semilunar valves

semilunar valves

atrioventricular valve

right ventricle

atrioventricular valve

left ventricle

septum

(inferior vena cava)

a

(apex of heart)

b

aorta

coronary arteries

c

**Figure 38.7** The human heart, partial view of the interior (**a**) and external view (**b**). Location of the coronary arteries (**c**).

### Blood Circulation Routes in Humans

The human heart is divided into two halves, which are the basis of two cardiovascular circuits through the body. These are the pulmonary and systemic circuits.

In the **pulmonary circuit**, blood from the right half of the heart is pumped to the lungs, where it picks up oxygen and gives up carbon dioxide. From the lungs, the freshly oxygenated blood flows to the left half of the heart. In the **systemic circuit**, the oxygen-enriched blood is pumped through the rest of the body. After passing through tissue regions (where oxygen is used and carbon dioxide is produced), the blood flows to the right half of the heart. In both circuits, blood travels through arteries, arterioles, capillaries, venules, and finally veins.

As Figure 38.6a suggests, a given volume of blood making either circuit generally passes through only one capillary bed. A notable exception is the blood passing through capillary beds in the digestive tract, where it picks up glucose and other substances absorbed from food. That blood moves on through another capillary bed, in the liver—an organ with a key role in nutrition. The slow flow of blood through this second bed gives the liver time to metabolize absorbed substances.

### The Human Heart

**Heart Structure.** During a seventy-year life span, the human heart beats some 2.5 billion times, and it rests only briefly between heartbeats. The heart's structure reflects its role as a durable pump. The heart is mostly cardiac muscle tissue surrounded and protected by a tough, fibrous sac (pericardium). Its inner chambers have a smooth lining (endocardium) composed of connective tissue and a single layer of epithelial cells. The epithelial cell layer is known as *endothelium*. It lines the inside of blood vessels as well as the heart.

Each half of the heart has two chambers—an **atrium** (plural, atria) and a **ventricle**. The flaps of membrane separating the two chambers serve as a one-way valve. As Figure 38.7 shows, the flaps are called an *AV valve* (short for "atrioventricular"). Each half of the heart also has a *semilunar valve*, located between the ventricle and the arteries leading away from it. Each time the heart beats, its valves open and close in ways that prevent backflow and keep blood moving in one direction.

Most cardiac muscle cells are not serviced by the blood moving inside the heart's chambers. The heart has its own "coronary circulation," with two main arteries leading into a large capillary bed (Figure 38.7c). Coronary arteries are the first to branch off the *aorta*, the major artery carrying oxygen-enriched blood away from the heart. Given their small diameter, coronary arteries can become clogged during some cardiovascular disorders (see the *Commentary* on page 671).

**Cardiac Cycle.** Each time the heart beats, its four chambers go through phases of contraction (systole) and relaxation (diastole). This sequence of muscle contraction and relaxation is a **cardiac cycle** (Figure 38.8).

When the relaxed atria are filling, fluid pressure inside them increases and forces the AV valves to open. Blood flows into the ventricles, which become completely filled when the atria contract. As the filled ventricles begin to contract, fluid pressure inside them increases,

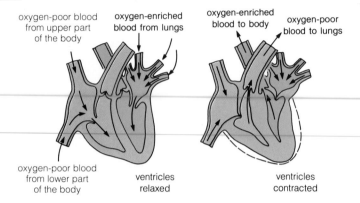

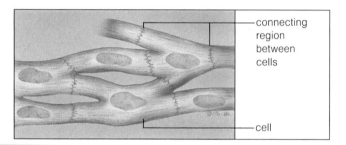

**Figure 38.8** Blood flow through the heart during part of a cardiac cycle.

**Figure 38.9** End-to-end regions between cardiac muscle cells. Communication junctions at these regions permit rapid signaling between cells that causes them to contract nearly in unison.

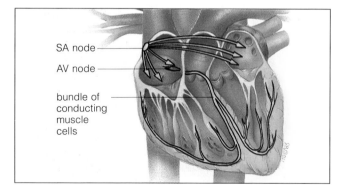

**Figure 38.10** Location of cardiac muscle cells that conduct signals for contraction through the heart.

and the AV valves snap shut. As the ventricles continue contracting, ventricular pressure rises sharply above that in blood vessels leading away from the heart. The semilunar valves are forced open by the increased pressure and blood flows out of the heart, into the aorta and pulmonary artery. After blood has been ejected, the ventricles relax, the semilunar valves close—and the already filling atria are ready to repeat the cycle.

Look back at the recording of Jimmie the bulldog's heartbeat (Figure 38.1). The atria of his heart were contracting between times P and Q. The ventricles began contracting forcefully through time QRS and were recovering from contraction at time T.

The blood and heart movements during the cardiac cycle generate vibrations that produce a "lub-dup" sound. You can hear the sound through a stethoscope positioned against the chest wall. At each "lub," the AV valves are closing as the ventricles contract. At each "dup" the semilunar valves are closing as the ventricles relax.

**During a cardiac cycle, atrial contraction simply helps fill the ventricles. *Ventricular contraction* is the driving force for blood circulation.**

**Mechanisms of Contraction.** In skeletal muscle tissue, the ends of individual cells are attached to bones. In cardiac muscle tissue, however, the cells branch and then connect with one another at their endings (Figure 38.9). Communication junctions occur where the plasma membranes of abutting cells are joined together. With each heartbeat, signals calling for contraction spread so rapidly across the junctions that cardiac muscle cells contract together as if they were a single unit.

Cardiac and skeletal muscle cells differ in another way. Signals from the nervous system bring about skeletal muscle contraction. But the nervous system can only *adjust* the rate and strength of cardiac muscle contraction. Even if all nerves leading to the heart are severed, the heart will keep on beating! How? Some cardiac muscle cells are self-excitatory; they produce and conduct the action potentials that initiate contraction. These cells are the basis of the **cardiac conduction system** shown in Figure 38.10.

Excitation begins with conducting cells in the *SA node* (short for "sinoatrial"), where major veins enter the right atrium. The SA node generates one wave of excitation after another, usually seventy or eighty times a minute. Each wave spreads over both atria, causes them to contract, then reaches the *AV node* (again, for "atrioventricular"). The wave spreads more slowly here. The delay gives the atria time to finish contracting before the wave of excitation spreads over the ventricles.

Although all cells of the system are self-excitatory, the SA node fires off action potentials at the highest frequency and comes to threshold first in each cardiac cycle. Thus, the SA node is the *cardiac pacemaker*, mentioned at the start of this chapter. Its rhythmic firing is the basis for the normal rate of heartbeat.

The SA node is the cardiac pacemaker. Its spontaneous, repetitive excitation spreads along a system of muscle cells that stimulate contractile tissue in the atria, then the ventricles, in a rhythmic cycle.

### Blood Pressure in the Vascular System

**Blood pressure**, the fluid pressure generated by heart contractions, is not the same throughout the circulatory system. Pressure normally is high to begin with in the aorta, then drops along the circuit away from and back to the heart. The pressure drops result from the loss of energy used to overcome resistance to flow as blood moves through blood vessels of the sort shown in Figure 38.11.

**Arterial Blood Pressure.** As we have seen, **arteries** conduct oxygen-poor blood into the lungs and oxygen-enriched blood to all body tissues. Arteries are pressure reservoirs that can "smooth out" the pulsations in blood pressure that are generated during each cardiac cycle. The thick, muscular wall of arteries bulges somewhat under the pressure surge caused by ventricular contraction, then the wall recoils and forces blood onward. With their relatively large diameters, arteries present little resistance to flow, so pressure does not drop much in the arterial portion of the blood circuits. Figure 38.12 shows how blood pressure is measured at large arteries of the upper arms.

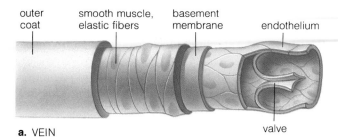

**a.** VEIN

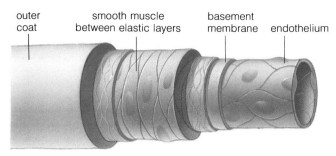

**b.** ARTERY

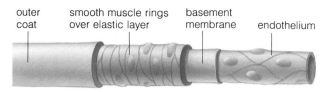

**c.** ARTERIOLE

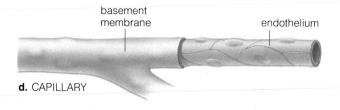

**d.** CAPILLARY

**Figure 38.11** Structure of blood vessels. The basement membrane is collagen-containing connective tissue.

**Figure 38.12** Measuring blood pressure with a device called a sphygmomanometer. A hollow cuff, attached to a pressure gauge, is wrapped around the upper arm and inflated with air to a pressure above the highest pressure of the cardiac cycle (at systole, when the ventricles contract). Above the systolic pressure, no sounds can be heard through a stethoscope positioned above the artery (because no blood is flowing through it).

Air in the cuff is slowly released, allowing some blood to flow into the artery. The turbulent flow causes soft tapping sounds, and when this first occurs, the value on the gauge is the systolic pressure—about 120mm Hg in young adults at rest. (This means the measured pressure would make a column of mercury rise a distance of 120 millimeters.)

More air is released until the sounds become dull and muffled. Just after this occurrence, blood flow is continuous; the turbulence and tapping sounds stop. The silence corresponds to the diastolic pressure (at the end of a cardiac cycle, just before the heart pumps out blood again). Generally the reading is about 80mm Hg. In this example, the *pulse pressure* (the difference between the highest and lowest pressure readings) is 120 − 80, or 40mm Hg.

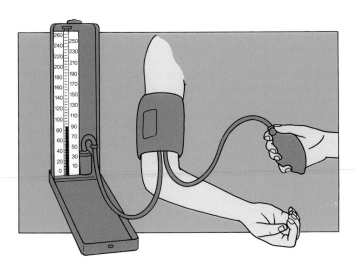

**Resistance at Arterioles.** Arteries branch into smaller diameter **arterioles**. By tracking the flow of blood along the systemic circuit, you can see that the greatest pressure drop occurs at arterioles (Figure 38.13). This indicates that arterioles offer the greatest resistance to blood flow in the circulation. By analogy, suppose you turn two open bottles of ketchup upside down. Both hold the same amount of ketchup, but one bottle has a very wide neck and the other has a narrow neck. Guess which bottle will be slowest to drain—that is, which will present more resistance to flow.

The diameter of arterioles can be made to increase or decrease by mechanisms that cause smooth muscle in their wall to contract or relax. As you will see shortly, adjustments are made in response to signals from the nervous and endocrine systems, as well as to changes in local chemical conditions. Blood is directed to a region of great metabolic activity when the diameters of arterioles in those regions enlarge. Blood is directed away from less active regions when the diameters of arterioles in those regions constrict. Thus, the more active the cells of a given region, the greater the blood flow to them.

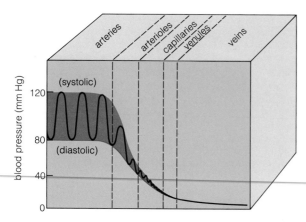

**Figure 38.13** Drops in blood pressure in the systemic circulation.

**Capillary Function.** Capillary beds are *diffusion zones* for exchanges between blood and interstitial fluid. Except for venules, all blood vessels are transport tubes. A **capillary** has the thinnest wall of any blood vessel. Its wall consists of a layer of flat endothelial cells, separated from one another by narrow clefts. The layer rests on a basement membrane (Figure 38.11).

Capillaries have such a small diameter, red blood cells must squeeze through them single file (Figure 38.3). Each capillary presents high resistance to blood flow. Yet there are so many of them in a capillary bed, their combined diameters are greater than the diameters of arterioles leading into them. Hence they present less *total* resistance to flow than arterioles. As a result, the drop in blood pressure here is not as great.

Capillaries thread through nearly every tissue in the body, coming within 0.01 centimeter of every living cell. Most solutes, including oxygen and carbon dioxide, move across the capillary wall by diffusion. Some proteins cross it by endocytosis or exocytosis (page 86), and certain ions probably pass through the clefts between endothelial cells. The clefts are wider in some capillary beds than in others, and those beds are functionally more "leaky" to solutes.

Finally, fluid also moves across the wall by bulk flow, with its water molecules being forced under pressure to move in the same direction. Such movements help maintain the proper fluid balance between the bloodstream and the surrounding tissues. This fluid distribution is important because blood pressure is maintained only when there is an adequate blood volume. Interstitial fluid is a reservoir that is tapped when blood volume drops to the point where there is a decrease in blood pressure, as during hemorrhage.

Figure 38.14 describes the two opposing forces (filtration and absorption) that bring about the fluid movements in capillary beds.

Arteries are pressure reservoirs that keep blood flowing away from the heart while the ventricles are relaxing. Their large diameters offer low resistance to flow, so there is little drop in blood pressure in arteries.

Arterioles are control points where adjustments can be made in the volume of blood flow to be delivered to different capillary beds. They offer great resistance to flow, so there is a major drop in pressure in arterioles.

Capillary beds are diffusion zones for exchanges between blood and interstitial fluid. Collectively, they have a greater cross-sectional area than that of arterioles leading into the beds, so they present less total resistance to flow. There is some drop in pressure here.

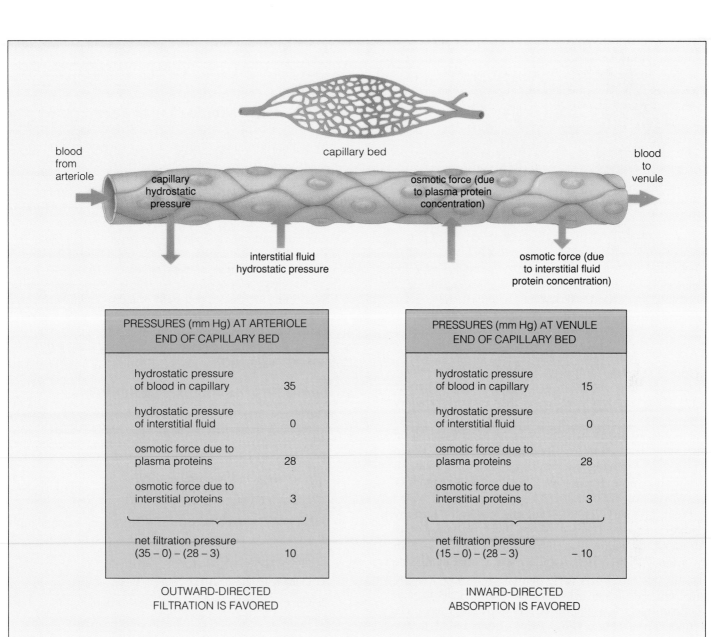

| PRESSURES (mm Hg) AT ARTERIOLE END OF CAPILLARY BED | | PRESSURES (mm Hg) AT VENULE END OF CAPILLARY BED | |
| --- | --- | --- | --- |
| hydrostatic pressure of blood in capillary | 35 | hydrostatic pressure of blood in capillary | 15 |
| hydrostatic pressure of interstitial fluid | 0 | hydrostatic pressure of interstitial fluid | 0 |
| osmotic force due to plasma proteins | 28 | osmotic force due to plasma proteins | 28 |
| osmotic force due to interstitial proteins | 3 | osmotic force due to interstitial proteins | 3 |
| net filtration pressure (35 – 0) – (28 – 3) | 10 | net filtration pressure (15 – 0) – (28 – 3) | – 10 |

OUTWARD-DIRECTED
FILTRATION IS FAVORED

INWARD-DIRECTED
ABSORPTION IS FAVORED

**Figure 38.14** Fluid movements in an idealized capillary bed. The movements play no significant role in diffusion. But they are important in maintaining the distribution of extracellular fluid between the bloodstream and interstitial fluid. The movements result from two opposing forces, called *filtration* and *absorption*.

At the arteriole end of a capillary, the difference between capillary blood pressure and interstitial fluid pressure leads to filtration. Because of the difference, some plasma (but very few plasma proteins) leaves the capillary. "Filtration" refers to this fluid movement out of the capillary.

In contrast, "absorption" refers to the movement of some interstitial fluid into the capillary. The difference in water concentration between plasma and interstitial fluid brings it about. Plasma has a greater solute concentration, with its protein components, and therefore a lower water concentration.

Fluid filtration at the arteriole end of a capillary bed tends to be balanced by absorption at the venule end. Normally, there is only a small *net* filtration of fluid, which the lymphatic system returns to the blood.

*Edema* is a condition in which excess fluid accumulates in interstitial spaces. This happens to some extent during exercise. As arterioles dilate in local tissue regions, capillary pressure increases and triggers increased filtration. Edema also results from an obstructed vein or from heart failure. Here again, capillary pressure and then filtration increase. Edema reaches its extreme during elephantiasis, which is brought on by a roundworm infection and subsequent obstruction of lymphatic vessels (page 417).

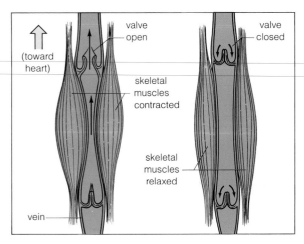

**Figure 38.15** Role of skeletal muscle contractions and venous valves in returning blood to the heart.

**Venous Pressure.** Capillaries merge into "little veins," or **venules**. Some diffusion also occurs across the venule wall, which is only a little thicker than that of a capillary (Figure 38.11). Venules merge into large-diameter **veins**, the transport tubes leading back to the heart. Blood movement is assisted by valves inside the veins. When blood starts moving backward because of gravity, it pushes the valves into a closed position. In this manner the valves prevent backflow.

Veins also serve as blood volume reservoirs, for they contain 50 to 60 percent of the total blood volume. Although the vein wall is thin and can bulge more than an arterial wall, it does contain some smooth muscle (Figure 38.11). When body activities increase, blood must circulate faster. The smooth muscle cells in vein walls contract; the walls stiffen and don't bulge as much. Venous pressure rises and drives more blood to the heart. The increased blood volume returned to the heart is now ejected and flows to active tissue regions.

Venous pressure rises when limbs move and skeletal muscles bulge against adjacent veins (Figure 38.15). It also is influenced by how rapidly you are breathing. When inhaled air pushes down on internal organs, it changes the pressure gradient between the heart and veins.

Venules overlap with capillaries in function; they afford some control over capillary pressure.

Veins are highly distensible blood volume reservoirs and help adjust flow volume back to the heart. They offer only low resistance to flow, so there is little drop in blood pressure here.

## On Cardiovascular Disorders

More than 40 million Americans have cardiovascular disorders, which claim about 750,000 lives every year. The most common cardiovascular disorders are *hypertension* (sustained high blood pressure) and *atherosclerosis* (a progressive narrowing of the arterial lumen). They are the major causes of most *heart attacks*—that is, the damage or death of heart muscle due to an interruption of its blood supply. (They also can cause *stroke*, or damage to the brain due to an interruption of blood circulation to it.)

Most heart attacks bring a "crushing" pain behind the breastbone that lasts a half-hour or more. Frequently, the pain radiates into the left arm, shoulder, or neck. The pain can be mild but usually is excruciating. Often it is accompanied by sweating, nausea, vomiting, and dizziness or loss of consciousness.

### Risk Factors in Cardiovascular Disorders

Cardiovascular disorders are the leading cause of death in the United States. Curiously, many factors associated with those disorders have been identified *and are controllable*. These are the known risk factors:

1. High level of cholesterol in the blood
2. High blood pressure
3. Obesity (page 649)
4. Lack of regular exercise
5. Smoking (page 710)
6. Diabetes mellitus (page 590)
7. Genetic predisposition to heart disorders
8. Age (the older you get, the greater the risk)
9. Gender (until age fifty, males are at much greater risk than are females)

The last four factors obviously cannot be avoided; but in most people, the first five can be. The risk associated with all five can be minimized simply by watching your diet, exercising, and not smoking.

For example, the fatter you become, the more your body develops additional blood capillaries to service the increased number of cells, and the harder the heart has to work to pump blood through the increasingly divided vascular circuit. As another example, the nicotine in tobacco stimulates the adrenal glands to secrete epinephrine, which constricts blood vessels and so triggers an accelerated heartbeat and a rise in blood pressure. The carbon monoxide present in cigarette smoke has a greater affinity for binding sites on hemoglobin than does carbon dioxide—and its action means that the heart has to pump harder to rid the body of carbon dioxide wastes. In short, smoking can destroy not only your lungs but also your heart.

Some examples of the tissue destruction resulting from cardiovascular disorders will now be described.

## Hypertension

Hypertension arises through a gradual increase in resistance to blood flow through the small arteries; eventually, blood pressure is sustained at elevated levels even when the person is at rest. Heredity may be a factor here (the disorder tends to run in families). Diet also is a factor; for example, high salt intake can raise the blood pressure in persons predisposed to the disorder. High blood pressure increases the workload of the heart, which in time can become enlarged and fail to pump blood effectively. High blood pressure also can cause arterial walls to "harden" and so influence the delivery of oxygen to the brain, heart, and other vital organs.

Hypertension has been called the silent killer because affected persons may show no outward symptoms; they often believe they are in the best of health. Even when their high blood pressure has been detected, some hypertensive persons tend to resist medication, corrective changes in diet, and regular exercise. Of 23 million Americans who are hypertensive, most are not undergoing treatment. About 180,000 will die each year.

## Atherosclerosis

"Arteriosclerosis" refers to a condition in which arteries thicken and lose their elasticity. In atherosclerosis, conditions worsen because lipid deposits also build up in the arterial walls and shrink the diameter of the arterial lumen. How does this occur?

Recall that lipids such as fats and cholesterol are insoluble in water (page 40). Lipids absorbed from the digestive tract are picked up by lymph vessels that empty into the bloodstream. There, the lipids become bound to protein carriers that keep them suspended in the blood plasma. In atherosclerosis, abnormal smooth muscle cells have multiplied and connective tissue components have increased in arterial walls. Lipids have been deposited within cells and extracellular spaces of the wall's endothelial lining. Calcium salts have been deposited on top of the lipids, and a fibrous net has formed over the whole mass. This *atherosclerotic plaque* sticks out into the lumen of the artery (Figures *a,b*).

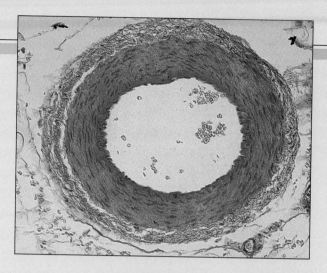

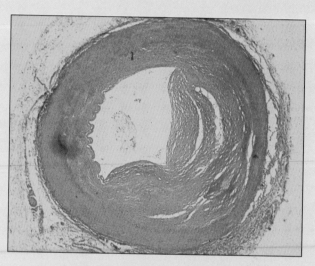

**a** Cross-section of a normal artery (above) and a partially obstructed one (below).

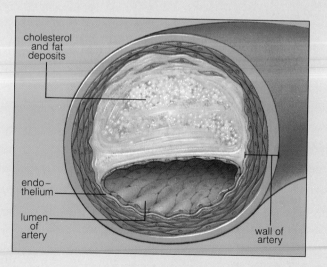

**b** Diagram of an atherosclerotic plaque.

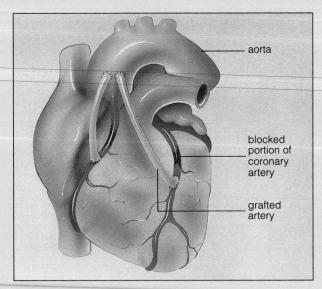

c Two coronary bypasses (green).

With their narrow diameter, the coronary arteries and their branches (Figure c) are extremely susceptible to clogging through plaque formation or occlusion by a clot. When such an artery becomes narrowed to one-quarter of its former diameter, the resulting symptoms can range from mild chest pain (angina pectoris) to a full-scale heart attack.

Atherosclerosis can be diagnosed on the basis of several procedures. These include stress electrocardiograms, or EKGs (recording the electrical activity of the cardiac cycle while a person is exercising on a treadmill) and *angiography* (injecting a dye that will stain plaques and then taking x-rays of the arteries; see page 42). Treatments of serious blockages include *coronary bypass surgery*. During this operation, a section of a vein taken from the arm or leg is stitched to the aorta and to the coronary artery below the narrowed or blocked region (Figure c). In another technique, called *laser angioplasty*, highly focused laser beams are used to vaporize the atherosclerotic plaques. *Balloon angioplasty* is more common. Here, a small balloon is inflated within a blocked artery to flatten a plaque and thereby increase the arterial diameter. All such procedures do not cure the underlying cardiovascular problem. They only buy time for the individual.

Plaque formation is related to cholesterol intake, but other factors are also at work here. For example, when cholesterol is transported through the bloodstream, it is bound to one of two kinds of protein carrier molecules: high-density lipoproteins (HDL) and low-density lipoproteins (LDL). High levels of LDL are related to a tendency toward heart trouble. LDLs, with their cholesterol cargo, have a penchant for infiltrating arterial walls. In contrast, the HDLs seem to attract cholesterol out of the walls and transport it to the liver, where it can be metabolized. (Atherosclerosis is uncommon in rats; rats have mostly HDLs. It is common in humans, who have mostly LDLs.) In addition, it appears that unsaturated fats, including olive oil and fish oil, can reduce the level of LDLs in the blood.

Sometimes platelets become caught on the rough edges of plaque and are stimulated into secreting some of their chemicals. When they do, they initiate clot formation. As the clot and plaque grow, the artery can become narrowed or blocked. Blood flow to the tissue that the artery supplies diminishes or may be blocked entirely. A clot that stays in place is called a *thrombus*. If it becomes dislodged and travels the bloodstream, it is called an *embolus*.

## Arrhythmia

Arrhythmias are irregular or abnormal heart rhythms. They can be detected by an EKG (Figure d). Some arrhythmias are normal. For example, the resting cardiac rate of many athletes who are trained for endurance is lower than average, a condition called *bradycardia*. Inhibition of their cardiac pacemaker by parasympathetic signals has increased as an adaptive response to ongoing strenuous exercise. A cardiac rate above 100 beats per minute (*tachycardia*) also occurs normally during exercise or stressful situations.

Serious tachycardia can be triggered by drugs (including caffeine, nicotine, and alcohol), hyperthyroidism, and other factors. Coronary artery disease also can cause arrhythmias.

A coronary occlusion and certain other disorders also may cause abnormal rhythms that can degenerate rapidly into a dangerous condition called *ventricular fibrillation*.

## Controls Over Blood Flow

**Maintaining Blood Pressure**.  Suppose you decide to measure your blood pressure every day while you are resting, in the manner shown in Figure 38.12. You will find that the resting value remains fairly constant over a few weeks, even months. Assuming you are a healthy adult, it will be somewhere around 120/80mm Hg.

This raises an interesting question. What mechanisms work to maintain the level of blood pressure over time and so ensure adequate blood flow to all regions of the body?

In the medulla oblongata of the brain are integrating centers that control blood pressure. The centers integrate information coming in from sensory receptors in cardiac muscle tissue and in certain arteries, such as the aorta and the carotid arteries in the neck. They use this information to coordinate the rate and strength of heartbeats with changes in the diameter of arterioles and, to some extent, of veins.

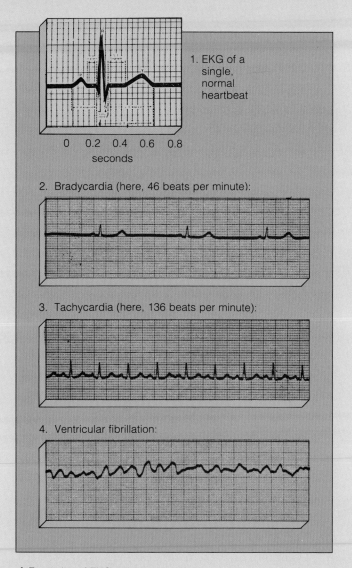

1. EKG of a single, normal heartbeat

    0    0.2   0.4   0.6   0.8
              seconds

2. Bradycardia (here, 46 beats per minute):

3. Tachycardia (here, 136 beats per minute):

4. Ventricular fibrillation:

**d** Examples of EKG readings.

Here, cardiac muscle in different parts of the ventricles contracts haphazardly, and the ventricles are unable to pump blood. Loss of consciousness occurs within a few seconds and may signify impending death. Sometimes a strong electric shock delivered to the chest can stop the fibrillation and may restore normal cardiac function.

When an *increase* in blood pressure is detected, the medulla commands the heart to beat more slowly and contract less forcefully. It also sends signals to smooth muscle cells in the wall of arterioles. The cells relax, the outcome being **vasodilation**—an enlargement (dilation) of arteriole diameter.

Conversely, when a *decrease* in blood pressure is detected, the medulla commands the heart to beat faster and contract more forcefully. It also stimulates the smooth muscle cells of arterioles to contract. In this case, the outcome is **vasoconstriction**—a decrease in arteriole diameter.

Hormones assist in maintaining blood pressure. Arterioles in various regions have different receptors that can be activated by epinephrine, angiotensin, and other hormones. (Epinephrine triggers vasoconstriction or vasodilation. Angiotensin triggers widespread vasoconstriction.)

**Integrating centers in the medulla oblongata of the brain function to keep the resting level of blood pressure fairly constant over time.**

**Control of Blood Distribution.** The nervous system, endocrine system, and changes in local chemical conditions interact to assure that blood circulation meets the metabolic demands of various tissues. Think about what happens after you have eaten a large meal. More blood is diverted to your digestive system, which swings into full gear as other systems more or less idle. When your body is exposed to cold wind or snow for an extended time, blood is diverted away from the skin to deeper tissue regions, so that the metabolically generated heat that warmed the blood in the first place can be conserved.

Chapter 33 indicated that the heart is serviced by parasympathetic nerves (which signal for decreases in heart rate) and sympathetic nerves (which signal for increases in heart rate). Adjustments in the signals traveling along those nerves lead to adjustments in blood circulation. Moreover, sympathetic nerves as well as hormones can stimulate vasoconstriction in many tissues, including the kidneys and muscle tissues that are not being called upon to contract. Thus they can cause blood to be diverted away from areas requiring less blood flow.

Finally, think about what happens when you swim or run. In your skeletal muscle tissue, the oxygen level decreases and the levels of carbon dioxide, hydrogen ions, potassium ions, and other substances increase. The changes in local chemical conditions cause smooth muscle cells of arterioles to relax, so arteriole diameter enlarges. Now more blood flows past the active muscles, delivering more raw materials and carrying away cell products and wastes. While this is occurring, arteriole diameter is decreasing in tissues of the digestive tract and kidneys.

**Adjustments in the distribution of blood flow are made in response to signals from the nervous system, endocrine system, and changes in local chemical conditions.**

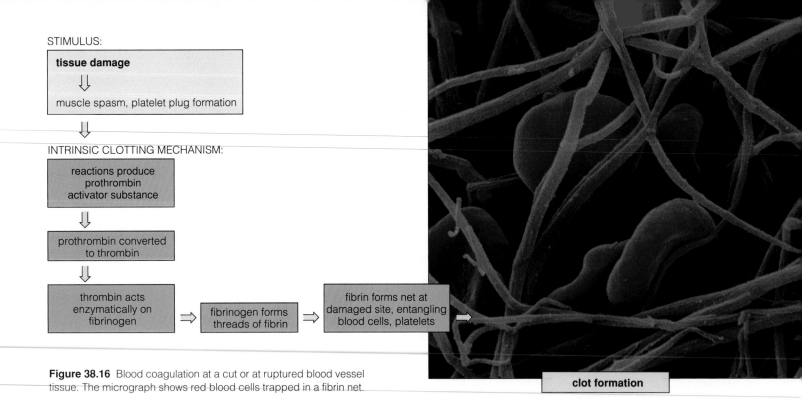

STIMULUS:

| **tissue damage** |
| ⇓ |
| muscle spasm, platelet plug formation |

⇓

INTRINSIC CLOTTING MECHANISM:

reactions produce prothrombin activator substance

⇓

prothrombin converted to thrombin

⇓

thrombin acts enzymatically on fibrinogen ⇒ fibrinogen forms threads of fibrin ⇒ fibrin forms net at damaged site, entangling blood cells, platelets ⇒

clot formation

**Figure 38.16** Blood coagulation at a cut or at ruptured blood vessel tissue. The micrograph shows red blood cells trapped in a fibrin net.

### Hemostasis

Don't even think about what would happen if the body could not repair breaks or cuts even in its small blood vessels. In **hemostasis**, blood vessel spasm, platelet plug formation, blood coagulation, and other mechanisms can stop bleeding.

First, smooth muscle in the wall of a damaged blood vessel contracts in a reflex response called a spasm. The blood vessel constricts, and the flow of blood is temporarily stopped. Second, platelets clump together, temporarily plugging the rupture. They also release substances that help prolong the spasm and attract more platelets. Third, blood *coagulates* (converts to a gel) and forms a clot. Finally, the clot retracts into a compact mass, drawing the walls of the vessel together.

Blood coagulates when damage exposes collagen fibers in blood vessel walls. The response is called the *intrinsic clotting mechanism*. A plasma protein becomes activated and triggers reactions that lead to the formation of an enzyme (thrombin), which acts on a large, rod-shaped plasma protein (fibrinogen). The rods adhere to one another, forming long, insoluble threads that stick to exposed collagen. The result is a net in which blood cells and platelets become entangled (Figure 38.16). The entire mass is a blood clot.

Blood also can coagulate through an *extrinsic clotting mechanism*. "Extrinsic" means that the series of reactions leading to blood clotting is triggered by the release of enzymes and other substances *outside* of the blood itself (that is, from damaged blood vessels or from the surrounding tissues). The substances lead to thrombin formation, and the remaining steps parallel those shown in Figure 38.16. Overall, fewer steps are involved than in the intrinsic clotting mechanism, and the reactions occur much more rapidly.

The contribution of the extrinsic clotting mechanism to hemostasis is unclear. However, it is definitely involved in walling off bacteria and preventing the spread of bacterial infection from invaded tissue regions.

### Blood Typing

All of your cells carry membrane proteins at their surface that serve as "self" markers; they identify the cells as being part of your own body. Your body also has proteins called **antibodies**, which can recognize markers on *foreign* cells (page 683). When the blood of two people mixes during transfusions, antibodies will act against any cells bearing the "wrong" marker. They can do the same thing during pregnancy, if antibodies diffuse from the mother's circulation system to that of her unborn child.

**ABO Blood Typing.** As we have seen, *ABO blood typing* is based on some of the surface markers on red blood cells (page 172). Type A blood has A markers on those cells, type B blood has B markers, type AB has both, and type O has neither one.

If you are type A, you do not carry antibodies against A markers—but you have antibodies against B markers. If you are type B, you have antibodies against A but not B markers. If you are type AB, you have no antibodies against A or B markers, so your body will tolerate donations of type A, B, or AB blood. If you are type O, however, you have antibodies against A and B mark-

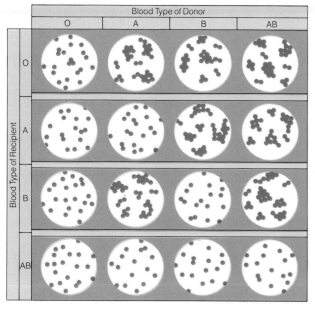

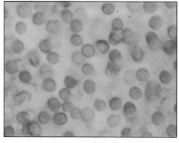

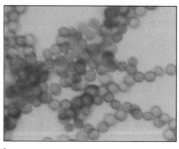

**Figure 38.17** (**a**) Agglutination responses in blood types O, A, B, and AB when mixed with blood samples of the same and different types. (**b**) Micrographs showing the absence of agglutination in a mixture of two different but compatible types (above) and agglutination in a mixture of incompatible blood types (below).

ers—and those antibodies will act against cells bearing one or both types.

Figure 38.17 shows what happens when blood from different types of donors and recipients is mixed together. In a response called **agglutination**, antibodies act against the "foreign" cells and cause them to clump. Such clumps can clog small blood vessels. They may lead to tissue damage and death. In looking at Figure 38.17, can you say what the agglutination responses will be to type AB blood? To type O blood?

**Rh Blood Typing**. Other surface markers on red blood cells also can cause agglutination responses. For example, *Rh blood typing* is based on the presence or absence of an Rh marker (so named because it was first identified in the blood of *rh*esus monkeys). Rh⁺ individuals have blood cells with this marker; Rh⁻ individuals do not. Ordinarily, people do not have antibodies that act against Rh markers. However, if someone has been given a transfusion of Rh⁺ blood, antibodies will be produced against it and will continue circulating in the bloodstream.

If an Rh⁻ female becomes pregnant by an Rh⁺ male, there is a chance the fetus will be Rh⁺. During pregnancy or childbirth, some red blood cells of the fetus may leak into the mother's bloodstream (Chapter 43). If they do, they will stimulate her body into producing antibodies against the Rh markers (Figure 38.18). If the woman becomes pregnant *again*, Rh antibodies will enter the fetal bloodstream. If this second fetus happens to have Rh⁺ blood, the antibodies will cause red blood cells to swell and then rupture, releasing hemoglobin into the bloodstream.

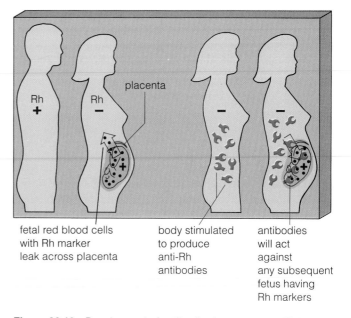

Rh +    Rh −    placenta    −    −

fetal red blood cells with Rh marker leak across placenta

body stimulated to produce anti-Rh antibodies

antibodies will act against any subsequent fetus having Rh markers

**Figure 38.18** Development of antibodies in response to Rh⁺ blood.

In extreme cases of this disorder, called *erythroblastosis fetalis*, too many cells are destroyed and the fetus dies before birth. If it is born alive, all the newborn's blood can be slowly replaced with blood free of the Rh antibodies. Currently, a known Rh⁻ female can be treated right after her first pregnancy with a drug, called Rho-Gam, that will protect her next fetus. The drug will inactivate any Rh⁺ fetal blood cells circulating in the mother's bloodstream before she can become sensitized and begin producing anti-Rh⁺ antibodies.

# LYMPHATIC SYSTEM

We conclude this chapter with a brief section on the lymphatic system, which supplements the circulatory system by returning excess tissue fluid to the bloodstream. But think of this section as a bridge to the next chapter, on immunity, for the lymphatic system is also vital to the body's defenses against injury and attack. As Figure 38.19 shows, the system's components include transport vessels and lymphoid organs. Tissue fluid that has moved into the transport vessels is called **lymph**.

## Lymph Vascular System

The **lymph vascular system** includes lymph capillaries, lymph vessels, and ducts. Collectively, these transport vessels serve the following functions:

1. Return of excess filtered fluid to the blood

2. Return of small amounts of proteins that leave the capillaries

3. Transport of fats absorbed from the digestive tract

4. Transport of foreign particles and cellular debris to disposal centers—that is, the lymph nodes

At one end of the lymph vascular system are *lymph capillaries*, no larger in diameter than blood capillaries. They occur in the tissues of almost all organs and serve as "blind-end" tubes. They have no entrance at the end located in tissues; their only "opening" merges with larger lymph vessels, as shown by Figure 38.20. Extracellular fluid simply diffuses into them through gaps in the capillary wall.

Like veins, *lymph vessels* have smooth muscle in their walls and flaplike valves that prevent backflow. When you breathe, movements of the rib cage and skeletal muscle adjacent to the lymph vessels help move fluid through lymph vessels, just as they do for veins. Lymph vessels converge into collecting ducts, which drain into veins in the lower neck. In this way, the lymph fluid is returned to the circulation.

## Lymphoid Organs

The **lymphoid organs** include the lymph nodes, spleen, thymus, tonsils, adenoids, and patches of tissue in the small intestine and appendix. These organs and tissue patches are production centers for infection-fighting cells, including lymphocytes. They also are sites for some defense responses.

Like all white blood cells, lymphocytes are derived from stem cells in bone marrow. The derivative cells enter the blood and take up residence in lymphoid organs. With proper stimulation, they divide by mitosis. In fact, most new lymphocytes are produced by divisions in the blood and lymphoid organs, not in bone marrow.

*Lymph nodes* are located at intervals along lymph vessels, as suggested by Figure 38.19. All lymph trickles through at least one node before being delivered to the bloodstream. Each node has several inner chambers. Lymphocytes and plasma cells (the progeny of certain lymphocytes) pack each chamber. Macrophages in the node help clear the lymph of bacteria, cellular debris, and other substances.

The largest lymphoid organ, the *spleen*, is a filtering station for blood and a holding station for lymphocytes. The spleen also has inner chambers, but these are filled with red and white "pulp." The red pulp contains large stores of red blood cells and macrophages. Red blood cells are produced here in developing human embryos.

The *thymus* secretes hormones concerned with the activity of lymphocytes. It also is a major organ where lymphocytes multiply, differentiate, and mature into fighters of specific types of disease agents. The thymus is central to immunity, the focus of the chapter to follow.

# SUMMARY

1. Animals ranging from worms to humans have a circulatory system consisting of a muscular pump (heart or heartlike structure), blood, and blood vessels.

   a. In closed circulatory systems, blood flows continuously inside the walls of these components; it exchanges substances with interstitial fluid only in diffusion zones.

   b. In open systems, a heart or heartlike structure pumps blood into tissues. The blood diffuses through tissue fluid and returns to the heart.

2. Blood is a transport fluid that carries oxygen and other substances to cells, and products and wastes (including carbon dioxide) from them. It helps maintain an internal environment favorable for cell activities.

3. Blood consists of red and white blood cells, platelets, and plasma.

   a. Plasma contains water, ions, nutrients, hormones, vitamins, dissolved gases, and the plasma proteins.

   b. Red blood cells transport oxygen (bound to hemoglobin) between the lungs and cells. They also transport some carbon dioxide.

   c. In all vertebrates, some white blood cells are scavengers of dead or worn-out cells and other debris. Others serve in the defense of the body against bacteria, viruses, and other foreign agents.

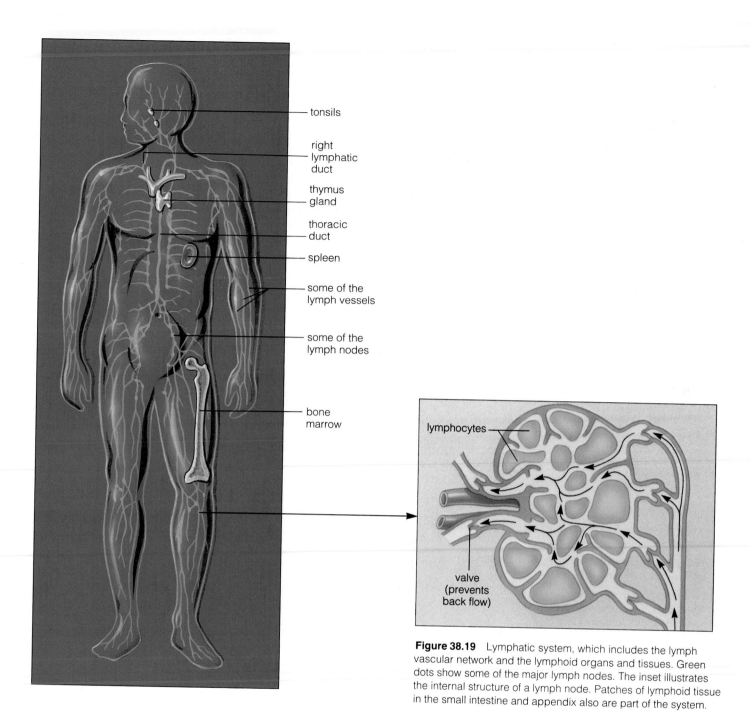

tonsils

right
lymphatic
duct

thymus
gland

thoracic
duct

spleen

some of the
lymph vessels

some of the
lymph nodes

bone
marrow

lymphocytes

valve
(prevents
back flow)

**Figure 38.19** Lymphatic system, which includes the lymph vascular network and the lymphoid organs and tissues. Green dots show some of the major lymph nodes. The inset illustrates the internal structure of a lymph node. Patches of lymphoid tissue in the small intestine and appendix also are part of the system.

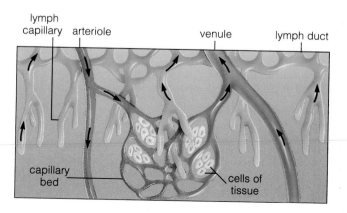

lymph
capillary    arteriole              venule        lymph duct

capillary
bed                                  cells of
                                     tissue

**Figure 38.20** Lymph vessels near a capillary bed.

4. In all vertebrates, the heart pumps blood into arteries. From there it flows into arterioles, capillaries, venules, veins, and back to the heart.

5. The human heart is divided into two halves, each with two chambers (an atrium and a ventricle). The division is the basis of two cardiovascular circuits. These are called the pulmonary and systemic circuits.

   a. In the pulmonary circuit, the right half of the heart pumps oxygen-poor blood to capillary beds inside the lungs, then oxygen-enriched blood flows back to the heart.

   b. In the systemic circuit, the left half of the heart pumps oxygen-enriched blood to all body regions, where it nourishes all tissues and organs. Then oxygen-poor blood flows from those regions back to the heart.

6. Heart contractions (specifically, the contracting ventricles) are the driving force for blood circulation. Fluid pressure is high at the start of a circuit, then drops in arteries, arterioles, capillaries, then veins. It is lowest in the relaxed atria.

7. Blood moves through several types of blood vessels in both the pulmonary and systemic circuits.

   a. Arteries are elastic pressure reservoirs that smooth out fluid pressure changes caused by heart contraction and relaxation.

   b. Arterioles are control points for the distribution of different volumes of blood to different body regions.

   c. Beds of capillaries are diffusion zones between the blood, interstitial fluid, and cells.

   d. Venules overlap capillaries and veins somewhat in function.

   e. Veins are blood volume reservoirs and help adjust volume flow back to the heart.

8. The lymphatic system supplements the circulatory system by returning excess fluid that seeps out of blood vessels back to the circulation. Some of its components have major roles in immune responses.

## Review Questions

1. What are some of the functions of blood? *660–661*

2. Describe the cellular components of blood. Describe the plasma portion of blood. *661–664*

3. Define the functions of the following: heart, cardiovascular system, and lymphatic system. *664–665, 676*

4. Distinguish between the following:
   a. open and closed circulation *660*
   b. systemic and pulmonary circuits *665*
   c. lymph vascular system and lymphoid organs *676*

5. Label the component parts of the human heart: *665*

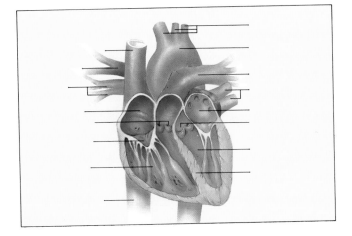

6. Explain how the medulla oblongata of the brain helps regulate blood flow to different body regions. *672*

7. State the main function of blood capillaries. What drives solutes out of and into capillaries in capillary beds? *668–669*

8. State the main function of venules and veins. What forces work together in returning venous blood to the heart? *670*

## Self-Quiz *(Answers in Appendix IV)*

1. In large, complex animals, a _____ system functions in the rapid exchange of substances to and from all living cells, and usually it is supplemented by a _____ system.

2. _____ and _____ are large-diameter blood vessels for fluid transport; _____ and _____ are fine-diameter, thin-walled blood vessels for diffusion; and _____ serve as control points over the distribution of different blood volumes to different body regions.

3. Which of the following are *not* components of blood?
   a. red and white blood cells
   b. platelets and plasma
   c. assorted solutes and dissolved gases
   d. all of the above are components of blood

4. Red blood cells are produced in the _____ and function in transporting _____ and some _____.
   a. liver; oxygen; mineral ions
   b. liver; oxygen; carbon dioxide
   c. bone marrow; oxygen; hormones
   d. bone marrow; oxygen; carbon dioxide

5. White blood cells are produced in the _____ and function in both _____ and _____.
    a. liver; oxygen transport; defense
    b. lymph glands; oxygen transport; pH stabilization
    c. bone marrow; day-to-day housekeeping; defense
    d. bone marrow; pH stabilization; defense

6. In the pulmonary circuit, the _____ half of the heart pumps _____ blood to capillary beds inside the lungs, then _____ blood flows back to the heart.
    a. left; oxygen-poor; oxygen-enriched
    b. right; oxygen-poor; oxygen-enriched
    c. left; oxygen-enriched; oxygen-poor
    d. right; oxygen-enriched; oxygen-poor

7. In the systemic circuit, the _____ half of the heart pumps _____ blood to all body regions, then _____ blood flows back to the heart.
    a. left; oxygen-poor; oxygen-enriched
    b. right; oxygen-poor; oxygen-enriched
    c. left; oxygen-enriched; oxygen-poor
    d. right; oxygen-enriched; oxygen-poor

8. Fluid pressure in the circulatory system is _____ at the beginning of a circuit, then _____ in arteries, arterioles, capillaries, and then veins. It is _____ in the relaxed atria.
    a. low; rises; highest     c. low; drops; lowest
    b. high; drops; lowest     d. high; rises; highest

9. Match the type of blood vessel with its major function.
    _____ arteries     a. diffusion
    _____ arterioles     b. control of blood volume
    _____ capillaries         distribution
    _____ venules     c. transport, blood volume
    _____ veins         reservoirs
        d. overlap of capillary
        function
        e. transport and pressure
        reservoirs

10. Match the circulation components with their descriptions.
    _____ capillary beds     a. two atria, two ventricles
    _____ lymph vascular     b. pressure reservoirs
        system     c. driving force for blood
    _____ heart chambers     d. zones of diffusion
    _____ veins     e. interstitial fluid
    _____ heart contractions     f. blood volume reservoirs
    _____ arteries

## Selected Key Terms

ABO blood typing 674
agglutination 675
antibody 674
aorta 665
arteriole 668
artery 667
atherosclerosis 671
atrium 665
AV valve 665
blood 660
blood pressure 667
capillary 668
capillary bed 668
cardiac conduction
    system 666
cardiac cycle 666
cardiac pacemaker 667
cell count 663
circulatory system 660
endothelium 665
heart 665
hemostasis 674
interstitial fluid 660

lymph 676
lymphatic system 660
lymph node 676
lymphoid organ 676
lymph vessel 676
plasma protein 661
platelet 664
pulmonary circuit 665
pulse pressure 667
red blood cell
    (erythrocyte) 662
Rh blood typing 675
semilunar valve 665
spleen 676
systemic circuit 665
thymus gland 676
vasoconstriction 673
vasodilation 673
vein 670
ventricle 665
venule 670
white blood cell
    (lymphocyte) 663

## Readings

Eisenberg, M. S., et al. May 1986. "Sudden Cardiac Death." *Scientific American* 254(5):37–43.

Golde, D. W., and J. C. Gasson. July 1988. "Hormones That Stimulate the Growth of Blood Cells." *Scientific American* 259(1):62–70.

Kapff, C. T., and J. H. Jandl. 1981. *Blood: Atlas and Sourcebook of Hematology*. Boston: Little, Brown. Beautiful micrographs of normal and abnormal blood and marrow cells.

Little, R., and W. Little. 1989. *Physiology of the Heart and Circulation*. Fourth edition. Chicago: Year Book Medical Publishers, Inc. Comprehensive coverage of cardiovascular physiology.

Robinson, T. F., et al. June 1986. "The Heart as a Suction Pump." *Scientific American* 254(6):84–91.

# 39 IMMUNITY

a

**Figure 39.1** Immunization past and present. (**a**) Statue honoring Edward Jenner's development of an immunization procedure against smallpox, one of the most dreaded diseases in human history. (**b**) Micrograph of an immune cell (T lymphocyte) being attacked by the virus (blue particles) that causes AIDS. Immunologists are working to develop weapons against this modern-day scourge.

## Russian Roulette, Immunological Style

Until about a century ago, smallpox swept repeatedly through the world's cities. Some outbreaks were so severe, only half of those stricken managed to survive. No one emerged unscathed. Even the survivors ended up with permanent scars on the face, neck, shoulders, and arms. Scarring was a small price to pay, however, because survivors seldom contracted the disease again—they were "immune" to smallpox.

No one knew what caused smallpox, but the possibility of acquiring immunity was dreadfully fascinating. In Asia, Africa, and then Europe, many who were in good health gambled with inoculations. They allowed themselves to be intentionally infected with material from the sores of diseased people. Thus Chinese of the twelfth century ground up crusts from smallpox sores and inhaled the powder. By the seventeenth century Mary Montagu, wife of the ambassador to Turkey, was championing inoculation. She went so far as to inject bits of smallpox scabs into the veins of her children; even the Prince of Wales did the same. Others soaked threads in the fluid from smallpox sores, then poked the threads into scratches on the body. The survivors of such practices acquired immunity to smallpox, but many came down with raging infections. As if the odds were not dangerous enough, those who were inoculated by such crude procedures also risked coming down with leprosy, syphilis, or hepatitis.

While this immunological version of Russian roulette was going on, Edward Jenner was growing up in the English countryside. At the time it was known that cowpox, a rather mild disease, could be transmitted from cattle to humans. Yet people who contracted cowpox never became ill with smallpox. No one thought much about this until 1796, when Jenner, by now a physician, took some material from a sore on a cowpox-infected person and injected it into the arm of a young boy (Figure 39.1). Six weeks later, after the reaction to cowpox subsided, Jenner inoculated the boy with fluid from smallpox sores. He hypothesized that the earlier inoculation would provoke immunity to smallpox—and he was right. The boy remained free of infection. The French mocked the procedure, calling it "vaccination" (which translates as "encowment").

Much later a French chemist, Louis Pasteur, devised similar procedures for other diseases. Pasteur also called his procedures vaccinations; only then did the term become respectable.

By Pasteur's time, improved microscopes were revealing a variety of bacteria, fungal spores, and other previously invisible forms of life. As Pasteur discovered, microorganisms abound in ordinary air. Did some of them cause contagious diseases? Probably. Could they settle into food or drink and cause it to spoil? Pasteur proved that they did. Boiling could kill them—Pasteur and others knew this. (Being a wine connoisseur, he also knew that you cannot simply boil wine—or beer or milk, for that matter—and end up with the same beverage. He devised a way to heat food or beverages at a temperature low enough not to ruin them but high enough to kill most of the microorganisms that cause spoilage. We still depend on his partial sterilization methods, which were named pasteurization in his honor.)

But it was a German physician, Robert Koch, who actually proved that a specific microorganism can cause a specific disease—namely, anthrax. In the late 1870s, Koch repeatedly transferred blood from infected animals to uninfected ones. Each time, the blood of the recipient ended up teeming with cells of a rather large bacterium (*Bacillus anthracis*). And each time, the animal developed symptoms of anthrax. More than

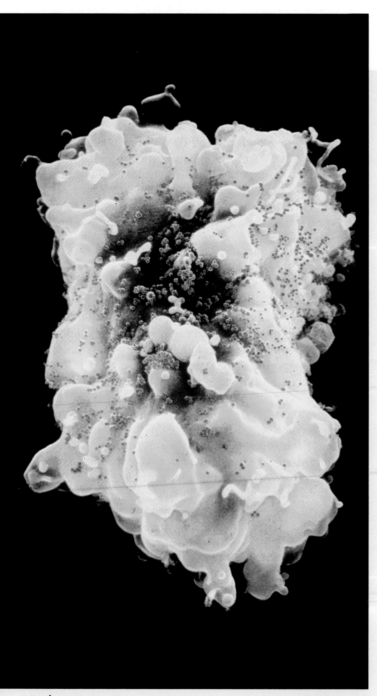

**b**

**1.** The vertebrate body defends itself against viruses, bacteria, and other foreign agents that enter the internal environment. Some defense responses are nonspecific, in that they occur when any kind of invasion is detected. Other responses are specific, with certain white blood cells being mobilized against a particular invader, not invaders in general.

**2.** White blood cells responsible for immune responses can distinguish between self-markers on the body's own cells and antigens. An antigen is any large molecule that white blood cells recognize as foreign and that triggers an immune response. Different foreign agents have different antigens on their surface.

**3.** Antibody-mediated immune responses are made against antigens circulating in the body's tissues or attached to the surface of an invader. Cell-mediated immune responses are made only against body cells already infected and against cancerous or mutated cells. In most cases, both responses proceed simultaneously.

this, bacterial cells cultured in nutrients outside the body could, when injected into an animal, cause the same disease!

Thus, by the beginning of the twentieth century, the promise of understanding the basis of immunity loomed large. The battle against infectious disease was about to begin in earnest. Those battles are the focus of this chapter.

When you suffer sneezes, a puffed-up dripping nose, and watery eyes, you have evidence that your body is being attacked by some type of cold virus. Yet you probably are not even aware that it is simultaneously fighting off attacks by many other pathogens—and does so every day of your life. **Pathogens** are a diverse assortment of viruses, bacteria, fungi, and protozoans, each able to infect and cause diseases in humans and other organisms. Examples of their infectious cycles were described earlier, in Chapters 22 and 25. With his pioneering procedure, Jenner was actually mobilizing cells to make an immune response to a specific virus. That type of response is one of the elegant defenses described in this chapter. Before we turn to the specific responses, however, let's start with the body's generalized defenses against attack.

## NONSPECIFIC DEFENSE RESPONSES

### Barriers to Invasion

When Julie Andrews sang "The hills are alive . . ." in the motion picture *The Sound of Music*, she might well have been referring to the microbial world. Hills, streams, plants, air, the animal body abound with invisible organisms, many harmless but some pathogenic. You and other vertebrates coevolved with most of them, however, so you need not lose sleep over this. Most of the time, the pathogens cannot even get past your body's physical and chemical barriers. Those barriers include the following:

1. Intact skin and mucous membranes. Few microorganisms can penetrate these.

2. Ciliated, mucous membranes in the respiratory tract. Like sticky brooms, the cilia trap and sweep out airborne bacteria.

3. Secretions from exocrine glands in the skin, mouth, and elsewhere. The enzyme lysozyme, for instance, destroys the cell wall of many bacteria. It is present in the fluid (tears) that bathes the eyes.

4. Gastric fluid. The acids in this fluid destroy many pathogens that are present in food when it enters the gut.

5. Normal bacterial inhabitants of the skin, gut, and vagina. They outcompete the pathogens for resources and so help keep them in check.

### Phagocytes: The Macrophages and Their Kin

What happens when physical barriers to invasion are breached by microorganisms, as when skin is cut or scraped? Then, the invasion mobilizes phagocytic white blood cells. These cells engulf and destroy foreign agents. As Figure 38.5 shows, these cells arise from stem cells in bone marrow. They include neutrophils, eosinophils, and monocytes that mature into macrophages, the "big eaters." Figure 39.2 shows a macrophage.

Phagocytes are strategically distributed cells. Some circulate within blood vessels, then enter damaged or invaded tissues by squeezing between endothelial cells making up the walls of capillaries. Some take up stations in lymph nodes and the spleen. (You may wish to review Figure 38.19, which shows the lymphatic system.) Other phagocytes are located in the liver, kidneys, lungs, joints, and the brain.

### Complement System

When certain microorganisms invade the body, about twenty plasma proteins interact as a system—the **complement system**. Those circulating proteins have roles in both nonspecific and specific defense responses. They are activated one after another in a "cascade" of reactions.

Once activated, each protein molecule helps activate many molecules of a different protein at the next reaction step. Each of these helps activate many molecules of a different protein at the next reaction step, and so on until huge numbers of complement proteins are mobilized. The reactions have these results:

1. Chemical gradients, created by the huge cascades of certain complement proteins, attract phagocytes to the scene (Figure 39.3).

2. Some complement proteins coat the surface of invading cells, and phagocytes zero in on the coat.

3. Other complement proteins help kill the pathogen by promoting lysis of its plasma membrane. (*Lysis* refers to gross induced leakage across the membrane that leads to cell death.)

### Inflammation

Many cells, the complement system, and other substances take part in the **inflammatory response**. This response is a series of events that destroy invaders and

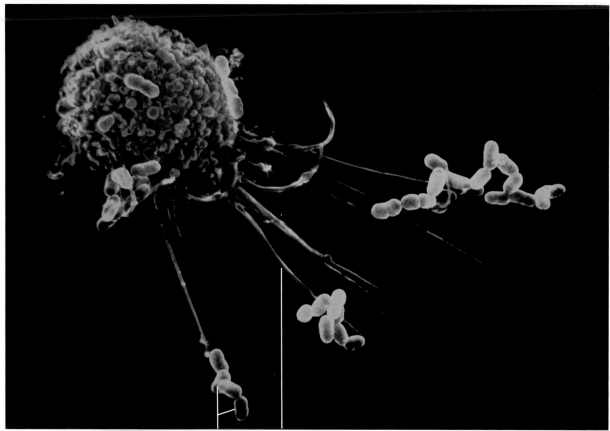

bacterial cells    cytoplasmic extension of macrophage

restore tissues to normal. The events are not limited to nonspecific defense responses. As you will see, they proceed also when the body acts against specific invaders.

For example, when the complement system is activated, circulating basophils (and mast cells, their counterparts in tissues) release *histamine*. This potent substance dilates capillaries and makes them "leaky," so fluid seeps out. The complement proteins and other substances used to fight an invasion are dissolved in this fluid and so gain access to tissues. Also, clotting mechanisms (page 674) are working to keep blood vessels intact and to wall off infected or damaged tissues. In short, the inflammatory response involves these events:

**1.** Localized warmth and redness occur in damaged or invaded tissues when capillaries dilate and become leaky.

**2.** Fluid seeping from capillaries causes local swelling and delivers infection-fighting proteins to the tissues.

**3.** Phagocytes, following chemical gradients to affected tissues, engulf foreign invaders and debris.

**4.** Clotting mechanisms help wall off the pathogen and help repair tissues.

**Figure 39.2** Scanning electron micrograph of a macrophage, probing its surroundings with cytoplasmic extensions. The macrophage engulfs bacterial cells that come in contact with it.

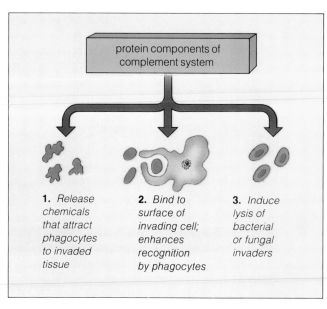

**Figure 39.3** Functions of proteins of the complement system. These proteins take part in specific as well as in nonspecific defense responses.

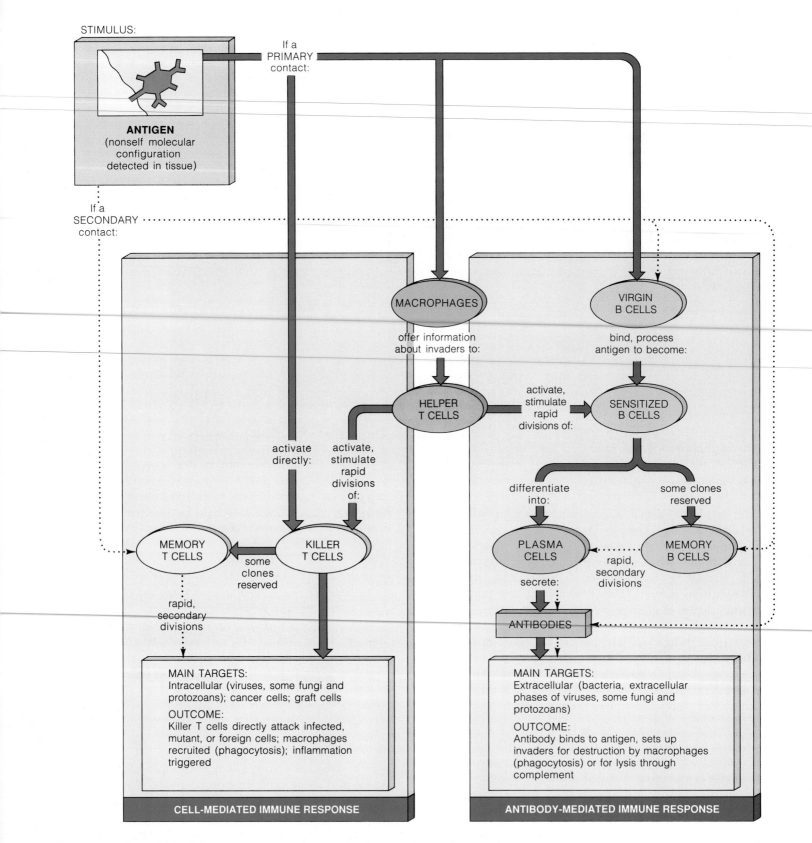

**Figure 39.4** Overview of the cell-mediated and antibody-mediated branches of the vertebrate immune system. Green arrows indicate a "primary" response, which follows a first-time encounter with a specific antigen. Dashed arrows indicate a "secondary" response to a subsequent encounter with the same kind of antigen. This illustration can be used as a road map as you make your way through the descriptions in the text. The details of the vertebrate immune system are astonishingly complex; even here, many events have been omitted so the main sequences can be seen clearly.

# SPECIFIC DEFENSE RESPONSES: THE IMMUNE SYSTEM

The body's nonspecific defenses—its physical barriers, complement system, and inflammatory response—are effective against many pathogens. Sometimes, however, the general attack responses are not enough to stop the spread of an invader, and illness follows. When that happens, three types of white blood cells—macrophages, T cells, and B cells—make precise counterattacks. Their interactions are the basis of the **immune system**.

The hallmarks of the immune system are *specificity* (its cells zero in on specific invaders) and *memory* (a portion of its cells can mount a rapid attack if the same type of invader returns).

## The Defenders: An Overview

Of every 100 cells in your body, one is a lymphocyte—a white blood cell. Here are the names and functions of the white blood cells responsible for immune responses:

**1. Macrophages**. Besides engulfing anything perceived as foreign, these phagocytic cells alert helper T cells to the presence of *specific* foreign agents.

**2. B cells**. The B cells and their progeny (plasma cells) produce antibodies. *Antibodies* are molecular weapons that lock onto specific targets and tag them for destruction by phagocytes or the complement system.

**3. Cytotoxic T cells**. These directly destroy body cells already infected by certain viruses or parasitic fungi.

**4. Helper T cells**. Helper T cells serve as master switches of the immune system. Among other things, they stimulate the rapid division of B cells and cytotoxic T cells.

**5. Suppressor T cells**. These "controller cells" slow down or prevent immune responses.

**6. Memory cells**. Memory cells are a portion of B cell and T cell populations produced during a first encounter with a specific invader but not used in battle. They circulate freely and respond rapidly to any subsequent attacks by the same type of invader.

The white blood cells just listed belong to two fighting branches of the immune system. Both are called into action during most battles. T cells dominate one branch; they carry out a "cell-mediated" response. B cells dominate the other branch; they carry out an "antibody-mediated" response. Figure 39.4 hints at how the two responses are interrelated.

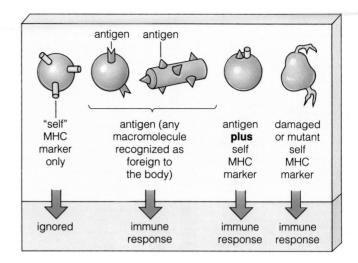

**Figure 39.5** Molecular cues that stimulate lymphocytes to make immune responses.

## Recognition of Self and Nonself

Before getting into the immunological battles, think about an important question. How do the defenders distinguish *self* (the body's own cells) from *nonself* (harmful foreign agents)? Such recognition is vital, for lymphocytes unleash extremely destructive immune reactions. We know this because on rare occasions the distinction is blurred—such that T and B cells make an autoimmune response. As you will see, this means that the cells turn on the body itself and irritate or damage tissues, sometimes with lethal consequences.

Among the surface proteins on your own cells are **MHC markers** (named after the genes coding for them). Your white blood cells recognize these as "self" markers and normally ignore the cells bearing them. MHC markers are unique to each individual. Except in the case of identical twins, no one has the same kinds.

But viruses, bacteria, fungi, ragweed pollen, bee venom, cells of organ transplants, and just about any other foreign agent have antigens on their surface, which lymphocytes do not ignore. An **antigen** is any large molecule with a distinct configuration that triggers an immune response. Most antigens are protein or oligosaccharide molecules.

B cells and T cells will not attack cells bearing MHC markers only. They will mount an immune response when they encounter an antigen. They will do so regardless of whether antigen is merely present in tissues or associated with MHC markers on cell surfaces (Figure 39.5).

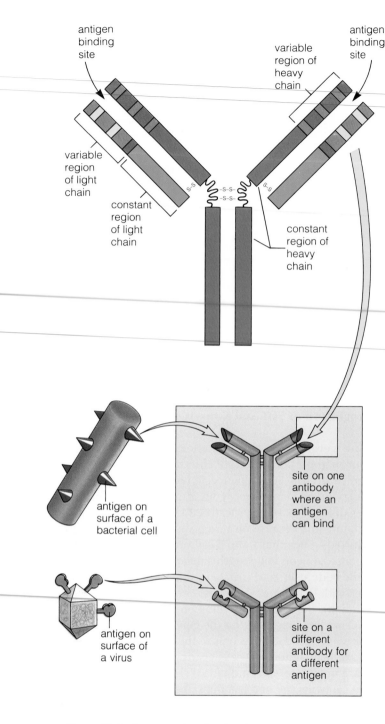

**Figure 39.6** Structure of antibodies. An antibody molecule has four polypeptide chains joined into a Y-shaped structure. Some regions are always the same in all antibody molecules. But the molecular configuration varies in one region; this is the antigen-binding site.

## Primary Immune Response

A *first-time* encounter with an antigen elicits a **primary immune response** from macrophages, T and B cells, and their products. Here we will consider an antibody-mediated response to such an encounter, then a cell-mediated one.

**Antibody-Mediated Immune Response**. An **antibody** is a Y-shaped protein molecule with binding sites for a specific antigen. Figure 39.6 is a diagram of its general structure. Only B cells and their progeny, called **plasma cells**, make antibodies.

B cells, recall, mature in bone marrow. While each B cell is differentiating, it makes many copies of just one kind of antibody. Some of these Y-shaped molecules become positioned at the cell surface, where they serve as receptors for a specific antigen. The tail of each "Y" is embedded in the plasma membrane, and the arms stick out above the surface. Now the cell is released into the circulation as a "virgin" B cell. This term signifies it has membrane-bound antibodies but has not yet made contact with antigen.

Suppose bacteria enter the body through a small cut (step *1* in Figure 39.7). The invasion triggers a general inflammatory response, and macrophages manage to engulf a few bacterial cells. The engulfed bacteria move into the cytoplasm inside endocytic vesicles, which fuse with other vesicles containing lysosomal enzymes (page 64). Although the enzymes digest the bacterial cells, they do not completely destroy their antigens. Antigen fragments are transported to the surface of the macrophage's plasma membrane. There, the fragments become bound to MHC markers (step *2* in Figure 39.7). *Each macrophage now displays antigen–MHC complexes at its surface.*

Some bacterial cells escape detection by macrophages, however. They multiply, and for a time they move undetected past a number of virgin B cells. Eventually they encounter the one B cell with antibodies able to bind to antigen on the bacterial cell surface (step *3*). Once the B cell binds with antigen, it becomes sensitive to stimulatory signals from macrophages and helper T cells.

What happens is this: When helper T cells make contact with the battling macrophages, some of their membrane receptors lock onto the antigen–MHC complexes at the macrophage surface (step *4* in Figure 39.7). Once the connection is made, macrophages secrete an *interleukin* that stimulates the helper T cells to secrete their own interleukins. The interleukins are communication signals. In their presence, *any B cell that has become sensitized to the antigen will start dividing.*

Rapid divisions among the stimulated B cell progeny give rise to a clone—a population of identical B cells.

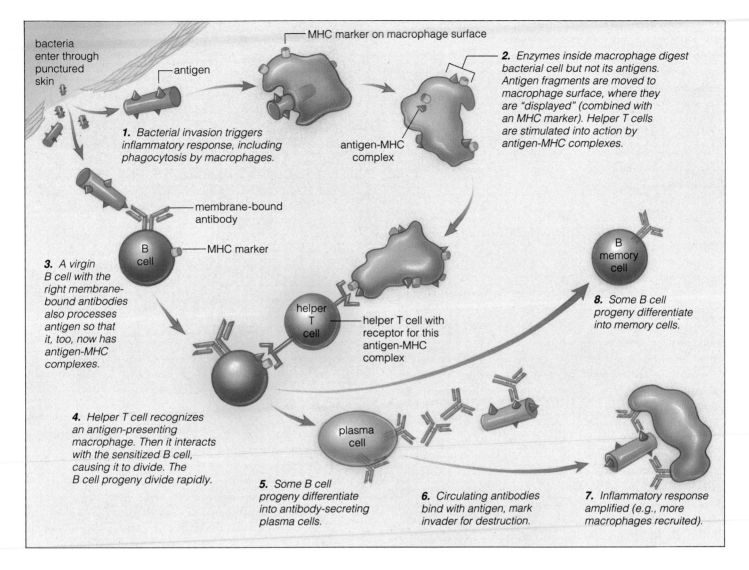

**Figure 39.7** Amplification of the inflammatory response by specific immune reactions. This example is of an *antibody-mediated response* to a bacterial invasion. Plasma cells (the progeny of activated B cells) release antibodies, which circulate and mark invaders for destruction by other defense agents, including more macrophages recruited to the battle scene.

Part of the population differentiates into plasma cells. The plasma cells are weapons factories. They make vast numbers of copies of the particular antibody that had been generated in the virgin B cell (step 5). For the next few days they secrete about 2,000 antibody molecules per second into their surroundings! The circulating antibodies do not destroy pathogens directly. They simply tag the invader for disposal by other means.

There are five classes of antibodies that serve different roles in defense. All belong to a group of plasma proteins, the **immunoglobulins** (Ig). When bound to antigen, the ones designated IgM and IgG enlist the aid of macrophages and complement proteins. IgG also can cross the placenta and help defend a developing fetus from pathogens. IgA, which is present in tears, saliva, and mucus, helps repel invaders at the start of the respiratory system, digestive system, urinary tract, and elsewhere. IgE calls histamine-secreting cells into action. Lastly, IgD and IgM work together to help bind antigen to B cells.

The main targets of an antibody-mediated response are bacteria and *extracellular phases* of viruses, some fungal parasites, and some protozoans. In other words, antibodies can't lock onto antigen if the invader has entered the cytoplasm of a host cell. The antigen must be circulating in tissues or at the cell surface.

**Cell-Mediated Immune Response.** This brings us to the viruses and other pathogens that have already penetrated host cells, where they remain hidden from antibodies. In a cell-mediated immune response, the host cells are killed before the pathogens can replicate and spread to other cells. Cytotoxic T cells serve as the executioners.

Stem cells in bone marrow give rise to the forerunners of cytotoxic T cells (Figure 38.5). These travel the circulatory highways to the thymus gland, where they mature into cytotoxic T cells. Each cell produces protein receptors that become positioned at its surface. (As you will see shortly, gene segments are shuffled into many different combinations that code for many different receptor proteins.) With these surface receptors, the cytotoxic T cell will be able to recognize any antigen encountered once it starts patrolling the body.

Cytotoxic T cells will patrol right on past circulating virus particles without even recognizing them. When a virus particle infects a cell, however, viral proteins become associated with MHC markers on the host's surface. The MHC markers identify the cell as belonging to the body. At the same time, the viral protein (in other words, the antigen) identifies the presence of something foreign. Cytotoxic T cells bind to this combination of MHC marker and antigen. Then they secrete *perforins*, which are proteins that effectively punch holes in the infected cell. It is possible that they also induce the infected cell to self-destruct, although the mechanism is not known. The remarkable scanning electron micrographs in the *Commentary* suggest that cytotoxic T cells have the same deadly effect on cancerous cells.

When the body rejects a tissue graft or an organ transplant, cytotoxic T cells are one of the reasons why. They recognize MHC markers on the grafted cells as being foreign, unless the donor is an identical twin. (Such twins have identical DNA, hence identical MHC markers.) Organ recipients take drugs to destroy cytotoxic T cells, but this compromises their ability to mount immune responses to pathogens. For example, pneumocystis infections are one of the leading causes of death among transplant recipients; the body cannot overcome the invading bacterium responsible for the disease.

Cytotoxic T cells may execute cancerous cells, but only when viruses have induced the cancer. More than 80 percent of all human cancers are not virus-induced. However, there may be other cells in the body that defend against cancer. Macrophages are candidates. *Natural killer cells*, designated NK cells, are others. NK cells, which are somewhat like lymphocytes, kill tumor cells and viral-infected cells. They do so in the absence of antibodies. They do so spontaneously, regardless of what type of cancerous cell they encounter. Like cytotoxic T cells, they punch holes in infected cells or, possibly, induce them to commit suicide.

## Commentary

## Cancer and the Immune System

Cancer is a disease in which cells have lost controls over cell division. It can arise when viral attack, chemical change, or irradiation induces mutation in genes that are central to the cell division cycle (page 232). The mutated cells start dividing relentlessly, and unless something stops them, they will destroy surrounding tissues and, ultimately, kill the individual.

Cytotoxic T cells and NK cells can destroy cancer cells—when they detect them. Typically, cancerous transformation involves alterations of glycoproteins positioned at the cell surface. The altered molecules are analogous to foreign antigens, in terms of how cytotoxic T cells and NK cells respond to them.

However, it may be that glycoproteins do not undergo alteration in all cases. It may be that they become chemically disguised or masked. Perhaps they are even released from the cell surface and begin circulating through the bloodstream and so lead NK cells down false trails. Whatever the reason, the transformed cells are free to divide uncontrollably and produce a tumor.

At present, surgery, drug treatment (chemotherapy), and irradiation are the only weapons against cancer. Surgery works when a tumor is fully accessible and has not spread, but it offers little hope when cancer cells have begun wandering. When used alone, chemotherapy and irradiation destroy good cells as well as bad.

*Immune therapy* is a promising prospect. The idea here is to mobilize cytotoxic T cells by deliberately introducing agents that will set off the immune alarm. *Interferons*, a group of small proteins, were early candidates for immune therapy. Most cells produce and release interferon following a viral attack. The interferon binds to the plasma membrane of other cells in the body and induces resistance to many viruses. So far, however, interferon has been useful only against some rare forms of cancer.

*Monoclonal antibodies* hold promise for immune therapy. It is difficult to get normal, antibody-secreting B cells to

**Control of Immune Responses.** Antibody-mediated and cell-mediated responses are regulated events. When the tide of battle turns, antibody molecules are "saturating" the binding sites on pathogens that have not yet been disposed of. With fewer exposed antigens, less antibody is secreted. Also, secretions from suppressor T cells call off the counterattack and keep the reactions from spiraling out of control.

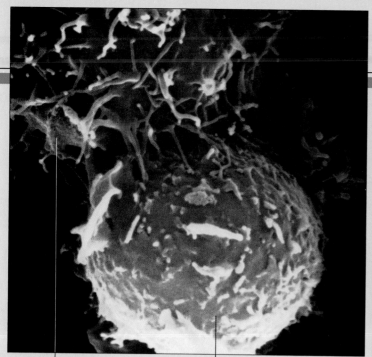

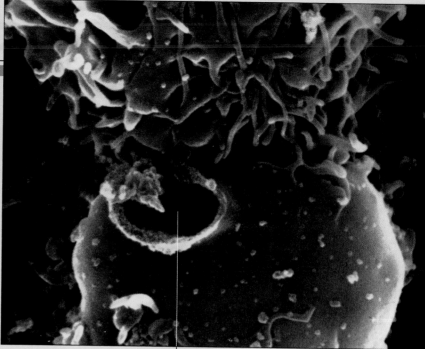

cytotoxic T cell        tumor cell

**a** A cytotoxic T cell recognizes and binds tightly to a tumor cell, then secretes pore-forming proteins that will destroy the integrity of the target cell membrane.

hole "punched" in tumor cell

**b** The target cell has become grossly leaky and has ballooned under an influx of the surrounding fluid; soon there will be nothing left of it.

grow indefinitely and mass-produce pure antibody in useful amounts. But Cesar Milstein and Georges Kohler discovered a way to do this. They immunized a mouse with a specific antigen. (The point was to allow lymphatic tissues in the mouse—the spleen especially—to become enriched with B cells specific for the immunizing antigen.) Later, B cells were extracted from the mouse spleen and were fused with a malignant B cell that showed indefinite growth. Some of the hybrid cells multiplied as rapidly as the malignant parent and produced quantities of the same antibodies as the parent B cells from the immunized mouse. Clones of such hybrid cells can be maintained indefinitely and they continue to make the same antibody. Hence the name "monoclonal antibodies." All the antibody molecules are identical, and all are derived from the same parent cell.

Monoclonal antibodies are being studied for use in passive immunization against malaria, flu viruses, and hepatitis B. They also are candidates for *cancer imaging*. This procedure uses scanning machines along with radioactively labeled monoclonal antibodies that are specific for certain types of cancer. It allows us to home in on the exact location of cancer in the body (compare page 22). Such scans indicate whether cancer is present, where it is located, and the size of the tumor.

Monoclonal antibodies might also help overcome one of the major drawbacks to chemotherapy. Such treatments have severe side effects because the drugs used are highly toxic and cannot discriminate between normal cells and cancerous ones. A current goal is to hook up drug molecules with a monoclonal antibody. As Milstein and Kohler speculated, "Once again the antibodies might be expected to home in on the cancer cells—only this time they would be dragging along with them a depth charge of monumental proportions." Such is the prospect of *targeted drug therapy*.

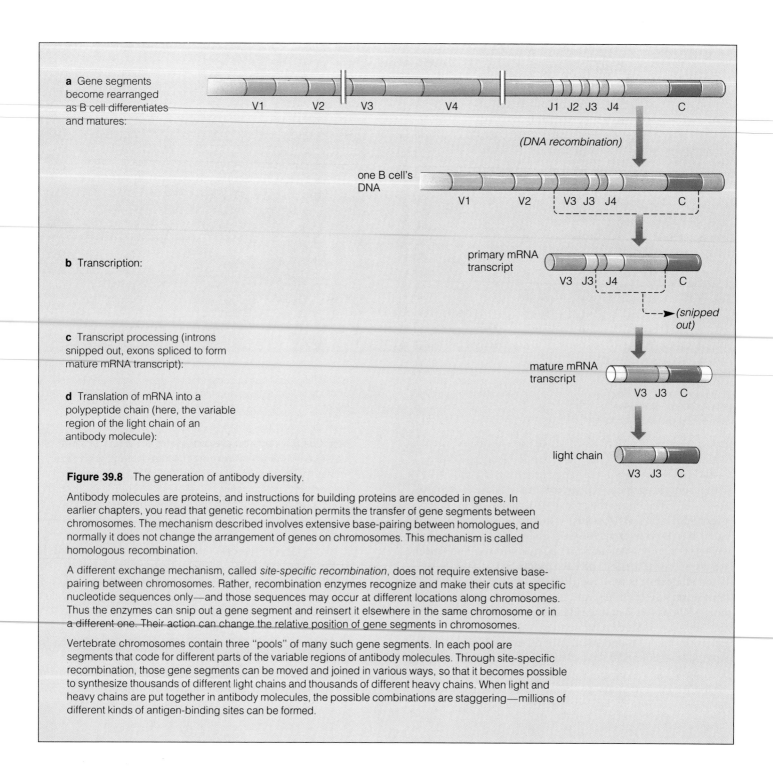

**a** Gene segments become rearranged as B cell differentiates and matures:

V1  V2  V3  V4  J1 J2 J3 J4  C

*(DNA recombination)*

one B cell's DNA

V1  V2  V3 J3 J4  C

**b** Transcription:

primary mRNA transcript

V3  J3  J4  C

*(snipped out)*

**c** Transcript processing (introns snipped out, exons spliced to form mature mRNA transcript):

mature mRNA transcript

V3  J3  C

**d** Translation of mRNA into a polypeptide chain (here, the variable region of the light chain of an antibody molecule):

light chain

V3  J3  C

**Figure 39.8** The generation of antibody diversity.

Antibody molecules are proteins, and instructions for building proteins are encoded in genes. In earlier chapters, you read that genetic recombination permits the transfer of gene segments between chromosomes. The mechanism described involves extensive base-pairing between homologues, and normally it does not change the arrangement of genes on chromosomes. This mechanism is called homologous recombination.

A different exchange mechanism, called *site-specific recombination*, does not require extensive base-pairing between chromosomes. Rather, recombination enzymes recognize and make their cuts at specific nucleotide sequences only—and those sequences may occur at different locations along chromosomes. Thus the enzymes can snip out a gene segment and reinsert it elsewhere in the same chromosome or in a different one. Their action can change the relative position of gene segments in chromosomes.

Vertebrate chromosomes contain three "pools" of many such gene segments. In each pool are segments that code for different parts of the variable regions of antibody molecules. Through site-specific recombination, those gene segments can be moved and joined in various ways, so that it becomes possible to synthesize thousands of different light chains and thousands of different heavy chains. When light and heavy chains are put together in antibody molecules, the possible combinations are staggering—millions of different kinds of antigen-binding sites can be formed.

## Antibody Diversity and the Clonal Selection Theory

Your body can be assaulted by an enormous variety of pathogens, each with a unique antigen. How do B cells produce the millions of different receptors (antibodies) required to detect all the potential threats? The answer lies with DNA recombinations occurring in the antibody genes of each B cell as it matures in bone marrow. Part of each arm of an antibody is a polypeptide chain, folded into a groove or cavity. All B cells have the same genes coding for the chain—but each shuffles the genes into one of millions of possible combinations, so they can give rise to virtually unlimited chain configurations (Figure 39.8).

Thus, it is not that you or any other individual inherited a limited genetic war chest from your ancestors, useful only against pathogens that were successfully

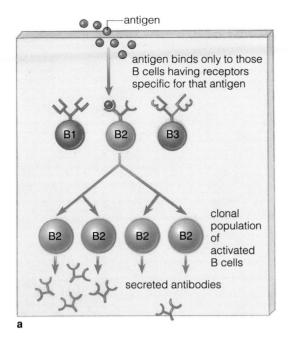

a

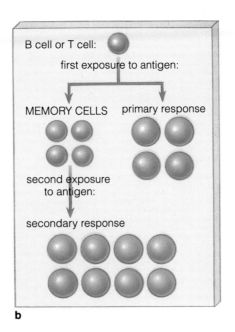

b

fought off in the past. Even if you encounter an entirely new antigen (as might occur when an influenza virus has mutated), your body may not be helpless against attack. It may be that DNA recombinations in one of your maturing B cells produced the exact chain configuration that can lock onto the invader. By happy accident, you have the precise weapon needed.

According to the **clonal selection theory**, proposed by Macfarlane Burnet, an activated B cell (or T cell) multiplies rapidly by mitotic cell division, and all of its descendants will retain specificity against the antigen causing the activation. They constitute a *clone* of cells, immunologically identical for the antigen that "selected" them (Figure 39.9).

**Figure 39.9** (**a**) Clonal selection of lymphocytes having receptors for specific antigens. The proteins from which the receptors are constructed are produced through random shufflings of DNA segments while lymphocytes are maturing. Only antigen-selected lymphocytes will become activated and give rise to a population of immunologically identical clones. (**b**) Immunological memory. Not all cells of the activated lymphocyte populations are used in the primary immune response to an antigen. Many continue to circulate as memory lymphocytes, which become activated during a secondary immune response.

### Secondary Immune Response

The clonal selection theory also explains how a person has "immunological memory" of a first-time response to an invasion. The term refers to the body's capacity to make a very rapid response to a subsequent invasion by the same type of pathogen. A **secondary immune response** to a previously encountered antigen can occur in two or three days. It is greater in magnitude than the primary response and of longer duration (Figure 39.10).

Why is this so? During a primary immune response, some B and T cells of the clonal populations do not engage in battle. They continue to circulate for years, even decades in some cases, as patrolling battalions of *memory cells*. When a memory cell encounters the same type of antigen, it divides at once. A large clone of active B cells or T cells can be unleashed, and it can be unleashed in a matter of days.

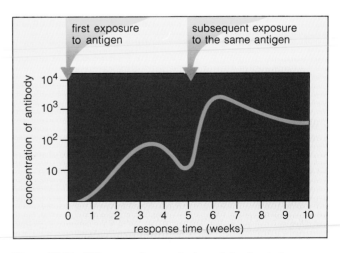

**Figure 39.10** Differences in magnitude and duration between a primary and a secondary immune response to the same antigen. (The secondary response starts at week 5.)

## AIDS—The Immune System Compromised

AIDS is a constellation of disorders that follow infection by the human immunodeficiency virus, or HIV. The virus cripples the immune system and leaves the body dangerously susceptible to opportunistic infections and some otherwise rare forms of cancer. Currently there is no vaccine against the known forms of this virus, which are called HIV-1, 2, and 3. And currently there is no cure for those already infected.

From 1981 to November of 1991, nearly 200,000 cases of AIDS had been reported in the United States alone. At that time the World Health Organization (WHO) had estimated that nearly half a million people had already died from AIDS and more than 1.5 million were infected with the virus. No one can say how many will be infected in the next decade. The number could be as high as 10 million.

### HIV Replication Cycle

HIV compromises the immune system by attacking helper T cells (also called CD4 or T4 lymphocytes) as well as macrophages. Sometimes the virus directly attacks the nervous system, causing mental impairment and loss of motor function.

HIV is a *retrovirus*; its genetic material is RNA rather than DNA. A protein core surrounds the RNA and several copies of an enzyme, a reverse transcriptase. The core itself is wrapped in a lipid envelope derived from the plasma membrane of a host helper T cell. Once inside a host, the enzyme uses the viral RNA as a template for making DNA, which then is inserted into a host chromosome (page 350).

## IMMUNIZATION

Jenner didn't know why his cowpox vaccine provided immunity against smallpox. Today we know that the viruses causing the two diseases are related, and they bear similar antigens at their surface. Let's express what goes on in modern terms.

**Immunization** means deliberately introducing an antigen into the body that can provoke an immune response and the production of memory cells. A **vaccine** (a preparation designed to stimulate the immune response) is injected into the body or taken orally. The first injection elicits a primary immune response. A second injection (the "booster shot") elicits a secondary response, which provokes the production of more antibodies and memory cells to provide long-lasting protection against the disease.

Many vaccines are made from killed or weakened pathogens. Sabin polio vaccine, for example, is a preparation of a weakened polio virus. Other vaccines are made from toxic but inactivated by-products of dangerous organisms, such as the bacteria causing tetanus.

Recently, selected antigen-encoding genes from pathogens were incorporated into the vaccinia virus. The virus was then used successfully to immunize laboratory animals against hepatitis B, influenza, rabies, and other serious diseases. A genetically engineered virus is not as potentially dangerous as a weakened but still-intact pathogen, which could revert to the virulent form.

For people already exposed to diphtheria, tetanus, botulism, and some other bacterial diseases, antibodies purified from some other source are injected directly to confer **passive immunity**. The effects are not lasting because the person's own B cells are not producing the antibodies. The injected antibodies may help counter the immediate attack.

## ABNORMAL OR DEFICIENT IMMUNE RESPONSES

### Allergies

Many of us suffer from **allergies**, in which the body makes a secondary immune response to a normally harmless substance. Exposure to dust, pollen, insect venom, drugs, certain foods, perfumes, cosmetics, and other substances triggers the abnormal response. Some allergic reactions occur within minutes; others are delayed. In either case, they can cause tissue damage.

Some individuals are genetically predisposed to allergies. But infections, emotional stress, even changes in air temperature can trigger or complicate reactions to dust and other substances that the body perceives as antigens. Every time allergic individuals are exposed to certain antigens, they produce IgE antibodies. The IgE initiates a local inflammatory response; it provokes cells into secreting histamine, prostaglandins, and other potent substances. Besides promoting fluid seepage from

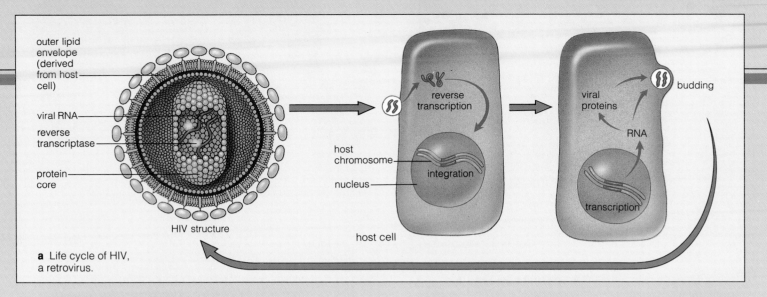

outer lipid envelope (derived from host cell)

viral RNA

reverse transcriptase

protein core

HIV structure

reverse transcription

host chromosome

nucleus

integration

host cell

viral proteins

RNA

budding

transcription

**a** Life cycle of HIV, a retrovirus.

It may take up to three years after infection before antibodies to several HIV proteins can be detected in the body. The antibodies do not inactivate the circulating virus particles or target infected cells for elimination. Those cells can harbor the foreign DNA for months, even years.

However, when the body is called upon to make an immune response, the infected cell may be activated. When that happens, it transcribes parts of its DNA—including the foreign insert. Transcription yields copies of viral RNA, which are translated into viral proteins. New virus particles

capillaries, histamine also stimulates exocrine glands to secrete mucus. Prostaglandins constrict smooth muscle in different organs and contribute to platelet clumping. In *asthma* and *hay fever*, the resulting symptoms of an allergic response include congestion, sneezing, a drippy nose, and labored breathing.

On rare occasions, inflammatory responses are explosive, and they can trigger a life-threatening condition called *anaphylactic shock*. For example, the few individuals who are hypersensitive to wasp or bee venom can die within minutes following a single sting. Air passages leading to their lungs undergo massive constriction. Fluid escapes too rapidly from grossly permeable capillaries. Blood pressure plummets and can lead to circulatory collapse.

Allergy-producing substances often can be identified by tests, and in some cases the body can be stimulated to make a different type of antibody (IgG) that can block the inflammatory response. Over an extended time, increasingly larger doses of the antigen are administered to allergy patients. Then, circulating IgG antibodies will be produced that bind with and mask molecules of the offending substance before they interact with IgE to produce the abnormal response.

### Autoimmune Disorders

In an **autoimmune response**, lymphocytes are unleashed against the body's cells. An example is *rheumatoid arthri-*

*tis*, in which movable joints especially are inflamed for long periods. Often, affected persons have high levels of an antibody (rheumatoid factor) that locks onto the body's IgG molecules as if they were antigens, then deposits them on membranes at the joints. The deposits trigger the complement cascade and inflammation. The membranes become prime targets for abnormal events, including increased fluid seepage from capillaries. The accumulating fluid separates the membrane from underlying tissues, membrane cells divide repeatedly in response, and the joint thickens. These and other events continue in cycles of inflammation that do not end until the joint is totally destroyed.

### Deficient Immune Responses

On rare occasions, cell-mediated immunity is weakened and the body becomes highly vulnerable to infections that might not otherwise be life-threatening. This is what happens in **AIDS** (acquired immune deficiency syndrome).

AIDS is caused by the "human immunodeficiency virus," or HIV. The *Commentary* above describes some immunological aspects of HIV infection—how the virus replicates inside a human host and what the prospects are for treating or curing infected persons. Social implications of the worldwide AIDS epidemic are described on page 780, in a *Commentary* on sexually transmitted diseases.

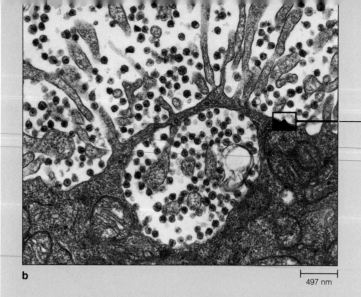

**b**

497 nm

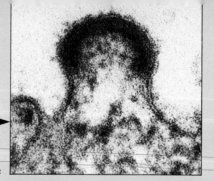

**c**

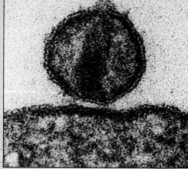

45 nm

**(b)** Transmission electron micrograph of HIV particleś (black specks) escaping from an infected cell. **(c)** Closer views of a virus particle budding from the host cell's plasma membrane.

are put together from the RNA and proteins. They bud from the plasma membrane of the host helper T cell or are released when the membrane ruptures (Figures *a–c*). With each new round of infection, more and more helper T cells are destroyed or impaired.

In time, the helper T cell population is depleted, and the body loses its ability to mount immune responses. Initially, infected persons may feel well or have a bout of flulike symptoms. In time, as the population of functional helper T cells undergoes serious depletion, a condition called ARC (AIDS-related complex) may develop. There may be persistent weight loss, joint and muscle pain, fatigue and malaise, nausea, bed-drenching night sweats, enlarged lymph nodes, various minor infections, and other symptoms. Eventually, when the body's ability to mount immune responses is entirely lost, full-blown AIDS appears. Often it is heralded by opportunistic infections, such as a form of pneumonia caused by a fungus (*Pneumocystis carinii*) and tuberculosis. Blue-violet or brown-colored spots may appear on the legs especially. These are signs of Kaposi's sarcoma, a deadly form of cancer that affects blood vessels in the skin and some internal organs.

## Modes of Transmission

Like any human virus, HIV requires a medium by which it can leave the body of its host, survive in the environment into which it is released, and enter a susceptible cell that can support its replication.

HIV is transmitted when bodily fluids of an infected person enter another person's tissues. Initially in the United States, transmission occurred most often among male homosexuals and among intravenous drug abusers who shared needles. As many as two-thirds of all drug abusers may now carry the virus. The incidence of HIV carriers also is increasing among heterosexuals in the general population. Besides this, HIV has been transmitted from infected mothers to their infants during pregnancy, birth, and breast-feeding. Contaminated blood supplies accounted for some cases before screening for HIV was implemented in 1985. In 1991 there were four cases of HIV transmission by way of donated tissues. In several developing countries, HIV has spread through contaminated transfusions and through reuse of unsterile needles by health care providers.

HIV generally cannot survive for more than about one or two hours outside the human body. Virus particles

## Case Study: The Silent, Unseen Struggles

Let us conclude this chapter with a case study of how the immune system helps *you* survive attack. Suppose on a warm spring day you are walking barefoot to class. Abruptly you stop: A thorn on the ground punctured one of your toes. Even though you remove the thorn at once, the next morning the punctured area is red, tender, and swollen. Yet a few days later, your foot is back to normal and you have forgotten the incident.

All that time your body had been struggling against an unseen enemy. Both the thorn and your bare foot carried some soil bacteria. When the thorn broke through your skin, it carried several thousand bacterial cells with it. Inside, the bacteria found conditions suitable for

growth. They soon doubled in number and were on their way to doubling again. Meanwhile, the products of bacterial metabolism were already starting to interfere with your own cell functions. If unchecked, the invasion would have threatened your life.

Yet as soon as your skin was punctured, defenses were being mobilized. Blood from ruptured blood vessels began to pool and clot around the wound. Histamine and other secretions from basophils and mast cells caused capillaries to dilate and become more permeable to plasma proteins, including complement. Now phagocytic white blood cells crawled through clefts between cells in capillary walls. Like bloodhounds on the trail, they moved toward higher complement concentrations. They began engulfing bacteria, dirt, and damaged cells.

on needles and other objects are readily destroyed by disinfectants, including household bleach. At this time, there is no evidence that HIV can be effectively transmitted by way of food, air, water, or casual contact. The virus *has* been isolated from blood, semen, vaginal secretions, saliva, tears, breast milk, amniotic fluid, cerebrospinal fluid, and urine. However, only infected blood, semen, vaginal secretions, and breast milk contain the virus in concentrations that seem high enough for successful transmission.

### Prospects for Treatment

At present, researchers may be close to developing vaccines. Among other things, they have isolated the genes for HIV proteins and are attempting to genetically engineer them in ways that might provoke effective immune responses. The task of developing an effective vaccine is formidable—HIV has the highest mutation rate of any known virus. It may be difficult to produce a vaccine that will work against all its mutated forms. Even if a vaccine can be developed that could coax the body into producing antibodies to HIV, the antibodies may not protect against AIDS. There is laboratory evidence that antibodies do not neutralize the virus.

The drug AZT (azidothymidine) is being used to prolong the life of AIDS patients. In combination with other drugs (such as interferon) or with bone marrow transplants, it may turn out to be useful in developing a cure. The search also is proceeding for compounds that might disrupt the ability of HIV to bind to the receptors by which it gains entry to cells. Other efforts involve looking for ways to interrupt the HIV replication cycle by inactivating a key protein-cutting enzyme that is necessary for viral replication. In the meantime, checking the spread of HIV depends absolutely on implementing behavioral controls through education on a massive scale. We return to this topic in Chapter 43.

If there had been no bacteria on the thorn or if they were unable to multiply rapidly in your tissues, then the inflammatory response would have cleaned things up. This time, bacterial cell divisions outpaced the nonspecific defenses—and B and T cells were called up.

If this had been your first exposure to the bacterial species that invaded your body, few B and T cells would have been present to respond to the call. The immune response would have been a primary one, and it would have been a week or more before B cells divided enough times to produce enough antibody. But when you were a child, your body did fight off this bacterial species, and it still carries vestiges of the struggle—memory cells. When the invader showed up again, it encountered an immune trap ready to spring.

As inflammation progressed, B and T cells were also leaving the bloodstream. Most were specific for other antigens and did not take part in the battle. But some memory cells locked onto antigens and became activated. They moved into lymph vessels with their cargo, tumbling along until they reached a lymph node and were filtered from the fluid. For the next few days, memory cells accumulated in the node, secreted communication signals and divided rapidly.

For the first two days the bacteria appeared to be winning; they were reproducing faster than phagocytes, antibody, and complement were destroying them. By the third day, antibody production peaked. The tide of battle turned. For two weeks or more, antibody production will continue until the bacteria are destroyed. After the response draws to a close, memory cells will go on circulating, prepared for some future struggle with this same invader.

## SUMMARY

1. The vertebrate body is equipped for these tasks:
   a. *Defense* against many viruses and bacteria, certain fungi, and some protozoans.
   b. *Defense* against mutant or cancerous cells.
   c. *Extracellular housekeeping* that eliminates dead cells and cellular debris from the internal environment.

2. "External" lines of defense against invasion include intact skin, exocrine gland secretions, gastric fluid, normal bacterial inhabitants of the body (which compete effectively against many invaders), and ciliated, mucous membranes of the respiratory tract.

3. The initial "internal" lines of defense include phagocytic cells (which engulf pathogens, dead body cells, and other debris) and the complement system. Complement proteins circulate in inactive form and act against bacteria and certain fungi. The ones generated during cascade reactions attract phagocytes. Some coat the surface of invading cells, enhancing its recognition by phagocytes; others cause the invading cells to lyse.

4. The body makes nonspecific (general) and specific responses to foreign agents that enter the internal environment. During the inflammatory response, phagocytes, complement proteins, and other factors are mobilized to destroy any agents detected as foreign, then to restore tissue conditions. They also are mobilized during immune responses.

5. White blood cells and their products are the basis of the immune system. Some of these cells show specificity (they attack only a particular pathogen, not invaders in general). They also show memory (they make rapid, secondary responses to the same pathogen whenever it is

**Table 39.1 Summary of White Blood Cells and Their Roles In Defense**

| Cell Types | Take Part In | Main Characteristics |
|---|---|---|
| Lymphocytes: | | |
| 1. Cytotoxic T cell | Cell-mediated immune response | Each cell equipped with membrane receptors specific for one type of antigen; each can directly destroy virus-infected cells (and possibly cancer cells) by punching holes in them |
| 2. Helper T cell | Cell-mediated and antibody-mediated immune responses | Master switch of immune system; stimulates rapid divisions of cytotoxic T cell and B cell populations |
| 3. Suppressor T cell | Same as above | Modulates degree of immune response (slows down or prevents activity by other lymphocytes) |
| 4. Virgin B cell | Antibody-mediated immune response | Not-yet-activated lymphocyte with *membrane-bound* antibodies (serving as antigen-specific receptors) |
| 5. Plasma cell | Same as above | *Antibody-secreting* descendant of an activated B cell |
| 6. Memory cell | Cell-mediated or antibody-mediated immune responses | One of a clonal population of T cells or B cells set aside during a primary immune response that can make a rapid, secondary immune response to another encounter with the same type of invader |
| Macrophages | Inflammatory, cell-mediated, and antibody-mediated immune responses | Phagocytic (engulfs foreign agents and infected, damaged, or aged cells); develop from circulating monocytes and take up stations in tissues; present antigens to immune cells; secretions trigger T cell and B cell proliferation |
| Neutrophil | Inflammatory response | Phagocytic; most abundant type of white blood cell; dominates early stage of inflammation |
| Eosinophil | Inflammatory response | Phagocytic (engulfs antigen-antibody complexes, kills certain parasites); combats effect of histamine in allergic reactions |
| Basophil and mast cell | Inflammatory response | Release histamine and other substances that contribute to vasodilation and a rapid inflammatory response |
| Natural killer cell (NK) | Nonspecific response | Directly destroy tumor cells, some virus-infected cells; distinct from T and B cells |

encountered again). Table 39.1 summarizes these cells and their functions.

6. Cells of the immune system communicate with one another by chemical secretions (notably interleukins), which stimulate rapid growth and division of certain lymphocytes (B cells, cytotoxic T cells, and helper T cells) into large armies against particular invaders.

7. Lymphocytes mount immune responses against circulating antigen or against cells bearing foreign antigen *in combination with* self-MHC markers.

8. An antigen is any large molecule that lymphocytes perceive as foreign and that triggers an immune response. Antigens occur at the surface of viruses, bacterial cells, fungal cells, and so on.

9. An antibody-mediated immune response is made against antigen circulating in the body's tissues or attached to the surface of an invading pathogen. First, macrophages and virgin B cells become sensitized to antigen (they display antigen–MHC complexes at their surface). When helper T cells encounter antigen–MHC complexes, they stimulate the virgin B cell to divide. Some B cell progeny develop into antibody-secreting plasma cells, others become memory B cells. Antibody molecules bind to specific antigens and mark the invaders for disposal (by macrophages or by complement).

10. A cell-mediated immune response is made against infected body cells and cancerous or mutant cells. Viral-infected cells have viral proteins combined with self-MHC markers on their surface. Cytotoxic T cells have receptors on their surface that may recognize the antigen–MHC complex. They directly destroy infected cells before the virus can replicate and spread to other cells. Natural killer cells, which are less specific about their targets, do the same.

11. After a primary (first-time) immune response, portions of the B and T cell populations produced continue to circulate as memory cells. They are available for a rapid, amplified response to subsequent encounters with the same antigen (a secondary immune response).

## Review Questions

1. The vertebrate body has physical and chemical barriers against invading pathogens. Name five such barriers. *682*

2. Which four events characterize an inflammatory response? *683*

3. The vertebrate immune system is characterized by *specificity* and *memory*. Describe what these terms mean. *685, 691*

4. Define the following types of white blood cells: macrophages, helper T cells, B cells, cytotoxic T cells, suppressor T cells, and memory cells. *685*

5. Are phagocytes deployed during nonspecific defense responses, immune responses, or both? *683, 685*

6. Antibodies and interleukins are central to immune responses. Define them and state their functions. *685, 686*

7. What is immunization? What is a vaccine? *692*

8. What is the difference between an allergy and an autoimmune response? What type of disease is AIDS? *692–693*

8. Which of the following is *not* a molecular cue that triggers a normal immune response?
   a. self-MHC marker alone
   b. antigen
   c. antigen combined with self-MHC marker
   d. damaged or mutant self-MHC marker
   e. all of the above serve as molecular cues

9. An antibody is _____.
   a. an activated plasma cell
   b. a receptor molecule with binding sites for virgin B cells
   c. a receptor molecule with binding sites for antigen
   d. an out-of-body experience

10. Match the immunity concepts.
   _____ cytotoxic T cells
   _____ helper T cells
   _____ macrophages
   _____ some B cell progeny
   _____ portions of B and T cell populations

   a. stimulate virgin B cells and cytotoxic T cells to divide
   b. destroy infected cells by punching holes in them
   c. circulate as memory cells
   d. origin of plasma cells
   e. phagocytosis; stimulate helper T cells

## Self-Quiz *(Answers in Appendix IV)*

1. _____ are any large molecules that white blood cells perceive as foreign and that elicit an immune response.

2. *Antibody-mediated* immune responses are made against _____; *cell-mediated* ones against _____.

3. External barriers to invasion include _____.
   a. unbroken skin
   b. lysozyme
   c. gastric fluid
   d. ciliated mucous membranes
   e. all of the above

4. Inflammatory responses require _____ and _____ as well as other factors to destroy foreign agents.
   a. complement, anticomplement
   b. phagocytes, antigens
   c. red blood cells, antigen
   d. phagocytes, complement

5. _____ are the fighting cells of the immune system.
   a. Red blood cells
   b. White blood cells
   c. Blue blood cells
   d. Antigens

6. The immune system shows _____ in that its cells are stimulated by specific antigens. It also exhibits _____, the ability to recognize the same invader upon subsequent attacks.
   a. communication; perception
   b. specificity; memory
   c. general responses; specific responses
   d. flexibility; recognition

7. The body's own uninfected cells are ignored by the immune response when they bear _____ at their surface.
   a. complement
   b. self-MHC markers
   c. antigen
   d. antigen plus self-MHC markers

## Selected Key Terms

allergy *692*
antibody *686*
antibody-mediated response *686*
antigen *685*
autoimmune response *693*
B cell *685*
clonal selection theory *691*
clone *691*
complement system *682*
cytotoxic T cell *685*
helper T cell *685*
histamine *683*
immune system *685*
immunization *692*
immunoglobulin *687*
inflammatory response *682*
interferon *688*
interleukin *686*
lysis, *682*
macrophage *685*
memory *685*
memory cell *685*
MHC marker *685*
natural killer cell *688*
passive immunity *692*
pathogen *682*
perforin *688*
plasma cell *686*
primary immune response *686*
secondary immune response *691*
suppressor T cell *685*
vaccine *692*

## Readings

Golub, E. 1987. *Immunology: A Synthesis*. Second edition. Sunderland, Massachusetts: Sinauer Associates.

Kimball, J. 1990. *Introduction to Immunology*. Third edition. New York: Macmillan.

Leder, P. May 1982. "The Genetics of Antibody Diversity." *Scientific American* 246(5):102–115.

Roitt, I., J. Brostoff, and D. Male. 1989. *Immunology*. St. Louis: Mosby. Second edition. Lavishly illustrated.

Tizard, I. 1988. *Immunology: An Introduction*. Second edition. Philadelphia: Saunders.

# 40 RESPIRATION

**Figure 40.1** A climber approaching the summit of Chomolungma, where oxygen is brutally scarce.

## Conquering Chomolungma

To experienced climbers, possibly the ultimate challenge is Chomolungma, a Himalayan mountain that also goes by the name Everest (Figure 40.1). Its summit, 9,700 meters (29,108 feet) above sea level, is the highest place on earth.

To conquer Chomolungma, climbers must be skilled enough to ascend vertical, iced-over rock in driving winds, smart enough to survive blinding blizzards, and lucky enough to escape heart-stopping avalanches. To conquer Chomolungma, they also must come to terms with a severe scarcity of oxygen that can result in long-lasting damage to the brain.

Of the air we breathe, only one molecule in five is oxygen. The earth's gravitational pull keeps oxygen molecules concentrated near sea level—or, as we say for gases, under pressure. We use so much oxygen (for aerobic respiration) that a pressure gradient exists between the outside air and the tissues inside our body. Red blood cells pick up and deliver oxygen along that gradient. Their hemoglobin becomes more or less saturated with oxygen in the lungs (the top of the gradient), then releases it in oxygen-depleted tissues (the bottom of the gradient).

Most of us live at low elevations. When we find ourselves in mountains taller than 3,300 meters (10,000 feet), the breathing game changes. Gravity's pull is less pronounced, gaseous molecules spread out—and the pressure gradient decreases. What would be a normal breath at sea level simply will not deliver enough oxygen into the lungs. The oxygen deficit can produce *altitude sickness*. Symptoms range from shortness of breath, headache, and heart palpitations, to loss of appetite, nausea, and vomiting.

The Chomolungma base camp is 6,300 meters (19,000 feet) above sea level. When climbers reach it, they have left behind more than half of the oxygen in the earth's atmosphere. Higher up, at 7,000 meters (23,000 feet), conditions start to get murderous. Apparently, when

air pressure drops dramatically and oxygen becomes extremely scarce, blood vessels become leaky. Tissues in the brain and lungs become swollen with excess fluid, a condition called edema. With severe edema, climbers become comatose and die.

Given the risks, experienced climbers keep themselves in top physical condition. Besides this, climbers live for several weeks at high elevations before their assault on the summit. They know that "thinner air" triggers the formation of billions of additional red blood cells. Thinner air also stimulates the production of more blood capillaries, mitochondria, and myoglobin. (Myoglobin, an oxygen-binding protein, is present in skeletal muscle cells that depend on blood flow for oxygen deliveries.) These and other physiological changes improve the odds of survival. They are transient changes only. After the descent, the capacity

698

**1.** Of all organisms, animals are the most active. The energy to drive their activities comes mainly from aerobic metabolism, which uses oxygen and produces carbon dioxide wastes. In a process called respiration, animals move oxygen into their internal environment and give up carbon dioxide to the external environment.

**2.** All respiratory systems make use of the tendency of any gas to diffuse down its pressure gradient. Such a gradient exists between oxygen in the atmosphere (high pressure) and the metabolically active cells in body tissues (where oxygen is used rapidly; pressure is lowest here). Another gradient exists between carbon dioxide in body tissues (high pressure) and the atmosphere (with its lower amount of carbon dioxide).

**3.** Respiratory systems are all alike in having a respiratory surface—a thin, moist layer of epithelium that gases can readily diffuse across. In most animals, the oxygen is picked up by the general circulation and transported to body tissues, where carbon dioxide is picked up and transported back to the respiratory surface.

**4.** Respiratory systems differ in their adaptations for increasing gas exchange efficiency. They differ also in how they match air flow to blood flow.

for oxygen transport and utilization will return to normal.

Even with extensive preparation, climbers still can become incapacitated. One treatment for acute mountain sickness is to inhale bottled oxygen. More recently, a few casualties on Chomolungma were zipped inside an experimental inflatable bag. For the next two hours, oxygen was pumped inside the airtight bag and carbon dioxide removed from it until the internal pressure was the equivalent of descending 6,000 to 9,000 feet. Those climbers were among the lucky; they survived.

Few of us will ever find ourselves on Chomolungma, the roof of the world, pushing our reliance on oxygen to the limits. Here in the lowlands, disease, smoking, and other environmental insults push it in more ordinary ways—although the risks can be just as great, as you will see by this chapter's end.

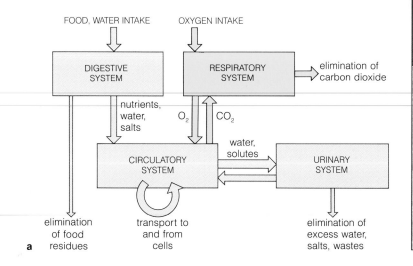

**Figure 40.2** (**a**) Roles of the respiratory system in complex animals. Unlike humans (**b**), flatworms (**c**) are small enough that a circulatory system is not required; oxygen can reach individual cells simply by diffusing across the body surface. Unlike flatworms, humans would never survive on the low concentrations of oxygen dissolved in water.

H igh in the mountains, in underground nooks and burrows, in shallow waters and deep in the oceans, you will find oxygen-dependent animals. They use oxygen for aerobic respiration—the only metabolic pathway that generates enough energy for their activities. And they give up carbon dioxide by-products of metabolism to the surroundings. In a process called **respiration**, oxygen moves into the internal environment of such animals and carbon dioxide is released to the external environment. The respiratory system works in conjunction with other organ systems in the body, most notably the circulatory system. Figure 40.2 is a diagram of their interrelationships.

## THE NATURE OF RESPIRATORY SYSTEMS

### Factors That Affect Gas Exchange

**Fick's Law**. Respiratory systems are diverse, but they are all alike in their reliance on *diffusion* of gases. Like other substances, oxygen and carbon dioxide diffuse down concentration gradients—or, as we say for gases, down pressure gradients. The more concentrated the molecules of a gas are outside the body, the higher the pressure and the greater the force available to drive individual molecules inside, and vice versa.

Oxygen and carbon dioxide are not the only gases in water or air. At sea level, a given volume of dry air is approximately 78 percent nitrogen, 21 percent oxygen, 0.04 percent carbon dioxide, and 0.96 percent other gases. Each gas obviously exerts only part of the total pressure exerted by the whole mix of gases. Said another way, each exerts a "partial pressure."

At sea level, atmospheric pressure is about 760mm Hg, as measured by a mercury barometer (Figure 40.3). Thus, the partial pressure of oxygen is (760 × 21/100), or about 160mm Hg. The partial pressure of carbon dioxide is about 0.3mm Hg.

Like any gas, oxygen and carbon dioxide tend to diffuse from areas of high to low partial pressure. Respiratory systems take advantage of this tendency, for they work with partial pressure gradients that exist between the internal and external environments. In all cases, gases diffuse across a thin, moist membrane called the **respiratory surface**. The surface must be kept moist at all times, for gases will diffuse into and out of an animal only when they are first dissolved in some fluid.

What determines the amount of oxygen or carbon dioxide diffusing across the respiratory surface in a given time? According to **Fick's law**, the amount depends on the surface area of the membrane and the differences in partial pressure across it. The more extensive the surface area and the larger the pressure gradient, the faster will be the diffusion rate.

**Surface-to-Volume Ratio**. An animal's surface-to-volume ratio affects diffusion rates. Imagine an animal growing in all directions, like an inflating balloon. As it expands, its surface area does not increase at the same rate as its volume. (As we saw in Figure 4.0, volume increases with the cube of its dimensions, but surface

c

**Figure 40.3** Atmospheric pressure as measured by a mercury barometer. At sea level, the level of mercury (Hg) in a glass column is about 760 millimeters (29.91 inches). At this level, the pressure exerted by the column of mercury equals atmospheric pressure outside the column.

760mm Hg

area only increases with the square.) Without further adaptations in the body plan, the animal would die once its diameter exceeded a single millimeter, for the diffusion distance between the respiratory membrane and all of its internal cells would be too great. That is why animals without respiratory organs have flattened or tubelike bodies; most internal cells are kept close to the respiratory surface. Flatworms and roundworms are examples.

**Ventilation**. There are many more adaptations to the constraints imposed by diffusion rates. When bony fishes move the "lids" over their gills, for example, they are stirring the water around them—and the gases dissolved in it. They are actively **ventilating** the body surface so that the fluid just outside the respiratory membrane does not become depleted of oxygen and loaded with carbon dioxide.

Similarly, the movements of microvilli on the collar cells of sponges help ventilate the body surface and so help improve diffusion rates. So does the action of ciliated cells that line certain cavities and body surfaces of sea stars. And when muscles move your rib cage as you breathe, the movement contributes to pressure gradients across the respiratory membrane of your lungs.

**Transport Pigments**. As we have seen, diffusion is faster when pressure gradients are steep. Hemoglobin and other pigments associated with the circulatory system are enormously important in this regard. In your own body, each hemoglobin molecule binds loosely with as many as four oxygen molecules in lungs (where oxygen concentrations are high), and it releases them in tissue regions where oxygen concentrations are low. By transporting oxygen away from the respiratory surface, hemoglobin plays a major role in maintaining the required pressure gradient.

**Aquatic Environments**. A liter of water that is "saturated" with dissolved oxygen still only holds 5 percent as much oxygen as a comparable volume of air. And water is much more dense and viscous than air. Aquatic animals work hard to maintain oxygen pressure gradients across their respiratory surfaces. For example, bony fishes busily ventilating their gills are doing so at great metabolic cost. Whereas a trout might devote 20 percent of its energy output to stirring up water around its gills, a buffalo staring out over the plains might devote a mere 2 percent to breathing.

Aquatic environments influence diffusion rates in other ways. The saltier the water or the higher its temperature, the less oxygen it can hold. The less sunlight there is, the lower the amount of oxygen released into the water by algae and other photosynthetic organisms. And the less the water circulates, the more depleted the oxygen becomes as aquatic animals and decomposers use up what is available. We will return to this topic in Chapter 47.

**Land Environments**. Compared to water, air has far more oxygen—but it also poses a far greater threat to respiratory systems. Any time the moist respiratory membranes dry out, they stick together and gases no longer can be exchanged across them. Conversely, earthworms and other inhabitants of the underground must contend with variable availability of oxygen in the soil. After a thunderstorm, for example, water may fill the spaces between soil particles. With the flooding, oxygen levels drop and carbon dioxide levels rise. That is why you often see earthworms thrashing about at the soil surface after a heavy rain.

Let's turn now to examples of respiratory systems. In the simplest of these systems, gases are exchanged across the body surface. In others, they are exchanged at specialized respiratory surfaces in gills, tracheas, and lungs.

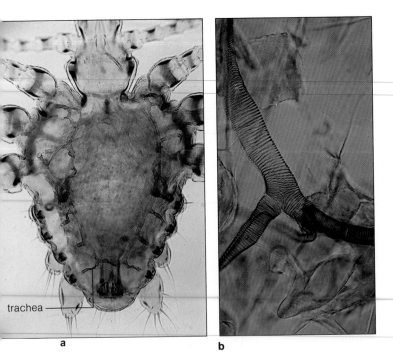

trachea

a        b

**Figure 40.4** (**a**) Respiratory system of an insect (a louse). (**b**) A closer view of some chitin-reinforced tracheas.

### Integumentary Exchange

Many animals do not have massive bodies or high metabolic rates, so their demands for respiration are not great. They rely on **integumentary exchange**, in which oxygen and carbon dioxide simply diffuse across a thin, vascularized layer of moist epidermis at the body surface. Most annelids, some small arthropods, and nudibranchs rely on integumentary exchange. To a large extent, so do frogs and other amphibians. For the water dwellers, the surroundings keep the respiratory surface moist. For land dwellers, mucus and other secretions provide the moisture.

For example, earthworms secrete mucus that helps moisten their integument, which is a single layer of epidermal cells. Oxygen molecules between soil particles dissolve in the mucus and diffuse across the integument. From there, oxygen diffuses into blood capillaries that project, fingerlike, between the epidermal cells. Pressure generated by muscular contractions of the body wall and by the pumping action of tiny "hearts" causes blood to circulate in the narrow, tubelike body. The bulk flow of blood enhances the transport of oxygen to individual cells.

### Specialized Respiratory Surfaces of Tracheas, Gills, and Lungs

The integument of many animals is too thick, too hardened, or too sparsely supplied with blood vessels to be a good respiratory surface. Also, animals larger than flatworms cannot depend on their integument alone to provide enough surface area for gas exchange. Without other adaptations in body plan, a larger animal would die. Gases could not diffuse across the integument fast enough to sustain the greater volume of interior cells. This is where tracheas, gills, lungs, and other specialized respiratory organs come in.

**Tracheas**. Insects and spiders are among the animals with air-conducting tubes called **tracheas**. Most insect tracheas are chitin-reinforced (Figure 40.4). They branch through the body and provide a rather self-contained system of gas conduction and exchange; assistance by a circulatory system is not required. Often a lid (spiracle) spans each opening at the body surface and helps keep the tubes moist by preventing evaporation.

Tracheas branch again and again until they become very fine dead-end tubes, and at their tips they are filled with the liquid that is obligatory for gas exchange. These blind tubes are most abundant in tissues with high oxygen demands. For example, many of them terminate against muscle cells.

Have you ever noticed how foraging bees stop every so often and pump the segments of their abdomen back and forth? The segments extend and retract like a telescope, forcing air into and out of the tracheal system. The stepped-up oxygen intake and carbon dioxide removal help support the high rate of metabolism required for insect flight.

**Gills**. A typical **gill** has a moist, thin, vascularized layer of epidermis that functions in gas exchange. External gills project from the body of some insects, the larval forms of a few fishes, and a few amphibians. The internal gills of adult fishes are rows of slits or pockets extending from the back of the mouth to the body surface. Water enters the mouth, moves down the pharynx, and flows out across the gills. As Figure 40.5 shows, the water moves *over* fish gills and blood circulates *through* them in opposite directions. Such movement of fluids in opposing directions is called **countercurrent flow**.

An extensive network of blood vessels is associated with the surface of the gills. Water passing over a fish gill first flows over the vessels leading back into the body. Blood inside the vessels contains less oxygen than the surrounding water, so oxygen diffuses inward. Next the water flows over blood vessels from deep body regions. This blood still has less oxygen than the (by now) oxygen-poor water. With the even greater difference in partial pressure, more oxygen diffuses inward. Through this opposing flow mechanism, fishes extract about 80 to 90 percent of the oxygen from water that flows over the gills. That is far more than they would get from a one-way flow mechanism, at far less energy cost.

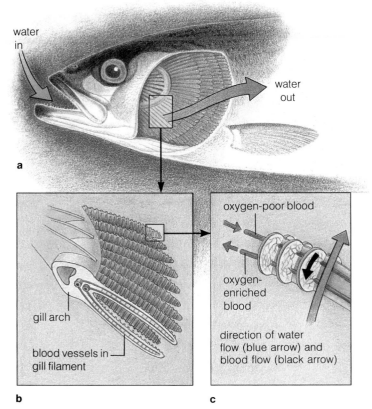

**a**

**b**            **c**

water in

water out

gill arch

blood vessels in gill filament

oxygen-poor blood

oxygen-enriched blood

direction of water flow (blue arrow) and blood flow (black arrow)

**Figure 40.5** Respiratory system of many fishes. (**a**) Location of gills. The bony covering over them has been removed for this sketch. (**b**, **c**) Each gill has extensive capillary beds between two blood vessels. One vessel carries oxygen-poor blood into the gills, the other carries oxygen-enriched blood back into the deeper body tissues. Blood flowing from one vessel to the other runs counter to the direction of water flowing over the gills. The arrangement favors the movement of oxygen (down its partial pressure gradient) into both vessels.

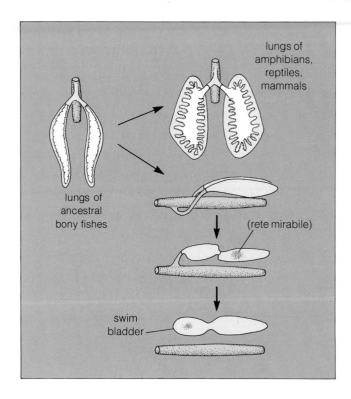

lungs of amphibians, reptiles, mammals

lungs of ancestral bony fishes

(rete mirabile)

swim bladder

**Figure 40.6** Evolution of vertebrate lungs and swim bladders. The esophagus (a tube leading to the stomach) is shaded gold; the respiratory tissues, pink.

Lungs originated as pockets off the anterior part of the gut; they increased the surface area for gas exchange in oxygen-poor habitats. In some lineages, lung sacs became modified into swim bladders: buoyancy devices that help keep the fish from sinking. Adjusting gas volume in the bladders allows fishes to remain at different depths.

Trout and other less specialized fishes have a duct between the swim bladder and esophagus; they replenish air in the bladder by surfacing and gulping air. Most bony fishes have no such duct; gases in the blood must diffuse into the swim bladder. Their swim bladder has a dense mesh of blood vessels (rete mirabile) in which arteries and veins run in opposite directions. Countercurrent flow through these vessels greatly increases gas concentrations in the bladder. Another region of the bladder allows reabsorption of gases by the body tissues.

**Lungs.** A **lung** is an internal respiratory surface in the shape of a cavity or sac. Simple lungs evolved more than 450 million years ago in fishes, and apparently they assisted respiration in oxygen-poor habitats. In some fish lineages, the lungs developed into moist, thin-walled organs called *swim bladders*. (Adjustments of gas volume in a swim bladder help maintain the body's position in the water. Some oxygen is also exchanged with blood and the surrounding tissues.) In other lineages, the lungs became complex respiratory organs, as indicated in Figure 40.6.

The evolution of lungs may be reflected in the respiratory systems of existing vertebrates. African lungfish have gills, but they also use lungs to supplement respiration. In fact, they will drown if they are kept from gulping air at the water's surface. Integumentary exchange still predominates in amphibians, but they, too, supplement respiration with a pair of small lungs. In reptiles, birds, and mammals, paired lungs are the major respiratory surfaces.

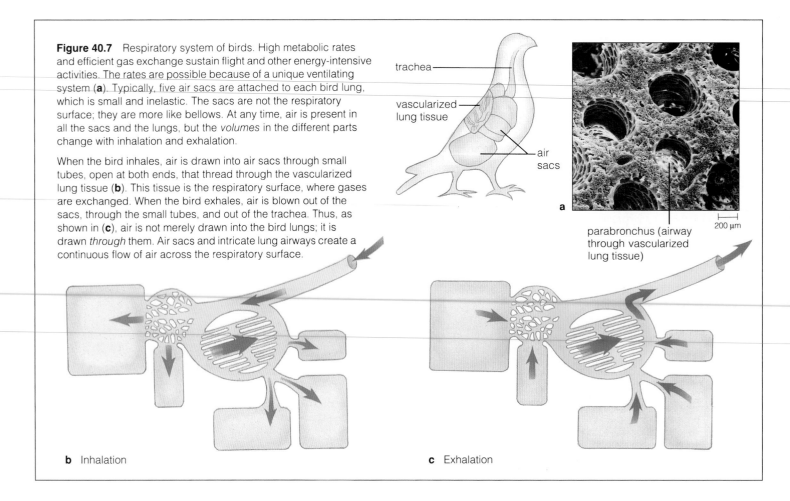

**Figure 40.7** Respiratory system of birds. High metabolic rates and efficient gas exchange sustain flight and other energy-intensive activities. The rates are possible because of a unique ventilating system (**a**). Typically, five air sacs are attached to each bird lung, which is small and inelastic. The sacs are not the respiratory surface; they are more like bellows. At any time, air is present in all the sacs and the lungs, but the *volumes* in the different parts change with inhalation and exhalation.

When the bird inhales, air is drawn into air sacs through small tubes, open at both ends, that thread through the vascularized lung tissue (**b**). This tissue is the respiratory surface, where gases are exchanged. When the bird exhales, air is blown out of the sacs, through the small tubes, and out of the trachea. Thus, as shown in (**c**), air is not merely drawn into the bird lungs; it is drawn *through* them. Air sacs and intricate lung airways create a continuous flow of air across the respiratory surface.

trachea

vascularized lung tissue

air sacs

a

200 μm

parabronchus (airway through vascularized lung tissue)

**b** Inhalation

**c** Exhalation

In all lungs, *airways* carry gas to and from one side of the respiratory surface, and *blood vessels* carry gas to and from the other side:

1. Air moves by bulk flow into and out of the lungs, and new air is delivered to the respiratory surface.

2. Gases diffuse across the respiratory surface of the lungs.

3. Pulmonary circulation (the bulk flow of blood to and from the lung tissues) enhances the diffusion of dissolved gases into and out of lung capillaries.

4. In other tissues of the body, gases diffuse between blood and interstitial fluid, then between interstitial fluid and individual cells.

Let's focus now on the human respiratory system; its operating principles are the same for most vertebrates. The major exception is the respiratory system of birds, shown in Figure 40.7. Birds possess air sacs that function in ventilating the lungs, where gas exchange occurs.

## HUMAN RESPIRATORY SYSTEM

### Air-Conducting Portion

The human respiratory system is shown in Figure 40.8. Air enters and leaves through the nose and, to a lesser extent, through the mouth. Hairs and ciliated epithelium lining the two nasal cavities filter out dust and other large particles. Also in the nose, incoming air becomes warmed and additionally picks up moisture from mucus.

The filtered, warmed, and moistened air moves into the **pharynx**, or throat; this is the entrance to both the **larynx** (an airway) and the esophagus (a tube leading to the stomach). When you breathe, a flaplike structure attached to the larynx points up. This is the *epiglottis*, shown in Figure 40.9.

*Vocal cords*, two thickened folds of the larynx wall, contain muscles that help produce the sound waves necessary for speech. Look again at Figure 40.9. Air forced through the space between the vocal cords (the glottis) gives rise to sound waves. The greater the air pressure on the vocal cords, the louder the sound. The greater the muscle tension on the cords, the higher the pitch of the sound.

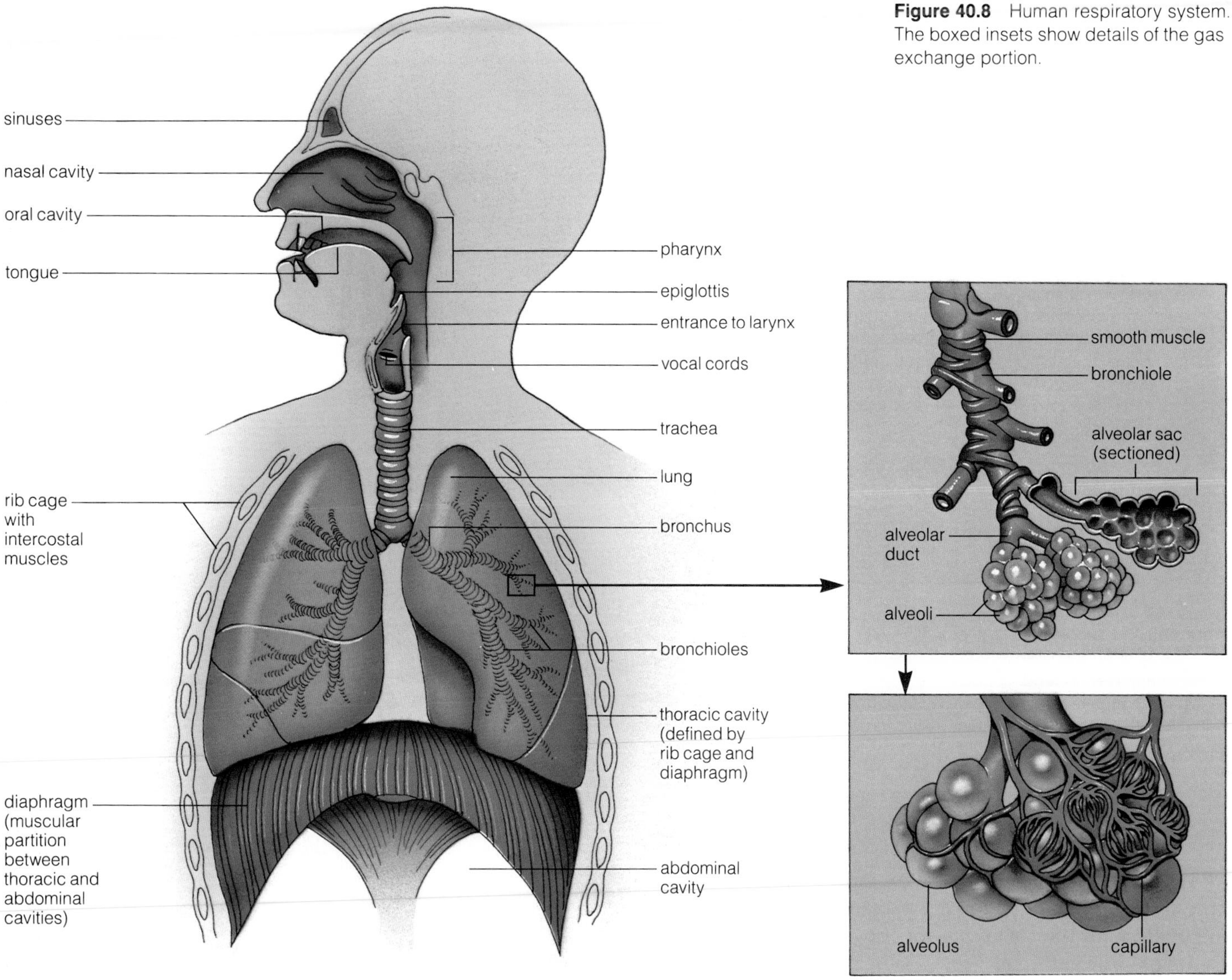

**Figure 40.8** Human respiratory system. The boxed insets show details of the gas exchange portion.

sinuses

nasal cavity

oral cavity

tongue

pharynx

epiglottis

entrance to larynx

vocal cords

trachea

lung

rib cage with intercostal muscles

bronchus

bronchioles

thoracic cavity (defined by rib cage and diaphragm)

diaphragm (muscular partition between thoracic and abdominal cavities)

abdominal cavity

smooth muscle

bronchiole

alveolar sac (sectioned)

alveolar duct

alveoli

alveolus

capillary

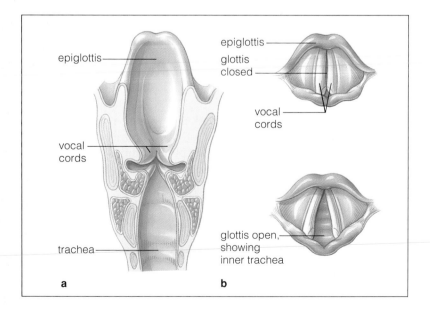

epiglottis

vocal cords

trachea

epiglottis

glottis closed

vocal cords

glottis open, showing inner trachea

a

b

**Figure 40.9** Where the sounds necessary for speech originate. (**a**) Front view of the larynx, showing the location of the vocal cords. (**b**) The two vocal cords as viewed from above when the glottis (the space between them) is closed or opened.

**Figure 40.10** The Heimlich maneuver. Each year, several thousand people choke to death when food enters the trachea instead of the esophagus (compare Figure 37.4). Strangulation can occur when the air flow is blocked for as little as four or five minutes. The Heimlich maneuver, an emergency procedure only, often can dislodge the misdirected chunks of food. The idea is to elevate the diaphragm forcibly, causing a sharp decrease in the chest cavity volume and a sudden increase in alveolar pressure. The increased pressure forces air up the trachea and may be enough to dislodge the obstruction.

To perform the Heimlich maneuver, stand behind the victim, make a fist with one hand, then press the fist, thumb-side in, against the victim's abdomen. The fist must be slightly above the navel and well below the rib cage. Next, press the fist into the abdomen with a sudden upward thrust. Repeat the thrust several times if needed. The maneuver can be performed on someone who is standing, sitting, or lying down.

Once the obstacle is dislodged, be sure the person is seen at once by a physician, for an inexperienced rescuer can inadvertently cause internal injuries or crack a rib. It could be argued that the risk is worth taking, given that the alternative is death.

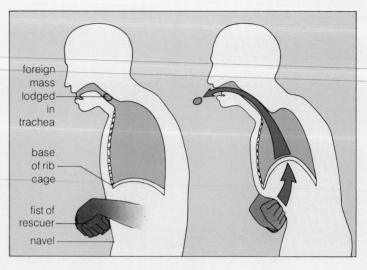

**Figure 40.11** Color-enhanced scanning electron micrograph of cilia (gold) in the respiratory tract. Mucus-secreting cells (rust-colored) are interspersed among the ciliated cells. Foreign material sticks to the mucus-coated microvilli at the free surface of these cells, then the cilia sweep the mucus-laden debris back toward the mouth.

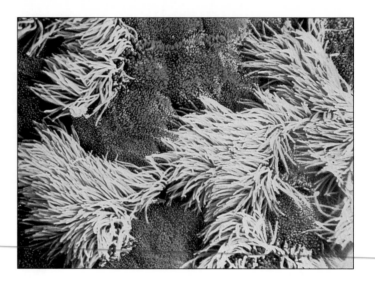

When you swallow, the larynx moves upward and presses the epiglottis down so that it partly covers the opening of the larynx. In this position, the epiglottis helps prevent food from going down the respiratory tract, the consequences of which can be fatal (Figure 40.10).

From the larynx, air moves into the **trachea**—the windpipe—which branches into the two airways leading into the lungs. Each airway is a **bronchus** (plural, bronchi). Its epithelial lining contains cilia and mucus-secreting cells, both with housekeeping roles (Figure 40.11). Bacteria and airborne particles stick in the mucus, then the upward-beating cilia sweep the debris-laden mucus toward the mouth.

### Gas Exchange Portion

Humans have two elastic, cone-shaped lungs, separated from each other by the heart. The lungs are located in the rib cage above the *diaphragm*, a muscular partition between the chest cavity and abdominal cavity. They are not attached directly to the wall of the chest cavity. Each is positioned within a *pleural sac*—a thin, doubled-over membrane of epithelium and loose connective tissue.

Imagine pushing a closed fist into a fluid-filled balloon (Figure 40.12). A lung occupies the same kind of position as your fist, and the pleural membrane folds back on itself, as does the balloon. Only an extremely narrow *intrapleural space* separates the two facing surfaces of the membrane, which are coated with a thin film of lubricating fluid. The fluid prevents friction between the pleural membranes while you breathe. When the pleural membrane becomes inflamed and swollen, friction initially occurs, and breathing can be quite painful. As mentioned in Chapter 35, this condition is called *pleurisy*.

Inside the lungs, airways become progressively shorter, narrower, and more numerous. The terminal air-

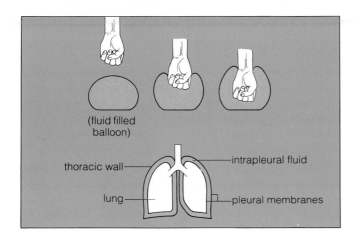

**Figure 40.12** Position of the lungs and pleural sac relative to the chest (thoracic) cavity. By analogy, when you push a closed fist into a fluid-filled balloon, the balloon completely surrounds the fist except at your arm. A lung is analogous to the fist; the balloon, to the pleural sac. Here, intrapleural fluid volume is enormously exaggerated for clarity.

ways, the **respiratory bronchioles**, have cup-shaped outpouchings from their walls. Each outpouching is an **alveolus** (plural, alveoli). Most often, alveoli are clustered together, forming a larger pouch called an **alveolar sac** (Figure 40.8). The alveolar sacs are the major sites of gas exchange.

A dense mesh of blood capillaries surrounds the 150 million or so alveoli in each lung. Together, the alveoli provide a tremendous surface area for exchanging gases with the bloodstream. If they were stretched out as a single layer, they would cover the floor of a racquetball court!

## AIR PRESSURE CHANGES IN THE LUNGS

### Ventilation

Every time you take a breath, you are ventilating the respiratory surfaces of your lungs. When you "breathe," air is inhaled (drawn into the airways), then exhaled (expelled from them). The air movements result from rhythmic increases and decreases in the chest cavity's volume. The changing volumes of the chest cavity reverse the pressure gradients between the lungs and the air outside the body—and gases in the respiratory tract follow those gradients.

Figure 40.13 shows what happens as you start to inhale. The dome-shaped diaphragm contracts and flattens, and muscles lift the ribs upward and outward. Then, as the chest cavity expands, the rib cage moves away slightly from the lung surface. Pressure in the

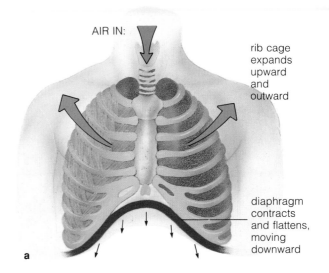

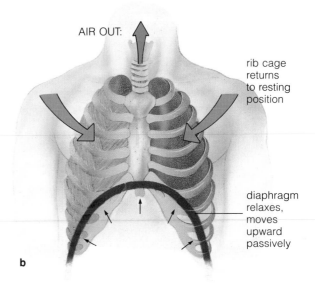

**Figure 40.13** Changes in the size of the chest cavity during breathing. Blue line indicates the position of the diaphragm during inhalation (**a**) and exhalation (**b**).

narrow space between each lung and the pleural sac becomes even lower than it was, compared to atmospheric pressure. The pressure difference causes the lung itself to expand more. The expansion allows fresh air to flow down the airways, almost to the respiratory bronchioles.

As you start to exhale, the elastic lung tissue recoils passively. The volume of the chest cavity decreases and compresses the air in alveolar sacs. The alveolar pressure becomes greater than the atmospheric pressure, so air follows the gradient and moves out from the lungs. When the demand for oxygen increases, as during exercise, muscles of the rib cage and abdomen are recruited to hasten airflow. Breathing is now "active."

## Lung Volumes

When you are resting, about 500 milliliters of air enter or leave your lungs in a normal breath. Although it takes a great leap of the imagination to compare the flow volume to the amount displaced by oceanic tides, this is nevertheless called the "tidal volume." The maximum volume of air that can move out of your lungs after a single, maximal inhalation is called the "vital capacity." You rarely use more than half the total vital capacity, even when you breathe deeply during strenuous exercise. To do so would exhaust the muscles used in respiration. Even at the end of your deepest exhalation, your lungs still cannot be completely emptied of air; about 1,000 milliliters would remain (Figure 40.14).

How much of the 500 milliliters of inhaled air is actually available for gas exchange? About 150 milliliters of exhaled air remain in the air-conducting tubes between breaths. Thus only (500 − 150), or 350 milliliters of fresh air reach the alveoli with each inhalation. When you breathe, say, 10 times a minute, you are supplying your alveoli with (350 × 10) or 3,500 milliliters of fresh air per minute.

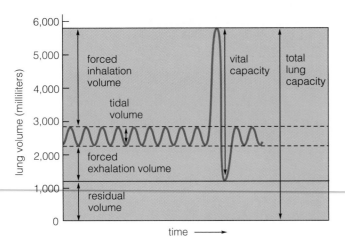

**Figure 40.14** Capacities of the human lung. During normal breathing, a tidal volume of air enters and leaves the lungs. Forced inhalation can bring much larger quantities into the lungs, and forced exhalation can release some of the air normally kept in the lungs. A residual volume of gas remains trapped in partially filled alveoli despite the strongest exhalation.

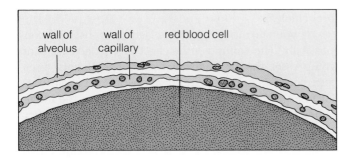

## GAS EXCHANGE AND TRANSPORT

### Gas Exchange in Alveoli

Each alveolus is only a single layer of epithelial cells, surrounded by a thin basement membrane. At most, a very thin film of interstitial fluid separates gas in the alveoli from blood in the lung capillaries. This is a narrow space, as Figure 40.15 suggests, and gases can diffuse rapidly across it.

Figure 40.16 shows the partial pressure gradients for oxygen and carbon dioxide throughout the human respiratory system. Passive diffusion alone is enough to move oxygen across the respiratory surface and into the bloodstream. And it is enough to move carbon dioxide in the reverse direction.

**Driven by its partial pressure gradient, oxygen diffuses from alveolar air spaces, through interstitial fluid, and into the lung capillaries.**

**Carbon dioxide, driven by its partial pressure gradient, diffuses in the reverse direction.**

### Gas Transport Between Lungs and Tissues

Blood can carry only so much oxygen and carbon dioxide in dissolved form. The transport of both gases must be enhanced to meet the requirements of humans and other large-bodied animals. Hemoglobin, a pigment in red blood cells, binds and transports oxygen and carbon dioxide. It increases oxygen transport by seventy times. It also increases carbon dioxide transport away from the tissues by seventeen times.

**Oxygen Transport.** There is plenty of oxygen and not much carbon dioxide in inhaled air that reaches the alveoli, but the opposite is true of blood entering the lung capillaries. So oxygen diffuses into the blood plasma, then into red blood cells. Once inside those cells, as many as four oxygen molecules can rapidly form a weak, reversible bond with each hemoglobin molecule. A hemoglobin molecule with oxygen bound to it is called **oxyhemoglobin**, or $HbO_2$. The amount of $HbO_2$ that forms depends on the partial pressure of oxygen. The higher the pressure, the more oxygen will be picked up, until all hemoglobin-binding sites are saturated.

**Figure 40.15** Diagram of a section through an alveolus and an adjacent blood capillary. By comparison to the diameter of the red blood cell, the diffusion distance across the capillary wall, the interstitial fluid, and the alveolar wall is exceedingly small.

HbO$_2$ holds onto its oxygen rather weakly and will give it up in tissues where the partial pressure of oxygen is lower than in the lungs. It especially does this in tissues where blood is warmer, has an increase in the partial pressure of carbon dioxide, and shows a decrease in pH. All four conditions occur in tissues having greater metabolic activity—hence greater demands for oxygen. That is why more oxygen is released from blood in vigorously contracting muscle tissues, for example.

**Carbon Dioxide Transport.** The partial pressure of carbon dioxide in metabolically active tissues is greater than it is in blood flowing through the capillaries threading through them. Carbon dioxide diffuses into the capillaries, then it is transported to the lungs. About 7 percent of the carbon dioxide remains dissolved in plasma. About 23 percent binds with hemoglobin, forming **carbaminohemoglobin** (HbCO$_2$). But most of it—approximately 70 percent—is transported in the form of bicarbonate (HCO$_3^-$).

The bicarbonate forms when carbon dioxide combines with water in plasma to form carbonic acid. The carbonic acid dissociates (separates) into bicarbonate and hydrogen ions:

$$CO_2 \; + \; H_2O \; \rightleftharpoons \; H_2CO_3 \; \rightleftharpoons \; HCO_3^- \; + \; H^+$$

The reactions proceed slowly in plasma, converting only 1 in every 1,000 molecules of carbon dioxide. Thus, the reaction rate is insignificant in plasma. But much of the carbon dioxide diffuses into red blood cells, which contain the enzyme **carbonic anhydrase**. With this enzyme, the reaction rate increases by 250 times. Inside red blood cells, most of the carbon dioxide that is not bound to hemoglobin is converted to carbonic acid and its dissociation products. As a result, the concentration of free carbon dioxide in the blood drops rapidly. Thus, carbonic anhydrase helps maintain the gradient that keeps carbon dioxide diffusing from interstitial fluid into the bloodstream.

What happens to the bicarbonate ions formed during the reaction? They tend to diffuse out of the red blood cells into the blood plasma. What about the hydrogen ions? Hemoglobin acts as a buffer for them and keeps the blood from becoming too acidic. A *buffer*, recall, is a molecule that combines with or releases hydrogen ions in response to changes in cellular pH.

The reactions are reversed in the alveoli, where the partial pressure of carbon dioxide is lower than it is in the surrounding capillaries. The water and carbon dioxide that form as a result of the reactions diffuse into the alveolar sacs. From there the carbon dioxide is exhaled from the body.

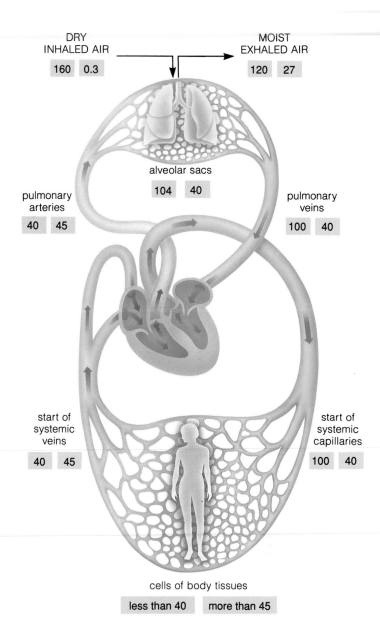

**Figure 40.16** Partial pressure gradients for oxygen (blue boxes) and carbon dioxide (pink boxes) through the respiratory tract.

The point to remember about the values shown is that *each gas moves from regions of higher to lower partial pressure.* That is why, for example, you become light-headed when you first visit places at high altitudes. The partial pressure of oxygen decreases with altitude, and your body does not function as well when the pressure gradient between the surrounding air and your lungs is lower than what you normally encounter.

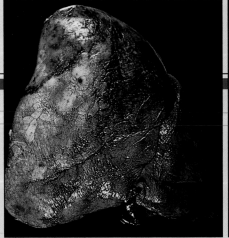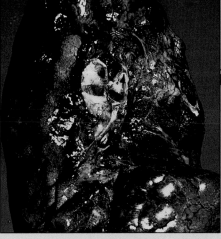

a

## Commentary

### When the Lungs Break Down

In large cities, in certain occupations, even near a cigarette smoker, airborne particles and irritant gases are present in abnormal amounts, and they put extra workloads on the lungs. Ciliated epithelium in the bronchioles is especially sensitive to cigarette smoke.

#### Bronchitis

A disorder called bronchitis can be brought on by smoking and other forms of air pollution that increase secretions of mucus and interfere with ciliary action in the lungs. Mucus and the particles it traps—including bacteria—accumulate in the trachea and bronchi, and this triggers coughing. The coughing persists as long as the irritation does and it aggravates the bronchial walls, which become inflamed. Bacteria or chemical agents start destroying the wall tissue. Cilia are lost from the lining, and mucus-secreting cells multiply as the body works to fight against the accumulating debris. Fibrous scar tissue forms and can obstruct parts of the respiratory tract.

#### Emphysema

An acute attack of bronchitis can be treated easily if the person is otherwise in good health. With continuing inflammation, however, fibrous scar tissue builds up and the bronchi become clogged with more and more mucus. Enzymes released from the inflammatory cells dissolve the elastic tissues of the alveoli, and the alveolar walls break down. Inelastic fibrous tissue comes to surround the alveoli. The remaining alveoli enlarge and the balance between air flow and blood flow is abnormal. The outcome is emphysema, in which the lungs are so distended and inelastic that gases cannot be exchanged efficiently. Running, walking, even exhaling can be difficult for those with emphysema.

(**a**) *Left:* Normal appearance of human lung tissue. *Right:* Appearance of lung tissue from someone affected by emphysema. (**b**) Cigarette smoke swirling into the human windpipe and down the two bronchial routes to the lungs.

Poor diet, smoking, and chronic colds and other respiratory ailments sometimes make a person susceptible to emphysema later in life. And many who suffer from emphysema do not have a functional gene coding for antitrypsin. This substance inhibits tissue-destroying enzymes produced by the inflammatory cells.

Emphysema can develop slowly, over twenty or thirty years. By the time it is detected, the damage to lung tissue cannot be repaired. On average, 1.3 million people in the United States alone suffer from the disorder.

#### Effects of Cigarette Smoke

The table below Figure *b* lists some effects of cigarette smoke on the lungs and other organs. Cilia in the bronchioles can be kept from beating for several hours by noxious particles in smoke from one cigarette. The particles also stimulate mucus secretions, which in time can clog the airways. They can kill the infection-fighting phagocytes that normally patrol the respiratory epithelium. "Smoker's cough" is not the only outcome; the coughing can pave the way for bronchitis and emphysema. Marijuana smoke also can cause extensive lung damage.

#### Controls Over Respiration

Gas exchange is most efficient when the rate of air flow is matched with the rate of blood flow. Both rates can be adjusted locally (in the alveoli of the lungs) and in the tissues of the body as a whole.

**Local Controls.** Local controls come into play in the lungs themselves when there are imbalances between air flow and blood flow. For example, when your heart is pounding and you don't breathe deeply enough, carbon dioxide levels increase in alveoli. The increase affects smooth muscle in the bronchiole walls. Bronchioles dilate, improving the match between the rates of air and blood flow.

Similarly, a decrease in carbon dioxide levels causes the bronchiole walls to constrict—thereby decreasing the air flow.

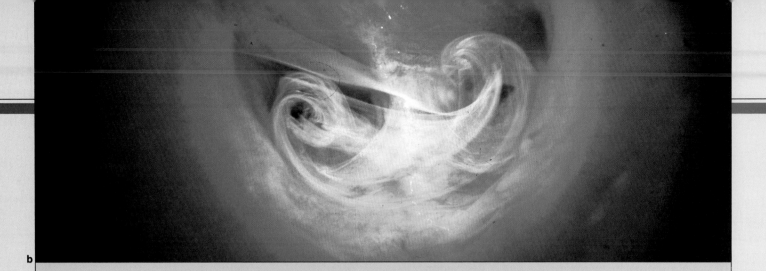

b

| **Risks Associated with Smoking** | **Benefits of Quitting** |
|---|---|
| *Shortened Life Expectancy:* Nonsmokers live 8.3 years longer on average than those who smoke two packs daily from the midtwenties on | Cumulative risk reduction; after 10 to 15 years, life expectancy of ex-smokers approaches that of nonsmokers |
| *Chronic Bronchitis, Emphysema:* Smokers have 4–25 times more risk of dying from these diseases than do nonsmokers | Greater chance of improving lung function and slowing down rate of deterioration |
| *Lung Cancer:* Cigarette smoking the major cause of lung cancer | After 10 to 15 years, risk approaches that of nonsmokers |
| *Cancer of Mouth:* 3–10 times greater risk among smokers | After 10 to 15 years, risk is reduced to that of nonsmokers |
| *Cancer of Larynx:* 2.9–17.7 times more frequent among smokers | After 10 years, risk is reduced to that of nonsmokers |
| *Cancer of Esophagus:* 2–9 times greater risk of dying from this | Risk proportional to amount smoked; quitting should reduce it |
| *Cancer of Pancreas:* 2–5 times greater risk of dying from this | Risk proportional to amount smoked; quitting should reduce it |
| *Cancer of Bladder:* 7–10 times greater risk for smokers | Risk decreases gradually over 7 years to that of nonsmokers |
| *Coronary Heart Disease:* Cigarette smoking a major contributing factor | Risk drops sharply after a year; after 10 years, risk reduced to that of nonsmokers |
| *Effects on Offspring:* Women who smoke during pregnancy have more stillbirths, and weight of liveborns averages less (hence, babies are more vulnerable to disease, death) | When smoking stops before fourth month of pregnancy, risk of stillbirth and lower birthweight eliminated |
| *Impaired Immune System Function:* Increase in allergic responses, destruction of macrophages in respiratory tract | Avoidable by not smoking |

Cigarette smoke contributes to lung cancer. Inside the body, certain compounds in coal tar and cigarette smoke become converted to highly reactive intermediates. These are the real carcinogens; they provoke uncontrolled cell divisions in lung tissues. In its terminal stage, the pain associated with lung cancer is agonizing.

Susceptibility to lung cancer is related to how many cigarettes the individual smokes daily and to how many times and how deeply smoke is inhaled. Cigarette smoking is responsible for at least 80 percent of all lung cancer deaths. It is a disorder that only 10 out of 100 smokers will survive.

Local changes also occur in the diameter of blood vessels that supply different lung regions. If air flow is too great relative to the blood flow, local concentrations of oxygen rise. The increase directly affects smooth muscle in the blood vessel walls, which undergo vasodilation, and thereby increase blood flow to the region. Similarly, if air flow is too small, vasoconstriction leads to a decrease in blood flow to the region, as described on page 673.

By now, it should be clear that gas exchange mechanisms at work in your body are like a fine Swiss watch—intricately coordinated and smooth in their operation. The *Commentary* describes a few respiratory disorders, some serious enough to stop the watch from ticking.

**Neural Controls**. The nervous system controls oxygen and carbon dioxide levels in arterial blood for the entire body. It does this by adjusting contractions of the diaphragm and muscles in the chest wall, and so adjusts the rate and depth of breathing.

The brain receives input from sensory receptors that can detect rising carbon dioxide levels in the blood. Such increases affect the $H^+$ concentration in cerebrospinal fluid, and the shift in pH stimulates the receptors. The brain also receives input from sensory receptors in the walls of arteries. Among the receptors are *carotid bodies* where the carotid arteries branch to the brain and the *aortic bodies* in arterial walls near the heart. Among other things, both types of receptors can detect changes in the partial pressure of oxygen dissolved in arterial blood. The brain responds by increasing the rate of ventilation, so more oxygen can be delivered to affected tissues.

Contraction of the diaphragm and muscles that move the rib cage are under the control of neurons in the brain's reticular formation (page 571). One cell cluster in this formation coordinates the signals calling for inhalation; another coordinates the signals for exhalation. The resulting rhythmic contractions are fine-tuned by respiratory centers in other parts of the brain, which can stimulate or inhibit both cell clusters.

## RESPIRATION IN UNUSUAL ENVIRONMENTS

### Decompression Sickness

In the world's oceans, the water pressure increases greatly with depth. Pressure in itself is not a problem, as long as it is dealt with. For example, to prevent their lungs from collapsing, deep-sea divers rely on tanks of compressed air (that is, air under pressure). Divers also must deal with the additional gaseous nitrogen ($N_2$) dissolved in their body fluids and tissues, especially adipose tissues.

As a diver ascends, the total pressure of the surrounding water decreases, so $N_2$ tends to move out of the tissues and into the bloodstream. If the ascent is too rapid, the $N_2$ comes out of solution faster than the lungs can dispose of it. When this happens, bubbles of $N_2$ form. Too many bubbles cause pain, especially at the joints. Hence the common name, "the bends," for what is otherwise known as *decompression sickness*. Bubbles that obstruct blood flow to the brain can lead to deafness, impaired vision, and paralysis. At depths of about 150 meters, $N_2$ poses still another threat, for at high partial pressures it produces feelings of euphoria, even drunkenness, and divers have been known to offer the mouthpiece of their airtank to a fish.

Weddell seals, fin whales, and other marine mammals have built-in respiratory adaptations that allow them to dive deeply with impunity. Members of a more ancient vertebrate lineage do the same, as the *Doing Science* essay describes.

## The Leatherback Sea Turtle

Late on a starlit night in May, biologist Molly Lutcavage and several colleagues leave their bungalows on an island in the Caribbean and set out on a turtle patrol. Picking their way down the beach, they keep a steady eye on the pale surfline. Finally a large, dark shape with flippers flailing appears in the surf. A female Atlantic leatherback sea turtle (*Dermochelys coriacea*) is emerging from the sea to excavate a nest in the sand.

Turtles often are portrayed as plodding representatives of the reptilian lineage, which extends 300 million years back in time. Yet sea turtles are graceful swimmers in the world where they spend most of their lives. To add grave injury to insult, all modern sea turtles—be they ridleys, loggerheads, green turtles, or leatherbacks—are now endangered or threatened species. They are on the brink of extinction.

The race is on to learn about the life histories and physiology of sea turtles—information that may help save them from extinction. The leatherbacks are the largest and least understood. Adult females may weigh more than 400 kilograms (880 pounds). An adult male may weigh nearly twice that amount.

Leatherbacks normally leave the water only to breed and lay their eggs. The rest of the time they migrate across vast stretches of the open ocean. And leatherbacks do something no other reptile on earth can do. They can dive as deep as 1,000 meters (3,000 feet) below sea level. Prior to tracking experiments in the mid-1980s, Weddell seals, fin whales, and some other marine mammals were the only natural deep-sea divers known.

By some estimates, a turtle would have to swim for nearly forty minutes to reach such depths—all without taking a breath! How do they dive so deep, for so long, and still have enough oxygen for aerobic metabolism?

Consider that underwater staying power has nothing to do with the lungs. (Even at depths between 80 and 160 meters, the pressure exerted by the surrounding water causes air-filled lungs to collapse.) However, marine mammals have an abundance of myoglobin, the oxygen-binding protein, in their muscle cells. They store plenty of oxygen in their muscles. Compared to other mammals, they also have more blood per unit of body weight, and their blood contains more red blood cells—hence more hemoglobin. Such adaptations allow marine mammals to engage in aerobic respiration without having to come up for air.

Could it be that the leatherback—a bona fide reptile—has mammal-like adaptations for diving?

**a** An endangered species—female leatherback turtles (*Dermochelys coriacea*), returning to the sea.

Imagine yourself with Lutcavage's team on the sandy beach. You want to draw blood from a leatherback so you can study the oxygen-binding capacity of its hemoglobin. You want to obtain samples of respiratory gases to learn more about how the leatherback actually uses oxygen. You cannot do either when the turtle is swimming in the open ocean. But for the twenty minutes or so when a nesting female is actually depositing eggs on a beach, she enters a trancelike state. During that time, she will not resist being gently handled.

The team gets to work as soon as the female begins her egg-laying behavior. You fit a specially constructed helmet equipped with an expandable plastic sleeve over her head. The fit is snug but not restrictive, so the turtle can breathe normally. With each exhalation, the expired gases flow through a one-way valve into a collection bag. At the same time, you count the number of breaths required to completely fill the bag. (As a baseline, you also tallied the breathing frequency before you placed the helmet over her head.) Then you store the gas sample according to established procedures and set it aside for laboratory analysis.

Working quickly, another member of the team draws blood from a superficial vein in the turtle's neck, then packs the sample in ice to preserve it. Back in the laboratory, the sample will be split into three portions—one for hemoglobin analysis, another for a red blood cell count, and the third for measuring the oxygen level, carbon dioxide level, and pH of the turtle's blood.

Lutcavage and her coworkers gathered samples of blood and respiratory gases from several turtles. They also analyzed skeletal muscle tissue, taken earlier from a drowned turtle. And they pieced together the following picture of how leatherbacks manage their spectacularly deep dives.

From calculations of the turtle's oxygen uptake and total tidal volume during breathing, it became clear that a leatherback cannot inhale and hold enough air in the lungs to sustain aerobic metabolism during a prolonged dive. Other oxygen-supplying mechanisms must be at work.

Analysis also showed that myoglobin levels are high in leatherbacks. In fact, when air in their lungs gives out (or when their lungs collapse) during a dive, large amounts of oxygen would still be available in the muscles. Those muscles therefore could work longer at swimming.

Besides this, leatherbacks have a notable abundance of red blood cells. And the hemoglobin of those cells has a truly high affinity for binding oxygen. (A previous study had revealed the presence of cofactors in leatherback hemoglobin. The cofactors seem to physically alter the positions of potential oxygen-binding sites in ways that maximize oxygen uptake.) As a result, a leatherback's blood may be a remarkable 21 percent oxygen by volume. That is an amount characteristic of a highly active human, not a "plodding turtle." In fact, the oxygen-carrying capacity of leatherbacks is the highest ever recorded for a reptile—and they are close to the capacity of deep-diving mammals. Add to this a myoglobin-based oxygen-storage system, a streamlined body, and massive front flippers, and you have a reptile uniquely adapted for diving.

Studies of leatherback sea turtles are only beginning. Measurements of oxygen consumption by a female nesting on a beach probably do not reveal much about the turtle's normal, day-to-day oxygen use when it swims in the ocean. Researchers are pursuing radio telemetry methods that will let them monitor a turtle's metabolism at sea and during dives. So far, it has proved extremely difficult and expensive to track particular turtles in the vastness of the seas.

### Physiological Comparison of a Few Diving Reptiles and Mammals

|  | Leatherback Turtle | Green Turtle | Crocodile | Killer Whale | Weddell Seal |
|---|---|---|---|---|---|
| Maximum diving depth (meters) | Greater than 1,000 | Less than 100 | Less than 30 | 260 | 600 |
| Red blood cell count (percent of total volume) | 39 | 30 | 28 | 44 | 58 |
| Hemoglobin level (grams/ deciliter of blood) | 15.6 | 8.8 | 8.7 | 16.0 | 17–22 |
| Myoglobin level (milligrams/ gram of muscle tissue) | 4.9 | —* | —* | —* | 44.6 |
| Oxygen-carrying capacity (volume percent) | 21 | 7.5–11.9 | 12.4 | 23.7 | 31.6 |

*Not known.

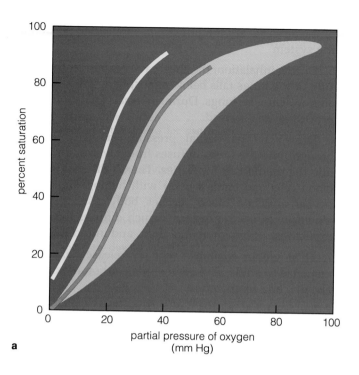

a

b

**Figure 40.17** (**a**) Comparison of the binding and releasing capacity of human hemoglobin (red line) and llama hemoglobin (yellow line). The area shaded light blue indicates the range that is typical for most mammals. (**b**) Llamas high up in the Peruvian Andes.

## Hypoxia

We conclude this chapter by coming full circle back to the example used to introduce it. The partial pressure of oxygen decreases with increasing altitude. Unlike the occasional climbers of Chomolungma, humans, llamas, and other animals accustomed to living at high altitudes have permanent adaptations to the thinner air. While they were growing up, more air sacs and blood vessels developed in their lungs. The ventricles in their heart became enlarged enough to pump larger volumes of blood. Llamas have an additional advantage. Compared to human hemoglobin, llama hemoglobin has a greater affinity for oxygen. It picks up oxygen more efficiently at the lower pressures characteristic of high altitudes (Figure 40.17).

Visitors who have not had time to adapt to the thinner air at high altitudes can suffer *hypoxia*, or cellular oxygen deficiency. Generally, at 2,400 meters (about 8,000 feet) above sea level, respiratory centers work to compensate for the oxygen deficiency by triggering *hyperventilation* (breathing much faster and more deeply than normal). As we saw at the start of this chapter, altitude sickness sets in at around 3,300 meters (10,000 feet).

Hypoxia also occurs when the partial pressure of oxygen in arterial blood falls because of *carbon monoxide poisoning*. Carbon monoxide, a colorless, odorless gas, is present in automobile exhaust fumes and in smoke from tobacco, coal, or wood burning. It binds to hemoglobin at least 200 times more strongly than oxygen does. Even very small amounts can tie up half of the body's hemoglobin and so impair oxygen delivery to tissues.

## Tale of the Desert Rat

Judging from the fossil record, about 375 million years ago some animals left the ancient, shallow seas behind and began invading the land. They had a demanding evolutionary legacy—their cells were geared to operating in a salty fluid of relatively stable temperature. Why did their invasion succeed? Largely because those animals carried water and solutes inside them—an "internal environment," so to speak—that could bathe and otherwise service their cells.

Living on land presented new challenges. Winds and radiant energy from the sun could dehydrate the body. Water was not always plentiful and most of it was fresh, not salty. Temperatures changed more dramatically from day to night and from one season to the next. Life on land required adaptations that could protect the volume, composition, and temperature of the internal environment.

Things are no different today. No matter which land-dwelling animal you observe, you discover fascinating aspects of body plan, body functions, and

behavior that counter threats to the stability of the internal environment.

Think about a kangaroo rat, living in an isolated desert of New Mexico (Figure 41.1). Winter rains are brief, then for months on end the sun bakes the sand and the only free water is the water sloshing in the canteens of researchers or tourists. Yet with nary a sip of free water, the kangaroo rat maintains its internal environment with exquisite precision until the next season's rains.

Kangaroo rats wait out the daytime heat in burrows, then forage in the cool of night for dry seeds or the occasional succulent. They are not sluggish about this. They hop rapidly over sizeable distances as they search for seeds and flee from snakes and other predators. All that hopping depends on a good supply of ATP energy *and* water for metabolic reactions.

Seeds provide the energy—and they replenish water in the internal environment. Seeds are rich in carbohydrates, which are chockful of energy. When living

**Figure 41.1**  A kangaroo rat, master of water conservation in the deserts.

| Table 41.1 | Normal Balance Between Water Gain and Water Loss in Humans and in Kangaroo Rats | | | |
|---|---|---|---|---|
| Organism | Water Gain (milliliters) | | Water Loss (milliliters) | |
| Adult human (measured on daily basis) | Ingested in solids: Ingested as liquids: Metabolically derived: | 850 1,400 350 ⎯⎯ 2,600 | Urine: Feces: Evaporation: | 1,500 200 900 ⎯⎯ 2,600 |
| Kangaroo rat (measured over 4 weeks) | Ingested in solids: Ingested as liquids: Metabolically derived: | 6.0 0 54.0 ⎯⎯ 60.0 | Urine: Feces: Evaporation: | 13.5 2.6 43.9 ⎯⎯ 60.0 |

cells break down carbohydrates, the reactions yield water. This "metabolic water" represents about 12 percent of the water that your own body gains each day. It represents a whopping *90 percent* of the total intake for kangaroo rats.

Beyond this, when kangaroo rats return to their burrow after a night of foraging, they empty their cheek pouches of seeds. There, in the cool burrow, the seeds soak up water vapor that is exhaled with each rat breath—and when the seeds are eaten, that water is recycled to the internal environment. In such ways, kangaroo rats use seeds to replenish their water stores.

Besides replenishing and recycling water, kangaroo rats are natural experts at conserving it. As the rat breathes, some water vapor condenses on the cool nasal epithelium, like beads of water condensing on a glass of iced tea, then diffuses back into the internal environment. Kangaroo rats have no sweat glands, so they don't lose water by perspiration. Like all other mammals, they do have kidneys. Water and solutes from the blood flow continuously through the kidneys, where adjustments are made with respect to how much water and which solutes are reabsorbed or disposed of (as urine). Of all the water your own body loses on an average day, nearly two-thirds of it exits by way of the kidneys. The kangaroo rat's kidneys are twice as efficient at reducing urinary water loss, for reasons that will become apparent in this chapter.

In such ways, the kangaroo rat gets and gives up water and solutes. Like other land animals, it survives only as long as the daily gains *balance* the daily losses, as Table 41.1 indicates. How the balancing acts are accomplished will be our initial focus in this chapter. Then we will consider how mammals survive in hot, cold, and sometimes capricious climates on land.

KEY CONCEPTS

1. To maintain a hospitable internal environment, all animals must make controlled adjustments to the external environment as well as obligatory exchanges with it.

2. Animals continually take in water, nutrients, and ions, and they produce metabolic wastes. In vertebrates, the kidneys help balance the intake and output of water and dissolved substances.

3. Urine forms in kidney nephrons. Its volume and composition result from the processes of filtration, reabsorption, and secretion.

4. Body temperatures of animals depend on the balance between heat produced through metabolism, heat absorbed from the environment, and heat lost to the environment.

5. Body temperature is maintained within a favorable range through controls over metabolic activity and adaptations in structure, physiology, and behavior.

## MAINTAINING THE EXTRACELLULAR FLUID

When we speak of the internal environment, we are talking about both blood plasma and interstitial fluid, the solute-rich solutions in the body's tissues. Together, they are the *extracellular fluid* that services all living cells of the body (page 542). Maintaining the volume and composition of this fluid is absolutely essential for cell survival. Humans and many other animals have a well-developed urinary system that serves this function. Figure 41.2 shows how the urinary system is linked with other organ systems in the human body.

### Water Gains and Losses

Ordinarily, and on a daily basis, humans and other mammals take in just as much water as they lose. Table 41.1 gives two examples of this balancing act. The body *gains* water through two processes:

1. Absorption of water from liquids and solid foods in the gut.

2. Metabolism—specifically, the breakdown of carbohydrates, fats, and other organic molecules in reactions that yield water as a by-product.

Among land-dwelling mammals, thirst behavior influences the gain of water. When water levels decrease, the brain compels the individual to seek out water holes, streams, cold drinks in the refrigerator, and so on. The mechanisms involved are described later in the chapter.

The mammalian body *loses* water mostly by the following processes:

1. Excretion by way of the urinary system.

2. Evaporation from the respiratory surface.

3. Evaporation through the skin.

4. Sweating.

5. Elimination by way of the gut.

The process of greatest importance in controlling water loss is **urinary excretion**. By this process, organs called kidneys help eliminate excess water and excess or harmful solutes from the internal environment. The evaporative processes listed are called "insensible water losses" because the individual is not consciously aware that they are taking place. Temperature-control centers in the brain govern sweating. Normally, the large intestine reabsorbs nearly all water in the gut, so very little water leaves the body in feces.

### Solute Gains and Losses

The preceding chapter described how oxygen enters the internal environment by crossing respiratory surfaces. Aside from oxygen, many diverse solutes enter the internal environment by three processes:

1. Absorption from the gut. The absorbed substances include *nutrients* such as glucose, which are used as energy sources and in biosynthesis reactions. They also include drugs, food additives, and *mineral ions*, such as sodium and potassium ions.

2. Secretion of hormones and other substances.

3. Metabolism, which produces carbon dioxide and other *waste products* of degradative reactions.

Carbon dioxide, the most abundant waste product of metabolism, is eliminated from the body at respiratory surfaces. Aside from carbon dioxide, these are the other major metabolic wastes that must be eliminated from the body:

1. *Ammonia*, formed in "deamination" reactions whereby amino groups are stripped from amino acids. If allowed to accumulate in the body, ammonia can be highly toxic.

2. *Urea*, produced in the liver in reactions that link two ammonia molecules to carbon dioxide. Urea is the main nitrogen-containing waste product of protein breakdown and is relatively harmless.

3. *Uric acid*, formed in reactions that degrade nucleic acids. If allowed to accumulate, uric acid can crystallize and sometimes collect in the joints.

4. *Phosphoric acid* and *sulfuric acid*, also produced during protein breakdown.

Usually, protein breakdown produces small amounts of other metabolic wastes besides the ones just listed. Some are highly toxic and may be responsible for many of the symptoms associated with kidney failure.

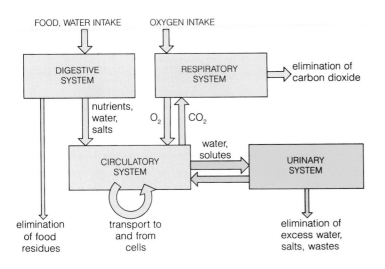

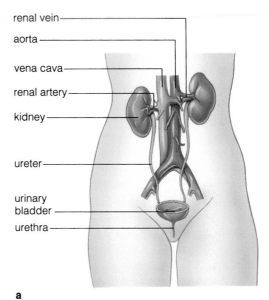

**Figure 41.2** Links between the urinary system and other organ systems that maintain operating conditions in the internal environment.

**a**

## Urinary System of Mammals

In all mammals, a pair of organs called **kidneys** continuously filter water, mineral ions, organic wastes, and other substances from the blood. Only a tiny portion of the water and solutes entering the kidneys leaves as a fluid called **urine**. In fact, all but about 1 percent is returned to the blood. But the composition of the fluid that *is* returned has been adjusted in vital ways. *Through their action, kidneys regulate the volume and solute concentrations of extracellular fluid.*

Each kidney has an outer cortex wrapped around a central region, the medulla. A tough coat of connective tissue encloses both. That coat is the renal capsule (from the Latin *renes*, meaning kidneys).

Internally, each kidney is divided into several lobes (Figure 41.3). Each lobe contains blood vessels and numerous slender tubes called **nephrons**. Water and solutes filtering out of the blood enter the nephrons. Most of the filtrate is reabsorbed from nephrons, but some continues on through tubelike collecting ducts and into the kidney's central cavity (renal pelvis). This fluid is the urine.

Urine flows from each kidney into a *ureter*, then into a storage organ, the *urinary bladder*. It leaves through a long tube, the *urethra*, which leads to the outside. The two kidneys, two ureters, urinary bladder, and urethra constitute the **urinary system** of mammals (Figure 41.3).

*Urination*, or urine flow from the body, is a reflex response. As a urinary bladder fills, tension increases in its strong, smooth-muscled walls. The increased tension causes muscles that prevent the flow of urine into the urethra to relax. At the same time, the bladder walls contract and force fluid through the urethra. The reflex response is involuntary but can be consciously blocked.

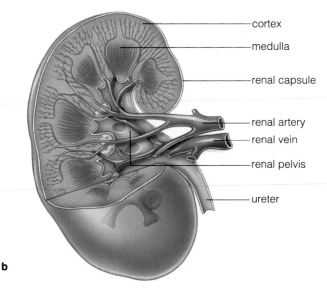

**b**

**Figure 41.3** (**a**) Components of the human urinary system. (**b**) Closer look at the kidney.

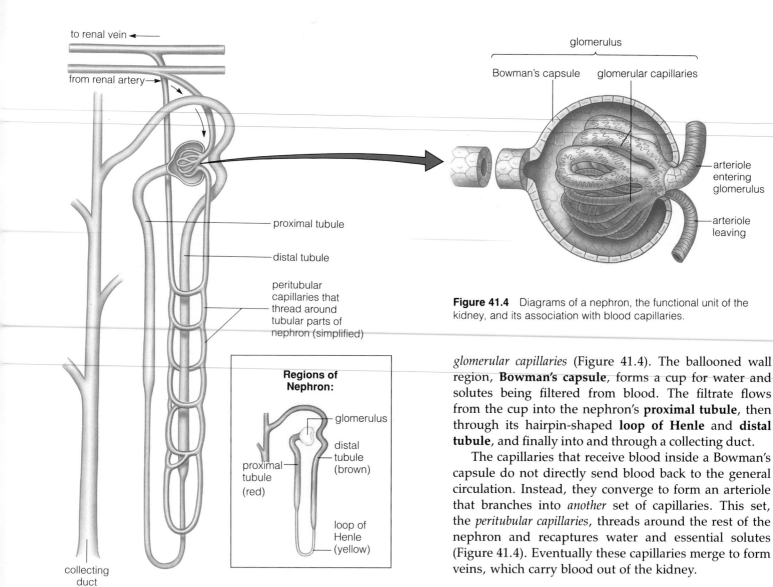

to renal vein

from renal artery

proximal tubule

distal tubule

peritubular capillaries that thread around tubular parts of nephron (simplified)

collecting duct

**Regions of Nephron:**

glomerulus

proximal tubule (red)

distal tubule (brown)

loop of Henle (yellow)

glomerulus

Bowman's capsule    glomerular capillaries

arteriole entering glomerulus

arteriole leaving

**Figure 41.4** Diagrams of a nephron, the functional unit of the kidney, and its association with blood capillaries.

You may have heard about *kidney stones*, these being deposits of uric acid, calcium salts, and other substances that settled out of urine and collected in the renal pelvis. At times a stone becomes lodged in the ureter or, on rare occasions, in the urethra. It can interfere with urine flow and cause pain. Kidney stones usually pass naturally from the body during urination. If they do not, they must be removed by medical or surgical procedures.

### Nephron Structure

Each fist-sized kidney of humans has more than a million nephrons. The nephron wall consists only of a layer of epithelial cells, but the cells and junctions between them are not all the same. Some wall regions are highly permeable to water and solutes. Still other regions bar the passage of solutes *except* at active transport systems built into the plasma membrane (page 85).

Filtration starts at the **glomerulus**, where the nephron wall balloons around a cluster of blood capillaries called

*glomerular capillaries* (Figure 41.4). The ballooned wall region, **Bowman's capsule**, forms a cup for water and solutes being filtered from blood. The filtrate flows from the cup into the nephron's **proximal tubule**, then through its hairpin-shaped **loop of Henle** and **distal tubule**, and finally into and through a collecting duct.

The capillaries that receive blood inside a Bowman's capsule do not directly send blood back to the general circulation. Instead, they converge to form an arteriole that branches into *another* set of capillaries. This set, the *peritubular capillaries*, threads around the rest of the nephron and recaptures water and essential solutes (Figure 41.4). Eventually these capillaries merge to form veins, which carry blood out of the kidney.

### URINE FORMATION

#### Urine-Forming Processes

By definition, urine is a fluid by which the body rids itself of water and solutes in excess of the amounts necessary to maintain the extracellular fluid. Figure 41.5 outlines the three processes required for urine formation. These are filtration, tubular reabsorption, and secretion.

**Filtration** starts and ends at the glomerulus. In this process, blood pressure (generated by heart contractions) forces water and solutes out of the glomerular capillaries and into the cupped region inside Bowman's capsule. The blood is said to be filtered here because blood cells, proteins, and other large solutes are left behind as water and smaller solutes (such as glucose, sodium, and urea) are forced out of the capillaries. The filtrate itself will flow on, into the proximal tubule.

**Reabsorption** takes place at tubular parts of the nephron. In this process, water and solutes move across the tubular wall and *out* of the nephron (by diffusion or active transport), then into adjacent capillaries. Most of the filtrate's water and usable solutes are reclaimed here and are returned to the general circulation. Table 41.2 gives some examples of daily reabsorption values.

**Secretion** also occurs across the tubular walls, *but in the opposite direction.* Excess hydrogen ions, potassium ions, and a few other substances move *out* of the capillaries and into cells making up the nephron walls. Then those cells secrete the substances into fluid within the tubule. Secretion, a highly regulated process, also rids the body of uric acid, some breakdown products of hemoglobin and other proteins, and some other metabolic wastes. Additionally, it can rid the body of many drugs, including penicillin, and other foreign substances.

**A concentrated or dilute urine forms through the processes of filtration, reabsorption, and secretion.**

**The urine contains waste products as well as water and solutes in excess of the amounts necessary to maintain the extracellular fluid.**

### Factors Influencing Filtration

Each day, more blood flows through the kidneys than through the tissues of any other organ except the lungs. Each minute, about 1-1/2 quarts of blood course through them! That is nearly one-fourth of the cardiac output. How can kidneys handle blood flowing through on such a massive scale? There are two mechanisms.

*First*, the arterioles delivering blood to a glomerulus have a wider diameter—and less resistance to flow—than most arterioles. The hydrostatic pressure caused by heart contractions therefore does not drop as much when blood flows through them. As a result, pressure in the glomerular capillaries is higher than in other capillaries. *Second*, glomerular capillaries are highly permeable. They do not allow blood cells or protein molecules to escape, but compared to other capillaries, they are 10 to 100 times more permeable to water and small solutes. Because of the higher hydrostatic pressure and greater capillary permeability, the kidneys can filter an average of 45 gallons (180 liters) per day.

At any given time, the *rate* at which kidneys filter a given volume of blood depends on the blood flow to them and on how fast their tubules are reabsorbing water. Reabsorption, as you will see shortly, is partly under hormonal control. Blood flow to the kidneys is influenced by neural controls over blood flow through the body as a whole. Recall that when you exercise, more blood than usual must be diverted to your heart and skeletal muscles to sustain the increased activity.

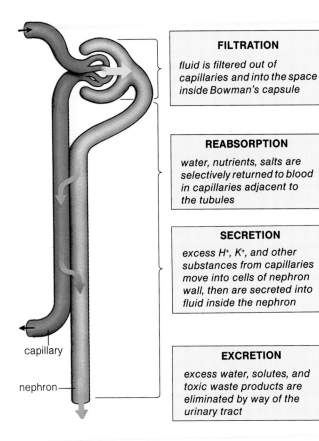

**FILTRATION**

*fluid is filtered out of capillaries and into the space inside Bowman's capsule*

**REABSORPTION**

*water, nutrients, salts are selectively returned to blood in capillaries adjacent to the tubules*

**SECRETION**

*excess H+, K+, and other substances from capillaries move into cells of nephron wall, then are secreted into fluid inside the nephron*

capillary

nephron

**EXCRETION**

*excess water, solutes, and toxic waste products are eliminated by way of the urinary tract*

**Figure 41.5**  Processes involved in urine formation and excretion.

| Table 41.2 | Average Daily Reabsorption Values for a Few Substances | | |
|---|---|---|---|
| | Filtered | Excreted | Proportion Reabsorbed |
| Water | 180 liters | 1.8 liters | 99% |
| Glucose | 180 grams | None, normally | 100% |
| Sodium ions | 630 grams | 3.2 grams | 99.5% |
| Urea | 54 grams | 30 grams | 44% |

Blood is diverted away from the kidneys when neural signals call for vasoconstriction of the arterioles leading into them. Because the flow volume to the kidneys is reduced, the filtration rate decreases.

Local chemical signals also influence filtration rates. When arterial blood pressure decreases, locally produced chemicals stimulate vasodilation of the arterioles leading into the glomeruli, so more blood flows in. When blood pressure rises, the arterioles are stimulated to constrict, and less blood flows in. This helps keep blood flow to the kidneys relatively constant, despite changes in blood pressure.

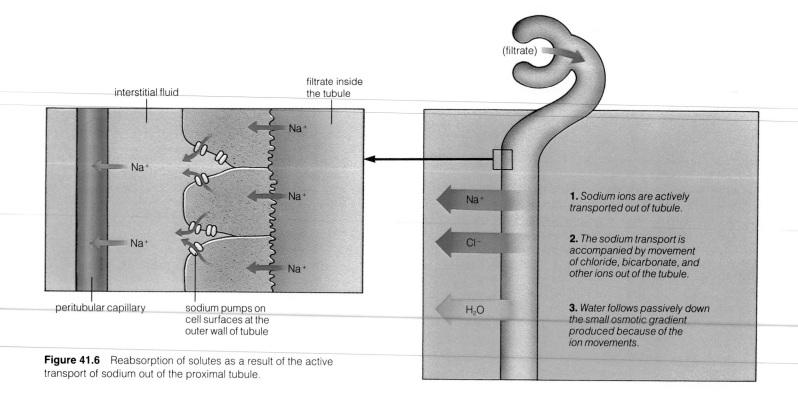

**Figure 41.6** Reabsorption of solutes as a result of the active transport of sodium out of the proximal tubule.

Labels in figure:

interstitial fluid

filtrate inside the tubule

Na⁺

Na⁺

Na⁺

Na⁺

Na⁺

peritubular capillary

sodium pumps on cell surfaces at the outer wall of tubule

(filtrate)

Na⁺

Cl⁻

H₂O

**1.** Sodium ions are actively transported out of tubule.

**2.** The sodium transport is accompanied by movement of chloride, bicarbonate, and other ions out of the tubule.

**3.** Water follows passively down the small osmotic gradient produced because of the ion movements.

### Reabsorption of Water and Sodium

Only tiny fractions of the water and sodium that enter the kidney are excreted in the urine. Yet even those minuscule amounts must be carefully regulated so that the internal environment remains constant.

Normally, for example, you cannot upset the balance for long by drinking too much water at lunch. The kidneys simply do not reclaim the excess; it will be excreted. Similarly, you cannot upset the balance by not taking in enough water, at least over the short term. The kidneys simply will conserve water already in the body; only a small volume of urine will be excreted.

Similarly, eating a large bag of salty potato chips for lunch will put excess sodium into your bloodstream, but the excess will be excreted. Conversely, if your doctor were to place you on a low-sodium diet, your body will work to conserve sodium. A larger fraction of the sodium in the filtrate passing through nephrons will be reabsorbed, and less will be excreted in the urine. The same thing happens when too much sodium is lost through diarrhea or excessive sweating. How are regulatory events such as this accomplished? Let's take a look.

**Proximal Tubule.** Of all the water and sodium filtered in the kidneys, more than half is promptly reabsorbed at the proximal tubule—the part of the nephron closest to the glomerulus. As Figure 41.6 shows, epithelial cells making up the tubule wall have transport proteins at their outer surface. Nearly all cells have pro-

teins of this sort, which function as sodium "pumps." In this case, the proteins actively transport sodium ions from the filtrate into the interstitial fluid. An outward movement of chloride, bicarbonate, and other ions accompanies the sodium transport.

The solute concentration inside the tubule drops only slightly because of this outward movement. At the same time, however, the movement affects the osmotic gradient between the filtrate and the interstitial fluid. The wall of the proximal tubule happens to be quite permeable to water. So now water follows the small osmotic gradient—it moves passively out of the tubule.

In this fashion, the volume of fluid remaining within the tubule decreases greatly, but the concentration of solutes—sodium especially—changes very little.

**Urine Concentration and Dilution.** The situation changes after fluid moves on through the proximal tubule and enters the loop of Henle. This hairpin-shaped structure plunges into the kidney medulla. In the interstitial fluid surrounding the hairpin, the solute concentration increases progressively with depth.

The descending limb of the loop is permeable to water. Water moves out, by osmosis, and the solute concentration in the fluid remaining inside increases until it matches that in the interstitial fluid. In the ascending limb, water cannot cross the tubule wall, but sodium is actively transported out. As sodium (and chloride) move out of the filtrate, the solute concentration rises outside the tubule and falls inside. This increase in sol-

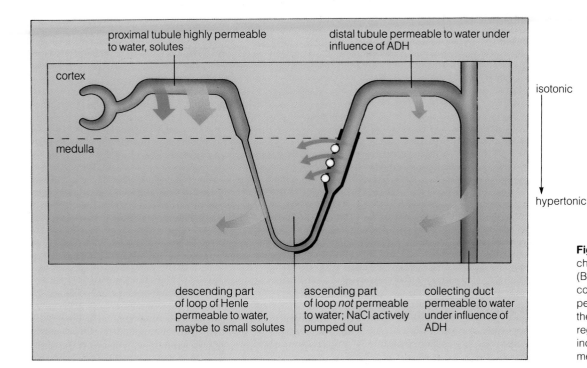

proximal tubule highly permeable to water, solutes

distal tubule permeable to water under influence of ADH

cortex

medulla

isotonic

hypertonic

descending part of loop of Henle permeable to water, maybe to small solutes

ascending part of loop *not* permeable to water; NaCl actively pumped out

collecting duct permeable to water under influence of ADH

**Figure 41.7** Permeability characteristics of the nephron. (Both the distal tubule and the collecting duct have very limited permeability to solutes; most of the solute movements in these regions are related directly or indirectly to active transport mechanisms.)

ute concentration outside the tubule favors reabsorption of water from the descending limb.

*Through this interaction between the ascending and descending limbs of the loop of Henle, a very high solute concentration develops in the deeper portions of the medulla.* At the same time, the interaction lowers the solute concentration in the fluid traveling up the ascending limb of the loop. So the tubular fluid finally delivered to the distal tubule in the kidney cortex is quite dilute, with a low sodium concentration. As you will now see, the stage is set for the excretion of either highly dilute or highly concentrated urine—or anywhere in between.

**Hormone-Induced Adjustments.** Because so much water and sodium are reabsorbed in the proximal tubule and loop of Henle, the volume of dilute urine reaching the start of the distal tubule has been greatly reduced. Yet if even that reduced volume were excreted without adjustments, the body would rapidly become depleted of both water and sodium. Controlled adjustments are made at cells located in the walls of distal tubules and collecting ducts. Two hormones serve as the agents of control. **ADH** (antidiuretic hormone) influences water reabsorption, and **aldosterone** influences sodium reabsorption.

Let's first consider the role of ADH in adjusting the rate of water reabsorption. The hypothalamus controls the release of ADH from the posterior pituitary gland. It triggers ADH secretion when the solute concentration of extracellular fluid rises above a set point (com-

pare page 542). This can happen when water intake is restricted or when the extracellular fluid volume is reduced, as by severe bleeding (hemorrhage). ADH acts on distal tubules and collecting ducts, making their walls more permeable to water (Figure 41.7). Thus, in the kidney cortex, water is reabsorbed from the dilute fluid inside the tubules. The volume of fluid inside is reduced somewhat, and now it passes down through the collecting ducts, which plunge down into the medulla. Remember that solute concentrations in the surrounding interstitial fluid are high in this region. Even more water is reabsorbed here—so only a small volume of very concentrated urine is excreted.

Conversely, when water intake is excessive, the solute concentration in extracellular fluid falls. ADH secretion is inhibited. Without ADH, the walls of the distal tubules and collecting ducts become impermeable to water. A large volume of dilute urine can be excreted, and in this way the body rids itself of the excess water.

ADH enhances water reabsorption at distal tubules and collecting ducts when the body must conserve water. When excess water must be excreted, ADH secretion is inhibited.

Let's now consider the role of aldosterone in adjusting the rate of sodium reabsorption. When the body loses more sodium than it takes in, the volume of extracellular fluid falls. Sensory receptors in the walls of blood vessels and the heart detect the decrease, and

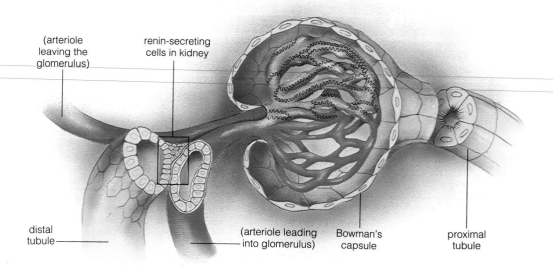

**Figure 41.8** Location of renin-secreting cells that play a role in sodium reabsorption.

(arteriole leaving the glomerulus)

renin-secreting cells in kidney

distal tubule

(arteriole leading into glomerulus)

Bowman's capsule

proximal tubule

renin-secreting cells in the kidneys are called into action. Those cells are part of the "juxtaglomerular apparatus." The name refers to a region of contact betwen the arterioles of the glomerulus and the distal tubule of the nephron (Figure 41.8).

Renin acts on molecules of an inactive protein that circulates in the bloodstream. In effect, enzyme action lops off part of the molecule. Then the fragment is converted to a hormone, angiotensin II. The hormone stimulates cells of the adrenal cortex, the outer portion of a gland perched on top of each kidney (page 588). In response to stimulation, target cells secrete aldosterone. This hormone causes cells of the distal tubules and collecting ducts to reabsorb sodium faster. The outcome? Less sodium is excreted in the urine.

Conversely, when the body contains too much sodium, aldosterone secretion is inhibited. Less sodium is reabsorbed, and more is excreted.

When the body cannot rid itself of excess sodium, it inevitably retains excess water, and this leads to a rise in blood pressure. Abnormally high blood pressure (hypertension) can adversely affect the kidneys as well as the vascular system and brain (page 671). One of the ways to control blood pressure is to restrict the intake of sodium chloride—table salt.

---

**Aldosterone enhances sodium reabsorption at distal tubules and collecting ducts when the body must conserve sodium.**

---

### Thirst Behavior

You are not entirely at the mercy of events in the kidney when your body is in need of water. The same stimuli that lead to ADH secretion and the stepped-up reabsorption of water in the kidneys can stimulate thirst behavior. Suppose you eat an entire box of salty popcorn at the movies. Soon the salt is moving into your bloodstream. The solute concentration in your extra-cellular fluid rises, the hypothalamus detects the increase, and it calls for ADH secretion. Besides this, your **thirst center** is stimulated. Signals from this cluser of nerve cells in the hypothalamus can inhibit saliva production. Your brain interprets the resulting sensation of dryness in your mouth as "thirst" and leads you to seek out drinking fluids.

In fact, a "cottony mouth" is one of the early signs that your body is becoming dehydrated. Dehydration commonly results after severe cases of hemorrhage, burns, or diarrhea. It also results after profuse sweating and after the body has been deprived of water for a prolonged time. At such times thirst behavior becomes exceptionally intense.

### Acid–Base Balance

So far, we have focused on the kidney's primary function—that is, how it controls water and sodium reabsorption, and so influences the total volume and distribution of body fluids. Another kidney function also has profound impact on the health of the individual. *Together with the respiratory system and other organ systems, the kidneys help keep the extracellular environment from becoming too acidic or too basic (alkaline).*

The overall acid–base balance is maintained by controls over ion concentrations, especially hydrogen ions ($H^+$). The controls are exerted through (1) buffer systems, (2) respiration, and (3) excretion by way of the kidneys.

Normally, the extracellular pH for the human body is between 7.35 and 7.45. Maintaining that value means neutralizing or eliminating a variety of acidic and basic

## Kidney Failure and Dialysis

An estimated 13 million people in the United States alone suffer from kidney disorders, as when diabetes or immune responses damage the nephron and interfere with urine formation.

When the kidneys malfunction, control of the volume and composition of the extracellular fluid is disturbed, and toxic by-products of protein breakdown can accumulate in the bloodstream. Nausea, fatigue, loss of memory and, in advanced cases, death may follow. A *kidney dialysis machine* can restore the proper solute balances. Like the kidney itself, the machine helps maintain extracellular fluid by selectively removing and adding solutes to the bloodstream.

"Dialysis" refers to the exchange of substances across a membrane between solutions of differing compositions. In *hemodialysis*, the machine is connected to an artery or a vein, then blood is pumped through tubes made of a mate-rial similar to sausage casing or cellophane. The tubes are submerged in a warm-water bath. The precise mix of salts, glucose, and other substances in the bath set up the correct gradients with the blood. In *peritoneal dialysis*, fluid of the proper composition is put into the abdominal cavity, left in place for a period of time, and then drained out. Here, the lining of the cavity (the peritoneum) serves as the dialysis membrane.

Hemodialysis generally takes about four hours; blood must circulate repeatedly before solute concentrations in the body are improved. The procedure must be performed three times a week. It is used as a temporary measure in patients with reversible kidney disorders. In chronic cases, the procedure must be used for the rest of the patient's life or until a functional kidney is transplanted. With treatment and controlled diets, many individuals are able to resume fairly normal activity.

---

substances entering the blood from the gut and from normal metabolism. Recall that acids lower the pH and bases raise it. In individuals on an ordinary diet, normal cell activities produce an excess of acids. The acids dissociate into $H^+$ and other fragments, and this lowers the pH. The effect is minimized when excess $H^+$ ions react with different kinds of buffers. The bicarbonate–carbon dioxide buffer system is an example. The reactions can be summarized this way:

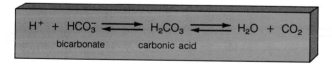

$$H^+ + HCO_3^- \rightleftharpoons H_2CO_3 \rightleftharpoons H_2O + CO_2$$

bicarbonate      carbonic acid

In this case, the $H^+$ is neutralized and the carbon dioxide that forms is excreted by the lungs.

Keep in mind that this buffer system and others do not *eliminate* acid; they only neutralize $H^+$ temporarily. *Only the urinary system can eliminate excess amounts of $H^+$ and restore the body's buffers.*

Notice, in the preceding equation, that the reaction arrows also run in reverse. The reverse reactions unfold in cells of the nephron wall. *First,* the $HCO_3^-$ that forms in those cells is moved into interstitial fluid, then into the capillaries around the nephrons. The capillaries deliver the $HCO_3^-$ to the general circulation, where it helps buffer excess acid. *Second,* the $H^+$ that forms in the

cells is secreted into the fluid inside the nephron. There, it can combine with bicarbonate ions to form $CO_2$, which is returned to the blood and excreted by the lungs. It also can combine with phosphate ions or ammonia ($NH_3$), which are excreted in urine. In such ways, hydrogen ions can be permanently removed from extracellular fluid.

Through such mechanisms, the kidneys help maintain the health of the individual. When such mechanisms fail, serious problems arise (see the *Commentary* above).

### On Fish, Frogs, and Kangaroo Rats

Now that you have an idea of how your own body maintains water and solute levels, you can gain insight into what goes on in some other vertebrates, including that kangaroo rat described at the start of the chapter.

The tissues of herring, snapper, and other bony fishes that live in the seas have about three times less solutes than seawater does. Such fishes continuously lose water (by osmosis) to their hypertonic environment, and continual drinking brings in replacements. (Experimentally prevent a fish from drinking and it will die from dehydration within a few days.) Marine fishes excrete ingested solutes against concentration gradients. Fish kidneys do not have loops of Henle, so it is impossible to excrete

**Figure 41.9** Salt-water balance by the salmon. Salmon are among the fishes that are able to live in both saltwater and freshwater. They hatch in streams and later move downstream to the seas, where they feed and mature. Then they return to their streams to spawn (page 136).

For most salmon, salt tolerance arises through changing concentrations of hormones, which seem to be triggered in some way by increasing daylength in spring. Prolactin, a pituitary hormone, plays a role in sodium retention in freshwater. When a freshwater fish has its pituitary gland removed, it will die from sodium loss—but that fish will live if prolactin is administered to it. By contrast, cortisol is a steroid hormone secreted by the adrenal cortex (page 588), and it is crucial in developing salt tolerance in salmon. Cortisol secretions are associated with increases in sodium excretion, in the sodium-potassium pumping activity by cells in the gills, and in absorption of ions and water in the gut. In young salmon, cortisol secretion increases prior to the seaward movement—and so does salt tolerance.

urine having a higher solute concentration than that of the body fluids. Most excess solutes are pumped out through membranes of fish gills, the cells of which actively transport sodium ions out of the blood.

In freshwater, a hypotonic medium, bony fishes and amphibians tend to gain water and lose solutes. They do not drink water. Rather, water moves by osmosis into the body, through thin gill membranes, or, in frogs and other adult amphibians, through the skin. Excess water leaves by way of well-developed kidneys, which excrete a large volume of dilute urine. Some solutes also are excreted, but the losses are balanced by solutes gained from food and by the active transport of sodium ions across the gills, into the body. Figure 41.9 describes the wonderful balancing act in a type of fish that makes its home in both freshwater *and* seawater.

And about that kangaroo rat! Its remarkable ability to restrict water loss results from its long loops of Henle. The solute concentration in the interstitial fluid around those loops becomes very high. The osmotic gradient between the fluid and the urine is so steep, most of the water reaching the equally long collecting ducts is reabsorbed. Thus kangaroo rats give up only a tiny volume of concentrated urine—three to five times more concentrated than that of humans.

## MAINTAINING BODY TEMPERATURE

Maintaining the volume and composition of the internal environment is serious business. Maintaining its temperature is equally serious. Consider that temperatures in the world just outside the body often change quickly and that exercise can send metabolic rates soaring. Such changes trigger slight increases or decreases in the normal **core temperature**. "Core" refers to the body's internal temperature, as opposed to temperatures of the tissues near its surface.

### Temperatures Suitable for Life

In a manner of speaking, the animal body runs on enzymes—and enzyme activity is affected by temperature. The enzymes of most animals commonly remain functional within the 0°–40°C range (Table 41.3). Above 41°C or so, they do not function as well because denaturation occurs. (Denaturation disrupts the chemical interactions holding a molecule in its required three-dimensional shape.) Also, the rate of enzyme activity generally decreases by at least half when the temperature drops by ten degrees. Clearly, then, metabolism can be upset if body temperatures exceed or fall below the proper range.

How do animals keep their body temperature fairly constant? They do so by balancing heat gains and heat losses. Many different physiological and behavioral responses can restore the normal temperature, as the following discussions will make clear.

### Heat Gains and Heat Losses

Enzyme-mediated reactions proceed simultaneously in the millions or billions of cells of a large-bodied animal. Heat is an inevitable by-product of all that metabolic activity. (Even as you sit quietly, reading this book, you are producing roughly 1 kilocalorie per hour per kilogram of body weight.) If that heat were to accumulate internally, the body temperature would steadily rise. You probably know that a warm body tends to lose heat to a cooler environment. Under such circumstances, its temperature will hold steady if the rate of heat loss strikes a balance with the rate of metabolic heat produc-

tion. Of course, the balance can tip, as when the body produces more heat than it loses. In general, the body's heat content depends on the balance between heat gains and losses, as summarized here:

| change in body heat | = | heat produced | + | heat gained | − | heat lost |

Here, the gains and losses refer to exchanges that take place at the outer surface of the body, which includes part of the respiratory tract.

Three processes—radiation, conduction, and convection—can move heat away from or to the body. A fourth process—evaporation—can only move heat away from the body.

**Radiation** is a process by which heat is gained when the body is exposed to intense radiant energy (as from the sun or a heat lamp) or to any surface warmer than its own surface temperature. Indoors, well over a third of your total heat loss typically occurs by way of radiation.

**Conduction** is the direct transfer of heat between objects that are in direct contact. Because heat moves down thermal gradients, we lose heat by conduction when we sit on cold ground, and we gain heat when we sit on warm sand at the beach.

**Convection** is the transfer of heat by way of moving fluid, such as air or water currents. The process involves conduction (heat moves down the thermal gradient between the body surface and the air or water next to it). It also involves mass transfer, with currents carrying heat away from or toward the body. When your skin temperature is higher than the air temperature, you lose heat by convection. Even when there is no breeze, your body loses heat by creating its own convective current. Air becomes less dense as it is heated and rises away from the body. Indoors, you typically lose as much heat by convection as you do by radiation.

**Figure 41.10** Evaporative water loss and sweating. Horses, humans, and some other mammals have sweat glands that move water and specific solutes through pores to the skin surface. In fact, an average-sized human can produce 1 to 2 liters of sweat per hour—for hours on end. For every liter of sweat that evaporates, the body loses 600 kilocalories of heat energy. During extreme exercise, this mechanism balances the high rates of heat production in skeletal muscle. In itself, sweat dripping from the skin does *not* dissipate body heat by evaporation. When you drip with sweat while exercising on a hot, humid day, the rate of evaporation does not keep pace with the rate of sweat secretion—the high water content of the surrounding air slows evaporation.

| Table 41.3 | Temperatures Favorable for Metabolism, Compared with Environmental Temperatures |
|---|---|
| Temperatures generally favorable for metabolism: | 0°C to 40°C (32°F to 104°F) |
| Air temperatures above land surfaces: | −70°C to +85°C (−94°F to +185°F) |
| Surface temperatures of open oceans: | −2°C to +30°C (+28.4°F to +86°F) |

**Evaporation** is the conversion of a liquid to a gaseous state. The conversion requires energy, which is supplied from the heat content of the liquid (page 28). Evaporation of water from the body surface has a cooling effect because the water molecules being released (as water vapor) carry away some of the energy with them. For land animals, some evaporative water loss occurs at the moist respiratory surfaces and across the skin. Animals that sweat, pant, or lick their fur also lose water by this process (Figure 41.10). The rate of evaporation depends on the humidity and on the rate of air movement. If air next to the body is already saturated with water (that is, when the local relative humidity is 100 percent), water will not evaporate. If air next to the body is hot and dry, evaporation may be the only means of countering heat production and heat gains (from radiation and convection).

---

**Heat can be brought into or carried away from the body by the processes of radiation, conduction, and convection. Evaporation carries heat away from it.**

**The factors that dictate how much heat is exchanged by these processes can be adjusted by behavioral and physiological means.**

---

## Classification of Animals Based on Temperature

**Ectotherms**. We humans have high metabolic rates that sustain our active way of life. But most animals have low metabolic rates, and on top of that, they are poorly insulated. This means they rapidly absorb and gain heat, especially when they have small bodies. They maintain their core temperature mostly by heat gains from the environment, not from metabolism. Such animals are **ectotherms**, which means "heat from outside." Lizards and other reptiles are examples.

Ectotherms are not entirely at the mercy of their environment, however, for they can make behavioral adjustments to changing external temperatures. We call this **behavioral temperature regulation**.

Lizards move about, putting themselves in places where they minimize heat or cold stress. To warm up, they move out of shade and keep orienting their body to expose the maximum surface area to the sun's infrared radiation. They gain heat by conduction when they bask on rocks that absorbed heat from the sun earlier in the day. In such ways, the lizard body can warm up as fast as 1°C per minute.

Lizards lose heat just as rapidly when the sun goes down and temperatures drop. Then, metabolic activity decreases and they become almost immobilized. Before that happens, they usually crawl into crevices or under rocks, where heat loss is not as great and where they are not as vulnerable to predators.

**Endotherms and Heterotherms**. Like most birds and mammals, we are **endotherms**, which means "heat from within." In endotherms, body temperature is controlled mainly by (1) metabolic activity and (2) controls over heat conservation and dissipation. We also make behavioral adjustments that supplement the physiological controls.

Most endotherms have an active life-style, made possible by high metabolic rates. It is a costly adaptation. A foraging mouse uses up to thirty times more energy than a foraging lizard of the same weight. Yet such energy outlays have advantages, for they are the main reason why endotherms can be active under a wide range of temperatures. Cold nights or cold seasons don't stop them from foraging, for example, or escaping from predators, or digging a burrow.

Endotherms conserve or dissipate heat associated with high metabolic rates by employing a variety of adaptations. Think about how fur, feathers, and layers of fat help reduce heat loss. Think about how clothing can reduce heat loss or heat gain. Or think about the ways that bodies are shaped and insulated. Some mammals in cold regions have more massive bodies than closely related species in warmer regions. Compared to the streamlined, thin-legged jackrabbits of the American Southwest, for example, you might think that the arctic hare is rather bulky. However, its more massive, rounder body has a greater volume of cells for generating heat and less surface area for losing it.

Like ectotherms, endotherms also adjust behaviorally to heat stress. During the day, the core temperatures of kangaroo rats and some other desert mammals of north temperate regions are much lower than temperatures of the air and the ground surface. Outside temperatures often are lower during the night as well as during winter. However, the soil well below the surface never heats up much, and it is here that most desert rodents and other mammals find refuge from the daytime heat. Typically those animals forage by night and spend the hottest part of the day in burrows or in the shade of bushes or rock outcroppings.

Some birds and mammals fall between the ectothermic and endothermic categories. Part of the time, these so-called **heterotherms** allow their body temperature to fluctuate as ectotherms do, and at other times they control heat exchanges as endotherms do. Hummingbirds have very high metabolic rates for their size, and they devote much of the day to locating and sipping nectar as an energy source for metabolism. Because hummingbirds do not forage at night, they could rapidly run out of energy unless their metabolic rates decreased considerably. At night, they may enter a sleeplike state and become almost as cool as their surroundings.

**Thermal Strategies Compared**. In general, ectotherms are at an advantage in the warm, humid tropics. They do not have to expend much energy to maintain body temperature, and more energy can be devoted to other tasks, including reproduction. Indeed, reptiles far exceed mammals in numbers and species diversity in the tropics. However, endotherms have the advantage and are more abundant in moderate to cold settings. High metabolic rates allow some endotherms to occupy even the polar regions, where you would never find a lizard.

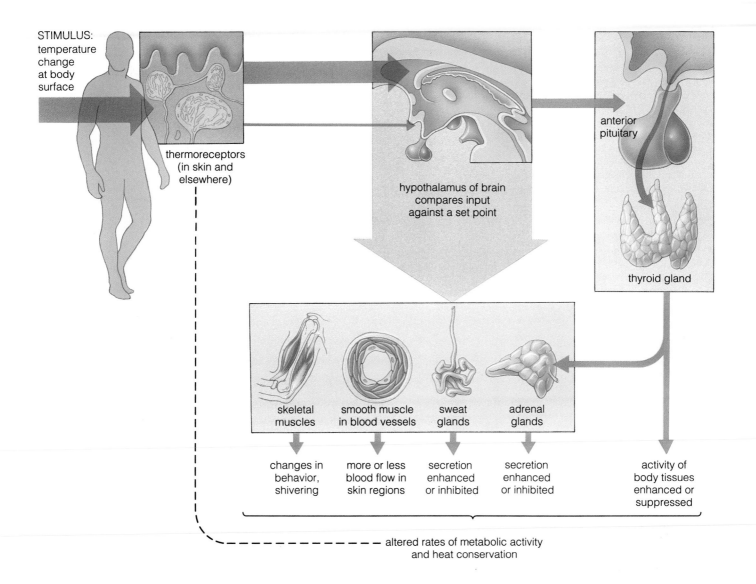

STIMULUS: temperature change at body surface

thermoreceptors (in skin and elsewhere)

hypothalamus of brain compares input against a set point

anterior pituitary

thyroid gland

| skeletal muscles | smooth muscle in blood vessels | sweat glands | adrenal glands |

changes in behavior, shivering

more or less blood flow in skin regions

secretion enhanced or inhibited

secretion enhanced or inhibited

activity of body tissues enhanced or suppressed

altered rates of metabolic activity and heat conservation

## Temperature Regulation in Birds and Mammals

**Responses to Cold Stress**. Table 41.4 lists the major responses to decreases in outside temperatures. They include peripheral vasoconstriction, the pilomotor response, shivering, and nonshivering heat production.

Among mammals, **peripheral vasoconstriction** is a normal response to a drop in outside temperature. As Figure 41.11 suggests, the hypothalamus governs this response. Thermoreceptors at the body surface detect the decrease in temperature and fire off signals to the hypothalamus. In turn, the hypothalamus sends out commands to smooth muscles in the walls of blood vessels in the skin. When the muscles contract, vasoconstriction occurs—and the bloodstream's convective delivery of heat to the body's surface is reduced. How effective is the response? To give an example, when your fingers or toes become cold, all but 1 percent of the blood that would otherwise flow to their skin is curtailed.

**Figure 41.11** Homeostatic controls over the internal temperature of the human body.

**Table 41.4 Responses to Cold Stress**

| Core Temperature | Responses Made |
|---|---|
| 36°–34°C (about 95°F) | Shivering response, increase in respiration. Increase in metabolic heat output. Constriction of peripheral blood vessels; blood is routed to deeper regions. Dizziness and nausea set in. |
| 33°–32°C (about 91°F) | Shivering response stops. Metabolic heat output drops. |
| 31°–30°C (about 86°F) | Capacity for voluntary motion is lost. Eye and tendon reflexes inhibited. Consciousness is lost. Cardiac muscle action becomes irregular. |
| 26°–24°C (about 77°F) | Ventricular fibrillation sets in (page 672). Death follows. |

**Figure 41.12** A tragic episode of hypothermia.

In 1912, the ocean liner *Titanic* set out from Europe on her maiden voyage across the cold Atlantic waters to America. In that same year, a huge chunk of the leading edge of a Greenland glacier broke off and began floating out to sea. Late at night on April 14, off the coast of Newfoundland, the iceberg and the *Titanic* made their ill-fated rendezvous. The *Titanic* was the largest ship afloat and was believed to be unsinkable. Survival drills had been neglected, and there were not enough lifeboats to hold even half the 2,200 passengers. The *Titanic* sank in about 2-1/2 hours.

Within two hours, rescue ships were on the scene—yet 1,517 bodies were recovered from a calm sea. All were wearing life jackets. None had drowned. Probably every one of those individuals had died from hypothermia—from a drop in body temperature below tolerance levels.

| Table 41.5 Summary of Mammalian Responses to Cold Stress and to Heat Stress | | |
|---|---|---|
| Environmental Stimulus | Main Responses | Outcome |
| Drop in temperature | Vasoconstriction of blood vessels in skin; changes in behavior (e.g., curling up the body to reduce surface area exposed to the environment) | Heat is conserved |
| | Increased muscle activity; shivering; nonshivering heat production | Heat production increases |
| Rise in temperature | Vasodilation of blood vessels in skin; sweating; changes in behavior; panting | Heat is dissipated from body |
| | Decreased muscle activity | Heat production decreases |

In another response to a drop in outside temperature, smooth muscle controlling the erection of hair or feathers is stimulated to contract. This is a **pilomotor response**. The plumage or pelt fluffs up, and this creates a layer of still air that reduces convective and radiative heat losses from the body. Heat loss can be further restricted by behavioral responses that reduce the amount of body surface exposed for heat exchange—as when cats curl up into a ball or when you hold both arms tightly against your body.

When other responses are not enough to counter cold stress, the hypothalamus calls for an increase in skeletal muscle activity that leads to **shivering**. The word refers to rhythmic tremors in which the muscles contract about ten to twenty times per second. Within a short time, heat production throughout the body increases several times over. Shivering comes at a high energy cost and is not effective for very long.

Heat production also can be increased without shivering. Prolonged or severe cold exposure can lead to a hormonal response that elevates the rate of metabolism (Figure 41.11). This **nonshivering heat production** is prominent in brown adipose tissue, a specialized tissue of hibernating animals and in rodents and other animals that become acclimatized to cold. Human infants have this tissue; adults have very little unless they are cold-adapted. Korean diving women, for example, who spend six hours a day in cold water, have well-developed brown adipose tissue.

When defenses against cold are not adequate, the result is *hypothermia*, a condition in which the core temperature falls below normal. In humans, a drop of only a few degrees affects brain function and leads to

**Figure 41.13** A jackrabbit (*Lepus californicus*) cooling off on a hot summer day in the mountains of Arizona. Notice the dilated blood vessels in its large ears. Both the large surface area of the ears and the extensive vascularization are useful for dissipating heat (by way of convection and radiation).

**Figure 41.14** (*Right*) Response to heat stress by mammals without sweat glands. Some animals resort to behavioral mechanisms, such as licking their fur or panting (breathing hard through the mouth). Panting enhances respiratory heat loss, the amount depending on the increases in ventilation and in watery secretions from salivary glands. The endurance of sled dogs during a hard race depends partly on their capacity to dissipate metabolic heat through evaporation from the tongue and respiratory tract.

confusion; further cooling can lead to coma and death (Figure 41.12). Some victims of extreme hypothermia, children particularly, have survived prolonged immersion in cold water of even lower temperatures. In fact, many animals can recover from profound hypothermia. However, cells that become frozen may be destroyed unless thawing is precisely controlled (this sometimes can be done in hospitals). Tissue destruction through localized freezing is called *frostbite*.

**Responses to Heat Stress**. Table 41.5 summarizes the main responses to heat stress as well as cold stress. Here again, the hypothalamus has roles in responses to increases in core temperature. In **peripheral vasodilation**, hypothalamic signals cause blood vessels in skin regions to dilate. More blood flows from deeper body regions to skin regions, where the excess heat it carries is dissipated (Figure 41.13).

Evaporative heat loss is another response that can be influenced by the hypothalamus, which can activate

sweat glands. Your skin has 2-1/2 million or more sweat glands, and considerable heat is dissipated when the water they give up to the skin surface evaporates. With extreme sweating, as might occur in a marathon race, the body loses an important salt—sodium chloride—as well as copious amounts of water. Such losses may change the character of the internal environment to the extent that the runner may collapse and faint.

What about mammals that sweat very little or not at all? Some of them, including dogs, pant. "Panting" refers to shallow, rapid breathing that increases evaporative water loss from the respiratory tract (Figure 41.14). Cooling occurs when the water evaporates from the nasal cavity, mouth, and tongue.

Sometimes peripheral blood flow and evaporative heat loss are not enough to counter heat stress, and *hyperthermia* results. This is a condition in which the core temperature increases above normal. For humans and other endotherms, an increase of only a few degrees above normal can be dangerous.

## From Frog to Frog and Other Mysteries

With a full-throated croak that only a female of its kind could find seductive, a male frog proclaims the onset of warm spring rains, of ponds, of sex in the night. By August the summer sun will have parched the earth, and his pond dominion will be gone. But tonight is the hour of the frog! Through the dark, a female moves toward the vocal male. They meet, they dally in behavioral patterns characteristic of their species. He clamps his forelegs about her swollen abdomen and gives it a prolonged squeeze. Out into the surrounding water streams a ribbon of hundreds of eggs. As the eggs are being released, the male expels a milky cloud of swimming sperm. Each sperm penetrates an egg, and soon afterward, their nuclei fuse. With this fusion, fertilization is completed. A single fertilized egg, the zygote, has formed.

For the leopard frog *Rana pipiens*, a drama now begins to unfold that has been reenacted each spring, with only minor variations, for many millions of years. Within a few hours after fertilization, the single-celled zygote begins dividing into two cells, then four, then many more to produce the early embryo. In less than twenty hours, it has become a ball of tiny cells, no larger than the zygote.

And now the cells begin to migrate, change shape, and interact. Some cells at the embryo's surface sink inward, forming a dimple. Their cellular descendants will give rise to internal tissue layers. Other cells at the surface lengthen or flatten out; together they form a groove. Their cellular descendants eventually will give rise to the nervous system. Through interactions between surface cells and interior cells, eyes start to develop. Within the embryo, a heart is forming and will soon start to beat rhythmically.

A tail takes shape; a mouth forms. These developments, appearing one after another, are signs of a process going on in *all* the cells that were so recently developed from a single zygote. The cells are becoming different from one another in both appearance and function!

Within twelve days after fertilization, the embryo has become a larval form—a tadpole—that swims and feeds on its own (Figure 42.1). After several

a

**Figure 42.1** Development of the leopard frog, *Rana pipiens*. (**a**) A male clasping a female in a behavior called amplexus. When the female releases her eggs into the water, the male releases his sperm over the eggs. (**b**) Frog embryos. (**c**) A larval form called a tadpole. (**d**) Transitional form between the tadpole and the young adult frog (**e**).

months, legs start to grow. The tail shortens, then disappears. The small mouth, once suitable for feeding on algae, develops jaws and now snaps shut on insects and worms. Eventually an adult frog leaves the water for life on land. With luck it will avoid predators, disease, and other threats in the months ahead. In time it may even find a pond filled by the new season's rains, and the cycle will begin again.

*How does the single-celled zygote of a frog or any other complex animal become transformed into all the specialized cells and structures of the adult?* With this question we turn to one of life's greatest mysteries—to the development of new individuals in the image of their parents. We are just starting to understand the underlying mechanisms.

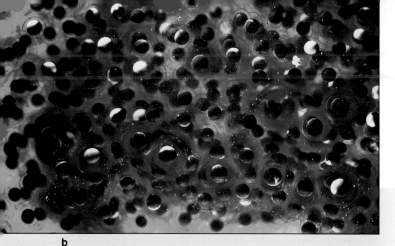

b

c

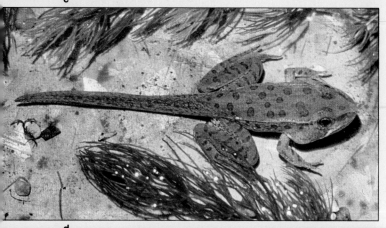

d

e

KEY CONCEPTS

**1.** Most animals reproduce sexually. Elaborate reproductive structures and forms of courtship and parental behavior have accompanied the separation into male and female sexes. The biological cost of those structures and behaviors is offset by the advantages afforded by diversity among the offspring. At least some of the diverse phenotypes should prove to be adaptive in changing or new environments.

**2.** Development commonly proceeds through six stages: gamete formation, fertilization, cleavage, gastrulation, organ formation, and growth and tissue specialization.

**3.** The fate of embryonic cells is determined partly at cleavage (when daughter cells inherit qualitatively different regions of cytoplasm) and partly by cell interactions in the developing embryo. Cell differentiation and morphogenesis depend on those events.

## THE BEGINNING: REPRODUCTIVE MODES

In earlier chapters we looked at the cellular basis of **sexual reproduction**, in which offspring are produced by way of meiosis, gamete formation, and fertilization. We also looked at **asexual reproduction**, in which offspring are produced by means other than gamete formation. Let's turn now to some structural, behavioral, and ecological aspects of these reproductive modes.

Think about a new sponge produced asexually, by budding from the parent body (page 409). Or think about a flatworm engaged in fission, in this case dividing lengthwise into two parts that give rise to two new flatworms. In both cases, the offspring are genetically identical to the parents. Having the same genes as parents is useful when parents are well adapted to the surroundings—and when the surroundings remain stable.

But most animals live under changing, unpredictable conditions. They rely mainly on sexual reproduction, with sperm from a male fertilizing eggs from a female.

a

b

Complete separation into male and female sexes is biologically costly. Getting sperm and eggs together depends on large energy investments in specialized reproductive structures and forms of behavior. Even so, the cost is offset by the variation in traits among the resulting offspring, at least some of which are likely to survive and reproduce in a changing environment.

Consider the question of *reproductive timing*. How do mature sperm become available exactly when eggs mature in a separate individual? Timing depends on energy outlays for sensory structures and rather involved hormonal controls in both parents. Both parents must produce mature gametes in response to the same cues, such as changes in daylength that mark the onset of the best season for reproduction. Moose, for instance, become sexually active in late summer or early fall—and their offspring are born the following spring, when the weather improves and food will be plentiful for many months.

Consider also the challenge of finding and recognizing a potential mate of the same species. Different

**Figure 42.2** Examples of where the animal embryo develops, how it is nourished, and how (if at all) it is protected. Snails (**a**) are *oviparous* (egg-producing) but are not doting parents; their fertilized eggs are unprotected. Egg-laying mammals, including the duck-billed platypus shown in Figure 26.1, are oviparous, yet they also secrete milk to nourish the juveniles. Birds, too, are oviparous. Their fertilized eggs, which have large yolk reserves, develop and hatch outside the mother's body.

(**b**) The fertilized egg of humans and most other mammals is retained inside the mother's body and nourished by her tissues until the time of birth. Such animals are *viviparous* (*viva-*, alive; *-parous*, to produce). The kangaroo (**c**) and opossum (**d**), both marsupials, are viviparous. But their young emerge in unfinished form and undergo further fetal development in a pouch on the ventral surface of the mother's body, where they are nourished from mammary glands.

Some fishes, lizards, and many snakes are *ovoviviparous*. Their fertilized eggs develop within the mother's body. Such eggs are not nourished by the mother's tissues; they are sustained by yolk reserves. (**e**) A copperhead is one of the ovoviviparous snakes. Her liveborn are still contained in the relics of egg sacs.

liveborn snake in egg sac

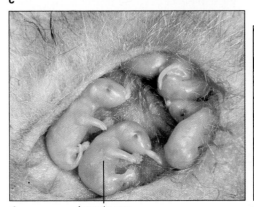

d    fetus in sac

e

species meet the challenge by investing energy in the production of chemical signals, structural signals such as feathers of certain colors and patterns, and sensory receptors able to pick up the signals that are being sent. Besides this, the males often expend considerable energy executing courtship routines (see Figure 50.3).

*Fertilization* also comes at a substantial cost. Many invertebrates and bony fishes simply release eggs and motile sperm into the water, and the chance of successful fertilization would not be good if adults produced only one sperm or one egg each season. Such species invest a great deal of energy in producing gametes by the hundreds of thousands. Nearly all land-dwelling animals depend on **internal fertilization**, the union of sperm and egg *inside* the body of the female parent. They invest energy on elaborate reproductive organs, such as a penis (by which a male deposits its sperm within the female) and a uterus (a chamber in the female where the embryo grows and develops).

Finally, energy is set aside for *nourishing some number of offspring.* Nearly all animal eggs contain **yolk**, a substance made up of proteins and lipids that can nourish the embryo. Eggs of some species have more yolk than others. Sea urchin eggs are small, they are released in large numbers, and the biochemical investment in yolk for each one is limited. There is a premium on rapid development, given the presence of predators that end up consuming most of the eggs. For example, sea stars engage in feeding frenzies when sea urchin eggs become available. No need of abundant yolk here; sea urchin offspring reach a self-feeding, free-moving larval stage in less than a day.

Bird eggs are also released from the mother, but they have large yolk reserves that nourish the embryo through a longer period of development, inside an eggshell that forms after fertilization. Human eggs have almost no yolk. After fertilization, the egg attaches to the uterus inside the mother's body, where the embryo is nourished and supported by physical exchanges with her tissues during an extended pregnancy (Figure 42.2).

As these examples suggest, animals show great diversity in reproduction and development. However, some patterns are widespread in the animal kingdom, and they will serve as a framework for our reading.

---

**Separation into male and female sexes is an energetically costly mode of reproduction. Specialized structures and forms of behavior are required for getting sperm together with eggs and for lending nutritional support to offspring.**

**This reproductive mode is advantageous in unpredictable environments, for at least some traits of the diverse offspring may prove adaptive under new or changing conditions.**

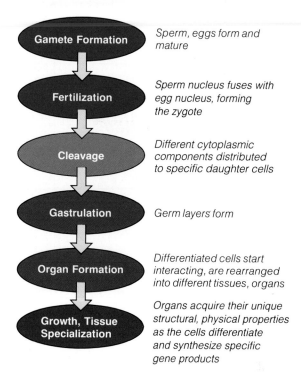

**Figure 42.3** Overview of stages of animal development.

## STAGES OF DEVELOPMENT

Figure 42.3 lists the stages of animal development. In the first stage, **gamete formation**, sperm or eggs form and mature within the parents. **Fertilization**, the second stage, starts when a sperm penetrates an egg. It ends when the sperm nucleus fuses with the egg nucleus and gives rise to the zygote, the first cell of the new individual. Next comes **cleavage**, when mitotic cell divisions typically convert the zygote to a ball of cells, the *blastula*. Cells increase in number but the embryo does not increase in size during this stage. The daughter cells collectively occupy the same volume as did the zygote.

As cleavage draws to a close, the pace of cell division slackens and gives way to **gastrulation**, a stage of major cell rearrangements. The organizational framework for the whole body is laid out as cells become arranged into two or three primary tissues, or *germ layers*. The human body arises from three such layers:

| | |
|---|---|
| **endoderm** | *inner layer; gives rise to inner lining of gut and organs derived from it* |
| **mesoderm** | *intermediate layer; gives rise to muscle, the organs of circulation, reproduction, and excretion, most of the internal skeleton, and connective tissue layers of the gut and integument* |
| **ectoderm** | *surface layer; gives rise to tissues of nervous system and outer layer of integument* |

| | **a**<br>Sea Urchin | **b**<br>Frog | **c**<br>Chick | **d**<br>Human |
|---|---|---|---|---|
| **FERTILIZATION:**<br><br>Fertilized egg<br>(outer membranes<br>shown here) | 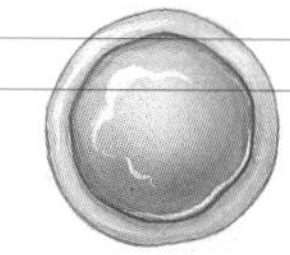 | 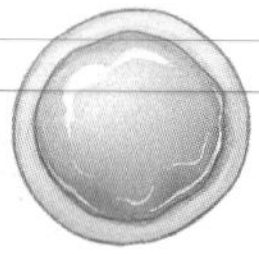 | 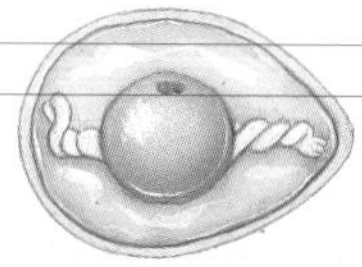 | 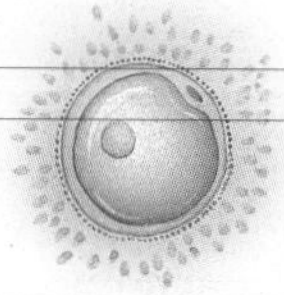 |
| **CLEAVAGE:**<br><br>First cleavage | 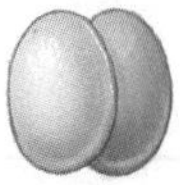 | 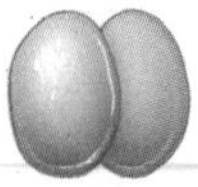 | (top<br>view) 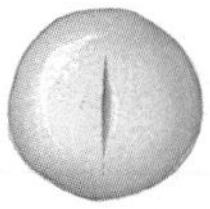 |  |
| Morula Stage | 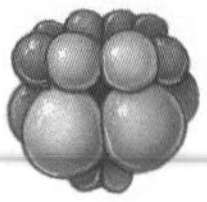 |  | (top<br>view) 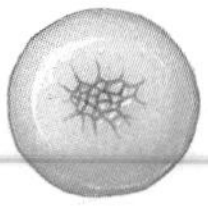 | morula<br><br>blastocyst 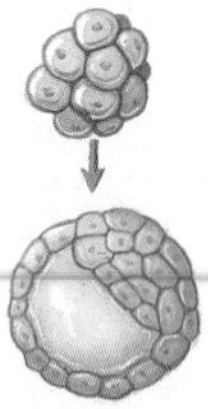 |
| Blastula Stage<br>(or blastodisk) | 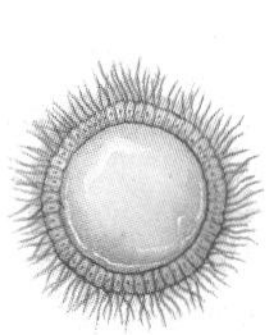 | 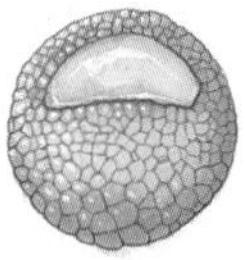 | blastocyst<br><br>(yolk) 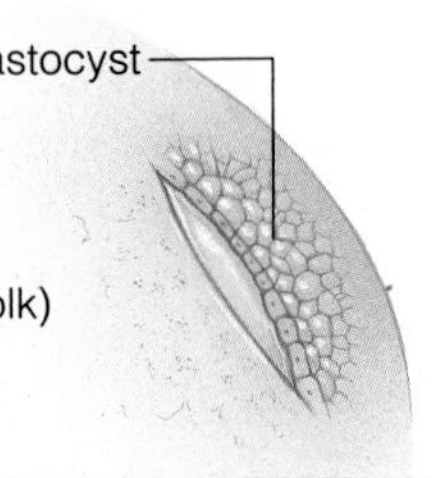 | blastodisk<br><br>(uterine<br>wall of<br>mother) 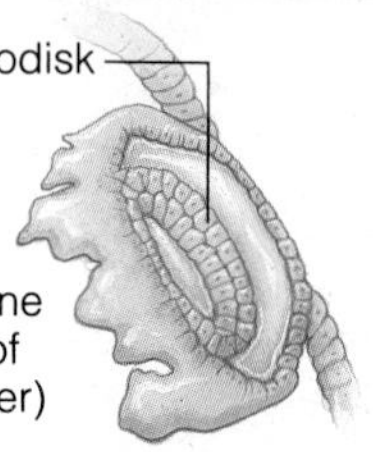 |
| **GASTRULATION:**<br><br>Gastrula<br>(germ layers<br>formed) | 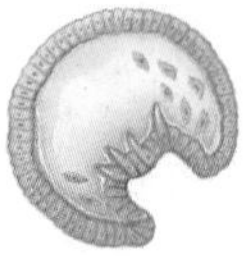 |  | 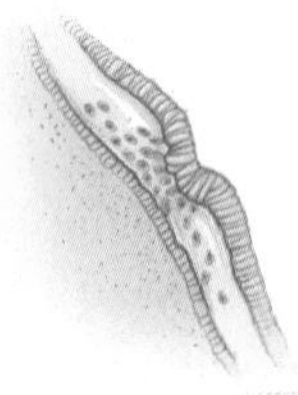 | 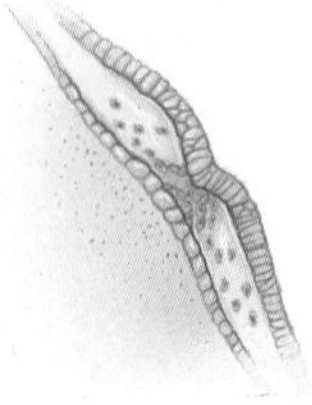 |
| **ORGANOGENESIS,<br>GROWTH, TISSUE<br>SPECIALIZATION:**<br><br>Some stages<br>of organ<br>formation | 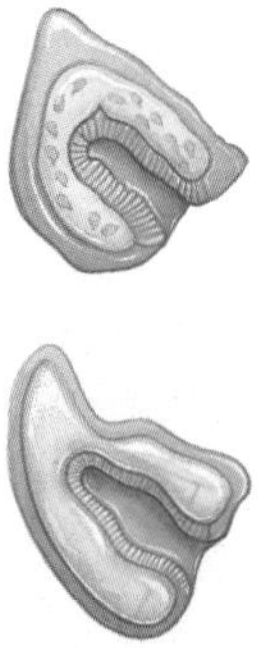 | (top<br>view)<br><br>(side<br>view) 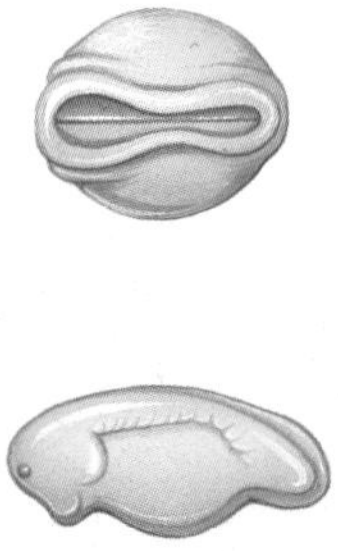 | (top<br>view)<br><br>(side<br>view) 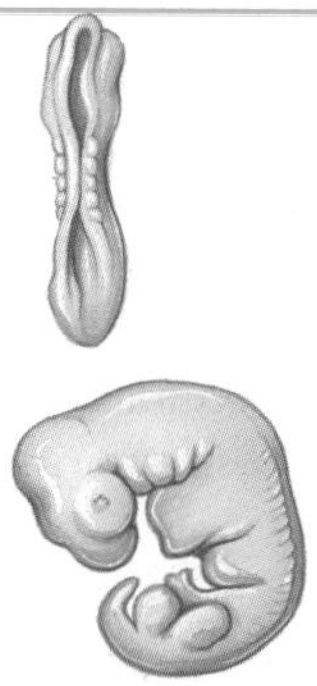 | (top<br>view)<br><br>(side<br>view) 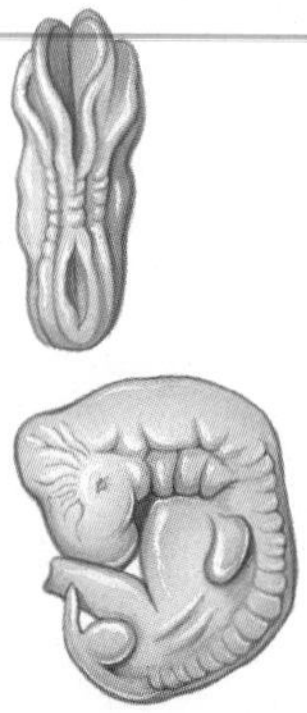 |
| Larval form<br>or advanced<br>embryo | 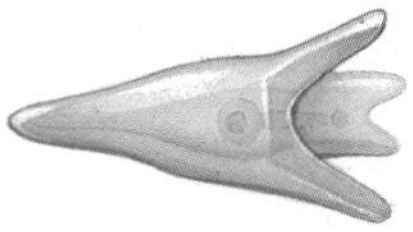 |  | 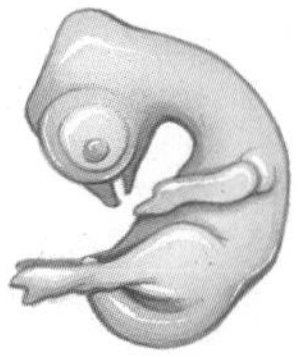 | 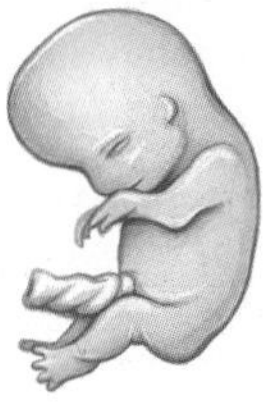 |

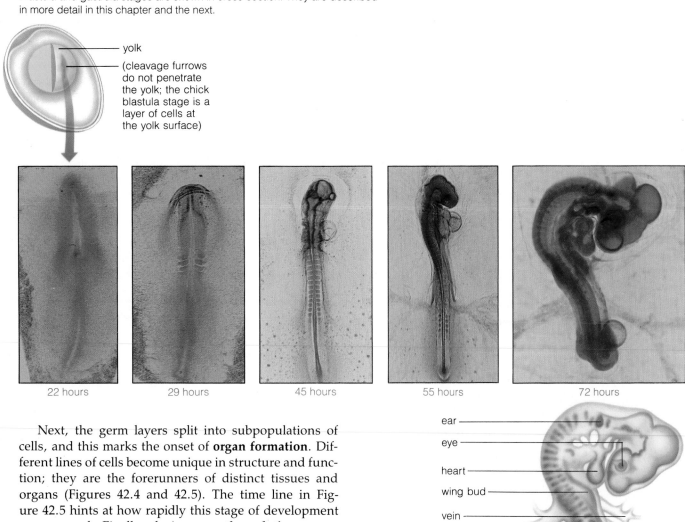

**Figure 42.4** (Left) Comparison of embryonic development in four different animals. The drawings are not to the same scale; however, they show the developmental patterns that are common to all four types. For clarity, the membranes surrounding the embryo are not shown from cleavage onward. Blastula and gastrula stages are shown in cross-section. They are described in more detail in this chapter and the next.

yolk

(cleavage furrows do not penetrate the yolk; the chick blastula stage is a layer of cells at the yolk surface)

22 hours    29 hours    45 hours    55 hours    72 hours

Next, the germ layers split into subpopulations of cells, and this marks the onset of **organ formation**. Different lines of cells become unique in structure and function; they are the forerunners of distinct tissues and organs (Figures 42.4 and 42.5). The time line in Figure 42.5 hints at how rapidly this stage of development can proceed. Finally, during **growth and tissue specialization**, organs acquire specialized properties. This stage continues into adulthood.

Adult animals of different species obviously do not all look alike. Neither do the embryonic forms, even though they commonly pass through all the stages just outlined. As Figure 42.4 suggests, by the end of each stage, the embryo has become more complex than it was before. This is important to think about, for structures that develop during one stage serve as the foundation for the stage following it.

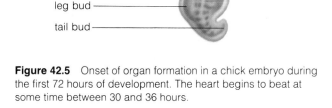

ear
eye
heart
wing bud
vein
artery

leg bud
tail bud

**Figure 42.5** Onset of organ formation in a chick embryo during the first 72 hours of development. The heart begins to beat at some time between 30 and 36 hours.

**Each stage of embryonic development builds on structures that were formed during the stage preceding it. Development cannot proceed properly unless each stage is successfully completed before the next begins.**

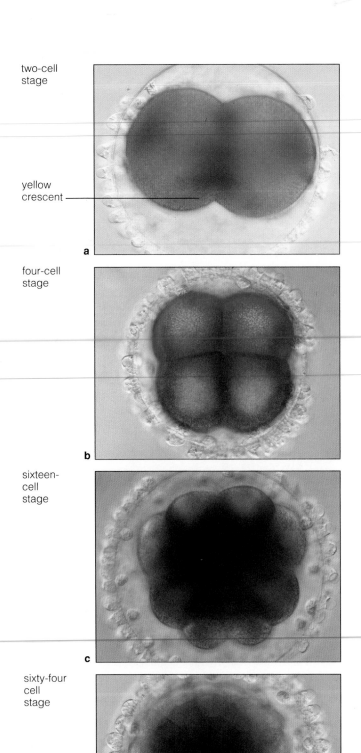

two-cell
stage

yellow
crescent

**a**

four-cell
stage

**b**

sixteen-
cell
stage

**c**

sixty-four
cell
stage

**d**

**Figure 42.6** Cytoplasmic localization in the embryo of a tunicate (*Styela*). Cleavage divides regions of cytoplasm into particular cells. The fertilized eggs of these animals have a pigmented region (the yellow crescent). This region becomes localized into a small group of cells that will give rise to the muscles of the tunicate larva.

## PATTERNS OF DEVELOPMENT

### Key Mechanisms: An Overview

In large part, developmental complexity is mapped out in the cytoplasm of an immature egg, or *oocyte*. Later on, as more cells form in the developing embryo, individual cells and groups of cells start to interact. Their interactions are a basis for **cell differentiation**, with cells in different locations in the embryo becoming specialized in prescribed ways at specific times. As the embryo continues its development, differentiated cells become organized into tissues and organs, a process called **morphogenesis**. Before we turn to specific examples, let's define two of the most important mechanisms underlying these developmental events. They are called cytoplasmic localization and embryonic induction.

**Cytoplasmic Localization.** The future shape and arrangement of body parts in an embryo depend largely on what goes on in a maturing oocyte, even before fertilization. An oocyte is larger and more complex than a sperm. As it matures, organelles, proteins, RNA, and other components accumulate and become distributed in prescribed locations in the cytoplasm. These "cytoplasmic determinants" are maternal instructions that will have roles in the way body parts become arranged in orderly fashion in the developing embryo.

During cleavage, daughter cells end up in specific locations in the early embryo. Simply by virtue of their location, those cells inherit different cytoplasmic determinants (Figure 42.6). This **cytoplasmic localization** helps seal the developmental fate of the descendants of those cells.

**Embryonic Induction.** As the embryo develops, one or more groups of cells interact physically or chemically in ways that affect the development of at least one of those groups. This mechanism is called **embryonic induction**. For example, certain groups of cells produce a hormone, growth factor, or some other substance that diffuses to other cell groups and triggers the synthesis of a particular protein or some other change in their activities. Such interactions give groups of cells their developmental marching orders, so to speak.

Development involves cell differentiation and morphogenesis. Both processes depend on the segregation of cytoplasmic determinants during cleavage and interactions among embryonic cells.

With these key mechanisms in mind, let's now look more closely at how early developmental events unfold.

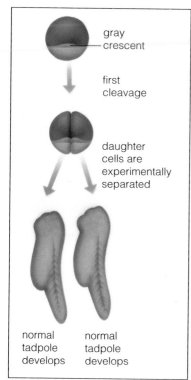

**a** Experiment 1

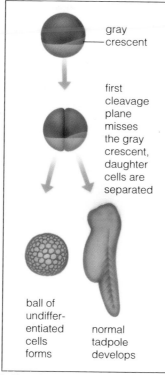

**b** Experiment 2

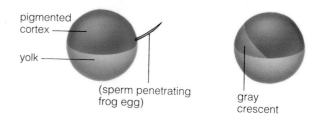

**Figure 42.7** Two experiments illustrating how qualitative differences in the cytoplasm of a fertilized egg help determine the fate of cells in a developing embryo. Frog eggs contain granules of dark pigment in their cortex (the plasma membrane and the cytoplasm just below it). The granules are concentrated near one pole of the egg; yolk is concentrated near the other pole. At fertilization, a portion of the granule-containing cortex shifts away from the yolk, and this exposes lighter colored cytoplasm in a crescent-shaped gray area:

Normally, the first cleavage divides the gray crescent between two daughter cells. (**a**) Even if the two daughter cells are separated from each other experimentally, each may still give rise to a complete tadpole. (**b**) But if a fertilized egg is manipulated so that the first cleavage plane misses the gray crescent entirely, only one of the two daughter cells gets the gray crescent and it alone will develop into a normal tadpole. The daughter cell deprived of substances in the gray crescent will only give rise to a ball of undifferentiated cells.

## Developmental Information in the Egg

A sperm cell produced during spermatogenesis is little more than paternal DNA, packaged with a few components that are necessary for moving the DNA to an egg (page 762). The oocyte is another story. It is the scene of dynamic activity, including increases in volume. Specialized proteins (including enzymes and yolk proteins), mRNA transcripts, and other molecules accumulate in its cytoplasm and will direct the development of the early embryo. For example, many of the mRNA molecules being transcribed from the maternal DNA are translated at once into enzymes, histones, and other proteins that will be used for chromosome replications in the early embryo. Ribosomal subunits and other cytoplasmic components necessary for protein synthesis are stockpiled. In addition, many mRNA transcripts accumulate in different regions of the egg cytoplasm. They will be allocated to different daughter cells during cleavage, then activated to direct the synthesis of specific sets of proteins.

Also present in the maturing oocyte are microtubules and other cytoskeletal elements, oriented in specific directions. These elements will influence the first cell divisions in the embryo. Whenever a cell divides in two, it does so at a prescribed angle relative to the adjacent cells, based partly on the orientation of microtubules of the mitotic spindle (page 141). The amount and distribution of yolk within the egg cytoplasm also will influence cleavage. Even the nucleus has a characteristic position in a frog egg and imparts "polarity" to it. The *animal pole* is simply the one closest to the nucleus. Opposite is the *vegetal pole*, where substances such as yolk accumulate. All animal eggs show some degree of polarity—that is, two identifiable poles—which will influence the structural patterns that emerge as the embryo develops.

## Fertilization

When a sperm penetrates an egg, it triggers structural reorganization in the egg cytoplasm. You can observe indirect signs of this reorganization in frog eggs, which contain granules of dark pigment in their cortex. (The "cortex" is the plasma membrane and the cytoplasm just beneath it.) The animal pole has more pigment granules and is darker than the vegetal pole. Sperm penetration causes microtubules to move the granules and then the cortex itself. The outcome is a **gray crescent**, an area of intermediate pigmentation near the equator, on the side of the egg *opposite* the sperm penetration site (Figures 42.7 and 42.8). This area establishes the body axis of the frog embryo, and gastrulation will begin there. This is another example of how the organization of the egg's cytoskeleton represents critical information for the embryo.

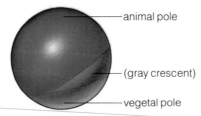

animal pole

(gray crescent)

vegetal pole

(gray crescent)

**Figure 42.8** Micrographs and diagrams of the early embryonic development of a frog. For these micrographs, the jellylike layer surrounding the egg has been removed, with the exception of (**i**). (**a**) Within about an hour after fertilization, the gray crescent establishes the body axis for the embryo, and gastrulation will begin here. (**b–f**) Cleavage leads to a blastula, a ball of cells in which a cavity (blastocoel) has appeared.

(**g, h**) Cells move about and become rearranged during gastrulation. Tissue layers form; a primitive gut cavity (archenteron) develops. (**i, j**) Neural developments now take place, and the fluid-filled body cavity in which vital organs will be suspended appears. (**k**) Differentiation proceeds, moving the embryo on its way to becoming a functional larval form.

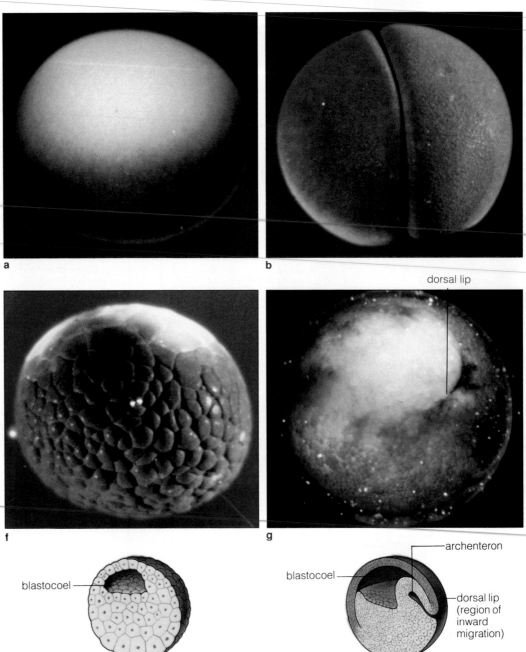

a

b

f

g

dorsal lip

blastocoel

blastocoel

archenteron

dorsal lip (region of inward migration)

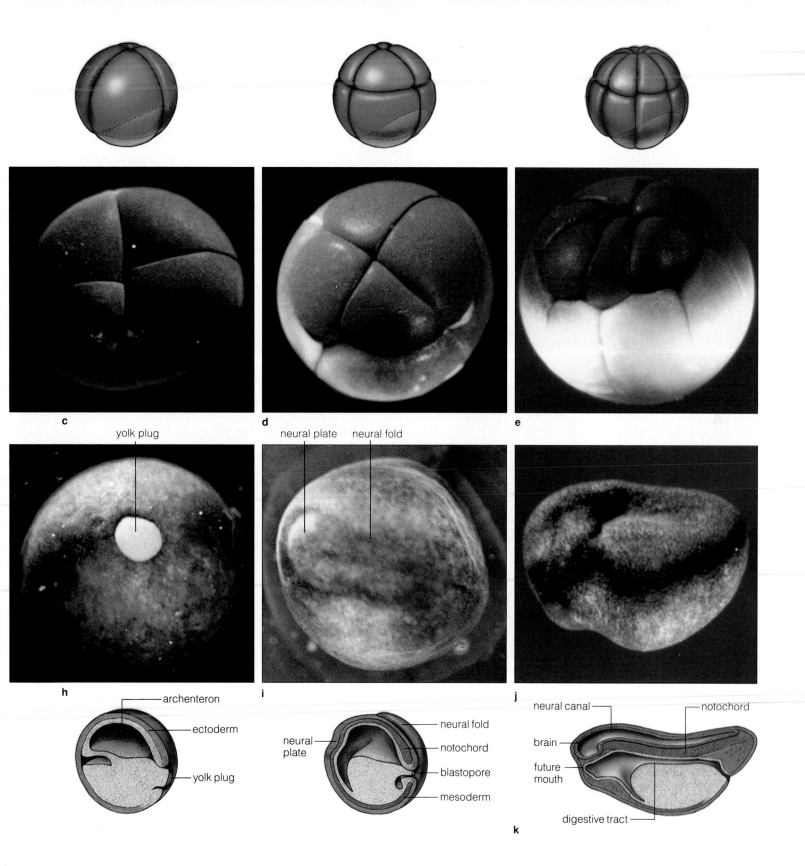

c

yolk plug

d

neural plate      neural fold

e

h

archenteron

ectoderm

yolk plug

i

neural
plate

neural fold

notochord

blastopore

mesoderm

j

k

neural canal

notochord

brain

future
mouth

digestive tract

## Cleavage

Fertilization is followed by mitotic cell divisions that convert the zygote into the early multicelled embryo. A cleavage furrow forms during each mitotic cell division (page 145), and it defines the plane where the cytoplasm will be pinched in two. Usually the embryo does not increase in size while the initial cleavages are proceed-ing. All the daughter cells collectively occupy the same volume as did the zygote, although they differ in size, shape, and activity.

Successive cleavages commonly produce a **blastula**, a ball of cells with a fluid-filled cavity (blastocoel) inside. The cleavage patterns themselves vary among different animal groups, however, often as a result of dramatic differences in the density and distribution of yolk. Cleavage of a sea urchin egg, which has little yolk, produces a hollow, single-layered sphere of cells (Figure 42.9). The concentrated yolk of an amphibian egg impedes cleavage near the vegetal pole, and a blastocoel forms near the animal pole. The abundant yolk of reptile, bird, and most fish eggs restricts cleavage to a tiny, caplike region at the animal pole. In these eggs, cleavage produces two flattened layers with a thin blastocoel between them, perched on the yolk surface (Figures 42.4c and 42.5c).

The abundant yolk of insect eggs influences cleavage in a truly distinctive way. Consider a *Drosophila* egg (Figure 42.10). At first the nucleus divides repeatedly and cleavage furrows do not form. Within hours, the fertilized egg is a bag of nuclei crowded together in the yolky cytoplasm. Then most of the nuclei migrate toward the egg surface, where a layer of cells forms. This layer (the blastoderm) is the embryo.

The blastula stage of mammals, called a blastocyst, differs from that of other animals in a key respect. By this stage, *two* distinct cell regions can be discerned. Some cells, which form a hollow sphere, are not part of the embryo. Their cellular descendants will form a portion of the placenta. (That structure, composed of embryonic and maternal tissues, will be vital for nurturing the developing embryo.) Other cells are clustered together and attached to the inner surface of the blastocyst wall. As described in the next chapter, the embryo proper develops from this "inner cell mass."

As these examples make clear, the particular cleavage planes that form in an animal zygote help dictate the size and spatial positions of the resulting cells. They also dictate which cytoplasmic determinants each cell will inherit. Together, these factors will affect how each cell will interact with others during the gastrula stage—and so on through subsequent stages of development.

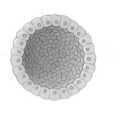

**a** Cross-section through blastula

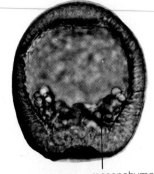

**b** Early gastrula

mesenchyme

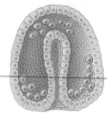
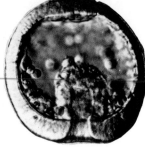

**c** Mature gastrula

**d** blastopore

**Figure 42.9** (**a**) Blastula of a sea urchin (*Lytechinus*). The inward migration of surface cells during gastrulation is apparent in the cross-sections in (**b**) and (**c**). Some of the cells, designated mesenchyme, will ultimately develop into a third germ layer, the mesoderm. (**d**) This scanning electron micrograph shows a surface view of individual cells and the inward cell migrations of an early gastrula.

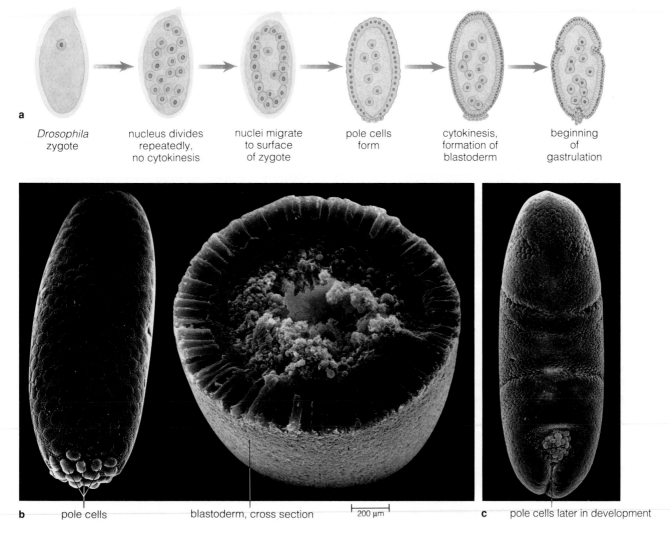

**a**

Drosophila zygote

nucleus divides repeatedly, no cytokinesis

nuclei migrate to surface of zygote

pole cells form

cytokinesis, formation of blastoderm

beginning of gastrulation

**b**    pole cells      blastoderm, cross section    ⊢ 200 µm ⊣      **c**    pole cells later in development

## Gastrulation

As cleavage draws to a close, the pace of cell division slackens. The cells begin to move about and change their positions relative to one another. Gastrulation, the stage of cell rearrangements, has begun.

There is little (if any) increase in size during gastrulation, but the cell rearrangements dramatically change the embryo's appearance. In sea urchins, for example, surface cells migrate inward, and some form the lining of an internal cavity (archenteron). That cavity will eventually become the gut (Figure 42.9). In other species, cell rearrangements establish a long axis for further development after gastrulation. For example, a tube will form along this axis in vertebrate embryos. As you will see shortly, the neural tube is the forerunner of the brain and spinal cord.

This last example brings us to the significance of gastrulation. Think about the structural organization of animals. With few exceptions, animals have an internal region of cells, tissues, and organs that function in digestion and absorption of nutrients. They have a surface

**Figure 42.10** (**a**) Delayed cleavage in a Drosophila zygote. The nucleus divides repeatedly, but cleavage is postponed until the nuclei migrate toward the zygote's surface. Then, a layer of single cells forms above the yolky cytoplasm. That layer, the blastoderm, is the early embryo (**b**).

(**c**) Cytoplasmic localization is known to affect a key aspect of Drosophila development. While the blastoderm is forming, cytoplasmic components called polar granules migrate to and become isolated in pole cells that form at one end of the zygote. In certain mutant embryos, the polar granules do not migrate properly—and reproductive cells never do form. The adult fly is sterile.

tissue region that protects internal parts and is equipped with sensory receptors for detecting changes in the outside world. Most animals have an intermediate region of tissues organized into many internal organs, such as those concerned with movement, support, and blood circulation. The three regions develop from the three germ layers—the endoderm, ectoderm, and mesoderm. This three-layered organization is typical of most animals, and it arises through gastrulation.

## CELL DIFFERENTIATION

Through cell differentiation, a single fertilized egg gives rise to diverse types of specialized cells. All cells produce a number of the same kinds of proteins. But each differentiated cell type also produces some proteins that are *not* found in other cell types. Those proteins are the basis of distinctive cell structures and functions. Differentiated cells have the same number and kind of genes (they are all descended from the same zygote). Through gene controls, however, restrictions are placed on *which* genes will be expressed in a given cell (page 236).

For example, while you were still an embryo and your eyes were developing, some cells started synthesizing quantities of crystallin, a family of proteins that would be used in the construction of transparent fibers in the lens. No other cell type in your developing body could activate the genes necessary to do this. The transparent fibers caused the lens cells to elongate and flatten and gave them their unique optical properties. The crystallin-producing cells are only one of many differentiated types.

By conservative estimates, adult mammals end up with populations of at least 150 different cell types, each with its distinctive structure, products, and functions. (It also has populations of *stem cells,* which can self-replicate, differentiate, or both. Descendants of these stem cells continually replace worn-out or dead cells in skin, blood, and the gut mucosa.)

With only rare exceptions, the fully differentiated state of a given cell type is reached without any loss of genetic information. How do we know this? Consider John Gurdon's experiments with the South African clawed frog, *Xenopus laevis.* Gurdon and his coworkers removed or inactivated the nucleus of an unfertilized frog egg. They also isolated intestinal cells from *Xenopus* tadpoles and carefully ruptured the plasma membrane, leaving the nucleus and most of the cytoplasm intact. Then they inserted the nucleus of the ruptured tadpole cell into the enucleated egg. In some cases, the transplanted nucleus—which was from a highly differentiated intestinal cell—directed the developmental program leading to a whole frog! Clearly the intestinal cell nucleus still contained the same genes as the original zygote nucleus.

As another example, consider what happens when a human embryo spontaneously splits during the first cleavage into two separate cells. The result is not two half-embryos but *identical twins,* or two complete, normal individuals having the same genetic makeup. (By contrast, nonidentical twins occur when two different eggs are fertilized at the same time by two different sperm.) Spontaneous splitting at cleavage is actually the normal pattern of development for armadillos. For those animals, the embryo splits at the four-cell stage to produce quadruplets, every time.

## MORPHOGENESIS

For each animal, tissues and organs become organized with great precision into patterns characteristic of the species. What kinds of events give rise to that organization? Considerable research in embryology has shown that the main ones are these:

1. Cell division
2. Cell migrations
3. Changes in cell size and shape
4. Localized growth
5. Controlled cell death

Let's consider a few examples of the morphogenetic changes that such cellular events can bring about.

### Cell Migrations

During morphogenesis, cells and entire tissues migrate from one site to another. In **active cell migration,** cells move about by means of pseudopods. Pseudopods are temporary projections from the main cell body, some rather bulbous, others fingerlike. The cells migrate over prescribed pathways, reaching a prescribed destination and establishing contact with cells already there. These movements must be extremely accurate. Through such migrations, for instance, the forerunners of nerve cells make billions of precise connections that enable the human nervous system to function.

How do the cells "know" where to move? In part, they move in response to chemical gradients, a behavior called *chemotaxis.* The gradients are probably created when specific substances are released from target tissues. This type of movement was shown in Figure 22.18, which described the development of the slime mold *Dictyostelium discoideum.*

Cells also move in response to *adhesive cues,* provided by recognition proteins at the surface of other cells and by molecules of the extracellular matrix. In vertebrate embryos, pigment cells are following such cues when they move along blood vessels but not along the axons of neurons. Similarly, Schwann cells are following adhesive cues when they migrate along the axons of neurons but not along blood vessels. It seems likely that the synthesis, secretion, deposition, and removal of specific extracellular substances help coordinate the active migration of cells in a developing embryo.

How do the migrating cells know when to stop moving? Again, adhesive cues appear to be involved. Migrating cells move to locations where adhesive interactions are strongest. Once they become arranged in a manner that maximizes their adhesion, further migration is impeded.

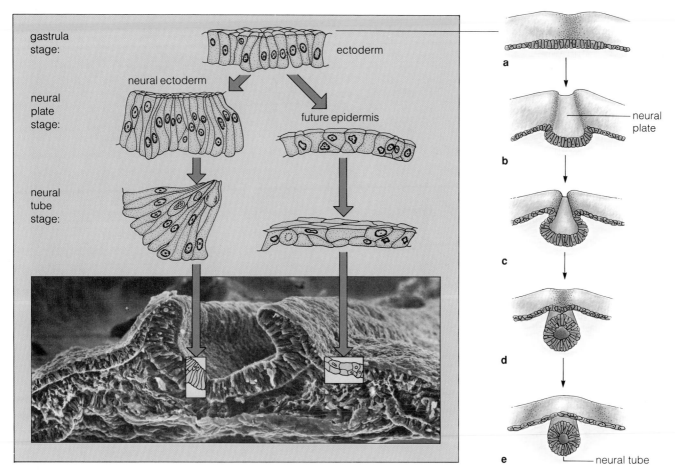

**Figure 42.11** Example of morphogenesis—the changes in cell shape that underlie the formation of a neural tube (the forerunner of the brain and spinal cord). As gastrulation draws to a close, the ectoderm is a uniform sheet of cells (**a**). Some ectodermal cells elongate, forming a neural plate, then they constrict at one end to become wedge-shaped. The changes in cell shape cause the ectodermal sheet to fold over the neural plate to form the neural tube (**b–e**). Other ectodermal cells flatten while these changes are occurring; they will become part of the epidermis. The scanning electron micrograph and diagrams show the neural tube and epidermis forming in a chick embryo.

## Changes in Cell Size and Shape

Another morphogenetic movement is the inward or outward folding of sheets of cells. Such folding is brought about by coordinated changes in cell shapes. Think about what happens after the three germ layers form in the embryos of amphibians, reptiles, birds, and mammals. As Figure 42.11 shows, ectodermal cells at the midline of the embryo elongate and form a **neural plate**, the first indication that a region of ectoderm is on its way to developing into nervous tissue. The change in each cell's shape is supported by the elongation of microtubules in its cytoplasm. Next, cells near the middle become wedge-shaped. The shape arises when a ring of microfilaments in the cytoplasm constricts one end of each elongated cell. Collectively, the changes in cell shape cause the neural plate to fold over and meet at the embryo's midline to form the **neural tube**. This tube is destined to become

the brain and spinal cord (page 566). Meanwhile, other ectodermal cells have flattened; they will become the epidermis above the tube.

## Localized Growth and Cell Death

Morphogenesis depends on **localized growth**, which contributes to changes in the sizes, shapes, and proportions of body parts. How localized growth occurs in some tissues more than others is not fully understood, but regulatory genes are almost certainly involved. Recall that there are few differences between chimpanzee and human DNA. Yet adult chimpanzees and humans differ quite a bit in the proportions of their body parts, including the skull, even though their fetuses develop at about the same rate, in parallel ways (page 296). Possibly during human evolution, there were mutations in regulatory genes that affected proportional changes in the skull.

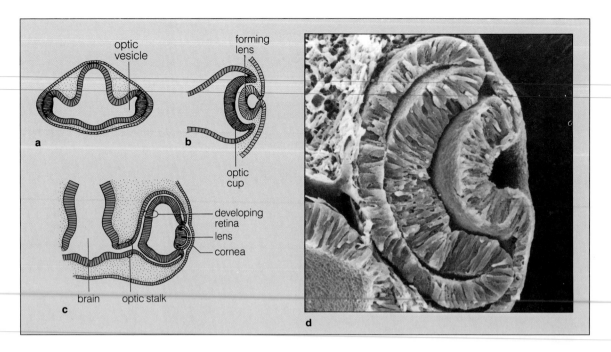

**Figure 42.12** Eye formation. The retina develops as an outgrowth of the brain; the lens, as an ingrowth of the ectoderm. (**a**) An optic vesicle grows out of the side of the brain. When it contacts the head ectoderm, it induces the elongation and inward folding of the ectodermal cells to form a lens vesicle (**b**). Meanwhile, the optic vesicle is induced to sink inward, forming the optic cup. (**c**) The cup's inner layer will form the retina. (**d**) Scanning electron micrograph of an optic cup and lens in a chick embryo.

Morphogenesis also depends on **controlled cell death**. The term refers to the elimination of tissues and cells that are used for only short periods in the embryo or the adult. Controlled cell death is genetically programmed, as the following examples suggest.

Perhaps you have noticed that kittens and puppies are born with their eyes sealed shut. The eyelids form as an unbroken layer of skin. Just after birth, the cells stretching in a thin line across the middle of each eyelid die on cue. As the dead cells degenerate, a slit forms in the skin, and then the upper and lower lids part company.

In a human embryo, hands and feet start out as paddle-shaped structures. Skin cells between the lobes of four "paddles" die on cue, leaving separate toes and fingers (Figure 9.1). This mechanism is genetically programmed. Between the time a death signal is sent and the actual time of death, protein synthesis declines dramatically in the doomed cells.

Finally, the embryos of ducks also have paddlelike appendages, but cell death normally does not occur in them; that is why ducks have webbed feet instead of separated toes. In some mice and some humans, a gene mutation blocks cell death in the paddles, and the digits remain webbed.

## Pattern Formation

For morphogenesis to proceed smoothly in a developing embryo, cells must sense their position relative to one another and use the information to produce ordered, spatial arrangements of differentiated tissues. The term **pattern formation** refers to the mechanisms responsible for the specialization of tissues and their positioning in space. As we have seen, the most important of those mechanisms are cytoplasmic localization and embryonic induction.

**Vertebrate Eye Formation.** Hans Spemann's research into the development of the vertebrate eye provides us with a classic example of embryonic induction. The retina of an eye originates from the forebrain but its lens, which focuses light onto the retina, originates from the epidermis (Figure 42.12). Spemann experimented with a salamander embryo in which optic cups had already begun to grow out of the forebrain. He surgically removed one of the optic cups and inserted it under the ectoderm of the belly region. Belly epidermal cells that came in contact with the transplanted optic cup were induced to form a lens—which fit perfectly into the transplanted optic cup!

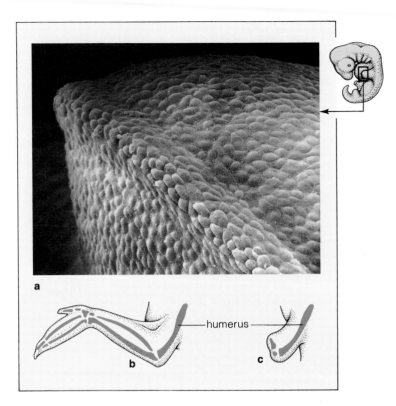

**Figure 42.13** (**a**) Scanning electron micrograph of the normal ectodermal ridge of a wing bud in a chick embryo. (**b**) Normal pattern of bone formation. (**c**) Pattern of bone formation when ectodermal cells at the ridge are surgically removed before the wing bud is fully grown. All new development ceases. Only the bones that have been determined by the time of the surgery will differentiate; notice the deficiencies in the wing skeleton.

## Chick Wing Formation.

The importance of embryonic induction is further illustrated by the precise positioning of bones that occurs when a chick wing develops. Through cell divisions in its mesoderm and the ectodermal covering, the wing bud grows out from the body. When certain groups of ectodermal cells are removed from the tip of a half-grown wing bud, terminal wing bones never develop (Figure 42.13). Apparently, as new cells are added to the outwardly growing tip, they "assess" which bones were formed previously, then they become the next bones in line. If the ectodermal ridge is surgically removed, there will be no new mesodermal cells to form the remaining bones.

## Pattern Formation in *Drosophila*.

As is the case for other arthropods, the *Drosophila* body is segmented. Segmentation is one consequence of cytoplasmic localization. Before cleavage, cytoplasmic components direct the migration of different nuclei in the *Drosophila* zygote toward different areas of the zygote's surface (Figure 42.10).

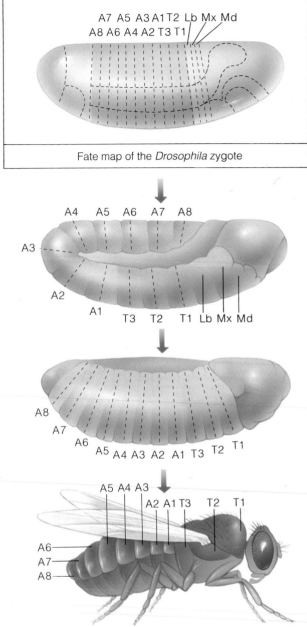

**Figure 42.14** Fate map of the *Drosophila* zygote. Dashed lines indicate the regions that will develop into specialized segments, many of which have highly specialized appendages. The developmental fate of the different segments begins with cytoplasmic localization in the zygote. The drawings are not to scale.

In certain experiments, small regions of the cytoplasm were destroyed in the zygote and segments failed to develop properly. Such experiments have helped developmental biologists draw a "fate map" on the surface of the *Drosophila* zygote. The map is used to determine the origin of each kind of differentiated cell in the different body segments (Figure 42.14).

**Figure 42.15** Imaginal disks of a *Drosophila* larva. Each disk will give rise to a specific part of the adult fly.

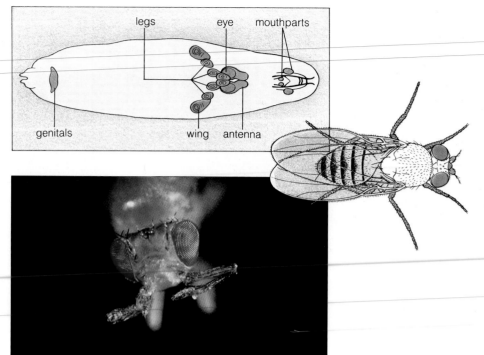

legs    eye    mouthparts

genitals    wing    antenna

**Figure 42.16** Experimental evidence of embryonic induction in *Drosophila*. A cluster of cells (imaginal disk) that gives rise to antennae was surgically removed and exposed to abnormal tissue environments. Then it was reinserted into a different location in a larva undergoing metamorphosis. Legs appeared on the head where antennae would normally be.

**Figure 42.17** Controls over the development of a silkworm moth, *Platysamia cecropia*. A hatched larva eats until it completes five near-doublings in size through rapid epidermal cell divisions and enlargements. The cells also secrete chitin to the epidermal surface, and this forms a cuticle that sets an upper limit on increases in mass. When the limit is reached, cell division idles and the larva molts (sheds its cuticle). Cell divisions resume, more chitin is secreted, and a larger cuticle is produced. The larva grows and molts repeatedly. Then chitin is deposited over the whole insect, legs and all; this is the pupal stage.

The insect spins a cocoon around itself for three days, then the body undergoes massive cell destruction and tissue reorganization. The contents of degraded cells form a nutrient-rich soup that sustains the growth of imaginal disks. The disks had been growing slowly, but now they rapidly give rise to what will become adult tissues and organs. The pupa lasts eight winter months. In spring, cell death and tissue changes transform the pupa into the adult.

Transformation of a larva into a moth is an example of metamorphosis. The adult "plan" is already laid out in immature tissues. Neural and endocrine controls dictate when the plan will be fulfilled. Cells in the larval brain secrete a hormone that acts on paired prothoracic glands. The glands produce and secrete precursors of *b-ecdysone*, a hormone that activates gene expression and stimulates molting. Hormones also act on two glands (the corpora allata) behind the brain. The glands secrete *juvenile hormone* (JH) which, at high levels, prolongs the juvenile state.

The *absence* of JH triggers cell differentiation into the adult. High levels of both JH and ecdysone promote larval growth and development. The amount of JH declines steadily and disappears by the fifth larval stage. The pupa forms when ecdysone levels are high and JH levels are low. Then, developmental restraints are lifted from cells of the imaginal disks.

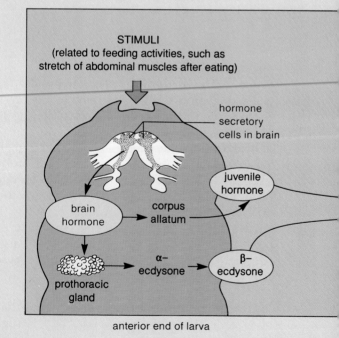

**STIMULI**
(related to feeding activities, such as stretch of abdominal muscles after eating)

hormone secretory cells in brain

brain hormone

corpus allatum

juvenile hormone

prothoracic gland

α–ecdysone

β–ecdysone

anterior end of larva

Other *Drosophila* experiments provide evidence of embryonic induction. Some cells in *Drosophila* larvae are arranged in clusters, each of which will give rise to specific body parts of the adult (Figure 42.15). The clusters are *imaginal disks*. Normally, the fate of the cells in each disk is sealed early in development. However, when an "antenna" disk is surgically removed and exposed to abnormal tissue environments before reinsertion into a larva undergoing metamorphosis, it may later differentiate into a leg (Figure 42.16). The experiment provides evidence that inducer signals influence which genes will be "read" when cells of the disks start the program of differentiation.

The same thing happens as a result of homeotic mutations. Apparently, a **homeotic mutation** affects a regulatory gene that activates sets of genes concerned with development. These single-gene mutations lead either to alterations in the inducer substance itself or to abnormal responses to it. One homeotic mutation activates the wrong set of genes in the cells of an "antenna" disk in *Drosophila*, causing those cells to differentiate into a leg.

**Inducer Signals.** What acts as an inducer during pattern formation? How is it transmitted from the inducer tissue to the responding tissue? Several experiments indicate that chemical signals diffuse from one tissue to the other. When the two tissues destined to become the epithelium and connective tissue of a pancreas are surgically separated from each other in a mouse embryo, the future epithelial tissue does not differentiate properly. However, suppose the two tissues are grown in a culture medium, separated only by a filter that is permeable to large molecules (but not to cells or cell structures). Then, the responding tissue differentiates properly. Although physical contact with the inducing tissue had been prohibited, signals still reached it.

This completes our discussion of the types of mechanisms underlying specific steps along the road leading from the embryo to the adult form. We conclude with the example in Figure 42.17, which shows how separate mechanisms can be integrated and controlled to bring about the transformation. This figure describes how controls govern the dramatic metamorphosis of a silkworm larva into a moth.

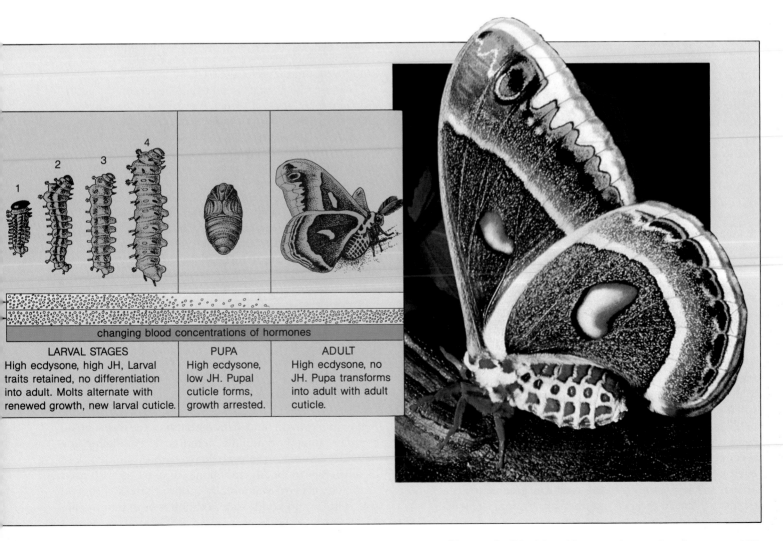

changing blood concentrations of hormones

| LARVAL STAGES | PUPA | ADULT |
|---|---|---|
| High ecdysone, high JH, Larval traits retained, no differentiation into adult. Molts alternate with renewed growth, new larval cuticle. | High ecdysone, low JH. Pupal cuticle forms, growth arrested. | High ecdysone, no JH. Pupa transforms into adult with adult cuticle. |

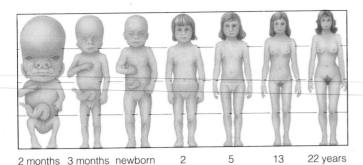

2 months  3 months  newborn    2      5       13    22 years

**Figure 42.18**  Diagram of changes in the proportions of the human body during prenatal and postnatal growth.

## POST-EMBRYONIC DEVELOPMENT

Once all the organs necessary for feeding and other vital activities have formed and are functioning, the new individual is ready to lead a more or less independent existence. Embryonic development is over, and now the young animal follows a prescribed course of further growth and development that leads to the **adult**, the sexually mature form of the species.

In nematodes and a few other animals, the transition from embryo to adult is straightforward. The young simply are miniatures of what is to come. All it takes to reach adulthood are increases in size and maturation of the gonads. The post-embryonic development of reptiles, birds, and mammals involves increases in size and changes in body proportions, as Figure 42.18 suggests. Other organs besides the gonads may not be fully developed. Newly hatched birds, for example, have only partially developed primary feathers just under the skin.

For insects and some other kinds of animals, the course of development is "indirect," for a larval stage intervenes between the embryo and the adult. First the embryo grows into a larva (a sexually immature, free-living and free-feeding animal), then the larva itself grows and changes into the sexually mature form. Extended larval stages are typical of animals that release small and relatively yolkless eggs into water. Short larval stages are common among animals that lay large, yolky eggs in water.

In some species, the transformation from larva to adult is gradual, with the immature form simply growing in size. In other species, the transformation involves massive tissue reorganization and drastic remodeling into the adult form, as in frogs (Figure 42.1). The reactivated growth and transformation of a larva into the sexually mature adult is called **metamorphosis**.

A different kind of reactivated growth occurs during **regeneration**: the replacement of body parts that have been lost by accident. For example, if a predator grasps one of the legs of a crab, the crab can give it up (the better to make an escape), and then grow a replacement.

## Death in the Open

By Lewis Thomas (*Printed by permission from the author and the* New England Journal of Medicine, *January 11, 1973, 288:92–93*)

Everything in the world dies, but we only know about it as a kind of abstraction. If you stand in a meadow, at the edge of a hillside, and look around carefully, almost everything you can catch sight of is in the process of dying, and most things will be dead long before you are. If it were not for the constant renewal and replacement going on before your eyes, the whole place would turn to stone and sand under your feet.

There are some creatures that do not seem to die at all; they simply vanish totally into their own progeny. Single cells do this. The cell becomes two, then four, and so on, and after a while the last trace is gone. It cannot be seen as death; barring mutation, the descendants are simply the first cell, living all over again. The cycles of the slime mold have episodes that seem as conclusive as death, but the withered slug, with its stalk and fruiting body, is plainly the transient tissue of a developing organism; the free-swimming amoebocytes use this mode collectively in order to produce more of themselves.

There are said to be a billion billion insects on the earth at any moment, most of them with very short life expectancies by our standards. Someone has estimated that there are 25 million assorted insects hanging in the air over every temperate square mile, in a column extending upward for thousands of feet, drifting through the layers of atmosphere like plankton. They are dying steadily, some by being eaten, some just dropping in their tracks, tons of them around the earth, disintegrating as they die, invisibly.

Who ever sees dead birds, in anything like the huge numbers stipulated by the certainty of the death of all birds? A dead bird is an incongruity, more startling than an unexpected live bird, sure evidence to the human mind that something has gone wrong. Birds do their dying off somewhere, behind things, under things, never on the wing.

## AGING AND DEATH

Following growth and differentiation, the cells of all complex animals gradually deteriorate. Paralleling the deterioration are structural changes and a gradual loss of efficiency in bodily functions, as well as increased sensitivity to environmentally induced stress. Progressive cellular and bodily deterioration is built into the life

Animals seem to have an instinct for performing death alone, hidden. Even the largest, most conspicuous ones find ways to conceal themselves in time. If an elephant missteps and dies in an open place, the herd will not leave him there; the others will pick him up and carry the body from place to place, finally putting it down in some inexplicably suitable location. When elephants encounter the skeleton of an elephant in the open, they methodically take up each of the bones and distribute them, in a ponderous ceremony, over neighboring acres.

It is a natural marvel. All of the life on earth dies, all of the time, in the same volume as the new life that dazzles us each morning, each spring. All we see of this is the odd stump, the fly struggling on the porch floor of the summer house in October, the fragment on the highway. I have lived all my life with an embarrassment of squirrels in my backyard, they are all over the place, all year long, and I have never seen, anywhere, a dead squirrel.

I suppose that is just as well. If the earth were otherwise, and all the dying were done in the open, with the dead there to be looked at, we would never have it out of our minds. We can forget about it much of the time, or think of it as an accident to be avoided, somehow. But it does make the process of dying seem more exceptional than it really is, and harder to engage in at the times when we must ourselves engage.

In our way, we conform as best we can to the rest of nature. The obituary pages tell us of the news that we are dying away, while birth announcements in finer print, off at the side of the page, inform us of our replacements, but we get no grasp from this of the enormity of the scale. There are 4 billion of us on the earth in this year, 1973, and all 4 billion must be dead, on a schedule, within this lifetime. The vast mortality, involving something over 50 million each year, takes place in relative secrecy. We can only really know of the deaths in our households, among our friends. These, detached in our minds from all the rest, we take to be unnatural events, anomalies, outrages. We speak of our own dead in low voices; struck down, we say, as though visible death can occur only for cause, by disease or violence, avoidably. We send off for flowers, grieve, make ceremonies, scatter bones, unaware of the rest of the 4 billion on the same schedule. All of that immense mass of flesh and bone and consciousness will disappear by absorption into the earth, without recognition by the transient survivors.

Less than half a century from now, our replacements will have more than doubled in numbers. It is hard to see how we can continue to keep the secret, with such multitudes doing the dying. We will have to give up the notion that death is a catastrophe, or detestable, or avoidable, or even strange. We will need to learn more about the cycling of life in the rest of the system, and about our connection in the process. Everything that comes alive seems to be in trade for everything that dies, cell for cell. There might be some comfort in the recognition of synchrony, in the information that we all go down together, in the best of company.

cycle of all organisms in which differentiated cells show extensive specialization. The process is called **aging**.

Aging in humans leads to structural changes such as loss of hair and teeth, increased skin wrinkling and fat deposition, and decreased muscle mass. Less obvious are gradual physiological changes. For example, metabolic rates decline in kidney cells, so the body cannot respond as effectively as it once did to changes in extracellular fluid volume and composition. Another change involves collagen, the fibrous protein that is present in the extracellular spaces of nearly all tissues. Collagen may represent as much as 40 percent of your body's proteins. With increasing age, new collagen fibers that are being produced are structurally altered—and such structural changes are bound to have widespread physical effects.

## The Journey Begins

At first, nothing spectacular happens to the fertilized egg. It doesn't burgeon abruptly in size, like an inflating balloon. It doesn't even grow. Then comes the first clue that a spectacular journey is about to begin. That single cell starts *carving itself up*. By self-contained machinery, the cell cleaves itself in one direction, then another, and another until it has transformed itself into a hollow ball of about sixty tiny cells. The miniature cells all look alike, even the small bunch of cells huddled together against part of the ball's inner surface. But now a space is opening up inside the huddle. Fluid moves in; the space widens. The fluid is lifting the cell mass away from the inner surface, like tissue fluid does to the skin above a blister.

By this time the cells of the "blister" are spread out as an oval-shaped disk, like a pancake that didn't quite make it into a circle when the batter was poured. Make that two stacked pancakes—the cells of the disk are arranged as two layers. The disk itself is tiny. At less than a millimeter across, it could stretch out across the head of a straightpin with room to spare. But that speck is an embryo, and it is already going places!

Now a third layer is forming between the original two. Within those three layers, cells are following commands to change shape, to get moving. Many adjacent cells of one sheetlike layer are all constricting on one side of the cell body but not the other. Their coordinated constriction makes part of the sheet curve up and fold over! Other cells are elongating, making their component layer thicker. And still other cells are migrating to new destinations! The effect is stunning— a mere pancake of an embryo has become a pale, translucent crescent with surface bumps, tucks, and hollows.

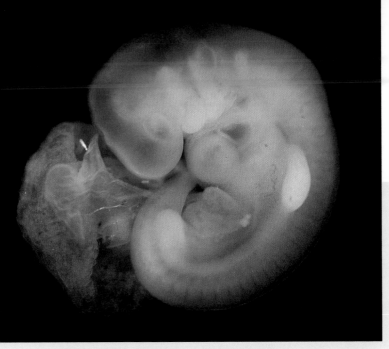

**Figure 43.1** *Below left*, the billowing entrance to an ovarian duct—the tubelike road to the uterus. There, on that road, a sperm traveling from the opposite direction encountered an egg, and a remarkable developmental journey began. *Above*, the embryo just four weeks after the moment of fertilization.

Day after day, the structural molding goes on. By the third week after fertilization, the tubelike forerunner of the nervous system appears. Sheets of cells roll up, forming tubes of a future circulatory system. By the fourth week, a patch of cells starts beating rhythmically in the thickening wall of one tube. They are descendants of the embryonic heart. From now on, through birth, adulthood, until the time of death, those cells will continue beating as the heart's pacemaker. On the embryo's sides, budding regions of tissue mark the beginning of upper and lower limbs.

At this stage it is clear that the embryo is a vertebrate. Already it has a blunt head end, an obvious bilateral symmetry, a series of tissue segments along its back, and an embryonic tail. But is it a human? A minnow? A duckling? At this stage only an expert could answer.

Five weeks into the journey, the limb buds have paddles—the start of hands and feet. Round dots appear under the transparent skin and gradually darken and grow larger; the eyes are forming. Around and below them, cells migrate in concert, interacting in ways that sculpt out a nose, cheeks, and a mouth. By now, the embryo is recognizably a human in the making.

This was your beginning. Later embryonic and fetal events filled in the details, rounded out contours, added flesh and fat and hair and nails to your peanut-sized body. For all of those events to unfold, gametes first had to form in your parents and meet in a moist tunnel leading away from one of your mother's ovaries (Figure 43.1). That beginning, and all the subsequent events that unfolded inside your mother's body, is the story of this chapter.

KEY CONCEPTS

**1.** Humans have a well-developed reproductive system, consisting of a pair of primary reproductive organs (testes in males, ovaries in females), accessory glands, and ducts. Testes produce sperm; ovaries produce eggs. Both types of gonads also produce sex hormones, the secretion of which is under the control of the hypothalamus and pituitary gland.

**2.** Human males produce sperm continuously from puberty onward. The hormones testosterone, LH, and FSH control male reproductive functions.

**3.** The reproductive capacity of human females is cyclic and intermittent, with eggs being released and the uterine lining being prepared for pregnancy on a monthly basis. The hormones estrogen, progesterone, FSH, and LH control this cyclic activity.

**4.** As for other vertebrates, human development proceeds through six stages: gamete formation, fertilization, cleavage, gastrulation, organ formation, and growth and tissue specialization.

In the preceding chapter, we looked at some general principles of animal reproduction and development. Here, we will focus on humans as a way of presenting an integrated picture of the structure and function of reproductive organs, gamete formation, and the stages of development from fertilization to birth. As part of this picture, we will also consider some of the mechanisms controlling reproduction.

## HUMAN REPRODUCTIVE SYSTEM

For both men and women, the reproductive system consists of a pair of primary reproductive organs (gonads), accessory glands, and ducts. Male gonads are **testes** (singular, testis), and female gonads are **ovaries**. Testes produce sperm; ovaries produce eggs. Both also secrete sex hormones, which influence reproductive functions and the development of secondary sexual traits. Such

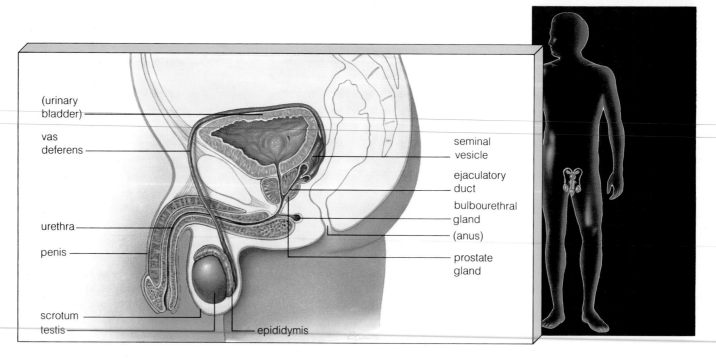

**Figure 43.2** Reproductive system of the human male.

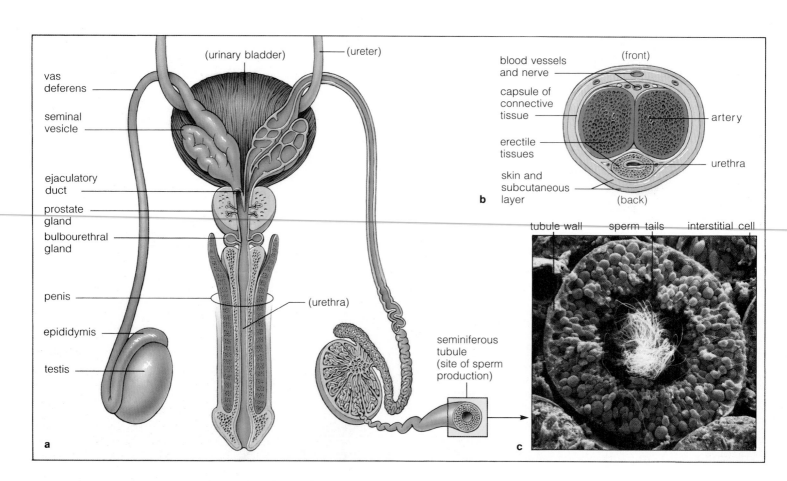

**Figure 43.3** (**a**) Posterior view of the male reproductive system. (**b**) Cross-section of the penis. (**c**) Scanning electron micrograph showing the cells inside a seminiferous tubule.

| Table 43.1 | Organs and Accessory Glands of the Male Reproductive Tract |
|---|---|
| **Organs:** | |
| Testis (2) | Production of sperm, sex hormones |
| Epididymis (2) | Sperm maturation site and storage |
| Vas deferens (2) | Rapid transport of sperm |
| Ejaculatory duct (2) | Conduction of sperm |
| Penis | Organ of sexual intercourse |
| **Accessory Glands:** | |
| Seminal vesicle (2) | Secretions large part of semen |
| Prostate gland | Secretions part of semen |
| Bulbourethral gland (2) | Production of lubricating mucus |

traits are distinctly associated with maleness and femaleness, although they do not play a direct role in reproduction. Examples are the amount and distribution of body fat, hair, and skeletal muscle.

Gonads look the same in all early human embryos. But after seven weeks of development, activation of genes on the sex chromosomes and hormone secretions trigger their development into testes *or* ovaries. The gonads and accessory organs are already formed at birth, but they do not reach full size and become functional until twelve to sixteen years later.

**Male Reproductive Organs**

**Where Sperm Form**. Figure 43.2 shows the male reproductive system and Table 43.1 lists its components. The testes start forming as buds from the wall of the embryo's abdominal cavity. Before birth they descend into the scrotum, an outpouching of skin below the pelvic region. Sperm develop properly when the scrotum's interior is kept a few degrees cooler than the body's normal temperature. Through controlled contractions of muscles in the scrotum, the temperature stays at 95°F or so. When it is cold outside, contractions draw the pouch closer to the (warmer) body. When it is warm outside, the muscles relax and lower the pouch.

Each testis is partitioned into as many as 300 wedge-shaped lobes. Each lobe contains two to three highly coiled tubes, the **seminiferous tubules**, and this is where sperm develop (Figure 43.3). Although a testis is only about 5 centimeters long, 125 meters of tubes are packed into it! Connective tissue between the tubes contains *Leydig cells* with endocrine functions. They secrete primarily the sex hormone testosterone.

Just inside the walls of these tubes we find undifferentiated diploid cells called spermatogonia. Ongoing cell divisions force the cells away from the walls, toward

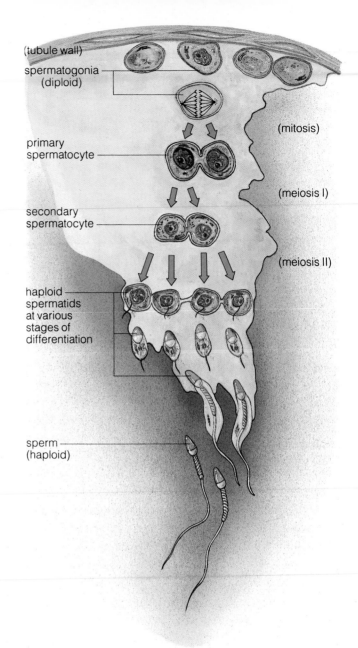

**Figure 43.4** (**a**) Sperm formation in a seminiferous tubule. Undifferentiated diploid cells (spermatogonia) are closest to the tubule walls. They are forced away from it by ongoing mitotic cell divisions and are transformed into primary spermatocytes. Following meiosis I, they become secondary spermatocytes. Each chromosome in these haploid cells still consists of two sister chromatids. Sister chromatids separate from each other during meiosis II. The resulting spermatids gradually develop into mature sperm. The entire process takes about nine to ten weeks.

the tube's interior. During their forced departure, the cells are gradually transformed into primary spermatocytes. As Figure 43.4 shows, they undergo meiosis I, the result being secondary spermatocytes. Although the cells are now haploid, keep in mind that each chromosome they contain is still in the duplicated state; it consists of two sister chromatids (page 153). The sister chromatids of each chromosome are separated from each other during meiosis II. The resulting cells are haploid spermatids,

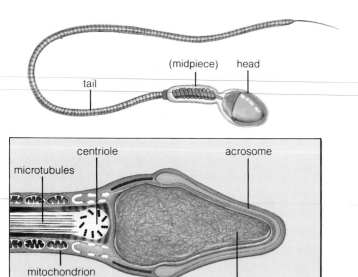

**Figure 43.5** Structure of a mature human sperm.

**Figure 43.6** (*Right*) Hormonal control of reproductive function in human males. The black dashed line indicates that increased testosterone secretion inhibits LH secretions through its negative effect on hypothalamic GnRH. The blue dashed line indicates that an inhibitory signal from Sertoli cells influences GnRH and FSH secretions. The "signal" is the hormone inhibin.

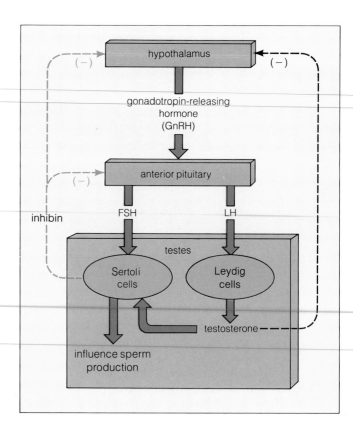

which gradually develop into **sperm**—the male gametes. The entire process takes about nine to ten weeks. All the while, the developing cells receive nourishment and chemical signals from adjacent *Sertoli cells*, which are the only other type of cell inside the tubule.

From puberty onward, sperm are produced continuously, with many millions in different stages of development on any given day. A mature sperm has a tail, a midpiece, and a head with a DNA-packed nucleus. An enzyme-containing cap (acrosome) covers most of the head. Its enzymes help the sperm penetrate an egg at fertilization. Mitochondria in the midpiece supply energy for the tail's whiplike movements (Figure 43.5).

**How Semen Forms.** Sperm become part of *semen*, a thick fluid that eventually is expelled from the penis. When sperm move out of a testis, they enter a long coiled duct, the epididymis. The sperm are not fully developed at this time, but secretions from the duct walls help them mature. Until sperm leave the body, they are stored in the last part of the epididymis. When they are about to leave, they pass through a thick-walled tube (vas deferens), ejaculatory ducts, then through the urethra—the channel leading outside, to the body's surface.

Secretions from glands along this route become mixed with sperm, forming the semen. Secretions from the seminal vesicles include the sugar fructose, which nourishes sperm. They also include prostaglandins, which may trigger contractions in the female reproductive tract and assist sperm movement. Secretions from the prostate gland probably help buffer acid conditions in the vagina. Vaginal pH is about 3.5–4, but sperm motility and fertility improve when it is about 6. Bulbourethral glands secrete some mucus-rich fluid into the urethra during sexual arousal. This fluid lubricates the penis, facilitating its penetration into the vagina. It also aids sperm movement.

### Hormonal Control of Male Reproductive Functions

Three hormones control male reproductive functions. One is **testosterone**, produced by endocrine cells in the testes. The others are **LH** (luteinizing hormone) and **FSH** (follicle-stimulating hormone). As we have seen, LH and FSH are produced by the anterior lobe of the pituitary gland (page 584). They were named for their effects in females, but we now know that their molecular structure is exactly the same in males.

Testosterone stimulates sperm production and controls the growth, form, and function of all parts of the male reproductive tract. This hormone has roles in normal sexual behavior and may tend to promote aggressive behavior. Growth of facial hair and pubic hair, lowering of the voice, and other secondary sexual traits depend on testosterone secretions.

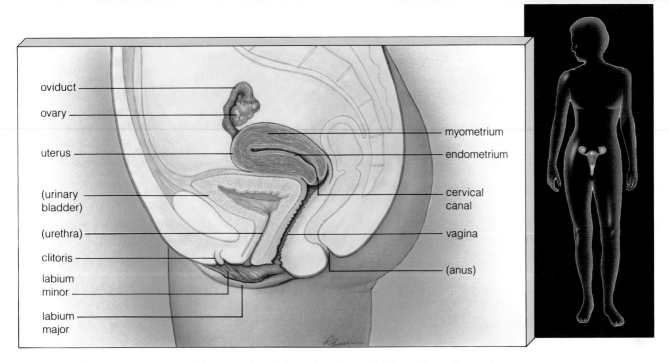

**Figure 43.7** Reproductive system of the human female. The uterus has a thick layer of smooth muscle (myometrium) and an inner lining (endometrium).

The hypothalamus governs testosterone secretion. When testosterone levels are low, the hypothalamus secretes **GnRH**, a releasing hormone. GnRH prods the anterior pituitary into secreting LH, which the bloodstream carries to the testes. There, LH triggers testosterone secretion by Leydig cells. The testosterone acts on the Sertoli cells in ways that stimulate sperm production. GnRH also prods the pituitary into secreting FSH, which acts directly on Sertoli cells. FSH stimulates the development of sperm during puberty.

Negative-feedback loops between the hypothalamus and testes control sperm production. These loops are shown in Figure 43.6. As the concentration of blood-borne testosterone increases, it exerts an inhibitory effect on the hypothalamus. GnRH secretion slows down as a result.

### Female Reproductive Organs

Figure 43.7 shows the female reproductive system and Table 43.2 lists its components. The two ovaries reside in the abdominal cavity. During a woman's reproductive years, they release eggs on a monthly basis, and they secrete the sex hormones **estrogen** and **progesterone**.

Take a look at Figure 10.12, the generalized picture of how meiosis I and II proceed in an immature egg, or *oocyte*. Even before a female is born, about 2 million oocytes start forming in her ovaries. They enter meio-

| Table 43.2 | Female Reproductive Organs |
|---|---|
| Ovaries | Oocyte production, sex hormone production |
| Oviducts | Conduction of oocyte from ovary to uterus |
| Uterus | Chamber in which new individual develops |
| Cervix | Secretion of mucus that enhances sperm movement into uterus and (after fertilization) reduces the embryo's risk of bacterial infection |
| Vagina | Organ of sexual intercourse; birth canal |

sis I, but in this case the division process is arrested! Meiosis will resume in one oocyte at a time—but not until puberty. By then, only 30,000 or 40,000 oocytes will still be around. And only about 400 of those will mature and escape from the ovary on a monthly basis, over the next three decades or so. Even then, meiosis II will not be completed unless fertilization occurs.

An oocyte released from an ovary enters a nearby channel, an **oviduct**. Fingerlike projections from the oviduct extend over part of the ovary, and they sweep the oocyte into the channel. From there, the oocyte moves into a hollow, pear-shaped organ, the **uterus**. Following fertilization, the new individual grows and develops here. The uterus is mostly a thick layer of smooth muscle (the myometrium). Its interior lining, the **endometrium**, consists of connective tissue, glands, and blood vessels. The lower portion of the uterus (the narrow part of the "pear") is the cervix. A muscular tube, the vagina, extends from the cervix to the body surface. This tube receives sperm and functions as part of the birth canal.

At the body surface are the external genitalia (vulva), which include the organs for sexual stimulation. The outermost structures are a pair of skin folds (the labia majora), which contain adipose tissue. Within the cleft formed by these folds are the labia minora—a smaller pair of skin folds that are highly vascularized but have no fatty tissue. At the anterior end of the vulva, the interior folds of skin partly enclose the *clitoris*, a small organ sensitive to sexual stimulation. The opening of the urethra is about midway between the clitoris and the vaginal opening.

## Menstrual Cycle

Most mammalian females follow an "estrous" cycle. They can become pregnant only during estrus. At that time, they are said to be in heat, or sexually receptive to males. Estrus occurs only at certain times of year, when oocytes mature and hormone action primes the endometrium to receive a fertilized egg.

The females of humans and other primates follow a **menstrual cycle**. For them, the release of oocytes and priming of the endometrium is cyclic and intermittent. A menstrual cycle differs from estrus, for there is no correspondence between heat and the time of fertility. In other words, all female primates can be physically and behaviorally receptive to sexual activity at any time.

Human menstrual cycles begin at about age thirteen. As Table 43.3 indicates, it takes about twenty-eight days to complete one cycle. Twenty-eight days simply is the average time span; the cycle runs longer for some women and shorter for others. Menstrual cycles continue until menopause, in the late forties or early fifties. By then, the egg supply is dwindling, hormone secretions slow down, and eventually menstruation (and fertility) is over. The eggs that are still to be released late in a woman's life are at some risk of acquiring chromosome abnormalities when meiosis finally resumes. A newborn with Down syndrome is one of the possible outcomes (page 198).

**Ovarian Function.** Figure 43.8 provides a closer look at how an oocyte develops in an ovary. Each oocyte becomes surrounded by a single layer of *granulosa cells*. Each "primary" oocyte, together with the surrounding cell layer, is a **follicle**. This is what FSH, the "follicle-stimulating hormone," will act upon. The granulosa cells gradually deposit a layer of material, the *zona pellucida*, around the oocyte.

Usually only one follicle reaches maturity during a menstrual cycle. Within that follicle, meiosis I resumes in the oocyte and two cells form. This division process is shown in Figure 10.12. One cell, the **secondary oocyte**, ends up with nearly all the cytoplasm. The other cell is the first **polar body**. It is a tiny cell that functions as a "dumping ground" for half the diploid number of chromosomes. The distribution of chromosomes between the secondary oocyte and the first polar body assures that both cells will be haploid. Neither cell will complete meiosis II until fertilization.

As the follicle develops, it secretes an estrogen-containing fluid. The fluid accumulates in the follicle and causes it to balloon outward from the ovary's surface, then rupture. The fluid escapes, carrying the secondary oocyte with it. The release of a secondary oocyte from an ovary is called **ovulation**. The granulosa cells left behind in the follicle differentiate into a glandular structure, the **corpus luteum**, which secretes progesterone and some estrogen.

A corpus luteum can persist for about twelve days if fertilization does not follow ovulation. During that time, the hypothalamus signals the anterior pituitary to decrease its FSH secretions. This prevents other follicles from developing until the menstrual cycle is over.

The corpus luteum degenerates during the last days of the cycle if fertilization does not occur. Apparently, it self-destructs by secreting prostaglandins, which interfere with its function. With the corpus luteum gone, progesterone and estrogen levels fall rapidly. Now FSH secretions can increase, another follicle can be stimulated to mature—and the cycle begins anew.

| Table 43.3 | Events of the Menstrual Cycle | |
|---|---|---|
| Phase | Events | Days of the Cycle* |
| Follicular phase | Menstruation; endometrium breaks down | 1–5 |
| | Follicle matures in ovary; endometrium rebuilds | 6–13 |
| Ovulation | Secondary oocyte released from ovary | 14 |
| Luteal phase | Corpus luteum forms; endometrium thickens and develops | 15–28 |

*Assuming a 28-day cycle.

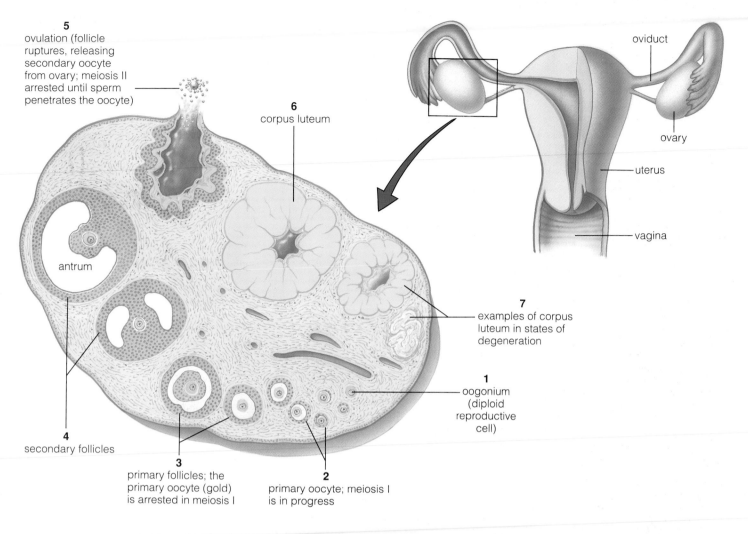

**5**
ovulation (follicle ruptures, releasing secondary oocyte from ovary; meiosis II arrested until sperm penetrates the oocyte)

antrum

**6**
corpus luteum

oviduct

ovary

uterus

vagina

**7**
examples of corpus luteum in states of degeneration

**1**
oogonium (diploid reproductive cell)

**4**
secondary follicles

**3**
primary follicles; the primary oocyte (gold) is arrested in meiosis I

**2**
primary oocyte; meiosis I is in progress

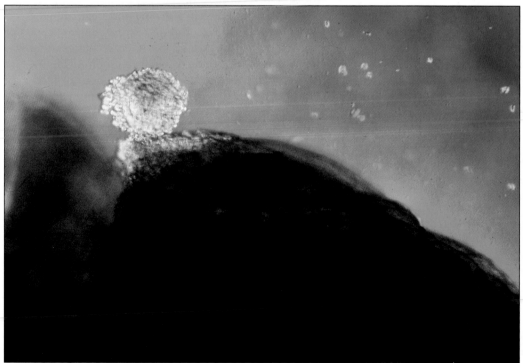

**Figure 43.8** A human ovary, drawn as if sliced lengthwise through its midsection. Events in the ovarian cycle proceed from the growth and maturation of follicles, through ovulation (rupturing of a mature follicle with a concurrent release of a secondary oocyte), through the formation and maintenance (or degeneration) of an endocrine structure called the corpus luteum. The positions of the oocyte and corpus luteum are varied for illustrative purposes only. The maturation of an oocyte occurs at the *same* site, from the beginning of the cycle to ovulation. The photograph shows a secondary oocyte at the moment of ovulation.

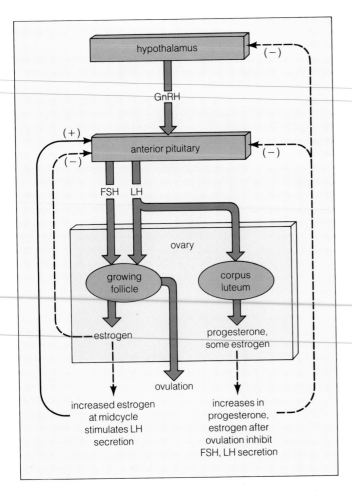

**Figure 43.9** Feedback loops among the hypothalamus, anterior pituitary, and ovaries during the menstrual cycle.

Feedback loops among the hypothalamus, pituitary, and ovaries control events in the ovary. Figure 43.9 shows the loops. When the menstrual cycle begins, the hypothalamus signals the anterior pituitary to release LH and FSH, which in turn signal the ovary to secrete estrogen. About midway through the cycle, the increased estrogen level in the blood causes a brief outpouring of LH from the pituitary. *It is this midcycle surge of LH that triggers ovulation.*

**Uterine Function**. The changing estrogen and progesterone levels just described cause profound changes that prepare the uterus for pregnancy. Estrogen stimulates the growth of the endometrium and its glands in the uterus. Progesterone causes blood vessels to grow rapidly in the thickened endometrium.

At ovulation, estrogen acts on the cervix, the narrow opening to the uterus from the vagina. It causes the cervix to secrete large amounts of a thin, clear mucus—an ideal medium through which sperm can travel. Right after ovulation, progesterone from the corpus luteum acts on the cervix. The mucus becomes thick and sticky, forming a barrier against vaginal bacteria that might enter the uterus through the cervix and endanger a new zygote.

When fertilization does not occur and the corpus luteum self-destructs, the endometrium starts to break down. Deprived of oxygen and nutrients, its blood vessels constrict, and its tissues die. Blood escapes from the ruptured walls of weakened capillaries. This menstrual flow consists of blood and sloughed endometrial tissues, and its appearance marks the first day of a new cycle. The menstrual sloughing continues for three to six days, until rising estrogen levels stimulate the repair and growth of the endometrium.

Each year, between 4 and 10 million American women are affected by *endometriosis*, the spread and growth of endometrial tissue outside the uterus. Estrogen acts on endometrial tissue wherever it occurs. Its action may lead to sensations of pain during menstruation, sexual relations, or urination. Also, endometrial scar tissue on the ovaries or oviduct can cause infertility. Endometriosis might arise when some menstrual flow backs up through the oviducts and spills into the pelvic cavity. Or perhaps some embryonic cells were positioned in the wrong place before birth and are stimulated to grow at puberty, when sex hormones become active.

Figure 43.10 summarizes the correlations between changing hormone levels and changes in the ovary and uterus during the menstrual cycle.

### Sexual Intercourse

Suppose a secondary oocyte happens to be on its way down the oviduct when a female and male are engaged in sexual intercourse, or *coitus*. Within mere seconds of sexual arousal, the penis can undergo changes that will help it penetrate into the vaginal channel. As Figure 43.3 shows, the penis contains three cylinders of spongy tissue. The mushroom-shaped tip of one cylinder (the glans penis) is loaded with sensory receptors that are activated by friction. Between times of sexual arousal, blood vessels leading into the three cylinders are constricted and the penis is limp. Upon arousal, blood flows into the cylinders faster than it flows out and collects in the spongy tissue, so the penis lengthens and stiffens.

During coitus, pelvic thrusts stimulate the penis as well as the female's vaginal walls and clitoral region. The mechanical stimulation causes rhythmic, involuntary contractions in the male reproductive tract. The contractions move sperm from their main storage site in the last portion of the epididymis. The contractions force the contents of seminal vesicles and the prostate into the urethra, then ejaculation of semen into the vagina. (During ejaculation, a sphincter closes and prevents urine from being excreted from the bladder.)

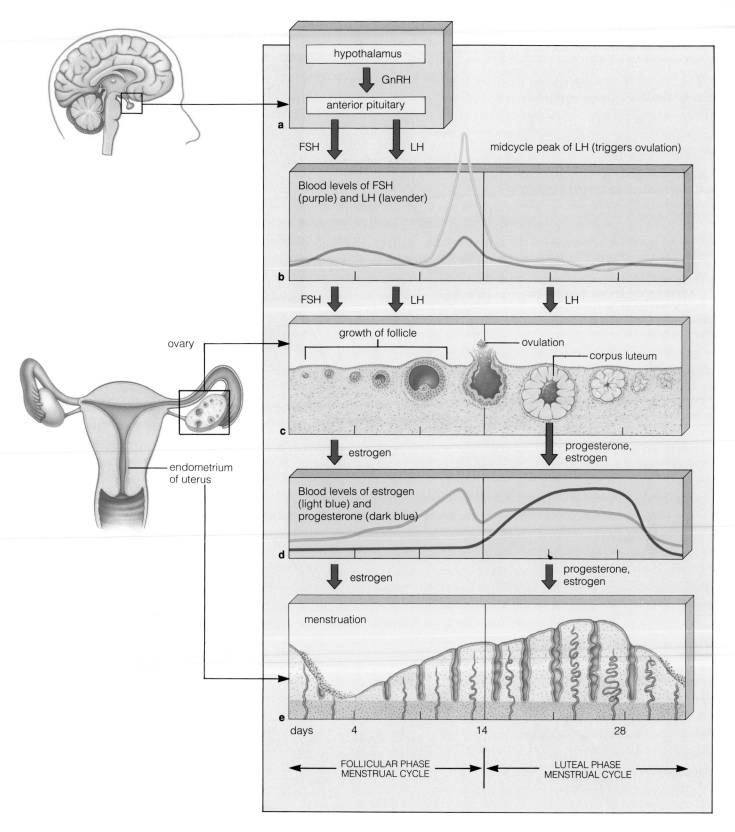

**Figure 43.10** Correlation between changes in the ovary and uterus with changing hormone levels during the menstrual cycle. Green arrows indicate which hormones dominate the follicular phase or the luteal phase of the cycle. A releasing hormone (GnRH) from the hypothalamus (**a**) controls the release of FSH and LH from the pituitary. The FSH and LH promote changes in ovarian structure and function (**b**, **c**), then estrogen and progesterone from the ovary promote changes in the endometrium (**d**, **e**).

**Figure 43.11** Fertilization and implantation in the uterus. The left photograph shows a human sperm about to penetrate the membranous covering (zona pellucida) around a secondary oocyte. The right photograph shows the three polar bodies above a mature ovum; these products of meiosis will degenerate shortly.

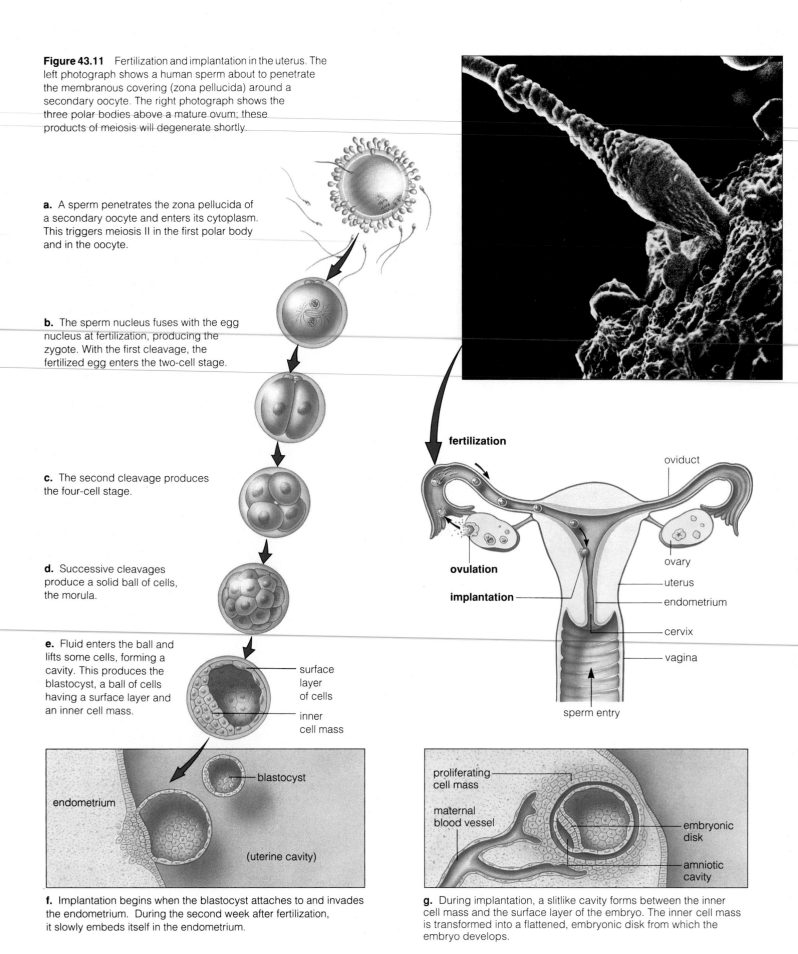

**a.** A sperm penetrates the zona pellucida of a secondary oocyte and enters its cytoplasm. This triggers meiosis II in the first polar body and in the oocyte.

**b.** The sperm nucleus fuses with the egg nucleus at fertilization, producing the zygote. With the first cleavage, the fertilized egg enters the two-cell stage.

**c.** The second cleavage produces the four-cell stage.

**d.** Successive cleavages produce a solid ball of cells, the morula.

**e.** Fluid enters the ball and lifts some cells, forming a cavity. This produces the blastocyst, a ball of cells having a surface layer and an inner cell mass.

surface layer of cells

inner cell mass

**fertilization**

oviduct

**ovulation**

ovary

**implantation**

uterus

endometrium

cervix

vagina

sperm entry

blastocyst

endometrium

(uterine cavity)

proliferating cell mass

maternal blood vessel

embryonic disk

amniotic cavity

**f.** Implantation begins when the blastocyst attaches to and invades the endometrium. During the second week after fertilization, it slowly embeds itself in the endometrium.

**g.** During implantation, a slitlike cavity forms between the inner cell mass and the surface layer of the embryo. The inner cell mass is transformed into a flattened, embryonic disk from which the embryo develops.

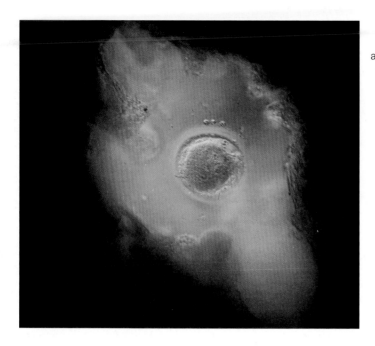

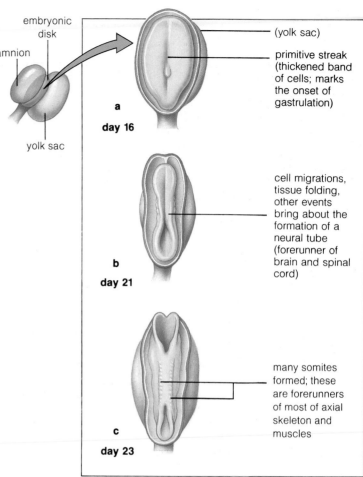

embryonic disk

amnion

yolk sac

(yolk sac)

**a**
**day 16**

primitive streak (thickened band of cells; marks the onset of gastrulation)

**b**
**day 21**

cell migrations, tissue folding, other events bring about the formation of a neural tube (forerunner of brain and spinal cord)

**c**
**day 23**

many somites formed; these are forerunners of most of axial skeleton and muscles

**Figure 43.12** Transformation of the pancake-shaped embryonic disk into the early embryo. Shown are three dorsal views (the embryo's back).

Together, the muscular contractions, ejaculation, and associated sensations of release, warmth, and relaxation are called *orgasm*. Female orgasm involves similar events, including an intense vaginal awareness, involuntary uterine and vaginal contractions, and sensations of relaxation and warmth. Even if the female does not reach this state of excitation, she can still get pregnant.

## FROM FERTILIZATION TO BIRTH

### Fertilization

Fertilization may occur if sperm enter the vagina any time between a few days before ovulation to a few days afterward. Within thirty minutes after ejaculation, muscle contractions move sperm deeper into the female reproductive tract. As many as 150 million to 350 million sperm may be deposited in the vagina during one ejaculation. Yet only a few hundred reach the upper region of the oviduct, where fertilization most commonly takes place.

When a sperm encounters a secondary oocyte, it releases digestive enzymes from its acrosome. Those enzymes clear a path through the zona pellucida (Figure 43.11a). Several sperm can reach the egg, but usually only one enters its cytoplasm. The arrival of that sperm stimulates both the secondary oocyte and the first polar body into completing meiosis II. As Figure 43.11 shows, there are now three polar bodies and a mature egg, or **ovum**. The sperm nucleus fuses with the nucleus of the ovum and their chromosomes intermingle, restoring the diploid number for the zygote.

### Implantation

For the first three or four days after fertilization, the zygote travels down the oviduct. It picks up required nutrients from maternal secretions and undergoes the first cleavages, in the manner described on page 746. By the time the cluster of dividing cells reaches the uterus, it is a solid ball of cells (the morula). The ball becomes transformed into a *blastocyst*, an embryonic stage consisting of a surface layer of cells and an inner cell mass (Figure 43.11e).

**Implantation** occurs before the end of the first week. By this process, the blastocyst adheres to the uterine lining, and some of its cells send out projections that invade the maternal tissues. While the invasion is proceeding, the inner cell mass becomes transformed into an **embryonic disk**. This is the oval, flattened pancake described in the chapter introduction (Figure 43.12). The disk will give rise to the embryo proper during the week following implantation, when all three germ layers will form.

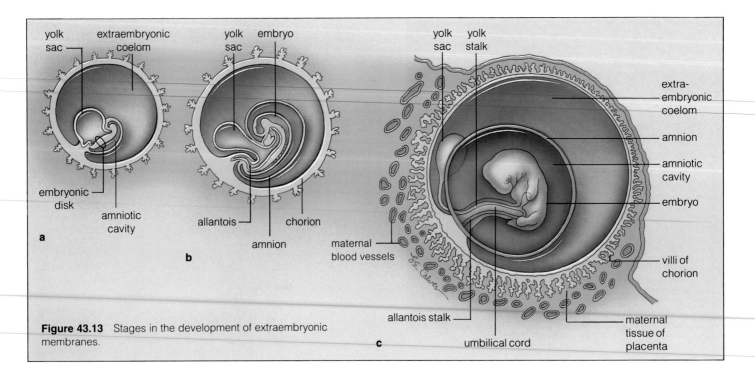

Figure 43.13 Stages in the development of extraembryonic membranes.

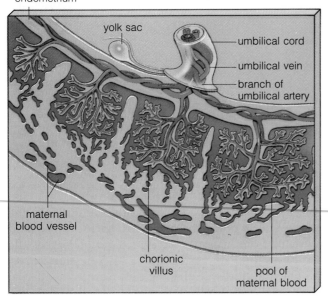

Figure 43.14 Relationship between fetal and maternal tissues in the placenta. The diagram shows how chorionic villi become progressively developed (from left to right across the illustration).

## Membranes Around the Embryo

To understand what happens after implantation, think back to the shelled egg, which figured in the vertebrate invasion of land (page 456). Inside most shelled eggs are four membranes, the **yolk sac**, **allantois**, **amnion**, and **chorion**. These "extraembryonic" membranes protect the embryo and serve in its nutrition, respiration, and excretion.

A human embryo is not housed in a shell or nourished by yolk, but it is still served by a yolk *sac* as part of its vertebrate heritage. The sac forms below the embryonic disk *as if* yolk were still there (Figures 43.12 and 43.13). The sac will play a role in the formation of a digestive tube.

In hard-shelled eggs, the allantois stores wastes from protein metabolism and its blood vessels supply the embryo with oxygen. In humans, the allantois is not involved in waste storage, but its blood vessels still function in oxygen and nutrient transport by the placenta.

The amnion of all land vertebrates is a fluid-filled sac that completely surrounds the embryo and keeps it from drying out. The fluid inside also absorbs shocks. Just before childbirth, "water" flows freely from the vagina. This is amniotic fluid, released when the amnion ruptures.

In time, only a thick cord connects the growing human embryo to parts of the yolk sac, allantois, and amnion. This **umbilical cord** is well endowed with blood vessels (Figure 43.14). The chorion develops as a protective membrane around the embryo and other structures. As you will see shortly, this membrane becomes part of the placenta. The chorion secretes a hormone (chorionic gonadotropin) that maintains the corpus luteum. And progesterone secreted from the corpus luteum in turn maintains the uterine lining during the first three months of pregnancy. After that, the placenta produces sufficient progesterone and estrogen.

gill arches    somites

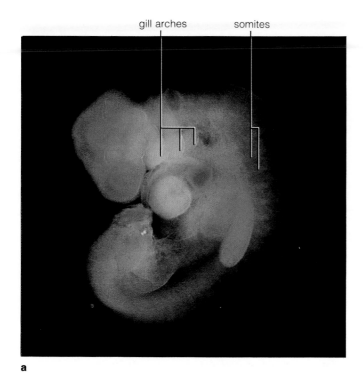

a

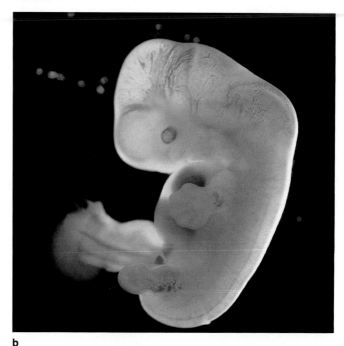

b

umbilical cord    amniotic sac

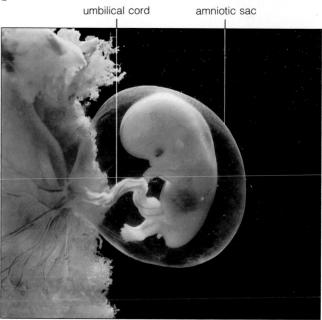

c

## The Placenta

Three weeks after fertilization, almost a fourth of the inner surface of the uterus has become a spongy tissue composed of endometrium *and* embryonic membranes, the chorion especially. By this tissue, the **placenta**, the embryo receives nutrients and oxygen from the mother and sends out wastes in return. The embryo's wastes are quickly disposed of through the mother's lungs and kidneys.

The tiny projections sent out from the blastocyst during implantation develop into many chorionic villi, each endowed with small blood vessels (Figure 43.14). When the embryo starts developing, its bloodstream will remain distinct from that of its mother. Substances simply will diffuse out of the mother's blood vessels, across the blood-filled spaces in the uterine lining, then into the embryo's blood vessels. At the same time, they diffuse in the opposite direction from the embryo.

## Embryonic and Fetal Development

**First Trimester**. The "first trimester" of the nine months of human development extends from fertilization to the end of the third month. The introduction to this chapter sketched out a picture of what the embryo looks like during this period. Figure 43.15 gives a more detailed picture of its shape. As indicated in the *Commentary* on the next page, the first trimester is a very critical period of embryonic development.

**Figure 43.15** (**a**) Another view of the embryo at four weeks, about 7 millimeters (0.3 inch) long. Notice the tail and gill arches, features that emerge in developing embryos of all vertebrates. Arm and leg buds are visible now. (**b**) Embryo at the end of five weeks, about 12 millimeters long. The head starts to enlarge and the trunk starts to straighten. Finger rays appear in the paddlelike forelimbs. A pigmented retina outlines the forming eyes. (**c**) An embryo poised at the boundary between the end of the first trimester and the start of the second, which will extend from the fourth month through sixth. Once past this boundary, the new individual is called a *fetus*. Here, the embryo floats in fluid within the amniotic sac. The chorion, which covers the amniotic sac, has been opened and pulled aside.

# Commentary

## Mother as Protector, Provider, Potential Threat

Many safeguards are built into the female reproductive system. The placenta, for example, is a highly selective filter that prevents many noxious substances in the mother's bloodstream from gaining access to the embryo or fetus. Even so, from fertilization to birth, the developing individual is at the mercy of the mother's diet, health habits, and life-style.

### Some Nutritional Considerations

During pregnancy, a balanced intake of carbohydrates, amino acids, and fats or oils usually provides all vitamins and minerals in amounts sufficient for normal development. The mother's vitamin needs are definitely increased, but the developing fetus is more resistant than she is to vitamin and mineral deficiencies. (The placenta preferentially absorbs vitamins and minerals from her blood.) In cases where the diet is marginal, the money spent on vitamin pills and other food supplements would usually do the mother (hence the fetus) more good if spent on wholesome, protein-rich food.

A few years ago, it was accepted medical practice for a pregnant woman to keep her total weight gain to ten or fifteen pounds. It is now clear that if the woman restricts her food intake too severely, especially during the last trimester, fetal development will be affected and the newborn will be underweight. Significantly underweight infants face more post-delivery complications than do infants of normal weight; in fact, they represent nearly half of all newborn deaths. They also will suffer a much higher incidence of mental retardation and other handicaps later in life. In most cases, a woman should gain somewhere between twenty and twenty-five pounds during pregnancy.

As birth approaches, the growing fetus demands more and more nutrients from the mother's body. During this last phase of pregnancy, the mother's diet profoundly influences the course of development. Poor nutrition damages most organs—particularly the brain, which undergoes its greatest growth in the weeks just before and after birth.

### Risk of Infections

Throughout pregnancy, antibodies transferred across the placenta protect the developing individual from all but the most severe bacterial infections. However, certain viral diseases can have damaging effects if they are contracted

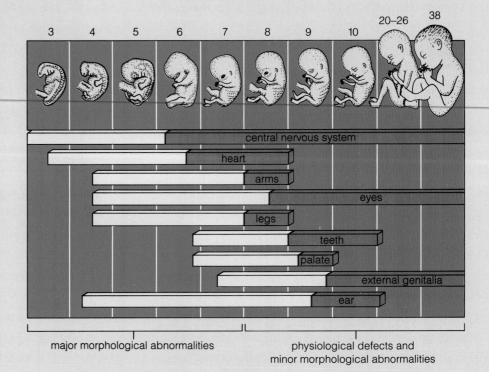

**a** Critical periods of embryonic and fetal development. Red indicates periods in which organs are most sensitive to damage from cigarette smoke, alcohol, viral infection, and so on. Numbers signify the week of development.

during the first six weeks after fertilization, the critical time of organ formation. For example, if the woman contracts German measles during this period, there is a 50 percent chance that her embryo will become malformed. If she contracts the measles virus when the embryo's ears are forming, her newborn may be deaf. (German measles can be avoided by vaccination *before* pregnancy.) The likelihood of damage to the embryo diminishes after the first six weeks. The same disease, contracted during the fourth month or thereafter, has no discernible effect on the fetus.

## Effects of Prescription Drugs

During the first trimester, the embryo is highly sensitive to drugs. A shocking example of drug effects came during the first two years after *thalidomide* was introduced in Europe. Women using this prescription tranquilizer during the first trimester gave birth to infants with missing or severely deformed arms and legs. Once the deformities were traced to thalidomide, the drug was withdrawn from the market. However, there is evidence that other tranquilizers (and sedatives and barbiturates) might cause similar, although less severe, damage. Even certain anti-acne drugs increase the risk of facial and cranial deformities. Tetracycline, a commonly prescribed antibiotic, causes yellowed teeth. Streptomycin causes hearing problems and may affect the nervous system.

At no stage of development is the embryo impervious to drugs in the maternal bloodstream. Clearly, the woman should take no drugs at all during pregnancy unless they are prescribed by a knowledgeable physician.

## Effects of Alcohol

As the fetus matures, its physiology becomes increasingly like that of the mother's. Alcohol passes freely across the placenta and has the same kind of effect on the fetus as on the woman who drinks it. *Fetal alcohol syndrome* (FAS) is a constellation of deformities that are thought to result from excessive use of alcohol by the mother during pregnancy. FAS is the third most common cause of mental retardation in the United States. It also is characterized by facial deformities, poor coordination and, sometimes, heart defects (see Figure *b*). Between 60 and 70 percent of alcoholic women give birth to infants with FAS. Some researchers now suspect that drinking any alcohol at all during pregnancy may be dangerous for the fetus. Increasingly, physicians are urging total or near-abstinence during pregnancy.

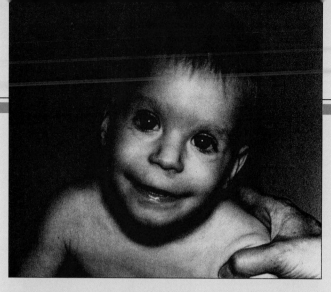

**b** An infant affected by FAS. Symptoms include a small head, low and prominent ears, poorly developed cheekbones, and a long, smooth upper lip. The child can expect to encounter growth problems and abnormalities of the nervous system. About 1 in 750 newborns in the United States are affected by this disorder.

## Effects of Cocaine

Cocaine, particularly crack cocaine, disrupts the function of the fetal nervous system as well as the mother's. The consequences can be devastating, and they extend beyond birth. Here you may wish to read again the introduction to Chapter 33.

## Effects of Cigarette Smoke

Cigarette smoking has an adverse effect on fetal growth and development. Newborns of women who have smoked every day throughout pregnancy have a low birth weight. That is true even when the woman's weight, nutritional status, and all other relevant variables are identical with those of pregnant women who do not smoke. Smoking has other effects as well. For example, for seven years in Great Britain, records were kept for all births during a particular week. The newborns of women who had smoked were not only smaller, they also had a 30 percent greater incidence of death shortly after delivery and a 50 percent greater incidence of heart abnormalities. More startling, at age seven, their average "reading age" was nearly half a year behind that of children born to nonsmokers.

In this last study, the critical period was shown to be the last half of pregnancy. Newborns of women who had stopped smoking by the middle of the second trimester were indistinguishable from those born to women who had never smoked. Although the mechanisms by which smoking exerts its effects on the fetus are not known, its demonstrated effects are further evidence that the placenta—marvelous structure that it is—cannot prevent all assaults on the fetus that the human mind can dream up.

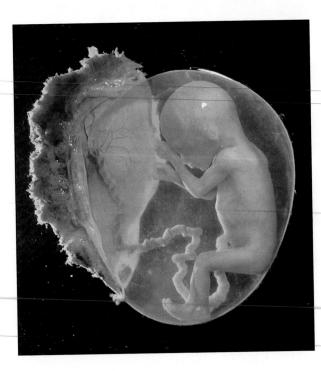

**Figure 43.16**  The fetus at sixteen weeks.

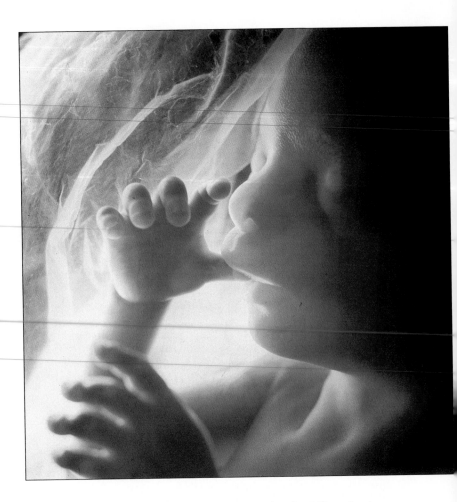

**Figure 43.17**  The fetus at eighteen weeks, about 18 centimeters (a little more than 7 inches) long. The sucking reflex begins during the earliest fetal stage, as soon as nerves establish functional connections with developing muscles. Legs kick, arms wave, fingers make grasping motions—all reflexes that will be vital skills in the world outside the uterus.

| Table 43.4 | Tissues and Organs Derived from the Three Germ Layers in Human Embryos |
|---|---|
| Germ Layer | Main Derivatives in the Adult |
| Endoderm | Various epithelia, as in the gut, respiratory tract, urinary bladder and urethra, and parts of the inner ear; also portions of the tonsils, thyroid and parathyroid glands, thymus, liver, and pancreas |
| Mesoderm | Cartilage, bone, muscle, and various connective tissues; gives rise to cardiovascular system (including blood), lymphatic system, spleen, and adrenal cortex |
| Ectoderm | Central and peripheral nervous systems; sensory epithelia of the eyes, ears, and nose; epidermis and its derivatives (including hair and nails), mammary glands, pituitary gland, subcutaneous glands, tooth enamel, and adrenal medulla |

Data from Keith Moore, *The Developing Human*.

Once the embryonic disk forms, development proceeds along the course described in Chapter 42. Gastrulation starts during the second week. It leads to the formation of three germ layers—ectoderm, mesoderm, and endoderm. The mesoderm forms when certain cells migrate inward from the surface of the embryo. Ectoderm remaining at the surface will give rise to the nervous system as well as to certain glands and other structures (Table 43.4). Later, endoderm will give rise to parts of the respiratory and digestive systems. Mesoderm will develop into the heart, muscles, bone, and many other internal organs. After the third week, an early, tubelike heart is beating.

By the end of the fourth week, the embryo has grown 500 times its original size. Its growth spurt gives way to four weeks in which the main organs develop rather slowly. The nerve cord and the four heart chambers form. Respiratory organs form but are not yet functional.

Finally, the segmentation characteristic of all vertebrate embryos becomes apparent during the first trimester. Figure 43.15 shows some of the paired segments, or *somites*—the start of connective tissues, bones, and muscles. Arms, legs, fingers, and toes now develop, along with the tail that also emerges in all vertebrate embryos. The human tail gradually disappears after the eighth week.

**Second Trimester.** The "second trimester" extends from the start of the fourth month to the end of the sixth. All major organs have formed, and the growing individual is now called a **fetus**. Figure 43.16 shows what the fetus looks like at the ninth and the sixteenth week of development. Movements of facial muscles produce frowns and squints. The sucking reflex also is evident (Figure 43.17). Before the second trimester draws to a close, the mother already can sense movements of the fetal arms and legs.

When the fetus is five months old, its heart can be heard through a stethoscope on the mother's abdomen. Soft, fuzzy hair (the lanugo) covers its body. Its skin is wrinkled, rather red, and protected from abrasion by a thick, cheesy coating. During the sixth month, eyelids and eyelashes form. During the seventh month, the eyes open.

**Third Trimester.** The "third trimester" extends from the seventh month until birth. Not until the middle of the third trimester will the fetus be able to survive on its own if born prematurely or removed surgically from the uterus. However, with intensive medical support, fetuses as young as 23–25 weeks have survived early delivery. Although development appears to be relatively complete by the seventh month, few fetuses would be able to breathe normally or maintain a normal body temperature, even with the best medical care. By the ninth month, survival chances increase to about 95 percent.

**Birth and Lactation**

Birth takes place about thirty-nine weeks after fertilization, give or take a few weeks. The birth process begins when the uterus starts to contract. For the next two to eighteen hours, the contractions become stronger and more frequent. The cervical canal dilates fully and the amniotic sac usually ruptures. Birth typically occurs less than an hour after full dilation. Immediately afterward, uterine contractions force fluid, blood, and the placenta from the body (Figure 43.18). The umbilical cord—the lifeline to the mother—is now severed, and the newborn embarks on its nurtured existence in the outside world.

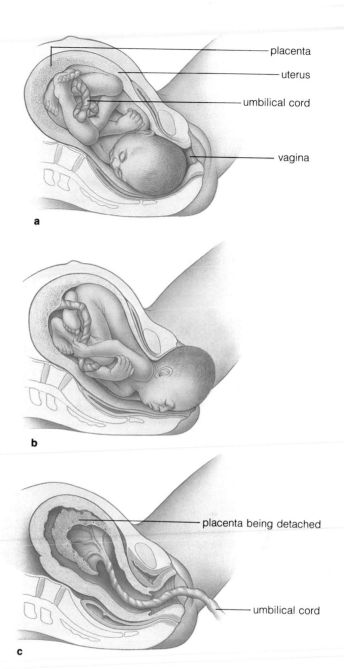

**Figure 43.18** Expulsion of the fetus during the birth process. The placenta, fluid, and blood are expelled shortly afterward (this is the "afterbirth").

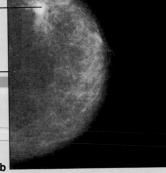

tumor

*Commentary*

# Cancer in the Human Reproductive System

For both men and women, reproductive function depends absolutely on hormonal controls of the sort described in this chapter. Hormonal imbalances contribute to a variety of disorders of the reproductive system, including many forms of cancer. Unless it is eradicated from the body, cancer kills—and it is not often eradicated easily. This makes cancer one of the most feared disorders of modern times. Here we focus on two types: cancer of the breast and of the testis. Both often can be detected through routine self-examination, a habit that may save your life.

## Breast Cancer

Of all cancers in women, breast cancer currently is second only to lung cancer in having the highest mortality. Despite intensive medical research, that rate has not been lowered by much over the past fifty years. Each year in the United States, well over 100,000 women develop breast cancer; more than a third die from it. Obesity, high blood cholesterol, and excessively high levels of estrogen and perhaps other hormones play roles in the development of cancer, but how they do this is not clear.

Chances for cure are excellent if breast cancer is detected early and treated promptly. That is why a woman should examine herself once a month, about a week after each menstrual period. The following steps have been recommended by the American Cancer Society:

1. Lie down and put a folded towel or pillow under the right shoulder, then put your right hand behind your head. With the left hand (fingers flat), begin the examination by following the outer circle of arrows shown in the diagram of Figure *a*. Gently press the fingers in small, circular motions to check for any lump, hard knot, or thickening. Next, follow the inner circle of arrows. Continue doing this for at least three more circles, one of which should include the nipple. Then repeat the procedure for the left breast.

2. For a complete examination, repeat the procedure of step 1 while standing in a shower or tub (hands glide more easily over wet skin).

3. Stand before a mirror, lift your arms over your head, and look for any unusual changes in the contour of your breasts, such as a swelling, dimpling, or retraction (inward sinking) of the nipple. Also check for any unusual discharge from the nipple.

If you discover a lump or any other change during a breast self-examination, it's important to see a physician at once.

Most changes are not cancerous, but let the physician make the diagnosis.

Currently, *mammography* (which uses low doses of x-rays) is the only imaging procedure with a proven record for detecting small cancers in the breast; it is 80 percent reliable. Figure *b* shows a mammogram that revealed a tumor, which a biopsy indicated to be cancerous. (The white patches at the front of the breast are milk ducts and fibrous tissue.)

Most often, cancerous tumors are removed through *modified radical mastectomy*. All the breast tissue, the overlying skin, and lymph nodes in adjacent tissues are removed, but muscles of the chest wall are left intact to permit more normal shoulder motions following surgery. In another procedure (*lumpectomy*), which is followed by radiation therapy, some of the breast tissue is left in place. In both cases, the removed lymph nodes are examined to determine the need for further treatment and to predict the prospects of a cure. If tumor cells are present in the nodes, there is a high risk of metastasis (cancer spread). Treatment then may include hormone therapy (aimed at shrinking any tumor masses) and radiation therapy. Treatment is most promising when cancer is detected at an early stage.

## Cancer of the Testis

You may be surprised to learn that cancer of the testis is a frequent cause of death in young men. About 5,000 cases are diagnosed in a given year in the United States alone. In its early stages, testicular cancer is painless. If not detected in time, however, it can spread to lymph nodes in the abdomen, chest, neck and, eventually, the lungs. Once it has metastasized, the cancer kills as many as half of its victims.

Once a month, from high school onward, men should examine each testis separately after a warm bath or shower (when the scrotum is relaxed). The testis should be rolled gently between the thumb and forefinger to check for any type of lump, enlargement, or hardening. Changes of that sort may or may not cause discomfort—but they must be reported to a physician, who can make a complete examination. Treatment of testicular cancer has one of the highest rates of success—when the cancer is caught before it can spread.

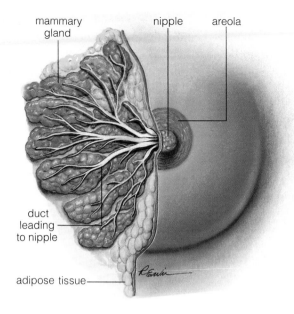

mammary gland · nipple · areola · duct leading to nipple · adipose tissue

**Figure 43.19** Breast of a lactating female. This cutaway view shows the mammary glands and ducts. The *Commentary* describes how to examine breast tissues for cancer.

During pregnancy, estrogen and progesterone were stimulating the growth of mammary glands and ducts in the mother's breasts (Figure 43.19). For the first few days after birth, those glands produce a fluid rich in proteins and lactose. Then prolactin secreted by the pituitary stimulates milk production (page 584).

When the newborn suckles, the pituitary also releases oxytocin, which causes breast tissues to contract and so force milk into the ducts. Oxytocin also triggers uterine contractions that "shrink" the uterus back to its normal size.

## POSTNATAL DEVELOPMENT, AGING, AND DEATH

Following birth, the new individual follows a prescribed course of further growth and development that leads to the **adult**, the mature form of the species. Table 43.5 summarizes all the prenatal and postnatal stages. (Prenatal means before birth; postnatal means after birth.) Figure 42.18 shows how the human body changes in proportions as the course is followed.

Late in life, the body gradually deteriorates through processes that are collectively called **aging**. Cell structure and function starts to break down, and this is accompanied by structural changes and gradual loss of body functions. As we saw in Chapter 42, all organisms with extensively differentiated cells undergo aging.

**Table 43.5  Stages of Human Development: A Summary**

**Prenatal Period:**

| | | |
|---|---|---|
| 1. Zygote | Single cell resulting from fusion of sperm nucleus and egg nucleus at fertilization |
| 2. Morula | Solid ball of cells produced by cleavages |
| 3. Blastocyst | Ball of cells with surface layer and inner cell mass |
| 4. Embryo | All developmental stages from two weeks after fertilization until end of eighth week |
| 5. Fetus | All developmental stages from the ninth week until birth (about thirty-nine weeks after fertilization) |

**Postnatal Period:**

| | | |
|---|---|---|
| 6. Newborn | Individual during the first two weeks after birth |
| 7. Infant | Individual from two weeks to about fifteen months after birth |
| 8. Child | Individual from infancy to about twelve or thirteen years |
| 9. Pubescent | Individual at puberty, when secondary sexual traits develop; girls between twelve and fifteen years, boys between thirteen and sixteen years |
| 10. Adolescent | Individual from puberty until about three or four years later; physical, mental, emotional maturation |
| 11. Adult | Early adulthood (between eighteen and twenty-five years), bone formation and growth completed. Changes proceed very slowly afterward. |
| 12. Old age | Aging follows late in life |

# CONTROL OF HUMAN FERTILITY

## Some Ethical Considerations

The transformation of a zygote into an intricately detailed adult raises profound questions. *When does development begin?* As we have seen, key developmental events occur even before fertilization. *When does life begin?* During her lifetime, a human female can produce as many as four hundred eggs, all of which are alive. During one ejaculation, a human male can release a quarter of a billion sperm, which also are alive. Even before sperm and egg merge by chance and establish the genetic makeup of a new individual, they are as much alive as any other form of life. It is scarcely tenable, then, to say "life begins" when they fuse. *Life began billions of years ago; and each gamete, each zygote, each mature individual is only a fleeting stage in the continuation of that beginning.*

This fact cannot diminish the meaning of conception, for it is no small thing to entrust a new individual with the gift of life, wrapped in the unique evolutionary threads of our species and handed down through an immense sweep of time.

Yet how can we reconcile the marvel of individual birth with the growing awareness of the astounding birth rate for our whole species? While this book is being written, an average of 10,700 newborns enter the world every single hour. By the time you go to bed tonight, there will be 257,000 more people on earth than there were last night at that hour. Within a week, the number will reach 1,800,000—about as many people as there are now in the entire state of Massachusetts. *Within one week.* Worldwide population growth has outstripped resources, and each year millions face the horrors of starvation. Living as we do on one of the most productive continents on earth, few of us can know what it means to give birth to a child, to give it the gift of life, and have no food to keep it alive.

And how can we reconcile the marvel of birth with the confusion surrounding unwanted pregnancies? Even highly developed countries have inadequate educational programs concerning fertility control, and a good number of their members are not inclined to exercise control. Each year in the United States alone, there are more than 100,000 "shotgun" marriages, about 200,000 unwed teenage mothers, and perhaps 1,500,000 abortions. Many parents encourage early boy–girl relationships, at the same time ignoring the risk of premarital intercourse and unplanned pregnancy. Advice is often condensed to a terse, "Don't do it. But if you do it, be careful!"

The motivation to engage in sex has been evolving for more than 500 million years. A few centuries of moral and ecological reasoning that call for its suppression have not prevented unwanted pregnancies. And complex social factors have contributed to a population growth rate that is out of control.

How will we reconcile our biological past and the need for a stabilized cultural present? Whether and how fertility is to be controlled is one of the most volatile issues of our time. We will return to this issue in the next chapter, in the context of principles governing the growth and stability of populations. Here, we can briefly consider some possible control options.

## Birth Control Options

The most effective method of birth control is complete *abstinence*, no sexual intercourse whatsoever. It is unrealistic to expect many people to practice it.

A modified form of abstinence is the *rhythm method.* The idea is to avoid intercourse during the woman's fertile period, beginning a few days before ovulation and ending a few days after. Her fertile period is identified and tracked either by keeping records of the length of her menstrual cycles or by taking her temperature each morning when she wakes up. (It rises by one-half to one degree just before the fertile period.) But ovulation can be irregular, and miscalculations are frequent. Also, sperm deposited in the vaginal tract a few days before ovulation may survive until ovulation. The method *is* inexpensive (it costs nothing after you buy the thermometer) and does not require fittings and periodic checkups by a physician. But its practitioners do run a large risk of pregnancy (Figure 43.20).

*Withdrawal*, or removing the penis from the vagina before ejaculation, dates back at least to biblical times. But withdrawal requires very strong willpower, and the method may fail anyway. Fluid released from the penis just before ejaculation may contain some sperm.

*Douching*, or rinsing out the vagina with a chemical right after intercourse, is next to useless. Sperm can move past the cervix and out of reach of the douche within ninety seconds after ejaculation.

Other methods involve physical or chemical barriers to prevent sperm from entering the uterus and moving to the ovarian ducts. *Spermicidal foam* and *spermicidal jelly* are toxic to sperm. They are packaged in an applicator and placed in the vagina just before intercourse. These products are not always reliable unless used with another device, such as a diaphragm or condom.

A *diaphragm* is a flexible, dome-shaped device, inserted into the vagina and positioned over the cervix before intercourse. A diaphragm is relatively effective when fitted initially by a doctor, used with foam or jelly before each sexual contact, and inserted correctly with each use.

*Condoms* are thin, tight-fitting sheaths of rubber or animal skin, worn over the penis during intercourse. They are about 85 to 93 percent reliable, and they help prevent the spread of sexually transmitted diseases (see the *Commentary* on page 780). However, condoms can tear and leak, in which case they are rendered useless.

The most widely used method of fertility control is *the Pill*, an oral contraceptive of synthetic estrogens and progesterones. It suppresses the normal release of these hormones from the pituitary and so stops eggs from maturing and being released at ovulation. The Pill is a prescription drug. Formulations vary and are selected to match each patient's needs. That is why it is not wise for a woman to borrow oral contraceptives from someone else.

When a woman does not forget to take her daily dosage, the Pill is one of the most reliable methods of controlling fertility. It does not interrupt sexual intercourse, and the method is easy to follow. Often the Pill corrects erratic menstrual cycles and decreases cramping. However, the Pill has some side effects for a small number of users. In the first month or so of use, it may cause nausea, weight gain, tissue swelling, and minor headaches. Its continued use may lead to blood clotting in the veins of a few women (3 out of 10,000) predisposed to this disorder. Some cases of elevated blood pressure and some abnormalities in fat metabolism might be linked to a growing number of gallbladder disorders in Pill users.

In *vasectomy*, a tiny incision is made in a man's scrotum, and each vas deferens is severed and tied off. The simple operation can be performed in twenty minutes in a physician's office, with only a local anesthetic. After vasectomy, sperm cannot leave the testes and so will not be present in semen. So far there is no firm evidence that vasectomy disrupts the male hormone system, and there seems to be no noticeable difference in sexual activity. Vasectomies can be reversed, but half the men who have had the surgery develop antibodies against sperm and may not be able to regain fertility.

For females, surgical intervention includes *tubal ligation*, in which the oviducts are cauterized or cut and tied off. Tubal ligation is usually performed in a hospital. A small number of women who have had the operation suffer recurring bouts of pain and inflammation of tissues in the pelvic region where the surgery was performed. The operation can be reversed, although major surgery is required and success is not always assured.

Once conception and implantation have occurred, the only way to terminate a pregnancy is *abortion*, the dislodging and removal of the embryo from the uterus. *RU-486*, the "morning-after Pill," can induce termination of pregnancy. It it administered under a physician's supervision, at least in Europe. Those opposed to abortion are currently fighting its use in the United States.

At one time, abortions were generally forbidden by law in the United States unless the pregnancy endangered the mother's life. The Supreme Court in 1973 ruled that the government does not have the right to forbid abortions during the early stages of pregnancy (typically up to five months). Before this ruling, there were

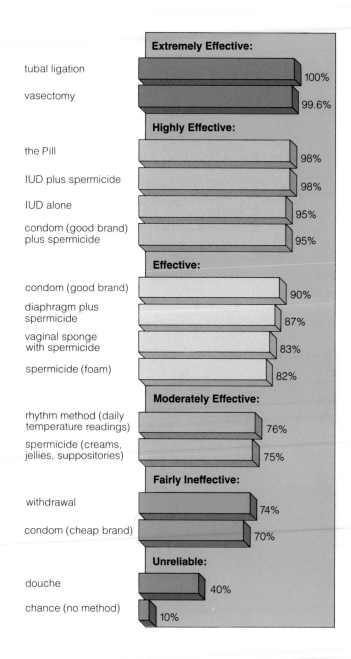

**Figure 43.20** Comparison of the effectiveness of some contraceptive methods.

# Commentary

## Sexually Transmitted Diseases

*(This Commentary is based on information from the Centers for Disease Control, Atlanta)*

Sexually transmitted diseases (STDs) have reached epidemic proportions, even in countries with the highest medical standards. The disease agents are mostly bacteria and viruses. Usually they are transmitted from infected to uninfected persons during sexual intercourse. In the United States alone, many millions of young adults have reported that they have some form of STD. No one can estimate the number of unreported cases.

The economics of this health problem are staggering. The cost of treatment far exceeds $2 billion a year—and this does not include the accelerating cost of treating AIDS patients. In many developing countries, AIDS alone threatens to overwhelm health-care delivery systems and to unravel decades of economic progress. The social consequences are sobering. Of every twenty babies born in the United States, one will have a chlamydial infection. Type II *Herpes* virus will infect as many as one of every 10,000 newborns, possibly killing half of them and leaving a fourth with serious neurological defects. Each year 1 million female Americans contract pelvic inflammatory disease, usually as a complication of gonorrhea and other STDs. Of every 200,000 who are hospitalized, over 100,000 become permanently sterile, and 900 die.

The examples just given only hint at the alarming complications of many sexually transmitted diseases.

### AIDS

Acquired immune deficiency syndrome (AIDS) is a set of chronic disorders that can follow infection by the human immunodeficiency virus (HIV). The virus cripples the immune system, in the manner described in Chapter 39. The body becomes highly vulnerable to illnesses, many of which would not otherwise be life-threatening. (Hence the description, "opportunistic" infections.)

AIDS is mainly a sexually transmitted disease, with most infections occurring through the transfer of bodily fluids during vaginal or anal intercourse. Such fluids include blood, semen, urine, and vaginal secretions. The virus enters the body through cuts or abrasions on the penis, vagina, or rectum. Mucous membranes in the mouth may be another point of entry. Once inside the body, the virus locks onto cells that are capable of sustaining its replication (page 692). Helper T cells (the T4 lymphocytes), macrophages, brain cells, and epithelial cells of the cervix are known targets.

At present there is no effective treatment or vaccine for AIDS. *There is no cure.* Infected persons may be symptom-free at first but as many as half develop AIDS within five to ten years. Others develop ARC (AIDS-related complex), which are milder symptoms than those characterizing AIDS (page 694). Will most or all of those infected by HIV eventually develop AIDS? We do not know enough about the natural history of the virus and the progression rates of the disease to discount that possibility.

HIV apparently existed in some parts of Central Africa for at least several decades. In the 1970s and early 1980s, it spread to different countries and was finally identified in 1981. Today it is spreading through African populations mainly by heterosexual contact. In the United States and other developed countries, HIV spread at first among male homosexuals. The pool of infection now includes a significant portion of the heterosexual population, largely as a result of needle-sharing among intravenous (IV) drug abusers. *During the next decade, as many as 10 million may be infected worldwide.*

As an example of the relative frequencies of the main modes of transmission, the following cases were reported to the Centers for Disease Control through July 1991:

| | White Males | Black Males | White Females | Black Females |
|---|---|---|---|---|
| Homosexual or bisexual contact: | 80% | 44% | — | — |
| IV drug abuse (heterosexual): | 7% | 36% | 41% | 57% |
| Both of the above: | 7% | 8% | — | — |
| Heterosexual contact: | 1% | 7% | 30% | 33% |
| Blood transfusion: | 2% | 1% | 21% | 3% |

Free or low-cost, confidential testing for AIDS is available through public health facilities and many physicians' offices. Keep in mind that there may be a time lag from a few weeks to six months or longer before detectable antibodies form in response to infection. The presence of antibodies indicates exposure to the virus, but this in itself does not mean that AIDS will develop. Even so, anyone who tests positive should be considered capable of spreading the virus.

The spread of HIV has led to massive public education programs, including the strongest possible advocacy for safe sex. Yet there is confusion about what constitutes "safe" sex. Proper use of high-quality, latex condoms, together with a spermicide that contains nonoxynol-9, is assumed to be highly effective in stopping transmission—but there is still a small risk of irreversible infection. Open-mouthed, intimate kissing with a person who tests positive for the virus should be avoided. Caressing carries no risk—if there are no lesions or cuts through which the virus can enter the body. Such lesions commonly accompany other sexually transmitted diseases, and they apparently are correlated with increased susceptibility to HIV infection.

In sum, AIDS has reached epidemic proportions mainly for three reasons. First, we did not know that the virus is transmitted by semen, blood, and vaginal fluid and that *behavioral* controls can limit its spread. Second, we did not have tests that could be used to identify symptom-free carriers who could unwittingly infect others; we do now. Third, we thought AIDS was a threat associated only with homosexual behavior. The medical, social, and economic consequences of its rapid spread throughout the world make it everyone's problem.

### Gonorrhea

Unlike AIDS, gonorrhea is a sexually transmitted disease that can be cured by prompt diagnosis and treatment. Gonorrhea ranks first among the reported communicable diseases in the United States, with 1 million new cases reported each year. There may be anywhere from 3 million to 10 million unreported cases.

Gonorrhea is caused by *Neisseria gonorrhoeae*. This bacterium can infect epithelial cells of the genital tract, eye membranes, and the throat. Since 1960 its incidence in the population has been rising at an alarming rate. The increase has coincided with the use of birth control pills and increased sexual permissivity.

Males have a greater chance than females do of detecting the disease in early stages. Within a week, yellow pus is discharged from the penis. Urination becomes more frequent and painful. Females may or may not experience a burning sensation while urinating. They may or may not have a slight vaginal discharge; even if they do, the

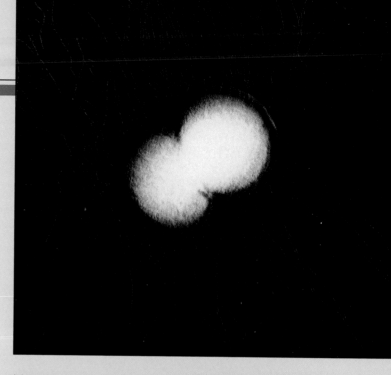

(a) *Neisseria gonorrhoeae*, a bacterium that typically is seen as paired cells, as shown here. The threadlike structures (pili) evident in this electron micrograph help the bacterium attach to its host, upon which it bestows gonorrhea.

discharge may not be perceived as abnormal. Thus, in the absence of worrisome symptoms, gonorrhea often goes untreated. The bacteria may spread into the oviducts. Eventually, there may be violent cramps, fever, vomiting and, often, sterility due to scarring and blocking.

Complications arising from gonorrheal infection can be avoided with prompt treatment. As a preventive measure, males who have multiple sexual partners can wear condoms to help prevent the spread of infection. Part of the problem is that the initial stages of the disease are so uneventful that the dangers are masked. Also, many infected persons wrongly believe that once cured of gonorrhea, they are safe from reinfection—which simply is not true. Multiple reinfections can and do occur.

### Syphilis

Syphilis is caused by a motile, corkscrew-shaped bacterium, *Treponema pallidum*. As many as 300,000 humans may become infected in a given year in the United States, but only about 30,000 are reported. In the past five years, its incidence has nearly doubled among females between ages fifteen and twenty-four. The bacterium is transmitted by sexual contact. After it has penetrated exposed tissues, it produces a chancre (localized ulcer) that teems with treponeme offspring. Usually the chancre is flat, not bumpy, and it is not painful. It becomes visible between one and eight weeks following infection—and it is a symptom of the primary stage of syphilis. By then, treponemes have already moved into the lymph vascular system and bloodstream.

During the second stage of infection, lesions can occur in mucous membranes, the eyes, bones, and the central nervous system. Afterward, the infection enters a latent stage, and there are no outward symptoms. Syphilis can be detected only by laboratory tests during the latent stage, which can last many years. All the while, the immune system works against the bacterium. Sometimes the body does cure itself, but this is not the usual outcome.

If untreated, syphilis in its tertiary stage can produce lesions of the skin and internal organs, including the liver, bones, and aorta. Scars form; the walls of the aorta can weaken. Treponemes also damage the brain and spinal cord in ways that lead to various forms of insanity and paralysis. Women who have been infected typically have miscarriages, stillbirths, or sickly and syphilitic infants.

### Chlamydial Infections

An intracellular parasite, *Chlamydia trachomatis*, is the culprit behind a variety of sexually transmitted diseases. Each year, anywhere from 3 million to 10 million Americans—college students particularly—are affected.

Among other things, the parasite infects cells of the genitals and urinary tract. Infected men may have a discharge from the penis and have a burning sensation when they urinate. Women may have a vaginal discharge as well as burning and itching sensations. Sometimes, however, there may be no apparent evidence of infection—yet the parasite can still be passed on to others.

Following infection, the parasites migrate to lymph nodes, which become enlarged and tender. The enlargement can impair lymph drainage and lead to pronounced tissue swelling. Chlamydial infections can be treated with tetracycline and sulfonamides. Most of the infections have

**(b)** *Treponema pallidum*, a bacterium that causes syphilis

no long-term complications. However, in some females the infection leads to pelvic inflammatory disease.

### Pelvic Inflammatory Disease

A condition called pelvic inflammatory disease (PID) affects about 14 million women each year. It is one of the serious complications of gonorrhea, chlamydial infections, and other STDs. It also can arise when microorganisms that normally inhabit the vagina ascend into the pelvic region. Most often, the uterus, oviducts, and ovaries are affected. Pain may be so severe, infected women often think they are having an attack of acute appendicitis. The oviducts may become scarred, and scarring can lead to abnormal pregnancies as well as to sterility.

### Genital Herpes

Genital herpes is an extremely contagious viral infection of the genitals. It is transmitted when any part of a person's body comes into direct contact with active *Herpes* viruses or sores that contain them. Mucous membranes of the mouth or genital area are especially susceptible to invasion, as is broken or damaged skin. Transmission seems to require direct contact; the virus does not survive for long outside the human body.

There are an estimated 5 million to 20 million persons with genital herpes in the United States alone. From 1965 to 1979, the number of reported cases increased by 830 percent. And 200,000 to 500,000 cases are still being reported annually.

Newborns of infected mothers are among those cases. Contact with the mother's active lesions during normal vaginal delivery can lead to a form of herpes that is often fatal. Lesions arising in the infant's eyes can cause blindness. Chronic herpes infection of the cervix also increases the risk of cervical cancer.

The many strains of *Herpes* viruses are classified as types I and II. The type I strains infect mainly the lips, tongue, mouth, and eyes. Type II strains cause most of the genital infections. Disease symptoms occur two to ten days after exposure to the virus, although sometimes symptoms are mild or absent. Among infected women, small, painful blisters appear on the vulva, cervix, urethra, or anal tissues. Among men, the blisters occur on the penis and anal tissues. Within three weeks, the sores crust over and heal without leaving scars.

After the first sores disappear, sporadic reactivation of the virus can produce new, painful sores at or near the original site of infection. Recurrent infections may be triggered by sexual intercourse, emotional stress, menstruation, or other infections. At present there is no cure for genital herpes. Acyclovir, an antiviral drug, decreases the healing time and often decreases the pain and viral shedding.

dangerous, traumatic, and often fatal attempts to abort embryos, either by pregnant women themselves or by quacks.

Newer methods have made abortion relatively rapid, painless, and free of complications when performed during the first trimester. Abortions in the second and third trimesters will probably remain extremely controversial unless the mother's life is clearly threatened. For both medical and humanitarian reasons, however, it is generally agreed in this country that the preferred route to birth control is not through abortion but through control of conception in the first place.

## In Vitro Fertilization

Controls over fertility also extend in the other direction—to help childless couples who are desperate to conceive a child. In the United States, about 15 percent of all couples cannot do so because of sterility or infertility. For example, hormonal imbalances may prevent ovulation in females, or the sperm count in the male may be too low to assure fertilization.

With *in vitro fertilization*, external conception is possible, provided sperm and oocytes obtained from the couple are normal. A hormone is administered that prepares the ovaries for ovulation. Then a physician locates and removes the preovulatory oocyte with a suction device. Before the oocyte is removed, sperm from the male is placed in a solution that simulates the fluid in oviducts. When the suctioned oocyte is placed with the sperm, fertilization may occur a few hours later. About twelve hours later, the newly dividing zygote is transferred to a solution that will support further development, and about two to four days after that, it is transferred to the female's uterus. Implantation occurs in about 20 percent of the cases, and each attempt costs several thousand dollars.

## SUMMARY

1. Most animals reproduce sexually. Separation into male and female sexes involves specialized reproductive structures and forms of behavior that help assure successful fertilization and that lend initial nutritional support to offspring.

2. Humans have a pair of primary reproductive organs (sperm-producing testes in males, egg-producing ovaries in females), accessory ducts, and glands. Testes and ovaries also produce hormones that influence reproductive functions and secondary sexual traits.

3. The hormones testosterone, LH, and FSH control sperm formation. They are part of feedback loops among the hypothalamus, anterior pituitary, and testes.

4. The hormones estrogen, progesterone, FSH, and LH control egg maturation and release, as well as changes in the lining of the uterus (endometrium). They are part of feedback loops involving the hypothalamus, anterior pituitary, and ovaries.

5. The following events occur during a menstrual cycle:

a. A follicle, which is an oocyte surrounded by a cell layer, matures in an ovary, and the endometrium starts to rebuild. (It breaks down at the end of each menstrual cycle when pregnancy does not occur.)

b. A midcycle peak of LH triggers the release of a secondary oocyte from the ovary. This event is called ovulation.

c. A corpus luteum forms from the remainder of the follicle. Its secretions prime the endometrium for fertilization. When fertilization occurs, the corpus luteum is maintained, and its secretions help maintain the endometrium.

6. Human development proceeds through gamete formation, fertilization, cleavage, gastrulation, organ formation, and growth and tissue specialization.

7. All tissues and organs in the developing embryo arise from three germ layers: the endoderm, ectoderm, and mesoderm of the early embryo.

8. Embryonic development depends on the formation of four extraembryonic membranes:

a. Yolk sac: parts give rise to the embryo's digestive tube.

b. Allantois: its blood vessels function in oxygen transport.

c. Amnion: a fluid-filled sac that surrounds and protects the embryo from mechanical shocks and keeps it from drying out.

d. Chorion: a protective membrane around the embryo and the other membranes; a primary component of the placenta.

9. The embryo and the mother exchange substances by way of the placenta (a spongy tissue of endometrium and extraembryonic membranes).

10. The placental barrier provides some protection for the fetus, but it may suffer harmful effects from the mother's nutritional deficiencies, infections, intake of prescription drugs, illegal drugs, alcohol, and smoking.

11. At delivery, contractions of the uterus dilate the cervical canal and expel the fetus and afterbirth. Estrogen and progesterone stimulate growth of the mammary glands. After delivery, nursing causes the release of hormones that stimulate milk production and release.

12. Control of human fertility raises important ethical questions. These questions extend to the physical, chemical, surgical, or behavioral interventions used in the control of unwanted pregnancies.

1. Study Table 43.1. Then list the main organs of the human male reproductive tract and identify their functions. *761*

2. Which hormones influence male reproductive function? *762*

3. Label the component parts of the female reproductive tract: *765*

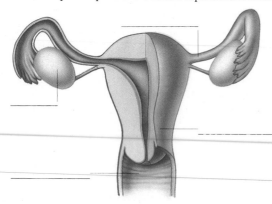

4. What is the menstrual cycle? Which four hormones influence this cycle? *764–766*

5. List four events that are triggered by the surge of LH at the midpoint of the menstrual cycle. *766–767*

6. What changes occur in the endometrium during the menstrual cycle? *766*

## Self-Quiz *(Answers in Appendix IV)*

1. Besides producing gametes, human male and female gonads also produce sex hormones. The _____ and the pituitary gland control secretion of both.

2. _____ production is continuous from puberty onward in males; _____ production is cyclic and intermittent in females.
   a. egg; sperm
   c. testosterone; sperm
   b. sperm; egg
   d. estrogen; egg

3. Sperm formation is controlled through _____ secretions.
   a. testosterone
   c. FSH
   b. LH
   d. all of the above are correct

4. During the menstrual cycle, a midcycle surge of _____ triggers ovulation.
   a. estrogen
   c. LH
   b. progesterone
   d. FHS

5. Which is the correct order for one turn of the menstrual cycle?
   a. corpus luteum forms, ovulation, follicle forms
   b. follicle forms, ovulation, corpus luteum forms

6. Parts of the _____, an extraembryonic membrane, give rise to the embryo's digestive tube.
   a. yolk sac
   c. amnion
   b. allantois
   d. chorion

7. The _____, a fluid-filled sac, surrounds and protects the embryo from mechanical shocks and keeps it from drying out.
   a. yolk sac
   c. amnion
   b. allantois
   d. chorion

8. Blood vessels of the _____, an extraembryonic membrane, transport oxygen and nutrients to the embryo.
   a. yolk sac
   c. amnion
   b. allantois
   d. chorion

9. Substances are exchanged between the embryo and mother through the _____, which is composed of maternal and embryonic tissues.
   a. yolk sac
   d. amnion
   b. allantois
   e. chorion
   c. placenta

10. Match the reproduction and development concepts.
   ____ extraembryonic membranes
   ____ corpus luteum
   ____ follicle
   ____ egg
   ____ germ layers

   a. endoderm, ectoderm, mesoderm
   b. immature oocyte and surrounding cell layer
   c. yolk sac, allantois, amnion, chorion
   d. mature ovum
   e. its secretions help prepare endometrium for fertilization

## Selected Key Terms

| | | |
|---|---|---|
| adult *777* | FSH *762* | polar body *764* |
| aging *777* | GnRH *763* | progesterone *763* |
| allantois *770* | granulosa cell *764* | secondary oocyte *764* |
| amnion *770* | implantation *769* | semen *762* |
| blastocyst *769* | Leydig cell *761* | seminiferous tubule *761* |
| chorion *770* | LH *762* | Sertoli cell *762* |
| clitoris *764* | menstrual cycle *764* | somite *775* |
| coitus *766* | oocyte *763* | sperm *762* |
| corpus luteum *764* | orgasm *769* | testis *759* |
| embryonic disk *769* | ovary *759* | testosterone *762* |
| endometrium *764* | oviduct *763* | umbilical cord *770* |
| estrogen *763* | ovulation *764* | uterus *763* |
| fetus *775* | ovum *769* | yolk sac *770* |
| follicle *764* | placenta *771* | zona pellucida *764* |

## Readings

Carlson, B. 1988. *Patten's Foundations of Embryology*. Fifth edition. New York: McGraw-Hill.

Gilbert, S. 1991. *Developmental Biology*. Sunderland, Massachusetts: Sinauer. Third edition. Acclaimed reference text.

Nilsson, L. et al. 1986. *A Child Is Born*. New York: Delacorte Press/ Seymour Lawrence. Extraordinary photographs of embryonic development.

Saunders, J. W. 1982. *Developmental Biology: Patterns, Problems, Principles*. New York: Macmillan.

Schatten, G. 1983. "Motility During Fertilization." *Endeavor* 7(4): 173–182.

FACING PAGE: *Two organisms—a fox in the shadows cast by a snow-dusted spruce tree. What are the nature and consequences of their interactions with one another, with other organisms, and with their environment? By the end of this last unit of the book, you possibly will see worlds within worlds in such photographs.*

# 44 POPULATION ECOLOGY

## A Tale of Nightmare Numbers

Suppose this year the federal government passes legislation to control population growth by limiting the size of each family to three children. Suppose they mandate sterilization of each father after his third child is born, and that if he refuses, he will be sterilized with or without his consent. *It would never happen here*, you might be thinking. Such an invasion of privacy would never be tolerated in our society. Besides, family size is not much of an issue in North America, where standards of food production, hygiene, and medical care are among the world's highest.

Most populations cannot take these things for granted, yet many are still growing at alarming rates. For example, there already are more people in India than in North and South America combined. Most do not have adequate food or medical care, living conditions are often appalling, and unemployment is a nightmare. Each *week*, 100,000 people enter the job market, for nonexistent jobs. Each *day*, 100 acres of the croplands necessary to support the population are being

removed from agriculture. Why? Too many salts from irrigation water have accumulated in the soil, and India does not get enough rain to flush them out.

The government of India has supported birth control programs for more than two decades, but these have not worked well. Most of the people live in remote villages, so the programs are difficult to administer. Illiteracy is widespread, so information must be conveyed by word of mouth. The very idea of limiting family size is met with resistance. Disease and starvation are so pervasive, many villagers believe that survival depends on having many children. Without large families, they ask, who will help a father tend fields? Who will go to cities and earn money to send back home? How can a father otherwise know he will be survived by a son, who must, by Hindu tradition, conduct the last rites so the soul of his dead father will rest in peace?

India's population may reach 1 billion in less than thirty years. It is not pleasant to think about how many then will face poverty, disease, and starvation. In 1976,

**Figure 44.1** A sampling of the more than 5.4 billion humans on earth. In this chapter we turn to the principles governing the growth and sustainability of populations, including our own.

out of desperation, the government passed a law calling for compulsory sterilization, although public outrage became so great the law eventually was rescinded.

Is there a way out of such dilemmas? Should other nations that are currently using the greatest share of the world's resources make concerted efforts to use fewer resources—and to use them more efficiently? Should they donate surplus food to the growing populations of less fortunate nations? Would donations help, or would they encourage greater increases in population size? Suppose the expanded populations came to depend on continuing support. What if the benefactor nations were hit by severe droughts year after year and had trouble meeting even the demands of their own populations? Whether we consider humans or any other kind of organism, *certain ecological principles govern the growth and sustainability of populations over time*. This chapter describes those principles, then shows how they apply to the past, present, and future growth of the human population.

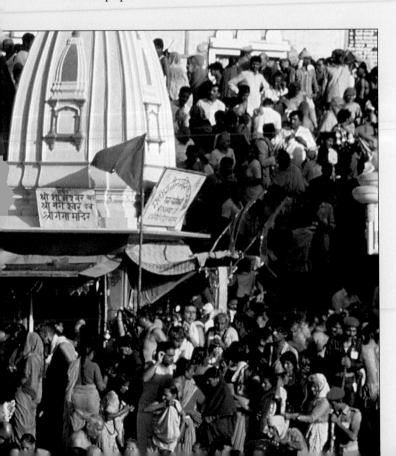

KEY CONCEPTS

**1.** A population is a group of individuals of the same species occupying a given area at the same time. As we have seen, the population—not the individual or the species—is the unit of evolution.

**2.** To gain more insight into the changing nature of the human population and to predict its likely future, we can consider ecological principles that govern the growth and sustainability of all populations over time.

**3.** Population growth generally follows certain patterns. When the birth rate remains even slightly above the death rate, and when both rates are constant, a population shows exponential growth. This means it increases in size by ever larger amounts per unit of time. When the environment imposes limits on growth, the pattern may become one of logistic growth. Here, a low-density population expands rapidly in numbers, then growth levels off as scarce resources limit further expansion or even cause declines.

**4.** All populations face limits to growth, for no environment can indefinitely sustain a continuously increasing number of individuals. Competition for resources, disease, predation, and other factors act as controls over population growth. The controls vary in their relative effects and they vary over time.

## FROM POPULATIONS TO THE BIOSPHERE

With this unit of the book, we turn to the interactions of organisms with one another and with their physical and chemical environment. **Ecology** means the study of those interactions, and those studies proceed through the following levels of biological organization:

1. The **population**: a group of individuals of the same species occupying a given place at the same time. The place where a population (or individual) lives is its *habitat*.

2. The **community**: the populations of *all* species occupying a habitat. The term also is used for groups of organisms with similar life-styles in the same habitat, such as the bird community and the plant community.

3. The **ecosystem**: a community and its environment. An ecosystem has a *biotic* component (all of its living members) and *abiotic* (nonliving) components, such as temperature, rainfall, atmospheric gases, and soil nutrients.

4. The **biosphere**: the entire realm in which organisms exist—the lower regions of the atmosphere, the waters of the earth, and the surface rocks, soils, and sediments of the earth's crust.

We begin now with the relationships that influence the size, structure, and distribution of populations. We will return to the nature of human population growth at the chapter's end. Communities, ecosystems, and the biosphere will be our focus in chapters to follow.

## POPULATION DYNAMICS

### Characteristics of Populations

Populations as a whole display certain characteristics, including size, density, distribution, and age structure. The *size* of a population is the number of individuals making up its gene pool. Its *density* is the number of individuals per unit area or volume, such as the number of guppies in each liter of water in a small stream. *Distribution* refers to the general pattern in which the population's members are dispersed through its habitat. Distribution varies in space and it varies with time, often in response to changing seasons or with changes in population density.

Finally, the *age structure* of a population is the relative proportion of individuals of each age. These are often divided into pre-reproductive, reproductive, and post-reproductive age categories. The middle category represents the *reproductive base* for the population.

### Population Density and Distribution

Population density generally varies greatly among species living in the same area. You probably know this if you have ever collected seashells or butterflies or kept a running list of birds visiting your backyard. For every common species you observed, you probably saw several rare ones.

Ecological relationships help shape population density. To study their effects, it helps to know what the density is at a particular point in space and time. The information can be used as a baseline for charting changes in density. For sparse populations of trees, snakes, or some other organism, a simple "head count" can be made in a defined area. For dense populations, counts in small, randomly selected sampling areas can yield an estimate of overall density.

Knowing the population density doesn't tell us how its members are distributed in space. They might be dispersed randomly through the habitat, clumped together in parts of it, or spread out rather uniformly.

**Random Dispersion**. Suppose that environmental conditions are the same throughout a habitat. Suppose further that members of the population are neither attracting nor repelling each other. Under such conditions, their dispersion may be random. Wolf spiders prowling about on a forest floor may be an example of random spacing among individuals.

Random dispersion is the exception rather than the rule. In population studies, it is a theoretical baseline for measuring the degree of clumping or uniformity among members of a population.

**Clumped Dispersion**. Most commonly, members of a population live in clumps in different parts of their habitat. There are three reasons for this. First, suitable physical, chemical, and biological conditions are usually patchy rather than uniform. In a pasture, certain plants grow luxuriantly in small patches of soil where cowpats fell weeks or months before—cowpats that enriched the soil with nitrogen. Some parts of the habitat offer more shade, more water, or better hiding places for prey organisms and hunting possibilities for predators. Second, many animals form social groups (Figure 44.2a). Such groups might provide mutual defense against predators, more efficient foraging for food, assistance for offspring, and better reproductive possibilities (page 912). Third, the dispersal of seeds, larvae, and other forms of each

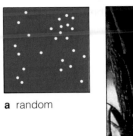

**a** random

**b** clumped

**c** nearly uniform

**Figure 44.2** Dispersion patterns for members of a population. (**a**) Random dispersion is uncommon over large areas. (**b**) Clumping occurs when living space, hiding places, and other aspects of the habitat are patchy, not uniform. It occurs also among social animals, such as baboons. (**c**) Competition for resources may foster nearly uniform distribution. Shown here, creosote bushes (*Larrea*) near Death Valley, California.

new generation is often limited. The offspring of sponges are not spectacularly motile, for example, and they simply settle near their parents and one another.

**Uniform Dispersion.** Trees in an orchard are evenly spaced, but nearly uniform distribution is rare in nature. Whenever a population does show some uniformity, this seems to be an outcome of competition among its members. Consider creosote bushes in dry scrub deserts of the American Southwest (Figure 44.2b). Large, mature plants deplete the soil around them of available water, and seed-eating ants and rodents concentrate their activities near the plants. Thus seeds and seedlings often cannot survive near mature plants, so clumps do not form.

**The Importance of Scale.** Whether the spatial pattern of a population appears to be clumped, random, or uniform usually depends on how wide a view we take. At the scale of a few square meters, the acorns under an oak tree are likely to be more randomly spaced. Yet, at the scale of a forest of mixed hardwood trees, the same acorns will appear to be clustered near the parent tree.

**Distribution over Time.** Population distribution varies with time, often in response to environmental rhythms. Few environments are so productive that they yield abundant resources all year long, and many animals often move from one local habitat to another with the seasons. For example, in regions of deciduous tropical forests, many birds and mammals crowd into narrow "gallery forests" along watercourses during the dry season. There, the trees are often evergreen and provide food and shelter. The animal density becomes much greater than it is during the wet season.

**Figure 44.4** (**a**) Idealized S-shaped curve characteristic of logistic growth. Following a rapid growth phase, growth slows and the curve flattens out as the carrying capacity is reached. Sometimes a population grows rapidly and overshoots the carrying capacity. This happened to the reindeer population introduced on one of the Pribilof Islands (Figure 44.5). At other times, the carrying capacity itself changes abruptly as environmental conditions change. This happened to the human population in Ireland before the turn of the century, when a disease (late blight, page 376) wiped out the potatoes that were the mainstay of the diet.

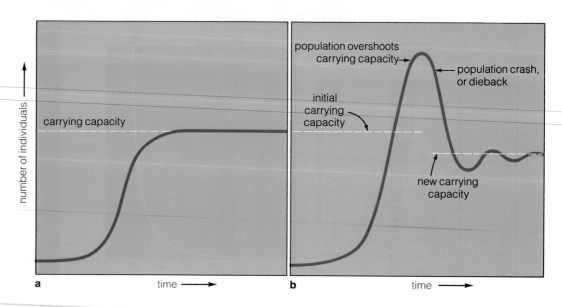

tically alter chemical conditions in the flask. The bacteria ended up polluting their environment. Their own activities put a stop to their growth.

All the limiting factors acting on a population collectively represent the environmental resistance to its growth. They all influence the number of individuals of a given species that can be sustained indefinitely in a given area.

**Carrying Capacity and Logistic Growth.** As a small population increases in size, the same nutrients, energy supplies, living space, and other resources must be shared by a greater and greater number of individuals. As the portion of resources available to each member declines, fewer individuals may be born or more may die (due to starvation or lack of nutrients). This situation will proceed until deaths equal births. The population size at which births are balanced by deaths is called the **carrying capacity** of the environment for that population. *A sustainable supply of resources defines the carrying capacity for a particular population in a particular environment.*

A rapidly growing population may temporarily exceed its carrying capacity. But eventually high death rates and low birth rates will bring population size down to the carrying capacity. They may even bring it down lower, as Figure 44.4 suggests. Often, the resource supply itself is damaged by burgeoning populations and the carrying capacity actually decreases.

For example, when mountain lions, wolves, and other natural predators of deer are killed off, herds of deer often grow too large for the supply of food that their habitat can provide. Before the deer population can decline to its former carrying capacity, the damage that their foraging inflicts on the plants that sustain them may dictate a new, lower carrying capacity. A similar

pattern can follow the introduction of grazing animals to islands where large predators are absent (Figure 44.5). Many biologists believe that the human population is now exceeding its own carrying capacity through soil erosion, deforestation, and changes in global climate (Chapter 48).

In nature, other population controls besides scarce resources often keep a population well below the carrying capacity set by resources. Predation and disease are effective this way. So are late frosts, storms, naturally occurring fires, and other disturbances. Later, you will see how we have sidestepped such controls over population size during recent human history.

What is the simplest model we can use for population growth when resources are limiting? It is this: A low-density population starts growing slowly, goes through a rapid growth phase, and then growth levels off once the carrying capacity is reached. This pattern is called **logistic growth**. It can be represented by this equation:

$$\begin{pmatrix}\text{population}\\\text{growth}\\\text{rate}\end{pmatrix} = \begin{pmatrix}\text{maximum net}\\\text{reproduction}\\\text{per individual}\end{pmatrix} \times \begin{pmatrix}\text{number}\\\text{of indi-}\\\text{viduals}\end{pmatrix} \times \begin{pmatrix}\text{portion of}\\\text{unexploited}\\\text{resources}\end{pmatrix}$$

In symbols, this equation becomes

$$G = r_{max} N\left(\frac{K - N}{K}\right)$$

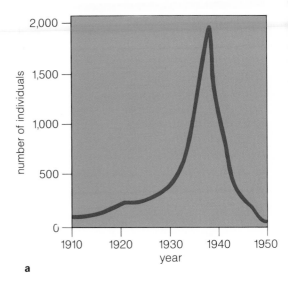

**b**

**Figure 44.5** Carrying capacity and a reindeer herd. In 1910, four male and twenty-two female reindeer were introduced on one of Alaska's Pribilof Islands. Within thirty years the herd increased to 2,000, and its members had to compete for dwindling vegetation. Overgrazing destroyed most of the vegetation, and in 1950 the herd plummeted to eight members. The growth pattern in (**a**) shows how it "overshot" the carrying capacity and then rapidly crashed.

**a**

where $K$ represents the carrying capacity and $N$ represents the number of individuals. The term within the parentheses is near 1 when the population is small. It falls to zero when the population reaches its carrying capacity.

A plot of logistic growth gives us an **S-shaped curve**, of the sort shown in Figure 44.4. This curve is only a simple approximation of what goes on in nature, however. Because environmental conditions vary, carrying capacity also can vary over time.

---

**Carrying capacity is the number of individuals of a species that can be sustained indefinitely by resources in a given area.**

**Resource availability is only one factor that limits population growth. Predation, competition, and other factors can do the same.**

**Because these factors vary in their effects over time, both the carrying capacity and the population size fluctuate for most populations.**

---

### Checks on Population Growth

**Density-Dependent Control.** When a population grows and its density increases, competition for resources, predation, parasitism, and disease tend to increase in intensity. These are **density-dependent controls**, and they have self-adjusting effects on population growth. Once density decreases, the pressures ease and population size may increase once more.

When prey or host populations become increasingly dense, for example, their members face greater risk of being killed by predators, colonized by parasites, or infected by contagious disease. A classic example is the *bubonic plague* that killed 25 million Europeans during the fourteenth century. *Yersinia pestis*, the bacterial agent of this disease, normally lives in wild rodents; fleas transmit it to new hosts. The bacterium multiplies in the flea gut and blocks digestion, the fleas attempt to feed more and more often, and so the disease spreads. It spread like wildfire through the cities of fourteenth-century Europe. Why? Human habitats were crowded together, sanitary conditions were poor, and the rats were abundant.

Subsequent outbreaks of plague have not been as dramatic. However, *Y. pestis* still lurks in many places—for example, in wild rodents that live in habitats from the Rocky Mountains to the West Coast of the United States.

**Density-Independent Control.** Some events tend to increase the death rate or decrease the birth rate more or less independently of population density. Such events are said to be **density-independent controls** over population growth. One member of a population might be more competitive than others when it comes to staking out part of a dwindling food supply. But if it is caught out in the open in a severe storm, that member is just as likely to be killed as any of the others.

As an example of a density-independent control, think about a summer snowstorm in the Colorado Rockies. Every so often, those freak snowstorms wipe out butterfly populations. Similarly, application of a potent pesticide in your backyard may kill most insects, mice, cats, birds, and other animals regardless of how dense their populations are.

**Table 44.1   Life Table for a Cohort of Annual Plants (*Phlox drummondii*)**

| Age Interval (days) | Survivorship (number surviving at start of interval) | Number Dying During Interval | Death Rate per Individual During Interval | "Birth" Rate (number of seeds produced per individual) During Interval |
|---|---|---|---|---|
| 0–63 | 996 | 328 | 0.329 | 0 |
| 63–124 | 668 | 373 | 0.558 | 0 |
| 124–184 | 295 | 105 | 0.356 | 0 |
| 184–215 | 190 | 14 | 0.074 | 0 |
| 215–264 | 176 | 4 | 0.023 | 0 |
| 264–278 | 172 | 5 | 0.029 | 0 |
| 278–292 | 167 | 8 | 0.048 | 0 |
| 292–306 | 159 | 5 | 0.031 | 0.33 |
| 306–320 | 154 | 7 | 0.045 | 3.13 |
| 320–334 | 147 | 42 | 0.286 | 5.42 |
| 334–348 | 105 | 83 | 0.790 | 9.26 |
| 348–362 | 22 | 22 | 1.000 | 4.31 |
| 362– | 0 | 0 | 0 | 0 |
| | | 996 | | |

Data from W. J. Leverich and D. A. Levin, *American Naturalist* 1979, 113:881–903.

**Table 44.2   Life Table for the United States Population, 1982**

| Age Interval | Survivorship (number alive at start of interval, per 100,000) | Number Dying During Interval |
|---|---|---|
| 0–1 | 100,000 | 1,107 |
| 1–5 | 98,893 | 269 |
| 5–10 | 98,624 | 175 |
| 10–15 | 98,449 | 181 |
| 15–20 | 98,268 | 497 |
| 20–25 | 98,771 | 673 |
| 25–30 | 97,098 | 663 |
| 30–35 | 96,435 | 725 |
| 35–40 | 95,710 | 986 |
| 40–45 | 94,724 | 1,483 |
| 45–50 | 93,241 | 2,352 |
| 50–55 | 90,889 | 3,483 |
| 55–60 | 87,406 | 5,063 |
| 60–65 | 82,343 | 7,281 |
| 65–70 | 75,062 | 9,005 |
| 70–75 | 66,057 | 12,214 |
| 75–80 | 53,843 | 14,455 |
| 80–85 | 39,388 | 14,467 |
| 85+ | 24,921 | 24,921 |

## LIFE HISTORY PATTERNS

So far, we have treated populations as if each were made up of identical members. Yet the members of most species pass through many stages of development, with each new stage of life having its own perils and rewards. In short, *the patterns of reproduction, death, and migration vary through the life span characteristic of the species.* Let's take a look at a few examples of these age-specific, **life history patterns** before considering the nature of some of the environmental variables that might have helped shape them.

### Life Tables

Each species has a characteristic life span, but few individuals reach the maximum age possible. Death is more probable at some ages and less so at others. Also, individuals are more likely to reproduce or leave the population at some ages than at others, and these ages vary from one species to the next. The study of such age-specific patterns is the subject of demography. Life insurance and health insurance companies were the first to study age-specific patterns, but ecologists now apply demographic methods to populations of plants and animals.

For example, **life tables** can be constructed to summarize the age-specific patterns of birth and death for a given population in a given area. Typically, ecologists follow the fate of a group of newborn individuals until the last one dies. Such a group is called a *cohort*. Besides

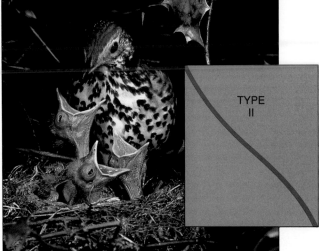

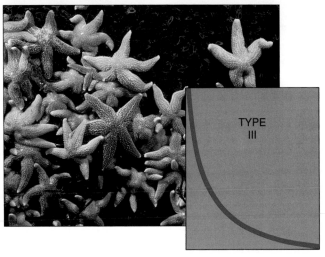

**Figure 44.6** Three generalized types of survivorship curves. For Type I populations, there is high survivorship until some age, then high mortality. Type II populations show a fairly constant death rate at all ages. For Type III populations, survivorship is low early in life.

recording the age at death of each individual, they keep track of how many offspring (if any) each surviving individual produces during each age interval of its life. The death rate and birth rate for each age interval are then easily computed, producing birth and death "schedules" for the cohort.

The death schedule is usually transformed to its more cheerful reflection, "survivorship." A survivorship schedule lists the number of individuals of the original cohort that survive long enough to reach age *x*. Table 44.1 presents a life table for a cohort of 996 phlox plants in Texas. Table 44.2 shows a life table that includes some of the same kinds of data for a human population.

## Survivorship Curves and Reproductive Patterns

**Survivorship curves** are plots of the age-specific survival of a cohort of individuals for a population in a given environment. Three types of survivorship curves are common in nature. Type I curves reflect high survivorship until fairly late in life, then a large increase in deaths (Figure 44.6). Such curves are typical of large mammals that produce only a few, large offspring but

provide them with extended parental care. Elephants do this. The females produce only four or five calves in a lifetime, and they devote several years of parental care to each one. Type I curves also are typical of human populations, especially when they are provided with good health care services. Historically, and where health care is poor today, infant deaths put a sharp drop at the start of the curve, which then levels off.

Type II curves reflect a fairly constant death rate at all ages. They are typical among organisms that are just as likely to be killed or die of disease at any age. This is the pattern for some songbirds, lizards, and small mammals.

Type III curves reflect a high death rate early in life. For example, the curve plummets precipitously for sea stars. Although sea stars produce mind-boggling numbers of tiny offspring, their young must feed and grow rapidly on their own, without nutritional support, protection, or guidance from the parents. The offspring are highly vulnerable, and most are quickly eaten by corals and other animals. Plummeting survivorship curves are typical of other marine invertebrates besides sea stars. They also are typical of most insects and many fishes, plants, and fungi—all of which produce many small offspring, with little or no parental care.

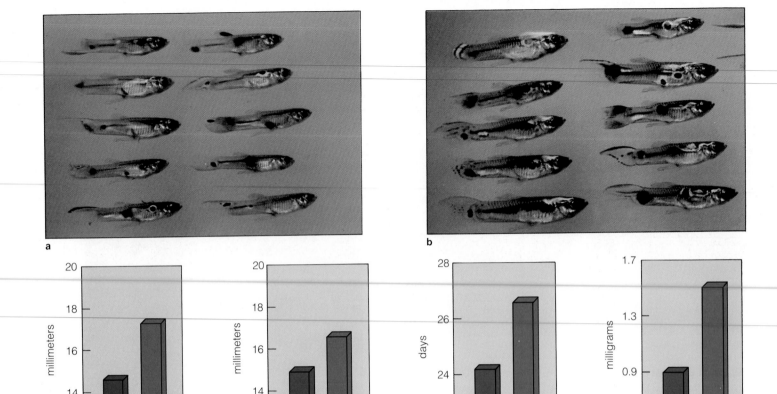

**Figure 44.7** Guppies (*Poecilia reticulata*) from the mountain streams of Trinidad. Guppies devote considerable time to courting and avoiding predators and must be adapted to do both. Depending on where they live, guppies might meet up with weak or dangerous predatory fishes.

(**a**) In places where predators are dangerous, guppies are smaller, more streamlined and duller in color patterning. (**b**) In places where predators eat guppies only occasionally, the guppies are larger, less streamlined, and more brightly colored. The differences are accompanied by pronounced differences in life history patterns (*see text*). (**c**) Differences in body size, interval between broods, and embryo weights for guppies from streams with pike-cichlids (green bars) or killifish (red bars) on the island of Trinidad. Pike-cichlids prey on larger guppies and killifish prey on smaller ones, selecting for the differences shown here.

## Evolution of Life History Patterns

Ecologists have been working to understand the connections between natural selection, environmental variables, and life history patterns. Early on, it seemed that selection processes favored two kinds of patterns—either rapid production of many relatively small offspring early in life, or production of only a few large offspring late in life. These two patterns are now known to be extremes, at opposite ends of a range of possible life histories. Also, *both* patterns as well as intermediate ones can sometimes characterize different populations of the same species!

Consider some elegant studies that David Reznick and John Endler conducted to identify which environmental variables influence the life history patterns of guppies. These small, live-bearing fish were studied first in the streams of Trinidad, an island in the southern Caribbean Sea. Then they were studied in the laboratory.

Male guppies are smaller than the females, which they attract with brightly colored patterns on the body and with complex courtship displays. Males stop growing once they are sexually mature, but the drab-colored females continue to grow larger as they reproduce.

In the mountains of Trinidad, guppies living in different streams—and even in different parts of the same streams—are subject to different dangers. In some streams, a small killifish preys heavily on immature guppies but cannot tackle the (larger) adults. In other streams, a larger pike-cichlid prefers mature (bigger) guppies and tends not to waste time hunting small ones.

As might be predicted from our understanding of selection processes, the individuals of guppy populations confronted with pike-cichlids (which favor large-bodied meals) mature sooner, are smaller at maturity, and reproduce at a younger age. They produce far more

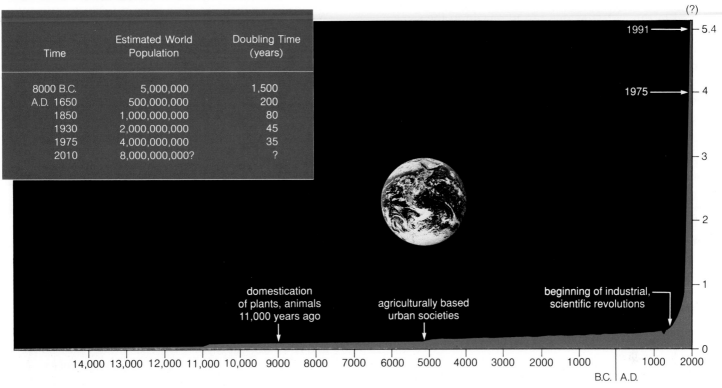

| Time | Estimated World Population | Doubling Time (years) |
|---|---|---|
| 8000 B.C. | 5,000,000 | 1,500 |
| A.D. 1650 | 500,000,000 | 200 |
| 1850 | 1,000,000,000 | 80 |
| 1930 | 2,000,000,000 | 45 |
| 1975 | 4,000,000,000 | 35 |
| 2010 | 8,000,000,000? | ? |

**Figure 44.8** The curve of global human population growth. The vertical axis of the graph represents world population, in billions. (The slight dip between the years 1347 and 1351 shows the time when 25 million people died in Europe as a result of bubonic plague.) The growth pattern over the past two centuries has been exponential, sustained by revolutions in agriculture, industrialization, and improvements in health care.

offspring and do so more often than their counterparts in killifish streams (Figure 44.7).

Could the differences in guppy populations be a result of some other, unknown differences between the streams? Reznick and Endler checked this out by raising guppies from each kind of stream in the laboratory for two generations. Both experimental populations were raised under identical conditions (with no predation). The same life history differences emerged. This provided experimental evidence of a genetic (heritable) basis for the differences.

The role of predators in the evolution of size differences among guppies was also checked. Guppies were grown for many generations in the laboratory—some alone, some with killifish, and some with pike-cichlids. As predicted, the guppy lineage subjected to predation over time by killifish became larger at maturity, whereas the lineage living with pike-cichlids showed a trend toward earlier maturity. Finally, Reznick and Endler introduced guppies from a pike-cichlid stream to another stream in Trinidad that contained killifish but no pike-cichlids or guppies. After eleven years, the guppies had evolved. As predicted, they had become larger in size and slower at reproducing—characteristics typical of natural guppy populations that live with killifish.

## HUMAN POPULATION GROWTH

In 1991, the human population reached 5.4 billion (Figure 44.8). In that year, almost 92 million more individuals were added to it. That amounted to an average of 1.8 million more per week, 257,000 per day, or 10,700 per hour. This staggering display of growth is occurring while at least one in five humans already on the planet is malnourished or starving, without clean drinking water, and without adequate shelter. It is occurring when health care delivery and sewage treatment facilities are nonexistent for a third of the population.

Suppose it were possible, by monumental efforts, to double the food supply to keep pace with growth. We would do little more than maintain marginal living conditions for most of the world, and death from starvation could still reach 20 to 40 million a year. Even this would come at great cost, for we are introducing serious new limiting factors into the very environment that must sustain us. Salted-out cropland, desertification, deforestation, global pollution—these are some of the factors you will be reading about in Chapter 48, and they do not bode well for our future.

For a while, it would be like the Red Queen's garden in Lewis Carroll's *Through the Looking Glass*, where one is forced to run as fast as one can to remain in the same place. But what happens when the human population doubles again? Can you brush this picture aside as being too far in the future to warrant your concern? It is no farther removed from you than the sons and daughters of the next generation.

### How We Began Sidestepping Controls

How did we get into this predicament? Human population growth has been slow for most of human history. But in the past two centuries, there have been astounding increases in the rate of population growth. Why has our growth rate increased so dramatically? There are three possible reasons:

1. We steadily developed the capacity to expand into new habitats and new climate zones.

2. Carrying capacities increased in the environments we already occupied.

3. We removed several limiting factors.

Let's consider the first possibility. Early humans apparently were restricted to savannas, and they were mainly vegetarians who added scavenged bits of meat to their diet when they could. By 200,000 years ago, small bands of hunters and gatherers had emerged (page 341). By 40,000 years ago, hunter-gatherers had spread through much of the world.

For most animal species, such extensive migrations could not have occurred as rapidly. Humans were able to migrate largely as a consequence of their very complex brains. They applied learning and memory to problems such as how to build fires, assemble shelters, create clothing and tools, and plan community hunts to exploit the abundance of wild game. Learned experiences were not confined to individuals but spread quickly from one band to another because of language—the ability for cultural communication. (It took less than seven decades from the time we first ventured into the air until we landed on the moon.) Thus, *the human population expanded into new environments, and it did so in an extremely short time span compared with the geographic spread of other organisms.*

What about the second possibility? About 11,000 years ago, people began to shift from the hunting and gathering way of life to agriculture. They shifted from risky, demanding moves following the game herds to a settled, more dependable basis for existence in more favorable settings. A milestone was the domestication of wild grasses, including the species ancestral to modern wheats and rice. Seeds were harvested, stored, and planted in one place; animals were domesticated and kept close to home for food and for pulling plows. Water was diverted into hand-dug ditches to irrigate crops.

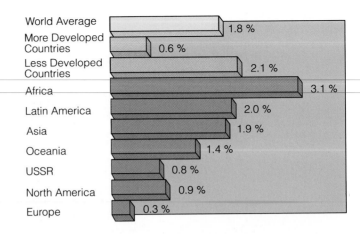

**Figure 44.9** Average annual population growth rate in various groups of countries in 1990.

The emerging agricultural practices increased productivity. With a larger, more dependable food supply, the rate of human population growth increased. Towns and cities developed, and with them came a social hierarchy that provided a labor base for more intensive agriculture. Much later, food supplies increased again with the use of fertilizers and pesticides. Thus, *even in its simplest form, management of food supplies through agriculture increased the carrying capacity for the human population.*

What about the third possibility—removing a series of limiting factors? Consider what happened when medical practices and sanitary conditions improved. Until about 300 years ago, malnutrition, contagious diseases, and poor hygiene kept the death rate relatively high, especially among infants. The death rate more or less balanced the birth rate. Contagious diseases are density-dependent factors, and they spread rapidly through crowded settlements and cities. Without proper hygiene and sewage disposal methods, and plagued with such disease carriers as fleas and rats, the human population increased only slowly at first. Then plumbing and sewage treatment methods were developed. Bacteria and viruses were recognized as disease agents. Vaccines, antitoxins, and drugs such as antibiotics were developed. The result was a sharp drop in the death rate. With births now exceeding deaths, the human population grew rapidly.

And consider what happened when humans discovered how to harness the energy stored in fossil fuels, beginning with coal. This discovery occurred in the mid-eighteenth century. Within a few decades, large industrialized societies emerged in Western Europe and North America. After World War I, more efficient technologies developed. Cars, tractors, and other economically affordable goods were now mass-produced in factories. The use of machines reduced the number of farmers needed to produce food, and those fewer farmers could support a larger population. Thus, *by bringing*

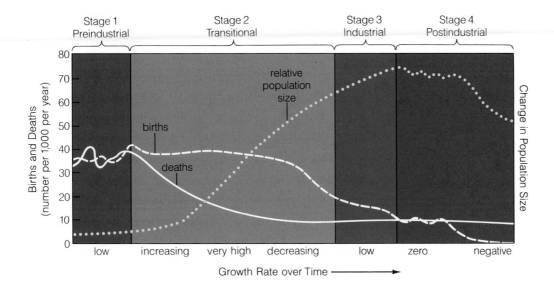

**Figure 44.10** The demographic transition model of changes in population size as correlated with changes in economic development.

*many disease agents under control and by tapping into concentrated, existing stores of energy, humans removed certain factors that had previously checked their population growth.*

## Present and Future Growth

What are the consequences of our farflung travels and advances in agriculture, industrialization, and health care? It took *2 million years* for the human population to reach the first billion. It took only 130 years to reach the second billion, 30 years to reach the third, 15 years to reach the fourth, *and only 12 years to reach the fifth!*

Figure 44.9 shows the annual growth rate for different parts of the world in 1990. If, as projected, the world average growth rate dips to 1.7 percent, we can expect the population to soar past 6 billion within the next decade. It may prove very difficult to achieve similar increases in food production, drinkable water, energy reserves, and all the wood, steel, and other materials we use to meet everyone's basic needs—something we are not even doing now. There is evidence that harmful by-products of our activities—pollutants—are changing the land, seas, and the atmosphere in ominous ways (Chapter 48). From what we know of the principles governing population growth, and unless there are spectacular technological breakthroughs, it is realistic to expect an increase in human death rates. *Although our stupendously accelerated growth continues, it cannot be sustained indefinitely.*

## Controlling Population Growth

Today, there is widespread awareness of the links between population growth, resource depletion, and increased pollution. Many governments attempt to control their population size by restricting immigration from other countries. Only the United States, Canada, Aus-

tralia, and a few others allow large annual increases. Some attempt to reduce population pressures by encouraging emigration to other countries. But most efforts focus on decreasing the birth rate so that death rates will not rise.

Two general approaches to decreasing birth rates are through economic development and family planning. The first involves providing more economic security and educational programs. Such programs may alleviate pressure on individuals to have large numbers of children to help them survive. Family planning involves educating individuals in ways to choose when and how many children they will have.

**Control Through Economics.** In the **demographic transition model**, changes in population growth are linked with changes that unfold during four stages of economic development. This model is shown in Figure 44.10.

In the *preindustrial stage*, living conditions are harsh and birth rates are high, but so are death rates. Thus, there is little population growth.

In the *transitional stage*, industrialization begins, food production rises, and health care improves. Death rates drop, but birth rates remain high, so the population grows rapidly. Growth continues at high rates (2.5 to 3 percent, on the average) over a long period. Then growth starts to level off as living conditions improve.

Population growth slows in the *industrial stage*, when industrialization is in full swing. The slowdown occurs mostly because urban couples regulate family size. Many decide that raising children is expensive and having too many puts them at an economic disadvantage.

In the *postindustrial stage*, zero population growth is reached. Then the birth rate falls below the death rate, and the population slowly decreases in size.

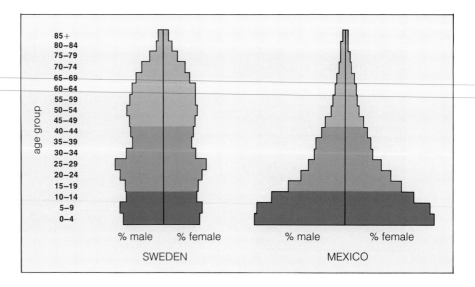

**Figure 44.11** Age structure diagrams for two countries in 1977. Dark green indicates the pre-reproductive base. Purple indicates reproductive years; light blue, the post-reproductive years. The portion of the population to the left of the vertical axis in each diagram represents males; the portion to the right represents females. Mexico has a very rapid rate of increase. In 1980, Sweden showed zero population growth.

Today, the United States, Canada, Australia, Japan, the Soviet Union, New Zealand, and most countries of Western Europe are in the industrial stage, and their growth rate is slowly decreasing. Eighteen countries, including Sweden, the United Kingdom, Germany, and Hungary, are close to, at, or slightly below zero population growth.

Mexico and other less-developed countries are in the transition stage. At current growth rates, Mexico's population may increase from 86 million to 120 million between 1990 and 2020. Like many countries in this stage, its large population does not have enough skilled workers to compete effectively in today's technological markets. Fossil fuels and other resources that drive industrialization are being used up there as well as in the industrialized countries—and fossil fuels are not renewable. Fuel costs may become too high for countries at the bottom of the economic ladder before they can enter the industrial stage. If population growth keeps outpacing economic growth, the death rate may increase. Thus many countries may now be stuck in the transitional stage. They may not stay there. In many cases they may return to the harsh conditions of the preceding stage.

**Control Through Family Planning**. Family planning programs that are thoughtfully developed and carefully administered may bring about a faster decline in birth rates, at less cost, than economic development alone. Such programs vary from country to country, but all provide information on methods of birth control, as described on page 778.

Suppose family planning programs were successful beyond our wildest imagination, so that each couple decided to have only two children to replace themselves. (Actually, the average "replacement rate" for zero population growth is slightly higher, for some female children

die before reaching reproductive age. It is about 2.5 children per woman in less-developed countries, and 2.1 in more-developed countries.) Even if the replacement rate for zero population growth were achieved globally, the human population would keep on growing for at least another sixty years! Why? An immense number of already existing children will themselves be reproducing.

Take a look at Figures 44.11 and 44.12, which show age structure diagrams for three populations growing at different rates. In these diagrams, ages 15 to 44 are used as the average range of childbearing years. The one for Mexico, a rapidly growing population, has a broad base. The base includes more than men and women of reproductive age. It also includes an even larger number of children who will move into that category during the next fifteen years. As Figure 44.13 indicates, *more than a third of the world population now falls in the broad reproductive base.* This gives us an idea of the magnitude of the effort it will take to control population growth.

One way to slow down the birth rate is to encourage delayed reproduction. For example, childbearing in the early thirties might be encouraged, rather than in the mid-teens or early twenties. This practice slows population growth by lengthening the generation time and by lowering the average number of children in each family.

In China, for example, the government has established the most extensive family planning program in the world. Couples are strongly urged to postpone the age at which they marry. Married couples have ready access to free contraceptives, abortion, and sterilization. Paramedics and mobile units ensure such access even in remote rural areas. Couples who pledge not to have more than one child are given extra food, better housing, free medical care, and salary bonuses; their child will be granted free tuition and preferential treatment when he or she enters the job market. Those who break the pledge forego all the benefits.

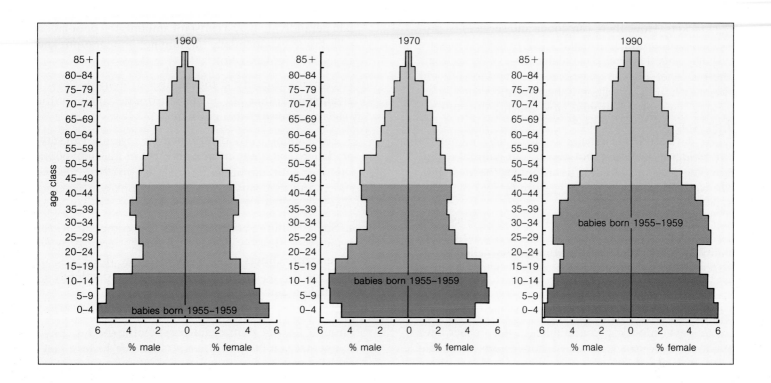

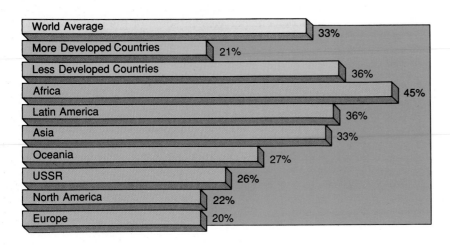

**Figure 44.12** *(Above)* Age structure of the U.S. population in 1960, 1970, and 1990. The population bulge of babies who were born between 1955 and 1959 has been moving up.

**Figure 44.13** Percentage of individuals under age fifteen in various regions in 1990.

These may seem like outrageously drastic measures unless you know that family planning has been China's alternative to mass starvation. Between 1958 and 1962 alone, an estimated 30 million Chinese died because of famine.

By 1987, the fertility rate in China had dropped from a previous high of 5.7 to 2.4 children per woman. Even so, the population time bomb has not stopped ticking. China's population now numbers 1.1 billion—and 340 million of its young women are about to move into the reproductive age category. By the year 2020, China is projected to have 1.5 billion people. Said another way, over the next thirty years, the population in China will have added to it *twice* as many individuals as are now living in the United States.

## Questions About Zero Population Growth

For the human population, as for all others, the biological implications of extremely rapid growth are staggering. Yet so are the social implications of achieving and maintaining zero population growth or of a population decline.

For instance, as you have seen, most members of an actively growing population fall in younger age brackets. Under conditions of constant growth, the age distribution means that there is a large work force. A large work force is capable of supporting older, nonproductive individuals with various programs, such as social security, low-cost housing, and health care. With zero population growth, far more people will fall in the older age brackets. How, then, can goods and services be provided for nonproductive members if productive ones are asked to carry a greater and greater share of the burden? These are not abstract questions. Put them to yourself. How much are you willing to bear for the sake of your parents, your grandparents? How much will your children be able to bear for you?

We have arrived at a major turning point, not only in our biological evolution but also in our cultural evolution. The decisions awaiting us are among the most difficult we will ever have to make, yet it is clear that they must be made, and soon.

All species face limits to growth. In one sense, we may think we are different from the rest, for our unique ability to undergo cultural evolution has allowed us to postpone the action of most of the factors limiting growth. But the key word here is *postpone*. No amount of cultural intervention can hold back the ultimate check of limited resources.

We have side-stepped a number of the smaller laws of nature. In doing so, we have become more vulnerable to those laws which cannot be repealed. Today there may be only two options available. Either we make a global effort to limit population growth in accordance with environmental carrying capacity, or we wait until the environment does it for us.

## SUMMARY

1. The growth rate of a population depends on the birth rate, death rate, and the rates of immigration and emigration. If the birth rate per individual exceeds the death rate per individual by a constant amount, the population will grow exponentially (assuming immigration and emigration remain zero).

2. Carrying capacity is the number of individuals of a given species that can be sustained indefinitely by the available resources in a given area.

3. Population size is determined by carrying capacity, predation, competition, and other factors that limit population growth. Limiting factors vary in their relative effects and they vary over time, so population size also can change over time.

4. Some limiting factors, such as competition for resources, disease, and predation, are density-dependent. Others tend to increase the death rate or decrease the death rate more or less independently of population density.

5. Patterns of reproduction, death, and migrations vary through the life span characteristic of a species. Environmental variables help shape their life history (age-specific) patterns.

6. Currently, human population growth varies from zero in some of the more-developed countries to more than 4 percent per year in some of the less-developed countries. In 1990 the world average growth rate was 1.8 percent per year.

7. Rapid growth of the human population during the past two centuries was possible largely because of our capacity to expand into new environments, and because of agricultural and technological developments that increased the carrying capacity.

### Review Questions

1. Why do populations that are not restricted in some way tend to grow exponentially? *790*

2. If the birth rate equals the death rate, what happens to the growth rate of a population? If the birth rate remains slightly higher than the death rate, what happens? *790*

3. What defines the carrying capacity for a particular environment? Can you diagram what happens when a low-density population shows a logistic growth pattern? *792–793*

4. At present growth rates, how many years will elapse before another billion are added to the human population? *799*

5. How have human populations developed the means to expand steadily into new environments? How have humans increased the carrying capacity of their environments? How have they avoided some of the limiting factors on population growth? Or is the avoidance an illusion? *798–799*

6. Write a short essay about a hypothetical population that shows either one of the following age structures. Describe what might happen to younger and older age groups when members move into new categories. *800*

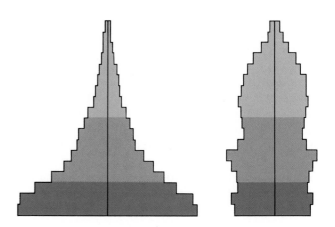

7. If a third of the world population is now below age fifteen, what effect will this age distribution have on the growth rate of the human population? What sorts of humane recommendations would you make that would encourage this age group to limit the number of children they plan to have? *800–801*

## Self-Quiz *(Answers in Appendix IV)*

1. _____ is the study of how organisms interact with one another as well as with their physical and chemical environment.

2. A _____ is a group of individuals of the same species that occupy a certain area at the same time.

3. The rate at which a population grows or declines depends upon the _____.
   a. birth rate
   b. death rate
   c. immigration rate
   d. emigration rate
   e. all of the above

4. Populations grow exponentially when _____.
   a. birth rate remains above death rate and neither changes
   b. death rate remains above birth rate
   c. immigration and emigration rates are equal (a zero value)
   d. emigration rates exceed immigration rates
   e. both a and c combined are correct

5. The number of individuals of a species that can be sustained indefinitely by the resources in a given region is the _____.
   a. biotic potential
   b. carrying capacity
   c. environmental resistance
   d. density control

6. Which of the following factors does *not* affect sustainable population size?
   a. predation
   b. competition
   c. available resources
   d. pollution
   e. each of the above can affect population size

7. Population growth controls such as resource competition, disease, and predation are said to be _____.
   a. density independent
   b. population sustaining
   c. population dynamics
   d. density dependent

8. At present, human population growth varies from about _____ percent in developed countries to more than _____ percent in some developing countries.
   a. four; zero
   b. zero; zero
   c. zero; four
   d. four; four

9. During the past two centuries, rapid growth of the human population has occurred largely because of _____.
   a. worldwide increased birth rate
   b. worldwide increased death rate
   c. carrying capacity reduction
   d. carrying capacity expansion

10. Match the population ecology terms.
    _____ carrying capacity
    _____ exponential growth
    _____ population growth rate
    _____ density-dependent controls
    _____ population

    a. examples are disease and predation
    b. group of individuals of the same species occupying a given area at the same time
    c. depends on birth, death, immigration, emigration rates
    d. number of individuals of a given species that can be sustained indefinitely by the resources in a given area
    e. increases in population size by ever larger amounts per unit of time

## Selected Key Terms

abiotic *788*
age structure *788*
biosphere *788*
biotic *788*
biotic potential *790*
carrying capacity *792*
cohort *795*
community *788*
demographic transition model *799*
density-dependent control *793*
density-independent control *793*
ecology *788*
ecosystem *788*
exponential growth *790*
habitat *788*
J-shaped curve *790*
life history pattern *794*
life table *794*
logistic growth *792*
population *788*
reproductive base *788*
S-shaped curve *793*
survivorship curve *795*
zero population growth *790*

## Readings

Begon, M., J. Harper, and C. Townsend. 1986. *Ecology.* Sunderland, Massachusetts: Sinauer.

Krebs, C. 1985. *Ecology.* Third edition. New York: Harper & Row.

Miller, G. T. 1990. *Living in the Environment.* Sixth edition. Belmont, California: Wadsworth. This author consistently pulls together information on human population growth into a coherent picture.

Polgar, S. 1972. "Population History and Population Policies from an Anthropological Perspective." *Current Anthropology* 13(2)203–241. Analyzes often-ignored cultural barriers to programs for population control.

Ricklefs, R. 1990. *Ecology.* Third edition. New York: Freeman. Chapters 15–19.

# 45

## COMMUNITY INTERACTIONS

### No Pigeon Is an Island

Flying through the rain forests of New Guinea is an extraordinary pigeon, with cobalt blue feathers and plumes on its head. It is about as big as a turkey, and it flaps so slowly and noisily that its flight has been likened to the sound of an idling truck. Like eight species of smaller pigeons living in the same forest, it perches on branches to eat fruit.

Why are there *nine* species of large and small pigeons in the same forest, all of which eat fruit? Wouldn't competition for food eventually leave one species the winner?

In fact, each species has its own role in the forest, as defined by its relations with other organisms and with its physical surroundings. The larger pigeons can perch on heavier branches when they feed, and they eat larger fruit. The smaller ones eat fruit hanging from branches too thin to support the weight of a turkey-size pigeon, and they have bills too small to open large fruit. The species of trees vary in terms of fruit size and thickness of fruit-bearing branches, so the different kinds of pigeons end up foraging on different trees. Parts of the food supply that are used less by one kind are used more by others in the same forest.

Trees in the forest do not give the pigeons something for nothing. Their pigeon-enticing fruits contain seeds. The seeds are adapted to passing through the pigeon gut unharmed—they have tough coats. And while the seeds are inside the pigeon gut, the pigeons are flying about. When the seeds are expelled from the pigeon body, chances are they will have been dispersed to new locations in the forest. When they sprout, then, the new seedlings may not be forced to compete directly with their parents for sunlight, water, and nutrients. In such competition the parent trees, with their extensive, well-developed roots and leafy crowns, might well win.

Leaf-eating, fruit-munching, and bud-nipping insects also have their own roles in the forest, as do nectar-drinking bats, birds, and insects that pollinate the trees. So do many decomposers living on the forest floor. The decomposers extract energy from the remains and wastes of other organisms and, by their metabolic activity, they recycle nutrients to the trees.

Like humans, then, no pigeon is an island, isolated from and unperturbed by the rest of the living world.

**Figure 45.1** Tropical rain forest of New Guinea, habitat of many diverse species, including the turkey-sized Victoria crowned pigeon and eight species of smaller pigeons. Within this habitat, each species has its own niche, as defined by the full range of its relations with other organisms and with its physical surroundings.

Given their differences in body size and bill size, the nine species of New Guinea pigeons eat fruit of different sizes, so they disperse seeds of different sorts. The differences in dispersal affect where different trees grow and decomposers flourish—and the tree distribution and decomposition activities affect how the entire community is organized. *Directly or indirectly, populations of each kind of organism are affected by interactions with populations of other kinds that are its neighbors.* With this chapter we turn to community interactions that influence the characteristics of populations over time and in the space of their environment.

KEY CONCEPTS

**1.** Communities are associations of different populations that occupy the same habitat. We characterize a community by the kinds and diversity of species, as well as by the numbers and dispersion of their individuals through the habitat.

**2.** Rainfall, temperature, and other physical aspects of the habitat influence the properties of each community. Resource availability, adaptations that enable individuals to exploit resources, and species interactions also influence those properties.

**3.** Each species has its own niche, defined by the full range of its relations with other organisms and with its physical surroundings. Coexistence in the same habitat is often an uneasy balance. That balance is maintained partly through a partitioning of resources among species with similar niches. And it is maintained partly by predation, parasitism, disease, and physical disturbances.

**4.** The most stable state for a community in a given habitat emerges through succession, which is the change in structure and composition leading toward the climax community. Most communities, however, are characterized by a mosaic of successional stages.

## CHARACTERISTICS OF COMMUNITIES

### Community Properties

A **community** is an association of populations, tied together directly or indirectly by way of predation, competition for resources, and other interactions. Each population in the community is adapted to living under the physical and biological conditions prevailing in a given habitat at a given point in time.

The **habitat** of an organism is the type of place where it normally lives. It is characterized by physical and chemical features, as well as by the presence of certain other species. Damselfish live in a coral reef habitat, muskrats live in a streambank habitat, moles in an underground habitat. Humans live in disturbed habi-

**Figure 45.3** Interference competition between male broadtailed hummingbirds (*above*) and rufous hummingbirds (*below*) in the Rocky Mountains. From the time spring wildflowers appear until they finish blooming in late summer, interference competition *within* species is evident. The male broadtails defend feeding territories around wildflower patches during this period. A male broadtail's aggressive aerial defense of his territory depends on the hot pursuit of intruders—both other males as well as females of his own species.

In August, however, interference competition *between* species is evident. Then, male rufous hummingbirds migrate from their breeding grounds in the Pacific Northwest to their wintering grounds in Mexico. Their migratory route takes them through the high Rockies. The migrating rufous males are even more effective at territorial defense behavior. They evict resident broadtails from their territories all along their migratory route!

to a resource by other individuals—or they exploit the resource more efficiently.

In *interference competition*, one species may limit another's access to some resource regardless of whether the resource is abundant or scarce. By analogy, even if you are sharing a ten-gallon milkshake with a friend, you could still pinch your friend's straw.

Nature provides spectacular examples of interference competition between species. Corals kill neighboring corals of other species by poisoning and growing over them. Hummingbirds chase hummingbirds of other species, even bees, away from defended clumps of flowers (Figure 45.3). Some limpets slowly "plow" competing species out of defended territories along the seashore. (Their shape, shown in Figure 10.1, lends itself to the plowing.) A strangler fig tree surrounds a victim tree and eventually kills it while growing its own massive canopy of leaves.

In pure *exploitation competition*, both species have equal access to a required resource but they differ in how fast or how efficiently they exploit it. Thus one hampers the growth, survival, or reproduction of the other indirectly, by reducing the common supply of resources. Such competition does not occur when shared resources are abundant. If you and a friend share a milkshake, each drinking from your own straw, you wouldn't care how fast your friend drank if it were a ten-gallon milkshake, replaced daily.

Said another way, even when two species exploit the same resource, they may continue to coexist provided that the resource is abundant.

## Competitive Exclusion

Any two species usually differ to some extent in their adaptations for securing food or avoiding enemies, so one usually has the competitive edge. The two are less likely to coexist in the same habitat when they are very similar in their use of scarce resources.

G. F. Gause demonstrated this idea by growing two species of *Paramecium* separately, then together (Figure 45.4). Because the two species exploited the same food (bacterial cells), there was strong competition between them. The test results suggested that complete competitors cannot coexist indefinitely, a concept now called **competitive exclusion.**

In other experiments, Gause used two other *Paramecium* species that did not overlap as much in their use of resources. When grown together, one species tended to feed on bacteria suspended in the liquid in the culture tube. The other species tended to feed on yeast cells at the bottom of the tube. The growth rate decreased for both populations—but not enough for either population to exclude the other. They continued to coexist and compete, with neither species excluding the other.

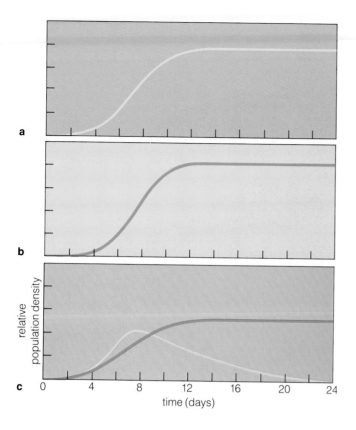

**Figure 45.4** Competition between two species of *Paramecium*. When grown separately, *P. caudatum* (**a**) and *P. aurelia* (**b**) established stable populations. (**c**) When grown together, *P. aurelia* (red curve) drove the other species (gold curve) toward extinction.

**Figure 45.5** Two species of salamanders that coexist in places where their habitats overlap: (**a**) *Plethodon glutinosus* complex and (**b**) *P. jordani*.

**The more two species differ in their use of scarce resources in the same habitat, the more likely they are to coexist.**

Salamanders provide evidence of competitive coexistence in natural habitats. In one field experiment, N. Hairston studied two species of salamanders in the Great Smoky Mountains and in the Balsam Mountains (Figure 45.5). *Plethodon glutinosus* generally lives at lower elevations than its relative, *P. jordani*, but at some sites the ranges overlap. Hairston removed one or the other species from different test plots in the overlap areas. He also left some plots untouched to serve as controls.

After five years, nothing had changed in the control plots; the two species continued to coexist. However, in plots that had been cleared of one salamander species (*P. jordani*), the other species (*P. glutinosus*) increased in numbers. In plots that had been cleared of *P. glutinosus*, there were proportionally more young *P. jordani* salamanders—evidence of a growing population. Taken together, the results suggest that where the two species coexist in nature, they suppress each other's population growth rate through competition for resources.

In another field experiment in Britain, A. Tansley showed that exploitation competition influenced the distribution of two species of bedstraw plants. One species normally grows in acidic soils and the other, in basic soils. When one species was transplanted to the opposite type of soil, it could grow—as long as the other species was not there with it! When the two were grown together on either an acidic or a basic soil, the surviving species was the one that normally grew there. In this case, each plant species excludes the other from that part of the habitat in which it grows best.

In short, the niches of two or more species living in the same habitat may overlap in several respects. But the mere existence of overlap is not a sure indicator of competition among them. Birds overlap in their need for oxygen, but they don't compete for it. Hairston's salamanders overlap in their use of the habitat, and although they *do* compete, they are able to coexist. Tansley's bedstraws also compete, but as a result of competition, they show *no* overlap in their use of the habitat.

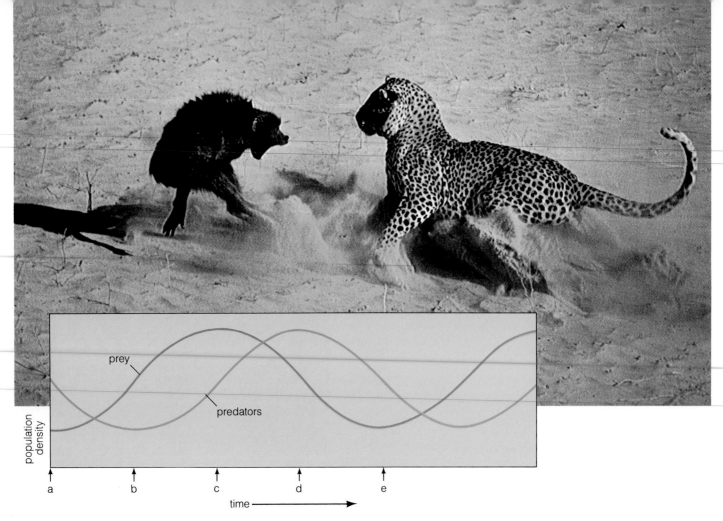

**Figure 45.6** Idealized cycling of predator and prey abundances. (The scale exaggerates predator density; predators usually are less common than their prey at all points in the cycle.) The pattern arises through time lags in predator responses to changes in prey abundance. Starting at time *a*, prey population density is low, so predators are hungry and their population is declining. In response to the decline, prey start increasing, but the predator population does not start increasing until reproduction gets under way (time *b*). Both populations grow until predation causes the prey population to decline (time *c*). Predators continue to increase and take out more prey animals, but the lower prey density leads to starvation among predators and their growth rate slows (time *d*). At time *e*, a new cycle begins.

## CONSUMER-VICTIM INTERACTIONS

### "Predator" Versus "Parasite"

Of all community interactions, predation is the most riveting of our attention—as well as the prey's. Take a look at the leopard in Figure 45.6 as it closes in for the kill. A goat pulling up a thistle plant for breakfast, although less dramatic, is also a predator. Its prey is a living organism, killed for food. And what about a horse grazing on but not killing plants? What about a mosquito taking blood from your arm before it flies off? What about ticks or fleas taking blood for long periods before they get off one host and lay their eggs else-

where? What about tapeworms, mistletoe, and other parasites that remain with their host?

For simplicity's sake, we will use only two broad definitions for all interactions between consumers and their victims. A **predator** gets food from other living organisms (its *prey*), which it may or may not kill, but it does not live on or in them. A **parasite** also gets food from other living organisms (its *hosts*), which it may or may not kill, but they do live on or in the host organism for a good part of their life.

### Dynamics of Predator-Prey Interactions

Predator and prey populations interact in diverse ways. Some interactions lead to stable coexistence at relatively steady population levels for both species. Others cause recurring cycles of abundance and population crashes, erratic population cycles, or even prey extinction. Three factors influence the outcome of the interaction:

1. Carrying capacity for the prey population, in the absence of predation.

2. Reproductive rates of the predator and prey.

3. Behavioral capacity of individual predators to respond to increases in prey density (by eating more or by moving to areas where prey are more abundant).

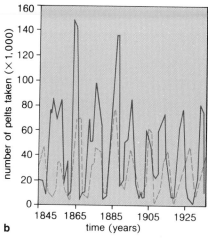

**b**

Recall that *carrying capacity* means the number of individuals of a species that can be sustained indefinitely by the available resources in a given environment. When predators keep the prey population from overshooting its carrying capacity, the two populations tend to coexist in a stable relationship. Predators can do this by reproducing promptly and by eating more when there are more prey organisms around to eat. Population densities tend to fluctuate when predators do not reproduce as fast as their prey, when they eat only so many prey organisms at a time no matter how many are around, and when the carrying capacity for the prey is high. Sometimes the fluctuations are extreme and irregular.

**Stable coexistence occurs when a predator population can keep a prey population from overshooting its carrying capacity.**

**Cyclic or irregular fluctuations in population density are likely when there are time lags in the predator's response to changes in prey abundance.**

The graph lines in Figure 45.6 represent a cyclic fluctuation of predator and prey abundances. In this idealized case, time lags in the predator's response to changes in prey abundance lead to the cycling. In nature, correspondence between the rise and fall of predator and prey populations is frequent. Other factors besides predation may also contribute to such cyclic changes.

**Figure 45.7** Predator-prey interactions between the Canadian lynx and snowshoe hare (**a**). The abundances of both populations, shown in (**b**), are based on counts of pelts that trappers sold to Hudson's Bay Company over a ninety-year period. The dashed line represents the abundance of lynx and the solid line, the abundance of hares.

This figure is a good test of how willing you are to accept conclusions without questioning their scientific basis. (Remember the discussion of scientific methods in Chapter 1?) What other factors could have influenced the relative abundances of lynx and hare? Did weather vary greatly, with more rigorous winters imposing greater demand for food (required to keep warm) and higher death rates? Did competition between lynx and other predators (owls, goshawks, coyotes, foxes) complicate the lynx cycle? Did predators turn to alternate prey species during low points of the hare cycle? Did trapping increase with rising fur prices in Europe, and did they decrease as pelt supply outstripped the demand?

For example, long-term studies of the snowshoe hare in Canada provide evidence that cyclic changes in hare population density occur every nine to ten years. The cycle tends to be synchronized across much of Canada and Alaska. Records of pelts taken by trappers and sold to the Hudson's Bay Company show that Canadian lynx populations rise and fall with about the same periodicity as the hares (Figure 45.7). Careful studies have revealed, however, that lynx are not the only predators involved in the cycle. Great horned owls, goshawks, coyotes, foxes, and other predators also feast on the hares when they are near their peak abundance. Predator populations recover only when the hares begin to increase in density once again.

**Figure 45.8** Mimicry. Many animals—especially those bite-sized morsels, the insects—avoid being eaten by having a bad taste, obnoxious secretion, or painful bite or sting. Among predators, knowledge of these traits is usually not inherited. Each young predator learns about them the hard way, by often unpleasant trials.

Among many prey species, dangerous or unpalatable individuals are easily recognized and remembered. If this were not the case, many individuals would be lost as inexperienced predators learned their lessons. Thus repugnant species tend to have distinctive, memorable appearances—bright colors (such as red, which predatory birds see so well), and bold markings (such as stripes, bands, and spots). These species make no effort to conceal themselves. Sometimes they even deliberately flash colors with an uplift of the body or the wings. Their coloration and patterning are called "aposematic" (*apo-*, meaning "away," and *sematic*, meaning "signal").

Each of the hundreds of dangerous or unpalatable species does not have a distinct warning signal. Too many signals probably would tax the learning capacity of predators. Instead there are whole groups of related species having nearly identical appearances, so the many benefit from a single taste trial. In turn, many less related

**a** A dangerous species (above) that serves as a model . . .
. . . and three of its mimics (below):

**b**    **c**    **d**

In field experiments, researchers provided extra food for the snowshoe hares when they were at peak densities, to see if food scarcity played a role in their subsequent decline. The hare populations declined anyway, suggesting that predators are the primary cause of the decline. To test this hypothesis, further experiments that keep predators out of experimental areas are under way. The food supply for hares probably does play an indirect role in population density. As food becomes scarce, hares are forced to take more risks to reach the remaining edible plants—exposing themselves to increasing risk of predation.

**Prey Defenses**

Predators and prey exert continual selection pressure on each other. When some new, heritable means of defense arises in a prey population, predators not equipped to counter the defense won't eat. *When the prey evolves, the predator also evolves to some extent because the change affects selection pressures operating between the two.* This is an example of **coevolution**. The word refers to the joint evolution of two (or more) species that are interacting in close ecological fashion. Let's take a look at some of the outcomes for consumers and their victims.

or totally unrelated species avoid predation by mimicking the appearance and behavior of the repugnant or dangerous "model."

Some mimics are as dangerous or unpalatable as their models; they are *Müllerian mimics*. Other mimics may be harmless or edible yet are still avoided; they are *Batesian mimics*. The stinging wasp in (**b**) is probably a Müllerian mimic for the yellowjacket in (**a**), *Vespula arenari*. Nonstinging insects such as beetles (**c**) and flies (**d**) may be Batesian mimics. (**e**) Certain butterflies of the New World tropics are frequent models for mimicry. Only a specialist can distinguish among the many lookalike species (such as the mimic *Dismorphia* shown in **f**).

There are other types of mimicry. In *aggressive mimicry*, parasites or predators resemble their hosts or prey. In *speed mimicry*, sluggish, easy-to-catch prey species resemble fast-running or fast-flying species that predators have given up trying to catch. This might well be termed "frustration" mimicry. For example, flesh flies have gray and black bodies, red eyes, and red tail ends (**g**). Birds soon give up trying to catch these fast-flying insects. In the American tropics, many sluggish insects, such as the weevil *Zygops rufitorquis* (**h**), closely resemble flesh flies and so reduce the chance of being eaten. (From Edward S. Ross, California Academy of Sciences)

A speedy model . . .

. . . and a slow moving mimic

**g**

**h**

**e** An unpalatable model . . .

**f** . . . and a palatable mimic

**Warning Coloration and Mimicry.** Predation pressure has favored the evolution of less-than-desirable prey species. Many such species are bad-tasting, toxic, or able to inflict pain (as by stingers) on their attackers. The conspicuous colors and bold patterns of many toxic prey species serve as warning signals to predators. Inexperienced predators might attack a prominently striped skunk, bright-orange monarch butterfly, or yellow-banded wasp—once. They quickly learn to associate the colors and patterning with pain or digestive upsets.

Weaponless prey species often have warning colors very similar to those of bad-tasting, toxic, or dangerous ones. The resemblance of an edible species to a relatively inedible one is called **mimicry**. Mimicry is a splendid demonstration of adaptive evolution. Some examples are shown in Figure 45.8.

**Moment-of-Truth Defenses.** When cornered, some prey animals defend themselves by startling or intimidating the predator with display behavior or chemical defense. Their behavior may create a moment of confusion, and a moment may be all it takes for a getaway. A bombardier beetle under attack sprays a noxious chemical. The adaptation works against many would-be pred-

a    b    c

**Figure 45.9** Moment-of-truth defensive behavior. (**a**) A cornered short-eared owl spreads its wings in a startling display that must have worked against some of its predators some of the time; it is part of the behavioral repertoire of the species. (**b**) As a last resort, some beetles spray noxious chemicals at their attackers, which works some of the time but not all of the time. (**c**) Grasshopper mice plunge the chemical-spraying tail end into the ground and feast on the head end.

**Figure 45.10** Camouflage. (**a**) Find the scorpion fish, a dangerous predator that lies motionless and camouflaged from prey. (**b**) Find the plants (*Lithops*) that hide in the open from herbivores; they have the form, pattern, and coloring of stones. (**c**) A yellow crab spider lurked motionless against a yellow background and so escaped detection by this prey tidbit. (**d**) A caterpillar looks like an unappetizing bird dropping by virtue of its coloration and body positioning. (**e**) What bird??? With the approach of a predator, the least bittern stretches its neck (colored much like the surrounding withered reeds), thrusts its beak upward, and sways gently like reeds in the wind.

a    c

b    d

ators but not against grasshopper mice, which have learned to pick up the beetle, shove its tail end into the earth, and munch the head end (Figure 45.9).

Chemicals produced by many plant and animal species serve as warning odors, repellants, and outright poisons. Earwigs, stink beetles, and skunks produce awful odors. Tannins in the foliage and seeds of certain plants taste bitter and make the plant tissues hard to digest. Nibble on a buttercup (*Ranunculus*) and you will badly irritate the lining of your mouth.

**Camouflage.** Predation pressure also has favored the evolution of prey that can **camouflage** themselves—that is, hide in the open. Adaptations in form, patterning, color, or behavior help the organism blend with its surroundings and escape detection. One desert plant looks like a small rock (Figure 45.10). It flowers only during a brief rainy season, when other plants and water are available for plant eaters. Camouflage, of course, is not the exclusive domain of prey. Stealthy predators also blend well with their backgrounds. Think about polar

e

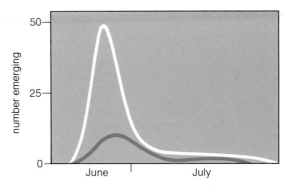

weakened by the attack that it succumbs to secondary infections. Generally, however, the attack causes death only when the parasite infects a new kind of host, one with no coevolved defenses against it. A host able to live longer may spread far more parasites about than a vulnerable, rapidly dying host ever could do. So parasite populations also tend to coevolve with their hosts, producing less-than-fatal effects.

**Parasitoids**. Unlike true parasites, **parasitoids** are insect larvae that always kill what they eat. While parasitoids are growing up, they completely consume the soft tissues of their hosts. This sounds horrendous, but fortunately their hosts are not humans but the larvae or pupae of other insect species. Actually, parasitoids serve as natural controls over other insects. Many species are raised commercially and released as an alternative to chemical pesticides.

Peter Price studied the evolutionary effects of a parasitoid wasp that lays its eggs on sawfly cocoons. The sawflies lay *their* eggs in trees. In time, the fly larvae drop to the forest floor. They spin cocoons after they have burrowed into the leaf litter, with some burrowing deeper than others. The first adult sawflies emerge from the shallowly buried cocoons. Later in the season, more sawflies emerge from the deeply buried ones (Figure 45.11).

The parasitoid wasps tend to lay eggs on cocoons near the surface of the leaf litter. They exert strong selection pressure on the sawfly population, for the deep-burrowing ones are more likely to escape detection. The sawflies exert selection pressure on the wasps. There are only so many flies near the surface, so the wasp able to locate cocoons deeper in the litter will be competitive in securing food for her larvae.

As Price points out, the host stays ahead in this coevolutionary contest. Each time the sawfly larvae bur-

bears against snow, tigers against tall-stalked and golden grasses, and pastel spiders against pastel flower petals.

### Parasitic Interactions

**True Parasites**. The blood flukes and tapeworms described in Chapter 25 are examples of true parasites, which live on or in a host organism and gain nourishment by tapping into its tissues. Sometimes parasites indirectly cause death, as when the host becomes so

row deeper, the female wasps have to spend more time searching for them. Fewer wasp eggs are laid—so the wasp population is held in check.

**Social Parasites.** Some parasites do not obtain nutrients directly from the tissues of a host. Rather, they depend on the social behavior of another species to complete their life cycle. We call such animals **social parasites**.

The North American brown-headed cowbird is an example. It never builds a nest, never incubates its eggs, and never cares for its offspring. The cowbird removes an egg from the nest of another kind of bird and lays one as a "replacement." Some birds can't tell the difference, so they end up hatching the egg and raising a young cowbird. The large, aggressive cowbird often pushes the rightful occupants out of the nest or gets most of the food, so the others starve to death.

## COMMUNITY ORGANIZATION, DEVELOPMENT, AND DIVERSITY

Community stability is the result of forces that have come into balance—sometimes an uneasy balance. Resources are sustainable, as long as populations do not start dancing dangerously around their carrying capacity. Predators and prey coexist only as long as neither wins. Competitors have no sense of fair play. Even mutualists are really antagonists. A flower gives as little nectar as necessary to attract a pollinator, and a pollinator takes as much nectar as it can for the least effort. Let's take a look at some community patterns arising from these conflicting forces.

### Resource Partitioning

Think back on those nine species of fruit-eating pigeons in the same forest in New Guinea. In any community, similar species generally share the same resource in different ways, in different areas, or at different times. This community pattern, called **resource partitioning**, arises in two ways. *First*, the ecological differences between established and competing populations may increase through natural selection. *Second*, species that are dissimilar from established ones often are more likely to succeed in joining an existing community.

Consider how three species of annual plants partitioned resources in a plowed, abandoned field. As for other plants, the resources were sunlight, water, and dissolved mineral salts. Each species exploited different parts of the habitat. Where soil moisture varied from day to day, foxtail grasses became established; their shallow, fibrous root systems absorb rainwater rapidly

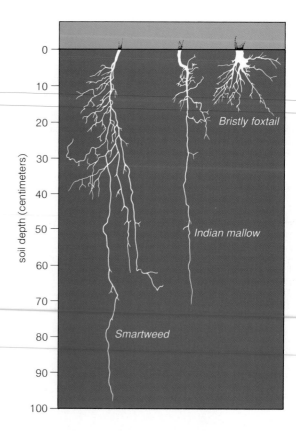

**Figure 45.12** Partitioning of a resource (soil, with its nutrients and water) by three annual plant species that became established in a field that had been plowed under the year before.

and help the plants recover from drought. Where deeper areas of soil were moist early in the growing season but drier later on, mallow plants took hold; their taproot system grows deeper in the soil. Where soil was continuously moist, smartweed prevailed; its taproot system branches in topsoil *and* in soil below the roots of the other species (Figure 45.12).

Resource partitioning may occur when ecological differences develop between similar species (as by natural selection) or when a dissimilar species joins an existing community.

### Effects of Predation and Disturbances on Competition

By reducing prey population densities, predation can reduce competition among prey species and promote their coexistence. Robert Paine kept sea stars out of experimental plots in a rocky intertidal zone for several years. He left sea stars and their prey—fifteen invertebrate species—in control plots. In the plots kept clear of

**a** Periwinkles in a tidepool

**b** *Enteromorpha*, a filamentous green alga

**Figure 45.13** Effect of grazing by periwinkles (*Littorina littorea*) on the number of algal species in tidepools. Jane Lubchenco's studies showed that the number of algal species is greatest in tidepools that have an intermediate number of algae-eating periwinkles. In tidepools having only a few periwinkles, the competitively dominant alga (*Enteromorpha*) eliminates all other algal species. In tidepools with very high densities of periwinkles, the grazing pressure eliminates all algal species except the tough, unpalatable *Chondrus*.

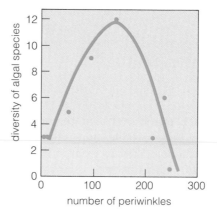

**c** *Chondrus*, another alga

predators, the number of prey species declined from fifteen to eight. The rest were crowded out by mussels, the main prey of sea stars and yet the strongest competitors in their absence.

In this experiment, predation on the dominant competitor helped maintain a high diversity of prey species by preventing competitive exclusion. In nature, storms, fire, landslides, floods, and other disturbances can have similar effect. In fact, communities subjected to moderate levels of predation or disturbance often support greater species diversity than either undisturbed or severely disturbed communities (Figure 45.13).

**Moderate levels of predation or disturbances to the habitat often encourage a greater diversity of species in a community.**

*Commentary*

## Hello Lake Victoria, Goodbye Cichlids

As the human population continues to grow exponentially, we keep looking for new ways to manage or control the populations of edible plants and animals that serve as our food production base. Although such efforts are well-intentioned, they can have disastrous consequences when ecological principles are not taken into account.

Consider what happened several years ago, when someone thought it would be a great idea to introduce the Nile perch into Lake Victoria in East Africa. People had been fishing there for thousands of years using simple, traditional methods, but now they were overfishing the lake. In time there would not be enough fish to feed the local populations and no excess catches to sell for profit. But Lake Victoria is a very big lake, the Nile perch is a very big fish (more than 2 meters long), and this seemed an ideal combination for commercial fishermen from the outside, with their big, elaborate nets—right? Wrong.

The native fishermen had been harvesting a variety of native fishes called cichlids. Cichlids eat mostly detritus and aquatic plants, but the Nile perch eats other fish—including cichlids. Although the cichlids were native to

the lake, the Nile perch was not. Having had no prior evolutionary experience with the perch, the cichlids simply had no defenses against it.

And so the Nile perch was able to eat its way through the cichlid populations and destroy the natural fishery. Dozens of cichlid species found nowhere else on earth became extinct. And the Nile perch, by wiping out its own food source, ended the basis for its own population growth and ceased to be a potentially large, exploitable food source for people living around the lake.

As if that weren't enough, the Nile perch is an oily fish. Unlike cichlids, which can be sun-dried, it required preservation by smoking—and smoking required firewood. And so the people started whacking away more rapidly at the trees in the local forests, which are not rapidly renewable resources. To add final insult to injury, the people living near the shores of Lake Victoria never liked to eat Nile perch anyway; they preferred the flavor and texture of cichlids.

What is the lesson? A little knowledge and some simple experiments in a contained setting could have prevented the whole mess at Lake Victoria.

### Species Introductions

Many species have been introduced to different geographic regions, some intentionally and others by accident. Most introductions of species from one continent to another probably fail to become established—although few records are kept of such nonevents. Nearly all of the species we know about are agricul-

turally useful or are pests. The useful ones include rice, wheat, corn, and potatoes grown far from their place of origin. The pests include water hyacinths.

In the 1880s, the water hyacinth from South America was put on display at the New Orleans Cotton Exposition. Flower fanciers from Florida and Louisiana car-

**Table 45.2  Detrimental Effects on Introduced Species in the United States**

| Species Introduced | Origin | Mode of Introduction | Outcome |
|---|---|---|---|
| Water hyacinth | South America | Intentionally introduced (1884) | Clogged waterways; shading out of other vegetation |
| Dutch elm disease: The fungus *Ophiostoma ulmi* (the disease agent) | Europe | Accidentally imported on infected elm timber used for veneers (1930) | Destruction of millions of elms; great disruption of forest ecology |
| Bark beetle (the disease carrier) | | Accidentally imported on unbarked elm timber (1909) | |
| Chestnut blight fungus | Asia | Accidentally imported on nursery plants (1900) | Destruction of nearly all eastern American chestnuts; disruption of forest ecology |
| Argentine fire ant | Argentina | In coffee shipments from Brazil? (1891) | Crop damage; destruction of native ant communities; mortality of ground-nesting birds |
| Camphor scale insect | Japan | Accidentally imported on nursery stock (1920s) | Damage to nearly 200 species of plants in Louisiana, Texas, and Alabama |
| Japanese beetle | Japan | Accidentally imported on irises or azaleas (1911) | Defoliation of more than 250 species of trees and other plants, including commercially important species such as citrus |
| Carp | Germany | Intentionally released (1887) | Displacement of native fish; uprooting of water plants with loss of waterfowl populations |
| Sea lamprey | North Atlantic Ocean | Through Erie Canal (1860s), then through Welland Canal (1921) | Destruction of lake trout and lake whitefish in Great Lakes |
| European starling | Europe | Released intentionally in New York City (1890) | Competition with native songbirds; crop damage; transmission of swine diseases; airport runway interference; noisy and messy in large flocks |
| House sparrow | England | Released intentionally (1853) | Crop damage; displacement of native songbirds; transmission of some diseases |
| European wild boar | Russia | Intentionally imported (1912); escaped captivity | Destruction of habitat by rooting; crop damage |
| Nutria (large rodent) | Argentina | Intentionally imported (1940); escaped captivity | Alteration of marsh ecology; damage to earth dams and levees; crop destruction |

After David W. Ehrenfeld, *Biological Conservation*, 1970, Holt, Rinehart and Winston and *Conserving Life on Earth*, 1972, Oxford University Press.

ried home clippings of the blue-flowered plants and set them out for ornamental display in ponds and streams. Unchecked by their natural predators, the fast-growing hyacinths spread through the nutrient-rich waters and displaced many native species. In time they choked off ponds and streams, then rivers and canals. They are still thriving—now as far west as San Francisco—and they are still bringing river traffic in many areas to a halt. The *Commentary* describes another intentional species introduction that had unforeseen and truly awful consequences.

Species successfully introduced into established communities do not always lead to such wholesale disasters, but few (if any) are without ecological consequences. On the one hand, "imported" honeybees have become part of existing communities in the United States, even though they have displaced native bees in many areas. On the other hand, remember that aggressive honeybees brought into South America from Africa were supposed to have stayed put for cross-breeding with local bees, which farmers thought weren't aggressive enough about pollinating crops. Descendants of the Africanized bees have been migrating and collectively stinging the occasional cow and person, and they have caused some deaths. Hence the name, "killer bees" (page 118). Table 45.2 lists other introduced species and their effects.

## Succession

The repercussions of newly introduced species in a community raise an interesting question. How do communities come to exist in the first place? New communities may arise in habitats initially devoid of life, such as newly forming volcanic islands, or in disturbed areas that have been previously inhabited, such as abandoned pastures. Through a process called **succession**, the first species in the habitat thrive, then are replaced by other species, which are replaced by others in reasonably orderly progression until the composition of species becomes steady under prevailing conditions. This more or less stable array of species is the **climax community**.

**Primary Succession**.   In **primary succession**, changes begin when pioneer species colonize a barren habitat. A new volcanic island is such a habitat; so is land exposed by the retreat of a glacier that had kept it buried for many thousands of years (Figure 45.14). *Pioneer species* are adapted to growing in exposed areas with intense sunlight, wide swings in air temperature, and soil that is deficient in nitrogen and other nutrients. Pioneers typically are small plants with short life cycles. Each year they produce an abundance of small seeds, which are quickly dispersed.

Once pioneers are established, they improve living conditions for other species—and commonly set the stage for their own replacement. For example, many are symbiotic with nitrogen-fixing soil microbes, and the gradual accumulation of plant litter adds nitrogen to the soil. Also, pioneers form low-growing mats that can shelter seeds of later species without shading out the seedlings. In time, later successional species crowd out the pioneers, whose seeds travel as fugitives on the wind or water—destined, perhaps, for a new but equally temporary habitat.

**Secondary Succession**.   In **secondary succession**, a community or patch of habitat progresses once again toward the climax state after being disturbed. This pattern of change occurs in ponds and shallow lakes, abandoned fields, and parts of established forests where falling trees or other disturbances have opened the canopy of leaves, letting sunlight reach the forest floor. Unlike primary succession, many plants arise from seeds or even seedlings already present when the process begins.

In secondary succession on land, both early and late species often are able to grow under prevailing conditions, but the later species simply are growing more slowly. In time they will exclude the others through competition. Also, the early successional species might inhibit the growth of later ones, which only prevail if some disturbance removes the established competitors.

a

b

c

d

**The Role of Disturbance**.   Even the most stable climax communities have successional patches brought about by major and minor disturbances. Winds, fires, insect infestations, and overgrazing all modify and shape the direction of succession by encouraging some species and eliminating others in different parts of the habitat. In fact, community persistence often requires episodes of disturbance that permit the regeneration of dominant species.

Giant sequoia trees grow in isolated groves in the Sierra Nevada in California. Some are more than 4,000

**Figure 45.14** Primary succession in the Glacier Bay region of Alaska (**a**), where changes in newly deglaciated regions have been carefully documented. A comparison of maps from 1794 onward shows that ice has been retreating at annual rates ranging from 3 meters (at the glacier's sides) to a phenomenal 600 meters at its tip over bays. (**b**) When a glacier retreats, the constant flow of meltwater tends to leach the newly exposed soil of minerals, including nitrogen. Less than ten years ago, the soil here was still buried below ice. (**c,d**) The first invaders of these nutrient-poor sites are the feathery seeds of mountain avens (*Dryas*), drifting over on the winds. Mountain avens is a pioneer species that benefits from the nitrogen-fixing activities of symbiotic microbes. It grows and spreads rapidly over glacial till.

(**e**) Within twenty years, young alders take hold. These deciduous shrubs also are symbiotic with nitrogen-fixing microbes. Young cottonwood and willows also become established (**f**). In time, alders form dense thickets (**g**). As the thickets mature, cottonwood and hemlock trees grow rapidly, as do a few evergreen spruce trees. (**h**) By eighty years, the spruce crowd out the mature alders.

(**i**) In areas deglaciated for more than a century, dense forests of Sitka spruce and western hemlock dominate. By this time, nitrogen reserves are depleted, and much of the biomass is tied up in peat: excessively moist, compressed organic matter that resists decomposition and forms a thick mat on the forest floor.

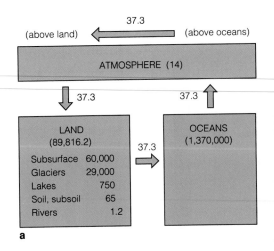

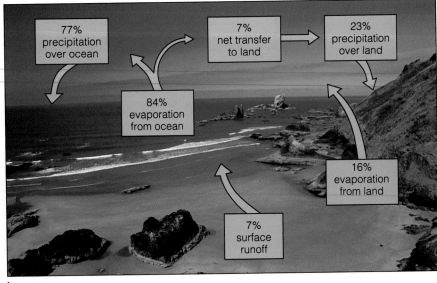

a

b

**Figure 46.10** (**a**) Simplified picture of the hydrologic cycle. Total quantities of water in different provinces and annual net rates of transfer from one province to another are shown (in thousands of cubic kilometers). The values in boxes indicate the quantities of water present at a given time. The values next to the arrows are the fluxes per year. For example, there is an annual net transfer of $37.3 \times 10^3$ cubic kilometers of water from the atmosphere to the land. This is balanced by a comparable net loss from the oceans to the atmosphere and a net gain by oceans of that amount (through runoff).

(**b**) Global water budget. The percentages indicate the annual movement of water into and out of the atmosphere. Of the water entering, 84 percent is by way of evaporation from oceans. Of that, 7 percent is carried horizontally to land, which returns it to oceans by way of rivers and streams. The remaining 77 percent (that is, $84 - 7$) leaves the atmosphere as precipitation over oceans.

**Figure 46.11** Movement of water through a watershed.

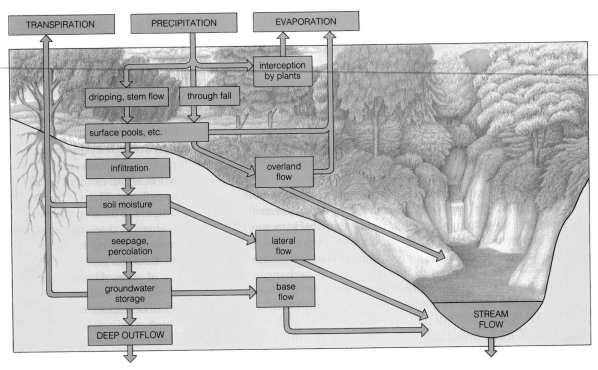

**a** View of experimental watersheds at Hubbard Brook Valley.

**Figure 46.12** Effects of different human activities on a forest ecosystem and its biogeochemistry. (**a**) These are experimental watersheds within the Hubbard Brook Valley of New Hampshire. Here, researchers have studied the effects of clear-cutting (foreground), progressive strip cutting (upper left), and deforestation (upper center). One of the gauging weirs used to collect all the water draining from an area under study is shown in (**b**). This watershed had been experimentally deforested, as shown in (**c**), then herbicides were applied to prevent regrowth during three years of studies. Two years after herbicide applications stopped, the vegetation recovered to the extent shown in (**b**).

**c**

## Hydrologic Cycle

Cold and warm ocean currents, clouds, winds, and rainfall are all part of the global hydrologic cycle. Driven by solar energy, the waters of the earth move slowly and on a vast scale through the atmosphere, on or through the uppermost layers of land masses, to the oceans, and back again (Figure 46.10). Water moves into the atmosphere by evaporation. There it remains aloft as vapor, clouds, and ice crystals. It falls back as precipitation—mostly rain or snow.

Water molecules do not stay aloft for more than ten days, on average, so the turnover rate is rapid for the airborne part of the cycle. Water released as rain or snow remains on land for an average of about 10 to 120 days, depending on the season and where it falls. Some of it evaporates, and rivers and streams carry the rest to the seas. With large-scale evaporation from the seas, the cycle begins again.

In itself, water is essential for life in any ecosystem. *But water is also an important medium by which nutrients move into and out of ecosystems.* This became clear through studies in **watersheds**, which are regions where all of the precipitation becomes funneled into a single stream or river. A watershed can be any size. The Mississippi River watershed extends across roughly one-third of the United States. Watersheds at Hubbard Brook Valley in the White Mountains of New Hampshire average about 14.6 hectares (36 acres).

Water enters a watershed mainly as rain or snow, and most filters into the soil or runs off along the surface. Some is lost by evaporation from surfaces. Plants absorb the water at their roots and lose it by transpiration from their leaves. As Figure 46.11 shows, water also seeps and percolates through soil. Some of it reaches the water table and is temporarily stored as groundwater, and some moves into a stream.

At Hubbard Brook watersheds, such as the ones shown in Figure 46.12, ecologists measured precipitation inputs and streamwater and groundwater outputs. They also estimated evaporation and transpiration outputs. Precipitation varied from year to year, as did the

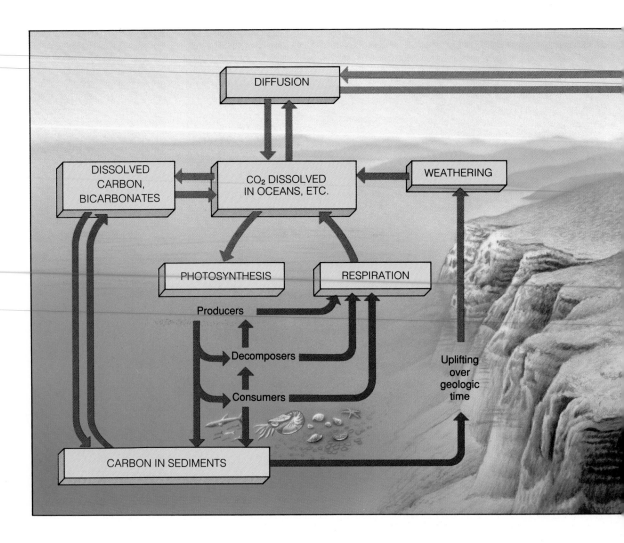

rate of stream flow—but evaporation and transpiration varied only slightly. This discovery has practical applications. It indicates how much water can be expected for use in cities that are downstream of this type of watershed.

Studies at Hubbard Brook and elsewhere also revealed that plants greatly influence how fast nutrients move through the ecosystem phase of biogeochemical cycles. For example, you might think that water draining a watershed would rapidly leach away calcium ions and other minerals. Yet in studies of watersheds with young, undisturbed forests, only about eight kilograms of calcium were lost from each hectare. Rainfall and the weathering of rocks brought calcium replacements into the watershed. Tree roots were also "mining" the soil, so that calcium was being stored in a growing biomass of tree tissue.

Nutrient outputs change when land is cleared. All the plants were removed from one watershed at Hubbard Brook, then herbicides were applied for three years to prevent regrowth. The soil itself was not disturbed; no organic material was removed. Yet the loss of calcium in the stream outflow was *six times* greater than in undisturbed watersheds. Given how slowly calcium and other nutrients move through geochemical cycles, *stripping the land of vegetation may have long-term disruptive effects on nutrient availability for the entire ecosystem.*

### Carbon Cycle

Carbon is an element that moves in an atmospheric cycle. In the **carbon cycle**, this element moves from reservoirs in the atmosphere and oceans, through organisms, then back to the reservoirs. Carbon enters the atmosphere by way of aerobic respiration, fossil fuel burning, and volcanic eruptions, which release carbon from rocks deep in the earth's crust. Carbon exists as a gas in the atmosphere—mostly as carbon dioxide ($CO_2$). About half of all the carbon entering the atmosphere each year will move into two large "holding stations"—

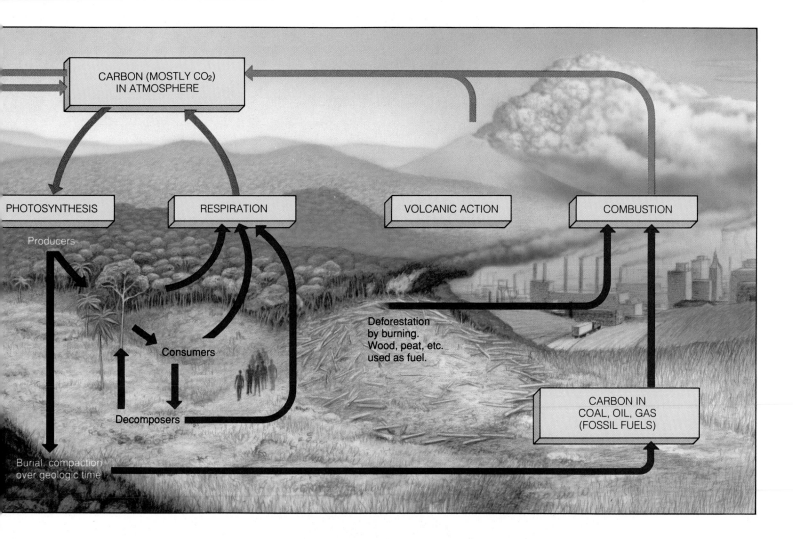

**CARBON (MOSTLY CO₂) IN ATMOSPHERE**

PHOTOSYNTHESIS

RESPIRATION

VOLCANIC ACTION

COMBUSTION

Producers

Consumers

Decomposers

Deforestation by burning. Wood, peat, etc. used as fuel.

Burial, compaction over geologic time

**CARBON IN COAL, OIL, GAS (FOSSIL FUELS)**

**Figure 46.13** Global carbon cycle. To the left, the movement of carbon through marine ecosystems; to the right, its movement through terrestrial ecosystems.

that is, into an accumulation of plant biomass and into the oceans.

Each year, photosynthesizers capture airborne or dissolved carbon dioxide and incorporate billions of metric tons of its carbon atoms into organic compounds. However, the average length of time that a carbon atom remains captured in any given ecosystem varies greatly.

In tropical forests, for example, decomposition and carbon uptake are rapid, so not much carbon is tied up in litter on the soil surface. In bogs, marshes, and other anaerobic settings, organic compounds are not broken down completely and carbon accumulates in forms such as peat (refer to Figure 45.14).

In aquatic food webs, carbon becomes incorporated into shells and other hard parts. When the shelled organisms die, they sink and become buried in bottom sediments of different depths. Carbon in the deep oceans can remain buried for millions of years until geologic movements bring it to the surface. Still more car-

bon is slowly converted to long-standing reserves of gas, petroleum, and coal deep in the earth—reserves we tap for use as fossil fuels. Figure 46.13 summarizes these and other aspects of the carbon cycle.

The worldwide burning of fossil fuels is putting more carbon into the atmosphere than can be returned to the global holding stations (oceans and plant biomass). This activity and others may be intensifying the greenhouse effect and may be triggering a global warming. The *Commentary* on the next page describes this effect and its consequences.

## Greenhouse Gases and a Global Warming Trend

The atmospheric concentrations of carbon dioxide, water, ozone, methane, nitrous oxide, and chlorofluorocarbons profoundly influence the average temperature near the earth's surface, and that temperature influences global climates. Collectively, molecules of these gases act somewhat like a pane of glass in a greenhouse (hence their name, "greenhouse gases"). They allow wavelengths of visible light to reach the earth's surface, but they impede the escape of longer, infrared wavelengths—that is, heat—from the earth into space. They absorb infrared wavelengths, much of which gets reradiated back toward the earth (Figure *a*). In short, the greenhouse gases cause heat to build up in the lower atmosphere, a warming action called the **greenhouse effect**.

If there were no greenhouse gases, the earth would be a cold and lifeless planet. But there can be too much of a good thing. Largely as a result of human activities, the levels of greenhouse gases have been increasing (Figure *b*), and they may be contributing to an alarming increase in the global warming.

What is so alarming about a warmer planet? Suppose the temperature of the lower atmosphere were to rise by only 4°C (7°F). Sea levels could rise by about 2 feet, or 0.6 meter. Why? Ocean surface temperatures would increase—and water expands when heated. Also, global warming could cause partial melting of glaciers and Antarctic ice sheets. Low coastal regions would flood.

Imagine what a long-term rise in sea level, combined with high tides and storm waves, would do to the waterfronts of Vancouver, Boston, San Diego, Galveston, and other coastal cities. Huge tracts of Florida and Louisiana would face saltwater intrusions. Agricultural lowlands and deltas in India, China, and Bangladesh—where much of the world's rice is grown—would be submerged.

Global warming could affect world agriculture in other ways. Regional patterns of precipitation and temperature would change. Crop yields would decline in currently productive regions, including parts of Canada and the United States, and increase in others. There is speculation

**a** The greenhouse effect

**1.** Sunlight penetrating the atmosphere warms the earth's surface.

**2.** The earth's surface radiates heat (infrared wavelengths) to the atmosphere, and some escapes into space. Greenhouse gases and water vapor absorb some infrared wavelengths and reradiate a portion of them toward the earth.

**3.** When greenhouse gases build up in the atmosphere, more heat is trapped near the earth's surface. Ocean surface temperatures rise, more water vapor enters the atmosphere, and the earth's surface temperature increases.

that warmer temperatures would promote insect breeding, with the increased population sizes of insect pests leading to more extensive crop losses.

In the late 1950s, a laboratory was set up on a mountaintop in the Hawaiian Islands to measure the concentrations of different greenhouse gases. The remote site was selected because it was free of local contamination and would represent average conditions for the Northern Hemisphere, and the monitoring activities are still going on. Consider what these studies tell us about carbon dioxide levels alone.

It turns out that the levels of atmospheric carbon dioxide follow the annual cycle of plant growth in the Northern Hemisphere. The levels are lower during summer, when plants are photosynthesizing most rapidly. They are higher in winter, when aerobic respiration continues even while photosynthetic activity declines. The lows and highs are represented by the peaks and troughs around the graph line in Figure *b* (part 1). *For the first time, scientists could see the integrated effects of the carbon balances of land and water ecosystems of a whole hemisphere.*

Disturbingly, the midline of the peaks and troughs in the cycle showed a continuous increase. Here was evidence that a buildup of carbon dioxide in the atmosphere may intensify the greenhouse effect over the next century.

Increasing carbon dioxide levels are attributed mostly to the burning of fossil fuels, coal especially, throughout the world. Deforestation is another contributing factor. Today, vast tracts of tropical forests are being cleared and burned at a rapid rate (see, for example, Figure 48.9). Carbon is being released during the wood burning. And more importantly, the number of plants that absorb carbon dioxide during photosynthesis is plummeting.

Atmospheric concentrations of greenhouse gases are expected to continue increasing into the middle of the twenty-first century, with global warming by several degrees occurring in the process. It is doubtful that we can sharply reduce fossil fuel burning and deforestation soon enough to prevent significant global warming. There is widespread agreement among scientists that we should begin preparing for the consequences. For example, research in genetic engineering could be intensified to develop drought-resistant and salt-resistant plants. Such plants may prove crucial in regions of saltwater intrusions and climatic change.

**b** (Right) Relative contributions of different greenhouse gases to the global warming trend, projected to the year 2020.

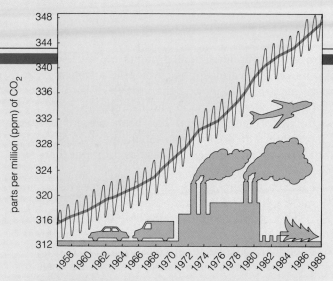

**1. Carbon Dioxide** ($CO_2$). By the year 2020, the relative contribution of the greenhouse gas $CO_2$ to the global warming trend is expected to be about 50 percent. Fossil fuel burning, factory emissions, car exhaust, and deforestation are all contributing to the increased concentration.

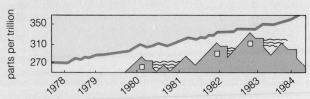

**2. Chlorofluorocarbons** (CFCs). By 2020, this gas will probably be responsible for about 25 percent of the greenhouse effect. CFCs are used in plastic foams, air conditioners, refrigerators, and industrial solvents.

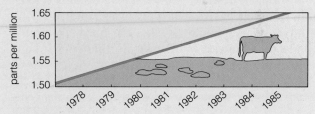

**3. Methane** ($CH_4$). By 2020, methane may be responsible for 15 percent of the greenhouse effect. Methane is produced by bacterial fermentation in anaerobic settings such as swamps, landfills, and the gut of cattle and other ruminants. It is also produced by termite activities.

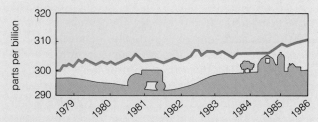

**4. Nitrous Oxide** ($N_2O$). By 2020, this gas may be responsible for about 10 percent of the greenhouse effect. It is a natural by-product of denitrifying bacteria; it is released from fertilizers and animal wastes, as in livestock feedlots.

## Nitrogen Cycle

Nitrogen, a component of all proteins and nucleic acids, also moves in an atmospheric cycle, the **nitrogen cycle**. Of all nutrients influencing the growth of land plants, nitrogen is often the one in shortest supply. Since the beginning of life, nitrogen has been present in the atmosphere and oceans but not in the earth's crust. Today, nearly all nitrogen in soils has been put there by nitrogen-fixing organisms.

The atmosphere is the largest nitrogen reservoir. About 80 percent of it is composed of gaseous nitrogen ($N_2$). Stable, triple covalent bonds hold the two nitrogen atoms together ($N \equiv N$), and few organisms can break those bonds. Some bacteria, volcanic action, and lightning can convert $N_2$ into forms that can be used in ecosystems.

Nitrogen is lost from ecosystems through metabolic activities of bacteria that "unfix" the fixed nitrogen. In the case of ecosystems on land, nitrogen is lost by leaching of soils. Thus, leaching also creates nitrogen inputs to aquatic ecosystems such as streams, lakes, and the oceans.

**The Cycling Processes**. Figure 46.14 shows the six major processes of the nitrogen cycle. These are called nitrogen fixation, assimilation and biosynthesis, decomposition, ammonification, nitrification, and denitrification.

In **nitrogen fixation**, a few kinds of bacteria convert $N_2$ to ammonia ($NH_3$), which dissolves rapidly in water to produce ammonium ($NH_4^+$). The fixed nitrogen is used in the synthesis of amino acids, then proteins and nucleic acids.

*Anabaena, Nostoc*, and other cyanobacteria are the nitrogen fixers of aquatic ecosystems. *Rhizobium* and *Azotobacter* are nitrogen fixers of many land ecosystems. *Rhizobium* is a symbiont with plants, and *Azotobacter* lives in soil. All of these bacteria are small in size but mighty in number. Collectively they fix about 200 million metric tons of nitrogen each year!

Fixed nitrogen becomes available to other organisms in the ecosystem. It moves into the tissues of many plants that have entered mutually beneficial interactions with the free-living or symbiotic nitrogen fixers (page 489). The most important of these plants are legumes such as peas and beans. Ammonium and other nitrogen-containing substances become available when the nitrogen fixers die and decompose. Such substances dissolve in soil water, from which they can be taken up by the roots of plants.

Plants are the only nitrogen source for animals, which feed directly or indirectly on them.

Later, in **ammonification**, bacteria and fungi break down the nitrogen-containing wastes and remains of plants and animals. The decomposers use the released amino acids and proteins for growth and give up the excess as ammonia or ammonium, some of which is picked up by plants.

Ammonia or ammonium in soil also gets the attention of nitrifying bacteria. In *nitrification*, these compounds are stripped of electrons, forming nitrite ($NO_2^-$). Then other nitrifying bacteria use nitrite in metabolism and produce nitrate ($NO_3^-$). Nitrification is an example of chemosynthesis (page 115).

**Nitrogen Scarcity**. The continual production of ammonia by nitrogen-fixing bacteria would seem to assure land plants of plenty of nitrogen. Yet soil nitrogen is scarce. Ammonium, nitrite, and nitrate are soluble and vulnerable to leaching. Some fixed nitrogen is lost to the air by *denitrification*. By this process, bacteria convert nitrate or nitrite to $N_2$ and a small amount of nitrous oxide ($N_2O$).

Most species of denitrifying bacteria ordinarily rely on aerobic respiration. But when soil is waterlogged and poorly aerated, they switch to anaerobic pathways in which nitrate, nitrite, or nitrous oxide is used as the final electron acceptor instead of oxygen. (This type of metabolic pathway is described in Chapter 8.) In doing so, the bacteria convert the fixed nitrogen to $N_2$.

In addition, nitrogen fixation comes at a high metabolic cost to the plants that are symbiotic with nitrogen-fixing bacteria. To gain nitrogen, the plants give up sugars and other photosynthetic products that can only be assembled with heavy investments of ATP and NADPH. Such plants are better off when soil nitrogen is scarce, but often they are displaced from nitrogen-rich soils by species that do not have to pay the metabolic price.

Nitrogen losses are great in agricultural regions. Some nitrogen departs from the fields in the harvested plant tissues. Losses also occur through soil erosion and leaching. European and North American farmers traditionally have rotated crops, as when they alternate wheat with legumes. Crop rotation has helped maintain soils in stable and productive condition, sometimes for thousands of years.

Today, intensive agriculture has come to depend instead on nitrogen-rich fertilizers. Plant varieties are bred for their ability to use these fertilizers, and crop yields per hectare have doubled and even quadrupled over the past forty years. Whether modern technologies of pest control and soil management can sustain high yields indefinitely remains uncertain, for reasons that will become apparent in Chapter 48.

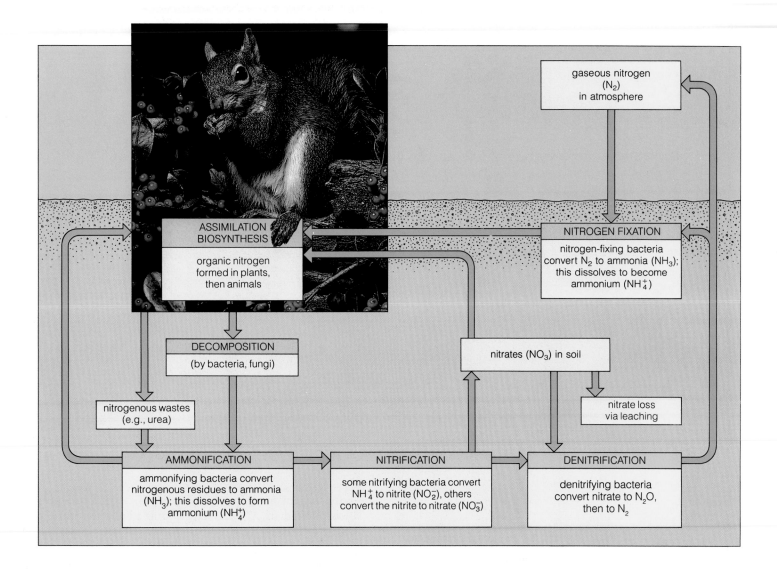

**Figure 46.14** Simplified picture of the nitrogen cycle for a terrestrial ecosystem.

Another catch is that we can't get something for nothing. Enormous amounts of energy go into fertilizer production—not energy from the unending stream of sunlight, but energy from fossil fuels. Fossil fuel supplies were once thought to be unending, so there was little concern about fertilizer costs. In fact, we still commonly pour more energy into the soil (in the form of fuels, fertilizers, and other chemicals) than we are getting out of it (in the form of food).

As any hungry person will tell you, food calories are more basic to survival than are gasoline calories or perhaps, even, than a car. As long as the human population continues to grow exponentially, farmers will be engaged in a constant race to supply food to as many individuals as possible. Soil enrichment with nitrogen-containing fertilizers is part of the race, as it is now being run.

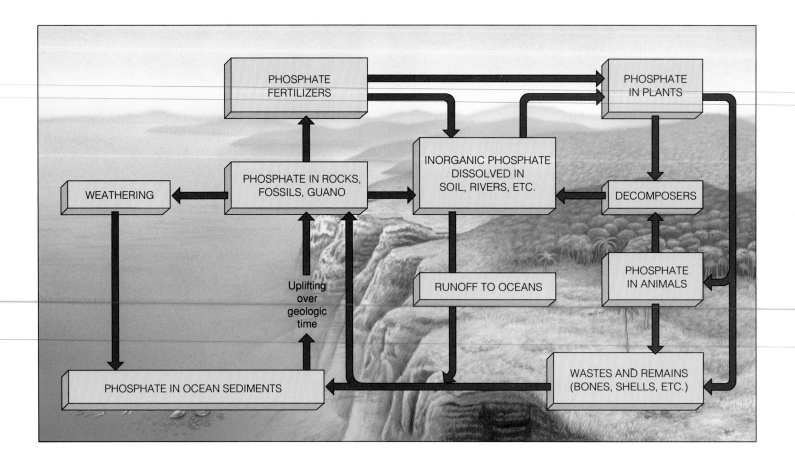

**Figure 46.15** The phosphorus cycle. This is an example of a sedimentary cycle.

## Phosphorus Cycle

Phosphorus is a mineral element that moves through a sedimentary cycle. In the **phosphorus cycle**, phosphorus moves from land to sediments in the seas and then back to the land (Figure 46.15). The earth's crust is the main storehouse for this and other minerals, such as calcium and potassium.

Phosphorus is present as phosphates in rock formations on land. Through weathering and erosion, it is washed into rivers and streams, then it moves to the oceans. There, largely on continental shelves, phosphorus accumulates with other minerals as insoluble deposits. Millions of years pass, and geologic forces thrust up the seafloor, which thereby becomes exposed as new land surfaces. Then weathering releases phosphorus from the rocks, and the geochemical phase of the cycle begins again.

The ecosystem phase of the phosphorus cycle is far more rapid than the long-term geochemical phase. All living organisms require phosphorus, which becomes incorporated in ATP, NADPH, phospholipids, nucleic acids, and other organic compounds. Plants have the metabolic means to take up dissolved, ionized forms of phosphorus. Actually, they do this so rapidly and efficiently that they often reduce soil concentrations of phosphorus to extremely low levels. Herbivores obtain phosphorus only by eating the plants; carnivores obtain it by eating herbivores. Herbivores and carnivores excrete phosphorus as a waste product in urine and feces. Phosphorus is also released to the soil by the decomposition of organic matter. The plants then take up phosphorus and so recycle it rapidly within the ecosystem.

## Transfer of Harmful Compounds Through Ecosystems

Human activities have major and minor effects on the functioning of ecosystems. We will consider some of these effects in Chapter 48, but for now a simple example will underscore the point.

Applications of DDT, the first of the synthetic organic pesticides, began during World War II. In mosquito-infested regions of the tropical Pacific, individuals were being stricken with a dangerous disease, malaria. DDT helped control the mosquitoes, which were transmitting the sporozoans responsible for the disease (*Plasmodium japonicum*). In the war-ravaged cities of Europe, individuals were being stricken with the crushing headache, fever, and rashes associated with typhus. DDT helped control the body lice that were transmitting the bacterium responsible for this terrible disease (*Rickettsia rickettsii*). After the war, it seemed like a good idea to use DDT in battles against insects that were agricultural or forest pests, transmitters of disease agents, or merely nuisances in homes and gardens.

DDT is a relatively stable hydrocarbon compound. It is almost insoluble in water, so you might think that it would stay put and exert its intended effect only where applied. But winds can carry DDT in vapor form; water can transport fine particles of it. DDT also is highly soluble in fats, so that when it comes in contact with organisms, it accumulates in their tissues. Thus, as we now know, DDT can show **biological magnification**—the increase in concentration of a nondegradable (or slowly degradable) substance in organisms as it is passed along food chains. Most of the DDT from all the organisms eaten by a consumer during its lifetime will become concentrated in its tissues. Besides this, many organisms have the means to partially metabolize DDT to modified compounds (such as DDE), which have different but still disruptive effects. Both DDT and those modified compounds are toxic or physiologically disruptive to *many* kinds of water-dwelling and land-dwelling animals.

After the war, DDT began to move through the global environment, infiltrate food webs, and affect organisms in ways that no one had predicted. In cities where DDT was sprayed to control Dutch elm disease, songbirds started dying. In streams flowing through forests where DDT was sprayed to control spruce budworms, salmon started dying. In croplands sprayed to control one kind of pest, new kinds of pests moved in—DDT indiscriminately was killing off the natural predators that had been keeping pest populations in check. It took no great leap of the imagination to make the connection, given that all those organisms were dying at the same time and place as the DDT applications.

Then the side effects of biological magnification started showing up in places far from the areas of DDT application—and much later in time. Most devastated were species at the end of food chains—peregrine falcons, ospreys, and bald eagles among them. One of the breakdown products of DDT disturbs the physiology of birds, causing eggshells to become thin and brittle. As a result, many birds of the new generation never did make it to hatching time. DDT actually brought some species to the brink of extinction.

Since the 1970s, DDT has been banned in the United States, except for some restricted applications where public health is endangered. Many hard-hit species have partially recovered in numbers. Even today, however, some birds breeding in the United States lay thin-shelled eggs. They pick up DDT on their winter ranges in Latin America. As recently as 1990, the California State Department of Health recommended that a fishery off the coast of Los Angeles be closed. Why? It shows persistent DDT pollution, from industrial waste discharges that ceased twenty years before.

This brief example brings us full circle to the story that opened this chapter. For it reinforces our understanding that disturbances to one part of an ecosystem can have unexpected effects on other, seemingly unrelated parts.

A recent approach to predicting such unforeseen effects is through **ecosystem modeling**. This is a method of identifying crucial bits of information about the different components of a system and combining the information, through computer programs and models, in order to predict the outcome of the next disturbance. For example, an analysis of which species feed on which others in the food web shown in Figure 46.3 can be turned into a series of equations describing how much of each species is consumed. The equations can be used to predict what the effect would be, say, of overharvesting whales or of greatly expanding the harvest of krill.

As we attempt to deal with larger and more complex ecosystems, it becomes more difficult and expensive to run desired experiments in the field. The temptation is to run them instead on the computer. This is a valid exercise if the computer model adequately represents the system. The danger is that we may not have identified all the important relationships in the ecosystem and incorporated them accurately into the model. The most important fact may be one that we do not yet know.

## SUMMARY

1. An ecosystem is a whole complex of producers, consumers, detritivores, and decomposers and their physical environment, all interacting through a flow of energy and a cycling of materials.

2. Ecosystems are open systems, with inputs and outputs of energy as well as nutrients. With few exceptions, photosynthetic autotrophs are the primary producers. They secure energy from sunlight and take up much of the nutrients used by other members of the systems. Ecosystems generally are most open for inputs and outputs of water, carbon, and energy. Nutrients such as nitrogen are mostly recycled within the ecosystem.

3. Energy fixed by photosynthesis passes through grazing food webs and detrital food webs. Both types of webs typically are interconnected in the same ecosystem. In both cases, energy is lost (as heat) through aerobic respiration and other metabolic activities.

4. Biogeochemical cycles include the movement of water, nutrients, and other elements and compounds from the physical environment to organisms, then back to the environment.

5. Ecosystems on land have predictable rates of nutrient losses that generally increase when the land is cleared or otherwise disturbed.

6. Fossil fuel burning and conversion of natural ecosystems to cropland or grazing land are contributing to increased atmospheric concentrations of carbon dioxide. The increase is believed to be causing a global warming trend.

7. Nitrogen availability is often a limiting factor for the total net primary productivity of land ecosystems. Gaseous nitrogen is abundant in the atmosphere, but it must be converted to ammonia and to nitrates that can be used by primary producers. A few species of bacteria as well as volcanic action and lightning can cause the conversion.

## Review Questions

1. Label the three categories of organisms necessary for the flow of energy and cycling of materials through ecosystems. Include examples of each: *827–828*

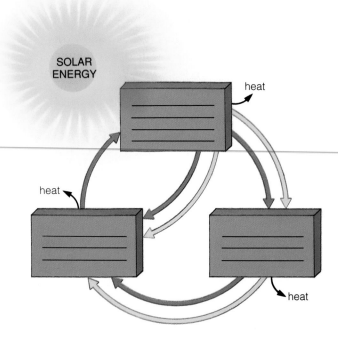

SOLAR ENERGY

heat

heat

heat

2. Define ecosystem. Why do autotrophs play such a central role in an ecosystem? *827–828*

3. Define trophic level. Can you name and give examples of some trophic levels in ecosystems? What is the energy source for each level? *828–829*

4. Distinguish between a food chain and a food web. Can you imagine an extreme situation whereby you would be a participant in a food chain? *829–830*

5. If you were growing a vegetable garden, what variables might affect its net primary production—that is, the amount of energy stored in the organic compounds of plant tissues? *830*

6. There are two major pathways of energy flow through ecosystems: grazing food webs and detrital food webs. How would you characterize each? How does energy leave each one? How are the two types of food webs interconnected? *830–831*

7. Describe the greenhouse effect. Make a list of the agricultural products and manufactured goods that you depend on yet are implicated in the amplification of the greenhouse effect. *838–839*

8. What is a biogeochemical cycle? Give an example of one and describe the reservoirs and organisms involved. *833–842*

9. Define these terms: nitrogen fixation, nitrification, ammonification, and denitrification. *840*

## Self-Quiz *(Answers in Appendix IV)*

1. An _____ is a complex of organisms and their physical environment, linked by a flow of _____ and a _____ of materials.

2. Energy flow through organisms in nearly all ecosystems begins with the metabolic pathway of _____, and it flows in _____ direction.

3. Ecosystems are open systems that are not self-sustaining; they require inputs of _____ and _____, and they have outputs of _____ and _____.

4. Global movements of water or nutrients between the physical environment and organisms and then back to the environment are _____ cycles.

5. Trophic levels can be described as _____ in an ecosystem.
   a. structured feeding relationships
   b. who eats whom
   c. a hierarchy of energy transfers
   d. any one of the above descriptions is appropriate

6. A feeding relationship that proceeds from algae to a fish, then to a fisherman and finally to a shark is best described as _____ .
   a. a food chain
   b. a food web
   c. bad luck for the fisherman
   d. a and c above

7. The primary productivity of an ecosystem is affected by _____ .
   a. the balance between photosynthesis and plant respiration
   b. how many plants avoid being eaten or decomposed
   c. the amount of rainfall
   d. the range of temperatures
   e. all of the above

8. Disturbances to land ecosystems have the predictable effect of _____ nutrient losses.
   a. lessening
   b. equalizing
   c. increasing
   d. stabilizing

9. An apparent global warming trend may be due to increases in greenhouse gases that are being brought about by _____ .
   a. fossil fuel burning
   b. gases released from fertilizers and livestock wastes
   c. deforestation, especially of vast tropical forests
   d. CFCs in refrigerators, etc.
   e. all of the above contribute to greenhouse gases

10. Match the ecosystem terms.
    _____ primary producers
    _____ consumers
    _____ decomposers
    _____ detritivores

    a. feed on partly decomposed particles of organic matter
    b. break down remains or products of other organisms
    c. herbivores, carnivores, omnivores, parasites
    d. photosynthetic autotrophs

11. Match the ecosystem terms.
    _____ nitrogen availability
    _____ ecosystem components
    _____ global warming
    _____ ammonium
    _____ biogeochemical cycles

    a. movement of water or nutrients from environment to organisms and back to environment
    b. form of nitrogen suitable for primary producers
    c. limiting factor for total net primary productivity of land ecosystems
    d. producers, consumers, detritivores, decomposers, and physical environment
    e. attributed to increases in greenhouse gases

## Selected Key Terms

ammonification *840*
atmospheric cycle *833*
biogeochemical cycle *833*
biological magnification *843*
biomass *831*
carbon cycle *836*
carnivore *828*
consumer *828*
decomposer *828*
denitrification *840*
detrital food web *830*
detritivore *828*
ecosystem *828*
ecosystem modeling *843*
energy input *828*
energy output *828*
energy pyramid *832*
food chain *829*

food web *829*
grazing food web *830*
greenhouse effect *838*
herbivore *828*
hydrologic cycle *833*
nitrification *840*
nitrogen cycle *840*
nitrogen fixation *840*
nutrient input *828*
nutrient output *828*
omnivore *828*
parasite *828*
phosphorus cycle *842*
primary producer *830*
primary productivity *830*
sedimentary cycle *833*
trophic level *828*
watershed *835*

## Readings

Berner, R., and A. Lasaga. March 1989. "Modeling the Geochemical Carbon Cycle." *Scientific American* 260(3):74–81.

Botkin, D. B. 1990. *Discordant Harmonies: A New Ecology for the Twenty-first Century.* New York: Oxford University Press.

Kerr, R. 1988. "Is the Greenhouse Here?" *Science* 239:559–561.

Post, W., et al. 1990. "The Global Carbon Cycle." *American Scientist* 78:310–326.

Rambler, M., L. Margulis, and R. Fester. 1989. *Global Ecology: Towards a Science of the Biosphere.* San Diego, California: Academic Press.

# 47 THE BIOSPHERE

## Does a Cactus Grow in Brooklyn?

When Charles Darwin and other naturalists made their grand explorations more than a century ago, they perceived broad patterns in the world distribution of plants and animals. If you had traveled with them, you would have discovered that many plants and animals are unique to some regions, yet strikingly similar to unrelated species in distant regions.

Suppose you had visited deserts of Africa and the American Southwest. Flowering plants with spines, extremely small leaves, and columnlike, fleshy stems grow in both places, yet they evolved from different ancestors (Figure 47.1). Suppose you also had explored the hills near the California coast. Woody, many-branched chaparral plants growing there are very much like the woody, many-branched, yet unrelated plants growing near the Mediterranean Sea, at the southern tip of Africa, and in central Chile.

By mapping the locations of those structurally similar but distant species, you find that the American and African desert plants grow about the same distance from the equator. The chaparral plants and their distant look-alikes tend to grow along western or southern coasts of continents between 30° and 40° latitudes.

Why do we find certain plants and animals in certain regions and not in others? In part, "accidents of history" put many species in particular places. Think about what happened after the supercontinent Pangea broke up more than 100 million years ago. When the huge fragments drifted apart, many species became geographically isolated from others, including species that evolved into eucalyptus trees, koalas, and kangaroos of Australia.

But the distribution of species also depends on climate, topography, and species interactions. With diligence, a member of the human species living in Brooklyn or some other New York borough can keep a cactus alive under artificial lights in an artificially heated room. Plant that cactus outside and it won't last one winter. By the same token, the glorious tropical plants of the Amazon basin wouldn't last one growing season outdoors in St. Louis, Missouri, although they flourish there inside an enclosed, climate-controlled botanical garden.

a

b

This example reminds us that we humans tinker with the distribution of species. Not all of our meddling is as harmless as growing a cactus or two in Brooklyn. Think about the impact of our collective interactions with other species—including our "predation" on forests and fishes or our pesticide battles with insects in the competition for food. We have already considered the general nature of predation, competition, and other species interactions in earlier chapters. Let's turn now to the physical forces shaping the character of the

**Figure 47.1** Convergent evolution of plants that are native to different geographic realms. (**a**) *Echinocereus*, a member of the cactus family (Cactaceae), grows in deserts of the American Southwest. (**b**) *Euphorbia*, a member of the spurge family (Euphorbiaceae), grows in deserts of southwestern Africa. Although the plants appear to be similar, they evolved from leafy plants that are not related.

biosphere itself. This will serve as a foundation for thinking about the impact of the human species on the biosphere—the topic of the chapter to follow.

KEY CONCEPTS

**1.** Besides being the primary energy source for ecosystems, solar radiation influences their distribution on the earth's surface. Heat energy derived from the sun warms the atmosphere and drives the earth's weather systems.

**2.** Global air circulation patterns, ocean currents, and topographic features interact to produce regional variations in temperature and rainfall. Patterns of temperature and rainfall influence the composition of soils and sediments. The patterns also influence the growth and distribution of primary producers—and through them, the distribution of ecosystems.

**3.** The world's major land regions are divided into six biogeographic realms. Each realm has an array of biomes. A biome is a particular type of regional ecosystem, dominated by certain plants and other organisms. Biomes include deserts, shrublands, woodlands, forests, and tundra. Their distribution corresponds roughly with regional variations in climate, topography, and soil type.

**4.** The water provinces cover more than 70 percent of the earth's surface. They include bodies of standing freshwater, moving freshwater, inland seas, and oceans. Each freshwater and marine ecosystem has gradients in light penetration, temperature, and dissolved gases. The gradients, which vary daily and seasonally, influence primary productivity and species diversity.

## CHARACTERISTICS OF THE BIOSPHERE

### Biosphere Defined

The **biosphere** is the entire realm in which organisms live—the waters of the earth, the surface rocks, soils, and sediments of its crust, and the lower atmosphere. All liquid and frozen water on or near the earth's surface constitutes the **hydrosphere**. This includes the oceans and smaller bodies of water, groundwater, the polar ice caps, and a small amount of airborne water. The **atmosphere** is a region of gases, airborne particles, and water vapor enveloping the earth. About 80 percent of its mass is distributed within 17 kilometers of the earth's surface.

## Global Patterns of Climate

Ecosystems of the biosphere range in size from vast continental forests to tiny ponds. Except for a few remote ecosystems in deep oceans, all are influenced profoundly by climate. **Climate** refers to prevailing weather conditions, including temperature, humidity, wind speed, cloud cover, and rainfall.

Of the many factors that shape climate, four are paramount: (1) variations in the amount of incoming solar radiation, (2) the earth's daily rotation and its path around the sun, (3) the world distribution of continents and oceans, and (4) the elevations of land masses. Interactions among these factors produce prevailing winds and ocean currents that influence global patterns of climate. Climate shapes the development of soils and sediments. The composition of soils and sediments influences the growth of primary producers—and through them, the distribution of entire ecosystems.

**Mediating Effects of the Atmosphere.** Only about half the solar radiation reaching the atmosphere actually gets to the earth's surface. Ozone ($O_3$) and oxygen molecules in the atmosphere absorb ultraviolet wavelengths. Ozone concentrations are greatest between 17 and 25 kilometers above sea level (Figure 47.2). This *ozone layer* is important, given that ultraviolet wavelengths can be lethal to most forms of life. Clouds, dust, and water vapor absorb more wavelengths from the sun or reflect them back into space.

The remaining radiation warms the earth's surface, which then gives up heat (by radiation and evaporation). Molecules of the lower atmosphere absorb some heat, then reradiate part of it back toward the earth. The effect is something like heat retention in a greenhouse. A greenhouse allows the sun's rays to penetrate but retains heat being lost from the plants and soil inside (page 838).

Why is this important? Heat energy derived from the sun warms the atmosphere—and *that energy drives the earth's weather systems*.

**Air Currents.** The sun's rays have different heating effects at different latitudes. The rays are less spread out at the equator than at the poles, so air is heated more at the equator. The global pattern of air circulation begins when warm equatorial air rises and spreads northward and southward. Then the earth's rotation modifies the circulation into worldwide belts of prevailing east and west winds (Figure 47.3). The differences in solar heating at different latitudes, together with the modified air circulation patterns, define the earth's major temperature zones (Figure 47.4).

Global air circulation patterns also cause differences in rainfall at different latitudes. At the equator, heated

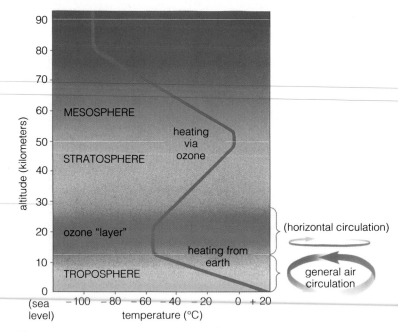

**Figure 47.2** Earth's atmosphere. Most of the global air circulation occurs in the troposphere, where temperatures decrease rapidly with altitude. Ultraviolet wavelengths from the sun are absorbed in the upper atmosphere primarily by ozone ($O_3$) molecules. Ozone is most concentrated between 17 and 25 kilometers above sea level; this is the ozone "layer."

air rises to cooler altitudes and gives up moisture as rain, and where the rain falls, it supports luxuriant forests. The air, which is now drier, moves away from the equator and then descends at latitudes of about 30°. It becomes warmer and drier as it descends, and there deserts tend to form. Farther from the equator, the air picks up moisture, ascends to high altitudes, then creates another moist belt at about 60° latitudes. Finally the air descends in polar regions, where low temperatures and almost nonexistent precipitation create cold, dry, polar deserts. Thus, *the latitudinal belts of temperature and rainfall influence where different ecosystems occur*.

**Seasonal Variations in Climate.** Throughout the year, the amount of solar radiation reaching the earth's surface changes in the Northern and Southern Hemispheres. The variation leads to seasonal changes in climate (Figure 47.5a). Many biological rhythms coincide with the seasons. In temperate regions, organisms respond most to seasonal changes in daylength and temperature. In deserts and tropical forests, they respond more to seasonal changes in rainfall. Plants, for example, show seasonal cycles of leafing out, flowering, fruiting, and leaf drop. Caribou, many birds, butterflies, and other animals follow cycles of breeding and migration. Turtles, seals, and whales are among those that migrate in the seas. Such movements correspond with seasonal bursts of primary productivity (compare Figure 4.1).

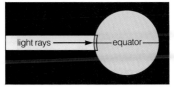

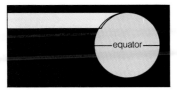

**a**

**Figure 47.3** Global air circulation patterns, brought about by three interrelated factors. *First*, the sun's rays are less spread out in equatorial than in polar regions (**a**). Warm equatorial air rises and spreads northward and southward to produce the initial pattern of air circulation (**b**).

*Second*, air pressure variations arise as a result of the nonuniform distribution of land masses. Land absorbs and gives up heat faster than oceans do, so some parcels of air rise (or sink) faster than others. Atmospheric pressure decreases where warm air rises (and increases where it sinks). The pressure differences give rise to winds, which disrupt the overall air movement from the equator to the poles.

*Third*, easterly and westerly deflections in wind directions arise as a result of the earth's rotation and overall shape. Each time the ball-shaped earth makes a full rotation, its surface turns faster beneath air masses at the equator and slower beneath those at the poles. Thus, a rising air mass can't really move "straight north" or "straight south." It is deflected to the east or west. This is the source of prevailing east and west winds (**c**).

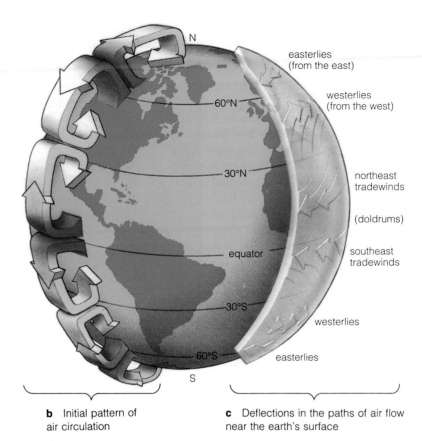

easterlies (from the east)

westerlies (from the west)

northeast tradewinds

(doldrums)

southeast tradewinds

westerlies

easterlies

**b** Initial pattern of air circulation

**c** Deflections in the paths of air flow near the earth's surface

**Figure 47.4** World temperature zones

cold
cool temperate
warm temperate
tropical
— — — — (equator) — — — —
tropical
warm temperate
cool temperate
cold

**Figure 47.5** (**a**) Annual variation in the amount of incoming solar radiation. Notice that the northern end of the earth's axis tilts toward the sun in June and away from it in December. Notice also the annual variation in the position of the equator relative to the boundary of illumination between day and night. Such variations in the intensity and duration of daylight lead to seasonal variations in temperature in the two hemispheres. Seasonal change becomes more pronounced with distance from the equator. It is greatest in the central regions of continents, where the moderating effects of oceans are minimal.

(**b**) Monarch butterflies, migrants that gather in winter in trees of California coastal regions and central Mexico. Monarchs typically travel hundreds of kilometers south to these regions, which are cool and humid in winter; if they stayed in their breeding grounds, they would risk being killed by more severe conditions.

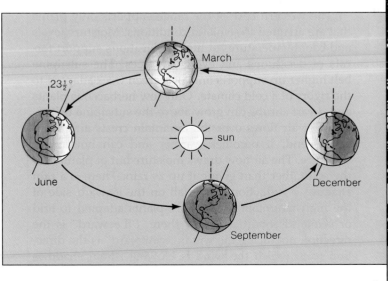

March

$23\frac{1}{2}°$

sun

June

December

September

**a**

**b**

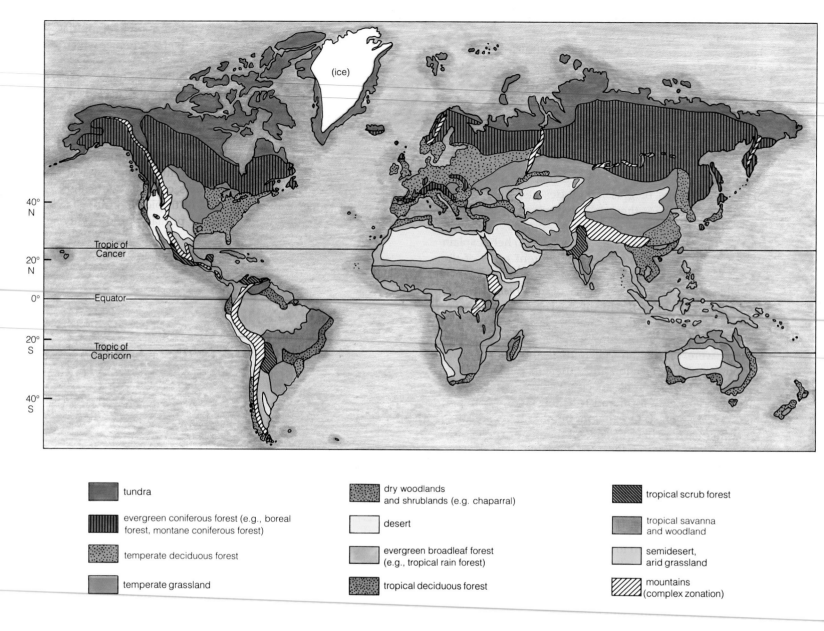

**Figure 47.10** Simplified picture of the world's major biomes. The overall pattern of biome distribution roughly corresponds with distribution patterns for climate and soil type. Compare this map with the two-page photograph of the earth's surface that precedes Chapter 1.

**Legend:**

- tundra
- evergreen coniferous forest (e.g., boreal forest, montane coniferous forest)
- temperate deciduous forest
- temperate grassland
- dry woodlands and shrublands (e.g. chaparral)
- desert
- evergreen broadleaf forest (e.g., tropical rain forest)
- tropical deciduous forest
- tropical scrub forest
- tropical savanna and woodland
- semidesert, arid grassland
- mountains (complex zonation)

Figure 47.10 shows the distribution of the major biomes. Overall, their distribution tends to correspond with climate, topography, and soil types. **Soil** is a mixture of rock, mineral ions, and organic matter in some state of physical and chemical breakdown. Water, air, and a variety of organisms are mixed with its components. As soils develop, they tend to take on a characteristic vertical structure, or "profile." Figure 47.11 shows soil profiles for some major types of biomes.

Let's now consider the major types of biomes—the deserts, shrublands, woodlands, grasslands, forests, and tundra. Keep in mind that the composition of species is not entirely uniform within any one of those biomes. Local variations in climate, landforms, and other physical features have favored patches of distinct communities within its boundaries.

**Figure 47.11** Soil characteristics, which influence the distribution of ecosystems on land. Soil is a mixture of rock, mineral ions, and organic matter in some state of physical and chemical breakdown. The rocks range from coarse-grained gravel to sand, silt, and finely grained clay. *Humus* (partly decomposed organic matter) helps retain water-soluble ions. *Loam*, a mixture of sand, silt, clay, and humus, retains nutrients and is well aerated. Most plants do not do well in poorly aerated, poorly draining soils.

During *leaching*, water percolates too rapidly through gravelly or sandy soils and depletes them of mineral ions. Clay soils with fine, closely packed particles are poorly aerated and do not drain well. Few plants can grow in waterlogged clay soils.

*Topsoil*, the most fertile soil layer, occurs just below any surface litter. It may be less than a centimeter deep on steep slopes to more than a meter deep in grasslands. Most crops are grown in grasslands. When tropical rain forests are cleared for agriculture, the heavy seasonal rains leach most nutrients from the exposed topsoil.

Crops usually cannot grow in deserts; there is not enough rain. With soil management and extensive irrigation, soils in some desert areas (such as Imperial Valley in California) can support crops. But they can become unproductive from waterlogging and salt buildup when they are irrigated without proper drainage.

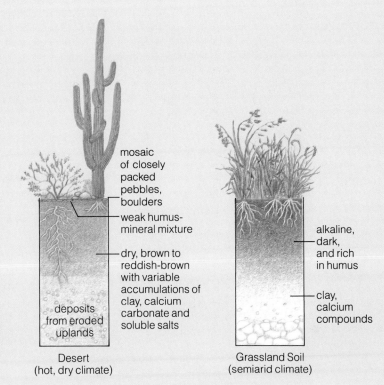

Desert
(hot, dry climate)

- mosaic of closely packed pebbles, boulders
- weak humus-mineral mixture
- dry, brown to reddish-brown with variable accumulations of clay, calcium carbonate and soluble salts
- deposits from eroded uplands

Grassland Soil
(semiarid climate)

- alkaline, dark, and rich in humus
- clay, calcium compounds

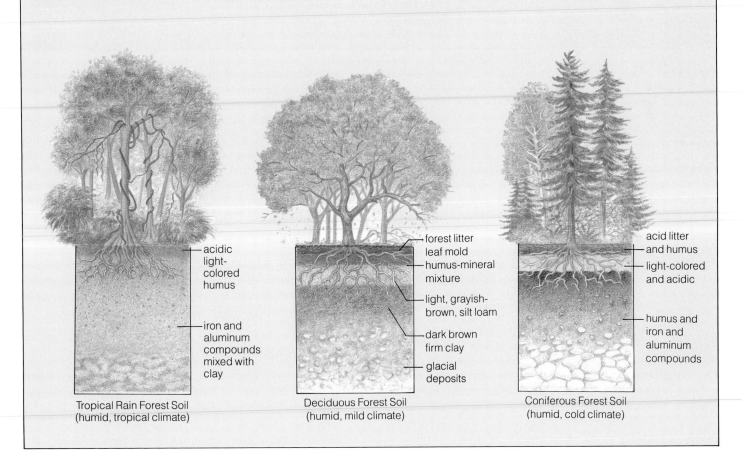

Tropical Rain Forest Soil
(humid, tropical climate)

- acidic light-colored humus
- iron and aluminum compounds mixed with clay

Deciduous Forest Soil
(humid, mild climate)

- forest litter leaf mold
- humus-mineral mixture
- light, grayish-brown, silt loam
- dark brown firm clay
- glacial deposits

Coniferous Forest Soil
(humid, cold climate)

- acid litter and humus
- light-colored and acidic
- humus and iron and aluminum compounds

**Figure 47.12** Warm desert near Tucson, Arizona. The vegetation includes creosote bushes; tall, multistemmed ocotillo; columnlike saguaro cacti; and prickly pear cacti with rounded pads.

## Deserts

*Deserts* occur where the potential for evaporation greatly exceeds rainfall. Such conditions prevail at about 30° north and south latitudes, where rainfall is scarce. There we find the deserts of the American Southwest, northern Chile, northern and southern Africa, Arabia, and Australia. Farther north are the vast Gobi of Asia, the Kyzyl-Kum east of the Caspian Sea, and the high deserts of eastern Oregon. The more northern deserts result largely from the rain shadow effect of extensive mountain ranges.

Deserts do not have much vegetation. The infrequent rains often fall in heavy, brief pulses that rapidly erode the unprotected soil. Because humidity is low, sunlight easily penetrates the atmosphere and the ground surface heats rapidly. The surface cools rapidly at night, when it radiates heat back to the sky.

Arid or semiarid conditions do not favor large, leafy plants, but deserts still show plenty of diversity (Figure 47.12). The same patch of Arizona desert may have creosote bushes and other deep-rooted evergreen shrubs as well as fleshy-stemmed, shallow-rooted cacti. It may have short plants such as prickly pear and tall plants such as saguaro. The patch also may have ocotillo, which can drop leaves more than once a year, then grow new ones within a week after each rain. And it may have annuals and perennials that flower briefly but spectacularly after a rain. Even deep-rooted species such as mesquite and cottonwood may grow along streambeds having a permanent underground water supply.

More than a third of the world's land area is already arid or semiarid. Yet in many parts of the world, there is an alarming trend toward *desertification*—the conversion of hundreds of thousands of hectares of grasslands and other ecosystems to dry wastelands. This trend is described on page 886.

## Dry Shrublands and Woodlands

The western or southern coastal regions of continents between 30° and 40° latitudes have a semiarid climate, like that around the Mediterranean Sea. They get more rain than deserts, but not much more. It rains mostly during mild winter months and summers are long, hot, and dry. The dominant plants often have hard, tough,

**Figure 47.13** (**a**) Closer view of a chaparral plant community in the California foothills. (**b**) Chaparral-covered hills in California. What appear to be dirt "roads" are firebreaks.

a

b

evergreen leaves adapted to dry conditions. Dry shrublands and woodlands dominate these regions and other areas with related climates.

*Dry shrublands* prevail when annual rainfall is less than 25 to 60 centimeters. They go by such exotic local names as maquis, fynbos, and chaparral (Figure 47.13). California alone has about 2.4 million hectares (6 million acres) of chaparral-covered hills. The dominant plants of this biome are woody, multibranched, and never more than a few meters tall. They can grow together into a nearly impenetrable mass. Every so often, lightning-sparked firestorms sweep through them. Many of the shrubs have highly flammable leaves and their aboveground parts burn rapidly. Yet they are exquisitely adapted to episodes of fire and quickly resprout from their root crowns. Trees (and suburban housing developments) do not survive firestorms nearly as well. The shrubs—which actually feed the fires—have the competitive edge.

*Dry woodlands* occur when annual rainfall is about 40 to 100 centimeters. The dominant trees do not form a dense, continuous canopy, but they can be quite tall. The eucalyptus woodlands of southwestern Australia and oak woodlands of the Pacific states are like this.

**Figure 47.14** Rolling shortgrass prairie to the east of the Rocky Mountains. Sixty million bison originally constituted the dominant large herbivore of North American grasslands.

## Grasslands and Savannas

Great grasslands and savannas sweep across South Africa, Australia, South America, and midcontinental regions of North America and the Soviet Union. The main types—shortgrass prairie, tallgrass prairie, and tropical savannas—have much in common. The land is usually flat or rolling. Warm temperatures prevail during summer in the temperate zones and throughout the year in the tropics. The 25 to 100 centimeters of annual rainfall is enough to keep the regions from turning into deserts but not enough to support forests. Grazing and burrowing species are the dominant forms of animal life. Grazing activities and periodic fires stop the encroachment of forests at the fringes of many grassland biomes.

Where winds are strong, rainfall light and infrequent, and evaporation rapid, we find *shortgrass prairie* (Figure 47.14). Plant roots above the permanently dry subsoil soak up the brief, seasonal rainfall. Much of the shortgrass prairie of the American Great Plains was overgrazed and plowed under for wheat, which requires more moisture than the region sometimes provides. During the 1930s, strong prevailing winds, drought, and poor farming practices turned much of the prairie into a Dust Bowl (Figure 48.11). John Steinbeck's *The Grapes*

**Figure 47.15** A patch of natural tallgrass prairie in eastern Kansas.

of *Wrath* and James Michener's *Centennial,* two historical novels, speak eloquently of the disruption of this biome and its consequences.

*Tallgrass prairie* of the sort shown in Figure 47.15 once extended west from the temperate deciduous forests of North America. Daisies, sunflowers, and other composites as well as legumes also were abundant. Humans have converted most tallgrass prairie to farmland. However, efforts are now being made to restore large areas of this ecosystem in several locations.

*Tropical savannas* cover broad belts of Africa, South America, and Australia. In savanna regions of low rainfall, the main plant species are rapidly growing grasses (Figure 47.16). Acacia and other shrubs grow in scattered patches where there is slightly more moisture. Where rainfall is higher, savannas grade into tropical woodlands with tall, coarse grasses, shrubs, and low trees.

Parts of southern Asia have *monsoon grasslands.* "Monsoon" refers to a season of heavy rainfall that corresponds to a shift in prevailing winds over the Indian Ocean. It alternates with a pronounced dry season. The climate favors dense stands of tall, coarse grasses, which provided frighteningly effective concealment for opposing forces during the Vietnam war.

**Figure 47.16** African savanna, a region of warm grasslands punctuated by stands of shrubs and trees. More large ungulates (hooved, plant-eating mammals) live here than anywhere else. They include migratory wildebeests (shown here); giraffes, which browse on leaves beyond the reach of other ungulates; and Cape buffalo, formidably horned animals that live in herds. Other ungulates are zebras and impalas, which are abundant and vulnerable to such predators as lions and cheetahs. Their remains (as well as the remains of lions and cheetahs) are picked over by hyenas, jackals, vultures, and other scavengers. The remains and wastes of all these organisms are broken down by detritivores and decomposers, which cycle nutrients back to the plant species.

**Figure 47.17** Tropical rain forest in Costa Rica.

## Forests

The world's major forest biomes have tall trees growing close enough together to form a fairly continuous canopy over a broad region. There are three general types of forest trees. Which type prevails in a given region depends partly on distance from the equator. *Evergreen broadleafs* are dominant between 20° north and south latitudes. *Deciduous broadleafs* are dominant at moist, temperate latitudes where winters are not severe. *Evergreen conifers* are most common at higher, colder latitudes and in mountainous regions of the temperate zone. Figures 47.17 through 47.20 show examples of some forest biomes.

**Evergreen Broadleaf Forests.** These forest biomes occur in tropical parts of Africa, southeastern Asia, the East Indies, the Malay Archipelago, South America, and Central America. Annual rainfall can exceed 200 centimeters and is never less than 130 centimeters.

Where regular, heavy rainfall coincides with high humidity (80 percent or more) and the annual mean temperature is about 25°C, you will find highly productive *tropical rain forests* (Figure 47.17). Here, evergreen trees produce new leaves and shed old ones throughout the year. Near the fringes of the biome, some trees are periodically bare during the driest season, but not for more than a few weeks. Because leaf production and leaf drop are generally continuous, tropical rain forests produce more litter than any other forest biome. However, decomposition and mineral cycling are rapid in the hot, humid climate. Also, the highly weathered soils, with little humus, are not a good reservoir of nutrients. We will look more closely at tropical rain forests in the next chapter.

**Deciduous Broadleaf Forests.** As we move out from the tropical rain forests, we enter regions where temperatures remain mild but rainfall dwindles during part of the year. Here we find *tropical deciduous forests*, in which many trees drop some or all of their leaves during the pronounced dry season. The *monsoon forests* of India and southeastern Asia also have such trees. Farther north, in the temperate zone, rainfall is even lower and temperatures become cold during the winter. Here are regions of *temperate deciduous forests* (Figure 47.18). Conditions do not favor decomposition as much as they do in the humid tropics, but nutrients are conserved in the accumulated litter on the forest floor.

At one time, deciduous broadleaf forests stretched across northeastern North America, Europe, and eastern Asia. Ash, beech, birch, chestnut, elm, and deciduous oak trees dominated the forests, which largely

Spring

Summer

Autumn

Winter

disappeared as land was cleared for farming. Some species introduced to North America brought diseases that wiped out nearly all chestnuts and many American elms (Table 45.2). Today, maple and beech forests predominate in the Northeast. They give way to oak-hickory forests farther west, which grade into oak woodlands and tallgrass prairies of the Midwest.

**Figure 47.18** The changing character of a temperate deciduous forest in spring, summer, autumn, and winter. The one shown here is south of Nashville, Tennessee.

**Figure 47.28** Vertical zonation in the intertidal zone of a rocky shore in the Pacific Northwest. The vertical difference between high and low tides varies from a few centimeters (in the Mediterranean Sea) to more than 15 meters (in the Bay of Fundy, next to Nova Scotia). It is about 3 meters for the area shown in this photograph.

In the sparsely populated upper littoral, the primary producers are cyanobacteria, green algae, and the algal part of lichens, all of which grow in mats or jellylike masses on the infrequently wet rocks. Small snails and limpets feed on the producers. Some large, aggressive shore crabs feed on the snails and limpets, and seabirds feed on snails, limpets, and smaller crabs. The upper littoral may also be populated by barnacles that can get by with a few hours of filter feeding each month, when the highest tides carry plankton to them. They also grow profusely in the mid-littoral, along with mussels, sea stars, and red, brown, and green algae. Diversity is greatest in the lower littoral, where abundant seaweeds form a leafy canopy for many organisms.

upper littoral

mid-littoral

lower littoral

**Figure 47.29** (*Below*) The kinds of residents of tide pools along the coasts of the Pacific Northwest.

**Life Along the Coasts.** Along the rocky and sandy shores of coastlines are ecosystems of the **intertidal zone**, which is not exactly renowned for its creature comforts. The resident organisms are battered by waves, fiercely so during storms. They are alternately submerged, then exposed by the tides. The higher up they are on the shore, the more they might dry out, freeze in winter, or bake in summer, and the less food comes their way. The lower they are on the shore, the more they must compete for the limited space available. At low tides, birds, rats, and raccoons move in to feed on them. High tides bring the predatory fishes.

It is nearly impossible to generalize about life here, because bewildering arrays of habitats are constantly being resculpted by waves and tides. About the only feature common to all rocky and sandy shores is a vertical zonation. *Rocky shores* often have three vertically arranged zones (Figure 47.28). The *upper littoral* is submerged only during the highest tide of the lunar cycle, and it is sparsely populated. The *mid-littoral* is submerged during the highest regular tide and exposed during the lowest tide of each day. Red, brown, and green algae, sea anemones, snails, nudibranchs, hermit crabs, and small fishes typically live in the tide pools characteristic of this zone (Figure 47.29). The *lower littoral* is exposed only during the lowest tide of the lunar cycle. Diversity is greatest here. Erosion prevents detritus from accumulating on rocky shores, so grazing food webs tend to predominate in all three zones.

*Sandy* and *muddy shores* are stretches of loose sediments, continually rearranged by waves and currents. Few large plants grow in these unstable places, so you won't find many grazing food webs. Organic debris from offshore or from the land nearby form the basis of detrital food webs. Vertical zonation is less obvious.

Along temperate coasts, blue crabs and sea cucumbers live below the low tide mark. Sand in between the high and low tide marks may house burrowing marine worms, crabs, and small isopods. Isopods prey on smaller animals. And some of those smaller animals graze on bacteria and diatoms on individual sand grains. Beach hoppers and ghost crabs burrow in at the high tide mark during the day, then bound or lurch about the beach at night.

**The Open Ocean.** Beyond the intertidal are two vast provinces of the open oceans (Figure 47.30). The **pelagic province** includes the entire volume of ocean water. It is subdivided into two zones. Its *neritic zone* consists of the relatively shallow waters overlying the continental

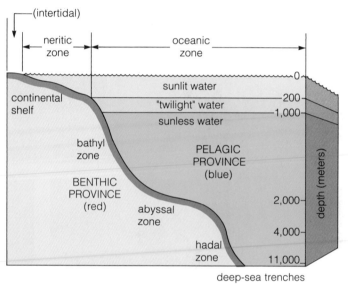

**Figure 47.30** Zones of the ocean.

sea anemone

sea star

sea anemone

crown-of-thorns sea star

moray eel

pillar coral

green tube coral

**Figure 47.31** A sampling of the diversity characteristic of tropical reefs. Long ago, corals began to grow and reproduce in the warm, nearshore waters off the islands shown. The skeletons they left behind served as a foundation for more corals to grow upon. As skeletons and residues accumulated, the reef grew, and tides and currents carved ledges and caverns in it. Today, the reef's spine may be decked out with as many as 750 species of corals, such as the ones shown here.

Red algae typically encrust the coral foundation. In shallow waters behind the reef, red algae give way to blue-green forms. Many small, transparent animals feed on algae and other plants and in turn are food for predators, including the fishes, sea star, and sea anemone shown. Fishes and other animals are food for the moray eel.

coral reef      island      lagoon      open ocean

shelves. Its *oceanic zone* consists of all the water over the ocean basins. Tropical reefs occur in the neritic zone and around islands in the oceanic zone. An example is shown in Figure 47.31. The **benthic province** includes all sediments and rocky formations of the ocean bottom. It begins with the *continental shelf* and extends down to the deep-sea trenches.

In the neritic and oceanic zones, photosynthetic activity is restricted to the upper surface waters. There,

phytoplankton often drift with the currents and form vast, suspended pastures for zooplankton. The zooplankton include copepods and shrimplike krill—all food for larger, strong-swimming carnivores such as squids and fishes.

Organic remains and wastes sink down from waters near the ocean's surface to the benthic province. They serve as the basis of detrital food webs for most communities in those zones.

**Figure 47.32** Looking out from shore to the open ocean, where the deeper you go, the less familiar the organisms become. (**a**) Whale leaping out of the water (breaching) near a coastline. (**b**) Sea lions commonly cavort around snorkelers off the California coast.

Hydrothermal vent ecosystems. In 1977, biologists discovered a distinct type of ecosystem deep in the Pacific Ocean, where sunlight never penetrates. John Corliss and his coworkers were exploring the Galápagos Rift, a volcanically active boundary between two of the earth's crustal plates. There, on the ocean floor, the near-freezing seawater seeps into fissures, becomes heated, and is spewed out through vents at high temperatures.

This hydrothermal outpouring results in deposits of zinc, iron, and copper sulfides as well as calcium and magnesium sulfates (all leached from rocks as pressure forces the heated water upward). In marked contrast to most of the deep ocean floor, these nutrient-rich, warm "oases" support diverse marine communities. Chemosynthetic bacteria use hydrogen sulfide ($H_2S$) as an energy source. They are the primary producers in a food web that includes tube worms, clams, sea anemones, crabs, and fishes (**c**, **d**).

So far, other hydrothermal vent ecosystems have been discovered near Easter Island (in the South Pacific Ocean); in the Gulf of California, about 150 miles south of the tip of Baja California, Mexico; near the Galápagos Islands; and in the Atlantic. In 1990, a team of United States and Soviet scientists discovered a unique hydrothermal vent ecosystem in Lake Baikal in Siberia which, at 1.7 kilometers, is the world's deepest lake. The tectonically formed lake basin seems to be splitting apart (hence the hydrothermal vents) and may mark the beginning of the formation of a new world ocean.

Although we have yet to discover huge populations in the abyssal zone, remarkable communities do thrive at **hydrothermal vents**. At these vents, water is heated and spewed from fissures between two crustal plates (Figure 47.32). The primary producers of these unique communities do not use sunlight as the primary energy source. The producers, chemosynthetic bacteria, obtain energy through reactions involving hydrogen sulfide. The bacterial populations are grazed upon by clams, mussels, marine worms, and other organisms.

**Upwelling**. Primary productivity increases whenever currents stir ocean water and keep nutrients circulating back to the surface. Consider the effects of **upwelling**, an upward movement of deep, nutrient-rich water along the margins of continents. Upwelling occurs when winds force surface waters along the coast to move away from the shore. When the surface water moves out, deep water moves in vertically to replace it.

For example, prevailing winds from the south and southeast force surface water away from the west coast of Peru. Cold, deeper water brought to the continental shelf by the Humboldt Current moves toward the surface and brings up tremendous amounts of nitrate and phosphate from below. Phytoplankton using these nutrients during growth are the basis of one of the world's richest fisheries, with schools of anchoveta and other fishes reaching huge numbers.

Periodically, warm surface waters of the western equatorial Pacific move eastward. The massive displacement of warm water affects the prevailing winds, which accelerate the eastward movement. The movement is enough to displace the cooler waters of the Humboldt Current and prevent upwelling. This phenomenon, which local fishermen named **El Niño**, causes productivity to decline. The decline has catastrophic effects on anchoveta-eating birds as well as on the anchoveta industry. The *Commentary* gives a closer look at the El Niño phenomenon. With it, we return to a concept presented at the start of this chapter—that interactions among the atmosphere, oceans, and land profoundly influence the world of life.

# El Niño and Oscillations in the World's Climates

Oceans cover more than 70 percent of the earth's surface, so it is not difficult to understand why variations in ocean surface temperatures affect climates around the world. But only within the past decade has numerical modeling begun to show how dramatic the effects may be.

In the winter of 1989–1990, for example, rainfall was below normal along the California coast, and by spring, strict water conservation and rationing programs were in the works. People were taking quick showers instead of baths, cutting back on flushing the toilet and running the dishwasher, and no longer deep-watering lawns and gardens because of the drought.

Drought is a relative term. If you live in Tucson, Arizona, or another desert city, "drought" is what you expect nearly year around. If you live along the western coasts or in the interior of continents, where semiarid conditions prevail, prolonged drought conditions can be disastrous.

Yet "abnormally" dry seasons may be part of a recurring feedback relationship between sea surface temperature, drought-related conditions on land, and drought-sustaining atmospheric circulation patterns.

Consider the global climate system called the *El Niño Southern Oscillation (ENSO)*. The "El Niño" part of this phenomenon is an irregular but episodic warming of surface waters in the eastern equatorial Pacific. The "Southern Oscillation" is a global-scale seesaw in atmospheric pressure at the earth's surface—specifically, at Indonesia, northern Australia, and the southeastern Pacific. This area of the Southern Hemisphere is the world's largest reservoir

of warm water (Figure *a*), and more warm, moisture-laden air rises here than anywhere else. Rainfall is also heavy here, and it releases much of the heat energy that drives the world's air circulation system.

Every two to seven years, the warm reservoir and the associated heavy rainfall move eastward (Figure *b*). This causes prevailing surface winds in the western equatorial Pacific to pick up speed. The stronger winds have a more pronounced effect on "dragging" the ocean surface waters eastward. Upper ocean currents are affected to the extent that the westward transport of water slows down and the eastward transport increases. The outcome? *More* warm water in the vast reservoir moves east—and so on in a feedback loop between the ocean and the atmosphere.

The rainfall pattern in the Pacific and Indian oceans is massively dislocated when ENSO warm episodes occur. During the 1982–1983 episode, the vital monsoon rains hardly materialized over India—and record droughts occurred in Australia and nearby regions as well as in the Hawaiian Islands. Month after month, record rainfall drenched the arid and semiarid coasts of Ecuador and Peru. The ENSO also prolonged a devastating drought that already was under way in Africa, with its consequent and appalling human starvation.

Numerical models are now being used to study these and other episodes of climatic change a few seasons in advance. More reliable forecasting should follow when more and better observations are made of the interrelated systems of the ocean, land, and atmosphere.

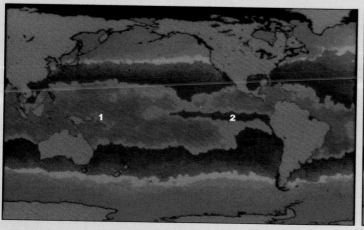

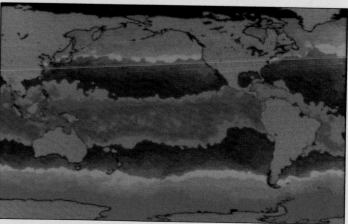

**a** Satellite images of the distribution of ocean surface temperatures in normal years, with the warmest waters found in the western equatorial Pacific (1), and a tongue of relatively cold water extending westward along the equator from South America (2).

**b** Distribution of ocean surface temperatures that were associated with the 1982–1983 ENSO episode.

## SUMMARY

1. The biosphere is the narrow zone of water, the lower atmosphere, and the fraction of the earth's crust in which organisms live. It is composed of ecosystems, each influenced by the flow of energy and the movement of materials on a global scale.

2. The world distribution of species is a result of "accidents of history," climate, topography, and species interactions.

3. Climate means prevailing weather conditions, including temperature, humidity, wind velocity, degree of cloud cover, and rainfall. It is an outcome of differences in the amount of solar radiation reaching equatorial and polar regions, the earth's daily rotation and its annual path around the sun, the distribution of continents and oceans, and the elevation of land masses.

4. Climatic factors interact to produce prevailing winds and ocean currents, which together influence global weather patterns. Weather affects soil composition, sedimentation, and water availability, which influence the growth and distribution of primary producers. Through these interactions, climate influences ecosystems.

5. The earth's land masses can be classified as six major biogeographic realms, each with characteristic types of plants and animals and each more or less isolated by oceans, mountain ranges, or desert barriers.

6. Biomes are particular types of major ecosystems. They are shaped by regional variations in climate, landforms, and soil composition. Each is dominated by plant species adapted to a particular set of conditions. The major types are deserts, dry shrublands and woodlands, grasslands, forests, and tundra.

7. The water provinces, which cover more than 70 percent of the earth's surfaces, include standing freshwater (such as lakes), running freshwater (such as streams and rivers), and oceans and seas. Their marine ecosystems include estuaries, intertidal zones, rocky and sandy shores, tropical reefs, and scattered ecosystems of the open oceans.

8. All freshwater and marine ecosystems have gradients in light penetration, temperature, salinity, and dissolved gases, features that vary daily and seasonally and that influence primary productivity.

9. In the open ocean, photosynthetic activity is greatest in shallow coastal waters and in regions of upwelling along the margins of continents. Upwelling is an upward movement of deep, cooler water that carries nutrients to the surface.

10. The interrelatedness of ocean surface temperatures, the atmosphere, and the land is especially clear through studies of the El Niño Southern Oscillation. This recurring phenomenon is accompanied by abnormal drought conditions in many parts of the world.

### Review Questions

1. Define climate. What interacting factors influence climate? What does climate in turn influence? *848–850*

2. How do prevailing air and ocean currents help dictate the distribution of different types of ecosystems? *848–851*

3. Describe the rain shadow effect. *850*

4. Distinguish between biogeographic realm and biome. In what type of biome would you say you live? *851*

5. How does the composition of regional soils affect ecosystem distribution? *852–853*

6. List some of the characteristics that all freshwater and marine ecosystems have in common. *862*

7. Spend some time outdoors observing the land or any freshwater or marine ecosystems around you. Write a short essay on some of the features you are able to identify.

8. Define upwelling, and give an example of a region where upwelling has a profound effect on primary productivity. *870*

### Self-Quiz *(Answers in Appendix IV)*

1. The primary energy source for ecosystems, and a determining factor in their distribution, is _____.

2. Heat energy derived from the sun warms the _____, and that energy drives the earth's _____.

3. Global air circulation patterns, ocean currents, and topography interact to create regional differences in _____.

4. The term "biosphere" includes _____.
   a. water zones
   b. the lower atmosphere
   c. the portion of the earth's crust inhabited by organisms
   d. all of the above are included in the term "biosphere"

5. "Accidents of history," climate, topography, and the interactions of species profoundly affect the world distribution of
   _____.
   a. water
   b. oxygen
   c. species
   d. weather
   e. solar radiation

6. Nearly all ecosystems are influenced by climate (prevailing weather conditions), which is shaped by _____.
   a. variations in amount of incoming solar radiation
   b. earth's daily rotation and annual path around sun
   c. world distribution of continents and oceans
   d. elevation of land masses
   e. all of the above are factors that shape climate

7. The earth's major _____ have characteristic types of plants and animals and are, for the most part, isolated by oceans, desert barriers, and mountain ranges.
   a. biomes
   b. ecosystems
   c. biogeographic realms
   d. water provinces

8. Examples of _____ are deserts, shrublands, grasslands, forests, and tundra.
   a. biomes
   b. ecosystems
   c. biogeographic realms
   d. plant populations

9. Gradients in light penetration, temperature, salinity, and dissolved gases are all features of _____ ecosystems.
   a. marine
   b. forest
   c. desert
   d. freshwater
   e. both a and d are correct

10. Match the biosphere terms appropriately.
   _____ biomes
   _____ biogeographic realms
   _____ climate
   _____ biosphere
   _____ solar radiation
   _____ water

   a. earth's six major land regions
   b. covers 70 percent of the earth's surface
   c. temperature, humidity, wind velocity, degree of cloud cover, and rainfall
   d. zones of the earth's crust, waters, and atmosphere that support life
   e. primary energy source for ecosystems
   f. broad subdivisions of biogeographic realms

## Selected Key Terms

atmosphere 847
benthic province 869
biogeographic realm 851
biome 851
biosphere 847
climate 848
convergent evolution 851
El Niño Southern Oscillation (ENSO) 871
estuary 865
eutrophication 864
fall overturn 862
hydrosphere 847
hydrothermal vent 870
intertidal zone 866

lake ecosystem 862
limnetic zone 862
littoral 862
oceanic zone 869
oligotrophic 863
ozone layer 848
pelagic province 867
permafrost 861
plankton 862
rain shadow 850
soil 852
spring overturn 862
stream 864
thermocline 862
upwelling 870

## Readings

Colinvaux, P. May 1989. "The Past and Future Amazon." *Scientific American* 260(3):102–108.

Gibbons, B. September 1984. "Do We Treat Our Soil Like Dirt?" *National Geographic* 166(3):350–388.

Ricklefs, R. 1990. *Ecology*. Third edition. New York: Freeman.

Smith, R. 1989. *Ecology and Field Biology*. Fourth edition. New York: Harper & Row. Good descriptions of biomes.

Sumich, J. 1988. *Biology of Marine Life*. Fourth edition. Dubuque, Iowa: Brown.

# 48 HUMAN IMPACT ON THE BIOSPHERE

b

## *Tropical Forests — Disappearing Biomes?*

On his first night out on an expedition into a tropical forest, a young graduate student carefully covered his body with a cotton mosquito net. He woke up before dawn with sudden, gut-wrenching awareness that things were crawling all over him. During the hot, humid night, a platoon of large, voracious insects had devoured most of the drenched net.

Tropical forests contain not only hungry insects but also the world's biggest ones. They are home to the greatest variety of spectacularly plumaged birds and to the plants with the largest flowers (Figure 48.1). Large animals are not as profuse as they are in other parts of the world, but they are splendidly varied. They range from the tapirs, monkeys, and jaguars of South America to the apes, okapi, and leopards of Africa. Specimens of trees from these forests may have made their way into your home or your dentist's waiting

room—*Ficus benjamina*, rubber plants, and tree ferns, to name a few. In the forests themselves, vines twist around tree trunks and grow up toward the sunlight. Orchids, mosses, lichens, and other plants grow on tree branches, absorbing minerals brought to them in rainwater. Entire communities of insects, spiders, and amphibians live, breed, and die in the small pools of water that collect in the leaves of aerial plants.

Developing countries in Latin America, Africa, and Southeast Asia have been clearing their tropical forests on a massive scale. Those countries have great and growing numbers of people and not enough food, fuel, or lumber. (One of every four humans lives in those countries which, in 1990, were responsible for about one-third of the world's human population growth.) At present rates of clearing, most tropical forests will disappear within your lifetime.

a

c

**Figure 48.1** (**a**) El Yunque rain forest, Puerto Rico. Among the diverse occupants of different tropical rain forests are (**b**) jaguars and (**c**) *Rafflesia*, a leafless, foul-smelling plant with a fly-pollinated flower that can grow 3 meters across.

Why does it matter? For purely ethical reasons, many condemn the total destruction of a major chunk of the biosphere. For practical reasons, consider how the destruction might affect your own life. The world's food supply depends on a very small number of crop and livestock species. That base of food production can be broadened and made less vulnerable if we can develop new or hybrid crop plants. By using genetic engineering and tissue culturing methods, we can tap the diverse organisms of tropical rain forests as genetic resources. They also can be tapped for developing new antibiotics and vaccines. Many tropical plants already give us the alkaloids for drug treatments of cardiovascular disorders, cancer, and other illnesses. Aspirin, the most widely used drug in the world, was formulated by using a chemical "blueprint" of a compound extracted from the leaves of tropical willow trees. Coffee, bananas, cocoa, cinnamon and other spices, sweeteners, Brazil nuts, and many other foods we take for granted originated in the tropics. So did latex, gums, resins, dyes, waxes, and many oils; these and other substances are used in such diverse products as ice cream, toothpaste, shampoo, condoms, cosmetics, perfumes, compact discs, tires, and shoes.

If aspirin or condoms don't catch your attention, think about this: Massive destruction of tropical forests is changing the mixture of gases in the atmosphere. Among other things, the burning releases carbon dioxide, carbon monoxide, hydrocarbons, nitrogen dioxide, and nitric oxide—and atmospheric concentrations of these gases are increasing. The increases are contributing to acid rain, smog, holes in the ozone layer, and probably increases in world temperatures.

As an individual, you might choose to cherish, brood about, or ignore any aspect of the world of life. Whatever the choice, the bottom line is that you and all other organisms are in this together. Our lives interconnect, to degrees that we are only now starting to comprehend.

KEY CONCEPTS

**1.** The human population has undergone rapid exponential growth since the mid-eighteenth century. Today we have the population size, the technology, and the cultural inclination to use energy and modify the environment at astonishing rates.

**2.** The world of life ultimately depends on energy from the sun, which drives the complex interactions between the atmosphere, oceans, and land. The accumulation of human-generated pollutants in the atmosphere especially is disrupting those interactions. Those disruptions may have alarming consequences in the near future.

**3.** We as a species must come to terms with the principles of energy flow and resource utilization that govern all systems of life on earth.

## ENVIRONMENTAL EFFECTS OF HUMAN POPULATION GROWTH

Of all the concepts introduced in the preceding chapter, the one that should be foremost in your mind is this: Complex interactions between the atmosphere, oceans, and the land are the engines of the biosphere. Driven by energy from the sun, they produce the worldwide temperatures and circulation patterns on which life ultimately depends. With this chapter, we turn to a related concept of equal importance. Simply put, the human population has been straining the global engines without fully comprehending that engines can crack.

To gain perspective on what is happening, think about something we all take for granted—the air around us. The composition of the present atmosphere is the outcome of geologic and metabolic events, including photosynthesis, that began billions of years ago. At some time between 6 and 4 million years ago, the first humans emerged. Like us, they breathed in oxygen from an atmosphere of ancient origins. Like us, they were protected from harmful ultraviolet radiation by an ozone shield in the stratosphere. Their population sizes were not much to speak of, and their interactions with the biosphere were not significant. About 10,000 years ago, however, agriculture began in earnest, and it laid the foundation for rapid population growth. With agriculture, and with the medical and industrial revolutions that followed, human population growth became exponential in a mere blip of evolutionary time (Figure 44.8).

Today, our burgeoning population may be placing demands on the biosphere that cannot be sustained. As we take energy and resources from it, we give back wastes in monumental amounts. In the process, we are destroying the stability of ecosystems on land, contaminating the hydrosphere, and changing the composition of the atmosphere. Our carbon dioxide wastes alone are contributing to an amplified "greenhouse effect," described on page 838, and this is expected to bring about a warming of the entire planet.

In a few developed countries of Europe, population growth has more or less stabilized, and rates of increase in the resource use per individual have slowed somewhat. But the resource utilization levels in those countries are already high. At the same time, population growth and demands for resources are increasing rapidly in the developing countries of Central America, South America, Asia, Africa, and elsewhere—even though millions there are already starving to death and hundreds of millions more suffer from malnutrition and inadequate health care.

Many of the problems we will consider in this chapter are not going to go away tomorrow. It will take decades, even centuries, to reverse some of the trends already in motion, and not everyone is ready to make the effort. A scattering of enlightened individuals in Michigan or Alberta or New South Wales can make good attempts at resource conservation—but scattered attempts will not be enough. Individuals of every nation will make a concerted effort to reverse global trends only when they perceive that the dangers of *not* doing so outweigh the personal benefits of ignoring them.

Does this seem pessimistic? Think about the exhaust fumes being released into the atmosphere every time you drive a car. Think about the oil refineries, paper mills, and food-processing plants that supply you with goods but also release chemical wastes into the nation's waterways. Think about Mexico and other developing countries that produce cheap food by using an unskilled labor force and dangerous pesticides—which poison the people who work the land, the land itself, perhaps even you. Who changes behavior first? We have no answer to the question. We suggest, however, that a strained biosphere can rapidly impose an answer upon us.

## CHANGES IN THE ATMOSPHERE

If you were to compare the earth to an apple from the supermarket, the atmosphere would be no thicker than the layer of shiny wax applied to that apple. Yet this thin, finite wrapping of air around the planet receives more than 700,000 metric tons of pollutants each day in the United States alone. **Pollutants** are substances with which ecosystems have had no prior evolutionary experience, in terms of kinds or amounts, and so have no mechanisms for dealing with them. From the human perspective, pollutants are substances that adversely affect our health, activities, or survival.

Table 48.1 lists the major classes of air pollutants. They include carbon dioxide, sulfur oxides, nitrogen oxides, and chlorofluorocarbons (CFCs). They also include photochemical oxidants, formed by interactions between sunlight and many of the chemicals that we release to the atmosphere.

### Local Air Pollution

Whether air pollutants are dispersed through the atmosphere or concentrated at their source in a given time period depends on local climate and topography. Consider what happens during a **thermal inversion**, when weather conditions trap a layer of dense, cool air beneath a layer of warm air (Figure 48.2). The pollutants cannot be dispersed by winds or rise higher in the atmosphere, so they accumulate to dangerous levels close to the ground. By intensifying a phenomenon known as smog, thermal inversions have contributed to some of the worst local air pollution disasters.

Two types of smog (gray air and brown air) can occur in major cities. Where winters are cold and wet, **industrial smog** forms as a gray haze over industrialized cities that burn coal and other fossil fuels for heating, manufacturing, and producing electric power. The burning

| Table 48.1 | Major Classes of Air Pollutants |
|---|---|
| Carbon oxides | Carbon monoxide (CO), carbon dioxide ($CO_2$) |
| Sulfur oxides | Sulfur dioxide ($SO_2$), sulfur trioxide ($SO_3$) |
| Nitrogen oxides | Nitric oxide (NO), nitrogen dioxide ($NO_2$), nitrous oxide ($N_2O$) |
| Volatile organic compounds | Methane ($CH_4$), benzene ($C_6H_6$), chlorofluorocarbons (CFCs) |
| Photochemical oxidants | Ozone ($O_3$), peroxyacyl nitrates (PANs), hydrogen peroxide ($H_2O_2$) |
| Suspended particles | Solid particles (dust, soot, asbestos, lead, etc.), liquid droplets (sulfuric acid, oils, dioxins, pesticides) |

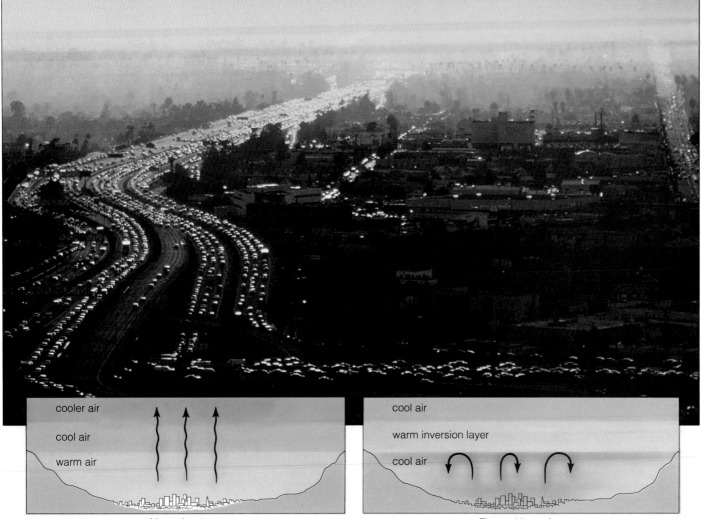

| | |
|---|---|
| cooler air | cool air |
| cool air | warm inversion layer |
| warm air | cool air |
| Normal pattern | Thermal inversion |

releases airborne pollutants, including dust, smoke, ashes, soot, asbestos, oil, bits of lead and other heavy metals, and sulfur oxides. When winds and rain do not disperse them, the pollutants may reach lethal concentrations. Industrial smog was the cause of London's 1952 air pollution disaster, in which 4,000 people died. Like London, New York, Pittsburgh, and Chicago once were gray air cities until they restricted coal burning. Now, most industrial smog forms in cities of developing countries, including China and India, as well as in Hungary, Poland, and other countries of eastern Europe.

In warm climates, **photochemical smog** forms as a brown, smelly haze over large cities. Where the surrounding land forms a natural basin, as it does around Los Angeles and Mexico City, photochemical smog can reach harmful levels. The main culprit is nitric oxide, produced mainly by cars and other vehicles with internal combustion engines. When nitric oxide reacts with oxygen in the air, nitrogen dioxide forms. When exposed to sunlight, nitrogen dioxide can react with hydrocarbons (spilled or partly burned gasoline, most often) to form photochemical oxidants. The main oxidants in smog are ozone and PANs (short for *peroxyacyl nitrates*). PANs are similar to tear gas; even traces can sting the eyes, irritate the lungs, and damage crops.

**Figure 48.2** Trapping of airborne pollutants by a thermal inversion layer. The photograph shows a Los Angeles freeway system under its self-generated blanket of smog at twilight.

## Acid Deposition

Oxides of sulfur and nitrogen are among the most dangerous air pollutants. Coal-burning power plants, factories, and metal smelters are the main sources of sulfur dioxides. Vehicles, power plants that burn fossil fuels, and nitrogen fertilizers are the main sources of nitrogen oxides.

Depending on climatic conditions, tiny particles of these substances may be airborne for a while and then fall to earth as **dry acid deposition**. Most sulfur and nitrogen dioxides dissolve in atmospheric water to form a weak solution of sulfuric acid and nitric acid. Winds can distribute them over great distances before they fall to earth in rain and snow. This is **wet acid deposition**. Normal rainwater has a pH of about 5. Acid rain can be ten to a hundred times more acidic than this, sometimes becoming as acidic as lemon juice. The acids chemically attack marble, metals, mortar, rubber, plastic, even nylon stockings. They also disrupt the physiology of organisms and the chemistry of ecosystems.

Because soils and vegetation are not identical in all watersheds, some regions are more sensitive to acid deposition than others (Figure 48.3). Highly alkaline soils neutralize the acids before runoff carries them into lakes and streams. Water with high concentrations of carbonates also will neutralize the acids. However, in watersheds throughout much of northern Europe, southeastern Canada, and in scattered regions of the United States, thin soils overlie solid granite—and such soils provide little buffer against the acids.

The precipitation in much of eastern North America is thirty to forty times more acidic than it was several decades ago, and croplands and forests are suffering (Figure 48.4). All fish populations have vanished from more than 200 lakes of the Adirondack Mountains of New York. Some Canadian biologists predict that within the next two decades, fish will disappear from 48,000 lakes in Ontario. Acidic pollutants originating in industrial regions of England and Germany are key factors in the destruction of large tracts of forests in northern Europe. Such pollutants are being linked to the decline of spruce and fir at high elevations in the Appalachian Mountains. They also are emerging as a serious problem in heavily industrialized parts of Asia, Latin America, and Africa.

Researchers confirmed years ago that power plants, factories, and vehicles are the main sources of acid pollutants, and that acid deposition is indeed damaging the environment. In fact, some responses to local air pollution problems that were recognized decades ago actually contributed to our current problems. For example, tall smokestacks were added to power plants and smel-

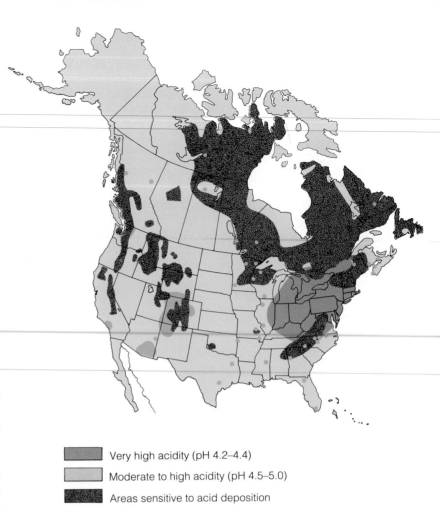

Very high acidity (pH 4.2–4.4)

Moderate to high acidity (pH 4.5–5.0)

Areas sensitive to acid deposition

**Figure 48.3** Average acidity of precipitation, and soil sensitivity to acid deposition, for regions of North America (1984).

**Figure 48.4** Dead spruce trees in the Whiteface Mountains of New York (**a**) and in Germany (**b**). Acid deposition, perhaps in combination with prolonged exposure to other pollutants, seems to be contributing to the rapid destruction of trees in these and other forests. In some cases, prolonged exposure to multiple air pollutants directly damages the trees, especially conifers. In many cases, the pollutants weaken the trees and make them more susceptible to drought, disease, and insect attacks.

a

b

ters. The idea was to dump the smoke high in the atmosphere so winds could distribute it elsewhere—which winds readily did. The winds helped transform local air pollution into regional acid deposition. Today the world's tallest smokestack, in Sudbury, Ontario, accounts for about 1 percent by weight of the annual worldwide emissions of sulfur dioxide.

But Canada cannot be singled out in this issue. Canada presently receives more acid deposition from industrialized regions of the midwestern United States than it sends across its southern border. Most of the acidic pollutants in Finland, Norway, Sweden, the Netherlands, Austria, and Switzerland are blown there from industrialized regions of western and eastern Europe. Prevailing winds do not stop at national boundaries; the problem is of global concern.

### Damage to the Ozone Layer

The ozone layer in the lower stratosphere has been thinning. Each year, from September through mid-October, an ozone "hole" appears over the Antarctic. It extends over an area about the size of the continental United States. Less pronounced thinning extends into the mid-latitudes. Why worry about this? The ozone layer absorbs most of the ultraviolet wavelengths from the sun. And those wavelengths harm organisms (page 848).

Satellites and high-altitude planes have been monitoring the ozone hole for more than a decade (Figure 48.5). Since 1987, the ozone layer over the Antarctic has been thinning by about half every year. Since 1969, it has decreased by 3 percent over heavily populated regions of North America, Europe, and Asia. By 2050, the ozone layer over those regions may decrease 10 to 25 percent.

With each reduction in the ozone layer, more harmful ultraviolet radiation reaches the earth's surface. The consequences are wide-ranging and potentially serious. Skin cancers are increasing dramatically, and these are almost certainly an outcome of increases in ultraviolet radiation (Figure 36.7). Cataracts, an eye disorder, may become more common. It appears that ultraviolet radiation can weaken the immune system, making individuals more vulnerable to some viral and parasitic infections. Reduction in the ozone layer also may adversely affect the world's populations of phytoplankton—the basis of food

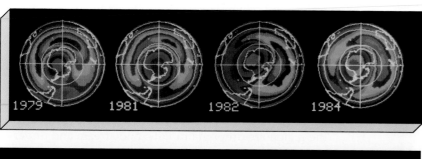

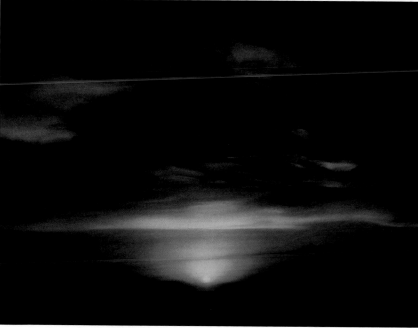

**Figure 48.5** Expansion of the ozone hole over Antarctica from 1979 to October 1987, as recorded by special high-altitude planes. The lowest ozone values (the "hole") are indicated by pink colors in the plots for 1979 through 1984. In the plot for 1987, the lowest value ever recorded by that year is indicated by the black area at the center of the plot. The photograph shows the ice clouds over Antarctica that play a role in the formation of an ozone hole each spring.

webs in freshwater and marine ecosystems and a factor in maintaining the composition of the atmosphere. Collectively, these photosynthesizers act to remove carbon dioxide from surface waters and release oxygen (page 50).

More than any other factor, chlorofluorocarbons (CFCs) are responsible for the ozone reduction. CFCs are odorless, invisible, and otherwise harmless compounds of chlorine, fluorine, and carbon. They are widely used as propellants in aerosol spray cans, coolants in refrigerators and air conditioners, industrial solvents, and plastic foams. CFCs enter the atmosphere slowly and resist breakdown. By some estimates, about 95 percent of the CFCs released between 1955 and 1990 are still making their way up to the stratosphere.

When a CFC molecule absorbs ultraviolet light, it gives up a chlorine atom. The chlorine can react with ozone to form an oxygen molecule and a chlorine monoxide molecule. When the chlorine monoxide reacts with a free oxygen atom, another chlorine atom is released that can attack another ozone molecule. Each chlorine atom released in the reactions can convert as many as 10,000 molecules of ozone to oxygen!

Recent studies show that chlorine monoxide levels above Antarctica are 100 to 500 times higher than at mid-latitudes. Why? High-altitude clouds of ice form there during the frigid winters, and they are isolated from other latitudes by winds that rotate around the south pole for most of the winter months (Figure 48.5). The same thing happens, to a lesser extent, in the Arctic. The ice provides a surface that facilitates the breakdown of chlorine compounds, so that chlorine is free to destroy ozone when the Antarctic air warms somewhat in the Antarctic spring. Hence the ozone hole.

Since 1978, the United States, Canada, and most Scandinavian countries have banned the use of CFCs in aerosol spray cans. Aerosol uses have risen sharply in western Europe, however, as have nonaerosol uses of CFCs throughout the world. In late 1987, an international group assembled by the United Nations Environment Program agreed to a draft treaty to halve CFC emissions by the year 1999. Most nations seem certain to ratify its provisions. The treaty is a step in the right direction, although some feel that it is too little and too late. CFCs already in the air will be there for over a century, before natural processes neutralize them. You, your children, and your grandchildren will be living with their destructive effects. Think about that, the next time you turn on the air conditioner in your car.

## CHANGES IN THE HYDROSPHERE

There is a tremendous amount of water in the world, yet three of every four humans do not have enough water or, if they do, it is contaminated. Most water is too salty for human consumption or agriculture. For every million liters of water, only about 6 liters are in a readily usable form.

### Consequences of Large-Scale Irrigation

Expansion of agricultural production is coupled with the exponential growth of the human population—and about one-third of the food being produced today grows on irrigated land (Figure 48.6). Water is piped into agricultural fields from groundwater or from lakes, reservoirs, and other sources of surface waters.

Irrigation can change the productivity of the land. Water available for irrigation often is loaded with mineral salts. In regions where the soil does not drain well, evaporation may cause **salination** (salt buildup) in the soil. Salination can stunt growth, decrease yields, and eventually kill crop plants.

Also, improperly drained irrigated lands can become waterlogged. Water accumulating underground can gradually raise the *water table* (the upper limit of ground that is fully saturated with water). When the water table is close to the surface, the soil around plant roots becomes saturated with toxic saline water. Salinity and waterlogging can be corrected with proper management of the water-soil system. However, the economic cost is high.

We use groundwater for many purposes, but irrigation is often paramount. Consider what is happening to the Ogallala aquifer in the United States (Figure 48.7). Farmers withdraw so much water from the aquifer that the annual overdraft (the amount of water not replenished) is nearly equal to the annual flow of the Colorado River! As a result, the already low water tables in much of the region are dropping rapidly, and stream and underground spring flows are dwindling. Where will the water come from when we deplete the aquifer?

Well, what about the oceans, which cover about three-fourths of the earth's surface? Why not "desalinate" seawater, since there is so much of it and since technologies for doing so already exist? The reason is cost. **Desalination** methods require great amounts of (expensive) energy to remove salts from seawater by distilling it or by forcing it through membranes (a method called reverse osmosis). To be sure, Saudi Arabia and some other Persian Gulf countries desalinate seawater for their urban populations. But those countries are among the very few with huge energy reserves and cash to spare. After suffering through prolonged droughts, cities along the California coast are now willing to consider desalination, as well as reclamation of wastewater to irrigate landscaping. But such efforts cannot solve the core problem. *Desalination may not ever be cost-effective enough for large-scale agriculture—which accounts for almost two-thirds of the human population's annual use of fresh water.*

## Maintaining Water Quality

Not having enough water is serious enough. Yet the problem is being compounded by pollution of the water that *is* available. Water becomes unfit to drink, even to swim in, once it contains human sewage, animal wastes, and various toxic chemicals. Those pollutants can encourage contamination by pathogenic microbes. Agricultural runoff pollutes water with sediments, pesticides, and plant nutrients. Power plants and factories pollute water with chemicals, radioactive materials, and excess heat (thermal pollution).

Pollutants accumulate in lakes, rivers, and bays before reaching their ultimate destination, the oceans. Many cities throughout the world dump untreated sewage into coastal waters. Cities along rivers and harbors maintain shipping channels by dredging the polluted muck and barging it out to sea. They barge out sewage sludge, also. ("Sludge" refers to coarse, settled solids. It contains bacteria, viruses, toxic metals, and possibly hospital wastes.) Sometimes, following storms, this black sludge washes ashore.

The United States has about 15,000 facilities for treating the liquid wastes from about 70 percent of the population and 87,000 industries. The remaining wastes, mostly from suburban and rural populations, are treated in lagoons or septic tanks or discharged—untreated—directly into waterways.

There are three levels of wastewater treatment. In *primary treatment*, screens and settling tanks remove the sludge, which is then dried, burned, dumped in landfills, or treated further. Chlorine often is used to kill pathogens in the water, but it does not kill them all. Also, chlorine may react with certain industrial chemicals to produce chlorinated organic compounds, some of which are carcinogens.

**Figure 48.6** Irrigation-dependent crops growing in the Sahara Desert of Algeria.

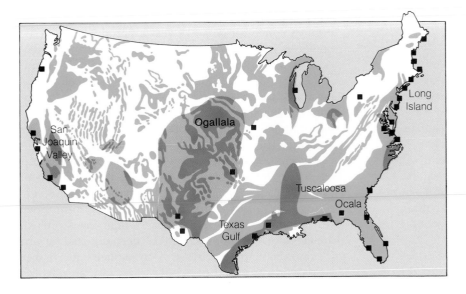

**Figure 48.7** Major underground aquifers containing 95 percent of all fresh water in the United States. These aquifers are being depleted in many areas and contaminated elsewhere through pollution and saltwater intrusion. Blue areas indicate major aquifers; gold, the areas of groundwater depletion; and the black boxes indicate areas of saltwater intrusion.

In *secondary treatment*, microbial populations are employed to degrade the organic matter. Secondary treatment occurs after primary treatment but before chlorination. The wastewater is either (1) sprayed and trickled through large beds of exposed gravel in which the microbes live or (2) aerated in tanks and seeded with microbes. Toxic solutes sometimes poison the microbial "employees," however. When that happens, the treatment facilities are shut down until the microbial populations are reestablished.

The combination of primary and secondary treatment does remove most oxygen-demanding wastes and suspended solids. But considerable nitrogen, phosphorus, and toxic substances, including heavy metals, pesticides, and industrial chemicals still remain. Usually the water gets chlorinated before being released into the waterways, but sometimes not.

*Tertiary treatment* is largely experimental. Tertiary treatment may adequately reduce pollution levels. Being expensive, it is used on only 5 percent of the nation's wastewater.

In short, most wastewater is not being properly treated. A typical pattern is repeated thousands of times along our waterways. Water for drinking is removed *upstream* from a city, and wastes from industry and sewage treatment are discharged *downstream*. It takes no great leap of the imagination to see that pollution intensifies as rivers flow toward the oceans. In Louisiana, where waters drained from the central states flow toward the Gulf of Mexico, pollution levels are high enough to be a threat to public health. Water destined for drinking does get treated to remove pathogens. But the treatment does not remove toxic wastes from numerous factories upstream. *You may find it illuminating to investigate where your own city's supply of water comes from and where it has been.*

## CHANGES IN THE LAND

### Solid Wastes

Resources are scarce in the developing countries, and very few materials are discarded. In the more affluent countries, the United States especially, a "throwaway" mentality prevails. Consumers use something once, discard it, then buy another.

Billions of metric tons of solid wastes are dumped, burned, and buried annually in the United States alone. About 60 billion beverage containers are part of it—50 billion of which are nonreturnable cans and bottles. Fully one half of the total volume of solid wastes are paper products.

It takes more than 500,000 trees to supply the Sunday paper to Americans. Every week. If everyone in the United States recycled even one out of ten newspapers, we could save 25 million trees a year. Besides this, using recycled paper reduces air pollution that results from paper manufacture by 95 percent. And recycling paper takes 30 to 50 percent less energy than making new paper.

Associated with the throwaway mentality is a problem that is unique in the world of life—what to do with the solid wastes. Instead of recycling materials, as is done in natural ecosystems, we bury them in landfills or burn them in incinerators. Incinerators add some toxic pollutants to the air and leave a highly toxic ash that must be disposed of safely. In any event, land available and acceptable for landfills is scarce. Because all landfills eventually "leak," they pose a threat to groundwater supplies.

A transition from a throwaway mentality to one based on recycling and reuse is feasible. It is affordable, and we have the technology to do it. Consumers can influence manufacturers by refusing to buy goods that are lavishly wrapped and boxed, packaged in indestructible containers, and designed for one-time use. Individuals can ask the local post office to turn off their daily flow of junk mail. Unsolicited mail wastes an astounding amount of paper, time, and energy—and means higher postage rates for everyone.

Finally, individuals can urge local governments to develop well-designed, large-scale resource recovery centers. Such a plant has been operating in Saugus, Massachusetts. With such systems, existing dumps and landfills would be urban "mines" from which usable materials can be sorted out and recovered.

### Conversion of Marginal Lands for Agriculture

Most of the world's prime agricultural land has long been exploited for crop or livestock production. Today, however, human population growth is forcing expansion of agriculture onto marginally productive land. Almost 21 percent of the earth's land is now being used for agriculture. Another 28 percent is said to be potentially suitable for cropland or grazing land. But its potential productivity is so low that conversion may not be worth the cost (Figure 48.8).

Asia and some other heavily populated regions suffer severe food shortages, yet more than 80 percent of their productive lands are already being intensively cultivated. Valiant efforts have been made to improve crop production on existing land. Under the banner of the **green revolution**, research has been directed toward (1) improving the varieties of crop plants for higher yields and (2) exporting modern agricultural practices

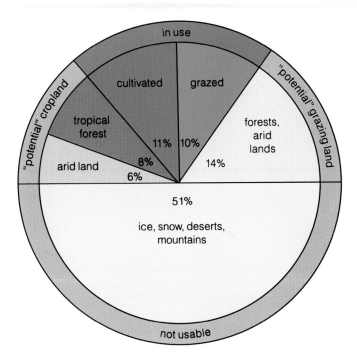

**Figure 48.8** Classification of the earth's land. Theoretically, the world's cropland could be more than doubled by clearing tropical forests and irrigating arid lands. But converting this marginal land to cropland would destroy valuable forest resources, cause serious environmental problems, and possibly cost more than it is worth.

And such importations add to the foreign debt of developing countries.

Pressures for increased food production are most intense in areas of Central and South America, Asia, the Middle East, and Africa. There, human populations are rapidly expanding into marginal lands. The repercussions will extend beyond national boundaries, as the following discussion will make clear.

### Deforestation

The world's great forests play major roles in the biosphere. As we saw in Chapter 46, forested watersheds are like giant sponges that absorb, hold, and release water gradually. By influencing the downstream flow of water, forests help control soil erosion, flooding, and sediment buildup in rivers, lakes, and reservoirs. **Deforestation**, the removal of all trees from large tracts of land, leads to loss of fragile surface layers of soil and disrupts watersheds. The disruption is especially pronounced in watersheds with steep slopes.

In the tropics, soil loss means long-term fertility loss as nutrients are quickly washed out of the system, leaving nutrient-poor soil behind. As the introduction to this chapter pointed out, tropical forests have great diversity. But despite that diversity, they are one of the worst places to grow crops and raise pasture animals. Because temperatures remain high and rain falls heavily and often, conditions favor rapid decomposition. This means that dead tissues and wastes decompose too fast for much litter to build up on the forest floor. Plant roots absorb the nutrients as they become available, and these become incorporated into plant tissues. Few of the nutrients released during the decomposition processes remain in the subsoil.

Through **shifting cultivation** (once called slash-and-burn agriculture), forests can be cut and burned, then the ashes tilled into the soil. Crops can be grown for one to several seasons after that. Nutrients are quickly leached from the exposed soils, however. The cleared plots soon become infertile and are abandoned. Shifting cultivation on small, widely scattered plots probably does not inflict much damage on forest ecosystems. When larger areas are cleared and when plots are cleared again at shorter intervals, this practice can degrade the fertility of the entire forest.

More than this, deforestation also can change regional patterns of rainfall as a result of altered rates of evaporation, transpiration, and runoff. Between 50 and 80 percent of the water vapor above tropical forests alone is released from the trees themselves. Without trees, annual precipitation declines and the region gets hotter and drier. Rain that does fall rapidly runs off the bare soil.

and equipment to developing countries. Many people in those countries rely on *subsistence agriculture*, with energy inputs from sunlight and from human labor. They rely also on *animal-assisted agriculture*, with energy inputs from draft animals, such as oxen.

By contrast, intensive, mechanized agriculture is based on massive inputs of fertilizers, pesticides, and ample irrigation to sustain high-yield crops. It is based also on fossil fuel energy to drive farm machines. Crop yields *are* four times as high. But the modern practices use up a hundred times more energy and mineral resources.

The plain truth is that many farmers in developing countries cannot afford to take widespread advantage of new, high-yield crop strains. Of necessity, the costs of fertilizers and machinery are reflected in market food prices—which are too high for much of the country's own population. The ones who can afford the investment come to depend on industrialized producers of fertilizers and machinery. These items often must be imported.

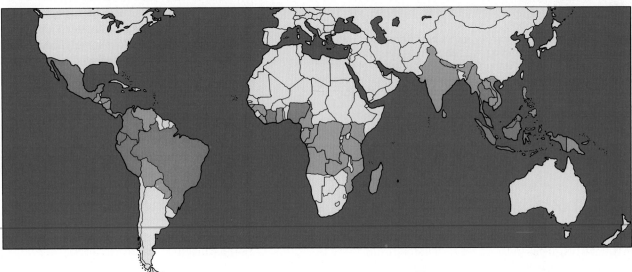

**Figure 48.9** Countries in which the largest destruction of tropical forests is occurring. Red shading signifies where 2,000 to 14,800 square kilometers are deforested annually; orange signifies regions of more moderate deforestation (100 to 1,900 square kilometers).

As the local climate gets hotter and drier, soil fertility and moisture levels decline even more. Eventually, sparse grassland or even desertlike conditions might prevail where there had once been a rich tropical forest.

Clearing large tracts of tropical forests also may have global repercussions. These forests absorb much of the solar radiation reaching the equatorial regions of the earth's surface. When they are cleared, the land becomes shinier, so to speak, and reflects more incoming energy back into space. Also, the trees of these vast forested regions help maintain the global cycling of carbon and oxygen through their photosynthetic activities.

When they are harvested or burned, the carbon stored in their biomass is released to the atmosphere in the form of carbon dioxide—and this may play a role in the amplified greenhouse effect.

Almost half of the world's expanses of tropical forests have already been cleared for cropland, grazing land, timber, and fuelwood. Deforestation is greatest in Brazil, Indonesia, Colombia, and Mexico (Figures 48.9 and 48.10). At present rates of clearing and degradation, only Brazil and Zaire will have large tracts of tropical forest in the year 2010. By 2035, most of those forests also will be gone.

**Figure 48.10**  A view from space of the vast Amazon River Basin of South America in September of 1988. All features of the land were obscured by smoke from the fires that had been set deliberately to clear tropical forests, pasturelands, and croplands during the dry season. The smoke extends to the Andes Mountains on the horizon to the west, approximately 650 miles (nearly 10,500 kilometers) away.

This smoke cover was the largest that astronauts had ever before observed. It extended 1,044,000 square miles over the Amazon Basin, an area comparable to more than one-third of the forty-eight contiguous United States. The smoke plume near the center of the photograph alone covered an area comparable to the huge forest fire in Yellowstone National Park in that same year.

**Figure 48.11** Dust storm approaching Prowers County, Colorado, in 1934. The Great Plains of the American Midwest are normally dry, windy, and subject to severe recurring droughts. Beginning in the 1870s, the land was converted to agriculture. Overgrazing destroyed large regions of natural grassland, leaving the ground bare. In May 1934, the entire eastern portion of the United States was blanketed with a massive dust cloud of topsoil blown off the land, giving the Great Plains a new name— the Dust Bowl. About 3.6 million hectares (about 9 million acres) of cropland were destroyed and over 32 million more severely damaged. Today, without massive irrigation and intensive conservation farming, desertlike conditions could prevail in this region.

**Figure 48.12** Desertification in the Sahel, a region of West Africa that forms a belt between the dry Sahara Desert and tropical forests. This is savanna country that is rapidly undergoing desertification as a result of overgrazing and overfarming.

## Desertification

**Desertification** refers to the conversion of grasslands, rain-fed cropland, or irrigated cropland to a more desertlike state, with a drop in agricultural productivity of 10 percent or more. Worldwide, about 9 million square kilometers have become desertified over the past fifty years. At least 200,000 square kilometers are still being affected each year. Prolonged drought may accelerate desertification, as it did in the American Great Plains many decades ago (Figure 48.11). Today, however, overgrazing on marginal lands is the main cause of large-scale desertification.

In Africa, for example, there are too many cattle in the wrong places. Cattle require more water than the wild herbivores that are native to the region. This means the cattle have to move back and forth between grazing areas and watering holes. As they do, they trample grasses and compact the soil surface (Figure 48.12). In contrast, gazelles, elands, and other native herbivores obtain most (if not all) of the water they require from the plants they eat. They also are better at conserving water; little is lost in feces, compared to cattle.

In 1978, the biologist David Holpcraft started a ranch composed of antelope, zebra, giraffe, ostrich, and other native herbivores. He is raising cattle as "control groups" in order to compare costs and meat yields on the same land. So far, results are exceeding expectations. Native herds are increasing steadily and yielding meat. And range conditions are improving, not deteriorating. There still are some vexing problems to overcome. African tribes have their own idea of what constitutes "good" meat, and some tribes view cattle as the symbols of wealth in their society.

# A QUESTION OF ENERGY INPUTS

Paralleling the J-shaped curve of human population growth is a steep rise in total and per capita energy consumption. The rise is due not only to increased numbers of energy users, but also to extravagant consumption and waste.

For example, in one of the most temperate of all climates, a major university constructed seven- and eight-story buildings with narrow, sealed windows. The windows cannot be opened to catch the prevailing ocean breezes. The buildings and their windows were not designed or aligned to take advantage of the abundant sunlight for passive solar heating and breezes for passive cooling. Massive energy-demanding cooling and heating systems are used instead.

When you hear talk of abundant energy supplies, keep in mind that there is an enormous difference between the total supply and the net amount available. **Net energy** is the energy left over after subtracting the energy used to locate, extract, transport, store, and deliver energy to consumers. Some sources of energy, such as direct solar energy, are renewable. Others, such as coal and petroleum, are not. Currently, 82 percent of the energy stores being tapped falls in the second category (Figure 48.13).

## Fossil Fuels

**Fossil fuels** are the carbon-containing remains of plants that lived hundreds of millions of years ago (page 34). The plants were buried and compressed in sediments and gradually transformed into coal, petroleum (oil), and natural gas. They are nonrenewable resources.

Even with stringent conservation efforts, known petroleum and natural gas reserves may be depleted during the next century. As petroleum and natural gas deposits become depleted in easily accessible areas, we begin to seek new sources. Our explorations have been taking us to wilderness areas in Alaska and to other fragile environments, such as the continental shelves. The *net* energy decreases as costs of extraction and transportation to and from remote areas increase. And the environmental costs of extraction and transportation escalate. At present, the long-term impact of the 11-million-gallon spill from the tanker *Valdez* in Alaska's coastal waters is still not understood.

Besides this, the known reserves are vulnerable to human folly and to what has become known as "environmental terrorism." There is no question that the long-term impact of the release of 460 million gallons of oil into the Persian Gulf during the war over Iraq's invasion of Kuwait will be appalling.

Colorado, Utah, and Wyoming probably have more potential oil than the entire Middle East. These states

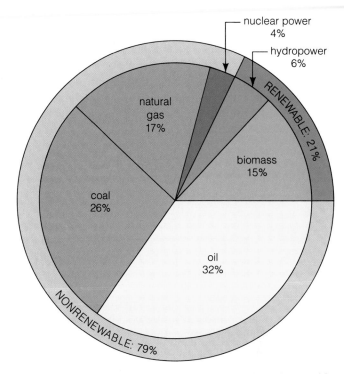

**Figure 48.13** World consumption of nonrenewable and renewable energy sources in 1986.

have vast deposits of oil shale, a type of buried rock containing the hydrocarbon kerogen. However, collecting, concentrating, heating, and converting kerogen into shale oil may cost so much that the net energy yield would be low. The extraction process would disfigure the land, increase water and air pollution, and tax existing water supplies in regions already facing water shortages.

What about coal? In theory, world reserves can meet the energy needs of the entire human population for at least several centuries. But coal burning has been the largest single source of air pollution. Most coal reserves contain low-quality, high-sulfur material. Unless the sulfur is removed before or after burning, sulfur dioxides are released into the air. They add to the global problem of acid deposition. Fossil fuel burning also puts carbon dioxide into the atmosphere and so adds to the greenhouse effect.

Pressure is on to permit widespread strip-mining of coal reserves close to the earth's surface. Strip mining limits the usefulness of the land for agriculture, grazing, and wildlife. Restoration of the land is difficult and expensive in arid and semiarid lands, where much strip-mining is proceeding.

## Nuclear Energy

As Hiroshima burned in 1945, the world recoiled in horror from the destructive potential of nuclear energy. Optimism replaced horror as nuclear energy became publicized during the 1950s as an instrument of progress. Today, nuclear power plants dot the landscape. France and other industrialized nations that are poor in energy resources depend heavily on nuclear power. Yet in most countries, plans to rely more on nuclear energy have been delayed or cancelled. The cost, efficiency, environmental impact, and safety of nuclear energy are being seriously questioned.

By 1990, the use of nuclear energy to generate electricity in the United States was costing slightly more than using coal—even when the coal-burning plants are equipped with expensive pollution control devices. At present, nearly all electricity-generating methods have average costs below those of new nuclear power plants. The use of solar energy is the exception, but by the year 2000, solar energy with natural gas backup is expected to be lower in cost also.

What about safety? Radioactivity escaping from a nuclear plant during normal operation is actually less than the amount released from a coal-burning plant of the same capacity. However, there is the potential danger of a **meltdown**. As nuclear fuel breaks down, it releases considerable heat. Typically, water that is circulating over the fuel absorbs the heat. The heated water produces steam, which drives electricity-generating turbines. Should a leak develop in the circulating water system, water levels around the fuel might plummet. The nuclear fuel would heat rapidly, past its melting point. The melting fuel would pour onto the floor of the generator, where it would come into contact with the remaining water and instantly convert it to steam. Formation of enough steam, along with other chemical reactions, could blow the system apart. Radioactive material would be released into the surroundings. Also, the overheated reactor core could melt through its thick concrete containment slab and into the earth, causing groundwater contamination.

In 1986, the potential dangers of nuclear power were brought into sharp focus when a meltdown occurred at the Chernobyl power station in the Soviet Union. Radiation was released into the atmosphere as the plant's containment structures were breached. A number of people died immediately and others died of radiation sickness during the weeks after the accident. Throughout Europe today, people are still concerned about the long-term consequences. How long will the environment be contaminated? How many individuals ultimately will get cancer from such exposure? We are in the midst of our first major, real-world experiment to find out.

The Chernobyl incident underscores the dangers of nuclear accidents. What about routine nuclear wastes? Nuclear fuel cannot be burned to harmless ashes, like coal. After about three years, the fuel elements of a reactor are spent. They still contain uranium fuel. They also contain hundreds of new radioactive isotopes produced during the reactor operation. Taken together, the wastes

## Commentary

### Biological Principles and the Human Imperative

Molecules, cells, tissues, organs, organ systems, multicelled organisms, populations, communities, ecosystems, the biosphere. These are the architectural systems of life, assembled in increasingly complex ways over the past 3.8 billion years. We are latecomers to this immense biological building program. Yet, during the relatively short span of 10,000 years, we have been restructuring the stuff of life at all levels—from recombining DNA of different species to changing the nature of the land, the oceans, and the atmosphere.

It would be presumptuous to think we are the only organisms that have ever changed the nature of living systems. Even during the Proterozoic Era, photosynthetic organisms were irrevocably changing the course of biological evolution by gradually enriching the atmosphere with oxygen. In the present as well as the past, competitive adaptations have assured the rise of some groups, whose dominance has assured the decline of others. Thus change is nothing new to this biological building program. What *is* new is the accelerated, potentially cataclysmic change being brought on by the human population. We now have the population size, the technology, and the cultural inclination to use energy and modify the environment at frightening rates.

Where will rampant, accelerated change lead us? Will feedback controls begin to operate as they do, for example, when population growth exceeds the carrying capacity of the environment? In other words, will negative feedback controls come into play and keep things from getting too far out of hand?

Feedback control will not be enough, for it operates only when deviation already exists. Our explosive population growth and patterns of resource consumption are founded on an illusion of unlimited resources and a

are an enormously radioactive, extremely dangerous collection of materials. As they undergo radioactive decay, they produce tremendous heat. They are immediately plunged into water-filled pools and stored for several months at the power plant. The water cools the wastes and keeps radioactive material from escaping. At the end of the holding period, the remaining isotopes are still lethal, and the decay rates of some of them mean that they must be isolated for at least 10,000 years. If one kind of plutonium isotope ($^{239}$Pu) is not removed, the wastes must be kept isolated for a quarter of a million years!

forgiving environment. A prolonged, global shortage of food or the passing of a critical threshold for the global engines can come too fast to be corrected. At some point, such deviations may have too great an impact to be reversed.

What about feedforward mechanisms? Many organisms have early warning systems. For example, skin receptors sense a drop in outside air temperature. Each sends messages to the nervous system, which responds by triggering mechanisms that raise the core temperature before the body itself becomes dangerously chilled. With feedforward control, corrective measures can begin before change in the external environment significantly alters the system.

Even feedforward controls are not enough for us, for they go into operation only when change in under way. Consider, by analogy, the DEW line—the Distant Early Warning system. This system is like a sensory receptor, one that detects intercontinental ballistic missiles that may be launched against North America. By the time the system detects what it is designed to detect, it may be far too late, not only for North America but for the entire biosphere.

It would be naive to assume we can ever reverse who we are at this point in evolutionary time, to de-evolve ourselves culturally and biologically into becoming less complex in the hope of averting disaster. However, there is reason to believe that we can avert disaster by using a third kind of control mechanism, one that is uniquely our own. We have the capacity to anticipate events *before* they happen. We are not locked into responding only after irreversible change has begun. We have the capacity to anticipate the future—it is the essence of our visions of utopia or of nightmarish hell. Thus we all have the capacity for adapting to a future that we can partly shape. We can, for example, learn to live with less. Far from being a return to primitive simplicity, it would be one of the most complex and intelligent behaviors of which we are capable.

Having that capacity and using it are not the same thing. We have already put the world of life on dangerous ground because we have not yet mobilized ourselves as a species to work toward self-control. Our survival depends on predicting possible futures. It depends on designing and constructing ecosystems that are in harmony not only with what we define as basic human values but also with the biological models available to us. Human values can change; our expectations can and must be adapted to biological reality. *For the principles of energy flow and resource utilization, which govern the survival of all systems of life, do not change.* It is our biological and cultural imperative that we come to terms at last with these principles, and ask ourselves what our long-term contribution will be to the world of life.

Recently the United States government approved a permanent underground storage program. Radioactive wastes are to be sealed in ceramic material, placed in steel cylinders, then buried deep underground in supposedly stable rock formations that are free from exposure to salt water. The first underground depository is not expected to open until 2010. Scientific and political difficulties may postpone the opening until much later.

The development of another type of reactor—the breeder reactor—is being considered. Such a reactor "breeds" fuel by converting an abundant isotope of uranium into a fissionable isotope of plutonium. Even though a conventional reactor cannot explode like an atomic bomb, a breeder reactor *could* undergo a very small nuclear explosion. Besides this, breeder reactors may cost several times more than conventional nuclear reactors. Also, it may take 100 to 200 years to produce enough plutonium to fuel a significant number of other breeder reactors.

Theoretically, a third nuclear power source is fusion power. The idea is to fuse hydrogen atoms to form helium atoms, as this would release considerable energy. The process is analogous to the reactions that produce heat energy in the sun. The scientific, technological, and economic problems associated with developing fusion power are great. Without major breakthroughs, fusion power is not expected to be available to produce electricity on a commercial basis until the last half of the next century, if ever.

Quite probably, nuclear energy has given us the means to cause a mass extinction equal to those of past geologic eras (page 307). According to one scenario, a nuclear exchange involving about one-third of the existing American and Soviet arsenals would probably kill between 40 and 65 percent of the human population, along with a good portion of most other forms of life. Those escaping rapid death would have to remain in shelters for a week to three months or more to avoid exposure to dangerous radiation levels. The nuclear detonations would inject a huge, dark cloud of soot and smoke over most of the earth. The cloud would be dense enough in the Northern Hemisphere to block out the sun. Much of the planet would be thrown into darkness and temperatures would fall below freezing for months—an effect called *nuclear winter*. If the freezing lasted for a shorter time, we might have a nuclear "autumn." But even then, the cold temperatures and darkness would be well beyond the tolerance limits of many plant and animal species.

## SUMMARY

1. For more than a century, the human population has been growing rapidly, with concurrent increases in energy demands and pollution of the atmosphere, the hydrosphere, and the land.

2. Pollutants are substances with which ecosystems have had no prior evolutionary experience (in terms of kinds and amounts) and so have no mechanisms for dealing with them. Many pollutants are substances that adversely affect the health, activities, or survival of human populations.

3. Industrial smog, photochemical smog, and acid deposition are examples of regional air pollution. Depletion of ozone in the lower stratosphere and amplification of the greenhouse effect are examples of global air pollution.

4. The rapid exponential growth of the human population has its foundations in the expansion of agriculture, which requires large-scale irrigation. Global supplies of freshwater are limited, and supplies are being polluted by agricultural runoff (which includes sediments, insecticides, herbicides, and fertilizers), industrial wastes, and human sewage.

5. Human populations are adversely affecting the land surface by the tremendous accumulation of solid wastes and by the conversion of marginal lands for agriculture. Millions of hectares are undergoing desertification every year. The destruction of tropical forest biomes is affecting regional soils and patterns of rainfall. The deforestation may also be amplifying the greenhouse effect.

6. Energy supplies in the form of fossil fuels are nonrenewable, they are dwindling, and their extraction and use come at high environmental cost. Nuclear energy normally does not pollute the environment as much as fossil fuels do, but the costs and risks associated with fuel containment and with storing radioactive wastes are enormous.

### Review Questions

1. Make a list of the advantages you personally enjoy as a member of an affluent, industrialized society. Then list some of the drawbacks. Do you believe that the benefits outweigh the costs? (This is not a trick question.)

2. Describe some of the potential global consequences of acid deposition, reduction of the ozone layer, and amplification of the greenhouse effect. *877–880; see also page 838*

3. What are some of the consequences of deforestation? Of desertification? *883–886*

4. Is it possible that massive amounts of money and technology can be used to put the biosphere on a more stable footing? What prospects and problems do you foresee in making a global effort in this regard?

5. List six activities you pursue each day that are harmful to the environment. Then list six ways in which you might "tread more gently" on the earth in your daily life.

6. How do you view our world and the effects we have on it? After reading this chapter, write your own caption for the photograph at the bottom of the next page.

### Self-Quiz *(Answers in Appendix IV)*

1. Since the mid-eighteenth century, the pattern of human population growth has been _____.

2. Human-generated _____ are disrupting the complex interactions of the sun, atmosphere, oceans, and land that are the engines of the biosphere.

3. For more than 100 years, the human population has been _____.
   a. growing exponentially
   b. increasing its energy demands
   c. polluting the atmosphere
   d. polluting the oceans and the land
   e. all of the above

4. Pollutants disrupt ecosystems because _____.
   a. they are composed of elements that differ from those of natural molecules
   b. only humans have uses for them
   c. ecosystems have not encountered them before and so do not have any evolved mechanisms that can handle them
   d. their only effect is on ecosystems but not humans

5. Industrial and photochemical smogs and acid deposition are examples of _____ air pollution; depletion of ozone in the lower stratosphere and the greenhouse effect are examples of _____ air pollution.
   a. local; local
   b. local; global
   c. global; global
   d. global; local

6. Rapid deforestation, especially of rain forests, may be amplifying the _____.
   a. pollution of water
   b. already serious loss of energy
   c. greenhouse effect
   d. dwindling of fossil fuels

7. Energy from fossil fuels is _____; their extraction and use come at _____ cost to the environment.
   a. renewable; low
   b. nonrenewable; low
   c. renewable; high
   d. nonrenewable; high

8. Nuclear energy normally pollutes the environment _____ than fossil fuels; the problems associated with it are _____ than with fossil fuels.
   a. less; lesser
   b. more; greater
   c. more; lesser
   d. less; greater

9. Match the human ecological concepts.
   _____ atmospheric pollutants
   _____ agricultural runoff
   _____ exponential
   _____ deforestation
   _____ factors that modify environment

   a. includes sediments, insecticides, herbicides, fertilizers
   b. amplifies greenhouse effect
   c. current pattern of human population growth
   d. population size, cultural habits, technology, energy use
   e. disrupts interactions between sun, atmosphere, oceans, land

## Selected Key Terms

acid deposition 877
chlorofluorocarbon (CFC) 880
deforestation 883
desalination 880
desertification 886
dry-acid deposition 877
fossil fuel 887
green revolution 882
industrial smog 876
meltdown 888
net energy 887
nuclear energy 887
photochemical smog 877
pollutant 876
salination 880
shifting cultivation 883
subsistence agriculture 883
thermal inversion 876
water table 880
wet-acid deposition 877

## Readings

Collins, M. 1990. *The Last Rain Forests*. New York: Oxford University Press.

Gribbin, J. 1988. *The Hole in the Sky*. New York: Bantam.

Gruber, D. 1989. "Biological Monitoring and Our Water Resources." *Endeavour* 13(3):135–140.

Miller, G. T. 1992. *Living in the Environment*. Seventh edition. Belmont, California: Wadsworth.

Mohnen, V. August 1988. "The Challenge of Acid Rain." *Scientific American* 259(2):30–38.

Schneider, S. 1990. *Global Warming*. New York: Random House.

Western, D., and M. Pearl. 1989. *Conservation for the Twenty-First Century*. New York: Oxford University Press.

Wilson, E. 1988. *Biodiversity*. Washington, D.C.: National Academy of Sciences.

## Deck the Nest with Sprigs of Green Stuff

Of all the birds in the United States, starlings are among the most familiar and least admired. Since their introduction into New York about a century ago, they have multiplied astonishingly, thanks in part to their ability to evict native North American birds from valuable nest sites in tree cavities. Once a male and female starling commandeer a tree hole, they build a nest. But theirs is no ordinary nest, constructed merely of dead, dried grasses and twigs. Starlings go on to *decorate* the nest bowl with many green sprigs, freshly plucked.

Why do starlings do this? Maybe they are camouflaging the nest from predators. Or maybe the greenery is insulative and will keep the forthcoming eggs warm. Or maybe it will repel or poison parasites that might otherwise attack the nestlings.

If the starlings were engaged in nest camouflaging, we would expect their nesting site to be vulnerable to keen-eyed predators. It isn't. Starlings build their nest in the dark cavity of a tree hole, where it is concealed from view. If the starlings were attempting to insulate the nest, we would expect them to use materials that would hold heat better than dry plant parts. But moist, green plant parts actually promote heat loss. Well, what about the third hypothesis? Do starlings have problems with parasites? They do.

Any birds that reuse old nest holes run the risk of encountering tiny mites that hunkered in after the previous nesting effort. Although each mite is tiny, a few individuals can quickly produce thousands upon thousands of descendants, all of which can suck the blood from a nestling. In a severely infested nest, a starling chick can lose 10 percent of its blood each day—a donation that reduces its capacity to grow and its chance of living.

But how might green plant parts reduce mite infestations in the starling nest? Biologists Larry Clark and Russell Mason found out. Starlings do not weave just any green plant material into their nests. They are selective, favoring the leaves of wild carrot and certain other plants. Clark and Mason built a set of experimental nests, some with fresh-cut wild carrot leaves and some without. They then removed the natural nests that pairs of starlings had constructed and had already started using. Half the nesting pairs received replacements decorated with sprigs of wild carrot. No sprigs decorated the replacement nests for the other pairs of starlings.

Figure 49.1 shows the results. The average number of mites in the greenery-free nests was consistently greater than in nests with greenery. In one experiment, nests with no carrot sprigs were teeming with an average of 750,000 mites—and the ones with the sprigs contained only 8,000.

Shoots of wild carrot happen to contain a highly aromatic steroid compound. Almost certainly the compound helps the plant by repelling herbivores. By coincidence, the compound also interferes with the development of immature bird mites. A starling that

Wild carrot growing in a meadow

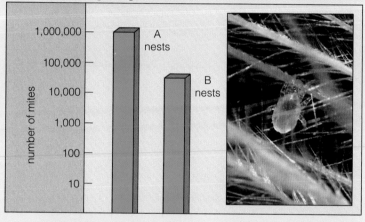

**Figure 49.1** A most excellent fumigator in nature—the European starling (*Sturnus vulgaris*), which combats mite infestations by decorating previously owned nests with fresh sprigs of aromatic green plants. The graph shows the effectiveness of its fumigation behavior. Experimental nests designated A were kept free of fresh green sprigs of wild carrot, which contains chemicals that prevent baby mites from reaching adulthood. Experimental nests designated B had fresh sprigs added every seven days. The head counts of mites (*Ornithonyssus sylviarum*) infesting the nests were made at the time the starling chicks left the nest, twenty-one days after the experiment began.

collects and adds fresh wild carrot sprigs to its nest will prevent mites in the nest cavity from reaching adulthood and starting a mite population explosion. "Decorating" with a bit of aromatic greenery turns out to be far from trivial after all. The behavior helps fumigate the nest and increases a starling's chance of producing healthy, surviving offspring.

And so the starlings lead us into the world of behavioral research. This research takes two forms, with some studies focusing on the mechanisms that enable individuals to behave as they do and other studies focusing on the adaptive value of some behavior to an individual's reproductive success. This chapter explores both aspects of animal behavior.

KEY CONCEPTS

**1.** "Behavior" refers to the observable, coordinated responses an animal makes to stimuli in its environment. Some aspects of behavior are instinctive, others are learned, and still others appear to be instincts modified to varying degrees by learning.

**2.** The mechanisms underlying instinctive behavior enable an animal to make complete responses to certain key stimuli the first time they are encountered. The mechanisms underlying learned behavior are different. They permit an animal to modify its behavior by acquiring information from certain experiences.

**3.** Behavior has a heritable basis, in that genes contain instructions for the development of the nervous and endocrine mechanisms necessary to detect, process, and respond to stimuli. Thus, like other traits having a genetic basis, forms of behavior are subject to evolution by natural selection.

**4.** If we assume that behavioral mechanisms have evolved as a result of natural selection, then those mechanisms must have functioned in ways that promoted the reproductive success of the individual. In other words, they promoted the ability of the individual to pass on its genes.

## MECHANISMS UNDERLYING BEHAVIOR

### Genetic Effects on Behavior

Through its sensory receptors and its integrators—that is, the nervous and endocrine systems—an animal detects and processes information about the environment, then issues commands for appropriate responses. Those responses are carried out by its effectors—muscles and glands. The observable responses to environmental stimuli are what we call **behavior**.

Let's reflect for a moment on the neural foundations of behavior. Response to a stimulus depends on the organization of neurons in the nervous system and on the patterns of neural activity. In this respect, behavior

Figure 49.2 (a) Banana slug, food for garter snakes of coastal
California (b). A newborn garter snake from a coastal population
(c), tongue-flicking at a cotton swab drenched with banana slug
fluids.

a

b

c

has a heritable basis. As an elegant example of this
point, consider the feeding behavior of garter snakes.
Stevan Arnold studied isolated populations of one species of garter snake in California. One population lived
in dry inland habitats, the other in coastal habitats. Individuals from the coastal population eat banana slugs
with great relish (Figure 49.2). Individuals from the
inland population prefer tadpoles and small fish. Offer
them a banana slug, and they will ignore it.

Arnold conducted experiments to discover what
causes the difference in feeding preferences. He offered
captive, recently born garter snakes a chunk of slug for
their very first meal. The offspring of coastal parents
almost always ate the slug. The offspring of inland parents rarely did. Inland baby snakes did not even flick their
tongue at cotton swabs drenched in essence of slug,
although coastal baby snakes did. (Snakes "smell" by
flicking their tongue in and out; this draws airborne
chemical scents into their mouth.)

The difference in response to slugs could not have
been learned through direct experience with slugs. It
showed up between captive snakes soon after birth. Possibly through some sensing mechanism, the offspring of
coastal snakes were able to detect a slug's odor and find
it attractive. Possibly the offspring of inland snakes had
no comparable mechanism.

To test this hypothesis, Arnold crossed coastal male
snakes with inland female snakes. He also crossed coastal
females with inland males. If there were a genetic difference between the two populations, then at least
some "hybrid" offspring would be intermediate in their
response to chunks and odors of banana slugs. The actual
observations matched the predicted results. Many baby
snakes of mixed parentage tongue-flicked at slug-scented
cotton swabs more often than the typical newborn inland
snakes—and they did so less often than the typical newborn coastal snakes. The results strongly suggest that the
differences in perception and feeding behavior stem from
differences in genes that affect how odor-sensing mechanisms are put together during embryonic development.

---

**Because genes influence how the nervous system develops,
they contribute in an indirect yet major way to behavioral
abilities.**

---

## Hormonal Effects on Behavior

Hormones contribute directly and indirectly to behavior. Imagine that it is early spring in a Canadian forest and a male white-throated sparrow is whistling a
song (Figure 49.3). It sounds rather like "Sam Peabody,
Peabody, Peabody." He repeats the song thousands of
times, with such clarity and consistency that you wonder how he does it.

Singing behavior depends indirectly on melatonin,
a hormone produced by the pineal gland (page 592).
Sunlight, which acts on photoreceptors in the pineal
gland, affects melatonin secretion. When the hours
of daylight (daylength) increase with the approach of
spring, melatonin output falls, and the growth of the
gonads (testes and ovaries) is freed from suppression by
melatonin. When gonadal activity steps up, hormones
are produced that are directly involved in the control of
singing behavior.

Among white-throated sparrows and some other
songbirds, the singers are primarily males. Through

**Figure 49.3** A male white-throated sparrow and a sound spectrogram of his territorial song. Sound spectrograms provide a visual record of the pitch (frequency) of each note, which is measured in kilohertz.

**Figure 49.4** *Right:* (**a**) Zebra finch, with sound spectrograms of the full song of (**b**) the male and of the song of (**c**) a female that had been exposed to estrogen as a nestling, then given a testosterone implant as an adult.

anatomical studies, we know that the males and females differ in the structure and size of brain regions that control the muscles of the syrinx, the vocal organ of songbirds. This brain region is the **song system**.

Two hormones, estrogen and progesterone, bring about the difference in this brain region between male and female birds of those species in which only males sing. In very young *males*, the estrogen level is higher than in nestling females. This is the hormone that initiates the development of a masculine song system. Zebra finches provide experimental evidence of this effect. When nestling *female* zebra finches received implants of tiny pellets of estrogen, they developed a malelike song system.

In itself, however, a masculinized brain is not enough to cause a bird to sing. At the start of the breeding season, the enlarged testes of a male bird manufacture more testosterone. Molecules of this hormone travel to and bind with receptors on cells in the song system. The cells respond by changing their metabolic activity. The end result is that, for many songbirds, the male begins to sing as it stakes out a territory and attracts a female.

As you might predict, female zebra finches that receive testosterone implants when they are adults will sing—but only if their brain had been masculinized through exposure to extra estrogen early in life (Figure 49.4).

Thus, estrogen *organizes* the development of the song system. Then testosterone *activates* the song system and prepares the bird to sing when properly stimulated. Coordinated interactions between hormones and nerve cells constitute part of the mechanisms that are necessary if a white-throated sparrow is ever to sing so much as a single "Sam Peabody, Peabody, Peabody."

Hormones influence the organization and activation of mechanisms required for particular forms of behavior.

# FORMS OF BEHAVIOR

## Instinct and Learning

By one traditional approach, we can divide behavior into two categories: "instinctive" or "learned." The categories are based on possible differences between physiological mechanisms that regulate an animal's instinctive and learned responses to environmental cues. In **instinctive behavior**, an animal has neural components that enable it to carry out complex, stereotyped responses to particular cues—often quite simple ones—in the environment.

Consider the cuckoo, a social parasite. Adult females slip their eggs into the nests of other kinds of birds. The young cuckoos will eliminate the natural-born offspring and then receive the undivided attention of their unsuspecting foster parents. Newly hatched cuckoos are blind and featherless, but they respond to contact with an egg in the nest. They maneuver the egg onto their back, then push it out of the nest (Figure 49.5). This is a **fixed action pattern**—an instinctive response that is triggered by a well-defined, simple stimulus and that is performed in its entirety, once it has been set in motion.

The tongue-flicking, orientation, and strike of a newborn garter snake are other examples of instinctive behavior. The snake is born with the neural capacity to perform these actions which, under natural conditions, usually help a snake get something nutritious to eat.

Similarly, when human infants are just two or three weeks old, they already have the neural components that allow them to smile when an adult's face comes close to theirs (Figure 49.6). The smiling response can be triggered by presenting a flat, face-sized mask with two dark spots where the eyes would be on a real human face. A mask with one "eye" will not do the trick. This suggests that an infant's early smiles are an instinctive response to simple stimuli.

By contrast, in **learned behavior**, information gained from the animal's specific experiences becomes incorporated into parts of its nervous system and is used to make *changes* in responses.

For example, a young toad will instinctively flip out its sticky-tipped tongue at any dark object moving across its field of vision. In the toad's world, most such objects are edible insects or similar kinds of prey. At some point in its life, the toad may capture a large bumblebee and get stung on its tongue. The lesson of this painful experience is "Bumblebee-sized objects with black and yellow bands can sting." The toad learns to leave bumblebees alone.

There is widespread belief that instincts are "genetically determined" and learned responses are "environmentally determined." But this simply is not true. Genes and environment play critical roles in the development of both kinds of behavior. Both kinds of behavior depend on special neural and often hormonal mechanisms, as we

**Figure 49.5** (**a**) An instinctive behavior. A newly hatched cuckoo is maneuvering to get the egg of its host, a member of another species, onto its back. Once in position, the cuckoo will eject the egg from the nest. (**b**) A foster parent, dwarfed by the cuckoo.

saw in the case of the zebra finch and its song system. No physiological mechanism can develop without an interaction between genes and environment, a matter discussed at length in Chapter 11. The toad could not "learn" to avoid bumblebees without a specially wired subsystem within its brain. The construction and organization of that subsystem requires genetic information of a particular sort. Thus, the toad's ability to learn, like its instinctive capacity to recognize possible prey, demands both genetic and environmental influences.

Instincts and learned behavior are regulated by different kinds of neural and hormonal systems.

Genes and the environment contribute to the mechanisms underlying both instinctive and learned behavior.

**Figure 49.6** Example of an instinctive response in humans. Smiling in very young infants is a fixed action pattern that can be triggered automatically by certain simple stimuli. Somewhat older infants also have the instinctive capacity to imitate facial expressions of adults.

## Forms of Learning

The role of heredity in the development of learning abilities became clear as researchers discovered many specialized forms of learning throughout the animal kingdom. Some of the traditional categories of learning may be defined this way:

1. *Classical conditioning.* An animal learns to associate an automatic, *unconditioned* response with a novel stimulus that does not normally trigger the response. Ivan Pavlov's experiments with dogs are a classic example. Dogs salivate just before eating. Pavlov's dogs were conditioned to salivate even in the absence of food. They did so in response to the sound of a bell or a flash of light that was initially associated with the presentation of food.

2. *Operant conditioning.* An animal learns to associate a voluntary activity with the consequences that follow. Thus toads learn to avoid stinging or bad-tasting insects after attempting voluntarily to eat these noxious prey.

3. *Habituation.* An animal learns through experience not to respond to a situation if a response has no positive or negative consequences. Many urban birds learn not to flee from humans or cars, which pose no real threat to them.

4. *Spatial or latent learning.* Through inspection of its environment, an animal acquires a mental map of a region, often by learning the position of local landmarks. Bluejays, for instance, can store information about the position of dozens, if not hundreds, of places where they have stashed food.

5. *Insight learning.* An animal abruptly solves a problem without trial-and-error practice at the solution. Chimpanzees, for instance, exhibit insight learning in cap-

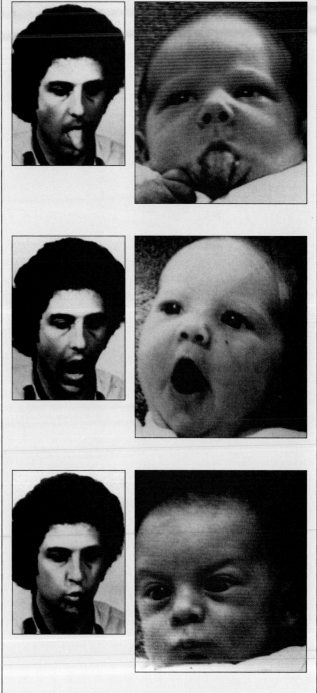

tivity when they suddenly solve a novel problem created for them by their captors (as when they abruptly stack together several boxes and use a stick to reach a bunch of bananas suspended out of their reach).

6. *Imprinting.* An animal that has been exposed to specific key stimuli, usually early in its behavioral development, forms an association with the object and may later show sexual behavior toward similar objects.

Here we will restrict our attention to the genetically guided nature of learning, using imprinting and song learning by birds as our examples.

a

## Imprinting

The ethologist Konrad Lorenz pioneered research on imprinting, which is a time-dependent form of learning among many animals, including geese. Lorenz discovered that baby geese (goslings) would form an attachment to him—or to *any* moving object—if they were separated from their mother shortly after hatching (Figure 49.7). But imprinting occurred only when the goslings were exposed to a moving object during a short, sensitive period early in life. During that period, the goslings are primed to learn a special piece of information: the identity of the individual they will follow in the months ahead. In nature, that individual is normally their mother or father. Thus the imprinting mechanism provides them with a bit of information that they could not know prior to hatching.

Lorenz discovered that imprinting further has long-term consequences on the sexual behavior of male greylag geese. When the males he had been observing became adults, they directed their sexual behavior toward members of whatever species they had been sexually imprinted upon in their youth. In nature, the males become imprinted on their mother and will later court female greylags. The ones that had been imprinted on Lorenz early in life courted human beings instead of members of their own species!

Clearly, greylag geese are born with neural wiring that enables them to learn something important. They can only learn it early in life, and only with respect to which individual to follow during a dependent phase and which kind of individual to court upon sexual maturity. Greylag geese could not exhibit this specialized learning without the genes that guided the development of the imprinting mechanism in their brain.

## Song Learning

Song acquisition by the white-crowned sparrow is a learning process that has been studied in detail. Male white-crowns have a distinctive, species-specific song. However, males in different regions belt out distinctive variations, or "dialects," of the standard song. Peter Marler suspected that male white-crowns learned parts of their song from other males in their population. To test his hypothesis, Marler took male nestlings to the laboratory and raised them in soundproof chambers. The isolated nestlings never heard adult males of their species singing. When they were mature, they sang. But

**b**

**Figure 49.7** (**a**) Human imprinting objects. No one can tell these goslings that Konrad Lorenz is not Mother Goose. (**b**) An imprinted rooster wading out to meet the objects of his affections. During a critical period of the rooster's life, he was exposed to a mallard duck. Although sexual behavior patterns were not yet developing during that period, the imprinting object became fixed in the rooster's mind for life. Then, with the maturation of sexual behavior, the rooster sought out ducks, forsaking birds of his own kind. These observations of imprinting help explain why birds of a feather do flock together.

# THE ADAPTIVE VALUE OF BEHAVIOR

Now that we have considered some of the mechanisms underlying the behavior of an individual animal, let's turn to evolutionary mechanisms by which diverse forms of behavior come about. We begin by defining a few terms currently used in animal behavior studies:

**1.** *Reproductive success:* survival and production of offspring.

**2.** *Adaptive behavior:* forms of behavior that promote the propagation of an individual's genes and that thereby tend to occur at increased frequency in successive generations.

**3.** *Selfish behavior:* forms of behavior by which an individual protects or increases its own chance of producing offspring, regardless of the consequences for the group to which it belongs.

**4.** *Altruistic behavior:* self-sacrificing behavior; the individual behaves in a way that helps others but, in so doing, decreases its own chance to produce offspring.

**5.** *Natural selection:* a measure of the difference in survival and reproduction that has occurred among individuals that differ from one another in their heritable traits.

their song was odd indeed, having none of the detailed structure of a typical adult's song.

Then Marler gave isolated, laboratory-reared males a chance to listen to tape recordings of songs by both white-crowned sparrows and song sparrows. When the isolated males reached adulthood, they sang a complete song—but only the white-crown song. More than this, they matched whatever dialect was on the tape. But a white-crown that hears only song sparrow tapes will not sing *those* songs. Evidently, male white-crowned sparrows have the capacity to learn some—but not all—of the acoustical cues to which they are exposed.

Marler also demonstrated that an isolated sparrow cannot acquire the full song of its species unless it hears tapes of that song between ten and fifty days after it has hatched. Thus young sparrows, too, have a sensitive period, when a learning mechanism is primed to acquire a restricted piece of information from the environment.

Recent research also shows that when young, hand-reared white-crowns have a chance to interact with a "social tutor" (as opposed to listening to taped songs), they are more likely to learn the song of the alien species, such as a song sparrow. In other words, their learning mechanisms are primed to be influenced by social experience as well as by certain acoustical cues.

When behavioral biologists speak of a "selfish" or an "altruistic" individual, they don't mean the individual is consciously aware of what it is doing or of the ultimate (reproductive) goal of its behavior. Referring to an animal as if it were trying to reproduce as much as possible is evolutionary shorthand for what would be a cumbersome mouthful. (To wit, "The animal's nervous and endocrine systems, the physiological foundation for behavior, have evolved through differences in individual reproductive success in the past, the outcome being behavioral mechanisms that tend to promote success in the competition to reproduce.") A lion does not have to know that eating zebras is good for its survival and reproductive success. It simply has the kind of nervous system that activates hunting behavior when it is hungry and sees a vulnerable zebra.

Before the 1960s, the idea that the ultimate function of animal behavior is to enable individuals to reproduce and pass on their genes was not widely accepted. Many biologists assumed that the members of a group or species would sacrifice their own chance of producing offspring, if such self-sacrificing behavior promoted the welfare of the group as a whole. In 1966, however, George Williams wrote a book based on a key question:

What would happen in a population composed of self-sacrificing types if mutation gave rise to a "selfish" individual able to leave more descendants than the "altruists"? If the selfish individual and its descendants out-reproduced the species-benefiting types, then "selfish" genes would become more and more frequent in subsequent generations. In time, altruists simply would disappear.

An example will help illustrate Williams' point. Norwegian lemmings disperse when population densities become extremely high, and food becomes scarce. Many die during their travels. According to one view, lemmings on the march are sacrificing themselves to prevent habitat destruction that could put the entire species on the brink of extinction. But are lemmings committing suicide to help their species—or merely dying by accidental drowning, starvation, and predation as they disperse to new areas, away from an overcrowded habitat?

There is a wonderfully instructive cartoon by Gary Larson that shows lemmings plunging into water, presumably in the act of suicide. However, one member of the group has come equipped with an inflated inner tube about its waist. Unlike the suicidal altruists, that obviously "selfish" member will leave copies of its genes in its descendants. Therefore, a more reasonable hypothesis is that overcrowding leads to the dispersal of lemmings to less crowded habitats, where they can reproduce.

In sum, the current consensus among behavioral researchers is this: *In developing a working hypothesis to explain some behavioral trait, it will almost always be more profitable to use an approach based on natural (individual) selection rather a species-benefiting approach.* A few examples will illustrate how hypotheses consistent with this concept of individual selection can be developed and tested. We will draw these examples from feeding and reproductive behavior.

### Feeding Behavior

In northern forests around the world, ravens scavenge carcasses of deer, elk, or moose, which are few and far between. When a raven comes across the rare carcass, even in winter when food is scarce, it often calls loudly, attracting a crowd of ravens. The birds then feed together on the dead animal.

Bernd Heinrich found the calling behavior puzzling. It seems to go against the interests of the caller. Wouldn't a raven that kept quiet have more food to eat, a better survival chance, and ultimately more descendants than an "unselfish" vocalizer? Isn't there a cost in terms of calories and nutrients lost to other diners? If the behavior is an outcome of natural selection, the cost must be offset by some reproductive benefit for the individual caller. But what benefit?

One possibility is that a single bird picking at a carcass is in danger. Maybe predators lie in wait for ravens at such "baits." Maybe the calls bring in others to provide vigilance against predators. Ravens, however, are large and agile birds with very few known enemies. In fact, after watching ravens that were feeding singly or in pairs at a carcass, Heinrich (who was hidden in a blind) never saw predators attack any one of them.

Then Heinrich noticed that lone ravens or a pair of them sometimes do *not* loudly advertise the presence of a carcass. He noticed this after he had hauled a cow carcass out into the woods in Maine. Were the silent birds adults that had previously staked out a large territory—which happened to include the spot where the carcass ended up? A **territory** is any area that one or more individuals defends against competitors.

Presumably, the resident territorial pair of ravens would have nothing to gain by attracting others to the bonanza. But what if the woods were subdivided into territories and defended by powerful adults? A wandering young bird would have no chance of feeding for long in an aggressive pair's territory. However, recruiting a gang of other, nonterritorial ravens to the spot might overwhelm the defensive capacity of the resident pair.

Heinrich predicted that wandering ravens give carcass-advertising calls but that territorial residents do not. He captured and put large, numbered tags on many ravens while also estimating their age (on the basis of their plumage). His prediction turned out to be correct. Nonterritorial ravens behave in their own self-interest by loudly advertising the presence of food that they would not, on their own, be able to exploit.

### Mating Behavior

Within sexually reproducing species, other members often create obstacles to an individual's reproductive success. When a male white-throated sparrow sings "Sam Peabody" in a forest, he faces constant challenges from other males. They may force him away from the breeding site he has selected and so reduce his chances to reproduce. Even if he succeeds in holding his chosen territory, females of his species may refuse to settle there, leaving him with no descendants to show for his effort. Thus, there may be both competition among the members of one sex for access to mates as well as selectivity in the choice of mates between the sexes. Competition for mates and selectivity among potential mates results in **sexual selection**, a process that leads to evolutionary change in the males and females of a species.

The effects of sexual selection appear to be widespread, for male competition for females and female choosiness are common. Each male typically produces a

**Figure 49.8** Sexual competition in bison. These males fight for access to a cluster of females in the area.

great many small gametes (sperm) compared to the relatively few large gametes ("eggs") made by each female. Therefore, if we measure a male's reproductive success in terms of the number of his descendants, it will usually be related to the number of females he mates with. Generally speaking, the more females he mates with, the more eggs he will fertilize and the more offspring he will produce.

What about a female's reproductive success? In general, this depends largely on how many eggs she can produce or how many offspring she can care for. For females, the *quality* of a mate, not the quantity of partners, should usually be the prime factor influencing her sexual preferences.

With this simple background, we can interpret mating tactics in terms of sexual selection theory.

Consider male bighorn sheep, which sometimes fight fiercely with one another in head-butting contests (Figure 1.7f). Fighting is costly in terms of time lost, energy expended, and the risk of injury. To counterbalance the costs, there should be reproductive benefits for successful fighters. And there are. Male bighorns only fight to control areas in which receptive females gather every year for about three months, during the winter rutting season. As predicted, winners do mate frequently with numerous females.

The losers of head-clashing bouts generally do not waste time in futile challenges to stronger, larger males. They employ another tactic. They gather in numbers within the area guarded by a prime fighter, thus overwhelming his capacity to drive them all away. The "loser" males attempt to mate with females—literally on the run, which they sometimes can do even though the females attempt to avoid them.

Competition between males for access to mates has evidently favored the evolution of backup options. With the pursuit tactic, males that otherwise would leave no descendants *do* mate with a few females and so may leave some surviving offspring, after all.

The sexually receptive females of bison, lions, elk, and many other species also cluster together, and here again, competing males engage in intense fights. The competition favors large males with formidable combative abilities—obvious in the strength of male lions, the antlers of elk, and the readiness of bison to gore each other (Figure 49.8). Here again the rewards can be great, with successful males having access to a ready-made harem.

**Figure 49.9** Flamboyant dancing display by a male sage grouse, performed at his own vigorously defended spot within a compact mating area called a lek. The females (the smaller brown birds) congregate at the lek, and observe the prancing males before choosing the one they will mate with.

What happens when females are *not* clustered in defendable groups? Female sage grouse, for example, are widely scattered through their prairie habitat in the western United States. Male grouse make no attempt to maintain a territory large enough to include the feeding or nesting sites of several females, presumably because such a territory would be too large to defend. Instead, males gather during the breeding season in a **lek**, a communal display ground. There, each male defends his own tiny territory, a few square meters in size. Females are attracted to the lek, but not to feed or nest. Instead they observe the male grouse prancing about in a truly astonishing display.

With tail feathers erect and splayed, held forward, and neck pouches inflated, a male booms out calls, all the while stamping about like a big wind-up toy on his display ground (Figure 49.9).

After inspecting the displaying males, females eventually select a mate from among the many. Then they go off to nest by themselves, with no assistance from their partner. Many females choose the same male, conferring exceptional reproductive success upon that individual. Most of the other males never mate.

What is the basis for this peculiar reproductive system? No well-tested hypothesis exists. Male sage grouse may form leks as a tactic of last resort because females cannot be directly or indirectly defended against rival males. Under the circumstances, choosy females require males to display their competitive ability at a communal display arena before accepting a partner.

In any event, male grouse compete for chances to mate and females choose among males. Choosiness carries a cost for females, particularly in the time invested to examine potential mates. What benefit might they gain by being selective? In this species, males provide their mates with the genes in their gametes, nothing more. Perhaps females assess male display behavior as a guide to his survival ability or endurance. To the extent that these behavioral attributes are heritable, a male able to put on a superior show might endow his offspring with useful abilities. Females that selected a male with more adaptive genes might therefore enjoy greater reproductive success.

In some species, such as white-throated sparrows, males offer their mate more than genes. When the males differ in the quality of their parental care or in the resources they transfer to their mates, females might well benefit from an investment in choosiness—provided they can identify males able to offer them superior material benefits.

Consider the hangingfly (*Harpobittacus apicalis*). Male hangingflies capture and kill a moth or some other insect, which they hold for a predatory female (Figure 49.10). Females come to males that have "nuptial gifts," attracted by the sex pheromone that the males release. The size of the male's offering varies, so we can predict that females should favor males able to provide larger, more calorie-rich nuptial gifts. In fact, females do exercise adaptive mate choice. They regulate the quantity of sperm they will accept from a partner on the basis of the size of the nuptial gift.

If a female hangingfly accepts a male's present, she permits him to mate with her—but she accepts no sperm until she has eaten for about five minutes. Thereafter she accepts a steady flow of sperm, which she stores in her reproductive tract—but only as long as the food holds out. At any point up to twenty minutes, she can break off the mating and leave without a full complement of sperm. If so, she will mate again and accept another male's gametes, diluting or replacing her first partner's sperm.

In most species, the eggs of females are a limited resource, from the male perspective. Thus female hangingflies, female bighorn sheep, and females of many other species dictate the rules of male competition, with males employing tactics that will help them fertilize as many eggs as possible.

**Figure 49.10** Female choice among the insects. Females of certain hangingflies choose sexual partners on the basis of the size of prey that males offer to them. Here a male dangles a moth as his nuptial present for a future mate.

## SUMMARY

1. Behavioral research can be divided into two broad categories. One category includes studies that focus on the internal genetic and physiological mechanisms that control behavioral responses. The other category includes studies that deal with the evolution and adaptive value of behavioral abilities.

2. Behavioral responses themselves have been categorized as either instinctive or learned. All behavioral responses depend on neural and hormonal mechanisms, which require a gene-environment interaction if they are to develop. However, the mechanisms underlying instincts trigger complete responses to key stimuli the first time that an animal reacts to those stimuli. In contrast, the mechanisms underlying learned behavior enable the individual to store information, which is used to modify the animal's response based on past experience.

3. Persons interested in the adaptive value of behavior, rather than behavioral mechanisms, usually use a research approach based on the theory of evolution by natural selection. A key premise of this approach is this: Traits that have evolved as a result of individual differences in reproductive success in the past should have

reproductive benefits that exceed their reproductive costs (or disadvantages).

4. Several alternative hypotheses can usually be proposed on why the benefits of a behavioral ability might exceed its costs. Discriminating among competing explanations requires using the hypotheses to generate predictions, which can then be tested by collecting new information and checking whether the actual results match the expected observations.

5. Behavioral biologists have tested many hypotheses on the possible adaptive value of various aspects of reproductive behavior. These biologists have been aided in the development of their hypotheses by recognizing that members of the same species often create obstacles to reproductive success for each other, either through competition for access to mates or through selective mate choice.

### Review Questions

1. Rephrase the statement "The singing behavior of white-throated sparrows is genetically determined." (The revision "The singing behavior of white-throated sparrows is environmentally determined" is not acceptable.) *894–899*

2. What role does the environment play in the development of an instinct, like the egg-ejecting behavior of baby cuckoos? *896*

3. How does song acquisition by white-crowned sparrows illustrate the principle that genetic mechanisms contribute to the development of learned responses? *898–899*

4. If a female zebra finch receives a testosterone implant as an adult, why doesn't she sing the courtship song of males of her species? *894–895*

5. Give a group selectionist and individual selectionist explanation for territorial behavior by white-throated sparrow males, and then criticize the theoretical basis of your group selectionist hypothesis. *899–900*

6. You find a very large moth caterpillar in the tropics that responds to being poked by dropping part way from the vine on which it had been resting and by puffing up the anterior portion of its body:

Provide at least one hypothesis on the underlying mechanism that enables the caterpillar to behave this way. Then provide at least one hypothesis on the possible adaptive value of the trait. How would you test both hypotheses? *896, 899*

7. What are the evolutionary costs and benefits to male hangingflies of offering nuptial gifts to their partners? How might a female's choosiness depend on these costs and benefits? *903*

8. Develop an adaptive value hypothesis for the observation that male lions kill the offspring of females they acquire after they chase away the males that had been the previous pride holders and mates of these females. How would you test your hypothesis? *900–902*

9. Suppose you are traveling in Africa and you see a black heron standing in the water with its wings held over its head like an umbrella, as shown below.

You might suppose that, like other herons, it is on the lookout for fish. But most other herons keep their wings down when they are hunting. Possibly the black heron is creating shade on the water, and perhaps minnows are being drawn to the shade because it appears to offer shelter. How would you go about testing this hypothesis?

1. Instincts differ from learned behavior in that _____.
   a. they are genetically determined rather than environmentally determined
   b. they are less environmentally determined than learned behaviors
   c. they are less advanced than learned responses
   d. they are produced in full and functional form the first time the animal responds to the appropriate sign stimulus

2. Studies of imprinting demonstrate that _____.
   a. only lower animals can be imprinted
   b. learned behavior depends on "innate" neural mechanisms that facilitate the acquisition of modified behaviors
   c. learning is genetically determined
   d. young ducks and geese instinctively know how to identify their parents

3. The statement "Starlings decorate their nests with sprigs of carrot in order to interfere with the development of nest mites" is a _____.
   a. hypothesis
   b. prediction
   c. test of an hypothesis
   d. proximate conclusion

4. Genes affect the behavior of individuals by _____.
   a. influencing the development of nervous systems
   b. affecting the kinds of hormones in individuals
   c. providing hereditary information that shapes the development of muscles and skeletons in some organisms
   d. all of the above

5. Which of the following environmental factors do not influence the acquisition of song by a white-crowned sparrow male reared in captivity?
   a. the experience of hearing the taped song of mature white-crowned sparrow males
   b. the experience of hearing the taped song of mature song sparrow males
   c. the age of the bird when it is exposed to songs of its species
   d. the experience of interacting socially with a singing male of its own species

6. When evolutionary biologists claim that a particular human activity is "adaptive," they mean that the activity _____.
   a. will promote the survival of our species
   b. increases the probability that an individual will behave altruistically
   c. increases the survival chances of the individual that behaves this way
   d. increases the number of genes that an individual will transmit to the next generation

7. The statement that lemmings, by dispersing at high population densities, reduce their population and bring it in line with available resources _____.
   a. has been proven to be true
   b. is consistent with Darwinian evolutionary theory
   c. is based on the theory of evolution by group selection
   d. is supported by the finding that most animals behave altruistically during their lives

8. A Darwinian biologist might be most surprised and puzzled by which of the following observations:

   a. A parent bird permits one of its offspring to kill another of its offspring.
   b. A moth beautifully mimics the appearance of a leaf.
   c. A raven is able to find food extremely efficiently.
   d. A white-throated sparrow's territorial song is different from a white-crowned sparrow's call.

9. In a species of mammal, females live in herds. Under these conditions we would predict to observe which of the following mating systems?
   a. a lek mating system
   b. competition among males to defend clusters of receptive females
   c. territorial defense of feeding sites by males
   d. males capturing food to lure females

10. Sexual selection theory differs from Darwinian natural selection by focusing on _____.
    a. how traits help species survive
    b. the evolutionary effects of differences in the survival abilities of individuals
    c. the evolution of traits that help individuals acquire mates
    d. why altruism has evolved in some species

## Selected Key Terms

| | |
|---|---|
| adaptive behavior *899* | lek *902* |
| altruistic behavior *899* | natural selection *899* |
| behavior *893* | reproductive success *899* |
| fixed action pattern *896* | selfish behavior *899* |
| imprinting *897* | sexual selection *900* |
| instinctive behavior *896* | song system *895* |
| learned behavior *896* | territory *900* |

## Readings

Alcock, J. 1989. *Animal Behavior: An Evolutionary Approach*. Fourth edition. Sunderland, Massachusetts: Sinauer. A broad-ranging survey of the many topics that make up the study of behavior, with a strong emphasis on evolutionary theory.

Dawkins, R. 1989. *The Selfish Gene*. New York: Oxford University Press. An updated, annotated edition of an entertainingly written book. Dawkins explains why natural selection favors behavioral traits that are associated with individual success in propagating the individual's genes.

Krebs, J., and N. Davies (editors). 1991. *Behavioral Ecology: An Evolutionary Approach*. Cambridge, Massachusetts: Blackwell Scientific. A more advanced text consisting of chapters written by experts on many of the issues covered in this chapter.

Welty, C., and L. F. Baptista. 1988. *The Life of Birds*. Fourth edition. Philadelphia, Pennsylvania: Saunders College Publishing. This edition contains a thorough review of many of the concepts discussed in this chapter as they relate to bird behavior.

Williams, G. C. 1966. *Adaptation and Natural Selection*. Princeton, New Jersey: Princeton University Press. The classic book that brought about a revolution in thinking about animal behavior in terms of individual selection.

## Consider the Termite

In a forest in southern Queensland, Australia, a brittle tube about a centimeter wide runs up the trunk of a dead eucalyptus tree. When you chip away a fragment from the tube, which is made of mud and tiny fecal pellets, some small, nearly white insects scurry away from the light. At the same time, a few brown ones rush in and make a defensive stand at the breach. Their swollen, eyeless head tapers into a long, pointed "nose" (Figure 50.1). Disturb them and they will shoot thin jets of silvery goo out of their nose!

Both kinds of individuals in the tunnel are members of a complex termite colony. The brown insects, called soldier termites, protect the colony from ants and other enemies. The white insects are worker termites. They build the tunnels that lead to places where they can safely gather fibers of wood. They carry the fibers down to an underground nest, and there the termites engage in a special sort of gardening. The wood fibers provide nourishment for a fungus. The termites eat portions of the fungus. But they also chew the wood into small

particles and swallow them. (Unlike most of the world's 2,000 or so species of termites, the Australian termites do not themselves produce the enzymes necessary to digest cellulose, a major component of wood. Instead, symbiotic microorganisms in their gut digest the cellulose for them and yield nutrients that the termites can absorb.)

Neither the workers nor the soldiers reproduce; they are sterile. The task of reproduction falls to at least one king and one immense queen with a huge, egg-producing abdomen. These are the "parents" for the entire society.

This brief glimpse into a termite society raises two fundamental questions that are the focus of this chapter.

*First*, how do the members of a termite society or any other social unit manage to cooperate with each other in ways that benefit the group and themselves? For example, if soldiers are to defend a column of workers effectively, they must be able to communicate with one another, locate a threat, and act together against it. Identifying the signals that are used and the manner in which individuals respond to signals is the subject of much research in social biology.

*Second*, what is the adaptive value of social life? In the preceding chapter, we considered the theoretical basis for the claim that animals behave in ways that increase the likelihood of reproductive success. Yet in many animal societies, individuals appear to help others. In some extreme cases, such as the termites, helpful individuals cannot reproduce; they are sterile. This is a major puzzle for evolutionary biologists, and its possible solution is a key part of this chapter.

**Figure 50.1** Soldier termites defending their social colony by guarding a break in a foraging tunnel. Members of this caste shoot out strands of glue from a pointed "nose." The glue entangles intruders, such as invading ants.

KEY CONCEPTS

**1.** Social interactions are based on modes of communication that have developed over evolutionary time. A communication signal is an action that has net benefits for both the sender and the receiver of the signal.

**2.** Social living has costs and benefits, which can be measured in terms of an individual's ability to pass on its genes to offspring. Not every environment favors the evolution of social life. Under most circumstances, solitary individuals can leave more descendants than social ones would be able to do.

**3.** Sociality can evolve through the increased reproductive success of individuals. In other words, social animals need not exhibit altruism. Altruism means helping others but, in doing so, sacrificing personal reproductive success.

**4.** The evolution of altruism requires circumstances that enable individuals to propagate their genes indirectly by helping *relatives* reproduce successfully.

## MECHANISMS OF SOCIAL LIFE

### Functions of Communication Signals

Most animals interact with others of their species by way of communication signals. Whether such an interaction is lifelong or fleeting, as during a one-time sexual encounter, we call this **social behavior**. The **communication signals** underlying social behavior may be defined as evolved actions or cues that have a net beneficial effect on a *signaler* (the member of the species sending the signal) and on a *receiver* (the member receiving it). In other words, we reserve "communication" for actions that transfer information to the benefit of both the signaler and the receiver.

When a termite tunnel is breached, for example, the pale workers bang their head against the ceiling and floor of the tunnel, creating vibrations that alert nearby soldiers and enable them to locate the disturbance. The soldiers run to the spot and point their nose in the direction of potential danger, as announced by the scent of ants or physical contact with intruders. When one soldier fires off his sticky strand, the volatile odors released from the gluey substance attract more soldiers to the spot and cause the release of more repellent.

The vibrations from the head-banging workers and the odor of the defensive chemicals apparently convey information from one termite to another. You probably would agree that both the sounds and the odor are communication signals. But what about the scent of an ant? It sends information from the ant to a soldier termite, which certainly responds to the information. Is the ant's scent also a communication signal for termites? In this case, no. By announcing the ant's presence to the soldier, the odor does the ant no good.

Natural selection obviously has not favored ants that issue cues *because* they cause termite soldiers to target them with an immobilizing glue. The scent from the surface of an ant's body may have some evolved function, but communicating with termites is not one of them. So the ant's scent is a *cue*, but not an evolved signal.

There are instances where communication signals are exploited by animals of other species. The scents on an ant's body probably help identify it as a member of a particular ant colony and trigger cooperative behavior from its colonymates. When termite soldiers detect the scent and kill their ant enemies, they are *illegitimate receivers* that are exploiting one of the ant's communication signals. Similarly, nature has its share of *illegitimate signalers*. Certain assassin bugs hook the dead and drained bodies of their termite prey on their dorsal surface and so acquire the odor of their victims. They send a false signal that they are themselves members of a termite colony. The deception enables the bugs to hunt additional termite victims more easily.

In general, however, a communication signal evolves or is maintained when it tends to increase the reproductive success of both the sender and the receiver. If it turns out that there are more disadvantages than advantages either to sending the signal or responding to it, natural selection will favor individuals that do not signal or do not respond.

---

**A communication signal is an action that has a net beneficial effect on both the animal sending the signal and the one receiving it.**

**Natural selection tends to favor communication signals that promote the reproductive success of both the signaler and the receiver.**

**Illegitimate signalers and receivers are members of one species that exploit the channels of communication of another species. The interaction is not social communication; benefits do not flow both ways.**

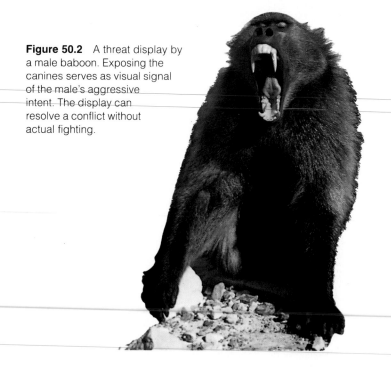

**Figure 50.2** A threat display by a male baboon. Exposing the canines serves as visual signal of the male's aggressive intent. The display can resolve a conflict without actual fighting.

### Types of Communication Signals

Communication signals involve different kinds of sensory stimuli, as described on page 600. Here we will mention some of the signals based on chemical, visual, tactile, and acoustical stimuli.

**Chemical Signals**. Here and elsewhere in the book, we have given examples of **chemical signals**, airborne molecules that travel from one individual to another of the same species. Chemical signals are perhaps the most widespread of all "channels" of communication in the animal kingdom. Termite alarm signals and the sex pheromones of hangingfly males (page 903) are examples. Compared to visual cues, chemical signals have the advantage of working well in complete darkness. Furthermore, the signal can be a highly specific one that only a legitimate receiver is likely to detect. Thus a sex pheromone often is unique to one species and can be effective at low concentrations—so low that other animals (especially the insect's predators) cannot easily detect it.

**Visual Signals**. Many animals that are active by day employ **visual signals**, which are observable actions or cues that have specific and sometimes very complex social meaning. Consider a male baboon that "yawns" conspicuously when he is confronted by a rival for the same receptive female (Figure 50.2). The yawn, which exposes the male's formidable canines, is indeed a threat display and may be followed by a physical attack. Suppose this time the rival (a signal receiver) backs away from the yawning male (the signaler). Benefits accrue to the signaler, who gains access to the female with-

a

b

c

d

out having to fight for it. Benefits accrue to the signal receiver if the threat display prevents him from battling with a male that can inflict a serious beating, which may lead to infection and death. Thus, even threat displays can evolve through natural selection; they potentially benefit both signaler and receiver.

Visual signals are a vital part of more obviously cooperative interactions. The courtship displays of lekking grouse, albatross, and some other birds are wonderfully elaborate (Figure 50.3). The visual displays are thought to be a means by which individuals assess the reproductive state and intentions of potential partners.

Certain visual signals work well at night. Some deep-sea fishes and certain nocturnal animals have the means to produce bioluminescent flashes (page 101). Fireflies, for example, have special light-generating organs at the tip of the abdomen. The typical male firefly emits a complex visual signal as he flies about. Perched female fireflies answer with a simple flash, and they do so a few seconds after the male's signal is completed. The male and female flash back and forth as the male approaches and locates her.

The females of certain predatory fireflies provide us with a classic example of an illegitimate signaler. They have broken the code of communication between males

**Figure 50.3** Courtship displays of albatross. (**a**) The male spreads his wings and points his head to the sky in a visual signal to the female. (**b-d**) Mutual displays between male and female that involve visual, acoustical, and tactile components. These displays precede copulation.

and females of other firefly species. When the predatory female spots one of the signaling males, she waits the appropriate interval, then flashes back. Eventually the male may be lured into landing within attack range. Instead of mating, the male is eaten. The possibility of being lured to a predator is an *evolutionary cost* associated with the come-hither signaling mechanism. Although a particular male might live a long time by ignoring that signal, he would have no surviving descendants. And *reproductive success*, remember, is the real measure of evolutionary success.

**Acoustical Signals.** Many animals communicate by **acoustical signals**, which are sounds having precise, species-specific information. The distinctive song of the male white-throated sparrow, described in the preceding chapter, is a splendid example. That song, remember, attracts mates and secures territory. The males of many frog species also send acoustical signals, frequently

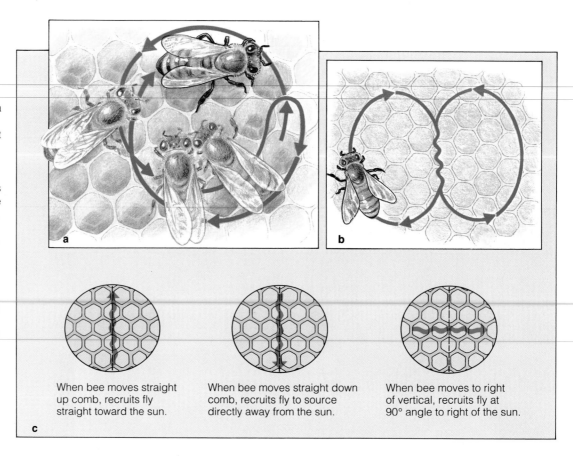

**Figure 50.4** Dances of honeybees, which convey information through tactile signals. **(a)** Bees trained to visit feeding stations close to a hive perform a round dance on the comb. Worker bees that remain in contact with the dancer search for food in the vicinity of the hive. **(b)** Bees trained to visit feeding stations more than 100 meters from the hive perform a waggle dance. During the portion of the dance when the bee moves in a straight line, it waggles its abdomen. The slower the waggles, the more distant the food source. **(c)** The angle of the straight run with respect to vertical provides information about the direction of the food source relative to the sun and hive.

When bee moves straight up comb, recruits fly straight toward the sun.

When bee moves straight down comb, recruits fly to source directly away from the sun.

When bee moves to right of vertical, recruits fly at 90° angle to right of the sun.

c

at night, that convey information to rival males and receptive females.

Among the calling frogs is a well-studied species that is exploited by an illegitimate receiver. The tungara frog of Central America has a two-part call, which is rather like a whine followed by a "chuck." The "chuck" component is highly attractive to females. It also is highly attractive to and easily pinpointed by a predator, the fringe-lipped bat. The bat tracks calls to their source and sweeps the male from the water (compare Figure 35.1).

**Tactile Signals**. Think about a handshake, a hug, a caress, or a shove, and you know how much can be conveyed by **tactile signals**, in which a signaler physically touches the receiver in socially significant ways. Or think about the highly social honeybees. Complex tactile signals figure prominently in their society. After finding a rich source of pollen or nectar, a foraging honeybee performs a complex dance upon returning to its hive. The dancing bee moves in circles, jostling in the dark through a crowded mass of workers on a honeycomb. Other bees may follow and maintain physical contact with the dancer. In so doing, they may acquire information about where the pollen or nectar is located.

The information is conveyed in the following way. When a forager has discovered food close to the hive, it performs a "round dance" of the sort shown in Figure 50.4a. When the forager has discovered a rich source far from the hive, it performs a "waggle dance" as shown in Figure 50.4b. Workers that have followed a forager during a round dance may be stimulated to fly out of the hive and search for the food source within 60 to 80 meters of the colony. They do not know the precise location of the nectar or pollen source. However, they smelled the scent of the particular flowers on the body of the dancer. They can search in the neighborhood for that scent and find food more quickly than they would by searching randomly.

Karl von Frisch, a pioneer in studies of social behavior, discovered that bees convey more precise information about the location of a food source when they do a waggle dance. He noted that the waggle dancer vibrates its abdomen only during a so-called straight run (Figure 50.4c). The orientation of the straight run varies from forager to forager. Von Frisch learned that the variations were related to differences in the location of food. When he set out a dish of honey on a direct line between the hive and the sun, the foragers that located it returned to the hive and oriented their straight runs right up the honeycomb. When the food source was placed at right angles to a line drawn between the hive and the sun, the foragers made their straight runs at 90 degrees to vertical on the honeycomb. Thus, a honeybee "recruited" into searching for food by a waggle-dancing forager could orient its flight *with respect to the sun and the hive*—and so waste less time and energy during the search.

**Figure 50.5** A colony of royal penguins on Macquarie Island, between New Zealand and Antarctica.

Besides this, waggle dancers vary the speed of their dance in relation to the distance of the food source from the hive. Thus, recruited bees got information about the distance between the hive and the food source. When a site is 150 meters from a hive, for example, the dance is much faster, with more waggles per straight run, compared to a dance concerning a food source that is 500 meters away.

You may be wondering, How was von Frisch able to distinguish one bee in a hive from another? He had painted a dot on the foragers that visited the honey dish. How was he able to observe their dances inside the hive? He had earlier arranged to have a special observation hive constructed with glass walls.

Von Frisch and others devised many experiments to test whether recruits actually used the information contained in the waggle dances of their experienced fellow foragers. Although controversy continues on this issue, many bee biologists believe that the evidence supports the hypothesis that the tactile component of the waggle dances conveys information that guides recruited bees toward a particular patch of flowers in the bees' environment.

## COSTS AND BENEFITS OF SOCIAL LIFE

### Disadvantages to Sociality

Communication is the glue that holds animals together in social groups and facilitates their cooperative, mutually beneficial actions. As you have just seen, a rich assortment of communication signals can convey a broad range of messages to other members of a social group. In fact, you may be thinking that all animal species are as sophisticated in their social life as termites, honeybees, and humans. But a survey of the animal kingdom reveals a considerable range of social behavior. The members of some species are largely solitary, and others live in small family groups. Others besides termites and honeybees live in huge groups of thousands of individuals. Still others, like modern humans, live in large social units containing many unrelated individuals. What is the basis for the diversity?

In answering this question, evolutionary biologists have employed a *cost-benefit approach*, as described in the preceding chapter. This approach is based on the hypothesis that all actions have both costs and benefits, which can be measured in terms of an individual's success in contributing genes to the next generation.

With this hypothesis in mind, consider that sociality clearly can have reproductive *disadvantages*. A nesting colony of herring gulls provides an example. Each breeding pair will cannibalize the eggs or young chicks of their neighbors in an instant if given the opportunity to do so. As another example, many herring gulls crowded together in the same habitat will deplete the available food more rapidly than they would if they were dispersed.

Another reproductive cost of sociality is greater vulnerability to parasites and contagious diseases. Think about a huge colony of cliff swallows, prairie dogs, or penguins (Figure 50.5). In such colonies, an individual or its offspring is more likely to be weakened by parasites, which are more readily transmitted from host to host in large groups. Our own history tells us

**Figure 50.6** Social defense by musk oxen. The pressure of predation can favor the evolution of living in groups when the members of the group are safer than are solitary animals.

a

b
**Figure 50.7** Social defense by Australian sawflies. The animals have smeared chemical secretions on each other in response to a disturbance.

that plagues spread like wildfire through the densely crowded cities of earlier times, before widespread reliance on plumbing, sewage treatment, and medical care (page 798).

Given the costs to individual reproductive success, why *do* the members of any species live close to others? As the following examples will indicate, *it may be that the benefits of social life are great enough to overcome the costs in certain environments.*

### Advantages to Sociality

**Predator Avoidance**. Predation is a risk factor that poses obvious disadvantages to individuals, but in many cases social behavior may mediate the risk. For example, simply having more pairs of eyes around can help individuals detect predators sooner. Or individuals can combine their defenses and attack a predator in ways that reduce the net risk to any given individual. Cooperative behavior that apparently reduces an individual's risk of being eaten also can be observed in many other species, including the musk oxen shown in Figure 50.6.

The biologist Birgitta Sillen-Tullberg studied the caterpillars of an Australian sawfly to gain insight into the benefits of sociality. Figure 50.7 shows how the caterpillars live together in a clump. When the caterpillars are disturbed, they collectively rear up, and writhe about, and regurgitate partially digested food. Food in this case is eucalyptus leaves, which contain chemical compounds that are toxic to most other animals—including predatory birds.

Sillen-Tullberg hypothesized that sawfly caterpillars benefit from the coordinated repulsion of predators. She used her hypothesis to predict that insect-eating song-

**a**   **b**

**Figure 50.8**   A dominant male and female member of a wolf pack (**a**). Typically, the dominant male is the only pack member likely to breed successfully. (**b**) Subordinate members of the pack greet a dominant male.

birds would be more likely to consume solitary caterpillars than a cluster of them. She tested her prediction by offering young, hand-reared Great Tits (*Parus major*) a chance to feed either on caterpillars offered up one by one or in a group of twenty per offering. She did this for a standard number of presentations. The ten birds that were offered one sawfly at a time consumed an average of 5.6 sawflies. The ten birds presented with clumped-together sawflies ate only 4.1, on average. Thus, in this experiment individuals of this species of sawfly were, as expected, somewhat safer in a group than on their own.

**The Selfish Herd**.   Cooperative defense against enemies is not the only factor that can favor the clumping of individuals. Some animals living in groups may benefit from the simple fact that some number of them will serve as living shields against predators, although not consciously so. Chinstrap penguins head out to sea in groups when they leave their rookeries, or communal nests. It may be that when they enter the water, at least some individuals benefit by being shielded by others of their group from sharks and killer whales.

If this scenario is correct, such a penguin group would qualify as a **selfish herd**, a simple society held together by reproductive self-interest. The selfish herd hypothesis has not been tested for chinstraps, but it has for bluegill sunfish. Male bluegills build their nests together at the bottom of lakes. Each male pushes muddy sediments aside with its fins, creating a small depression where his mate or mates will deposit eggs. If the colony of bluegill males is a selfish herd, we can predict competition for the "safe" sites—at the center of the colony. Compared to the periphery, eggs laid in nests at the center are less likely to be attacked by snails and largemouth bass. Competition

of this sort does indeed exist. The largest, most powerful males tend to claim the central locations. Other, smaller males must assemble around them. Although males at the periphery bear the brunt of predatory attacks, they still are better off in a group than on their own, fending off a bass singlehandedly, so to speak.

## SOCIAL LIFE AND SELF-SACRIFICE

In selfish herds, individuals make no personal sacrifice for other members of their group. Apparently, the personal benefits of living together simply outweigh the costs. By contrast, the members of some social groups show **self-sacrificing behavior**. That is, they behave in ways that help *other* individuals survive and reproduce—at personal cost.

How might we explain behavior that "sacrifices" some or all of an individual's own chance of reproductive success? One hypothesis is that making the sacrifice is really a cost of belonging to the group. Bluegills are a case in point. The peripheral males help the males in the center of the colony. But they do so only because they cannot displace their stronger rivals and because solitary life would expose them to even greater dangers.

Does the hypothesis hold up when we examine other social groups, such as baboon troops and wolf packs? A great deal of helpful behavior goes on in such groups, as when individuals cooperate and fend off predators. Wolves even share food. Nevertheless, the opportunities to reproduce are distributed in a most unequal fashion. For example, only *one* male and *one* female of a wolf pack typically produce pups (Figure 50.8). Other members of the pack do not breed, even though they sometimes bring back food to members that have stayed at the den, guarding the pups. Similarly, some baboons quickly give up safe sleeping places, choice bits of food, and even receptive females to others upon receiving a threat signal from another troop member. In both cases, some individuals have adopted a subordinate status to others of their social group (Figure 50.9). That is, a **dominance hierarchy** exists. A more detailed example of such hierarchies is described on page 580.

Why should nonbreeding adults in a baboon troop or wolf pack remain in the group and make sacrifices to their dominant peers? Like small bluegill sunfish, they may derive sufficient personal benefits to make acceptance of some sacrifices worthwhile. Maybe subordinates that challenge stronger individuals receive a life-shortening injury. Maybe those that leave the group have difficulty surviving long enough to form a group of their own. A solitary baboon surely quickens the pulse of the first leopard that sees it.

And perhaps self-sacrificing behavior gives them the chance to assume reproductive status, *if* they live long enough and *if* predation, weakness of old age, or some other occurrence removes dominant peers in the hierarchy. If this is true, then we can predict that some sub-

ordinate wolves and baboons *do* move up the social ladder as dominant members slip down a rung or fall off. Repeated and long-term field observations of wolves, baboons, chimpanzees, and similar animals do indeed indicate that acceptance of subordinate status can have long-term reproductive payoffs for the patient individual.

**Self-sacrificing behavior may be more apparent than real. Although individuals seem to be giving up their chance of reproduction while helping others, they may be better off with the group than without it—and they may be presented with an opportunity to reproduce later on.**

## THE EVOLUTION OF ALTRUISM

A termite soldier and a honeybee worker are sterile, incapable of reproducing at any time in their life. For them, we cannot hypothesize that an eventual chance at reproduction is a potential benefit of their helpful behavior. Rather, we must explain their **altruistic behavior**. An altruistic individual is one that helps increase the reproductive success of others but pays a cost, for the behavior decreases the number of offspring it can produce in its own lifetime.

How can the genetic basis for altruism persist over evolutionary time? As explained in Chapter 49, it is unlikely that individuals give up their chance to reproduce in order to help their group survive. To account for altruism, we need a different explanation, which evolu-

**Figure 50.9** Appeasement behavior among baboons. Notice the assured position of the dominant animal—and the abject stare and groveling posture of the subordinate one, who is intent on making little conciliatory smacking noises with its lips.

tionary biologists call the theory of **indirect selection**. The key point of this theory is that individuals can indirectly pass on their genes by helping relatives survive and reproduce. This theory gives us insight into the reproductive advantages of parental behavior and the self-sacrificing behavior of sterile workers in insect societies.

## Parental Behavior and Caring for Relatives

The parents of some species take care of their offspring until they are capable of surviving on their own. For example, adult Caspian terns (Figure 50.10) incubate the eggs, shelter the nestlings and feed them, even accompany and protect them after they begin flying. Such activities use up time and energy that might otherwise be spent on improving the parents' own chance of living to reproduce another time. There is a cost associated with parenting—a loss of future reproduction. Strictly speaking, parental care is not altruistic behavior as evolutionary biologists use the term. Under some circumstances, parenting improves the likelihood that the current offspring will survive, and this immediate reproductive success can outweigh the cost of reduced reproductive success at some future time.

When a sexually reproducing parent helps one of its offspring, it is not helping an exact genetic copy of itself. The new individual has received only half of that parent's genes. Other individuals besides parents and offspring share common genes as a result of having the same ancestors. Genetically, two siblings are as similar as a parent and one of its offspring. An uncle and a nephew or a niece will be alike in about one-fourth of their genes.

William Hamilton proposed in 1964 that caring for nondescendant *relatives* might perpetuate the genetic basis for a helpful behavior in much the same way that *parental* behavior can do. If an uncle helps a nephew or a niece survive long enough to reproduce, then the uncle has effectively made a genetic contribution to the next generation. His contribution can be measured in terms of the genes that he and his relative share. The genetic cost of his altruism may be lost opportunities to reproduce himself. But if the cost is less than the benefit, the action will propagate the uncle's genes and favor the spread of his kind of altruism in the species. This example encapsulates the logic of the theory of indirect selection. If an uncle saves two nieces, this is equivalent to saving his own daughter.

## Indirect Selection and Social Insects

Hamilton applied his theory of indirect selection to social insects, with their sterile castes of self-sacrificing workers. He hypothesized that the workers are indirectly promoting their "self-sacrifice" genes through altruistic behavior. If true, then their altruism should be directed toward relatives, not toward all bees indiscriminately.

This prediction is correct. Honeybee colonies, as well as those of ants and termites, are actually large families. The families include a worker force that labors on behalf of siblings, some of which are future kings and queens. When a guard bee drives her stinger into a raccoon, she dies when the stinger pulls away from her own body.

For reasons that need not be explored here, worker bees, ants, and wasps may show notable genetic similarity to their sisters. The similarity may promote the

**Figure 50.10** Male and female Caspian terns, protecting their chicks. Parental care carries costs as well as benefits.

**Figure 50.11** Life in a honeybee colony. (**a**) The only fertile female in the hive is the queen bee, shown here surrounded by her court of sterile worker daughters. These individuals feed her and relay her pheromone throughout the hive, regulating the activity of all of its members.

(**b**) Transfer of food from bee to bee, one of the helpful actions within honeybee society. (**c**) Bee dance. The central bee is in the midst of a complex dance maneuver, which contains information about the direction and distance to a food source. Floral odors on the dancer's body are also useful to the recruits, which follow her movements before flying out in search of the food. (**d**) Guard bees. Worker females assume a typical stance at the colony entrance. They are quick to repel intruders from the hive. (**e**) The queen is much larger than the workers, in part because her ovaries are fully developed (unlike those of her sterile daughters). (**f**) Stingless drones are produced at certain times of the year. They do not work for the colony but instead attempt to mate with queens of other hives. (**g**) Worker bees forage, feed larvae, guard the colony, construct honeycomb,

and clean and maintain the nest. Between 30,000 and 50,000 are present in a colony. They live about six weeks in the spring and summer, and can survive about four months in an overwintering colony. (**h**) Scent-fanning, another cooperative action by the worker. As air is fanned, it passes over the exposed scent gland of the bee. The pheromones released from the gland help other bees orient to the colony entrance. (**i**) Worker bees constructing new honeycomb from wax secretions. Here, honey or pollen may be stored or new generations may be cared for from the egg stage to the emergence of the adult.

(**j**) The initial stages in the life cycle of a bee. The brood cell caps have been removed, revealing eggs and larvae of various ages. Larvae are fed by young worker bees. (**k**) Worker pupae. Once again the cell caps have been removed, exposing pupae that will metamorphose into future workers. (**l**) Complete sequence of developmental stages of a worker bee, from the egg to a six-day-old larva (fourth from left), to a twenty-one-day-old pupa about to become an adult.

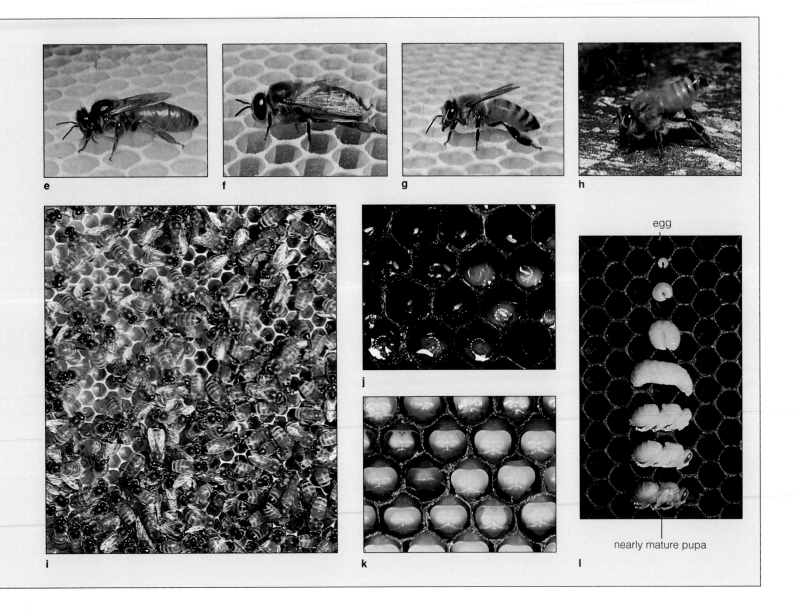

e f g h

i j k l

egg

nearly mature pupa

evolution of altruism among these insects. In such cases, the genetic benefits for self-sacrificing worker females of the sort shown in Figure 50.11 are unusually high.

Sterility and extreme self-sacrifice appear to be much more rare among social vertebrates than among social insects. The *Doing Science* essay describes one of the exceptions.

**The existence of sterile and self-sacrificing individuals in insect societies may be explained by assuming that they pass on their genes indirectly by helping relatives survive and reproduce. This point is the basis of the theory of indirect selection.**

## Naked Mole-Rats and the Evolution of Self-Sacrificing Behavior

Naked mole-rats look like bucktoothed sausages with wrinkled, pink skin from which a few hairs sprout out, here and there (Figure *a*). Although the appearance of these mammals is rather bizarre, their behavior is extremely fascinating. The fascination stems from the discovery that, unlike any other known vertebrate, many naked mole-rat individuals appear to spend their entire lives as non-breeding helpers in the groups in which they live. How can helpful behavior persist in mole-rat populations if helpers fail to reproduce—and so fail to have offspring that carry on the genetic basis for helping?

Before we try to solve this puzzle, let's examine what it is that helper mole-rats do. These highly social mammals live underground in the arid plains of northeastern Africa. They are always found in colonies that have anywhere from 25 to about 300 members. But each colony has only a single reproducing female that mates with one to three males. All other members of the colony devote themselves to caring for the "queen" and the "king" or "kings" and their offspring.

The nonreproducers dig an extensive network of sub-terranean tunnels with special chambers that serve as living rooms or waste-disposal centers. The tunnel diggers work like a chain gang, so to speak, with the lead animal gnawing into the cement-like soil of their environment. The followers push the loosened dirt under them and eventually up to the surface.

Helpers do more than the arduous work of tunnel con-struction. They also locate and chop up large, underground tubers that serve as food for the colony. Workers carry back bits of root to feed the queen and her retinue of males and offspring. They also deliver food to some other helpers that spend time loafing about, shoulder to shoulder (and belly to back), with the reproductives. The "loafers," which are usually larger than the diggers, get busy when a snake or some other enemy threatens the colony. At great personal risk, they launch an attack against the predator, chasing it away or killing it. Thus, different kinds of nonreproducing helpers occur in naked mole-rat colonies, and they work for a tiny minority of reproducing colony members. Is their altruism adaptive?

Helpers might *increase* the number of offspring produced by other individuals. If the additional individuals were related to them, the helpers could increase the number of genetically similar relatives that exist in their population.

Therefore, the loss of offspring production experienced by a self-sacrificing altruist could be genetically offset by the extra relatives that exist, thanks to the helper's assistance. Under some conditions, an animal might actually make a greater genetic contribution to the next generation by helping relatives survive to reproduce than by reproducing itself.

The theory of indirect selection applied to naked mole-rats produces a hypothesis and the following testable prediction:

*Hypothesis*. If helping is genetically advantageous to some individuals in a naked mole-rat colony,

*Prediction*. then it follows that helpers will be related to the reproductive members of the colony that benefit from their altruism.

*Test*. The prediction can be tested by determining the genetic relationships between the subordinate, sterile helpers and the dominant, reproducing queen and king(s).

One way to conduct this test would be to study a naked mole-rat colony in its natural environment, marking individuals and gradually establishing a genealogy for the group. Such a test would be highly impractical, however, given the hidden, subterranean life-style of these rodents.

A research group led by H. Kern Reeve took advantage of *DNA fingerprinting*, a method that can be used to estab-

### HUMAN SOCIAL BEHAVIOR

Is it possible to analyze human behavior in the same way that biologists have analyzed the behavior of ter-mites and naked mole-rats? There is resistance to the idea. Apparently, many people believe that attempting to identify the adaptive value of a particular human trait is the same thing as attempting to define its moral or social advantage. But there is a clear difference between trying to explain something by reference to its evolution-

lish degrees of genetic relatedness among individuals. DNA fingerprinting makes use of restriction fragments, of the sort described on page 253. In this case, restriction enzymes cut a DNA molecule wherever a particular base sequence (for example, adenine-cytosine-thymine-adenine) occurs. If individuals differ in their DNA base sequences, they may differ in the locations and number of times that particular series (A-C-T-A) appears among the millions of bases in their DNA. If so, they will differ in the number and length of the DNA fragments created.

It is possible to construct a visual record of the different sets of fragments derived from the DNA of different individuals. These are the DNA "fingerprints." Identical twins will have essentially identical DNA fingerprints because they possess DNA with exactly the same base sequences. Individuals with the same father and mother will have similar but not identical DNA and thus similar but not identical DNA fingerprints. On the average, genetically unrelated individuals should differ much more in their "fingerprints" than siblings or other kinds of relatives.

Inspection of DNA fingerprints taken from samples of tissue collected from naked mole-rats reveals that individuals from any given colony are remarkably similar. On the average, they are much more similar than you would expect a group of siblings with the same parents to be. This finding strongly supports the prediction that helper mole-rats are helping *very* close relatives.

Members of different colonies differ a great deal in their DNA fingerprints. This result, coupled with the genetic similarity within colonies, indicates that naked mole-rat colonies are highly inbred. This is the result of generations of brother-sister, mother-son, or father-daughter matings. Lineages created in this fashion have very reduced genetic variability. Each colony apparently has its own long history of inbreeding and so is genetically distinct from most other colonies.

Therefore, when a naked mole-rat helps a member of its own colony produce more offspring, the "extra" individuals produced by its help carry a very high proportion of the specific genes (alleles) that the helper possesses. Their genotypes may be perhaps 90 percent identical. Giving up reproduction under such circumstances does not condemn the individual's genes to extinction. Instead, the helper can help spread its genes in precisely the manner theorized by William Hamilton, with helpful individuals indirectly creating more copies of their genes by helping rear siblings.

It remains to be discovered how new colonies come into being in naked mole-rats. But at least we now have a better understanding of how sterility can persist in this species, thanks to a creative theory and the testable hypothesis that it produced.

ary history and trying to justify it. "Adaptive" does not mean "moral." It only means valuable in the transmission of an individual's genes.

Few people are bothered with the concept of adaptive value when they discuss the behavior of animals other than ourselves. For example, no one has complained to Doug Mock about his investigation of **siblicide** among baby egrets, the larger of which make lethal assaults on their smaller nestmates. Mock tested the hypothesis that the lethal assaults are adaptive for both the victorious chick and the nestlings' parents. He found that by eliminating the competition for food, large siblings had a better chance of fledging. The parent egrets also gained benefits, for if food becomes scarce, their attempt to rear too many chicks is doomed to failure. Mock was *not* attempting to justify siblicide by the egrets or to claim that the attacks were moral in some sense.

By the same token, evolutionary hypotheses about the adaptive value of "selfish" and "altruistic" actions of humans can be tested. Consider the following example, which illustrates how one might study the evolutionary basis of our behavior with no intent to justify any particular characteristic.

Humans engage in many helpful acts toward others. Adopting a baby is a particularly dramatic example. If we seek to explain why some people adopt children, we might choose an evolutionary approach to generate testable hypotheses. Whether adoption is moral or socially desirable is an entirely separate issue, about which biologists have no more to say than anybody else.

At first glance, it is difficult to see how adults might gain genetic representation in the next generation by adopting someone else's offspring. The theories of natural selection and indirect selection are based on the premise that individuals act in ways that promote their genetic self-interest. Therefore, we might argue that adoption is not adaptive for most individuals. We can predict that parents with some dependent, care-requiring children of their own will be much less likely to adopt, compared to adults who have no children or who have already raised their children and are now living by themselves. The prediction could be tested by securing a sufficient sample of data, ideally from a variety of human cultures, on the relationship between childlessness and adoption.

However, some conditions might make adoption adaptive for an individual. Under special circumstances, indirect selection could favor adults who directed their parenting assistance to relatives, thereby indirectly promoting the success of their shared genes. A prediction based on this hypothesis would be that such adults are more likely to be related to the adopted child than would be expected by chance alone. Joan Silk has tested this prediction. She showed that in some traditional societies, children are indeed adopted overwhelmingly by relatives. In modern societies in which adoption agencies and other means of facilitating adoption exist, relative adoption is not predominant. The point is that evolutionary hypotheses can be tested. In so doing, we gain additional understanding about the evolution of our behavior.

## SUMMARY

1. Social behavior depends on communication among animals living together. Communication signals are a form of cooperation, with a signaler and a receiver both benefiting from the transfer of information between them. Illegitimate signalers and receivers are members of one species that exploit the communication signals of another species.

2. Signaling animals employ many different sensory modes as channels of communication. Among these are chemical, visual, acoustical, and tactile signals.

3. Social species, those that live in groups, are not more highly evolved than solitary species whose members live primarily in isolation from one another.

4. Sociality carries with it high risks of contagious disease and competition for limited resources. But the costs of social life can be outweighed under some circumstances, particularly if predation pressure is severe.

5. Most animal societies are composed of individuals that do not sacrifice their reproductive chances to help others in their group.

   a. Even when apparently helpful acts occur, they may be the result of dominant individuals forcing subordinates to relinquish useful resources.

   b. When subordinates give in to dominant members of their group, they may receive compensatory benefits of group living, particularly safety from predators.

   c. In some species, subordinates may eventually reproduce if they live long enough.

6. It is theoretically possible for individuals to reduce their net lifetime production of offspring in order to help others reproduce, provided that the others are relatives.

7. Indirect selection can lead to the extreme altruism shown by workers of some species of social insects and by the naked mole-rat. Although workers of these groups normally do not reproduce, they help their reproducing siblings survive, passing on "by proxy" the genes underlying the development of altruism.

### Review Questions

1. A hyena places scent marks on vegetation in its territory by releasing specific chemicals from certain glands. What evidence would you need to demonstrate that this action is an evolved communication signal? *908–910*

2. Explain how a threat display can be considered an example of cooperation. *908–909*

3. Why don't the members of a "selfish herd" live apart if each member is "trying to take advantage" of the others? *913*

4. How can a self-sacrificing individual of a social group still pass on more of its genes than a noncooperative individual? *913–914*

5. Is parental behavior always adaptive? *915*

6. How can an individual propagate its genes even if it is sterile? *914–917*

## Self-Quiz *(Answers in Appendix IV)*

1. An *evolved* communication signal is one that _____ .
   a. benefits the signaler, not the receiver
   b. benefits the receiver, not the signaler
   c. benefits both signaler and receiver
   d. benefits the species as a whole by promoting cooperation

2. Social behavior evolves because _____ .
   a. social species are more advanced evolutionarily than solitary ones
   b. under some ecological conditions, the costs of social life are less than its benefits to the group
   c. under some ecological conditions, the costs of social life are less than its benefits to the individual
   d. predator pressures always favor social living

3. The behavior of subordinate animals in groups with dominance hierarchies poses a puzzle to Darwinian biologists because _____ .
   a. subordinates use valuable resources that the more adaptive dominants should receive if the group is to persist
   b. dominants often physically attack subordinates
   c. subordinates are usually younger or weaker individuals
   d. subordinates often do not even attempt to reproduce when dominants are present

4. Altruism is (in terms of evolutionary biology) helpful _____ .
   a. behavior that cannot evolve
   b. behavior that can spread through a species only if it raises the altruist's reproductive success
   c. behavior that reduces an individual's reproductive success
   d. behavior that always helps spread the altruist's genes

5. Which of the following represents an evolutionary cost of giving a loud acoustical signal?
   a. the energy expended by the caller
   b. the risk that an illegitimate receiver will detect and exploit the call
   c. the time spent calling that cannot be spent feeding
   d. all of the above

6. The discovery that members in large colonies of cliff swallows are more likely to have nests infested with nest parasites provides evidence that _____ .
   a. sociality has disadvantages as well as possible benefits
   b. cliff swallows will never nest solitarily
   c. the costs of sociality are greater than its benefits
   d. social life forms are more advanced evolutionarily than solitary ones

7. The finding that predators eat more sawflies when the prey are offered to them experimentally one by one rather than in groups of ten is _____ .
   a. a hypothesis          c. a test of a prediction
   b. a prediction          d. a proven hypothesis

8. The theory of evolution by indirect selection differs from the Darwinian theory of evolution by natural selection in _____ .
   a. focusing on how evolution for group benefit could occur
   b. explaining that altruism is always adaptive
   c. examining the consequences of differences among individuals in their effects on the reproductive success of relatives
   d. showing how families that have many descendants are superior to families that exhibit self-sacrificing behavior

9. The degree of genetic similarity between an uncle and his nephew is _____ .
   a. the same as between a parent and his offspring
   b. greater than between two full siblings
   c. dependent on how many other nephews the uncle has
   d. less than that between a mother and her daughter

10. You discover a new species of insect in which groups consist of sterile males and females that cooperate to help rear the young of four pairs of reproducing males and females. You might use DNA fingerprinting to test which of the following predictions based on the theory of indirect selection:
    a. The sterile males will not be closely related to the sterile females.
    b. The sterile individuals will be more closely related to each other than to the reproducing males and females.
    c. The reproducing individuals will be cousins.
    d. All the members of one group will be much more closely related genetically to each other than to members of any other group.

## Selected Key Terms

| | |
|---|---|
| acoustical signal *909* | indirect selection *915* |
| altruistic behavior *914* | receiver *907* |
| chemical signal *908* | selfish herd *913* |
| communication signal *907* | self-sacrificing behavior *913* |
| cost-benefit approach *911* | siblicide *919* |
| DNA fingerprinting *918* | signaler *907* |
| dominance hierarchy *914* | social behavior *907* |
| illegitimate receiver *908* | tactile signal *910* |
| illegitimate signaler *908* | visual signal *908* |

## Readings

In addition to the readings outlined at the end of the preceding chapter, all of which deal with animal behavior, a number of excellent texts focus primarily on material covered in this chapter, including:

Daly, M., and M. Wilson. 1983. *Sex, Evolution, and Behavior*. Second edition. Boston: Willard Grant Press. This book has two particularly good chapters on the evolutionary analysis of human behavior.

Frisch, K. von. 1961. *The Dancing Bees*. New York: Harcourt Brace Jovanovich. A classic on the natural history and behavior of honeybees.

Trivers, R. 1985. *Social Evolution*. Menlo Park, California: Benjamin/Cummings. A complete and readable account of all the topics that make up an evolutionary approach to social behavior.

Wilson, E. O. 1975. *Sociobiology: The New Synthesis*. Cambridge, Massachusetts: Harvard University Press. The pivotal book in presenting a new approach to the study of social behavior.

Wittenberger, J. F. 1981. *Animal Social Behavior*. Boston: Willard Grant Press. A book that covers much the same ground as *Social Evolution* by Robert Trivers but at a somewhat more advanced level. The book does an excellent job of laying out competing hypotheses on aspects of social behavior and evaluating the evidence for them.

# APPENDIX II
# Brief Classification Scheme

This classification scheme is a composite of several used in microbiology, botany, and zoology. Although major groupings are more or less agreed upon, what to call them and (sometimes) where to place them in the overall hierarchy are not. As Chapter 20 indicated, there are several reasons for this. First, the fossil record varies in its quality and completeness, so certain evolutionary relationships are open to interpretation. Comparative studies at the molecular level are firming up the picture, but this work is still under way.

Second, since the time of Linnaeus, classification schemes have been based on perceived morphological similarities and differences among organisms. Although some original interpretations are now open to question, we are so used to thinking about organisms in certain ways that reclassification proceeds slowly. For example, birds and reptiles traditionally are considered separate classes (Reptilia and Aves)—even though there now are compelling arguments for grouping lizards and snakes as one class, and crocodilians, dinosaurs, and birds as another.

Finally, botanists as well as zoologists have inherited a wealth of literature based on schemes that are peculiar to their fields; and most see no good reason to give up established terminology and so disrupt access to the past. Thus botanists continue to use Division as a major taxon in the hierarchical schemes and zoologists use Phylum in theirs. Opinions are notably divergent with respect to an entire Kingdom (the Protista), certain members of which could just as easily be called single-celled forms of plants, fungi, or animals. Indeed, the term protozoan is a holdover from earlier schemes that ranked the amoebas and some other forms as simple animals.

Given the problems, why do we bother imposing hierarchical schemes on the natural history of life on earth? We do this for the same reason that a writer might decide to break up the history of civilization into several volumes, many chapters, and a multitude of paragraphs. Both efforts are attempts to impart structure to what might otherwise be an overwhelming body of information.

One more point to keep in mind: The classification scheme in this Appendix is primarily for reference purposes, and it is by no means complete (numerous phyla of existing and extinct organisms are not represented). Our strategy is to focus mainly on the organisms mentioned in the text, with numerals referring to some of the pages on which representatives are illustrated or described. A few examples of organisms are also listed under the entries.

---

**SUPERKINGDOM PROKARYOTA.** Prokaryotes (single-celled organisms with no nucleus or other membrane-bound organelles in the cytoplasm).

---

**KINGDOM MONERA.** Bacteria, either single cells or simple associations of cells; autotrophic and heterotrophic forms. *Bergey's Manual of Systematic Bacteriology*, the authoritative reference in the field, calls this "a time of taxonomic transition" and groups bacteria mainly on the basis of form, physiology, and behavior, not on phylogeny (Table 22.4 gives examples). The scheme presented here does reflect the growing evidence of evolutionary relationships for at least some bacterial groups.

**SUBKINGDOM ARCHAEBACTERIA.** Methanogens, halophiles, thermophiles. Strict anaerobes, distinct from other bacteria in their cell wall, membrane lipids, ribosomes, and RNA sequences. 356, 357

**SUBKINGDOM EUBACTERIA.** Gram-negative and Gram-positive forms. Peptidoglycan in cell walls. Photosynthetic autotrophs, chemosynthetic autotrophs, and heterotrophs. 358–359

DIVISION GRACILICUTES. Typical Gram-negative, thin wall. Autotrophs (photosynthetic and chemosynthetic) and heterotrophs. *Anabaena, Chlorobium, Escherichia, Shigella, Desulfovibrio, Agrobacterium, Pseudomonas, Neisseria.* 57, 354, 356, 781
DIVISION FIRMICUTES. Typical Gram-positive, thick wall. Heterotrophs. *Staphylococcus, Streptococcus, Clostridium, Bacillus, Actinomyces.* 346, 359, 557, 680
DIVISION TENERICUTES. Gram-negative, wall absent. Heterotrophs (saprobes, parasites, pathogens). *Mycoplasma.* 356

---

**SUPERKINGDOM EUKARYOTA.** Eukaryotes (single-celled and multicelled organisms; cells typically have a nucleus and other organelles).

---

**KINGDOM PROTISTA.** Mostly single-celled eukaryotes. Some colonial forms.

PHYLUM GYMNOMYCOTA. Heterotrophs.
  Class Acrasiomycota. Cellular slime molds. *Dictyostelium.* 362, 363
  Class Myxomycota. Plasmodial slime molds. *Physarum.* 362
PHYLUM EUGLENOPHYTA. Euglenoids. Mostly heterotrophic, some photosynthetic. Flagellated. *Euglena.* 364
PHYLUM CHRYSOPHYTA. Golden algae, yellow-green algae, diatoms. Photosynthetic. Some flagellated, others not. *Vaucheria.* 50, 364–368
PHYLUM PYRRHOPHYTA. Dinoflagellates. Mostly photosynthetic, some heterotrophs. *Gonyaulax.* 365
PHYLUM MASTIGOPHORA. Flagellated protozoans. Heterotrophs. *Trypanosoma, Trichomonas.* 366–367

PHYLUM SARCODINA.    Amoeboid protozoans.
Heterotrophs. Amoebas, foraminiferans, heliozoans, radiolarians. 366–367
PHYLUM CILIOPHORA.    Ciliated protozoans.
Heterotrophs. *Paramecium, Didinium.* 367–368
SPOROZOANS.    Parasitic protozoans, many intracellular.
("Sporozoans" is the common name for these diverse organisms; it has no formal taxonomic status.) *Plasmodium.* 368–369

**KINGDOM FUNGI.**    Mostly multicelled eukaryotes.
Heterotrophs (mostly saprobes, some parasites). All rely on extracellular digestion and absorption of nutrients.

DIVISION MASTIGOMYCOTA.    All produce flagellated spores.
Class Chytridiomycetes. Chytrids. 375
Class Oomycetes. Water molds and related forms. *Plasmopora, Phytophthora, Saprolegnia.* 375, 376
DIVISION AMASTIGOMYCOTA.    All produce nonmotile spores.
Class Zygomycetes. Bread molds and related forms. *Rhizopus, Pilobolus.* 376, 377
Class Ascomycetes. Sac fungi. Most yeasts and molds; morels, truffles. *Saccharomyces, Morchella.* 376, 378
Class Basidiomycetes. Club fungi. Mushrooms, shelf fungi, bird's nest fungi, stinkhorns. *Agaricus, Amanita.* 376, 378–380
FORM-DIVISION DEUTEROMYCOTA.    Imperfect fungi. All with undetermined affiliations because sexual stage unknown; if better known they would be grouped with sac fungi or club fungi. *Verticillium, Candida.* 376, 380–381

**KINGDOM PLANTAE.**    Nearly all multicelled eukaryotes.
Photosynthetic autotrophs, except for a few saprobes and parasites. 385, 396

DIVISION RHODOPHYTA.    Red algae. *Porphyra.* 308, 388, 389, 869
DIVISION PHAEOPHYTA.    Brown algae. *Fucus, Laminaria.* 388–390
DIVISION CHLOROPHYTA.    Green algae. *Ulva, Spirogyra.* 16, 55, 308, 390–391, 817
DIVISION CHAROPHYTA.    Stoneworts.
DIVISION BRYOPHTYA.    Liverworts, hornworts, mosses. *Marchantia, Sphagnum.* 392–393
DIVISION RHYNIOPHYTA.    Earliest known vascular plants; extinct. *Cooksonia, Rhynia.* 310, 393
DIVISION PSILOPHYTA.    Whisk ferns. 310, 393–394
DIVISION LYCOPHYTA.    Lycopods, club mosses. *Lycopodium, Selaginella.* 394
DIVISION SPHENOPHYTA.    Horsetails. *Equisetum.* 280, 281, 394, 396
DIVISION PTEROPHYTA.    Ferns. 34, 395, 874
DIVISION PROGYMNOSPERMOPHYTA.    Progymnosperms. Ancestral to early seed-bearing plants; extinct. *Archaeopteris.* 294, 397
DIVISION PTERIDOSPERMOPHYTA.    Seed ferns (extinct fernlike gymnosperms).
DIVISION CYCADOPHYTA.    Cycads. *Zamia.* 311, 397
DIVISION GINKGOPHYTA.    Ginkgo. *Ginkgo.* 311, 398
DIVISION GNETOPHYTA.    Gnetophytes. *Ephedra, Welwitschia, Gnetum.* 398
DIVISION CONIFEROPHYTA.    Conifers. 309, 398, 852, 858, 860
Family Pinaceae. Pines, firs, spruces, hemlock, larches, Douglas firs, true cedars. *Pinus.* 399, 483, 784, 878
Family Cupressaceae. Junipers, cypresses, false cedars. 382
Family Taxodiaceae. Bald cypress, redwood, Sierra bigtree, dawn redwood. *Sequoia.* 10, 820, 821
Family Taxaceae. Yews.
DIVISION ANTHOPHYTA.    Flowering plants. 400, 469ff.
Class Dicotyledonae. Dicotyledons (dicots). Some families of several different orders are listed: 402, 474
Family Magnoliaceae. Magnolias, tulip trees. 478
Family Ranunculaceae. Buttercups, delphinium. 177, 480, 814

Family Nymphaeaceae. Water lilies. 400, 402
Family Papaveraceae. Poppies, including opium poppy. 516
Family Brassicaceae. Mustards, cabbages, radishes, turnips.
Family Malvaceae. Mallows, cotton, okra, hibiscus. 461
Family Solanaceae. Potatoes, eggplant, petunias. 114, 376, 818
Family Salicaceae. Willows, poplars. 477
Family Rosaceae. Roses, peaches, apples, almonds, strawberries. 503, 510, 511
Family Fabaceae. Peas, beans, lupines, mesquite, locust. 166, 477, 489, 519, 854
Family Cactaceae. Cacti. 512, 846, 854
Family Euphorbiaceae. Spurges, poinsettia. 847
Family Cucurbitaceae. Gourds, melons, cucumbers, squashes.
Family Apiaceae. Parsleys, carrots, poison hemlock. 254, 479, 514
Family Aceraceae. Maples. 109, 477, 511
Family Asteraceae. Composites. Chrysanthemums, sunflowers, lettuces, dandelions. 10, 511, 384, 857
Class Monocotyledonae. Monocotyledons (monocots). Some families of several different orders are listed: 402, 474
Family Liliaceae. Lilies, hyacinths, tulips, onions, garlic. 401, 816
Family Iridaceae. Irises, gladioli, crocuses. 474
Family Orchidaceae. Orchids. 326, 400, 874
Family Arecaceae. Date palms, coconut palms. 397, 477
Family Cyperaceae. Sedges.
Family Poaceae. Grasses, bamboos, corn, wheat, sugarcane. 115, 401, 479, 518, 856–857
Family Bromeliaceae. Bromeliads, pineapple, Spanish moss. 511

**KINGDOM ANIMALIA.**    Multicelled eukaryotes. Heterotrophs (herbivores, carnivores, omnivores, parasites, decomposers, detritivores). 310, 407, 440

PHYLUM PLACOZOA.    Small, organless marine animal. *Trichoplax.* 369, 440
PHYLUM MESOZOA.    Ciliated, wormlike parasites, about the same level of complexity as *Trichoplax.*
PHYLUM PORIFERA.    Sponges. 408–409
PHYLUM CNIDARIA    410
Class Hydrozoa. Hydrozoans. *Hydra, Obelia, Physalia.* 410, 412
Class Scyphozoa. Jellyfishes. *Aurelia.* 410, 411
Class Anthozoa. Sea anemones, corals. *Telesto.* 410, 565, 624, 867, 868
PHYLUM CTENOPHORA.    Comb jellies. *Pleurobrachia.* 413
PHYLUM PLATYHELMINTHES.    Flatworms. 414
Class Turbellaria. Triclads (planarians), polyclads. *Dugesia.* 414, 642, 700
Class Trematoda. Flukes. *Schistosoma.* 414–415, 416
Class Cestoda. Tapeworms. *Taenia.* 414–415, 416, 417
PHYLUM NEMERTEA.    Ribbon worms. 418
PHYLUM NEMATODA.    Roundworms. *Ascaris, Trichinella.* 416–417, 418–419
PHYLUM ROTIFERA.    Rotifers. 419
PHYLUM MOLLUSCA.    Mollusks. 420
Class Polyplacophora. Chitons. 421
Class Gastropoda. Snails (periwinkles, whelks, limpets, abalones, cowries, conches, nudibranchs, tree snails, garden snail), sea slugs, land slugs. 150, 421, 422, 817
Class Bivalvia. Clams, mussels, scallops, cockles, oysters, shipworms. 216, 423
Class Cephalopoda. Squids, octopuses, cuttlefish, nautiluses. *Loligo.* 282, 310, 424–425, 553
PHYLUM BRYOZOA.    Bryozoans (moss animals).
PHYLUM BRACHIOPODA.    Lampshells.
PHYLUM ANNELIDA.    Segmented worms. 425
Class Polychaeta. Mostly marine worms. 425, 426

# APPENDIX III
# Answers to Genetics Problems

## Chapter Eleven

1. a. *AB*
   b. *AB* and *aB*
   c. *Ab* and *ab*
   d. *AB, aB, Ab,* and *ab*

2. a. *AaBB* will occur in all the offspring.
   b. 25% *AABB*; 25% *AaBB*; 25% *AABb*; 25% *AaBb*.
   c. 25% *AaBb*; 25% *Aabb*; 25% *aaBb*; 25% *aabb*.
   d. $\frac{1}{16}$ *AABB* ( 6.25%)

   $\frac{1}{8}$ *AaBB* (12.5% )

   $\frac{1}{16}$ *aaBB* ( 6.25%)

   $\frac{1}{8}$ *AABb* (12.5% )

   $\frac{1}{4}$ *AaBb* (25% )

   $\frac{1}{8}$ *aaBb* (12.5% )

   $\frac{1}{16}$ *AAbb* ( 6.25%)

   $\frac{1}{8}$ *Aabb* (12.5% )

   $\frac{1}{16}$ *aabb* ( 6.25%)

3. Yellow is recessive. Because the first-generation plants must be heterozygous and had a green phenotype, green must be dominant over the recessive yellow.

4. a. *ABC*
   b. *ABc* and *aBc*
   c. *ABC, aBC, ABc,* and *aBc*
   d. *ABC, aBC, AbC, abC, ABc, aBc, Abc,* and *abc*

5. Because the man can only produce one type of allele for each of his ten genes, he can only produce one type of sperm. The woman, on the other hand, can produce two types of alleles for each of her two heterozygous genes; she can produce 2 × 2 or 4 different kinds of eggs. As can be observed, as the number of heterozygous genes increases, more and more different types of gametes can be produced.

6. The first-generation plants must all be double heterozygotes. When these plants are self-pollinated, $\frac{1}{4}$ (25%) of the second-generation plants will be doubly heterozygous.

7. The most direct way to accomplish this would be to allow a true-breeding canary having yellow feathers to mate with a true-breeding canary having brown feathers. Such true-breeding strains could be obtained by repeated inbreeding (mating of related individuals; for example, a male and a female of the same nest) of yellow and brown strains. In this way, it should be possible to obtain homozygous yellow and homozygous brown canaries. When true-breeding yellow and true-breeding brown canaries are crossed, the progeny should all be heterozygous. If the progeny phenotype is either yellow or brown, then the dominance is simple or complete, and the phenotype reflects the dominant allele. If the phenotype is intermediate between yellow and brown, there is incomplete dominance. If the phenotype shows both yellow and brown, there is codominance.

8. a. Mother must be heterozygous for both genes; father is homozygous recessive for both genes. The first child is also homozygous recessive for both genes.
   b. The probability that the second child will not be able to roll the tongue and will have free earlobes is $\frac{1}{4}$ (25%).

9. a. The mother must be heterozygous ($I^A i$). The man having type B blood could have fathered the child if he were also heterozygous ($I^B i$).
   b. If the man is heterozygous, then he *could be* the father. However, because any other type B heterozygous male also could be the father, one cannot say that this particular man absolutely must be. Actually, any male who could contribute an O allele (*i*) could have fathered the child. This would include males with type O blood (*ii*) or type A blood who are heterozygous ($I^A i$).

10. a. $F_1$ genotypes and phenotypes: 100% *Bb Cc*, brown progeny. $F_2$ phenotypes: $\frac{9}{16}$ brown + $\frac{3}{16}$ tan + $\frac{4}{16}$ albino.

    $F_2$ genotypes: $\begin{cases} \frac{1}{16} BB\ CC + \frac{2}{16} BB\ Cc + \frac{2}{16} Bb\ CC + \\ \frac{4}{16} Bb\ Cc;\ \left(\frac{9}{16}\text{ brown}\right) \\ \frac{1}{16} bb\ CC + \frac{2}{16} bb\ Cc;\ \left(\frac{3}{16}\text{ tan}\right) \\ \frac{1}{16} BB\ cc + \frac{2}{16} Bb\ cc + \frac{1}{16} bb\ cc;\ \left(\frac{4}{16}\text{ albino}\right) \end{cases}$

    b. Backcross phenotypes: $\frac{1}{4}$ brown + $\frac{1}{4}$ tan + $\frac{2}{4}$ albino.

    Backcross genotypes: $\begin{cases} \frac{1}{4} Bb\ Cc;\ \left(\frac{1}{4}\text{ brown}\right) \\ \frac{1}{4} bb\ Cc;\ \left(\frac{1}{4}\text{ tan}\right) \\ \frac{1}{4} Bb\ cc + \frac{1}{4} bb\ cc;\ \left(\frac{1}{2}\text{ albino}\right) \end{cases}$

11. The mating is *Ll* × *Ll*.
    Progeny genotypes: $\frac{1}{4}$ *LL* + $\frac{1}{2}$ *Ll* + $\frac{1}{4}$ *ll*

    Phenotypes: $\begin{cases} \frac{1}{4}\text{ homozygous survivors (}LL\text{) +} \\ \frac{1}{2}\text{ heterozygous survivors (}Ll\text{) +} \\ \frac{1}{4}\text{ lethal (}ll\text{) nonsurvivors.} \end{cases}$

    Thus, among the survivors, there is a $\frac{2}{3}$ probability that any individual will be heterozygous.

12. To work this problem, consider the effect of each pair of genes separately.
    a. *Aa* × *aa*: The resulting progeny genotypes are $\frac{1}{2}$ *Aa* and $\frac{1}{2}$ *aa*. Thus, half of the progeny will receive an *A* allele, which permits kernel pigmentation.

b. $cc \times Cc$: Here, too, half of the progeny will receive a dominant $C$ allele, which permits kernel pigmentation.

c. $Rr \times Rr$: In this case, $\frac{3}{4}$ of the progeny will receive at least one $R$ allele, which permits kernel pigmentation. Remember that in order to have pigmented kernels, at least one dominant allele of each and every one of these three gene loci must be simultaneously present. Since the $A$, $C$, and $R$ loci assort independently, to find the overall fraction of kernels that are pigmented, you must multiply together the fraction of pigmented kernels produced separately by the $A$, $C$, and $R$ loci, and that is: $\frac{1}{2} \times \frac{1}{2} \times \frac{3}{4} = \frac{3}{16}$.

## Chapter Twelve

1. a. Males inherit their X chromosome from their mothers.

b. A male can produce two types of gametes with respect to an X-linked gene. One type will lack this gene and possess a Y chromosome. The other will have an X chromosome and the linked gene.

c. A female homozygous for an X-linked gene will produce just one type of gamete containing an X chromosome with the gene.

d. A female heterozygous for an X-linked gene will produce two types of gametes. One will contain an X chromosome with the dominant allele, and the other type will contain an X chromosome with the recessive allele.

2. a. Because this gene is only carried on Y chromosomes, females would not be expected to have hairy pinnae because they normally do not have Y chromosomes.

b. Because sons always inherit a Y chromosome from their fathers and because daughters never do, a man having hairy pinnae will always transmit this trait to his sons and never to his daughters.

3. A 0% crossover frequency means that 50% of the gametes will be $AB$ and 50% will be $ab$.

4. The first-generation females must be heterozygous for both genes. The 42 red-eyed, vestigial-winged and the 30 purple-eyed, long-winged progeny represent recombinant gametes from these females. Because the first-generation females must have produced 600 gametes to give these 600 progeny, and because $42 + 30$ of these were recombinant, the percentage of recombinant gametes is 72/600, or 12%, which implies that 12 map units separate the two genes.

5. The rare vestigial-winged flies could be explained by a deletion of the dominant allele from one of the chromosomes, due to the action of the x-rays. Alternatively, the radiation may have induced a mutation in the dominant allele.

6. If the longer-than-normal chromosome 14 represented the translocation of most of chromosome 21 to the end of a normal chromosome 14, then this individual would be afflicted with Down syndrome due to the presence of this attached chromosome 21 as well as two normal chromosomes 21. The total chromosome number, however, would be 46.

7. Using $c$ as the symbol for color blindness and $C$ for normal vision, then the cross can be diagrammed as follows:

$$C(Y) \text{ female} \times Cc \text{ male}$$

In mugwumps, a son receives one sex-linked allele from each of his parents, but a daughter inherits her unpaired sex-linked allele solely from her father. In this cross, half of the sons will be $CC$ and half $Cc$, but none will be color blind. Of the daughters, half will be $C(Y)$ and half $c(Y)$. There is a 50% chance that a daughter will be color blind. Note: this answer is backward from the way it would be in humans; but it is correct not only for mugwumps but also for all birds and Lepidoptera (moths and butterflies), as well as a few other forms.

8. In order to produce a female suffering from childhood muscular dystrophy, not only must her mother be a carrier of the disease, but her father must have it. Few, if any, such males who survive to adulthood are capable of having children.

# APPENDIX IV
# Answers to Self-Quizzes

**CHAPTER 1**
1. DNA
2. energy
3. Metabolism
4. Homeostasis
5. adaptations
6. mutations
7. reproductive capacity
8. c
9. c
10. c

**CHAPTER 2**
1. electrons
2. d
3. electrons
4. Isotopes
5. c
6. b
7. d
8. b
9. b
10. d
11. b, e, a, c, d

**CHAPTER 3**
1. carbon
2. a
3. polysaccharides, lipids, proteins, nucleic acids
4. c
5. c
6. d
7. b
8. b
9. c
10. c, e, b, d, a

**CHAPTER 4**
1. Cells
2. e
3. c
4. c
5. d
6. d
7. b
8. d
9. cytoskeleton
10. c, h, g, e, d, b, a, f

**CHAPTER 5**
1. plasma membrane
2. c
3. a
4. a
5. b
6. transport, receptor, and recognition proteins
7. b
8. d
9. c
10. g, f, d, a, e, b, c

**CHAPTER 6**
1. metabolism
2. thermodynamics
3. c
4. d
5. e
6. d
7. d
8. d
9. c
10. c, e, d, a, b

**CHAPTER 7**
1. carbon
2. carbon dioxide; sunlight
3. d
4. c
5. b
6. e
7. c
8. c
9. c, d, e, b, a

**CHAPTER 8**
1. glucose, other organic compounds
2. fermentation; anaerobic electron transport
3. pyruvate; carbon dioxide; water
4. d
5. c
6. d
7. c
8. b
9. b
10. b, c, a, d

**CHAPTER 9**
1. mitosis; meiosis
2. chromosomes; DNA
3. c
4. d
5. c
6. a
7. c
8. d
9. b
10. b
11. d, b, c, a

**CHAPTER 10**
1. diploid; two
2. c
3. c
4. c
5. d
6. c
7. d
8. c
9. b

**CHAPTER 11**
1. a
2. c
3. a
4. c
5. b
6. c
7. b
8. d
9. c
10. c, d, e, b, a

**CHAPTER 12**
1. c
2. e
3. e
4. c
5. e
6. c
7. d
8. d
9. d
10. c, e, d, b, a

**CHAPTER 13**
1. Hydrogen
2. e
3. d
4. d
5. c
6. a
7. a
8. d
9. d
10. d, e, b, c, a

**CHAPTER 14**
1. three
2. d
3. b
4. b
5. d
6. c
7. a
8. b
9. a
10. a

**CHAPTER 15**
1. chemical
2. differentiation
3. d
4. e
5. d
6. b
7. c
8. d
9. a
10. b, e, c, d, a

**CHAPTER 16**
1. diversity
2. genetic engineering
3. Plasmids
4. cloning vectors
5. c
6. d
7. b
8. c
9. e
10. d, e, a, c, b

**CHAPTER 17**
1. body
2. evolve
3. fossils
4. population
5. b
6. c
7. e
8. e
9. c, d, e, b, a

**CHAPTER 18**
1. population
2. differences
3. e
4. a
5. e
6. d
7. c
8. c
9. d
10. d, c, e, a, b

## CHAPTER 19

1. macroevolution
2. mass extinctions; adaptive radiations
3. gradual; bursts
4. e
5. e
6. c
7. b
8. e
9. b, d, e, a, c

## CHAPTER 20

1. d
2. e
3. c
4. c
5. b
6. a
7. d
8. b
9. e, c, g, a, d, f, b

## CHAPTER 21

1. e
2. b
3. c
4. c
5. a
6. b
7. d
8. c
9. d, a, e, c, b

## CHAPTER 22

1. a living host cell
2. e
3. d
4. d
5. c
6. c
7. e
8. b
9. c
10. c, e, d, a, b

## CHAPTER 23

1. decomposers
2. mycelium; hyphae
3. a
4. c
5. b
6. c
7. b
8. f
9. d
10. c, e, d, b, a

## CHAPTER 24

1. autotrophs
2. green algae
3. b
4. d
5. a
6. b
7. c
8. d
9. c
10. d, e, c, f, b, a

## CHAPTER 25

1. body symmetry, cephalization, type of gut, type of body cavity, segmentation
2. a
3. b
4. d
5. a
6. a
7. b
8. c

## CHAPTER 26

1. bilateral
2. c
3. b
4. brain
5. c
6. c
7. b
8. c
9. e

## CHAPTER 27

1. dermal, ground, vascular
2. apical; lateral
3. b
4. a
5. a
6. b
7. c
8. a
9. d
10. c, d, f, a, e, b

## CHAPTER 28

1. hydrogen bonds
2. stomata
3. d
4. e
5. c
6. b
7. d
8. d
9. d
10. e, d, c, a, b

## CHAPTER 29

1. a
2. pollinators
3. d
4. c
5. b
6. c
7. b
8. a
9. a, c, d, e, b

## CHAPTER 30

1. c
2. c
3. e
4. d
5. d
6. a
7. c
8. c
9. a, e, c, d, b

## CHAPTER 31

1. epithelial, connective, muscle, nervous
2. e
3. d
4. c
5. b
6. a
7. d
8. c
9. a
10. Receptors, integrator, effectors
11. d, e, c, b, a

## CHAPTER 32

1. neurons
2. sensory neurons, interneurons, motor neurons
3. a
4. c
5. a
6. d
7. b
8. d
9. d
10. c, d, a, b, e

## CHAPTER 33

1. nerve nets
2. cephalization; bilateral
3. a
4. b
5. c
6. a
7. d
8. c
9. a
10. b
11. e, d, b, c, a

## CHAPTER 34

1. Hormones; transmitter substances; local signaling molecules; pheromones
2. hypothalamus; pituitary
3. negative feedback
4. receptors
5. a
6. d
7. e
8. b
9. d
10. d, b, e, c, a

## CHAPTER 35

1. stimulus
2. sensation
3. perception
4. d
5. b
6. d
7. c
8. b
9. a
10. f, d, c, e, a, b

## CHAPTER 36

1. integumentary
2. Skeletal; muscular
3. smooth; cardiac; skeletal
4. d
5. d
6. c
7. b
8. c
9. d, e, b, a, c

## CHAPTER 37

1. digestive; circulatory; respiratory and urinary
2. digesting; absorbing; eliminating
3. caloric; energy
4. carbohydrates
5. essential amino acids; essential fatty acids
6. b
7. c
8. d
9. c
10. e, d, a, c, b

## CHAPTER 38

1. circulatory; lymphatic
2. Arteries; veins; capillaries; venules; arterioles
3. d
4. d
5. c
6. b
7. c
8. b
9. e, b, a, d, c
10. d, e, a, f, c, b

## CHAPTER 39

1. Antigens
2. antigen that is circulating or attached to pathogen; infected, cancerous, or mutated cells
3. e
4. d
5. b
6. b
7. b
8. a
9. c
10. b, a, e, d, c

## CHAPTER 40

1. Oxygen; carbon dioxide
2. gas; partial pressure gradient
3. epithelium; diffuse
4. adaptations; air flow
5. d
6. e
7. b
8. d
9. d
10. e, a, d, b, c

## CHAPTER 41

1. filtration; absorption; secretion
2. c
3. c
4. d
5. d
6. a
7. a
8. c
9. d
10. b, d, e, c, a

## CHAPTER 42

1. d
2. a
3. a
4. a
5. c
6. morphogenesis
7. b
8. c
9. e
10. d, a, f, e, c, b

## CHAPTER 43

1. hypothalamus
2. b
3. d
4. c
5. b
6. a
7. c
8. b
9. c
10. c, e, b, d, a

## CHAPTER 44

1. Ecology
2. population
3. e
4. a
5. b
6. e
7. d
8. c
9. d
10. d, e, c, a, b

## CHAPTER 45

1. Communities
2. niche
3. habitat
4. e
5. d
6. d
7. b
8. d
9. c
10. e, c, a, d, b

## CHAPTER 46

1. ecosystem; energy; cycling
2. photosynthesis; one
3. energy; nutrients; energy; nutrients
4. biogeochemical cycles
5. d
6. d
7. e
8. c
9. e
10. d, c, b, a
11. c, d, e, b, a

## CHAPTER 47

1. sunlight
2. oceans; weather systems
3. rainfall
4. d
5. c
6. e
7. a
8. a
9. e
10. f, a, c, d, e, b

## CHAPTER 48

1. exponential
2. pollutants
3. e
4. c
5. b
6. c
7. d
8. d
9. e, a, c, b, d

## CHAPTER 49

1. d
2. b
3. a
4. d
5. b
6. d
7. c
8. a
9. b
10. c

## CHAPTER 50

1. c
2. c
3. d
4. c
5. d
6. a
7. c
8. c
9. d
10. d

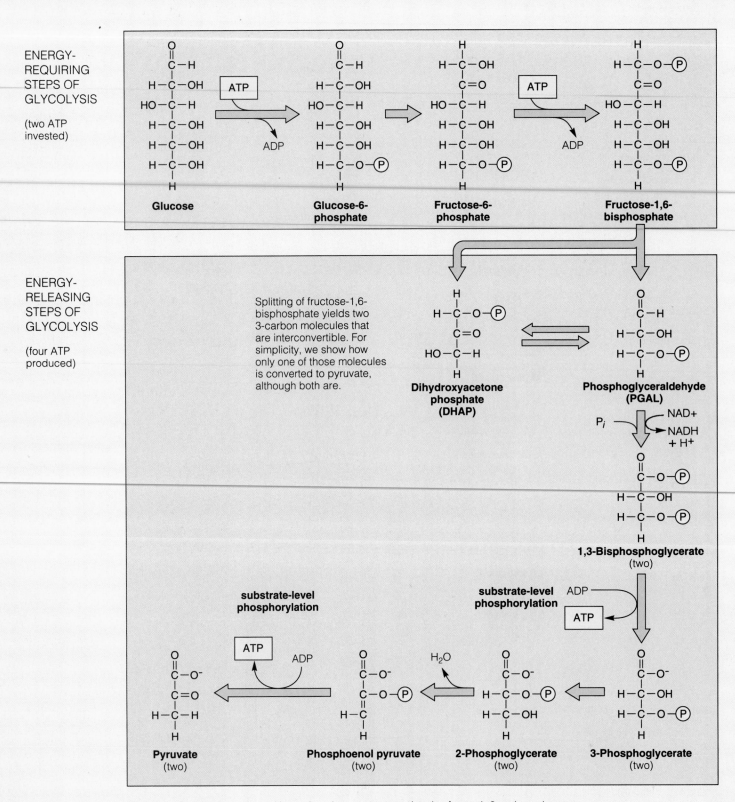

**Figure a**  Glycolysis, ending with two 3-carbon pyruvate molecules for each 6-carbon glucose entering the reactions. The *net* energy yield is two ATP molecules (two invested, four produced).

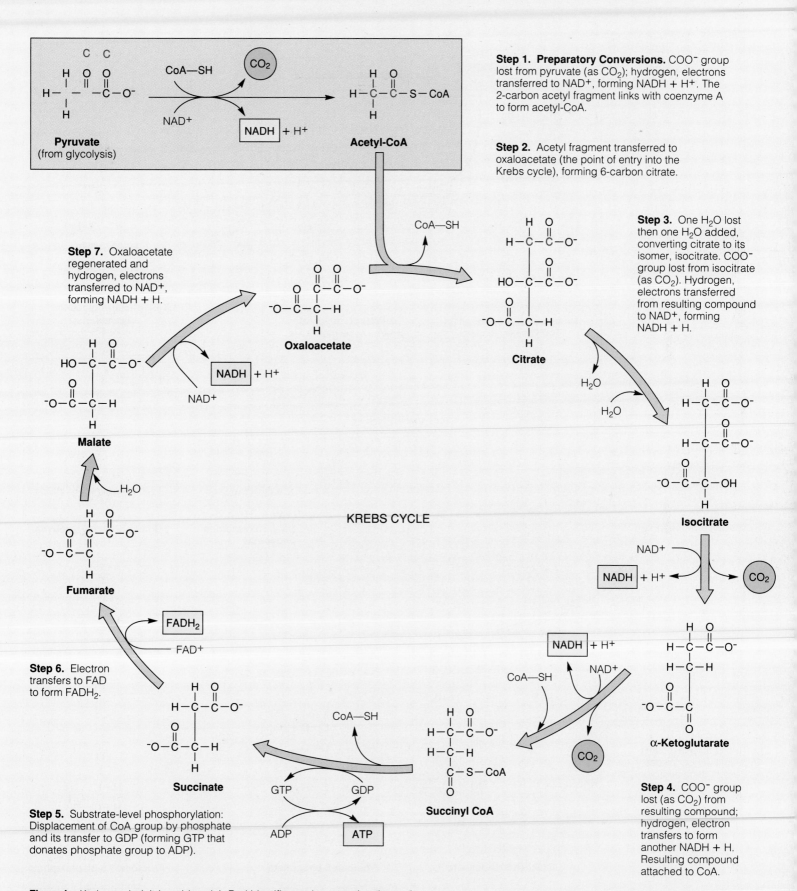

**Step 1. Preparatory Conversions.** $COO^-$ group lost from pyruvate (as $CO_2$); hydrogen, electrons transferred to $NAD^+$, forming $NADH + H^+$. The 2-carbon acetyl fragment links with coenzyme A to form acetyl-CoA.

**Step 2.** Acetyl fragment transferred to oxaloacetate (the point of entry into the Krebs cycle), forming 6-carbon citrate.

**Step 3.** One $H_2O$ lost then one $H_2O$ added, converting citrate to its isomer, isocitrate. $COO^-$ group lost from isocitrate (as $CO_2$). Hydrogen, electrons transferred from resulting compound to $NAD^+$, forming $NADH + H$.

**Step 7.** Oxaloacetate regenerated and hydrogen, electrons transferred to $NAD^+$, forming $NADH + H$.

**KREBS CYCLE**

**Step 6.** Electron transfers to FAD to form $FADH_2$.

**Step 5.** Substrate-level phosphorylation: Displacement of CoA group by phosphate and its transfer to GDP (forming GTP that donates phosphate group to ADP).

**Step 4.** $COO^-$ group lost (as $CO_2$) from resulting compound; hydrogen, electron transfers to form another $NADH + H$. Resulting compound attached to CoA.

**Figure b** Krebs cycle (citric acid cycle). Red identifies carbon entering the cycle by way of acetyl-CoA. Blue identifies carbon destined to leave the substrates (as carbon dioxide molecules).

A11

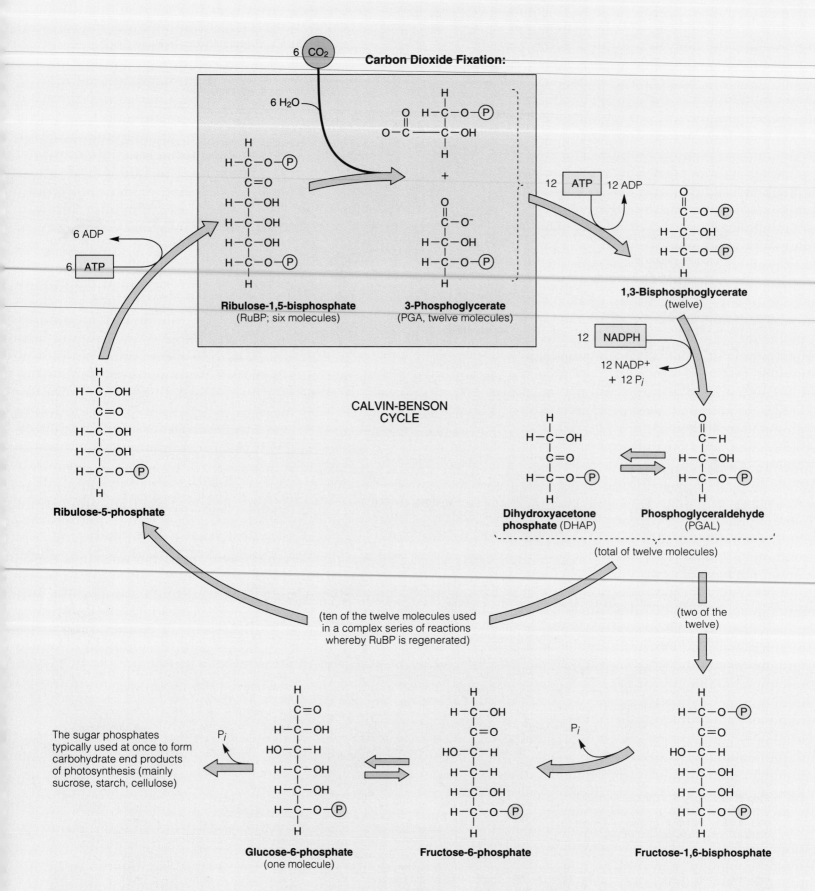

**Figure c** Calvin-Benson cycle of the light-independent reactions of photosynthesis.

# GLOSSARY OF BIOLOGICAL TERMS

**ABO blood typing** Method of characterizing blood according to particular proteins at the surface of red blood cells.

**abortion** Spontaneous or induced expulsion of the embryo or fetus from the uterus.

**abscisic acid** (ab-SISS-ik) Plant hormone that promotes stomatal closure, bud dormancy, and seed dormancy.

**abscission** (ab-SIH-zhun) [L. *abscissus*, to cut off] The dropping of leaves, flowers, fruits, or other plant parts due to hormonal action.

**absorption** For complex animals, the movement of nutrients, fluid, and ions across the gut lining and into the internal environment.

**acid** [L. *acidus*, sour] A substance that releases hydrogen ions ($H^+$) in solution.

**acid deposition, dry** The falling to earth of airborne particles of sulfur and nitrogen oxides.

**acid deposition, wet** The falling to earth of snow or rain that contains sulfur and nitrogen oxides.

**acoelomate** (ay-SEE-la-mate) Type of animal that has no fluid-filled cavity between the gut and body wall.

**actin** (AK-tin) A globular contractile protein. Within sarcomeres, the basic units of muscle contraction, actin molecules form two beaded strands, twisted together into filaments, that are pulled by myosin molecules during contraction.

**action potential** An abrupt but brief reversal in the steady voltage difference across the plasma membrane (that is, the resting membrane potential) of a neuron and some other cells.

**activation energy** The minimum amount of collision energy required to bring reactant molecules to an activated condition (the transition state) at which a reaction will proceed spontaneously. Enzymes enhance reaction rates by lowering the activation energy.

**active site** A crevice on the surface of an enzyme molecule where a specific reaction is catalyzed.

**active transport** The pumping of specific solutes through transport proteins that span the lipid bilayer of a plasma membrane, most often against their concentration gradient. The proteins act when they receive an energy boost, as from ATP.

**adaptation** [L. *adaptare*, to fit] In evolutionary biology, the process of becoming adapted (or more adapted) to a given set of environmental conditions. Of sensory neurons, a decrease in the frequency of action potentials (or their cessation) even when a stimulus is being maintained at constant strength.

**adaptive radiation** A burst of speciation events, with lineages branching away from one another as they partition the existing environment or invade new ones.

**adaptive trait** Any aspect of form, function, or behavior that helps an organism survive and reproduce under a given set of environmental conditions.

**adaptive zone** A way of life, such as "catching insects in the air at night." A lineage must have physical, ecological, and evolutionary access to an adaptive zone to become a successful occupant of it.

**adenine** (AH-de-neen) A purine; a nitrogen-containing base found in nucleotides.

**adenosine diphosphate** (ah-DEN-uh-seen die-FOSS-fate) ADP, a molecule involved in cellular energy transfers; typically formed by hydrolysis of ATP.

**adenosine phosphate** Any of several relatively small molecules, some of which function as chemical messengers within and between cells, and others as energy carriers.

**adenosine triphosphate** *See* ATP.

**ADH** Antidiuretic hormone, produced by the hypothalamus and released from the posterior pituitary; stimulates reabsorption in the kidneys and so reduces urine volume.

**adrenal cortex** (ah-DREE-nul) Outer portion of either of two adrenal glands; its hormones have roles in metabolism, inflammation, maintaining extracellular fluid volume, and other functions.

**adrenal medulla** Inner region of the adrenal gland; its hormones help control blood circulation and carbohydrate metabolism.

**aerobic respiration** (air-OH-bik) [Gk. *aer*, air, + *bios*, life] Degradative, oxygen-requiring pathway of ATP formation. Oxygen serves as the final acceptor of electrons stripped from glucose or some other organic compound. The pathway proceeds from glycolysis, then through the Krebs cycle and electron transport phosphorylation. Of all degradative pathways, aerobic respiration has the greatest energy yield, with 36 ATP typically formed for each glucose molecule.

**agglutination** (ah-glue-tin-AY-shun) Clumping of foreign cells, induced by the cross-linking of antigen-antibody complexes at their surface.

**aging** A range of processes, including the breakdown of cell structure and function, by which the body gradually deteriorates. Characteristic of all organisms showing extensive cell differentiation.

**AIDS** Acquired immune deficiency syndrome, a set of chronic disorders following infection by the human immunodeficiency virus (HIV), which destroys key cells of the immune system.

**alcoholic fermentation** Anaerobic pathway of ATP formation in which pyruvate from glycolysis is broken down to acetaldehyde; the acetaldehyde then accepts electrons from NADH to become ethanol.

**aldosterone** (al-DOSS-tuh-rohn) Hormone secreted by the adrenal cortex that helps regulate sodium reabsorption.

**allantois** (ah-LAN-twahz) [Gk. *allas*, sausage] A vascularized extraembryonic membrane; in reptiles and birds, it functions in excretion and respiration; in placental mammals, it functions in oxygen transport by way of the umbilical cord.

**allele** (uh-LEEL) One of two or more alternative forms of a gene at a given locus on a chromosome.

**allele frequency** The relative abundance of each kind of allele that can occur at a given gene locus for all individuals in a population.

**allergy** An abnormal, secondary immune response to a normally harmless substance.

**allopatric speciation** [Gk. *allos*, different, + *patria*, native land] Formation of new species as a result of geographic isolation.

**allosteric control** (AL-oh-STARE-ik) Control of enzyme functioning through the binding of a specific substance at a control site on the enzyme molecule.

**altruistic behavior** (al-true-ISS-tik) Self-sacrificing behavior; the individual behaves in a way that helps others but, in so doing, decreases its own chance to produce offspring.

**alveolar sac** (al-VEE-uh-lar) Any of the pouchlike clusters of alveoli in lungs; the major sites of gas exchange.

**alveolus** (ahl-VEE-uh-lus), plural **alveoli** [L. *alveus*, small cavity] Any of the many cup-shaped, thin-walled outpouchings of the respiratory bronchioles; a site where oxygen from air in the lungs diffuses into the bloodstream and where carbon dioxide from the bloodstream diffuses into the lungs.

**amino acid** (uh-MEE-no) A small organic molecule having a hydrogen atom, an amino group, an acid group, and an R group covalently bonded to a central carbon atom; amino acids strung together as polypeptide chains represent the primary structure of proteins.

**ammonification** (uh-moan-ih-fih-KAY-shun) A decomposition process by which certain bacteria and fungi break down nitrogen-containing wastes and remains of other organisms.

**amnion** (AM-nee-on) In land vertebrates, an extraembryonic membrane in the form of a fluid-filled sac around the embryo; it absorbs shocks and keeps the embryo from drying out.

**anaerobic pathway** (an-uh-ROW-bok) [Gk. *an*, without, + *aer*, air] Degradative metabolic pathway in which a substance other than oxygen serves as the final electron acceptor for the reactions.

**analogous structures** Occurrence of similar body parts being used for similar functions in evolutionarily remote lineages; evolutionary outcome of morphological convergence.

**anaphase** (AN-uh-faze) The stage of mitosis when sister chromatids of each chromosome separate and move to opposite poles of the spindle.

**anaphase I and II** Stages of meiosis when each chromosome separates from its homologous partner (anaphase I) and, later, when sister chromatids of each chromosome separate and move to opposite poles of the spindle (anaphase II).

**angiosperm** (AN-gee-oh-spurm) [Gk. *angeion*, vessel, + *spermia*, seed] A flowering plant.

**animal** A heterotroph that eats or absorbs nutrients from other organisms; is multicelled, usually with tissues arranged in organs and organ systems; is usually motile during at least part of the life cycle; and goes through a period of embryonic development.

**Animalia** The kingdom of animals.

**annual plant** Vascular plant that completes its life cycle in one growing season.

**anther** [Gk. *anthos*, flower] In flowering plants, the pollen-bearing part of the male reproductive structure (stamen).

**antibody** [Gk. *anti*, against] Any of a variety of Y-shaped receptor molecules with binding sites for a specific antigen (molecule that triggers an immune response); produced by B cells of the immune system.

**anticodon** In a tRNA molecule, a sequence of three nucleotide bases that can pair with an mRNA codon.

**antigen** (AN-tih-jen) [Gk. *anti*, against, + *genos*, race, kind] Any large molecule (usually a protein or polysaccharide) with a distinct configuration that triggers an immune response.

**aorta** (ay-OR-tah) [Gk. *airein*, to lift, heave] Main artery of systemic circulation; carries oxygenated blood away from the heart.

**apical dominance** The influence exerted by a terminal bud in inhibiting the growth of lateral buds.

**apical meristem** (AY-pih-kul MARE-ih-stem) [L. *apex*, top, + Gk. *meristos*, divisible] In most plants, a mass of self-perpetuating cells at a root or shoot tip that is responsible for primary growth, or elongation, of plant parts.

**appendicular skeleton** (ap-en-DIK-you-lahr) In vertebrates, bones of the limbs, pelvic girdle (at the hips), and pectoral girdle (at the shoulders).

**arteriole** (ar-TEER-ee-ole) Any of the blood vessels between arteries and capillaries; arterioles serve as control points where the volume of blood delivered to different body regions can be adjusted.

**artery** Any of the large-diameter blood vessels that conduct oxygen-poor blood to the lungs and oxygen-enriched blood to all body tissues; with their thick, muscular wall, arter-

ies are pressure reservoirs that smooth out pulsations in blood pressure caused by heart contractions.

**asexual reproduction** Mode of reproduction in which offspring arise from a single parent, and inherit the genes of that parent only.

**atmosphere** A region of gases, airborne particles, and water vapor enveloping the earth; 80 percent of its mass is distributed within seventeen miles of the earth's surface.

**atmospheric cycle** A biogeochemical cycle in which a large portion of the element being cycled between the physical environment and ecosystems occurs in a gaseous phase in the atmosphere; examples are the carbon cycle and nitrogen cycle.

**atom** The smallest unit of matter that is unique to a particular kind of element.

**atomic number** The number of protons in the nucleus of each atom of an element; differs for each element.

**ATP** Adenosine triphosphate, an energy carrier composed of adenine, ribose, and three phosphate groups; directly or indirectly transfers energy to or from nearly all metabolic pathways; produced during photosynthesis, aerobic respiration, fermentation, and other pathways.

**australopith** (OHSS-trah-low-pith) [L. *australis*, southern, + Gk. *pithekos*, ape] Any of the earliest known species of hominids; that is, the first species on the evolutionary branch leading to humans.

**autoimmune response** Abnormal immune response in which lymphocytes mount an attack against the body's own cells.

**autonomic nervous system** (auto-NOM-ik) Those nerves leading from the central nervous system to the smooth muscle, cardiac muscle, and glands of internal organs and structures; that is, to the visceral portion of the body.

**autosomal dominant inheritance** Condition in which a dominant allele on an autosome (not a sex chromosome) is always expressed to some extent.

**autosomal recessive inheritance** Condition in which a mutation produces a recessive allele on an autosome (not a sex chromosome); only recessive homozygotes show the resulting phenotype.

**autosome** Any of those chromosomes that are of the same number and kind in both males and females of the species.

**autotroph** (AH-toe-trofe) [Gk. *autos*, self, + *trophos*, feeder] An organism able to build all the organic molecules it requires using carbon dioxide (present in air and in water) and energy from the physical environment. Photosynthetic autotrophs use sunlight energy; chemosynthetic autotrophs extract energy from chemical reactions involving inorganic substances. Compare *heterotroph*.

**auxin** (AWK-sin) Any of a class of plant growth-regulating hormones; auxins promote stem elongation as one effect.

**axial skeleton** In vertebrates, the skull, backbone, ribs, and breastbone (sternum).

**axon** A long, cylindrical extension of a neuron with finely branched endings. Action potentials move rapidly, without alteration,

along an axon; their arrival at the axon endings can trigger the release of transmitter substances that may affect the activity of an adjacent cell.

**bacterial flagellum** Whiplike motile structure of many bacterial cells; unlike other flagella, it does not contain a core of microtubules.

**bacteriophage** (bak-TEER-ee-oh-fahj) [Gk. *baktērion*, small staff, rod, + *phagein*, to eat] Category of viruses that infect bacterial cells.

**balanced polymorphism** Of a population, the maintenance of two or more forms of a trait in fairly stable proportions over the generations.

**Barr body** In the cells of female mammals, a condensed X chromosome that was inactivated during embryonic development.

**basal body** A centriole which, after having given rise to the microtubules of a flagellum or cilium, remains attached to the base of either motile structure.

**base** A substance that, in solution, releases ions that can combine with hydrogen ions.

**base pair** A pair of hydrogen-bonded nucleotide bases; in the two strands of a DNA double helix, either A–T (adenine with thymine), or G–C (guanine with cytosine). When an mRNA strand forms on a DNA strand during transcription, uracil (U) base-pairs with the DNA's adenine.

**behavior, animal** A response to external and internal stimuli, following integration of sensory, neural, endocrine, and effector components. Behavior has a genetic basis, hence is subject to natural selection, and it commonly can be modified through experience.

**benthic province** All of the sediments and rocky formations of the ocean bottom; begins with the continental shelf and extends down through deep-sea trenches.

**bilateral symmetry** Of animals, a body plan with left and right halves that are basically mirror-images of each other.

**biennial** Flowering plant that lives through two growing seasons.

**binary fission** Of bacteria, a mode of asexual reproduction in which the parent cell replicates its single chromosome, then divides into two genetically identical daughter cells.

**biogeochemical cycle** The movement of carbon, oxygen, hydrogen, or some mineral element necessary for life from the environment to organisms, then back to the environment.

**biogeographic realm** [Gk. *bios*, life, + *geographein*, to describe the surface of the earth] In one scheme, one of six major land regions, each having distinguishing types of plants and animals.

**biological clocks** In many organisms, internal time-measuring mechanisms that have roles in adjusting daily and often seasonal activities in response to environmental cues.

**biological magnification** The increasing concentration of a nondegradable or slowly degradable substance in body tissues as it is passed along food chains.

**biological systematics** Branch of biology that assesses patterns of diversity based on information from taxonomy, phylogenetic reconstruction, and classification.

**biomass** The combined weight of all the organisms at a particular trophic (feeding) level in an ecosystem.

**biome** A broad, vegetational subdivision of some biogeographic realm, shaped by climate, topography, and composition of regional soils.

**biosphere** [Gk. *bios*, life, + *sphaira*, globe] the entire realm in which organisms exist; the lower regions of the atmosphere, the earth's waters, and the surface rocks, soils, and sediments of the earth's crust; the most inclusive level of biological organization.

**biosynthetic pathway** A metabolic pathway in which small molecules are assembled into lipids, proteins, and other large organic molecules.

**biotic potential** Of a population, the maximum rate of increase, per individual, under ideal conditions.

**bipedalism** A habitual standing and walking on two feet. Humans and ostriches are examples of bipedal animals.

**blastocyst** (BLASS-tuh-sist) [Gk. *blastos*, sprout, + *kystis*, pouch] In mammalian development, a modified blastula stage consisting of a hollow ball of surface cells and an inner cell mass.

**blastula** (BLASS-chew-lah) Among animals, an embryonic stage consisting of a ball of cells produced by cleavage.

**blood** A fluid connective tissue composed of water, solutes, and formed elements (blood cells and platelets).

**blood-brain barrier** Set of mechanisms that help control which blood-borne substances reach neurons in the brain.

**blood pressure** Fluid pressure, generated by heart contractions, that keeps blood circulating.

**brainstem** The vertebrate midbrain, pons, and medulla oblongata, the core of which contains the reticular formation that helps govern activity of the nervous system as a whole.

**bronchus**, plural **bronchi** (BRONG-cuss, BRONG-kee) [Gk. *bronchos*, windpipe] Tubelike branchings of the trachea that lead to the lungs.

**bud** An undeveloped shoot of mostly meristematic tissue; often protected by a covering of modified leaves.

**buffer** A substance that can combine with hydrogen ions, release them, or both, in response to changes in pH.

**bulk flow** In response to a pressure gradient, a movement of more than one kind of molecule in the same direction in the same medium (as in blood, sap, or air).

**C4 pathway** Alternative light-independent reactions of photosynthesis in which carbon dioxide is fixed twice, in two different cell types. Carbon dioxide accumulates in the leaf and helps counter photorespiration (a "wasteful" process that reduces synthesis of sugar phosphates). The first compound formed is the 4-carbon oxaloacetate.

**calorie** (KAL-uh-ree) [L. *calor*, heat] The amount of heat needed to raise the temperature of 1 gram of water by 1°C. Nutritionists sometimes use "calorie" to mean kilocalorie (1,000 calories), which is a source of confusion.

**Calvin-Benson cycle** Cyclic, light-independent reactions, the "synthesis" part of photosynthesis. For every six carbon dioxide molecules that become affixed to six RuBP molecules, twelve 3-carbon PGA molecules form; ten are used to regenerate RuBP and the other two to form a sugar phosphate. The cycle runs on ATP and NADPH formed in the light-dependent reactions.

**CAM plant** A plant that conserves water by opening stomata only at night, when carbon dioxide is fixed by way of the C4 pathway.

**cambium**, plural **cambia** (KAM-bee-um) In vascular plants, one of two types of meristems that are responsible for secondary growth (increase in stem or root diameter). Vascular cambium gives rise to secondary xylem and phloem; cork cambium gives rise to periderm.

**camouflage** An outcome of form, patterning, color, or behavior that helps an organism blend with its surroundings and escape detection.

**cancer** A type of malignant tumor, the cells of which show profound abnormalities in the plasma membrane and cytoplasm, abnormal growth and division, and weakened capacity for adhesion within the parent tissue (leading to metastasis), and, unless eradicated, lethality.

**capillary** [L. *capillus*, hair] A thin-walled blood vessel; component of capillary beds, the diffusion zones for exchanges of gases and materials between blood and interstitial fluid.

**carbohydrate** [L. *carbo*, charcoal, + *hydro*, water] A simple sugar or a polymer composed of sugar units, and used universally by cells for energy and as structural materials. A monosaccharide, oligosaccharide, or polysaccharide.

**carbon cycle** Biogeochemical cycle in which carbon moves from reservoirs in the land, atmosphere, and oceans, through organisms, then back to the reservoirs.

**carbon dioxide fixation** Initial step of the light-independent reactions of photosynthesis. Carbon dioxide becomes affixed to a specific carbon compound (such as RuBP) that undergoes rearrangements leading to regeneration of that carbon compound *and* to a sugar phosphate.

**carcinogen** (kar-SIN-uh-jen) Ultraviolet radiation and many other agents that can trigger cancer.

**cardiac cycle** [Gk. *kardia*, heart, + *kyklos*, circle] The sequence of muscle contractions and relaxation constituting one heartbeat.

**cardiovascular system** Of animals, an organ system composed of blood, one or more hearts, and blood vessels.

**carnivore** [L. *caro, carnis*, flesh, + *vovare*, to devour] An animal that eats other animals; a type of heterotroph.

**carotenoid** (kare-OTT-en-oyd) A light-sensitive pigment that absorbs violet and blue wavelengths but reflects yellow, orange, and red.

**carpel** (KAR-pul) One or more closed vessels that serve as the female reproductive parts of a flower. The chamber within a carpel is the ovary where eggs develop and are fertilized, and seeds mature.

**carrier protein** Type of transport protein that binds specific substances and changes shape in ways that shunt the substances across a plasma membrane. Some carrier proteins function passively, others require an energy input.

**carrying capacity** The number of individuals of a given species that can be sustained indefinitely by the available resources in a given environment; births are balanced by deaths in a population at its carrying capacity.

**Casparian strip** In plant roots, a waxy band that acts as an impermeable barrier between the walls of abutting cells of the endodermis or exodermis.

**cDNA** Any DNA molecule copied from a mature mRNA transcript by way of reverse transcription.

**cell** [L. *cella*, small room] The basic *living* unit. A cell has the capacity to maintain itself as an independent unit and to reproduce, given appropriate conditions and resources.

**cell cycle** A sequence of events by which a cell increases in mass, roughly doubles its number of cytoplasmic components, duplicates its DNA, then undergoes nuclear and cytoplasmic division. The cycle extends from the time the cell forms until its own daughter cells form.

**cell junction** Of multicelled organisms, a point of contact that physically links two cells or that provides functional links between their cytoplasm.

**cell plate** Of plant cells undergoing mitotic cell division, a partition that forms at the spindle equator, between the two newly forming daughter cells.

**cell theory** A theory in biology, the key points of which are that (1) all organisms are composed of one or more cells, (2) the cell is the smallest unit that still retains a capacity for independent life, and (3) all cells arise from preexisting cells.

**cell wall** A rigid or semirigid supportive wall outside the plasma membrane; a cellular feature of plants, fungi, protistans, and most bacteria.

**central nervous system** Of vertebrates, the brain and spinal cord.

**central vacuole** Of living plant cells, a fluid-filled organelle that stores nutrients, ions, and wastes; its enlargement during cell growth has the effect of improving the cell's surface-to-volume ratio.

**centriole** (SEN-tree-ohl) A small cylinder of triplet microtubules near the nucleus in most animal cells. Centrioles occur in pairs; some give rise to the microtubular core of flagella and cilia, and some may govern the plane of cell division.

**centromere** (SEN-troh-meer) [Gk. *kentron*, center, + *meros*, a part] A small, constricted region of a chromosome having attachment sites for microtubules that help move the chromosome during nuclear division.

**cephalization** (sef-ah-lah-ZAY-shun) [Gk. *kephalikos*, head] Of an animal body, having sensory structures and nerve cells concentrated in the head.

**cerebellum** (ser-ah-BELL-um) [L. diminutive of *cerebrum*, brain] Hindbrain region that coordinates motor activity for refined limb movements, appropriate posture, and spatial orientation.

**cerebral cortex** Thin surface layer of the cerebral hemispheres. Some regions of the cortex receive sensory input, others integrate information and coordinate appropriate motor responses.

**cerebrospinal fluid** Clear extracellular fluid that surrounds and cushions the brain and spinal cord. The bloodstream exchanges substances with the fluid, which exchanges substances with neurons.

**cerebrum** (suh-REE-bruhm) Part of the vertebrate forebrain governing responses to olfactory input and, in mammals, the most complex information-encoding and information-processing center. Divided into two cerebral hemispheres; in humans, the left hemisphere deals generally with spoken language skills and the right, with abstract, nonverbal skills.

**channel protein** Type of transport protein that serves as a pore through which ions or other water-soluble substances move across the plasma membrane. Some channels remain open; others are gated, and open and close in controlled ways.

**chemical bond** A union between the electron structures of two or more atoms or ions.

**chemical synapse** (SIN-aps) [Gk. *synapsis*, union] A junction where a small gap, the synaptic cleft, separates two neurons (or a neuron and a muscle cell or gland cell). The presynaptic neuron releases a transmitter substance into the cleft, and this may have an excitatory or inhibitory effect on the postsynaptic cell.

**chemiosmotic theory** (kim-ee-OZ-MOT-ik) Theory that an electrochemical gradient across a cell membrane drives ATP synthesis. Metabolic reactions cause hydrogen ions ($H^+$) to accumulate in some type of membrane-bound compartment. The combined force of the resulting concentration and electric gradients propels hydrogen ions down the gradient, through channel proteins. Enzyme action at those proteins links ADP with inorganic phosphate to form ATP.

**chemoreceptor** (KEE-moe-ree-sep-tur) Sensory receptor that detects chemical energy (ions or molecules) dissolved in body fluids next to the cell.

**chemosynthetic autotroph** (KEE-moe-sin-THET-ik) One of a few kinds of bacteria able to synthesize all the organic molecules it requires using carbon dioxide as the carbon source and certain inorganic substances (such as sulfur) as the energy source.

**chlorophyll** (KLOR-uh-fill) [Gk. *chloros*, green, + *phyllon*, leaf] Photosynthetic pigment molecule that absorbs light of blue and red wavelengths and transmits green. Special chlorophyll pigments of photosystems give up electrons used in photosynthesis.

**chloroplast** (KLOR-uh-plast) Of plants and certain protistans, a membrane-bound organelle that specializes in photosynthesis.

**chordate** An animal having a notochord, a dorsal hollow nerve cord, a pharynx, and gill slits in the pharynx wall for at least part of its life cycle.

**chorion** (CORE-ee-on) Of land vertebrates, a protective membrane surrounding an embryo and other extraembryonic membranes; in placental mammals it becomes part of the placenta.

**chromatid** The name applied to each of the two parts of a duplicated eukaryotic chromosome for as long as the two parts remain attached at the centromere. Each chromatid consists of a DNA double helix and associated proteins, and it has the same gene sequence as its "sister" chromatid.

**chromosome** (CROW-moe-some) [Gk. *chroma*, color, + *soma*, body] In eukaryotes, a DNA molecule and many associated proteins. A chromosome that has undergone duplication prior to nuclear division consists of two DNA molecules and associated proteins; the two are called *sister chromatids*. A bacterial chromosome does not have a comparable profusion of proteins associated with the DNA.

**cilium** (SILL-ee-um), plural **cilia** [L. *cilium*, eyelid] Short, hairlike process extending from the plasma membrane and containing a regular array of microtubules. Cilia are typically more profuse than flagella. Cilia serve as motile structures, help create currents of fluids, or are part of sensory structures.

**circadian rhythm** (ser-KAYD-ee-un) [L. *circa*, about, + *dies*, day] Of many organisms, a cycle of physiological events that is completed every 24 hours or so, even when environmental conditions remain constant.

**circulatory system** Of multicelled animals, an organ system consisting of a muscular pump (heart, most often), blood vessels, and blood; the system transports materials to and from cells and often helps stabilize body temperature and pH.

**cladistics** An approach to biological systematics in which organisms are grouped according to similarities that are derived from a common ancestor.

**cladogram** Branching diagram that represents patterns of relative relationships between organisms based on discrete morphological, physiological, and behavioral traits that vary among taxa being studied.

**cleavage** Stage of animal development when mitotic cell divisions convert a zygote to a ball of cells, the *blastula*. Different cytoplasmic components end up in different daughter cells, and this *cytoplasmic localization* helps seal the developmental fate of their descendants.

**cleavage furrow** Of animal cells undergoing cytokinesis, a depression that forms at the cell surface as contractile microfilaments pull the plasma membrane inward; defines where the cell will be cut in two.

**climate** Prevailing weather conditions for an ecosystem, including temperature, humidity, wind speed, cloud cover, and rainfall.

**climax community** Of an ecosystem, a more or less stable array of species that results from the process of succession.

**clonal selection theory** Theory of immune system function stating that lymphocytes activated by a specific antigen rapidly multiply, giving rise to descendants (clones) that all retain the parent cell's specificity against that antigen.

**cloned DNA** Multiple, identical copies of DNA fragments contained within plasmids or some other cloning vector.

**codominance** Condition in which two alleles of a pair are not identical yet the expression of both can be discerned in heterozygotes. Each gives rise to a different phenotype.

**codon** One of a series of base triplets in an mRNA molecule that code for a series of amino acids that will be strung together during protein synthesis. Different codons specify different amino acids; a few serve as a stop signal and one type as a start signal.

**coelom** (SEE-lum) [Gk. *koilos*, hollow] Of many animals, a type of body cavity that occurs between the gut and body wall and that has a lining, the *peritoneum*.

**coenzyme** An organic molecule that serves as a carrier of electrons or atoms in metabolic reactions and that is necessary for proper functioning of many enzymes. $NAD^+$ is an example.

**coevolution** The joint evolution of two or more closely interacting species; when one evolves, the change affects selection pressures operating between the two species so the other also evolves.

**cofactor** A metal ion or coenzyme that either helps catalyze a reaction or serves briefly as an agent that transfers electrons, atoms, or functional groups from one substrate to another.

**cohesion** Condition in which molecular bonds resist rupturing when under tension.

**cohesion theory of water transport** Theory that water moves up through vascular plants due to hydrogen bonding among water molecules confined inside the xylem pipelines. The collective cohesive strength of those bonds allows water to be pulled up as columns in response to transpiration (evaporation from leaves).

**collenchyma** Of vascular plants, a ground tissue that helps strengthen the plant body.

**colon** (CO-lun) The large intestine.

**commensalism** [L. *com*, together, + *mensa*, table] Two-species interaction in which one species benefits significantly while the other is neither helped nor harmed to any notable extent.

**communication signal** Of social animals, an evolved action or cue that transfers information, to the benefit of both the member of the species sending the signal *and* the member receiving it.

**community** The populations of all species occupying a habitat; also applies to groups of organisms with similar life-styles in a habitat (such as the bird community).

**comparative morphology** [Gk. *morph*, form] Detailed study of differences and similarities in body form and structural patterns among major taxa.

**competition, exploitation** Interaction in which both species have equal access to a required resource, but differ in how fast or how efficiently they exploit it.

**competition, interference** Interaction in which one species may limit another species' access to some resource regardless of whether the resource is abundant or scarce.

**competition, interspecific** Two-species interaction in which both species can be harmed due to overlapping niches.

**competition, intraspecific** Interaction among individuals of the same species that are competing for the same resources.

**competitive exclusion** The theory that populations of two species competing for a limited resource cannot coexist indefinitely in the same habitat; the population better adapted to exploit the resource will enjoy a competitive (hence reproductive) edge and will eventually exclude the other species from the habitat.

**complement system** A group of about twenty proteins circulating in blood plasma that are activated during both general responses and immune responses to a foreign agent in the body; part of the *inflammatory response.*

**concentration gradient** A difference in the number of molecules or ions of a substance between one region and another, as in a given volume of fluid. In the absence of other forces, the molecules tend to move down their gradient.

**condensation reaction** Enzyme-mediated reaction leading to the covalent linkage of small molecules and, often, the formation of water.

**cone cell** In the vertebrate eye, a type of photoreceptor that responds to intense light and contributes to sharp daytime vision and color perception.

**conjugation** [L. *conjugatio*, a joining] Of some bacterial species, the transfer of DNA between two different mating strains that have made cell-to-cell contact.

**consumer** [L. *consumere*, to take completely] A heterotrophic organism that obtains energy and raw materials by feeding on the tissues of other organisms. Herbivores, carnivores, omnivores, and parasites are examples.

**continuous variation** For many traits, small degrees of phenotypic variation that occur over a more or less continuous range.

**contractile vacuole** (kun-TRAK-till VAK-you-ohl) [L. *contractus*, to draw together] In some protistans, a membranous chamber that takes up excess water in the cell body, then contracts, expelling the water through a pore to the outside.

**control group** In a scientific experiment, a group used to evaluate possible side effects of the manipulation of the experimental group. Ideally, the experimental group differs from the control group only with respect to the key factor, or variable, being studied.

**convergence, morphological** Resemblance of body parts between dissimilar and only distantly related species, an evolutionary outcome of their ancestors having adopted a similar way of life and having used those body parts for similar functions.

**cork cambium** Of woody plants, a type of lateral meristem that produces a tough, corky replacement for the epidermis on older plant parts.

**corpus callosum** (CORE-pus ka-LOW-sum) In the human brain, a band of 200 million axons that functionally links the two cerebral hemispheres.

**corpus luteum** (CORE-pus LOO-tee-um) A glandular structure; it develops from cells of a ruptured ovarian follicle and secretes progesterone and some estrogen, both of which maintain the lining of the uterus (endometrium)

**cortex** [L. *cortex*, bark] In general, a rindlike layer; the kidney cortex is an example. In vascular plants, ground tissue that makes up most of the primary plant body, supports plant parts, and stores food.

**cotyledon** A so-called seed leaf that develops as part of a plant embryo; cotyledons provide nourishment for the germinating seedling.

**covalent bond** (koe-VAY-lunt) [L. *con*, together, + *valere*, to be strong] A sharing of one or more electrons between atoms or groups of atoms. When electrons are shared equally, the bond is *nonpolar*. When electrons are shared unequally, the bond is *polar*—slightly positive at one end and slightly negative at the other.

**crossing over** During prophase I of meiosis, an event in which nonsister chromatids of a pair of homologous chromosomes break at one or more sites along their length and exchange corresponding segments at the breakage points. As a result, new combinations of alleles replace old ones in a chromosome.

**culture** The sum total of behavior patterns of a social group, passed between generations by learning and by symbolic behavior, especially language.

**cuticle** (KEW-tih-kull) A body covering. In plants, a cuticle consisting of waxes and lipid-rich cutin is deposited on the outer surface of epidermal cell walls. Annelids have a thin, flexible cuticle. Arthropods have a thick, protein- and chitin-containing cuticle that is flexible, lightweight, and protective.

**cyclic AMP,** cyclic adenosine monophosphate (SIK-lik ah-DEN-uh-seen mon-oh-FOSS-fate) A nucleotide present in cytoplasm that serves as a mediator of the cell's response to hormonal signals; a type of second messenger.

**cyclic photophosphorylation** (SIK-lik foe-toe-FOSS-for-ih-LAY-shun) Photosynthetic pathway in which electrons excited by sunlight energy move from a photosystem to a transport chain, then back to the photosystem. Operation of the transport chain helps produce electrochemical gradients that lead to ATP formation. (Compare *chemiosmotic theory.*)

**cytochrome** (SIGH-toe-krome) [Gk. *kytos*, hollow vessel, + *chrōma*, color] Iron-containing protein molecule present in the electron transport systems used in photosynthesis and aerobic respiration.

**cytokinesis** (SIGH-toe-kih-NEE-sis) [Gk. *kinesis*, motion] The actual splitting of a parental cell into two daughter cells; also called cytoplasmic division.

**cytokinin** (SIGH-tow-KY-nun) Any of the class of plant hormones that stimulate cell division, promote leaf expansion, and retard leaf aging.

**cytomembrane system** [Gk. *kytos*, hollow vessel] The membranous system in the cytoplasm in which proteins and lipids take on their final form and are distributed. Components of the system include the endoplasmic reticulum, Golgi bodies, lysosomes, and a variety of vesicles.

**cytoplasm** (SIGH-toe-plaz-um) [Gk. *plassein*, to mold] All cellular parts, particles, and semifluid substances enclosed by the plasma membrane *except* the nucleus (in eukaryotes) or the nucleoid (in prokaryotes).

**cytosine** (SIGH-toe-seen) A pyrimidine; one of the nitrogen-containing bases in nucleotides.

**cytoskeleton** Of eukaryotic cells, an internal "skeleton" that structurally supports the cell, organizes its components, and often moves components about. Microtubules, microfilaments, and intermediate filaments are the most common cytoskeletal elements.

**decomposer** [L. *de-*, down, away, + *companere*, to put together] Generally, any of the heterotrophic bacteria or fungi that obtain energy by chemically breaking down the remains, products, or wastes of other organisms. Their activities help cycle nutrients back to producers.

**degradative pathway** A metabolic pathway by which molecules are broken down in stepwise reactions that lead to products of lower energy.

**deletion** Loss of a chromosome segment, nearly always resulting in a genetic disorder.

**demographic transition model** Model of human population growth in which changes in the growth pattern correspond to different stages of economic development. These are a preindustrial stage, when birth and death rates are both high, a transitional stage, an industrial stage, and a postindustrial stage, when the death rate exceeds the birth rate.

**denaturation** (deh-NAY-chur-AY-shun) Disruption of bonds holding a protein in its three-dimensional form, such that its polypeptide chain(s) unfolds partially or completely.

**dendrite** (DEN-drite) [Gk. *dendron*, tree] A short, slender extension from the cell body of a neuron.

**denitrification** (DEE-nite-rih-fih-KAY-shun) The conversion of nitrate or nitrite, by certain bacteria, to gaseous nitrogen ($N_2$) and a small amount of nitrous oxide ($N_2O$).

**density-dependent controls** Factors such as predation, parasitism, disease, and competition for resources, which limit population growth by reducing the birth rate, increasing the rates of death and dispersal, or all of these.

**density-independent controls** Factors such as storms or floods that increase a population's death rate more or less independently of its density.

**dentition** (den-TIH-shun) The type, size, and number of an animal's teeth.

**dermis** The layer of skin underlying the epidermis, consisting mostly of dense connective tissue.

**desertification** (dez-urt-ih-fih-KAY-shun) Conversion of grasslands, rain-fed cropland, or irrigated cropland to desertlike conditions, with a drop of agricultural productivity of 10 percent or more.

**detrital food web** A network of interlinked food chains in which energy flows from plants through decomposers and detritivores.

**detritivore** (dih-TRY-tih-vore) [L. *detritus*; after *deterere*, to wear down] An earthworm, crab, nematode, or other heterotroph that obtains energy by feeding on partly decomposed particles of organic matter.

**deuterostome** (DUE-ter-oh-stome) [Gk. *deuteros*, second, + *stoma*, mouth] Any of the bilateral animals, including echinoderms and chordates, in which the first indentation in the early embryo develops into the anus.

**diaphragm** (DIE-uh-fram) [Gk. *diaphragma*, to partition] Muscular partition between the thoracic and abdominal cavities, the contraction and relaxation of which contribute to breathing. Also, a contraceptive device used temporarily to prevent sperm from entering the uterus during sexual intercourse.

**dicot** (DIE-kot) [Gk. *di*, two, + *kotylēdōn*, cup-shaped vessel] Short for dicotyledon; class of flowering plants characterized generally by seeds having embryos with two cotyledons (seed leaves); net-veined leaves; and floral parts arranged in fours, fives, or multiples of these.

**differentiation** Of the cells of multicelled organisms, differences in composition, structure, and function that arise through selective gene expression. All the cells inherit the same genes but become specialized by activating or suppressing some fraction of those genes in different ways.

**diffusion** Tendency of molecules or ions of the same substance to move from a region of greater concentration to a region where they are less concentrated. See *concentration gradient*.

**digestive system** Of most animals, an internal tube or cavity in which ingested food is reduced to molecules small enough to be absorbed into the internal environment; often divided into regions specialized for food transport, processing, and storage.

**dihybrid cross** An experimental cross between two organisms, each of which breeds true (is homozygous) for forms of *two* traits that are distinctly different from those displayed by the other organism. For each trait, the first-generation offspring inherit a pair of nonidentical alleles.

**diploid** (DIP-loyd) Of sexually reproducing species, having two chromosomes of each type (that is, homologous chromosomes) in somatic cells. Except for sex chromosomes, the two homologues of a pair resemble each other in length, shape, and which genes they carry. Compare *haploid*.

**directional selection** A shift in allele frequencies in a population in a steady, consistent direction in response to a new environment (or a directional change in the old one), so that forms of traits at one end of a range of phenotypic variation become more common than the intermediate forms.

**disaccharide** (die-SAK-uh-ride) [Gk. *di*, two, + *sakcharon*, sugar] A type of simple carbohydrate, of the class called oligosaccharides; two monosaccharides covalently bonded.

**disruptive selection** Selection that favors forms of traits at both ends of a range of phenotypic variation in a population, and operates against intermediate forms.

**distal tubule** The tubular section of a nephron most distant from the glomerulus; a major site of water and sodium reabsorption.

**divergence** Accumulation of differences in allele frequencies between reproductively isolated populations of a species.

**divergence, morphological** Among similar and evolutionarily related species, decreased resemblance in one or more aspects of body patterning or function, usually corresponding to divergences in life-styles.

**diversity, organismic** Sum total of variations in form, function, and behavior that have accumulated in different lineages. Those variations generally are adaptive to prevailing conditions or were adaptive to conditions that existed in the past.

**DNA** Deoxyribonucleic acid (dee-OX-ee-RYE-bow-new-CLAY-ik) Usually, two strands of nucleotides twisted together in the shape of a double helix. The nucleotides differ only in their nitrogen-containing bases (adenine, thymine, guanine, cytosine), but *which* ones follow others in a DNA strand represents instructions for assembling proteins, and, ultimately, new organisms.

**DNA library** A collection of DNA fragments produced by restriction enzymes and incorporated into plasmids.

**DNA ligase** (LYE-gaze) Enzyme that links together short stretches of nucleotides on a parent DNA strand during replication; also used by recombinant DNA technologists to join base-paired DNA fragments to cut plasmid DNA.

**DNA polymerase** (poe-LIM-uh-raze) Enzyme that assembles a new strand on a parent DNA strand during replication; also "proofreads" for mismatched base pairs, which are replaced with correct bases.

**dominance hierarchy** Form of social organization in which some members of the group are subordinate to other members, which in turn are dominated by others.

**dominant allele** In a diploid cell, an allele whose expression masks the expression of its partner on the homologous chromosome.

**dormancy** [L. *dormire*, to sleep] Of plants, the temporary, hormone-mediated cessation of growth under conditions that might appear to be quite suitable for growth.

**double fertilization** Of flowering plants only, the fusion of one sperm nucleus with the egg nucleus (to produce a zygote), *and* fusion of a second sperm nucleus with the nuclei of the endosperm mother cell, which gives rise to nutritive tissue.

**duplication** Type of chromosome rearrangement in which a gene sequence occurs in

excess of its normal amount in a chromosome.

**ecology** [Gk. *oikos*, home, + *logos*, reason] Study of the interactions of organisms with one another and with their physical and chemical environment.

**ecosystem** [Gk. *oikos*, home] A whole complex of organisms and their environment, all of which interact through a one-way flow of energy and a cycling of materials.

**ecosystem modeling** Method of identifying pieces of information about different components of an ecosystem, then combining that information with computer programs and models in order to predict the outcome of a disturbance to the system.

**ectoderm** [Gk. *ecto*, outside, + *derma*, skin] Of animal embryos, the outermost primary tissue layer, or *germ layer*, that gives rise to the outer portion of the integument and to tissues of the nervous system.

**effector** A muscle (or gland) that responds to nerve signals by producing movement (or chemical change) that helps adjust the body to changing conditions.

**egg** A female gamete; of complex animals, a mature ovum.

**electron** Negatively charged particle occupying one of the orbitals around the nucleus of an atom.

**electron transport phosphorylation** (FOSS-for-ih-LAY-shun) Final stage of aerobic respiration, in which ATP forms after hydrogen ions and electrons (from the Krebs cycle) are sent through a transport system that gives up the electrons to oxygen. (Compare *chemiosmotic theory* and *electron transport system*).

**electron transport system** An organized array of membrane-bound enzymes and cofactors that accept and donate electrons in sequence. Operation of such systems leads to the flow of hydrogen ions ($H^+$) across a cell membrane, and this flow results in ATP formation and other reactions.

**element** Any substance that cannot be decomposed into substances with different properties.

**embryo** (EM-bree-oh) [Gk. *en*, in, + probably *bryein*, to swell] Of animals generally, the early stages of development (cleavage, gastrulation, organogenesis, and morphogenesis). In most plants, a young sporophyte, from the first cell divisions after fertilization until germination.

**embryo sac** In flowering plants, the female gametophyte.

**endergonic reaction** (en-dur-GONE-ik) Chemical reaction showing a net gain in energy.

**endocrine gland** Ductless gland that secretes hormones into interstitial fluid, after which they are distributed by way of the bloodstream.

**endocrine system** System of cells, tissues, and organs that is functionally linked to the nervous system and that helps control body functions with its hormones and other chemical secretions.

**endocytosis** (EN-doe-sigh-TOE-sis) A process by which part of the plasma membrane encloses substances (or cells, in the case of phagocytes) at or near the cell surface, then pinches off to form a vesicle that transports the substance into the cytoplasm.

**endoderm** [Gk. *endon*, within, + *derma*, skin] Of animal embryos, the inner primary tissue layer, or *germ layer*, that gives rise to the inner lining of the gut and organs derived from it.

**endodermis** In roots, a sheetlike wrapping of single cells around the vascular cylinder; functions in controlling the uptake of water and dissolved nutrients. An impermeable barrier (*Casparian strip*) prevents water from passing between the walls of abutting endodermal cells.

**endometrium** (EN-doh-MEET-ree-um) [Gk. *metrios*, of the womb] Inner lining of the uterus, consisting of connective tissues, glands, and blood vessels.

**endoplasmic reticulum** or **ER** (EN-doe-PLAZ-mik reh-TIK-yoo-lum) System of membranous channels, tubes, and sacs in the cytoplasm in which many newly formed proteins become modified and the protein and lipid components of most organelles are manufactured. Rough ER has ribosomes on the surface facing the cytoplasm; smooth ER does not.

**endoskeleton** [Gk. *endon*, within, + *sklēros*, hard, stiff] In chordates, the internal framework of bone, cartilage, or both. Together with skeletal muscle, supports and protects other body parts, helps maintain posture, and moves the body.

**endosperm** (EN-doe-sperm) Nutritive tissue that surrounds and serves as food for a flowering plant embryo and, later, for the germinating seedling.

**endospore** Of certain bacteria, a resistant body that forms around the DNA and some cytoplasm under unfavorable conditions; it germinates and gives rise to new bacterial cells when conditions become favorable.

**energy** The capacity to make things happen, to do work.

**energy pyramid** A pyramid-shaped representation of the trophic structure of an ecosystem, based on the decreasing energy flow at each upward transfer to a different trophic level.

**entropy** (EN-trow-pee) A measure of the degree of disorder in a system—that is, how much energy in the system has become so dispersed (usually as low-quality heat) that it is no longer available to do work.

**enzyme** (EN-zime) One of a special class of proteins that greatly speed up (catalyze) reactions involving specific substrates.

**epidermis** The outermost tissue layer of a multicelled plant or animal.

**epistasis** (eh-PISS-tih-sis) An absence of an expected phenotype owing to a masking of the expression of one gene pair by another gene pair.

**epithelium** (EP-ih-THEE-lee-um) Of multicelled animals, one or more layers of adhering cells having one free surface; the opposite surface rests on a basement membrane that intervenes between it and an underlying connective tissue. Epithelia cover external body surfaces and line internal cavities and tubes.

**equilibrium, dynamic** [Gk. *aequus*, equal, + *libra*, balance] The point at which a chemical reaction runs forward as fast as it runs in reverse, so that there is no net change in the concentrations of products or reactants.

**erythrocyte** (eh-RITH-row-site) [Gk. *erythros*, red, + *kytos*, vessel] Red blood cell.

**esophagus** (ee-SOF-uh-gus) Tubular portion of a digestive system that receives swallowed food and leads to the stomach.

**essential amino acid** Any of eight amino acids that human cells cannot synthesize and must obtain from food.

**essential fatty acid** Any of the fatty acids that the human body cannot synthesize and must obtain from food.

**estrogen** (ESS-trow-jun) A sex hormone required in egg formation, preparing the uterine lining for pregnancy, and maintaining secondary sexual traits; also influences growth and development.

**estrus** (ESS-truss) [Gk. *oistrus*, frenzy] For mammals generally, the cyclic period of a female's sexual receptivity to the male.

**estuary** (EST-you-ary) A partly enclosed coastal region where seawater mixes with freshwater from rivers or streams and runoff from the land.

**ethylene** (ETH-il-een) Plant hormone that stimulates fruit ripening and triggers abscission.

**eukaryotic cell** (yoo-CARRY-oht-ik) [Gk. *eu*, good, + *karyon*, kernel] A cell that has a "true nucleus" and many other membrane-bound organelles; any cell except bacteria.

**evaporation** [L. *e-*, out, + *vapor*, steam] Changes by which a substance is converted from a liquid state into vapor.

**evolution** [L. *evolutio*, act of unrolling] Change within a line of descent over time; entails successive changes in allele frequencies in a population as brought about by mutation, natural selection, genetic drift, and gene flow.

**excitatory postsynaptic potential** or **EPSP** One of two competing signals at an input zone of a neuron; a graded potential that brings the neuron's plasma membrane closer to threshold.

**excretion** Any of several processes by which excess water, excess or harmful solutes, or waste materials are passed out of the body. Compare *secretion*.

**exergonic reaction** (EX-ur-GONE-ik) A chemical reaction that shows a net loss in energy.

**exocrine gland** (EK-suh-krin) [Gk. *es*, out of, + *krinein*, to separate] Glandular structure that secretes products, usually through ducts or tubes, to a free epithelial surface.

**exocytosis** (EK-so-sigh-TOE-sis) A process by which substances are moved out of cells. A vesicle forms inside the cytoplasm, moves to the plasma membrane, and fuses with it, so that the vesicle's contents are released outside.

**exodermis** Layer of cells just inside the root epidermis in most flowering plants; functions in controlling the uptake of water and dissolved nutrients.

**exon** Any of the portions of a newly formed mRNA molecule that are spliced together to form the mature mRNA transcript and that are ultimately translated into protein.

**exoskeleton** [Gk. *exo*, out, + *sklēros*, hard, stiff] An external skeleton, as in arthropods.

**exponential growth** (EX-po-NEN-shul) Pattern of population growth that occurs when *r* (the net reproduction per individual) holds constant; then, the number of individuals increases in doubling increments (from 2 to 4, then 8, 16, 32, 64, 128, and so on).

**extinction, background** A steady rate of species turnover that characterizes lineages through most of their histories.

**extinction, mass** An abrupt increase in the rate at which major taxa disappear, with several taxa being wiped out simultaneously.

**extracellular fluid** In animals generally, all the fluid not inside cells; includes blood plasma and interstitial fluid, which occupies the spaces between cells and tissues.

**extracellular matrix** Of animals, a meshwork of fibrous proteins and other components in a ground substance that helps hold many tissues together in certain shapes and that influences cell metabolism by virtue of its composition.

**facilitated diffusion** The passive transport of specific solutes through the inside of a channel protein or carrier protein that spans the lipid bilayer of a cell membrane; the solutes simply move in the direction that diffusion would take them.

**family pedigree** A chart of the genetic relationships of the individuals in a family through a number of generations.

**fat** A lipid with one, two, or three fatty acid tails attached to a glycerol backbone.

**fatty acid** A compound having a long, unbranched carbon backbone (a hydrocarbon) with a —COOH group at the end.

**feedback inhibition** A mechanism of enzyme control in which the output of the reaction (such as a particular molecule) works in a way that inhibits further output.

**fermentation** [L. *fermentum*, yeast] Type of anaerobic pathway of ATP formation that begins with glycolysis and ends with electrons being transferred back to one of the breakdown products or intermediates. Glycolysis yields two ATP; the rest of the pathway serves to regenerate $NAD^+$.

**fertilization** [L. *fertilis*, to carry, to bear] Fusion of sperm nucleus with egg nucleus. See also *double fertilization*.

**fibrous root system** Adventitious roots and their branchings.

**filtration** In urine formation, the process by which blood pressure forces water and solutes out of glomerular capillaries and into the cupped portion of a nephron wall (Bowman's capsule).

**first law of thermodynamics** [Gk. *therme*, heat, + *dynamikos*, powerful] Law stating that the total amount of energy in the universe remains constant. Energy cannot be created or destroyed, but can only be converted from one form to another.

**flagellum** (fluh-jell-um) plural **flagella**, [L. whip] Motile structure of many free-living eukaryotic cells; has a 9 + 2 microtubule array.

**flower** The often showy reproductive structure that distinguishes angiosperms from other seed plants.

**fluid mosaic model** Model of membrane structure in which proteins are embedded in a lipid bilayer or attached to one of its surfaces. Lipids impart structure to the membrane as well as impermeability to water-soluble molecules. Packing variations and movements of lipids impart fluidity to the membrane. Proteins carry out most membrane functions, such as transport, enzyme action, and reception of chemical signals.

**follicle** (FOLL-ih-kul) In a mammalian ovary, a primary oocyte (immature egg) together with the surrounding layer of cells.

**food chain** A linear sequence of who eats whom in an ecosystem.

**food web** A network of crossing, interlinked food chains, encompassing primary producers and an array of consumers, detritivores, and decomposers.

**forebrain** Brain region that includes the cerebrum and cerebral cortex, the olfactory lobes, and the hypothalamus.

**fossil** Recognizable evidence of an organism that lived in the distant past. Most fossils are skeletons, shells, leaves, seeds, and tracks that were buried in rock layers before they could be decomposed.

**fossil fuel** Coal, petroleum, or natural gas; formed in sediments by the compression of carbon-containing plant remains over hundreds of millions of years.

**founder effect** An extreme case of genetic drift in which a few individuals leave a population and establish a new one. Simply by chance, allele frequencies for many traits may differ from those in the original population.

**fruit** [L. after *frui*, to enjoy] In flowering plants, the ripened ovary of one or more carpels, sometimes with accessory structures incorporated.

**functional group** An atom or group of atoms covalently bonded to the carbon backbone of an organic compound, contributing to its structure and properties.

**Fungi** The kingdom of fungi.

**fungus** A heterotroph that secretes enzymes able to break down an external food source into molecules small enough to be absorbed by cells (extracellular digestion and absorption). Saprobic types feed on nonliving organic matter; parasitic types feed on living organisms. Fungi as a group are major decomposers.

**gall bladder** Organ of the digestive system that stores bile secreted from the liver.

**gamete** (GAM-eet) [Gk. *gametēs*, husband, and *gametē*, wife] Haploid cell (sperm or egg) that functions in sexual reproduction.

**gametophyte** (gam-EET-oh-fite) [Gk. *phyton*, plant] The haploid, multicelled, gamete-producing phase in the life cycle of most plants.

**ganglion** (GANG-lee-un), plural **ganglia** [Gk. *ganglion*, a swelling] A clustering of cell bodies of neurons into a distinct structure in regions other than the brain or spinal cord.

**gastrulation** (gas-tru-LAY-shun) Stage of embryonic development in which cells become arranged into two or three primary tissue layers (germ layers); in humans, the layers are an inner endoderm, an intermediate mesoderm, and a surface ectoderm.

**gene** [short for German *pangan*, after Gk. *pan*, all + *genes*, to be born] Any of the units of instruction for heritable traits. Each gene is a linear sequence of nucleotides that calls for the assembly of a sequence of specific amino acids into a polypeptide chain.

**gene flow** Microevolutionary process whereby allele frequencies in a population change due to immigration, emigration, or both.

**gene frequency** More precisely, allele frequency: the relative abundances of different alleles carried by the individuals of a population.

**gene locus** Particular location on a chromosome for a given gene.

**gene mutation** [L. *mutatus*, a change] Change in DNA due to the deletion, addition, or substitution of one to several bases in the nucleotide sequence.

**gene pair** In diploid cells, the two alleles at a given gene locus on a pair of homologous chromosomes.

**gene pool** Sum total of all genotypes in a population. More accurately, allele pool.

**gene therapy** Inserting one or more normal genes into existing cells of an organism as a way to correct some genetic defect.

**genetic code** [After L. *genesis*, to be born] The correspondence between nucleotide triplets in DNA (then in mRNA) and a specific sequence of amino acids in the resulting polypeptide chains; the basic language of protein synthesis.

**genetic drift** Microevolutionary process whereby allele frequencies in a population change randomly over time, as a result of chance events.

**genetic engineering** Altering the information content of DNA through use of recombinant DNA technology.

**genetic equilibrium** Hypothetical state in a population in which allele frequencies for a trait remain stable through the generations; a reference point for measuring rates of evolutionary change.

**genetic recombination** Presence of a new combination of alleles in a DNA molecule compared to the parental genotype; the result of processes such as crossing over at meiosis, chromosome rearrangements, gene mutation, and recombinant DNA technology.

**genome** All the DNA in a haploid number of chromosomes of a species.

**genotype** (JEEN-oh-type) Genetic constitution of an individual. Can mean a single gene pair or the sum total of the individual's genes. Compare *phenotype*.

**genus**, plural **genera** (JEEN-us, JEN-er-ah) [L. *genus*, race, origin] A taxon into which all species exhibiting certain phenotypic similarities and evolutionary relationship are grouped.

**germ cell** Animal cell that may give rise to gametes. Compare *somatic cell*.

**germ layer** Of animal embryos, one of two or three primary tissue layers that form during gastrulation and that gives rise to certain tissues of the adult body. Compare *ectoderm*, *mesoderm*, and *endoderm*.

**germination** (jur-min-AY-shun) Of plants, the time at which an embryo sporophyte breaks through its seed coat and resumes growth.

**gibberellin** (JIB-er-ELL-un) Any of a class of plant hormones that promote stem elongation.

**gill** A respiratory organ, typically with a moist, thin, vascularized layer of epidermis that functions in gas exchange.

**glomerulus** (glow-MARE-you-luss) [L. *glomus*, ball] Region where a nephron wall balloons around a cluster of capillaries and where water and solutes are filtered from blood.

**glucagon** (GLUE-kuh-gone) Type of hormone, secreted by alpha cells of the pancreas, that stimulates conversion of glycogen and amino acids to glucose.

**glyceride** (GLISS-er-eyed) A molecule having one, two, or three fatty acid tails attached to a backbone of glycerol. Glycerides—fats and oils—are the body's most abundant lipids and its richest source of energy.

**glycerol** (GLISS-er-ol) [Gk. *glykys*, sweet, + L. *oleum*, oil] A three-carbon molecule with three hydroxyl groups attached; combines with fatty acids to form fat or oil.

**glycogen** (GLY-kuh-jen) In animals, a storage polysaccharide that is a main food reserve; can be readily broken down into glucose subunits.

**glycolysis** (gly-CALL-ih-sis) [Gk. *glykys*, sweet, + *lysis*, loosening or breaking apart] Initial stage of both aerobic and anaerobic pathways by which glucose (or some other organic compound) is partially broken down to pyruvate, with a net yield of two ATP.

**Golgi body** (GOHL-gee) Organelle in which many newly forming proteins and lipids undergo final processing, then are sorted and packaged in vesicles.

**gonad** (GO-nad) Primary reproductive organ in which gametes are produced.

**graded potential** Of neurons, a local signal that slightly changes the voltage difference across a small patch of the plasma membrane. Such signals vary in magnitude, depending on the stimulus. With prolonged or intense stimulation, graded potentials may spread to a trigger zone of the membrane and initiate an *action potential*.

**granum**, plural **grana** Within many chloroplasts, any of the stacks of flattened membranous compartments that incorporate chlorophyll and other light-trapping pigments and reaction sites for ATP formation.

**gravitropism** (GRAV-i-TROPE-izm) [L. *gravis*, heavy, + Gk. *trepein*, to turn] The tendency of a plant to grow directionally in response to the earth's gravitational force.

**gray matter** Of vertebrates, the dendrites, neuron cell bodies, and neuroglial cells of the spinal cord and cerebral cortex.

**grazing food web** A network of interlinked food chains in which energy flows from plants to herbivores, then through some array of carnivores.

**greenhouse effect** Warming of the lower atmosphere due to the buildup of so-called greenhouse gases—carbon dioxide, methane, nitrous oxide, ozone, water vapor, and chlorofluorocarbons.

**green revolution** In developing countries, the use of improved crop varieties, modern agricultural practices (including massive inputs of fertilizers and pesticides), and equipment to increase crop yields.

**ground meristem** (MARE-ih-stem) [Gk. *meristos*, divisible] Of vascular plants, a primary meristem that produces ground tissue, hence the bulk of the plant body.

**guard cell** Either of two adjacent cells having roles in the movement of gases and water vapor across leaf or stem epidermis. An opening (stoma) forms when both cells swell with water and move apart; it closes when they lose water and collapse against each other.

**gut** A body region where food is digested and absorbed; of complete digestive systems, the portions from the stomach onward.

**gymnosperm** (JIM-noe-sperm) [Gk. *gymnos*, naked, + *sperma*, seed] A plant that bears seeds at exposed surfaces of reproductive structures, such as cone scales. Pine trees are examples.

**habitat** [L. *habitare*, to live in] The type of place where an organism normally lives, characterized by physical features, chemical features, and the presence of certain other species.

**hair cell** Type of mechanoreceptor that may give rise to action potentials when bent or tilted.

**haploid** (HAP-loyd) Of sexually reproducing species, having only one of each pair of homologous chromosomes that were present in the nucleus of a parent cell; an outcome of meiosis. Compare *diploid*.

**heart** Muscular pump that keeps blood circulating through the animal body.

**hemoglobin** (HEEM-oh-glow-bin) [Gk. *haima*, blood, + L. *globus*, ball] Iron-containing, oxygen-transporting protein that gives red blood cells their color.

**hemostasis** (HEE-mow-STAY-iss) [Gk. *haima*, blood, + *stasis*, standing] Stopping of blood loss from a damaged blood vessel through coagulation, blood vessel spasm, platelet plug formation, and other mechanisms.

**herbivore** [L. *herba*, grass, + *vovare*, to devour] Plant-eating animal.

**heterocyst** (HET-er-oh-sist) Of some filamentous cyanobacteria, a type of thick-walled, nitrogen-fixing cell that forms when nitrogen is scarce.

**heterotroph** (HET-er-oh-trofe) [Gk. *heteros*, other, + *trophos*, feeder] Organism that cannot synthesize its own organic compounds and must obtain them by feeding on plants or other autotrophs. Animals, fungi, many protistans, and most bacteria are heterotrophs.

**heterozygous condition** (HET-er-oh-ZYE-guss) [Gk. *zygoun*, join together] For a given trait, having nonidentical alleles at a particular locus on homologous chromosomes.

**hindbrain** Of the vertebrate brain, the medulla oblongata, cerebellum, and pons; includes reflex centers for respiration, blood circulation, and other basic functions; also coordinates motor responses and many complex reflexes.

**histone** Of eukaryotic chromosomes, any of a class of structural proteins intimately associated with the DNA.

**homeostasis** (HOE-me-oh-STAY-sis) [Gk. *homo*, same, + *stasis*, standing] Of multicelled organisms, a physiological state in which the physical and chemical conditions of the internal environment are stabilized within tolerable ranges.

**hominid** [L. *homo*, man] All species on the evolutionary branch leading to modern humans. *Homo sapiens* is the only living representative.

**hominoid** Apes, humans, and their recent ancestors.

**homologous chromosome** (huh-MOLL-uh-gus) [Gk. *homologia*, correspondence] One of a pair of chromosomes that resemble each other in length, shape, and the genes they carry, and that pair with each other at meiosis. X and Y chromosomes differ in these respects but still function as homologues.

**homologous structures** Similarity in some aspect of body form or patterning between different species; evolutionary outcome of descent from a common ancestor.

**homozygous condition** (HOE-moe-ZYE-guss) For a given trait, having two identical alleles at a particular locus on homologous chromosomes.

**homozygous dominant** An individual having two dominant alleles at a given gene locus (on a pair of homologous chromosomes).

**homozygous recessive** An individual having two recessive alleles at a given gene locus (on a pair of homologous chromosomes).

**hormone** [Gk. *hormon*, to stir up, set in motion] Any of the signaling molecules secreted from endocrine glands, endocrine cells, and some neurons and that travel the bloodstream to nonadjacent target cells.

**hydrogen bond** Type of chemical bond in which an atom of a molecule interacts weakly with a hydrogen atom already taking part in a polar covalent bond.

**hydrogen ion** A hydrogen atom that has lost its electron and so bears a positive charge ($H^+$); a "naked" proton.

**hydrologic cycle** A biogeochemical cycle in which hydrogen and oxygen move, in the form of water molecules, through the atmosphere, on or through the uppermost layers of land masses, to the oceans, and back again; driven by solar energy.

**hydrolysis** (high-DRAWL-ih-sis) [L. *hydro*, water, + Gk. *lysis*, loosening or breaking apart] Enzyme-mediated reaction that breaks covalent bonds in a molecule, which splits into two or more parts; at the same time, $H^+$ and $OH^-$ (derived from a water molecule) become attached to the exposed bonding sites.

**hydrophilic substance** [Gk. *philos*, loving] A polar substance that is attracted to water molecules and so dissolves easily in water.

**hydrophobic substance** [Gk. *phobos*, dreading] A nonpolar substance that is repelled by water molecules and so does not readily dissolve in water. Oil is an example.

**hydrosphere** All liquid or frozen water on or near the earth's surface.

**hypha**, plural **hyphae** (HIGH-fuh) [Gk. *hyphe*, web] Of fungi, a generally tube-shaped filament with chitin-reinforced walls and, often, reinforcing cross-walls; component of the mycelium.

**hypodermis** A subcutaneous layer having stored fat that helps insulate the body; although not part of skin, it anchors skin while allowing it some freedom of movement.

**hypothalamus** [Gk. *hypo*, under, + *thalamos*, inner chamber or possibly *tholos*, rotunda] Of vertebrate forebrains, a brain center that monitors visceral activities (such as salt-water balance, temperature control, and reproduction) and that influences related forms of behavior (as in hunger, thirst, and sex).

**hypothesis** A plausible answer, or "educated guess," concerning a question or problem. In science, predictions drawn from hypotheses are tested by making observations, developing models, and performing repeatable experiments.

**immune system** White blood cells (macrophages, T lymphocytes, and B lymphocytes) and their interactions and products; the system shows specificity in response to a particular foreign agent, and memory—the ability to mount a more rapid attack if that specific agent returns.

**immunization** Deliberate introduction into the body of an antigen that can provoke an immune response and the production of memory lymphocytes.

**imprinting** Category of learning in which an animal that has been exposed to specific key stimuli early in its behavioral development forms an association with the object.

**incomplete dominance** Of heterozygotes, a condition in which one allele of a pair only partially dominates expression of its partner.

**independent assortment** Mendelian principle that each gene pair tends to assort into gametes independently of other gene pairs located on nonhomologous chromosomes.

**indirect selection** A theory in evolutionary biology that self-sacrificing individuals can indirectly pass on their genes by helping relatives survive and reproduce.

**induced-fit model** Model of enzyme action whereby a bound substrate induces changes in the shape of the enzyme's active site, resulting in a more precise molecular fit between the enzyme and its substrate.

**inflammatory response** A series of events involving many cells, complement proteins, and other substances that destroy foreign agents in the body and that restore tissues and internal operating conditions to normal. Occurs during both nonspecific defense responses and immune responses.

**inheritance** The transmission, from parents to offspring, of structural and functional patterns that have a genetic basis and are characteristic of each species.

**inhibiting hormone** A signaling molecule produced and secreted by the hypothalamus that controls secretions by the anterior lobe of the pituitary gland.

**inhibitor** A substance that can bind with an enzyme and interfere with its functioning.

**inhibitory postsynaptic potential**, or **IPSP** Of neurons, one of two competing types of graded potentials at an input zone; tends to drive the resting membrane potential away from threshold.

**instinctive behavior** The capacity of an animal to complete fairly complex, stereotyped responses to particular environmental cues without having had prior experience with those cues.

**insulin** Hormone that lowers the glucose level in blood; it is secreted from beta cells of the pancreas and stimulates cells to take up glucose; also promotes protein and fat synthesis and inhibits protein conversion to glucose.

**integration, neural** [L. *integrare*, to coordinate] Moment-by-moment summation of all excitatory and inhibitory synapses acting on a neuron; occurs at each level of synapsing in a nervous system.

**integument** Of animals, a protective body covering such as skin. Of flowering plants, a protective layer around the developing ovule; when the ovule becomes a seed, its integument(s) harden and thicken into a seed coat.

**integumentary exchange** (in-teg-you-MEN-tuh-ree) Of some animals, a mode of respiration in which oxygen and carbon dioxide diffuse across a thin, vascularized layer of moist epidermis at the body surface.

**interleukin** A type of communication signal that sensitizes specific cells of the immune system to the presence of a foreign agent and that stimulates them into action.

**interneuron** Any of the neurons in the vertebrate brain and spinal cord that integrate information arriving from sensory neurons and that influence other neurons in turn.

**internode** In vascular plants, the stem region between two successive nodes.

**interphase** Of cell cycles, the time interval between nuclear divisions in which a cell increases its mass, roughly doubles the number of its structures and organelles, and replicates its DNA. The interval is different for different species.

**interstitial fluid** (IN-ter-STISH-ul) [L. *interstitus*, to stand in the middle of something] In multicelled animals, that portion of the extracellular fluid occupying spaces between cells and tissues.

**intertidal zone** Generally, the area on a rocky or sandy shoreline that is above the low water mark and below the high water mark; organisms inhabiting it are alternately submerged, then exposed, by tides.

**intervertebral disk** One of a number of disk-shaped structures containing cartilage that serve as shock absorbers and flex points between bony segments of the vertebral column.

**intron** A noncoding portion of a newly formed mRNA molecule.

**inversion** Type of chromosome rearrangement in which a segment that has become separated from the chromosome is reinserted at the same place but in reverse, so the position and sequence of genes are altered.

**invertebrate** Animal without a backbone.

**ion, negatively charged** (EYE-on) An atom or a compound that has gained one or more electrons, hence has acquired an overall negative charge.

**ion, positively charged** An atom or a compound that has lost one or more electrons, hence has acquired an overall positive charge.

**ionic bond** An association between ions of opposite charge.

**isotonic condition** Equality in the relative concentrations of solutes in two fluids; for two fluids separated by a cell membrane, there is no net osmotic (water) movement across the membrane.

**isotope** (EYE-so-tope) An atom that contains the same number of protons as other atoms of the same element, but that has a different number of neutrons.

**karyotype** (CARRY-oh-type) Cut-and-paste micrograph of a cell's metaphase chromosomes, arranged according to length, shape, banding patterns, and other features.

**keratin** A tough, water-insoluble protein manufactured by most epidermal cells.

**kidney** In vertebrates, one of a pair of organs that filter mineral ions, organic wastes, and other substances from the blood, and help regulate the volume and solute concentrations of extracellular fluid.

**kinetochore** At the centromere of a chromosome, a specialized group of proteins and DNA that serves as an attachment point for several spindle microtubules during mitosis or meiosis. Each chromatid of a duplicated chromosome has its own kinetochore.

**Krebs cycle** Stage of aerobic respiration in which pyruvate is completely broken down to carbon dioxide and water. Resulting hydrogen ions and electrons are shunted to the next stage, which yields most of the ATP produced in aerobic respiration.

**lactate fermentation** Anaerobic pathway of ATP formation in which pyruvate from glycolysis is converted to the three-carbon compound lactate, with a net yield of two ATP.

**large intestine** The colon; a region of the gut that receives unabsorbed food residues from the small intestine and concentrates and stores feces until they are expelled from the body.

**larva**, plural **larvae** Of animals, a sexually immature, free-living stage between the embryo and the adult.

**larynx** (LARE-inks) A tubular airway that leads to the lungs. In humans, contains vocal cords, where sound waves used in speech are produced.

**lateral meristem** Of vascular plants, a type of meristem responsible for secondary growth; either vascular cambium or cork cambium.

**leaf** For most vascular plants, a structure having chlorophyll-containing tissue that is the major region of photosynthesis.

**learning** The adaptive modification of behavior in response to neural processing of information that has been gained from specific experiences.

**lichen** (LY-kun) A symbiotic association between a fungus and a captive photosynthetic partner such as a green alga.

**life cycle** A recurring, genetically programmed frame of events in which individuals grow, develop, maintain themselves, and reproduce.

**light-dependent reactions** First stage of photosynthesis, in which sunlight energy is absorbed and converted to the chemical energy of ATP alone (by the cyclic pathway) or to ATP and NADPH (by the noncyclic pathway).

**light-independent reactions** Second stage of photosynthesis, in which sugar phosphates are assembled with the help of the ATP and NADPH produced during the first stage.

**lignification** Of mature land plants, a process by which lignin is deposited in secondary cell walls. The deposits impart strength and rigidity, stabilize and protect other wall components, and form a waterproof barrier around the cellulose. Probably a key factor in the evolution of vascular plants.

**lignin** An inert substance containing different sugar alcohols in amounts that vary among plant species.

**limbic system** Brain regions that, along with the cerebral cortex, collectively govern emotions.

**lineage** (LIN-ee-age) A line of descent.

**linkage** Tendency of genes located on the same chromosome to stay together during meiosis and to end up together in the same gamete.

**lipid** A compound of mostly carbon and hydrogen that generally does not dissolve in water, but that does dissolve in nonpolar substances. Some lipids serve as energy reserves; others are components of membranes and other cell structures.

**lipid bilayer** Of cell membranes, two layers of mostly phospholipid molecules, with all the fatty acid tails sandwiched between the hydrophilic heads as a result of hydrophobic interactions.

**liver** Glandular organ with roles in storing and interconverting carbohydrates, lipids, and proteins absorbed from the gut; maintaining blood; disposing of nitrogen-containing wastes; and other tasks.

**local signaling molecules** Secretions from cells in many different tissues that alter chemical conditions in the immediate vicinity where they are secreted, then are swiftly degraded.

**locus** (LOW-cuss) The specific location of a particular gene on a chromosome.

**logistic growth** (low-JIS-tik) Pattern of population growth in which the growth rate of a low-density population goes through a rapid growth phase and then levels off.

**loop of Henle** The hairpin-shaped, tubular region of a nephron that functions in reabsorption of water and solutes.

**lung** An internal respiratory surface in the shape of a cavity or sac.

**lymph** (LIMF) [L. *lympha*, water] Tissue fluid that has moved into the vessels of the lymphatic system.

**lymphatic system** System of lymphoid organs (which function in defense responses) and lymph vessels (which return excess tissue fluid to the bloodstream and also transport absorbed fats to it).

**lymphocyte** Any of various white blood cells that take part in vertebrate immune responses.

**lymphoid organs** The lymph nodes, spleen, thymus, tonsils, adenoids, and patches of tissue in the small intestine and appendix.

**lysis** [Gk. *lysis*, a loosening] Gross induced leakage across a plasma membrane that leads to cell death. Examples are lysis of a virus-infected cell or of a pathogen under chemical attack by complement proteins.

**lysosome** (LYE-so-sohm) In eukaryotic cells, an organelle containing digestive enzymes that can break down polysaccharides, proteins, nucleic acids, and some lipids.

**lytic pathway** A mode of viral replication; the virus quickly takes over a host cell's metabolic machinery, the viral genetic material is replicated, and new virus particles are produced and then released as the cell undergoes lysis.

**macroevolution** The large-scale patterns, trends, and rates of change among major taxa.

**macrophage** A phagocytic white blood cell that develops from circulating monocytes and defends tissues; takes part in inflammatory responses and in both cell-mediated and antibody-mediated immune responses.

**mass extinction** An abrupt rise in extinction rates above the background level; a catastrophic, global event in which major taxa are wiped out simultaneously.

**mass number** The total number of protons and neutrons in an atom's nucleus. The relative masses of atoms are also called atomic weights.

**mechanoreceptor** Sensory cell or cell part that detects mechanical energy associated with changes in pressure, position, or acceleration.

**medulla oblongata** Part of the vertebrate brainstem with reflex centers for respiration, blood circulation, and other vital functions.

**medusa** (meh-DOO-sah) [Gk. *Medousa*, one of three sisters in Greek mythology having snake-entwined hair; this image probably evoked by the tentacles and oral arms extending from the medusa] Free-swimming, bell-shaped stage in cnidarian life cycles.

**megaspore** Of seed-bearing plants, a type of spore that develops into a female gametophyte.

**meiosis** (my-OH-sis) [Gk. *meioun*, to diminish] Two-stage nuclear division process in which the parental number of chromosomes in each daughter nucleus becomes haploid (with one of each type of chromosome that was present in the parent nucleus). Basis of gamete formation, also of spore formation in plants. Compare *mitosis*.

**membrane excitability** A membrane property of any cell that can produce action potentials in response to appropriate stimulation.

**memory** The storage and retrieval of information about previous experiences; underlies the capacity for learning.

**memory lymphocyte** Any of the various B or T lymphocytes of the immune system that are formed in response to invasion by a foreign agent and that circulate for some period, available to mount a rapid attack if the same type of invader reappears.

**menopause** (MEN-uh-pozz) [L. *mensis*, month, + *pausa*, stop] Physiological changes that mark the end of a human female's potential to bear children.

**menstrual cycle** The cyclic release of oocytes and priming of the endometrium (lining of the uterus) to receive a fertilized egg; the complete cycle averages about 28 days in female humans.

**menstruation** Periodic sloughing of the blood-enriched lining of the uterus when pregnancy does not occur.

**mesoderm** (MEH-so-derm) [Gk. *mesos*, middle, + *derm*, skin] In most animal embryos, a primary tissue layer (germ layer) between ectoderm and endoderm; gives rise to muscle, organs of circulation, reproduction, and excretion, most of the internal skeleton (when present), and connective tissue layers of the gut and body covering.

**messenger RNA** A linear sequence of ribonucleotides transcribed from DNA and translated into a polypeptide chain; the only type of RNA that carries protein-building instructions.

**metabolic pathway** A linear or cyclic series of breakdown or synthesis reactions in cells, the steps of which are catalyzed by the action of specific enzymes.

**metabolism** (meh-TAB-oh-lizm) [Gk. *meta*, change] All those chemical reactions by which cells acquire and use energy as they synthesize, accumulate, break apart, and eliminate substances in ways that contribute to growth, maintenance, and reproduction.

**metamorphosis** (met-uh-MOR-foe-sis) [Gk. *meta*, change, + *morphe*, form] Transformation of a larva into an adult form by way of major tissue reorganization.

**metaphase** Stage of mitosis when spindle has fully formed, sister chromatids of each chromosome become attached to opposite spindle poles, and all chromosomes lie at the spindle equator.

**metaphase I** Stage of meiosis when all pairs of homologous chromosomes are aligned with their partners at the spindle equator.

**metaphase II** Stage of meiosis when the chromosomes, already separated from their homologous partner but still in the duplicated state, are aligned at the spindle equator.

**metazoan** Any multicelled animal.

**MHC marker** Any of the surface receptors that mark an individual's cells as self; except for identical twins, the markers are unique to each individual.

**microevolution** Changes in allele frequencies brought about by mutation, genetic drift, gene flow, and natural selection.

**microfilament** [Gk. *mikros*, small, + L. *filum*, thread] Component of the cytoskeleton; involved in cell shape, motion, and growth.

**microspore** Of seed-bearing plants, a type of spore that develops into an immature male gametophyte called a pollen grain.

**microtubular spindle** Of eukaryotic cells, a bipolar structure composed of organized arrays of microtubules; it forms during mitosis or meiosis and moves the chromosomes.

**microtubule** Hollow cylinder of mainly tubulin subunits; a cytoskeletal element with roles in cell shape, motion, and growth and in the structure of cilia and flagella.

**microtubule organizing center**, or **MTOC** Small mass of proteins and other substances in the cytoplasm; the number, type, and location of MTOCs determine the organization and orientation of microtubules.

**microvillus** (MY-crow-VILL-us) [L. *villus*, shaggy hair] A slender, cylindrical extension of the animal cell surface that functions in absorption or secretion.

**midbrain** Of vertebrates, a brain region that evolved as a coordinating center for reflex responses to visual and auditory input; together with the pons and medulla oblongata, part of the brainstem, which includes the reticular formation.

**migration** Of certain animals, a cyclic movement between two distant regions at times of year corresponding to seasonal change.

**mimicry** (MIM-ik-ree) Situation in which one species (the mimic) bears deceptive resemblance in color, form, and/or behavior to another species (the model) that enjoys some survival advantage.

**mineral** Any of a number of small inorganic substances required for the normal functioning of body cells.

**mitochondrion**, plural **mitochondria** (MY-toe-KON-dree-on) Organelle in which the second and third stages of aerobic respiration occur; those stages are the Krebs cycle and preparatory conversions for it, as well as electron transport phosphorylation.

**mitosis** (my-TOE-sis) [Gk. *mitos*, thread] Type of nuclear division that maintains the parental number of chromosomes for daughter cells. It is the basis of bodily growth and, in some cases, of asexual reproduction of eukaryotes.

**molecule** A unit of two or more atoms of the same or different elements, bonded together.

**molting** The shedding of hair, feathers, horns, epidermis, or exoskeleton in a process of growth or periodic renewal.

**Monera** The kingdom of single-celled prokaryotes; bacteria.

**monocot** (MON-oh-kot) Short for monocotyledon; a flowering plant in which seeds have only one cotyledon, whose floral parts generally occur in threes (or multiples of threes), and whose leaves typically are parallel-veined. Compare *dicot*.

**monohybrid cross** [Gk. *monos*, alone] An experimental cross between two parent organisms that breed true for distinctly different forms of a single trait; heterozygous offspring result.

**monomer** A simple sugar or some other small organic compound that can serve as one of the individual units of *polymers*.

**monophyletic group** A set of independently evolving lineages that share a common evolutionary heritage.

**monosaccharide** (MON-oh-SAK-ah-ride) [Gk. *monos*, along, single, + *sakharon*, sugar] A sugar monomer; the simplest carbohydrate. Glucose is an example.

**monosomy** Abnormal condition in which one chromosome of diploid cells has no homologue.

**morphogenesis** (MORE-foe-JEN-ih-sis) [Gk. *morphe*, form, + *genesis*, origin] Processes by which differentiated cells in an embryo become organized into tissues and organs, under genetic controls and environmental influences.

**motor neuron** Nerve cell that relays information away from the brain and spinal cord to the body's effectors (muscles, glands, or both), which carry out responses.

**mouth** An oral cavity; in human digestion, the site where polysaccharide breakdown begins.

**multicelled organism** An organism that has differentiated cells arranged into tissues, and often into organs and organ systems.

**multiple allele system** Three or more alternative molecular forms of a gene (alleles), any of which may occur at the gene's locus on a chromosome.

**muscle tissue** Tissue having cells able to contract in response to stimulation, then passively lengthen and so return to their resting state.

**mutagen** (MEW-tuh-jen) An environmental agent that can permanently modify the structure of a DNA molecule. Certain viruses and ultraviolet radiation are examples.

**mutation** [L. *mutatus*, a change, + *-ion*, result of a process or an act] A heritable change in the molecular structure of DNA.

**mutualism** [L. *mutuus*, reciprocal] A type of community interaction in which members of two species each receive benefits from the association. When the interaction is intimate and involves a permanent dependency, it is called *symbiosis*.

**mycelium** (my-SEE-lee-um), plural **mycelia** [Gk. *mykes*, fungus, mushroom, + *helos*, callus] A mesh of tiny, branching filaments (hyphae) that is the food-absorbing part of a multicelled fungus.

**mycorrhiza** (MY-coe-RISE-uh) "Fungus-root," a symbiotic arrangement between fungal hyphae and young roots of many vascular plants, in which the fungus obtains carbohydrates from the plant and in turn releases dissolved mineral ions to the plant roots.

**myelin sheath** Of many sensory and motor neurons, an axonal sheath that affects how fast action potentials travel; formed from the plasma membranes of Schwann cells that are wrapped repeatedly around the axon and are separated from each other by a small node.

**myofibril** (MY-oh-FY-brill) One of many threadlike structures inside a muscle cell; composed of actin and myosin molecules arranged as sarcomeres, the fundamental units of contraction.

**myosin** (MY-uh-sin) One of two types of protein filaments that make up sarcomeres, the contractile units of a muscle cell; the other is actin.

**NAD+** Nicotinamide adenine dinucleotide, a large organic molecule that serves as a cofactor in enzyme reactions. When carrying electrons and protons (H+) from one reaction site to another, it is abbreviated NADH.

**NADP+** Nicotinamide adenine dinucleotide phosphate. When carrying electrons and protons (H+) from one reaction site to another, it is abbreviated NADPH.

**natural selection** A microevolutionary process; a measure of the differences in survival and reproduction that have occurred among individuals of a population that differ from one another in one or more traits.

**negative feedback mechanism** A homeostatic feedback mechanism in which an activity changes some condition in the internal environment and so triggers a response that reverses the changed condition.

**nematocyst** (NEM-ad-uh-sist) [Gk. *nema*, thread, + *kystis*, pouch] Of cnidarians only, a stinging capsule that assists in prey capture and possibly protection.

**nephridium**, plural, **nephridia** (neh-FRID-ee-um) Of earthworms and some other invertebrates, a system of regulating water and solute levels.

**nephron** (NEFF-ron) [Gk. *nephros*, kidney] Of the vertebrate kidney, a slender tubule in which water and solutes filtered from blood are selectively reabsorbed and in which urine forms.

**nerve** Cordlike communication line of the peripheral nervous system, composed of axons of sensory neurons, motor neurons, or both packed within connective tissue. In the brain and spinal cord, similar cordlike bundles are called nerve pathways or tracts.

**nerve cord** Of many animals, a cordlike communication line consisting of axons of neurons.

**nerve impulse** See *action potential*.

**nerve net** Cnidarian nervous system.

**nervous system** System of neurons oriented relative to one another in precise message-conducting and information-processing pathways.

**neuroendocrine control center** Those portions of the hypothalamus and pituitary gland that interact to control many body functions.

**neuroglial cell** (NUR-oh-GLEE-uhl) Of vertebrates, one of the cells that provide structural and metabolic support for neurons and that collectively represent about half the volume of the nervous system.

**neuromodulator** Type of signaling molecule that influences the effects of transmitter substances by enhancing or reducing membrane responses in target neurons.

**neuromuscular junction** Chemical synapses between the axon terminals of a motor neuron and a muscle cell.

**neuron** A nerve cell; the basic unit of communication in nervous systems. Neurons collectively sense environmental change, integrate sensory inputs, then activate muscles or glands that initiate or carry out responses.

**neutral mutation** Mutation in which the altered allele has no more measurable effect on survival and reproduction than do other alleles for the trait.

**neutron** Subatomic particle of about the same size and mass as a proton but having no electric charge.

**niche** (NITCH) [L. *nidas*, nest] Of a species, the full range of physical and biological conditions under which its members can live and reproduce.

**nitrification** (nye-trih-fih-KAY-shun) Process by which certain soil bacteria strip electrons from ammonia or ammonium, releasing nitrite ($NO_2$) that other soil bacteria break down, releasing nitrate ($NO_3$).

**nitrogen cycle** Cycling of nitrogen atoms between living organisms and the environment, through nitrogen fixation, assimilation and biosynthesis of nitrogen-containing compounds, decomposition, ammonification, nitrification, and denitrification.

**nitrogen fixation** Process by which a few kinds of bacteria convert gaseous nitrogen ($N_2$) to ammonia, which dissolves rapidly in water to produce ammonium. Other organisms as well as the bacteria use the fixed nitrogen.

**nociceptor** A sensory receptor, such as a free nerve ending, that detects any stimulus causing tissue damage.

**node** In vascular plants, a point on a stem where one or more leaves are attached.

**noncyclic photophosphorylation** (non-SIK-lik foe-toe-FOSS-for-ih-LAY-shun) [L. *non*, not, + Gk. *kylos*, circle] Photosynthetic pathway in which new electrons derived from water molecules flow through two photosystems and two transport chains, and ATP and NADPH form.

**nondisjunction** Failure of one or more chromosomes to separate during meiosis.

**notochord** (KNOW-toe-kord) Of chordates, a rod of stiffened tissue (not cartilage or bone) that serves as a supporting structure for the body.

**nuclear envelope** A double membrane (two lipid bilayers and associated proteins) that is the outermost portion of a cell nucleus.

**nucleic acid** (new-CLAY-ik) A single- or double-stranded chain of nucleotide units; DNA and RNA are examples.

**nucleoid** Of bacteria, a region in which DNA is physically organized apart from other cytoplasmic components.

**nucleolus** (new-KLEE-oh-lus) [L. *nucleolus*, a little kernel] Within the nucleus of a nondividing cell, a mass of proteins, RNA, and other material used in ribosome synthesis.

**nucleosome** (NEW-klee-oh-sohm) Of eukaryotic chromosomes, an organizational unit consisting of a segment of DNA looped twice around a core of histone molecules.

**nucleotide** (NEW-klee-oh-tide) A small organic compound having a five-carbon sugar (deoxyribose), nitrogen-containing base, and phosphate group. Nucleotides are the structural units of adenosine phosphates, nucleotide coenzymes, and nucleic acids.

**nucleotide coenzyme** A protein that transports hydrogen atoms (free protons) and electrons from one reaction site to another in cells.

**nucleus** (NEW-klee-us) [L. *nucleus*, a kernel] In atoms, the central core of one or more positively charged protons and (in all but hydrogen) electrically neutral neutrons. In eukaryotic cells, a membranous organelle containing the DNA.

**obesity** An excess of fat in the body's adipose tissues, caused by imbalances between caloric intake and energy output.

**oligosaccharide** A carbohydrate consisting of a small number of covalently linked sugar monomers. One subclass, disaccharides, consists of two sugar monomers. Compare *monosaccharide* and *polysaccharide*.

**omnivore** [L. *omnis*, all, + *vorare*, to devour] An organism able to obtain energy from more than one source rather than being limited to one trophic level.

**oncogene** (ON-coe-jeen) Any gene having the potential to induce cancerous transformations in a cell.

**oogenesis** (oo-oh-JEN-uh-sis) Formation of a female gamete, from a germ cell to a mature haploid ovum (egg).

**operator** A short base sequence between a promoter and the start of a gene; interacts with regulatory proteins to control transcription.

**operon** Of transcription, a promoter-operator sequence that services more than a single gene. The lactose operon of *E. coli* is an example.

**orbitals** Volumes of space around the nucleus of an atom in which electrons are likely to be at any instant.

**organ** A structure of definite form and function that is composed of more than one tissue.

**organ formation** Stage of development in which primary tissue layers (germ layers) split into subpopulations of cells, and different lines of cells become unique in structure and function; foundation for growth and tissue specialization, when organs acquire specialized chemical and physical properties.

**organ system** Two or more organs that interact chemically, physically, or both in performing a common task.

**organelle** Any of various membranous sacs, envelopes, and other compartmented portions of cytoplasm. Organelles separate different, often incompatible metabolic reactions in the space of the cytoplasm and in time (by allowing certain reactions to proceed only in controlled sequences).

**organic compound** In biology, a compound assembled in cells and having a carbon backbone, often with carbon atoms arranged as a chain or ring structure.

**osmosis** (oss-MOE-sis) [Gk. *osmos*, act of pushing] Of cell membranes, the passive movement of water through the interior of membrane-spanning proteins in response to solute concentration gradients, a pressure gradient, or both.

**ovary** (OH-vuh-ree) In female animals, the primary reproductive organ in which eggs form. In seed-bearing plants, the portion of the carpel where eggs develop, fertilization takes place, and seeds mature. A mature ovary and sometimes other plant parts is a fruit.

**oviduct** (OH-vih-dukt) Duct through which eggs travel from the ovary to the uterus. Formerly called Fallopian tube.

**ovulation** (AHV-you-LAY-shun) During each turn of the menstrual cycle, the release of a secondary oocyte (immature egg) from an ovary.

**ovule** (OHV-youl) [L. *ovum*, egg] Any of one or more structures that form on the inner wall of the ovary of seed-bearing plants and that, at maturity, are seeds. An ovule contains the female gametophyte with its egg, surrounded by nutritive and protective tissues.

**ovum** (OH-vum) A mature female gamete (egg).

**oxidation-reduction reaction** An electron transfer from one atom or molecule to another. Often hydrogen is transferred along with the electron or electrons.

**pancreas** (PAN-cree-us) Glandular organ that secretes enzymes and bicarbonate into the small intestine during digestion, and that also secretes the hormones insulin and glucagon.

**pancreatic islets** Any of the 2 million clusters of endocrine cells in the pancreas, including alpha cells, beta cells, and delta cells.

**parasite** [Gk. *para*, alongside, + *sitos*, food] An organism that obtains nutrients directly from the tissues of a living host, which it lives on or in and may or may not kill.

**parasitoid** An insect larva that grows and develops inside a host organism (usually another insect), eventually consuming the soft tissues and killing it.

**parasympathetic nerve** Of the autonomic nervous system, any of the nerves carrying signals that tend to slow the body down overall and divert energy to basic tasks; such nerves also work continually in opposition with *sympathetic nerves* to bring about minor adjustments in internal organs.

**parathyroid glands** (PARE-uh-THY-royd) In vertebrates, endocrine glands embedded in the thyroid gland that secrete parathyroid hormone, which helps restore blood calcium levels.

**parenchyma** Most abundant ground tissue of root and shoot systems. Its cells function in photosynthesis, storage, secretion, and other tasks.

**parthenogenesis** Development of an embryo from an unfertilized egg.

**passive immunity** Temporary immunity conferred by deliberately introducing antibodies into the body.

**passive transport** Movement of a solute across a cell membrane in response to its concentration gradient, through the interior of proteins that span the membrane. No energy expenditure is required.

**pathogen** (PATH-oh-jen) [Gk. *pathos*, suffering, + *-genēs*, origin] Disease-causing organism.

**pattern formation** Of animals, mechanisms responsible for specialization and positioning of tissues during embryonic development.

**PCR** *See* polymerase chain reaction.

**pelagic province** The entire volume of ocean water; subdivided into *neritic zone* (relatively shallow waters overlying continental shelves) and *oceanic zone* (water over ocean basins).

**penis** A male organ that deposits sperm into a female reproductive tract.

**perennial** [L. *per-*, throughout, + *annus*, year] A plant that lives for three or more growing seasons.

**pericycle** (PARE-ih-sigh-kul) [Gk. *peri-*, around, + *kyklos*, circle] Of a root vascular cylinder, one or more layers just inside the endodermis that give rise to lateral roots and contribute to secondary growth.

**periderm** Of vascular plants showing secondary growth, a protective covering that replaces epidermis.

**peripheral nervous system** (per-IF-ur-uhl) [Gk. *peripherein*, to carry around] Of vertebrates, the nerves leading into and out from the spinal cord and brain, and the ganglia along those communication lines.

**peristalsis** (pare-ih-STAL-sis) A rhythmic contraction of muscles that moves food forward through the animal gut.

**peritoneum** A lining of the coelom that also covers and helps maintain the position of internal organs.

**permafrost** A permanently frozen, water-impenetrable layer beneath the soil surface in arctic tundra.

**PGA** Phosphoglycerate (FOSS-foe-GLISS-er-ate); a key intermediate in glycolysis as well as the Calvin-Benson cycle.

**PGAL** Phosphoglyceraldehyde; a key intermediate in glycolysis as well as the Calvin-Benson cycle.

**pH scale** A scale used in measuring the concentration of free (unbound) hydrogen ions in solutions; pH 0 is the most acidic, 14 the most basic, and 7, neutral.

**phagocyte** (FAG-uh-sight) [Gk. *phagein*, to eat, + *kytos*, hollow vessel] A macrophage or certain other white blood cells that engulf and destroy foreign agents in body tissues.

**phagocytosis** (FAG-uh-sigh-TOE-sis) [Gk. *phagein*, to eat, + *kytos*, hollow vessel] Engulfment of foreign cells or substances by amoebas and some white blood cells, by means of endocytosis.

**pharynx** (FARE-inks) A muscular tube by which food enters the gut and, in land vertebrates, the windpipe (trachea).

**phenotype** (FEE-no-type) [Gk. *phainein*, to show, + *typos*, image] Observable trait or traits of an individual; arises from interactions between genes, and between genes and the environment.

**pheromone** (FARE-oh-moan) [Gk. *phero*, to carry, + *-mone*, as in hormone] A type of signaling molecule secreted by exocrine glands that serves as a communication signal between individuals of the same species.

**phloem** (FLOW-um) Of vascular plants, a tissue with living cells that interconnect and form the tubes through which sugars and other solutes are conducted.

**phospholipid** A type of lipid with a glycerol backbone, two fatty acid tails, and a phosphate group to which an alcohol is attached; the main lipid of plant and animal cell membranes.

**phosphorylation** (FOSS-for-ih-LAY-shun) The attachment of inorganic phosphate to a molecule; also the transfer of a phosphate group from one molecule to another, as when ATP phosphorylates glucose.

**photolysis** (foe-TALL-ih-sis) [Gk. *photos*, light, + *-lysis*, breaking apart] First step in noncyclic photophosphorylation, when water is split into oxygen, hydrogen, and associated electrons; photon energy indirectly drives the reaction.

**photoreceptor** Light-sensitive sensory cell.

**photosynthesis** The trapping and conversion of sunlight energy to chemical energy (ATP, NADPH, or both), followed by synthesis of sugar phosphates that become converted to sucrose, cellulose, starch, and other end products.

**photosynthetic autotroph** An organism able to synthesize all organic molecules it requires using carbon dioxide as the carbon source and sunlight as the energy source. All plants, some protistans, and a few bacteria are photosynthetic autotrophs.

**photosystem** Of photosynthetic membranes, a light-trapping unit having organized arrays of pigment molecules and enzymes.

**photosystem I** A type of photosystem that operates during the cyclic pathway of photosynthesis.

**photosystem II** A type of photosystem that operates during both the cyclic and noncyclic pathways of photosynthesis.

**phototropism** [Gk. *photos*, light, + *trope*, turning, direction] Adjustment in the direction and rate of plant growth in response to light.

**phylogeny** Evolutionary relationships among species, starting with the most ancestral forms and including all the branches leading to all their descendants.

**phytochrome** Light-sensitive pigment molecule, the activation and inactivation of which trigger plant hormone activities governing leaf expansion, stem branching, stem lengthening, and often seed germination and flowering.

**phytoplankton** (FIE-toe-PLANK-tun) [Gk. *phyton*, plant, + *planktos*, wandering] A freshwater or marine community of floating or weakly swimming photosynthetic autotrophs, such as diatoms, green algae, and cyanobacteria.

**pineal gland** (py-NEEL) Of vertebrates, a light-sensitive endocrine gland that secretes melatonin, a hormone that influences the development of reproductive organs and reproductive cycles.

**pioneer species** Typically small plants with short life cycles that are adapted to growing in exposed, often windy areas with intense sunlight, wide swings in air temperature, and soils deficient in nitrogen and other nutrients. By improving conditions in areas they colonize, pioneers invite their replacement by other species.

**pituitary gland** Of vertebrate endocrine systems, a gland that interacts with the hypothalamus to coordinate and control many physiological functions, including the activity of many other endocrine glands. Its *posterior lobe* stores and secretes hypothalamic hormones; the *anterior lobe* produces and secretes its own hormones.

**placenta** (play-SEN-tuh) Of a uterus, an organ composed of maternal tissues and extraembryonic membranes (chorion especially); delivers nutrients to and carries away wastes from the embryo.

**plankton** [Gk. *planktos*, wandering] Any community of floating or weakly swimming organisms, mostly microscopic, in freshwater and saltwater environments. See *phytoplankton* and *zooplankton*.

**plant** Most often, multicelled autotroph able to build its own food molecules through photosynthesis.

**Plantae** The kingdom of plants.

**plasma** (PLAZ-muh) Liquid component of blood; consists of water, various proteins, ions, sugars, dissolved gases, and other substances.

**plasma cell** Of immune systems, any of the antibody-secreting daughter cells of a rapidly dividing population of B cells.

**plasma membrane** Of cells, the outermost membrane that separates internal metabolic events from the environment but selectively permits passage of various substances. Composed of a lipid bilayer and proteins that carry out most functions, including transport across the membrane and reception of outside signals.

**plasmid** Of many bacteria, a small, circular DNA molecule that carries some genes and replicates independently of the bacterial chromosome.

**plasmodesma** (PLAZ-moe-DEZ-muh) Of multicelled plants, a junction between linked walls of adjacent cells through which nutrients and other substances flow.

**plasticity** Of the human species, the ability to remain flexible and adapt to a wide range of environments, rather than becoming narrowly adapted to one specific environment.

**plate tectonics** Arrangement of the earth's outer layer (lithosphere) in slablike plates, all in motion and floating on a hot, plastic layer of the underlying mantle.

**platelet** (PLAYT-let) Any of the cell fragments in blood that release substances necesary for clot formation.

**pleiotropy** (PLEE-oh-troh-pee) [Gk. *pleon*, more, + *trope*, direction] Form of gene expression in which a single gene exerts multiple effects on seemingly unrelated aspects of an individual's phenotype.

**pollen grain** [L. *pollen*, fine dust] Of gymnosperms and flowering plants, an immature male gametophyte (gamete-producing body).

**pollen sac** In anthers of flowers, any of the chambers in which pollen grains develop.

**pollen tube** A tube formed after a pollen grain germinates; grows down through carpel tissues and carries sperm to the ovule.

**pollination** Of flowering plants, the arrival of a pollen grain on the landing platform (stigma) of a carpel.

**pollutant** Any substance with which an ecosystem has had no prior evolutionary experience, in terms of kinds or amounts, and that can accumulate to disruptive or harmful levels. Can be naturally occurring or synthetic.

**polymer** (POH-lih-mur) [Gk. *polus*, many, + *meris*, part] A molecule composed of three to millions of small subunits that may or may not be identical.

**polymerase chain reaction** DNA amplification method; DNA having a gene of interest is split into single strands, which enzymes (polymerases) copy; the enzymes also act on the accumulating copies, multiplying the gene sequence by the millions.

**polymorphism** (poly-MORE-fizz-um) [Gk. *polus*, many, + *morphe*, form] Of a population, the persistence through the generations of two or more forms of a trait, at a frequency greater than can be maintained by new mutations alone.

**polyp** (POH-lip) Vase-shaped, sedentary stage of cnidarian life cycles.

**polypeptide chain** Three or more amino acids joined by peptide bonds.

**polyploidy** (POL-ee-PLOYD-ee) A condition in which offspring end up with three or more of each type of chromosome characteristic of the parental stock.

**polysaccharide** [Gk. *polus*, many, + *sakcharon*, sugar] A straight or branched chain of hundreds of thousands of covalently linked sugar monomers, of the same or different kinds. The most common polysaccharides are starch, cellulose, and glycogen.

**polysome** Of protein synthesis, several ribosomes all translating the same messenger RNA molecule, one after the other.

**population** A group of individuals of the same species occupying a given area.

**positive feedback mechanism** Homeostatic mechanism by which a chain of events is set in motion and intensifies an original condition.

**post-translational controls** Of eukaryotes, controls that govern modification of newly formed polypeptide chains into functional enzymes and other proteins.

**predator** [L. *prehendere*, to grasp, seize] An organism that feeds on and may or may not kill other living organisms (its *prey*); unlike parasites, predators do not live on or in their prey.

**pressure flow theory** Of vascular plants, a theory that organic compounds move through phloem because of gradients in solute concentrations and pressure between source regions (such as photosynthetically active leaves) and sink regions (such as growing plant parts).

**primary growth** Plant growth originating at root tips and shoot tips.

**primary immune response** Actions by white blood cells and their products, elicited by a first-time encounter with an antigen; includes both antibody-mediated and cell-mediated responses.

**primary productivity, gross** Of an ecosystem, the total rate at which the producers capture and store a given amount of energy, as by photosynthesis, during a specified interval.

**primary productivity, net** Of an ecosystem, the rate of energy storage in the tissues of producers in excess of their rate of aerobic respiration.

**procambium** (pro-KAM-bee-um) Of vascular plants, a primary meristem that gives rise to the primary vascular tissues.

**producer, primary** An organism such as a plant that directly or indirectly nourishes consumers, decomposers, and detritivores.

**progesterone** (pro-JESS-tuh-rown) Female sex hormone secreted by the ovaries.

**prokaryote** (pro-CARRY-oht) [L. *pro*, before, + Gk. *karyon*, kernel] Single-celled organism that has no nucleus or other membrane-bound organelles; only bacteria are prokaryotes.

**promoter** Of transcription, a base sequence that signals the start of a gene; the site where RNA polymerase initially binds.

**prophase** First stage of mitosis, when each duplicated chromosome becomes condensed into a thicker, rodlike form.

**prophase I** Stage of meiosis when each duplicated chromosome condenses and pairs with its homologous partner, followed by crossing over and genetic recombination among nonsister chromatids.

**prophase II** Brief stage of meiosis after interkinesis during which each chromosome still consists of two chromatids.

**protein** Organic compound composed of one or more chains of amino acids (polypeptide chains).

**Protista** The kingdom of protistans.

**protistan** (pro-TISS-tun) [Gk. *prōtistos*, primal, very first] Single-celled eukaryote.

**proton** Positively charged unit of energy in the atomic nucleus.

**proto-oncogene** A gene sequence similar to an oncogene but that codes for a protein required in normal cell function; may trigger cancer, generally when specific mutations alter its structure or function.

**protostome** (PRO-toe-stome) [Gk. *proto*, first, + *stoma*, mouth] A bilateral animal in which the first indentation in the early embryo develops into the mouth. Includes mollusks, annelids, and arthropods.

**proximal tubule** Of a nephron, the tubular region that receives water and solutes filtered from the blood.

**pulmonary circuit** Blood circulation route leading to and from the lungs.

**Punnett-square method** A diagramming technique for predicting the possible outcome of a mating or an experimental cross.

**purine** Nucleotide base having a double ring structure. Adenine and guanine are two examples.

**pyrimidine** (pih-RIM-ih-deen) Nucleotide base having a single ring structure. Cytosine and thymine are examples.

**pyruvate** (PIE-roo-vate) Three-carbon compound produced by the initial breakdown of a glucose molecule during glycolysis.

**radial symmetry** Body plan having four or more roughly equivalent parts arranged around a central axis.

**rain shadow** A reduction in rainfall on the leeward side of high mountains, resulting in arid or semiarid conditions.

**reabsorption** Of urine formation, the diffusion or active transport of water and usable solutes out of a nephron and into capillaries leading back to the general circulation; regulated by ADH and aldosterone.

**receptor** Of cells, a molecule at the cell surface or within the cytoplasm that may be activated by a specific hormone, virus, or some other outside agent. Of nervous systems, a sensory cell or cell part that may be activated by a specific stimulus.

**receptor protein** Protein that binds a signaling molecule such as a hormone, then triggers alterations in cell behavior or metabolism.

**recessive allele** [L. *recedere*, to recede] In heterozygotes, an allele whose expression is fully or partially masked by expression of its partner; fully expressed only in the homozygous recessive condition.

**recognition protein** Protein at cell surface recognized by cells of like type; helps guide the ordering of cells into tissues during development and functions in cell-to-cell interactions.

**recombinant technology** Procedures by which DNA (genes) from different species may be isolated, cut, spliced together, and the new recombinant molecules multiplied in quantity.

**red blood cell** Erythrocyte; an oxygen-transporting cell in blood.

**red marrow** A substance in the spongy tissue of many bones that serves as a major site of blood cell formation.

**reflex** [L. *reflectere*, to bend back] A simple, stereotyped, and repeatable movement elicited by a sensory stimulus.

**reflex arc** [L. *reflectere*, to bend back] Type of neural pathway in which signals from sensory neurons can be sent directly to motor neurons, without intervention by an interneuron.

**refractory period** Of neurons, the period following an action potential at a given patch of membrane when sodium gates are shut and potassium gates are open, so that the patch is insensitive to stimulation.

**releasing hormone** A hypothalamic signaling molecule that stimulates or slows down secretion by target cells in the anterior lobe of the pituitary gland.

**repressor protein** Regulatory protein that provides negative control of gene activity by preventing RNA polymerase from binding to DNA.

**reproduction, asexual** Production of new individuals by any mode that does not involve gametes.

**reproduction, sexual** Mode of reproduction that begins with meiosis, proceeds through gamete formation, and ends at fertilization.

**reproductive isolating mechanism** Any aspect of structure, functioning, or behavior that prevents successful interbreeding (hence gene flow) between populations or between local breeding units within a population.

**resource partitioning** A community pattern in which similar species generally share the same kind of resource in different ways, in different areas, or at different times.

**respiration** [L. *respirare*, to breathe] In most animals, the overall exchange of oxygen from the environment and carbon dioxide wastes from cells by way of circulating blood. Compare *aerobic respiration*.

**resting membrane potential** Of neurons and other excitable cells that are not being stimulated, the steady voltage difference across the plasma membrane.

**restriction enzymes** Class of enzymes that cut apart foreign DNA that enters a cell, as by viral infection; also used in recombinant DNA technology.

**reticular formation** Of the vertebrate brainstem, a major network of interneurons that helps govern activity of the whole nervous system.

**reverse transcriptase** Viral enzyme required for reverse transcription of mRNA into DNA; also used in recombinant DNA technology.

**reverse transcription** Assembly of DNA on a single-stranded mRNA molecule by viral enzymes.

**RFLPs** (restriction fragment length polymorphisms) Slight but unique differences in the banding pattern of DNA fragments from different individuals of a species; result from individual differences in the number and location of DNA sites that restriction enzymes can recognize and cut.

**ribosomal RNA** (rRNA) Type of RNA molecule that combines with proteins to form ribosomes, on which polypeptide chains are assembled.

**ribosome** Of cells, a structure having two subunits, each composed of RNA and protein molecules; the site of protein synthesis.

**RNA** Ribonucleic acid; a category of nucleotides used in translating the genetic message of DNA into protein.

**rod cell** A vertebrate photoreceptor sensitive to very dim light and that contributes to coarse perception of movement.

**root hair** Of vascular plants, an extension of a specialized root epidermal cell; root hairs collectively enhance the surface area available for absorbing water and solutes.

**RuBP** Ribulose bisphosphate, a five-carbon compound required for carbon fixation in the Calvin-Benson cycle of photosynthesis.

**salivary gland** Any of the glands that secrete saliva, a fluid that initially mixes with food in the mouth and starts the breakdown of starch.

**salt** An ionic compound formed when an acid reacts with a base.

**saltatory conduction** In myelinated neurons, rapid, node-to-node hopping of action potentials.

**saprobe** Heterotroph that obtains its nutrients from nonliving organic matter. Most fungi are saprobes.

**sarcomere** (SAR-koe-meer) Of skeletal and cardiac muscles, the basic unit of contraction; a region of organized myosin and actin filaments between two Z lines of a myofibril inside a muscle cell.

**sarcoplasmic reticulum** (sar-koe-PLAZ-mik reh-TIK-you-lum) A calcium-storing membrane system surrounding myofibrils of a muscle cell.

**Schwann cells** Specialized neuroglial cells that grow around neuron axons, forming a *myelin sheath*.

**sclerenchyma** Of vascular plants, a ground tissue that provides mechanical support and protection in mature plant parts.

**second law of thermodynamics** The spontaneous direction of energy flow is from high-quality to low-quality forms. With each conversion, some energy is randomly dispersed in a form that is not as readily available to do work.

**second messenger** A molecule inside a cell that mediates and generally triggers amplified response to a hormone.

**secondary immune response** Rapid, prolonged immune response by white blood cells, memory cells especially, to a previously encountered antigen.

**secretion** Generally, the release of a substance for use by the organism producing it. (Not the same as *excretion*, the expulsion of excess or waste material.) Of kidneys, a regulated stage in urine formation, in which ions and other substances move from capillaries into nephrons.

**sedimentary cycle** A biogeochemical cycle without a gaseous phase; the element moves from land to the seafloor, then returns only through long-term geological uplifting.

**seed** Of gymnosperms and flowering plants, a fully mature ovule (contains the plant embryo), with its integuments forming the seed coat.

**segmentation** Of earthworms and many other animals, a series of body units that may be externally similar to or quite different from one another.

**segregation** (Mendelian principle of) [L. *se-*, apart, + *grex*, herd] The principle that diploid organisms inherit a pair of genes for each trait (on a pair of homologous chromosomes) and that the two genes segregate during meiosis and end up in separate gametes.

**selective gene expression** Of multicelled organisms, activation or suppression of a fraction of the genes in unique ways in different cells, leading to pronounced differences in structure and function among different cell lineages.

**selfish behavior** Form of behavior by which an individual protects or increases its own chance of producing offspring, regardless of the consequences for the group to which it belongs.

**semen** (SEE-mun) [L. *serere*, to sow] Sperm-bearing fluid expelled from a penis during male orgasm.

**semiconservative replication** [Gk. *hēmi*, half, + L. *conservare*, to keep] Reproduction of a DNA molecule when a complementary strand forms on each of the unzipping strands of an existing DNA double helix, the outcome being two "half-old, half-new" molecules.

**senescence** (sen-ESS-cents) [L. *senescere*, to grow old] Sum total of processes leading to the natural death of an organism or some of its parts.

**sensory neuron** Any of the nerve cells that act as sensory receptors, detecting specific stimuli (such as light energy) and relaying signals to the brain and spinal cord.

**sessile animal** Animal that remains attached to a substrate during some stage (often the adult) of its life cycle.

**sex chromosomes** Of most animals and some plants, chromosomes that differ in number or kind between males and females but that still function as homologues during meiosis. Compare *autosomes*.

**sexual dimorphism** Phenotypic differences between males and females of a species.

**sexual reproduction** Production of offspring by way of meiosis, gamete formation, and fertilization.

**sexual selection** Natural selection based on a trait that provides a competitive edge in mating and producing offspring.

**shoot system** Stems and leaves of vascular plants.

**sieve tube member** Of flowering plants, a cellular component of the interconnecting conducting tubes in phloem.

**sink region** In a vascular plant, any region using or stockpiling organic compounds for growth and development.

**sliding filament model** Model of muscle contraction, in which actin filaments physically slide over myosin filaments toward the center of the sarcomere. The sliding requires ATP energy and the formation of cross-bridges between actin and myosin filaments.

**small intestine** Of vertebrates, the portion of the digestive system where digestion is completed and most nutrients absorbed.

**smog, industrial** Gray-colored air pollution that predominates in industrialized cities with cold, wet winters.

**smog, photochemical** Form of brown, smelly air pollution occurring in large cities in warm climates.

**social behavior** Tendency of individual animals to enter into cooperative, interdependent relationships with others of their kind; based on the ability to use communication signals.

**social parasite** Animal that depends on the social behavior of another species to gain food, care for young, or some other factor to complete its life cycle.

**sodium-potassium pump** A transport protein spanning the lipid bilayer of the plasma membrane. When activated by ATP, its shape changes and it selectively transports sodium ions out of the cell and potassium ions in.

**solute** (SOL-yoot) [L. *solvere*, to loosen] Any substance dissolved in a solution. In water, this means its individual molecules are surrounded by spheres of hydration that keep their charged parts from interacting, so the molecules remain dispersed.

**solvent** Fluid in which one or more substances are dissolved.

**somatic cell** (so-MAT-ik) [Gk. *somā*, body] Of animals, any cell that is not a germ cell (which develops by meiosis into sperm or eggs).

**somatic nervous system** Those nerves leading from the central nervous system to skeletal muscles.

**source region** Of vascular plants, any of the sites of photosynthesis.

**speciation** (spee-cee-AY-shun) The time at which a new species emerges, as by divergence or polyploidy.

**species** (SPEE-sheez) [L. *species*, a kind] Of sexually reproducing species in nature, one or more populations composed of individuals that are interbreeding and producing fertile offspring, and that are reproductively isolated from other such groups.

**sperm** [Gk. *sperma*, seed] Mature male gamete.

**spermatogenesis** (sperm-AT-oh-JEN-ih-sis) Formation of mature sperm following meiosis in a germ cell.

**sphere of hydration** Through positive or negative interactions, a clustering of water molecules around the individual molecules of a substance placed in water. Compare *solute*.

**sphincter** (SFINK-tur) Ring of muscle between regions of a tubelike system (as between the stomach and small intestine).

**spinal cord** Of central nervous systems, the portion threading through a canal inside the vertebral column and providing direct reflex connections between sensory and motor neurons as well as communication lines to and from the brain.

**spindle apparatus** A bipolar structure composed of microtubules that forms during mitosis or meiosis and that moves the chromosomes.

**sporangium**, plural **sporangia** (spore-AN-gee-um) [Gk. *spora*, seed] The protective tissue layer that surrounds haploid spores in a sporophyte.

**spore** Of fungi, a walled, resistant cell or multicelled structure, produced by mitosis or meiosis, that can germinate and give rise to a new mycelium. Of land plants, a reproductive cell formed by meiosis that can develop into a gametophyte (gamete-producing body).

**sporophyte** [Gk. *phyton*, plant] Diploid, spore-producing stage of plant life cycles.

**stabilizing selection** Mode of natural selection in which the most common phenotypes in a population are favored, and the underlying allele frequencies persist over time.

**stamen** (STAY-mun) Of flowering plants, a male reproductive structure; commonly consists of pollen-bearing structures (anthers) on single stalks (filaments).

**start codon** Of protein synthesis, a base triplet in a strand of mRNA that serves as the start signal for mRNA translation.

**steroid** (STAIR-oid) A lipid with a backbone of four carbon rings. Steroids differ in the number and location of double bonds in the backbone and in the number, position, and type of functional groups.

**stimulus** [L. *stimulus*, goad] A specific form of energy, such as light, heat, and mechanical pressure, that the body can detect through sensory receptors.

**stoma** (STOW-muh), plural **stomata** [Gk. *stoma*, mouth] A controllable gap between two guard cells in stems and leaves; any of the small passageways across the epidermis through which carbon dioxide moves into the plant and water vapor moves out.

**stomach** A muscular, stretchable sac that receives ingested food; of vertebrates, an organ between the esophagus and intestine in which considerable protein digestion occurs.

**stop codon** Of protein synthesis, a base triplet in a strand of mRNA that serves as the stop signal for translation, so that no more amino acids are added to the polypeptide chain.

**stroma** [Gk. *strōma*, bed] Of chloroplasts, the semifluid matrix surrounding the thylakoid membrane system; the zone where sucrose, starch, and other end products of photosynthesis are assembled.

**substrate** Specific molecule or molecules that an enzyme can chemically recognize, bind briefly to itself, and modify in a specific way.

**substrate-level phosphorylation** Enzyme-mediated reaction in which a substrate gives up a phosphate group to another molecule, as when an intermediate of glycolysis donates phosphate to ADP, producing ATP.

**succession, primary** (suk-SESH-un) [L. *succedere*, to follow] Orderly changes from the time pioneer species colonize a barren habitat through replacements after replacements by various species; the changes lead to a *climax community*, when the composition of species remains steady under prevailing conditions.

**succession, secondary** Orderly changes in a community or patch of habitat toward the climax state after having been disturbed, as by fire.

**surface-to-volume ratio** A mathematical relationship in which volume increases with the cube of the diameter, but surface area increases only with the square. Of growing cells, the volume of cytoplasm increases more rapidly than the surface area of the plasma membrane that must service the cytoplasm. Because of this constraint, cells generally remain small or elongated, or have elaborate membrane foldings.

**symbiosis** (sim-by-OH-sis) [Gk. *sym*, together, + *bios*, life, mode of life] A form of mutualism in which interacting species have become intimately and permanently dependent on each other for survival and reproduction.

**sympathetic nerve** Of the autonomic nervous system, any of the nerves generally concerned with increasing overall body activities during times of heightened awareness, excitement, or danger; such nerves also work continually in opposition with *parasympathetic nerves* to bring about minor adjustments in internal organs.

**sympatric speciation** [Gk. *syn*, together, + *patria*, native land] Speciation that follows after ecological, behavioral, or genetic barriers arise within the geographical boundaries of a single population.

**synaptic integration** (sin-AP-tik) Moment-by-moment combining of excitatory and inhibitory signals arriving at a trigger zone of a neuron.

**systematics** Branch of biology that deals with patterns of diversity among organisms in an evolutionary context; its three approaches include taxonomy, phylogenetic reconstruction, and classification.

**systemic circuit** (sis-TEM-ik) Circulation route in which oxygen-enriched blood flows from the lungs to the left half of the heart, through the rest of the body (where it gives up oxygen and picks up carbon dioxide), then back to the right side of the heart.

**taproot system** A primary root and its lateral branchings.

**taxonomy** (tax-ON-uh-mee) Approach in biological systematics that involves identifying organisms and assigning names to them.

**telophase** (TEE-low-faze) Of mitosis, final stage when chromosomes decondense into threadlike structures and two daughter nuclei form.

**telophase I** Of meiosis, stage when one of each type of duplicated chromosome has arrived at one or the other end of the spindle pole.

**telophase II** Of meiosis, final stage when four daughter nuclei form.

**temperate pathway** Mode of viral replication in which the virus enters latency instead of killing the host cell outright; viral genes remain inactive and, if integrated into the bacterial chromosome, may be passed on to any of the cell's descendants—which will be destroyed when the viral genes do become activated.

**testcross** Experimental cross in which hybrids of the first generation of offspring ($F_1$) are crossed with an individual known to be true-breeding for the same recessive trait as the recessive parent.

**testis**, plural **testes** Male gonad; primary reproductive organ in which male gametes and sex hormones are produced.

**testosterone** (tess-TOSS-tuh-rown) In male mammals, a major sex hormone that helps control male reproductive functions.

**theory** A related set of hypotheses that, taken together, form a broad-ranging explanation about some aspect of the natural world; differs from a scientific hypothesis in its breadth of application. In modern science, only explanations that have been extensively tested and can be relied upon with a very high degree of confidence are accorded the status of theory.

**thermal inversion** Situation in which a layer of dense, cool air becomes trapped beneath a layer of warm air; can cause air pollutants to accumulate to dangerous levels close to the ground.

**thermoreceptor** Sensory cell that can detect radiant energy associated with temperature.

**thigmotropism** (thig-MOTE-ruh-pizm) [Gk. *thigm*, touch] Of vascular plants, growth oriented in response to physical contact with a solid object, as when a vine curls around a fencepost.

**threshold** Of neurons and other excitable cells, a certain minimum amount by which the steady voltage difference across the plasma membrane must change to produce an action potential.

**thylakoid membrane** Of chloroplasts, an internal membrane commonly folded into flattened channels and stacked disks (*grana*); contains light-absorbing pigments and enzymes used in the formation of ATP, NADPH, or both during photosynthesis.

**thymine** Nitrogen-containing base in some nucleotides.

**thymus gland** Of endocrine systems, a gland in which certain white blood cells multiply, differentiate, and mature, and which secretes hormones that affect their functions.

**thyroid gland** Of endocrine systems, a gland that produces hormones that affect overall metabolic rates, growth, and development.

**tissue** Of multicelled organisms, a group of cells and intercellular substances that function together in one or more specialized tasks.

**tonicity** The relative concentrations of solutes in two fluids, such as inside and outside a cell. When solute concentrations are *isotonic* (equal in both fluids), water shows no net osmotic movement in either direction. When one fluid is *hypotonic* (has less solutes than the other), the other is *hypertonic* (has more solutes) and is the direction in which water tends to move.

**trachea** (TRAY-kee-uh), plural **tracheae** Of insects, spiders, and some other animals, a finely branching air-conducting tube that functions in respiration; of land vertebrates, the windpipe that carries air between the larynx and bronchi.

**tracheid** (TRAY-kid) Of flowering plants, one of two types of cells in xylem that conduct water and dissolved minerals.

**transcript-processing controls** Of eukaryotic cells, controls that govern modification of new mRNA molecules into mature transcripts before shipment from the nucleus.

**transcription** [L. *trans*, across, + *scribere*, to write] Of protein synthesis, the assembly of an RNA strand on one of the two strands of a DNA double helix; the base sequence of the resulting transcript is complementary to the DNA region on which it is assembled.

**transcriptional controls** Controls influencing when and to what degree a particular gene will be transcribed.

**transfer RNA** (tRNA) Of protein synthesis, any of the type of RNA molecules that bind and deliver specific amino acids to ribosomes *and* pair with mRNA code words for those amino acids.

**translation** Of protein synthesis, the conversion of the coded sequence of information in mRNA into a particular sequence of amino acids to form a polypeptide chain; depends on interactions of rRNA, tRNA, and mRNA.

**translational controls** Of eukaryotic cells, controls governing the rates at which mRNA transcripts that reach the cytoplasm will be translated into polypeptide chains at ribosomes.

**translocation** Of cells, the transfer of part of one chromosome to a nonhomologous chromosome. Of vascular plants, the conduction of organic compounds through the plant body by way of the phloem.

**transmitter substance** Any of the class of signaling molecules that are secreted from neurons, act on immediately adjacent cells, and are then rapidly degraded or recycled.

**transpiration** Evaporative water loss from stems and leaves.

**transport control** Of eukaryotic cells, controls governing when mature mRNA transcripts are shipped from the nucleus into the cytoplasm.

**transposable element** DNA element that can spontaneously "jump" to new locations in the same DNA molecule or a different one. Such elements often inactivate the genes into which they become inserted and give rise to observable changes in phenotype.

**trisomy** (TRY-so-mee) Of diploid cells, the abnormal presence of three of one type of chromosome.

**trophic level** (TROE-fik) [Gk. *trophos*, feeder] All organisms in an ecosystem that are the same number of transfer steps away from the energy input into the system.

**tropism** (TROE-pizm) Of vascular plants, a growth response to an environmental factor, such as growth toward light.

**tumor** A tissue mass composed of cells that are dividing at an abnormally high rate.

**turgor pressure** (TUR-gore) [L. *turgere*, to swell] Internal pressure applied to a cell wall when water moves by osmosis into the cell.

**upwelling** An upward movement of deep, nutrient-rich water along coasts to replace surface waters that winds move away from shore.

**uracil** (YUR-uh-sill) Nitrogen-containing base found in RNA molecules; can base-pair with adenine.

**urinary system** Of vertebrates, an organ system that regulates water and solute levels.

**urine** Fluid formed by filtration, reabsorption, and secretion in kidneys; consists of wastes and excess water and solutes.

**uterus** (YOU-tur-us) [L. *uterus*, womb] Chamber in which the developing embryo is contained and nurtured during pregnancy.

**vagina** Part of a female reproductive system that receives sperm, forms part of the birth canal, and channels menstrual flow to the exterior.

**variable** Of the factors characterizing or influencing an experimental group under study, the only one (ideally) that is *not* identical to those of a control group.

**vascular bundle** One of several to many strandlike arrangements of primary xylem and phloem embedded in the ground tissue of roots, stems, and leaves.

**vascular cambium** A lateral meristem that increases stem or root diameter of vascular plants showing secondary growth.

**vascular cylinder** Of plant roots, the arrangement of vascular tissues as a central cylinder.

**vascular plant** Plant having tissues that transport water and solutes through well-developed roots, stems, and leaves.

**vein** Of the circulatory system, any of the large-diameter vessels that lead back to the heart; of leaves, one of the vascular bundles that thread lacily through photosynthetic tissues.

**vernalization** Of flowering plants, stimulation of flowering by exposure to low temperatures.

**vertebra**, plural **vertebrae** Of vertebrate animals, one of a series of hard bones, arranged with intervertebral disks, into a backbone.

**vertebrate** Animal having a backbone of bony segments, the *vertebrae*.

**vesicle** (VESS-ih-kul) [L. *vesicula*, little bladder] Of cells, a small membranous sac that transports or stores substances in the cytoplasm.

**villus** (VIL-us), plural **villi** Any of several types of absorptive structures projecting from the free surface of an epithelium.

**viroid** An infectious nucleic acid that has no protein coat; a tiny rod or circle of single-stranded RNA.

**virus** A noncellular infectious agent, consisting of DNA or RNA and a protein coat; can replicate only after its genetic material enters a host cell and subverts that cell's metabolic machinery.

**vision** Precise light focusing onto a layer of photoreceptive cells that is dense enough to sample details concerning a given light stimulus, followed by image formation in the brain.

**vitamin** Any of more than a dozen organic substances that animals require in small amounts for normal cell metabolism but generally cannot synthesize for themselves.

**water potential** The sum of two opposing forces (osmosis and turgor pressure) that can cause the directional movement of water into or out of a walled cell.

**watershed** A region where all precipitation becomes funneled into a single stream or river.

**wax** A type of lipid with long-chain fatty acid tails; waxes help form protective, lubricating, or water-repellent coatings.

**white blood cell** Leukocyte; of vertebrates, any of the macrophages, eosinophils, neutrophils, and other cells which, together with their products, comprise the immune system.

**white matter** Of spinal cords, major nerve tracts so named because of the glistening myelin sheaths of their axons.

**wild-type allele** Of a population, the allele that occurs normally or with greatest frequency at a given gene locus.

**wing** Of birds, a forelimb of feathers, powerful muscles, and lightweight bones that functions in flight. Of insects, a structure that develops as a lateral fold of the exoskeleton and functions in flight.

**X-linked gene** Any gene on an X chromosome.

**X-linked recessive inheritance** Recessive condition in which the responsible, mutated gene occurs on the X chromosome.

**Y-linked gene** Any gene on a Y chromosome.

**xylem** (ZYE-lum) [Gk. *xylon*, wood] Of vascular plants, a tissue that transports water and solutes through the plant body.

**yolk sac** Of many vertebrates, an extraembryonic membrane that provides nourishment (from yolk) to the developing embryo; of humans, the sac does not include yolk but helps give rise to a digestive tube.

**zooplankton** A freshwater or marine community of floating or weakly swimming heterotrophs, mostly microscopic, such as rotifers and copepods.

**zygote** (ZYE-goat) The first diploid cell formed after fertilization (fusion of nuclei from a male and a female gamete).

# CREDITS AND ACKNOWLEDGMENTS

**Front Matter**

**Pages xiv–xv** Stock Imagery / **Pages xvi–xvii** James M. Bell/Photo Researchers / **Pages xviii–xix** S. Stammers/SPL/Photo Researchers / **Pages xx–xxi** © 1990 Arthur M. Greene / **Pages xxii–xxiii** Thomas D. Mangelsen / **Pages xxiv–xxv** Lennart Nilsson from *A Child Is Born*, © 1966, 1977 Dell Publishing Company, Inc. / **Pages xxvi–xxvii** Jim Doran

**Page 1** Tom Van Sant/The GeoSphere Project, Santa Monica, CA

**Chapter 1**

**1.1** Frank Kaczmarek / **1.4** (left) Paul DeGreve/FPG; (right) Norman Meyers/Bruce Coleman, Inc. / **1.5** (a) Walt Anderson/Visuals Unlimited; (b) Edward S. Ross; (c) Gregory Dimijian/Photo Researchers; (d) Alan Weaving/Ardea, London / **1.6** Jack deConingh / **1.7** (a) Tony Brain/SPL/Photo Researchers; (b) M. Abbey/Visuals Unlimited; (c) Edward S. Ross; (d) Dennis Brokaw; (e) Edward S. Ross; (f) Pat & Tom Leeson/Photo Researchers / **1.8** Levi Publishing Company

**Page 17** James M. Bell/Photo Researchers

**Chapter 2**

**2.1** Martin Rogers/FPG / **2.2** Jack Carey / **Page 21** (a) Kingsley R. Stern; (b) Chip Clark/ **Page 22** (c) Stanford Medical Center; (d) (left) Hank Morgan/Rainbow; (right) Dr. Harry T. Chugani, M.D., UCLA School of Medicine / **2.8** Art by Palay/Beaubois / **2.9** (a) Richard Riley/FPG / **2.10** Colin Monteath, Hedgehog House New Zealand; art by Palay/Beaubois / **2.11** (a), (b) Paolo Fioratti; (c) H. Eisenbeiss/Frank Lane Picture Agency / **2.14** Michael Grecco/Picture Group

**Chapter 3**

**3.1** (a) Field Museum of Natural History, Neg. #75400C, Chicago; (b) Brian Parker/Tom Stack & Associates / **3.2, 3.3, 3.5** Art by Palay/Beaubois / **3.8** Biophoto Associates/SPL/Photo Researchers / **3.9** Art by Jeanne Schreiber / **3.13** Clem Haagner/Ardea, London / **3.14** Larry Lefever/Grant Heilman / **Page 42** Lewis L. Lainey / **3.19** Art by Palay/Beaubois / **3.20** (a) CNRI/SPL/Photo Researchers; (b), (c) art by Robert Demarest / **3.22** (b) A. Lesk/SPL/Photo Researchers

**Chapter 4**

**4.1** (a) Jan Hinsch/SPL/Photo Researchers; (b), (c) NASA / **4.2** (a) (left) National Library of Medicine; (right) Armed Forces Institute of Pathology; (b) The Bettmann Archive; (c) George Musil/Visuals Unlimited / **4.5** (a–d) Jeremy Pickett-Heaps, School of Botany, University of Melbourne / **4.7** (a) Gary Gaard and Arthur Kelman; (b) micrograph G. Cohen-Bazire; art by Palay/Beaubois / **4.8–4.9** Art by Leonard Morgan / **4.10** Micrograph M.C. Ledbetter, Brookhaven National Laboratory; art by D. & V. Hennings / **4.11** Micrograph G.L. Decker; art by D. & V. Hennings / **4.12** Micrograph D. Fawcett, *The Cell*, Philadelphia: W.B. Saunders Co.; art by D. & V. Hennings / **4.13** (a) Don W. Fawcett/Visuals Unlimited; (b) A.C. Faberge, *Cell and Tissue Research*, 151:403–415, 1974 / **4.14** Art by A. Kasnot / **4.16** (a), (b) Micrographs Don W. Fawcett/Visuals Unlimited; (below, right) art by Robert Demarest / **4.17** (left) Art by Robert Demarest after a model by J. Kephart; micrograph Gary W. Grimes / **4.18** Gary W. Grimes / **4.19** Micrograph Keith R. Porter / **4.20** Micrograph L.K. Shumway; (below) art by Palay/Beaubois / **4.21** (a–c) Mark McNiven and Keith R. Porter; (d) Andrew S. Bajer; (e) J. Victor Small and Gottfried Rinnerthaler / **4.23** (a) Sidney L. Tamm; (b) art by D. & V. Hennings / **4.24** (b) U.W. Goodenough and J.W. Heuser / **4.25** (a) After Alberts et al., *Molecular Biology of the Cell*, Garland Publishing Co., 1983; (b) Dianne T. Woodrum and Richard W. Linck / **4.26** Sketch by D. & V. Hennings after P. Raven et al., *Biology of Plants*, third edition, Worth Publishers, 1981; micrograph P.A. Roelofsen

**Chapter 5**

**5.1** (a) Runk/Schoenberger/Grant Heilman; (b) Inigo Everson/Bruce Coleman Ltd. / **5.3** Art by Leonard Morgan; micrograph H.C. Aldrich / **Pages 78–79** Art by Palay/Beaubois; micrograph P. Pinto da Silva and D. Branton, *Journal of Cell Biology*, 45:598, by copyright permission of The Rockefeller University Press / **5.5** Micrograph M. Sheetz, R. Painter, and S. Singer, *Journal of Cell Biology*, 70:193, by copyright permission of The Rockefeller University Press / **5.6** Frieder Sauer/Bruce Coleman Ltd. / **5.7** Photographs Frank B. Salisbury / **5.8** Art by Leonard Morgan after Alberts et al., *Molecular Biology of the Cell*, second edition, Garland Publishing Co., 1989 / **5.9–5.11** Art by Leonard Morgan / **5.12** From *Molecular Cell Biology* by James Darnell, Harvey Lodish, and David Baltimore. Copyright © 1986 by Scientific American Books. Reprinted with permission of W.H. Freeman and Company; art by Palay/Beaubois / **5.13** M.M. Perry and A.B. Gilbert / **5.14** Art by Palay/Beaubois

**Chapter 6**

**6.1** Robert C. Simpson/Nature Stock / **6.3** (left) NASA; (right) Manfred Kage/Peter Arnold, Inc. / **6.4** Art by Palay/Beaubois / **6.8** Thomas A. Steitz / **6.9** Art by Palay/Beaubois / **6.12** Douglas Faulkner/Sally Faulkner Collection / **6.15** Art by L. Calver after B. Alberts et al., *Molecular Biology of the Cell*, Garland Publishing Co., 1983 / **Page 101** (a) Kathie Atkinson/Oxford Scientific Films; (b) Keith V. Wood

**Chapter 7**

**7.1** Photograph Sam Zarember/Image Bank / **7.2** (a) Hans Reinhard/Bruce Coleman Ltd.; (b), (c) micrographs David Fisher; (b) art by Palay/Beaubois; (c), (d) art by K. Kasnot / **7.3** Photograph Barker-Blakenship/FPG; art by Victor Royer / **7.4** Larry West/FPG / **7.5** Photograph E.R. Degginger / **7.6** Art by Illustrious, Inc. / **7.10** Art by L. Calver

**Chapter 8**

**8.1** Stephen Dalton/Photo Researchers / **8.2** (a), (b) NASA; (c) Janeart/Image Bank / **8.4** (right) Art by Palay/Beaubois / **8.5** (a) Keith R. Porter; (b), (c) art by L. Calver / **8.9** Photograph Adrian Warren/Ardea, London / **8.11** David M. Phillips/Visuals Unlimited / **Page 130** Ralph Pleasant/FPG / **Page 133** R. Llewellyn/Superstock, Inc.

**Page 135** © Lennart Nilsson

**Chapter 9**

**Page 136** (left) Chris Huss; (right) Tony Dawson / **Page 137** (above) Chris Huss; (below) Tony Dawson / **9.4** Andrew S. Bajer, University of Oregon / **9.5** Photographs Andrew S. Bajer, University of Oregon / **9.6** Micrographs Ed Reschke; art by K. Kasnot / **9.7** Art by Palay/Beaubois / **9.8** Micrographs H. Beams and R.G. Kessel, *American Scientist*, 64:279–290, 1976 / **9.9** B.A. Palevitz and E.H. Newcomb, University of Wisconsin/ BPS/Tom Stack & Associates / **9.10** (a–c), (e) Lennart Nilsson from *A Child Is Born*, © 1966, 1967 Dell Publishing Company, Inc.; (d) Lennart Nilsson from *Behold Man*, © 1974 by Albert Bonniers Forlag and Little, Brown and Company, Boston

**Chapter 10**

**10.1** (a) Jane Burton/Bruce Coleman Ltd.; (b) Dan Kline/Visuals Unlimited / **10.2** Courtesy of Kirk Douglas/The Bryna Company / **10.3** Art by L. Calver / **10.4** Art by K. Kasnot / **10.5** Micrograph B. John / **10.6** CNRI/SPL/Photo Researchers / **10.7** Art by K. Kasnot / **10.12** (b) © David M. Phillips/Visuals Unlimited / **10.14** Art by K. Kasnot

**Chapter 11**

**11.1** (a) David M. Phillips/Visuals Unlimited; (b) Bill Longcore/Photo Researchers; (c) Moravian Museum, Brno / **11.2** Photograph Jean M. Labat/Ardea, London; art by Jennifer Wardrip / **11.11** Photographs William E. Ferguson / **11.13** Tedd Somes / **11.14** (a), (b) Michael Stuckey/Comstock Inc.; (c) Russ Kinne/Comstock Inc. / **11.15** David Hosking / **11.16** Photograph Bill Longcore/Photo Researchers / **11.17** Photograph Jane Burton/Bruce Coleman Ltd.; art by D. & V. Hennings / **11.18** After John G. Torrey, *Development in Flowering Plants*, by permission of Macmillan Publishing Company, copyright © 1967 by John G. Torrey / **11.19** (top to bottom) Frank Cezus; Frank Cezus; Michael Keller; Ted Beaudin; Stan Sholik/all FPG / **11.20** (a) F. Blakeslee, *Journal of Heredity*, 1914 / **Page 181** Evan Cerasoli

**Chapter 12**

**12.1** Eddie Adams/AP Photo / **12.3** Art by Palay/Beaubois; photograph Omikron/Photo Researchers / **Page 186** (a) Reprinted by permission from page 109 of *Human Heredity: Principles and Issues*, second edition. Copyright © 1991 by West Publishing Company. All rights reserved. Art by Robert Demarest / **Page 187** Art by Robert Demarest after Patten, Carlson, and others / **12.5** Photographs Carolina Biological Supply Company / **12.8** Photograph Dr. Victor A. McKusick / **12.12** After Victor A. McKusick, *Human Genetics*, second edition, copyright © 1969. Reprinted by permission of Prentice-Hall, Inc., Englewood Cliffs, NJ; photograph The Bettmann Archive / **12.13** (a) Courtesy of David D. Weaver, M.D. / **12.15** Photograph courtesy of B.R. Brinkley from D.E. Merry et al., *American Journal of Human Genetics*, 37:425–430, 1985. © 1985 by The American Society of Human Genetics. All rights reserved. Used by permission of The University of Chicago Press / **12.16** Art by K. Kasnot / **12.17** (a) Cytogenetics Laboratory, University of California, San Francisco; (b) after Collman and Stoller, *American Journal of Public Health*, 52, 1962 / **12.18** (top left) Used by permission of Carole Lafrate; (top right and below) courtesy of Peninsula Association for Retarded Children and Adults, San Mateo Special Olympics, Burlingame, CA / **Page 200** Art by Palay/Beaubois / **Page 203** (above) Bonnie Kamin/Stuart Kenter Associates; (below) Carolina Biological Supply Company


## Chapter 13

**13.1** Photograph A.C. Barrington Brown © 1968 J.D. Watson; model A. Lesk/SPL/Photo Researchers / **13.3** Micrograph Lee D. Simon, Waksman Institute of Microbiology / **13.5** Micrograph Biophoto Associates/SPL/Photo Researchers / **13.19** (a) Photograph W.C. Earnshaw. From the *Journal of Cell Biology*, 1985, 100:1716–1725 by copyright permission of The Rockefeller University Press; (b), (c) photographs U.K. Laemmli from *Cell*, 12:817–828, copyright 1977 by MIT/Cell Press / **13.10** (a) C.J. Harrison et al., *Cytogenetics and Cell Genetics* 35:21–27, copyright 1983 S. Karger A.G., Basel; (c) U.K. Laemmli; (d) B. Hamkalo; (e) O.L. Miller, Jr., and Steve L. McKnight; art by Palay/Beaubois


## Chapter 14

**14.1** (above) Kevin Magee/Tom Stack & Associates; (below) Dennis Hallinan/FPG / **14.4** Art by Palay/Beaubois / **14.7** Art by Palay/Beaubois / **14.8** Photograph courtesy of Thomas A. Steitz from *Science*, 246:1135–1142, December 1, 1989 / **14.10** Art by L. Calver / **14.11** Micrograph Dr. John E. Heuser, Washington University School of Medicine, St. Louis, MO; art by Palay/Beaubois / **14.12** Art by L. Calver / **14.14** Peter Starlinger


## Chapter 15

**15.1** (a) Lennart Nilsson © Boehringer Ingelheim International GmbH / **15.2** Art by Palay/Beaubois / **15.4** Art by Palay/Beaubois / **15.5** (a) M. Roth and J. Gall; (e) W. Beerman / **15.6** Stuart Kenter Associates / **15.7** Jack Carey


## Chapter 16

**16.1** Secchi-Lecague/Roussel–UCLAF/CNRI/SPL/ Photo Researchers / **16.2** (a) Dr. Huntington Potter and Dr. David Dressler; (b) C.C. Brinton, Jr., and J. Carnahan / **16.6** Art by Jeanne Schreiber / **Page 252** Photograph Damon Biotech, Inc. / **Pages 252–253** Art by Palay/Beaubois / **16.9** Michael Maloney/ San Francisco Chronicle / **16.10** Runk/Schoenberger/ Grant Heilman / **16.11** W. Merrill / **16.12** Keith V. Wood / **16.13** (a), (b) Monsanto Company; (c), (d) Calgene, Inc. / **16.14** R. Brinster and R.E. Hammer, School of Veterinary Medicine, University of Pennsylvania


**Page 259** S. Stammers/SPL/Photo Researchers

## Chapter 17

**17.1** (a) Art by Leonard Morgan after R.H. Dott, Jr., and R.L. Batten, *Evolution of the Earth*, third edition, McGraw-Hill, 1981. Reproduced by permission of McGraw-Hill Inc.; (b) Werner Stoy/ Bruce Coleman Ltd.; / **17.2** (a) Jen & Des Bartlett /Bruce Coleman Ltd.; (b) Kenneth W. Fink/Photo Researchers; (c) Dave Watts/A.N.T. Photo Library / **17.4** (a) Courtesy George P. Darwin, Darwin Museum, Down House; (c) Christopher Ralling / **17.5** (a) D. Barrett/Planet Earth Pictures; (b) (above) Field Museum of Natural History (Neg. No. CK21T), Chicago, and the artist Charles R. Knight; (below) Lee Kuhn/FPG / **17.6** (a) Photograph Dieter & Mary Plage/Survival Anglia; art by Leonard Morgan; (b) C.P. Hickman, Jr.; (c), (d) Heather Angel / **17.7** (a) George W. Cox; (b) David Cavagnaro; (c) Heather Angel; (d) Alan Root/Bruce Coleman Ltd.; (e) David Steinberg / **17.8** Down House and The Royal College of Surgeons of England / **17.9** Photograph John H. Ostrom, Yale University


## Chapter 18

**18.1** Elliott Erwitt/Magnum Photos, Inc. / **18.2** Alan Solem / **18.3** (left) Eric Crichton/Bruce Coleman Ltd.; (right) William E. Ferguson / **18.5** After D. Futuyma, *Evolutionary Biology*, Sinauer, 1979 / **18.6** David Cavagnaro / **18.7** (above) David Neal Parks; (below) W. Carter Johnson / **18.9** (a) Thomas N. Taylor; (b) Edward S. Ross / **18.10** (c), (d) Alex Kerstitch / **18.11** After M. Karns and L. Penrose, *Annals of Eugenics*, 15:206–233, 1951 / **18.12** J.A. Bishop and L.M. Cook / **18.13** (a) Bruce Beehler; (b) Charles W. Fowler/National Marine Fisheries; (c) D. Avon/Ardea, London / **18.14** After F. Ayala and J. Valentine, *Evolving*, Benjamin-Cummings, 1979 / **18.15** After V. Grant, *Organismic Evolution*, W.H. Freeman and Co., 1977 / **18.16** Jen & Des Bartlett/ Bruce Coleman Ltd. / **Page 289** (a) Nancy Sefton/ Photo Researchers; (b) Daniel W. Gotshall / **18.17** After W. Jensen and F.B. Salisbury, *Botany: An Ecological Approach*, Wadsworth, 1972


## Chapter 19

**19.1** (left) Vatican Museums; (right) Martin Dohrn /SPL/Photo Researchers / **19.2** (a) Patricia G. Gensel; (b) A. Feduccia, *The Age of Birds*, Harvard University Press, 1980; (c) Jonathan Blair/Woodfin Camp & Associates; (d) Donald Baird, Princeton Museum of Natural History / **19.3** David Noble/FPG / **19.4** From T. Storer et al., *General Zoology*, sixth edition, McGraw-Hill, 1979. Reproduced by permission of McGraw-Hill, Inc. / **19.5** Art by Victor Royer / **19.6** Art by Joel Ito / **19.7** (top) Douglas P. Wilson/Eric & David Hosking; (center) Superstock, Inc.; (bottom) E.R. Degginger / **19.8** After "Reconstructing Bird Phylogeny by Comparing DNA's" by C.G. Sibley and J.E. Ahlquist, *Scientific American*, February 1986. Copyright © 1986 by Scientific American, Inc. All rights reserved / **19.9** Chesley Bonestell / **19.10** Gary Byerly, LSU / **19.12** (a) Sidney W. Fox; (b) W. Hargreaves and D. Deamer / **19.13** Art by Leonard Morgan / **19.14** (below) After S.M. Stanley, *Macroevolution: Pattern and Process*, W.H. Freeman and Co., 1979 / **19.15** After P. Dodson, *Evolution: Process and Product*, third edition, Prindle, Weber & Schmidt / **19.16** Art by Leonard Morgan / **19.17** Data from J.J. Sepkoski, Jr., *Paleobiology*, 7(1):36–53 and J.J. Sepkoski, Jr., and M.L. Hulver in Valentine, ed., *Phanerozoic Diversity Patterns: Profiles in Macroevolution*, Princeton University Press, 1985 / **19.18** (a) Stanley W. Awramik; (b) M.R. Walter / **19.19** (a), (b) Neville Pledge/South Australian Museum; (c), (d) Chip Clark / **19.21** (a) H.P. Banks; (b) Patricia G. Gensel / **19.22** (a) From *Evolution of Life*, Linda Gamlin and Gail Vines (Eds.), Oxford University Press, 1987; art by D. & V. Hennings; (b) Rod Salm/Planet Earth Pictures / **Pages 312–313** (a) NASA; (b) © John Gurche 1989; (c) William K. Hartmann / **19.23** Jack Carey / **19.24** (a), (b) Field Museum of Natural History (Neg. Nos. CK46T & CK8T), Chicago, and the artist Charles R. Knight


## Chapter 20

**20.1** (top) Thomas D. Mangelsen/Images of Nature; (center) Jeffrey Sylvester/FPG; (bottom) Kjell Sandved/Visuals Unlimited / **20.2** Edward S. Ross / **20.3** Art by Raychel Ciemma after C.T. Regan and E. Trewavas, 1932 / **20.4** From F. Salisbury and C. Ross, *Plant Physiology*, fourth edition, Wadsworth, 1991 / **20.6** (a) Suzanne L. Collins and Joseph T. Collins; (b) from *The Amphibians and Reptiles of Missouri* by Tom R. Johnson. Copyright © 1987 by the Conservation Commission of the State of Missouri. Reprinted by permission / **20.7** Kevin Schafer/Tom Stack / **Page 324** Art by D. & V. Hennings / **Pages 325–327** Art by John & Judy Waller / **Page 326** (left to right) Larry Lefever/Grant Heilman; R.I.M. Campbell/Bruce Coleman Ltd.; Runk /Schoenberger/Grant Heilman; Bruce Coleman Ltd.


## Chapter 21

**21.1** (left) FPG; (right) Douglas Mazonowicz /Gallery of Prehistoric Art / **21.3** (a) Bruce Coleman Ltd.; (b) Tom McHugh/Photo Researchers; (c) Larry Burrows/Aspect Picture Library / **21.4** Art by D. & V. Hennings / **21.6** © Time Inc. 1965/Larry Burrows Collection / **21.8** Art by D. & V. Hennings / **21.9** (a) Louise M. Robbins; (b) Dr. Donald Johanson, Institute of Human Origins / **21.10–21.11** Art by D. & V. Hennings / **21.12** Photographs by John Reader copyright 1981 / **Page 342** Art by Palay/Beaubois after "Emergence of Modern Humans" by Christopher B. Stringer, *Scientific American*, December 1990. Copyright © 1990 by Scientific American, Inc. All rights reserved


**Page 345** © 1990 Arthur M. Greene

## Chapter 22

**22.1** Tony Brain and David Parker/SPL/Photo Researchers / **22.2** (b) Art by L. Calver / **22.3** Art by Palay/Beaubois / **22.4** (a) George Musil/Visuals Unlimited; (b) K.G. Murti/Visuals Unlimited / **22.5** Art by Palay/Beaubois / **22.6** Kenneth M. Corbett / **Page 352** Kent Wood/Photo Researchers / **22.7** Art by L. Calver / **22.8** L.J. LeBeau, University of Illinois Hospital/BPS / **22.9** (a) CNRI/SPL/Photo Researchers; (b) Stanley Flegler/Visuals Unlimited / **22.10** Micrograph J.J. Cardamone, Jr./BPS / **22.11** Paul A. Zahl / © 1967 National Geographic Society / **22.12** (a) John D. Cunningham/Visuals Unlimited; (b) Tony Brain/SPL/Photo Researchers; (c) P.W. Johnson and J. McN. Sieburth, University of Rhode Island/BPS / **22.13** Stanley W. Watson, *International Journal of Systematic Bacteriology*, 21:254–270, 1971 / **22.14** T.J. Beveridge, University of Guelph/BPS / **22.15** Richard Blakemore / **22.16** Hans Reichenbach, Gesellschaft für Biotechnologische Forschung, Braunschweig, Germany / **Page 361** (a) Art by Palay/Beaubois; (b) (above) H. Stolp; (below) L. Margulis / **22.17** (a) Edward S. Ross; (b) John Shaw/Bruce Coleman Ltd. / **22.18** (a) Art by Leonard Morgan; (b), (c) M. Claviez, G. Gerish, and R. Guggenheim; (d) London Scientific Films; (e–g) Carolina Biological Supply Company; (h) photograph courtesy Robert R. Kay from R.R. Kay et al., *Development*, 1989 Supplement, pp. 81–90, © The Company of Biologists Ltd. 1989 / **22.19** (a) P.L. Walne and J.H. Arnott, *Planta*, 77:325–354, 1967; (b) T.E. Adams/Visuals Unlimited; art by Palay /Beaubois / **22.20** (a–c, e) Ronald W. Hoham, Dept. of Biology, Colgate University; (d) (above) G.A. Fryell; (below) G. Shih and R.G. Kessel, *Living Images*, Jones and Bartlett Publishers, Inc., Boston © 1982 / **22.21** (left) C.C. Lockwood; (right) Florida Department of Natural Resources, Bureau of Marine Research / **22.22** (a) John D. Cunningham/Visuals Unlimited; (b) Jerome Paulin/Visuals Unlimited; (c) David M. Phillips/Visuals Unlimited / **22.23** (a) M. Abbey/Visuals Unlimited; (b) John Clegg/Ardea, London; (c) Manfred Kage/Bruce Coleman Ltd.; (d) T.E. Adams/Visuals Unlimited / **22.24** Micrograph Gary W. Grimes and Steven L'Hernault; art by Palay/Beaubois / **22.25** Gary W. Grimes and Steven L'Hernault / **22.26** Art by Leonard Morgan; micrograph Steven L'Hernault / **Page 369** (a) Richard W. Greene; (b) Laszlo Meszoly in L. Margulis, *Early Life*, Jones and Bartlett Publishers, Inc., Boston, © 1982


## Chapter 23

**23.1** Robert C. Simpson/Nature Stock / **23.2** Philip Springham from A.D.M. Rayner, *New Scientist*, November 19, 1988 / **23.3** (b) G.T. Cole, University of Texas, Austin/BPS / **23.5** M.S. Fuller, *Zoosporic Fungi in Teaching and Research*, M.S. Fuller and A. Jaworski (Eds.), 1987, Southeastern Publishing Company, Athens, GA / **23.6** Heather Angel/ **Page 242** W. Merrill / **23.7** (a) John D. Cunningham/Visuals Unlimited; (b) David M. Phillips/Visuals Unlimited / **23.8** John Hodgin / **23.9** (a) After T. Rost et al., *Botany*, Wiley, 1979; (c), (b) Robert C. Simpson/Nature Stock; (d) Eric Crichton/Bruce Coleman Ltd. / **23.10** (a), (b) Robert C. Simpson/Nature Stock; (c) Victor Duran; (d) Jane Burton/Bruce Coleman Ltd.; (e), (f) Thomas J. Duffy / **23.11** (b) Martyn Ainsworth from A.D.M. Rayner, *New Scientist*, November 19, 1988; (c–e) G. Shih and R.G. Kessel, *Living Images*, Jones and Bartlett Publishers, Inc., Boston, © 1982 /

**23.12** (a) G.T. Cole, University of Texas, Austin/BPS; (b) N. Allin and G.L. Barron; (c) G.L. Barron, University of Guelph / **23.13** (above) Mark Mattock/Planet Earth Pictures; (below) Ken Davis/Tom Stack & Associates / **23.14** After Raven, Evert, and Eichhorn, *Biology of Plants*, fourth edition, Worth Publishers, New York, 1986 / **23.15** © 1990 Gary Braasch / **23.16** F.B. Reeves

## Chapter 24

**24.1** (left) Raymond A. Mendez/Animals Animals; (above) Ken Lewis/Earth Scenes / **24.2** Art by Jeanne Schrieber / **24.3** (a) D.P. Wilson/Eric & David Hosking; (b) Douglas Faulkner/Sally Faulkner Collection / **24.4** (left) Steven C. Wilson/Entheos; (right) J.R. Waaland, University of Washington/BPS / **24.5** (a) Dennis Brokaw; (b) art by Jennifer Wardrip based on Gilbert M. Smith, *Marine Algae of the Monterey Peninsula*, Stanford University Press / **24.6** (a) Hervé Chaumeton/Agence Nature; (b) Alex Kerstitch/Tom Stack & Associates; (c), (d) Ronald W. Hoham, Dept. of Biology, Colgate University / **24.7** Carolina Biological Supply Company / **24.8** Photograph D.J. Patterson/Seaphot Limited: Planet Earth Pictures; art by D. & V. Hennings / **24.9** Hervé Chaumeton/Agence Nature / **24.10** Photograph Jane Burton/Bruce Coleman Ltd.; art by D. & V. Hennings / **24.11** (a), (b) Kingsley R. Stern; (c) John D. Cunningham/Visuals Unlimited / **24.12** Kingsley R. Stern / **24.13** (a) Edward S. Ross / **24.14** (left) W.H. Hodge; (right) Kratz/ZEFA / **24.15** (a) Art by D. & V. Hennings; photograph A. & E. Bomford/Ardea, London; (b) Lee Casebere; (c) Jean Paul Ferrero/Ardea, London / **24.16** Art by Jennifer Wardrip / **24.17** Ed Reschke / **24.18** (a) John H. Gerard; (b) Kingsley R. Stern; (c) Edward S. Ross; (d) F.J. Odendaal, Duke University/BPS / **24.19** Photograph Edward S. Ross; art by D. & V. Hennings / **24.20** (a) Martin Grosnick/Ardea, London; (b) Hans Reinhard/Bruce Coleman Ltd.; (c) Edward S. Ross; (d) Heather Angel; (e) Dick Davis/Photo Researchers (f) Peter F. Zika/Visuals Unlimited; (g) L. Mellichamp/Visuals Unlimited / **24.21** Art by D. & V. Hennings

## Chapter 25

**25.1** (a) Chip Clark; (b) (above) Jim Stewart/Scripps Institution of Oceanography; (below) Chip Clark / **25.5** Art by D. & V. Hennings / **25.6** (a) (above) Douglas Faulkner/Sally Faulkner Collection; (below) David C. Haas/Tom Stack & Associates; (b) Marty Snyderman/Planet Earth Pictures / **25.7** Art by Palay/Beaubois / **25.8** (b), (c) Kim Taylor/Bruce Coleman Ltd. / **25.9** (a) Frieder Sauer/Bruce Coleman Ltd.; (b) Walter Deas/Seaphot Limited: Planet Earth Pictures; (c) Bill Wood/Seaphot Limited: Planet Earth Pictures; (d) Douglas Faulkner /Sally Faulkner Collection; (e) F. Stuart Westmorland /Tom Stack & Associates / **25.11** Photograph Andrew Mounter/Seaphot Limited: Planet Earth Pictures; art by Raychel Ciemma / **25.12** Photograph E.R. Degginger; art by Joan Carol after T. Storer et al., *General Zoology*, sixth edition, © 1979 McGraw-Hill / **25.13** Larry Madin/Planet Earth Pictures / **25.14** Photograph Kim Taylor/Bruce Coleman Ltd.; art by K. Kasnot / **25.15** (above) Cath Ellis, University of Hull/SPL/Photo Researchers; (below) Robert & Linda Mitchell / **Page 416** (a) Photograph Robert L. Calentine / **Page 417** (b) Photograph Carolina Biological Supply Company; art by K. Kasnot; (c) Lorus J. and Margery Milne; (d) Dianora Niccolini / **25.16** Kjell B. Sandved / **25.18** Photograph J. Solliday/BPS; art by Raychel Ciemma / **25.19** Art by Palay/Beaubois / **25.21** (c) Anthony & Elizabeth Bomford/Ardea, London; (d) Kjell B. Sandved / **25.22** Jeff Foott/Tom Stack & Associates / **25.24** (a) Rick M. Harbo; (b) Alex Kerstitch; (c) Hervé Chaumeton/Agence Nature / **25.25** Art by Laszlo Meszoly and D. & V. Hennings / **25.26** Hervé Chaumeton/Agence Nature / **25.27** J. Grossauer/ZEFA / **25.28** Photograph Douglas Faulkner/Sally Faulkner Collection; art by Laszlo Meszoly and D. & V. Hennings / **25.29** © Cabisco/Visuals Unlimited / **25.30** J.A.L. Cooke/Oxford Scientific Films / **25.31**

(a) Hervé Chaumeton/Agence Nature; (b) Jon Kenfield/Bruce Coleman Ltd. / **25.32** Art by Raychel Ciemma / **25.33** C.B. & D.W. Frith/Bruce Coleman Ltd. / **25.34** (a) Peter Green/Ardea, London; (b) Angelo Giampiccolo/FPG; (c) Jane Burton/Bruce Coleman Ltd. / **25.35** (a) John H. Gerard; (b) Ken Lucas/Seaphot Limited: Planet Earth Pictures; (c) P.J. Bryant, University of California, Irvine/BPS / **25.37** (a) Frans Lanting/Bruce Coleman Ltd.; (b) photograph Hervé Chaumeton/Agence Nature; art by Laszlo Meszoly / **25.38** Fred Bavendam/Peter Arnold, Inc. / **25.39** Agence Nature / **25.40** (a) Z. Leszczynski/Animals Animals; (b) Steve Martin/Tom Stack & Associates / **25.41** Art by D. & V. Hennings / **25.42** (a) David Maitland/Seaphot Limited: Planet Earth Pictures; (b–e), (g–i), (k) Edward S. Ross; (f) Ralph A. Reinhold/FPG; (j) C.P. Hickman, Jr. / **25.43** (a) Ian Took/Biofotos; (b) Kjell B. Sandved; (c) John Mason/Ardea, London; (d) Chris Huss/The Wildlife Collection / **25.44** (a) Hervé Chaumeton/Agence Nature; (b), (c) art by L. Calver / **25.45** (a) Kjell B. Sandved; (b) Jane Burton/Bruce Coleman Ltd.

## Chapter 26

**26.1** (a) Tom McHugh/Photo Researchers; (b) Jean Phillipe Varin/Jacana/Photo Researchers / **26.2** Art by D. & V. Hennings / **26.3** Photographs (left) Rick M. Harbo; (above) Peter Parks/Oxford Scientific Films/Animals Animals; (a–d) from *Living Invertebrates*, V. & J. Pearse and M. & R. Buchsbaum, The Boxwood Press, 1987. Used by permission / **26.4** (a) C.R. Wyttenbach, University of Kansas/BPS; art by D. & V. Hennings / **26.5** Photograph Hervé Chaumeton/Agence Nature; art by Laszlo Meszoly and D. & V. Hennings / **26.6** Art by D. & V. Hennings / **26.7** After A.S. Romer and T.S. Parsons, *The Vertebrate Body*, sixth edition, Saunders College Publishing, © 1986 CBS College Publishing; art by Laszlo Meszoly and D. & V. Hennings / **26.8** After C.P. Hickman, Jr., and L.S. Roberts, *Integrated Principles of Zoology*, seventh edition, St. Louis: Times Mirror/Mosby College Publishing, 1984; art by Palay/ Beaubois / **26.9** Heather Angel / **26.10** (a) Allan Power/Bruce Coleman Ltd.; (b) Erwin Christian/ZEFA; (c) Tom McHugh/Photo Researchers; (d) Patrice Ceisel/© 1986 John G. Shedd Aquarium; (e) Douglas Faulkner/Sally Faulkner Collection; (f) Robert & Linda Mitchell; (g) William H. Amos / **26.11** Photograph Bill Wood/ Bruce Coleman Ltd.; art by Raychel Ciemma / **26.12** Peter Scoones/Seaphot Limited: Planet Earth Pictures / **26.13** Art by Laszlo Meszoly and D. & V. Hennings / **26.14** Art by D. & V. Hennings after Romer and others / **26.15** (a) From *The Vertebrate Body*, sixth edition, by A. Romer and T. Parsons, copyright © 1986 by Saunders College Publishing, reprinted by permission of the publisher; art by Leonard Morgan; (b) Jerry W. Nagel; (c) Stephen Dalton/Photo Researchers; (d) © John Serraro/Visuals Unlimited; (e) Juan M. Renjifo/Animals Animals / **26.17** (a) Zig Leszczynski/Animals Animals / **26.18** Art by D. & V. Hennings / **26.19** (a) D. Kaleth/Image Bank; (b) Peter Scoones/Seaphot Limited: Planet Earth Pictures / **26.20** (a) Andrew Dennis/A.N.T. Photo Library; (b) Kim Taylor/ Bruce Coleman Ltd.; (c) Stephen Dalton/Photo Researchers; art by Raychel Ciemma; (d) Bob McKeever/Tom Stack & Associates; (e) C.B. & D.W. Frith/Bruce Coleman Ltd.; (f) W.J. Weber/Visuals Unlimited / **26.21** (a) Heather Angel; (b) Kevin Schafer/Tom Stack & Associates; (c) W.A. Banaszewski/Visuals Unlimited / **26.22** (a) Robert A. Tyrrell; (b) Rajesh Bedi; (c) J.L.G. Grande/Bruce Coleman Ltd.; (d) Thomas D. Mangelsen/Images of Nature / **26.23** (a) Gerard Lacz/A.N.T. Photo Library; (b) art by D. & V. Hennings / **26.24** D. & V. Blagden/A.N.T. Photo Library / **26.25** (a) Jack Dermid; (b) Douglas Faulkner/Photo Researchers; (c) Clem Haagner/Ardea, London; (d), (e) Leonard Lee Rue III/FPG; (f) Sandy Roessler/FPG

**Page 467** Bonnie Rauch/Photo Researchers

## Chapter 27

**27.1** (a) Roger Werth; (b) © 1980 Gary Braasch; (c) © 1989 Gary Braasch / **27.3** (left) Micrograph James D. Mauseth, *Plant Anatomy*, Benjamin-Cummings, 1988; (a–c) Biophoto Associates / **27.4** Thomas Eisner, Cornell University / **27.5** Art by Jennifer Wardrip / **27.6** Micrographs H.A. Core, W.A. Coté, and A.C. Day, *Wood Structure and Identification*, second edition, Syracuse University Press, 1979 / **27.7** (a) Chuck Brown; (b) G. Shih and R.G. Kessel, *Living Images*, Jones and Bartlett Publishers, Inc., Boston, © 1982 / **27.8** Art by D. & V. Hennings / **27.9** (a), (b) Edward S. Ross / **27.10** Art by D. & V. Hennings; (center) Carolina Biological Supply Company; (right) James W. Perry / **27.11** (left) Art by D. & V. Hennings: (center) Ray F. Evert; (right) James W. Perry / **27.12** (a) Robert & Linda Mitchell; (b), (c) Roland R. Dute / **27.13** (b–d) E.R. Degginger / **27.14** Art by D. & V. Hennings / **27.15** Heather Angel / **27.16** (a) C.E. Jeffree et al., *Planta*, 172(1):20–37, 1987. Reprinted by permission of C.E. Jeffree and Springer-Verlag; (b) art by D. & V. Hennings / **27.17** John E. Hodgin / **27.18** Micrograph E.R. Degginger / **27.19** Sketch after T. Rost et al., *Botany: A Brief Introduction to Plant Biology*, second edition, © 1984, John Wiley & Sons; micrographs Chuck Brown / **27.20** Carolina Biological Supply Company / **27.21** Art by Palay/Beaubois / **27.22** Ripon Microslides, Inc. / **27.25** After Marian Reeve / **27.26** (b) Jerry D. Davis / **27.27** (a) Biophoto Associates; (b) H.A. Core, W.A. Coté, and A.C. Day, *Wood Structure and Identification*, second edition, Syracuse University Press, 1979

## Chapter 28

**28.1** (a) Robert & Linda Mitchell; micrograph John N.A. Lott, *Scanning Electron Microscope Study of Green Plants*, St. Louis: C.V. Mosby Company, 1976; (b) Robert C. Simpson/Nature Stock / **28.2** (a) Art by Jennifer Wardrip; (c) Mark E. Dudley and Sharon R. Long; (d) Adrian P. Davies/Bruce Coleman Ltd.; (e) NifTAL Project, University of Hawaii, Maui / **28.3** Micrograph Jean Paul Revel / **28.4** Art by Leonard Morgan / **28.6** Art by Leonard Morgan / **28.7** (a) John Troughton and L.A. Donaldson; (b) W. Thomson, *American Journal of Botany*, 57(3):316, 1970 / **28.8** Micrograph Jeremy Burgess/SPL/Photo Researchers; art by Palay/Beaubois / **28.9** T.A. Mansfield / **28.11** Martin Zimmerman, *Science*, 133:73–79, © AAAS 1961 / **28.12** (b) David Fisher; (c–d) micrographs David Fisher; art by Palay/ Beaubois

## Chapter 29

**29.1** (a) (above) Edward S. Ross; (below) Thomas D.W. Friedmann/Photo Researchers; (b) Jeffry Myers/FPG / **29.3** Photograph Bonnie Rauch/ Photo Researchers / **29.4** John Shaw/Bruce Coleman Ltd. / **29.5** (a), (b) David M. Phillips/Visuals Unlimited; (c) David Scharf/Peter Arnold, Inc. / **29.7** Art by Leonard Morgan / **29.8** Art by D. & V. Hennings / **Page 506** (a) Peter Steyn/Ardea, London/ (c) Thomas Eisner, Cornell University / **Page 507** (b) M.P.L. Fogden/Bruce Coleman Ltd.; (d) Edward S. Ross / **Page 508** (e) Ted Schwartz / **29.9** (a), (b) Patricia Schulz; (c), (d) Ray F. Evert; (e), (f) Ripon Microslides; (far right) Kingsley R. Stern / **29.10** F. Bracegirdle and P. Miles, *An Atlas of Plant Structure*, Heinemann Educational Books, 1977 / **29.11** Janet Jones / **29.12** (a) B.J. Miller, Fairfax, VA/BPS; (b) R. Carr/Bruce Coleman Ltd.; (c) Richard H. Gross, Motlow State Community College / **29.13** (a) Grant Heilman; (b) Kjell Sandved / **Pages 512–513** John Alcock

## Chapter 30

**30.1** (a) R. Lyons/Visuals Unlimited; (b) Michael A. Keller/FPG / **30.2** Photograph Carolina Biological Supply Company / **30.3** Photograph Hervé Chaumeton/Agence Nature / **30.4** Art by Palay/ Beaubois / **30.5** (a) Kingsley R. Stern / **30.6** Frank B. Salisbury /

30.7 John Digby and Richard Firn / 30.8 B.E. Juniper / 30.9 Cary Mitchell / 30.10 Frank B. Salisbury / 30.12 Frank B. Salisbury / 30.14 Jan Zeevart / 30.15 Photograph N.R. Lersten / 30.16 A.C. Leopold et al., *Plant Physiology*, 34:570, 1958 / 30.17 A.C. Leopold and M. Kawase, *American Journal of Botany*, 51:294–298, 1964 / 30.18 R.J. Downs in T.T. Kozlowski, ed., *Tree Growth*, The Ronald Press, 1962 / **Page 528** Edward S. Ross / **Page 529** Dennis Brokaw

**Page 531** © Kevin Schafer

**Chapter 31**

31.1 David Macdonald / 31.2 Photographs (a) Lennart Nilsson from *Behold Man*, © 1974 Albert Bonniers Forlag and Little, Brown and Company, Boston; (b) Manfred Kage/Bruce Coleman Ltd.; (c) Ed Reschke/Peter Arnold Inc. / 31.3 (a) Art by Palay/Beaubois; (b) Focus on Sports; (inset) Manfred Kage/Bruce Coleman Ltd. / 31.4 Art by Palay/Beaubois / 31.5 Photographs Ed Reschke / 31.6 (left) Art by L. Calver / 31.7 Photographs Ed Reschke / 31.8 Lennart Nilsson from *Behold Man*, © 1974 Albert Bonniers Forlag and Little, Brown and Company, Boston / 31.9 Art by L. Calver / 31.10 Art by Palay/Beaubois / **Page 544** (a) Manfred Kage/Bruce Coleman Ltd.; (b–d) Ed Reschke

**Chapter 32**

32.1 Adrian Warren/Ardea, London / 32.2 Manfred Kage/Peter Arnold, Inc. / 32.3 (left) Art by Kevin Somerville; (right) art by L. Calver / 32.4, 32.6 Art by Leonard Morgan / 32.7 (a) A.L. Hodgkin, *Journal of Physiology*, vol. 131, 1956 / 32.8 (top) Art by Leonard Morgan; (bottom) art by Jeanne Schreiber / 32.9 (a) Carolina Biological Supply Company; art by Leonard Morgan / 32.10 Art by D. & V. Hennings; (c) J.E. Heuser and T.S. Reese / **Page 558** Painting by Sir Charles Bell, 1809, courtesy of Royal College of Surgeons, Edinburgh / 32.12 (a) Art by Robert Demarest; (b) from *Tissues and Organs: A Text-Atlas of Scanning Electron Microscopy*, by R.G. Kessel and R.H. Kardon. Copyright © 1979 by W.H. Freeman and Company. Reprinted with permission / 32.13 Art by K. Kasnot

**Chapter 33**

33.1 Comstock/Comstock Inc. / **Page 564** Art by D. & V. Hennings / 33.2 Photograph Francois Gohier/Photo Researchers; art by Raychel Ciemma / 33.3 Art by Palay/Beaubois / 33.4–33.5 Art by Kevin Somerville / 33.6 (b) Art by Kevin Somerville / 33.8 Art by Robert Demarest / 33.9 (a) Art by L. Calver; (b) Manfred Kage/Peter Arnold, Inc. / 33.11 Art by Kevin Somerville / 33.12 C. Yokochi and J. Rohen, *Photographic Anatomy of the Human Body*, second edition, Igaku-Shoin Ltd., 1979 / **Page 573** Art by Palay/Beaubois / 33.13 Art by Joel Ito / 33.14 Art by Palay/Beaubois after Penfield and Rasmussen, *The Cerebral Cortex of Man*, copyright © 1950 Macmillan Publishing Company, Inc. Renewed 1978 by Theodore Rasmussen / 33.15 Art by Robert Demarest / 33.16 After H. Jasper, 1941

**Chapter 34**

34.1 Hugo van Lawick / 34.2–34.3 Art by Kevin Somerville / 34.4–34.7 Art by Robert Demarest / 34.8 (a) Mitchell Layton; (b) Syndication International (1986) Ltd. / 34.9 Photographs courtesy of Dr. William H. Daughaday, Washington University School of Medicine. From A.I. Mendelhoff and D.E. Smith, eds., *American Journal of Medicine*, 20:133 (1956) / 34.12 The Bettmann Archive / 34.13 Biophoto Associates /SPL/Photo Researchers / 34.14 Art by Leonard Morgan / **Page 592** Evan Cerasoli

**Chapter 35**

35.1 Merlin D. Tuttle, Bat Conservation International / 35.2 Eric A. Newman / 35.3 Art by Kevin Somerville / 35.4 Art by Palay/Beaubois after Penfield and Rasmussen, *The Cerebral Cortex of Man*, copyright © 1950 Macmillan Publishing Company, Inc. Renewed 1978 by Theodore Rasmussen / 35.5 From Hensel and Bowman, *Journal of Physiology*, 23:564–568, 1960 / 35.6 Art by Ron Ervin; photograph Ed Reschke / 35.7 Art by D. & V. Hennings / 35.8 Art by Robert Demarest; micrograph Omikron/SPL/Photo Researchers / 35.9 Art by Kevin Somerville / 35.10 Art by Robert Demarest / 35.11 (a), (b) Robert E. Preston, courtesy Joseph E. Hawkins, Kresge Hearing Research Institute, University of Michigan Medical School / 35.12 Photograph Edward W. Bower/ © 1991 TIB/West; art by Kevin Somerville / 35.13 (a) Keith Gillett/Tom Stack & Associates; (b–d) after M. Gardiner, *The Biology of Vertebrates*, McGraw-Hill, 1972 / 35.14 (a) E.R. Degginger / 35.15 G.A. Mazohkin-Porshnykov (1958). Reprinted with permission from *Insect Vision*, © 1969 Plenum Press / 35.16 Art by Robert Demarest / 35.17–35.18 Art by Kevin Somerville / 35.19 Micrograph Lennart Nilsson © Boehringer Ingelheim International GmbH / **Page 613** Photographs Gerry Ellis/The Wildlife Collection; art by Kevin Somerville / 35.20 Art by Robert Demarest / 35.21 Art by Palay/Beaubois after S. Kuffler and J. Nicholls, *From Neuron to Brain*, Sinauer, 1977 / **Page 616** Art by Robert Demarest

**Chapter 36**

36.1 Jeff Schultz/AlaskaStock Images / 36.2 Art by L. Calver / 36.3 (a) Robert & Linda Mitchell; (b) Jane Burton/Bruce Coleman Ltd. / 36.4 Chaumeton-Lanceau/Agence Nature / 36.5 Art by Robert Demarest / 36.6 Ed Reschke / 36.7 Michael Keller/FPG / 36.8 CNRI/SPL/Photo Researchers / 36.9 Linda Pitkin/Planet Earth Pictures / 36.10 Photograph Stephen Dalton/Photo Researchers; art by Raychel Ciemma / 36.11 D.A. Parry, *Journal of Experimental Biology*, 36:654, 1959 / 36.12 Art by D. & V. Hennings / 36.13 Art by Joel Ito; micrograph Ed Reschke / 36.14 Art by K. Kasnot / 36.15 National Osteoporosis Foundation / 36.16 (b) Art by Ron Ervin / **Page 629** Photograph C. Yokochi and J. Rohen, *Photographic Anatomy of the Human Body*, second edition, Igaku-Shoin Ltd., 1979 / 36.17 (b), (c) Art by L. Calver / 36.18 (a) Ed Reschke; (b) D. Fawcett, *The Cell*, Philadelphia: W.B. Saunders Co., 1966 / 36.21 Art by R.M. Jensen / 36.22 Adapted from R. Eckert and D. Randall, *Animal Physiology: Mechanisms and Adaptations*, second edition, W.H. Freeman and Co., 1983 / 36.23 (a) Art by Kevin Somerville; (b) Ed Reschke / **Page 637** Photograph Michael Neveux / 36.25 N.H.P.A./A.N.T. Photo Library

**Chapter 37**

37.1 (a) D. Robert Franz/Planet Earth Pictures; (b) art by D. & V. Hennings adapted from *Mammalogy*, third edition, by Terry Vaughan, copyright © 1986 by Saunders College Publishing. Used by permission of the publisher / 37.2 (a) Kim Taylor/Bruce Coleman Ltd.; (b) Wardene Weisser/Ardea, London / 37.4 Art by Robert Demarest / 37.6 Art by Raychel Ciemma; (b) after A. Vander et al., *Human Physiology: Mechanisms of Body Function*, fifth edition, McGraw-Hill, 1990. Used by permission / 37.8 (a), (c) Lennart Nilsson © Boehringer Ingelheim International GmbH; (b) Biophoto Associates/SPL/Photo Researchers; art by Victor Royer / 37.9 Art by Robert Demarest / 37.10 Art by L. Calver / 37.11 (b) Steven Jones/FPG / **Page 650** Photograph CNRI/Phototake / 37.12 Photograph Ralph Pleasant/FPG / 37.13 Modified after A. Vander et al. *Human Physiology*, fourth edition, McGraw-Hill, 1985 / **Page 655** Photograph courtesy of David Steinberg

**Chapter 38**

38.1 (a) From A.D. Waller, *Physiology, The Servant of Medicine*, Hitchcock Lectures, University of London Press, 1910; (b) photograph courtesy of The New York Academy of Medicine Library / 38.3 (b) (below) After M. Labarbera and S. Vogel, *American Scientist*, 70:54–60, 1982 / **Page 662** Art by Palay/Beaubois / 38.4 (a) CNRI/SPL/Photo Researchers; (b) Lennart Nilsson from *Behold Man*, © 1974 by Albert Bonniers Forlag and Little, Brown and Company, Boston / 38.5 (left) Art by L. Calver and Victor Royer; (right) art by Victor Royer / 38.6 (a) Art by Leonard Morgan; (b) art by Kevin Somerville / 38.7 (a) Art by Joel Ito; (b) C. Yokochi and J. Rohen, *Photographic Anatomy of the Human Body*, second edition, Igaku-Shoin Ltd., 1979 / 38.11 Art by Robert Demarest based on A. Spence, *Basic Human Anatomy*, Benjamin-Cummings, 1982 / 38.14 Art by Raychel Ciemma / 38.15 After J. A. Gosling et al., *Atlas of Human Anatomy with Integrated Text*, copyright © 1985 by Gower Medical Publishing Ltd. **Page 671** (a) (above) Ed Reschke; (below) F. Sloop and W. Ober/Visuals Unlimited / 38.16 Photograph Lennart Nilsson © Boehringer Ingelheim International GmbH / 38.17 (a) After F. Ayala and J. Kiger, *Modern Genetics*, © 1980 Benjamin-Cummings; (b) Lester V. Bergman & Associates, Inc. / 38.18 After Gerard J. Tortora and Nicholas P. Anagnostakos, *Principles of Anatomy and Physiology*, sixth edition, copyright © 1990 by Biological Sciences Textbooks, Inc., A & P Textbooks, Inc. and Elia-Sparta, Inc. Reprinted by permission of Harper Collins Publishers / 38.19 Art by Kevin Somerville

**Chapter 39**

39.1 (a) The Granger Collection, New York; (b) Lennart Nilsson © Boehringer Ingelheim International GmbH / 39.2 Lennart Nilsson © Boehringer Ingelheim International GmbH / 39.5 Art by Palay/Beaubois / 39.6 Art by L. Calver and Victor Royer / 39.7 Art by Palay/Beaubois after S. Tonegawa, *Scientific Ameican*, October 1965 / **Page 689** Photographs Dr. Gilla Kaplan / 39.8 Art by Palay/Beaubois / 39.9 Art by Palay/Beaubois after B. Alberts et al., *Molecualr Biology of the Cell*, Garland Publishing Company, 1983 / **Page 693** (a) Art by L. Calver / **Page 694** (b), (c) Micrographs Z. Salahuddin, National Institutes of Health

**Chapter 40**

40.1 Galen Rowell/Peter Arnold, Inc. / 40.2 (b) Steve Lissau/Rainbow; (c) Peter Parks/Oxford Scientific Films / 40.4 Ed Reschke / 40.5 Art by D. & V. Hennings after C. P. Hickman et al., *Integrated Principles of Zoology*, sixth edition, St. Louis: C. V. Mosby Co., 1979 / 40.6 After C. P. Hickman et al., *Integrated Principles of Zoology*, sixth edition, St. Louis: C. V. Mosby Co., 1979 / 40.7 Art by Palay/Beaubois adapted from H. Scharnke, *Z. vergl. Physiol.*, 25:548–583 (1938) in *Form and Function in Birds*, Vol. 4, A. King and J. McLelland, Eds., Academic Press, 1989; micrograph H.R. Duncker, Justus-Liebig University, Giessen, Germany / 40.8 Art by L. Calver / 40.9 Art by Kevin Somerville / 40.11 CNRI/SPL/Photo Researchers / 40.12 After A. Vander et al., *Human Physiology*, third edition, McGraw-Hill, 1980 / 40.13 Art by K. Kasnot / 40.14 From L.G. Mitchell, J.A. Mutchmor, and W.D. Dolphin, *Zoology*, © 1988 Benjamin-Cummings Publishing Company / 40.16 Art by Leonard Morgan / **Page 710** (a) Gerard D. McLane / **Page 711** (b) Lennart Nilsson from *Behold Man*, © 1974 by Albert Bonniers Forlag and Little, Brown and Company, Boston / **Page 713** Christian Zuber/Bruce Coleman Ltd. / 40.17 (b) Giorgio Gualco/Bruce Coleman Ltd.

**Chapter 41**

41.1 Claude Steelman/Tom Stack & Associates / 41.3 Art by Kevin Somerville / 41.4 Art by Robert Demarest / 41.8 Art by Joel Ito / 41.9 Thomas D. Mangelsen/Images of Nature / 41.10 (left) David Jennings/Image Works; (right) Evan Cerasoli / 41.11 Art by Kevin Somerville / 41.12 The Bettmann Archive / 41.13 Terry Vaughan / 41.14 Fred Bruemmer

## Chapter 42

**42.1** (a) Hans Pfletschinger; (b) Carolina Biological Supply Company; (c–e) John H. Gerard / **42.2** (a) Frieder Sauer/Bruce Coleman Ltd.; (b) Evan Cerasoli; (c) Fred McKinney/FPG; (d) Carolina Biological Supply Company; (e) Leonard Lee Rue III / **42.4** Art by Palay/Beaubois adapted from R.G. Ham and M.J. Veomett, *Mechanisms of Development*, St. Louis: C.V. Mosby Co., 1980 / **42.5** Photographs Carolina Biological Supply Company; sketch after M.B. Patten, *Early Embryology of the Chick*, fifth edition, McGraw-Hill, 1971 / **42.6** J.R. Whittaker / **42.8** Photographs Carolina Biological Supply Company / **42.9** (a), (b) Micrographs J.B. Morrill and N. Ruediger; (c), (d) Micrographs J.B. Morrill; art by Raychel Ciemma after V.E. Foe and B.M. Alberts, *Journal of Cell Science*, 61:32, © The Company of Biologists 1983 / **42.10** Micrographs F.R. Turner; art by Raychel Ciemma / **42.11** Sketches after B. Burnside, *Developmental Biology*, 26:416–441, 1971; micrograph K.W. Tosney / **42.12** (a–c) Adapted by permission of Macmillan Publishing Company from *Developmental Biology: Patterns, Problems, Principles* by John W. Saunders, Jr., Copyright © 1982 by John W. Saunders, Jr.; (d) S.R. Hilfer and J.W. Yang, *The Anatomical Record*, 197:423–433, 1980 / **42.13** (a) K.W. Tosney / **42.14** Art by Palay/Beaubois after Robert F. Weaver and Philip W. Hedrick, *Genetics*. Copyright © 1989 Wm. C. Brown Publishers, Dubuque, Iowa. All rights reserved. Reprinted by permission / **42.15** After J.W. Fristrom et al., in E.W. Hanly, ed., *Problems in Biology: RNA Development*, University of Utah Press / **42.16** Carolina Biological Supply Company / **42.17** Sketches after Willier, Weiss, and Hamburger, *Analysis of Development*, Philadelphia: W.B. Saunders Co., 1955; photograph Roger K. Burnard / **42.18** Art by Raychel Ciemma adapted from L.B. Arey, *Developmental Anatomy*, Philadelphia: W.B. Saunders Co., 1965 / **Page 755** Dennis Green/Bruce Coleman Ltd.

## Chapter 43

**43.1** Lennart Nilsson from *A Child Is Born*, © 1966, 1977 Dell Publishing Company, Inc. / **43.2** (left) Art by Ron Ervin; (right) art by L. Calver / **43.3** Art by L. Calver; (c) R.G. Kessel and R.H. Kardon, *Tissues and Organs: A Text-Atlas of Scanning Electron Microscopy*, W.H. Freeman and Co., copyright © 1979 / **43.4–43.5** Art by Ron Ervin / **43.7** (left) Art by Ron Ervin; (right) art by L. Calver / **43.8** (top) Art by Robert Demarest; photograph Lennart Nilsson from *A Child Is Born*, © 1966, 1977 Dell Publishing Company, Inc. / **43.10** Art by Robert Demarest / **43.11** Art by Robert Demarest; (left) micrograph from Lennart Nilsson, *A Child Is Born*, © 1966, 1977 Dell Publishing Company, Inc.; (right) from Lennart Nilsson, *Behold Man*, © 1974 by Albert Bonniers Forlag and Little, Brown and Co., Boston / **43.12** Art by Robert Demarest / **43.13** Art by L. Calver; (c) after A.S. Romer and T.S. Parsons, *The Vertebrate Body*, sixth edition, Saunders College Publishing, © CBS College Publishing / **43.14** Art by L. Calver after Bruce Carlson, *Patten's Foundations of Embryology*, fourth edition, McGraw-Hill, 1981 / **43.15** From Lennart Nilsson, *A Child Is Born*, © 1966, 1977 Dell Publishing Company, Inc. / **Page 772** Modified from Keith L. Moore, *The Developing Human: Clinically Oriented Embryology*, fourth edition, Philadelphia: W.B. Saunders Co., 1988 / **Page 773** James W. Hanson, M.D. / **43.16–43.17** From Lennart Nilsson, *A Child Is Born*, © 1966, 1977 Dell Publishing Company, Inc. / **43.18** Art by Robert Demarest / **Page 776** Mills-Peninsula Hospitals / **43.19** Art by Ron Ervin / **Page 781** (a) Cheun-mo To and C.C. Brinton / **Page 782** (b) Joel B. Baseman

## Page 785 Alan and Sandy Carey

## Chapter 44

**44.1** Antoinette Jongen/FPG / **44.2** (above) Fran Allan/Animals Animals; (below) E.R. Degginger / **44.3** (c) Stanley Flegler/Visuals Unlimited / **44.5** (b) E. Vetter/ZEFA / **Page 794** Photograph Eric Crichton /Bruce Coleman Ltd. / **44.6** (left) Jonathan Scott/ Planet Earth Pictures; (right) (above) Wisniewski/ ZEFA; (below) Fred Bavendam/Peter Arnold, Inc. / **44.7** (a), (b) John Endler; (c) art by Raychel Ciemma / **44.8** Photograph NASA / **44.11** After G.T. Miller, *Living in the Environment*, sixth edition, Wadsworth, 1990 / **44.12** Data from Population Reference Bureau

## Chapter 45

**45.1** (left) Edward S. Ross; (right) Dona Hutchins / **45.2** (a), (c) Harlo H. Hadow; (b) Bob and Miriam Francis/Tom Stack & Associates / **45.3** Edward S. Ross; (b) Roger T. Petersen/NAS/Photo Researchers; (b), (c) Thomas Eisner, Cornell University / **45.10** (a) Douglas Faulkner/Sally Faulkner Collection; (b) W.M. Laetsch; (c), (d) Edward S. Ross; (e) James H. Carmichael / **45.11** Data from P. Price and H. Tripp, *Canadian Entymology*, 104:1003–1016, 1972 / **45.12** After N. Weland and F. Bazazz, *Ecology*, 56:681–688, © 1975 Ecological Society of America / **45.13** (a), (b) Jane Burton/Bruce Coleman Ltd.; (c) Heather Angel; graph from Jane Lubchenco, *American Naturalist*, 112:23–29, © 1978 by The University of Chicago Press / **Page 818** R. Slavin/FPG / **45.14** (a–f), (i) Roger K. Bernard; (g), (h) E.R. Degginger / **45.15** Photograph Dr. Harold Simon/Tom Stack & Associates; (below) after S. Fridriksson, *Evolution of Life on a Volcanic Island*, Butterworth, London, 1975 / **45.16** After J.M. Diamond, *Proceedings of the National Academy of Sciences*, 69:3199–3201, 1972 / **45.17** After M.H. Williamson, *Island Population*, Oxford University Press, 1981 / **45.18** (a) After F.G. Stehli et al., *Geological Society of America Bulletin*, 78:455–466, 1967; (b) after M. Kusenov, *Evolution*, 11:298–299, 1957; (c) after T. Dobzhansky, *American Scientist*, 38:209–221, 1950

## Chapter 46

**46.1** Wolfgang Kaehler / **46.3** Photograph Sharon R. Chester / **46.7** (b) Photograph Steven D. Bach / **46.10** Photograph © 1991 Gary Braasch / **46.11** Art by Raychel Ciemma / **46.12** (a) Photograph by Gene E. Likens from G.E. Likens and F.H. Bormann, *Proceedings First International Congress of Ecology*, pp. 330–335, September 1974, Centre Agric. Publ. Doc. Wagenigen, The Hague, the Netherlands; (b), (c) photographs by Gene E. Likens from G.E. Likens et al., *Ecology Monograph*, 40(1):23–47, 1970 / **Pages 838–839** Art by Raychel Ciemma; photograph NASA / **46.14** Photograph William J. Weber/Visuals Unlimited

## Chapter 47

**47.1** (a) (left) Edward S. Ross; (right) David Noble/ FPG; (b) (left) Edward S. Ross; (right) Richard Coomber/Planet Earth Pictures / **47.3** (b) Art by L. Calver / **47.5** (b) Edward S. Ross / **47.7** Art by Raychel Ciemma / **47.9** Art by Joan Carol / **47.10** Art by D. & V. Hennings after G.T. Miller, *Environmental Science: An Introduction*, Wadsworth, 1986 / **47.11** After Whittaker; Bland; and Tilman / **47.12** Harlo H. Hadow / **47.13** (a) Jack Wilburn/ Earth Scenes; (b) John D. Cunningham/Visuals Unlimited / **47.14** Kenneth W. Fink/Ardea, London / **47.15** Ray Wagner/Save the Tall Grass Prairie, Inc. / **47.16** Jonathan Scott/Planet Earth Pictures / **47.17** © 1991 Gary Braasch / **47.18** Thomas E. Hemmerly / **47.19** Dennis Brokaw / **47.20** Jack Carey / **47.21** Fred Bruemmer / **47.22** D.W. MacManiman / **47.24** Modified after Edward S. Deevy, Jr., *Scientific American*, October 1951 / **47.25** D.W. Schindler, *Science*, 184:897–899 / **47.27** (a) E.R. Degginger; (b) art by D. & V. Hennings / **47.28** Courtesy of J.L. Sumich, *Biology of Marine Life*, fourth edition, William C. Brown, 1988 / **47.29** (left and center) © 1991 Gary Braasch; (right) Phil Degginger / **47.31** (top right) Jim Doran; all other photographs Douglas Faulkner/Sally Faulkner Collection / **47.32** (a) McCutcheon/ZEFA; (b) Chuck Niklin; (c) Fred Grassle, Woods Hole Institution of Oceanography; (d)

Robert Hessler / **Page 871** Photographs R. Legeckis/NOAA

## Chapter 48

**48.1** (a) Gerry Ellis/The Wildlife Collection; (b) Adolf Schmidecker/FPG; (c) Edward S. Ross / **48.2** Photograph John Lawlor/FPG / **48.3** After G.T. Miller, *Environmental Science: An Introduction*, Wadsworth, 1986, and the Environmental Protection Agency / **48.4** (a) USDA Forest Service; (b) Heather Angel / **48.5** (bottom left) National Science Foundation; (top left; right) NASA / **48.6** Dr. Charles Henneghien/Bruce Coleman Ltd. / **48.7** From Water Resources Council / **48.8** Data from G.T. Miller / **48.9** (above) R. Bieregaard/Photo Researchers; (below) after G.T. Miller, *Living in the Environment*, sixth edition, Wadsworth, 1990 / **48.10** NASA / **48.11** USDA Soil Conservation Service/ Thomas G. Meier / **48.12** Agency for International Development / **48.13** Data from G.T. Miller / **Page 889** J. McLoughlin/FPG / **Page 891** © 1983 Billy Grimes

## Chapter 49

**49.1** (left) Robert Maier/Animals Animals; (right) (above) John Bova/Photo Researchers; (below) photograph Jack Clark/Comstock Inc.; graph L. Clark, *Parasitology Today*, 6(11), 1990, Elsevier Trends Journals, Cambridge, U.K. / **49.2** (a) Eugene Kozloff; (b), (c) Stevan Arnold / **49.3** Photograph John S. Dunning/Ardea, London; sonogram J. Bruce Falls and Tom Dickson, University of Toronto / **49.4** Photograph Hans Reinhard/Bruce Coleman Ltd.; sonogram G. Pohl-Apel and R. Sussinka, *Journal for Ornithologie*, 123:211–214 / **49.5** (a) Eric Hosking; (b) Stephen Dalton/Photo Researchers / **49.6** (left) Evan Cerasoli; (right) from A.N. Meltzoff and M.K. Moore, "Imitation of Facial and Manual Gestures by Human Neonate," *Science*, 198:75–78. Copyright 1977 by the AAAS / **49.7** (a) Nina Leen in *Animal Behavior*, Life Nature Library; (b) F. Schultz / **49.8** Michael Francis/The Wildlife Collection / **49.9** (left) David C. Fritts/Animals Animals; (right) Ray Richardson/Animals Animals / **49.10** John Alcock / **Page 904** (left) Lincoln P. Brower; (right) Eric Hosking

## Chapter 50

**50.1** John Alcock / **50.2** Edward S. Ross / **50.3** (a) E. Mickleburgh/Ardea, London; (b–d) G. Ziesler/ ZEFA / **50.4** Art by D. & V. Hennings / **50.5** A.E. Zuckerman/Tom Stack & Associates / **50.6** Fred Bruemmer / **50.7** John Alcock / **50.8** Patricia Caulfield / **50.9** Timothy Ransom / **50.10** Frank Lane Agency/Bruce Coleman Inc. / **50.11** Kenneth Lorenzen / **Page 919** Gregory D. Dimijian/Photo Researchers

# INDEX

# U

# V